Left:
Bunchgrass terrain degraded by sheep-grazing on Meseta de las Viscachas in Santa Cruz in southern Argentina. Fox *Dusicyon griseus* and White-bellied Seedsnipes *Attagis malouinus*.

Right:
Interior of undisturbed *Polylepis* forest at 4300 meters elevation in Quebrada Balcón in the mountains southeast of Abancay in Apurímac, Peru. The flowering creeper is *Salpichroa* sp. Giant Hummingbird *Patagona gigas* chasing White-tufted Sunbeam *Aglaeactis castelnaudii*. Above an Ash-breasted Tit-tyrant *Anairetes alpinus*.

Birds of the High Andes

To Else and Judith

BIRDS OF
THE HIGH ANDES

by

Jon Fjeldså and Niels Krabbe

illustrated by

Jon Fjeldså

A Manual to the Birds of the Temperate Zone
of the Andes and Patagonia, South America

Published by
Zoological Museum,
University of Copenhagen
and
Apollo Books, Svendborg, Denmark

© Copyright 1990 by Jon Fjeldså and Niels Krabbe

All rights reserved. No part of this book may be reproduced or translated in any form, by print, photoprint, microfilm, microfiche or any other means without written permission from the publisher.

Published by:
Zoological Museum, University of Copenhagen
DK-2100 Copenhagen, Denmark

Distributed by:
Apollo Books, Lundbyvej 36
DK-5700 Svendborg, Denmark

Technical editor:
Leif Lyneborg, Fauna Books
DK-3600 Stenløse, Denmark

Colour reproduction:
Lito Gården Fyn A/S
DK-5230 Odense M

Printed in Denmark by:
Nørhaven A/S
DK-8800 Viborg, Denmark

ISBN 87-88757-16-1, standard edition (5000 copies)
ISBN 87-88757-18-8, collector's edition (200 copies)
ISBN 87-88757-22-6, unique edition (26 copies)

Contents

Introduction	9
Acknowledgements	9
Abbreviations	11
The compilation of this book	11
Our field-work	11
Other sources of information	12
Reference of sources	13
Andean natural history	14
Topography	14
Climate	16
Vegetation zones	18
The tropical Andes	18
The subtropical Andes	21
The southern temperate zone	22
Wetlands	22
Pristine or disturbed habitats?	23
Habitat definitions	24
The Andean avifauna	27
The exploration of the Andean avifauna	27
General characterization of landbird communities	29
Conservation and the future	35
Practical hints for field-work in the Andes	38
Equipment	38
Travelling in the High Andes	39
Useful addresses for research and conservation	40
Introduction to the main text	41
How to use the book	41
Which species are included?	41
Completeness	42
Kind of treatment	42
How are the birds classified?	43
The meaning of classification	43
Currently used species concepts	45
Microtaxonomic categories	47
Family accounts	47
Species accounts	47
Names	47
Reference to illustrations	48
Description	48
Habits	49
Voice	49
Breeding	50
Habitat	50
Range and numerical status	51

The maps	51
Note	51
The plates	52
Rheiformes	53
Rheas – Family Rheidae	53
Tinamiformes	55
Tinamous – Family Tinamidae	55
Podicipediformes	65
Grebes – Family Podicipedidae	65
Pelecaniformes	72
Cormorants – Family Phalacrocoracidae	72
Darters – Family Anhingidae	74
Pelicans – Family Pelecanidae	74
Ciconiiformes	75
Herons, egrets and bitterns – Family Ardeidae	75
Storks – Family Ciconiidae	81
Ibises – Family Threskiornithidae	82
Flamingos – Family Phoenicopteridae	84
Falconiformes	88
New world vultures – Family Cathartidae	88
Ospreys – Family Pandionidae	91
Hawks – Family Accipitridae	91
Falcons – Family Falconidae	106
Anseriformes	115
Waterfowl – Family Anatidae	115
Galliformes	135
Cracids – Family Cracidae	135
Pheasants and quail – Family Phasianidae	139
Gruiformes	143
Rails – Family Rallidae	143
Finfoots – Family Heliornithidae	154
Charadriiformes	155
Jacanas – Family Jacanidae	155
Oystercatchers – Family Haematopodidae	155
Stilts and avocets – Family Recurvirostridae	157
Lapwings and plovers – Family Charadriidae	158
Sandpipers – Family Scolopacidae	166
Seedsnipes – Family Thinocoridae	180
Gulls, skimmers and terns – Family Laridae	184
Columbiformes	189
Doves and pigeons – Family Columbidae	189
Psittaciformes	200
Parrots – Family Psittacidae	200
Cuculiformes	217
Cuckoos and anis – Family Cuculidae	217
Strigiformes	221
Barn owls – Family Tytonidae	221
Typical owls – Family Strigidae	222

Caprimulgiformes	232
Oilsbirds – Family Steatornithidae	232
Potoos – Family Nyctibiidae	232
Nightjars – Family Caprimulgidae	233
Apodiformes	237
Swifts – Family Apodidae	237
Hummingbirds – Family Trochilidae	241
Coraciiformes	299
Trogons – Family Trogonidae	299
Kingfishers – Family Alcedinidae	301
Motmots – Family Momotidae	303
Piciformes	304
Puffbirds – Family Bucconidae	304
Barbets – Family Capitonidae	304
Toucans – Family Ramphastidae	305
Woodpeckers – Family Picidae	310
Passeriformes	319
Woodcreepers – Family Dendrocolaptidae	319
Ovenbirds – Family Furnariidae	323
Antbirds – Family Formicariidae	397
Tapaculos – Family Rhinocryptidae	418
Cotingas – Family Cotingidae	444
Plantcutters – Family Phytotomidae	451
Tyrant flycatchers – Family Tyrannidae	453
Larks – Family Alaudidae	527
Swallows and martins – Family Hirundinidae	528
Crows, jays and allies – Family Corvidae	535
Dippers – Family Cinclidae	537
Wrens – Family Troglodytidae	539
Mockingsbirds – Family Mimidae	548
Thrushes – Family Turdidae	552
Waxwings and allies – Family Bombycillidae	563
Wagtails, pipits and allies – Family Motacillidae	563
Gnatcatchers and allies – Family Polioptilidae	566
Vireos and allies – Family Vireonidae	567
Icterids – Family Icteridae	570
Wood-warblers – Family Parulidae	581
Honeycreepers – Family Coerebidae	594
Tanagers – Family Thraupidae	607
Finches – Family Fringillidae	642
Weaverbirds – Family Ploceidae	695
General ornithological literature for the Andean zone	697
Color plates I to LXIV	703
Addenda	846
Index	848
Plants illustrated on the color plates	876
Maps	877

Introduction

The Andean zone of South America has a higher diversity of animal and plant life than any other part of the World. 17% of all the World's species of birds have been recorded within the Andean zone, or only 1.3% of the World's land surface. The humid temperate-elevation forests of the tropical Andes, which cover only 0.2% of the World's land surface, house 6.3% of the World's bird species. However, many habitats are disappearing at an alarming rate. The development of action plans for preservation of the habitats and their enormous genetic resources requires, first of all, scientific studies.

The motive for writing this book was our feeling that the needed upsurge of ornithological activity in the area would not take place unless modern fieldguides became available. Many previous texts were suited mainly for identifying birds 'in the hand' (i.e., in museum collections) but difficult to use in the field due to few illustrations and informations about habits, calls, habitat requirements, etc. Since we started, Colombia has been well covered, but most parts of the Andes still lack a modern treatment. We hope the present book will meet the needs of a genuine fieldguide.

This book does not cover the Andean states as such. Political boundaries have been determined by man and not by nature. Thus, they do not often conform to the limits of faunal areas. We found it more useful to describe a biogeographic region. The book covers the temperate and arctic zones of South America through the Andean Cordilleras and Patagonia, with the exception of strictly coastal and marine species of the southern temperate zone. Altogether 2180 forms, of which 1100 are currently treated as full species, are included; 2200 birds (different taxa and plumages) are illustrated, 2038 of these in color. There are 937 distribution maps. However, forms which are not well established within the zone are described only briefly. This means that the main bulk of the text covers typical high Andean birds.

JF wrote most of the introductory chapters and the text for the non-passerine birds, and made the line-drawings and paintings. NK was responsible for the text for the passerine birds. However, throughout the work we cooperated over the collection of data, and mutually commented on each other's achievements. Despite a wide agreement, some differences in our respective writing styles have been unavoidable.

Acknowledgements

Our efforts could hardly have been successful without the cooperation of many other people, especially those other ornithologists working in the Andes, who never kept their unpublished records secret. (**Abbreviations following persons' names below will be used in the subsequent text**).

For assistance and companionship in the field we thank Peter Arctander, Viviana Babtista, Else Bering, David M Boertmann, John Brandbyge, Esteban and Patricia Brehmer, Steven Cardiff, Tristan J Davis (**TD**), Gustavo Dreyfus, Kirsten Fahnøe, Daniel Gerzon, Skip Glenn (**SG**), Paul Greenfield (**PG**), Robert A Hughes (**RH**), Ole F Jakobsen, Steve James, Andres Johnson (**AJ**), Ole Karsholt, Kenn Kaufmann (**KK**), Judith Krabbe, Juan Carlos Matheus P (**JM**), Benjamin of Nahempaimi, Gary L Nuechterlein, Bosque

Pati, Ole Høegh Post, Mark B Robbins (**MR**), Reyes Rivera A, Donna Schmitt, Thomas S Schulenberg (**TS**), Bruce Sorrie, Francisco and Justo Tueros A, Natividad V Urquiso L, Barry Walker, Jan Erling Wasmuth, and Enrique Zerda.

For help with formalities for the field work we thank Rosario Acero, Francisco Erize, Eliana Flores, Irma Franke, Hernando de Macedo R, Fausto Sarmiento E, and the Danish Foreign Ministery and embassies and consulates in the Andean states.

Throughout the Andes we recieved generous help and hospitality from a great number of people. We are particularly indebted to Helia Macedo M in Lima, the Tueros family of Ondores in Junín, Judy Tennant in La Paz, and Jorge Jaramillo in Quito.

For various kinds of information we thank Hanne Bloch (**HB**), Lene Brøndal, TD, G Egli (sound recordings of Chilean birds), Stig Englund, Eliana Flores, Irma Franke, Ole Frimer, Pierre Devillers, Gary R Graves (**GG**), PG, JW Hardy, Jorge Hernandez C, Steven Hilty, Pedro Hocking, RH, AJ, Michael Kessler, Thomas Læssøe, KK, Hernando de Macedo R, JM, Erik Mølgaard, Fernando Ortiz C, Manuel A Plenge, Michael Køie Poulsen (**MKP**), Carsten Rahbek (**CR**), Jan Fisher Rasmussen (**JFR**), J Van Remsen Jr (**VR**), Robert S Ridgely (**RR**), MR, Ken Rosenberg (**KR**), RA Rowlett, TS, Derek Scott (**DS**), Helmuth Sick, Doug Stotz, Melvin A Traylor Jr., Barry Walker, Brett Whitney (**BW**), David E Willard, and RW Woods. A number of new data were obtained through the headquarters of the International Council of Bird Preservation in Cambridge, where NK worked for 9 months as compilor on the Red Data Book Programme. This allowed him to use all collecting data for museum specimens of red-listed species assembled by Nigel Collar (**NC**).

In particular we thank Theodore A Parker III (**TP**) for an enormous amount of life history data. As he is by far the greatest capacity on the life histories of neotropical birds there ever was, his contribution to our knowledge of neotropical birds can not be stressed strongly enough.

Comments to manuscript drafts were offered by Robert Bleiweiss (hummingbirds), GG, PG, RH, Stuart Hurlburt (flamingos), Hans Meltofte (shorebirds), Manuel Nores (**MN**), TP, RR, Karl-L Schuchmann (**KS**, hummingbirds), TS, Lester L Short (Woodpeckers), and Tjitte de Vries (raptors). Part of the introductory chapters have been commented on by David Boertmann, John Brandbyge, NC, Jürgen Haffer, Hans Meltofte and Jens B Rasmussen. We are especially indepted to Nigel Collar making linguistic corrections to part of the manuscript. Permissions to describe and depict yet unnamed taxa were given by GG, VR, TS and John W Terborgh (**JT**). Jan Haugum made the final preparations of most of the maps.

For courtesy and hospitality during museum visits and with loans of specimens we thank Robert Bleiweiss, Wesley E Lanyon, Mary LeCroy, Lester L Short and others of the American Museum of Natural History in New York (**AMNH**, visited on 7 occasions); Frank B Gill, Mark and Kathy Robbins, and RR of the Academy of Natural Sciences of Philadelphia (**ANSP**, 3 visits); Peter Colston of the British Museum of Natural History (ornithology) in Tring (**BMNH**, 2 visits); Ken Parkes of the Carnegie Museum of Pittsburgh (**CMP**, 1 visit); John Fitzpatrick and David E Willard of the Field Museum of Natural History in Chicago (**FMNH**, 2 visits); JR, TP and wife Carol, TS and wife Kathreen, John O'Neill, TD, KR, and others of the Louisiana State University Museum of Zoology, Baton Rouge (**LSUMZ**, 3 visits); E Costa and P Devillers of L'Institute Royal des Sciences Naturelles in Bruxelles (**IRSNB**, 1 visit); Jorge Navas of Museo Argentino de Ciencias Naturales in Buenos Aires (**MACN**, 2 visits); G Mauersberger of Museum Alexander Humboldt in East Berlin (**MAH**, 2 visits); F Sarmiento of Museo Ecuatoriana de Sciencias Naturales (**MESN**, 2 visits); F Gehringer of the Musée d'Histoire Naturelle Neuchatel (**MHNN**, 1 visit); Irma Franke and Hernando de Macedo R of the Museo de Historia Natural 'Javier Prado' de la UNMSM, Lima (**MJPL**, c. 10 visits); Hernando Romero Z of Museo Nacional de Ciencias Naturales, Bogotá (**MNCN**, 1 visit); Eliana Flores of Museo Nacional de Historia Natural, La Paz (**MNHN LP**, 1 visit); Chr. Erard of Musee National d'Histoire Naturelle de Paris (**NHNP**, 1 visit); S Mathiasson of the Natural History Museum of Gothenburgh (**NHMG**, 1 visit); GF Mees of Rijksmuseum van Natuurlijke Historie in Leiden (**RNH**, 1 visit); JA Dick of the Royal Ontario Museum in Toronto (**ROM**, 1 visit); Carl Edelstam and E Åhlander of the Swedish Museum of Natural History in Stockholm (**SMNH**, 1 visit); GG and Richard Zusi of the United States National Museum of Natural History in Washington, D.C. (**USNM**, 3 visits); Karl-L Schuchmann of the

Zoologischer Forschungsinstitut und Museum Alexander König in Bonn (**MAK**, 1 visit).

The work was possible due to generous funds from the Danish National Science Research Council (grants 511-8136, 11-2250, 11-4143, 11-5958, 11-7073) and from the Carlsberg Foundation, and further the Frank M Chapman Foundation, Bøje Benzons Foundation, the Danish Education Ministry, GEC Gads Foundation, Knud Højgaards Foundation, Queen Margrethe and Prince Henriks Foundation, the Saxild Family Foundation, and William Wetts Foundation.

An interest-free loan from the Aage V Jensen Charity Foundation made it possible for the Zoological Museum, University of Copenhagen, to produce the book and use the income as funds for further field studies in the Andes.

Abbreviations

The following abbreviations are used throughout the species accounts:

ad.	adult (bird or plumage)
acc. to	according to
C	central
c.	*circa* = approximatively
cord.	cordillera
esp.	especially
etc	*et cetera* = and so on
excl.	exclusive(ly)
imm.	immature (bird or plumage)
incl.	including, included
isl(s)	island(s)
juv.	juvenile = a young bird in its first feathers
km	kilometer (**km²** = square kilometer)
mts	mountains
occ.	occasional(ly)
prob.	probabl(e/ly)
rec.	record(s), or recorded
resp.	respective, respectively
s	second, south
sp	species (**spp** in plural form)
ssp	subspecies (**sspp** in plural form, and in the combination **megassp**)
Sta	Santa (in Sta Marta and Sta Cruz)
subad.	subadult
subtrop.	subtropical
temp.	temperate (climate or zone)
trop.	tropical
vs.	*versus* = compared to, against

Directions are abbreviated as **e** (east), **n** (north), **s** (south), and **w** (west), or as combinations, **nw** (northwest), etc. These can be combined with **c** (central), e.g., **wc** (western central). Also used in combinations, as **ewards** (east-wards) etc. Capital letter is used when part of a geographical name (**C Andes** = the Central Andean Cordillera; **nC Andes** = northern Central Andes).

Frequently mentioned names of geographic areas are abbreviated, viz. **N, C,** and **S Am.** (North, Central, and South America, but The Americas spelled out), **Arg.** (Argentina), **Bol.** (Bolivia), **Col.** (Colombia), **Ecu.** (Ecuador), **Parag.** (Paraguay), **Urug.** (Uruguay), and **Ven.** (Venezuela), but the short country names Chile and Peru are spelled out.

The months of the year are abbreviated with three letters: **Jan** (January), **Feb** (February) etc., but June and July are spelled out.

Abbreviations of persons' names and institutions are mentioned in the Acknowledgements. The names of the present authors are abbreviated **JF** and **NK** in the text.

Measurements are metric, usually **cm** (centimeters).

In comparative remarks on very similar species, the group name is usually abbreviated (e.g., Andean Tinamou vs. Ornate **T.**).

The Compilation of this Book

OUR FIELD-WORK

The background for writing this book was altogether 28 months of intensive field studies in the Andes by NK, 17 months by JF.

JF made detailed biological studies of grebes in 1977/78, 1981, and Feb 1984. A number of other species, notably waterbirds, were also covered well during these studies, the synthesis of which are published by Fjeldså (1981, 1985) and in papers referred to in the chapters on grebes and rails; other data were incorporated in Scott & Carbonell (1986). NK made a general birding trip in 1978/79.

The initial planning of this book was made in 1982. We used the next expedition (1983/84, and a smaller trip in 1985) almost exclusively for improving our personal field experience, and for making sound recordings etc of as many species as possible. Some records from this study were compiled in Fjeldså & Krabbe (1986). We also tried to identify neglected problems and needs for conservation management, and decided to pay special attention to the relict high Andean woodlands. On a 1987 expedition, with ornithologists, botanists, and an entomologist, we visited numerous such forests in Peru and Bolivia, to increase the general knowledge of these habitats. We now analyze the evolution and biogeography of this fauna in order to lay a sound scientific basis for future conservation management (Fjeldså 1987). During this expedition we also obtained data of the same kind as in 1983/84, NK paying special attention to tapaculos.

Localities visited by us are indicated on the map.

OTHER SOURCES OF INFORMATION

The main sources used in the early phase of planning were existing manuals (e.g. Blake 1977, Meyer de Schauensee 1966, 1970, as well as the *Catalogues of Birds in the British Museum* by RB Sharpe and others, *Catalogue of Birds of the Americas* (Nos.1-15, 1918-49) by CE Hellmayr (CB Cory and B Conover in part) and *Check-List of Birds of the World* (16 vols., 1931-87) by JL Peters and others. On this basis we made the first tentative lists of species for inclusion, compiled basic data about identification and distribution, and defined problems worth closer study. This was gradually supplemented with extensive surveys of the literature.

Much of the older literature rests on studies by group specialists and biogeographic compilers of the large amounts of 19th century trade skins and of specimens taken by collecting expeditions in

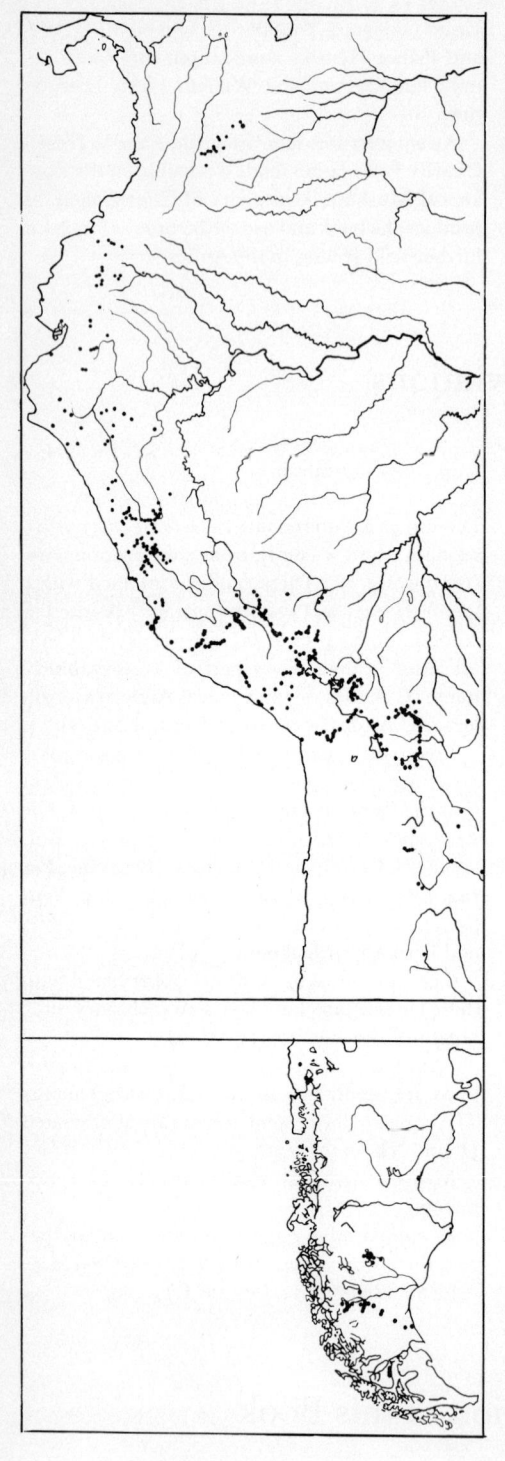

Map showing our own study sites in western South America.

the early part of our century, now lodged in the major museums (see The History of Andean Ornithology). Some books give diagnoses for identification, but are less suited for field use. The reason is the novelty of the description of habits, vocalizations etc., or simply that the compilers mainly had seen museum specimens. A few books, like Koepcke (1969) and some very recent works (especially Hilty & Brown 1986) are exceptions, but these publications cover only certain parts of the Andes.

Owing to these shortcomings of the past literature, we payed more attention to practical field identification, habits, calls etc., and on feeding habits, habitat requirements etc. This was done both through our fieldwork, and in discussion with other people working in the Andes.

In order to understand plumage variation (both geographic variation and plumage sequences), we performed additional, selective specimen collection. We also re-examined the collections on which previous writers had based their views, as well as the important recent collections by J Fairon, GG, P Hocking, John P O'Neill, T Orellano R, TP, VR, H Romero Z, TS, JT, F Vuilleumier, JS Weske, and others (museum visits specified under Acknowledgements). We thus checked, at least superficially, over 100,000 bird specimens, and studied the variation in some species in detail. The color plates were painted on basis of specimens kept in the Zoological Museum University of Copenhagen; those lacking there were borrowed, or painted during a visit in 1985 to museums in the US.

REFERENCE OF SOURCES

When we started to work on this book, we had in mind an ordinary bird-watcher's fieldguide without references. Even when the manuscript increased beyond this level, we felt that putting references into the main text would make it unreadable. Furthermore, it would often be hard to define the source. Most life history accounts are in fact mixtures of our own experiences, published data and commentaries from readers of manuscript drafts. The spellings of vocalizations may be compromises between our own and other independent attempts. It seems hopeless (without greatly enlarging the text) to specify how different information contributed to the final concise wording.

On the other hand, scientist readers would certainly appreciate statements about our sources of evidence. Such information reduces the risks of introducing errors or misinterpretations into the ornithological literature. In case of unpublished data, it would assure credit to the observers. We therefore finally included some references in the main text, but only in case of significant, unpublished data, or very recently published data not incorporated in other handbooks.

Reference to published information is given in three different ways:

A **basic reference list** gives some main continent-wide or national, or regional compilations, papers describing local faunas, and biogeographic syntheses. We regard any information from publications in this list as 'basic knowledge' that is not usually referred further, except in the introductory chapters. The basic reference list is found on p. 697.

Literature covering individual species or families is compiled at the end of the family accounts, and only rarely elsewhere in the text. We limit ourselves to significant recent contributions, especialy those not already incorporated in other regional handbooks. Systematic revisions are only referred to if they include a broad analysis, or are more recent than the family revision in Peters' ***Check-List of Birds of the World***. These references (and a few given in the text) are ultra-short, with only the data needed for locating the source through a research library: abbreviated journal name, volume, year, and pages. These lists will give students of a particular family or smaller group a key to the recent writings over this group, which in turn refer to older sources. Under the species account we refer the original description of taxa published after the appearance of Peters' Check-list.

It can normally be assumed that information not found through the literature cited are our own. As far as distributional data and taxonomy are concerned, many of our records have now (or will soon be) published elsewhere. Unpublished information received from other people are in most cases adjustments or specifications of e.g. behavior or habitat, and we generally restrict ourselves to thanking these persons in the Acknowledgements. However, in cases where an unpublished taxon is referred to, or where rather noteworthy unpublished records are mentioned, we give the informers' name or initials directly in the text (abbreviations: see under Acknowledgements).

Andean Natural History

TOPOGRAPHY

The Andes (**Cordilleras de los Andes**) comprise the mountain chains along the west coast of South America, from the Caribbean Sea to Cape Horn. Geologically, they are connected with the large mountain systems of western North America. Both systems were folded in recent geological times. The uplift of the Patagonian part had started in the early tertiary, but the northernmost massif (Santa Marta mountains) were uplifted during the last 1 million years. Volcanic activity and earthquakes reveal that the upheaval has not yet come to an end.

The Andes span 10,000 km and cover altogether 1.8 million square kilometers representing the most impressive magnitudinal grandeur and diversity of sceneries. In fact, several more or less distinct parallel ranges (**cordilleras**) are involved, and the main ridge is rarely below 2000 m, and exceeds 5000 m locally. The highest peaks in the Illampu massif in La Paz in Bolivia, and Aconcagua in central Chile, exceed 7000 m. However, steep slopes, gorges, and wild peaks characterize only part of the zone, other parts being rolling tablelands and flat plains (**altiplanos**), mainly around 4000 m.

The snowline is at 4500-5000 m near the Equator, rising to above 5000 m at 15 degrees south, and then dropping to near 1000 m furthest south in the continent.

The northern parts of the Andes are fairly low, with a number of higher mountain massifs, the **páramos**. In this part, then, the alpine habitat is broken up into smaller areas separated by passes of subtropical habitat.

The south of Venezuela and the eastern half of Colombia are lowlands, with large savannas in the Orinoco basin, and with 'seas' of tropical forest stretching south across the Amazon basin. The low areas isolate the Andean zone from the broken-up and old **Tepui** mountains of the Guiana shield. Low cordilleras follow the coast of Venezuela, and form higher páramos (from Lara through the high Mérida mountains to Táchira, with the highest peak slightly over 5000 m). This eastern branch of the Andes and the low Perijá mountains of the northern Colombian/Venezuelan frontier section merge into the East Andes of Colombia. Near the border to Colombia, the **Táchira depression** forms a 40 km gap down to 8-900 m elevation, with a dry subtropical climate (see Vuilleumier 1984). **Sierra Nevada de Santa Marta** lies isolated near the Caribbean coast of Colombia, rising like a large pyramid from the lowland plains to nearly 5800 m. However, the montane habitat is continuous in the western half of Colombia, which has three cordilleras, the **East**, **Central**, and **West Andes**, merging into a single range near the border to Ecuador. These cordilleras are separated by two large valleys draining north: the broad tropical **Magdalena Valley** between the East and Central Andes, and the narrower and less deep **Cauca Valley** between the Central and West Andes. The East Andes slope steeply to the east, but is wide (100-200 km over most of its length), with plateaus and gently rolling hills at 2500-3500 m, and a few higher massifs with snowfields, as Sierra Nevada del Cocuy (5493 m). The **Macarena** mountains, which rise to 1800 m shortly east of this cordillera, is an erosion remnant that is geologically isolated from the Andes. The Central Andes has a prominent ridge line at 3000-3500 m and páramos and nevados mainly in the Caldas-Quindío region (5399 m) and Nevado de Huila (5750 m). The Western Andes is the lowest range, with the ridge mainly at 2000 m and only 4-5 peaks above 4000 m, little páramo, and no snowcaps.

The Andes of southern Colombia (Nariño) and Ecuador comprise two long volcanic ranges connected by some transverse ridges (nudos), dividing the tablelands into a number of broad interandine valleys (**Central Valley** system). Some of the valley segments drain west (Río Mira, Guaillabamba, Chimbo), others east (Pastaza, Paute, Zamora), causing deep gaps in the cordilleras. Thus, a number of larger and smaller páramos are formed: Páramo de Angel, Imbabura, Cayambe-Antisana-Cotopaxi massif, Pichincha-Atacazo-Illiniza-Chimborazo páramos, and Tungurahua-Sangay massif, with some volcanoes rising to over 5000 m. The southern parts are generally lower.

A major gap in the highland habitats exist in Cajamarca in northwest Peru. This disjunction, the **North Peru Low**, has two components: the low valley of the Amazon tributary Marañón forms a broad gap into the Andes, and turns south and deeply separates the Central and West Cordilleras (**Marañón valley**) and sends another branch north (**Huancabamba**); the **Porculla pass** in the West Andes is at 2000 m (Vuilleumier 1984).

Also south of the North Peru Low the West Andes is broken up into several massifs, but the highland habitat becomes continuous from southern Cajamarca, with pictoresque snow-capped peaks to well above 6000 m in **Cordilleras Blanca**, **Huayhuash**, and **Raura**. The Central Andes are continuous, but with a rather low ridge, from Amazonas to Huánuco. In Junín in central Peru the cordilleras encircle the altiplano of Lake Junín, and from here and south to Aconcagua in central Chile the highland is continuous, without a single pass below 4000 m. The highlands of central Peru are intersected from the east by several deep zigzagging valleys, and several low foothill ranges run parallel in the inner Amazon area. From Cuzco to Central Bolivia the East Andes (**Cordillera Real**) forms a continuous chain interrupted only by two deep canyons in La Paz, and with a tremendous altitudinal gradient towards the lowlands (the **Yungas**). The highest parts are clearly shaped by glaciation, with alpine habitats locally, and with u-shaped or cirque-shaped valleys with small lakes. The largest peaks, in the Illampu massif in La Paz, have been measured to more than 7000 m. The area near the junction of Bolivia, Chile and Peru in the West Andes has several large volcanoes (to 6542 m). However, most of the middle Andes, between the cordilleras, is morraine slopes, rolling hills and plains, including the enormous **Peruvian/Bolivian Altiplano** stretching from right north of Lake Titicaca across western Bolivia. The Andes reaches its maximum width of 920 km at 19 degrees south (between Madeira and Pilcomayo).

The Altiplano is an area of inner drainage with enormous lake basins. The whole plain was once covered by one huge lake, 'Ballivian', now fragmented into the smaller present lakes: In the north the 8,300 sq km large deep Lake Titicaca, draining south into the saline Lake Poopó, and in the southern part of the altiplano enormous salt-plains, flooded in the rainy season. The large lakes Uyuni and Coipasa cover 10,000 and 2,220 sq km, respectively.

The East Andes ends in the Tunari range in

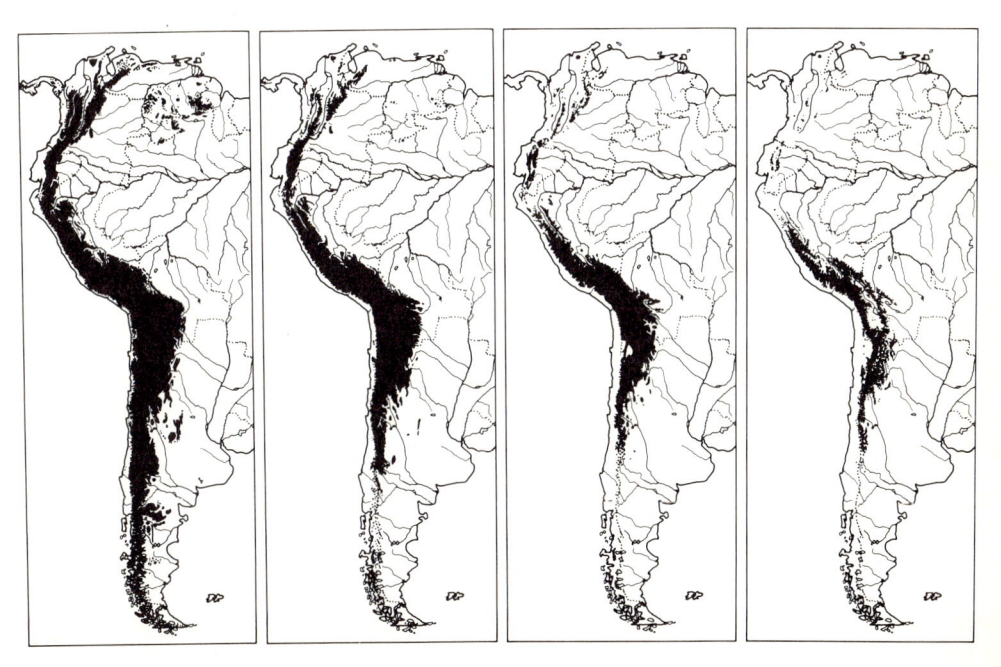

The topography of western South America.
From the left, land over 1000, 2000, 3000, and 4000 m elevation, respectively.

Cochabamba, Bolivia. South of this range, deep valleys penetrate the east Andean slope (the **Valles**), causing a somewhat irregular edge of the highland. From Tucumán to La Rioja, Sierras de Aconquija and Famatina form projecting branches of the Andes, and Sierras de Velasco, Valle Fértil, la Huerta, Pie de Palo, San Luis, and Córdoba form small isolated 'punas' surrounded by montane shrub and low, mainly closed basins (**bolsones**) east of the Andes in northern Argentina. **Sierras de Córdoba** (2880 m) are isolated as much as 350 km from the Andes.

The cordilleras narrow near the Aconcagua massif at 32 degrees south, and the southern Andes (**Patagonian Cordillera**) forms a narrow ribbon of mountains, which includes the wildest peaks in the Continent (e.g., Cerro de Fitzroy, Torres de Paine). The mountains are intersected by deep u-shaped valleys, with narrow, ribbon-like lakes (especially the 'lake district' at 39-42 degrees south, and adjacent to the two large ice-caps that cover the cordillera between 46 and 51 degrees south).

Through most of Chile, a lowland area (**Central Valley**) separates a low **Coastal Cordillera** from the Andes. South of Puerto Montt the Central Valley sinks below the ocean, and through southern Chile outcrops of the coastal range form myriads of offshore islands, starting with Chiloé and continuing to Cape Horn.

Most of southern Argentina is undulating, with some large plateaus and lowland plains especially in Sta Cruz. The plains are strewn with shallow claypan lakes formed by deflation of soil and sand. Near the base of the Andes, wide flat valleys separate basaltic foothill plateaus of c. 1000 m elevation. These 'mesetas' are densely dotted by small lakes in rift valleys and in circular craters ('maars').

Often, the entire southern South America south of Río Maule in central Chile and Río Colorado in central Argentina is referred to as **Patagonia**. For convenience we use the word **Patagonian plains** for the plains and foothills east of the Andes, and call the western part the **South Andean zone**.

CLIMATE

The life zones of the Andes were defined by Chapman (1917, 1926), and most later biologists maintain his subdivision. There are no rigid limits to his zones, however, as they intergrade and are modified by local climates. In humid zones near the Equator, the temperate zone normally means the stunted and epiphyt-clad woodlands that form the transition from the tall premontane forest to the barren highlands. In Venezuela, the temperate zone is from 2300-2600 to 2900-3500 m, lowest in the rainy eastern parts. The lower limit is normally at 2500 m throughout the Andes from Colombia to Bolivia, but is lower in western Peru and in very rainy areas, and slightly higher in some dry valleys. From northern Argentina, the limit of the zone falls gradually, to reach sea level in Patagonia. We do not follow the diagrams by G Mann (pp. 180-181 in Fittkau et al. 1968), which sets the limit between the warm and cold zones at 2000 m, and at 2500 m on the eastern slope from northern Ecuador to northwestern Argentina.

The Andes is a zone of climatic extremes (see review by G Sarmiento in Vuilleumier & Monasterio 1986). Theoretically, the temperature should fall 0.77 degrees Celsius per 100 m ascent, but the topography produces other climatic changes that locally, and at different times of the day, cause marked differences over few hundred meters distance. The elevational gradients alter prevailing atmospheric flows, and cause elevated heat sources (e.g., over the high plains and in rain-shadow valleys; by their inertia cold airmasses 'falling' into these valleys tend to fall past the equilibrium level and thus become warm and dry, rise, and cause even more drying). Also the vegetation is an important regional factor, as the vast Amazonian and montane forests loose moisture through evaporation. This creates daily cycles of cooling, with formation of fog and afternoon rain as the air rises up the Andean slopes.

The local Andean climates are influenced by those of the surrounding lowlands. The Caribbean coasts of South America are affected by a southwards shift of the northern trade winds from mid-December to mid-April. This produces a severe dry season. Pacific Colombia is very different, as the dry winds give way southwards to a hot humid, non-seasonal and very rainy climate (locally with well above 10,000 mm precipitation in the premontane zone). Further south the cold Peruvian current, a branch of the Antarctic current, makes the rain fall far offshore. A carpet of mists covering this coastal zone April-October is the only source of moisture, except in the few years when the Equatorial Current shifts south (el Niño years). The desert is baking hot in the sunny period November-March. The eastern lowlands are ever-humid, and this climate influences the steep eastern Andean slopes from Venezuela to the Yungas of Bolivia.

The climates are also modified by the movements of the sun and the general atmospheric circulation of the tropics. This causes seasonal differences in rainfall. In the northern Andes, the rain falls mainly in mid-March to mid-June and October to mid-December, with the longer 'dry season' from June through September. Regionally, the topography and vegetation drastically alters this shift, and the eastern Andean slope of Colombia only has one marked peak of rainfall (March-August). From Popayán southwards through Ecuador, Peru, and Bolivia there are two seasons, a rainy season November-April, and a dry southern winter. This pattern is modified locally.

Depending on the topography, the rainfall varies from 800 to 4000 mm through the highlands of the tropical Andes. Thus, there may be much rain on the high parts of the mountain slopes, but rarely on the hot dry valley floors below. In the Bogotá area (2700 m) in the East Andes of Colombia, high rainfall characterises the hills right above the city, while the outer edge of the Bogotá plateau towards the Magdalena valley is semi-desert. Right below the edge there is humid montane forest, but the bottom of the Magdalena Valley has a dry tropical climate.

Certain ridges block incoming moisture-filled clouds, causing local 'rain shadows'. This is seen especially in valleys that zigzag between sharp mountain ranges of the rain-drenched slopes. This causes gaps in the humid treeline habitats in the upper parts of the Río Sucio, Dagua, and Patía valleys in western Colombia; Chota, Guayllabamba, Pastaza, Chanchán, Guamote, and Paute river valleys in Ecuador; the North Peru Low and Marañón Valley, the upper Huallaga valley, Huánuco, the Mantaro-Apurímac, Urubamba and Paucartambo systems of central Peru, the headwaters of Imbaburi, Puno, and Mapíri and La Paz canyons in the East Andes of Bolivia.

While the precipitation follows a clear seasonal pattern, the mean temperatures vary little (usually 1-2 degrees, even as far south as northern Bolivia). This is above all due to the slight (13%) seasonal variation in the daily solar radiation near the Ecuator. Thus, on high elevations, the temperature changes more in the course of the day than between months. One may talk of day-summer and night-winter conditions. On the altiplanos of Peru and Bolivia, the dry season has warm midday sunshine but strong nightly frost, the day/night amplitude reaching 40-50 degrees Celsius in some places. The rainy season may start with violent afternoon hailstorms, but the peak of this season is characterized by cool and cloudy weather with only occasional frosts. Despite the nightly frost, high Andean lakes (north of 15 degrees south) are constantly icefree, usually with temperatures of 9-12 degrees Celsius. By day, they are heated well by intense solar radiation, and the water conserves the heat overnight.

The Tunari range in Cochabamba, Bolivia, causes a large rain shadow area from central Cochabamba and south to the Argentinean frontier. In this area many valleys have subtropical heat up to 2300-2600 m. Only a few ridges have frequent mists.

Further south, the climate becomes increasingly dry all the way across the continent. Thus an arid diagonal stretches from the Peruvian coastal desert across the southern part of the altiplanos and south to the Patagonian plains. Within this zone, the highlands have from 350 to less than 100 mm annual precipitation. Rain mainly falls October-April, but is everywhere unpredictable, and most wetlands in this zone dry out in certain years. In winter, lakes freeze regularly, except in places with hot water springs. However, a zone of seasonally humid climate is formed where the air rises from the Argentinean chaco up the steep premontane slopes in southernmost Bolivia and northwestern Argentina.

On the west slope of northern Chile, rain falls only once or twice a year (in certain places virtually never). However, the transition in central Chile to the humid southern parts has a pleasant 'mediterranean' climate with 500-1000 mm rain, mostly at high elevations.

The southern part of the continent is under the influence of the subantarctic western winds. Swept by persistent gales, the zones south of 40 degrees south are referred to as the 'roaring forties' and 'thundering fifties'. The west coast has a typical oceanic climate, with rain all year round (over 3000 mm locally in southern Chile). The summer is cool, the winter mild, but with occasional sleet and snow. After passing the glaciers and peaks of the southern Andes, the storms cause an extremely inhospitable climate of cold winds. Some snow may fall at any season, and large areas have permanent snow in winter. The yearly precipitation is 100-150 mm. Only the deep, sheltered valleys through the Andes and between the eastern foothills have a pleasant summer.

VEGETATION ZONES

The major vegetational formations correspond closely to prevailing climates. Discrepancies are mainly caused by human activity which today prevents development of climactic vegetation in many parts.

The plant list at the end of the book functions, by using the references to plants shown on the color plates, as an identification guide to some plants often mentioned in the text.

THE TROPICAL ANDES

Detailed descriptions of the vegetation of the tropical Andean zone can be found in Cabrera, AL (1968) *Ecologia vegetal de la puna*. Colloq. Geogr. Bonn 9:91-116; Cleef, AM (1981) *The vegetation of the páramos of the Colombian Cordillera Oriental*. Diss. Bot. 61; Garcia B, et al. (1966-79) *Catalogo illustrado de las plantas de Cundinamarca*. 7 vols. Bogotá; Diels, L (1937) *Beitrag zur Kenntnis der Vegetation und Flora von Ecuador*. Stuttgart; Gentry, AH (1980) The flora of Peru: A conspectus. *Fieldiana, Bot. New Ser.* 5:1-73; Hueck, K (1966) *Die Wälder Südamerikas*. Stuttgart; Lasser, T (1964) *Flora de Venezuela*. Inst. Bot. Caracas; Monasterio, M (ed.) (1980) Estudios ecológicos en los páramos andinos, Mérida, Venezuela. Univ. de los Andes; Vareschi, V (1970) *Flora de los paramos de Venezuela*. Mérida; Weberbauer, A (1945) *El Mundo vegetal de los Andes Peruanos*. Lima. See also several papers in Vuilleumier & Monasterio (1986).

THE HUMID MONTANE SLOPES: Humid tropical forests surround the Andean foothills in Venezuela, Colombia, and northwestern Ecuador, and cover the Amazon lowlands and adjacent foothill ranges of eastern Ecuador and Peru to northwestern Bolivia (interrupted by savannas in Venezuela and adjacent Colombia and locally in the Amazon basin). The adjacent Andean slopes are covered by **humid forests** varying from moderately to very wet types, which, according to temperature, can be classified as **subtropical (premontane)** and **temperate (montane)**. The upper parts are often referred to as **ceja de selva** ('eyebrows of the jungle').

The vegetation is generally mature with little clearing by man (except in parts of Ecuador and Colombia). The generally low level of human activity is caused by the unpleasantly misty and wet climate, the tremendous topography and impenetrable vegetation. In the premontane zone, most trees are tall and canopied. The understory is cluttered, and landslides and windfall gaps quickly fill with mountain bamboo (*Chusquea*) (unlike the dominance of large-leaved monocots in disturbed habitats of the tropical zone). Tree ferns are common. The vegetation changes aspect as we pass into the montane zone. While the trees lower down are fast-growing (e.g., tall *Cecropias*) with relatively few epiphytes, often just a thin, slimy layer of algae and moss on the trunks, they now become shorter, gnarled, and heavily burdened by spongy mosses, bromeliads, ferns etc. Dominant trees in the lower temperate zone are Lauraceae, Moraceae, and Myrtaceae. Oaks (*Quercus humboldti* and *colombiana*) are locally common in Colombia. Common trees in the higher zone are *Clethra, Clusia, Drimys, Hesperomeles, Miconia, Myrica, Persea, Schefflera, Tecoma, Trema, Vallea*, and *Weinmannia*, and a few areas have forest dominated by the conifer *Podocarpus*. *Rubus* brambles and vines with tubular hummingbird flowers (*Bomarea, Passiflora*) are common near edges. Near the treeline, the vegetation becomes even more cluttered, but with some larger, flat-canopied trees, the whole vegetation clad with ferns and white or black lichens. Common trees here are *Escallonia, Hedyasmum, Gynoxys, Polylepis*, and *Weinmannia*. Open glades often have dense brush (*Baccharis, Brachyotum, Senecio, Vaccinium, Viburnum*), and the ground is spongy, with thick layers of mosses and ferns.

THE ARID MOUNTAIN SLOPES: Desert scrub and dry forest, thorny woodlands etc. cover the lowlands near the Caribbean coast, in the broad Magdalena valley, in western Ecuador, and northwestern Peru, the arid floodplain at the Marañón bend, and the chaco of eastern Bolivia. Leguminaceous trees (*Acacia, Prosopis, Mimosa, Inga*) and large cacti are also important. However, most of the original semi-deciduous forests have been destroyed now, except in the chaco.

The coastal deserts of Peru and northern Chile are almost totally devoid of vegetation. The dull pinkish gray landscapes are broken only by green strips of dense riparian vegetation along the few rivers that bring water from the Andes. This vegetation is mainly scrub, pepper trees (*Schinus molle*), willows (*Salix chilensis*), *Parkinsonia aculeata*, leguminaceous trees, and the giant grass ('caña brava') *Gynerium sagittatum*. Today, virtually all level parts in these valleys are cultivated, and additional fields are formed by irrigation. Due to precipitation from fog, some vegetation (**loma formation**) of annuals, shrubs, and scattered, *Tillandsia*-clad trees and columnar cacti develop on the mountain slopes around 1000 m. This vegetation is very rich in endemic species. The slopes towards the lower temperate zone have some low scrub of *Jatropa*, scattered trees of the thick large-leaved *Carica candicans*, and dispersed 'woods' of columnar cacti. As we pass into the temperate zone, the landscape becomes greener. In addition to typical xerophytes, mesophytic shrub becomes increasingly abundant (especially *Eupatorium*), and in the rainy season the ground is covered by short grass and annuals. In some areas, taller herbs, bush-like lupins (*Lupinus*), and *Mutisia* bushes bloom densely, and some areas have thorny thickets (*Cassia, Colletia, Hesperomeles*). In places with frequent mists at 2400-3300 m, there are woodlands (**algarrobal formation**) of *Eugenia* and *Oreopanax*. However, these habitats are strongly degraded, and remain only locally as small patches, mainly in northwestern Peru, but with scattered patches south to Arequipa.

Less extremely drought-influenced formations exist in the tablelands of the East Andes of Colombia, in the tablelands of Nariño and Ecuador, and the Andean valleys of Peru and northern Bolivia. The temperate zone of these valleys (at 2500-3500 m) has been densely populated for millenia, and human activity is an obvious determinant of the vegetation. Slopes which are not presently cultivated are often terraced as a sign of former cultures (e.g. Limbani and Sandia valleys of Puno Peru), and the usual vegetation is low mesophytic shrub. The low bush *Dodonea viscosa* covers large areas, and other characteristic plants are bushy *Eupatorium*, agaves, cacti, euphorbiads, and the introduced *Spartium junceum*. Pepper-trees (*Schinus molle*) and thorny leguminaceous trees form a dense riparian vegetation. The small fields are often separated by hedgerows and small stands of *Schinus* and introduced *Eucalyptus*, and the ravines often have some coppice with *Alnus acuminata*. However, steep terrain of difficult access may have dense scrub, often thorny bushes (*Cassia, Colletia, Barnadesia*) or trees (balsas as *Ochroma* and *Bombax lagopus*, and *Escallonia* in high parts). A few relict patches of true cloud forest can sometimes be seen on the ridges 1000-1500 m above the valley bottoms. Such woodland habitats are still well developed in the large valleys of Ayacucho and Apurímac, central Peru.

PARAMO and PUNA: The timberline, at 3000-3400 m (higher in Peru), gives way to grasslands. These are sometimes collectively referred to as **páramo** (which means cold and desolate tracts), but we use this term in the narrow sense to cover habitats above the treeline in the more humid parts of the northern Andes (Venezuela, Colombia, Ecuador). Along the eastern slopes of north and central Peru the corresponding vegetation is called **jalca**.

The humid lower páramo vegetation is mire-like, with spongy and thick layers of mosses and lichens, coarse bunchgrass, and some short bamboo (*Swallenochloa*), sedges (*Carex, Luzula*), ferns (*Blechnum, Pteridium*), heather-like small bushes (*Baccharis, Befaria, Brachyotum, Cavendishia, Hypericum, Libonathamus, Pernettya, Senecio, Vaccinium*), some columnar ferns, and *Puya* bromeliads. Acid bogs are common, and some areas have richer fen-like *Werneria* bogs. There are occasional patches of stunted, mossclad wood (*Buddleia incana, Escallonia, Gynoxys, Polylepis*). The higher parts become very barren, often stony 'arctic' desert with scattered cushion and rosette plants (*Mona, Arenaria, Azorella, Calandrina, Draba, Senecio*). *Espeletia*s are huge caulescent rosette plants that give a distinctive physiognomic appearance to many such areas in Venezuela and Colombia.

The grasslands of the drier plateaus of Peru to northwest Argentina we call **puna**. This habitat is dominated by grasses, especially tussocky bunchgrass (*Agrostis, Calamagrostis*, and *Festuca*) and short and matted *Aciacne* filling gaps between the tussocks.

The puna differs from páramo and jalca by having a dry season with strong nightly frost. Yet the rolling grassy terrain of the Peruvian Andes and east of the Bolivian altiplano is called **humid puna**, to contrast the semi-deserts to the south and southwest, called **arid puna**. The humid puna is dominated by uniform bunchgrass vegetation, above all *Stipa ichu* and *Festuca dolichophylla*. Karst terrain and rocky slopes have some heather-like composite brush, as *Baccharis* and the orange-flowered, thorny *Chuquiragua spinosa*. Drier areas have scattered grass, low thorny scrubs (e.g. *Margarocarpus, Tetraglochin*) and the heather-like bush *Lepidophyllum quadrangulare*.

The *Polylepis* woods, which typically fringe streams or form small patches in gorges and on rocky slopes and cliff-ledges, vary from dense, humid types with lichens, many mistletoes, and with rich soil and a varied undergrowth, to scattered shrub on almost bare desert soil.

In very high parts, the vegetation is short, densely matted in the boggy parts, adapted to withstand radiation, drought, wind and severe frost. Many plants are closely related to well-known lowland forms of the northern temperate zones, but exceedingly different by aspect. Many

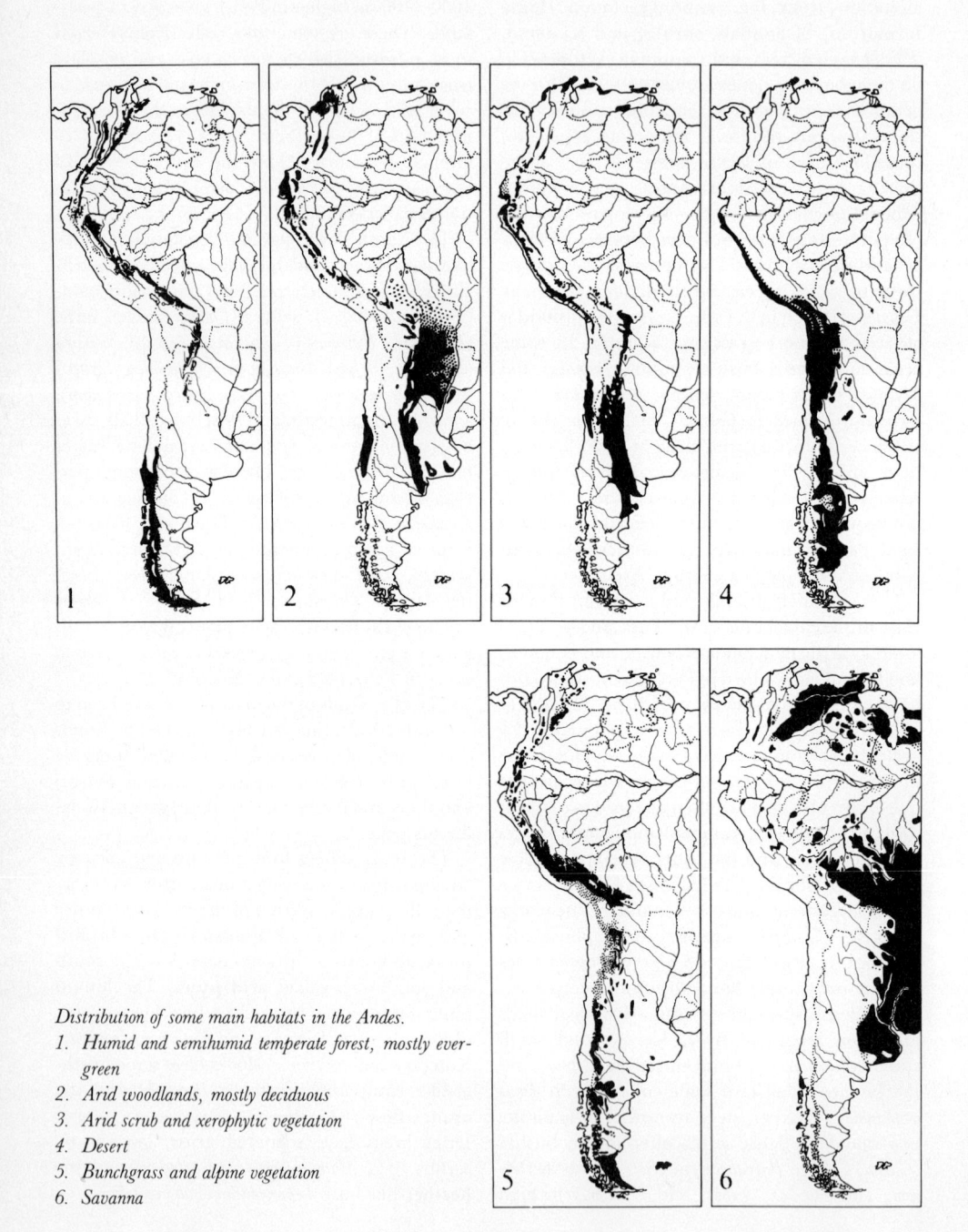

Distribution of some main habitats in the Andes.
1. *Humid and semihumid temperate forest, mostly evergreen*
2. *Arid woodlands, mostly deciduous*
3. *Arid scrub and xerophytic vegetation*
4. *Desert*
5. *Bunchgrass and alpine vegetation*
6. *Savanna*

have rosettes of leaves compressed on the ground, and very deep 'tap roots'. Many plants have beautiful flowers, almost placed on the ground (*Acaulimalva, Gentiana, Hypochoeris, Liabum, Werneria*). Others form compact and often remarkable cushions (*Aciachne pulvinata, Azorella, Larretia compacta, Pycnophyllum*, and woolly-haired cushion-cacti *Opuntia floccosa* and *lagopus*).

The flat bottoms of the glacial valleys often have bogs with a densely matted vegetation of compact cushion-plants (*Plantago rigida, Distichia muscoides*). Above 4600-4800 m the slopes are bare and rocky with very scattered plants, excepting patches of bog in waterlogged places.

THE SUBTROPICAL ANDES

Principal botanical sources are Adolfo, H (1962) *Plantas del valle de Cochabamba*. 2 vols. Cochabamba; Cabrera, A (1957) La Vegetatión de la Puna Argentina. *Rev. Invest. Agric. Buenos Aires* 11:317-512; Cabrera, A ed. (1972-) *Flora de la Provincia de Jujuy, Republica Argentina*. Buenos Aires; Herzog, T (1923) *Die Pflanzenwelt der bolivischen Anden und ihres östliche Vorlandes*. Leipzig; Hueck, K (1966) *Die Wälder Südamerikas*. Stuttgart.

This part of the Andes is surrounded mainly by arid or semi-arid lowlands:

The Pacific slope has lifeless desert. The temperate zone of the north Chilean mountains has very little scrub and herbs, as the sterile subtropical desert usually grades directly into true alpine habitat. The Atacama desert grades southwards through the **pampa del tamarugal** (with *Prosopis tamarugo*) into the deciduous '**matoral**' scrub in the Central Valley of mid-Chile.

The lowlands east of the Andes are known as the **chaco**. This is a mixture of more or less xerophilous forest and savanna, with many halophytic and swamp-like vegetation formations. In the southern parts, deciduous thickets grade into the subdesert scrublands of Patagonia.

The east Andean slope south of the Tunari range has mainly semi-arid wood and scrub, with aspect of savanna in some valleys. *Prosopis* woodlands cover large areas of rolling hills, but other slopes have deciduous woodlands with large trees. Such zones alternate with small punas, but certain high slopes have patches of humidity-demanding wood (*Alnus acuminata, Podocarpus, Vallea stipularis, Weinmannia*, bamboo in deep ravines, and *Polylepis* near 4000 m).

Furthest south in Bolivia and in northwestern Argentina the premontane and lower temperate zone has humid semi-deciduous forest and some evergreen forest with aspect of cloud forest highest up (**Tucumán-Bolivian forest**). A few relict forest patches occur as far south as in the arid zone of Catamarca.

The transition from the wooded zone to alpine habitat is called the **prepuna**. This comprises large desolate 'badlands' of thick masses of finely withered rock material, eroded by casual rainshovers, and lacking vegetation. Other parts are bush-steppes with composites (*Gochnatia glutinosa* or *Aphyllocladus spartioides*), leguminaceous bushes (*Cassia crassiramea*), and 'woods' of large columnar cacti (*Tephrocactus pasacana* and *terschekii*), and occasional patches of *Polylepis tomentella* and *australis* in rocky parts. Some *Salix chilensis* fringes the few streams. A few *Prosopis* woods lie isolated in central Potosí.

The high plateaus of the **arid puna** are characterized by the silence of the desert, blazing sun, and strong nightly frost, and a very sparse and low vegetation. As in the prepuna, many areas have been eroded and withered for millenia by frost and thawing, and are covered by fine rock debris. A rather closed short vegetation fringes the wells at the foot of the mountain ranges and around the saltpans ('**bofedales**') of the areas of inner drainage. In the less extreme northern parts of the arid puna, enormous plains are heath-like with scattered bunchgrass and up to meter-tall brush, mainly the heather-like composite *Lepidophyllum quadrangulare* ('**tola**' or '**thola**'). Rocky places have some low spiny scrubs (*Adesmia* species, *Fabiana densa, Prosopis ferox* and *Tetraglochin cristatum*) and some low rosette plants and scattered bunchgrass. The very high parts have many strongly specialized cushion-plants. The pumice slopes of the large volcanoes of western Bolivia represent frostfree soils. Here, scattered bushes of *Polylepis tomentella* form open heath-like 'woods' over large areas, almost to the snowline at 5000 m.

The premontane forest of northwestern Argentina grades southwards into the arid zone of central Argentina, the **Monte** or ***Larrea* semidesert**, characterized by grassland and scattered evergreen creosote scrubs or **jarilla** (*Larrea divaricata, cuneifolia*, and *nitida*, and *Bougainvillea spinosa, Chuquiragua arinacea, Prosopis torquata*) and others that loose their leaves in the dry season (*Cassia aphylla, Monttea aphylla*), and locally small xerophilous thorn forests (**algarrobal**, with *Celtis, Jodina*, and *Prosopis*). The low parts of this zone has salt and saltpetre plains with low succulent vegetation (*Atriplex lampa, Salicornia, Suaeda divaricata*).

The arid subtropical zone isolates some mountains with small areas of dry puna-like grassland from the Andes.

THE SOUTHERN TEMPERATE ZONE

Principal botanical references are Cabrera, AL (1953) Esciema fitogeográfica de la República Argentina. *Rev. Mus. La Plata, Bot.* 8:87-168; Dimitri, MJ (1962) La Flora de los Bosques Andinopatagonica. *An. Parques Nac.* 9:1-130; Lewis, JP & Collantes, MB (1973) El Espinal Periestépico. *Ciencia e Investigación* 29:345-408; Hueck, K (1966) *Die Wälder Südamerikas*. Stuttgart; Morello, J (1958) La provincia fitogeografica del Monte. *Opera Lilloana* 2:11-155.

In the humid southern Chile, subantarctic wet forests stretch from 35 degrees south to the southern tip of the continent (a few forest fragments also further north, at high elevations). Even the harsh zone near Cape Horn has evergreen forest of tropical aspect in sheltered places. This forest zone, surrounded by mountains, arid steppes, and the mattoral of central Chile, is 1100 km away from the nearest montane forest in northwestern Argentina and 1400 km from the nearest lowland forest in northeastern Argentina.

One fact that has whetted the curiosity of biogeographers for a century is that similar forest occurs only in Tasmania and New Zealand.

These forests are dominated by 'southern beech' *Nothofagus* (its Spanish name 'roble' does not mean oak, as in Colombia or Spain). The very rainy central part, the **Valdivian rain forest**, is dominated by large dense forests of tall evergreen *Nothofagus dombeyi*, and conifers (*Austrocedrus, Fitzroya, Pilgerodendron, Podocarpus, Saxegothea*), and has lianas (*Hydrangea*) and a dense bushy stratum of *Abutillon, Campsidium, Embothrium, Fuchsia*s, and *Chusquea* bamboo. The deciduous trees *Nothofagus pumilo* and *antarctica* form the treeline north to 40° south. Here, large tracts at 1000-1600 m have stands of the coniferous 'monkey-puzzle tree' *Araucaria araucana*. Further south the forest is a mixture of *Nothofagus pumilo, antarctica*, and the evergreen *betuloides*, which locally resembles the Valdivian forest, but normally is low and gnarled, mossy, or densely clad with pale lichens (*Usnea*), and the peculiar *Myzodendron* mistletoe. *Empetrum* heaths and moorland dominate in the exposed parts of the archipelago.

The forest continues through the many low passes of the South Andes to form rather dry woodland pockets (*Nothofagus antarctica* and sometimes *Australocedrus*) on the east side, and a narrow ecotone of bushy heathland towards the steppes of south Argentina (*Berberis, Colletia, Jurellia, Maytenus boaria, Mutisia*).

The **Patagonian semidesert** has eroded bunchgrass plains and areas of low shrub and brush (e.g., low *Chuquiragua avellanedae, Nassauvia glomerulosa*, the 'mata negra' *Verbena tridentata*, and the big thorny cushion-plant *Mulinum spinosum*). Previously, lush rushy grassland fringed the rivers. The basaltic upland plateaus have strongly wind influenced and eroded bunchgrass areas and some alpine vegetation.

The ecotone between the semidesert and the more humid Fuegian zone has bunchgrass vegetation of similar aspect as the humid puna.

WETLANDS

The Andean highlands have some of the finest wetlands in South America. However, the wetlands of the high parts are well isolated from the swamps and várzea habitats of the humid tropics, or from the enormous swampy savannas of the chaco and pampas.

There are few lakes in the northern Andes. Acid, unproductive ponds and tarns occur near the snowfields, as signs of past glacial erosion. Such poor wetlands also occur in northern Peru, clustered together in the highest parts.

Only a few lakes with high plant production exist in the Andes of Colombia. Until the recent disturbance by man, the Ubaté and Bogotá plateaus in the temperate zone of the East Andes of Colombia had enormous marshes and swamps. The wetlands here are fringed by tall and dense reeds (*Scirpus, Typha*, and some *Cortadera*) and some *Alnus* swamp, and the open shallows are full of *Elodea, Myriophyllum brasiliense, Potamogeton* etc. Only few such habitats remain outside Lake Tota (3040 m) today. Those of the savanna plains are strongly influenced by pollution and siltation: here, the submergents disappear, and carpets of *Azolla, Ludwigia peploides*, and *Limnobium stoloniferum* spread over the surface.

A few páramo lakes with dense floating *Myriophyllum elatinoides* exist in the East and Central Andes and in Nariño (Laguna Cocha). Also the tablelands of Ecuador had many fine lakes of reed-fringed as well as open, weedy types. Except for the lakes of the Cajas plateau in Azuay, they have become more or less valueless due to siltation.

The best Andean wetlands are in the puna zone. A wide variety of types is represented. Lake Junín in central Peru is surrounded by 300 sq km of reed-marsh and seasonally inundated meadows. Lake Titicaca is very deep, most coasts steep and rocky, but other coasts are flat, with enormous zones of aquatic vegetation, and the surrounding plains are dotted with rushy lakes. Other significant reedy lakes are near Jauja in Junín, and in the Tungasuca area in Cuzco. Larger, shallow freshwater lakes with only submergent (and floating) plants are Conococha in Ancash, and the Yaurihuiri lakes and Parinacocha in Ayachucho, and Lagunillas in western Puno. Alkaline, slightly salt lakes with abundant submergents (*Chara, Myriophyllum, Potamogeton, Ruppia*) are widespread in the arid puna, but the few strongly alkaline lakes are often turbid ('white', green etc), lacking plants other than large unicellular algae. Finally, the high alpine parts often have clusters of sterile tarns.

The upland plateaus of the Patagonian steppe have similar wetlands as the arid puna, but with exceeding limnological differences between neighbouring lakes, and with many strongly alkaline ponds. The lakes of the southern Andean zone are typically clear (oligotrophic), with little vegetation, except some reeds fringing sheltered bays.

PRISTINE OR DISTURBED HABITATS?

The high parts of the Andes are normally classified as steppic habitat.

Recent botanical studies in Ecuador have led to the view, that the páramos once were dominated by elfin forest and *Polylepis* woods up to the edge of true alpine vegetation. Grasslands may have occurred locally as an early successional stage following occasional fires. The present wide occurrence of grassland apparently are results of centuries of slashburning, cutting for firewood and grazing. Regular burning is sometimes considered a threat to the grassland, but in fact secures the continued dominance of hardy grasses. Except in the steepest and most moisture-influenced zones, cultivations, pasture, and second growth now spreads rapidly. Thus the entire Andean zone of Colombia and Ecuador, except the wet slopes towards the Pacific and to the eastern lowlands, are cultural landscapes with pristine habitat left only locally.

Similar changes may have taken place in the puna zone. This is less obvious, though, as the changes started very long time ago. H Ellenberg, in *Die Umschau* 21(1958):645-681, was the first to suggest that the grassland vegetation of the puna was the result of hundreds if not thousands of years of human activity. The role of man from prehistoric to recent times is further described by J Ansion (1986, *El arbol y el bosque en la sociedad Andina*. Inst. nac. For. Fauna – FAO, Lima). His view, based on ethnobotanical considerations, is corroborated by some botanical data. The original vegetation was assumed to be extensive *Polylepis* woodlands and *Puya* and *Buddleia* stands. Some palynological studies suggest that the *Polylepis* disappeared suddenly from some cordilleras 3-5000 years ago.

Several botanists are still sceptical to this view (see, e.g. BB Simpson in *Smithsonian Contr. Bot.* 43(1977)). They claim that the widely scattered patches of wood depend on local conditions creating 'lower elevation conditions'. It is true that the woods occurring above 4400 m (e.g. in Cordillera Blanca, and along the Bolivian/Chilean border) are restricted to rocky and mostly frostfree soils, and that *Polylepis* often grows in places with nightly mists or where the cold air easily flows down the slope. However, our studies in Peru and Bolivia in 1987 (Fjeldså 1987), and an analysis by E Jordan (in *Tuexenia* 3(1983):101-112, based on aerial fotos of western Bolivia) document that certain *Polylepis* forms grow under quite diverse conditions. By lack of grazing the forest regenerates well in dense and tall grass as well as on extremely exposed barren ground. This shows that the distribution of forest patches cannot be explained from present ecological conditions alone. Instead, we believe that large parts of the presently steppic habitats were once densely dotted with woodlands.

Possibly, much of the virgin vegetation was destroyed already by early Andean cultures. The mountain slopes were terraced and cultivated, and the high parts were cleared to improve the grazing. Thus the woodlands became restricted to rocky places unsuitable for other use, but were maintained here as renewable firewood resources. This may at least have been the case during the Incac period, when forests were saved for the nobelty to hunt in, and the area under cultivation was minimized by transporting soil and developing sophisticated irrigation channels from the rain-drenched slopes into the drier valleys. Later, under the mestizo descendents of the Conquistadores, large areas were depopulated, but this permitted the new governors of the Andes to monopolize the best land. When the indian population started to recover by 1780 AD, even the steepest slopes were turned into a quilt of tiny farms. Unfortunately, the land reforms that started when the military deposed the aristocracy in parts of the Andes led to chaos: break-down of well-functioning agriculture in some areas, uncontrolled clearing of forest in other areas. Today, the sheep-grazing and the burning done to improve the pastures almost inhibits forest regeneration. Therefore, the woodland survives only as widely isolated, tiny patches.

The deforestation has been exceptionally intensive in the mining areas in the high Andes.

This interpretation may apply at least through the mountain ranges of Peru into northern Bolivia, but *Polylepis* may always have been lacking on the dilluvial plains and in truly alpine habitats, and was rare in southern Bolivia and northwestern Argentina.

While the puna is used mainly for grazing, the valleys and semiarid slopes have been terraced and cultivated for thousands of years. The destruction of the closed primary vegetation results in a drier climate with less regular rainfall. This in turn leads to soil erosion. In areas now abandoned for cultivation, the poor soil quality and continued grazing prevents regeneration (other than *Dodonea* shrub). Unfortunately, the reforestation is mainly with introduced *Eucalyptus globosus*, pines, and cypresses. The high phenolic contents of the *Eucalyptus* leads to further impoverishment of the soil. Furthermore, with its deep roots, the *Eucalyptus* removes much of the surface water. The planting therefore fails to solve the problems, except that the firewood produced lowers the pressure on the relict patches of indigenous forest (see Poore, MED & C Fries (1985) *The ecological effects of eucalyptus*. FAO Forestry Paper 59, Rome).

Also in western Peru, and on the dry foothills of central Argentina, the 'algarrobo' woodlands have become very restricted.

In Patagonia, introduced junipers and roses spread rapidly. The grassy river-plains have been turned into orchards, or are dominated by introduced grasses and herbs. A very grave change has happened on the southern Patagonian plains.

The broad river plains, which few decades ago had tall grass and rushes, have changed into short-grass semi-desert, owing to hay-making and sheep-grazing. The blowing soil fills up thousands of shallow lakes, which gradually become arid land.

Also in the Andean highlands, the partial replacement of the native grazers (rheas and camelids) with sheep has led to considerable degradation of the grasslands. Still, however, tall grassy vegetation is widespread outside the driest parts of the puna.

HABITAT DEFINITIONS (Vegetation and landscapes)

Habitats with trees

ARID SCRUB (or **DESERT SCRUB**): Low, often thorny bushes and small drought-resistant trees and often several cacti. Permanent scrub vegetation cover the arid zones near the Caribbean coast and in some rain shadow valleys.

BORDER: The edge of forest habitat towards open terrains, as streams, roads and clearings. May be divided in **forest** and **woodland borders**. Similar habitat occurs where a tree has fallen, taking down some of its neighbours. A border frequently represents an impenetrable wall of dense and shrubby vegetation, often with vines and epiphytes. In humid zones, large-leaved plants (especially monocots) dominate these habitats. Borders are fine habitats for studying birds, because canopy species come low here, and because observation is easy.

CEJA: See under elfin forest.

CLOUD FOREST: A term of popular although ambiguous usage. In the strict sense, cloud forests occur in otherwise dry regions in places where the terrain is often wrapped in humidity-giving mists. Such forests, which exist above the coastal desert in Peru, and in some dry intermontane valleys, mostly at 2500-3200 m, we call **dry cloud forest**. Montane forests which receive considerable rainfall are often called humid montane forest or montane rainforest. We talk of **humid cloud forest** in order to tell parts usually enwrapped in clouds from forest which is usually below cloud cover.

DRY FOREST and **DRY WOODLAND**: Forest and woodland with rather well spaced trees growing where the dry season is long and often severe. The trees are often small and thorny as in savanna woodland, but usually there are some large trees with thick trunks, and such trees may be conspicuous elements, e.g. in the balsa forests. The trees often have 'spanish moss' (*Tillandsia usneoides*) hanging from the branches. Many trees loose their leaves for part of the year. If these trees dominate, the habitat is called **deciduous forest**. Of the balsa forests, the true balsa (*Bombax*) is evergreen, while *Ochroma* is deciduous. These habitats are widespread in the valleys of central Peru, but are widely cleared in other parts.

DWARF FOREST: See under Elfin Forest.

ELFIN FOREST (and **ELFIN WOODLAND**): A stunted humid cloud forest on spongy ground at high elevation. We restrict the term to patches of humid arboreal vegetation that occur in fairly protected places above the limit of continous forest. These habitats often have an inpenetrable vegetation, with much bamboo along the borders. We call the particularly gnarled types formed in places with constant exposure to cold wind and fog **dwarf forest**. All humid treeline habitats, both continuous cloud forest, elfin woodland, and dwarf forest may collectively be called 'ceja' or 'ceja de selva'.

FOREST: Extensive tracts of dense arboreal vegetation (unlike **wood**, which is often lighter and forms smaller patches).

GALLERY FOREST: A strip of trees confined to the edge of a watercourse. Most common in tropical savanna. We prefer to call the lower riverside vegetation of scrub and gigantic grasses (*Cortaderia*, *Gynerium*) in the Andean zone **Riparian thickets**.

HUMID FOREST: Usually at higher elevations and with less extreme rainfall than wet forest. Often, the weather is quite sunny in the dry season (e.g. in the yungas of Bolivia).

POLYLEPIS WOODLAND: High Andean woodland or shrub dominated by *Polylepis* trees, usually isolated patches of few hectares size, or rarely larger areas (in mosaic with grass- or heathland). Sometimes occurs near the treeline, as elfin forest (mixed with *Buddleia*, *Escallonia*, and *Gynoxis*), but pure *Polylepis* stands occur high above the general treeline, at 4000-4500 m. Bushy *Polylepis* vegetation sometimes occurs on cliff-ledges or on pumice slopes of the desert puna to near 5000 m.

SAVANNA WOODLAND: Often rather open or well spaced scrubby forest on poor (sandy) soil, with a low canopy. The trees are usually thorny or contain tannins, phenols, and other defensive chemicals in their leaves to resist grazing.

SCRUBBY AREA: Unlike shrubby areas this is a permanent plant community of chiefly drier regions. Some are 'shrubby' by origin, but become so permanently altered by overgrazing, erosion, firewood cutting etc that forest regeneration is unlikely. The vegetation is characterized by scrubs of retarded growth, often thorny, and small leguminaceous trees (*Acacia*, *Prosopis*, often also large cacti). *Dodonea viscosa* bushes may cover large areas. Areas dominated by canopied small trees are classified as **scrubby woodland** (e.g. *Prosopis* wood). Widespread in the Central Valley of Ecuador and deep montane valleys of Peru and Bolivia.

SECOND GROWTH: A term used to define a larger spectrum of plant communities from the initial rapid regrowth after deforestation, with tall grasses, herbs, and bamboo to the development of a fairly mature forest. A relatively mature regrowth stage (sub-climax) often found over large areas which in time will develop into true forest (if the pressure from humans permits). In the most disturbed Andean valleys the second growth is often allowed to develop freely in steep ravines, until the trees grow so large that they can be used for construction.

SHRUBBY AREA (SHRUB): A regrowth stage where the terrain is dominated by shrubs (e.g. many composites), thickets, and some fast-growing trees (e.g. *Cecropia*s in the humid zone). The habitat may be called **coppice** where woody stems are cut on a rotational basis, often in association with a dispersed silviculture.

WET FOREST: Forest tracts with heavy rainfall year round. Mature, canopied, tall, with some very high emergent trees. Occurs in the premontane zone of Pacific Colombia and northern Ecuador and on certain ridges in the eastern Andean zone of Ecuador and Peru. We avoid the word rainforest, which is ambiguous due to popular usage: any tropical lowland 'jungle'. Due to the very different species composition from the tropics, the forest of southern Chile is called **subantarctic wet forest**.

Steppic and alpine habitats

BRUSH STEPPE: Plains and gently rolling arid land, often sandy or gravelly, with 0.5-1 m tall microphyllic and often thorny bushes but little ground cover. One common type in the puna zone is the **tola** or **thola heaths** with *Lepidophyllum quadrangulare* and scattered grass.

CRAG: An abrupt rock face, not necessarily high.

DESERT: A large tract of waterless land, with insignificant or no vegetation.

ESPELETIA STAND: Páramo habitats with a scattered vegetation of large (up to 10 m tall) composites of the genus *Espeletia* (locally called frailejónes = fat munks). They represent the so-called caulescent rosette plants, which have evolved in several high tropical mountains, and are characterized by a thick column mainly of dead leaf structure that elevates a large rosette of pale felt-like leaves and yellow flowers up above the extreme microclimate near the ground.

GORGE: A steep-sided narrow deep valley, often flanked with crags.

GRASSLAND: A term of broad usage, here used for level or gently rolling terrain dominated by grassy vegetation, except temporarily flooded parts (rushy meadows, shore meadows), and arid types (brush steppe).

PAJONAL: Spanish word for pasture. Often used for grassy (cleared) slopes in the cloud forest and lower páramo zones.

PARAMO: Barren tracts above the treeline. Used here for the usually rather humid grassy habitats, often with some heather-like vegetation, ferns etc. in Venezuela, Colombia, and Ecuador. The equivalent habitat in northern Peru is called **jalca**.

PUNA: Barren, seasonally dry highlands mainly with grassy vegetation. Used for the high Andean zone of Peru, Bolivia, and northern Argentina and Chile. Subdivided into **humid puna**, with predictable rainfall October-March, and **arid puna**, with a severe dry season and unpredictable rainfall.

PUYA STAND: Puna terrain with giant caulescent bromeliads *Puya raimondii*, which grow 8-12 m tall when flowering. A number of birds nest in the safety of the dense leaves, which are stiff and hard, and have hook-shaped spines. Other birds assemble to feed from the flowers. Very local in the highlands of Peru and Bolivia, but some smaller *Puya* species are more widespread.

RAVINE (**quebrada** in Spanish): Narrow valley, deep in proportion to its width, with rocky precipitous sides or abrupt grassy slopes. A miniature ravine is a **gulley**, a very large one is a **canyon**.

SAVANNA: A tropical grassland with short bunchgrass or tall grass and scattered trees or palms (or no trees; **grassland savanna**). They are caused by severe environmental factors such as fire or poor sandy soils that deter forest development, and not by the climate. The precipitation is not necessarily lower than in the forested parts.

Wetland habitats

DELTA: A complex river mouth where much silt is deposited, splitting the flow into several, often meandering channels, and sometimes having lagoons (a single inlet is called **estuary**).

LAGOON: Shallow lake enclosed by sandy ridges, marshy vegetation etc. near a coast or shore of a larger lake.

LAKE: Any fairly large natural body of standing fresh water. Subject to wind and wave action on margins. Deep lakes in high montane (glaciated) areas often are **oligotrophic** (infertile), often without marginal vegetation. If shallow, with fringing reedbeds, often **eutrophic** (fertile). Lakes with slightly alkaline water often have rich submergent vegetation. These lakes are often the best bird lakes, but the few typical **alkaline lakes** (with pH over 9, often salt or bittersalt) have high numbers of only a few species. The usually small lakes in humid páramo and jalca have boggy margins, and brown acid water (**dystrophic**).

MARSH: Low areas with more or less permanently standing water and emerging vegetation, but not trees. In the higher parts of the Andes, the dominent marsh plant is *Scirpus californicus tatora*, which is up to 3-4 m tall. Also lower *Juncus* species occur, in Colombia also *Typha, Cortadera* etc.

MIRE: Peaty, wet and sometimes flooded land with sedges, grasses, and occasional shrubs. If alkaline or receiving nutritious water from outside, called a **fen**, if derived from rain or condensation from fog (thus poor in nutrients), called **bog**. We use the latter term also for any waterlogged habitat at very high elevation, where the vegetation is very short and matted, often dominated by cushion-plants. This habitat is typically found on the flat bottoms of once glaciated valleys. Similar habitats, where fresh water seeps in along the margins of salt plains of the arid puna, is often called **bofedales**.

POND: Small body of standing water (compare lake). An oligo- or dystrophic pond in barren montane habitat is called a **tarn**.

RIVER: Major flowing watercourse, more than 5 m broad. Slow-flowing rivers with rather low banks may have auxiliary backwaters, oxbows, and small wetlands, in a riverine zone.

RUSHY MEADOW: Sometimes inundated grassland with moderate grazing pressure. Tall grasses (as tussocky *Festuca*) and rushes (*Juncus*) are dominent plants, but open spaces with short matted vegetation or standing water with floating plants may occur.

SALINE: Saltpan, salt pond, or lagoon which may seasonally dry up as large salt-encrusted flats.

SHORE MEADOW: Low areas often with stagnant water, but unlike rushy meadows strongly grazed and lacking tall vegetation. The dominant plants are short matted grass, and the tiny *Lachemilla pinnata* covering the surface in inundated creeks. A typical plant in shallow pools is the peculiar small sedge-like umbelliferous *Lilaeopsis*. If not maintained by strong grazing the habitat soon changes into rushy meadow or marsh. Sometimes called **floodland**.

STREAM: Small watercourse. Stretches with chronic turbulence are called **torrents**. Here, **rapids** and **waterfalls** are frequent.

SWAMP: Areas with more or less permanently stagnant water and trees occurring. The trees may occur rather scattered in open water or in tall grassy vegetation or among reeds, or may grow more densely (**swamp forest; várzea forest** in case of seasonally flooded tropical forest). These are mainly lowland habitats, but *Alnus acuminata* and *Escallonia myrtilloides* trees may once have been important elements of the wetlands in the East Andes of Colombia.

The Andean Avifauna

THE EXPLORATION OF THE ANDEAN AVIFAUNA

The first natural history from the Andean zone, of Chile, was written as early as 1782 by Father Molina. Of other early explorers, Don Felix de Azara described the natural history of the chaco 1802-1805. Alexander Humboldt, known mainly for his explorations in the tropics, mounted Volcan Chimborazo. Early in the 1800s, the famous naturalist A d'Orbigny collected birds in many parts of Argentina and Bolivia 1826-32. Darwin visited Patagonia and Chile on the Beagle expedition 1832-6. Lesson, Jameson and Bourcier were other prominent explorers of this epoch. Gould, Gray, and Lafresnaye described many of the birds brought to Europe by the early collectors.

Most birds of the northern Andes became known to science through the trade skins exported from around 1830 to the end of the century from 'Bogotá' or 'Nouvelle Grenada'. Such skins, made by natives, were sent to Europe in hundreds of thousands. Most 'Bogotá specimens' may have been collected in the hills surrounding the city or fringing the Magdalena valley, but some certainly came from other areas. 'New Granada' sometimes has no more significance than South America.

A browsing of the older systematic literature reveals an extreme confusion due to the very unprecise or misleading label data of the early collections. In fact, there were very few records from definite localities before 1850.

The first descriptions of Venezuelan birds were based on Baron NJ Jacquin's travel in 1784, and the collection by R Schomburgk in Roraima 1840-4, and by A la Sallé near Caracas 1846-8.

Principal collections in the coastal mountains of Venezuela were by A Goering (1866-72), SM Klages (1909-14), CK Cherrie (1918-19), HJ Clement and NG Netty (1929-30), ER Blake (1932) and A Wetmore (1937). In Mérida, JJ Linden collected hummingbirds 1841-4, and A Goering collected here in 1869. From 1872, the professional collectors of the Briceño family provided birds from the Mérida mountains to several museums. Collecting in the Venezuelan Andes was done by MA Carriker in 1909-11 and 1922. Modern studies of páramo birds were made by D Ewert and F Vuilleumier.

The birdlife of the Tepuis of the southern highlands was explored in the late 1800's by H Whitely and FV McConnel, and in the early half of our century by GHH Tate, A Olalla, ET Gillard, EG Holt, by PS Peberdy and AS Pinker, and by WH Phelps Jr and KD Phelps.

In our century, the Phelps family patronized bird collecting throughout the country, took part in several of the expeditions, and established the 'Phelps' collections' in Caracas. The exploration of the Venezuelan avifauna was outlined by Phelps (1944), and a modern description of the country's birdlife was given by Meyer de Schauensee & Phelps (1978).

Important collections from Colombia were G Wyatt's from Santander 1870, TK Salmon's from Antioquia 1872-8, MG Palmer's from Buenaventura Cali and upper Río San Juan 1907-9, and by Wheeler and Detwiler, count von Berlepsch, and W Robinson, in the Magdalena Valley. In the Santa Marta mountains, collecting was done in the late 1800s by FA Simons, WW Brown and H Smith, and 1911-20 by MA Carriker (see Todd & Carriker 1922). Carriker, from his coffee plantation in Santa Marta, continued working throughout the western half of Colombia 1941-53. His activity, spanning the whole tropical Andes, resulted in 53,000 bird skins for North American museums.

8 expeditions by the AMNH 1910-5 in the Colombian Andes resulted in the early zoogeographic synthesis by Chapman (1917). K von Sneidern collected large series in various parts of the country 1938-52 (and again in the early 1970s). This material, together with that of A Dugand, JI Borrero, Brother Nicéforo, and the Mena brothers, formed the basis of Meyer de Schauensee's catalogues (1948-52). Brother Olivares, FC Lehman, J Haffer, and J Hernandez C were central in the more recent ornithological exploration, and H Romero Z made a great effort, until his early death, to develop the national bird collections (MNCN).

The knowledge of Colombian birds has now permitted compilation of modern bird-books and broad synthesis (Haffer 1967, 1970, 1974, 1975, Meyer de Schauensee 1964, 1966, 1970, Hilty & Brown 1986).

The exploration of Ecuador started late, and through the late 1800s into our century most material was obtained through native collectors. Their main patron, L Söderstrom, secured large series (of poorly labelled birds) for the SMNH (Lönnberg & Rendal 1922). More modern collecting in the late 1800s was by Fraser, Siemiradski and Stolzmann (Sztolcmann), Rosenberg, Festa, Goodfellow, and Hamilton.

SN Rhoads made significant collections in 1911, and 1913-25 the AMNH had 8 expeditions, and hired the famous native collectors of the Olalla family in 1922-5. After this remarkable geographic coverage, a thorough synthesis of the avifauna of Ecuador was possible (Chapman 1926). Since then, the expedition activity stagnated, until very recently, when RS Ridgely, MB Robbins, TS Schulenberg, FB Gill and others started an exploration of the premontane forests. P Greenfield has made general observations throughout the country.

Curiously, native collectors played no great role further south in the Andes. The early ornithological exploration of Peru was by trained collectors as von Tschudi (1844-6), Sclater and Salvin, Jelski, Taczanowski, count von Berlepsch, Kalinowski and Baron. They mainly explored the highlands and Andean valleys, but the humid montane forest was studied only in the Utcubamba area in Amazonas and around Maraynioc in Junín. Kalinowski and Whitely spent several years in the Urubamba region, apparently in the valleys within few miles of humid zones, but never visited this difficult habitat. Thus, the humid treeline habitat of Cuzco was first studied in 1915 by E Heller. The Urubamba region was extensively sampled during the great scientific activity that followed the archaeological expeditions by H Bingham (see Chapman 1921). Carriker made 4 collecting expeditions in Peru. For the BMNH, PO Simons collected birds 1898-1901 from Ecuador through Peru and Bolivia to northwestern Argentina (Chubb 1919), and A Morrison collected from Huánuco to Apurímac in 1936-37. DA Griswold collected in Maraynioc in 1939.

Much of the 'writing-desk work' was done by JT Zimmer, in the many systematic revisions in *Amer. Mus. Novit.*

The exploration of Peru progressed further as M and HW Koepcke studied the distribution of birds on the Pacific slope in detail from 1951 until M's tragic death in 1971. H de Macedo's hacienda in Puno was important for the study of puna birds (e.g. by J Dorst). P Hocking (and in part M Villar) collected for various museums in the relict forests of the poorly known central provinces. The last decades have witnessed an upsurge of expedition activity from North America, starting with the studies by J Terborgh & JS Weske in Cordilleras Sira and Vilcabamba, and by GH Lowery, and continuing with collecting by GR Graves, JP O'Neill, JV Remsen, TS Schulenberg, and others, and especially life-history observations by TA Parker. Through the initiative of I Franke, Peruvian students are now also efficiently engaged in recording the country's avifauna.

The birds of Bolivia became known through the work of d'Orbigny in the early 1800's, and some collecting by Buckley and Rusby. G Garlepp, after years in tropical Peru and the chaco, collected in the early 1890s in several places in the cordilleras of Bolivia. His younger brother O Garlepp continued the work into the early 1900s (but taking only insects in his later years).

MA Carriker's collected for the ANSP through Bolivia in 1934-38, providing the main material for Bond & Meyer de Schauensee (1941, 1946). Currently, the Steinbach family of Cochabamba exported material to various museums in North America. AM Olalla undertook collecting expeditions in Brazil and Bolivia for the SMNH, including in the highlands of La Paz and Cochabamba (Fjeldså & Krabbe in press). In 1951-52, G Niethammer collected through La Paz and Cochabamba (Niethammer 1953-6). Yet, Bolivia remained poorly sampled, although recently R Crossin, JV Remsen, TS Schulenberg, CG and DC Schmitt, J Fitzpatrick, and D Willard, have filled in many gaps.

In the later half of the 1800s, various parts of Argentina were visited by H Burmeister, A Doering (Córdoba and Río Negro), WH Hudson (the pampas and Río Negro), and F Schulz (Tucumán) (Sclater & Hudson 1888). In 1920-1, A Wetmore explored wintering quarters of northern shorebirds from the chaco to central Argentina and Chile.

The late C Olrog, Swedish, lived in Tucumán from 1948, and was a central figure in the development of Argentinean ornithology. He gained a prodigious first-hand insight into the country's bird-life. Other noteworthy explorers of Argentine birdlife were R Dabbene, WH Partridge, JA Pereyra, NA Bó, G Hoy, H Krieg, and WJ Mogensen. More lately, JR Contreras studied birds of central Argentina, M Nores the birdlife of Córdoba and of habitat islands along the base of the Andes of northern Argentina and into Bolivia.

In Chile, birds were collected on several occasions from the early 1800's by naturalists of the naval expeditions, such as Lesson, Garnot, von Kittlitz, Darwin, FJF Meyer, and Bridges. Ornithological data from Gay's studies (1830-42) were compiled by O des Murs. Chilean ornithology entered its scientific phase in the 1850's, with the arrival of L Landbeck and RA Philippi. From this period, a number of residents collected birds. The accumulated information was compiled by Hellmayr (1932), who studied the birds brought to Chicago by the Marshall Field Chilean Expedition (1922-4).

Detailed field studies were started in the 1920's by JD Goodall and AW Johnson, and in part F Behn, WR Millie, and LE Peña. This epoch ended with the publication of the *Birds of Chile*, first in Spanish, then in English (Johnson 1965, 1967).

Isla Grande (Tierra del Fuego) was visited by several naval expeditions, and the birds became known also through R Crawshay, RH Beck, and the families Reynolds and Bridges at Viamonte. Important studies in recent decades were by Olrog, Philippi, Johnson, Goodall & Behn, H Johansen, and an expedition for the Yale Peabody Museum of Natural History (Humphrey et al. 1970). Today, C Venegas C is active in this area.

In all Andean states, the ornithological interest now turns increasingly towards ecological studies and conservation management. In connection with this work, a modern bird-watching interest begins to develop (actually, the tradition is quite old in Argentina. Owing to the efforts of M Rumboll, the Argentinean national park rangers have a remarkable insight into natural history). Also the organization of a network has started, so that registrations can be coordinated (as in much of the Northern Hemisphere).

GENERAL CHARACTERIZATION OF LANDBIRD COMMUNITIES

The high-altitude habitats of the Andes are extremely interesting from the viewpoint of diversification and evolution of faunas, as the high-elevation conditions appeared mainly in the last two million years or so. Thus, it is still feasible to reconstruct scenarios of colonization and diversification of the complex fauna. A number of attempts have been presented in Vuilleumier & Monasterio (1987), and illustrate a complex picture with some long-distance colonizations from other cool areas, others from adjacent lowlands.

Since the Andes are young, compared with the surrounding lowlands, one must expect lowland birds to predate Andean ones. The few analyses made so far also support the view. The montane forests were colonized especially by premontane forest birds; the steppic highlands were colonized from the Patagonian steppe and the chaco and cerrado. A noteworthy exception is the Andean element of the avifauna of the low mountains of southern Brazil. This element was apparently established as Andean birds spread across the chaco during the glacial periods (see H Sick on pp. 233-7 in Buckley et al., eds. (1985)). There are also instances of colonizations going from the Andes to the Patagonian lowlands and to habitats near the Pacific coast.

The evidence does not indicate that the high Andes were colonized by birds from the Tepuis of Venezuela, but rather suggests that the old and broken-up Tepuis received emigrants from the geologically younger Andes.

The avifauna of the region is complex, but can be divided into two main components: one of the humid slopes mainly of the tropical Andes, one of the steppic zones. Smaller communities inhabit arid scrub and woodlands of the Andean valleys, and the Valdivian forest.

The Andes have very few birds without near relatives: the small seedsnipe family (Thinocoridae), 2 peculiar long-billed plovers, Torrent Duck, Tit-like Dacnis, White-capped Tanager, Plush-capped Finch, the Pardusco, and a few peculiar species of the Valdivian forest zone. The other Andean birds have close relatives among the birds of the adjacent lowlands. Adaptations to highland life seem mainly to be metabolic adaptations to the harsh climate (see e.g. *Comp. Biochem. Physiol.* 82A (1985):847-850). Due to their efficient respiratory system, birds do not strongly need specializations to the low oxygen pressure (*J. Orn.* 129(1988):217-226, but modifications of eggshell characteristics are important − see *J. exper. Zool.* suppl.1 (1987):247-252). Also the morphological adaptation is usually slight, with tendencies for larger size and dull or often dark colors, the latter possibly a protection against harmful radiation.

Many bird groups have shown a significant adaptive radiation after the colonization of the highlands. The Andes have many endemic waterbirds. The large synallaxine subfamily of ovenbirds and the large fluviatiline subfamily of tyrant flycatchers occur almost exclusively in southern lowlands and north in the high Andes. Other groups with many (or all) species in the highlands are *Nothoprocta* tinamous, *Metriopelia* ground-doves, *Bolborhynchus* parakeets, *Scytalopus* tapaculos, *Grallaria* antpittas, and sierra-finches. Among hummingbirds, typical highland genera are *Oreotrochilus, Aglaeactis, Coeligena, Heliangelus, Eriocnemis, Metallura,* and *Chalcostigma*. Tanagers are mainly tropical, but include distinctive assemblies of highland forms: colorful mountain-tanagers (*Anisognathus, Buthraupis, Dubusia, Iridosornis*), hemispinguses (*Hemispingus*), and flowerpiercers (*Diglossa*).

Birds of the humid montane forest

The humid montane and premontane forests of the tropical Andes compete with the Amazon forest in species richness. Of the 2,900 species of non-marine birds of South America, at least 70% are recorded in the montane forest of the tropical Andean region. This species richness is caused by a dense vertical segregation of allied species (Terborgh 1971, 1977) and the presence of many very local forms. In isolated mountains (e.g., Cordillera Sira in Peru) not reached by typical Andean temperate-zone birds, premontane species expand their ranges to the summit. This illustrates the significance of competition in causing vertical segregation (Terborgh & Weske 1975).

The birds of the montane forest are mainly small and colorful forms of the canopy and bamboo foliage, and a number of brown or sooty forms in the understory.

Some birds move around a good deal within suitable habitat, and crossing smaller barriers. Throughout the humid tropical zones, many birds rely on insect foods or fruits that are distributed rather spottily or unpredictably. Such foods are not easy to monopolize, and instead of defending territories many species therefore move around within a larger area. Hummingbirds fly around alone, 'traplining' for special kinds of flowers. Many other birds move around in mixed-species flocks, which can follow very fixed routes and time schedules day after day. The observer will often, after hours of silence, experience a short period when the air is full of fine call notes, as a feeding party passes through the terrain, searching all strata from understory to treetops. The party members naturally use the same macrohabitat, but the species segregate finely by microhabitat, fouraging techniques, and food selection. The feeding habits and interactions of the cloud-forest birds are described by Moynihan (1979) and Remsen (1985).

A number of these fairly dispersive forest birds are distributed through most of the forest from Venezuela or Colombia to Peru or Bolivia: Andean Guan, Band-tailed Pigeon, hummers as Collared Inca, Greater Sapphirewing, Swordbilled Hummingbird, Amethyst-throated Sunangel, and Tyrian Metaltail, furthermore Pearled Treerunner, Barbtails, certain tapaculos, fruiteaters, Red-crested Cotinga, various tyrannulets, Golden-crowned, Cinnamon, and Dusky-capped Flycatchers, Black-throated Tody-tyrant, some Chat-tyrants, Collared Jay, Great and Glossy-black Thrushes, Mountain Wren, certain Mountain-tanagers, Masked Flowerpiercer, a few Brushfinches etc. Widespread species of bamboo thickets are Speckled Hummingbird, *Thryothorus* wrens, *Basileuterus* warblers, and hemispinguses, Mountain Cacique, and Plush-capped Finch. Specialists in seeding bamboo are Maroon-chested Ground-dove, Barred Parakeet, Paramo Seedeater, and Slaty Finch, which are scarce but maybe widespread.

Most birds of the forest floor are insectivores

and omnivores that defend large territories. In the rather stable understory environment, any potential territory is permanently occupied, which eliminates the advantage of emigrating. For this reason, most understory birds evolve extremely sedentary habits. Although able to fly, they develop a 'fear of flying' across open land. Even small strips of cleared land thus act as dispersal barriers for many species. Many such species have extremely restricted ranges. However, most of these 'endemic' forms have 'counterparts' in similar habitat in the neighbouring mountains. Such patterns are found among metaltail hummingbirds, thistletails, spinetails, antpittas, *Scytalopus* tapaculos, flowerpiercers, and several brush-finches.

The deep dry valleys that intersect the humid mountain slopes represent dispersal barriers for many forest birds. The resulting differentiation of populations often results in a '**mosaic pattern**' of variation, where different characters vary independently along the Andes. In the very linear distributions shown by Andean forest birds, a very aberrant-looking population is often inserted between populations of very similar appearance (**leap-frog pattern**, see Remsen 1984). Because of the sedentary habits of many forest birds, such patterns are unlikely to arise by long-distance dispersal. More probably, the patterns result from temporary fragmentation of the ranges, and rapid divergent evolution in some population segments. Certain species show a widely disjunct range, which suggests that many segments of a once continuous linear range went extinct. A few species occur only in one small area, lacking near counterparts at all.

There are particularly many endemic forms in the northern Andes: in the Mérida mountains of Venezuela, the Santa Marta mountains, and in the East and Central Andes of Colombia. Further south, local endemics cluster together in some cloud-forest areas in southwestern Ecuador and adjacent northwestern Peru, in the northern and the southern parts of the Central Andes, Cordilleras Vilcabamba and Vilcanota in Peru, and in the yungas of Bolivia, with other taxa centered around Jujuy in northwestern Argentina.

Obviously, the endemism of the Sta Marta and Mérida mountains reflects the isolation and very casual colonization across the surrounding gaps of lowland. The Marañón gap in north Peru efficiently prevents dispersal of forest birds along the east Andean slopes, but the Porcula Pass does not stop all páramo and elfin forest birds (Parker & al. 1985). Other barriers are less easy to define. Graves (1985, 1988) documents that the tendency for disjunct species ranges and differentiation of local populations increases from the lowland forest to the treeline. The geographic variation is negatively correlated with the amplitude of the species' elevational distribution, presumably because discontinuities are more likely to occur in habitat types restricted to narrow elevational zones. Particularly patchy distributions, with several forms replacing each other along the Andean slope, are found in the narrow zone of humid treeline habitat.

Terborgh (1980), Haffer (1974) and others invoked historical factors to explain the species richness of the humid forests of South America. Probably the diversification is caused by a dynamic process, as opportunities for dispersal alternate with periods of isolation. However, founding of small and isolated new populations by dispersal across existing barriers, and 'vicariance', where once widespread forms become divided in small population segments represent two competing explanations. Isolating barriers may often be the deep, dry gorges that intersect the cloud forest zone, but the barriers may have become much wider during the dry and cold glacial periods, when the montane forests shrinked to local refuges. A similar **refuge theory** was used by Haffer (1987) to explain the differentiation of birds of lowland forest. Little direct evidence exists for the location of specific refuges, but enough is known about past climates to strongly suggest that wet forest had very restricted distribution in certain time periods.

Despite these opportunities for differentiation of new species, competition between the species may maintain rather stable equilibria. Thus, the exceptionally high total number of species in the Andes is caused more by the differentiation of local faunas (endemism) than by exceptional number of sympatric species (Remsen 1985).

Endemism is not just manifestations of present ecological conditions, but have historic explanations. However, we disagree with Cracraft (1985) that species with different requirements follow congruent historic patterns. According to our experience with Andean bird distributions, different species groups are very differently affected by the features of the landscape. Thus, landscape structures that represent barriers to some species serve as dispersal corridors for others. Accordingly, species with different requirements cluster together in different areas. These may, however, form mosaics, and clusters of interlocking areas of endemism are easily lumped if this supports the preconceived vicariance model.

Birds of the drier woodland

The drier slopes and valleys have Andean and Curvebilled Tinamous, Bare-faced Ground-dove, Andean Swift, Sparkling Violetear, Giant Hummingbird, trainbearers, comets, and several other hummers, certain canasteros, tit-tyrants, Golden-billed Saltator, warbling-finches etc., and a number of seedeating finches. Most such birds are common and widespread, at least in Peru and Bol. Some birds of prey and owls, White-tipped Dove, certain parakeets, and seed-eating passerines, occur in the savanna patches all the way across the Amazon lowlands and through the chaco and cerrado zones, and high up in some Andean valleys. Particularly many lowland forms (parakeets, earthcreepers, spine-tails, thornbirds, antshrikes, warbling-finches) move high up in the 'Valles' of Bolivia, sometimes to the temperate zone. This area also has several endemic species, some with counterparts in the lowlands (a macaw, an earthcreeper, a mockingbird, Bolivian Blackbird, a saltator, a warbling-finch), in other montane woodlands (Cochabamba Mountain-finch), and in the puna zone (Wedge-tailed Hillstar, Citron-headed Yellow-finch). Other species are more widespread in the algarroba woodlands of the Andean slope of Bolivia and northwestern Argentina.

The only other zone of dry scrub or woodland having many endemic forms is on the Pacific slope of Peru and southwestern Ecuador. A few very local forms occur also in the subtropical parts of the Magdalena valley, the Marañón valleys or the valleys of central Peru (Kalinowski's and Taczanowski's Tinamous, Bearded Mountaineer, inca-finches, and certain warbling-finches). Some of these endemics actually belong to rather isolated tracts of humid habitat within arid zones: Bearded Guan, sunbeam hummingbirds, Purple-throated Sunangel, Rainbow Starfrontlet, Gray-greasted Comet, and a few spinetails and finches. Some of these move up into *Polylepis* forests seasonally.

Birds of the steppic zones

The highlands mainly have birds of dull brown and gray colors. The ground-doves and yellow-finches wander around in flocks. Sierra-finches occur more dispersed, and mixed-species flocks are rare here (except wandering flocks of several species of vividly colored siskins). Most bird species of the steppic habitats are common and widespread. Adapted to wet-dry cycles they show low site tenacity and high pioneering abilities, and quite opportunistic feeding habits. This may mean that closely related species segregate mainly according to physical aspects of their habitats.

Typical birds of the grasslands are tinamous (*Nothoprocta, Nothura*), seedsnipes, ground-doves, earth-creepers, some canasteros, and pipits. The short-grass habitat of grazed shore meadows have Andean Lapwing, miners, many ground-tyrants, yellow-finches, and white-winged Negrito. Stony slopes often have ground-tyrants, Bar-winged Cinclodes, and sierra-finches. The desert puna has very few species, but some miners and finches, Puna hawk, and Mountain Caracara can be seen in the very high and rugged parts, and Rufous-breasted Seedsnipe and White-winged Diuca-finch are partial to habitats near the snow-line.

Only very few puna birds have very small total ranges (Darkwinged Miner, White-bellied Cinclodes, White-throated Finch, and a few birds of the Fuegian zone). The rarest forms live in very bleak habitats where no birds can be expected to have high densities (Diademed Sandpiper-plover, Olivaceous Thornbill, White-bellied Cinclodes, Short-tailed Finch, Creamy-rumped Miner).

Whereas the puna seems quite uniform upon casual observation, it is sufficiently complex to restrict the distribution of many birds. Vuilleumier & Simberloff (1980) claim a patchy distribution of many puna birds. However, local distributions could reflect requirements that are only satisfied locally. In our experience, very few birds of open habitats are more patchily distributed within the puna zone than what habitat gradients and presence of related forms will account for.

Only 115 of the 290 species living above the general Andean treeline and on the Patagonian plains decidedly prefer the open boggy, grassy, and desert-like habitats. This applies especially in the northern Andes, where very few birds are partial to open grassland (2 canasteros, a ground-tyrant, pipits, a lapwing, and an ibis).

Of the remaining 175 species, 66 inhabit bushy or wooded canyons or slopes at high elevations, 38 mainly elfin forest and their borders, 33 scrubby areas bordering arable land. So, despite the present dominance of grassland and semi-desert, the highest species richness is in the small patches of taller vegetation. It is among these birds that we find the patchy or disjunct distributions.

Birds of relict high elevation woods often have very small or disjunct ranges. Sure enough, some of them are as continuously distributed as their patchy habitat permits (e.g. Rufous-webbed Tyrant, D'orbignys Chat-tyrant, Giant Conebill) but others have widely separated populations (Tawny Tit-spinetail, Junin and Line-fronted Canasteros, Ash-breasted Tit-tyrant, Stout-billed Cinclodes, Blue-mantled Thornbill, Tit-like Dacnis). Many species show more or less well marked geographic variation, with leap-frog patterns suggesting rapid divergent evolution in some isolated populations. Royal Cinclodes, White-browed Tit-spinetail and Berlepsch' Canastero have extremely small ranges. Similarly, many inhabitants of páramo shrub in the northern Andes are very local and often well differentiated.

Some high Andean forest birds have very specialized feeding habits. The Giant Conebill is strictly dependent on the *Polylepis* trees, where it searches insects by flaking bark. Blue-mantled Thornbill, Tit-like Dacnis, and Plain-tailed Warbling-finch feed on sugary secretions on the undersides of *Gynoxys* leaves. A close dependence (mutualism) also exists between the White-cheeked Cotinga and mistletoes.

Many birds of the isolated high Andean woodlands also have populations in the humid treeline habitats lower down. The speciation may have occurred as treeline birds spread into the highlands in periods with extensive *Polylepis* tracts, and subsequently became isolated (e.g., in the West Cordillera of Peru) as the forest became fragmented in cold and dry climatic periods. While the nominate subspecies of the Tit-like Dacnis is fairly widespread (but scarce) in humid treeline habitats in Peru (feeding in the interspersed *Gynoxys* bushes), the subspecies (or species) *petersi* occurs very locally (but sometimes abundantly) in dense *Gynoxys* stands associated with isolated *Polylepis* woods in the Western Cordillera.

The arid zone of central Argentina has a poor fauna, mainly widespread openland and chaco forms. Burrowing Parakeet, certain tinamous, canasteros, and finches, Crested and Sandy Gallitos are distinctive birds of this zone. Some isolated patches of grassy highland have small relict populations of puna birds. This isolation has favored the speciation process for Argentinian highland birds (Nores & Yzurieta 1983). Most noteworthy are the Córdoba mountains, with three distinctive ovenbirds.

The drier zones of central Chile have a similar but poorer fauna than central Argentina. Crag Chilia and the two *Scelorchilus* tapaculos are endemic.

The most characteristic birds of the grasslands of Patagonia are the geese, especially Upland Goose, which has a densely barred type inhabiting Isla Grande and boggy glades of the forest zone, and a scarcely barred type inhabiting the open plains. Despite large-scale persecution by man, owing to claimed competition with sheep, these geese are still abundant in the south of the continent. Preferring river plains in the lowlands and near lakes, they also occur far from water, whereever there is boggy ground or low green vegetation associated with melting snow or wells.

Otherwise, the fauna resembles that of the arid puna, with a few specialties, e.g. Band-tailed Earthcreeper, White-bellied Seedsnipe, Patagonian Tinamou, and *Melanodera* finches.

Birds of the Valdivian forest zone

A large proportion of the birds of the southern forest zone are actually also widespread in bushy parts of the surrounding arid zones. However, some are partial to mature humid forest, and have their nearest counterparts in remote areas (Rufous-tailed Hawk, Chilean Pigeon, a hummer, two owls, woodpeckers, two *Scytalopus* and two large *Pteroptochos* tapaculos). A few lack close relatives (two conures, Des Murs' Wiretail, Rayadito, and White-throated Treerunner), but at least belong in typical South American groups. Unlike plants and insects, no birds are related to forms of Australia's or New Zealand's faunas. Due to the lack of true barriers within the forest zone today and during the glacial periods, very few Valdivian birds have been differentiated into species pairs (Vuilleumier 1985).

Waterbirds

Data about the main bird-lakes of South America are compiled in Scott & Carbonell (1985). Maybe the most spectacular sites are the weakly alkaline lakes of the puna zone. Most of these have few bird species, but sometimes large numbers, e.g., 10,000s of flamingos, and sometimes 100,000s of Wilson's Phalaropes. Lakes with abundant submergents have large numbers of dabling ducks, Giant Coots, Silvery Grebes, some Ruddy Ducks, and often Andean Gull, Puna Ibis, Puna Plovers, and shorebirds on the surrounding plains. The few lakes which, despite intensive harvesting of waterplants, have reed-beds, may show an even larger selection of species. Lake Junín in central Peru has the record with 60 species of waterbirds recorded, some of them in 10,000s (but fewer in recent years).

Most waterbird species are widespread from Patagonia north in the Andes to Peru, or even further. A few are strictly Andean (e.g. the Andean Goose, Andean Gull, and the peculiar Diademed Sandpiper-plover), and a few birds are endemic to smaller parts, e.g., endemic (flightless) grebes in Lakes Junín and Titicaca, Horned Coot, and two flamingos in the desert puna.

As most Andean slopes lack wetlands, the highland and lowland populations are often well isolated. This explains a certain differentiation into subspecies or species pairs. Isolation in the large glacial lake that once covered the Bolivian altiplano may have played a major role for the evolution of endemic Andean waterbirds (Fjeldså 1985).

The lake districts of the South Andes have moderate densities of birds, but some fine wetlands exist along the Río Cruces in Valdivia and in the Fuegian zone. These lakes are close to marine wetlands that hold large numbers of wintering waterfowl: the tidal estuaries and salt marshes in Ancud, on the Chiloë island, near Puerto Natales, and Porvenir on Isla Grande. The claypan lakes of the Patagonian steppe have become almost unsuited for waterbirds, owing to soil erosion. The main concentrations of Patagonian waterbirds are found on the small volcanic lakes that dot the volcanic foothill mesetas to the east. Thus the foothill at 800-1200 m between Lakes Cardiel and Strobel had more than 60,000 waterfowl in February 1984. Endemic forms of the southern region are the remarkable Hooded Grebe, some geese and steamer-ducks, Spectacled Duck, Magellanic Oystercatcher, Two-banded Plover, Rufous-chested Dotterel, and the peculiar Magellanic Plover, and some distinct subspecies.

The páramos of the northern Andes generally have only small local populations of rather adaptable waterbirds. Several marsh-birds from the adjacent lowlands ascend to the temperate zone here. A few of the latter have evolved endemic subspecies in the wetlands of the East Andes of Colombia. According to the analysis by Fjeldså (1985), the avifauna of these wetlands is not saturated, as certain species guilds are very incomplete. It was suggested that time had not been adequate for adaptive shifts in response to 'vacant' niches in open lakes. More probably, open lake habitats are generally too small or isolated to support populations of species adapted to such habitats over longer time periods.

Migrants

Bird migration in the Andean zone can be divided into three main components: **North American migrants** visiting during the northern winter (October-April or longer), **southern migrants**, which move north during the southern winter (March-October), and **local migrants**, which inhabit high Andean habitats in the rainy season, and move down the slopes at other times.

The northern migrants naturally dominate in Colombia and Venezuela. Most of the migrant passerines winter in the lowlands, but some cross the northern Andes on their way south. A number of passerines establish winter territories, often in the same spot year after year. Others wander. The migrants compete for space with local species in forest habitats that currently shrink, because of human activity.

The most conspicuous northern migrants are the waterbirds – both waterfowl that mainly visit Colombia, and shorebirds, of which several fly all the way to Patagonia. Yellowlegs, Baird's Sandpipers, and Wilson's Phalaropes winter in large numbers in the puna zone. Some remain in South America also in their normal breeding months. These are mainly second-year birds, not yet in breeding condition.

Most birds of the South Andes have to move down to low elevations in winter (e.g., seedsnipes), or migrate north in the Andes (e.g. seven species of ground-tyrants). Many Valdivian forest birds are sedentary, but some move to central Chile. Among the Fuegian waterbirds, many remain in estuaries and coastal waters far south, but others move to the latitude of Buenos Aires, with a minor current travelling north to central Chile. Also most birds of the inhospitable Patagonian steppes move north.

Most lowland birds of central Argentina are partly migratory, and in widespread forms, some populations migrate into the range of resident populations (e.g. Picui Ground-dove, Dark-billed Cuckoo, Elaenias, Swainson's, Scrub, and Euler's Flycatchers, Forktailed and Dusky-capped Flycatchers, Tropical Kingbird, Brown-chested Martin, Blue-and-white Swallow, and maybe American Kestrel). The extent of such movements are poorly known because of the difficulty of recognizing the populations.

A number of birds of chaco and savanna show rather irregular wanderings. Thus, storks, whistling ducks, Masked Duck, Purple and Azure Gallinules, and Jacanas, appear casually in the high Andes and other strange places. Multidirectional postbreeding dispersal of ibises and herons may also be mentioned.

Although many forest birds are unwilling to pass gaps of unsuitable terrain, they can move some distance through unbroken habitat. This usually permits movements up and down the Andean slopes, which causes pronounced seasonal differences in the local faunas. It is widely known that hummingbirds move seasonally up and down mountain slopes as the blooming shifts, but the amplitude is poorly known. The shifts can be very swift, as entire populations virtually disappear overnight. The parrots are sometimes quite erratic in occurrence, because of shifting food supplies, and any bird adapted to seeding bamboo is thought to show erratic movements. Oilbirds sometimes move (post-breeding) from their traditional breeding caves to new potential supplies, often far away, wandering seemingly aimlessly.

The birds of the highlands are more mobile in general, but even here most movements are local, e.g. seasonal flocking and habitat shifts among yellow-finches, seed-eaters, Ash-breasted Sierra-finch, and siskins, or movements of hummingbirds between high-elevation woodlands in the rainy season and Andean valleys in 'winter'. Highland birds can be seen to rush down into adjacent valleys during snowstorms, but a few species (e.g., Tawny Tit-spinetail, Stripe-faced Antpitta, chat-tyrants) remain on their territories despite some snow.

CONSERVATION AND THE FUTURE

The human activity – although it has strongly altered many Andean habitats – has caused very few total extinctions of bird species so far. However, numerous forms have declined strongly, and may soon become endangered unless protective measures are taken.

Very little has been done so far to preserve Andean habitats and birds. This is first of all due to lack of information, for no proper planning of conservation actions is possible before the problems have been documented. Fortunately the increase of modern biological investigations in the Andes has made it possible to develop some conservation priorities. Data about endangered and vulnerable birds have been compiled in the IUCN/ICBP Red Data Books, of which a strongly revised volume for South America is in progress. 'Blue lists' of species of indeterminate status have also been given (Hilty & Brown 1985). These are potentially endangered birds which, subjected to current research, may serve as good 'early warning systems'.

A considerable proportion of the threatened birds of South America belongs in the Andean zone. Of the blue-listed birds of Colombia, 48% occur primarily in the subtropics (9% of the land surface) and 20% in the temperate zone (6% of the land surface). This ratio also seems to apply further south.

The primary cause of the problems are forest destruction and inappropriate landuse leading to massive soil degradation, erosion and poor water quality, and heavy use of insecticides locally. Such problems are today global, and cause interminable problems due to the growth of the human population, and the failure to view nature as a renewable resource. The locals are forced to seek short-term and inappropriate solutions to long-term problems.

One might think that the birds of the most disturbed parts are most endangered. This is not necessarily true. On the contrary - because man's impact in the puna and in the Andean valleys has already been severe for thousands of years – we feel that most of those species that are still present in these disturbed areas are in no danger. The centers of the old Andean cultures were mainly in semiarid regions, where most wild species have opportunistic habits. They possess considerable potential for adapting to altered conditions, and for moving around. Most granivores, and species adapted to short grass, may actually benefit from the clearing of woodland and cultivation. In many areas a rotational use maintains mosaic vegetation, and thus habitat for many species.

Even very local forms, vulnerable due to their small ranges, are often fairly numerous. Worries exist, though, for woodland birds, since their habitat is in general proximity to high-density human populations, and have shrinked to extremely small and scattered patches. Among birds inhabiting the valleys of the central Andes, the following may be in trouble: Taczanowski's and Kalinowski's Tinamous, Purple-backed Sunbeam, Gray-breasted Comet, Little Woodstar, Great Spinetail and Rufous-breasted Warbling-finch. The endemics of the remaining patches of humid forest on the Pacific slope of southwestern Ecuador into northwestern Peru are in general trouble. The situation is also serious for some birds of the relict woods at very high elevation (Fjeldså in press). The Royal Cinclodes *Cinclodes (excelsior) aricomae* and the Ash-breasted Tit-tyrant are extremely scarce.

In the grasslands and semideserts – although severely altered by grazing – most birds maintain good populations. Only a few may have suffered greatly. The Puna Rhea has been hunted nearly to extinction in some parts, and the overgrazing of the plains may have harmed such species as the Short-eared Owl, which is extremely scarce in the Andes. The severe decline of the tiny Ruddy-headed Goose of Patagonia is normally attributed to persecution and predation by foxes. We find it more probable that it suffered from the disappearance of tall lush grass. This could also account for the apparent disappearence of the Antarctic Rail, and maybe for the present rarity of Austral Canastero and Black-throated Finch.

The wet montane forests are intact in large parts, notably on the steep slopes facing the Amazon lowlands. The elfin forest zone is particularly unsuited for cultivation. Thus, in the upper parts of the cloud forests of Peru and the Yungas of Bolivia even the very local forms seem to be safe. However, the habitat is strongly fragmented in other parts. The premontane forest is particularly threatened because this is an optimal climatic zone for growing coffee, coca, citrus fruits etc. But in the northern part of the Andes the forest destruction extends up into the montane zone. In Colombia, the forest destruction is widespread all the way between the rain-drenched slopes towards the Pacific side and the eastern lowlands (Hilty & Brown 1986), and the situation seems even worse in Ecuador. Large portions of once forested regions now consist of eroded, semi-cultivated slopes with small patches of shrub, and this process seems to accelerate. As the forest fragments shrink and become increasingly isolated, they can no longer hold viable populations of all species.

The effect of forest loss is evident from the examination of the 'Bogota trade skins' from the 1800s. Brown Tinamou, Andean Potoo, Banded Snipe, Turquoise-throated Puffleg, and Rufous-browed Tyrannulet, have not been recorded in Colombia since. The Gorgeted Woodquail occurs only very locally in *Quercus* woods. The large canopy fruiteaters, guans, trogons, toucans, and notably parrots, must have declined seriously (see Yellow-eared and Rusty-faced Parrots). The rich hummingbird fauna has changed dramatically: More than 20 'Bogotá' hummingbirds once thought to be species are now regarded as hybrids. Yet, curiously, hybrid hummingbirds lack in collections made in the present century. 'Bogota' hybrids may reflect collecting done when rapid land clearing permitted contact between previously isolated siblings, but some of these birds may represent forms that have now disappeared (Hilty 1985). (Unfortunately, the ornithological exploration of Ecuador started too late to permit a similar comparison here).

While the frugivores and probably also several birds of the forest floor have become rare or local in the remaining forest patches, a number of unspecialized small insectivores have become very abundant. Thus, the communities of birds in strongly fragmented forest tracts have changed from a high species diversity to impoverished communities dominated strongly by some few species. Unfortunately, detailed analyses of this process still lack in the Andes, although the University of Copenhagen has initiated some projects.

Although the ecological impact of reforestation with *Eucalyptus* has hardly been studied in the Andes, it seems that these plantations represent valid habitat for only the most tolerant forms. Small patches may give some ecotone effects, but larger plantations are exceedingly poor for birds.

The soil erosion following forest clearing causes siltation of the wetlands. Most bird-lakes in the highlands of Ecuador are now of very low value. The Southern Pochard, which requires clear, shallow water with dense submergents, has virtually disappeared. A large-scale decline of marshbirds in the temperate savannas in the East Andes of Colombia is mainly caused by soil erosion and draining. 500 years ago the Bogotá savanna was an enormous swamp; today it is pastureland and greenhouse areas with some tiny marshes. This habitat destruction has already led to the extinction of the Colombian Grebe and Niceforo's Pintail. Other endemic waterbirds have become very local, but may still be saved if adequate measures are taken to secure marsh habitats.

Owing to siltation of Andean rivers, the Torrent Duck has disappeared from several drainage systems.

The Junín Flightless Grebe declines rapidly towards extinction, because of pollution from mine-washing and the regulation of its lake for hydroelectric power. Regulation schemes, mining etc now also threaten some of the spectacular flamingo lakes in Bolivia. The waterbird fauna of the Bolivian altiplano may also have suffered from the egg-culling traditions of the Aymara Indians. This may be the reason for the lack of ibises in large areas with apparently suitable habitat. The exploitation has become a severe threat to the flamingos, as people now can reach remote salt lakes by truck.

That wildlife is threatened by human activity is now so frequently announced in the mass media that one could fear that extinction is apathetically perceived as compulsory. Yet, the future may not look altogether bleak. There is finally some political understanding of the imperative needs for long-sighted use of the World's resources. This understanding is exemplified by the adoption, in the United Nations, of a World Conservation Strategy, by the recent policy change in the World Bank, and by the recommendations of the 1986 World Commission for Environment and Development. Although nature conservation as such is not part of the strategy for 'sustainable development', habitat conservation has certainly become more feasible. 'Biological diversity' is gradually accepted as a valuable resource for the future generations.

Still, the regulating local authorities are often unable or unwilling to see the problems, or to enforce environmental politics. A strict opposition against exploitation may be viewed as an elitarian view pressed upon them by experts of the highly developed countries. Their view is particularly difficult to understand for populations existing at subsistence levels. Although exploitation for the cage-bird market threatens certain species by extinction, a total exportation ban on wildlife is not necessarily the best way to protect wildlife, because it removes some local incitements for preserving wild habitats. Also projects directed narrowly at single endangered species may be bad investments, by diluting efforts from the conservation of critical habitats. Thus, a modern conservation action plan must represent a balanced approach, where the natural habitats are incorporated in the local economies. By focusing on the ecosystem rather than on its species we make advantage of the self-regulating property of the ecosystem – if the ecosystem remains intact the survival of its species comes for free!

Conscious and conservation-oriented biologists can now be found in several institutions in the Andean states. Their training inproves, but they are constantly starved for resources for field work. Although we believe that only South Americans can save their natural environments, the participation of 'experts' from scientifically advanced countries is still necessary. However, they should not only explore, but also engage themselves in the funding and training of the locals, and in the development of new sustainable methods for landuse. Funding institutions in developed countries have favored projects with purely scientific aims, or for their own museum collections. However, it is indeed possible to formulate projects that combine analysis of general biological theory with engagement in the local conservation education. Furthermore, a number of national agencies for development aid finally begin to realize the relevance of supporting conservation management. Also the non-professional birder can contribute by collecting basic data, and by providing external contacts for often quite isolated local naturalists.

The ornithologists of South America received a tremendous source of contact by the development of a continental coordination for bird preservation through ICBP/IWRB. This started during the registration of important waterbird habitats in the Neotropics (Scott & Carbonell 1986). The network established for this project has been reinforced. Education projects, workshops, and exchange of information are flourishing activities (see the *Boletín Panamericana*).

Several governmental and non-governmental institutions in the US and Canada have shown a strong interest in threats to habitats for their migratory birds in the Neotropics. Conservation efforts in North America would be in vain if species in fact are controlled by factors in South America. Therefore, a number of projects for monitoring flyways of waterbirds, making ecological studies in the winter quarters, and training South American ornithologists, have now started in the US and Canada.

Organizations in these countries are even more concerned with the indigenous tropical fauna. While the voluntary organizations previously focused on individual species, often animals of public appeal, they now realize the fundamental needs for preserving ecosystems. Thus, the Pan-American section of World Wide Fund for Nature (WWF) recognized the tropical Andes as a major conservation priority 'primarily because of the tremendous diversity of plant and animal species in the Andean montane and premontane forests, and the accelerated pressures of habitat conversion that are affecting them'.

Despite the large amounts of intact forest, WWF considers it imperative to preserve (in time) key areas which can 'help maintain ecological and evolutionary processes to support a wealth of species, and ensure the conservation of natural genetic reservoirs of great economic value.' If this network of reserves shall function, key areas must be selected on basis of biological knowledge and not just from political convenience. One main criterion is biodiversity; but here most emphasis should be on the 'load' of endemic taxa. The endemism reflects the history and future potential of each area. Areas for protection should ideally encompass altitudinal gradients through zones with high rainfall and high species diversity. They should also be large and defendable. 16 areas already protected and 15 unprotected areas are part of this plan.

An equally important facet of the campaign is conservation education at all levels from school children to upper level students, politicians, administrators, and journalists.

Much research remains to be done in order to know the adequacy of such reserves, and to state the needs for additional (smaller) reserves. Several endemic forms live outside actual and proposed reserves, and undiscovered birds are still expected to exist in poorly sampled areas. Even more is needed before the local authorities can develop and reinforce management plans for a rational and sustainable use of the areas.

Practical Hints for Field-work in the Andes

EQUIPMENT

Binoculars: 7x35 to 10x50 is best, and for fieldwork in the humid parts airtight (preferably holding pure nitrogen rather than normal air with water-vapor) models are recommendable. For studies of waterbirds, telescope 20x magnification is most useful. In the cool clear weather dominating in the best Andean waterbird areas magnifications up to 40x can be used.

Notebook and pencils: Never rely on memory, but take frequent notes. Since the notebooks for sale locally are often of poor quality, we recommend that an adequate supply of a 'favorite model' is brought from home. Bring pencils as they will write in cold and rain. Large pockets and a rainproof bag is important, both for notebooks and fieldguides.

Maps: In most Andean states, excellent maps (road maps, 1:100,000 or 1:50,000 topographic maps) can now be bought in the *Instituto Geographical Nacional* (or *Militar*) in the Capital.

Photographing: A telelens is normally required for bird photographing. It is safest to bring all the films you expect to use from home. Many of the Andean birds have never been photographed or tape-recorded. Especially scarce are photos of forest birds which are often hidden to view or stay in deep shade, where a flash-light is needed. Also photos of raptors in flight are in great need.

Tape-recording: A directional microphone (e.g., Sennheisser ME 80 or even better ME 88) will give good results even with a small casette recorder, but a parabola gives a louder recording. A recorder with a loudspeaker (rather than a walkman) is preferable, as the birds can be attracted with playback for safe identification. Always note, when you make recordings, whether the bird was taped after playback (many birds tend to get unusually excited after playback), and try to guess the function of the vocalization. Talk onto the tape, as written notes tend to get lost. Use playback only when strictly necessary (during research), and don't make playback with endangered species, which may be induced to leave their territories. Tape-recordings are essential for surveys of forest, as many species will be detected in no other way. TA Parker suggests the following: pick a good spot for recording, and be there before the first light. Choose a different spot each morning, preferably areas at least 500 m apart, and let the recorder run for 15 minutes or more. Point the microphone in the directions of louder sounds for at least 60 s, and try to record (at least briefly) in all directions, and from the undergrowth to the canopy. Also find areas where mixed-species flocks are forming at dawn and record for at least 10-15 minutes, and get 5-10 minutes of sounds from any large flock at any time of day. Place recordings with one or more of the collections of natural sounds: Library of Natural Sounds, Cornell University; Bioacoustical Laboratory (Gainesville, Florida); Arquivo Sonoro Neotropical (Universidade Estadual de Campinas, Brazil); Laboratorio de Sonidos Naturales (Museo de Ciencias Naturales, Buenos Aires); British Library of Wildlife Sounds (London); Bioakustisk Laboratorium (Aarhus University, Denmark).

Mistnets: Recording birds in dense and entangled forest require long experience with voices, and that the observations are supplemented by capturing of understory birds in fine nets stretched out between poles. Mistnets must usually be ordered through a bird-ringing institution. It would be irresponsible of us to describe the use of mistnets further here, since no-one should mistnet birds without first having worked together with a trained bird-netter.

Information: Local people can often give valuable information about the birdlife of an area. But be critical. Try to test the reliability by asking specific questions. 'Yes' or 'no' questions are of little use. They give no basis for judging whether the informant understands. And remember that in some cultures it is 'polite' to answer 'yes' (or persons will answer 'yes' to hide their ignorance). Don't put words into the person's mouth.

TRAVELLING IN THE HIGH ANDES

Travel information: The 'Bible' for travelling in South America is the **South American Handbook** from Trade & Travel Publications, Bath, England. Revised annual volumes (c. 1300 pages) give an immense amount of practical information about travelling, hotels, sights, shopping etc. (but next to nothing for the special interest of biologists). Today, a number of more local guides, especially trekking guides, have also appeared.

Permits: Although visas are not normally needed in Andean states, it is advisable to check with the Foreign Ministery or with an embassy well in advance of the departure (especially if you plan to stay long). For scientific missions and especially collecting or exportating of material, special permits are needed. Precise information should be obtained through national museums and the CITES administration (see adresses below).

Transport: National air flights are generally cheap within the Andean states, and most larger towns are connected by bus routes. However, the time schedules may be unreliable (especially in the rainy season, when the roads are often blocked by landslides). Furthermore, it may be difficult to obtain tickets, the same seat is sometimes sold to more than one person, and buses are usually overcrowded. Be aware of pickpockets on the crowded bus- (and railway) stations, and as you fight your way to your seat.

The locals mainly travel by truck. This transportation is very cheap, but unpredictable and uncomfortable. However, on the minor sideroads there are few alternatives, unless you have your own car. Car rental is generally very expensive, and it is often cheaper to hire a car with a driver.

Backpacking is tough at high elevations. Above 3000 m, you easily loose your breath when walking uphill with a heavy burden. To bring the equipment up to a high-elevation camp it is wise (and not expensive, if you know how to bargain) to hire a mule-driver and some animals.

Camping and cooking: The arid puna - although day temperatures are pleasant – often is bitterly cold at night. A sleeping pad and a good sleeping bag (or two, one inside the other) is necessary. On the humid slopes, your tent must be waterproof, but as the temperatures are not so extreme, the sleeping-bag quality is less important. Wellington boots are useful on the puna bogs, and in the humid zones good rainclothes are indispensible. It is impossible to buy rainclothes in the Andes, as local people use ponchos or plastic bags etc.

A foldable plastic watercontainer and a small portable alcohol- or kerosine-stove need not take much space. Water is usually drinkable in the high parts of the Andes and in uninhabited tracts of humid forest, but in other areas it should be boiled or sterilized chemically. Kerosene is easy to obtain, while alcohol (*alcohol industrial*) often can be obtained only from special vendors in the larger towns (in some areas only sold in pharmacies). Only alcohol more than 90% is efficient for cooking at high elevations. Remember that the cooking time for potatoes, beans, rice etc. is very long at high elevations, although it can be somewhat shortened by adding the salt as late as possible.

Salami or dried meat is a lasting proteinsource (even in humid climates), but generally is sold only in larger towns, as is also the case with good coffee and yellow cheese. Eggs, fresh meat (sheep, alpaca, cow, guinea-pig etc.) are available on the markets throughout the highlands. Bread is often harder to come by, and usually must be obtained in the morning. Local cheese is found in many places, but may give digestive problems. Fresh milk is generally hard to find, but powdered milk is widely sold. Sweets, beer, and bottled softdrinks are ubiquitous, as are cigarettes. Don't ex-

pect to be able to live from wild fruits and berries.
Insects, health: There are generally very few poisonous snakes in the highlands. Insect repellant is rarely necessary in the high Andes. Malaria pills need not be taken, and most other tropical diseases are also absent. Beware of dogs, which may carry rabies. **Hepatitis** and **diarrhoea** are the commonest problems. Local pharmacies sell various pills that cure most intestinal infections. But use them as little as possible, as some may have serious secondary effects. Lasting diarrhoea involves a severe loss of water, and it is important to drink much water (water with a spoonfull of sugar and a teaspoonful of salt per liter).

Altitude problems: While most people can travel from sea level to 3000 m in a few hours without complications, care must be taken with greater elevational movements. Going straight to 4500 m will cause complications in at least one out of three persons. Dizziness, headache, vomiting, and anaemeia are common symptoms. If you turn blue or get lung oedemia (coughs with excessive amounts of pink mucus), you must immediately return to lower elevations. Eat light, or not at all if you are anaemic, as blood flow to the intestines increases during the digestive process. Be especially careful with alcohol. Do not run, and avoid going uphill. Most people can manage 4500 m after one night at 3000-3500 m, providing they do not engage in strenuous efforts. After 3-5 days above 4000 m, normal conditions are mostly restored. Coca tea can often help the first few days, and also some pills against altitude sickness are for sale (they must, however, be taken before the ascent, and they are often without effect).

Sun: In the tropics, hat and clothes is necessary as protection against the sun during most of the day. **This becomes increasingly important at high elevations**, where especially the U.V.-radiation is less filtered by the atmosphere. In snow-covered landscapes with a bright glare, wear masks with tiny slits to see through. Snow blindness may not start until 4-6 hours after the exposure, symptoms being very painful eyes or even temporary blindness for a day or more.

The mid-day temperatures may be pleasant, but be prepared for a drastic fall in temperature after sunset.

USEFUL ADDRESSES FOR RESEARCH AND CONSERVATION

International Council for Bird Preservation (ICBP), 32 Cambridge Road, Girton, Cambridge, CB3 OPJ, UK. International Union for Conservation of Nature and Natural Resources (IUCN), Avenue du Mont Blanc, 1196 Gland, Switzerland (data monitoring center at same adress as ICBP). International Waterfowl Research Bureau (IWRB), Slimbridge, Gloucester GL2 7BX, UK. International Affairs Staff, US Fish and Wildlife Service, Department of the Interior, Washington DC, 20240 USA. World Wide Fund for Nature (WWF) US, 1601 Connecticut Avenue NW, Washington D C, 20009, USA.

Argentina: Administración de Parques Nacionales, Santa Fe 690, 1059 Capital Federal. Dirección Nacional de Fauna Silvestre, Paseo Colon 922-2° piso, oficina 201, 1063 Buenos Aires (CITES administration). Museo Argentino de Ciencias Naturales 'Bernadino Rivadavia', Avenida Angel Gallardo 470, 1405 Buenos Aires. Museo de la Plata, Paseo del Bosque, 1900 La Plata. Universidad Nacional de Buenos Aires, Facultad de Ciencias Exactas y Naturales, Ciudad Universitaria, Pabellon 2, Capital Federal. Universidad Nacional de Córdoba, Pabellon Peru, Ciudad Universitaria, 5000 Córdoba. Universidad Nacional de Salta, Museo de Ciencias Naturales, 4400 Salta. Universidad Nacional de Tucuman, Inst. 'Miguel Lillo', Miguel Lillo 205, 4000 San Miguel de Tucuman, Tucuman. – Voluntary: Fundación Vida Silvestre Argentino, Leandro N Alem 968, 1000 Capital Federal. Asociación Natura, 25 de Mayo 749, 1°Piso, Buenos Aires. Centro de Ecología y Recursos Naturales Renovables, Univ. Nacional de Córdoba, C.C. 395, 5000 Córdoba. Comité Argentino de Conservación de la Naturaleza. Avenida Santa Fe 1145, Buenos Aires.

Bolivia: Museo Nacional de Historia Natural, Casilla 5829, La Páz. Centro de Desarollo Forestal. Av. Camacho 1471 6to Piso. Casilla de Correo No 23398, La Páz (CITES administration). Universidad Mayor de San Andres, Instituto de Ecologia, Casilla 201 27, La Paz.

Chile: Centro Administrativa de Chile para CITES, Servicio Agricola y Ganadero. Avda. Bulnes 285-5° Piso, Casilla 4088, Santiago. Corporación Nacional Forestal (CONAF), Avenida Bulnes 285-5° Piso, Santiago. Instituto de Ecología y Evolución, Universidad Austral de Chile, Casilla 567, Valdivia. Facultad de Ciencias Biologicas y de Recursos Naturales, Univ. de Concepción, Casilla 2407, Apdo 10, Concepción. Museo Nacional de Historia Natural, Casilla 787. Santiago. Pontifica Universidad Catolica de Chile, De-

partamento de Biologia Ambiental, Casilla 114-D, Santiago. Universidad Catolica de Valparaiso, Inst. de Biologia, Casilla 4059, Valparaiso. Universidad de Magellanes, Instituto de la Patagonia, Casilla 102D, Punta Arenas. – Voluntary: Comité Nacional pro Defensa de la Fauna y Flora, Casilla 3675, Huerfanos 972, Oficina 508, Santiago. Consejo Internacional Protección de Aves – Chile. Calle del Arzobispo 0605, Santiago.
Colombia: Instituto Nacional de los Recursos Naturales Renovables y del Ambiente (INDERENA), Gerente General, Diagonal 34, Numero 5-18, Apdo Aéreo 13458, Bogotá (CITES administration). Instituto de Ciencias Naturales, Museo de Historia Natural. Apdo Aéreo 7495, Bogotá. Dept. Biología, Universidad del Valle, Apdo 2188, Cali. Museo de Historia Natural, Universidad del Cauca, Calle 5A No. 3-38, Apartado Aereo 742, Popayan, Cauca. Universidad de Tolima, Departamento de Biologia, Apdo. Aereo 25360, Cali. – Voluntary: Asociación National para la Defensa de la Naturaleza, Apdo Aéreo 6227, Cali. Sociedad Colombiana de Ecología, Calle 59 N° 13-83, Of. 302, Bogotá. Sociedad Vallecauca de Ornitologia, Apartado Aeria 26538, Cali.
Ecuador: Director Ejecutivo del Programa Nacional Forestal, Ministerio de Agricultura y Ganaderia, Casilla 2919, Quito (CITES management). Museo Ecuatoriano de Ciencias Naturales, Tamayo 516 y Carrión, Casilla 8976, Suc. 7, Quito. Dept. de Biología, Univ. Católica, Apdo. 2184, Quito. Pontificio Universidad Católica del Ecuador. Dept. de Ciencias Biologicas. 12 de Octubre y Roca. Apto. 2184 Quito. – Voluntary: Fundación Natura. Jorge Juan 481, Casilla 243, Quito.
Peru: Dirección General Forestal y de Fauna, Ministerio de Agricultura. Jiron Natalio Sanchez 220, 3°piso, Jesus Maria, Lima (CITES management). Dirección General de Conservación, INFOR, Santa Cruz 734, Jesus María, Lima 11. Museo de Historia Natural 'Javier Prado', Ave. Arenales 1256, Apdo Postal 11010, Lima 14. Centro de Datos para la Conservación (CDC), Dept. Manejo Forestal, Univ. Agraria La Molina, Apdo 456, Lima. Departamento de Biologia, Univ. Nacional de San Antonio Abad del Cusco, Urbanización Kennedy A-19, Huanchac, Cusco. - Voluntary: Asociación Peruana para la Conservación de la Naturaleza (APECO), Atahualpa 335, Lima 18. Pro Defensa de la Naturaleza (PRODERA), Avenida Nicolas de Pierola, 742, Of. 703, Edificio Int., Lima.
Venezuela: Ministerio del Ambiente y de los Recursos Naturales Renovables, Torre Sur, Piso 19, Centro Simón Bolivar-El Silencio, Caracas 1010 (CITES management). Ministerio del Ambiente ..., Apartado 184, Maracay, Aragua. Depto. de Concervación, Univ. de los Andes, Facultad de Ciencias Forestales, Mérida. Gerencia de Parques Nacionales, Apdo. 76471, Caracas 1062. Servivio Navional de Fauna Silvestre, Apdo. 184, Maracay-Estado Aragua. Universidad de los Andes, Facultad de Ciencias, Departamento de Biologia, Apdo. 256, Merida. Universidad Central de Venezuela, Instituto de Zoologia Tropical, Apdo. 59058, Caracas. – Voluntary: Coleccion Ornitólogical W.H. Phelps, Apdo. 2009, Caracas 1010-A. Fundación para la Defensa de la Naturaleza (FUDENA), Apdo 70376, Caracas 1070A. Sociedad Conservacionista Audobon de Venezuela, Apto 80450, Caracas 1080A. Sociedad Venezolana de Ciencias Naturales, Apdo 1521, Calle Arichuana con Cumaco, El Marqués, Caracas 1010A.

Introduction to the Main Text

HOW TO USE THE BOOK

Almost every species treated in this book is illustrated, often in all main plumages, and with the most distinctive subspecies shown. We therefore provide no identification key, but instead suggest that the observer look through the plates to find what best resembles the bird seen. This tentative identification can then be tested by checking the diagnosis and distribution data on the facing text page, and if necessary by reading also the species account in the main text. For difficult groups it is advisable to read not only the description, but also information about habits and voice.

WHICH SPECIES ARE INCLUDED?

Of c. 1750 bird species known from the Andean

region, over 600 are well established breeding birds above 2500 m elevation. These deserve particular attention in this book. As several Andean forms have close counterparts in the Patagonian lowlands, an inclusion of this area would give a complete or fairly complete review of certain groups of birds. We therefore also include the c. 100 bird species recorded in the low passes of the southern Andes and the large foothill plateaus east of the Patagonian Cordillera and the Córdoba mountains.

Besides the typical Andean-Patagonian birds, many premontane birds reach the lower fringe of the temperate zone locally or seasonally. Many birds of the chaco woodlands go high up in the warm 'valles' of southern and central Bolivia. Some tropical lowland forms may wander to barren uplands, and North American birds migrate through the area. This brings the total number of species recorded in the temperate zone up above 1000, with 100 more being borderline cases.

As the work proceeded and we realized how 'contaminated' the high Andean fauna is by lowland forms, we approached the limits of what we could manage to survey. However, the 'point of no return' was reached quite early, as JF had painted all species recorded in the temperate zone on the waterbird plates made in the optimistic first phase of the work in 1983. The limit at the lower edge of the temperate zone was therefore maintained.

The inclusion of lowland birds, although in a way unbalancing, presents valuable lessons for the full interpretation of Andean biogeography. Some such forms represent candidates for future colonization of the highlands, and their vagrancy illustrates how the high Andes could have been colonized in the past.

The temperate zone, as defined above, under CLIMATE, lies above 2300-2600 m elevation in the tropical Andes (with some local variation), and falls gradually southwards to reach sea level at 40 degrees south. However, we do not include southern temperate-zone birds that are limited to the littoral zone (Kelp Goose, flightless species of steamerduck, sheathbills, marine *Cinclodes* species), and we omit typical seabirds.

COMPLETENESS

A coverage of the temperate zone of the Andes and Patagonia implies a complete or nearly complete review of certain families, tribes, or large genera. Our book thus represents a nearly complete illustrated guide to miners, earthcreepers, tit-spinetails, canasteros, *Scytalopus* tapaculos, ground-tyrants, yellow-finches, sierra-finches, warbling-finches, and some other difficult groups.

Besides being a treatise of the Andean/Patagonian avifauna, the book can be used as a guide to Chilean landbirds (except 2 coastal *Cinclodes* species). For Argentina it gives a complete coverage of all decided landbirds occurring south of Río Colorado. Most birds of the coastal desert of western Peru (all except the endemics of the north-west) also ascend high enough for inclusion.

KIND OF TREATMENT

In order to emphasize the typical highland fauna, we decided to treat species in three alternative ways:

(1) **Casual visitors and hypothetical or very marginal forms** are mentioned only briefly: Name, concise statement of their typical distribution, and to what extent they reach the temperate zone, followed by a diagnosis. Such presentations fill only one paragraph.

(2) **Lowland forms which are well established in the lower temperate zone or sometimes wandering higher up**, receive a broader treatment. Here, the heading, with name and reference to the illustration, is followed by a description with some biological data which may aid identification. A second paragraph describes habitat and distribution.

(3) **Birds established well into the temperate zone or visiting the zone in some numbers receive 'full treatment'**, with separate paragraphs on appearance, habits, vocalizations, breeding, habitat, and distribution. The genuine highland birds are treated in most detail, while ubiquitous species ranging from low to high elevations often are described more superficially, as they are covered well in other books (e.g. Hilty & Brown 1986).

Distribution maps are given in the margin adjacent to the species accounts of the third and second categories, but only occasionally for first-category species.

HOW ARE THE BIRDS CLASSIFIED?

The sequence of families and species, the grouping of species in genera, and the decisions about what to recognize as full species, follow in general the opinions of Meyer de Schauensee (1966, 1970). A few exceptions are due to recent systematic revisions. The first families (tinamous through waterfowl) were revised in a 2nd edition of Peters' *Checklist of Birds of the World* (Mayr & Cottrell 1979), and we have accepted most decisions of this revision. This involves a new sequence of the species of several families, and waterfowl and birds of prey changing positions. The birds of prey then come after the wading birds (to whom they seem to be related), and the waterfowl come before gallinaceous birds (with whom they are traditionally thought to be allied). Other exceptions from the Meyer de Schauensee sequence is in each case mentioned in the family accounts. We do not follow the re-classification of ovenbirds by C Vaurie (*Bull. Am. Mus. nat. Hist* 1(1980):1-357). As the discovery of some new forms of tapaculos renders Zimmer's revision (*Am. Mus. Novitates* 1044(1939):1-18) incomplete and partly misleading, we give a new classification (*viz.*, the current view of NK and TS).

We accept that the massive effort to directly compare the genetic material DNA of birds (Sibley & Ahlquist 1985ab and unpublished results) have revealed that the current classification is a poor reflection of the evolution of birds, and has shown several plausible alternatives. However, we stick to the more traditional classification until these methods have become refined and the results corroborated by other evidence, and the new classifications have become widely accepted. The same also applies to a number of other well-founded recent suggestions of changes to the arrangement of bird groups.

We do not find it appropriate to formally publish and name new-discovered taxa in a guide like this. Thus new forms not yet named are referred to here as '**unnamed species**' and '**unnamed subspecies**'.

As the classification involves personal judgements and taste, the arrangement of species and the number of recognized species in a group to some extent reflects who classified them. This is a real problem for those analysing regional variations in species richness (biodiversity). Meyer de Schauensee made a formidable progress towards a consistent classification of the birds of South America. However, there are still controversies. We therefore comment on the classification in **NOTEs** following individual species accounts.

To make the notes meaningful to non-experts, we give, below, a thorough outline of how species evolve, how the problems in delimiting species arise, and what the technical terms mean.

For describing and analysing the patterns of evolution of species and of faunas, it is useful to have a large set of categorical levels, specifying the degree of differentiation of populations. Such **microtaxonomic categories** (their history and meaning reviewed by J Haffer in *Z. zool. Syst. Evolut.* 24(1986):169-90) encourage and facilitate analysis of the internal structuring of related forms, and we therefore define these categories, and comment on the microtaxonomic status of some forms in the above-mentioned NOTEs.

This review covers a difficult topic, and readers not particularly interested in the basis of classification should skip this chapter.

THE MEANING OF CLASSIFICATION

Birds are classified and listed in a hierarchical system: species are grouped into genera, which in turn are assembled in families, these in turn in orders (some 30 orders composing the class of birds). This is a fixed convention, but the practical application involves subjective judgement.

Biological facts make extra sense in the light of biological evolution, and it is therefore generally agreed that the basis for the hierarchical grouping should be how species evolved from common ancestors. Thus the system should reflect the various levels of differentiation reached in the evolutionary process, and the pattern of descent (**phylogeny**), as far as this can be expressed in a linear sequence of names. Unfortunately, it is extremely difficult to reconstruct phylogeny. Thus, the classification continually changes (improves?) as new interpretations find support.

The science of clarifying the evolutionary relationships of organisms is called **systematics**; the description of units of living beings (**taxa**, or **taxon** in singular form), their ranking by assignment to taxonomic categories, as well as their practical classification, is called **taxonomy**.

In a classification which reflects phylogeny, any taxon, be it a family, genus, or species, should represent all the descendants from one common ancestor (**monophyly**). Unrelated forms evolving under very similar conditions can sometimes become confusingly similar. Thus, it is sometimes revealed that a currently accepted taxon in fact involves two or more independent evolutionary histories (**polyphyly**). The original taxon is then **polyphyletic**; it is not a 'natural' (evolutionary) unit, and must be divided into **monophyletic** components.

The **scientific name** (called latin or latinized, though just as often greek) of a bird consists of two parts, first the **generic name** (with capital first letter), then the **specific name** (epithet), e.g., *Attagis gayi* for Rufous-bellied Seedsnipe. This double naming is called the **binominal nomenclature**. If birds from a certain geographical provenance can be diagnosed, this local population is referred to as a subspecies, which is named by adding a third name – the **subspecies name**, e.g., *Attagis gayi latreilli* for the dark rufous Ecuadorian population of Rufous-bellied Seedsnipe. We now have a **trinominal nomenclature**. Subspecies are often referred to as 'races'. The population on which the original description of the species was based is the **nominate subspecies**, and has identical species and subspecies names, e.g. *Attagis gayi gayi* (abbreviated *Attagis g. gayi*) for the pale Rufous-bellied Seedsnipe of the southern Andes).

When a new species or subspecies is described, one collected specimen is designated as a **type specimen** (**holotype**, or simply **the type**). This is not necessarily a typical representative, but serves as a 'bearer of the name'. In this way, any dispute about which form the name applies to can be settled by re-examining the type. It is sometimes revealed that two given names refer to the same form (e.g. if different-looking males and females once were thought to be different species). Then the first published name should be valid (it has **'priority'**), and the later is suppressed as a **synonym** (it is **'synonymized'**).

Within local populations, males, females, young, and old birds often look different, but if individuals of the same such category fall into distinct types (e.g., color phases or **morphs**) we talk of **polymorphic forms** (**polymorphism**). Such forms never receive separate scientific names. The subspecies concept refers **only** to differences over geographic distances. These may be (1) gradual over large distances in continuously distributed populations (**clinal**, or **stepped clinal** if the gradient steepens locally), or (2) they may refer to isolated populations. The variation may be trivial, detectable only by direct comparison of many specimens, or so striking that it becomes difficult to tell what is subspecies, and what species. Species which have two or more subspecies are **polytypic**. While some species look alike all over their range, others are strongly polytypic, and may show several levels of variation.

According to current practice, biologically similar forms which interbreed are ranked as subspecies. The resulting intergradation is often seen as rather narrow zones of increased individual variation (**hybrid zones**). They are geographical representatives of the same biological species. Forms which do not hybridize are considered distinct species. Thus a **species** can be defined as **that most inclusive population of individual bisexual organisms which has a common fertility system**. If the forms live in different geographic areas, the status must be inferred from the degree of difference in appearance, voice, and behavior, reaction to playback of tape recordings etc.

Species of birds arise in two or three different ways:
- (**1**) by **phyletic speciation**, which represents the gradual change of a species in time, until so many genetic changes have accumulated in the population that it becomes practical to introduce another name (not all authors consider this change in time, without a split, as 'speciation').
- (**2**) as an ancestral species is split into two or more forms, which finally become so different that hybridization is precluded. This process, **speciation proper**, seems almost invariably to start by a geographical separation of two or more populations, which for a long time prevents mating contact and exchange of genetic material between them (**gene flow**). The separation may be caused by habitat changes leading to fragmentation of a once continuous range (speciation by **vicariance**). The speciation may also take place after active dispersal of individuals across a pre-existing barrier – e.g., a gap of sea, or deep valley – and the resulting establishment of a population in the new area (**peripatric speciation**). As the isolated populations evolve independently over a long time, becoming 'acclimatized' to their respective local condition, they gradually acquire a distinctive appearance. If contact is re-established after a short geological period, hybridization normally leads to a stepped cline; if they remain isolated longer time they finally become too different to interbreed. They now act as distinct species if they were to meet. (There may, however,

be differences between different contact zones, as the forms may intergrade locally, but act as separate species in other contact zones – so-called **ring species**). What prevents hybridization of forms is rarely known. Their independent evolution may have resulted in such genetic differences that matings result in sterile or inferior offspring, or no offspring at all (**genetic incompatibility**). With other taxa, although fertile hybrids can be produced in captivity, different vocalizations, courtship displays, or habitat preferences preclude hybridization under natural conditions. The forms may in the end become so different that they can live together without serious competition. **Ecological (or full) compatibility** is reached, and the speciation process is complete.

- (3) **by reductive speciation**, which involves the hybridization of two forms and establishment of a stabilized hybrid population. This is certainly a rare event (except in plants).

Forms inhabiting different geographic areas are called **allopatric** (the phenomenon: **allopatry**); fully compatible forms living in the same area **sympatric** (**sympatry**). There exists also an intermediate level of **parapatric** forms (**parapatry**). These forms exclude each other geographically in fairly uniform habitat zones. Their distributions thus resemble large-scale mosaics of neatly interlocking patches formed by the ranges of component forms (**paraspecies**). These forms have reached the differentiation where they do not hybridize, but coexistence is not yet possible. Instead the forms exclude each other. The whole group of not yet fully compatible related forms is called a **superspecies**.

Although recent studies have shown that parapatry has wide occurrence, the underlying biological causes remain poorly known. The nicest examples are described from the Amazon lowlands, and from the Afrotropics, but such patterns are apparent also in the distribution of Andean forms. In some cases physical barriers are involved (e.g., large rivers, mountain ranges or deep arid valleys which reduce actual contact). In other cases the forms may compete ecologically or actually expel each other (**competition parapatry**). It seems that small barriers are involved in most cases where species replace each other along the Andes (see e.g. Graves 1985). However, abrupt species replacement with actual contact is common along elevational gradients on the Andean slopes (see Terborgh 1971, 1977, Terborgh & Weske 1975). Such species form a pseudo-overlap as they live in the same general area, and only close study reveals how slight the contact is (sometimes replacement over 100 m elevational distance).

Hybridization along contact zones between forms is traditionally used as evidence for considering them subspecies of the same species. However, hybridization does not necessarily mean a breakdown of the integrity of populations. If hybrid offspring have reduced viability, parental types continue to dominate, and intermediate types hardly spread outside the narrow contact zone, as in the Carbonated Flowerpiercer (*Diglossa carbonaria*) superspecies. Despite the lack of mating isolation, the discrete reproductive entities are in fact maintained. Few decades ago, a record of an intermediate ('hybrid') specimen between two 'species' normally lead to a lumping under one species name; today, both species are considered valid, as long as the 'parental' types dominate in the area of contact (see *Auk* 86 (1969):84-105).

As species evolve gradually, there will at any moment exist some forms which are just about to become full species. The many intermediate levels between subtle forms and fully compatible species is hardly a biological mystery, but it causes controversies about the limitation of the species category. Various proposed species concepts place the species limit at nearly every conceivable level of differentiation: see figure on next page.

CURRENTLY USED SPECIES CONCEPTS

PHYLOGENETIC SPECIES: 'The smallest diagnosable cluster of individual organisms within which there is a parental pattern of ancestry and descent' (J Cracraft in *Current Ornithology* 1(1983):159-87). This means that any population characterized by an evolutionary novelty (apomorphy) is recognized as a species. Then, any local variant population that is not just a 'subtle form' receives species rank, and subspecies may become superfluous. In effect this 'new' concept corresponds to the species concept used a century ago.

As long as the concept defines taxa solely by morphological characters, the 'smallest diagnosable clusters' will not be evolutionarily comparable across different animal groups (e.g., groups of birds using vocal versus visual signals in their pair formation). The concept thus precludes

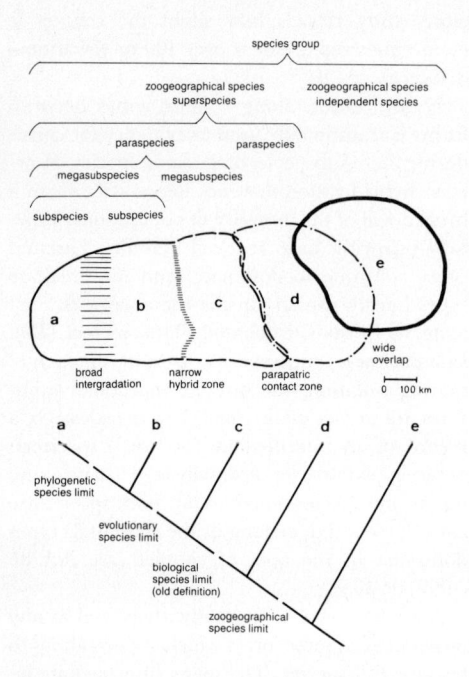

Species limits under different species concepts. In this schematic representation of an imaginary field situation, increased reproductive isolation between geographical forms along their contact zone (upper part) coincides with increasingly distant genealogical relationships (lower part). In this example, 5, 4, 3, or 2 taxonomic species are recognized, depending on which species concept is applied, and where the corresponding species limit is placed (after Haffer).

broad synthesis. Another practical problem is that many 'species' will 'disappear' by reductive speciation. The concept has a clear logical potential for the analysis of speciation events. But to be practical also for other purposes, it may require the introduction of a higher categorical level. It is therefore tempting to suggest that the speciation analysis is made directly on the data describing the population structure, without changing the classification.

EVOLUTIONARY SPECIES: A single lineage of ancestor-descendant populations which maintains its identity from other such lineages and which has its own evolutionary tendencies and historical fates. This concept admits specific rank to forms which hybridize strongly, as long as the hybridization occurs within well-defined contact zones and does not break down the genetic integrity of the component forms. Also this concept is useful for biogeographic analysis. It acknowledges subspecies and 'subtle forms' (EO Willis in *Syst. Zool.* 27(1978), and LS Stepanyan in *Zool. Zhurnal* 57(1978):1461-1471).

BIOLOGICAL SPECIES (or **BIOSPECIES**): A group of interbreeding natural populations that are reproductively isolated from other such groups. Often a reference is given to 'potential interbreeding' in order to cover forms which live so far apart, that the inability to interbreed can only be inferred from morphological as well as biological differences. This is the traditional concept developed in the 1930s by E Stresemann, B Rensch, and especially E Mayr, and the genetician T Dobzhansky. A revised definition (in Mayr 1982: *The growth of biological thought*. Cambridge Mass.) also includes the ecological differentiation as a criterion: '...a reproductive community of populations ... that occupies a specific niche in nature'. For a long period, biospecies were considered as real entities in nature, a consensus now challenged by other species concept. The biospecies concept operates with species (which may be semispecies, paraspecies, allospecies, or synspecies; definitions below), megasubspecies, subspecies and 'subtle forms'.

ZOOGEOGRAPHICAL SPECIES: Forms which have reached reproductive isolation as well as ecological compatibility with other related forms. The component forms are called the 'sectors' of the species, except intergrading forms which have status as subspecies and 'subtle forms'. From the point of view of the biological species, this is a concept of communities of descent rather than of taxonomic units.

Currently, the biospecies limit, at an intermediate level of differentiation, is the most widely used compromise. However, this unit 'conceals' many conspicuously differentiated entities. Classified as subspecies, such forms tend to be neglected (both by students of variation in life history traits, and in connection with conservation). Other difficulties with using the biospecies concept are pointed out by MC McKitrick and RM Zink in *Condor* 90(1988):1-14.

MICROTAXONOMIC CATEGORIES.

SUBTLE FORMS: Local populations that differ from other populations only in a statistical sense, as the individual variation gives extensive overlap. If continuously distributed, these forms intergrade broadly (clinally).

SUBSPECIES: Distinct geographical representatives of the same species. The meaning has shifted somewhat, as previous authors gave subspecies names even to the most subtle local variants. Today, names are normally applied when all or most specimens from a certain geographic area differ from individuals from other areas. Subspecies should ideally be named only after a thorough analysis of the population structure within the entire species. Numerous Andean species need revision of this kind.

MEGASUBSPECIES: Distinct taxa (subspecies or clusters of related subspecies) which do not intergrade with other such taxa in a gradual way, but form well marked and often narrow hybrid zones if they meet. Some megasubspecies thus fall within the evolutionary species concept. Megasubspecies have also been called taxo-evolutionary or biological subspecies, ecophysiological subspecies, or 'Sippen' (in German literature). A biospecies composed of two or more megasubspecies has been designated a **megaspecies**.

SEMISPECIES: Taxa that are even more differentiated than megasubspecies and actually or potentially form a 'zone of overlap and hybridization', or some parapatric hybridization (quasispecies and vicespecies are synonymous terms). The semispecies has the taxonomic rank of a species, like paraspecies and allospecies.

PARASPECIES: A distinct component taxon (biospecies) of a superspecies whose geographical range is in contact with that of other paraspecies, but does not overlap with it.

ALLOSPECIES: A distinct component taxon (biospecies) of a superspecies, which lives without geographical contact with other component forms.

SUPERSPECIES: A group composed of two or several semispecies, paraspecies, or allospecies. If the component forms are monophyletic, and were once the conspecific representatives of one ancestral form, we talk of a **first-order superspecies**; if derived from two or more directly related ancestors (presumably paraspecies) we talk of **second-order superspecies (mega-superspecies)**. Superspecies are called Artenkreis in German literature.

EX-CONSPECIES: Term used for two or more semispecies or megasubspecies under the narrow evolutionary species concept.

SYNSPECIES: Reproductively and ecologically isolated species with geographically overlapping ranges.

SPECIES GROUP: Two or more closely related synspecies (called ex-superspecies by Vuilleumier 1985). The term is also used for groups of closely related but ecologically different species inhabiting different vegetation zones or biogeographic regions, as they are regarded as potentially compatible.

FAMILY ACCOUNTS

A family account – which is extensive in case of a large group well represented in the high Andes – starts with a statement of the total number of species and distribution, and of the assumed systematic relationship of the family. A general description of the birds' appearance and biology follows. Such generalizations may be helpful by emphasizing distinctive family characters. Since members of a family usually show fairly uniform life history patterns and habits we incorporate such general information in the family account instead of repeating it under each individual species.

The family accounts end with short listings of important publications about the family or some family member.

Within larger families we often give special accounts of subfamilies, tribes, or genera, emphasizing how these differ from other such subgroups of the family.

SPECIES ACCOUNTS

NAMES

English names chosen are those we should like to see become standard usage. In general we support the decisions of Meyer de Schauensee, thus maintaining names already in common use. Although not all are equally appropriate, we try to avoid confusion by changing names. However, a few exceptions are made: It is practical to name species with extremely limited distributions after geographical sites. Such species are at least potentially endangered by extinction, and international conservation bodies already use this way of naming in order to call attention to local needs for conservation management. A few such changes have been adopted here, e.g. for the two very local flightless grebes. In any case, we mention the name used by Meyer de Schauensee (and other commonly used names) in the NOTEs.

Scientific names chosen are according to Meyer de Schauensee, except for those families where a more recent classification is referred to in the family account.

REFERENCE TO ILLUSTRATIONS

The reference to line-drawings and color plates (plate number in italics, and reference to the numbers and letters used on the plate) is given immediately after the scientific name.

DESCRIPTION

BOLDFACE TEXT: Details of importance for the practical identification are emphasized by the use of bold-face letters. Categories, as ad., male, female, imm., juv. and pull. are also written in bold types to emphasize where in the text each category is described.

MEASUREMENTS: All accounts start with total lengths in centimeters. Other measurements are given when appropriate (hummingbird bills, long tail-feathers, wingspan, etc). The length is usually taken from a number of study skins, more rarely from intact birds. The total length is not an exact measurement, but should be adequate for comparative purposes.

UNMISTAKABLE: Our use of this word signifies that a first tentative identification from the plates is likely to be correct, since the species is so characteristic that confusion is improbable.

SEQUENCE AND DICTION: To facilitate comparisons of species, we have tried to standardize the descriptions. The plumage is as a rule described from the forehead and back, upperparts first, then the underparts. Colors of soft parts are given afterwards. Striking color patterns sometimes make another sequence more convenient, but normally the same sequence is used for very similar species.

TERMS FOR THE TOPOGRAPHY OF BIRDS: The naming of the different external parts of a bird is shown on the drawing. Such terms are standard usage, although for convenience we sometimes deviate from the strict technical conventions. Thus, we use the word **wing-coverts** (or just **coverts**) as a collective term for the lesser, middle, and greater upper wing-coverts, and **wing-linings** for the under wing-covert plus axillaries. When describing characteristics of a flying bird, a distinctively colored area formed by the lesser and middle secondary coverts is called **forewing**. A pale area on the closed wing formed by pale outer webs of some flight-feathers is called **wing panel** (or just **panel**).

COLORS: Generally, we use rather broad and hopefully self-explanatory terms, except in groups where a very exact statement is needed for identification. The color **hue** represents the position of a color between any contiguous pair of spectral colors, the **tone** the darkness, ranging from pallid (whitish) to very dark (blackish). A **bright** color is spectrally pure, versus a saturated color. The colors are illustrated and explained in FB Smithes' (1975, 1981) *Naturalists Color Guide* (AMNH, New York); or if a more precise information is needed, in Ridgway's (1912) *Color Standard and Color Nomenclature*. Washington DC (we give such color terms with capital initial letters).

PATTERNS: Some important terms of color patterns are: **hooded** = with an aberrant (often dark) color covering most of the head; **capped** = dark top of the head; with **crown-cap** = cap small, not including forehead and superciliary. A **barred** pattern is transverse, **striped** or **streaked** pattern longitudinal. Terms describing the markings of the individual feathers are: Dark edges of round feather-tips cause a **scaled** pattern, or **scalloped** pattern if several concentric lines are involved. **Mottling** is a very fine, vague, or irregular transverse pattern. **Edged** = with aberrant (pale) lateral sides of the feather, **fringed** (or **bordered**) = with the whole feather margin deviating (pale); **notched** = with several small marks (usually pale) evenly spaced along the feather-margins (often causing a serrated egde if the feather is worn); **streaked** = with deviating (usually dark) mark along the feather-shaft; regular longitudinal markings which do not necessarily follow the shaft is called **stripes**; **tipped** = with aberrant color at the feather-tip, near the shaft.

PLUMAGES: Adult males in breeding dress are described first, as they normally represent the most distinctive appearance of the species. This is followed by female, immature, juvenile and downy young (with other possible aberrant plumages inserted). All too often previous students neglected age variation and described only adult breeding plumage, or uncritically called all drabber birds juveniles. Modern collectors therefore state the ossification of the skull and condition of gonads, molt data etc. on the specimen labels, enabling critical analysis of the plumage sequences. We paid special attention to drab-looking and molting specimens, and therefore hope to have avoided basic misinterpretations.

Pull. = pulli, are young in their early downy plumage. **Juv.** = juvenile, are young birds in the first plumage of true feathers. **Imm.** = immature, represents a distinctive plumage inserted between juvenile and adult plumages (though most birds retain juv. flight- and tail-feathers). Larger birds often have one, two, or more immature plumages before they become sexually mature. A bird which is breeding, but has not yet acquired the definite plumage, is called **subad.** = subadult. **Ad.** = adult, is a sexually mature bird in definite plumage. Some species alternate between adult breeding and non-breeding (winter) plumages. It should be noted that, at certain times of the year, most birds in a population molt, and have very checkered plumages. These are normally not described, but a few examples are shown on the plates.

SUBSPECIES VARIATION: As a rule, we describe all subspecies recorded in the Andean zone. It is often convenient to start by describing a 'central' or widespread subspecies, and then add how the others deviate. In other cases we start by describing the species as such, and then briefly outline the variation. It is difficult to avoid describing subspecies in relative terms ('darker than' etc.); this could be regarded as a weakness of our description, but actually stresses that a safe identification requires direct comparison (often of large numbers of specimens). Many subspecies are described from single specimens, which may thus be aberrant individuals, or represent points on clines. They may have to be synonymized when a thorough analysis of the population is made. For such reasons, we sometimes find it most responsible to just state that 'the subspecies are poorly differentiated and doubtfully valid' or 'the variation is trivial'.

For sedentary birds, the subspecies name will normally be given by the geographic location.

HABITS

In recent years, the field techniques for identification have become sophisticated far beyond what the traditional bird guide conveys. The trained birder is often able to recognize a bird out on the horizon merely by its wing action. Such clues for identification can be almost as reliable as colors. They combine with other features to give each species a special feel or 'jizz' (a term derived from the fighter pilots' acronym 'GIS' – General Impression and Shape). This 'holistic method' of recognition is hard to convey in a text. Yet we try to describe attitudes, shapes, movements etc. which characterize the species' general habits, and at the same time facilitate its identification.

We hope that the terminology used to describe various feeding methods is self-explanatory.

Sociability is also described. Many birds move together in flocks, either erratically (mainly in the steppic habitats), or in accordance with fixed group territories, routes, and daily time schedules (mainly in humid forest). Birds in a flock are less vulnerable to predation, and the good warning system in the group gives the individual bird more time to feed. However, individuals in **pure-species flocks** may suffer from food competition. Many forest birds therefore form **mixed-species flocks**, often with one pair or family group of each species, which ensures that most flock members select different foods. A related phenomenon is **bird waves**, which is an activity wave elicited by a wide-ranging 'core species' moving through the territories of sedentary species. Whilst these latter normally stay within dense vegetation cover, they utilize the warning system offered by passing birds for a burst of intensive feeding in the open parts of the habitat. Thus, the otherwise very silent habitat suddenly is full of life. The participants of these flocks generally have very similar calls. Many species also join mixed 'feeding aggregations' in places with a temporary superabundance of food, e.g., a fruiting tree.

VOICE

Many birds will first be detected by their vocalizations, and familiarity with the sounds of a variety of species is therefore a valuable aid to rapid identification, and it is indispensable for studies in forest. For some species, the voice alone enables a positive identification even of an unseen bird, and a tape recording may be as good a documentation of its presence as a collected specimen.

The only way to learn birds' voices is by listening to them in the field or on tape. Some readers find spellings of bird songs absurd. We agree that it is impossible to convey the right impressions of the quality of a vocalization. Yet we believe that such descriptions can aid identification at least by ruling out certain species.

Some important aspects of sounds are: **1, frequency = pitch** (measured in Hz): A 'fine' or 'high' sound here denotes a high-pitched sound, as produced by touching a small, tightly spanned, light-material string or membrane. The terms 'low' or 'deep' are here used for low-frequency sounds. **2, amplitude = loudness**, sometimes called volume or strength (measured in decibel): A 'loud' sound has great amplitude, a 'weak' sound small amplitude (thus we never use the term 'low' in respect to loudness). An accented sound (written with bold letters) has greater amplitude than the sounds immediately before and after it. **3, quality** or 'sound': All natural sounds are complex, each 'note' consisting of a fundamental 'note' and a number of higher-pitched harmonics or overtones. The pitch of the 1st overtone is the double frequency of the fundamental, that of the 2nd overtone three times the frequency of the fundamental, and so on. The 1st harmonic is synonymous with the fundamental, so 1st overtone is the same as the 2nd harmonic. The various harmonics are dampened to different degrees (those finding resonance being almost completely undampened), and the resulting number of audible harmonics and their relative loudness produces the **timbre**, which is the main aspect of quality. The term **sonance** has been used to denote changes (in amplitude, frequency, or timbre) within a single 'note' (e.g. a rising or falling note), but it is often impossible to define 'notes' in bird sounds (despite our wide use of the word in the text). Also birds have two independent sound-producing membranes, so they can emit two different sounds simultaneously, and interference between the two sounds may further complicate the pattern. Sometimes the word 'note' has been replaced by other terms that conveys one aspect or another, e.g. scream, squeak, chirp.

Bird song may attract mates, stimulate breeding behavior (via hormonal production), and serve a territorial function. A song is usually composed of phrases or bouts, i.e. a series of 'notes' followed by a pause before the next **bout or phrase**. An unvaried song, where a 1-second bout is followed by a **pause** of 2 seconds before the next bout, is here described as 'song lasts 1 s and repeated every 3 s', or 'song lasts 1 s and given at 2 s intervals'. We have thus used the term 'interval' synonymous with 'pause'. The bout usually consists of several 'notes'. The **speed** of the song here indicates how many 'notes' are given per second. A song-bout lasting 1 second and having 3 'notes' is a slow song, if having 10 'notes' it is fast. If the 'notes' in a bout are more or less alike, and given faster than 10 per second, the bout is here called a **churr** (short bout) or a **trill** (long bout). We thus use the word 'trill' somewhat differently from music, where it means a rapid alternating between two notes. The words **ascending** and **descending** we only use to refer to rising and falling pitch, while **increasing** and **decreasing** may refer to either amplitude or speed.

BREEDING

The paragraph gives the time of egg-laying, and points out whether the breeding habits deviate from what is said in the family account. A designation such as 'Feb-Mar (Jan-May)' means laying in February and March, with occasional clutches from January to May. A difference between parts of the range may be indicated (to the extent that this is known). Only for a few species do we feel qualified to clearly define the breeding season, but by stating our general impression we may convey at least an idea of the best season for a field study. Where our evidence is meagre (especially for small birds), we present it in full: the available record (based on publication, our records from the labels of museum specimens, and from our fieldwork) of nests with eggs, nestlings (pull.), fledglings, and young birds (juv.), which permits an estimation of when the eggs were laid.

HABITAT

Habitat categories are defined in the 'Andean natural history', and a crude guideline for recognizing plants mentioned in the text is offered at the end of the book by the reference list to some plants shown on the color plates.

The information about the bird's habitat is dangerous to use for identification, but may be valuable as a suggestion of where to search for

a certain species. The data may also form a basis for biogeographers analysing community structures.

RANGE AND NUMERICAL STATUS

An account usually starts with a very concise statement of total range. Occurrence within the Andes is outlined in more detail. For fairly common species we give the extreme points of the range (usually starting in the north, except for southern forms migrating northwards). For very local or rare forms we may refer the precise localities. Department names are followed by the abbreviated country name. We specify the ranges of all subspecies, and mention subspecies in adjacent lowlands if only a few are involved.

The paragraph ends with a comment on abundance. **Abundant** means that a species is widespread, and often occurs in high numbers, **very common** that it is almost invariably recorded in the right habitat, often at fair density. **Common** birds are recorded on most trips in suitable terrain, but normally in low numbers (**fairly common** is a less strong statement). These terms thus describe wide distribution within proper habitat. Some readers may be surprised to find the terms applied to forms of which extremely few records are published. Many such forms inhabit areas of difficult access, or they have retiring behavior. However, they can be found fairly easily once the calls or precise habitats are known.

Uncommon means that a species is recorded only on few trips in proper habitat. The term **local** has nothing to do with numerical status, but reflects that the species lives only in certain places within broadly suitable habitat, and is also used in case of very patchy (relict) habitats. The term **rare** is used when a species is decidedly difficult to find even in the right habitat and season.

The following terms refer to the regularity of occurrence rather than numbers. **Casual**: few records, and not expected to occur except at long or unpredictable intervals. **Accidental**: probably very rare and unpredictable. **Hypothetical**: no record, but assumed to occur judging from its requirements or mobility. Inhabitants of lowlands east of the Andes are hypothetical visitors to the high Andes if recorded accidentally on the Pacific side.

Only for very few species are estimates of the total population of adult birds given.

THE MAPS

As a supplement to the text, distribution maps are shown in the adjacent margin. The maps show the total occurrence in western South America, but in general only the occurrence within the Andes is shown in detail. For species inhabiting the vast lowland savannas and forests east of the Andes we found it vastefull to make a full survey of the record, and instead just give our general impression of the extension and regularity of occurrence.

BLACK AREAS show areas of known or very probable regular occurrence. Some large areas are covered by very few observations/collections, but most fairly common species are assumed to occur throughout the area between the known distribution extremes, except in obviously unsuitable terrain. However, for species which are generally assumed to be rare, local, or disjunctly distributed we often show only positive records. For poorly known species a stippled line connects the records (dots) along the Andean slopes or along the cordilleras, indicating that the gaps are probably collecting gaps.

SHADING indicates (1) larger areas where the species is resident but very rare, or (2) larger areas with scattered records of vagrants, or where the species is assumed to pass through unnoticed, and (3, small patch) a stray record.

QUESTION-MARKS are used for a record which cannot be regarded as satisfactorily documented, and to indicate that we have not found out whether the species occurs in the area.

SINGLE-HEADED ARROW is used to emphasize a very local record.

DOUBLE-HEADED ARROW is used to indicate that a species migrates back and forth through the area. Ideally, they connect breeding and non-breeding grounds (consult the text).

WHITE BROKEN LINE through black areas separate breeding and non-breeding ranges.

NOTE

Under this heading we mention any current dispute or debate about the naming or the taxonomic status of a species. Here we also express our personal doubts about the soundness of the current classification. The meaning of the terminology was discussed in a previous paragraph.

In case the species in question is sometimes split into two or more full species, we give both scientific and proposed vernacular names, capitalizing names for component species occurring in the high Andes.

THE PLATES

The plates were painted by JF in water color from field sketches (supported by photos), and with study skins at hand. As a rule, birds kept in the Zoological Museum of the University of Copenhagen are shown, but plumages lacking there were painted from specimens borrowed from other collections, or painted during museum visits.

All major plumages are shown for typical high Andean forms, and in a few cases birds in molt are illustrated. For most marginal forms, only a main plumage is shown. Distinct subspecies are also painted.

The birds on a plate often differ in attitude, some birds resting, others active, some flying, although identical attitudes would make the comparison of similar-looking forms easier. The artist's motive was not only to make the plates 'alive', but also to convey information about the characteristic attitudes, activities etc., which combine with the colors and pattern to give the species its special feel or 'jizz'.

Details regarding the arrangement and scales and the text pages facing the plates are given in the introduction to the plates.

Puna Rhea

Rheiformes

Rheas – Family Rheidae.

2-3 species of large and flightless ostrich-like birds with 3 toes endemic to S Am., where they inhabit grasslands and savanna. The wings are weak, but can be used for balance in running, or spread as a display signal. Detailed distributional data for the Andean taxa are published in *Ibis* 124(1982):168-172; *Neotropica* 28(1982):47-50; diet described in *Idia* Nos 429-432(1985):63-73.

LESSER RHEA *Pterocnemia pennata*

95 cm, with 15-25 kg weight. Plumper and less long-legged than better-known Greater R. (*Rhea americana*) of the pampa and chaco, sspp *garleppi* and *tarapacensis* particularly short-legged and with only 8-10 scutes in front of the tarsus (vs. 16-18 in *pennata*). Plumage downy, with long plume-like feathers of the upperparts hanging down over the flanks, but small head and long neck have short down. **Ad.** with top of head and entire neck light gray-brown, face and throat more fulvous. Upperparts of body gray-brown, mantle and forewings with warmer hue, varying from tawny in *pennata* to dark chestnut-brown in *garleppi* and *tarapacensis*. Belly and vent white. Strong legs light gray. **Most dorsal feathers have white tips**, which vary from fine streaks on mantle to big spots on the larger wing- and flank-feathers, sometimes giving the impression of large lateral splotches and bands. Females generally have smaller and fewer white spots than males, and look duller and more unicolored (but males in worn plumage also have small spots). Statements that *garleppi* and *tarapacensis* have smaller marks than *pennata* seem to rest on comparison of birds of different ages. Birds in their 2nd and 3rd years are more fulvous, esp. on the upper back, and hardly show white feather-tips. **Juv.** are quite fulvous also on top of head, and have uniform brown body, except for white underparts. **Pull.** pale gray-brown with 3 blackish bands on back.

Habits: Usually in bands of 5-30 and rarely even more birds, often 1 male with 2 females or with several young birds of sometimes quite variable size. Sometimes grazes in loose company of lamas or other camelids. When undisturbed, walks slowly with head low, neck in deep U-curve. In most places very wary. When disturbed, usually runs with upright neck, but leans the neck partly forwards at full speed. May suddenly throw itself down and hide very efficiently by lying down, stretched on the ground.

Voice: Although chicks are very vocal, ad. seem usually to be mute. In the mating season male gives loud booming call (like foghorn), raising wings and inflated swollen lower neck.

Breeding: Males simultaneously polygynous, females serially polyandrous. Several females lay 10-30 eggs (olive-buff) in large scrape. Male incubates and conducts large brood of nidifugous young. Eggs in Sep-Jan in the n, from July in Rio Negro, in Nov furthest s.

Habitat: Northern populations on desert-like salt puna or pumice flats, usually large plains with some *Lepidophyllum* brush and some bog. In the s usually on shrub steppe and lush grass of the flood-plains, but seems mostly to breed on undisturbed upland with bunchgrass heath. **Range**: Very locally at 3500-4500 m in desert puna from Moquegua and w Puno, s Peru (vanishing) through w Bol., commoner Jujuy to San Juan (reserva Guillermo), nw Arg. (*garleppi*), and Aríca to Atacama, n Chile (*tarapacensis*). Magellanic Chile through Patagonian steppes of s Arg. and to 2000 m on adjacent Andean foothills n to southern Mendoza, Neuquen and Río Negro; introduced and established on northern Isla Grande (*pennata*).

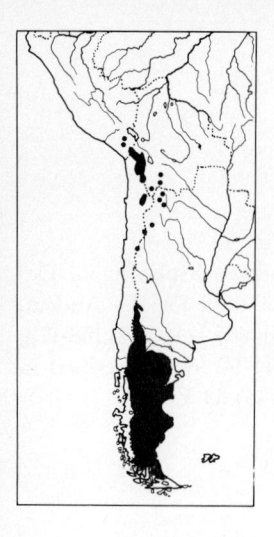

NOTE: Highland forms *garleppi* and *tarapacensis* are doubtfully different, but stand out sufficiently from *pennata* to be considered a distinct (allo)species by some: **PUNA RHEA (*Pterocnemia tarapacensis*).**

Darwin's Nothura

Tinamiformes

Tinamous – Family Tinamidae

Tinamous represent a group of terrestrial birds related to the flightless ratite birds, possibly the most ancient lineage of extant birds. The family is restricted to the Neotropics, and comprises 33 species. These chicken-like birds are much sought for game, and are kept captive in some parts of the Andes. The body is compact, round, and apparently tail-less, as the small tail-feathers are directed down- or forwards and may be hidden by partly disintegrated fur-like coverts. Wings are short and rounded. The neck is slender, head small (sometimes crested) with rather thin, often somewhat curved bill and very deep corners of the mouth. The colors are cryptic: Open-land species show intricate vermiculated and striped patterns of buff, black, gray, and white, while forest species are more uniform deep brown.

Tinamous are furtive and avoid observation by running and crouching. They are flushed at close range like an explosion, with whirring, noisy wing-beats (in bursts of violent flaps and alternate glides) and accompanied by loud whistled screams. Being weak fliers, they prefer to fly downhill, and soon drop back into cover and disappear. A useful way to detect tinamous is by listening, morning or evening, for their loud, tremulous or whistled, monotonously repeated calls. Unfortunately, the vocal differences between species are insufficiently known.

The diet comprises seeds, roots, bulbs, leaves, and insects. Usually, females are larger and more aggressive than males. Most species are polyandrous, a female supplying 2 or more males with eggs. *Nothocercus*, *Nothoprocta*, and *Eudromias*, however, are polygamous, 2 or more females laying in a common scrape. In any case, only males incubate and care for young. Tinamou eggs are unicolored with a characteristic porcelain-like gloss. The young are highly precocial, and resemble miniature ad. with decomposed feather structure. They soon grow typical feathers and become capable of short fluttering flight.

Information about Andean species are found in Todd (1942a), *Auk* 72(1955):113-27; *Bonn. Zool. Beitr.* 19(1968):225-34; *Fieldiana (Zool.)* (Chicago) 31(1950): 339-74.; *J. Orn.* 95(1954): 219-32; *Rev. Univ. Nac. Col.* 23(1958):245-304; *Trans. San Diego Soc. Nat. Hist.* 16(1971):291-302; *Wildlife* 120(1969): 88-138; *Zoologia* 58(1973):13-40, Blake (1977), and in the general faunas.

Ornate Tinamou

Rufous-breasted Seedsnipe

HIGHLAND TINAMOU *Nothocercus bonapartei* – Plate I 6a-b

38 cm. Rather uniform deep **rufous brown**. Told from related species by combination of slaty crown, **rufous throat** and c underparts, entire body marked with profuse dusky vermiculations, and rump and wing-coverts with small buffy-white spots. Flight-feathers with dusky and buff barring. Sspp *bonapartei* and *plumbeiceps* very similar, the latter dark throughout, *discrepans* most uniform rufescent, lacking light spots except on the wings, *intercedens* deviating by much less rufescent underparts, and pale ochraceous buff throat. **Pull**. similar, but most of head mottled gray, with interrupted black stripes through eye and along crown and nape. Very retiring, living singly or in small groups. Wide-ranging, somewhat guttural call *cooyooh* is repeated over and over again in sequences of each 4-5 calls. When flushed, gives loud galline cackling *quok quok quok*.... Turquoise eggs in Jan-June (Ven.), Feb-June or Nov (Col., Ecu.).

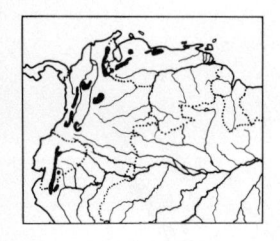

Locally distributed and uncommon in dense humid forest and near forest glades, esp. in areas with seeding bamboo, mainly subtrop., but locally to above 2500 m. Southern C Am. (*frantzii*), and in the n Andes as follows: W Andes of Col. s to the Patía canyon (*intercedens*); locally in C and E Andes and Perijá mts, Col., and Andes of n Ven. to Lara (*bonapartei*), at base of Andes in Tolima and Meta, Col. (*discrepans*); and e slope of Andes through Ecu. to Chaupe in Cajamarca, extreme northern Peru (*plumbeiceps*).

TAWNY-BREASTED TINAMOU *Nothocercus julius* – Plate I 5

38 cm. **Rich chestnut**-brown. Told from related species by the warm chestnut forecrown and face, and combination of well-marked **white chin and throat**, and body conspicuously and densely barred black except for **clear cinnamon-rufous c breast** and belly. Inner remiges show cinnamon notches. **Juv**. has paler head and more extensively rufous underparts. **Pull**. similar, but with tawny crown and mottled gray nape demarcated with black against paler headsides.
Habits: Shy. Solitary or in small groups.
Voice: '2-note whistle' reported.
Breeding: June-Aug.
Habitat: Forest edge, low, fairly open semi-humid woodland and second growth.
Range: At 1700-3250 m. Páramo de Tamá in extreme w Ven. through all Andean ranges of Col., Pichincha, and e Andean slopes of Ecu., and in Peru from Piura to Vilcabamba mts, Cuzco. Now local.

HOODED TINAMOU *Nothocercus nigrocapillus* – Plate I 8a-b

33 cm. **Rich chestnut**-brown, very dark, owing to dense fine black vermiculations. Told from related species by combination of **blackish top of head**, well-marked **white chin and throat**, and conspicuous black

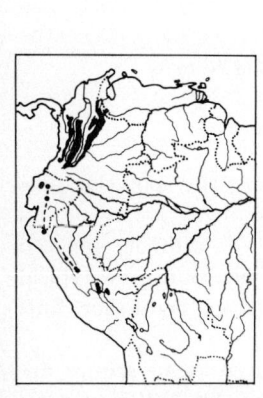

and buff spots forming **bars on wings**. Chest ochraceous gold to buffy brown with black vermiculations (*nigrocapillus*) or darker rufescent with obsolete vermiculations (*cadwaladeri*). **Pull**. unknown.
Habits and **voice**: Shy and difficult to see, and usually reveals itself only by wide-ranging single-note calls resembling those of the allopatric Highland T. Sometimes congregates in places with seeding bamboo, to eat seeds on the ground.
Breeding: No information.
Habitat: Wet tall montane forest, in dark places with little undergrowth and thick leaf litter, or around bamboo stands.
Range: In subtrop. to lower temp. zones (2000-3000 m). Along e Andean slopes from Amazonas s of Marañón to Chaupe, Cajamarca, n Peru (*cadwaladeri*), and from Junín, ec Peru, to yungas of Cochabamba, Bol. (*nigrocapillus*). Probably local and uncommon. Graves (1985) suggests a continuous range, but very few rec. are known.

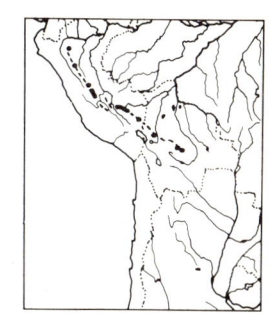

BROWN TINAMOU *Crypturellus obsoletus* – Plate I 7

27 cm. Somewhat like a quail-dove in shape and coloration. **Head dark gray**, crown slaty or black. **Body dark chestnut**, almost uniform, very dark in *castaneus*, or grading to brighter rufous below (*punensis, ochraceiventris*, the latter quite rufous throughout); belly somewhat barred dusky, and vent sooty scaled with whitish feather-margins. Eyes orange. Feet yellow. **Juv**. with 2 white dots on most back-feathers. **Pull**. tawny with crown dark black with tawny mottles centrally, and contrasting with pale headsides; black dot behind eye.
Habits and **Voice**: Usually seen running along vegetation borders, e.g. roadsides through forest. Morning and evening gives **whistled outbursts** *wuee-dwydwy* every 5-10 s, or accelerating series of hysterical, rolling *trrry-rhee-rhee-rhee…*, swelling and falling slightly at the end.
Breeding: Eggs deep pink. Pull. Oct (Cuzco).
Habitat: Along glades of humid montane forest, at forest edge, in copses of *Alnus*, and in other second growth.
Range: Several sspp locally e of Andes, some with small isolated populations in subtrop. mts of n Ven., others being widespread in e Andean foothills, ascending to near lower limit of temp. zone in e Ecu. (and Guayabamba valley, if correct) and n Peru (*castaneus*), and reaching 2900 m between Huánuco and Marcapata Valley in Cuzco, Peru (*ochraceiventris*), and in yungas of La Paz to Sta Cruz, Bol. (*punensis*).

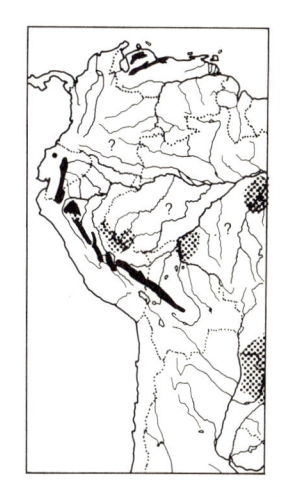

RED-WINGED TINAMOU *Rhynchotus rufescens* – Plate I 1a-b

40 cm, with rather long, curved bill. **Head and neck buffy cinnamon** with dark barred cap and some streaks on cheek and creamy throat, in *maculicollis* with some black streaks on foreneck. Otherwise light gray-brown **boldly barred black and pale buff above**, more faintly barred on flanks, and more cinnamon below. In **flight** shows **strongly cinnamon-rufous outer parts of wings**. 'uv. apparently not distinguishable. **Pull**. colored like ad., but head distinctly striped (crown with 3 heavy black stripes), and body streaked rather than barred, and with

numerous projecting, bristle-like cilia. Usually well hidden in tall grass, often sitting partly under overhanging grass. The rare flight is very noisy, with constantly whirring wings and no glides inserted. Call beautiful flutelike whistles, single *weeut*, or *weeut tu tu*, the initial *wee-* lower pitched than the rest, often several birds calling in unison. Vinaceous eggs Sep-Dec.

Widespread, but declining, with several sspp in grassland and semi-arid woodland savanna in lowlands from c Arg. to extreme e Peru. Ssp *maculicollis* lives from 1000 m to at least 3050 m locally on semi-arid grassy mountain tops and Andean slopes from Catamarca and Tucumán, nw Arg., to Cochabamba and around the La Paz and Mapiri valleys in La Paz, Bol.

Nothoprocta tinamous

8 species inhabiting grassland and scrub of chaco, puna, and páramo habitats. All are round-bodied, with somewhat elongate crown-feathers, which are raised as a bushy crest in excitement. The colors are ochraceous to more brown or gray with intricate mottles and often whitish streaks above. The rear of the body has an exceptionally fur-like plumage.

TACZANOWSKI'S TINAMOU *Nothoprocta taczanowskii* – Plate I 14a-b

36 cm, with rather long, **curved bill**. Appears very **dark gray-brown with pale streaks and spots**. Above black with dark gray-brown bars, conspicuously striped with pale buff near the drab lateral feather-edges. Wing-coverts dark drab brown with irregular black and buff bars and spots, tertials and all flight-feathers with pale cinnamon bars on both webs. Side of head, neck, and breast dark drab gray with small pale buff spots with dark margins, and with whitish throat. Below ochraceous buff with obscure drab gray bars on sides. **Juv.** richer brown above and on breast. **Pull.** densely streaked, darker than in related species, and with almost black crown.
Habits: Wary. Often runs rapidly along borders, edges of fields etc., and difficult to flush. 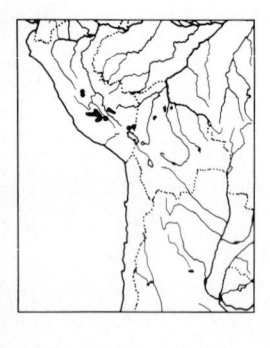
Voice: A loud cackling *cuyy-cuyy...* noted from a flushed bird.
Breeding: Eggs Apr.-May, pull. May (Junín), Oct (Puno).
Habitat: Rocky or grassy slopes with puyas or low copses, partly cleared, burned, or cultivated parts of montane shrub forest, open *Polylepis* woodland and at fringe of humid wood. May frequent small fields with tuber crops.
Range: Peru. At 2775-3700 m, and possibly higher. Rec. in Maraynioc in Junín, Chincheros, Naupallagta, Nevado Ampay area, and Pomayaco in Apurímac, and locally from Cuzco through E Andes of Puno. Uncommon, and may vanish as cultivation threatens to transform its habitat.

NOTE: *Nothoprocta godmani* Taczanowski, 1886, may be an immature *taczanowskii*.

KALINOWSKI'S TINAMOU *Nothoprocta kalinowskii* – Plate I 13

34 cm. **Resembles Ornate T.**, but generally **darker**, and with more rufous hue, more coarsely spotted black above. **Wing-coverts very densely barred** with a deeper gray tone, and **secondaries and tertials with rufous-brown bars**.
Habits, voice, and **breeding**: No data.
Habitat: Possibly on shrubby slopes and in small fields. Information in Walker & Ricalde (1988) that it inhabits forest is a misprint (B Walker).
Range: Peru. 2 specimens. Rec. from Tulpo (3000 m) se of Huamachuco, La Libertad, and in Licamachay (4575 m, may be wrong) s of Cuzco. Probably vanishing.

NOTE: This form has been suggested to be a well-marked ssp of Ornate T., but is more probably a distinct lower-elevation species.

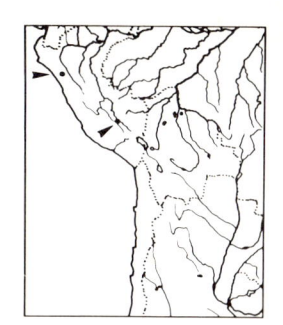

ORNATE TINAMOU *Nothoprocta ornata* – Plate I 10a-b

34 cm. Bill most curved in male. **Speckled grayish ochraceous**, with some variation in tone and hue. **Head whitish profusely dotted with black**. Upperparts of body mainly ochraceous gray, often rather olivaceous towards rump, densely vermiculated, mottled and barred with black and gray, and with gray or brown, or occ. white feather-edges (but never striped as Taczanowski's and Andean T.). Wing-coverts profusely barred buff, black, and gray; inner secondaries coarsely barred buff, black, and gray, but primaries only with white notches. **Neck and chest rather uniform drab-gray**, lower underparts more pinkish buff or ochraceous, sometimes faintly barred. Northern *branickii* small, dark, with heavy dorsal spotting, and little pale barring; *ornata* with prominent tawny feather-margins above; *rostrata* more decidedly buffy above, tinged brown on breast, and longer-billed. **Juv.** with less spotted face, but malar stripe indicated, and all underparts faintly barred.
Pull. as for Andean T. (Plate I 11d).
Habits: Usually single or 'pair' together. Sometimes seen running through the grass with raised neck and erect crown-feathers (looks tufted).
Voice: When flushed, gives a very thin, high-pitched, but slightly descending *weeeu weeu weeu...*, clearly different from Andean T. Also has a single whistle. On the ground may give a thin *weet-kweet* of squealing quality, and a low *wuck?* in response to playback (BW). Song high-pitched *zee-ee-ee-e-e-r*.
Breeding: Eggs in Dec-Apr and June-Aug (Peru) and Jan (Tarapacá, Chile); pull. Apr (Cochabamba and Oruro Bol).
Habitat: Mainly in monotonous grassland, usually places with rather tall bunchgrass tussocks, and sometimes in tola heath or open bushy country, or neighbouring slopes with some puyas or scattered *Polylepis* trees.
Range: Through large parts of puna zone at (2500-) 3500-4800 m. From Ancash to uplands of Ayacucho and Apurímac, Peru (*branickii*); from Cuzco across the altiplanos to Potosí and Chuquisaca, Bol., and

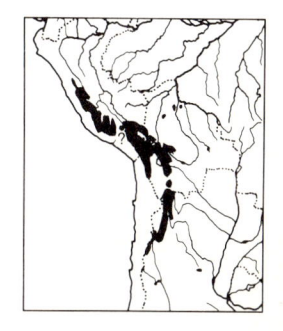

to c Tarapacá, Chile (*ornata*), and Jujuy to La Rioja, nw Arg. (*rostrata*). Widespread but in most places in moderate densities.

CHILEAN TINAMOU *Nothoprocta perdicaria* – Plate I 12

30 cm, with curved bill. Above and on sides of breast distinctly **striped**, as each feather has broad, mottled black-and-tawny median zone separated by white line from light gray (*perdicaria*) or gray-brown (*sanborni*) feather-edges. **Wing-coverts coarsely barred cinnamon** and dusky, secondaries with rufous brown, primaries with white notches. Crown barred black and brown, rest of head, neck, and underparts light drab gray (*perdicaria*) or ochraceous (*sanborni*), with some pale spots on c breast. **Juv.** with some dark spots on breast. **Pull.** as in Andean T.
Habits: Usually in small coveys.
Voice: Territorial call a whistled *tweewít*. Flushed birds give shrill whistles.
Breeding: Chocolate-brown eggs in Dec.
Habitat: Semiarid grassland and wheat fields. Often on roadsides.
Range: C Chile. Mainly lowlands, from Huasco Valley, Atacama, to Nuble (*perdicaria*), and from Maule s to Llanquihue and through low Andean passes to adjacent sw Arg. (*sanborni*). Strongly persecuted, now scarce in the n, but commoner in the s.

NOTE: Forms a species group with Andean and Curve-billed T.

ANDEAN TINAMOU *Nothoprocta pentlandii* – Plate I 11a-d

25-30 cm, with slender, curved bill. Sspp **pentlandii, doeringi,** and **mendozae** of the e Andes quite drab gray (*pentlandii* with brown rather than gray dorsal feather-edges, *mendozae* decidedly longer-billed), head with dark spots on crown and behind eye, and upperparts of body with several broad, tawny **stripes**, most of which include black bars and vermiculations, and with whitish streaks demarcating **gray lateral feather-edges**. Rump rather obscurely mottled compared with the preceding and following species. Flight-feathers look dark in flight, despite regular pale notches on the outer webs. **Breast bluish gray with white spots**, sides paler, more pinkish drab, sometimes vaguely barred; belly whitish. **Western sspp *fulvescens*, *oustaleti* and *ambigua* browner, rather fulvous below**, *ambigua* most ochraceous, breast sometimes with coarse brown bands laterally, but rather gray and with whitish spots centrally. Legs yellow or orange. **Juv.** more coarsely marked, lacking gray feather-edges above, and neck and breast showing profuse black spots. **Pull.** densely streaked brown, blackish and buff, crown tawny with less black spots than in Curve-billed T.
Habits: Usually walks in cover of dense bushes, roosting in sheltered place below rocks.
Voice: Clearly different from Ornate T. In W Andes, territorial call sharp *cheeleep*. When flushed gives long accelerating and descending series of melodic (sandpiper-like) calls **pyouc-pyuc-pyuc...yucyucyu.**. E

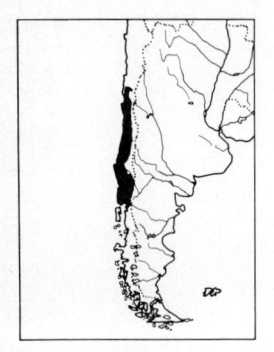

Andean birds noted to give a single whine, and a shorter series (compared with w Andean birds) of calls from flushed birds.
Breeding: Chocolate-brown eggs in July-Aug (Jan?) (w Peru).
Habitat: May prefer edges of cloud forest patches and thickets in humid ravines in semi-arid zones, e.g., bushy slopes with scattered *Carica* trees in w Peru, but also frequents hillsides with shrub-steppe and often coarse herbs (*Lupinus*, composite brush, or alfalfa and potato crops), columnar cacti, *Jatropa* or other thorny scrub.
Range: Disjunct. In mts of n Neuquén and Mendoza (*mendozae*), San Luís and Córdoba, Arg., (*doeringi*), at 1500-3600 m on Andean slopes from La Rioja and Santiago del Estero (Sierra de Guasayán), Arg., n to Potosi in s Bol., ascending to 4000 m in semiarid parts of Cochabamba and still higher in La Paz and Puno, e Peru (*pentlandii*; also birds from Putre and Socoroma in n Chile said to be this form). Around Cuzco, Peru (*fulvescens*). At 1500-3600 (3900) m along w slope of Andes in s and c Peru (*oustaleti*) and at similar elevations in nw Peru and at least previously above Zaruma on w Andean slope of s Ecu. (*ambigua*). Isolated local populations occur near coast of Peru (*niethammeri* and unnamed ssp.). Fairly common locally.

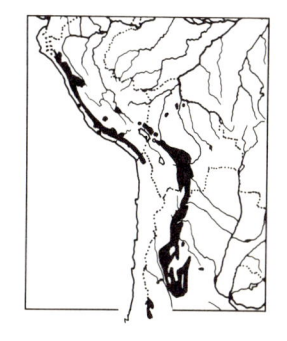

CURVE-BILLED TINAMOU *Nothoprocta curvirostris* – Plate I 9a-b

28 cm, with slender, **decurved bill**. Above tawny with black vermiculations and spots (largest in *curvirostris*, giving dark effect) and with **white streaks** laterally on most feathers (i.e. **lacking stripes of gray dorsal feather-edges of the preceding 2 species**). **Wing-coverts and secondaries barred black-and-rufous**, primaries notched white. Sides of head pale buff, rather streaked with brown and blackish. All **underparts light rufous** brown with some dusky and buff mottling on breast and sides. **Juv.** with black spots on breast. **Pull.** densely streaked and mottled brown, black, and whitish.
Habits: No data.
Voice: Song 3 whistles (1st note lower than next 2) repeated every 5 sec. When flushed screams *pee-pee-pee...*, much as in Andean T.
Breeding: Chocolate-brown eggs in Jan-Aug.
Habitat: Sandy bunchgrass slopes with evergreen ericads, *Hypericum*, and composite shrub (jalca), sometimes with slash-and-burn cultivation.
Range: At 2800-3700 m in Ecu. from Mt Pichincha to Condor mts in n Cajamarca, Peru (*curvirostris*); Andean valleys of n Peru from Cuervo in Cajamarca and Amazonas and through Marañón Valley s to Huánuco (*peruviana*). Generally common.

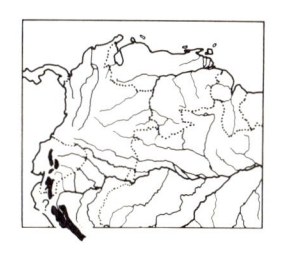

DARWIN'S NOTHURA *Nothura darwini* – Plate I 2

26 cm (*peruviana* small, wing 12.65-13.3 cm, *agassizii* large, wing 13.3-15.1 cm), with **small bill**, and rather short neck. Normally with the tail bent less strongly down between the legs than in a *Nothoprocta*, and therefore looks more flat-backed and less rotund. Sspp *agassizii* and *boliviana* above **tawny, the c zone of each feather showing several**

black bars (**up to 1 cm wide**), most of these interrupted by buff or white lines demarcating thin gray lateral feather-edges. Wing-coverts and flight-feathers dark spotted and barred cinnamon, but outer primaries with light notches only on outer webs, (**hand thus appearing very dark in flight**). Face and neck buff with dusky dots, throat white, rest of underparts warm buff with several coarse brown bands edged with black and occ. white lines, more barred towards posterior sides. **Juv.** shows tendency for more dense black barring within the brown zones of the back-feathers, and spotted rather than striped breast. **Pull.** resembles a *Nothoprocta*, but has small bill and inconspicuous post-orbital, auricular and malar stripes.

Habits and **Voice**: Hides in grass rather than under shrub. Said to give sweet, somewhat melancholy whistle as it runs. Rarely seen, except when crossing roads, or when flushed at close range. Then flies short distance only, and usually emits a **single, stressed scream *kyyit***. Whistled song *pi-pi-pi-pi-**pi** pi pi pi*.

Breeding : Grayish-violet eggs.

Habitat: Mainly in grassland, both in bunchgrass and esp. in small grain-fields, but also in savanna brush and in the highlands also on *Lepidophyllum* heaths.

Range: In semiarid subtrop. Arg. (*darwinii, salvadorii*), ascending to 2000-2600 m from Tarija and Oruro to Cochabamba, Bol. (*boliviani*), locally ranging to 4300 m on altiplano of Bol./Peru from Lake Poopó to the Puno/Cuzco border and in the E Andes of Cuzco, Peru (*agassizii*); and rec. at Sta Ana in the dry trop. zone of Cuzco (*peruviana*). Generally common.

NOTE: Darwin's Nothura has by some been considered sspp of Spotted N., although they overlap, and have different songs.

SPOTTED NOTHURA *Nothura maculosa* – Plate I 3

23 cm. Strikingly similar to Darwin's N., but normally more long-legged (tarsus 3.4-4 cm, vs. 2.6-3.4 cm, but Patagonian Spotted N. has tarsus down to 3 cm). **Primaries barred across both webs**. The gray-brown median zone of each back-feather generally has denser arrangement of black bars, in the upland forms *pallida* and *submontana* actually a dense mottling, demarcated with thin black-and-white stripes from the gray feather-edges, so that the overall impression is drab brown with fine white streaks. **Juv.** of these sspp are very obscurely and densely mottled drab above. Gives low whistles when running. The song is a series of whistles accelerating towards end (unlike in Darwin's N.).

Widespread with several sspp in chaco, grassland savanna, shrub-steppe, and *Larrea* semidesert (partial to places abounding in thin scrub) from s Brazil to sc Arg. Ascends to 2300 m in Córdoba mts (*pallida?*) and to c. 2000 m on the Andean foothills of n Neuquén and the large upland plateaus in c Río Negro into n Chubut, Arg. (*submontana*).

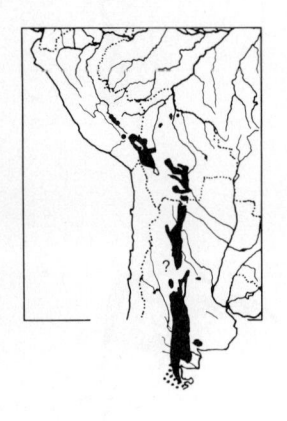

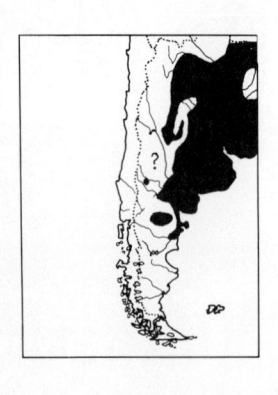

ELEGANT CRESTED-TINAMOU *Eudromias elegans* – Plate I 4a-b

39 cm. Heavy-bodied, with **thin, forward-curving black crest**. Sides of head finely streaked dusky and dull gray with whitish bands passing above and below eye. Otherwise **looks rather uniform olive-gray** at a distance. Above ochraceous gray, densely dotted with small round black-and-white marks, flight-feathers regularly barred fuscous-and-buff across both webs, tail large (for a tinamou), heavily barred black-and-white. Underparts ochraceous with black bars and streaks, densest on breast. Owing to independent variation in several plumage characters, the species has been subdivided into numerous broadly intergrading sspp: *intermedia*, *riojana*, and *magnistriata* are generally pale, fairly large-spotted, esp. the c double-spot on each feather, the former ssp buffy mixed with gray, the 2nd grayish and black, the latter more uniform dark gray-brown; *riojana* and *albida* are segments of a cline s-wards to *elegans*, *devia* and *patagonica*, which are generally dark olive-gray with abundant tiny spots, *devia* with the belly heavily barred. **Juv.** and **pull.** have broadly striped head, slight drab brown crest, and drab gray plumage with some dusky and whitish streaks (not dots).
Habits and **voice**: Generally in groups of 10-20 birds. Very fond of dustbathing. When scared, utters shrill squeaky cries as they run. Rarely flies, and then with bursts of whirring wingbeats – and soft wailing notes – with glides inserted.
Breeding: Yellowish green eggs.
Habitat: Arid and semi-arid open woodland, *Larrea* shrub, and steppe, in the temp. zone mainly sandy areas with thorny scrub and low evergreen bushes, e.g., 'mata negra' (*Verbena tridentata*).
Range: Widespread with several sspp in semi-arid parts of Arg. Locally to 2500 m (information about 3000-4000 m seems doubtful) on Andean slopes from Salta to Catamarca (*intermedia*), sw to Tucumán and n Córdoba (*magnistriata*) and s to La Rioja and San Juan (*riojana*), other forms on the lower subtrop. plains, and in the temp. zone of Patagonia: *elegans* in Río Negro and Neuquén, *devia* at base of Andes in Neuquén, *patagonica* from s Neuquén to Sta Cruz) enters low, sheltered valleys between the foothill plateaus and passes over to the Chilean side at 46°30's in Aysén.

NOTE: Sspp *numida* and *wetmorei* published by RC Banks in *Proc. Biol. Soc. Wash.* 89(1977):529-44 was synonymized in the revision by J Navas in *Rev. Mus. Arg. Cienc. Nat. 'Bernadino Rivadavia'*, 11(2)(1981): 33-59.

PUNA TINAMOU *Tinamotis pentlandii* – Plate I 15a-b

42 cm, **really bulky compared with other highland tinamous. Head and neck buffy white with unclear, blackish bands** on crown, through eye and on cheek. Otherwise warm buff or olive-buff towards the rump, profusely mottled with drab gray to olive-gray herringbone marks to give a rather uniform **olive-gray overall effect. Belly and vent uniform orange-buff. Juv.** has less rufous belly and less olive rump. **Pull.** colored much as juv., with striped black-and-white head, and finely streaked dusky and ochraceous body.

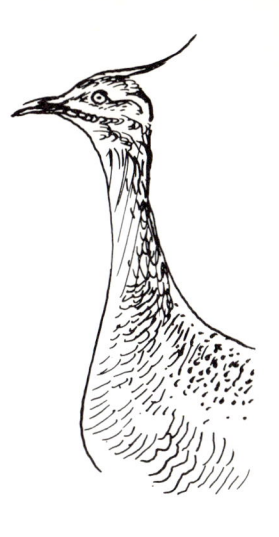

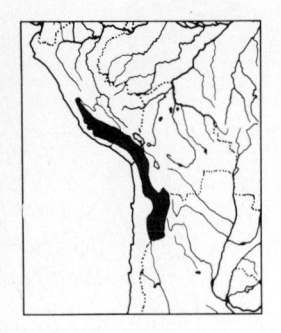

Habits and **voice**: Often in small groups. Song melodious, flute-like *kewla-kewla-kewla...* or *kiula-kiula-kiula...*, apparently in chorus. When flushed, gives a descending sequence of thin wailing whines. Glides with wings horizontal (vs. bowed in other highland tinamous).
Breeding: Yellowish green eggs in June-Aug (Peru) or probably mainly Sep-Oct (n Chile).
Habitat: Prefers tola heath (*Lepidophyllum*), but also inhabits other sandy or stony habitats or pumice with scattered *Larretia compacta* and other cushion-plants, and tussocks of coarse grass (*Festuca orthophylla*), and sometimes scattered arid *Polylepis* shrub.
Range: At 4000-5300 m from Junín, c Peru, along W Andes s to Antofagasta, Chile, extreme w Bol. and n Catamarca, Arg. Generally scarce, but fairly common near junction of Bol./Chile/Peru territories. A rec. from Cord. Real in La Paz (Ribera & Hanagart 1982) needs confirmation.

PATAGONIAN TINAMOU *Tinamotis ingoufi* – Plate I 16a-b

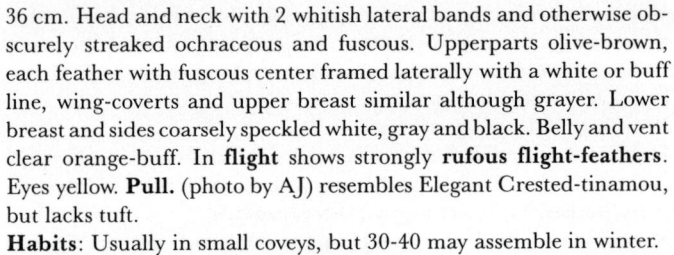

36 cm. Head and neck with 2 whitish lateral bands and otherwise obscurely streaked ochraceous and fuscous. Upperparts olive-brown, each feather with fuscous center framed laterally with a white or buff line, wing-coverts and upper breast similar although grayer. Lower breast and sides coarsely speckled white, gray and black. Belly and vent clear orange-buff. In **flight** shows strongly **rufous flight-feathers**. Eyes yellow. **Pull.** (photo by AJ) resembles Elegant Crested-tinamou, but lacks tuft.
Habits: Usually in small coveys, but 30-40 may assemble in winter.
Voice: Flute-like notes reported.
Breeding: Dark olive-buff eggs Nov-Dec.
Habitat: Grassland steppes, in sheltered valleys with patches of dense, low brush (*Berberis, Pernetyia, Verbena*). Sometimes near sea-coasts, esp. in winter.
Range: Mainly at 200-800 m in sheltered valleys between the barren and windy plateaus of Arg. Patagonia from w Río Negro to Sta Cruz and into Aysén and Magallanes, Chile. Generally uncommon.

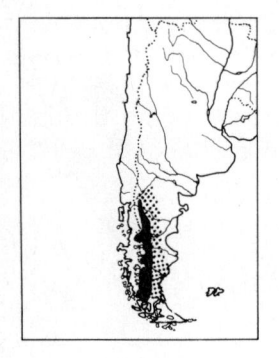

Patagonian Tinamou

Podicipediformes

Grebes – Family Podicipedidae

A group of uncertain systematic affinity, possibly distantly related to other waterbirds. The 20 species are distributed almost worldwide, mainly in freshwater lakes and marshes, but also in saline environments, and many species winter at sea.

Grebes are foot-propelled diving birds recognized by pointed bill and downy, almost tail-less rear. The folded wings are normally not discernible. The peculiar feet, in which each toe has a separate swimming lobe, are placed at the extreme rear end, and are not well suited for walking. Grebes normally leave the water only to ascend their nests, which are soggy floating structures of water-weeds anchored among reeds, or open to view among floating plants. Most Andean grebes lay only 2 chalky white eggs, which gradually become stained buff. The downy young can dive almost at once, but are vulnerable to chilling, and therefore carried under the parents' wings most of the time. They are fed for a long period by both parents.

Small round-bodied grebes often sunbathe, expanding their loose plumage as they turn their high sterns towards the sun, looking like floating feather-balls (6a on Plate II). Grebes characteristically preen rolled over in the water. When alarmed, they become slim and dive swiftly, hardly leaving a ripple. In emergency, esp. near nests, they often use a 'fold-in-middle' dive (2b on Plate II), sometimes with distinct splash (instead of normal diving with fore- or downwards thrust of neck). Certain species are experts in hiding under water, and come up for air hidden among floating plants, with only head visible. Typical open-water species are, on the other hand, easy to watch. The diet comprises small fish, water insects, and crustaceans, taken by pursuit, by gleaning from water-weeds, and picking from the surface.

Grebes need a long running start to become airborne. Birds from southern S Am. are migratory. They move between lakes at night, and often make short upwind flights by day. Grebes living near the Equator fly very rarely, and 2 Andean species are flightless. Grebes are also flightless during the period of wing-feather molt. Some open-water species congregate in good feeding habitats for this event.

The classification follows RW Storer (in Mayr & Cottrell 1979). Detailed studies of the biology of Andean grebes are published in *Ardea* 74(1986):40-58; *Bull. Brit. Orn. Club* 95 (1975):148-151; *Condor* 84(1982):370-380; *Dansk orn. Foren. Tidsskr.* 76(1982):37-68; *Hornero* 10(1967):339-350, 11(1977):377-379; *Ibis* 125(1983):463-481; *Living Bird* 20(1982)51-67; *Steenstrupia* 7(1981):237-259, 11(1985):133-155, *Trans. San Diego Soc. Nat. Hist.* 18(1976):113-126 and *Vid. Meddr. dansk Naturhist. Foren.* 143(1981):125-249.

WHITE-TUFTED GREBE *Rollandia rolland* –
Plate II 3a-c

24-30 cm, with ragged, somewhat square head, plumes of head longest in Patagonian birds. **Ad**. with head, neck, and back **black** with green sheen, save for large triangular patch of **white, black-streaked plumes on headside.** Sides tawny with obscure spots, rest of underparts rich chestnut, but stern pale. Towards end of breeding season black parts become dark brown, often with buff feather-edges, and underparts become cinnamon. **Non-breeders** have throat and belly almost white, but neck dull rufous. **Juv**. resembles non-breeder, but has 2 broad black bands (instead of thin streaks) across the buff head-side. In rare **flight** shows large white wing-patch. **Pull**. striped black and rich buff. **Sexes** separable by very stubby bill in female; however, bill size varies geographically, being smallest in Lake Titicaca area, largest in the stocky ssp *morrisoni*.

Habits: High-sterned and with fluffy plumage when calm. Often tame and inquisitive, but may skulk in the breeding season. Territorial or loosely social. Often seen threatening near reed-borders, deep in water with head low; or chasing one another out on open water, pattering across surface. Feeding dives rather clumsy, and often feeds in surface position with immersed head. Prefers fish for food, but also takes a variety of mainly larger arthropods.

Voice: **Usually silent**, but may give low, cracking and growling sounds, as *jarrh* when alarmed, and growling *chourrh* or *hrrr* during aggression.

Breeding: Multiple-brooded. Egg-laying mainly Oct-Dec, but in Peru at all seasons.

Habitat: Marshes, ponds and lake shallows. Prefers mosaics of aquatic vegetation and open spaces, and channels into reed-marshes, but also frequents reed-fringed bays of open lakes, and lakes and ponds with no other vegetation than dense floaring carpets of water-weeds.

Range: Lowlands throughout Arg. to Parag., Chile, except in the northern desert (southern populations migratory or wintering along adjacent sea-coasts) and in coastal Peru. Lake districts of the s Andes, and at 3500-4500 m in puna zone from Catamarca, wc Arg., through w Bol. and ne Chile to Ancash, c Peru (*chilensis*). Lake Junín, c Peru (*morrisoni*). Islas Malvinas (*rolland*). Generally common.

NOTE: Gigantic ssp *rolland* of the Malvinas is currently classified as a megassp. If species rank should be assigned (Rolland's Grebe), the name *R. chilensis* applies to the mainland forms.

TITICACA FLIGHTLESS GREBE *Rollandia microptera* – Plate II 4a-d

40 cm. Rather **slim, with long yellow-striped bill**. **Ragged chestnut cap streaked black**, nape uniform chestnut. **Sides of head below eye and throat white**, except for black streaks on the ear-plumes. **Ad**. has back dark brown with tawny feather-edges, and dull rufous neck, sides and underparts. **Subad**. and **imm**. are lighter, with dull gray back with fine dusky streaks, brown sides, but more or less white underparts. Flightless, but during chases sometimes exposes small wings with white patch. **Juv**. and **pull**. pale, drab gray above, head and neck striped dusky brown, rufous, and white.

Habits: Territorial, but often in family-groups, occ. 3 with pull. and juv. from an earlier breeding episode together. Sometimes assemble outside potential breeding habitats for social displays with much calling, pattering, and racing in more or less upright postures. Often threatens with lowered head, occ. raising back-feathers. Normally looks slim and never high-sterned. Crown-feathers often raised into

double transverse fan, but scared birds look sleek, and often raise small forehead tuft. When alarmed, swims offshore rather than into cover. Fast diver, mainly taking fish far offshore.
Voice: Repeated loud *gaeck* or *gjaong* calls. **Widely audible squeals** and mewing and whining notes, *chee chee chee...*, *eeer err aeeer*, or lower *djurrrh idjurrrh...*, often as duets.
Breeding: Multiple-brooded. Eggs most of year, but chiefly Nov-Dec.
Habitat: Open lake habitats, often feeding far offshore, as long out as some vegetation exists at the bottom. Breeds in rather open mosaic-like parts of wide reed-marshes in places with easy access to open water, or open to view in floating water-weeds.
Range: Peru/Bol. Lakes Titicaca, Arapa, Umayo, River Desaquadero, and lakes Uru-Uru and Poopó. Common, but some local populations may disappear in drought years, and become re-established in years of extreme flooding.

NOTE: Previously in genus *Centropelma*. Called Short-winged Grebe by Meyer de Schauensee.

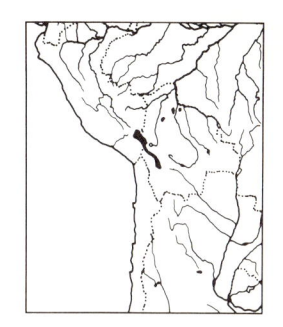

LEAST GREBE *Tachybaptus dominicus* – Plate II 1a-c

25 cm. A **tiny, round, dull-colored grebe**. Ad. sooty brown with pale stern, head and neck plumbeous with black cap and throat-bib. Eyes yellow. **Non-breeding** and **imm.** with pale throat, **juv.** striped head, **pull.** boldly striped also on neck and back. Unlike in White-tufted G. the species shows **absolutely no buff or rufous colors**, apart from a rufous crown-patch in the pull. In rare **flight** shows large white area on secondaries. This skulking species is most often revealed by its voice, either loud **trills** *dye-dye-ye-ye-e-e...*, a low *kirrr-r-r...*, or a somewhat nasal (gallinule-like) high-pitched note.

Inhabits swamps, ponds and shallow lakes, often with abundant floating vegetation. Very local mainly in subtrop. zones of the Americas (in S Am. *speciosus*, and tiny *eisenmanni* – described by RW Storer on pp. 31-39 in Buckley *et al.* (1985) – common in trop. w Ecu.). Breeds in Cauca and Patía valleys, w Col., at 2000-2700 m locally in E Andes, Col., and to similar elevations at Chachapoyas, n Peru, and Sta Cruz, Bol. Accidental in Andes of Ven.

NOTE: Previously in genus *Podiceps*.

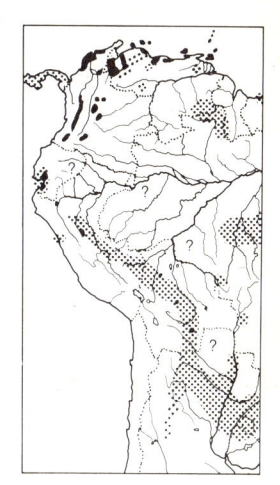

PIED-BILLED GREBE *Podilymbus podiceps* – Plate II 2a-c

32 cm. A gray-brown grebe with **thick chicken-like bill** and rather flat forehead. Ad. with sides of head gray, **throat black, bill white with black transverse band. Non-breeders** with cheeks and foreneck dull cinnamon-buff (not streaked cheek as in White-tufted G.), throat whitish, bill horn. **Juv.** similar, but side of head whitish with fuscous stripes. In **flight** lacks white in wing. **Pull.** striped mainly in black, gray, and white, unlike Least G. with rufous color restricted to nape, distinctly striped forehead, and black band on cheek interrupted.

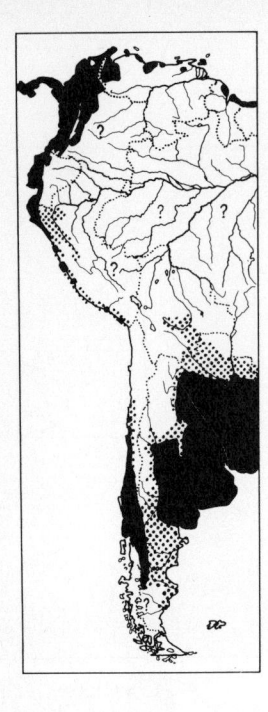

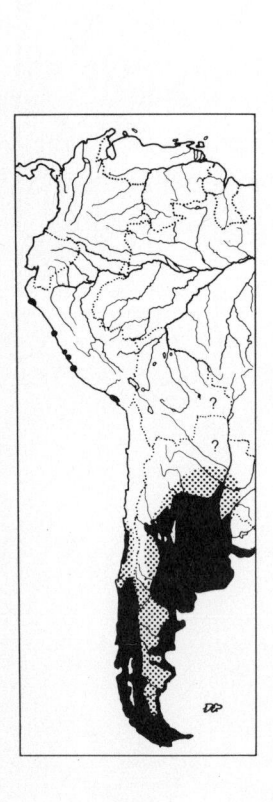

Habits: Territorial or loosely social. Often in crouched pose near vegetation border, and never in high-sterned sunbathing pose. Skulks when alarmed, disappearing in fold-in-middle dive, and hiding. Feeds mainly on fish and larger arthropods, incl. crayfish and freshwater crabs.
Voice: Usually silent, but in breeding season gives wailing and grunting sounds, and a **vibrant chatter** starting with whinnying calls *eeow-eeow-eeow, keeow*, then slowing down to a sequence of *cow* calls.
Breeding: Multiple-brooded. Breeds all year, but in Andes of Col. the egg-laying peaks Jan-Mar and Sep-Oct.
Habitat: Lakes, marshes, and ponds, usually with abundant reeds, floaters (*Azolla, Limnobium, Ludwigia*) and submergents, and not necessarily very much open water. However, some of its best sites in the northern Andes are large lakes with vegetation only along the margins.
Range: Large parts of the Americas. Mainly in lowlands, but breeds commonly at 2600-3100 m in E Andes, Col., and at 2200-2600 m in n Ecu. (common on Yaguarcocha and San Pablo). Otherwise accidental in high parts of Andes, but rec. to 2570 m in Cochabamba, Bol., and 2100 m in Jujuy, nw Arg. (*antarcticus*).

GREAT GREBE *Podiceps major* – Plate VIII 14a-c

70 cm, slim, with very long sinuous neck, and long, slightly up-turned bill. **Ad**. rufous-necked with plumbeous head and small occipital crest; back blackish, sides pale or cinnamon-rufous with obscure spots. Ssp *navasi* recognized by entirely black hind-neck without any gray laterally, face blackish slaty, bill black (not greenish). In winter more dull with throat whitish, neck grayish washed more or less with rufous. **Juv**. has black stripes on head. In **flight** shows extensively white flight-feathers. **Pull**. striped black-and-white all over and with conspicuous bare red spots on crown and lore.
Habits: Usually on open water, and easy to watch. Often curious. Loosely social, frequently nesting in colonies. Long neck stretched up or sinuous (low over water in threat). **Bill never raised** as in a cormorant. When alert or alarmed often raises tail to show buff vent. Feeding dives swift, often with elegant arched leap, but uneasy birds disappear in a fold-in-middle dive.
Voice: Call a far-carrying, **long, plaintive** *oooo* or *ooo-oo*. Also irregular sequence of soft *a* calls. When alarmed gives a loud, staccatto *ap ap ap....* When aggressive a drawn-out, accented *ao^{ow}*, encounters sometimes followed by machine-gun-like *a-a-a-a...*
Breeding: Season irregular, but sometimes protracted, with several successive clutches: eggs mainly Oct-Jan, but sometimes at other seasons (s Chile, Peru).
Habitat: Mainly estuarine marshes and reed-fringed bays of larger lakes. Often 'winters' in marine kelp-zones (*Macrocystis*), and occ. roams far offshore.
Range: Widespread in lowlands of southern S Am. Plains of Arg. from Isla Grande to se Brazil and to base of Andes in San Juan and in c Chile (*major*). Very common to 1200 m in lake districts from Tierra del Fuego to Bío Bío in s Chile and adjacent sw Arg. (*navasi*, see *Com. Mus. Arg. Ci. nat. Bernardino Rivadavia* 14(1985):115-119). Peruvian coastal zone (unnamed ssp).

COLOMBIAN GREBE *Podiceps andinus* – Plate X 4

33 cm. Head mainly black, crown-feathers slaty, somewhat elongated as a helmet-like crest, **auricular plumes chestnut**, the upper ones sometimes golden. Rest of upperparts blackish, **foreneck, breast, and sides chestnut**. Eyes red. Most of year, chin, cheeks and flanks more or less admixed with white feathers, but some chestnut always retained. **Juv**. and **pull**. unknown.
Habits: Social. When in danger said to swim offshore rather than into cover.
Voice: Possibly thin whistles.
Breeding: Probably Aug-Sep.
Habitat: Freshwater marshes and lakes with tall marginal reeds, and extensive shallows full of submergent water-weeds.
Range: Col. in E Andes. Up to the 1940s on Bogotá Savanna and Lakes Fuquene and Cucunuba. At least to the late 1960s in Lake Tota. Most probably became extinct in the 1970s (Varty et al. 1986, and JF).

NOTE: Often considered a megassp of Black-necked G. (*Podiceps nigricollis*) of N Am., Eurasia and Africa. Owing to apparently more primitive characters we treat it as an (allo)species. These taxa form a species group with Silvery and Junín Flightless G.s.

SILVERY GREBE *Podiceps occipitalis* – Plate II 6a-d

27 cm. A **small-billed, silvery gray grebe with red eyes and white foreneck**, belly, and stern. **Ad**. with **large black area on nape** continuous with dark stripe down hind-neck. Ssp *juninensis* is rather dark gray with gray-brown ear-plumes, sometimes with brass-like tinge, and with white throat. Ssp *occipitalis* has lighter crown in sharp contrast to black nape, lustrous straw-yellow ear-plumes, and light gray chin and throat. On some lakes stained ferrugineous on breast. **Juv**. lacks ear-plumes and has slaty gray nape. In **flight** shows white or gray-shaded secondaries and inner primaries. **Pull**. mainly gray above with white-striped forehead, small chicks of esp. *occipitalis* are quite distinctly striped on all upperparts, but large chicks appear uniform light gray.
Habits: Usually open to view, in pairs, groups, dispersed flocks, or sometimes in dense rafts. Often confident and easy to watch. Often sunbathes with fluffy plumage, stern high. In breeding season shows elaborate displays, when pairs 'dance', or groups show parallel races, in upright attitudes with bodies almost out of the water. Often 3 birds make a common barge-and-dive display. Feeds mainly on tiny arthropods by picking from the surface, and by rather clumsy, stationary diving.
Voice: Bouts of frail whistled calls, in *juninensis* mainly thin *dooi'th* or *vit*, in *occipitalis* less melodic *chook* or *djec*. In barge-and-dive displays gives *dzi-dzeee* calls between incessant violent dives.
Breeding: Colonial, often with the small nests packed closely together almost to form rafts. In highland lakes often nests open to view in places with waterweeds on the surface. Eggs mainly Nov-Jan (Patagonia), Nov-Jan and rarely Mar (puna zone) or Feb (Col.).

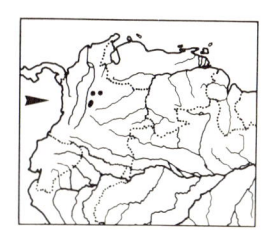

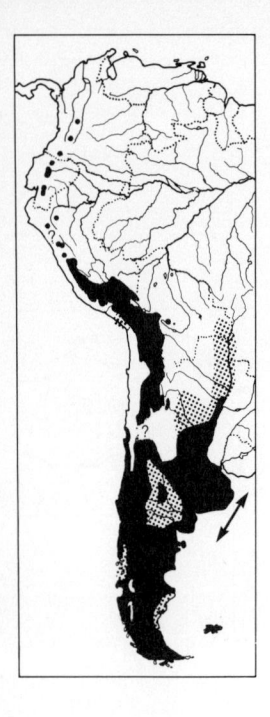

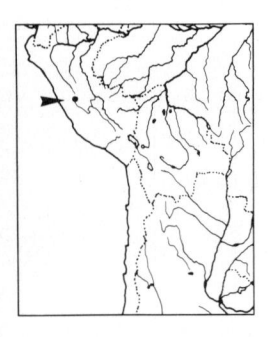

Habitat: Open lake habitats with or without reedbeds. Mostly shallow, weakly alkaline lakes with widespread but not too dense submergent vegetation, and rich insect life. Outside the breeding season sometimes congregate on salt-lakes.
Range: Lowlands of s and c Chile and Arg. n to Córdoba and Sta Fé, wintering to n Arg.; rec. to 2600 m in Aconcagua, 2800 m in Catamarca, but normally below 1300 m (*occipitalis*). At 3000-5000 m in puna zone from Catamarca, Arg., through Bol. and ne Chile to Ancash, Peru (descending to the Pacific coast in Arequipa), and in some páramos through n Peru and Ecu. to Nariño, and in lakes San Rafael and Otun in C Andes, Col. (*juninensis*). Abundant on Patagonian uplands and near junction of Bol./Chile/Peru territories, but rare and vanishing in n parts of range.

JUNIN FLIGHTLESS GREBE *Podiceps taczanowskii* – Plate II 7a-c

35 cm. Plumage as in sympatric Silvery G., although with whiter flanks. Somewhat larger, **slimmer and longer-necked**, with more pointed head, and longer bill (2.6-3.6 cm, vs. 1.6-2 cm in sympatric Silvery G.). **Bill mostly light gray** (not blackish); feet buff (not blackish). **Juv**. and **pull**. as in Silvery G., but with paler bill.
Habits: Usually in small parties. When calm and sunbathing hard to tell from Silvery G. However, distinctly longer-necked when alert. Feeds mainly on tiny fish (*Orestias*). During social feeding, several birds swim on a line and dive synchronously. Similar displays to Silvery G., sometimes both species dancing together.
Voice: Melodic whistles *dooi'th* and *wit*.
Breeding: Among tall reeds, usually in reed-isles with partly broken-down vegetation. Egg-laying mainly Dec-Jan.
Habitat: Open lake with widespread submergent vegetation of *Chara*. In the breeding season enters bays and channels in the outer edge of wide reed-marshes surrounding the lake, but stays far offshore part of the year.
Range: C Peru in Lake Junín 4080 m. Population falling rapidly towards extinction (thousands in the 1960s, over 300 1977, not much over 100 1987), owing to pollution and regulation for hydroelectric power.

NOTE: Called Puna Grebe by Meyer de Schauensee; the alternative name proposed here emphasizes its flightlessness, and its restriction to a single locality near the periphery of the puna zone.

HOODED GREBE *Podiceps gallardoi* – Plate II 5a-b

32 cm. More long-bodied than Silvery G., never so high-sterned, and looks **shiny white** at a distance. **Head black with white forehead grading into rufous of erectile crown-cap**, this hood contrasting with white neck with thin black line behind. Eyes red bordered with yellow wattle. Back dark slaty, sides white with minute dark spots. In **flight** shows extensively white wings with black tips and wrist spots. **Imm**. (2nd year) similar with a few white feathers on cheeks. **Juv**. with pale

lower forehead and black crown-cap contrasting with white lower parts of head and nape (vs. Silvery G., where dark cap and nape gradually tapers into the dark line on hind-neck). **Pull.** white-throated and black-capped; in the first week boldly striped on the body, later with light gray body.
Habits: Gregarious and very confident. Imm. and non-breeders may assemble on other lakes than those used for breeding, occ. hundreds together. In spring shows elaborate displays, e.g. dancing and barging parallel with bodies raised almost out of the water. Feeds gleaning small invertebrates from the waterweeds, and depend strongly on snails and amphipods when feeding young.
Voice: Melodious whistles. Call an extended whistled phrase *te-whreee-errr* of rolling quality. Near colony-site incessant *kee-kee-kee...* and *ki-wee ki-wee....*
Breeding: Nesting colonies always open to view in places where waterweeds (*Myriophyllum*) form dense floating carpets on the surface. Eggs in Dec-Jan (Feb). Never tries to raise more than 1 young.
Habitat: In extremely wind-blown upland habitats where lakes totally lack marginal reeds. Prefers clear, slightly alkaline lakes with submerged water-weeds. In the breeding lakes it feeds esp. along the edges of dense floating weed-carpets. The best breeding habitats are volcanic lakes surrounded by sheltering cliff-walls.
Range: S Arg. in inland of Sta Cruz. At 500-1200 m from 47° to 51°s, but common only on the basaltic plateau between Lakes Strobel, Quiroga, and Cardiel, the smaller plateau near Lake San Martín to the sw, and the Asador plateaus to the n. Possibly winters near the breeding area, on the large, permanently icefree clear lakes Islote, Quiroga, and Strobel. Casual near Payne, Magellanic Chile. Population c. 5000.

NOTE: Described by M Rumboll in *Com. Mus. Arg. Cienc. nat. Bernardino Rivadavia* 4(1974):33-35.

Junín Flightless Grebes, in front one Silvery Grebe

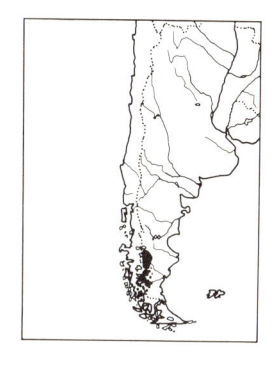

Pelecaniformes

Waterbirds probably related to wading birds. The order is apparently paraphyletic, as the cormorant line may have given rise to herons and ibises, the pelican line to storks and New World vultures (*Sci. Amer.* 1986:68-78). As traditionally defined, the order comprises birds with swimming-webs connecting all four toes and a more or less developed throat pouch.

Cormorants – Family Phalacrocoracidae

30 species of worldwide total distribution, mainly at sea, but several species inland.

Cormorants are rather large, foot-propelled diving birds, recognized by long strong neck and fairly long hooked bill, and long stiff tail used as 'hydroplane' under water. Usually swim low in the water, sometimes with no more than head and neck visible, bill tilting upwards. Never cock tail while swimming. Chasing fish, they proceed far under water, sometimes 100 m distance.

Unlike in other diving birds, the plumage soon becomes watersoaked. Cormorants (and darters) therefore spend much of the day perched conspicuously in trees overhanging water or on snags or rocks, standing upright with spread wings to dry the feathers.

Cormorants have some difficulty in becoming airborne, but have a strong flight. They are somewhat goose-like in flight, but many species slightly raise their head, and may glide and soar on arched wings. Being social, they often fly in lines or V-formation.

Cormorants breed in colonies, placing their coarse stick- or seaweed nests on cliffs, rocky skerries, trees, or reed-beds. The blackish, rather grotesque-looking young are fed for a long time in the nest.

No detailed studies have been made in the Andes.

NEOTROPIC CORMORANT *Phalacrocorax olivaceus* – Plate VIII 15a-c

70 cm, rather slim. **Ad. bronzy black** all over except for narrow white line bordering bare, dull yellow facial and gular skin. Eyes normally green, but light blue eyes may characterize birds of the Andean zone of s Peru to c Arg. (*Bull. Brit. Orn. Club* 108(1988):147-150). In breeding dress aquires small white plumes near ear, and a more prominent white border of the gular skin. **Imm. fuscous** above and pale drab brown below with almost white breast (breast spotted black during molt). **Juv**. dull brown at first, and without white border of the gular skin.
Habits: Alone, in pairs, or loose groups, sometimes feeding socially, many birds swimming on a line with immersed head, diving synchronously. In flight, neck withdrawn with a distinct kink, so that the bill is slightly raised.
Voice: Usually silent, but guttural, pig-like grunts and croaks are heard frequently in breeding colonies and at roosts.

Neotropic Cormorant

Breeding: Egg-laying in Jan-Mar (Apr) (puna zone) or Nov-Jan (Fuegian zone).
Habitat: Sea-coasts, estuaries, rivers, lakes, and rather open marsh habitats, but may prefer wide estuarine shallows at sea and in lakes. In the Andes, favored by introduction of trout (*Salmo gairdneri*) and pejereyes (*Odonthestes*).
Range: Widespread from s USA through S Am., from trop. to sub-antarctic zones. Rare straggler in the northern Andes (mainly imm.; many soaring flocks seen Jan 1983 at Lake Mucubayi, 3500 m, Ven.; K Malling Olsen). Commom at 3500-4200 m (and locally to 5000 m) s-wards from Ayacucho c Peru through puna zone of Bol., n Chile, and nw Arg., but further s in the continent mainly inhabits low-lying fjord lakes of the Andean zone (*olivaceus*). Common.

NOTE: Also called Olivaceous C.

IMPERIAL SHAG *Phalacrocorax atriceps* – Plate VIII 16

72 cm. Above black with white shoulder patch and mid-back patch. Below white from chin to vent. Eyes blue, legs pinkish brown. In *atriceps* white parts include cheeks, ear-coverts, and sides of neck. **Ad**. with yellow caruncle on forehead, and small curled crest. **Juv**. dark brown above, and lacking caruncle. Gregarious. In flight, neck stretched horizontally (unlike Neotropic Cormorant).

Breeds along subantarctic coasts. Common along coasts of s Arg. and Chile, and in large lakes around 800 m near foot of s Andes in Neuquen and Rio Negro, Arg., nesting on rocky isl. Victoria in Lake Nahuel Haupí at least since 1942 (*atriceps*).

NOTE: The name Imperial Shag includes Blue-eyed Cormorant (*P. atriceps*) and (Black-cheeked) King Cormorant (*P. albiventer*), lumped after the recent discovery of intermediate individuals and lack of behavioral isolation (*Gerfaut* 68(1978):53-86, *Wilson Bull.* 98(1986):571-580).

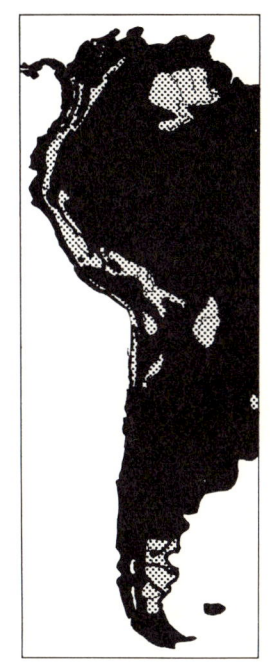

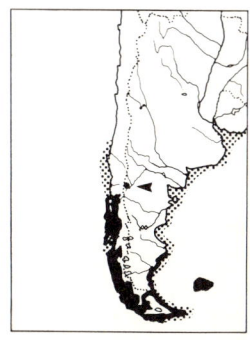

Flying Anhinga and Neotropic Cormorant

Darters – Family Anhingidae

4 species inhabit Australia, S Asia, Africa, and the Americas. Darters often swim submerged with only snake-like neck above water (hence often called snakebird). In flight, rapid flaps alternate with glides. Sometimes soars in wide circles to great heights.

ANHINGA *Anhinga anhinga* an inhabitant of trop. and subtrop. swamps from s USA to n Arg. – is accidental to the Bogotá savanna (2600 m) in E Andes and upper Cauca Valley, Col. 80 cm. Differs from cormorants by longer and very thin, sinuous neck, tiny head, thin pointed bill, and much longer, fan-shaped tail. **Male** mainly glossy black, wing-coverts and scapulars streaked silvery white, tail pale-tipped. In breeding dress side of upper neck has pale, elongate plumes, and hind-neck has black 'mane'. **Female** browner on back, and with head, neck, and upper breast buffy brown, lower breast chestnut. **Imm.** like non-breeding female, but duller, and without chestnut breast band. **Juv.** paler below, light marks on shoulder vague.

Pelicans – Family Pelecanidae.

7-8 species inhabit warm parts of all continents. Huge birds with enormous bills. Recognized in flight by majestic aspect as they plan with consummate ease on horizontal wings, often in line formation just over the waves. Also soar to great heights in wide circles. The head is retracted and tucked into the body, which gives a hunched appearance. The tail is short. 2 Am. species differ from other pelicans in catching fish by plunge-diving from the air.

BROWN PELICAN *Pelecanus occidentalis* – an inhabitant of coasts and lowlands of Pacific USA, the Caribbean area, and s to Galapagos and nw Peru – is accidental visitor at 2600 m on the Bogotá and Ubaté savannas in E Andes, and near Popayañ, Col. (*occidentalis*). 120 cm. **Ad.** gray-brown with pale silvery gray panel in wing, and white to straw-yellow head (except purplish black skin of face). Unlike larger Peruvian P. (*P. thagus*) underparts lack white streaks. In breeding dress chestnut color of lower neck extends up hind-neck to crested nape. **Juv.** dull brown all over, merging into white on belly.

Ciconiiformes

A polyphyletic order of wading birds. However, all families seem to have evolved from within the ancient Gruiform-Charadriiform-Pelicaniform complex of waterbirds.

Herons, egrets and bitterns – Family Ardeidae.

Herons are distributed worldwide, with altogether 62 species. They are medium-sized to large and lanky wading birds with straight pointed bills, and long kinked necks. The long legs and toes are suited for wading in shallow water, and also permit climbing in swamp vegetation. The plumage is coarse and loose, often with long delicate plumes on crown, foreneck, and mantle. In flight, retracted neck gives thick-chested appearance. The neck is extended only during brief flights and certain display flights. The wing-beats are rhythmic, calm, and quite shallow. The wings are broad, blunt, but not conspicuosly fingered, and are arched.

Herons live along shores of lakes, lagoons, and streams, and in marshy or swampy terrain. They often stand motionless near reed-borders, in characteristic, upright but rather hunched posture, with retracted neck (head resting on 'chest'). Most species stalk fish, amphibians, and insects solitarily, with slow strides and much motionless waiting. A few species are more social and the majority are gregarious when roosting and nesting. Roosts and breeding colonies are usually in inaccessible places in riverine woods, swamps, tall trees, cliffs, or islets. The nests are coarse platforms of sticks or reeds. Both sexes build, incubate, and feed young, which are nidicolous, covered with long hair-like down.

The classification is according to R B Payne (in Mayr & Cottrell 1979). Information about species occurring in the Andes is given in Hancock, J & H Elliot (1978) *The Herons of the World*, London; Olivares, A (1973) *Las Ciconiiformes Colombianos*. Bogotá; *Ardea* 59(1971): 1-16; *Auk* 83(1966):304-306, 85(1968):457-440, 92(1975):590-592; *Beitr. Vogelk.* 13 (19xx):397-454; *Bull. Brit. Orn. Cl.* 94(1976): 81-88; *Colonial Waterbirds* 11(1988):252-262; *Condor* 72(1970):407-416, 73(1971): 107-11; *Hornero* 10(1967):225-234; *Ibis* 109(1967):168-179; *Living Bird* 8(1970):95-111, 9(1971):176-194; *Misc. Publ. Mus. Zool. Univ. Mich.* 150 (1975); *Mit. Zool. Mus. Berlin* 53(1977):1-78; *Ostrich* 40(1969):75-127; *Wilson Bull.* 83 (1971): 435-438.

GREAT WHITE EGRET *Casmerodius albus* – Plate III 3

110 cm. A **large and slim white** heron, in breeding dress with long and fine upwards-curving plumes on lower back. **Bill** of *egretta* **yellow** in all plumages. Legs and feet blackish. When no other herons available for size comparison, look for extension of black line from gape to well beyond eye (see drawing accompanying Plate III). Note that Great Blue Heron has an extremely rare white color phase.
Habits: Shy. Usually feeds alone or in loose groups. Stalks prey slowly, methodically, freezing for long periods. Often with straight neck so that it stands high above other herons.

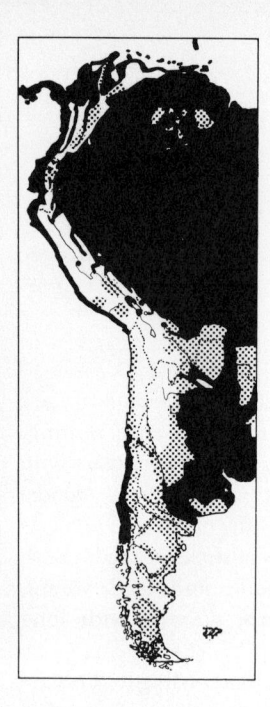

Voice: Usually silent. Coarse *krrahk* mostly heard when startled and near roosts.
Breeding: In reeds. Eggs probably in Sep in highlands.
Habitat: Rivers and marshes, mudflats, and lakes, partial to open habitats for feeding. Roosts in trees or reeds.
Range: Distributed worldwide. Widespread in lowlands of S Am. Migrates and occ. wanders to above 3000 m in the n Andes, regularly to 3000-4100 m in Ecu. and in puna zone of Peru and Bol. (*egretta*). Breeding attempts at 4080 m in Junín, c Peru, and possibly nests on tablelands of Ecu.

NOTE: Previously in genus *Egretta*, and transferred to *Ardea* by RB Payne. As long as the phylogeny is unresolved we prefer to keep it in the monotypic genus *Casmerodius*.

Cattle Egret   *Snowy Egret*

CATTLE EGRET *Bubulcus ibis* – Plate III 1

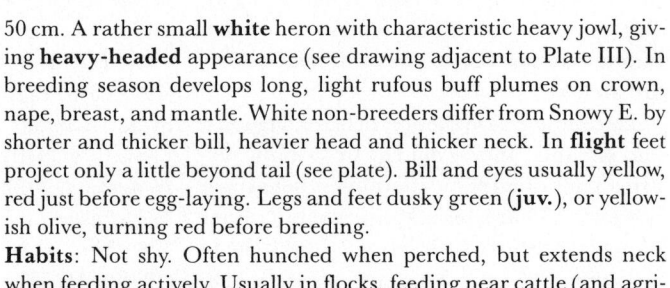

50 cm. A rather small **white** heron with characteristic heavy jowl, giving **heavy-headed** appearance (see drawing adjacent to Plate III). In breeding season develops long, light rufous buff plumes on crown, nape, breast, and mantle. White non-breeders differ from Snowy E. by shorter and thicker bill, heavier head and thicker neck. In **flight** feet project only a little beyond tail (see plate). Bill and eyes usually yellow, red just before egg-laying. Legs and feet dusky green (**juv.**), or yellowish olive, turning red before breeding.
Habits: Not shy. Often hunched when perched, but extends neck when feeding actively. Usually in flocks, feeding near cattle (and agricultural machines) for larger insects. Flies in long lines to and from roosts. Faster wingbeats than in the following 2 species.
Voice: Usually silent, but gives various croaking calls in breeding colony.
Breeding: Egg-laying in July in highlands of c Peru.
Habitat: Pastureland, often rather humid habitat of wide extension. In the New World often feeds on garbage dumps. Roosts and nests in thickets or tall reeds.
Range: Distributed worldwide after explosive recent expansion. Established in northern S Am. in the 1940s. Now widespread in all warm parts, and occurs regularly along the Atlantic side to Isla Grande (flocks in fall). In the Amazon area, however, only a seasonal migrant from N Am. Now seasonally common on wetlands around 2600 m on Bogotá and Ubaté savannas of E Andes, Col., and in high plateaus of Ecu. In puna zone occurs scattered at 2500-4600 m from nc Peru to nw

Arg., mainly in the rainy season, but is resident and nests at 4080 m in Junín, c Peru, and perhaps in some other places (*ibis*).

NOTE: Sometimes placed in genus *Ardeola*, and transferred to *Egretta* by RB Payne. As long as the phylogeny is unresolved we prefer to keep it in the monotypic genus *Bubulcus*.

LITTLE BLUE HERON *Egretta caerulea* – Plate III 4a-b

60 cm. A slim, **uniformly deep gray** heron, **ad.** with maroon head and neck. Eyes yellow, bill dark-tipped gray, legs greenish. **Juv. white**, differing from Snowy Egret only by **gray primary-tips**, particolored bill, and greenish legs and feet. This dress succeeded by intermediate 'piebald' dress. Generally shy. Singly or in small groups, stalking quietly on open mudflats.

Inhabits marshes, flooded grassland, and mangrove flats, mainly in trop. and subtrop. parts of the Americas. Regularly wanders to wetlands at 2500-3020 m in Boyacá and Cundinamarca, Col. Accidental at 3750 m in Cuzco, se Peru.

NOTE: Often in its own genus *Florida*.

SNOWY EGRET *Egretta thula* – Plate III 2

60 cm. A small and dainty **white** heron. In breeding plumage with graceful plumes on head, neck, and back, where they curve upwards. **Slender bill black** contrasting with yellow loral skin (red in breeding season). Legs black contrasting with yellow pads of feet (imm. and some ad. with yellow stripe up back of legs).
Habits: Usually in parties. More active than most other herons when feeding, rushing about and shuffling its feet in shallow water to flush small fish. Flight (in all egrets) elegant and buoyant, with elastic wingbeats.
Voice: Non-breeders (which visit highlands) silent.
Habitat: Flooded grassland and extensive shallows of salt, brackish, or fresh water. Visits marine rock-pools. Roosts and nests in trees or reeds.
Range: Widespread in warm parts of the Americas. In Col. casually to 2600 m. Regular at temp. zone lakes in n Ecu. and around 4000 m in Junín, Cuzco, and Puno, Peru, and occ. large flocks at Lake Alalay (2550 m) in Cochabamba, Bol. (*thula*).

STRIATED HERON *Ardeola striata* – Plate III 10 a-b

40 cm. **Ad.** with erectile crown blackish green. Wing-coverts and scapulars blackish green edged yellow, mantle dark grayish green. Cheeks, neck, and underparts gray with rufous streaks on mid-throat. Often looks more blue than green, and **all dark at a distance**. Bill black, lores yellow. Legs greenish and in breeding season flushing red. **Juv.** browner with diffuse buff streaks above (less conspicuous than in Night H.), but very coarse streaks below; legs and bill yellow.

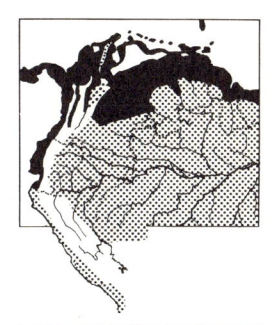

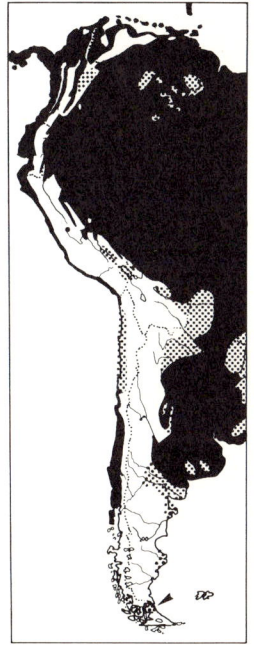

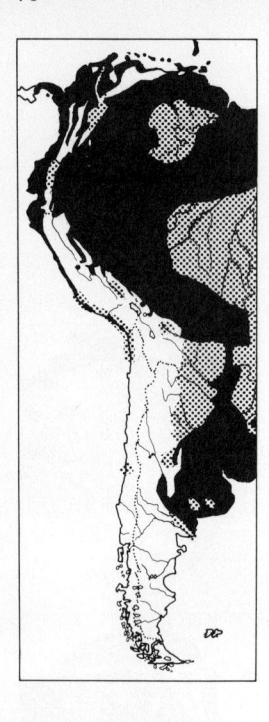

Habits: Usually solitary. Shy and crepuscular. Mostly seen in short, deliberate flight low over the marsh. Feeds from perch just out of water. Has a distinct neckless appearance when perched, and may freeze in horizontal attitude, resembling a stone in the water. At other times extends neck and raises shaggy crest when alarmed.
Voice: Sharp descending calls, *kyow* or *keoup*.
Habitat: Reed marsh and dense, woody vegetation fringing lakes and streams (*várzea* forest).
Range: Several sspp in warm parts of Australia, Africa, Asia, and S Am., where widespread and common in trop. and subtrop. zones. Regular visitor at 2600 m near Bogotá, and at 3020 m to Lake Tota in E Andes, Col., lakes of the c valley of Ecu., and apparently regular in Sep at 4080 m in Junín and near Cuzco, c Peru, and casual in other temp. areas of the n and c Andes, and with a stray rec. to 3700 m even in La Paz, Bol. (*striata*).

The very similar **GREEN HERON** *Ardeola v. virescens* of N Am. is a casual visitor to Ecu., at 2600 m in Bogotá area Col., and in Andes of Mérida Ven. It is greener (less gray) on the back, and has chestnut (vs. buffy or gray) sides of neck. Juvs. are probably indistinguishable in the field.

NOTE: The Green H. is variously treated as a semispecies or as a ssp group under Striated H. Previously a separate genus name *Butorides* was used.

AGAMI (CHESTNUT-BELLIED) HERON *Agamia agami* – a rare inhabitant of trop. forest swamps of C and S Am. – is rec. accidentally on Bogotá and Ubaté savannas (2600 m) in E Andes, Col. Recognized by **very long and slender bill and neck, and rather short legs**. **Dark**, ad. with back oily black, neck and belly chestnut, chest gray. Juv. brown with white streaks below.

NIGHT HERON *Nycticorax nycticorax* – Plate III 9a-c

62 cm. A **stocky** large-headed and apparently neckless heron with rather stout bill and short legs. Bill, **crown, and mantle glossy slaty black**, crown with 3 long slender white plumes. Eyes red. Frontal band, side of head, neck, and entire underparts white to light gray, possibly polymorphic (*hoactli*, gray birds most frequent in highlands), or dark smoky gray (*obscurus*). Wings, rump, and tail dull gray to sooty. Yellowish green legs turn bright salmon-pink in the breeding season. **Juv. dull brown, above with drop-shaped white spots,** below and on wing-linings light buff streaked brown. Bill yellow with dusky tip. 3/4-year old, crown and mantle become uniform dull brown, wings and underparts diffusely streaked. After next molt, acquires ad-like pattern, but in shades of dusky brown rather than black and gray. With increasing age, eye-color changes from yellow to red. **Ad.** plumage acquired in 3rd year. In **flight** shows slaty gray underwings, and toes barely visible beyond tail.
Habits: Usually single or few together, but sometimes forms large roosts. Crepuscular or partly nocturnal. By day, usually hidden, but

highland birds often active. Stands hunched by a reed-border, 'hunting' by stand-and-wait strategy. Seems chiefly to take frogs. When flushed, often flies a considerable distance, with deliberate wingbeats.
Voice: Harsh barked *quar*k when flushed.
Breeding: In reeds, trees, or on rocky isles. Eggs in Aug-Sep and Jan-Feb (Junín c Peru), around June and Jan (Bol.), or Nov (furthest s).
Habitat: Shores of any kind, in lakes, streams, or along coasts. Often along rocky shorelines, but may prefer edges of reed-beds.
Range: Nearly worldwide. Widespread in S Am. lowlands. In northern Andes only casual in temp. zone, hardly breeding except in Ecu., but common in puna zone from Ancash, nc Peru, to Cochabamba, Bol., and Atacama Chile, in some places breeding to 4800 m (*hoactli*). Forest zone and lake districts of s Chile and adjacent sw Arg., this population migrating to n Arg. and Chile (*obscurus*). Islas Malvinas (*falklandicus*).

BOAT-BILLED HERON *Cochlearius cochlearius* – Plate III 6

50 cm. Resembles Night H., but recognized by broad, **slipper-like black bill**. In ssp *cochlearius*, cap and long nape-feathers black (can be raised to a broad transverse fan), rest of upperparts pearly gray. Face, foreneck, and c breast white, lower breast and belly light cinnamon with distinct black lateral areas. **Juv.** mainly light brown with small black cap and white-striped sides. **Imm.** (on plate) more cinnamon, in 2nd year with some gray on wings and sides, but still retains some cinnamon on wing in 3rd year. A lethargic but shy, crepuscular and nocturnal bird. Mostly seen hunched in water or on branch in deep shade (often roosting with other herons). Gives heavy *qua* when flushed. Rattles bill.

Inhabits thick vegetation in swamps and along creeks, in trop. and subtrop. C and S Am. Occ. to 2600 m in Andes of Col., and casually to páramo zone of Mérida, Ven. (*cochlearius*).

FASCIATED TIGER-HERON *Tigrisoma fasciatum* – Plate III 8

65 cm. Bittern-like. **Ad.** has cap black, neck and back **very densely barred slaty and buff** (appearing gray at a distance); belly graybrown. **Juv.** (only depicted) **broadly barred black-and-tawny** all over, except for some white below, and for black-and-white barring on flanks and tail. Ad. differs from Rufescent T-h. (*Tigrosoma lineatum*) of lower elevations by lack of rich chestnut on head and neck. Juv. barely distinguishable (smaller bill, whiter underwing, narrower barring above, and pale cinnamon (vs. white) primary-tips.
Habits: Secretive and solitary. Sometimes freezes with neck extended and bill pointing up or stretched somewhat forwards when approached. When flushed, flies short distance up or down river, and lands on boulder, or low in a tree.
Voice and **breeding** : No data.

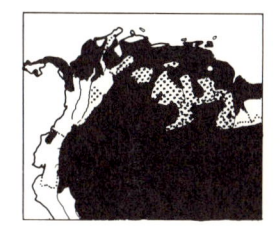

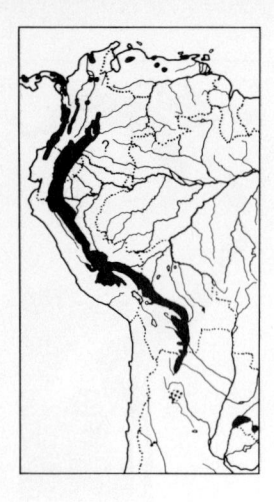

Habitat: Hilly country, along creeks and fast-flowing rivers with boulders, gravel bars, and wooded edges.
Range: Mainly in premontane areas. Scattered in C Am., n Ven., n and w Col. to Pichincha, nw Ecu. Widespread at 600-1800 m in e Andean foothills s to extreme nw Arg., and in se Brazil. Esp. juv. birds occur regularly in the temp. zone, sometimes to 3300 m in semi-arid valleys in Cuzco, and possibly Ayacucho and Apurímac, c Peru (*salmoni*). Possibly also reaches lower temp. zone in nw Arg. (*pallescens*).

LEAST BITTERN *Ixobrychus exilis* – Plate X 9a-b

A **tiny** heron with apparently very thick neck due to long, coarse plumage; larger *bogotensis* 34 cm. Buffy brown adorned with **black (male) or dark brown (female) cap, back, and tail.** Whitish border of scapular area. Cheeks ochraceous, neck tawny, streaked in front and near wing-bend, underparts rich buff. **Wing-coverts form large warm buff and chestnut area. Juv.** resembles female, but dorsal feathers have pale edges, the underparts more streaks. Could be confused with sympatric Bogotá Rail (*Rallus semiplumbeus*), which has all-chestnut wings. Striated Heron is dark-winged.
Habits: Usually single. Secretive and difficult to flush, but yet most often seen when startled into short flight with dangling legs and extended neck. Periods of vigorous wingflaps alternate with short glides. Sometimes freezes with bill up when approached.
Voice: Calls include series of low *kwuh* calls. Cackles when flushed.
Breeding: Builds small reed-nest. Season in Col. unknown. 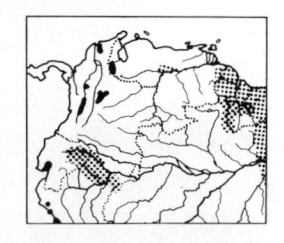
Habitat: Reeds, tall *Scirpus californicus* and *Typha* bordering lakes, but ssp *bogotensis* often feeds along ditches on rushy adjacent pastureland.
Range: Several sspp discontinuously in lowlands of N, C and S Am. Bogotá and Ubaté savannas (2600 m) and around Lake Tota (3020 m) in E Andes, and in Antioquia C Andes, Col., apparently vanishing owing to habitat destruction (*bogotensis*).

PINNATED BITTERN *Botaurus pinnatus* – Plate III 7

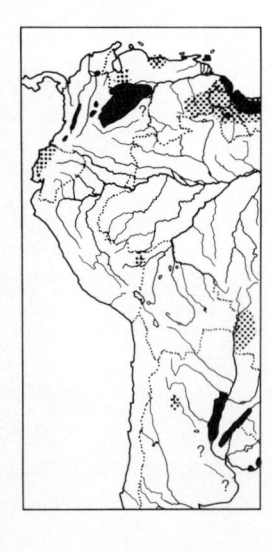

65 cm. **Stocky, with bulky neck**, and short legs, compared with other herons. Pale buff to **ochraceous with dense dark brown blotches, streaks and mottling**, and some tawny spots. Top of head mainly black. Remiges blackish vaguely tipped tawny. **Throat and underparts of body white centrally**, and foreneck broadly striped ochraceous or rufous brown. Differs from potentially visiting N Am. Bittern (*B. lentiginosus*) by lacking large black moustache, but otherwise has more contrasting colors, by black blotches on back, and black rather than dark gray flight-feathers. Differs from juv. tiger-herons by less rufescent hue, and streaked rather than barred pattern, and by **lack of white tail-bars**.
Habits: Solitary. Secretive nocturnal marsh bird, which moves quietly through the vegetation in hunched attitude with low head. Normally hides when disturbed, and at close range 'freezes' (often for very long periods) with bill straight up, and sways with the surrounding reeds. Most often seen rising clumsily above the reeds with legs dangling, in

owl-like flight, and soon dropping back into cover. In flight resembles a large owl.
Voice: Nasal *kwaak-kwaak* when flushed. Gives booming *oong-ka-choonk* at night.
Breeding: Builds large reed-nest in the marsh. Egg-laying in the Andes (Col.) possibly in July-Oct.
Habitat: Extensive marshes with *Scirpus californicus* and other tall rushes, and cattails.
Range: In trop. to up-country savanna in Mexico and S Am., possibly wandering. On Bogotá and Ubaté savannas (2600 m) in E Andes, and middle Cauca and upper Patía valleys, Col. Locally in trop. and subtrop. Ecu., e Col., Ven., Guyanas and se Brazil to n Arg. (*pinnatus*). Apparently rare.

WHITE-NECKED HERON *Ardea cocoi* – **Plate III 5** - widespread in S Am. lowlands – straggles at least to 2550 m along the larger valleys of Cochabamba and Tarija, Bol. – 110 cm. A **large blue-backed** heron. **Black cap down to eye-level**. Neck white with black streaks in front, and coarse plumes towards chest, belly black, thighs white. In flight shows black-and-white marks near wrist. **Juv.** with underside light gray conspicuously streaked, neck and back with brownish tinge. Non-breeders probably have blackish base of bill and black legs and feet, unlike depicted breeding bird. Solitary or in loose groups. Shy, but feeds open to view, often motionless for long periods. A harsh squawk given in flight.

GREAT BLUE HERON *Ardea herodias* from N Am. and the Caribbean Isls has occ. turned up on migration in the Bogotá area (2600 m) and upper Cauca Valley, Col., and near Quito, c Ecu. (*herodias*). Differs from White-necked H. by smaller crown-cap, buffy white to violaceous gray neck, and chestnut thighs. Its call is lower-pitched than that of White-necked H.

Storks – Family Ciconiidae

A worldwide group of 17-18 species of large or huge long-legged grassland birds. Resemble large herons, but have smoother plumage and generally larger and heavier bills. Stride sedately. Fly with neck extended (not tucked into the shoulders), and with deliberate wing-beats. Flight profile vulture-like, with large and rectangular, fingered wings. Often soar high in thermals, but unlike in vultures the individual birds in a flock tend to turn in the same direction. The food is invertebrates and smaller vertebrates of grassland and shallow water.

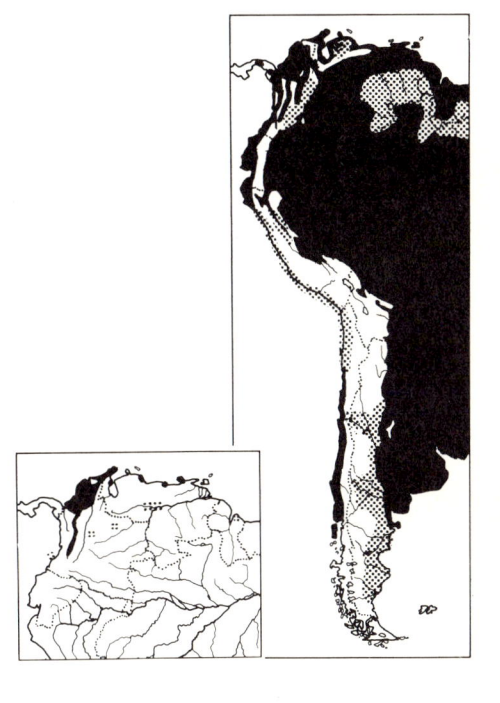

Further information about species visiting the Andes can be found in *Acta Zool. Lilloana* 25(1969):19-42, *Condor* 73(1971):220-229, *Ecol. Boliv.* 6(1985):73-81, *Living Bird* 10(1971):151-170, pp. 921-931 in Buckley *et al.* (1985), and *Rep. XXXI Annual Meeting IWRB* (1986):129-131.

MAGUARI STORK *Ciconia maguari* of trop. and subtrop. savanna swamps e of Andes wanders to 2500 m in Cochabamba, Bol., and probably also goes fairly high up in w Arg., and occ. crosses to c Chile. 100 cm. **White with black remiges, greater wing-coverts, and posterior humerals**. Long bill gray with red tip, bare skin around pale eyes, throat, and legs red. **Juv. sooty-brown** with white tail and throat. **Imm.** coarsely spotted.

WOOD STORK *Mycteria americana* of wet subtrop. and trop. savannas of the Americas is reported to cross between upper Magdalena valley and w Caquetá, Col., in June-July, and crossing the Andes in s Ecu. Andean crossings are also suggested by stray rec. near Quito, and on w coast of Ecu., in Junín and Cuzco, Peru, and in n Chile. 90 cm, with **long dusky bill with thick base and curved tip**. Head and neck dusky, naked, and much wrinkled. Plumage white with black flight- and tail-feathers. **Juv.** with head partly covered in short brownish down. Soaring birds recall King Vultures. Gregarious.

JABIRU *Jabiru mycteria* of the extensive marshy savannas e of Andes may cross the Andes, judging from stray rec. near Cuzco and in sw Peru. 130 cm, wingspan 230 m. **Gigantic stork** with very powerful, slightly upturned black bill. All **white except for naked black head,** and swollen black neck with red base. **Juv. brownish gray**, white feathers first appearing on back.

The three above mentioned species are shown in line-drawing on p. 710.

Ibises – Family Threskiornithidae.

30 species, distributed almost worldwide. Mainly medium-sized birds of grassy and marshy plains, with specialized bills (thin and decurved in most species). Plumage usually uniform glossy black, or boldly patterned. Unlike herons, fly with outstretched neck, resembling cormorants, with similar wing-shape and shallow rapid flapping alternating with gliding on horizontal wings. Often fly in long wavering lines.

Ibises normally feed socially in grasslands ranging from inundated or muddy areas to dry steppe. They usually walk slowly in loose flocks, picking and probing with small quick movements in grassy tussocks, or into wet soil. Andean species feed on frogs and a variety of insects and larvae, incl. hairy caterpillars.

Breeding is normally colonial, in swamps, reed-beds, or on cliffs, often with herons, and with similar coarse stick-nests. Both sexes incubate and feed the downy young. These are nidicolous, but soon capable of climbing around outside nests.

There are no detailed published accounts of Andean species. The classification follows RB Payne (in Mayr & Cottrell 1979).

SCARLET IBIS *Eudocimus ruber* of trop. coastal swamps of n and e S Am. is rec. accidentally at 2500 m in Ven. Jizz like White-faced I. **Ad. entirely scarlet** (except for black wingtips). **Juv.** brown and partly white below. May be conspecific with the White I. (*E. albus*).

WHISPERING (BARE-FACED) IBIS *Phimosus infuscatus* of trop. and subtrop. areas e of Andes and in n Col. has been sighted at 2600 m near Bogotá, Col. **Black with pinkish bill, face, and feet**.

WHITE-FACED IBIS *Plegadis chihi* – Plate IV 5

55 cm. With **dusky curved bill**. **Dark brown**, with humerals, wings, and tail bronzy with purplish sheen. In breeding dress with distinct white border of red facial skin and bill, and with deep chestnut fore- and underparts. Off-season duller, with fine white streaks on head and neck. **Juv.** dull brown below, and with head and neck heavily streaked white. Underside of body never blackish, as is usually the case in the Puna I.

Inhabits subtrop. marshes and rushy pastureland, breeding in n

Am, n Ven., and from se Brazil and e Bol. to c Arg. and c Chile. Winters casually to C Am. Reaches temp. valleys at foot of Andes of Arg., straggles to Patagonia, and has accidentally been rec. wandering to 3220 and 4300 m in Bol.

PUNA IBIS *Plegadis ridgwayi* – Plate IV 6 a-c

60 cm. With **dark red, curved bill**. Stockier, less slim and long-legged than White-faced I. Dark purplish brown, **looking black** at a distance, but in breeding dress head and neck are deep chestnut, and the back is always deep purple with green sheen, lesser wing-coverts more strongly glossed green, underparts fuscous-black. Legs black. Off-season with fine white streaks on dark chestnut head and neck. **Juv.** duller, with coarser white streaks, some brown below, and lighter and browner bill. **Pull.** dark brown with bands on bill. Albinistic individuals occur in Junín c Peru.
Habits: Usually in flocks, sometimes in large numbers. When feeding, often looks hunched-backed owing to raised back-feathers. In some areas very confident. Flies in clumps or in wavy lines, often low over the terrain.
Voice: Usually silent, but bewildered chorus of nasal *wut* and *cwurk* notes given when flushed. Juv. give whispering calls.
Breeding: In tall reeds. Variable season. Eggs mainly in Apr-July, with some laying in Jan-Feb (Junín), Jan-Mar (w Ayacucho), and Nov (Cochabamba Bol.). Fledglings Mar-Apr (La Paz and Cuzco/Puno border).
Habitat: Large marshy areas and rushy pastureland, mostly in places with mudflats, or short matted grass, feeding mainly along muddy creeks or in flooded places, but also in coarse bunchgrass on the hills, sometimes far from water.
Range: At 3500-4800 m from s Ancash, nc Peru, s to Arica and Tarapacá, n Chile, and La Paz, Bol., and isolated near Cochabamba town, Bol., and on altiplano of Jujuy, nw Arg. Visits Peruvian coast May-Sep, and accidental to La Libertad, n Peru. Common.

BUFF-NECKED IBIS *Theristicus caudatus* of trop. and subtrop. S Am. savannas from Ven. to n Arg. has a small population breeding above 2000 m in Córdoba mts, Córdoba, c Arg. (*hyperorius*). Can be expected occ. to cross some temp.-zone passes in Col. (*caudatus*) Resembles the following species, but lacks the breast-band, and instead has dark color of belly extending to lower breast.

BLACK-FACED IBIS *Theristicus melanopis* – Plate IV 4 a-d

75 cm. Stocky, with long, **very broad, slightly fingered wings, and very short tail**. Crown chestnut, rest of **head, neck, and breast pale buff** (*branickii*), or usually richer buff (*melanopis*), with gray breastband. Upperparts of body light gray mottled with dark brown (esp. in *melanopis*), wing-coverts light gray contrasting with black flight-feathers.

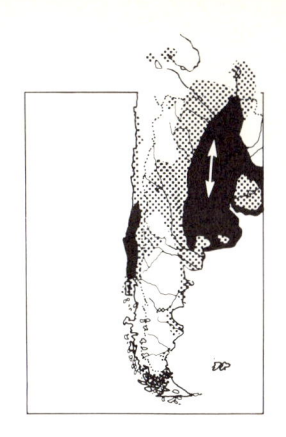

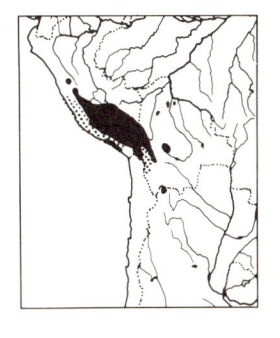

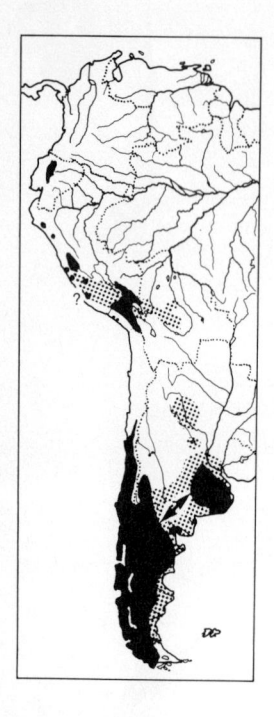

Belly black, but in *branickii* whitish buff of breast continues to c belly. **Underwings and tail region black**. Bare stripes at base of bill may in *melanopis* fuse to form a conspicuous median gular pouch. Pink legs short, feet not projecting beyond short tail in flight. **Juv.** with faint dusky streaks on neck, and diffuse buff feather-edges giving scalloped effect esp. on middle wing-coverts. **Pull.** buff.
Habits: A conspicuous but often shy bird. Single, in pairs, or small flocks, noisy when together. Walks slowly when feeding, probing into dense grassy tussocks. Flies in lines low over the terrain, but sometimes soars to great heights during long daily wanderings. May perch in trees.
Voice: Metallic clanking call *clack clack* ('quetal quetal') grading to incessant barked *quap* when flushed.
Breeding: Colonially, in rocky gullies or cliffs, or in reedbeds, occ. in trees. May nest together with Night Herons, or, in the s, with cormorants. Eggs in Sep-Mar (Junín c Peru) to Sep-Dec (s Chile).
Habitat: Feeds in open terrain, from boggy glades in Valdivian forest through fields, meadows, and damp rushy river valleys to arid rangeland and upland bunchgrass heaths, sometimes sandy habitats with very little vegetation. Mainly roosts in rocky outcrops.
Range: Disjunct. In páramos Antisana and Cotopaxi, c Ecu., and at 4000-5000 m in puna zone in Ancash, Junín, Huancavelica, w Ayacucho, and E Andes of Cuzco and Puno, Peru, to La Paz and Oruro (and casually Cochabamba), Bol., and to Arica, n Chile (*branickii*). S Arg./Chile, migrating to pampas of n Arg. Mainly in the subantarctic forest zone and adjacent Patagonian steppes, below 1000 m, and continuing at 2000-3000 m along Chilean Andes n to Antofagasta, and (vanishing) in some hills near coast of Peru (*melanopis*). Common in the s, otherwise sparse or very local.

NOTE: Previously treated as sspp of Buff-necked I. Now classified as a separate (allo)species.

ROSEATE SPOONBILL *Ajaja ajaja* is widespread in lowland marshes and swamps of S Am. s to c Arg. Andean crossings can be inferred from 3 stray rec. in c Chile. Unmistakeable. 70 cm. Long bill, expanded as a flat distal spoon. Plumage white (juv.), or mostly pink, with rosy crimson shoulder-bar and rump.

Flamingos – Family Phoenicopteridae

Highly specialized birds related to wading birds or shorebirds. 5 species inhabit Africa and adjacent parts of Europe and Asia, and the Neotropics.

A flamingo is immediately recognized by large size and lanky build, long sinuous neck, angled-down bill, and long legs with small webbed feet, and by the pink colors (ad.). Owing to often large observation distances the diagnostic details of bill and feet may be difficult to see. The identification then requires knowledge of proportions and habits. Flamingos are often seen together in large numbers, wading in extensive shallows of very productive saline lakes with barren surroundings. They also up-end and swim well. They feed with the bill held 'upside-down', filtering tiny organisms from water and sediment. The 'shallow-keeled' *Phoenicopterus* species feed on invertebrates like waterfleas, copepods, brine-shrimps, or brine-flies. 'Deep-keeled' *Phoenicoparrus* species feed mainly on large unicellular organisms, as diatoms and protozoans.

Flamingos are among the most social of birds, sometimes with hundreds engaged in communal displays with head-flagging and wing-flashing. Breeding takes place in sometimes huge colonies on miry clay or isls of difficult access. The nests, defended vigorously against neighbours, are usually 2 neck-lengths apart, but densely packed in some colonies in the Andes. They are normally 10-40 cm high conical clay-mounds, but on dry sites (on isles) they are much less high. A single, white egg is laid. The chick has straight bill at first, and is dressed in whitish down; after some time this down is replaced by a darker gray down and then gradually by brownish, somewhat streaked juv. plumage. Both parents incubate and feed the chick on secretions formed by glands in the oesophagus. The young assemble in large packs ('creches'), but apparently the parents manage to find their own chick in the group, and feed it alone. Flamingos do not breed until 3-5 years old, and often have very poor breeding success. Owing to extensive illegal egg-culling only very few young are now raised in the important breeding areas in Bol. At the same time, many flamingo habitats are becoming unsuitable owing to food competition from introduced fish. Still, the populations may remain relatively stable because of very high ad. survival. Currently the populations become threatened by saltmining and pollution even in once desolate parts of the altiplano of Bol.

Further information about Andean species can be found in Johnson (1965, 1972); Scott & Carbonell (1985); Kear, J & N Duplaix-Hall eds. (1975) *Flamingos*. Berkhamsted; and Ogilvie, M & C (1986) *Flamingos*. Gloucester. See also *Auk* 96(1979):328-342; *Condor* 60(1962):228-342, *Limnol. Oceanogr.* 31 (3)(1986):457-468; and *Proc. Natn. Acad. Sci. USA* 80(1983):4766-69.

CHILEAN FLAMINGO *Phoenicopterus chilensis* – Plate IV 3 a-d

Male 105, female 95 cm. More **long-legged** than the 2 other species, the heel ('knee') normally reaching beyond tip of tail (but note that 1st-year birds have shorter legs). **Ad. pinkish white**, sometimes with slightly streaked rosy breast. **Light vermilion wing-coverts and tertials** nearly cover black primaries and secondaries (when standing). **Lore and basal half of bill pinkish white**, bill-tip black. **Legs greenish gray with red joints and toes**. In flight, shows entirely red wing-linings. **Pull.** white with pink bill and feet; after 2 weeks adopts a coat of thick gray down. **Juv.** mainly light fawn-colored, upperparts with **bold black streaks, and pale feather-edges**; primary-coverts pink with black streaks, but greater secondary-coverts mainly fuscous. Bill mostly gray (as in juv. of other flamingos with less extensive black tip than in ad.), legs dusky. In subsequent plumages becomes gradually whiter, less striped, subad. being mainly white with some rosy tinge on wing.
Habits: Social, but flocks usually disperse when feeding. Walk steadily, sometimes rapidly, but may also stop and trample the lake bottom. Occ. feed swimming.
Voice: Goose-like *churr-hrr-hrr....* In aggression *dzhydzhydzhy....*
Breeding: Colonies on mud, gravel, and in Bol. also on the margins of large, sediment-covered icebergs. Eggs in Jan-Mar (altiplano), or Dec-Jan (s Arg).
Habitat: Extensive shallows of open, saline to moderately alkaline fresh-water lakes, estuaries and calm marine bays. Most numerous on

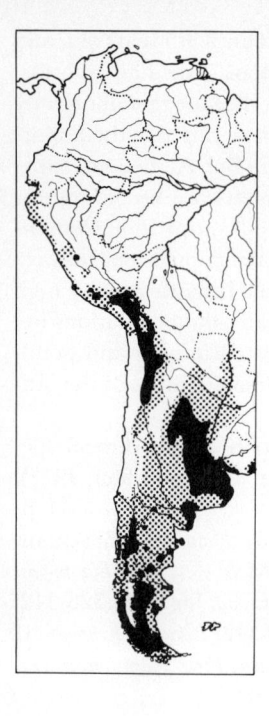

fish-free lakes, which often have many zooplankton organisms of fair size.
Range: Magellanic Chile and throughout c and s Arg. to s Brazil and Parag., but breeds only on few lakes in the Magellanic zone, on uplands of Patagonia, and in lowlands of Córdoba and Sta Fé. Continues n across altiplanos of n Arg., Chile, Bol., and Peru, mainly in salt lakes of Chalviri, Coipasa, and Poopó, but also breeds in s Ayacucho, and Junín, c Peru. 'Winters' in suitable places along Pacific coast, and straggles to Ecu. Population c. 500,000.

NOTE: Sometimes threated as a (mega)ssp of *Phoenicopterus ruber* of the Caribbean and Palaearctic regions.

ANDEAN FLAMINGO *Phoenicoparrus andinus* – Plate IV 1 a-d

C. 110 cm. **Ad.** white with a faint rosy tinge, **darker purplish rosy on lower neck, breast**, and wing-coverts, and with occ. purplish streaks on humerals. Humerals not long enough to cover **black tertials** and flight-feathers, and therefore form **large black triangle near the rear** (in standing birds, vs. thin strip of exposed black flight-feathers in the 2 other species). Wing-linings pink with purplish red axillaries forming triangular patch. **Bill rather short and deep, black, with pale yellow base** narrowly demarcated by violet on lore. **Legs pale yellow**. **Pull.** white with pink bill and feet, later gray. **Juv.** drab-colored with bold fuscous streaks above, and often indications of streaks on flanks; bill gray at base. In succeeding plumages increasingly white, and then pink.
Habits: Social, usually feeding patchily distributed within the shallows of a lake. While feeding meanders slowly, occ. stopping.
Voice: Conversation weak, high-pitched *chy-chy-chy*.... Also gives a loud vibrating call.
Breeding: Colonies on soft, miry clay. Eggs in Dec-Jan (Feb?).
Habitat: Extensive shallows in saline lakes with high production of diatoms.
Range: From Catamarca, nw Arg., to Arequipa, s Peru, mainly at 2300-4300 m, with main breeding concentration in Atacama salt-lake and some nearby lakes in Antofagasta, n Chile, but important staging sites in w Potosí, Bol. Visits lakes in Ayacucho and accidentally Conococha in Ancash, Peru. In winter sometimes common in lowlands of Córdoba, Arg. Population c. 150,000.

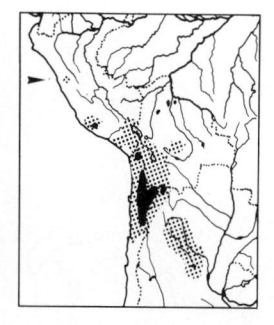

PUNA FLAMINGO *Phoenicoparrus jamesi* – Plate IV 2 a-c

90 cm. **Ad.** pinkish white, usually with distinct purplish streaks on lower neck and mantle, and sometimes solidly light purplish-rosy around lower neck. Light purplish rosy and quite long and narrow **plume-like humerals** are often raised, and may virtually hide black primaries and secondaries. In breeding season head flushes purplish pink. **Bill short and swollen, basal 2/3 deep yellow demarcated by**

red lores. Legs pinkish buff to brick-red, rather short (heel never beyond tail). Wing-linings pink with triangular red patch formed by axilliaries. **Pull.** light gray with orange-pink bill and feet, later dark gray. **Juv.** rather uniform fawn-colored with very thin black streaks above (looks uniform at a distance); in flight shows white primary-coverts and pink axillaries.
Habitat: Feeds more dispersed than Andean F., walking sedately or standing still.
Voice: Low, high-pitched conversation *bry-bry-bry...* to *bzyrr-bzyrr....* Also loud *gua-gua-gua...* and *arrrh* rec.
Breeding: Colonies on soft miry clay. Eggs in Dec-Feb.
Habitat: Extensive shallows with high production of diatoms in open, saline Andean lakes.
Range: C desert puna, mainly well above 4000 m in lakes and salt-lakes near Lake Poopó and Coipasa and esp. in Laguna Colorada in sw Potosi, Bol., Lake Pozuelos in Jujuy and in Catamarca, nw Arg., and Salar de Pedernales in Antofagasta, Chile. Sometimes large numbers in Lake Salinas in Arequipa, and also visits Lake Parinacochas in Ayacucho, Peru, and occ. from 15° to 42°s in the Andes and to the Pacific coast. Population 50,000 or somewhat more.

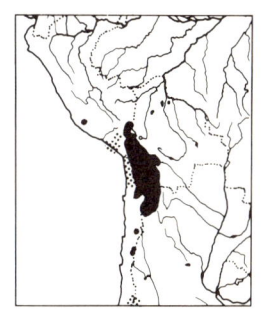

NOTE: Often called James's Flamingo.

Chilean Flamingoes

Falconiformes

A polyphyletic order, as New World Vultures undoubtedly are related to storks and pelicans, while the other families probably show more remote relationships with the waterbird assembly.

New world vultures – Family Cathartidae

The 7 species are all American, with 6 in S Am. Large to gigantic scavenging birds with ugly, more or less naked heads. Apart from their short bills, necks, and legs, they resemble storks in the air. The slightly hooked weak bills are best suited for eating fermented meat, and the chicken-like big feet do not permit carrying of prey. They therefore feed on carcasses, garbage etc. Lacking a syrinx, they are usually silent, but capable of producing hissing or grunting sounds.

Most often seen soaring high overhead or along rocky mountain sides. All species are broad-winged, but the flight varies between species from laborious flapping alternating with short glides to effortless soaring. Vultures may also be seen perched on poles, top branches, or cliffs. On the ground they walk sedately, but jump with raised wings when hurried, esp. during group dances. Like pelecaniform birds and storks they often stand with spread wings (sunbathing). Breeding is solitary, the nest placed in a cave in inaccessible rocky terrain, in concealed site on ground, or occ. in a big hollow tree. Dark gray, down-clad nidicolous young develop very slowly.

More information can be found in *Auk* 68(1951):120-126; *Condor* 75(1973): 60-8, *Ibis* 126(1984): 253-6, 131(1989):301-3; *Rev. Acad. Col. Cienc. exactas, Fisicas Nat.* 12(19xx): 21-8 and 261-7; *Smiths. Misc. Coll.* 146(1964) 6:1-18, Wallace, MP (1985) *Ecological studies of Andean Condors in Peru.* Ph.D. thesis Univ. Wisconsin – Madison, and pp 69-76 in Stiles & Aguilar (1985).

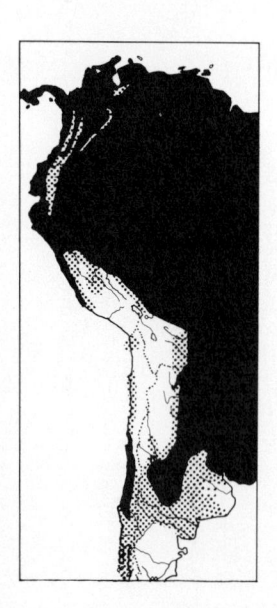

BLACK VULTURE *Coragyps atratus* – Plate XI 7a-b

63 cm, wingspan 150 cm. **Black except for whitish area at base of primaries, below** (restricted in *foetens*). Corrugated naked head and upper neck blackish, bill slender, quite long, pale distally. Feet pale gray. In **juv.** head less corrugated and partly feathered. Wings with 6 long 'fingers', broad and rectangular, rear edge curved. **Tail short** and square. In a perched bird, tail hardly reaches below footsole level.
Habits: Often gathers on carrion or garbage dumps, where sometimes assertive and bold. In flight, the long pointed head is directed forwards. Alternates between vigorous wing-flapping and short gliding with horizontal wings, but may also soar, sometimes to great height, with distinctly raised wings and widely spread 'fingers' with up-turned tips.
Voice: Normally silent, but snarling and croaking sounds when together.
Breeding: Eggs Sep-Oct (Chile); no rec. from other Andean regions.
Habitat: All kinds of open land, sometimes in forest, but most often in villages, on garbage dumps, and dry river beds.

Range: Southern N Am., C and S Am., except furthest s. Mainly in lowlands, but common to 2700 m (and sometimes higher) in semi-arid tablelands in Andes of Ven. and Col. (*brasiliensis*), and casually to 2900 m in Ecu. and n Peru (*foetens*). Because of disease (1960s) it has disappeared from most of s Peru.

TURKEY VULTURE *Cathartes aura* – Plate XI 8a-b

75 cm, wingspan 180 cm. Dark brown; gray underside of remiges contrasts with black wing-linings, to give a **2-toned black-and-gray appearance** (esp. in *falklandicus* and the large *jota*). Small naked head red (*ruficollis* with yellowish wrinkles on nape). Bill tipped white, feet pale flesh. **Juv.** has partly down-covered head, at first with slight white nuchal collar. In **flight** shows long wings with 6 not very widely spread 'fingers', **tail fairly long** and rounded.
Habits: Usually shy and solitary, but bold and often many together along coast of Peru. Often perched on poles, tall trees etc. Elegant flier, soaring with **wings in shallow V**, often tilting or rocking from side to side, and only occ. showing deep labored wing-beats. Sometimes glides with crooked wings. In this species, olfaction is important for locating food.
Voice: Usually mute.
Breeding: Egg-laying Aug-Sep (n Chile) or later (s Chile) (not rec. elsewhere in Andes).
Habitat: Ubiquitous, excepting only unbroken forest. Often in fishing villages, but unlike Black V. not in towns (except in s Peru, where Black V. has disappeared, and furthest s).
Range: Throughout most of the Americas. N Am. *meridionalis* (c. 300,000) partly migratory, occurring s to Ecu. and Parag. Mainly lowlands, but reaches páramos of Ven. (*ruficollis*), and is fairly widespread on semi-arid temp. tablelands of Col. and Ecu., local or rare in Andean

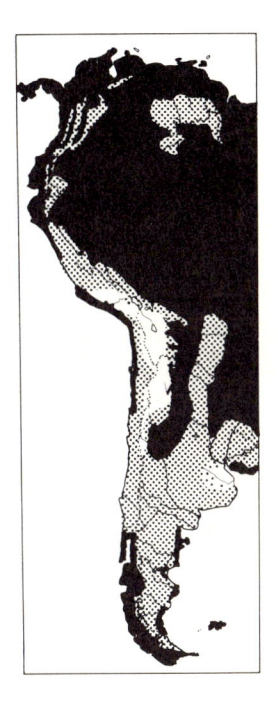

Profile and relative lengths of tail- and wing-tips identify perched vultures. From the left Andean Condor, Black Vulture, Turkey Vulture, and juv. King Vulture.

valleys and slopes from n Peru to La Paz Bol. (accidentally to 4300 m), rather common below 2500 m on e Andean slopes from Cochabamba, Bol., to Catamarca, wc Arg., and on Patagonian steppe (*jota*). Islas Malvinas, Isla Grande, and along Pacific coast n to Ecu., wandering regularly to 3500 m in s Peru (*falklandicus*).

GREATER YELLOW-HEADED VULTURE *Cathartes melambrotos* has been seen to at least 2000 m in yungas of La Paz, Bol., and at Calilegua in Jujuy, nw Arg., and can thus be expected sometimes to rise to temp.-zone elevation. Differs from Turkey V. in having yellowish head, blacker plumage (incl. remiges), and white primary-quills (from above). Looks slightly more broad-winged and shorter-tailed, and soars with less raised wings.

ANDEAN CONDOR *Vultur gryphus* – Plate XI 9a-c

120 cm, wingspan over 300 cm, weight 11 kg. In flight recognized by **long rectangular wings with 8 very long distended 'fingers'**, the inner 'finger' usually forming angle with rear edge of arm. Tail fairly short. With less broad-based wings than Black-chested Buzzard-eagle and shorter tail than Turkey Vulture. **Ad. black with silvery-white panel on upperside of wings** formed by outer webs of secondaries and their greater and medium coverts. Neck with prominent white collar of down, naked skin of head dusky flesh, crinkled, and in **male** with large median wattle and comb. **Juv.** entirely dusky brown with short brown down on head. **Imm.** blackish, and by age showing increasingly pale panel on wings, but lacking white collar of ad. However, details of plumage sequence have not been worked out.
Habits: Usually seen singly, in pairs, or 3 together, but sometimes gather near carrion. Often soars for hours over steep mountainsides on horizontal or slightly raised wings, without tilting of Turkey Vulture (and never showing vigorous flapping of Black V.). 'Fingers' often strongly up-curved. When gliding, sometimes strongly crooks wings. After a meal the distended crop may give the bird a grotesquely deep-chested appearance.
Voice: Normally silent, except for hissing whistle produced by action of wind upon wings.
Breeding: On cliffs, often steep rock-faces at tremendous heights above ground. Pull. July (Col.). Eggs Sep-Oct (c Chile). Owing to prolonged period of parental care, Condors only breed every other year.
Habitat: Mostly desolate montane areas with canyons or high, steep slopes and cliffs, both in wooded and barren landscapes. Also visits open grassland. Many spend their day near seabird colonies on Peruvian and southern coasts.
Range: From Sta Marta mts., n Col., and Mérida, Ven., through Andes s to Isla Grande, both low and high up, but more around deep valleys than on altiplanos, and descends to sea-level in Peru and further s. Locally extirpated and generally rare and local in the n parts, but widespread in Peru, and locally common further s.

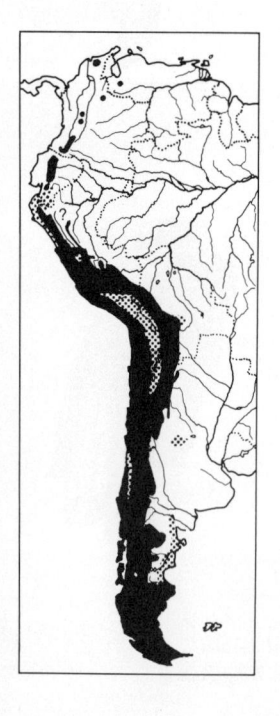

KING VULTURE *Sarcoramphus papa*

75 cm. **Ad. unmistakable, pinkish-white with many-colored bare head**, gray neck, and black flight- and tail-feathers. Likely to be confused (in flight) only with Wood Stork. **Juv.** blackish brown throughout, except for orange neck and base of bill, but progressively whiter with age. In **flight** recognized by very broad wings with 7-8 very long thin 'fingers' (5-6 'fingers' in eagles), and very short broad tail. Soars on flat wings, often at high altitudes in order to observe lower flying vultures, which guide them to carcasses in the forest. Usually alone, but occ. several together, or associated with other vultures around thermals. Perched in a tree, it often sinks down as if sitting on its belly.

Widespread, but generally sparse, in dry or humid unbroken lowland and esp. premontane forest of C and S Am, occ. reaching lower fringe of temp. zone (2300-2500 m, and seen accidentally to 3000 m) in Andes of Col., Peru, and Bol. Breeds (10-15 pairs) from Jujuy to Tucumán, nw Arg., and is regularly seen at 2300 m in forest on Calilegua mts in Jujuy.

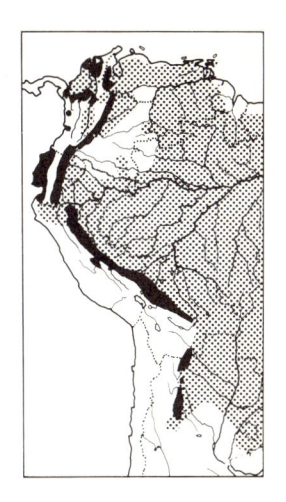

Ospreys – Family Pandionidae

One species. Migration in S Am. described in *J. Wildl. Mgmt.* 36(1972):1133-1141 and 51(1987):148-155.

OSPREY *Pandion haliaetus* – Plate XII 5

58 cm, wingspan 140-180 cm. **White with dark brown upperparts of body** and with broad **black mask** through eye. **Female** with streaked zone across breast. **Juv.** with white feather-tips above, and more streaked breast. In **flight** shows long, narrow, but slightly 'fingered' wings, and rather short tail. **From below mainly white** except for large **black patch on wrist**, and lightly barred flight- and tail-feathers.
Habits: Except when soaring in thermals, **wings always more or less crooked to form open M** (even when seen head-on). Hovers heavily high over water to plunge-dive for large fish, feet first. Spends much time on high, exposed perch.
Voice: Frail whistle rarely heard in winter quarters.
Habitat: On migration mainly along sea-coasts and larger rivers.
Range: Distributed worldwide. N Am. birds migrate to the Neotropics (Oct-Apr, but imm. also at other seasons, and Taczanowski (1884) reported a nesting rec. in nw Peru and seen displaying in sw Ecu. (HB, MKP, CR, JFR). Mainly along Pacific coast and in the w part of the Amazon basin, but passes over mountains on migration, and occ. feeds along shores and rivers to 4100 m from Ven. to Bol. (*carolinensis*).

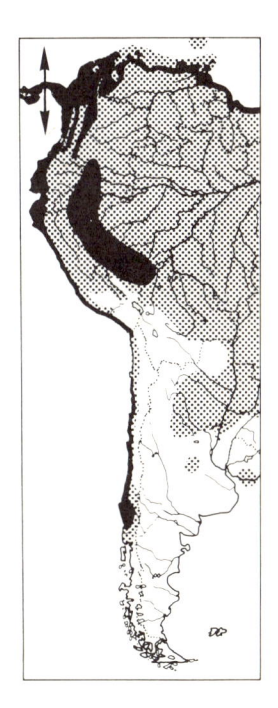

Hawks – Family Accipitridae

A worldwide family of more than 200 species of predatory birds ranging from small to very large forms. All have distinctly hooked bills surrounded by naked 'cere', and strong talons with long curved claws. The plumage is usually dark brown above, barred or streaked below. Many species have semi-concealed white occipital spot, which may be exposed as a social signal. Flight- and tail-feathers are often characteristically barred. Identification can nevertheless be difficult owing to age and individual variation. Flight styles are therefore main clues for field identification. Because of the heterogeneity of the group, we give separate diagnoses for each smaller species group, except for the very heterogeneous kite group.

Hawks usually take live prey, ranging from insects to sizeable mammals. Unlike in falcons, avian prey is rarely taken in the air. Most mid-sized species are generalists that feed opportunistically on what is available, but many kites have specialized feeding habits. As expected of predators, most hawks live at relatively low densities, and hold large territories. In much of the temp. Andes (even in regions with little persecution by man) hawks are astonishingly scarce, and certain species are definitely rare. The best method for monitoring populations is by watching from a good look-out early in the breeding season, when most species show aerial displays high over the terrain.

Hawks breed as monogamous pairs. There is a tendency for females to be larger than males, and in most species the sexes have somewhat different roles: the male often hunting near the nest-site, feeding mate and small young, the female hunting larger prey over a wide area, but remaining on the nest in the incubation period and as long as the young are small. Most species make substantial twig-nests in large trees, or sometimes on cliff-ledges. They usually have a small clutch of white or brown-spotted eggs, and nidicolous young dressed in white or gray down. The young are fed a long time in the nest, and often some additional time after fledging.

Further information about Andean species is published in Brown, LH & D Amadon (1968) *Eagles, Hawks and Falcons of the World*. London and Weick, F (1980) *Birds of prey of the World*. Hamburg-Berlin. See also *Amer. Mus. Novit.* 2166(1964); *Auk* 70(1953):470-8, 72(1955):204-5, 92(1975):380-82, 97(1980): 897-8; *Bol. Soc. Biol. Concepción* 22:3-5, *Bull. Brit. Orn. Club* 106(1986):170-3: *Condor* 59(1957) :156-65, 64(1962):277-90, 67(1965):235-41, 86(1984):221-2; *Ibis* 102(1960):362; *Novedades Colomb.* 1(1959-60):169-95 and 242-55.; Newton & Chancellor, eds. *ICBP Techn. Publ.* 5(1982):271-90; *Revista Geogr.* 21 I.G.M., Quito Apr. 1985; theses by R Sierra (1985, *Geronaetus*) in Univ.Católica, Quito, Ecu.; and in regional handbooks.

HOOK-BILLED KITE *Chondrohierax uncinatus* – Plate XII 11a-d

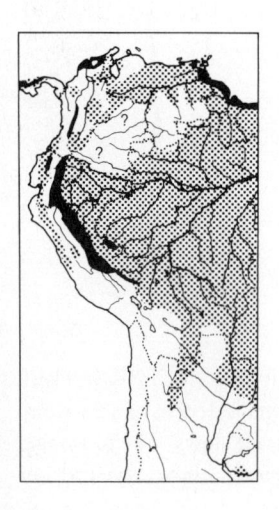

40 cm, wingspan 80 cm. Colors vary: **Gray phase** (mostly **males**) all gray, or with underparts lighter, finely barred, remiges vividly barred. **Brown phase** (mostly **females**) normally with hood dark gray, nuchal collar rufous, body dark brown above, barred rufous and white below, with underside of primaries barred gray and white. Tail normally slate-colored with 1 broad and 2 narrow (basal) whitish bars, but rare **black phase** shows only 1 bar. **Juv.** above black, except for narrow pale nuchal collar, below white with scattered bars, tail pinkish with 4 dusky bars. The species has an odd visage enhanced by pale iris and yellow patch in front of eye, and it has a **strongly hooked (parrot-like) bill** (of variable size). **In flight resembles a small, rather long-tailed *Buteo*, but wings generally blunter, more rectangular, distinctly narrowing near body.** Generally a secretive and shy forest bird perching inside the canopy foliage, but sometimes soaring high. Occ. migrates in flocks. In active flight alternate flaps and glides on horizontal wings.

Like many other kites, searches prey hovering, then glides down, seizing it in feet and swooping up again. Feeds on tree snails, frogs, and birds. Gives rapid, musical *wi-i-i-i-eeh* when soaring, screams, and gives harsh chatter *hah tetetete* when alarmed.

Inhabits lower canopy, chiefly of swampy forest, forest edge and gallery woodland, in all warm parts of the Americas. Mainly along base of Andes but rarely reaches temp. forest, 2500 m in Ven., 3100 m in Col., 2700 m in Ecu., 1600 m (nesting) in Salta, nw Arg. (*uncinatus*).

SWALLOW-TAILED KITE *Elanoides forficatus* – Plate XIII 1

60 cm, wingspan 130 cm. An elegant flier with pointed, gull-like, but more narrow-based wings, and **deeply forked tail**. **Clearcut pattern**: black (green-glossed on mantle, but *forficatus* purplish) with white head, underparts of body, wing-linings, and tertials. **Juv.** somewhat streaked on head. Flies with swallow-like grace, with much circling on horizontal wings. Seems constantly in the air, often in flocks, feeding mainly on flying insects. Voice a high-pitched *kee-kee-kee...* not unlike White-collared Swift, and a hissing whistle.

Inhabits swamps at forest edge, savannas, and marshland, almost throughout warm parts of the Americas, but migrates high, and occ. wanders to 2600 m in Andes of Col. and Ecu. (groups of 8 and 12 in Azuay and Loja), occ. to Lake Junín (4080 m), and rec. crossing Milloc Pass (5000 m) to Lima, Peru (*yetapa*, but *forficatus* from N Am. is a potential visitor).

WHITE-TAILED KITE *Elanus leucurus* – Plate XIII 2

38 cm, wingspan 100 cm. **White** to light blue-gray above, and with extensively **black shoulders**. In flight black wrist-spot visible below. **Juv.** has some orange-brown streaking. **Flight profile** like small gull, with rather long white tail. Active flight falcon-like, with angled wings. Often hovers like Am. Kestrel, sometimes flies slowly with raised wings like a miniature harrier, or soars and glides like a gull. Often on conspicuous perches. Partly crepuscular, feeding mainly on mice and insects. Voice a chirped or whistled *kewp*.

Inhabits open or widely cleared lowlands, bushy or sparsely wooded savanna, and marshes, throughout warm parts of the Americas. Sometimes wanders to 2600 m in Col. and to 2000 m in Tucumán, nw Arg. and c Chile, and rec. once at 4200 m in Cuzco, Peru (*leucurus*).

NOTE: Forms a superspecies with the Blue K. (*E. caeruleus*) of the Old World.

SNAIL KITE *Rostrhamus sociabilis* – Plate XIII 3a-b

40 cm, wingspan 112 cm. With slender, **sickle-shaped bill**, and with rectangular, very broad, but somewhat narrow-based wings. **White base of medium-long, squared-off tail**. **Ad. male** slaty black with

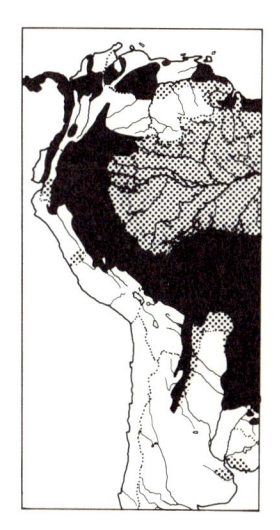

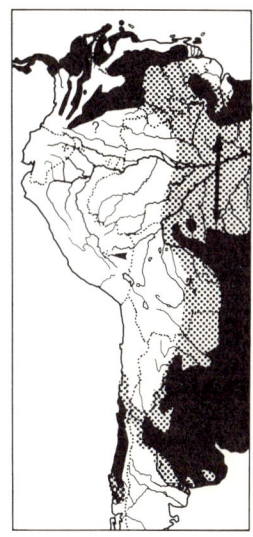

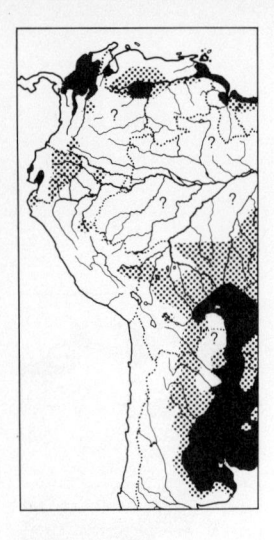

scarlet lores. **Female** dark brown with whitish forehead and eyebrow, and rusty and white marks below, and with barred rufous wing-linings. **Subad. male** browner than ad., streaked with cinnamon to chestnut below. **Subad. female** with white to pinkish buff streaks, and cinnamon thighs and wing-linings. **Juv.** with buff spots above, almost white face, and pale buff boldly streaked underparts. When quartering and soaring low over marshes somewhat harrier-like, but with floppier flight. Constantly moves tail. Often several together. Voice a rasping bleating *whe-he-he-he*....

Inhabits marshes and open swamps with *Pomacea* snails in warm parts of the Americas, but occ. straggles to lower temp. zone in E Andes Col. and in Ecu. (*sociabilis*).

NOTE: Called Everglade Kite in N Am.

PLUMBEOUS KITE *Ictinia plumbea* – an inhabitant of forest edge and gallery forest through trop. and subtrop. C and S Am. – is rec. at 2300 and 2600 m in Col. 36 cm. Uniform deep slaty gray with white band on middle of tail (and semi-concealed at base). When perched, wing-tips reach beyong tail-tip. In **flight** shows very long curving wings with large **chestnut area on primaries**. **Juv.** with underside streaked dusky and white, wings with restricted rufous patch, tail with 3 white bands.

Harriers – Genus *Circus*

Generally associated with marshes and rushy meadows, and do not require elevated perches. With rather small head and slight bill, and somewhat owl-like face. Readily distinguished by long wings with 4 distended 'fingers', narrow tail, and buoyant soaring flight low back and forth over the ground, with distinctly raised wings and much tilting and quartering. They sometimes hover briefly. *Buteo* hawks may also soar on raised wings, but have short and usually more spread tails, and broader, more rounded wings. Live mainly on water-bird chicks, rodents, and frogs.

NORTHERN HARRIER *Circus cyaneus* – Plate XXI 2a-b

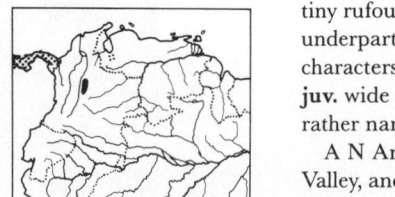

Resembles Cinereous H., but gray **male** paler, underparts whiter with tiny rufous dots; brown **female** resembles juv. Cinereous with buffier underparts, and **juv.** has almost fox-red main color below. Best field characters are, in **male**, the faintly barred off-white tail; in **female** and **juv.** wide and well spaced bars on flight-feathers, and 3 broad (vs. 4-5 rather narrow) dark tail-bars.

A N Am. migrant reaching the Bogotá wetlands (2600 m), Cauca Valley, and Antioquia Col., and Mérida mts (2500 m) Ven. (*hudsonius*). Male spends little time in the winter quarters.

NOTE: Forms a superspecies with the Cinereous H.

CINEREOUS HARRIER *Circus cinereus* – Plate XXI 1a-d

45 or 50 cm, wingspan 120 cm. **White-rumped. Ad. male ashy gray**, mantle darker and sometimes suffused with brown; underparts below gray, breast white densely barred rufous. In **flight**, 5 black 'fingers' contrast with white underside of wing, tail shows narrow, incomplete, but still conspicuous dark bars. **Ad. female chestnut to grayish brown** with numerous white or buff dots and bars, esp. below, below the dark breast, and on the wing-linings, and with prominent dusky gray and pale buff or whitish bars on flight- and tail-feathers. **Imm.** of both sexes have female-like wings and tail, but are coarsely streaked dark brown and white below (not dotted), and male soon becomes gray on foreparts. Also **Juv.** streaked below, but sometimes very dark fuscous above, and rich buff below, and always has cinnamon spots on wing-coverts.
Habits: Glides and soars with raised wings, but in active flight sometimes glides with flat or slightly bowed wings between series of 5-10 heavy wingbeats. Usually seen quartering back and forth, low over open plains or rolling grassland hills.
Voice: Not often heard. Repeated *chu-chu-chu...*, and a nasal, plaintive *pee peepee*.
Breeding: On ground in rushy vegetation. Eggs in Nov (Arg.).
Habitat: Associated with rushy fields, reeds, and open marshes, usually in wide uniform habitat, but may nest in small area of rushy vegetation surrounded by extensive heaths and bunchgrass terrain.
Range: Mainly lowlands from Tierra del Fuego to Parag. and Antofagasta, n Chile, and along entire coast of Peru, and in puna zone at 2500-4500 m from nw Arg. through w Bol. to Ancash, nc Peru, furthermore on tablelands of Ecu. and Nariño, s Col. (and at least previously at 1700-2600 m in Boyacá and Cundinamarca). In most parts the breeding sites may be quite widely scattered.

LONG-WINGED HARRIER *Circus buffoni* - a widespread inhabitant of lowland savannas and grasslands in S Am. – may straggle to the base of the Andes in San Juan to Tucumán, Arg. (normally below 690 m, but one spec. from 2000 m). Andean crossings are suggested by rec. in Cauca Valley, Col., and in c Chile. Lankier and distinctly longer-winged and longer-tailed than the 2 other harriers, and flies more buoyantly with more lifted wings. Upperparts slaty (**male**) or sooty brown (**female**); dark, somewhat white-streaked face and chest contrasting with white lower underparts; or all blackish. Winglinings buff or black, according to phase. In any case, flight- and tail-feathers are gray with several well-spaced black bars above and below. **Juv.** streaked below with dark or light, according to phase.

Hawks of Genus *Accipiter*

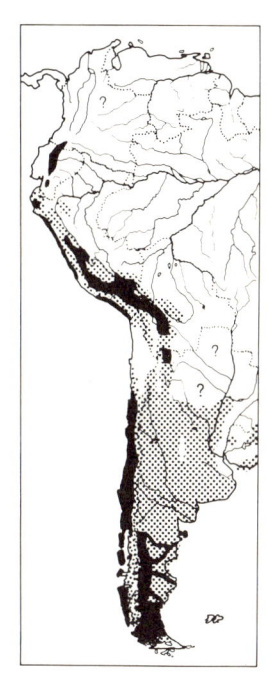

Small to medium-sized, shy, secretive, and very rapid forest raptors with rather short, broad-based and blunt wings, and long tails. Small bill and rather long thin legs. Active flight comprises series of rapid wingflaps alternating with gliding, where wing-tips become more pointed. When sometimes soaring high up, wings and tail are fully spread. When hunting, the bird weaves through the forest or along its edges to make surprise attacks, mainly on birds. Rarely seen perched, as they usually hide well in the vegetation. Females are much larger than males.

SHARP-SHINNED HAWK *Accipiter striatus* – Plate XIII 7a-g

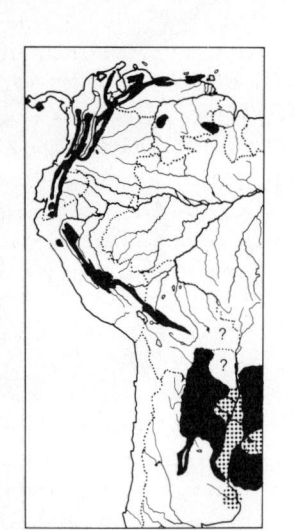

26 or 35 cm, wingspan 55 or 65 cm. A **tiny hawk with tail squared off** (slightly notched when folded). **Ad.** above bluish slate to fuscous, side of head lighter, sometimes with slight white eyebrow. Ssp *ventralis* has **white to russet, rarely dark smoky gray underparts and wing-linings**, finely streaked throat, and some short bars further down, but some birds have tawny or even darker underparts. Ssp *erythronemius* almost white below, with thin bars, rufous 'thighs' and midflank. Underside of flight- and tail-feathers distinctly barred, fuscous with 4-5 grayish white to cinnamon bars, exposed even in the darkest birds. Eyes rather big, red to yellow. **Juv.** usually with russet feather-edges and occ. white spots above, and mainly buff or russet with dark stripes below, but uniform rufous on 'thighs' (dark brown *ventralis* juv. have only faint rusty streaks below).
Habits: Glides and soars in circles on horizontal wings.
Voice: Shrill cackling *qui-qui-qui...* near nest.
Breeding: No information.
Habitat: Forest, mainly of semi-arid types, but ranges from woodland savanna through deciduous forest to humid or even wet montane forest.
Range: Most of the Americas. Sea level to 3000 (3400) m, almost throughout Ven. (but high up only n of the Orinoco) and throughout Col., Ecu. and n Peru, and on the e Andean slopes to Cochabamba, Bol. (rec. to 3540 m in Peru) (*ventralis*). Lowlands throughout the chaco of s Brazil to c Arg. and e Bol., ascending to lower temp. zone in s Bol. and casually to 2700 m in Jujuy (*erythronemius*). Fairly common.

NOTE: The megasspp *ventralis* and *erythronemius* are sometimes listed as full species.

BICOLORED HAWK *Accipiter bicolor* – Plate XIII 6a-d

35 or 42 cm, wingspan 65 or 80 cm. Tail rather long with bluntly rounded tip. General impression is usually **dark gray above, light gray or cinnamon below, with chestnut thighs**. Ssp *chilensis* above dark slaty gray, below dull rufous mottled with some gray, and with numerous large white spots and bars usually edged with dusky; 'thighs' and wing-linings rufous-chestnut. Tail gray with not very wide but prominent dark bars. Ssp *guttifer* is almost uniform cinnamon below (incl. on wing-linings), with chestnut 'thighs', resembling rufous-bellied individuals of Sharp-shinned H. (ssp *ventralis*), but larger, with less contrasting tail with only 3 light bars exposed. In ssp *bicolor* (largely sympatric with Sharp-shinned H.) the **underparts are uniform light**

gray, 'thighs' rufous, and wing-linings whitish; tail blackish with only 3 narrow but distinct pale bars visible. **Juv.** above blackish brown with occ. white or buff spots. Below buffy white to dark ochraceous, **in** *chilensis* **and** *guttifer* **with heavy dark streaks and barred 'thighs', in** *bicolor* **uniform creamy or cinnamon-buff.**
Habits: Shy and secretive. Soars on horizontal wings, occ. high up, and glides with pointed, slightly lowered wing-tips.
Voice: Scolding *khow-khow-khow...* or woodpeckerlike *kek-kek-kek....*
Habitat: Forest, forest edge, and parkland, *chilensis* mostly in mosaics of open land and deciduous forest mainly of Chilean oak *Nothofagus obliqua*, but also seen in open scrub. Little influenced by forest clearance.
Range: C and S Am. Chiefly lowlands or subtrop. hills, but can possibly reach lower limit of temp. zone in Ven., Col., and Ecu. (*bicolor*), and in Bol., and reaches 2500 m in nw Arg., with one nesting rec. at 2700 m (chaco form *guttifer*). Below 1000 m in Chile from Tierra del Fuego to O'Higgins, and in adjacent sw Arg. n to Neuquen, and migrating n to Tucumán (*chilensis*).

NOTE: Forms a superspecies with Cooper's H. and Gundlach's H. *A. gundlachi*. Ssp *chilensis* has sometimes been listed as a full species, either monotypic, or with *guttifer* as a ssp.

COOPER'S HAWK *Accipiter cooperii* – Plate XIII 5 – a N Am. migrant – is rec. once in E Andes of Col. 30 or 50 cm, wingspan 70 or 90 cm. Essentially a **larger version of Sharp-shinned H.**, but tail somewhat longer, with wedge-shaped or rounded tip. Also usually with more truncate wing-tips, head less tucked into the shoulders, and with less pop-eyed expression, and (esp. in juv.) with coloring of underparts more gradually reduced from breast towards belly.

Eagles

Vernacular name for a rather heterogeneous group of large (1-7 kg) raptors. Some are soaring open-land species and others inhabit forest, but all have large broad wings with 5-6 free 'fingers', and rather short broad tails.

SOLITARY EAGLE *Harpyhaliaetus solitarius* occurs very locally in C Am., and (*solitarius*) from Sta Marta mts, n Col. and e of Andes to Salta, nw Arg., mainly in heavy premontane forest, but possibly sometimes to lower temp. zone. A very **large (70 cm), dark** slaty gray to deep fuscous brown hawk with decidedly **short tail with broad white median band**, and an incomplete, partly concealed basal bar. Head with short bushy crest. **Juv.** with heavy buff streaks across nape, on sides of head (incl. broad eyebrow), wing-linings, and underside of body (exc. almost solid black thighs and usually black splotches forming black chest-bar). Tail gray-brown, paler near base, dotted, but never distinctly barred; less short than in ad. Wings with 6 rather short fingers. Larger, and with relatively longer wings, heavier head and slatier plumage (ad.) than common Great Black Hawk (*Buteogallus anthracinus*) of low elevations. Call *pipipipip* and arresting *yeeep yeeep yeeep....*

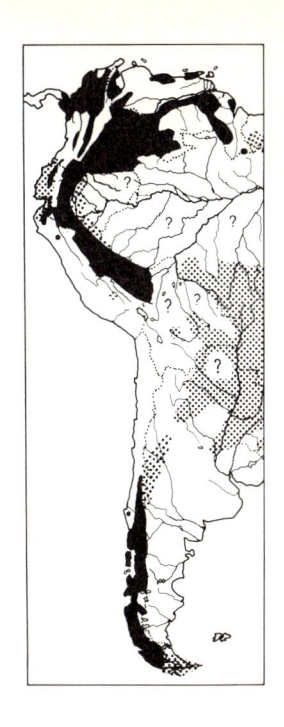

BLACK-AND-CHESTNUT EAGLE *Oroaetus isidori* – Plate XII 9ab

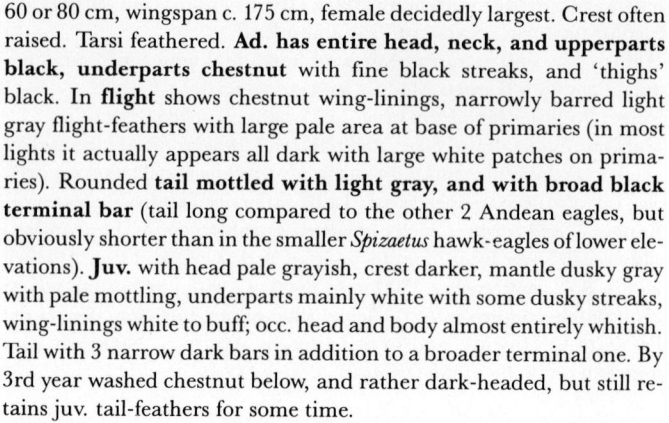

60 or 80 cm, wingspan c. 175 cm, female decidedly largest. Crest often raised. Tarsi feathered. **Ad. has entire head, neck, and upperparts black, underparts chestnut** with fine black streaks, and 'thighs' black. In **flight** shows chestnut wing-linings, narrowly barred light gray flight-feathers with large pale area at base of primaries (in most lights it actually appears all dark with large white patches on primaries). Rounded **tail mottled with light gray, and with broad black terminal bar** (tail long compared to the other 2 Andean eagles, but obviously shorter than in the smaller *Spizaetus* hawk-eagles of lower elevations). **Juv.** with head pale grayish, crest darker, mantle dusky gray with pale mottling, underparts mainly white with some dusky streaks, wing-linings white to buff; occ. head and body almost entirely whitish. Tail with 3 narrow dark bars in addition to a broader terminal one. By 3rd year washed chestnut below, and rather dark-headed, but still retains juv. tail-feathers for some time. 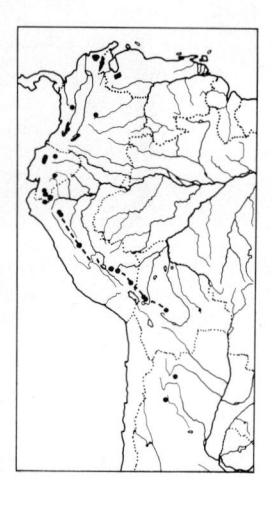
Habits: A secretive forest bird. Sometimes on exposed perch, and often seen soaring. Glides on horizontal wings. Takes larger birds and arboreal mammals.
Voice: Near nest loud *peeeeo*, alarm *chee-chee-chee*....
Breeding: In large tree-top nest. Breeds in Aug (Bol.). Recently fledged Mar-July (Ven.).
Habitat: Undisturbed heavy montane forest, probably mainly in larger valleys.
Range: Upper trop. to lower temp. zone (1750-2500 m, locally to 3500 m). From coastal mts of Carabobo and Aragua, Ven., Sta Marta mts, Col., and Andes from Perijá mts and Mérida, Ven. through Col. and Ecu. to nw Peru (s Piura), and along e Andean slope s to Jujuy and (1956-57) Tucumán, nw Arg. Local and rare.

BLACK-CHESTED BUZZARD-EAGLE *Geranoaetus melanoleucus* – Plate XI 6a-c

65-80 cm (juv. longest), wingspan 175 or 200 cm (female decidedly largest). **Flight profile of ad. triangular**, owing to **extremely broad-based wings and very short wedge-shaped tail** (Black Vulture also has short tail, but more rectangular wings with very long 'fingers'). **Juv. has longer tail**; it looks more like a Puna Hawk, but is decidedly larger, more broad-winged, with longer 'fingers', and heavier bill and head. In all plumages the flight-feathers are densely barred grayish with a very uniform general effect, except for the free 'fingers' which are uniform black. **Ad.** above dark slaty gray, with large finely barred **ashy-gray area formed by the upper wing-coverts. Blackish chest** contrasts with the pale lower underparts, which are white (*melanoleucos*), or finely dark-barred (*meridensis* and esp. *australis*; 1 specimen totally sooty black with gray 'thighs' may be a rare dark phase). Younger birds have a plumage strongly variegated in buff and black. A typical sequence is as follows: **1st year** head and nape blackish strongly streaked with white feather-bases (and some cinnamon), rest of upperparts very uniform blackish brown without pale area on wing-coverts or tawny

juvenile   *early 2nd year*   *late 2nd year*

Flying Black-chested Buzzard-eagles

checkering (of juv. Puna Hawks), but usually with small pale tips to fresh wing-coverts. Tail dark gray, densely mottled. Throat and breast buff to tawny, with few black streaks or splotches, but across belly almost black, and 'thighs' densely barred. **After molt** late in 1st year with some cinnamon and gray mottling on wing-coverts, throat and breast becoming almost covered with black splotches, but lower underparts are more orange-rufous with dense but fine black barring. Right after the body molt starts the development of a black ad.-like tail; however, a few of the conspicuously longer juv. rectrices may be retained for another year. **From late 2nd year** ad.-like, but with more irregular mottling on the wing-coverts, and more gradual transition from black chest to barred belly, and with some cinnamon color in the light parts. Even in the **4th year** with occ. dark smudges on gray wing-coverts and whitish belly.

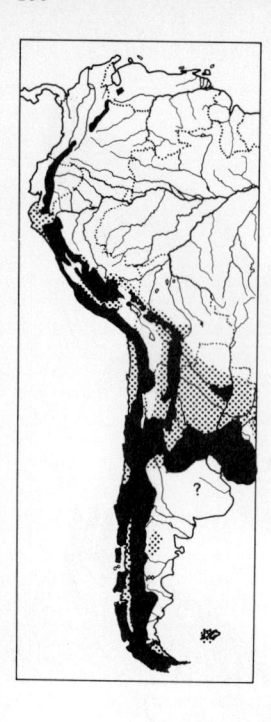

Habits: Alone or in pairs, occ. a few together. Often perched in trees or on posts, but most often seen soaring effortlessly, on horizontal to somewhat raised wings with wingtips up-turned. When gliding, often with pointed wing-tips curved back- and upwards. Occ. hovers heavily, or uses hanging hover. Usually hunts snakes and small to viscacha-sized mammals, but also takes birds, carrion, and sometimes insects, by picking from the ground.
Voice: Most commonly, both sexes give drawnout calls *kiiua kiiua*, sometimes repeated several times. Female at nest gives *pfiiu pfiiup*, and a very rapid *kiu-kiu-kiu...*, possibly to attrack mate (T de Vries).
Breeding: In large trees or on cliff-ledges. Eggs Sep-Feb (Ecu.). Nestlings Dec-Jan in the s, Aug (c and n Peru), Mar-July (Ven.). Juv. Nov-Feb (w Peru).
Habitat: Semi-arid open or sparsely wooded country, in the Andes mainly in rugged terrain with steep hillsides with some rock formations.
Range: Lowlands from Parag. to Córdoba, Arg. (*melanoleucus*), and throughout Andes from Mérida, Ven., to Isla Grande, with *meridanus* in the n, *australis* s of 25°s. Commonest on Andean slopes and up to 3500 m in the valleys, and absent from most altiplanos, although occ. seen to 4600 m. Fairly common, except furthest n.

NOTE: Previously called *Buteo fuscescens*.

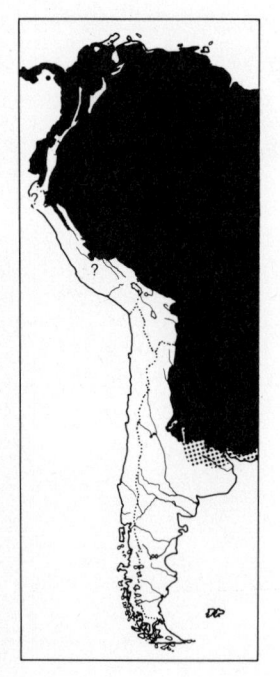

Hawks – Genus *Buteo*

Known as buzzards in the Old World. Mostly medium-sized raptors. Compact with rounded heads with rather slight bills, and with feathers of thighs forming 'flags'. The wings are broad with 3-5 short 'fingers' and somewhat curved trailing edge, tail rather short to medium long, fan-shaped. Most species are strongly polymorphic by color, and field identification requires experience. Often they soar high up in updraughts and thermals, but hunt mostly by silent waiting and watching from tree-tops, poles, or cliff-ledges, and suddenly fly and dive on the prey. Prey mainly terrestrial reptiles and rodents.

ROADSIDE HAWK *Buteo magnirostris* – Plate XIII 4a-b

33-40 cm, with strong bill and relatively heavy head. Fierce look caused by big yellow eyes and yellow lore and cere. Most sspp gray above, varying from light blue-gray (e.g., *insidiatrix*) through brownish gray (*occiduus*) to dusky brown with almost black head (*saturatus*). Breast light gray to rusty, lower underparts white with dense gray-brown to rusty bars. **Tail gray to rufous with 4 black bars.** In **flight** shows **large, brick-red area in wing**. **Juv.** colored much as juv. Broad-winged H., but with 4-banded tail, and far from as large-winged. Has relatively smaller wings than most *Buteo* hawks (somewhat *Accipiter*-like, although with rectangular and not particularly broad-based wings; tail = wing width). Rather sluggish. Mostly perched on pole or exposed dead branch, mainly hunting reptiles and insects. As in *Accipiter* hawks, series of shallow choppy wingbeats alternate with glides (with bowed wings). Rarely soars. 'Song' a long gliding (declining) wheezy *zweeeee-eey*'. Also *wee-wee-wee...* and *cla-cla....*

In savanna, bushy country, clearings, and at forest edge nearly throughout warm parts of C and S Am. Owing to clearing of montane forest locally ascends to 2500 m in Col., Ecu., and n Ven. (*insidiatrix*), nearly to 3000 m locally on e Andean slope of Peru (*occiduus*), and at least to 2350 m in Bol. (*saturatus*).

NOTE: Often in a separate genus *Rupornis*.

WHITE-RUMPED HAWK *Buteo leucorrhous* – Plate XII 1a-c

37 cm, wingspan 85 cm. **A mainly black small chunky soaring hawk**. In ad. black tail has 2 white bars (only the outer visible from above). Black with white rump and under tail-coverts, and somewhat rufous thighs, and (opposite dark Short-tailed H.) with **light buff or white wing-linings contrasting with black flight-feathers** with faint pale barring (Brown & Amadon suggests existence of an ashy brown phase with forehead, mid-throat, and all underparts white, thighs rufous, but we have never seen such a specimen). **Imm.** above fuscous, mottled and streaked with rufous; head and all underparts buff to rufous with coarse fuscous streaks, belly often barred black, thighs mainly brown. Tail as in ad. but with white tip. **Flight** profile with rather large tail, as in Broad-winged H., but easily separated by dark flight-feathers. Usually seen circling low over the forest. Sometimes tirelessly repeats thin, high-pitched, short *pyee* calls.

Inhabits lowland forest edge, chaco, and hills with dense montane forest, locally to 1500-2500 m (occ. to 3650 m) through n Ven and most of Andes of Col. and Ecu., and in e Andean premontane zone of Peru and Bol., to 2000 m in Jujuy, nw Arg.

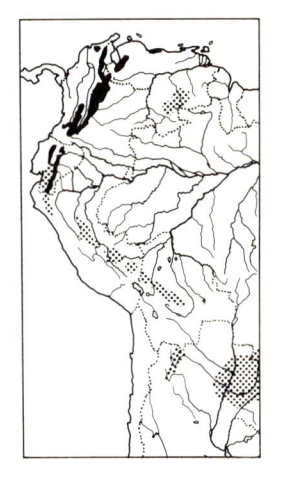

BROAD-WINGED HAWK *Buteo platypterus* – Plate XII 4a-b

34-45 cm, wingspan 80-100 cm. **A mainly gray-brown, small chunky soaring hawk**. Ad. above dark gray-brown, head often with white streaks and with white forehead and dark mustache. Throat white, somewhat streaked, breast mainly chestnut with white spots, lower underparts white barred with brown. **Tail distinctive with 2 whitish bars** (plus 1 concealed at base). Sooty-brown phase very rare. **Juv.** below white with dark streaks, flight-feathers distincly barred, tail with 5-6 bars. At all ages, **underside of wings mainly white**, and faintly to lightly barred, except for rather narrow but **distinct dark rear edge**. **Flight** profile with very broad rounded wings and rather long tail, approaching a soaring *Accipiter*, but wings much larger.
Habits: On migration often assembles to form large 'kettles' in thermals, then glides to new thermal to repeat soaring. Soars and glides with horizontal or slightly arched wings. Hunts by waiting on rather low perch at forest edge.
Voice: Shrill, high-pitched but rather weak *pwee* or *ker-wee-wee-eee*.

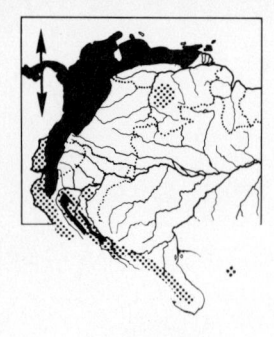

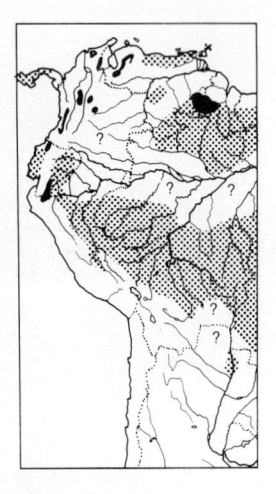

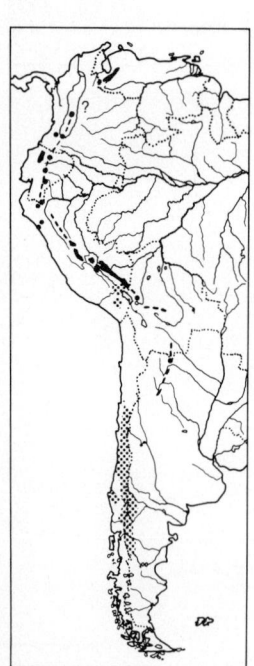

Habitat: Second growth and wooded hillsides, not within forest.
Range: 3-400.000 N Am. migrants visit forested parts of northern S Am., where common Oct-Mar at 500-3000 m (occ. to 4000 m) in Andes of Ven., Col. and Ecu., more rarely in Peru and Bol. (*platypterus*).

SHORT-TAILED HAWK *Buteo brachyurus* – Plate XII 3a-b

C. 36 cm, wingspan 90 cm. Small and compact, ample-winged and decidedly **short-tailed** compared with other **small** *Buteo* hawks. Always with **white at base of bill**. **Light phase**: with upperparts and sides of head and neck blackish, underparts white, incl. throat and wing-linings. Rare **dark phase** resembles black phase of Broad-winged H., but with black wing-linings contrasting with light flight-feathers. **Juv.** pale, with head streaked white, and underside and wing-linings pale buff with occ. black streaks, or dark somewhat spotted with buff. In all plumages, light but rather densely barred undersides of **flight-feathers become progressively darker towards rear edge of wing** (not distinct dark rear edge of Broad-winged H.). Tail distinctly barred, in ad. with 4-5 thin and a broader subterminal dusky bar, in juv. with c. 8 bars. Glides and soars in criss-crossing circles with wings held flat, but with pronounced up-turn of primaries. Rather active, diving down on prey spotted from high in the air. Takes rodents, reptiles, insects and birds. Rarely seen perched.

Chiefly seen high over soggy meadows and lake shores or open country near tall wood. Mainly inhabits warm parts of the Americas from s USA to n Arg., but occ. reaches 2500 m in Col., and rec. to 2200 m in Ven.

WHITE-THROATED HAWK *Buteo albigula* – Plate XII 2a-b

C. 38-48 cm, wingspan c. 95 cm. Differs from light Short-tailed H. as **rufous brown on neckside extends to breastside**, and rufous to dark brown streaks and splotches spread over much of underparts and wing-linings, and form **solid patch on each flank**. **Thighs barred** rufous. Tail conspicuously longer than in Short-tailed H., and with 8-10 very faint bars (appears uniform light gray below, dusky above). Occurrence of a rare dark phase suspected. **Juv.** has distinct but scattered black streaks below, and thighs and wing-linings have buff tinge. In flight resembles a tiny, light Red-backed H.

Habits: Usually seen in short glide close to hillside. Sometimes soars in criss-crossing circles. Feeds on rodents and birds.
Voice: Normally silent, but near nest gives squeal at high (but falling) pitch ^{kee}ea.
Breeding: On cliffs or in shrub. Season unknown.
Habitat: Dwarf forest patches and cloud forest adjoining open land, and regularly in *Eucalyptus* areas in Ecu., and in *Araucaria* woodlands in the s.
Range: At 2100-3500 m, locally from Mérida nw Ven. through Col. and Ecu. and along e Andean slopes of Peru (w slope in n Peru, occ. to

sea-level and in the puna) and Bol. to Salta, nw Arg. At lower elevations in the Andes of w Neuquen, Río Negro and Chubut, c Arg. (possibly migrating n to Salta and Tucuman) and Atacama and Valdivia, c Chile (to sea level). Rare.

NOTE: Sometimes listed as a ssp of Short-tailed H., but more probably a distinct (semi)species.

SWAINSON'S HAWK *Buteo swainsoni* – Plate XII 10

52 cm. Above dark brown. Common **light phase** buffy white below (incl. wing-linings), with brown **bib-like band on chest**. Flight- and tail-feathers lightly barred gray-brown with dark terminal bar. **Dark phase** blackish brown excepting light throat, forehead, under tail-coverts and some light under wing (but not contrasting pattern of dark Short-tailed H.). **Rufous phase** similar, with rusty bars and splotches below. **Juv.** somewhat streaked and spotted, esp. on chest. **Flight** outline characteristic, with rather narrow wings with long tips formed by 2nd-4th primaries. When perched, wingtips reach slightly beyond tail-tip. Soars in criss-crossing circles with stiff, slightly raised wings. Teetering vulture-like flight. Often migrates in flocks. Regularly walks on ground and perches on posts, banks etc.

N Am. migrant to La Plata region, Arg. C. 300,000 birds pass high up (4-6000 m), but occ. rec. when passing the C and E Andes of Col. and Ven. (e.g., 1 rec. of several thousand near Bogotá) Aug-Oct and Mar.

WHITE-TAILED HAWK *Buteo albicaudatus* of chiefly savanna country from southern N Am. to northern S Am. (*hypospodius*, *colonus*) and s of the Amazon area to c Arg. (*albicaudatus*), ascends to near limit of temp. zone in Bol. (rec. 2400 m). A large long-winged hawk, with wingtips reaching well beyond tip of tail when perched. Head blackish, above slaty gray or medium gray, with shoulders rufous, underparts white; tail white with single black band near tip. Juv. heavily streaked below.

RED-BACKED HAWK *Buteo polyosoma* – Plate XII 7a-c

45-53 cm, wingspan c. 120 cm (but *peruvianus* small). 3rd primary (counting inwards) slightly longer than 5th (visible in the field). When perched, wing-tips fall 5-7 cm short of tailtip. Similar color variation as in Puna H., but in most areas with light birds dominating, and few dark-phase birds. Dark birds with chestnut mantle and vest are rare, being replaced by a type with almost totally chestnut underparts with slate color only on throat, thighs and lower abdomen. Lighter, and in flight appears less large-winged and relatively longer-tailed (tail 3/4 of wing width).
Habits: Often hunts actively, hovering (with very floppy wingbeats compared with Puna H., the wings inverted on the backwards strokes), or using hanging hover in suitable wind. Mainly takes rodents and larger insects. Flies with rather shallow and stiff wingbeats, and soars with flatter wings than Puna H.

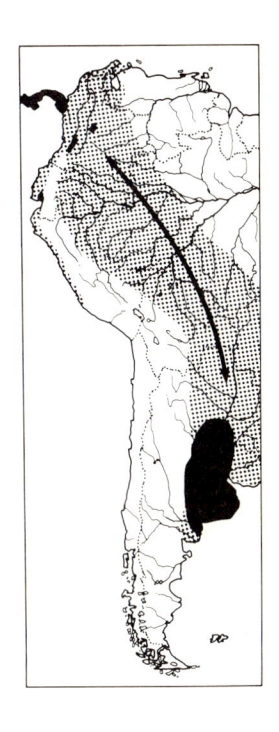

*Colour phases of adult Red-backed Hawks;
horizontal shading indicates rufous color.*

Voice: Sometimes calls when hunting. Loud *keeeow-kyow-kyow...* noted from pair near nest.
Breeding: On cliff-ledges or in large trees; eggs Oct-Nov (in the s) or Dec-Apr (Ecu.).
Habitat: Mainly open hillsides with some groves, shrub, and bracken, and sometimes on partly cultivated slopes, but in Arg. and Chile also lowland plains.
Range: From s tip of continent through beech forest zone and interior of the Patagonian steppe (migrating to plains of nc Arg.) and along Andes of Arg., Chile, Bol. (with 1 rec. also in the lowlands), and Peru to páramos of Ecu. and in C Andes, Col. (as migrant?), esp. on semi-arid slopes and valleys up to the tree-line (viz. mainly at 1800-3200 m, but locally higher, overlapping with Puna H.) (*polyosoma*). W Andean slope of Peru and sw Ecu. (*peruvianus*).

PUNA HAWK *Buteo poecilochrous* – Plate XII 8a-c

50 or 70 cm, wingspan 125 or 150 cm. A **stocky and large-winged version of Red-backed H.**, with 3rd primary (counting inwards) usually shorter than 5th. When perched, wingtips almost reach tip of tail. **Ad. with white tail** with several very thin dark bars and **broad black subterminal bar**; flight-feathers gray, pale below, finely barred and darker towards edge of wing. **Body plumage highly variable**: Light phase blue-gray above, occ. with some dense dark barring, and often (usually in females) with rufous mantle, and white below (incl. wing-linings) with some fine gray or rufous barring. Dark phase (which seems to be the commonest) deep slaty gray to sooty brown all over (incl. wing-linings, but with the above-mentioned pale flight- and tail- feathers); female with more or less chestnut mantle and 'vest' or sometimes almost black 'vest', with some white checkering on belly. **Juv. has gray flight- and tail-feathers with faint, very dense barring** and no broad distal bar, although feathers with medium-wide subterminal bar appear in

2nd year. Body plumage extremely variable, from white below and on head, with scattered black spots, and with dark brown upperparts with much gray or cinnamon checkering or feather-margins, to almost uniform dark sooty brown. Many birds are intermediate with dark whisker on rusty or pale face, and clear rusty to white breast contrasting densely and obscurely spotted and mottled or solid black 'vest'. **Flight** profile with large and broad-based wings with a quite rounded outline, and proportionally small head and short and apparently broad tail (2/3 of wing width, more eagle-like profile than Red-backed H.).
Habits: Often soars high above mountain ridges and valleys, single or in pairs, or a few together in the mating period. Wings normally raised when soaring, horizontal when gliding. Wingstrokes soft. Often hunts using 'hanging hover', with very slow or no wing-flaps on updraught winds, few meters over the terrain, or sometimes high up.
Voice: Long-drawn *peeeoh* ranging from copious cry to thin whistle, hoarse in juv. In display flight harsh *hee hee hee....*
Breeding: On rock ledges. Eggs in Feb and Sep in Ecu.
Habitat: Rugged mountainous country and rolling grassland, occ. with *Puya*s or some *Polylepis* woodland.
Range: At 2800-5000 m from Jujuy and Salta, nw Arg., throughout w Bol. (rare in adjacent Chile), Peru, and Ecu., and (as migrant?) in Upper Cauca and Patía Valleys, sw Col. Generally common.

NOTE: Sometimes regarded as a ssp of Red-backed H., but more probably a (semi)species replacing the other at high elevations. Sometimes called Variable H.

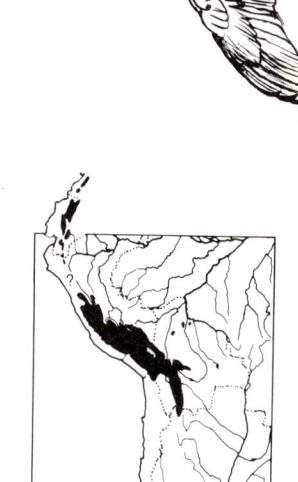

Juvenile Puna Hawks

RUFOUS-TAILED HAWK *Buteo ventralis* – Plate XII 6

C. 55 cm, wingspan 130 cm. Common **light phase** fuscous above with occ. buff checkering, side of head rather ferruginous, underparts changing from white on c throat to russet on 'thighs', with some black streaks and spots esp. on flanks and across lower breast. **Tail rufous, white-tipped, and with 8 distinct black bars**, terminal broadest. Underside of wings generally light, darkest towards trailing edge, and with black tip, and probably always with dark lesser under wing-coverts forming a **dark patch near leading edge of wing. Black phase** sooty black, with gray mottling below and some light barring on flight- and tail-feathers (*viz.*, not the black-tipped white tail of Red-backed H.). **Imm.** and **juv.** above black or fuscous, with occ. white spots, some birds with many cinnamon marks, dark malar area prominent. Below largely white with black blotches and streaks except on c breast. Underside of wings light buff (vs. white). Wing-linings white or buffy. Upper tail-coverts with white barring, tail gray-brown (not rufous) with c. 8 prominent black bars in imm., or 10 thin, but still conspicuous bars in juv. **Flight** profile not so unlike Variable H., but with more square wing-tips.
Habits: Soars and glides like Red-backed H., but main hunting mode sit-and-wait.
Voice: Harsh prolonged *kee-ahrr*.
Breeding: In tall tree. Egg-laying in Oct-Nov.
Habitat: Forest edge, open forest, and park-land, but we have seen imm. birds on ridges and plateaus of Patagonian brush-steppe.
Range: Chile to 1200 m from n Isla Grande to Ñuble and in adjacent sw Arg. Everywhere rare, but may not be at risk.

NOTE: By some regarded a ssp of the Red-tailed H. (*B. jamaicensis*; see *Birds of Prey Bull.* 3(1986):115-118).

BAY-WINGED HAWK *Parabuteo unicinctus* – widespread in open country in the warm parts of the Americas - is rec. accidentally in temp. zone of Chile and Lima, Peru, and (to 3300 m) in Guayllabamba valley n of Quito, Ecu. 50 cm. Resembles sympatric Red-backed and White-throated H.s, but has slenderer build, longer legs, tail and wings, 5 long 'fingers', and a more rapid dashing flight. **Ad. is black with chestnut wing-coverts**, and with broad white base and narrow white tip of tail. **Juv.** has pale eyebrow, checkered wing-coverts, buff coarsely streaked underparts, chestnut-barred thighs, and numerous thin tail-bars.

Falcons – Family Falconidae

A rather heterogeneous group of altogether c. 60 species united by skeletal characters and sequence of moult of the primaries, and by some biochemical characters. S Am. has some primitive groups of caracaras, forest-falcons (*Micrastur*), Spot-winged Falconet (*Spiziapteryx circumcinctus*), and Laughing Falcon (*Herpethotheres cachinnans*), and some species of the large cosmopolitan group of true falcons (*Falco*).

References concerning Andean caracaras are *Acta Zool. Lillo.* 18(1962):112-13, *Bol. Mus. nac. Hist. nat. Chile* 37(1980):113-16, *Breviora* 355(1970):1-29; and *Hornero* 12(1986) 223-9.

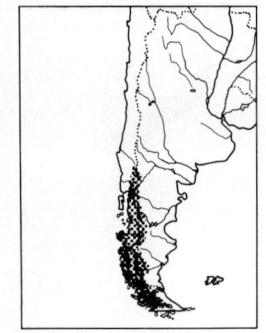

Caracaras – Subfamily Polyborinae

This New World group comprises 9 mediumsized scavengers, pirates, and insect-eaters (one lowland form eats fruit and preys on wasps and bees). They are lanky with long rectangular wings with distended 'fingers', and rather long tails. Most species have a rather weak 'chicken-bill', and partly naked face usually red, but changeable to yellow or white color in few s (when scared or excited). The neck is quite long. Large species have colored pouch on breast, protruding when distended by food. The legs are quite long, and caracaras run well. Much of the time is, however, spent perched or sedately walking. The flight is direct, with rather stiff, shallow, rapid wingbeats, some gliding on the wind, but little soaring. Caracaras build stick-nests in trees or on rock-ledges and lay 2-3 eggs.

CARUNCULATED CARACARA
Phalcoboenus carunculatus – Plate XI 5a-c

53 cm, wingspan 110 cm. With long, rectangular wings with broad tip of 5 long, distended 'fingers'. **Ad. mainly black**, but **entire underparts of body has coarse white streaking**, and lower abdomen and vent is white. Large area on rump and base of tail white, and also tail-tip, wing-linings, and bases and tips of primaries are white. Bare face and upper throat red, bill pale grayish, legs yellow. **Juv. sepia-brown**, usually with more cinnamon head and underparts, and normally with numerous, small buff feather-tips. Rump and upper tail-coverts pale buff with some thin fuscous bars. Also vent somewhat barred, tail dark brown, often with faint light bars and cinnamon underside. Bare face and feet bluish-white at first. In **flight** juv. shows **broad whitish buff zone across base of primaries**, and profusely barred dark wing-linings. **Imm.** has partly ad. plumage, but retains the barred tail-coverts.

Habits: Often seen walking and running on the ground, sometimes several together near cattle. Flight straight, with rapid, stiff, and very shallow wing-beats interspersed with glides on level wings, but said also to glide with half-closed winds on gusty mountain winds.

Voice: Rarely emits harsh barking calls.

Breeding: On rock ledges. Eggs Sep-Oct and fledglings by Jan, but one spec. close to fledging size is labelled 20 May.

Habitat: Open grassy and bushy pasture and páramo terrain.

Range: At 3-4000 m in páramos from c Loja, s Ecu. to Nariño and spreading into Cauca Valley of sw Col.

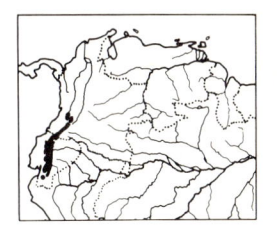

NOTE: Carunculated C. forms a superspecies with Mountain and White-throated C.s. Based on *carunculatus* traits in occ. Mountain C.s from n Peru, and signs of hybridization between Mountain and White-throated C.s in Río Negro and Chubut, Arg., the 3 were once considered sspp under *megalopterus*.

NOTE: *Phalcoboenus* is often included in *Polyborus*.

MOUNTAIN CARACARA *Phalcoboenus megalopterus* – Plate XI 2a-b

52 cm. **Ad. glossy black with white belly and wing-linings**, white bases and tips of primaries, **white rump** and tail-coverts, and broad white tail-tip. Bare face orange or red. **Juv. dark sepia-brown** throughout, except often for pale feather-tips on wing-coverts and belly, and for uniform pale vinaceous buff rump. Bare face and feet bluish white at first, later yellow. In **flight** juv. shows **broad whitish buff zone across base of primaries**, and extensive buff areas mainly on the outer tail-feathers. **Imm.** rather variegated, owing to a mixture of brown and black feathers, and with some thin dark bars on white base of tail.
Habits: Frequently seen walking, often a pair or ad. with one juv. together, or more birds, occ. hundreds on plowed fields. Often roost many together. Flight with stiff, shallow wingbeats. Glides on horizontal and sometimes strongly crooked wings.
Voice: Series of rasping, barked *kjah* or *ahk* calls.
Breeding: On rock-ledges. Eggs Oct-Nov in the s.
Habitat: Open puna grassland, feeding esp. on plains, shore meadows of lakes, and other strongly grazed areas, and on recently plowed land, but breeding and roosting in cliffs in rugged country.
Range: In puna zone at 3500-5000 m, sometimes to 760 m, and regularly to the Pacific coast in s Peru. From páramos on the Peru/Ecu. border through Peru and w Bol. to Talca, Chile, and Mendoza and Nahuel Haupí in Chubut, w Arg. Abundant esp. in n part of Bol./Peru altiplano.

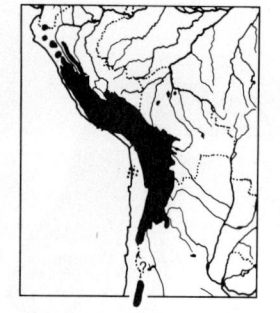

WHITE-THROATED CARACARA
Phalcoboenus albogularis – Plate XI 3a-b

55 cm. Differs from Mountain C. by yellow or orange facial skin, less glossy black upperparts, and **white underparts from chin to base of tail**, although with black spots and streaks laterally. **Juv.** seems often

Mountain Caracara

to be darker than in Mountain C., almost black above, with black ear-coverts and forehead, but may not be safely distinguished. Clearly larger and darker than sympatric Chimango C., and with more restricted pale zones on wings and tail.
Habits and **Voice**: Probably much as in Mountain C. The call is said to be a deeper version of that of Chimango C. On Isla Grande assembles on garbage dumps and sheep-slaughtering sites.
Breeding: On rock ledges. Eggs probably in Oct-Nov.
Habitat: Wild mountainous areas to low woodland.
Range: To 3000 m in Andean zone from s tip of continent to Neuquen, c Arg. and Ñuble, Chile, probably commonest in the s part of Isla Grande and in Magallanes and Aysén, Chile, but also breeds on rocky edges of basalt plateaus of the Patagonian steppe.

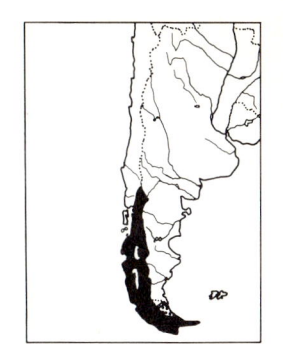

STRIATED CARACARA *Phalcoboenus australis* – of Islas Malvinas and the rocky Fuegian seacoasts, and recently found n to 54°s in Arg. and Chile – can be expected to move some way up in the coastal mts. It resembles juv. White-throated C., but is much more stockily built, broad-winged, and relatively shorter-tailed. **Juv.** uniform smoky brown with buff zone across base of primaries and on tail; **ad.** black with some white at base of primaries, thin white streaks on breast and around neck, tawny thighs, and white tail-tip.

CRESTED CARACARA *Polyborus plancus* – Plate XI 1a-b

49-65 cm, wingspan c. 120 cm; *plancus* largest. With stout and **very deep pale bill**, bare red face, and **flat crown with black cap ending in slight backwards pointed crest. Sides of head and throat whitish**. In **ad.** pale buff neck and breast densely barred blackish brown, rest of body almost uniform fuscous black (but with fine dorsal barring in *plancus*). **Juv.** generally browner and dingier, with ochraceous cheeks grading to drab gray on nape, and with coarsely streaked (not barred) underparts. Facial skin whitish at first. In **flight** shows barred wing-

Immature (molting) Crested and Mountain Caracaras

linings, **large white patch at base of primaries**, and broad black distal zone of otherwise whitish tail.
Habits: Sometimes several together. Often seen walking or running, or perched on posts or tree-tops. Flight irregular or zigzagging, rapid, usually with shallow stiff wingbeats alternating with sailing with somewhat arched and slightly drooping wings. In flight, hand more backwards deflected, and the free fingers less spread than in *Phalcoboenus*. Occ. soars. Feeds on garbage, road-kills, and larger carrion, sometimes with vultures.
Voice: Harsh bass rattle *hrrap-hrap-hrap...* or *chorr*. From perch a grating *crric* or *crruc*, sometimes repeated. Purring call given with head thrown back. In aerial courtship, pairs swoop at each other, giving shrill screaming *keeer*.
Breeding: Atop trees; eggs in Aug-Oct in c Arg., or Nov-Dec further s.
Habitat: In sparsely wooded lowlands (or sometimes forest and marsh) and shrub steppe, esp. ranch-land.
Range: Widespread in S, C, and southern N Am., mainly in lowlands, but after recent deforestation ascents to semiarid temp. regions of Col., and tablelands of Ecu. (*cheriway*). To 2000 m at base of Andes in wc Arg. and c Chile, and regularly visits uplands of the sheep districts of the Patagonian steppe (*plancus*).

NOTE: Genus name *Caracara* sometimes used.

CHIMANGO CARACARA *Milvago chimango* – Plate XI 4

40 cm. Finely mottled **rufous brown** with whitish, faintly mottled base of tail, and **broad whitish zone across hand**, esp. above (unlike in juv. of larger caracaras, this zone includes greater hand-coverts). Ssp *chimango* rufous buff below, *temucoensis* distinctly browner. **Juv.** more rufous and with paler tail and flight-feathers. Resembles a miniature light juv. *Phalcoboenus*, with smaller chicken-bill and feathered (not brightly colored) face; in flight, shows less prominent 'fingers'. Sometimes in parties, probably mainly family-groups, and may nest in colonies. Feeds esp. on orthopteroid insects and small vertebrates. Voice a petulant *chew-chew-chew...* and *ki-en, ki-en, ki-en...*, or single *eeeeee-eh*.

Widespread on heathlands, marshes, open wood, fields, and suburban areas in lowlands of southern S Am. Forested zone of s Chile and low valleys in the s Andes into adjacent sw Arg. (*temucoensis*); c Chile n to Tarapacá and throughout Arg. to Parag., and straggling to temp. Andean slopes in wc Arg. and c Chile (*chimango*).

YELLOW-HEADED CARACARA *Milvago chimachima* of trop.
and subtrop. zones from Urug./Parag. n-wards, is rec. near lower limit of temp. zone in Popayan and Bogotá areas Col. (*chimachima*). - **Ad. light buff with black upperside of body**. Wings with broad white zone across base of primaries. Tail with broad black terminal zone and some thin bars. **Juv.** has coarse fuscous streaks on all buff parts.

NOTE: Chimango and Yellow-headed C.s are paraspecies.

Falcons – Subfamily Falconinae

A cosmopolitan group of 37 small to medium-sized streamlined raptors. Apparently neckless, with small short head. Long pointed wings, often with quite dark but profusely white-spotted linings. Medium-long or long narrow tails. The short bill has distinctly notched cutting edges. Eyes dark (unlike in most *Accipiter* hawks). The Andean species all frequent open country, and are swift fliers. The flight is direct, choppy, and powerful, although the birds sometimes soar high up with spread tail. The hunting also involves silent waiting from treetops, posts, or cliffs. The prey is (except in kestrels) habitually killed in flight, either by striking it dead with a blow, or seizing it with the foot.

Falcons do not build nests, but use abandoned stick-nests in trees or cliffs, or nest directly on rock-ledges, or in holes. The eggs are round, thickly speckled red-brown. The nidicolous chicks are dressed in white down. The young are often attended considerable time after fledging.

Detailed information on Andean falcons can be found in Cade, TJ (1982) *The Falcons of the World*. London. See also Porter, RD et al. (1987) *Working bibliography of the peregrine falcon*. Nat. Wildlife Federation Washington DC; Weick, F (1980) *Birds of Prey of the World*. Hamburg-Berlin; *Ann. Inst. Pat.* 12(1981):221-8; *Auk* 74(1957):1-19, 100(1983):269-71; *Condor* 82 (1980):350-1; *Wilson Bull.* 67(1955):194-9, and thèsis by C Pacheo (1987) (*Falco sparverius*) in Univ. Católica, Quito, Ecu.

AMERICAN KESTREL *Falco sparverius* – Plate XIII 11a-h

25 cm, wingspan 55 cm. **A tiny falcon identified by russet back and double black bands on white face**, 2 black spots on buff nape (face-like pattern from behind) and blue-gray crown usually with brick-red spot centrally. In **male**, mantle rufous with some black bars, **wings blue-gray**, tail rufous with wide black terminal bar. Ssp. *aequatorialis* and *caucae* are intensely cinnamon-colored below, *ochraceus* slightly more ferruginous, but *peruvianus* and *cinnamominus* pale buff with some black dots below, *cinnamominus* also characterized by heavily barred mantle. Ssp *caerae* resembles *cinnamominus*, but male has more or less white outer tail-feathers with several black bars. In **female**, deep rufous brown upperparts are densely barred fuscous, lighter underparts have brown streaking, tail several thin bars in addition to broad terminal one. **Juv.** resembles ad., with similar sexual dimorphism, but more subdued pattern, and broader spots and bars (thin bars in tail, even the distal bar quite narrow).
Habits: Alone or in pairs, often quite confident. Flight hurried, often with rather stiff wingbeats alternating with gliding and soaring. **Mainly hunts hovering**. When dropping on terrestrial prey (insects, scorpions, lizards), wings often folded into sickle-shape. Perches conspicuously in treetops, larger cacti, wires, and posts.
Voice: Rapid high-pitched *kilee-kilee-kilee...* or *pitee-pitee....*
Breeding: On rock-ledges or in cavities. Jan-May (Ven., Col.), June-July (Ecu.) to Oct-Nov (furthest s).
Habitat: Forest edge and xerophytic areas to any kind of open terrain, provided presence of scattered trees, electric wires, or rocky bluffs. Often in suburban areas.

Range: Most of the Americas, excepting only trop. humid forest. In nw Ven. and E Andes of Col., mainly in drier areas at 2000-3200 m (*ochraceus*); hills bordering Cauca and Patía valleys, w Col. (*caucae*), and s Col. to Chanchan Valley,cw Ecu., to 4000 m on Cotopaxi (*aequatorialis*), and Pacific slope from s Ecu. to n Chile (*peruvianus*). All highlands and Andean valleys of c and e Peru through Bol. (to 4500 m) and at lower elevation through Arg. and Chile s to Tierra del Fuego (*cinnamominus*). A bird corresponding to the chaco form *caerae* is rec. at 2887 m in Cochabamba, Bol.

ORANGE-BREASTED FALCON *Falco deiroleucus* – a very rare inhabitant of mature trop. premontane forest of C Am and s to Bol., n Arg., and s Brazil - can possibly reach lower fringe of temp. zone in Col. and Cuzco, Peru. Resembles a compact and short-tailed Aplomado F. without white eyebrows and with broader buff barring on the black 'vest'.

APLOMADO FALCON *Falco femoralis* – Plate XIII 10a-c

36 or 45 cm, wingspan c. 110 cm. **Slender and long-tailed**, when perched tail projects well beyond wingtips. Characteristic pattern: above deep gray with prominent **white eyebrow continuing as buff nuchal collar**, and with prominent but narrow mustache. Cheeks, throat, and upper breast buff; lower breast and sides form **black 'vest'** with fine pale barring; belly and vent tawny (*femoralis*), or deep ochraceous (*pichinchae*). Tail blackish with narrow whitish bars. **Juv.** darker with brown feather-margins above, darker buff light parts, and coarse black streaks on breast. Flight profile distinctive due to long tail and long narrow wings (compare Peregrine).
Habits: Often pair together. Usually not very shy. Often perched very upright on post or low exposed branch. Flies swiftly and fast, with elegant turns. Hunting methods partly *Accipiter*-like, by surprise attacks, but may also chase a bird for long periods, zigzagging low over the terrain, and usually kills in the air.
Voice: High-pitched *cueek*s several s apart (resembles Andean Flicker). Repeated *ee ee ee...* (not as sharp as other falcons).
Breeding: Sep-Oct (Arg.).
Habitat: Open country ranging from savanna and bushland to rolling puna country or rugged mts, in the highlands sometimes associated with woodlands or *Eucalyptus* groves.
Range: S, C and southern N Am., commonest in drier parts. Andes at 2000-4600 m through puna zone of Peru to Catamarca, nw Arg. (except slopes with humid montane forest, but seasonally descends to the coast in Peru and in the s to Curico, Chile) (*pichinchae*). Lowlands of n Col. and e of Andes, and spreading across the continent in the s parts of Arg. and into Magallanes, Chile (*femoralis*).

MERLIN *Falco columbarius* – Plate XIII 8a-b

C 30 cm, wingspan 60 cm. Somewhat more compact and with shorter tail than Am. Kestrel. **Obscurely colored, mustache and other**

markings on face indistinct, and lacks pattern on nape. **Ad. male** above bluish slate, below pale buff with black stripes and often washed rusty, esp. on thighs. In **flight** shows white notches on flight-feathers, and broad black terminal bar and several thin bars on gray tail. **Ad. female** above deep gray-brown, below more cinnamon-buff with fuscous stripes, tail with 4 regularly spaced light bars. **Juv.** female-like, somewhat browner. Behavior somewhat *Accipiter*-like: flight dashing, with fast wingbeats interrupted by glides. Rarely soars, and often flies low over the ground, and perches low and inconspicuously.

Circumpolar. Breeders from subarctic N Am. winter Oct-May (occ. other seasons) s to Bogotá savanna and the nw of Col. and Ecu., sometimes to 3400 m, and casually to coast of Peru (*columbarius*, *bendirei*).

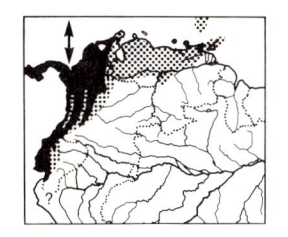

PEREGRINE FALCON *Falco peregrinus* – Plate XIII 9a-e

38 or 50 cm, wingspan 95 or 117 cm; female markedly larger than male. Powerful, appearing broad-shouldered, with backwards tapering and **rather short tail** (tail not beyond wingtips when perched). In all plumages with **dark hood** covering almost entire head, except for chin and mid-throat (*cassini*), or reaching well below eye and sending a broad moustache down between clear white throat cheek (*tundrius*; juv. of latter ssp with much buff streaking on sides of crown, and sometimes looking quite buff-crowned. **Ad.** above slaty black with some blue-gray checkering, below buffy white with some dark barring on lower parts (*tundrius* males may have almost clear white breast and c belly, but females have drop-shaped marks on these parts; *cassini* has strongly buffy breast). **Juv.** above fuscous with thin buff feather-margins, below buff with bold black streaks. Feet (and cere and eyering of ad.) yellow. In **flight** wings pointed, but arm appearing short and broad, underwing densely barred and spotted; tail length = width of wings (vs. Aplomodo F.).
Habits: Alone or in pairs. Usually flies with calm shallow wingbeats, but uses powerful wingbeats when hunting, and reaches incredible speed when swooping at prey with wings folded close to the body. Mainly takes flying birds, esp. waterbirds. Sometimes tries to flush crouching birds, by flying very close to the ground. Soars with horizontal or slightly drooping wings. Perches high on trees, posts etc.
Voice: Fairly high-pitched, slightly hoarse, weak and upslurred *hieks* repeated c. 7 times in 5 s. Near eyrie *witchew-witchew*....
Breeding: Mainly on rock-ledges. Eggs in May, June, July and Sep (Ecu.) or Sep-Oct (s Arg. and Chile).
Habitat: Open terrain, preferring coasts and extensive shore meadows with waterbirds, near sparsely wooded hills, or cliff-faces, where it may roost or breed.
Range: Distributed worldwide, and can be seen almost throughout S Am. N Am. migrants reach 40°s, but are common only along Pacific coast and on tablelands of Ecu. and Peru to c. 4100 m Oct-Apr (*tundrius*). Breeds in coastal and upland areas in s and c Arg. and Chile, and very locally further n (rec. breeding or probably breeding at 500-2700 m in Cochabamba, Bol. (*Bol. Doñana Acta Vert.* 13(1986): 170-173), Cuzco, Junín, Lambayeque, and Piura, Peru, and near Quito, Ecu.), and migrates to Col. (*cassini*). Common in the s.

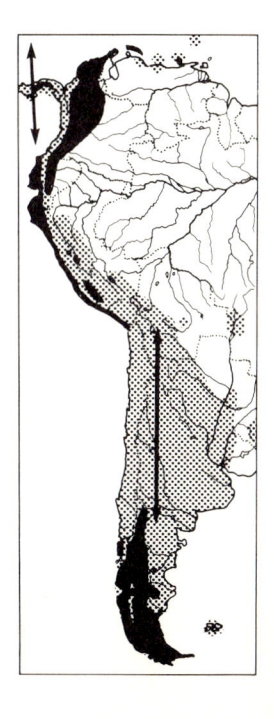

NOTE: Northern migrants were previously referred to ssp *anatum*, but apparently only the arctic *tundrius* reaches S Am.

NOTE: A pale color phase (recessive gene?) of *cassini* (see *Auk* 100(1983):269-71; *Raptors Res.* 18(1984):123-30; and *Natn. Geogr. Res.* 1(1985):338-94) was previously classified as a separate species: **KLEINSCHMIDT'S or PALLID FALCON (*Falco kreyenborgi*) – Plate XIII 9b**. This has much white on head, and rather thin mustache, upperparts of body dark gray with pale blotches and short bars (**ad.**), or fuscous checkered and margined with pale cinnamon or white (**juv.**). Underparts, incl. wing-linings, creamy white with faint streaks. Known from Isla Grande and Magallanes, Chile to the upland plateaus of Patagonia.

Puna Teals

Anseriformes

Waterfowl – Family Anatidae.

A group of disputed systematic affinity, although most evidence points to affinity with Galliformes. Distributed worldwide, with 145 species. Ecologically, waterfowl range from diving species which feed on invertebrates and fish to vegetarian grassland birds, but all can swim, and all are able to dive in emergency. The group is rather homogeneous morphologically. All species have rather short legs with 3 anterior toes connected with swimming webs, and a rather broad bill with corneous lamellas along the edges, and on the fleshy tongue, serving as a sieve (grazing species have more conical bills with the lamellas modified into tooth-like structures). The eyes are small. The plumage is dense with smooth surface, the tail more or less graduated, and usually short. Many waterfowl have an iridescent patch, the **speculum**, on the secondaries.

The gaith is waddling, in certain species decidedly awkward. The flight is fast, often somewhat noisy, and certain waterfowl need a running start to get airborne. The wing- and tail-feathers are molted simultaneously, and the birds therefore become flightless for a month's time, usually right after the breeding. In this critical period they lead a secluded life in thick marsh vegetation, or assemble in undisturbed lakes with plenty of food. Many species have a special migration before this molt. However, such movements, and those associated with seasonal drought or cold, are scarcely known for Andean species. Males of brightly colored species often have a female-like body-plumage in the period of wing-molt (**eclipse** plumage).

Most waterfowl are highly gregarious, but disperse in the breeding season, as the males (**drakes**) defend their mates (**ducks**) (very few species are strictly territorial). The breeding usually occurs scattered in marshland or other vegetation cover, but colony-like nesting sometimes occurs on isles. The nests are simple scrapes lined with the female's own down. The normally 4-10 eggs are pale and unspotted. The young are downy, highly active, and obtain their own food. The female incubates alone (except in whistling ducks and swans). The drake usually has a guarding role, both prior to egg-laying, when it secures undisturbed feeding for the duck, and after the young hatch. In most dabbling ducks, the drake leaves the duck (for the season being) before the hatching, but in most Andean species the pairs remain intact. This is a general tendency in southern waterfowl, and this difference from the better-known N Hemisphere relatives is associated with a more permanent pairing and less intense seasonal courtship in the southern life-zones.

The most detailed recent treatments of Andean waterfowl are given by Delacour, J (1954-64) *Waterfowl of the World*, London; Johnsgard, P (1965) *Handbook of waterfowl behavior*, Ithaca; and Johnsgard, P (1978) *Ducks, geese, and swans of the World*, Nebraska; Woods, RW (1975) *Birds of the Falkland Islands*, Oswestry; Livezey, BC (1985) *Systematics and flightlessness of Steamer-ducks (Anatidae, Tachyeres)*, Ph.D. thesis, Univ. Kansas. See further *Auk* 93(1976):560-70; *Condor* 87(1985):87-91; *Ibis* 125(1983):524-44; *Living Bird* 9(1970):5-27; *Spec. Publ. Mus. Texas Techn. Univ.* 20(1983); *Waterfowl* 23(1982):25-44; *Wildfowl* 17(1966): 66-74, 18(1967):98-107, 19(1968):33-40, 27(1976):45-53, 30(1979):5-15, 37(1986):113-22; *Wilson Bull.* 80(1968):189-212 and 87 (1975):83-90. Quantitative distribution data are summarised in Scott & Carbonell (1986).

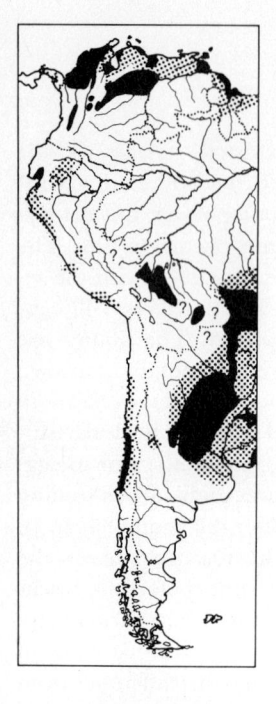

Whistling Ducks or Tree-ducks – Genus *Dendrocygna*

Gregarious, but quite quarrelsome. Highly vocal, their whistled calls making their presence evident even by night. All species are **slim**, rather goose-like and lanky, with **large feet** that project well beyond tail in flight. Also recognized in flight by black underside of wings, slightly drooping necks, and rather slow wing-beats (for a duck). The sexes are alike. **Pull.** boldly patterned, and identified by bands passing **across** nape. Most species have declined because of pesticides applied to rice.

FULVOUS WHISTLING DUCK *Dendrocygna bicolor* – Plate VIII 10

48 cm. Mainly **fulvous tawny**, back dark brown with obscure tawny bars, flank-feathers form creamy **white lateral stripe**. **Area around base of tail white** (spotted brown in **juv.**). Feet blue-gray. May resemble Yellow-billed Pintail (Plate VII), but distinguished by jizz, sidestripe, and **uniform dark wings**. Stands with more horizontally aligned body than Black-bellied W.-d. Voice a loud, whistled *pi-cheea*. Also a rapid *twitwitwi*....

Inhabits marshy parts of grasslands, ricefields, swamps, and lakes with rich waterside vegetation and usually some fringing wood. Distributed over the warm parts of India, Africa, and the Americas. Casual (but common until 1940s) at 2600 m on Bogotá and Ubaté savannas in E Andes, Col. Accidental in highland Peru, nesting once (Dec) at 4080 m in Junín, and accidental at 2550 m in Cochabamba, Bol..

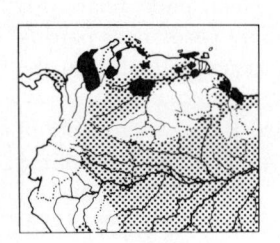

WHITE-FACED WHISTLING DUCK *Dendrocygna viduata*

45 cm. Dark with contrasting **white face** and throat-spot. Rear part of head black, **neck dark chestnut** grading to faintly streaked dark brown back, and densely barred black-and-white sides. Belly, wings and tail region blackish. Feet blue-gray. Younger birds show light grayish (**imm.**) or more ochraceous (**juv.**) head and underparts of body, but have chestnut lower neck. Often in large flocks. Call a characteristic very fine *vee-see-dee*.

Inhabits lakes and marshes, with no particular affinity to wooded areas. Usually roosts on mud or sandbars. Widespread in trop. and subtrop. Africa and S Am., wandering to temp. zone of E Andes, Col. and Ecu. (less in recent years), and occ. also in highlands of Peru.

BLACK-BELLIED WHISTLING DUCK *Dendrocygna autumnalis*

50 cm. Distinguished by **black underparts** of body and **large white patch on upperside of wings** (median and greater coverts, and base of primaries). Otherwise rich brown with contrasting light gray throat and sides of head. Bill and feet pink. **Juv.** duller with dusky bill. Rather terrestrial, with slender and gangling look, and erect posture. Often

perches in trees. Voice reedy whistled *che-che-reea* or *pitche wee reea*, and prolonged low *didrididlidi*....

Inhabits lagoons, streams, marshes, rice-fields, and neighbouring woodland. Widespread in trop. parts of the Americas, casually visiting temp. zone in E Andes, Col., tablelands near Quito, Ecu., Junín, c Peru, and Cochabamba, Bol. (*autumnalis*).

COSCOROBA SWAN *Coscoroba coscoroba* –
Plate VIII 1a-b

100 cm. A **huge, white 'duck-swan' with rosy bill**. In flight shows **black primary-tips**, and large, pink feet. **Juv.** has blackish cap and obscure gray-brown feather-tips on much of body. **Pull.** similar, with black cap and three broad drab gray dorsal bands. Swims with somewhat raised rear (unlike swimming flamingo), raising wingtips slightly in aggression. Usually in small groups, grazing, swimming, or wading in shallow water. Distinctive bugling call **cusgu**-*worr* ('cosco-roba'), the last part lower pitched.

Inhabits shallow lagoons, lakes and fresh-water marshes with abundant submergent vegetation. Widespread in Arg. and s Chile, esp. in the Magellanic and Pampean regions. Partly migratory, but important winter quarters exsist on n Isla Grande. Mainly lowlands, but not infrequently visits (and occ. nests on) weed-covered lakes and ponds near 1000 m on Andean foothills in Sta Cruz, Arg.

NOTE: Often classified as a swan, although relationship with whistling ducks seems more probable.

BLACK-NECKED SWAN *Cygnus melanocoryphus* –
Plate VIII 2a-b

125 cm. **Huge, white, with head and long neck black**, except for white 'spectacles'. Bill plumbeous with large red fleshy caruncle. **Imm.** has small caruncle and drab gray head and neck. **Juv.** is white obscurely spotted with rusty gray on head, neck, and rest of upperparts. **Pull.** grayish white.
Habits: Social, but aggressive and territorial in breeding season. Vegetarian, feeding mainly on stonewarts (*Chara*), pondweeds (*Potamogeton*), and other waterweeds, up-ending in shallow parts of lakes.
Voice: Usually mute, but in breeding season often gives repeated, hollow, metallic, whinnying *whyh-her-her-her*, and (male) a musical *hooee-hoo-hoo*.
Breeding: First breeds when several years old. Usually with large partly floating plant-nest in reed-beds. Eggs from July, but furthest s mainly in Sep-Nov. Both ad.s attend the young, and sometimes carry them on their backs.
Habitat: Shallow coastal areas, lakes, lagoons, and marshes, preferring waters with rich submergent vegetation.
Range: Southern S Am. n to Santiago, Chile and Urug., mainly breeding in La Plata area and along s Andes, incl. in low valleys within the chain. Non-breeders assemble to molt in large flocks on some large, clear lakes on Andean foothills of w Sta Cruz and Chubut. Common.

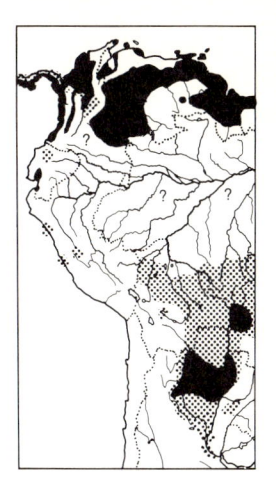

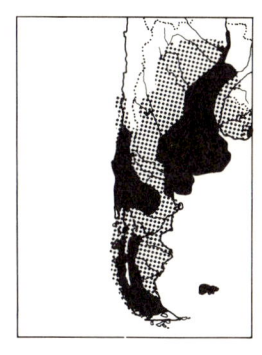

Sheld-geese of Genus *Chloephaga*

Quite large, long-necked waterfowl mainly with terrestrial habits. All species have very small conical bills, and graze on shore meadows and river plains. Probably first breed in 3rd year. Both sexes attend the young.

ANDEAN GOOSE *Chloephaga melanoptera* – Plate V 4a-c

70-90 cm; male largest, appearing very compact and thick-necked. **Both sexes mainly white**, with rosy bill and feet. Mantle and anterior scapulars with elongate sooty brown spots, **posterior scapulars and inner secondary-coverts purplish black** with some green sheen, tertials deep iridescent green. Primaries and tail black. **Juv.** less purely white, with mostly brownish gray scapulars. **Pull.** white, with black ear-spot, blackish dorsal band from forehead to tail, and another along side of back, blackish fading to fawn brown.
Habits: In loose flocks or pairs, but molting birds assemble in dense flocks. Usually on boggy shore meadows, mainly feeding on fleshy semi-aquatic plants (*Chara, Lilaeopsis, Myriophyllum, Nostoc*). In courtship, female walks in high-stepped upright posture, quacking repeatedly, while drake shows diagonally stretched neck, and gives whistled calls, or struts about with protruding chest, and shows head-rolling movements or distinct wing-raising display. During fights, strikes with wrist. The shyness varies according to the amount of local persecution.
Voice: Weak. Male gives soft *huit-wit-wit...*, *crip*, *quirop*, and low, grunting *kwuwuwu...*, single-syllable threat whistle, and double-syllable sexual calls. Female gives somewhat grating *kwa-kwak*.
Breeding: Usually on grassy or rocky slope overlooking bogs and water. Eggs Nov-Jan (often also later) in the n, Nov in the s.
Habitat: Open terrain with areas of very short matted grass. On bogs in well-watered mountain valleys, 'bofedales', and on river plains and shore meadows of lakes and lagoons.
Range: Through puna zone from Ancash w Peru through w Bol. to Mendoza (occ. Río Negro), Arg. and Ñuble, Chile, mainly breeding at 4-5000 m, but in the s down to 3500 m, and descending to 2000 m in winter (casually to coast). Common.

MAGELLAN GOOSE *Chloephaga picta* – Plate V 3a-f

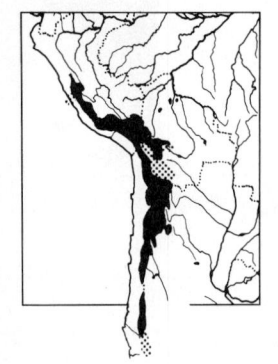

60-70 cm. **Sexually dimorphic at all ages. Male white, more or less extensively black-barred** on fore- and under-parts of body. Back gray posteriorly, with some fine black bars. Rump white, tail black. Bill and legs blackish. **Female has more sandy cinnamon head, neck**, and breast grading to gray-brown on posterior back, and chestnut on belly. Rump black, vent dusky cinnamon. Legs yellow. **Both sexes have partly white wings**, with blackish primaries, and dark band formed by iridescent green greater secondary-coverts. **Juv.** duller than ad., right after fledging, white parts of the male's plumage look dingy due to retained down, and female shows more drab brown neck, and less chestnut belly. Recognized through 1st year by dull speculum and dark-tipped median secondary-coverts. **Pull.** has blackish mask and otherwise varies from olive-gray with white bands and white underparts to almost uniform drab brown.

Habits: Quite tame where not persecuted. Gregarious, sometimes in enormous flocks said to compete with sheep. Principally grazes on short grasses (*Poa annua* and *pratensis*). Quite aggressive in spring, when male struts around in high-stepped, upright postures, sometimes with partly spread wings.

Voice: Male gives rather sad soft *wi-wi-wi...*, *wuee wuee...*, or whispering *whuit*; female low rattling *a-rrr*, but in courtship males give breathy multiple-toned whistles, females rather loud, harsh *hrc*, grunting *gu-rohr*, *gu gu ... gu rohr*, or a shorter *choonk-choonk*.

Breeding: Among grass or shrubs. Eggs from mid Oct, with some clutches as late as Dec.

Habitat: Grassland, esp. in well-watered valleys and near lakes, but also far from open water, and high up on bunchgrass hills and mountain plateaus, wherever there is some boggy terrain or low green vegetation associated with melting snow or wells.

Range: Despite strong persecution, occurs anywhere near the s end of the continent, breeding n to Neuquen, Arg., wintering in Magellanic zone, or migrating to ec Arg. or Colchagua, Chile (*picta*). Islas Malvinas (*leucoptera*). Abundant.

NOTE: The barring of the plumage varies. Males with barring only on flanks seem to be typical of unforested parts (but has invaded forested parts after recent clearing of forest), and migrates. Extensively barred males should be typical of the forested Andean zone and Isla Grande, and winters in the s. Proposed sspp (or species) *picta* and *dispar* are not currently recognized, owing to claimed intergradation. Our observations indicate that at least some areas have pure populations of one type or the other. Sometimes called Upland Goose.

ASHY-HEADED GOOSE *Chloephaga poliocephala* – Plate V 2a-d

52 cm. Essentially **tricolored gray/brown/white in both sexes**. **Head and upper neck gray**; mantle gray-brown; rump and tail black. Breast chestnut (male) or reddish brown finely barred (female), contrasting with posterior **underparts, which are white except for vertical black barring on sides**. Vent light chestnut. Legs and feet orange and black. **Wings partly white**, with blackish primaries and **dark band** formed by iridescent green greater secondary-coverts. **Juv.** duller, with cinnamon distinctly barred breast. Whole 1st year shows fuscous middle secondary-coverts. **Pull.** like light Magellan Goose pull., but with whiter face.

Habits: As in other sheldgeese. Usually in small flocks or pairs.

Voice: Wheezy wigeon-like whistles *wiwi-wi...* noted from male.

Breeding: In tall grass or hollow tree-trunks. Eggs late Oct-Nov.

Habitat: Sometimes in open land with Magellan Geese, but most typical of lake shores and swampy tracts in open bushy or wooded country.

Range: S Chile and adjacent Arg., wintering to Arg. pampas. Furthest n in the breeding range (Llanquihue and Bío Bío, Chile) only at 500-1300 m or even higher, in the mountains. Common.

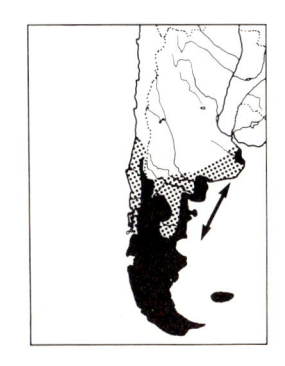

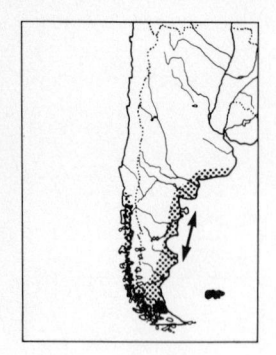

RUDDY-HEADED GOOSE *Chloephaga rubidiceps* – Plate V 1a-b – a vanishing species of lush grassland on northern Isla Grande and Islas Malvinas – visits valleys in the s Andes on migration. Resembles a tiny female Magellan Goose, but has white eye-ring, and head and neck are bright chestnut to faded cinnamon, contrasting with gray lower neck. Body densely barred with black, pale gray, and cinnamon, brown belly grading to **chestnut vent** (conspicuous when swimming).

ORINOCO GOOSE *Neochen jubata* of trop. lowlands e of Andes is rec. once on Ubaté savanna (2600 m) E Andes, Col. Large, with head, neck, and chest creamy, otherwise chestnut. Black wings with white patch on secondaries.

FLYING STEAMER-DUCK *Tachyeres patachonicus* – Plates V 5a-e and VIII 5

68 cm. A **heavy grayish duck with white wing-patch** (secondaries), and white belly. Pointed tail often raised to reveal white vent. Strong bill orange-yellow with some gray, esp. in female, or totally dark gray (juv.). **Ad.** with whitish 'spectacles' and somewhat scaled pattern. Considerable variation in color, with gray, gray-brown, or even dark chocolate color phases discernible, at least in some areas: Male usually plumbeous gray with vinous hue, esp. on breast, and with distinctly scaled pattern with pale feather-centers on back and flanks. Males of coastal populations of Arg. show white-headed plumage in Oct-Dec, and a very dark (juv.-like) plumage in Feb., but this may not apply to all upland populations. **Female** and **imm.** more gray, less scaled. **Juv.** rather uniform gray-brown. **Pull.** drab gray with long, broad, but discontinuous whitish supercilium.
Habits: Flight heavy, but rapid; at least in some coastal populations many birds are temporarily flightless (Apr) because of weight gained. When alarmed, dives rather than flies, and only feeds diving (taking molluscs and crustaceans). Territorial, and very pugnacious in breeding season, but imm. are gregarious. When patrolling territory, often swims in flat posture with head along surface, often raising tail high. Also 'steams' rapidly over lake surface, using wings simultaneously as oars (unlike in take-off run). During courtship regularly attacks and kills smaller waterfowl.
Voice: Male vocal in breeding season, esp. when patrolling early morning and evening: long series *titidididi...*, *huirr huirr huirr...*, *br-br-br...* and engine-like *toc-toc-toc...* alternating with loud whistled *bzheeeo*. Also low cackling *mrrr* and *kek kek*. Calls *bzee*.
Breeding: Usually nests in some vegetation cover, often far from water. Eggs mainly Oct-Dec.
Habitat: Seacoasts and open lakes of any kind. On uplands breeds (usually with one pair on each small lake) mainly on clear lakes with or without water-weeds, but non-breeders assemble on larger often turbid lakes sometimes of alkaline types.
Range: S Arg. and Chile, both along coasts and inland n to Ñuble, Chile. Common at 700-1200 m on barren upland plateaus of inland Sta Cruz and on the Somuncurá plateau of Río Negro, Arg., and in Ñuble ascends to 1800 m.

Perching Ducks – Tribe Cairinini

MUSCOVY DUCK *Cairina moschata* of swamps, rivers, and lagoons of trop. C and S Am. mainly e of Andes, accidentally visits temp. zone of E Andes, Col. Male 84, female 65 cm. Male with characteristic profile caused by knob at base of bill, humped crown, and hanging nape-tuft. Bill dark gray with 2 pink bars. **Black** with bronzy and purple reflections. **Large white forewing patch** (except in **juv.**). Voice weak hissing (male), or simple quack (female).

COMB DUCK *Sarkidiornis melanotos* – Plate VIII 8a-c

Male 75 cm, female 55 cm. Goose-like, male with large black **frontal comb**. **Head white** (yellowish when breeding) **with dense black dots**. Upperparts of body steel-blue, shining blue-green to purplish on wings. Below white with flanks light gray (female) to mostly black (male). Tail and **underside of wings black**. **Imm.** duller, barred on anterior mantle. **Juv.** brownish, obscurely marked; above dark with thin white supercilium, black eye-line, gray-brown cheeks and flanks grade to buff c underparts. Shy, usually in small groups, sometimes in large flocks, or associated with whistling ducks. Much of the year the sexes appear separate, ducks avoiding company of drakes. Gives faint, whistled *churr* (male) or weak quacks or grunts (female).

In open grassy savanna woodlands with inundated areas or lagoons in the tropics of se Asia, Africa, and S Am. Mainly on Llanos of Ven. and Pantanal of sw Brazil to c Arg., but owing to erratic movements can be seen in most warm parts of S Am., esp. in rice-fields. Rec. in mid- and upper Cauca valley (at least formerly), s Nariño, in wetlands at 2600 m in E Andes, Col., 2 highland rec. from Pichincha, Ecu., and repeatedly at 2550 m near Cochabamba town, Bol. Accidental in highlands of Ecu., and at 4080 m in Junín, c Peru (*sylvicola*).

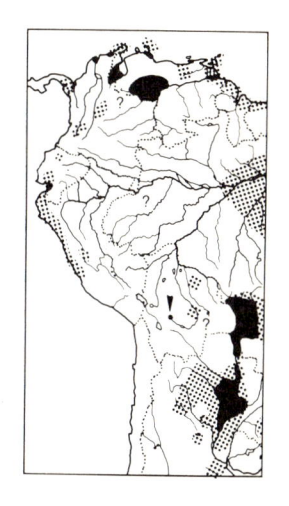

TORRENT DUCK *Merganetta armata* – Plates VII 1a-h and VIII 11a-b

40 cm. A slender duck with red bill and rather long stiff tail. **Male with head and neck white with black lines** along crown and hind-neck, and from eye down side of neck (and vertical bar below eye in *armata*). Body variably streaked white, gray, brown, and black. Ssp *colombiana* is light, gray and blackish, back-feathers white-edged, underparts white with gray streaks. *Leucogenys* complex shows great individual and local variation in mantle edging and underpart melanism: *leucogenys* with mantle mostly black, underparts buffy white with black streaks esp. on breast; *turneri* with some individuals almost black below, others streaked rusty; *garleppi* similar, but with whitish feather-edges above; *berlepschi* black-bodied with broad white streaks above, and some rufous on flanks, but certain individuals with similar body colors as ssp *armata*. The latter is streaked black-and-white above, and is black below, occ. with some ochraceous streaks on belly. **Female above plumbeous gray**, finely mottled to side of neck (except in *colombiana*), and with

black-and-white streaks on back, **below from chin to vent cinnamon-rufous** (extending to neck-side in *colombiana*). **Juv. is gray**, streaked above, barred on flanks, wings, and tail; head white with deep gray cap and hind-neck. In **flight** plumbeous wings show bronzy speculum with thin white demarcations. **Pull.** boldly striped and spotted black and white.

Habits: In pairs or family groups. Strong pair bond and cooperative defence of a considerable stretch of river. Although the sexes share a similar threat display repetoire, territorial birds primarily confront intruders of the same sex. Usually seen in rather upright stance on rocks in streams. Frequently bobs head up and down, or back and forth, and hops from rock to rock. Shows incredible ability to negotiate the tumultuous rapids, but mostly keeps close to rocks, often in eddies, and in shade of overhanging rocks, or vegetation of edge. Feeds by dabbling near the water-edge, and diving in the ripples. Swims deep in water, with neck inclined forwards, sometimes with body submerged. Rarely flies, with effort, and low over the river, and usually dives downstream when in danger. Tail used for maneuvering, both under water, and when swimming through rapids, and is decurved for support while climbing on rocks. Tail-cocking display is shown on rocks, never on water.

Voice: Rough guttural *dzhurr*. Warning note a sharp *wheek wheek*.

Breeding: Nests in holes or dense vegetation. Eggs Nov (Col.), Apr (Ecu.) or June-Sep (c Peru), Nov (Bol.), and Sep-Nov (in the s). Drake participates in brood-raising.

Habitat: Clear streams with projecting rocks and rapids interspersed with stretches of placid water. In open or forested areas, and often in canyons. Elevational distribution 300-4600 m, lowest in the s; usually ascending to where streams become too small to provide adequate foraging, and descending to where rivers become slow and turbid without rocks.

Range: Along the entire Andean chain, with distribution gaps separating the main sspp. From Mérida, Ven. and spottily in all Andean ranges of Col. and through Andes of Ecu. (*colombiana*). *Leucogenys* complex: Molinopampa, Amazonas Peru s to río Cañete in Lima, Peru (*leucogenys*), Junín to Cuzco (transitory), and sw through Arequipa, Peru to Arica, n Chile, and to Limbani in Puno, e Peru (*turneri*); La Paz to Tarija, Bol. (*garleppi*), and Tarija to Catamarca and La Rioja, nw Arg. (*berlepschi*). Andes s-wards from Mendoza Arg./Atacama Chile to Tierra del Fuego, and isolated in Nahuelbuta massif, wc Chile (*armata*). Common only locally, may have declined due to food competition from introduced trout, and has disappeared from many rivers, owing to siltation following deforestation.

NOTE: The main sspp *colombiana*, *leucogenys* and *armata* were previously treated as full (allo)species.

Dabbling Ducks – Tribe Anatini

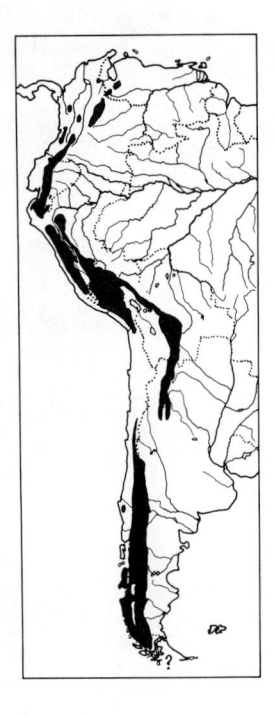

Dabbling ducks are surface-feeding or feed by up-ending in shallow water. They stain invertebrates, seeds, bulbs etc., but a few species also graze. On the water, they usually **slightly lift the rear**. Unlike most diving ducks they **alight steeply**, without running start. Most species (esp. those of N Hemisphere) are dimorphic: the drakes have brightly colored nuptial dresses most of year, and a female-like **eclipse plumage**. In most S Hemisphere species the sexes are alike; unlike in dimorphic species, drake usually accompanies duck (with a guarding function) throughout breeding period.

AMERICAN WIGEON *Anas americana* – Plate IX 11a-b

48 cm. **Male** with pale **gray head with white forehead**, and metallic green patch behind eye. Otherwise mainly pinkish brown with white band demarcating black tail region. **Female** and **juv.** gray-brown rather faintly mottled, and with gray head. In **flight** shows narrow wings with partly gray wing-linings, white belly, and pointed tail. The **forewing is white in the drake**, but mottled gray in **female** and esp. in **juv.** Voice of drake whistled.

N Am. migrant. Previously frequented wetlands of Bogotá and Ubaté savannas at 2600 m in E Andes, Col. (Oct-Apr), and may still visit Cauca Valley.

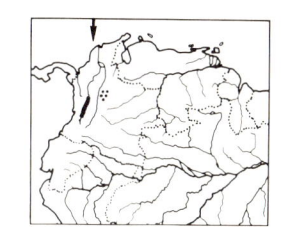

CHILOE WIGEON *Anas sibilatrix* – Plates VI 10a-c and IX 10

48 cm. Rather short-necked, and with triangular head and a short, tapering bill. Recognized by **sooty general coloration, and white face, forewing and surroundings of tail**. Head dark metallic green and magenta behind eye, and white on face and near ear. On dark upperparts, white marks form obscure bars on mantle and stripes further back. Underparts white, breast with dense dusky bars, and flanks more or less rusty. **Male** most glossy, **female** duller, with slight mottling on forewing. **Juv.** lacks metallic colors, and shows slightly barred flanks and mainly gray forewing. In **flight**, ad. recognized by large white forewing, black speculum demarcated towards body by white inner secondaries, pale wing-linings, white belly, and sharp tail. **Pull.** fuscous and rufous brown, lacking clearcut pattern.
Habits: Shy. Except when breeding, often in massive flocks. Usually swims with head low, bill near water. Sometimes grazes on shore. During infectious displays, drakes call and show distinct chin-lifting and raising of the rear.
Voice: Highly vocal, also at night. Male gives rapid chattering whistles and sometimes loud rolling *wee whee-err* (1st note lowest pitched). These calls create a wild chorous where many birds congregate. Short *cwi* notes are given in flight. Female gives short nasal quacks.
Breeding: Egg-laying in the s mainly in Oct-Dec.
Habitat: Lakes, lagoons, marshes, and broad rivers, preferring shore meadows with short grass, and wide lake shallows with dense submergents (esp. *Potamogeton, Ruppia,* and *Zanichellia*), or water with floating carpets of *Myriophyllum*.
Range: Widespread in southern S Am. n to Córdoba, Arg., wintering to Parag. and Atacama, n Chile, and accidental to Junín, c Peru. Mainly lowlands, but important staging and molting areas exist on the foothill plateaus e of the s Andes. Widespread and abundant.

GREEN-WINGED TEAL *Anas crecca* – **Plate IX 7** - a N Am. migrant – is rec. accidentally on Ubaté savanna of E Andes, Col. (*carolinensis*). 38 cm. **Male** with head dark chestnut with broad green band behind eye; body finely mottled gray with horizontal black stripe formed by the humerals; breast yellow stippled with black, and with vertical white demarcation towards gray side. On each side of vent, a large pale yellow patch framed with black. In **flight**, broad buffy wing-bar in front of and white line behind green speculum. **Female** resembles female Cinnamon T., but has smaller bill and dull gray forewing. Voice of male a clear *kry-yk*, of female hard *trr*.

SPECKLED TEAL *Anas flavirostris* – Plates VI 5a-d and IX 6

C. 40 cm. Rather compact and short-necked and somewhat square-headed, and with grayish general coloration. **In most of range with yellow bill** with black ridge (sspp *flavirostris* and *oxyptera*); **in the n, with gray bill** (*andium* and *altipetens*). Head and neck looks dark, owing to dense dusky stipling. Upperparts of body mainly black with tawny to grayish buff feather-margins, underparts and **rear of body light** gray, more or less dotted with fuscous on breast and anterior sides. In **flight**, shows gray-brown wings with short whitish band below; speculum black with metallic-green inner part, bordered tawny in front, and with white or buff rear edge. Ssp *andium* **is very dark**, *altipetens* less extreme, both showing narrow light feather-margins above, and coarsely spotted breast; *oxyptera* **is very pale-bodied** (consequently appearing dark-headed), with small black spots on light brown back, and with immaculate flanks. Ssp *flavirostris* may resemble Yellow-billed Pintail when resting with head tucked in, so that the shape is difficult to judge; however, it is safely distinguished by dark zone through eye. **Juv.** has rather obscurely mottled upperparts, with the black spots vaguely outlined, and obscurely dabbled rather than spotted breast. **Pull.** dark brown and honey-yellow, with more obscure light supercilium and broader band on cheek than Yellow-billed Pintail. Characteristic bill-colors develop after 2 weeks.
Habits: In pairs or in small groups, often with Puna Teal or Yellow-billed Pintail, but more than these it feeds in muddy creeks and along banks, often walking at the water-edge. Displays inconspicuous. Flight swift and erratic.
Voice: Male gives short high-pitched *cricket* or *cryc*; female a low *quec*, *rack*, or *cah*.
Breeding: Often in some cover, e.g., in rocky outcrops, holes in banks, or under roofs of houses, in dense vegetation, locally mainly in trees (e.g., in nests of Monk Parakeets). Several birds may nest close together. Eggs in Oct-Nov (furthest s), in Nov-Mar, with extra clutches June-July (Peru), Nov (Ecu.), Oct-Feb (Col.), or Aug and Dec (Ven.).
Habitat: All kinds of wetlands: muddy estuaries, shore meadows, and bogs with ponds and creeks, tarns, small acid lakes in elfin forest, rivers (incl. almost dry, gravelly or muddy beds). Sometimes along steep rocky lake-shores.
Range: All of southern S Am. to c Chile and Córdoba, Arg., incl. Patagonian uplands, and lakes of s Andes, migrating n to Parag. (*flaviro-*

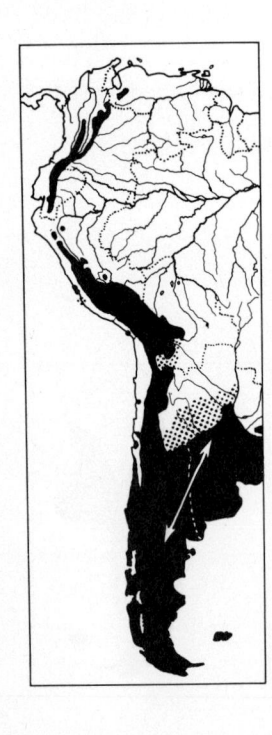

stris). Throughout highlands from Catamarca, Arg. to Cajamarca, n Peru at 2500-4500 m (locally higher), descending to Pacific lowlands along Lluta and Huasco rivers, n Chile, and in s Peru, in July-Oct (*oxyptera*). Virtually all páramos of Ecu. into Col. in s and c part of E Andes (*andium*), and continuing at 3200-4300 m (seasonally down to 2600 m) from Bogotá to sw Trujillo, nw Ven. (*altipetens*). Common, but abundant only in few areas.

NOTE: The 2 pairs of sspp have sometimes been listed as separate (allo)species: **YELLOW-BILLED TEAL** *Anas flavirostris* (incl. *oxyptera*), and **ANDEAN TEAL** *A. andium* (incl. *altipetens*).

CRESTED DUCK *Anas specularioides* – Plates VI 1a-d and VIII 4

60 cm, rather long-bodied, with long, pointed tail often raised. At a distance looks **gray-brown with dark tail**. Head and neck dull grayish buff, dark on crown and blackish around eyes, and with hanging occipital crest (best developed in male, which therefore looks heavy-headed). Eyes red (*specularioides*) or yellow (*alticola*). Body light gray-brown, somewhat rusty on breast, humerals pale scaled with fuscous feather-edges (esp. in *specularioides*), flanks with pale feather-tips. Wings dark earthy brown with glossy speculum (purplish/green against light in *alticola*, bronzy with some rosy in *specularioides*) with black and broad white posterior border; **white secondary-tips conspicuous in flight**. **Juv.** lacks crest, and is less black around eye. **Pull.** at first blackish marked whitish (unlike in most other dabbling ducks the pale supercilium is vestigial), but soon becomes fawn-colored. 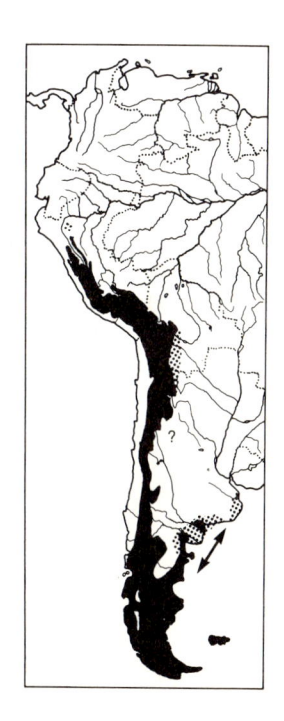
Habits: Usually in territorial pairs or family groups, but imm. and molting birds assemble in large flocks on some lakes. Flight with measured wing-beats and somewhat hanging neck.
Voice: Male gives short, harsh croaking *whrt*, *whorr*, or buzzy *wheeeoo*, often repeated; female a low, barked *whurrw*.
Breeding: On ground, sometimes far from water. Egg-laying in the s mainly Oct-Dec, in Andes mainly Jan-Mar, with occ. clutches in 'winter' (double-brooded?).
Habitat: Wide spectrum, from shallow coastal bays and open marshes to high-altitude lakes and lagoons, tarns, and bogs with small ponds, but prefers larger lakes with barren shores. Sometimes in hundreds on turbid, alkaline lakes with large concentrations of zooplankton.
Range: Southern S Am. Furthest s throughout from coast to high mts, but most common in shallow marine bays and on the volcanic foothills of Sta Cruz, s Arg. Some birds migrate to La Plata region of n Arg. In Talca, Chile, southern *specularioides* grades into *alticola*, which continues throughout puna and desert zones of n Arg. and Chile (down to 2000 m in winter), w Bol. and Peru n to Ancash and Amazonas, mainly breeding well above 4000 m. Common.

NOTE: Previously in a separate genus *Lophonetta*.

SPECTACLED DUCK *Anas specularis* –
Plates VI 2a-b and VIII 3

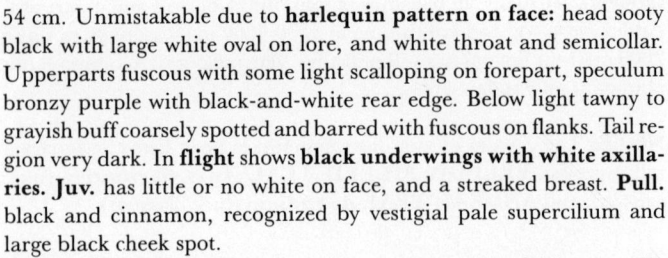

54 cm. Unmistakable due to **harlequin pattern on face:** head sooty black with large white oval on lore, and white throat and semicollar. Upperparts fuscous with some light scalloping on forepart, speculum bronzy purple with black-and-white rear edge. Below light tawny to grayish buff coarsely spotted and barred with fuscous on flanks. Tail region very dark. In **flight** shows **black underwings with white axillaries. Juv.** has little or no white on face, and a streaked breast. **Pull.** black and cinnamon, recognized by vestigial pale supercilium and large black cheek spot.
Habits: Usually in territorial pairs or family groups. Flying, usually follows course of river.
Voice: Drake said to have a trilled whistle, duck dog-like barks.
Breeding: Mainly on isles in rivers; eggs in Oct-Nov.
Habitat: Mostly in lakes and fast-flowing streams, sometimes lagoons, chiefly in forested regions.
Range: From s Chile n to Cautín and in adjacent Arg., ranging n to Neuquén and accidentally to Mendoza outside breeding season. Mostly in valleys in Andean zone, to 1500 m. Some wintering on ice-free streams on Isla Grande. Nowhere numerous.

NOTE: Sometimes called Bronze-winged Duck.

NORTHERN PINTAIL *Anas acuta* –
Plates VI 3 and IX 2a-b

60 cm. Very **slim**, with still longer and slenderer neck and smaller head than Yellow-billed P. Bill gray. **Male** with head and upper neck dark brown with white of lower neck extending as white line up side of nape. Body mainly light gray with black band along scapulars. Black base of needle-thin tail. **Female** and **juv.** much as slim Yellow-billed P., except for **gray color of bill**, and more densely vermiculated pattern of the individual body feathers. Also similar in flight, but rear demarcation of speculum whiter. In courtship flight the drake gives a distinctive pizicatto *dreep-eep* (often heard in southern winter quarters). Duck quacks.

N Am. migrant Oct-Apr to C Am. and Caribbean coast and Cauca Valley, Col. Up to l950s abundant on marshes of Bogotá and Ubaté savannas in E Andes, Col. (*acuta*).

YELLOW-BILLED PINTAIL *Anas georgica* –
Plates VI 4a-c, IX 1a-b and X 1

49-57 cm. **Slim-looking**, with slender bill, round head, thin neck, and sharply pointed tail. **Yellow bill** with black ridge. Light buffy brown all over, except for whitish throat. Head with fine dark stipples, body spotted with fuscous c parts of each feather, leaving only narrow buff feather-margins above. At a distance, **looks rather uniform milk-coffee-colored**, with quite uniform, light head. Ssp *niceforoi* is small, much

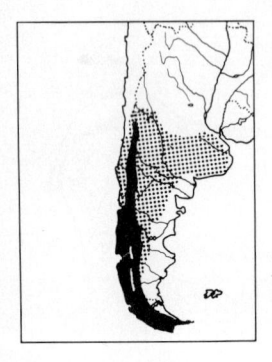

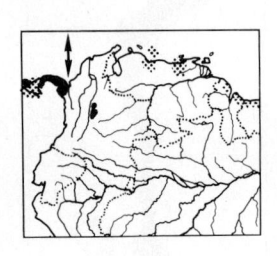

darker, with heavier fuscous markings, top of head black, tail less pointed. **Juv.** not distinctly spotted below, but instead obscurely smudged with light drab brown. In **flight** readily distinguished by slim jizz and sharp tail, dusky brown wing-linings, and gray forewings. Speculum blue-black in male, dull fuscous in female and juv., broadly framed with buff in front and behind (thin demarcations in juv.). **Pull.** fuscous with buff pattern; note that scapular spot is drawn out as a streak.
Habits: Gregarious. Sometimes in large flocks. When active, thin neck readily distinguishes it from Speckled Teal. Walks and runs well. Often feeds far offshore, upending, and may dive for food if vegetation is more than 40 cm below surface. In nuptial flight, drake looks hump-backed due to declining neck and tail.
Voice: Drake gives short *tric* and double-toned whistle in nuptial flight. Duck has short, grunting *rrah*.
Breeding: Mostly in tall grassy or rushy vegetation. Eggs in Oct-Dec far s, Aug-Mar and in Peru and Col., but all year in coastal Peru.
Habitat: All kinds of wetlands, but mostly lakes with wide shallows with submergents or floating waterweeds. Mainly breeds in wide reed-marshes or partially inundated rushy meadows.
Range: Southern S Am. from coast to Andes, some southern populations migrating (short period) to Parag. and e Bol. Through puna zone of Arg., Chile, Bol., and Peru, mainly at 3500-4600 m, but descends to coast of Peru. More patchily from n Peru through Andes of Ecu. and Nariño to w Putumayo, s Col. Generally widespread and abundant, except n of Junín, c Peru (*spinicauda*). Previously on Bogotá and Ubaté savannas and lake Tota in E Andes, Col. (and 1 rec. in Cauca valley); common to 1950s, but now probably extinct (*niceforoi*).

NOTE: Ssp *niceforoi* has by some authors been listed as a full (allo)species **NICEFORO'S PINTAIL**.

WHITE-CHEEKED PINTAIL *Anas bahamensis* – Plates VI 6 and IX 3

44 cm. Shaped much as Yellow-billed Pintail. Bill blue-gray with black ridge and red spot at base. **Brown-capped** with **white cheeks and throat**. **Body tawny** with dense black dots, pale fawn-colored towards pointed tail. Bronzy speculum with **broad** cinnamon demarcations in front and behind. Voice as in Yellow-billed P.
   In tide-water areas and shallow, weedy lagoons. Spottily distributed mainly in W Indies, along coasts of trop. and subtrop. S Am., and inland from e Bol. to c Arg. Occ. crosses the southern Andes to Chile. In recent years in several places in the Magellanic Region, where seen with juv. in Feb (Clark 1986). Accidental at 4080 m in Junín, c Peru, and on Bol. altiplano, and regular at 2550 m near Cochabamba, Bol. (*rubrirostris*).

NOTE: Sometimes called Bahama Pintail.

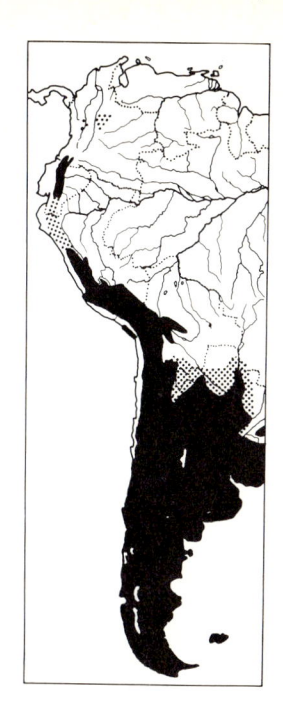

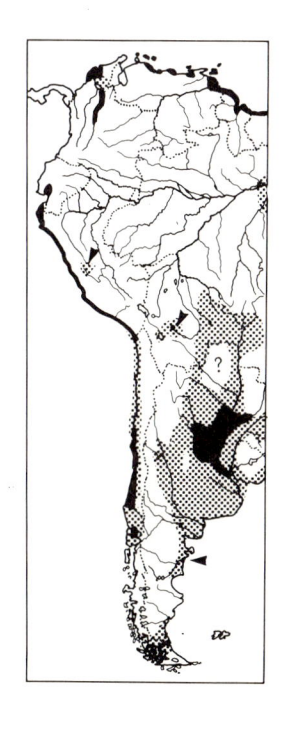

PUNA TEAL *Anas puna* – Plates VI 9a-d and IX 5

47 cm. Resembles Silvery T., but grayer, and with long and straight blue bill. **Black-capped** (to eyes) with **white cheeks and throat**. Neck and breast light buff with blackish dots. Back dusky brown with buff feather-margins, humerals uniform gray-brown, sides densely barred black-and-white (**male**), more coarsely barred black-and-buff (**female**), or obscurely mottled (**juv.**). **Rear of body pale gray, faintly mottled.** Juv. shows distinctive buff tinge on cheek. In **flight** shows dull metallic green speculum with white stripes in front and behind, and gray-banded under-wings. **Pull.** patterned sepia and whitish, recognized by pale face with short eye-line and small ear-spot.
Habits: Gregarious, in small and sometimes large, dispersed flocks. Displays inconspicuous. Normally feeds offshore, on floating vegetation, esp. *Chara*.
Voice: Low chatting *hueer, pt pt pt ...*, mecanic *trrrrrrrr* (rising pitch) and *dr-r-r....* Alarmed low *whr* and *errr*.
Breeding: Often in small colonies on isls. Eggs rec. Sep-Mar and July, probably peaking in July (c Peru).
Habitat: Prefers open, weakly alkaline lakes with much submergent or floating vegetation, esp. of *Chara*. Requires isls or reed-marsh for nesting.
Range: Puna zone mainly at 3500-4600 m from Jujuy, nw Arg. through w Bol. and Arica in extreme n Chile to Ancash, nc Peru. Casual to Pacific coast. Somewhat local, but generally common.

NOTE: Sometimes listed as a ssp of Silver T., but the different bill suggests a significant ecological difference.

SILVER TEAL *Anas versicolor* – Plates VI 8 and IX 4

43 cm. Small **black-capped duck with creamy buff cheeks**. Bill blue with black ridge and yellow spot at base (at least in male). Neck and breast light buff to fulvous with black spots. Upperparts of body dusky with buff feather-margins giving spotted appearance; underparts white with heavy black barring on flanks, whole rear finely barred (light gray at a distance). In **flight** shows bluish bronzy speculum with thin white demarcations, and gray-banded underwings. Female duller than male, browner, less distincly barred. Often in large flocks.

Inhabits marshes and grassy ponds on lowlands of southern S Am. Lowlands of Patagonian steppe and s Chile n to Aysén and low valleys of the s Andes (and straggling to 1200 m on upland plateaus (*fretensis*), and in lowlands further n, to Valparaiso, c Chile and se Bol. (*versicolor*). Very numerous in pampas marshes.

BLUE-WINGED TEAL *Anas discors* – Plates VII 3a-b and IX 8a-b

38 cm. **Male** has **head dull blue-gray with large white crescent in front of eye**. Entire body buffy brown finely dotted and striped with black. Behind thigh, broad white band separates black tail region.

Male in **eclipse** (Jul-Oct), **duck** and **juv.** grayish ocher mottled and spotted fuscous, except on whitish belly. Then resembles Cinnamon T. in corresponding plumages, but looks grayer with rather well-defined facial pattern of pale eye-brow, **whitish loral spot** and throat, and dark eye-line. Dark (not red) eye. Also separated by less heavy bill and by the profile, ridge of bill, forehead, and crown forming a smooth curve. In **flight**, with large chalky-blue forewing (light gray in juv.) with white line (faint and double in female) in front of dull green speculum; under-wings white and gray.
Habits: Gregarious. By day often asleep near vegetation cover, or among floating water-weeds, and may mainly feed by night (at least in areas with strong hunting). Flight swift and erratic, in tight flocks.
Voice: On wing, male gives *keck keck keck*, or a finer, whistled note, female a weak quack.
Habitat: Flooded fields, marshes, and lakes with reedy margins or dense carpets of floating waterferns or similar vegetation.
Range: N Am. migrant visitor Sep-May to coastal areas and to 3600 m in Andes of Ven., Col., Ecu., in Peru locally along the coast and in Lake Junín 4080 m, and accidental in c Arg. (*discors*). Common.

CINNAMON TEAL *Anas cyanoptera* –
Plates VII 2a-d, IX 9a-b, X 3

35-40 cm (*borreroi* and *orinomus* 45 cm). Differs from other teals by profile: longer, somewhat spatulate bill, and step-like forehead. **Male chestnut**, back-feathers black with rusty edges. Sspp *cyanoptera* and *orinomus* are uniform chestnut-rufous below, with few or no black spots, but birds in transitional plumages may be densely spotted black below. Ssp *septentrionalum* normally becomes darker below, and *tropicus* and *borreroi* are very dark below, nearly black on belly, and often have black spots on neck, breast, and flanks. Tail region looks black. Male in **eclipse**, **duck** and **juv.** buffy with definite cinnamon tinge, obscurely spotted and mottled with fuscous, but unlike Blue-winged T. in corresponding plumages the **facial pattern is less contrasting**, the pale loral spot sullied with dark spots. Bill gray, **eyes red** (from 8th week). In **flight** shows large pale blue forewing area, green speculum with white line in front, and white and gray wing-linings. **Pull.** fuscous with honey-yellow markings and underparts, distinct ear-spot (unlike *Netta* species).
Habits: Social, but rarely in large numbers.
Voice: Quiet. Male may give a low rattling chatter (like shoveler), female a weak *karr karr*.
Breeding: Mainly Oct (far s); Jan (puna zone); Feb, Mar, May, Oct (Col.).
Habitat: Lakes, ponds, and marshes, usually weakly alkaline with some reeds, and with abundant floating waterweeds.
Range: Disjunct. N Am. migrant to C Am., Col., Mérida, Ven. and Ecu. (*septentrionalum*). Trop. n Col. to subtrop. Cauca Valley (*tropicus*); wetlands of Bogotá savanna in E Andes, Col. (common in 1950s, now vanishing), Nariño, w Putumayo, s Col., and previously to n Ecu. (*borreroi*). Widespread in lowlands of southern S Am. n to Parag. and locally on coast of Peru (*cyanoptera*), and patchily at 2550-4300 m in puna

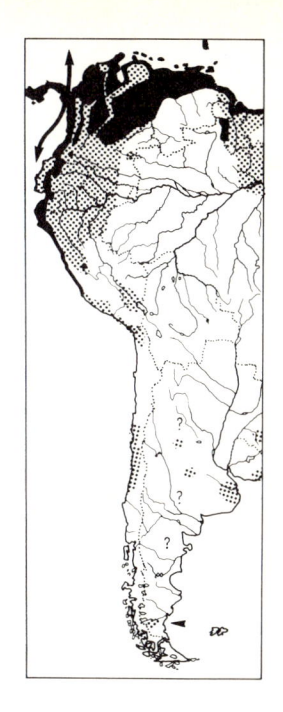

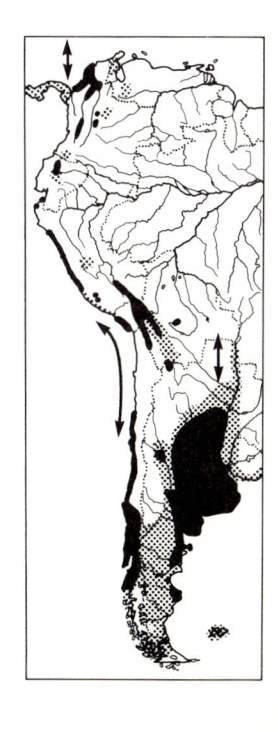

zone from Cajamarca to Jujuy, nw Arg., in Potosí and Cochabamba, Bol., and in the Titicaca region, Cuzco, and Ayacucho, Peru, and to 5000 m from Arica to Antofagasta, n Chile (*orinomus*). Common locally.

RED SHOVELER *Anas platalea* – Plates VI 7a-c and IX 12a-b

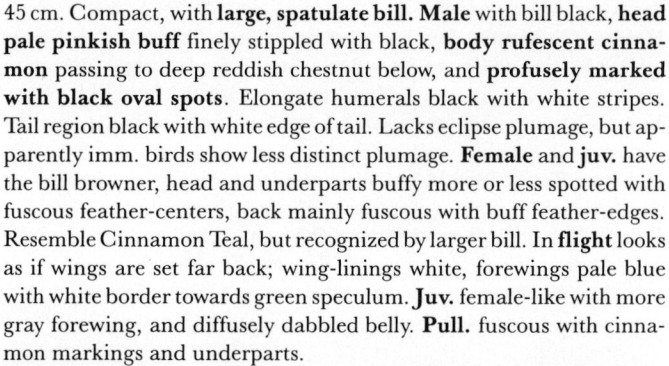

45 cm. Compact, with **large, spatulate bill. Male** with bill black, **head pale pinkish buff** finely stippled with black, **body rufescent cinnamon** passing to deep reddish chestnut below, and **profusely marked with black oval spots**. Elongate humerals black with white stripes. Tail region black with white edge of tail. Lacks eclipse plumage, but apparently imm. birds show less distinct plumage. **Female** and **juv.** have the bill browner, head and underparts buffy more or less spotted with fuscous feather-centers, back mainly fuscous with buff feather-edges. Resemble Cinnamon Teal, but recognized by larger bill. In **flight** looks as if wings are set far back; wing-linings white, forewings pale blue with white border towards green speculum. **Juv.** female-like with more gray forewing, and diffusely dabbled belly. **Pull.** fuscous with cinnamon markings and underparts.
Habits: Gregarious, sometimes hundreds together. Poor walker. Usually holds the head low, working weedy or shallow, muddy water by running bill along surface.
Voive: Quiet. Male may give low rattle, female rasping *whrrrt*. In flight deep *tuc tuc*.
Breeding: Rarely breeds on southern upland lakes, but one pull. rec. here in Dec. Pull. in Mar (Puno Peru).
Habitat: Lakes and marshes. Mainly feeds where water-weeds cover surface, but sometimes also assembles on turbid, alkaline lakes with dense zooplankton.
Range: Common in southern S Am. n to Aconcagua, Chile, and n Arg., migrating to lowlands of Bol. Sparse within Patagonian Andes, but enormous numbers stage and molt around 1000 m on upland plateaus of inland Sta Cruz. Possibly breeds in Jujuy (Lag. Nuntuyoc 3400 m), and small resident population exists in highlands of Puno and Cuzco, Peru.

Red Shoveler

NORTHERN SHOVELER *Anas clypeata* – Plate IX 13a-b

48 cm, with even longer bill than Red S. **Male** boldly patterned, with head black glossed green, eyes yellow; white-bodied with back streaked with black, belly and flanks chestnut, and tail black-based. Eclipse plumage never shown while in S Am., but a transitional plumage shown Sep-Oct somewhat resembles Blue-winged Teal male with indications of a pale facial crescent. Imm. males are similar. **Female** and **juv.** dressed as Red S. Habits and voice as in Red S.

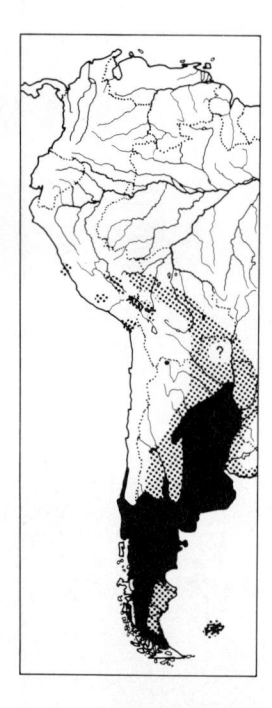

N Hemisphere. N Am. migrants reach northern S Am. in Sep-Mar: previously common (today rare but regular) on Bogotá and Ubaté savannas at 2600 m in E Andes Col.

Fresh-water Diving Ducks – Tribe Aythyini

Many members of this group dive for food and have the feet set somewhat back. They are thus awkward on land. On water, they usually show rounded back without lifted rear. However, the *Netta* species are intermediate towards dabbling ducks in this respect, and are mainly vegetarian, while *Aythya* takes more invertebrates, selecting individual prey visually. Patter feet on water during take-off. The species are generally silent, and do not show remarkable visual displays.

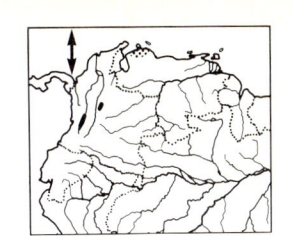

ROSY-BILLED POCHARD *Netta peposaca* –
Plates VII 5a-b and VIII 9

55 cm. Stocky, rather long-bodied. **Ad.** with somewhat swollen base of bill (distinct knob in male). **Male** with head, neck, and breast **purplish black**, back black with minute gray mottling; rest of **body appears drab-gray** owing to fine pale vermiculations. Vent white. **Bill and feet rosy. Female** generally dark brown (warmer than in Southern P.), with pale throat, foreneck and belly. Often white on lore and lower eyelid, and occ. around ear, but rarely approaching the distinct facial pattern of female Southern P. **Vent white** (usually as large oval spot). **Juv.** similar, but browner below. In **flight** shows white wing-linings and **a broad white bar on secondaries becoming more creamy on primaries**. Somewhat resembles a dabbling duck, owing to buoyant body and somewhat lifted tail. Grazes much on land, and feeds up-ending (hardly ever dives). Usually in small groups. Gives low growling and cracking sounds, and displaying male gives faint *whee-ow*.

Inhabits marshes, lakes, and ponds with dense waterweeds in lowlands of c Chile and c and e Arg., migrating to Bol. and coast of s Peru. After probably recent expansion now inhabits lakes in the s Andes, and is not uncommon in the Magellanic zone to southern Isla Grande. On spring migration (Nov) visits weedy ponds to 1000 m on upland plateaus of the Patagonian steppe.

NOTE: Sometimes called Rosybill, or Peposaca Duck.

SOUTHERN POCHARD *Netta erythrophthalma* –
Plates VII 4a-c and VIII 7a-b

50 cm. With slim jizz, and rather triangular head. **Male** with head, neck, and breast glossy **purplish black**, and tail region black, otherwise dark chestnut-brown. Bill light gray, eye red. **Female dusky brown**, somewhat lighter and more fulvous rusty below. **White areas at base of bill, on throat, and ear region**, interrupted by gray cheek. Some individuals have partly white vent. **Juv.** lighter brown with less white facial pattern. In **flight** shows **completely dark wing-linings**,

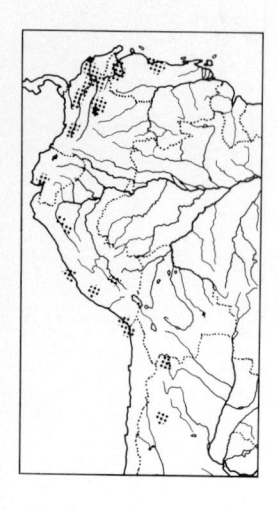

and a **short white wingbar** at base of secondaries. **Pull.** olive-brown above, yellow on sides of head and below.

Habits: Usually singly, sometimes social, but does not associate with other waterfowl. Feeds at the surface, by up-ending, and by diving.

Voice: Quiet. During courtship soft, mechanical *eerooow* and harsh *rrr-rrr*. In threat *quarrk*.

Breeding: Probably variable season, but no data from S Am.

Habitat: Marshes, shallow but permanent ponds, and lakes. Requires widespread submergent vegetation, and is therefore vulnerable to siltation caused by soil erosion.

Range: Africa and locally in S Am. Se Brazil. Up to l940s common in at least 2 lakes at 2600 m near Bogotá, E Andes, in Cauca valley, Col., and in c Ecu. May persist in trop. w Ecu. and was seen 1976 at 3650 m in Papallacta, ne Ecu., and 1977 and 1986 near Bogotá. Scattered rec. in Ven., in e and n Col., in San Martín, s Ayacucho and Cuzco, Peru, at coast of Peru (last seen 1962), and Arica, n Chile, in Jujuy, and recently in some highland lakes in Catamarca, nw Arg. (*erythrophthalma*). Apparently vanishing.

LESSER SCAUP *Aythya affinis* – Plate VIII 6a-b

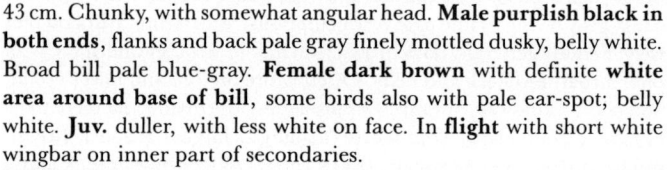

43 cm. Chunky, with somewhat angular head. **Male purplish black in both ends**, flanks and back pale gray finely mottled dusky, belly white. Broad bill pale blue-gray. **Female dark brown** with definite **white area around base of bill**, some birds also with pale ear-spot; belly white. **Juv.** duller, with less white on face. In **flight** with short white wingbar on inner part of secondaries.

Habits: Gregarious. Feeds diving. Shy.

Voice: Usually mute when in S Am. Male has short whistle, female grating *err-err*.

Habitat: In larger shallow marshes rather than the barren páramo lakes.

Range: N Am. migrant, which in Jan-Feb visits lakes in Surinam, n Ven. and to 3000 m in Col. (mainly E Andes) and Ecu.

Stiff-tailed Ducks – Tribe Oxyurini

Dumpy and short-necked ducks with large feet and distinctive stiff tails held parallel with the water, or cocked. The tail is used as a brake upon landing and for underwater maneuvering. 'Ruddies' are efficient divers, and may remain submerged up to 30 s, feeding on submergents (esp. *Chara*), and straining midge larvae etc. from the waterweeds or the mud. Not very social. Courtship often occurs in groups, with little active involvement from females, and there does not seem to be sustained pair-bonds. Young may be independent after few days. One S Am. lowland species is decided nest parasite, and also other species are prone to dump their eggs in other duck-nests.

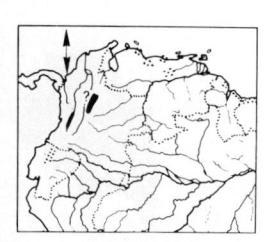

MASKED DUCK *Oxyura dominica* –
Plates VII 6a-b and VIII 12

33 cm. With small round head on thick neck. **Male ferruginous, with blue bill, black facial mask**, and black spots all over body. **Eclipse male**, **female**, and **juv.** more buffy brown, with pale head with 3 broad black parallel lines along crown, through eye, and along cheek; body profusely barred dusky brown and pale buff. Jizz, **light supercilium,** and bolder barring separates them from female Lake D. In **flight** shows dark wings with white axillaries and white patch on upperside formed by outer secondaries and adjoining coverts. Also fuscous **pull.** shows striped face pattern. Usually singly or in pairs. Low in the water, and never raises tail. Secretive. By day, rests quietly among floating plants. When disturbed dives quietly and hides with only head visible (blue bill of drake helps to detect it). Unlike other stifftails, launches vertically into flight without running start. Usually silent, but may give hen-like clucking, short *kuri-kirro* or *kirri-birroo*, and mechanic squawk.

   Inhabits marshes, swamps, and river backwaters, esp. places with extensive areas of floating-leaf vegetation (*Eychhornia*, *Limnobium*, and *Bidens*). In most trop. and subtrop. parts of the Americas, but seems to be fairly common only in n Arg. Regular and probably nesting at 2600 m near Bogotá, Col., and rec. at 2870 m near Quito.

NOTE: Sometimes in a separate genus *Erismatura*, or *Nomonyx*.

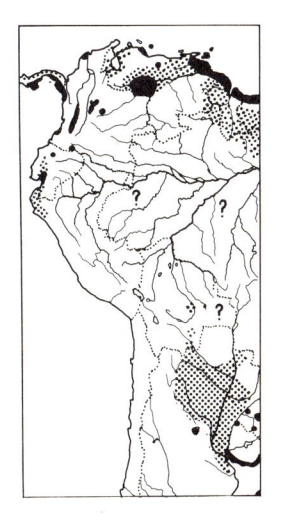

RUDDY DUCK *Oxyura jamaicensis* –
Plates VII 7a-d, VIII 13a-b and X 5a-b

45 cm. Stocky, with thick flabby neck, and large shovel-shaped bill. **Male with cobalt-blue bill**. Profile angular, owing to small erectile 'horns' over eyes. **Head and upper hind-neck black** with tiny white spot on chin, and casual white feathers on cheeks (*ferruginea*), or with white feathers regularly on lores and ears, or cheeks mostly white with some black spots and lines mainly on the middle (dimorphic ssp *andina*). Neck and **body reddish chestnut** with blackish rump and tail, belly and vent with some silvery white mottling (except if stained by rust). **Eclipse** plumage more dull, the black being replaced with dark gray-brown, the rufous parts with dusky mottling. **Female** with gray bill, generally dark brown with slight mottling, and faint dusky dots on face, and a tendency to lighter zone below eye and on throat (*ferruginea*, somewhat stronger pattern in *andina*, but far from as distinct as in N Am. birds). **Juv.** more distinctly mottled, and face below eye level lighter, with well-marked dark horizontal band across cheek from gape to ear. In **flight**, dark wings show short whitish stripe below. **Pull.** almost black, with faint light band below eye, and white shoulder-spot.
Habits: Normally in loose flocks away from other waterbirds. By day, usually seen sleeping well offshore (often in lake center), with bill buried in scapulars. Normally escapes in swift dive, and also feeds diving. Partly vegetarian (preferring *Chara*?). When undisturbed, holds tail at an angle above water. In display groups, drakes cock tails right up or even forwards, occ. inflates neck strongly, and hurriedly beats bill against breast. Also shows rapid forwards rush. Running take-off.

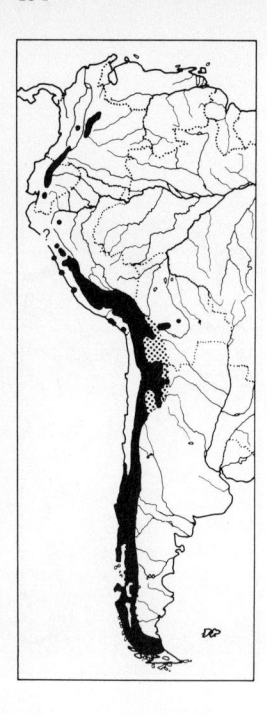

Rare flight low over water with shallow buzzy wing-beats.
Voice: Mechanic drumming or bubbling sounds during courtship. Displaying males also give nasal *aarp*.
Breeding: Nests in reeds, often over water, in roofed-over vegetation. Seemingly nests most of year (Col. and coastal Peru). Eggs Apr (Ecu.), Oct-Feb (Junín, Peru), Mar (Cochabamba, Bol.) or Dec (far s).
Habitat: Lakes and open marshes. Requires fairly deep areas in clear weakly alkaline lakes with plenty of waterweeds. Retreats and nests in fringing reed-marsh, but outside nesting season chiefly offshore.
Range: Through the Americas. Small populations at 2000-4000 m in E Andes, and Puracé in southern C Andes, Col. (*andina*). At 2500-4500 m from Nariño s Col. through Ecu., Peru (locally also near coast), Bol. (also at 875 m near Gutierrez, Sta Cruz) and n Arg./Chile, and (continuously?) at lower elevations from lake districts of s Andes to Isla Grande (*ferruginea*). Common locally.

NOTE: Ssp *ferruginea* was considered a separate species until the discovery of ssp *andina*, which resembles hybrids of *ferruginea* and *jamaicensis* of N Am.

LAKE DUCK *Oxyura vittata* – Plate VII 8a-b

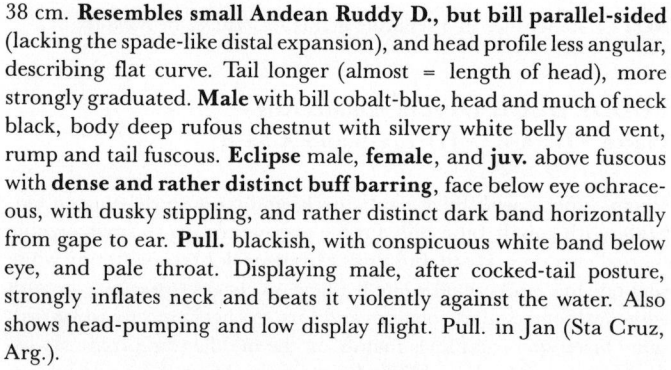

38 cm. Resembles small Andean Ruddy D., but bill parallel-sided (lacking the spade-like distal expansion), and head profile less angular, describing flat curve. Tail longer (almost = length of head), more strongly graduated. **Male** with bill cobalt-blue, head and much of neck black, body deep rufous chestnut with silvery white belly and vent, rump and tail fuscous. **Eclipse** male, **female**, and **juv.** above fuscous with **dense and rather distinct buff barring**, face below eye ochraceous, with dusky stippling, and rather distinct dark band horizontally from gape to ear. **Pull.** blackish, with conspicuous white band below eye, and pale throat. Displaying male, after cocked-tail posture, strongly inflates neck and beats it violently against the water. Also shows head-pumping and low display flight. Pull. in Jan (Sta Cruz, Arg.).

Inhabits open marshes, ponds and deep roadside ditches, sometimes in *Azolla*-covered water, in southern S Am. rather locally from Isla Grande to Atacama, Chile, and s Parag. (but furthest n only as migrant). Mainly lowlands, incl. low valleys in S Andes, but locally nests to 800 m (Bariloche).

INTRODUCED WATERFOWL: Some waterfowl are held captive in towns and villages, and are sometimes seen intermingling with wild waterfowl along creeks and on lake shores some distance away from human settlements.

MUTE SWAN *Cygnus olor* is huge (150 cm, very long-necked), white, with red bill with black basal knob. **Juv.** light gray-brown with gray bill. A few pairs semi-wild in Boyacá and Cundinamarca, Col.

GRAYLAG GOOSE *Anser anser*. 75-100 cm, large domestic birds flightless, very deep-vented. Gray-brown, obscurely barred on body, and with pale gray wing-coverts, domestic birds often more or less completely white. Big orange bill. In most of the Andes, and naturalized on Islas Malvinas.

MALLARD *Anas platyrhynchos*. 55-80 cm, large domestic birds usually flightless, very deep-vented, with cylindrical bodies. Typically, **male** has head metallic green separated by thin white neck-ring from dark brown breast, body mainly light gray, speculum blue, rump and vent black. Bill yellowish green. **Female** buffy brown spotted and mottled fuscous; speculum blue, bill dusky and partly orange. Domestic variants range from black (often with white throat) to all white with orange bill and feet. Common throughout the Andes, and naturalized on Islas Malvinas.

MUSKOVY DUCK *Cairina moscata*. Domestic birds much less gracefully shaped than wild (see above), and often more or less white, and with red bare face. Common throughout the Andes.

Galliformes

Probably one of the most ancient groups of landbirds. The two most primitive families are presently restricted to the Australasian region (Megapodidae) and the Neotropics (Cracidae), where other families are poorly represented.

Cracids – Family Cracidae

Neotropical family of c. 40 chicken- to turkey-sized gallinaceous birds much sought after for game. All highland forms are **guans – Tribe Penelopini** - which comprise bronze-colored, rather light and small species with unspecialized chicken-like bills. Their wings are short and strongly rounded, tails long, somewhat graduated.

Guans usually live in pairs or small parties, foraging on fruit, buds, and flowers in the canopy. They conduct a secluded life in virgin rain and cloud forest areas. They often sit motionless, sheltered in the foliage, run swiftly among the branches when disturbed, or dive straight down from the canopy to disappear. The flight is noisy with inserted glides.

Guans give a characteristic wing-whirring display in the breeding season, usually at the earliest break of daylight. It consists in a gliding flight on sloping wings from one perch to another. The wing-whirring – a dry rattling sound, like ripping a piece of canvas – is usually divided in 2 sections.

Nests are small simple platforms of twigs and green leaves, in a bush or on top of a broken snag, or some way up in a tree. The clutch usually comprises 3 pale eggs. The downy young are capable tree climbers, and can flutter away after 4-5 days. Despite this precocity, the parents are very attentive, presenting food in the bill-tip and also regurgitating food to them.

Detailed information about Andean guans are found in Delacour, J & D Amadon (1973) *Curassows and related Birds*. New York; *Amer. Mus. Novit.* 2251(1966); *Auk* 93(1976):194-5; *Bull. Amer. Mus. Nat. Hist.* 138 art.4(1968); and *Bull. Mus. Comp. Zool.* 134 art.1(1965).

CRESTED GUAN *Penelope purpurascens*, widespread in trop. zone from n Ven. through most of Col. and w Ecu. (possibly to Tumbes, Peru: TP), has been sighted to 2500 m in Loja (Celica mts), sw Ecu. (HB). C. 90 cm. With short bushy crest, the feathers broad without pale margins. Dark olive-brown, feathers of upper mantle indistinctly margined with gray, breast-feathers broadly edged laterally with white. Belly and rump chestnut. C tail-feathers like back, the outer blackish. Bare facial skin blue, large dewlap red, legs rosy.

ANDEAN GUAN *Penelope montagnii* – Plate XIV 4a-c

60 cm. With long, erectile crown-feathers. Small dull orange dewlap on throat partly feathered on anterior part (except in *sclateri*), bare ocular skin blue-gray. Dark bronzy olive with **feather-edges on anterior half margined silvery gray**, rump and belly chestnut, **tail uniform blackish**. Some geographic variation in tones and hues: the intergrading sspp *montagnii*, *atrogularis*, and *brooki* are rather dull-colored, less rufescent below (least developed in *montagnii*, *brooki* dullest and darkest), and conspicuously scalloped whitish on breast, foreneck, and anterior mantle, but with obscure, almost obsolete markings on head. Ssp *plumosa* and esp. *sclateri* are less scalloped, but have silvery white malar and superciliary stripes. Legs salmon red. **Juv.** has lighter, indistinctly vermiculated underparts of body. **Pull.** chestnut with blackish head with long, irregular white band along side of crown, and with 2 buff bands on back.
Habits and **Voice**: Small groups generally conduct an elusive and silent life in the canopy, but give low *dock* notes increasing to *quit-quit-quit* at alarm, sometimes repeated for minutes. By dusk, sometimes sings *chaah-choah-cha-cha-choam cha-cha-cha...*, somewhat like Chachalachas (*Ortalis*), or soft *cluir cluir lui lui luir*. In aerial display, a soft whistle anticipates double bout of muffled wing-drummings (single rattle according to Hilty & Brown (1986)). Responds to immitations of the whistles.
Breeding: Pull. rec. ult. Mar (Col.).
Habitat: Mainly hills of thick, epiphyt-laden humid forest, feeding in mid- and top strata of fruiting trees.
Range: 1800-3500 m. Along Andean slopes from Trujillo and Perijá mts, Ven., and part of E and C Andes, Col. (*montagnii*), sw Col. along w slopes of Ecu. to Azuay (w of Cuenca) (*atrogularis*), and e Andean slopes of Ecu. (at least s to Zapote Najda mts in Morona-Santiago) (*brooki*), e slope of Peru s-wards from Bagua in Amazonas (*plumosa*), and La Paz to Sta Cruz, Bol., and Oran in Salta, nw Arg. (*sclateri*). Fairly common, surviving in small forest patches unless persecuted.

RED-FACED GUAN *Penelope dabbenei* – Plate XIV 5

65 cm. With long crown-feathers, and **orange-red facial skin and lappet**. Rich brownish olive, anterior parts streaked with white feather-edges. Rump and belly dull rufescent, tail uniform dark brown. Lacks white bands on side of head (vs. southern Andean G.), but shows clear contrast between pale-streaked crown and **black forehead**. Wary and prudent. Low throaty warning notes increase to loud *kroa, kroa...* as it flies.

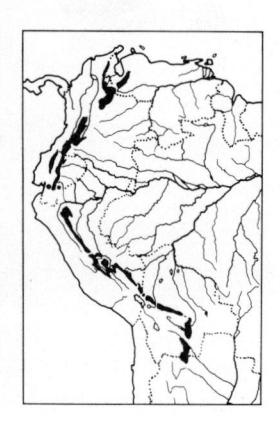

Andean Guan

Bearded Guan

At (900) 1800-2500 m in Chuquisaca and Tarija, Bol., and Cerro Calilegua in Jujuy and Salta, nw Arg., inhabiting large forest tracts with *Cedrela lilloi*, *Eugenia pungens*, or alders *Alnus acuminata* in the highest parts. Rare and local, and suffering from habitat loss in Bol.

BAND-TAILED GUAN *Penelope argyrotis* – Plates XIV 3 and LXIII 7

65 cm. Has long crown-feathers, conspicuous bare dull red throat dewlap, and blue-gray ocular skin. Dark bronzy olive with short, but distinct, **white** lateral feather-edges on anterior parts, incl. on wing-coverts. Rump and belly chestnut. **Lateral tail-feathers slaty black above, deep gray below, and with broad cinnamon tip** (pale buff in *albicauda*). In *argyrotis* and *albicauda* white feather-edges form light zones along supercilium and cheek. Ssp *colombiana* (Plate LXIII) is more grayish throughout, crest larger, composed of blunt feathers with white edges. Face darker, slightly white-streaked, but throat and breast with more developed white scalloping. C tail-feathers coppery. Juv. plumage virtually indistinguishable. **Pull.** very uniform dark brown, lighter below, with white spots near eye and on chin.

Habits: Said to be quite confident. Sometimes several territories close together. In drumming flights, often displays immediately upon take-off, without initial sail.

Voice: Noisy in the breeding season. A rolling *gurr wrrr urrru* increases to loud grating *gi gi gigggigik* at alarm. Low-pitched, melancholic *kuah* is given early morning and on moonlit nights.

Breeding: Probably nests Feb-Apr (May) until the torrential rains start.

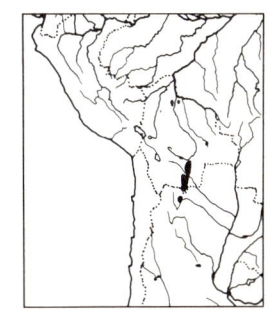

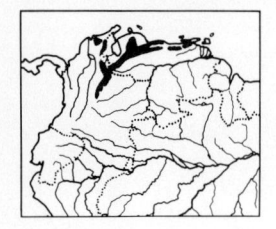

Habitat: Dense wet virgin forest, but sometimes in second growth. Usually at mid-levels, seldom on ground.
Range: Sta Marta mts, n Col. (*colombiana*), upper Río Negro of Perijá mts (*albicauda*), n Boyacá, Col., through Andes and coastal mts of Ven. to Falcón, mainly at 800-2400, locally to 3050 m, but low furthest e (*argyrotis*). Common locally.

BEARDED GUAN *Penelope barbata* – Plate XIV 2a-b

55 cm. Resembles Band-tailed G., but darker, more sparingly streaked with white. Chin and upper throat fully feathered. Legs feathered to well below heel, the bare tarsi only 3.5-4 cm long.
Habits: Apparently as in other *Penelope* guans, but reported often to feed on the ground. Confident where not hunted.
Voice: A peculiar whinnying call noted before drumming display. When disturbed, gives a faint ascending whistle, and persistently repeated, very loud yelping calls *clu clu clu*..., accelerated and hysterical just before take-off.
Breeding: Chicks known from Dec-Feb and July.
Habitat and range: Inhabits both humid and dry cloud forests at (1200) 1500-3000 m, sometimes even small relict forest patches, on w slope of Andes: Cord. de Chilla, San Lucas, Malacatos, Huaico (?) and P.N. Podocarpus (to the e slope) in Loja, s Ecu. (*barbata*), and at Cerro Chinguela and Cruz Blanca, Palambla, Abra de Porcula, Tambillo, Llama, Chota, and Taulis area in Cajamarca and Piura, nw Peru (*inexpectata*). It is not known whether the guans in Cord. Condor is this species or Andean G. Declining, owing to hunting and deforestation, rec. only on 5 sites in the past 50 years (but quite common in suitable sites in e Loja).

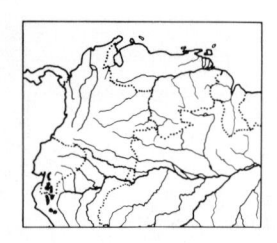

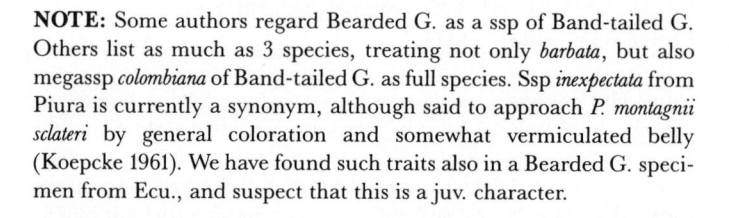

NOTE: Some authors regard Bearded G. as a ssp of Band-tailed G. Others list as much as 3 species, treating not only *barbata*, but also megassp *colombiana* of Band-tailed G. as full species. Ssp *inexpectata* from Piura is currently a synonym, although said to approach *P. montagnii sclateri* by general coloration and somewhat vermiculated belly (Koepcke 1961). We have found such traits also in a Bearded G. specimen from Ecu., and suspect that this is a juv. character.

WATTLED GUAN *Aburria aburri* of premontane zones from Sta Marta mts and Andes from Ven. to s Peru may sometimes (seasonally?) ascend to lower limit of temp. zone. 70 cm. **Blackish** with bill light blue tipped black; long throat wattle and legs yellow (ad.).

SICKLE-WINGED GUAN *Chamaepetes goudotii* – Plate XIV 1

55-65 cm (*tschudii* and *rufiventris* largest). More slender than a *Penelope*, and with tail slightly shorter, but dull red legs longer. No crest or throat lappet, but bare **facial skin light cobalt-blue**. Slaty or fuscous bronzy, somewhat duller fuscous on head and neck, **underparts chestnut to**

Wattled Guan

bright ferruginous. Ssp *sanctaemartae* is less dark above (brownish olive), but deep rufescent below; *goudotii* paler rufous below, with faint gray feather-edges; *fagani* dark (bronzy black) above, and with the dark sooty foreparts grading into uniform dark rufous lower breast and belly; *tschudii* similar, with only slightly more olive back and brighter rufous underparts; *rufiventris* still paler olive above, and with gray feather-margins on the foreparts. **Pull.** described as chocolate-brown with sooty black cap and throat, and rusty buff supercilia joining on nape.

Habits: In pairs or small groups. Although statements about terrestrial habits have been given, the species seems to be mainly arboreal, feeding in fruiting trees at dawn and dusk. Wary. When alarmed hides in the foliage, often after hopping or flying up into the canopy.

Voice: Clucking calls, and loud *kee-uck* at alarm. The specialized (narrow) outer primaries produce a 2 s long, very distinctive wing-rattle in the short display flight.

Breeding: Fledglings June (Sta Marta, n Col.), Sep (Pichincha, Ecu.), Feb-May (Peru).

Habitat: Steep forested hills with difficult access, mainly in zones of very high precipitation. Generally in tall forest, but may descend to coffee plantations and second growth outside breeding season.

Range: On subtrop. and lower temp. levels, with seasonal elevational movements (mainly at 1100-2500, locally to 3000 m). In Sta Marta mts. (*sanctaemartae*), all Andean ranges of Col. (*goudotii*), w Nariño, s Col., through w Ecu. (*fagani*), w Caquetá, Col. (ssp?), along e slope of Nariño through Ecu. to Yurinaqui Alto in Junín, c Peru (*tschudii*); and from Valcón in Puno, e Peru to Chuspipata in La Paz, Bol. (*rufiventris*). May be fairly common locally.

SPECKLED CHACHALACHA *Ortalis guttata* – a member of the widespread *O. motmot* superspecies – is common in interfaces of forest and grassy slopes and in second growth in subtrop. zone bordering the inner Amazon basin, ascending to near transition to temp. zone (at least to 2275 m in Bol.). A small and slender dull brown guan with small red throat-lappet and conspicuous white spots on throat and breast, and cinnamon lower underparts. Easily detected by loud chattering duets *hou-dou-á-ra-cou*.

Pheasants and quail – Family Phasianidae

The 36 native Am. species constitute the Toothed Quail subfamily Odontophorinae. Other subfamilies are widespread in the E Hemisphere. Toothed Quail are small compact birds with short and **very stout bills** (unlike tinamous), rounded wings, and rather short tails (but decidedly longer than in tinamous). They are terrestrial, and mostly live under the cover of vegetation, eating seeds, berries, and some insects. Esp. the forest-dwelling forms are detected mainly by far-carrying whistled calls. When approached, they usually run away, or stand motionless, relying on cryptic colors. At close range they take to sudden flight, with whirring wingbeats alternating with glides on bowed wings, and quickly drop back into cover.

The nest is a simple scrape on the ground, or (*Odontophorus*) a domed construction. The precocial young are able to flutter away within a week.

Further information about species covered below is given in Long (1981) and in *Caldasia* 13(1986):777-86.

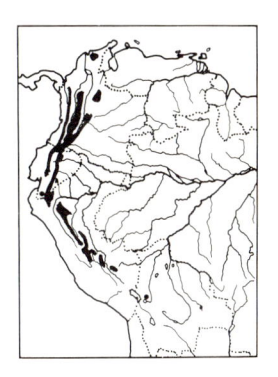

CALIFORNIA QUAIL *Lophortyx californicus*

23 cm. **Male** grayish with a long forward-curving drop-shaped black **head-plume**. Pale forehead continues as long white supercilium. Chin and throat black with white outline, the white lines demarcated black against rest of body, which is mainly gray, or buff on belly, with nicely scaled pattern on hindneck and belly. **Female** with lighter throat, no head markings, and small crest. Calls *cow-wkey* and *cuca cow*.

A N Am. species introduced to c Chile for game. Now very common on arid land in the coastal mts from Atacama to Concepción, and continuing into the Andes, passing over to w Arg. from San Juan to Río Negro and much of Neuquén.

Crested Bobwhite

CRESTED BOBWHITE *Colinus cristatus* – Plate XVII 1a-c

22 cm. With **pointed, light sandy brown crest**. Brownish, densely variegated with black, and dotted with white, lower neck and underparts mainly black **profusely dotted with white oval spots** which grade to coarser buff or tawny spots on belly. In *bogotensis*, **male** has eyebrow, cheek, and throat tawny with dusky feather-bases; lower forehead, line below eye, and chin white. Bill black in male, horn in female. **Female** has partly dusky crest, rather scaled throat, and indistinctly dotted chest. **Juv.** similar, with whitish streaks above and on chest. **Pull.** above fuscous, head with small chestnut cap bordered black, broad buffy superciliaries and black postocular line, back with 2 buff stripes, underparts dusky buff.
Habits: In pairs or 'coveys'. Generally shy. When flushed disperse in different directions, and soon drop into cover.

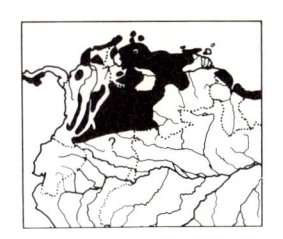

Voice: Male gives whistled *quoit, bob-white*, faster than in N Am. Bob-white (*C. virginianus*). In groups *hoy* or *hoy-hoy-poo*.
Breeding: No information from the Bogotá area.
Habitat: Xerophytic and grassy hills and bushy pastureland, usually near cover.
Range: Several sspp mainly in arid lowlands of northern S Am., locally to the subtrop. zone, and *bogotensis* breeding at 2600-3200 m in semi-arid parts of Boyacá and Cundinamarca, Col. Common locally.

BLACK-FRONTED WOOD-QUAIL
Odontophorus atrifrons

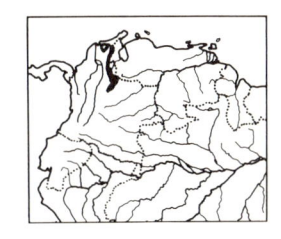

28 cm. Very thick-billed. **Head black with rufous brown crown**, and with occipital crest. Upperparts of body gray vermiculated with black, browner towards rump. Wings mottled and barred cinnamon and black, with white streaks on tertials. Underparts cinnamon-buff with some black and white spots. Wary and secretive, foraging in dense undergrowth, and seen only when flushed at close range. Call a whistled *bob-white*.

In humid subtrop. forest at n end of E Andes of Col. (*variegatus*), and at 1650-2700 m in Perijá mts (*navai*). Rec. to 2750 m in Sta Marta mts (*atrifrons*).

VENEZUELAN WOOD-QUAIL *Odontophorus columbianus* replaces Black-fronted W-q. in sw Táchira and coastal mts from Carabobo to w Miranda, Ven. Known from the humid premontane forest (1300-2400 m), and may reach the lower temp. zone. Resembles Black-fronted W-q., but lacks black on head, and shows inconspicuous buff-mottled supercilium, and **white throat** indistinctly bordered black.

CHESTNUT WOOD-QUAIL *Odontophorus hyperythrus* of the premontane forest of w and n Andes of Col., has been rec. to 2700 m. 26 cm. Bare area around and behind eyes white. Rich rufous brown, crown with short chestnut crest, back with some black bars and spots. Female with greyish belly. Song a rollicking, rapidly repeated *orrit-killyit*..., or either part separately.

GORGETED WOOD-QUAIL *Odontophorus strophium* – **Plate XVII 3** – of subtrop. oak forest in E Andes, Col. – could possibly reach lower edge of the temp. zone. Known from 8 old specimens from Río Beura, Subía, San Juan de Ríoseco, and Agualarga near Bogotá, but recently rec. in good numbers in the extensive area of intact habitat near Virolín in Santander. 24 cm. Very thick-billed. **Characteristic head pattern**: crown and ear-coverts blackish brown, sides of head otherwise white with black speckles, throat black crossed by distinct white crescent. Body **mainly chestnut**, above with fine black vermiculations, and some buff streaks, below with large buff spots (1 specimen uniform dusky brown below). In **Juv.**, underparts have small diamond-shaped buff spots outlined in black.

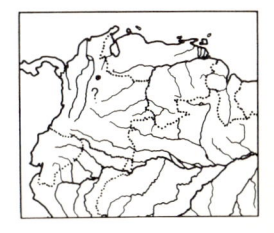

STRIPE-FACED WOOD-QUAIL *Odontophorus balliviani*
– Plate XVII 2

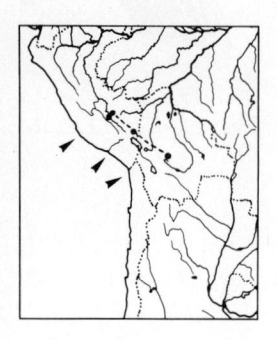

26 cm. Stocky and very thick-billed. **Head with 2 long rusty stripes** behind eye and along cheek separating crest-like rufous crown, black ear-coverts, and fuscous throat. Otherwise chestnut-brown with slight vermiculations, the humerals with black patches, underparts with diamond-shaped white spots. Voice a loud bubbling, rapidly repeated *whydly-i, whydly-i…*, given at early dawn and late dusk.

Inhabits heavy but seasonally somewhat dry montane forest with bamboo, epiphytes, and tree ferns, as well as very humid cloud forest, usually in very steep places of difficult access. At 2000-3000 m in Cordillera Vilcabamba and e Andean slope of Cuzco and Puno, Peru, and in Yungas of Cochabamba, Bol. Maybe continuous, but easily overlooked.

Stripe-faced Wood-quail

Black-fronted Wood-quail

Gorgeted Wood-quail

Chestnut Wood-quail

Gruiformes

An ancient and quite heterogeneous order of mainly ground-dwelling birds of forest, swamp, marsh, and grassland. Only the large rail family represented in the Andes.

Rails – Family Rallidae.

Distributed almost worldwide, even on remote oceanic islands, with altogether 129 species. Rails vary from the size of a sparrow to a big chicken. With their compressed bodies, they are well adapted to slip through dense vegetation. Species inhabiting quaggy places have very long toes, and the water-adapted coots (*Fulica*) have swimming-lobes along the toe margins. The bills vary from small and conical to long, slightly curved types. The plumage may be rather loose in texture, sometimes decidedly decomposed, the colors mainly olive-brown to buff, chestnut, bluish or black, unpatterned, or with bars or coarse streaks. Rails are weak fliers. The short and round wings develop at a late ontogenetic state, and rails therefore easily evolve flightlessness. A considerable proportion of the World's species are flightless or virtually so.

Although some rails inhabit grassland or forest, most live in waterside vegetation, where they climb well. They walk with bobbing heads and flirting tails, and lift their feet high, folding the toes for each step. Most rails have very secretive habits and are heard more than seen. Many are nocturnal, and even the easily observed coots fly between lakes mainly in the dark of the night. The flight-feathers are molted simultaneously, and in the flightless period coots sometimes assemble in thousands in undisturbed lakes with abundant food. However, the population movements associated with molt, breeding, or seasonal climates are scarcely known for Andean species. The diet is varied, mainly worms, slugs etc., but coots prefer waterweeds and other plants. Many species occ. rob eggs and chicks from nests.

Rails are pugnacious in the breeding season, and normally nest as single pairs, or sometimes in small polygamous groups. The nests are well concealed cups of reeds and sedges. Many species lay large clutches of buffy eggs with small dark spots. The nidicolous chicks are normally black, sometimes with brightly colored bills or heads. Most species feed their young a considerable time, and occ. juv. help feeding siblings of a later brood.

Detailed information is given in Ripley, DS (1977) *Rails of the World*. Boston; *Smithsonian Contr. Zool.* 417(1985), some quantitative data in Scott & Carbonell (1985). Detailed data about Andean coots are given in *Bull. Brit. Orn. Club* 103(1983):18-22; *Condor* 66(1968):209-11, 77(1975):324-7; *Hornero* 10(1984):119-25; *Ibis* 109(1967):409, 123 (1981):423-37; *J. Orn.* 100(1971):119-31; *Rev. Mus. Argent. Cienc. nat. (Zool.)* 6(1959):103-29; *Steenstrupia* 8(1982):1-21, and 9(1983):209-15.

PLUMBEOUS RAIL *Rallus sanguinolentus* – Plate XV 2a-d

32-38 cm (*sanguinolentus* and *simonsi* small, *luridus* very large). A stocky, **long-billed**, dark rail. **Ad. bluish slate**, back and wings browner with obscure blackish spots centrally on humerals and secondaries (lacking

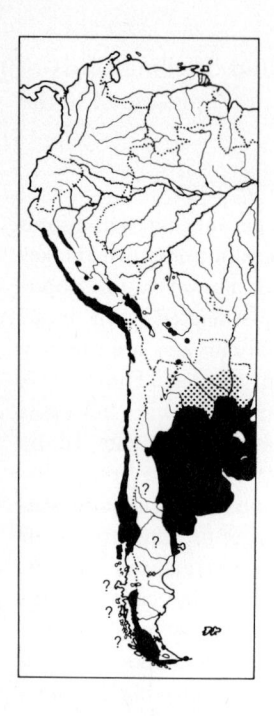

in *landbecki* and *luridus*, the latter may show faint buff feather-tips above, and is distinctly gray-brown on the posterior underparts). **Bill grass-green with bright blue base**, lower mandible red basally (only small spot in *landbecki*, vestigial or missing in *luridus*). Feet dark red. **Imm.** more dusky brown, and **juv.** all dusky brown without any gray, but with buffy white throat and belly, bill and feet black. **Pull.** fuscous.
Habits: Alone or in pairs. Fairly often out of cover compared with other rails. Sometimes curious. Frequently flicks tail.
Voice: Although mainly crepuscular, sometimes sings by day, with rolling squeals *rrueet'e rrueet'e rrueet'e...*, *pu-rueet*, or *huyr huyr huyr...*, etc., often with low, deep *hoos* sounding simultaneously (duet). Calls *gjyp gjyp gjyp...* or *wit wit....*
Breeding: Eggs and pull. in Oct., pull. Jan (Junín, c Peru). Breeds Nov-Dec (c Arg.).
Habitat: Reed-marsh, sometimes of small extension, esp. around muddy creeks and ponds with much floating vegetation (*Ranunculus*, *Hydrocotyle*). Locally also in irrigated areas, like alfalfa crops, or in ditches through rushy pasture in cultivated parts of semiarid valleys.
Range: Mainly southern S Am. Lowlands near Río de Janeiro (*zelebori*); from s Brazil and Parag. through n and c Arg. to base of Andes in Mendoza, and in Sta Cruz, Bol. (*sanguinolentus*). C Chile from Tarapacá to Aysén and adjacent forest refuges of w Chubut to Sta Cruz, sw Arg. (*landbecki*); s part of Isla Grande (common) and maybe further n (*luridus*). Along Pacific coast of n Chile through Peru, and noted to 2500 m in Moquegua, and nearly to the border of the puna in sw Ayacucho (*simonsi*); locally at 2500 m from Catamarca to Jujuy, nw Arg., in Tarija and Cochabamba, Bol., and patchily around 4000 m from Río Desaguadero in La Paz, n Bol. to Pisac Cuzco, in Huancavelica, Junín altiplano (where abundant), and in Cajamarca, n Peru (*tschudii*).

NOTE: In older literature often called *Rallus rytirhynchos*.

There is 1 stray rec. of **BLACKISH RAIL *Rallus nigricans*** from 4080 m, Junín, c Peru (*humilis*?). The species is patchily distributed in trop. e Brazil, Parag., and ne Arg., e Peru and Ecu., Zulia Ven., and subtrop. Cauca Valley, Col. A smaller version of Plumbeous R., with white chin and straighter bill yellowish-green to base.

NOTE: It has been proposed to transfer Plumbeous and Blackish Rails to the genus *Pardirallus*.

LESSER (or VIRGINIA) RAIL *Rallus limicola* – Plate XV 4a-c

22-25 cm. A **long-billed** rail with **deep rufous wings**. Crown dusky brown, rest of upperparts olive-brown obscurely marked with fuscous feather-centers. Sides of head gray, darkest on lore, which is demarcated above by pale line. Throat usually whitish, rest of **underparts fawn-colored or darker vinaceous brown** in *aequatorialis*, grading to black with thin whitish bars on flanks. Vent white spotted black. **Juv. blackish** all over, with only a few brown edges on scapulars and wings, and with some white on mid-throat and mid-belly (thus much blacker

than N Am. juv. birds). **Pull.** black with pink bill with black bar, and red skin visible on crown.
Habits: Skulks, but may feed open to view at dusk. Flicks tail.
Voice: Single *kuc* notes. Descending and accellerating 1.5-2 s trill of c. 10 notes of either raspy or hollow quality. Also rolling *yrll* and descending series of pig-like grunts.
Breeding: Pull/juv Aug (w Ecu.)
Habitat: Reed-marsh adjoining more open wetland, fens, rushy areas, and páramo bog.
Range: N and C Am. Rare and local in lowlands of w Ecu. and along coast of Peru (unnamed ssp), and at 1800-3660 m in the Andes of Ecu., esp. in the C Valley near Ibarra and Quito, and on both slopes of Nariño, s Col. (*aequatorialis*). Little known, and probably declining.

NOTE: A similar rail, with mainly gray neck and underparts, is described under the name ***Rallus peruvianus*** on basis of 1 specimen (apparently lost) probably from Peruvian highlands. Depiction (**Plate XV 5**) is based on the original description by Taczanowski (1884-86). Although currently recognized as a ssp of Lesser Rail, this bird most probably represents a ssp of the Bogotá R.

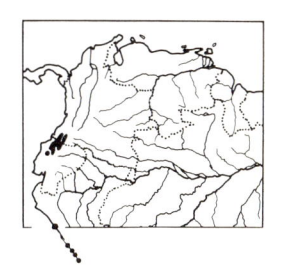

AUSTRAL RAIL *Rallus limicola antarcticus* – Plate XV 6

20 cm. Resembles Lesser and Bogotá R., but decidedly smaller. Dorsal feather-edges more sandy buff, all underparts plumbeous gray (not cinnamon-brown, but in fresh plumage washed olive-brown on breast and sides), and black flanks with broad regularly spaced white bars. **Juv.** differs from ad. only by slight sooty feather-tips on breast, and whitish throat. **Pull.** of probably this form black with whitish bill with black distal half of mandible.
Habits and **voice**: No data.
Breeding: Egg clutch from Oct (Chile).
Habitat: Probably fens and rushy parts of meadows.
Range: Known from several 19th Century specimens from Buenos Aires and c Chile s to Isla Grande. 1901 and 1909 in Valle de Lago Blanca, Chubut, Arg., later (year?) in Llanquihue, Chile, and 1959 in Andes of Bolsón, sw Río Negro, Arg. May vanish because of overgrazing of the habitat and haymaking.

NOTE: Sometimes listed as a full species (allospecies), and the small size (for a subantarctic counterpart of the Lesser R.) suggests a considerable niche shift. The juv. plumage in fact suggests closer relationship with Bogotá than with Lesser R.

BOGOTÁ RAIL *Rallus semiplumbeus* – Plate X 11

24 cm. **Long-billed**. Upperparts olive-brown obscurely spotted with fuscous. **Wing chestnut**. Face, **neck, breast, and anterior sides uniform plumbeous gray**, posterior sides, flanks, and wing-linings black with thin white bars. Vent whitish with the median zone buff barred black. **Juv.** similar, with slight sooty feather-tips on breast, and whitish

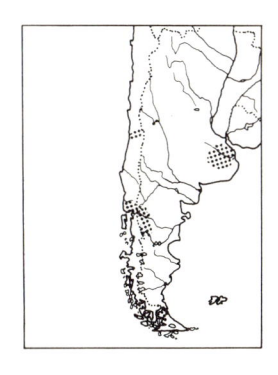

throat. **Pull.** black, bill white with black bar on middle, and entirely black mandible.
Habits: Skulks, but is not infrequently flushed right before one's feet, flying a short distance only.
Voice: Variety of squeaks, grunts, whistling and piping notes, usually starting low, growing in intensity, wavering and trailing of rather abruptly. Brief, rapid *titititirr*. Apparently does not call at night.
Breeding: Pull. in July and Sep. New nest Sep. 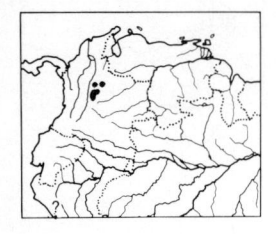
Habitat: Rushy fields, reed-beds with open (burnt) areas, reed-filled ditches, but often feeding in flooded grass, wet fen, or patches of dead waterlogged vegetation nearby. Sometimes also fens fringed with dwarf bamboo (*Swallenochloa*) in páramos.
Range: Col. in E Andes (and an unconfirmed rec. from Ecu.). On Bogotá and Ubaté savannas (2600 m) and at Lake Tota (3020 m), and to 3400 in nearby páramos. Despite much habitat destruction, still common in certain places (see Varty et al. 1986). Listed as vulnerable.

BLACK CRAKE *Laterallus jamaicensis* – Plate XV 3a-d

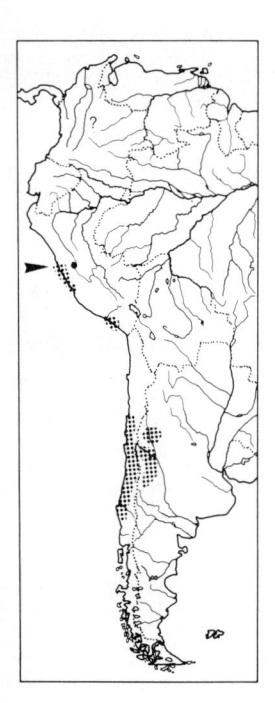

13 cm. A **tiny, slaty-black rail** (but note that other rails have black young). Mantle dull brown (more russet in *salinasi*), lower back, wings, flanks, and tail finely spotted white, more or less as short bars in all S Am. populations, and forming broad white bars across each feather in *tuerosi*. Under tail-coverts barred black and buff, or almost uniform buff (*tuerosi*). Bill black. **Juv.** with pale throat (short period) and sooty mantle. **Pull.** black with mostly white bill.
Habits: Crepuscular and extremely secretive. Hard to flush, and looks like a little black mouse as it runs through the grass right before ones feet.
Voice: Single *chirr*s. Male gives short succession of *chic* notes ending by a falling slurr. Female gives low *croo-croo-o*.
Breeding: In the rainy season (Junín, c Peru); Nov-Dec (Chile).
Habitat: Rushy areas with open spaces with partly flooded moss or short matted grass (Junín, c Peru). Saltmarsh and inundated fields (Arg. and Chile).
Range: Discontinuous in the Americas. Old Bogotá trade skin (doubtful ssp *pygmaeus*). A few marshes along coast of Peru (*murivagans*); at 4080 m in lake Junín, c Peru (*tuerosi*, described by JF in *Steenstrupia* 8(1983):277-282); Mejía lagoons, s Peru and Atacama to Malleco, c Chile to adjacent Catamarca (1200 m), Mendoza, San Juan, and La Rioja, c Arg. (*salinasi*). Rare.

NOTE: Ssp *tuerosi* differs as much from the other forms as does the Galapagos Rail (*L. spilonotus*), and may deserve species rank.

SORA or CAROLINA RAIL *Porzana carolina* – Plate XV 7

20 cm. Plump with small pointed bill. Upperparts **mainly olivaceous**, crown with black stripe, back with obscure fuscous stripes and **many white dots**. Underparts gray with dusky and white bars on flanks, **vent white**. Characteristic black 'mask' often blurred by light feather-margins outside breeding season. Bill yellow tipped dusky, feet yellow-

ish green. Young birds arrive to S Am. in post-juv. molt, with some buff feathers in gray parts of plumage.
Habits: Alone or in small groups. Skulking, but fairly easy to flush. Shows broken shifting flight. Easily takes to water, and dives if alarmed.
Voice: Alarm a cackling *cuck-cuck*, *kee*, and sharp, descending *vee-ker*. A distinctive descending whinny is rarely given in the winter quarters.
Habitat: Mosaics of tall marsh vegetation and open muddy areas, drainage ditches, and inundated fields. Mudflats.
Range: N Am. migrant visiting northern S Am. Sep-May (but may remain June-Aug). Mainly lowlands, but common to 2500 m in Ven. and in some wetlands at 2-3000 m in Andes of Col. and Ecu., and occ. in lake Junín (4080 m), c Peru.

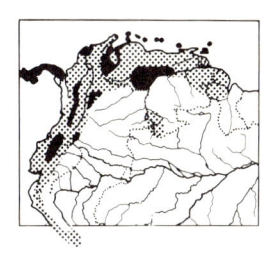

PAINT-BILLED CRAKE *Neocrex erythrops* – Plate XV 8

19 cm. Deep **plumbeous with crown, hind-neck, and back brown**; flanks, wing-linings, and **vent barred black-and-white**. Bill small, yellowish green, in **ad.** with conspicuous red base. Legs rosy. **Juv.** with pale gray faintly barred belly. Furtive, but often flies short distances when disturbed (whereas Black C. runs). Gives staccato froglike note followed by churring trill. Unlike other rails, the song is a long gradually accelerating and descending series of up to 36 staccato notes followed by 3-4 short churring notes that drop noticeably in pitch, last note usually as 3 s flat trill.

May prefer dense swamp vegetation and damp secondary woodland in flooded savanna. Local in trop. Ven., Col., Brazil, Parag. and se Bol. to Tucumán, nw Arg. Possibly with vagrant habits, judging from scattered rec. in the Amazon area and several rec. on the Bogotá and Ubaté savannas (2600 m) in E Andes, Col., mainly in Mar-Apr. Accidental in La Paz City, Bol. (*olivascens*). Coastal Peru from Lima to Lambayeque, and breeding on the Galapagos Isls. at least since 1953 (*erythrops*).

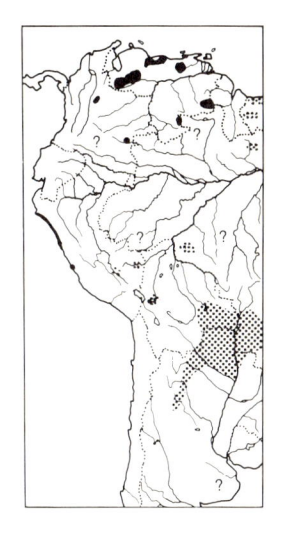

NOTE: This and the following species sometimes placed in *Porzana*.

COLOMBIAN CRAKE *Neocrex colombianus* – Plate XV 9 – is listed by Butler (1979) for the temp. zone of w Ecu. Scattered rec. exist for trop. and subtrop. Panama and n Chocó, Col., (*ripleyi*), Sta Marta mts and w of E Andes, Col., and nw Ecu. (*columbianus*), but we know no definite rec. from above 2100 m. Differs from Paint-billed C. by greenish bill with black tip (and sometimes red base), uniform dull brownish flanks, and **buff under tail-coverts**.

PURPLE GALLINULE *Porphyrula martinica* – Plate XV 11

30 cm. Rather slim, very long-legged and long-toed. Head, neck, and underparts deep purplish blue, back and wings blue-green. **Under tail-coverts all white** (vs. 2 white stripes of Common G.). Bill red with yellow tip, frontal shield light blue. Feet yellow. **Imm.** brownish, but

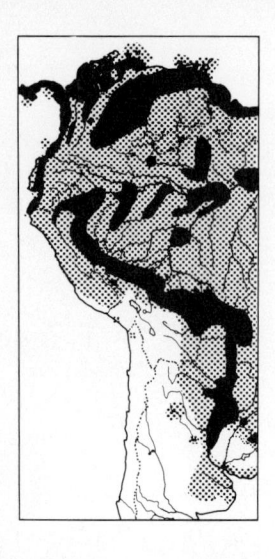

with bronzy blue wings and mostly white belly. **Juv.** mainly buffy brown except for white vent. Singles or pairs (occ. flocks in lowlands), most often in upright gait with hint of a stoop, or climbing in the reeds. Sometimes swims with highly raised rear and wing-tips. Runs across floating-leaf plants. Guttural, cackling calls, and *kek-kek-kek...* notes.

Frequents flooded savanna and ricefields (where considered a pest), marshes, and reed-borders throughout trop. and subtrop. parts of the Americas. During seasonal or irregular movements frequents wetlands of lower temp. zone in E Andes Col. (and maybe stationary in Lake Tota, 3020 m), and regularly moves up Urubamba Valley, Cuzco, sc Peru. Casual on tablelands near Quito, Ecu., and at 4080 m in Junín, c Peru.

AZURE GALLINULE *Porphyrula flavirostris* of the trop. lowlands e of Andes is rec. casually on the Bogotá savanna (2600 m) of E Andes, Col. Recognized by washed-out appearance: above olive-brown with bronzy blue wings; sides of head, neck, and breast pale turquoise grading to white on throat and lower underparts. Bill greenish yellow, feet yellow.

SPOT-FLANKED GALLINULE *Gallinula melanops* – Plates X 8a-b and XV 10a-b

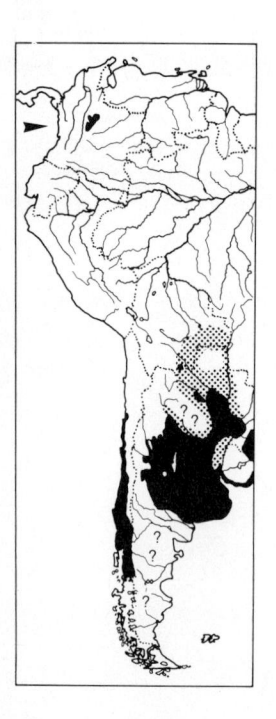

22-28 cm (*melanops* small, *crassirostris* largest). Slate-gray with blackish face; upperparts of body olive-brown, often more chestnut on shoulders and wing-coverts (esp. in *bogotensis*). **Sides profusely dotted white**. Under tail-coverts with 2 white stripes. Chicken-like bill and frontal shield lime-green. **Juv.** rather uniform olive-brown, lighter below, and with faint side-spots. **Pull.** black with almost bare pink and blue top of head, bill banded black and white.

Habits: Not shy, and often open to view. Alone, in pairs or family parties, and sometimes in flocks. Quarrelsome, often chasing each other pattering across the water. Feeds mostly swimming, picking from floating vegetation or sometimes from the water. On water appears flat-backed with scarcely raised rear (vs. Common and Purple G.).
Voice: Variety of clucking calls and *tik-tik-tik...*, in aggressive situations *tap-tap-tap...* growing to loud and laugh-like *huh-huh-huh...* that ends abruptly.
Breeding: Eggs Feb and June, but apparently also breeds at other seasons (Col.). Breeds Oct-Jan (Córdoba) and at least Oct-Nov (Chile).
Habitat: Ponds, ditches, marshes, and lake margins with often extensive floating-leaf vegetation.
Range: Disjunct. Widespread in sw Brazil, Parag., e Bol., and across lowlands of n and c Arg., to 750 m near base of Andes from Tucumán to Mendoza, and probably spreading s (*melanops*). C Chile from Copiapó to Aysén, and through low Andean valleys to adjacent w Río Negro and Chubut, Arg. (*crassirostris*). Common at 2000-3020 m in Cundinamarca, Boyacá, and Santander in E Andes, Col. (*bogotensis*).

NOTE: Sometimes in a separate genus *Porphyriops*.

COMMON GALLINULE *Gallinula chloropus* – Plate XVI 7a-e

30-38 cm (*garmani* largest). **Slate-colored**, becoming black on head and neck, the color of the back varying from neutral slate-gray (*hypomelaena*, some *garmani*) to definitely olive-brown (*pauxilla*), but the differences between sspp are not at all constant. Broken white line along upper edge of flank. **Under tail-coverts white, forming 2 broad stripes**. Bill red, tipped yellow, continuous with square red frontal shield. Legs olive-yellow with orange ring above heel. **Juv.** dark drab gray, pale below, becoming whitish on throat and belly, and showing white side-line and pattern below tail. Bill dusky pink. **Pull.** black with almost bare pink and blue crown, bill red tipped yellow.

Habits: Usually alone, in pairs, or family parties. Quarrelsome, but will sometimes assemble in flocks (100s or even 1000s in certain wetlands in Peru). Mostly feeds on open water, picking from surface, but quickly retreats to cover when disturbed. Walks and swims in jerking fashion, pumping head and bobbing tail with each step. More buoyant on the water than a coot, and often with distinctly raised wing-tips and rear. When threatening, often flares its plumage and spreads white of tail.

Voice: Variety of clucking and chattering calls *gu-cu-cu-cu...*, *qua-qua-qua...*, *quee-quecece...* etc., sometimes repeated for long periods (in the American sspp, which in this respect differ clearly from sspp of the E Hemisphere). Also gives gargling *kurrrl* and hard snarling *kr-r-rk*. Warns with trumpet-like *gut*.

Breeding: Both in July-Sep and in the rainy season (Peruvian highlands).

Habitat: Ponds, marshes, and lakes. Usually in interface of thick marsh vegetation and open water with submerged or floating weeds. On muddy shore meadows or partly dried-up waterweed zones often feeds some distance from cover.

Range: Belongs to a worldwide superspecies. In Panama, trop. Cauca and Magdalena Valleys, and locally in lower temp. zone in E Andes, Col., in the trop. w and locally on temp. tablelands of Ecu., and along coast of Peru (*pauxilla*). Locally in n Ven. and Guianas and widespread on savannas s of the Amazon area from e Brazil and se Peru to n and c Arg., reaching lower temp. zone (2100 m) from Mendoza to Catamarca (*galeata*). Vacas in Cochabamba, Bol. (*hypomelaena*, maybe a *galeata/garmani* intergrade), and apparently patchily in the puna zone at 2000-4200 m in Jujuy and Salta, nw Arg., from Lake Uru-Uru in La Paz, Bol., and Tarapacá, n Chile, to Arequipa, Cuzco, Huancavelica, and Junín to Ancash, Peru (*garmani*). Very numerous in certain wetlands.

NOTE: Called Moorhen in the Old World.

Coots – Genus *Fulica*

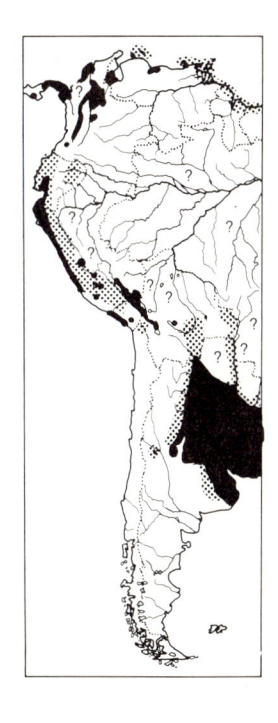

Stocky rails with slate-black plumage. Adapted to open water and usually easy to watch. Distinguished from ducks by small head with chicken-bill, and rounded back. Alights with running start. The systematic sequence of coots does not follow the traditional sequence, but is adapted here to reflect the phylogeny, starting with the most gallinule-like species (see Fjeldså 1985).

RED-FRONTED COOT *Fulica rufifrons* – Plate XVI 3a-b

40 cm. **Characteristic profile** with straight line from bill-tip to knob-like top of long frontal shield. Dark slaty gray inclining to black on head and neck, and with **under tail-coverts forming large white patch** (like an inverted heart, vs. 2 white lines of Common Gallinule). Bill yellow with ridge and frontal shield dark chestnut-red, legs olive. **Juv.** rather uniform gray-brown as in Common G., but has the diagnostic pointed shield and large white under-tail patch, and lacks sideline. **Pull.** black, crown partly naked pink and blue, throat with some broad, orange bristles; bill red with 2-3 black bands. Swims with somewhat less nodding than Common G., and with flatter back but usually raised tail. Mainly surface-feeds, but can dive. Long chattering series at deeper pitch than in Common G., and often slower towards end: *togo togo togo...*, *cu° cu° cu°...*, *puhuh puhuh puhuh...* etc. Alarm note *tuc*. Eggs mainly in Sep-Oct (Chile), May-Nov (Córdoba) or Sep-Jan (s Peru).

Prefers semi-open marshland, often with much floating duck-weeds and waterferns, and is rarely far from cover. Inhabits lowlands from Chubut to Chaco and to 650 m at foot of Andes in San Juan and Mendoza, Arg., around 800 m in lake districts on the border to Chile, also in Chilean lowlands from Osorno to Atacama, and in coastal s Peru (Mejía). Casual in lagoon Volcán (2100 m) in Jujuy, nw Arg., and in the Magellanic region.

RED-GARTERED COOT *Fulica armillata* – Plate XVI 5a-c

44 cm, with somewhat angular profile caused by elevated tip of frontal shield (White-winged C. is round-headed). Dark slate-gray inclining to black on head and neck. Under tail-coverts white forming 2 broad white stripes under short tail. Bill and pointed frontal shield sulfur-yellow (almost white from a distance), with dark red streak at base of culmen, so that **bill and shield seem disconnected**. Legs of ad. orange-yellow (but duller, more olive in 2nd and 3rd years), with red 'garter' above heel. **Imm.** with horn-colored bill, but soon aquires yellow shield; feet olive. **Juv.** with whole body dark drab gray, head and neck white with dusky mottling above, but with distinct whitish demarcation towards frontal shield. **Pull.** black, crown partly naked pink and blue, throat with small orange bristles, bill mainly black with orange band and red shield.
Habits: Gregarious, sometimes swimming or loafing in large rafts, but territorial and pugnacious when breeding. When threatening, proudly raises neck and tilts wing rear-edge up. Feeds up-ending and diving.
Voice: Male gives whistled *huit* at alarm, and explosive *pit* and repeated *wuw* in aggression. Female gives *yec* at alarm, and repeated hard *terr*.

Breeding: Egg-laying mainly Oct-Nov, or Nov-Jan furthest s.
Habitat: Open marshes and lakes with abundant water-weeds. On Patagonian uplands breeds mainly on exposed shallow lakes with extensive carpets of floating *Myriophyllum*.
Range: Southern S Am. n to Coquimbo, Chile and Jujuy, Arg., but migrating to Parag. and s Brazil. Mainly lowlands, but ascends at least to 1200 m on barren plateaus of inland Patagonia and to 1000 m in the lake districts of the s Andes, and breeds at 2100 m in lagoon Volcán in Jujuy. Common.

GIANT COOT *Fulica gigantea* – Plate XVI 1a-c

55 cm. Young birds not remarkably larger than Slate-colored C., but ad. **very heavy-bodied** and with relatively small head and characteristic profile with **concave forehead and 2 high knobs** formed by enlarged orbital rims. **Ad.** deep slaty, inclining to black on head and neck. Only slight indications of white streaks on vent. Bill deep red distally, white on ridge, yellow on sides continuing onto shield. **Feet dark red. Imm.** with dark gray breast and belly, less strong colors of bill, and feet light grayish-red. **Juv.** dark drab gray, foreneck and side of head below eye-level white, bill and feet dusky. **Pull.** black, crown downy, with hardly visible pink skin; bill buff with pale yellow tip, and magenta base separated by black bands; throat with few, small orange bristles.
Habits: Normally quite confident, but shy where persecuted. Usually in pairs or families (often 2 generations of young together). Vigorously defend territories, but social when out in middle of lake. Often seen attacking, running across water with heavy wingbeats. Threat posture with protruding breast and proudly raised neck, but not tilting wings. Imm. disperse by nightly flight, but ad. are too heavy for flight, and are normally attached to their territory and huge nest for life. When small breeding ponds sometimes freeze, ad. coots probably walk to larger ice-free lakes. Feeds on short grass on the shore (on foot), and on water-weeds (mainly on surface).
Voice: Male gives a widely audible gobbling, which alternates with growling sounds: *hrr horrr horrrr*, **houehouhouhouhoou**, *hrr*, **houehouhouhouhoou** etc. Female has cracking *chee-jrrrh* and soft *hi-hirr hirrr hirrr*....
Breeding: Nest often maintained over several years, and may form 3 m long raft that, when prepared for a breeding season, projects 0.5 m above water. Some territories also have small auxiliary platforms. Breeds at any season, but laying peaks in winter, June-July, often with 2nd clutch in Nov-Feb. During bad weather, pull. tend to remain inside nest, feeding on the nest material.
Habitat: Ponds and lakes in barren highlands. Nests in meter-deep water with dense water-weeds (esp. *Myriophyllum*, *Potamogeton* and *Ruppia*) to the surface, or sometimes on miry clay. Large populations in lakes with extensive weedy shallows. Imm. visit lower lakes with reeds, but always stay outside vegetation cover.
Range: Mainly breeds at 3600-5000 m (locally down to 3100 m) in puna zone from Lake Pozuelos in Jujuy, nw Arg., and Tarapacá, n Chile, through w Bol. to lake Paron in Ancash, nc Peru. Generally scarce, but large populations exist in Lauca Natn. Park in Arica (c. 12,000), n Arequipa, Yaurihuiri lakes in Ayacucho and Conococha in

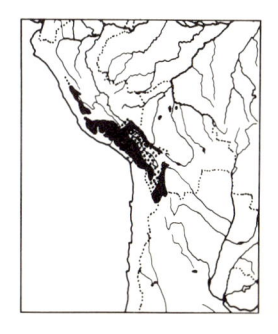

Ancash. Once considered rare, the population seems now to have 'exploded' in Chile and Peru, owing to control of the use of fire-arms. Straggles to lower elevations, accidentally to Pacific coast.

HORNED COOT *Fulica cornuta* – Plate XVI 2

50 cm. Shaped like Giant C. Uniformly slate-gray, head darkest. 2 prominent white stripes under the small tail. Bill greenish yellow, dull orange basally, with black ridge, forehead with black fringes and long **black proboscis** that normally rests on ridge of bill. **Feet olive** with dark joints. **Juv.** seems to resemble Giant C. Age of development of proboscis unknown. **Pull.** is (judging from photos) black with feathered crown, bill light gray and pale yellow at base and tip.
Habits and **voice**: Much as Giant C., although the loud braying 'song' has not been heard. In interaction, low grunting series of 3-5 syllables and loud sharp grunts noted (BW). Has been found nesting in loose colony. Ad. are capable of flight. Proboscis can be raised during display.
Breeding: Nest resembles that of Giant C., but sometimes rests on a pile of up to 1.5 tons of stones, apparently assembled by the coots themselves. Nest-building Nov., eggs Jan, but some breeding must occur at other seasons. Nest often used over many years, and also serves as a nest-site for other water-birds.
Habitat: Barren highland lakes, fresh and brackish, with dense submergents (esp. *Potamogeton* and *Ruppia*).
Range: At 3000-5200 m in desert puna from Tarapacá to Atacama, n Chile, and Potosí, Bol., and Jujuy to w Tucumán (and maybe San Juan), nw Arg. Rare, but good numbers around Sierra Aconquija (Vides-Almonacid in press), Pozuelos, in Maricunga salt lake, and a concentration of 2500 once rec. on lake Pelado of the Chalvíri area, Potosí. Local populations fluctuate greatly between periods of drought and inundation.

ANDEAN COOT *Fulica ardesiaca* – Plate XVI 6a-e

43 cm. Stocky, but much less heavy-bodied than ad. Giant C., and with rounder head. Slate-gray inclining to black on head and neck. Usually with white secondary-tips. Under tail with 2 white lines (weakly developed or missing in *atrura*). **Red-fronted phase**: With large rounded frontal shield deep chestnut, bill chrome-yellow becoming pale yellow and green, or sometimes white, towards pale blue tip; feet green. **Pale-fronted phase**: bill white, frontal shield white or sometimes orange-yellow; feet slaty. Some additional variants exist. **Juv.** dark drab gray with somewhat paler underparts, and mainly white face. **Pull.** blackish, top of head with some red and blue skin shining through, throat with rather short orange bristles, bill red with dark bar demarcating orange tip.
Habits: Often quite shy, pattering into cover if approached in a boat. Generally gregarious, sometimes assembling in 1000s, feeding in shallow water areas with dense submergent or floating vegetation in sometimes mixed flocks, or loafing or swimming in deep open water in pure flocks. Pugnacious when nesting. Threatens with head low, or raises neck proudly, and tilts wings over the back.

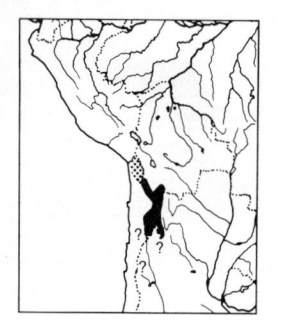

Voice: Usual call low *churr* or harder *hrrp*, often repeated. Females give low chitter.
Breeding: Among reeds, or open to view in floating water-weeds. Egg-laying peaks in July-Aug, but some clutches can be found at any season.
Habitat: Lakes and marshes. Mostly in fairly large lakes with extensive shallows with dense submergent vegetation, fringed by reeds, but also breeds in barren lakes lacking reeds. Except furthest s, there is a tendency for red-fronted birds to dominate in well vegetated lakes, and white-fronted birds to dominate in barren high-elevation lakes with *Chara* as the principal submergent.
Range: At 3500-4700 m from Catamarca, nw Arg. (down to 2100 m in Jujuy), and Tarapacá, n Chile, through highlands at least to Ancash, nc Peru (*ardesiaca*, with pure red-fronted populations s from Titicaca region and in Cuzco, white-fronted dominating in high parts of Junín). Coastal Peru, Guayaquil area, and all tablelands of Ecu. and e and w Nariño and Cauca, s Col. (*atrura*, described by JF in *Bull. Brit. Orn. Club* 103(1983):18-22). Seasonal population movements occur. Common, abundant in some areas.

NOTE: Chestnut- and white-fronted birds were previously treated as separate species **American C. *Fulica americana peruviana*** and **Slate-colored C. *F. ardesiaca***, and were later on lumped under the name *F. americana ardesiaca*. Although the color phases keep apart to a certain extent the interbreeding is too extensive for separation as species. New biological data and documentation of sympatry with *F. americana columbiana* support species rank for *ardesiaca*. The new definition of the taxa is emphasized by introducing a new vernacular name.

WHITE-WINGED COOT *Fulica leucoptera* – Plate XVI 4a-c

40 cm. Slate-colored inclining to black on head and neck. Secondaries with broad, white tips (conspicuous in flight). Under tail-coverts white forming 2 bold white stripes below short tail. Bill bluish pink to lemon-yellow, occ. with dark spot near tip; **small round frontal shield yellow or orange**; legs green, yellow above heel. **Imm.** and **juv.** resemble Red-gartered C., but show rounder head, and small round shield of rather pinkish color; juv. is also paler below. **Pull.** differs from other southern coots by denser yellow bristles around neck, and red bill without dark bands.
Habits: Gregarious, normally in large flocks, but territorial when breeding. Always threatens with head low over water, and often with expanded body plumage (tilted wings). Chiefly feeds by pecking from surface, and by up-ending.
Voice: Loquacious, with a variety of hollow cackling calls. On breeding sites mainly *wurt*, *wyt* etc., and aggressive male giving explosive *huc*.
Breeding: Egg-laying mainly Apr-Nov (lowlands of c Arg.) or Nov-Jan (all uplands).
Habitat: Prefers ponds, lagoons, river backwaters etc., with grassy shores and much submergents. Upland breeding sites are usually barren, without fringing reeds, but with floating *Myriophyllum* on much of surface.

Range: Southern S Am. n to Arica, n Chile, and Parag., migrating to e Bol. In Arg. mainly in lowlands, but nests scattered on upland plateaus of Patagonia, and in Chile reported to be a highland bird that furthest n ascends to 4900 m. Occ. in lake Alalay (2550 m) in Cochabamba, Bol. (HEM Dott). Common.

NOTE: May possibly be a megassp of American C.

AMERICAN COOT *Fulica americana* – Plate X 2

Ssp *columbiana* 43 cm. Slate-colored inclining to black on head and neck. Secondaries with small white tips, under tail-coverts white forming 2 bold stripes. Frontal shield rather small, blackish chestnut; bill white with yellow base, and **broken chestnut ring near tip**; legs and feet greenish gray, with much yellow or orange in old birds. **Juv.** with white face and foreneck, and light drab gray underparts of body (apparently lighter than Andean C.). **Pull.** differs from Andean C. in having denser yellow bristles around neck, and by red bill (pink distally, with tiny black tip, but no dark bar).
Habits: Gregarious, but territorial when breeding. Always threatens with head low, occ. expanding plumage, and tilting wings somewhat. Often feeds from surface, but dives in offshore areas.
Voice: Loquacious, with variety of cackling notes, *punk*, *hrk*, *kwa* etc., often in series. Female gives low nasal *kju-kju-kju...*, aggressive males explosive *hic*.
Breeding: Egg-laying peaks June-July, but some laying stated in Sep (Col.).
Habitat: Open marshes and reed-fringed lakes, with peak abundance in areas with mosaics of low *Bidens* and *Limnobium*, and open water with much submergents outside reed-borders.
Range: N Am. to Panama (*americana*). At 2100-3250 m in E Andes, Col., with main population (800 birds) in Lake Tota. At least previously nested in Otún in Caldas, and in sympatry with Andean C. in Páramo de Puracé in Cauca, and also rec. near Ibarra, n Ecu., and possibly occurs near Cotopaxi (*columbiana*) (statements in Hilty & Brown (1986) about occurrence in Nariño may be caused by confusion with red-fronted *ardesiaca*).

NOTE: May include the Caribbean C. (*F. caribea*) and possibly also the White-winged C. as (mega)sspp. See also the note under Andean C.

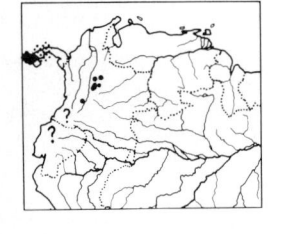

Finfoots – Family Heliornithidae

SUNGREBE *Heliornis fulica* of trop. swamp and river habitats in C and S Am. is rec. once at 2600 m near Bogotá, Col. 28 cm, grebe-like, but with rather long and broad tail. Head and neck boldly striped black and white (female with buffy cheeks), otherwise brown to white on c underparts.

Charadriiformes

An ancient group related to gruiform birds and possibly to certain other waterbirds. The order, as defined presently, comprises three main lineages: the sandpiper line (incl. jacanas and seedsnipes) and the plover complex, from which the gull/auk line diverged. The classification follows Johnsgaard, P (1981) *Plovers, sandpipers and snipes of the World*. Lincoln-London.

Jacanas – Family Jacanidae.

Altogether 7 species in warm parts of both hemispheres. Rail-like waterbirds with extremely long toes and claws permiting walking on waterferns or larger floating leaves. Unlike rails, they often appear open to view. For further details about Am. forms, see *Animal Behavior* 26(1978):207-18, *Auk* 89(1972): 743-76, and *Condor* 79(1977):98-105.

WATTLED JACANA *Jacana jacana* – Plate XV 1a-b

24 cm (excl. trailing feet). Looks black with delicately **greenish yellow (transparent-looking) flight-feathers** tipped black. 2-lobed frontal shield and rictal lap, and base of bill red, tip yellow. Head and neck black. In *melanopygia* most of body and wing coverts blackish maroon; in *intermedia* back, flanks, and wing-linings brighter rufous; *jacana* similar, but darker and with partly black scapulars. **Juv.** has upperparts and wing-linings bronzy or gray-brown, underparts and sides of head buffy white with black eyeline expanding as broad band on ear-coverts, frontal shield small. Territorial and aggressive towards other waterbirds (but chased off by coots). Usually seen walking deliberately on floating vegetation, and occ. raises the green wings. When approached, flies instead of skulking, and raises wings for a while after alighting. Has high-pitched squawking and cackling notes, as *wheek-wheek-wheek....*

Characteristic of ponds and river backwaters with abundant floating plants in trop. and subtrop. S Am. s to Córdoba, Arg. Casual at 2600 m in Boyacá, and possibly regular in Lake Tota (3020 m) in E Andes, Col. (*intermedia* or *melanopygia*). Regular in Nov at 2550 m in Cochabamba, Bol., and accidentally to temp. zone in nw Arg. (*jacana*).

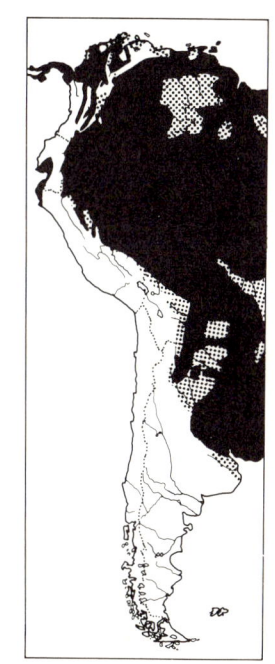

NOTE: Considered by some a ssp group of Northern J. (*J. spinosa*).

Oystercatchers – Family Haematopodidae

A worldwide group of c. 6 species related to stilts and plovers. Very conspicuous, fairly large, black-and-white shorebirds with long, straight, and laterally compressed bills, and not very long legs lacking hind toe. Feed mainly on mussels and other marine littoral invertebrates, but take insects in inland habitats. The nest is a poorly lined scrape in gravel or short vegetation. Unlike in most shorebirds, the nidifugous young are fed some time by both parents. Details about the Magellanic Oystercatcher are given in *Wilson Bull.* 92(1980):149-168.

MAGELLANIC OYSTERCATCHER
Haematopus leucopodus – Plate XVIII 9a-b

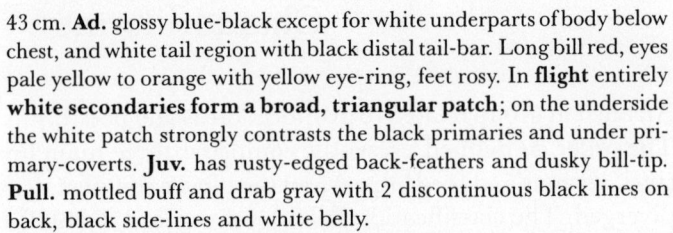

43 cm. **Ad.** glossy blue-black except for white underparts of body below chest, and white tail region with black distal tail-bar. Long bill red, eyes pale yellow to orange with yellow eye-ring, feet rosy. In **flight** entirely **white secondaries form a broad, triangular patch**; on the underside the white patch strongly contrasts the black primaries and under primary-coverts. **Juv.** has rusty-edged back-feathers and dusky bill-tip. **Pull.** mottled buff and drab gray with 2 discontinuous black lines on back, black side-lines and white belly.
Habits: Gregarious, but disperses to breed. Then agonistic and very vocal, displaying on the ground or flying around, usually low over the landscape, often 2-3 neighboring pairs together. Flight strong, with rapid, shallow wingbeats. When disturbed on the nesting terrain shows characteristic display with tail held vertically, and sometimes feigns injury.
Voice: Often heard. Has distinctly weaker and higher-pitched calls than other oystercatchers: drawn-out plaintive *peeeyi* or even higher-pitched, very thin *pee pee pee....*
Breeding: Eggs Sep-Dec, latest inland.
Habitat: Marine beaches and gravely or stony shores of inland lakes and rivers, but often nests far from water in short grass.
Range: Islas Malvinas and coasts of southern S Am. n to Chubut, Arg., and Llanquihue, Chile. Nests at least to 1000 m on the pre-andean foothills in Sta Cruz, s Arg. Winters esp. at Cabo Blanco and Río Deseado estuary and casually to s Buenos Aires and Llanquihue, but some pairs remain in the s. Common.

AMERICAN OYSTERCATCHER *Haematopus palliatus* inhabits sandy and rocky beaches of N, C, and S Am. s to c Chile and s Arg., and has been rec. near base of Andes at La Paz in Mendoza, wc Arg. (*pitanai*). 42 cm. Differs from Magellanic O. by slaty black head and neck, deep gray-brown upperparts, and by **wing-bar** formed by white greater coverts and base of secondaries. Eyelids red. No display with strongly raised tail. Voice a rapidly repeated piercing *wheep*.

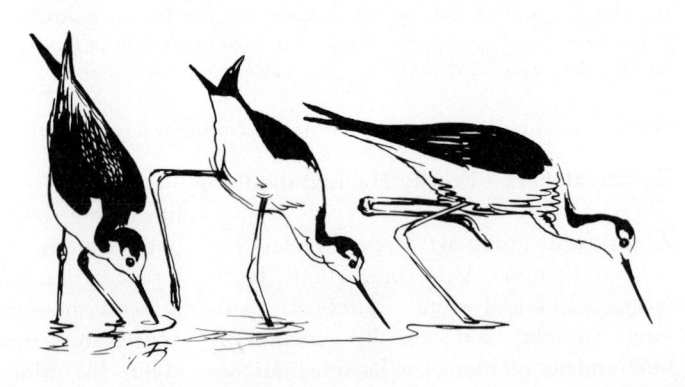

Black-necked Stilts

Stilts and avocets –
Family Recurvirostridae

A group of 7-10 species inhabiting all continents. Related to oystercatchers and plovers, but easily distinguished by smaller heads, slenderer necks, long and thin bills, and longer legs. Adapted to extensive shallows with high densities of invertebrates in fresh, brackish, or salt lakes. Feeding habits often bring them in company with Chilean Flamingos.

Nest in loose colonies, on mud or tussocks in shallow water. Substantial nests (for shorebirds) of straws and mud if the nesting habitat is very wet. The nidifugous young find their own food and run and swim well at an early age.

Further details in Hayman, P, Marchant, J & Prater, T (1986) *Shorebirds. An identification guide to the waders of the World.* London – Sydney.

BLACK-NECKED STILT *Himantopus mexicanus* – Plates VIII 2a-c and XX 15

37 cm (excl. legs). Very lanky, with needle-thin straight bill and **extremely long and slender pink legs** which in flight project 15 cm beyond tail. **Pied**: cap to well below eyes and hind-neck black except for white forehead and spot above eye (*mexicanus*), or only nape with a ramification to each eye black (*melanurus*). Mantle black (male) or partly dusky brown (female), sometimes with white patch; **pointed wings black above and below**. Tail pale gray. Otherwise white. **Juv.** usually has gray (and never white) crown, and the back browner with rusty feather-margins that soon wear off. **Pull.** above buffy gray with some black spots, below white.
Habits: Social. Usually quite shy. Strides deliberately and gracefully, lifting the feet high. Neck partly withdrawn. Feeds by pecking and snatching in mud below water and sometimes by scything movements in soft mud. Sometimes sits on ground, as if nesting.
Voice: Noisy when breeding. Alarm rapidly repeated tern-like *yip* notes, the mobbing continuous but irregular, puppy-like yelping *woyp woyp-woyp..woyp....* Young birds give feeble whistles.
Breeding: Eggs in Oct in Junín (Peru) and Sep-Mar (wc Arg.).
Habitat: Shallow, muddy, and often weakly alkaline coastal and inland waters surrounded by level grassland, often seasonally inundated meadows with some interrupting rushes or reeds (open marsh).
Range: Widespread in the Americas, mostly in lowlands. Main concentrations on llanos of Col. and Ven. (*mexicanus*) and on plains of n and c Arg. and c Chile (*melanurus*), but due to erratic habits it can nest wherever suitable habitat is formed, except furthest s. Casual in highlands of Col. and Ecu., but seen rather commonly in several wetlands at 2500-4200 m in the puna zone from c Peru to nw Arg. (nesting confirmed in Lake Junín, near Cuzco, near Cochabamba town, Bol. and Lagunas Pozuelos and Rontuyoc in Jujuy, nw Arg.). Most highland birds resemble *mexicanus*, but birds with *melanurus* traits are seen seasonally, and dominate in the s part.

NOTE: Sometimes treated as conspecific with the Old World forms; then named as sspp of Common S. *H. himantopus*.

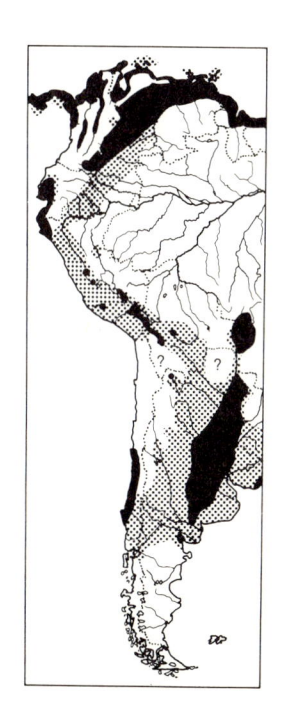

ANDEAN AVOCET *Recurvirostra andina* – Plate XVIII 1

46 cm. With strongly **upcurved bill**. Heavy-bodied and rather compactly built and short-legged compared with the stilt, and fairly broad-winged (like Andean Lapwing) in flight. **Mainly white** (occ. stained ferruginous), with back and wings (except linings) and tail black. Lower back white. Eyes orange, legs light blue. **Juv.** with inconspicuous rusty feather-edges above soon wearing off. **Pull.** white, above washed gray with a few dusky spots; webbed feet blue.
Habits: Social, but sometimes single. Usually seen wading in shallow water, feeding on free-swimming invertebrates with characteristic vigorous sideways sweeps of the bill in the water surface. Occ. swims and up-ends. Often associates with Chilean Flamingo.
Voice: A sharp di-syllabic barking *kluu-kluut* rapidly repeated.
Breeding: Egg-laying Sep-Jan, depending on time of rainfall.
Habitat: Salines and alkaline lakes with wide beaches, seasonally inundated saline meadows and in open parts of alkaline marshes.
Range: At 3100-4600 m in Lake Junín, c Peru, Lake Parinacochas in W Ayacucho and from Arequipa and Puno, s Peru, through w Bol. and adjacent Chile to Jujuy, Salta, and Catamarca, nw Arg. Casually to Pacific coast. Generally sparse.

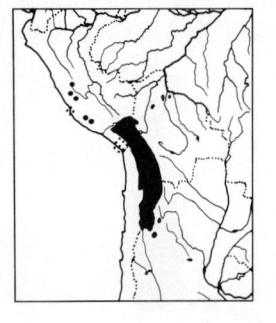

Lapwings and plovers – Family Charadriidae

A worldwide group of altogether c. 50 species. Small to medium-sized, compact and often 'neckless' shorebirds. The plumage is usually boldly patterned, often with pectoral bars, but species adapted to pale sandy beaches have reduced patterns, and many species show a winter plumage with little contrast. The bill is usually short (swollen tip shown in many field-guides characterize only dry museum specimens), adapted for picking, shallow probing, and 'soil biting'. The legs are moderately long, and many species adapted to dry or firm substrates lack hind-toe.

Plovers are less social than many other shorebirds, and often occur singly or in pairs. However, lapwing territories are often clumped on uniform grassland, and such loose colony-breeding probably facilitates communal mobbing of predators. Most species live in small flocks outside the breeding season. Plovers inhabit open terrain with little or very short vegetation: margins of desert ponds, beaches, rather dry parts of mudflats, well grazed meadows, and young crops. They run well, smaller species being remarkably quick. As they halt, they often make characteristic bobbing movements. Feeding birds typically use a run-and-wait technique, with quick runs alternating with a few s of motionless watching. They may also gently pat the ground with an outstretched foot to induce the prey to move. The diet comprises arthropods and some worms that live on or just beneath the surface. Most feeding occurs at night.

The nest is usually a shallow scrape in the ground, with a few straws, pebbles, or shell fragments used as lining. The 4 (or fewer) pear-shaped eggs and the downy young are extremely well camouflaged. The young find their own food and run well, but mostly avoid danger by motionless crouching and camouflage.

Information about Andean species is compiled in Hayman, P, Marchant, J & Prater, T (1986) *Shorebirds. An identification guide of the waders of the World*. London – Sydney. See also Johnsgard, P (1981) *Plovers, Sandpipers and Snipes of the World*. Lincoln – London; Woods, RW (1975) *Birds of the Falkland Islands*. Oswe-

Flying Southern and Andean Lapwings show slightly different wing patterns. From the left sspp fretensis *and* lampronotus *of Southern L., and Andean L., which has the largest white band.*

stry; *Auk* 84(1965):130-131 and 89(1972):299-324; *Hornero* 12(1986):150-155; *Mem. S Diego Soc. Nat. Hist.* 3(1968):1-54; *Trans. S Diego Soc. nat. Hist.* 18(1976):29-73 and *Wader Study Group Bull. 49, Suppl./IWRB Special Publ.* 7:(1987):57-69.

SOUTHERN LAPWING *Vanellus chilensis* – Plate XVIII 3a-b

C. 35 cm, *chilensis* and *fretensis* particularly stocky. Small wing-spur and thread-like crest on nape. Easily recognized by **black forehead and black midline on chin and throat** demarcated by white against light ashy gray main color of head and neck (pale blue-gray in *chilensis* and *fretensis*), and **black chest** contrasting white lower underparts. The black pattern is least developed (interrupted on throat and giving place to much white on face) in *cayennensis*, most developed (with very large black breast 'shield') in *chilensis* and *fretensis*. Upperparts of body bronzy brown, anterior humerals and wing-coverts glossed rosy, purplish, and green. In **flight** shows broad blunt wings, white greater and middle upper wing-coverts (unlike in Andean L. poorly developed on inner part of wing), and white wing-linings in contrast to black remiges. Tail region white with broad black distal bar. **Juv.** shows cinnamon feather-edges above, and virtually lacks throat line. **Pull.** is variegated grayish cinnamon and black above, with heavy black demarcation of white nape; underparts white with fuscous pectoral bar.
Habits: Very wary and noisy. Several pairs often assemble to mob intruders. Unhesitangly 'dive bomb' near nests and young. Upon landing, usually folds wings together immediately (unlike Andean L.). During courtship shows rising and falling aerial displays (wingbeats slower and heavier than in Andean L., and interspersed with soaring on spread wings).

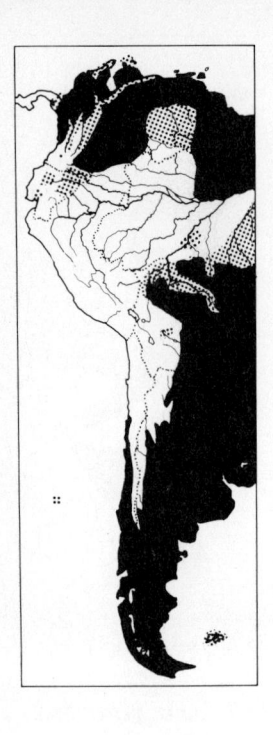

Voice: **Very noisy**. At any or no provocation utters harsh *parp-parp-parp...*, *keck-keck-keck...*, *que-que-que...*, or *chero-chero-chero...*, sometimes a sustained chorus audible kilometers away. Sspp *fretensis* and *chilensis* give a distinctive, much harsher, and parakeet-like shrill *tjirr-tjirr-tjirr...teree teree teree....*
Breeding: Eggs in July-Nov (s part of continent). Sometimes cooperative breeding with 1-2 extra ad. per nest.
Habitat: Fields, meadows, river plains, and open boggy habitat with short matted vegetation. Breeds mainly near moist or partly inundated parts of the terrain.
Range: Throughout lowlands of S Am., except for the w and c Amazon basin, and range of Andean L. Accidental visitor to temp. savannas in E Andes and Cauca and at 3800 m on Nevado del Ruiz in Caldas (GG pers comm.) in C Andes, Col., and one rec. in highlands of Ecu. (Cotopaxi) (*cayennensis*). Casual to 2550 m in Cochabamba, Bol., and common as high as 2100 m at Tafí del Valle in Tucumán and in Córdoba, nw Arg. (*lampronotus*). Common across the Patagonian steppes, ascending to 1200 m on the basaltic plateaus from Somuncura in Chubut to Sta Cruz (*fretensis*); s and c Chile and passing through Andean valleys to foothills of Mendoza to Chubut, w Arg. (*chilensis*). Common, apparently spreading due to forest clearance and cultivation.

NOTE: Previously in genus *Belonopterus*. Judging from different proportions and vocalizations 2 species might be involved (**SOUTHERN L. V. cayennensis**, incl. *lampronotus*, and **CHILEAN L. V. chilensis**, incl. *fretensis*). However, intergradation is known from one (man-influenced) locality (see *Neotropica* (La Plata) 32 No 88(1986):157-165).

ANDEAN LAPWING *Vanellus resplendens* – Plate XVIII 4a-d

33 cm. A large plover with **head, neck, and breast** ashy **gray** contrasting white lower underparts. Top of head looks creamy white, setting off dusky lores. Upperparts of body bronzy-green, the lesser wing-coverts dark metallic purple, contrasting white wing-bar. Rosy bill and legs. In **flight** shows pied wings with white linings, black remiges, and **a large white diagonal on the upper side** formed by white carpal joint, middle and greater wing-coverts, and inner secondaries. Black distal bar of white tail becomes narrow on the outer feathers. Unlike in Southern L., feet do not protrude beyond tail-tip. **Juv.** with cinnamon feather-edges above. **Pull.** lighter buff than Southern L., with thin and incomplete black demarcations of the white nuchal collar.
Habits: Wary and noisy as the Southern L. Wings blunt, but in most display-flights the wing-tips look pointed, and the wing-beats are rapid and shallow. Upon alighting, always stands some time with raised wings, exposing the striking color pattern.
Voice: **Very noisy**. Mobs with a harsh *criee-criee-cri...*, staccato *cwi-cwi-cwi...*, or more mellow and melodic *dididi---celeec-celeec-celeec-ce....* Sometimes a low, tremulous *kwiwiwiwirrr*.
Breeding: Eggs mainly in Oct-Dec, occ. later.
Habitat: Wide shore meadows, open parts of marshes and rushy pasture and boggy terrain, but avoiding saline marshes. Sometimes also dry fields and hill-sides with short grass. Main nesting habitat is

partly inundated and hummocky, but dry grassland and heaths are frequented outside the breeding season.
Range: Continuously through puna and páramo zones from Catamarca, nw Arg. through w Bol., n Chile, Peru, and Ecu., to Nariño and Cauca, s Col., and casually to E Andes, Col. Locally very common at 3000-4500 m (down to 1500 m furthest s, overlapping slightly with Southern L. in nw Arg.). During austral winter accidentally descends to Pacific coast and the inner Amazon area.

NOTE: Previously in genus *Ptiloscelis*.

AMERICAN GOLDEN PLOVER *Pluvialis dominica* – Plates XVIII 10 and XX 7

25 cm. A medium-sized plover densely mottled and dotted yellow and dusky gray on all upperparts (**looking uniform olivaceous at a distance**), but grayish on nape, and much paler below (partly white in **ad.**). During the migration in Mar-Apr, black feathers emerge on all underparts from chin to vent (incl. cheeks and flanks). In **flight** shows buffy gray wing-linings, and only faint traces of a wing-bar, rump buff, tail-feathers dark with dense light barring (but outer feathers almost uniform grayish buff in **juv.**).
Habits: Social, sometimes migrating in large flocks. Wary, unlike the smaller plovers often adopting an upright stance. Flight fast with measured, deep wing-beats.
Voice: A range of loud, clear single or 2-note whistles as *kl-ee* or *chee-it*, unlike plaintive 3-note call of Gray P.
Habitat: At seacoasts and on open marshes, fields, and uncultivated shortgrass habitats, in the Andes mainly wide shore meadows near lakes.
Range: Migrant from Arctic N Am., mainly to the coastal plains from s Brazil to nw Arg., but regularly reaches Isla Grande. On s migration flies right across the Caribbean and Amazon areas, with only small numbers visiting the Andes. The n migration slants across the Andes to C Am., and large numbers can be seen around 4000 m on altiplanos of nw Arg., Bol., and Peru in Mar-Apr.

NOTE: The Siberian Golden P. *Pluvialis fulva* has been considered a ssp of *P. dominica* (in which case the vernacular name Lesser Golden P. is used). Although nesting e to w Alaska it **always** migrates w of the Pacific Ocean.

GRAY (or BLACK-BELLIED) PLOVER
Pluvialis squatarola – Plates XVIII 11 and XX 6

29 cm. Resembles Am. Golden P., but more chunky, **heavy-billed, and paler**, the upperparts speckled dusky gray and white, underparts mostly white. In **flight** shows conspicuous white wing-bar, underwing whitish contrasting **black axillaries**, rump white, tail white barred dusky. Not very social, and inland rec. are usually of single birds associated with Am. Golden P. Call a drawn-out, plaintive, 3-note *plee-oo-eee* that drops in pitch on the middle.

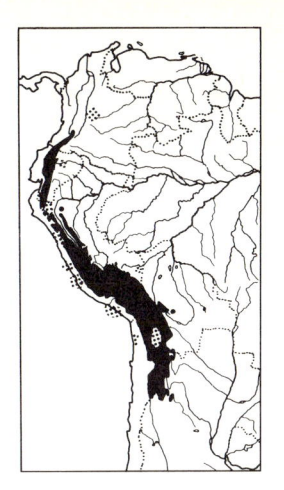

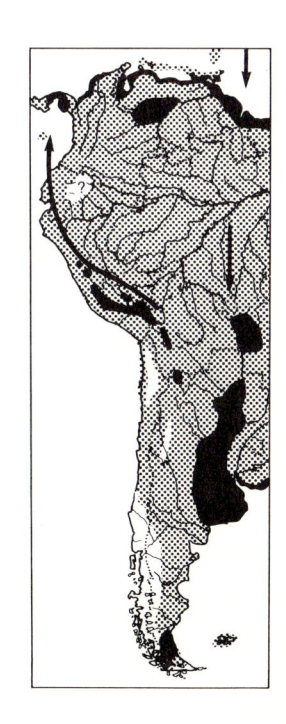

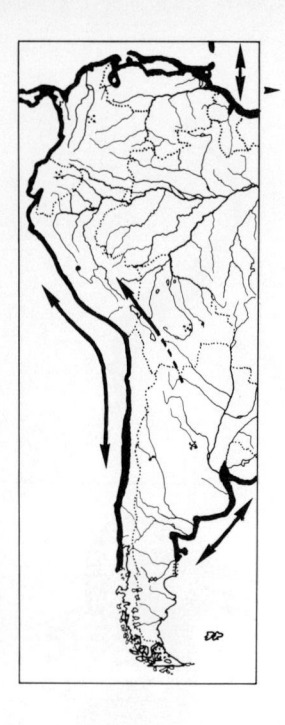

Migrant from arctic N Am. (and Asia?) to beaches, mudflats, and shore meadows mainly at the coasts of Peru, n Chile, and the La Plata region of Arg. Few rec. inland in C and E Andes, Col., e Ecu. and to 4100 m in c Peru (we suspect it to be regular here on n-wards migration).

SEMIPALMATED PLOVER *Charadrius semipalmatus* – **Plates XVIII 8 and XX 8** – a N Am. migrant to S Am. coasts (esp. Guianas) – is rec. accidentally around 4000 m on altiplanos of s Peru and nw Arg., and on Ubaté savanna (2600 m) in E Andes, Col. 17 cm. A small plover with stubby bill pale tipped black. Crown gray-brown with dark demarcation towards white frontal band, which continues as a white supercilium. Side of head with broad dark band from bill to occiput, white of throat continuing as **white nuchal collar**. **Breast with dark bar** continuing as dark anterior demarcation of the gray-brown back. Legs pink. The darkest zones are black in the breeding season, but diffuse, mainly dark gray-brown, when in the winter quarters. **Juv.** finely scalloped above. In **flight shows long white wing-stripe** and dark tail narrowly edged with white. Call a plaintive rising *cheweeet*.

NOTE: Sometimes considered a ssp of the Ringed P. (*Charadrius hiaticula*) of the Old World and Greenland.

KILLDEER *Charadrius vociferus* – **Plate XX 3** – of N, C and w S Am. – is rec. occ. during migration in Nov-Dec at 2500-4000 m in Andes of Col. and Ven. (*vociferus*). Frequents grassland with patches of bare soil. 24 cm. A medium-sized, slim plover patterned as Semipalmated P., except for having **2 black chest bars** in all plumages. Also characteristic **cinnamon-rufous rump** and base of the **long** and wedge-shaped, white-tipped **tail**. Very prominent white wingstripe. Singly or in loose groups. Calls comprise repeated, loud, and insistent *kill-deeah*, or plaintive *dee-ee*.

PUNA PLOVER *Charadrius alticola* –
Plates XVIII 6a-c and XX 10

17 cm, compact and 'neckless'. Above light gray-brown with some rufous peripherally on crown and nape (esp. in **male**). **Ad. with face white, demarcated by black bar across upper forehead** and through eye to side of neck. Chest with a faint drab gray bar, and another bar indicated on side of upper breast. Outside breeding season partly looses black pattern. Legs black. Birds from Catamarca, Arg. are very pale and almost lack traces of a chest-bar (see Plate XIX). In **flight**, primaries show white shafts, but **wing-bar** is indicated only on inner primaries; tail white laterally. **Juv.** lacks black pattern on head, but has mottled loral stripe, upperparts scalloped with pinkish buff, pectoral bar often conspicuous and quite rufous. **Pull.** buffy white with black speckles above.
Habits: Loosely social, but disperses to breed. When uneasy runs extremely fast, or flies low over the terrain. Feigns injury when disturbed with eggs or young.

Voice: Not very vocal. Thin *tseet* when running. Short *prit* in flight.
Breeding: Egg-laying on short, matted grass; mainly Sep-Oct, rarely Jan.
Habitat: Wide expanses of rather firm clay or mud, or partly flooded, strongly grazed shore meadows around salt and freshwater lakes.
Range: Puna zone from Junín, c Peru through ne Chile and w Bol. to Jujuy, Salta, and (isolated?) Catamarca, nw Arg. Mainly at 3000-4500 m, or even higher, but regularly descending to Pacific coast in Ica. Rather common in s and c parts of range.

NOTE: Sometimes regarded as a ssp of Two-banded P.

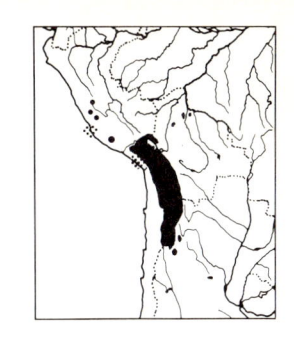

TWO-BANDED PLOVER *Charadrius falklandicus* – Plates XVIII 5a-b and XX 9

19 cm. Resembles Puna P., but darker and more boldly patterned. **Ad.** with black zone through eye continuing down side of neck to the black upper **pectoral bar, and another bold bar across lower chest** (on the Malvinas, many birds have interrupted upper bar). **Male** with extensive rufous color on crown and nape, **female** with more gray crown. **Off-season** with little rufous, side of head plain brown, chest-bars dark gray-brown. **Juv.** similar, but with buff feather-edges above and on the chest-bars. **Pull.** pale buff with black speckles and lines above.
Habits: As in Puna P., but usually very tame on the breeding grounds.
Voice: Thin *wheet* when running. Short *prit* or more liquid *wheet* in flight.
Breeding: Egg-laying in Sep-Dec, or even later in upland habitats.
Habitat: Gravelly and stony sea-coasts, but also breeds on beaches and short grass with patches of sand or gravel near streams and lakes in unforested inland areas.
Range: Islas Malvinas and southern S Am. from Cape Horn to c Arg. (breeding rec. in Córdoba) and Santiago, c Chile, and migrating to n Chile and se Brazil. Mainly along coasts, but breeds at least to 1200 m on the Andean foothills in Sta Cruz, s Arg. Common.

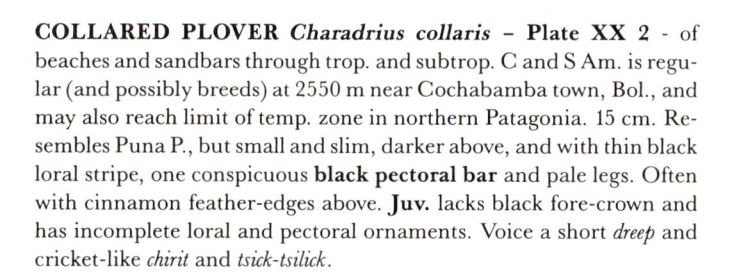

COLLARED PLOVER *Charadrius collaris* – **Plate XX 2** - of beaches and sandbars through trop. and subtrop. C and S Am. is regular (and possibly breeds) at 2550 m near Cochabamba town, Bol., and may also reach limit of temp. zone in northern Patagonia. 15 cm. Resembles Puna P., but small and slim, darker above, and with thin black loral stripe, one conspicuous **black pectoral bar** and pale legs. Often with cinnamon feather-edges above. **Juv.** lacks black fore-crown and has incomplete loral and pectoral ornaments. Voice a short *dreep* and cricket-like *chirit* and *tsick-tsilick*.

RUFOUS-CHESTED DOTTEREL *Charadrius modestus* – Plates XVIII 7a-d and XX 5

20 cm. **Foreparts dark**. **Ad.** unmistakable, with **white 'diadem' encompassing the fuscous crown**, face and throat deep gray, breast

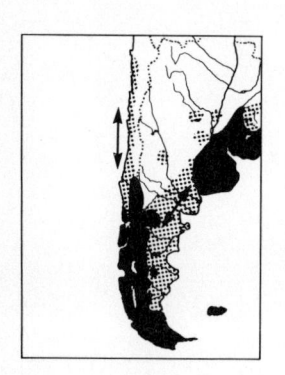

chestnut broadly bordered black against white belly. Upperparts of body dark gray-brown. **Off-season** diffusely gray-brown on entire head, neck and chest, except for buffy white supercilium and pale throat. **Juv.** has blackish upperparts with buffy yellow notches along the feather-margins as in Am. Golden Plover, supercilium whitish, chest brownish buff, normally finely barred with dusky feather-tips. This plumage is soon replaced by 1st winter plumage, with more uniform gray-brown foreparts. In rapid **flight** lacks wing-bar and shows black rump and **broad white sides of dusky brown tail**. **Pull.** golden buff conspicuously marked with black dots and lines, nape rufous.
Habits: In small groups, but disperse to breed. Active and often alert, bobbing on a summit. However, some birds are very confident.
Voice: Call a double, tremulous, slightly falling *peeetrr*, which is sometimes shortened and repeated. In courtship flight a *tik-tik-tik* alternating with rattling call and melodious *dreede-leedel-leedel*....
Breeding: Mainly in short matted vegetation. Eggs Oct-Nov, rarely Jan.
Habitat: Coastal shingle through eroded inland grassland, bogs, and heaths, often on high parts of the terrain. Sometimes breeds in stony terrain with alpine vegetation surrounding upland lakes.
Range: Islas Malvinas and southern S Am. from Cape Horn to Llanquihue, c Chile, and Sta Cruz, s Arg, migrating to n Chile and se Brazil. From coast to 2000 m on the windy foothills of inland Sta Cruz. Not uncommon.

NOTE: Previously placed in a separate genus *Zonibyx*.

TAWNY-THROATED DOTTEREL *Oreopholus ruficollis* – Plates XVII 8a-c and XX 4

27 cm. With rather **long and thin bill**, but typical plover jizz. Top of head, hind-neck, chest, and rump light brownish gray, rest of **upperparts buff heavily striped with fuscous**. Face light buff with black eye-line, **throat orange-rufous**, sides and belly light buff **with prominent black spot centrally**. Ssp *pallidus* pale. **Juv.** with hardly any ru-

Tawny-throated Dotterels often camouflage themselves by standing upright, turning the striped back towards the observer.

fous on throat, faintly scalloped upperparts, and small abdominal patch; legs gray (vs. dull pink in ad.). In **flight** resembles Am. Golden Plover, but has white wing-linings. Wing-stripe indicated on the primaries, and secondaries tipped white, tail mainly pale gray. **Pull.** cinnamon intricately patterned with black lines and zones of dense white 'powder-puffs'; *pallidus* chicks very pale.

Behavior: In flocks outside breeding season. Usually shy. When alert often bobs or stands very upright between short runs. When further approached will stand with striped back towards the observer, and becomes very well camouflaged. Flight high and direct, as in Am. Golden Plover.

Voice: In flight sad, somewhat reedy, falling *tryyy-y*, sometimes a long tremulous version, or shorter *deeyy*, occ. combined *deey terryyy-y-y*, and often in series.

Breeding: Eggs maybe in June-Sep (lomas near Peru coast), June-Sep (puna zone), Jan (Arequipa), Oct (Catamarca), Nov (Potosí), or Oct-Dec (furthest s).

Habitat: Mainly on windy semi-arid ridges of tola heathland and overgrazed puna grassland, or sandy areas with sparse vegetation. On migration also on fields and meadows in lowlands.

Range: Common throughout the Patagonian steppe, at 2000 m in Córdoba mts, c Arg., and on open heathlands of s Chile (migrates to se Brazil and occ. to s Ecu.). Continues to the plateaus of the s Andes and to 3500-4600 m in the puna zone of n Arg., ne Chile and w Bol. to se Peru, spottily at least to Junín, and seasonally in 'lomas' and irrigated parts of the coastal desert n to Lima, c Peru (*ruficollis*). Resident near coast of Lambayeque and Piura, n Peru (*pallidus*). Generally common in most of range, in the puna zone mainly during the southern winter, but probably resident locally.

NOTE: Sometimes placed in the genus *Eudromias*.

DIADEMED SANDPIPER-PLOVER *Phegornis mitchellii* – Plates XVII 10a-c and XX 20

18 cm. A **small**, distinctly colored plover with **long, thin bill**. **Ad.** with head dusky brown except for a **white 'diadem'** around front and sides of crown. Hind-neck rufous, rest of upperparts dark gray-brown. Lower throat with white bar, breast white with dense thin dusky bars disappearing gradually toward white belly. Legs bright yellowish orange. In **flight** the rather round wings and short tail are dark, except for white barring on inner secondaries and outer tail-feathers. **Juv.** and **pull.** lack the distinctive head-pattern, and have all upper- and foreparts coarsely spotted and barred fuscous and cinnamon-buff. Juv. soon molt to a plumage with grayer foreparts, but buff-edged wing-coverts may be retained through 1st year.

Habits: Alone or in pairs, or rarely a few together. Usually confident, but difficult to find, owing to its quiet behavior. Sometimes feeds hidden in eroded holes and creeks in the bogs, where it also may stand still for very long periods. Probes with its bill vertically. When active makes short fast runs, like a typical plover, and frequently bobs like a Spotted Sandpiper. The short flight is weak and undulating as in a passerine bird.

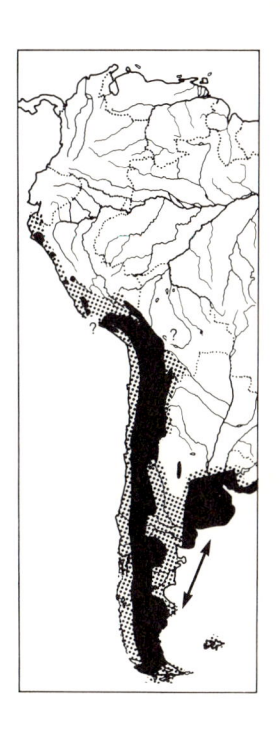

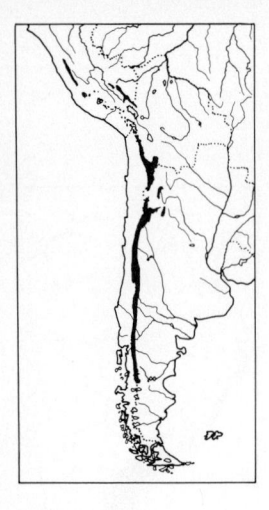

Voice: Usually silent, but a clear, plaintive, plover-like *pyeet* has been noted.
Breeding: Egg-laying in Oct-Dec (Chile) or Jan (Bol.).
Habitat: Waterlogged mossy tundra and bog with matted cushion-plant vegetation (esp. *Distichia* bogs) and gravel on river plains and near lakes.
Range: Andes from Chubut, s Arg., through n Chile and w Bol. to Ancash, nc Peru, breeding at 4-5000 m, but furthest s descending to 2000 m in winter. Poorly known, possibly rare or very local, although easily overlooked.

MAGELLANIC PLOVER *Pluvianellus socialis* –
Plates XVII 9a-c and XX 11

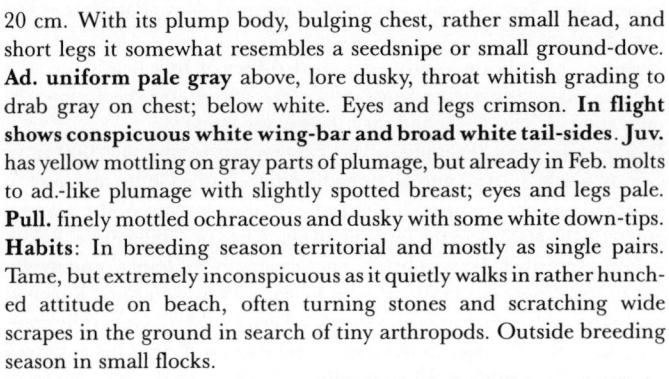

20 cm. With its plump body, bulging chest, rather small head, and short legs it somewhat resembles a seedsnipe or small ground-dove. **Ad. uniform pale gray** above, lore dusky, throat whitish grading to drab gray on chest; below white. Eyes and legs crimson. **In flight shows conspicuous white wing-bar and broad white tail-sides**. **Juv.** has yellow mottling on gray parts of plumage, but already in Feb. molts to ad.-like plumage with slightly spotted breast; eyes and legs pale. **Pull.** finely mottled ochraceous and dusky with some white down-tips.
Habits: In breeding season territorial and mostly as single pairs. Tame, but extremely inconspicuous as it quietly walks in rather hunched attitude on beach, often turning stones and scratching wide scrapes in the ground in search of tiny arthropods. Outside breeding season in small flocks. 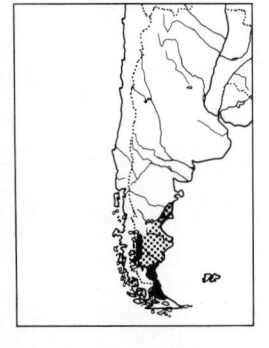
Voice: In spring, gives wide array of calls, incl. dove-like notes and *pip-wheet* at alarm. Otherwise quiet, except for sharp accentuated *ruee-ew* in flight.
Breeding: Eggs in Sep-Nov, to Dec in the highlands. Unlike in other waders the single chick chiefly stays in one spot, is fed for a long period, and develops very slowly.
Habitat: Near alkaline, fresh or salt, clear or clayey lakes with unstable water levels and wide clayey or pebbly shores.
Range: N Isla Grande, s Chile to Sta Cruz, s Arg., c. 1000 birds in the lowlands, and a few hundred at 700-1200 m on the upland plateaus in inland Sta Cruz. Winters on Patagonian coast n to Valdes peninsula.

NOTE: The relationships of this aberrant species is uncertain. Relationships with sheathbills Chionididae seems possible, and the species may be best placed in its own family.

Sandpipers – Family Scolopacidae

Altogether 85 species breed mainly in the N Hemisphere, but many occur in S Am. as migrant visitors. Most sandpipers are fairly small, slimmer than plovers, and with distinct necks and longer bills, which are flexible distally. Many long-billed species probe soft substrates for worms, arthropods, and molluscs, but also pick prey and seeds from the surface. The legs and toes are often long, but the hind-toe small. The migratory species are usually in winter plumage when in S Am., and then have grayish plumages.

Juv. are usually recognized as such by white or buff feather-margins above, which give a scaly effect. This plumage is replaced by a winter plumage in the course of the s-wards migration, but the inner median coverts are not molted, and identify 1st-winter birds. The untrained observer may find sandpipers confusing. However, differences in wing and tail patterns and calls permit safe field identification.

Except when breeding, sandpipers usually appear in flocks. The flight is rapid. Flying flocks often perform coordinated aerial movements, where all birds change direction simultaneously, with remarkable precision. Most species breed on moorland or tundra, but otherwise live on shores and mudflats.

Only snipes (*Gallinago*) breed in the Andes. They are monogamous and reputed for their aerial 'bleating'. All sandpipers lay 2-4 perfectly camouflaged, pear-shaped eggs in a simple scrape. The downy young are highly developed nidifugous chicks, which find their own food and soon become independent. The adults feign injury to distract enemies.

Further information can be found in Hayman, P, J Marchant & T Prater (1986) *Shorebirds - an identification guide to the waders of the World*. London - Sydney. See also Johnsgard, P (1981) *Plovers, sandpipers and snipes of the World*. Lincoln - London, and *Mem. S. Diego Soc. nat. Hist.* 3(1968):1-54. *Wader Study Group Bull.* 49 suppl./*IWRB Special Publ.* 7(1987): 57-69 summarises results of the comprehensive recent attempts to gather integrated data on shorebird populations of the Americas throughout the migration routes, e.g., by aerial counts. However, such studies have not yet covered the Andes. Papers based on observations in the Andes are *Rev. Chilena Hist. nat.* 57(1984):47-57, *Auk* 92(1975):442-451, and *Studies in Avian Biology* 12(1988):1-74 (on Wilson's Phalarope), and *Auk* 89(1972): 497-505 (on Banded Snipe).

Stints (peeps) or Calidridine sandpipers - Tribe Calidridini

26 arctic species, of which 11 regularly winter in the Neotropics. Most species are small, compact, not unlike small plovers, but less boldly patterned (except Ruddy Turnstone), and with distinctly longer, tapering bills. The taxonomy has varied, as the present genus *Calidris* is sometimes divided in *Calidris, Erolia, Ereunetes, Eurynorhynchus*, and *Crocethia*. The turnstones (*Arenaria*, 2 species) have had a complex taxonomic history.

RUDDY TURNSTONE *Arenaria interpres* – **Plate XX 1** – N Am. migrant to coasts of S Am. (esp. nc Brazil) – has been sighted at 4080 m in c Peru, and in extreme n Chile (4500 m, D Scott) (*morinellus*). 23 cm, stockily built, with short legs and short, conical and slightly upturned bill. In breeding plumage unmistakable, owing to black-and-white harlequin pattern of head, neck, and breast, and variegated rufous and black upperparts of body. In the non-breeding season all these parts are darker or lighter gray-brown with rather obscure pattern. **Juv.** similar, with more white on head, and scapulars and wing-coverts fringed with tawny buff. Readily recognized in **flight** by 2 white wingbars formed by lesser coverts and base of secondaries, and by **white mid-back separated by black rump from white tail with black bar.** Call *tuc-a-tuc* and sharper *kittik*, lower pitched than in most shorebirds.

RED KNOT *Calidris canutus* breeds locally in the northern high Arctic, and migrates along the coasts of S Am. to winter in s Arg. and Chile. Sighted at La Cumbre in La Paz (ssp?) (*American Birds* 40(1986):224). A bulky calidrid with almost straight conical bill, rather short greenish legs, clear but narrow wing-bar, and pale grayish rump and tail. **Ad.** non-breeding light gray above, white below, with some gray suffusion and marks on throat, breast, and sides. **Juv.** with buffy tinge and scaly pattern above. During spring molt plumage admixed with brick-red feathers.

PECTORAL SANDPIPER *Calidris melanotos* – Plates XIX 14 and XX 27

19-23 cm, male largest. Head rather small, bill = head, slightly drooping. Above richly hued brown with dense black streaks, and with 2 pale stripes indicated along each side of back. In all plumages with strongly **streaked brown breast ending abruptly, and contrasting white belly**. The breast is particularly developed (dark, with pale mottling, and somewhat bulging) in males in breeding plumage, and is lighter, more buffy in juv. Juv. is brighter all over, owing to more buffy feather-edges. White eyebrow broadest in front of eye (most distinct in juv., streaked in ad.). Feet ochraceous or olive. In **flight** upperside of wings look uniform dark. Black on lower back and centrally on rump and otherwise gray tail, black zone contrasting long white patch on either side of rump.
Habits: Normally not very social. Often wary, with a tendency to stretch upwards. Takes off in zigzag, much as a snipe, but sustained flight somewhat sluggish.
Voice: Call rich, but low *chuck*, *trrik*, or *chree-eep*.
Habitat: Fairly open parts of rushy shore meadows and marshes, and sometimes on bogs or open grassland. Rarely on open beaches or mudflats.
Range: Discontinuously holarctic, N Am. birds wintering in s half of S Am. from coast to inland. Fairly common on upland plateaus of Patagonia and at 3500-4500 m through puna zone of nw Arg, n Chile, w Bol., and s and c Peru, and on passage (mainly s-wards) also in highlands of Ecu. and Col.

WHITE-RUMPED SANDPIPER *Calidris fuscicollis* – Plates XIX 13 and XX 24

16 cm; quite thick, slightly drooping bill = head. Like Baird's S., with rather long wings (reach beyond tail). In winter quarters drab gray becoming white on lower underparts, and recognized in **flight** by thin short wing-stripe, and **white rump** contrasting dark tail. Legs black. Shows brown-tinged crown and ear-coverts, short white supercilium, fine dark stripes and spots on breast and flanks, and very faint shaft-streaks above. **Juv.** has rusty feather-margins above, or white margins along edge of mantle and scapulars. In **Breeding plumage** (partly adopted from Mar) has black back-feathers with irregular rusty margins, and distinct dusky streaking on foreparts and flanks.

Habits: Usually in small single-species flocks (rarely large flocks), which feed along the shoreline or in shallow water, picking and probing. Quite tame.
Voice: Call a characteristic thin mouse-like *tzeet*, or low *tip-tip*.
Habitat: Winters on coastal mud-flats and inland along lake margins with clay, pebbles, or short grass, often with tundra-like surroundings.
Range: Breeds in arctic N Am. and winters esp. in s Brazil and s Arg. Common on the Patagonian plains from coast to the numerous lakes and ponds on the foothills e of the Andes to at least 1200 m. Migrates e of Andes, but a few can be seen in highlands of Ecu. and Peru, and on Pacific coast of Peru and Chile.

LEAST SANDPIPER *Calidris minutilla* – Plate XX 29 - N Am. migrant to muddy S Am. sea-coasts s to n Chile (Arica) and se Brazil – is rec. casually in highlands near Quito, Ecu., at 4080 m in c Peru, and at 2550 m in Cochabamba, Bol. 14 cm, with **needle-thin bill** with slight droop. Above quite dark gray-brown with inconspicuous pale V on mantle. Throat and breast well streaked dusky brown, but lower underparts white. Legs pale (yellowish green, but may appear dark). **Juv.** has scaly rufous pattern above, with whitish stripes bordering the scapulars. In **flight** shows thin white wingstripe and whitish sides of rump and tail. Distinctive crouched appearance when feeding or running. Mainly feeds by pecking. Call high-pitched, drawn-out *three-eet*, and lower *prrit*.

BAIRD'S SANDPIPER *Calidris bairdii* –
Plates XIX 12 and XX 28

18 cm, with thin, slightly drooping bill. Long-winged as White-rumped S., but more elongate. Above buffy gray-brown obscurely blotched but never striped. **Foreparts buffy brown**, but eyebrow and throat whitish, and becoming white on lower underparts. Legs black. **Juv.** slightly scaled on back. In **breeding plumage** (partly attained from Mar) the back has black splotches. In **flight** shows very faint wingstripe, and dark median zone of lower back, rump, and tail.
Habits: Often seen in small groups or singly, and large flocks are often dispersed. Usually quite tame. Feeds mainly on land by pecking rather than probing.
Voice: Call low and raspy *tzurr*, *kreet*, or doubled *tjirrirr*.
Habitat: By lakes and marshes, but may be most typical of wide, damp, strongly grazed shore meadows with muddy, partly dry ponds. Often also far from water, on plains and slopes with short grass.
Range: Breeds in arctic N Am. and mainly migrates through the Andes to inland Patagonia. May pass rapidly through Col. (Aug-Oct), but small numbers occur Aug-Mar in páramos of Ecu., and it is common at 3500-4700 m through the puna of Peru, n Chile, w Bol., and nw Arg., with high numbers present in Sep-Nov, and fair numbers also Jan-Mar.

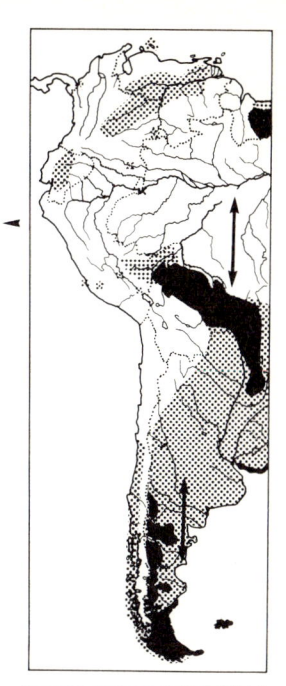

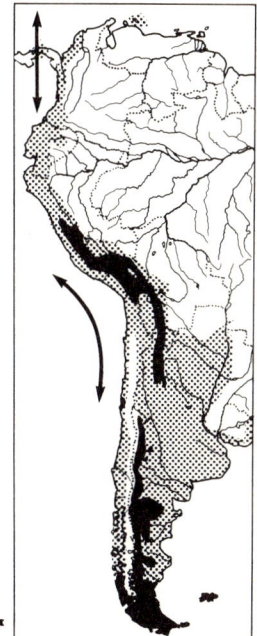

SANDERLING *Calidris alba* – Plate XX 25

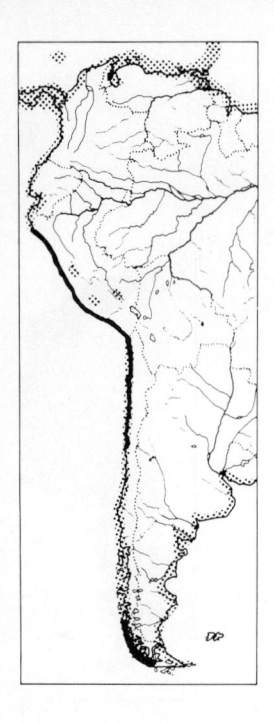

19 cm, incl. 2.5 cm long, **decidedly thick black bill.** In **winter plumage** white with crown and upperparts of body pale gray with large sooty 'shoulders'. **Juv.** checkered black-and-white above, with some pinkish buff on head and breast. **Breeding plumage** (attained from Mar) with mixed black and cinnamon foreparts and back. In **flight** shows long and **flashy white stripe on dusky wings**, and white sides of gray tail. Runs exceedingly fast, like a small plover. Call a short *trick*.

Holarctic. N Am. birds winter (and remain 1st year) on Pacific beaches of Peru and n and c Chile. On s-wards migration across the Caribbean and Col. singles and small flocks appear in Cauca valley, Col., and at 3500-4000 m in c and se Peru, but the n-wards migration may be restricted to the Pacific coast. Smaller numbers migrate across the Amazon area and winter along coast of Arg.

NOTE: Previously in a separate genus *Crocethia*.

STILT SANDPIPER *Micropalama himantopus* – Plate XX 26

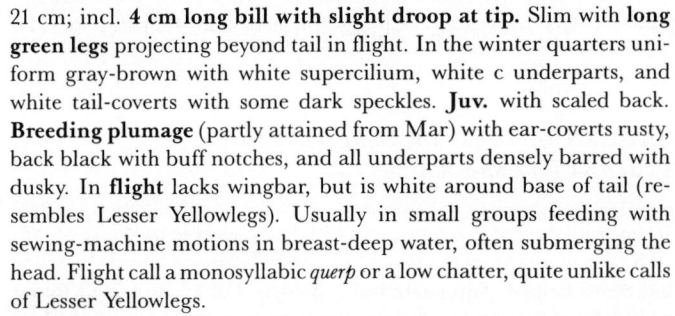

21 cm; incl. **4 cm long bill with slight droop at tip.** Slim with **long green legs** projecting beyond tail in flight. In the winter quarters uniform gray-brown with white supercilium, white c underparts, and white tail-coverts with some dark speckles. **Juv.** with scaled back. **Breeding plumage** (partly attained from Mar) with ear-coverts rusty, back black with buff notches, and all underparts densely barred with dusky. In **flight** lacks wingbar, but is white around base of tail (resembles Lesser Yellowlegs). Usually in small groups feeding with sewing-machine motions in breast-deep water, often submerging the head. Flight call a monosyllabic *querp* or a low chatter, quite unlike calls of Lesser Yellowlegs.

Migrant from subarctic N Am. to ponds and flooded grassland of the Pantanal from s Bol. to se Brazil and c Arg. Migrates e of Andes and along Pacific coast from Col. to Peru, and passes the altiplanos (sometimes common in lake Uru-uru in Oruro, Bol., and Lake Pozuelos in Jujuy, nw Arg.).

NOTE: Should possibly be included in *Calidris*, despite its lanky build.

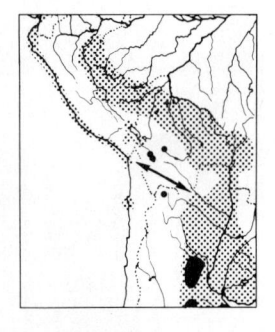

SHORT-BILLED DOWITCHER *Limnodromus griseus* – N Am. migrant to the coasts of northern S Am. – is sometimes seen at 3700 m near Cuzco, Peru (B. Walker). A fairly large long-billed species (**snipe-like by shape and action**), with a prominent blaze of **white up c back visible in flight**. Winter plumage rather plain grayish, with eyebrow, throat, and belly white. Juv. as well as ad. in breeding plumage extensively light rufous. Voice a mellow *tu-tu-tu* or *tuu*.

BUFF-BREASTED SANDPIPER *Tryngites subruficollis* – **Plate XX 30** – N Am. migrant to the short-grass habitats of the pampas of

Parag. and c Arg., migrating e of Andes – is rec. casually in E Andes, Col., and in n Ecu. Possibly overlooked in the highlands. 20 cm, with quite small head and small thin bill. Rather uniform **light buff with pale face and white eyering** ('innocent look'); crown and upperparts of body with black feather-centers and buff edges, which give a uniform scaly effect. **Legs yellow**. In **flight** shows a peculiar mottling on underside of primaries, outside the wrist. Tame and gregarious, often with Am. Golden Plover, and feeding with plover-like action. Calls include harsh *prrik* and sharp *tik* notes.

Phalaropes – Tribe Phalaropodini

3 tiny species related to tringine sandpipers, but previously often classified with avocets, with which they share partly webbed feet. Nest by grassland and tundra ponds in the N Hemisphere. All may winter in S Am., 2 along coasts, 1 inland.

WILSON'S PHALAROPE *Phalaropus tricolor* – Plates XIX 15ab and XX 19

23 cm, incl. 3 cm needle-thin bill. Elegant, with rather small head and slender neck, but peculiarly pot-bellied. In winter plumage white, with crown, eyestripe, hindneck, and upperparts of body light gray. **No wingbar**. Square **white rump separates gray tail**. **Juv.** has fuscous back with buff fringes, but **all** juv. feathers are replaced before end of year. Breeding plumage (partly attained in the migratory periods) has black eyestripe continuing as broad chestnut band down side of neck (esp. in female) and along c back, and blackish wing-coverts. Legs yellow (black in breeding plumage, this color developing from Jan).
Habits: Floats buoyantly on the water. Social, sometimes in enormous flocks associated at times with feeding Chilean Flamingo, which stirr up small invertebrates from the lake bottom. The phalaropes also stirr up prey themselves by spinning movements. In medium-large groups, the feeding is well organized, most birds swimming parallel to one another back and forth within the group. They feed swimming forward with the opened bill in the water, or up-ending. Sometimes feeds wading in belly-deep water, or walking near the shore.
Voice: Calls grunting *wurk*, *ca-work*, or louder *wah*, sometimes combined *wahgagaga* (like a subdued Chilean Flamingo, and very unlike the thin *kik* notes of other Phalaropes).
Habitat: Shallow parts of open alkaline lakes and lagoons with fresh, brackish or salt water, and on inundated parts of wide shore meadows.
Range: Breeds in c N Am. At other seasons it may appear anywhere in southern N Am. and the Neotropics, but with significant numbers only at a few highly productive localities: staging and molting in July-Aug esp. in Mono Lake, California USA, migrating non-stop over the e Pacific to land in Peru, and wintering on altiplano of w Bol., and in Arg. Juv. birds migrate more slowly, and over a broader front. 1000s can be seen in several lakes from Puno, Peru, into Bol., and 100,000s on lakes Hedionda, Kalina, and Loromayu. Also numerous at times in the chaco/pampa transition of Arg. (100,000s in Mar Chiquita in Córdoba), and on uplands of the Patagonian steppe, s Arg. Common in the migratory periods in some lakes in c Peru.

NOTE: Intermediate between a tringine sandpiper and the 2 other phalaropes, it is sometimes placed in a separate genus *Steganopus*.

Tringine sandpipers – Tribe Tringini

16 N Hemisphere species, of which 6 regularly visit S Am., migrating on broad fronts. Small to medium-sized slim waders, usually with rather lanky build. Often bob head and tail and readily take flight when alarmed. Usually single or a few together.

LESSER YELLOWLEGS *Tringa flavipes* – Plates XIX 9 and XX 16

26 cm, incl. **3.5 cm long straight bill. Legs yellow**. In **winter plumage** above drab gray only slightly broken with dusky and pale spotting; head, neck, and breast light gray, white on upper lore and on throat, and very vaguely streaked. Underparts of body white with some gray mottling on flanks. **Juv.** very uniform above, with regular pattern of small white dots evenly along all feather-margins (partly retained 1st winter). **Breeding plumage** (adopted from Mar) with prominent whitish and black speckles above. In **flight** shows **uniform dark upper wing and broad white base of the white-barred tail**.
Habits: Although often in large flocks on migration, not very social on the wintering grounds, where often defending feeding territories. Sometimes associates with calidridine sandpipers. Bobs 'nervously'. Wades in shallow or belly-deep water, often walks on floating waterweeds, with elegant high steps, and sometimes swims (unlike Greater Y.). Feeds by pecking at the surface and seldom by probing.
Voice: Flight call 1-3 rather flat **soft notes *tew* or *tew-tew***. When disturbed sometimes gives long series of evenly spaced *cew* notes, bobbing for every note.
Habitat: Mud-flats, shores, margins of grassy ponds, or on inundated grassland, preferring more sheltered locations than Greater Y.
Range: Breeds in boreal zone of N Am., wintering s throughout C and S Am., with peak abundance in warm savanna and grassland zones. Also throughout the puna of Peru, Bol., and nw Arg. and n Chile, but sporadic in highlands of Col. and Ecu. A few birds 'oversummer'.

GREATER YELLOWLEGS *Tringa melanoleuca* – Plates XIX 8 and XX 12

35 cm, incl. **5.5-6 cm long, heavy and slightly upturned bill** (light basally). Slim, and with **long, strongly yellow legs**. **Winter plumage** drab-gray with some white notches and dark speckles along the dorsal feather-margins, and quite streaked on head, neck, and breast; throat and underparts white with slightly gray-mottled flanks. **Juv.** has very regular pattern of small white notches along all dorsal feather-margins, and light drab gray, very vaguely marked foreparts. **Breeding plumage** (partly adopted from Mar) with irregular black dorsal splotches. In **flight** shows **uniform dark upper wing, whitish rump and light barred tail**.

Habits: Vary. Normally not very social, sometimes defending feeding territories in winter. Feeds in shallow water, probing or skimming the water in avocet manner, and often dashing after small fish.
Voice: In flight gives series of **3-5 loud ringing notes** *tew-tew-tew...* ('Dear! Dear! Dear!').
Habitat: Mudflats and rather wide open margins of lakes, lagoons, and ponds. Also open inundated parts of grasslands.
Range: Breeds in boreal zone of N Am., and winters s through all parts of the Neotropics, both coasts and inland, with main occurrence on warm savanna and grassland zones. Also to above 4000 m in Andes from Ven. s to nw Arg., being fairly common in highlands of Peru, and in Jujuy.

SOLITARY SANDPIPER *Tringa solitaria* – Plates XIX 10 and XX 22

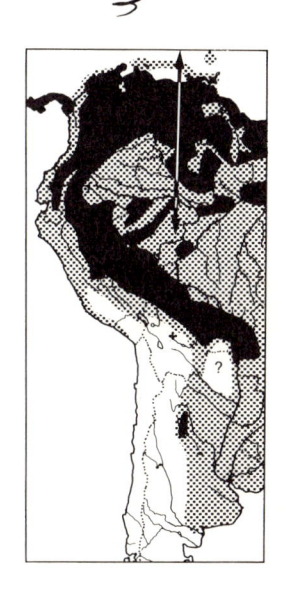

20 cm; incl. 3 cm long, thin bill. Like a diminutive Lesser Yellowlegs, but darker brown (looks black above in breeding plumage), with tiny pale spots on the feather-margins (buffy in *cinnamomea*, more whitish in *solitaria*, *cinnamomea* also distinguished by finely dotted lore without distinct dark bar), and conspicuous **white eyering** and upper lore. **Legs olive-gray. In flight** dark-winged (with densely barred under-wings), and also recognized by **dark rump and tail** with white-barred tail-sides.
Habits: Usually single. Flight darting, with quick almost swallow-like wing-strokes. Feeds wading in shallow water, nodding head as it walks slowly, and sometimes vibrating the leading foot to disturb insects.
Voice: When flushed gives thin *peet* or *peet-weet-weet*.
Habitat: Muddy or vegetation-fringed margins of freshwater lakes, inundated parts of grassland, or stagnant pools, forest ponds, ditches, and even puddles.
Range: Breeds in boreal zone of N Am., wintering throughout n and c parts of S Am. to s Peru and Río Negro, c Arg. Peak abundance on trop. savannas, but occurs sparsely July-Apr up to 2600 m, rarely straggling to near 4000 m through Andes of Ven., Col., Ecu., and Peru (*solitaria*, *cinnamomea*).

WANDERING TATTLER *Tringa incana* – Plate XX 21 – of Alaska winters throughout the trop. Pacific areas, and has been rec. once in the puna of Peru. 28 cm, incl. 4 cm long, rather thick bill. Pale yellow legs much shorter than in the yellowlegs. In winter plumage uniform drab gray, except for white stripe above dark lore, and white c underparts. Breeding plumage (adopted from Mar) with dense dark bars below. **Juv.** with whitish scaling above. **In flight looks totally unpatterned**. Not very social. Walks with bobbing action. Flight call an accelerating ringing trill *peew-ti-ti-ti-ti*.

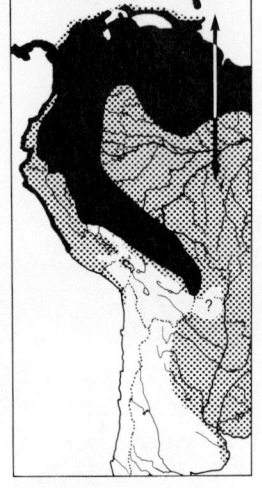

SPOTTED SANDPIPER *Tringa macularia* – Plates XIX 11 and XX 31

19 cm. **Winter plumage** above quite uniform olive-brown, with only faint dusky streaks and bars. Thin whitish eyebrow. Underparts white, with **color of upperparts spreading over the breast-sides**, and contrasting a white 'peak' in front of wing. **Juv.** more barred with dusky brown and buff feather-margins above. **Breeding plumage** (partly adopted in the migratory periods) with black-spotted underside. In **flight** shows white wingstripe and white-barred tail-sides.
Habits: Not very social, often defending feeding territories on migration. Recognized at once by its 'teetering' behavior with incessant bobbing movements, and by **flying with shallow and vibrating stiff wingstrokes below the horizontal plane**, and short glides (entirely unlike the deeper wingstrokes of other small shorebirds). Feeds on insects, almost exclusively along water-margins.
Voice: When flushed, gives characteristic high-pitched *weet* or *peet-weet-weet*.
Habitat: From sea level to alpine tarns, probably preferring lakes with wave-exposed wood-fringed shores and shaded margins of watercourses, but also common on floating-leaf vegetation along the outer edge of reed-beds.
Range: N Am. migrant wintering s to c S Am. (c Chile and c Arg.). Very common Aug-May at 2500-3100 m and sometimes higher in Andes of Col., more sparse in Ecu., and with small numbers wintering in the puna s-wards to c Peru and occ. Cuzco and Arequipa, s Peru.

NOTE: Often placed in the genus *Actitis*.

WILLET *Catoptrophorus semipalmatus* – N Am. migrant to the coasts of mainly the n half of S Am. – is sighted accidentally in the temp. zone in Junín and Cuzco, Peru. 35 cm. A gray-brown sandpiper, non-descript except for bold wing-pattern with **broad brilliant white wingbar separating** blackish wing-linings, primary-coverts, and flight-feather tips. Bill thick, straight, and dark. Tail gray. Resembles Hudsonian Godwit. The 3-note flight-call is much harsher than that of Greater Yellowlegs.

Curlews and godwits – Tribe Numenini

13 N Hemisphere species, of which 5 regularly winter in the Neotropics. Medium-sized to quite large shorebirds with long neck and usually long bill.

UPLAND SANDPIPER *Bartramia longicauda* – Plate XX 18 – N Am. migrant to the grasslands from Parag. to c Arg. – is rec. accidentally in Andes of Ven., Col., Ecu. (15 in flock, PG), Bol., and c Chile. 28 cm, with only 3 cm bill, but **long tail for a shorebird**. 'Pigeon-headed' with big eyes and slender neck. **Buffy brown densely streaked** dusky throughout, albeit lightly on the underparts. Legs yellow. In **flight** shows strongly barred underwings and sides, and black rump.

Flies with shallow wingbeats. Often holds wings elevated upon alighting. Calls *kip-ip-ip-ip*, or more rolling calls.

WHIMBREL *Numenius phaeopus* – **Plate XX 14** – N Am. migrant to C and S Am. seashores and tidal swamps (esp. ne Brazil and Chile) – is rec. accidentally in temp. zone of E Andes, Col., and around 4000 m in c Peru and n Bol. 43 cm, incl. **6.5-9.5 cm long bill with curved distal portion**. **Head striped**: crown dusky with buff mid-stripe, broad supercilium buffy, and eyeline dark. Upperparts dull brown with diffuse pale notches along all feather-margins, rest of plumage pale buff with dusky streaks and 'arrowheads', grading to white on belly. Shows mottled wing-linings and somewhat barred flight- and tail-feathers. Flocks often fly in line formation. Voice a sharp, even tittering *ky-ky-ky-ky-ky-ky*.

HUDSONIAN GODWIT *Limosa haemastica* – **Plate XX 13**

38 cm, incl. 7-9.5 cm **long, straight or slightly upturned bill. Winter plumage** light drab gray, except for white throat, belly, vent, and rump. In **breeding plumage** (partly adopted during migration) the underparts are chestnut with some dusky barring. In **flight** shows **long white wingbar**, dusky wing-linings and **black tail**. The flight-call is a double, slightly rolling *toe-wit*.

Breeds in subarctic N Am. Migrates in long non-stop flights, mainly staging in Surinam, and mainly wintering on ne Isla Grande, and on Chiloé Isl., Chile. Visits upland plateaus of the Patagonian steppe in small numbers, and rec. accidentally on altiplano of s Peru and nw Bol., and at 2680 m in n Ecu.

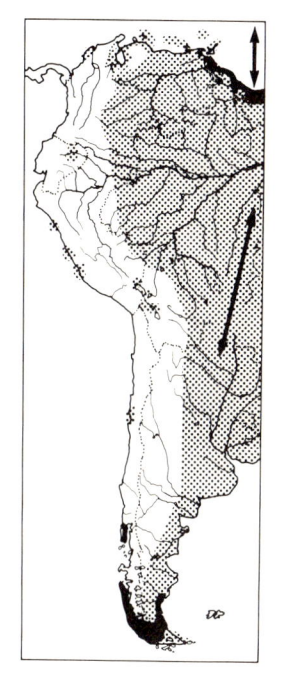

Snipes and Woodcocks – Subfamily Scolopacinae

14-16 species of snipe have a world-wide total distribution, while the 2 woodcock-snipes are Andean, and the 4 woodcocks inhabits the N Hemisphere and Indomalayan archipelago. All species have long and flexible soft-tipped bills, a peculiar visage owing to an elevated position of the eyes, and short legs. The bill is held partly downwards both on ground and in flight. The plumage is intricately mottled buffy brown, often with buff stripes demarcating the dorsal feather-tracts, and blends perfectly with the vegetation. All species are secretive and tight-sitting, and are rarely seen except when suddenly flushed at close range. Their presence is usually revealed during the evening twilight by characteristic calls and aerial displays in which the outer tail-feathers are spread and vibrate (influenced by modulations of the airflow made by the wings), producing a hollow, vibrating sound ('**winnowing**' or '**bleating**').

CORDILLERAN SNIPE *Gallinago stricklandii* –
Plate XIX 5a-b, 6a-b

30 cm, incl. 8-9 cm long, quite deep bill. **Heavy-bodied**. Above dense-

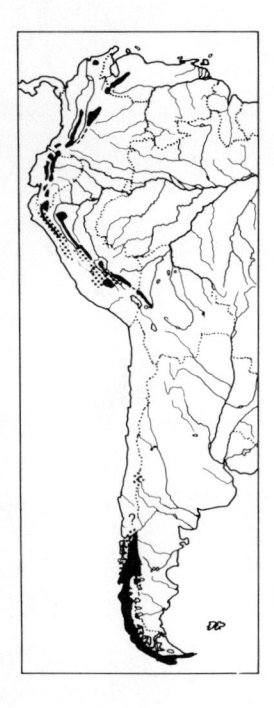

ly variegated fuscous with buff and cinnamon-brown feather-edges, but **not prominent light stripes** of smaller snipes. Broad dark zone along crown has diffuse buff mid-line. Face to foreneck warm buff variously stippled and streaked with fuscous, and with indications of eyeline and ear-stripe. Breast and lower underparts rather uniform warm buff (*stricklandii*), or pale, more grayish buff with dense dusky brown barring (*jamesoni*). **Juv.** has more prominent pale feather-edges bordering mantle and scapulars. In **flight** shows broad blunt wings with barred undersides, and unlike smaller snipes **no rufous or white in tail**. Feet do not project beyond tail. **Pull**. *jamesoni* generally tawny buff with bold fuscous pattern, and some buff-white stripes on head and down back; *stricklandii* brighter, with some indications of white 'powder-puffs' in the pale zones.

Habits and **voice**: Hidden in vegetation by day. When flushed, gives explosive *tzhyc*, and soon drops into cover. Most often heard shortly after sunset: From the ground gives long, sometimes interminable sequences of piercing *djyc* calls with rising and falling pitch (*jamesoni*), or *cheep cheep cheep...* (*stricklandii*). Also rodes in wide circles, with shallow wingbeats, calling loudly *witcheeuh witcheeuh...* for long periods, interrupted only by shallow stoop with a 2-3 s 'winnowing' *wzzzzzzhrrrrrrrr* (wheezy, with deep undertone in 2nd part, resembling a distant jet-plane).

Breeding: Eggs and pull. in May, Oct. (Col.), and Nov., Feb (Ecu.). Pull. Dec (s Chile).

Habitat: Ssp *jamesoni* inhabits ecotones of boggy páramo grassland, incl. *Espeletia* areas towards wet treeline forest, often a mosaic of grassy bog, bamboo (*Chusquea, Swallenochloa*) and lichen-clad dwarf forest, and sometimes cushion-plant bogs. Ssp *stricklandii* occurs from open swampy forest to coastal moorland with dwarf shrub and miry parts, and in wet grassy places above the treeline.

Range: At 2100-3400 m (sometimes to 4300 m in rainy season) in most páramos from Trujillo to Tamá, nw Ven., and Sta Marta mts through Col. and Ecu., very locally on w Andean slopes of n Peru and in some montane valleys, but more widespread along e Andean slopes through Peru to La Paz, Bol (*jamesoni*). Breeds esp. on isls around Cape Horn, with decreasing numbers in the fjordlands and deep valleys of the s Andes and n to Bío Bío (and accidentally Concepción) s Chile, and probably into adjacent extreme w Arg., but now may be extinct in the n parts of this range (*stricklandii*). Not uncommon, but difficult to find.

NOTE: Sspp *stricklandii* and *jamesoni* are sometimes treated as distinct species, **CORDILLERAN** and **ANDEAN SNIPES**, resp. These 2 forms and the Banded Snipe are intermediate between the smaller snipes and the stocky woodcocks (*Scolopax*) of other continents; *stricklandii* being most snipe-like, the Banded Snipe very similar to the Dusky Woodcock (*S. saturata*) of Java and New Guinea. The 'woodcock-snipes' are sometimes placed in a separate genus *Chubbia*.

BANDED SNIPE *Gallinago imperialis* – Plate XIX 7

30 cm, incl. 9 cm long, quite deep bill. A stocky **dark rufescent snipe boldly barred with black** throughout, except for crown (black with rufous midstripe), face, and neck (streaked), and abdomen (barred

Flight profiles of Noble, Cordilleran, and Banded Snipes

black-and-white). The uniform sepia-brown tail is covered by barred tail-coverts. In **flight** shows very broad rounded wings, and very short tail.
Habiats and **voice**: Crepuscular or nocturnal. Rodes in wide circles at dawn and dusk well into the darkness, giving c. 10 s long series of rough staccato notes that rapidly increase, then decrease in volumen, with double and triple notes in the climactic section, and then single notes decreasing in volume. Then makes a sharp dive, producing a clearly audible rush of air. Uniform sequences of 5-8 harsh notes are possibly given from the ground.
Breeding: Display rec. in July and Aug (Cuzco).
Habitat: Mountain ridges with dense, well-watered elfin forest with tree-ferns, and bamboo-fringed glades with marshy grassland that have thick layers of *Sphagnum* mosses.
Range: 19th Century rec. from E Andes, Col. (probably from páramos de Choachí and Laguna de Siecha). Recently found at 2745-3500 m along e Andean slopes of Peru in Piura (Cerro Chinguela), Amazonas (Cord. de Colán), e La Libertad (near Tayabamba), and Cuzco (Cord. Vilcabamba, Machu-Picchu (?), and Abra Malaga in Cord. Vilcanota). May be widespread, but very local, hard to collect, and difficult to find without being familiar with its song.

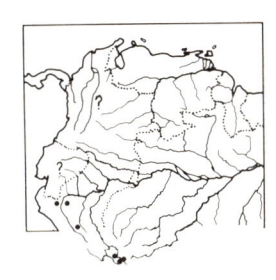

NOTE: Also called Imperial Snipe.

NOBLE SNIPE *Gallinago nobilis* – Plate XIX 4a-b

32 cm, incl. **8-10 cm long clearly 2-toned bill**. With rather **rich brown general hue**. Crown fuscous with buff midstripe, rest of head warm buff with dusky streaks suggesting eyeline and earstripe. Upperparts of body densely spotted and vermiculated with black, and with whitish lines demarcating the feather-tracts. Humerals and wing-coverts mostly tawny, the smaller and median wing-coverts often distinctly barred with white feather-tips. Neck and breast light brown

with obscure dusky streaks and mottling, flanks white with coarse dusky barring, belly white. Tail light rufous, with white sides barred black, the rufous hue extending to rump. Differs from Woodcock-snipes by lighter bill and less heavy build, **striped back**, unbarred c underparts, and above all the **rufous tail**. Differs from the following forms by **large size**, more tawny chest, more extensively rufous tail region, and densely barred wing-linings (underside of wings appearing mottled dark gray), and furthermore **lack of white trailing edge of wing**. The flight profile is intermediate between that of the round-winged woodcock-snipes and narrow-winged small forms. **Pull.** rich chestnut throughout, with black pattern, and rows of quite large and fluffy white 'powder-puffs'.

Habits and **voice**: Poorly known. Nasal grating *dzhit* and *tzhi-tzhi-tzhi...* when flushed. Flight rather heavy, slow, and direct (not twisting). Said to give a whistling winnowing sound in circling aerial display. Sometimes several together, and what may have been an assembly of 30-40 of this species was heard at late dusk from the rush-bed at Laguna del Limpio (3850 m) in Cotopaxi, Ecu. on 9 Oct (NK) (we suspect that this could be a lek behavior, as in the Palaearctic Great S. *Gallinago media*).

Breeding: Eggs known from July (Ven.), and Sep (Col.), pull. Sep (Ecu.).

Habitat: Said to inhabit grassy páramo bogs, and boggy glades in cloud forest, although less associated with peripheral bamboo and shrubbery than Cordilleran Snipe. In our experience most typical of rich habitats as rushy pasture and reed-marsh surrounding eutrophic lakes.

Range: At 2500-4000 m from páramo de Tamá, w Ven., through E Andes, in n end of W Andes and s part of C Andes of Col., and in most páramos of Ecu. s to Cerro Chinguela in Piura, n Peru. Locally common.

NOTE: Also called Páramo Snipe.

NOTE: This and the following species were previously placed in the genus *Capella*.

NORTH AMERICAN SNIPE *Gallinago gallinago* – Plate XIX 1

26 cm, incl. 6.5 cm long bill. Intricately speckled and vermiculated buff and black. Crown black with buff mid-stripe and supercilium, face buffy with dusky streaks suggesting eyeline and earstripe. Mantle mostly black, humerals more vermiculated with orange-tawny, and all dorsal feather-tracts demarcated with buff stripes. Wing-coverts mottled dusky and buff. Foreparts buff with rather obscure fuscous streaks changing to bars on the flanks. Belly white. **Tail rufous with white sides** barred with black, but **with mottled olive-buff rump** (vs. rufous hue extending to rump in Noble S.). **Juv.** virtually identical, but the wing-coverts more neatly fringed with whitish (scaly pattern).

Habits and **voice**: Take-off explosive, with distinctive rasping *dzhyt* or *tzhat* call with slightly rising inflection. Rises fast, then adopts zig-zag flight. The aerial 'winnowing' (sometimes heard on the wintering

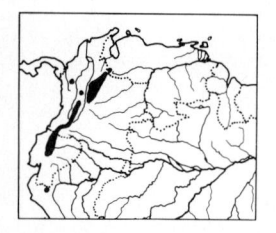

grounds) is a rich tremulous *whowhowhowhowho...*, sometimes alternating with vocal *yack*-ing.
Habitat: Bogs and mires, often with scattered shrubs, and rushy pasture, riparian meadows, and rather open parts of reedswamp, sometimes also muddy shores.
Range: N Am. migrant which in July-Mar visits S Am. s to e Ecu. and the Orinoco basin. Mostly lowlands, but also in temp. and páramo zones to 3900 m in Sta Marta mts and C and E Andes, Col., and to 3500 m in mts of n Ven. (*delicata*).

NOTE: Forms a superspecies with Magellanic and Puna Snipes. All forms are often lumped as 1 species, but differences in size (viz., the occurrence of large-sized birds in the trop. zone, and tiny birds in the high Andean climate), outer tail-feathers, and quality of the 'winnowing' suggest 3-4 species: Magellanic S. (*magellanica*, with *paraguaiae*), Puna S. (*andina*), and maybe also Eurasian S. (*gallinago*) and N Am. S. (*delicata*) separated.

MAGELLANIC SNIPE *Gallinago magellanica* –
Plate XIX 2a-b

28 cm, incl. 7.5 cm long bill. As N Am. Snipe, but larger, in ssp *magellanica* with upperparts more extensively marked buff and tawny, foreneck richer buff. Vent buff (not white). Axillaries with black bars narrower than white ones. Outer rectrix whiter, and only c. 0.5 cm wide. **Pull.** dark chestnut, entire head, neck, and chest with complex pattern of white bands outlined in black; white powder-puffs tiny but very dense.
Habits and **voice**: When flushed rises steeply, then proceeds with small jumps (not zigzag), and soon drops into cover. Upflight calls *dzhyt*, *atch*, or *creck-crack*, sometimes given 2-3 times. An excited *kek kek kek...* is given both after being flushed, in display flight, or from the ground (here also a *ke-kek ke-kek...*). The 'winnowing' is described as a deep rumble, more prolonged, pulsating, rising and falling than in related forms.
Breeding: Eggs in Nov (Malvinas), Aug. (C Arg.).
Habitat: Marshy bogs, mires, and moors, rushy shore meadows, and sometimes in open swampy wood. In s Patagonia often in tussock-grasses.
Range: Widespread in savanna and chaco habitats from Ven. and e Col., esp. s of the Amazon basin through tablelands of Brazil and the chaco to Buenos Aires and Mendoza, c Arg. (*paraguaiae*). From Neuquén and Río Negro, Arg., and Atacama, Chile, to s tip of continent, and Islas Malvinas (*magellanica*). Common in all low valleys of the s Andes and to 1000 m in Valdivian lake district. Not uncommon on upland plateaus of the Patagonian steppe in the migratory periods.

PUNA SNIPE *Gallinago andina* –
Plates XIX 3ab and XX 17

23 cm, incl. 5-6 cm bill. **Small** version of N Am. and Magellanic S.

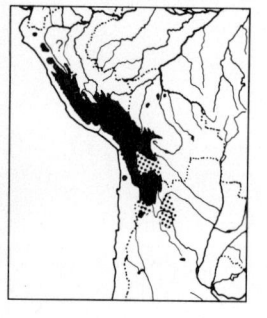

Usually more contrasting, with blackish mantle, primaries and primary coverts, and with white slightly barred wing-linings (except in *inornata*, where the under wing-coverts are densely barred). White tips of all flight-feathers. **Legs bright yellow** (vs. olivaceous or bluish in related forms; rather short, and probably not trailing behind tail in flight). **Pull.** has a much less contrasting pattern than Magellanic S., the powder-puffs not forming prominent white zones, and the belly very pale. **Habits** and **voice**: When flushed rises steeply, then proceeds ahead with small 'jumps' (but not zigzag), and usually soon drops into cover. Upflight call a rather high *dzeetch*. Calls from the ground *dyak dyak dyak...*, *dyuc dyuc dyuc...* or sometimes *dji-dji-dji...*. Display flight in wide circles with shallow dives; the 'winnowing', given in the horizontal phase after a dive, has a very hoarse and wheezy quality: *shushushushu....*

Breeding: Eggs mainly in Oct-Dec (c Peru), or Sep (n Chile).

Habitat: Boggy parts of the puna zone, ranging from cushion-plant bogs and boggy margins of creeks through mires and rushy parts of puna grassland to rather open reed-marsh.

Range: Mainly at 3000-4500 m, from Cruz Blanca n of the n Peru low and Cutervo in Cajamarca through W Andes, n Peru, and the puna of c and s Peru, w Bol., extreme n Chile (Arica), and nw Arg. s to Catamarca, furthest s wintering with Magellanic S. in adjacent lowlands (*andina*). Along Loa River in Antofagasta, n Chile (*inornata*). Common, except in the arid s parts.

Seedsnipes – Family Thinocoridae

A strictly Patagonian/Andean group of 4 species. These aberrant birds strikingly resemble the sandgrouse of the Old World by their chicken-like bills, plump shapes, short legs, and densely vermiculated plumages, as well as flight-calls. However, the rapid, erratic, often zigzagging flight when flushed is snipe-like. Indeed, the 2 smaller species can be confused with calidridine sandpipers at a first glance, and recent studies suggest that they are related to sandpipers. The nostrils are protected by a shield-like covering.

The name is somewhat misleading, as seedsnipes mainly browse directly on the vegetation, eating buds and small green leaves rather than seeds (but Cabot (1988) reports that Least S. eats more seeds). In arid climates, succulent leaves are eaten as a source of water. None of the species appear to drink under natural conditions.

In the breeding season, seedsnipes are monogamous, and occur in pairs or family-groups, but at other times gather in nomadic or migratory flocks. At least the *Thinocorus* species have characteristic aerial displays with a mellow song of monotonously repeated notes, which can also be given from traditional songposts on tops of rocks, fence-poles, bushes etc. Such perches are often situated near nests, and may serve as outlook posts. The nest is a scrape in the ground, in the open, or close to a clump of vegetation. The 4 pear-shaped (snipe-like) eggs, and the nidifugous young, have perfect camouflage color-pattern. Only the female incubates and she carefully covers the eggs with plant debris whenever she leaves. The birds feign injury, running with drooping and partly spread wings and tail when disturbed with eggs or young.

Further information can be found in *Bonn. Zool. Beitr.* 19(1968):235-248 and *Living Bird* 8(1970):33-80.

RUFOUS-BELLIED SEEDSNIPE *Attagis gayi* – Plate XVII 7a-c

29 cm. Intricately mottled rufous brown it resembles a tinamou until it is flushed, and shows its pointed wings. Above, each feather is black, scalloped with 1-2 buff and cinnamon lines (*latreilli*), or chiefly tawny, whitish along the feather edges, and with several concentric dusky lines on each feather (*simonsi, gayi*). Head densely speckled, neck and breast scalloped in same colors as upperparts, rest of underparts deep cinnamon-rufous (*latreilli*), pinkish cinnamon (*simonsi*) or light pinkish cinnamon (*gayi*), vent densely barred (*latreilli*), or nearly plain-colored. In **flight** shows no prominent wing or tail pattern, wing-linings colored like underparts. **Pull.** buff densely streaked and spotted black, this coat soon being replaced by some gray down, and then by feathers, which are more finely mottled than the ad. type (Plate XVII shows a molting *latreilli* chick with 3 feather generations present; juv. *gayi* has extremely finely mottled dorsal feathers, and appears almost uniform sand-colored).
Habits: Usually in territorial pairs or small groups, but can assemble in large (up to 80) flocks in winter. Usually tame and confiding. Rather upright when running (like a tinamou), but once flushed it flies fast, in zigzag, and sometimes flies far.
Voice: Calls continuously in flight (and sometimes while running), a melodic *gly-gly-gly...* or *cul-cul-cul...* (bewildered chorus when many birds are together). (Harsh *tchaa* when flushed, mentioned in the literature, is certainly not normal!).
Breeding: Egg-laying in Sep-Nov (Chile); pull. Oct (Ecu.).
Habitat: Inhabits rocky slopes, scree, and other bleak alpine terrain, but when feeding usually descends to the nearest bog or other area with cushion-plants or other matted vegetation. In the rainy season stays very near to the snowline.
Range: Andes of Arg. and Chile from s tip of continent to Salta and Antofagasta, above 1000 m furthest s, else above 2000 m (*gayi*). At 4000-5500 m in the highest massifs from Jujuy, nw Arg., through w Bol. and Tarapacá, ne Chile, to Cord. Huayhuash in n Lima, c Peru (*simonsi*). Highest parts of páramos of Ecu. (*latreilli*). Common locally, but poorly known.

WHITE-BELLIED SEEDSNIPE *Attagis malouinus* – Plate XVII 4a-b

28 cm. Above blackish cryptically scalloped with buff and cinnamon feather-edges, rump densely barred pale buff and black. Head speckled, neck and breast scalloped rich buff and black. Chin and belly white. **Male** seems normally to have very clear demarcation between scalloped breast shield and white belly. In **flight** white greater and middle under wing-coverts form a conspicuous band. Also shows narrow white tips to most tail-feathers. **Pull.** resembles Rufous-bellied S. closely.
Habits: In pairs or small groups, in winter in large flocks. Sometimes very confident; yet difficult to find, because of its quiet behavior, cryptic colors, and habit of hiding behind tussocks and turning its back to

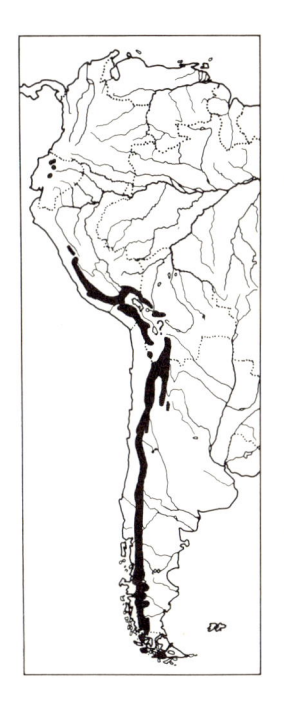

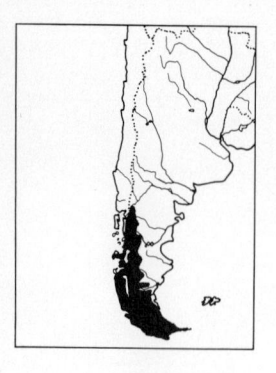

the observer. Flight not as fast as in Rufous-bellied S. Sometimes flies only short distance, but at other times dashes away in wild twisting flight, and flies far.
Voice: In flight calls continually, an excited *tu-whit tu-whit*... or *too-ee too-ee*....
Breeding: Eggs known from Jan.
Habitat: Above tree-line, mainly windswept places with *Azorella* cushions on bleak alpine moorland and stony slopes. Outside breeding season on stony parts of lower grassland, e.g., wide shores of partly dried-up lakes and dry river-beds.
Range: S Arg. and Chile from Río Negro to isls off Cape Horn, breeding from 650 to 2000 m, but usually descending in winter, e.g., to the steppe of northern Isla Grande. Apparently uncommon.

GRAY-BREASTED SEEDSNIPE *Thinocorus orbignyianus* – Plate XVII 5a-d

23 cm. Upperparts cryptically patterned cinnamon-buff, with more or less extensive fuscous to black marks which, together with buff feather-margins, give a scalloped effect, humerals vermiculated and barred. **Male** with face, **neck, and breast uniform plumbeous gray** bordered black against white upper throat and (narrowly) against the white lower underparts. **Female** and **juv.** with neck and breast buff boldly streaked with dusky. There is some variation in general color tone, but the ssp diagnosis rests only on size differences, *ingae* being smallest. In **flight** shows faint white wingbar above, broad white bar below contrasting black (male) or fuscous (female) wing-linings. Tail-feathers with prominent white tips. **Pull.** pale vinaceous to cinnamon drab gray, densely marked with black bands and white down-tips, the latter forming a complex pattern on back.
Habits: Not very social. In breeding season disperses on territories, in pairs or family groups (male guarding). At dusk and daybreak male often flies in wide circles, calling continuously. Characteristically raises tail and stiffly curves wings down during parachute-like landing after a display flight. Ground concerts are given from hummocks or rocks.
Voice: When flushed, gives a grating snipe-like *chrp*. Song a soft cooing *pooko-pooko-pooko*... or *poocuy-poocuy-poocuy*..., or a more grating *pooraake pooraak pooraak*..., often in interminable sequences. If given in flight, the song has rising and falling pitch. Easiest way to find this and the following species is by imitating the song, and listening for replies.
Breeding: From early to very late in the rainy season or summer, thus possibly multiple-brooded.
Habitat: A characteristic bird of the puna grassland, sometimes in dense bunchgrass, but usually in stony or rocky places with some cushion-plants and low herbs, or in short grass bordering bogs.
Range: From s tip of continent through Andes to La Rioja, Arg., and Tacna, Chile (*orbignyianus*) and puna zone from Catamarca, nw Arg., and Tarapacá, n Chile, through Bol. and n Peru to La Libertad (*ingae*). Furthest s breeds above 1000 m, but descends in winter. In Peru common at 3400-5000 m, highest up in the rainy season.

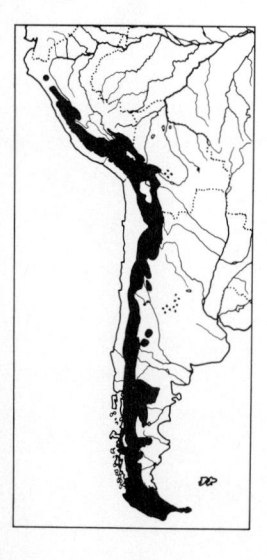

LEAST SEEDSNIPE *Thinocorus rumicivorus* – Plates XVII 6a-b and XX 23

17 cm (*bolivianus* 20 cm, size difference from Gray-breasted S. not always evident in the field). Above brownish buff thickly mottled with fuscous (*cuneicauda* paler than the depicted *rumicivorus*, and *bolivianus* distinctly pinkish cinnamon by hue). **Male** gray from face to breast, with broad **black band demarcating white upper throat, and continuing down c throat** to expand as a transverse bar separating gray chest from white lower underparts. **Female** with foreneck and chest sandy buff diffusely streaked and dotted. Resembles female Gray-breasted S., but white throat more boldly demarcated with fuscous, back less distinctly scalloped, and appearing more spotted (except in *bolivianus*), tertials darker with fewer light vermiculations and bars. **Juv.** like female, but back still more obscurely marked with thin black and white feather-edges, white throat not distinctly demarcated, and breast quite diffusely spotted rather than streaked or scalloped. Soon attains ad.-like plumage. In **flight** like Gray-breasted S., but tail distinctly wedge-shaped. **Pull.** as Gray-breasted S., but white down-tips form simpler pattern, usually with 2 parallel lines down back.

Habits: Commonly in flocks, sometimes in large numbers outside breeding season or, in deserts, on the long daily feeding excursions. Often very confident. When approached frequently runs with horizontal attitude, crouching flat each time it stops (but larger chicks run with upright stance). Display flight as in Gray-breasted S.

Voice: When flushed gives short snipe-like *djuc*, *djiric*, or low grunting calls, sometimes followed by low *tuc-tucketuck*.... The song may be 'interminable', but is generally less melodic and more variable than in Gray-breasted S.: in flight *vut-vut-vut...*, *puc-curr puc-curr...*, *pehooy pehooy...*, or accelerating *tjupo-cooo tjupo-cooo...*; from ground or fence-pole *krii-oko, krro, kro... pucui pucui-pucui*, or snipe-like *djek-djuk-djuk-djuk....*

Breeding: Eggs Aug-Feb (Patagonia, probably with several broods per season). We suspect that juv. become sexually mature 1st summer, at c. 4 months age.

Habitat: Mainly sandy areas with rather scattered bunchgrass or low herbs. Frequents roads and wide, gravelly shores with tiny annual herbs around partly dry claypan lakes. Often sings from fence-pole or bush-top.

Range: Southern temp. zone n to Mendoza, Arg., esp. on Patagonian steppe, below 1200 m, but migrating to plains of ne Arg., e.g. to above 2000 m in mts of Córdoba, and to Atacama, Chile (*rumicivorus*). Altiplano from Jujuy, nw Arg., to La Paz, Bol., and cordillera of Tarapacá and Atacama, n Chile, and expected in extreme s Peru (few rec., but common May-June at Ulla-Ulla in Cord. Real according to Cabot 1988) (*bolivianus*). From sea-level to over 1000 m in the deserts from Tarapacá, n Chile, along coast of Peru (*cuneicauda*), and in lowlands of extreme nw Peru and sw Ecu. (*pallidus*). Very common furthest s.

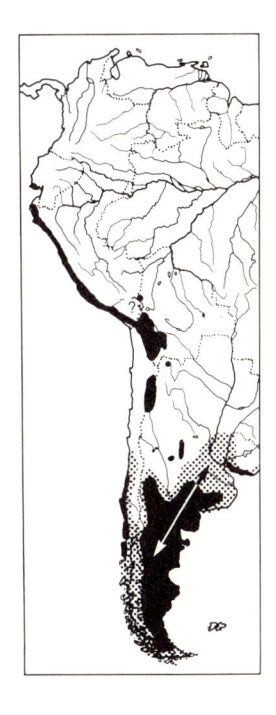

Gulls, skimmers and terns – Family Laridae

Gulls and terns, altogether 87 species, are familiar coast- and lake birds worldwide. Gulls (**subfamily Larinae**) dominate in northern seas, terns (**Sterninae**) in trop. seas. Several species, incl. the aberrant skimmers (**Rhynchopinae**), also occur inland.

Most species are white below, pale gray to black above, often with darker pattern on head and wing-tips, but a few gulls and terns are mainly dark. The feet are webbed. The subfamilies are readily told apart: Gulls are generally larger, heavily built, with slightly hooked bills, and usually square tails. Skimmers are also fairly large, with very short tails, and a peculiar laterally compressed bill with strongly protruding mandible. Most terns are smaller elegant birds with pointed bills, narrow and very pointed wings, and forked tails. Skimmers and terns have very short legs.

Gulls often soar high up. Skimmers usually fly low, rapidly, with stiff wingbeats interspersed with gliding on raised wings. Terns fly with deliberate, remarkably elastic wingbeats. They usually turn their bill down at an angle to the body.

Gulls often land on water, but mostly float passively, and use the webbed feet only to stir up prey when wading. They feed by scavenging along the shore, seizing floating items, and (smaller species) taking insects off the ground, the water surface, or in the air. Terns hardly ever land on water, but feed on the wing, picking insects from the surface or plunge-diving for small fish. Skimmers 'plough' the water with their knife-like mandible to snatch fish that come near the surface.

All species usually breed in colonies, on skerries, sandbars, or marshy places of difficult access. The nests vary from scrapes in dry substrates to substantial grass-cups in flooded places. The birds scold loudly or show diving attacks towards intruders near the nests (however, the Andean Gull is relatively silent and retiring when disturbed at its breeding colonies, which thus can be difficult to detect). The 2-3 eggs and the downy young have cryptic colors. Although the young can walk and swim, they usually stay near the nest and are fed by both parents (some terns feed their young long after fledging).

Further data on Andean species can be found in Harrison (1983); *Bull. Am. Mus. nat. Hist.* 52(1925):63-402; *Colonial Waterbirds* 8(1985):74-78, 11(1988):170-175; *Commun. Inst. Nac. Invest. Cienc. Nat., Cienc. Zool.* 2(1951):113-128; *Cormorant* 11(1984):65-67; *Estudios Oceanologicos* 3(1983):21-40; and *Univ. Calif. Publ. Zool.* 104(1974).

LAUGHING GULL *Larus atricilla* – a N Am. migrant to Caribbean and Pacific coasts of S Am. – is rec. once at 3020 m in E Andes, Col. Resembles Franklin's G., but larger, less graceful, with lighter semi-hood in the non-breeding plumages and black primaries lacking white pattern. Juv. has heavier tail-bar covering also the lateral feathers.

FRANKLIN'S GULL *Larus pipixcan* – **Plate XXI 3a-b**

36 cm. **Ad.** white (sometimes washed pink) with rather deep ashy gray upperside of body and wings. Black hood in the breeding season, but Sep-Mar shows semi-hood of streaked gray crown, and dusky patch from eye to nape. In graceful **flight**, black wing-tips are separated from the gray parts by a **white (transparent-looking) transverse bar**. In **1st winter** with plain grayish upperparts, **primaries lacking white**

pattern (except along trailing edge of inner feathers), tail pale gray with broad black bar except on outer feathers. In 2nd winter with intermediate wing-pattern and partial tail-band. Calls gliding *coo^i coo^i coo^i...* and lower *kugugugu* or nasal *karrr*.

N Am. migrant wintering on sandy beaches along the Pacific coast of Peru and n Chile. In recent years rec. to the Magellanic Strait and in c Arg. (e.g., many in Mar Chiquita in Córdoba), and occ. flocks are seen in Oct-Dec and May (imm.?) around 4000 m in puna zone of Junín c Peru, and altiplano of s Peru and Bol.

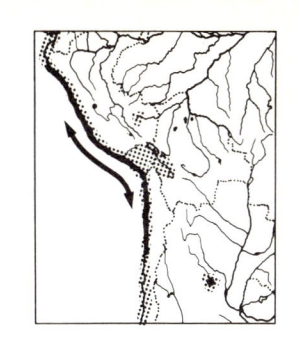

GRAY GULL *Larus modestus* breeds (Nov-Mar) in large colonies on stony hills at least to 1800 m in the lifeless 'nitrate desert' of Antofagasta, n Chile. Since only a few colonies of this common coastal bird have been located so far, undiscovered colonies might occur up to temp. zone. The species feeds along the coast of n Chile and Peru, arriving at, and leaving, its inland colonies in the dark of the night. **Sooty gray** throughout, except for a white rear edge of the wings, and white head in ad. plumage. Chicks are pale gray with some blurry dark gray spots above.

BROWN-HOODED GULL *Larus maculipennis* – Plate XXI 5a-c

40 cm. **Ad.** white with **very pale pearly gray mantle and upperside of wings. Chocolate-brown hood** in the breeding season, but in winter only dusky ear-spot and crescent near eye. In **flight** shows mainly black underside of the primaries except for a **broad white zone along the leading edge of the hand** (unnamed southern ssp; northern birds with more restricted white and black subapical zone, pattern approaching that of Andean G.). **Juv.** differs from winter birds by pale brown upperparts with buff feather-margins, white 'mirror spots' on 2 outer black primaries, and mostly gray inner primaries, and black tail-bar. The upperparts soon become whitish with scattered mottling, but in 2nd summer there are still some pale brown wing-coverts, and a partial tail-bar, and the hood is incomplete. **Pull.** is warm buff with black spots.
Habits: Gregarious, often in large flocks. Mainly eats insects, but uses various feeding tactics, incl. kleptoparasitism.
Voice: Thin and high-pitched, shrill and tern-like *zreee* and short *kip-kip-kip....* Harsh guttural *kwarr* in nest defence.
Breeding: Mainly on isls and rather open parts of reed-beds. Eggs in Oct-Dec.
Habitat: Seacoasts and marshy lakes. Mainly feeds over grassy river-plains and cultivated land.
Range: In lowlands throughout Arg., Islas Malvinas and c and s Chile, incl. low-lying lakes in the s Andes (southern populations represent an unnamed ssp). Often visits lakes on the foothills e of the Andes and nests to 800 m in w Chubut. Migrates to n Chile and e Brazil.

ANDEAN GULL *Larus serranus* – Plate XXI 8a-e

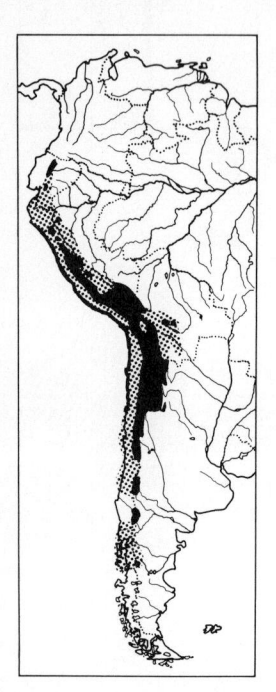

46 cm. **Ad.** white (sometimes tinged pink), with pearly gray mantle and upperside of wings. **Black hood** in breeding season; in winter only with dusky ear-spot and crescent near eye. Bill dark red. In **flight** shows **large white patch ('mirror')** inside tip of the otherwise mainly black outer 4-5 primaries, and black outer webs on remaining primaries. **Juv.** differs from winter birds by some light brown mottling on mantle and wing-coverts, restricted 'mirrors' only on 3 outer primaries, and black tail-bar. In 2nd summer still shows partial tail-bar, and some pale brown spots on wing-coverts. The hood is incomplete. **Pull.** light gray with black spots.
Habits: Social, but rarely in dense flocks. Mainly feeds on insects, over grassland, but may prey on eggs and bird-chicks in the breeding season.
Voice: Agitated, sometimes tremulous *yeeer* calls. Hoarse *raggh-aggh kee-aagh* and other low raspy notes.
Breeding: Breeds in usually very dispersed colonies in fairly open parts of reed-marshes, on inaccessible mudflats or isls and on abandoned coot-nests, or builds floating nests open to view among water-weeds. Eggs Sep-Jan, depending on when the rainy season starts (puna zone); fledglings Aug (Ecu.).
Habitat: Open high Andean marshes and lake-shores, bogs with small ponds, or on partly flooded plains, but often feeds over grassland, along rivers etc.
Range: 3000-4600 m, sometimes higher, and down to c. 2000 m in Jujuy, nw Arg., and furthest s. Throughout the puna zone of n Arg. and Chile, Bol., and Peru, less continuously on páramos of n Peru and Ecu., and in Andes of Chile s to Aysén. Winters on adjacent Pacific coast. Common in c parts of range.

Andean Gull

KELP GULL *Larus dominicanus* – Plate XXI 4a-b

62 cm. **Ad.** white with slaty **black mantle and upperside of wings**, except for broad white rear edge and spots on wing-tips. Bill strong, yellow with red spot on mandible. **Juv.** dark gray-brown mottled with whitish feather-edges all over, and with fuscous flight-feathers and wing-coverts, and white-mottled inner part of dark tail. Bill blackish. Ad. plumage acquired after 3 years, the tail becoming completely

white in 3rd winter (unlike in the Band-tailed Gulls *Larus belcheri* and *atlanticus* which maintain the tail-bar). The calls are copious *kiauw* or *kaukau*, mewing *kweeeu*, or a rolling descending whistle in juv. Alarm note a coarse *uk-uk-uk*.

Circumpolar at subantarctic latitudes. In S Am. mainly along the coasts of Chile and Arg., but straggles inland to feed on garbage and carrion, and to prey on goslings. Now breeds (Oct-Jan) near some low-lying lakes at the e base of the s Andes, and frequently visits upland plateaus at c. 1000 m and Andean lake districts at 40-44°s (may nest near Nahuel Huapí).

SOUTH AMERICAN TERN *Sterna hirundinacea* – Plate XXI 7a-c

42 cm (long-tailed ad.). **Looks almost white** (the Arctic and Antarctic T.s *Sterna paradisaea* and *vittata*, seen at southern coasts of S Am., are considerably grayer, and have finer bills). **Ad.** very pale gray with **black cap**, white cheeks and tail region, and with rather strong, orange-red bill with prominent angle on the mandible. Thin gray line on inner web borders the white shaft of the outer primary. When perched, streamers of forked tail almost reach wing-tips. **Imm.** is whiter, forehead and forecrown white, bill black. **Juv.** similar with dusky spots on wing-coverts. Flies with deliberate elastic wingbeats. Call a metallic *kyick* and a screeching *kee-err* or *cree-eer*.

Widespread along coasts of the s half of S Am., and may breed inland at some lakes and rivers in lowland Patagonia, and both ad. and imm. sometimes visit lakes around 1000 m on Andean foothills in Sta Cruz.

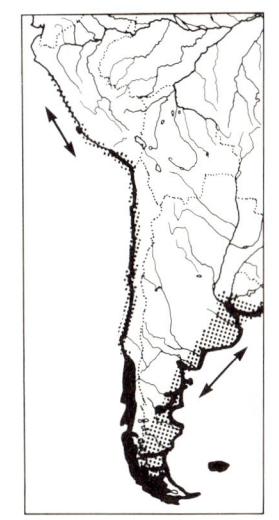

According to local fishermen, a tiny tern casually visits lake Junín at 4080 m, c Peru. The description suggests **PERUVIAN TERN** *Sterna lorata*: size like a small shorebird. With forked tail. Pale gray with black cap, white forehead; yellow base of bill.

LARGE-BILLED TERN *Phaetusa simplex* – Plate XXI 6

37 cm. A stocky tern with **large, very thick, yellow bill, and short tail with slight fork**. **Ad.** white with black cap (with a loop down below eye), back, lesser wing-coverts, and tail light gray. In **flight** shows characteristic **large white triangle** of carpal coverts, greater wing-coverts and secondaries. **Off-season** with crown partly white. **Juv.** with crown gray, ear-coverts black, carpal coverts and secondaries gray. Seen singly or in small parties in estuaries, over large rivers and lake shores. Gives 2-note *ga ga* call (*diwi diwi*, *sque-ee* and *ink-onk* also noted).

Widespread in trop. and subtrop. lowlands e of Andes. Apparently a regular visitor to Lake Tota (3020 m) and other wetlands in E Andes and Cauca (3450 m), Col., and accidental in highlands of Ecu. (lake Colta), c Peru (lake Junín), to Peruvian coast, and in Bol.

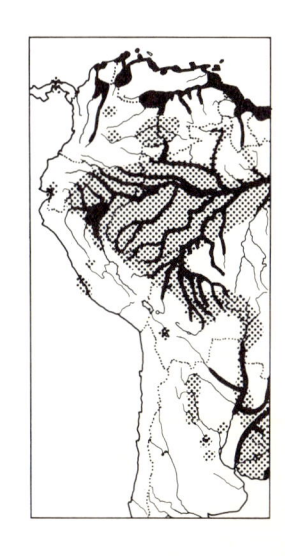

BLACK SKIMMER *Rhynchops niger*

45 cm. Grotesque profile due to the **huge blade-like bill with projecting mandible**. **Ad. sooty black** with orange-red feet and base of bill, and **white face and underparts**. Secondaries and outer rectrices edged white (*cinerascens* with reduced edges, and with less gray winglinings; however, imm. not to be referred to ssp). **Off-season** browner, with pale nuchal collar. **Juv.** rather brown with white streaks above, and with black bill. In **flight** seems to 'stoop', because of the large declining bill, and the small tail. The wings are very long, pointed, and with narrow base, the wing-beats rather shallow and stiff. Usually in dense flocks. Shy. Feeds in the evening, at night, and in the early morning. Call a 1-note ***gaaa*** (vs. 2-note in Large-billed Tern).

Breeds along all warm coasts and large rivers of the Americas. Northern migrant to C Am. and Ven. (*nigra*). Northern S Am., breeding s to c Brazil and w Ecu., suspected to cross Andes of s Col. (by night?) on its migration to the Pacific coast of Peru and Chile (*cinerascens*). S Brazil to n Arg., straggling to Patagonia (*intercedens*). Stray rec. at 2550 m in Cochabamba and at 3900 m on the altiplano, Bol.

Picui Ground-doves

Columbiformes

An order of uncertain systematic position; affinities suggested with gallinaceous and cuculiform birds, parrots, shorebirds, and passerine birds.

Doves and pigeons – Family Columbidae

Distributed worldwide, except in the coldest climates. The 225 species range from sparrow-size to near turkey-size. Smaller species are generally called doves, larger species pigeons. Pigeons are much sought-after for game. All are plump-bodied with small heads. The plumage is smooth, usually quite uniform pinkish or purplish gray-brown. Small ground-feeding species are well camouflaged. Wings and tail vary from short and broad to quite long and pointed. Usually, when flushed from the ground, doves first rise steeply with extremely deep wing-beats, the wings meeting over and under the body to produce distinct clapping or whirring sounds. Then they dart of in swift straight flight. Color patterns of underwings and tail, exposed as a bird takes of, are useful for identification. An alarmed bird may freeze in upright posture (or horizontal at the sight of an aerial predator).

Doves are social and unaggressive, except when acquiring or maintaining nest-sites. They live in pairs in the breeding season, but in some species several pairs may nest together. The nests are flimsy twig platforms (but the tiny ground-doves have quite well-built cup-nests, exceptional in the family). The nests are usually placed in trees or bushes or among dense tangles, but a few species are ground-nesting. The clutch is 1-2 white or buff eggs. The nidicolous young are ugly, nearly naked with some coarse buff hair-like down. They are fed by regurgitated 'crop-milk'. Some species have several annual clutches.

Further information about Andean species can be found in Goodwin, D (1983) *Pigeons and doves of the World*. 3rd ed. Ithaca-London. See also *Condor* 51(1959):3-19, 76(1974): 80-8, *Ecosur* 4(1944):157-85; *Hornero* 4(1978): 273-7, 6(1980):37-47, *Ibis* 99(1957):594-9, 110(1968):102-4; *J. Orn.* 98(1957):124; and *Wilson Bull.* 76(1964):211-47.

Columba pigeons

A world-wide genus of altogether 53 species which all resemble the feral pigeon in shape.

BAND-TAILED PIGEON *Columba fasciata* – Plate XXII 3a-b

35 cm, somewhat heavier than a feral pigeon. Ssp *albilinea* has head, neck, and all underparts dark grayish vinaceous, males generally most purplish. **White nuchal collar** separates vinaceous head and dark metallic green hind-neck. Mantle and wing-coverts dark bronze-brown grading to light blue-gray on lower back and rump, **tail mainly slaty with light gray distal zone**. Eyes, bill, and feet yellow. **In flight,**

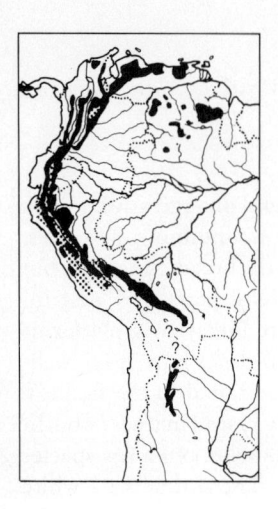

wings look uniform dark. **Juv.** almost uniform gray-brown with some rusty feather-margins above; eyes, bill and feet dark.

Habits: Gregarious. Usually seen in pairs or small groups, but sometimes form large feeding flocks, or roost in hundreds. Seen almost exclusively in the canopy, sometimes clambering and hanging upside down as it feeds on berries, young leaves, and blossoms. In display-flight ascends with slow but deep wing-beats or descends with stretched neck and spread tail and wings (sequence somewhat varied). Pair is sometimes seen swooping down-hill at great speed.

Voice: Weak owl-like 'hoo's usually combined in short sequences. Chirping *dzurrr* in display-flight. Also a croaking *graak* reported. Wing-sound when flushed *clap-clap-clap*....

Breeding: Eggs mainly in the rainy season, but season protracted, probably with several annual clutches, depending on food supply.

Habitat: Mainly humid heavily wooded hillsides and gullies, but locally in small semi-arid cloud forests, alder woods, second growth, or savanna.

Range: Mainly in the w parts of the Americas. Tepuis of s Ven., and at 2000-3000 m (500-3600, a straggler to 3900 m) on most Andean slopes of n Ven., Col., Ecu. and n and c Peru, and in the e Andean zone from Cuzco through Bol. and nw Arg. to Catamarca (*albilinea*). May show seasonal movements. Common.

NOTE: The Neotropical form *albilinea* has been listed by some as a full species. Forms a superspecies with the Chilean and Caribbean Pigeons (*C. araucana* and *caribaea*).

CHILEAN PIGEON *Columba araucana* – Plate XXII 4a-b

35 cm. As previous species, but head more purple, mantle and scapulars **dark vinaceous purple**, and all underparts more strongly vinaceous tawny, hind-neck below narrow white collar silvery green. Bill black, feet rosy. In **flight** shows gray tail with distinct black band on middle. **Juv.** unpatterned vinous gray-brown, but with wing-coverts, rump, and tail blue-gray.

Habits: Highly gregarious, previously sometimes in enormous flocks, and breeding in large, loose colonies. Arboreal. Particularly fond of *Araucaria* fruits.

Voice: Song, lasting 1 s, is a series of deep *hoo*s, the final syllable slightly drawn-out (BW).

Breeding: Eggs in Dec-Mar (May).

Habitat: Heavily wooded land, esp. large *Nothofagus dombeyi* trees and areas with monkeypuzzle trees *Araucaria*, esp. nesting in bamboo thickets. However, it is also frequently seen at forest borders and flying over cultivated areas.

Range: Breeds to 1000 m in forest zone of Chile from the c fiordlands n to 37°s and into adjacent sw Arg., but wintering mainly in c Chile (but some s to Peninsula de Taitao). Almost disappeared after disastrous epidemic in the 1950's (see *Zool. Vet. Med. B* 26(1979):430-432), but now rather common again in Chile, and has recently been seen in w Arg. at Lago Roca in sw Sta Cruz (A Johnson), and in Neuquén and Río Negro (*Nótul. Faun.* 8(1987):1-2).

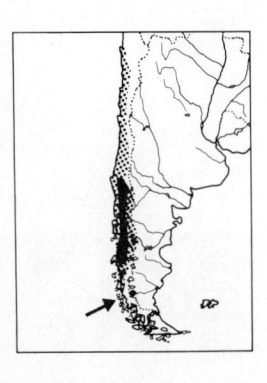

RUDDY PIGEON *Columba subvinacea* of humid trop. and subtrop. forest zones of C and S Am., is reported to 2800 m in Col. (but is rare above 1500 m). 30 cm. Dark ruddy or vinaceous brown, lightest below. Bill black, feet red. Voice (best difference from Plumbeous P. *C. plumbea*) rhythmic fairly high-pitched *wut wood-woooo ho*.

SPOT-WINGED PIGEON *Columba maculosa* – Plate XXII 2

32 cm. Built like a feral pigeon, but more broad-winged. Ssp *albipennis* with head, neck, rump, and all underparts **blue-gray**, somewhat tinged purplish (esp. **male**); mantle and wing-coverts gray-brown with white feather-edges, broad white outer webs of outer greater secondary-coverts forming **conspicuous white band slanting backwards from the wing-bend**, and contrasting with the black hand. Tail dark gray with black distal zone. Ssp *maculosa* lacks white diagonal in the wing. **Juv.** plumage unknown.
Habits: Social, sometimes in large flocks, but most often seen in small groups or as scattered breeding pairs. Often coos from tree-tops, but feeds on the ground. In ground display fans and raises tail and holds wings somewhat out. Display-flight with slow wing-beats, and in normal flight has calmer wingbeats than a Feral P.
Voice: Very faint soft coos followed by loud hoarse *cooouh-cuh-cuuuh*, 1st note low pitched. Short *corw* and low, growl-like sounds when displaying on the ground. Often gives heavy wing-clapping when startled.
Breeding: In trees; probably nesting most of the year.
Habitat: In the Andean zone in semi-humid to semi-arid fairly open woodlands and scrub, often in small patches, and sometimes around small *Eucalyptus* stands and near villages in otherwise unforested land. In Arg. considered a pest, esp. in sunflower crops.
Range: Widespread in lowlands from extreme s Bol. through Parag. to Chubut, c Arg., ascending at least to 1000 m at base of Andes in Catamarca and La Rioja (*maculosa*). At 2000-4200 m on Andean slopes from n Jujuy nw Arg., through s and c Bol. and locally in La Paz, on plateaus and in Andean valleys of e and c Peru and very locally on the w slope from Arequipa (RH) to Lima (*albilinea*). Common.

The highly variable **FERAL PIGEON** *Columba livia* is introduced from Europe and now widespread in urban settlements, villages, and sometimes near isolated farms, even in the highest parts of the Andes (but nowhere naturalized). In the wild original form blue-gray with rather light mantle and wing-coverts, and with 2 black wing-bars (on inner secondaries and greater coverts), some white on rump, and with blackish end of tail. Black-spotted, deep blue-gray, black, or more or less rufous or white forms are also common. Calls comprise soft *coo-cu-kuck-cooo*.

EARED DOVE *Zenaida auriculata* – Plate XXII 1a-d

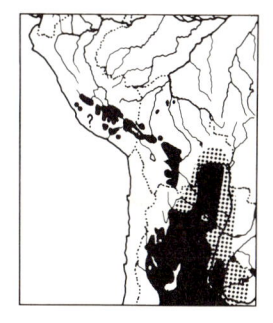

23-28 cm. Gracefully shaped. Upperparts olive-brown, blue-gray on crown, and grading to gray on wings and tail; side of neck with pink reflections and some golden-green. Black streak behind eye and near ear, and black spots on tertials. Underside more tinged cinnamon, or vinaceous on the foreparts (esp. in **male**), sides usually more blue-gray. Feet red. Sspp intergrade; *penthera* is generally very uniform cinnamon-rufous below, *caucae* also deep cinnamon, the southern *auriculata* less dark, but extensively vinaceous, and very gray on breast and sides, but *chrysauchenia* has white vent, and *hypoleuca* has pale pinkish buff belly grading to white on vent. In **flight** shows **graduated fairly short tail**, where the c feathers are colored like back, the rest blue-gray with **broad white tips** (**tawny in *penthera***) contrasting with black median bar (compare White-tipped and Large-tailed D.). **Juv.** is for a short period after fledging quite uniform earthy brown, locally with numerous small buff feather-tips on part of the plumage.

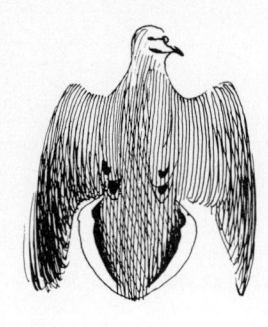

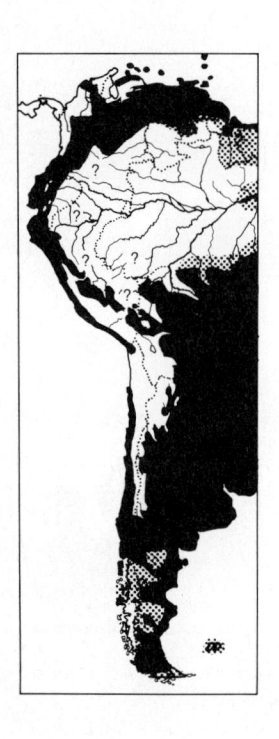

Habits: Social, and roosts and nests communally in trees or rocks, sometimes in large numbers. Often perches on electric wires. Feeds on the ground. In display-flight glides with set wings in narrow circles, throat inflated. Normal flight very fast.

Voice: A low, often slightly falling *whoo* or *ooh-whoo*, 2nd note accented and sometimes rolling or hoarse. Calls often repeated in long series. When startled, produces a weak wing-clapping sound.

Breeding: In most areas seems to breed most of year, but chiefly in Nov-Mar (c Arg.) or Oct (s Arg.). Eggs rec. Dec-Feb (Bol.) and Aug (n Peru). Fledglings Oct (Ecu.), Sep-Oct (Nariño), and Feb (E Andes, Col.).

Habitat: Mainly semi-arid open land or 'agricultural savanna', usually not far from groves, in the high parts usually associated with towns and villages. Considered a pest in many areas.

Range: With several ill-defined sspp throughout S Am., omitting only the most desolate parts of the puna, and the c Amazon area. Mainly in lowlands, but locally high up: at 2000-3200 m from Trujillo to Táchira, Ven., and in E Andes of Col., except humid and wooded parts (*penthera*). To temp. zone in Cauca valley, w Col. (*caucae*), grading towards *hypoleuca* in the Dagua and Patía valleys and Nariño, sw Col. (*'vulcania'*), and in arid temp. zone valleys and on Pacific slope of Ecu. and in n and c Peru, and locally to 4400 m on altiplanos of se Peru to La Paz, Bol. (*hypoleuca*). To at least 3000 m in semi-arid valleys from Cochabamba, Bol. to Jujuy, nw Arg. (*chrysauchenia*). S Brazil to c and s Arg. and to 2200 m in Chile from Llanquihue to Atacama (*auriculata*). Common.

NOTE: Previously in the genus *Zenaidura*.

COMMON GROUND-DOVE *Columbina passerina* – Plate XXII 11

16 cm, short-necked, and with rounded wings and short tail. Light gray-brown, **male** with a decided pinkish tinge all over. Face pale, plumage of neck and **breast with scalloped appearance**, owing to dark centers and pale edges of the feathers. Some wing-coverts have purple bars and spots. Bill light, tipped black. In **flight shows bright rufous, black-tipped wing-feathers**, rufous wing-linings, and white tips to the black outer tail-feathers. **Juv.** with slight pale feather-edges above. Most of time in flocks on ground, walking in jerky manner.

Usually unafraid. Flies in zig-zag course, and soon lands. Raises wing in threat. Gives a low soft *woot*, *coo^{wa}*, or bouts of *wut* notes. Sometimes nests on ground; eggs most of year.

In the highlands inhabits open dry range-land with *Acacia*-like bushes and agaves, fields, and often villages. Several vague sspp range through arid and semi-arid parts of the Americas from s US to e Brazil. Mainly trop., but ascends to the transition to the temp. zone in Tunja in E Andes, upper Cauca, Dagua, and Patía valleys, Col. (*nana*), and to 2850 m in the c valley of Ecu. from Guayllabamba canyon to Riobamba (*quitensis*).

NOTE: Has also been called Scaly-breasted G-d. This and the following species were previously placed in the genus *Columbigallina*.

PLAIN-BREASTED GROUND-DOVE *Columbina minuta* of sandy arid areas with some scrub in the lowlands of C and S Am., possibly comes near lower limit of temp. zone above Peruvian coast, and in the Marañón and Huánuco valleys of n Peru (*minuta*). 13 cm. Uniform light gray-brown with rows of black spots on the wing-coverts, **below uniform** light. Male with blue-gray head and nape, and a vinous tinge below; female more brown throughout. Bill blackish. In **flight** shows chestnut underside of wings, and black distal part of the outer tail-feathers. Gives low, insistent coos *woo-ahk, woo-ahk*....

RUDDY GROUND-DOVE *Columbina talpacoti* (incl. Ecuadorian Ground-dove *C. buckleyi*), widespread in drier trop. and premontane zones of C Am. and S Am. s to n Arg., has been rec. at 2400 m in E Andes, Col., and Andean crossing can be inferred from 2 rec. in c Chile. 16.5 cm. Head pale gray, otherwise cinnamon-rufous (**male**), or duller brown, although ruddy brown on back (**female**). Wing-coverts with some black spots. Outer feathers of short tail black tipped cinnamon.

PICUI GROUND-DOVE *Columbina picui*

18 cm, fairly slim and long-tailed. Sandy gray-brown, somewhat blue on nape, and male with vinous breast, underparts buffy white. Black stripe across lesser wing-coverts and another along tertials. A white band is visible in the closed wing. **In flight shows prominent white band on wing (greater secondary coverts), and white sides of tail, and wing-linings black contrasting with white patch at base of primaries.** Feet violet. **Juv.** birds are mottled with buffy feather-edges.
Habits: Feeds on the ground, on seeds and grain, occ. in flocks on fields. Often seen on roads, flying up into surrounding bushes when disturbed.
Voice: *go-^{up} go-^{up} go-^{up}...* or *woo^{loow}*.
Breeding: Rec. in Jan-Mar (Arg.), nestling May (Cochabamba).
Habitat: Mainly arid. At forest edge, open *Prosopis* scrub-forest, grassland with scattered trees and scrub, cultivations, gardens, suburban areas etc, in places with short or scattered cover of grasses and annuals.
Range: Widespread in lowlands in nc Chile, and from Chubut, c Arg.

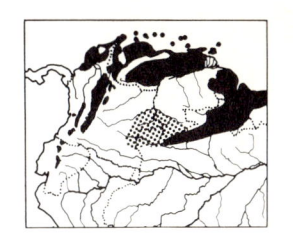

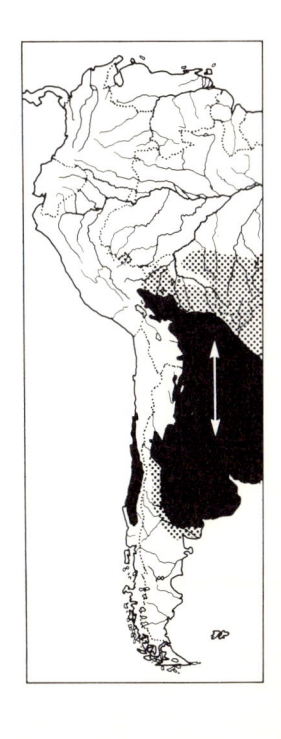

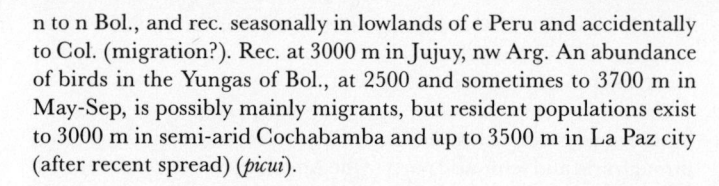

n to n Bol., and rec. seasonally in lowlands of e Peru and accidentally to Col. (migration?). Rec. at 3000 m in Jujuy, nw Arg. An abundance of birds in the Yungas of Bol., at 2500 and sometimes to 3700 m in May-Sep, is possibly mainly migrants, but resident populations exist to 3000 m in semi-arid Cochabamba and up to 3500 m in La Paz city (after recent spread) (*picui*).

CROAKING GROUND-DOVE *Columbina cruziana* – Plate XXII 12

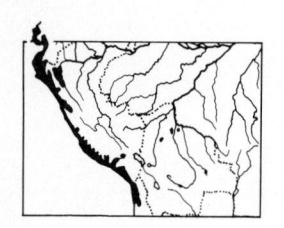

15 cm. Light gray-brown with pale blue-gray head, and light pinkish brown underparts. Purplish streak on lesser wing-coverts, and (unlike in Common and Plain-breasted G-d.) **primaries black**, wing-linings partly black, and **bill longer, bright yellow with black tip**. Voice a characteristic croaking *qworr* or *tweorr*. Dust-bathes. Breeds Sep-Nov.

In gardens and arid trop. and subtrop. scrub and riparian thickets of arid Pacific zone of s Ecu. and Peru to n Chile. Although normally replaced by Bare-faced G-d. in the higher parts, it sometimes ascends to 2400 (accidentally 2900) m both in Loja, n Peru, and in the Arequipa area.

NOTE: Previously in a separate genus *Eupelia*.

MAROON-CHESTED GROUND-DOVE *Claravis mondetoura* – Plate XXIV 13a-b

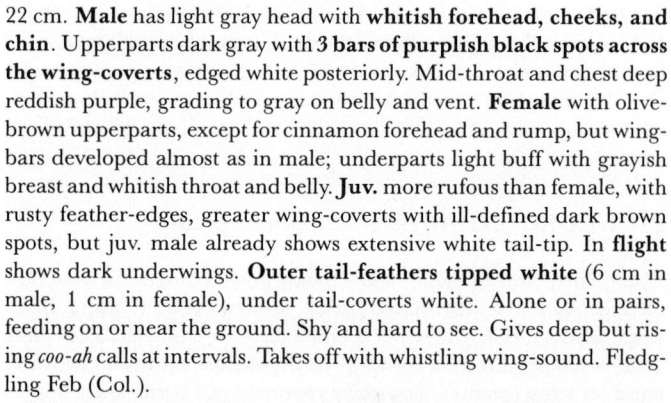

22 cm. **Male** has light gray head with **whitish forehead, cheeks, and chin**. Upperparts dark gray with **3 bars of purplish black spots across the wing-coverts**, edged white posteriorly. Mid-throat and chest deep reddish purple, grading to gray on belly and vent. **Female** with olive-brown upperparts, except for cinnamon forehead and rump, but wing-bars developed almost as in male; underparts light buff with grayish breast and whitish throat and belly. **Juv.** more rufous than female, with rusty feather-edges, greater wing-coverts with ill-defined dark brown spots, but juv. male already shows extensive white tail-tip. In **flight** shows dark underwings. **Outer tail-feathers tipped white** (6 cm in male, 1 cm in female), under tail-coverts white. Alone or in pairs, feeding on or near the ground. Shy and hard to see. Gives deep but rising *coo-ah* calls at intervals. Takes off with whistling wing-sound. Fledgling Feb (Col.).

Heavy undergrowth in wet montane forest. Strongly associated with bamboo of certain growth stages. At 1300-3000 m through most of Andes of n Ven., all Andean ranges of Col., and along e Andean zone through Ecu. and Peru (to W Andes at the Porculla pass, and accidentally at Caraz in Ancash, O Frimer), and to yungas of Cochabamba Bol. (*mondetoura*). Rare, local, or unpredictable because of nomadic habits. Other sspp in C Am.

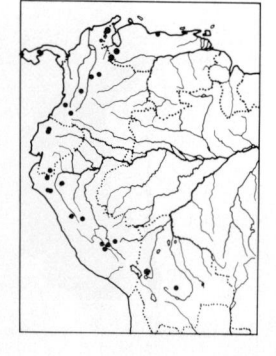

Ground-doves – Genus *Metriopelia*

A typical high Andean group of 4 small, drab-colored ground-doves.

BARE-FACED GROUND-DOVE *Metropelia ceciliae* – Plate XXII 9a-c

17 cm, compact. **Naked orange-yellow orbital area** surrounded by black line. Earthy brown with gray to ruddy cinnamon hue (very dull and grayish in *ceciliae* and *obsoleta*, warm ruddy in *gymnops*, and with rosy tinge on breast esp. in male). Marked with **pale buff tips on most back-feathers and wing-coverts**. Belly and vent creamy to cinnamon-buff. In **flight** shows dark sooty gray under-side of wings, and dull rufous tinge on inner webs of primaries; on upper side of wings, buff-tipped coverts form a light zone contrasting the black hand. **Tail-feathers black with broad white outer corners**. **Juv.** with duller soft-part colors and slightly looser feather-structure.
Habits and **voice**: Usually in small groups, feeding on ground. Dust-bathes. After being flushed, often lands on projecting cliffs. Roosts and nests on sheltered rock-ledges or in holes in houses. When startled, the wings produce a frail, metallic rattle. We have never detected vocalizations from this species.
Breeding: Ground- and cliff-nesting, or using holes in buildings. Eggs in Mar (Chile), occ. in winter (Lima). Juv. Apr-July (Cochabamba, Bol.).
Habitat: Arid and semi-arid habitats, always sandy ground with some stones and rocks, and sparse plant cover. Both on bushy or lightly wooded slopes and in cactus land. Also in highland areas totally lacking arboreal vegetation, but then always associated with villages or houses.
Range: On Andean slopes and altiplanos, usually at 2500-4500 m, but lower in w Peru, seasonally down to 700 m in some valleys, and in s Peru in the coastal range, and occ. to sea-level. From the Porculla pass n Peru up the Marañón valley, and locally in valleys of Huancavelica to Apurímac (doubtful ssp *obsoleta*), and Pacific slope and W Andes of Peru (*ceciliae*). Altiplanos of s Peru and Tarapacá, n Chile through w Bol., and occ. to Jujuy and 1 old specimen from Salta, nw Arg. (*gymnops*).

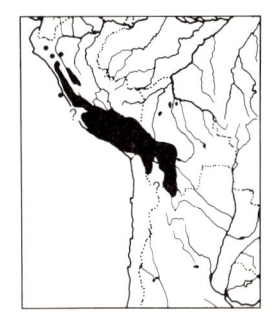

MORENO'S BARE-FACED GROUND-DOVE *Metriopelia morenoi* – Plate XXII 10a-b

17 cm. As previous species, but **bare ocular region more strikingly orange-red**, and all **upperparts uniformly gray-brown without pale spots**; rump slightly ruddier, underparts uniformly light gray grading to dull cinnamon on vent. **Juv.** shows indication of buff fringes of the dorsal feathers. In **flight** shows slaty underside of wings and small white corners of the tail compared with Bare-faced g.-d.
Habits: Social ground-feeder. Hard to find.
Voice: When startled, wings produce a frail rattling sound. Otherwise silent.
Breeding: No information.
Habitat: Sandy and rocky hills with scattered trees, scrub, and columnar cacti (pre-puna vegetation).
Range: Above 2000 m. Locally in Jujuy and Salta, and from Tucumán to La Rioja, nw Arg. Not uncommon.

NOTE: Called Bare-eyed Ground-dove by Meyer de Schauensee. Since the paraspecies *M. ceciliae* and *morenoi* are equally 'bare-faced', we will not support calling one 'bare-faced' and the other 'bare-eyed'.

BLACK-WINGED GROUND-DOVE
Metriopelia melanoptera – Plate XXII 7

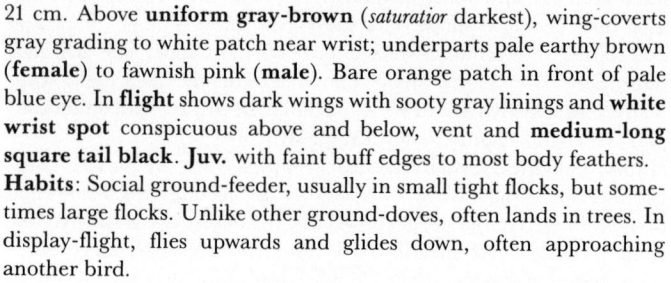

21 cm. Above **uniform gray-brown** (*saturatior* darkest), wing-coverts gray grading to white patch near wrist; underparts pale earthy brown (**female**) to fawnish pink (**male**). Bare orange patch in front of pale blue eye. In **flight** shows dark wings with sooty gray linings and **white wrist spot** conspicuous above and below, vent and **medium-long square tail black**. **Juv.** with faint buff edges to most body feathers.
Habits: Social ground-feeder, usually in small tight flocks, but sometimes large flocks. Unlike other ground-doves, often lands in trees. In display-flight, flies upwards and glides down, often approaching another bird.
Voice: Usually silent, but in breeding season males coo incessantly from a tree morning and evening, a rolling *trrē͞eooi*, like a brief version of the call of the abundant highland toad *Bufo spinulosus*. When startled, wings produce a weak rattling *why-why-why....*
Breeding: In thick bushes, cacti, and puyas (occ. on buildings or the ground), sometimes several nests close together. Eggs in Aug-Oct (furthest n), Mar-May (Peru) to (Nov)Dec-Feb in the s.
Habitat: Mainly near the tree-line in arid and semi-arid regions. By mid-day feeds on grassy plains often far from trees, but morning and evening birds assemble on bushy slopes or in *Polylepis* woodlands. Also on sheltered hillsides with cultivation terraces and leguminaceous scrub, often near houses, and sometimes associated with columnar cacti or puyas.
Range: Most of Andes, at 2000-4300 m, but to 4800 m in the n and lower in the s, esp. in winter, and accidentally to near sea level in Peru. From ne Isla Grande through Andes of Arg., Chile, Bol., Peru n to Junín, and at least locally n to Cajamarca in the W Andes (*melanoptera*). In the larger páramos of Ecu. and into Pasto in Nariño, sw Col. (*saturatior*). Common, locally numerous.

GOLDEN-SPOTTED GROUND-DOVE
Metriopelia aymara – Plate XXII 8a-b

18 cm, very compactly built, and with small bill and very short tail and legs. **Pale fawn-brown** grading to pink below (esp. in male). Lesser wing-coverts with short row of golden spots (conspicuous in certain lights only), tertials with 2-3 small purplish black spots. In **flight**, shows long but blunt wings, underside of black primaries with large chestnut area (hardly visible on the upperside, and even on the underside the brown color is apparent only in good light). Small blackish **tail concealed by fawn-brown coverts, except for the outer corners** (compare larger black tail of Black-winged G-d.). **Juv.** has whitish feather-edges above, and lacks pink tinge; golden spots absent or present.

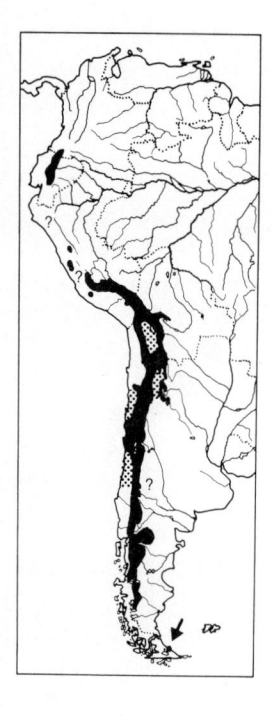

Habits and **sounds**: Social ground-feeder, usually in small groups, but sometimes in considerable flocks. Usually in horizontal attitude with body very close to the ground, and tail lowered (looks legless). Very difficult to see on the semi-desert soil. When startled, produces whistled wing-sound, which (unlike in other ground-doves) is discernible also in sustained flight. Wing-sound well marked during the display-flight, which involves a wide circle followed by a long dive. No vocalizations noted.
Breeding: In grass and maybe also in trees or rocks. Eggs in Apr-Aug (Bol., n Chile) or May-June (c Peru), juv. rec. July-Sep and Dec (Bol.).
Habitat: Usually on **level** areas with bare semi-desert soil and scattered *Lepidophyllum* brush, cushion-plants, tussocks, or low annuals, both on plains, wide lake-shores, or dried-up claypans. When nesting, assembles in *Polylepis* shrub or rocky terrain, sometimes at very high elevations.
Range: At 3500-5000 m (lower furthest s, and occ. down to 300 m) from Mendoza, w Arg., and Atacama, Chile, through altiplanos of Bol. to Puno and Arequipa, and locally in w Ayacucho and (casually?) Junín, Peru. Common locally at certain times, but possibly wanders.

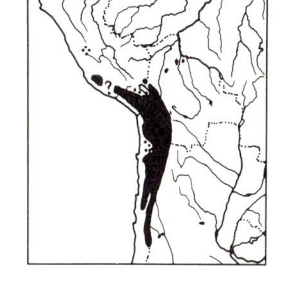

TOLIMA DOVE *Leptotila conoveri* – a very local bird of subtrop. e slope of C Andes of Col., from Ibique to head of Magdalena valley, reaches lower limit of temp. zone (2500 m). Differs from sympatric White-tipped D. by light blue-gray top of head, buffy belly, and less broad white tail-tip.

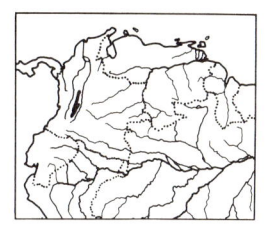

WHITE-TIPPED DOVE *Leptotila verreauxi* – Plate XXII 6a-b

29 cm. Head pale pinkish drab gray, otherwise light grayish (*decolor*) or grayish olive-brown, darkest above, and with light violet hindneck (quite rosy in *verreauxi* to deeper purplish blue in *chalcauchenia*), with the feathertips on the transition to the upper back always glossed with blue-green. Below paler, becoming white on belly and vent. Bill thin, black; bare loral stripe and eye-ring blue, or red in *decolor* and *chalcauchenia*. In **flight** shows bluntly rounded **wings with chestnut-rufous underside** (only tips of remiges dusky; outer primary strongly attenuated) and rather large tail, which is gray-brown above, blackish below, the 3-4 outer feathers-pairs black with broad white tips. Tail slightly graduated, but when spread wide (in a startled bird) looks much more fan-shaped than in the Eared D. **Juv.** duller, olive-brown above, with faint light 'scaling', slightly tinged chestnut on greater wing-coverts, and outer primary only slightly attenuated.
Habits: Feeds on the ground, on seeds and berries, usually singly or in pairs, but sometimes in loose groups. Quite wary. When approached walks away quietly, or flies to perch in a tree, where it nervously nods head and opens and closes tail with a twitching motion.
Voice: Repeated drawn-out deep *woob-wooooo* of male sounds like when blowing across the top of a bottle. Wing-sound when flushed *wiwiwi....*
Breeding: Nest in bush or vines. Breeding season protracted, but may peak June-July (Col.).

Habitat: Dry bushy slopes, thickets, open scrubby wood, often with scattered large trees (*Bombax, Carica, Ochroma*), and frequently with a mid-story of thorny leguminaceous trees. Usually feeds along edges, in cover below trees or shrubs.

Range: Widespread, with several sspp in trop. and subtrop. parts of C and S Am. Mainly in lowlands, but distribution probably determined by aridity rather than elevation, as it ascends to 2500 or 3000 m wherever thick humid montane forest in missing, e.g. in n Ven. and around Magdalena valley Col. (*verreauxi*), in Cauca and upper Dagua and Patía valleys and Nariño, sw Col., and up the Guayllabamba valley to Quito, nw Ecu., and in the Andean valleys of n and c Peru (*decolor*). May also reach temp. zone in the valles of Bol. and nw Arg. (*chalcauchenia*). Common.

NOTE: Forms with red and blue orbital skin could be different species.

GRAY-HEADED DOVE *Leptotila plumbeiceps* of subtrop. woodland in C Am. and Cauca valley, Col. is rec. to 2600 m near Popayán. - Differs from sympatric White-tipped D. by light gray top of head and nape, buffy cheeks, and more pinkish tinged breast. Eye-ring red (vs. blue in White-tipped D. of Col.). Calls apparently very similar.

LARGE-TAILED DOVE *Leptotila megalura* – Plate XXII 5

29 cm. Confusingly similar to White-tipped D. Normally described as having almost white forehead contrasting with blue-gray crown, but this is often not true. The general coloration is darker, tinged more ruddy brown above (esp. in *saturata*), with purplish gloss on the hindneck and upper mantle (**never greenish glossed feathertips of White-tipped D.**). Tail slightly less graduated, the white terminal bar slightly narrower than in White-tipped D., and the brown dorsal color covering the 4(2-6) c tail-feathers to the tip (vs. gray 2(4) c feathers in White-tipped D.). Rufous of under-wings less extensive (broad drab gray rear edge). Attenuated tip of outer primary slightly expanded subapically, unlike in White-tipped D. **Juv.** duller, with slightly scaled wing-coverts, as in White-tipped D. Song deep, as in White-tipped D., but of 3-4 syllables, of which the 1st is accented and dropping at the end *whoooo, hoo-hoo-hooo* (BW).

In alder woods, coppice, and second growth in humid parts as well as semi-arid zones with *Prosopis* woodland, *Schinus*, and *Molina*, often near cultivation. Subtrop. zone and occ. to temp. zone (2800 m) in the yungas of nw Bol. (*megalura*) and s to Tucumán, nw Arg. (*saturata*). Common in the Yungas and e.g. at Vila-Vila in Cochabamba.

Quail-doves – Genus *Geotrygon*

7 species of ground-living bulky short-tailed doves with strong feet. Characteristic stripes on head-sides and peculiar expression caused by strongly inclining angle of bill. Inhabits dark litter-covered floor of forest and shrub, walking with waggling rear end. More often heard than seen (usually glimpsed only when flushed from the ground).

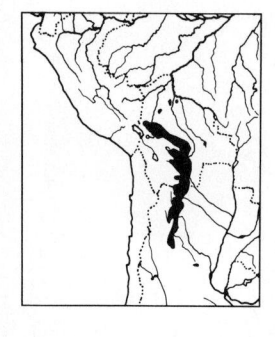

NOTE: The below-mentioned taxa were previously placed in the genus *Oreopeleia*.

RUDDY QUAIL-DOVE *Geotrygon montana* – Plate XXII 13a-b
of humid trop. and subtrop. forest of C and S Am., ascends to 2600 m at least near Bogotá and maybe elsewhere in Col. (*montana*). 25 cm. **Male** with all upperparts glossy rich purplish brown, breast pinkish cinnamon contrasting with rich cinnamon-buff lower underparts. Cream-colored band under eye separated by purplish horizontal stripe from cream-colored throat. **Female** olive-brown to fuscous where male is purplish, quite dark cinnamon on breast and sides, grading to white on c belly and vent. **Juv.** duller than ad. of same sex, and has small rusty tips on wing-coverts, some **thin** rusty bars on many dorsal feathers, breast somewhat freckled, sides quite gray. Solitary. Call a low-pitched (and falling) far-reaching sad moan *hoooooooooo*, or low, *mmmmm* (like a cow). Distress call series of short, sharp *cu cu cu...* notes.

WHITE-THROATED QUAIL-DOVE *Geotrygon frenata* – Plate XXII 14a-b

34 cm. A very large quail-dove, gray-brown (*frenata*), or darker and more reddish (*boucieri*). **Ad.** somewhat glossed purplish; **crown and nape pinkish to bluish gray** demarcated by thin dark eyeline; cheek pale pinkish cinnamon demarcated by bold horizontal black bar from white upper throat. Eyes orange. **Juv.** coarsely barred tawny and fuscous above and below. Solitary or 2-3 together, shy. Ground-living, but flies to low branches when disturbed. Call a deep *hoohooo* from midstory or high in a tree.

Entirely within humid forest with some undergrowth or in tall second growth in steep terrain. At (900) 1500-2500 m in W and C Andes of Col., on both slopes of Ecu. (to 2900 m) s to Zaruma (*bourcieri*), and to 2950 m on the e Andean slopes and in some humid 'pockets' in Andean valleys of Peru to yungas of Cochabamba, Bol. (*frenata*), and locally to Jujuy and n Salta, nw Arg. (*margaritae*).

LINED QUAIL-DOVE *Geotrygon linearis*, which replaces the White-throated Q-d. in the Sta Marta mts (*infusca*), Perijá mts, along e slope of E Andes and n in the C Andes, Col. (*linearis*), may locally reach lower limit of temp. zone (2400-2500 m). 27 cm. Differs from a small White-throated Q-d. by the forecrown being cinnamon-buff grading through vinous to gray on nape. Horizontal line between cheek and throat narrow. Underparts tawny-buff.

NOTE: Form a superspecies with White-faced and Rufous-breasted Q-d. (*G. albifacies* and *chiriquensis*) of C Am. Sometimes included in *G. frenata*.

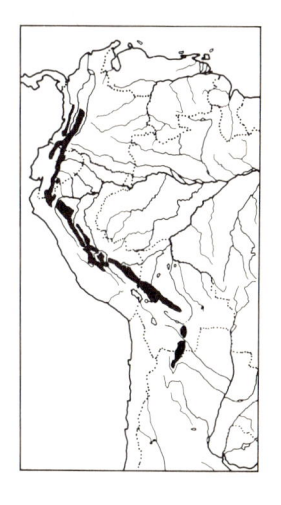

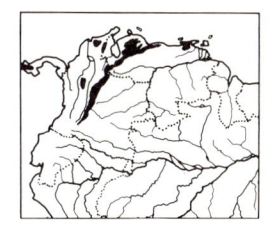

Psittaciformes

An ancient order of arboreal landbirds possibly related to doves.

Parrots – Family Psittacidae

Altogether 333 species in warm parts of both hemispheres, with the highest species richness in Australia and S Am. Easily recognized by their noisy behavior, bright mostly green colors, and peculiar very deep bill with the highly moveable maxilla strongly down-curved and fitting over the short mandible. The bill is specialized for peeling seeds, and is used as an extra claw when climbing. The legs are short, with 2 toes forwards, 2 back (zygodactyl). The species range from the size of a sparrow to over a metre. The sexes are usually alike, but juv. often have duller colors.

Most parrots are arboreal, but some feed on the ground. The group seems primarily to be adapted to eating grass-seeds, but many species eat a variety of seeds, nuts, berries, fruit, buds, and flowers, and an Australian group eats pollen and nectar. All seeds are husked before being swallowed. Most species are gregarious, and their social behavior is complex and includes play behavior. Mating is usually for life, and is maintained by mutual preening and feeding. Most species are sedentary, except for daily wandering between feeding places and roosts. However, some species show regular or irregular movements associated with climate or fruiting seasons. Some species raid crops, and are persecuted as pests. The nestlings of many species are exploited as human food and for the commercial pet market, and some species are now threatened.

Parrots usually nest in tree holes, sometimes in rocks, or with a crude nest in dense vegetation (and see under Burrowing Parrot and Monk Parakeet). The 2-5 eggs are white. The nidicolous chicks hatch blind and naked, or with sparse down. The female incubates, and in the beginning also feeds the young, while the male brings food which is regurgitated.

Information about Andean species is compiled in Arndt, T (1981) Südamerikanische Sittiche 3: Kilschwanzsittiche, (1983) 4: Rotschwanzsittiche/*Pyrrhura*,(1986) Schmalsnabelsittiche – Maronenstirnsittich, Walsrode; Forshaw, JM & Cooper (1978) *Parrots of the World* (2nd ed.), Melbourne; Pasquier, RF, ed. (1981) *Conservation of New World Parrots*. ICBP (Techn. Publ. 1); Low, R (1984) *Endangered Parrots. A conservation study*. See also Lautermann, W (1986) *Die Papageien Mittel- und Südamerikas*. M & H Schlaper, Hannover; *Bol. Tech. Corp. Nac. Forest. Chile* 11(1984):1-55; *Papageien* 1988:9-11, 86-88, 1989:16-17; and *Trochilus* 8(1987): 105-6.

Macaws – Genus *Ara*

50-100 cm long parrots with very strong maxillas, quite extensive naked areas on face, and long, pointed tails. They usually live in the canopy of trop. forest, and fly high. Eat fruit pulp and seeds, incl. considerable amounts that are distasteful or toxic to humans.

MILITARY MACAW *Ara militaris*

70 cm. With strong black bill and bare **rosy face** with black feather-stripes. **Grass-green with red forehead**. Remiges and their greater coverts blue, tail blue above with red base of the 4 c feathers, vent blue. In **flight** shows golden olive underside of flight- and tail-feathers. Voice a raucous *craaak*.

Very local in gallery forest and rather dry deciduous forest slopes, often with canyons, in Mexico (*mexicana*), n Col. from Sta Marta mts and n Ven. to Dagua Valley w of Andes, and locally e of Andes through Col. to Cajamarca and Huánuco (in humid forest), e Peru (*militaris*), and Sta Cruz, Tarija, and Chuquisaca, Bol., and Salta, nw Arg. (doubtful ssp *boliviana*). Mainly below 2000 m, but possibly wanders higher up in Bol., moves up to 2400 m in the Marañón valley in n Peru, and may pass over the n Peru low in Sep-Oct (feather found at 3100 m), and maybe also Nariño, s Col. Variable characters in the small populations of Great Green Macaws (*Ara ambigua*) in w Ecu. suggests hybridization with visiting Military M. (*Bull. Brit. Orn. Club* 107(1987):28-31).

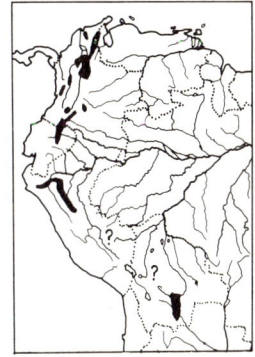

RED-FRONTED MACAW *Ara rubrogenys*

55-60 cm. **Resembles a giant Mitred Conure**: olive-green with orange-red forehead and crown, ear-spots, 'shoulders', **wing-linings** and thighs. Bare facial skin rosy, bill blackish. Flight-feathers and long tail blue-green above, olive-yellow below. **Juv.** shows little red color. Voice a loud *cra cra cra...*, clearly deeper, more raucous, than in the Mitred Conures of the same area.

Inhabits dry subtrop. woodland with columnar cacti and esp. leguminaceous trees, and nests colonially on cliffs. In **Bol.**, at 1100-2500 m from s Cochabamba and w Sta Cruz through Chuquisaca to e Potosí, esp. in río Grande valley and near Mizque, but also along río Pilcomayo. Population 3-5000 birds (D Lanning), heavily tolled by trapping, and persecuted as it raids maize and peanut crops.

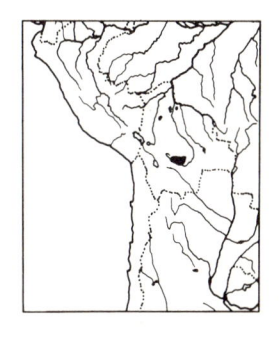

Conures – Genera *Aratinga*, *Pyrrhura* and others

Called **parakeets** by Meyer de Schauensee. Miniature versions of macaws, with relatively smaller beaks, and restricted bare eyerings. Like macaws, they have long graduated tails, pointed in *Aratinga*, blunt-tipped in the smaller *Pyrrhura*s. The *Aratinga*s seem particularly often to become smudged by juices from the fleshy fruits on which they feed. *Pyrrhura*s eat more seeds, incl. cereals, and may find insect larvae in decaying wood. A vegetarian diet is supplemented by eating mineral-rich soil. The flight is direct, with rapid shallow wingbeats. Flying birds are often revealed at great distance by incessant high-pitched shrieking.

BLUE-CROWNED CONURE *Aratinga acuticaudata* – Plate XXIII 5a-b

40 cm (but northern sspp smaller). Grass-green with dark **azure-blue**

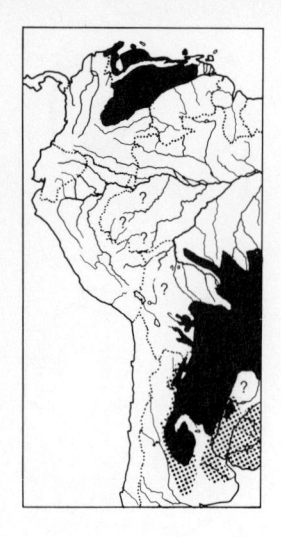

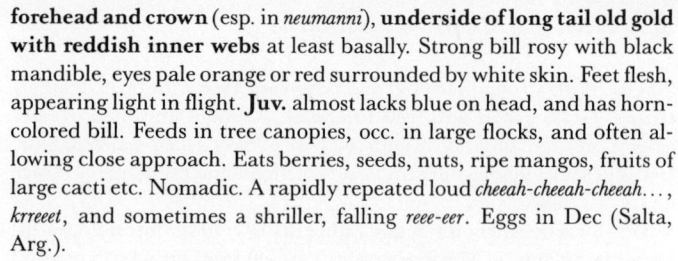

forehead and crown (esp. in *neumanni*), **underside of long tail old gold with reddish inner webs** at least basally. Strong bill rosy with black mandible, eyes pale orange or red surrounded by white skin. Feet flesh, appearing light in flight. **Juv.** almost lacks blue on head, and has horn-colored bill. Feeds in tree canopies, occ. in large flocks, and often allowing close approach. Eats berries, seeds, nuts, ripe mangos, fruits of large cacti etc. Nomadic. A rapidly repeated loud *cheeah-cheeah-cheeah...*, *krreeet*, and sometimes a shriller, falling *reee-eer*. Eggs in Dec (Salta, Arg.).

In rather dry savanna and deciduous woodland, esp. thorny leguminaceous wood with columnar cacti. Arid trop. and subtrop. areas from Sta Marta region of n Col., e through Ven. (*neoxena*), on Margarita isl. (ssp), and s-wards, on some savannas of the inner Amazon area (*haemorrhous*); in e Brazil and through chaco of e Bol. to the pampa, and at moderate elevations to Tucumán and San Juan, Arg. (*acuticaudata*). At 1500-2650 m in Cochabamba, Sta Cruz, Chuquisaca, and Tarija, Bol. (*neumanni*). Generally common.

SCARLET-FRONTED CONURE *Aratinga wagleri* – Plate XXIII 2a-d

32-42 cm (*minor* smallest, *frontata* largest). **Green with scarlet forehead** and usually a few red feathers elsewhere. Wing-linings green (but some orange or red may occur near wing-bend), underside of flight- and tail-feathers old gold. Bill light horn, eyes orange-brown surrounded by white skin. Red of forehead slightly developed in **juv.**, and also restricted in *transilis* and *wagleri*; extensive and reaching eye and lore in *frontata* and *minor*, but not to cheeks, as in Mitred C. Ssp *frontata* is strongly washed with bronzy yellow (moss-colored), *minor* quite dark dull green with much pale orange-red on the tibiae. Tail-tip slightly blunter than in other *Aratinga*s (see plate).

Habits: Social, sometimes hundreds together. Very obvious on the high daily flights between roosts on rock faces or very steep hillsides and the feeding grounds. Usually feed high in canopy. Often raid crops and fruit plantations.

Voice: Loud strident and high-pitched *chee-ey*. The flocks often maintain a loud screechy chatter.

Breeding: In holes in high rock-faces; eggs Apr-June (Ven.) or Dec-June (Col.).

Habitat: Humid forest with fruiting trees. In Peru in light semi-arid cloud forest (*Acacia*, *Prosopis*, and *Ochroma*, roosting in large *Oreopanax* trees) and *Ochroma* woodlands of the c valleys. May feed in cactus locations, orchards, and fields.

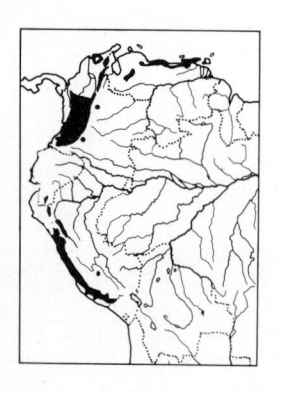

Range: Mainly in premontane zone, but sometimes to sea level, and to 2800 m in Col., 2700 m in Marañón valley, and furthest s sometimes to above 3000 m. Coastal mts of Ven. from Monagas to Aragua (*transilis*), Andes of Ven. from Lara and Perijá mts continuing to Sta Marta mts, and w of E Andes of Col. s to n Nariño (*wagleri*). 3 specimens rec. from Loja (San José, San Lucas, Lunamá) s Ecu., and distributed in n and w Peru s to Lima and locally to w Arequipa, and seen also in Tacna (*frontata*). Inter-Andean valleys from Cajamarca and La Libertad (Marañón valley) to Apurímac (Ninabamba), possibly Cuzco, Peru (*minor*). Common, esp. in the n parts.

MITRED CONURE *Aratinga mitrata* – Plate XXIII 1a-c

38 cm. Green, light and somewhat yellowish below. **Forehead dark purplish red grading to pale red on forecrown**, where contrasting with the green crown (no such contrast in Scarlet-fronted C., which is the commoner of the 2 in the zone of overlap). Wing-linings olive, undersides of flight- and tail-feathers old gold. Bill and feet whitish horn, eyes grayish buff or light gray surrounded by white skin. **Ssp *mitrata* has red lores and cheeks**, and some red around the orbital skin, on carpal joints and thighs, and scattered red feathers also elsewhere; ***alticola* has very little red** except on forehead/crown, and has **glaucous green upperparts**. **Juv.** has little or no red, and has brown eyes and light underside of body.
Habits and **voice**: Calls rather deep and harsh *chrrree* and *queer queer*, and a snarling *whee-eee eeeh...* never heard in Red-fronted C.
Breeding: On cliffs or in tree hollows; eggs in Dec (Arg.).
Habitat: Montane deciduous forest and fairly dry cloud forest, often with many leguminaceous trees. Often along very steep hills and tall rock-faces.
Range: The large valleys of c Peru from Huánuco to Cuzco, and locally in valleys in the E Andes to Bol., and widespread from s of the Tunari range in Cochabamba into nw Arg., to La Rioja (Córdoba?). Mainly in subtrop. zone, but often to well above 2500 m, and seen to 3400 m in Apurímac (*mitrata*). Rec. 1914 at 3400 m near Cuzco town (*alticola*; possibly this form seen 1983 at 4000 m in nearby Abra Málaga). Generally common, but now uncommon ('pre-critical situation') in Arg.

NOTE: Since the nominate form occurs in subtrop. habitat very near the type locality of *alticola*, this latter may well prove to be a distinct higher-elevation species. Its breeding habitat has not been located.

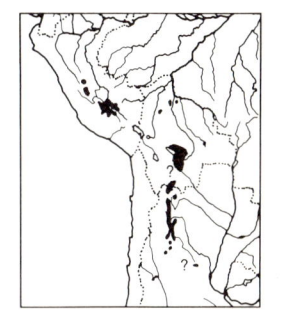

GOLDEN-PLUMED CONURE *Leptosittaca branickii* – Plate XXIII 3

34 cm. Green, lower forehead orange-brown, **yellow line below eye ends in small ear-plume**. Belly variously suffused with orange to dull red. Undersides of flight-feathers old gold, **underside of tail with dull red inner webs**. Bill gray, eyes red with white surrounding skin (**Juv.** with dark eyes).
Habits and **voice**: Poorly known. In small chattering groups, occ. in larger flocks, voice resembling Red-fronted C., but more nasal (macaw-like, although softer). Usually high in the canopies, often partly hidden, and rarely stay long in each tree. Sometimes feed low, in shrub and corn.
Breeding: Breeding Feb (Puracé, Col.). Mating Aug (Loja, Ecu.).
Habitat: Mossy cloud forest, often in tracts of difficult access, but occ. in small remnant woodlands. At least in Cajas mts in Azuay said to be strongly associated with the coniferous tree *Podocarpus*. Quite seasonal and nomadic.
Range: Known from 1300-3500 m (mainly in the upper part) from Caldas to Cauca in C Andes, in the Volcan Puracé region and at Cerro Munchique in W Andes, and Llorente in Nariño, Col.; Imbabura

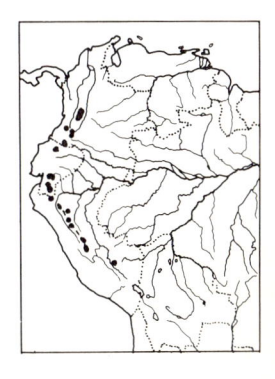

(1931), Chilla mts in El Oro and in Cajas mts, Azuay, Zapote-Najda mts in adjacent Morono-Santiago, and San Lucas, Cajanuma and Yangana in Loja, Ecu. More widespread in Peru, in Cajamarca, San Martín, Amazonas, e La Libertad, Huánuco, Junín, and Cuzco, Peru (TP). Certainly sparse and very local in the heavily deforested northern parts of the range.

NOTE: By some placed in the genus *Aratinga*.

YELLOW-EARED CONURE *Ognorhynchus icterotis* – Plate XXIII 4

42 cm, macaw-like in shape. Green with quite contrasting pattern: above and on neck-side to below cheek dark green; forehead, lores, and **wide fan of ear-plumes yellow**, underparts light yellowish green turning dark green towards vent. Pale yellowish green wing-linings contrast with drab gray underside of flight-feathers. Underside of tail dusky orange. Bill very large, maxilla blackish, mandible gray demarcated by white skin at base; eyes red surrounded by gray skin.
Habits and **voice**: Flocks probably wander seasonally. Seen feeding on fruits of *Sapium*. Give disyllabic nasal (goose-like) calls, resembling musical babble at a distance.
Breeding: Reported from holes in wax-palms in Mar-May.
Habitat: Wet montane forest and partially cleared terrain. Said to be associated with *Ceroxylon* palms, at least for breeding and roosting, but wanders seasonally.
Range: At (1200-) 2000-3400 m. Previously locally in W and E Andes, nearly throughout C Andes, and in Carchi, Imbabura and Pichincha, nw Ecu. Recent observations only at Cerro Munchique and San Rafael páramo in Cauca, in Puracé (stragglers?), and P. N. Los Guácharos in Huila, P. N. La Planada in w Nariño, Col., and on w slope in Carchi, nw Ecu. Vanishing owing to deforestation.

Burrowing Parrots

BURROWING PARROT *Cyanoliseus patagonus* – Plate XXIII 12a-d

45 cm, stockily built, square-headed, **with very long tail** (half the total length, unlike Slenderbilled and Austral Conures, which furthermore have smaller and rounder heads). **Dark olive-gray conure with yellow rump** (olive-yellow rump in *andinus*) and dark olive-gray underwings. Head, neck, and mantle deep olive-gray, wing-coverts and back more glossy olive, flight-feathers blue above. Anterior underparts and undersides of wings and tail, mostly drab gray tinged olive (or violet-gray breast in *conlara*); whitish half-collar on each side of breast charac-

terizes *patagonus*, and expands as a broad bar across chest in *byroni*. Belly yellow with red c part (*patagonus, byroni*), more restricted and greenish yellow (*conlara*), or deep drab-colored, with only traces of olive and maroon color (*andinus*). Bill black (light horn in **juv.**), eyes pale yellow surrounded by white skin.

Habits: Gregarious, previously sometimes in very large flocks. Feeds on the ground, and in the vegetation (in *Schinus*, *Lycium*, and *Discaria* trees, where the plumage blends well with the foliage, and a pest on sunflower and corn crops locally). When flying around locally by day, usually flies low, but high up on the way to and from roosts. In flight sleeked and streamlined, but spreads tail just before alighting.

Voice: Shrill rollicking screeches in flight, *crryi*, *cuirryy* etc., rising or falling at the end. Guttural cries and low creaking conversation when perched.

Breeding: In burrows in high sandy edges and limestone cliffs. Eggs in Sep(-Nov) (c Chile) or Dec (Arg.).

Habitat: Open arid country with thorny scrub and columnar cacti, esp. near streams. Roosts in groves or on cliff-faces.

Range: Lowlands from Chubut and s Neuquen to s Buenos Aires and Córdoba (and migrating to Urug.) (*patagonus*), intergrading in s Mendoza and San Luis to the population inhabiting the pre-Andean tablelands, to 2000 m, n to Cachi area in Salta, c Arg. (*andinus*); n San Luis and neighboring area in Córdoba (*conlara*). To 1900 m (in summer) in the Andean foothills and in the coastal range of Atacama to Valdivia, Chile (*byroni*, once common, but now (1984) restricted by persecution to 2800 birds with 12 breeding sites in the c provinces of Chile). It has disappeared from most of Córdoba and Buenos Aires, Arg.

GREEN-CHEEKED CONURE *Pyrrhura molinae* – Plate XXIII 9

26 cm. Mainly deep green; crown dusky brown, ear-coverts forming light gray patch; primaries and their coverts dark blue, **tail brownish red above and below**; throat, neck-side, and breast dusky, broadly barred with pale buff feather-edges (*molinae*; *australis* slightly darker, often with more olivaceous pectoral banding, but doubtfully valid). C belly with diffuse maroon patch. Bill and feet dark; dark eyes surrounded by white skin. Undersides of flight- and tail-feather dusky, winglinings green. **Juv.** has light orange-yellow breast barred with black feather-tips (but lacking gray bases of the feathers).

Habits: Usually in flocks of 10-20 birds, sometimes considerable numbers. Mostly seen passing through forest glades in low, swift, somewhat erratic flight. When alighting, glides past a tree, then swoops back and down into it. Feeds in treetops, where the plumage blends perfectly with the foliage.

Voice: Incessant, shrill *kree-cyt*, 2nd note lower, or sharp *kreet* or *quee*, but quiet when feeding.

Breeding: In tree-holes, eggs in Feb (nw Arg.).

Habitat: Dense rather low forest with open glades from chaco to hills with second growth and wet mossy cloud forest.

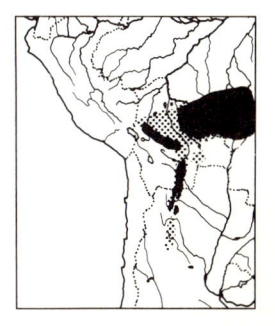

Range: Chaco of s Brazil, Parag., and Bol. (2 or 3 sspp). In yungas of La Paz and Cochabamba, on subtrop. elevation, and in the rainy season to at least 2900 m (possibly continuing to the tree-line), and locally in semi-arid zone of Cochabamba and Chuquisaca, Bol. (*molinae*). To 2000 m in Tarija, s Bol. and 1500 m in Salta and Jujuy (and occ. Tucumán and Catamarca), nw Arg. (*australis*). Generally common.

SANTA MARTA CONURE *Pyrrhura viridicata* – Plate LXIII 1

25 cm. Dark green. Lower forehead red, ear-coverts maroon, edge of wing near bend and **wing-linings orange**, primaries blue above. Belly variably patterned with orange-red feathers. Tail green above, reddish brown below. Bill horn, eyes dark surrounded by white skin. 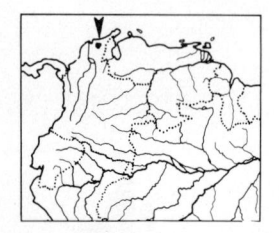
Habits and voice: In flocks, screeching like other *Pyrrhura*s. Said to be very wild.
Breeding: Breeding indications in June and Sep.
Habitat: Humid montane forest and slopes of grass and bracken with some montane shrub.
Range: N Col.: at 1700-2500 m and probably higher in Sta Marta mts. Common at San Lorenzo.

MAROON-TAILED CONURE *Pyrrhura melanura* – Plate XXIII 6a-c

24 cm. Dark green. Part or all of crown fuscous (*pacifica* with forecrown green, lower forehead rusty). Chest dusky green barred very narrowly with buffy white (*pacifica*), more broadly with grayish white to pale cinnamon (in male) (*souancei*), or with very broad whitish barring continuing around neck (*chapmani*). **Wing-bend and primary-coverts red**, primaries dark greenish blue above, black below, wing-linings green. Tail maroon (more red in *pacifica*) changing to green near base, and dusky below. C belly often with brown patch (*souancei* and *chapmani*). Bill gray (light horn in **juv.**), dark eyes surrounded by white skin (gray in *pacifica*).
Habits and **voice**: Usually in flocks of 6-12 birds. When passing open terrain flies very close to the ground. Restless and loquacious, shrieking harshly like most *Pyrrhura*s, but at approach of danger freeze and suddenly become silent.
Breeding: Eggs in Apr-June (Ecu.).
Habitat: Medium tall wet and humid forest and cloud forest, sometimes partially cleared. 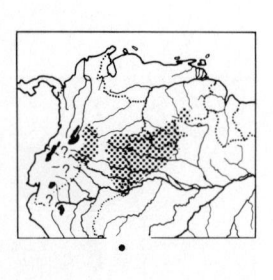
Range: Trop. zone locally in the inner Amazon basin (*melanura*); Huallaga Valley, c Peru (*berlepschi*); locally on lower Andean slopes in n Peru and e Ecu. (occ. to 3200 m), and lowlands of Caquetá to Macarena mts, s Col. (*souancei*); from 1600 m and well up in the temp. zone on e slope of C Andes, Col. (*chapmani*); in wet premontane zone on Pacific slope of Nariño, s Col. to Esmeraldas, nw Ecu. (*pacifica*). Generally common.

NOTE: The highland form *chapmani* may possibly represent a distinct species. Ssp. *berlepschi* previously considered a distinct species: Berlepsch's Conure.

FLAME-WINGED (BROWN-BREASTED) CONURE
Pyrrhura calliptera – Plate XXIII 7

22 cm. Dark green, crown tinged dusky blue; ear-coverts, throat, neck-sides, and chest dark rufous brown, and also belly more or less brown. **Wing-bend and upper primary-coverts strikingly yellow or orange** (except in **juv.**), primaries dark blue-green above, black below; wing-linings green. Tail brownish red. Bill horn, eyes buffy brown surrounded by white skin.
Habits and **voice**: In small or sometimes large flocks. Seems to resemble the Maroon-tailed C. closely.
Breeding: Breeding condition Oct.
Habitat: Cloud forest and regenerating treeline shrubbery, sometimes in very wet boggy tracts, but may descend to humid montane forest, and sometimes feed in second growth and in clearings.
Range: E Andes of Col. At (1500?) 1850-3400 m on e slope, and rec. on w slope recently (and prior to 1914). Although still numerous in some undisturbed areas, it is clearly now threatened by deforestation.

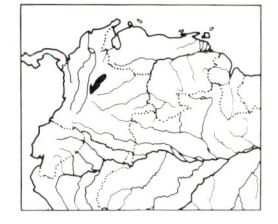

ROSE-CROWNED CONURE *Pyrrhura rhodocephala* – Plate LXIV 1

24 cm. Dark green with **large rose-red crown-cap** sometimes also spreading to the ear-coverts. **Primary-coverts white**, primaries dark blue above, wing-linings green. Tail light brownish red. Bill horn, eye-ring white. **Juv.** with blue-green crown with scattered light rosy feathers, primary-coverts blue, tail greenish at base.
Habits and **voice**: Apparently as in other *Pyrrhura*s, but clearly more quiet.
Breeding: Said to breed May-June.
Habitat: Humid forest and elfin woodlands, but also reported from second growth.
Range: Ven., at 800-3050 m in Andes from Táchira to n Trujillo. Fairly common, but now apparently very local owing to deforestation (see *Papageien* 1988:88).

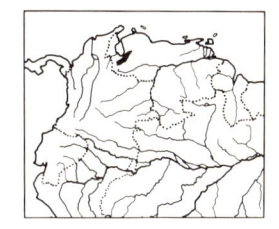

AUSTRAL CONURE *Enicognathus ferrugineus* – Plate XXIII 11

33 cm. **Olive-green faintly barred** with dusky feather-tips; forehead chestnut (looks black from a distance). Wings quite dull green, or somewhat glaucous, wing-linings green; **tail sharply pointed, brownish red**. Belly somewhat rufous, sometimes as a large dull red shield (esp. in the large and quite yellowish *ferruginea*). Small bill deep gray, bare orbital skin brown. **Juv.** with restricted red on forehead and belly.
Habits: In pairs or small groups, and wandering in large flocks. Flight swift and direct, often near ground through forest glades. Usually confident, and sometimes curious. While feeding in treetops, often climbs with body close to the branches. In feeding pauses often perches on dead uppermost branches, or hangs, head down, from the thinnest

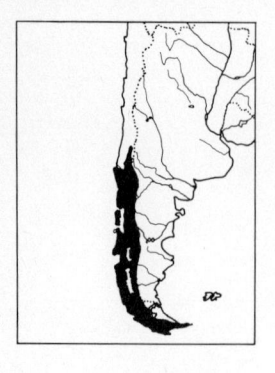

twigs. Eats seeds, esp. of *Nothofagus* and *Araucaria*, grain, berries, leaf buds of trees, and seeds of *Chusquea* bamboo, but also often feeds on the ground.
Voice: Series of raucous and grating metallic shrieks *queerc*, not unlike those of Southern Lapwing.
Breeding: In tree-holes, eggs in Dec.
Habitat: Inside woods, mainly *Nothofagus* forest, but also favors *Araucaria* and *Drymis winteri*, and sometimes in open park-like vegetation.
Range: To 1200 m in the s Andean zone from Isla Grande and isls of Cabo del Hornos to Aysén and into adjacent Arg. (*ferruginea*), and continuing n to Bío-Bío (and straggling to Colchagua) Chile, and from w Sta Cruz to Neuquén, Arg., the Argentinean birds wintering in Chile (*minor*). Common.

SLENDER-BILLED CONURE
Enicognathus leptorhynchus – Plate XXIII 10

40 cm. As previous species, but larger, with deeper mandible, and 2.7-3.7 cm **long, thin maxilla**. Generally more strongly grass-green, lacking barring below, and with red lores and forehead. Orbital skin gray (whitish in **Juv.**, which also has pale billtip).
Habits: Highly gregarious at all times of the year, sometimes in large flocks (up to 2000 in roosts). One 'chief' bird is said to lead each group. Usually flies high. Feeds on buds, saplings, seeds, and berries, in treetops and on the ground. Concentrate in *Araucaria* trees in Mar-Apr to feed on their seeds.
Voice: Almost continuous screeching. Call notes raucous, widely audible. Flight calls *clleek*, not as trilled as those of Austral C. Conversation while feeding a shrill chatter.
Breeding: In terrain of difficult access, nesting in tree-holes (often several pairs in one tree) or occ. in rock crevices. Eggs in Nov-Dec.
Habitat: Forested hills (but not the wettest types), mainly with large *Nothofagus* and *Araucaria*. Esp. in winter often in semi-open and cultivated lowlands, provided that groups of tall trees exist for roosting.
Range: C Chile from n Aysén to Aconcagua, but mainly between Chiloé and Cautín. To 2000 m in summer, mainly along the Andes but also isolated in Nahuelbuta massif. Has declined because of deforestation, persecution and disease, but still seems to be fairly common locally.

MONK PARAKEET *Myiopsitta monachus* – Plate XXIII 8

29 cm. **Face, forecrown, and throat light gray**, breast normally scaled, but uniform light gray in *luchsi*. Nape strongly grass-green, but back duller green or more drab-gray, greater wing-coverts and **flight-feathers dark blue (appearing black in flight)**, wing-linings mainly duller blue. Tail sharply pointed, blue-green. Lower chest olive-yellow, belly green. **Juv.** with green tinge on forehead. (Color phase with pale blue instead of green occurs in Neuquén, Arg.). Highly gregarious, sometimes in large flocks. Usually flies low and straight. Feeds on seeds, esp. of *Celtis tala* and leguminaceous trees, palms and corn, sun-

flower and sorghum crops. A notorious pest. In flight continually gives loud raucous guttural shrieks at various pitches. Often feeds on the ground, and then usually silent, but sometimes gives high-pitched chatter. Builds huge twig-nest (up to 200 kg), with several compartments, in trees. Eggs Oct-Dec. The nest is also used by Speckled Teal, Am. Kestrel, Firewoodgatherer (*Anumbius annumbi*), Spot-winged Falconet (*Spiziapteryx circumcinctus*), and many other birds.

Mainly in savanna woodland with low rainfall, thorny leguminaceous scrub, xerophytic areas, and farmland with groves and orchards. Common in subtrop. regions below 1000 m. Through lowlands from se Brazil to C Arg. and Tarija, Bol. (3 sspp), and locally at 1430-2700 m, in arid valleys of s Cochabamba and adjacent Chuquisaca, Bol. (*luchsi*).

Parakeets – Genus *Bolborhynchus*

Small, with long or fairly short pointed and strongly graduated tails. Recognized from other highland parakeets and conures by twittering calls. The flight is swift and direct. When disturbed during feeding, they usually fly only a short distance to alight in the top of a bush or tree.

GRAY-HOODED PARAKEET *Bolborhynchus aymara* – Plate XXIV 10

20 cm, incl. **10 cm long, narrow, and pointed tail**. Slim. Head with **dark gray cap** to below eyes, rest of upperparts green. **Cheeks, throat and breast pale gray** grading to yellowish green on sides, and light blue-green on belly and underside of tail. Underwings yellowish gray, with yellowish green lesser coverts. Bill fleshy pink to light gray.
Habits: Gregarious, usually in pairs or small groups, but sometimes 10-40 together. Feeds in seeding grass, herbs (e.g., *Viguera* and other composites), and bushes. Flies fast, in long waves, with series of rapid wing-beats, and elegantly swings tail when maneuvering.
Voice: High-pitched agitated *tjic* calls combined 3 (2-5) together.
Breeding: In holes in earth banks. Eggs in Nov (Tucumán, Arg.).
Habitat: Arid and semi-arid shrubby hillsides, incl. strongly farmed areas. Often in dense scrub and congregated in places with seeding herbs.
Range: 2000-4200 m (usually lower than Mountain P., and down to 600 m in the dry season). In the deep inter-montane valleys of La Paz, n Bol., and on the dry Andean slopes from Cochabamba s to Mendoza and Córdoba mts, Arg. (reports from Tarapacá in Chile probably erronous). Common.

NOTE: Previously placed in its own genus *Amoropsittaca*.

MOUNTAIN PARAKEET *Bolborhynchus aurifrons* – Plate XXIV 8a-e

18 cm, with **8-9 cm long slender tail** (6.6-8 cm in *margaritae*, this form

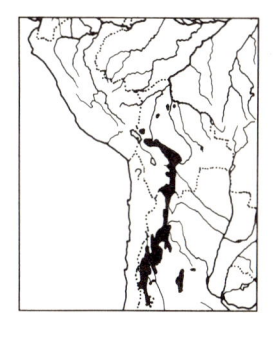

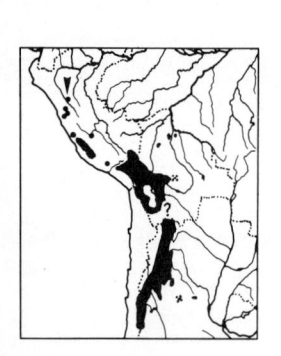

and *rubrirostris* more compactly built than the northern forms). Grass-green, lightest below, and with outer webs of secondaries blue-green, of primaries violet-blue, forming **distinct blue panel in the closed wing**. The coloration varies from rather pale dull green in *margaritae* to richer grass-green in *aurifrons* and *robertsi*, and nearly emerald-green above in *rubrirostris* (but never approaches the rich dark green of Andean P.). **Males of *aurifrons* and *robertsi* have forehead, lores, and foreparts of cheeks bright yellow**, rest of cheeks emerald-green, in *aurifrons* also with lower breast and belly almost yellow. **Female** and **juv.** of *aurifrons* and *robertsi*, and all plumages of *margaritae* and *rubrirostris*, lack yellow on face, and the 2 latter hardly have yellowish tinge at all. Underside of wings neutral gray with blue-green linings. Bill light or darker horn, in *rubrirostris* fleshy pink in male, dark gray in female; feet rosy in the northern forms, grayish flesh in *margaritae* and *rubrirostris*.
Habits: Gregarious, usually 5-10 together. Feeds on ground and in bushes. In the desert puna feeds much on buds and seeds of *Lepidophyllum*, *Fabiana densa*, and *Adesmia* and other leguminaceous scrub.
Voice: A high-pitched rather unmusical twitter *trreet*, *tzirr-zirr*, or *zrit*.
Breeding: In burrows in steep cliffs and canyons, eggs in Oct-Dec (n Chile).
Habitat: From small riparian thickets to wooded and shrubby places with fog vegetation in the coastal hills of Peru, to slopes with composite brush, thorny scrub, or cacti in the puna zone, and sometimes in open puna grassland with very scattered brush. Often in gardens and other cultivated areas.
Range: Soquián in Marañón valley, La Libertad, nw Peru (*robertsi*). Pacific slopes of Peru at least from Ancash (also Cordillera Blanca) and Lima to Arequipa, usually at 1000-2900 m, but sometimes higher, or to sea level (*aurifrons*). At 3000-4500 m on altiplanos from Puno, s Peru through Bol. to Tucumán and ne Catamarca, nw Arg., and in adjacent n Chile from Arica Tarapacá to Antofagasta (*margaritae*); and mainly around 2500 m from Catamarca to Mendoza and occ. in w Córdoba, c Arg., and across the Andes to Santiago, c Chile (*rubrirostris*). Common wherever natural scrubby vegetation occurs in the highlands.

NOTE: Previously in its own genus *Psilopogon*.

ANDEAN PARAKEET *Bolborhynchus orbygnesius* – Plate XXIV 9

16 cm, incl. **6 cm long rather broad-based but pointed tail**. Somewhat plumper than Mountain P. **Rich dark green**, with only slightly more yellowish face and underparts. Outer webs of primaries blue-green rarely approaching violet-blue, underside of wings dull blue-green. Bill pale greenish yellow (looks white except at the base), feet flesh. **Juv.** lacks yellowish tinge.
Habits: In pairs or small to sometimes fairly large flocks, feeding in bamboo, bushes, brambles, and leguminaceous trees, or on the ground.
Voice: Of a richer quality than in Mountain P. Both in trees and in flight give chattering *dydydy gy, dydydy gy...* or series of *gurk* notes, when flushed a rolling *rrueet'e rrueet'e rrueet'e....* While feeding occ. gives weaker calls.

Breeding: In burrows in steep banks.
Habitat: Mainly semi-arid zones often with *Tillandsia*-clad trees, e.g. semi-arid cloud forest and *Polylepis* wood, and sometimes in bushy ravines in more open land. Apparently prefers woods of a rather complex botanical composition.
Range: Mainly at 3000-4000 m, but occ. much higher, and seasonally lower down in montane valleys. Andes from Cajamarca and La Libertad in n Peru s in the valleys in C and E Andes to Cochabamba, Bol., and in Lima on the w slope. Common locally, but generally rarer than Mountain P.

NOTE: Previously called *B. andicola*, and sometimes included in *B. aurifrons*. However, its nearest relative is probably the Rufous-fronted P., with which it forms a superspecies.

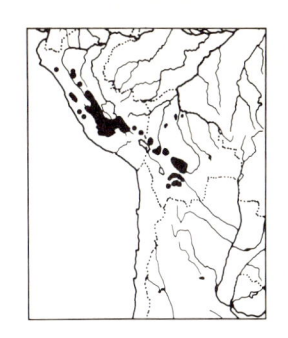

RUFOUS-FRONTED PARAKEET
Bolborhynchus ferrugineifrons – Plate XXIV 11

19 cm, incl. 6-7 cm long, rather short pointed tail, and appears stocky both perched and in flight. **Very rich dark grass-green**, slightly tinged olive-yellow below. Washed ochraceous around base of bill, with lower forehead and lores rufous. Outer web of primaries blue-green, underside of wings glaucous. Tail-feathers with rufous shafts. Bill gray to dull white with horn-colored base.
Habits and **voice**: Usually in flocks of 10-15, feeding on grass seeds. Calls moderately pitched, recalling those of Andean P.
Breeding: Probably in burrow. Large gonads Jan (Cauca).
Habitat: Grassy slopes punctuated with ravines with stunted wood, scattered shrubs, and *Espeletia*s.
Range: Col. At (2835) 3200-3900 m in C Andes: Volcano complex of Nevados del Tolima and del Ruiz, and the slopes of Volcán Puracé in Cauca. Recent rec. comprise 1 specimen from 1957 in Cauca, importations to Germany in the 1970s, probable sighting on Nevado del Ruiz 1975, and found to be fairly common near Lake Otún in Risaralda 1985, and seen on the n and nw flanks of Nevado del Ruiz 1986 (*Gerfaut* 77(1987):89-92). May be fairly common locally (recent inf., GG).

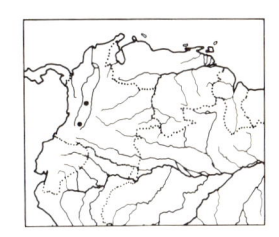

BARRED PARAKEET *Bolborhynchus lineola* –
Plate XXIV 12a-b

17 cm, with 5-6 cm pointed tail. Green varying from emerald-green on forehead and crown to dark olivaceous on upperparts of body, sides of breast, and flanks, and yellowish green below. Flight- and tail-feathers dark blue-green. Except on face and c breast and belly, **barred** with black terminal feather-edges (but other parakeets may also appear barred in worn plumage). These markings form 2 bars across the wings, and develop into **drop-shaped spots on the tail-coverts**. Tail tipped black. The pattern is best developed in **male**, poorest in **juv**. Bill pale pinkish horn.
Habits: In small parties and sometimes large flocks, which fly high, very fast, with direct, buzzy flight. Tame, but sluggish and difficult to

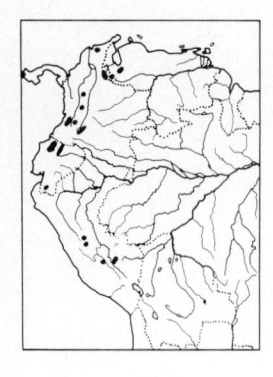

detect when feeding. Feeds on bamboo seeds, buds, blossoms, *Cecropia* catkins, corn, berries etc.

Voice: In flight high-pitched musical twittering, very different from calls of parrotlets.

Breeding: Possibly Dec (Panama) and July-Aug (Col.).

Habitat: Groves and humid forest in misty premontane and montane habitats. Partial to clearings and landslides with dense bamboo.

Range: Occurrence rather unpredictable because of the association with seeding bamboo, and erratic habits. C Am. (*lineola*). At 900-1500 m from Distr. Federal to Táchira, n Ven., known from scattered localities at 1100-2600 m in all 3 ranges of Col., and in nw and ne Ecu. (recently rec. Papallacta), Loreto, sw Ecu., and at 1600-3300 m in Amazonas, Huánuco, Ayacucho, and Cuzco, Peru, but possibly more widespread (*tigrina*, incl. *maculata* of Ecu.). Generally uncommon.

SPECTACLED PARROTLET *Forpus conspicillatus* of dry woodland of Andean valleys and llanos of Col. and Ven. is rec. occ. on Bogotá savanna (2600 m) in the E Andes of Col. (*caucae*), and possibly this species (certainly not *F. coelestis*) seen in 3 temp. zone forest localities (to 3000 m) in Loja, sw Ecu. (HB, MKP, CR, JFR) (ssp?). 13 cm, with short tail. Light green with pale bill, male pale blue around eye, and with purplish blue wing-coverts, wing-linings, and rump. Has finch-like twitter and an undulating passerine-like flight.

CANARY-WINGED PARAKEET *Brotogeris versicolorus* –

22 cm, with medium-long strongly graduated and pointed tail. Dull green, darkest above, *chiriri* and *behni* rather pale, tinged yellowish below, and showing a conspicuous patch of **yellow secondary-coverts** (**juv.** may have green edges on these feathers) (ssp *versicolorus* has a larger wing-patch, incl. white secondaries and inner primaries.) Wing-linings dull green. Bill pale horn, orbital skin gray. Social, sometimes in large flocks, which give shrill metallic notes in flight.

Widespread in the Amazon area (*versicolorus*). Tablelands of s Brazil through e Bol. and Parag. to n Arg. (*chiriri*) and s Bol. into Salta, nw Arg. (*behni*). Common in lowlands, but *behni* ascends to 2700 m in arid zones, in places with scattered trees (*Molinus, Prosopis, Schinus*) and fig cactus (*Opuntia*) in Cochabamba, Bol.

NOTE: Ssp *chiriri* possibly represents a distinct species **ORANGE-WINGED PARAKEET** (with *behni*)..

BLACK-WINGED PARROT *Hapalopsittaca melanotis* – Plate XXIV 2a-b

24 cm, with rather short graduated tail. Green with forecrown, lore and neck more or less inclining towards blue-gray (esp. in *melanotis*). Ear-coverts black (*melanotis*) or buff (*peruviana*). **Wing-coverts form large black area**, primaries and tail-tip dark purplish blue, under-

wings blue-green. In **juv.** greater and median secondary coverts have wide green edges. Bill gray.
Habits: In sometimes fairly large flocks. Feeds conspicuously in treetops, mainly on berries. Flies with rapid fairly deep wing-beats, often high.
Voice: Rapid, undistinctive, but quite melodic *shrit shrit…* and *ylp ylp…* noted.
Breeding: No information.
Habitat: Humid forest with fruiting trees, from rather tall types to boggy elfin forest type, sometimes to the border of cultivated areas.
Range: At 2800-3450 m in Carpish mts in Huánuco, e Pasco and Auquimarca, and Chilpes in Junín, Peru (*peruviana*), and at 1740-2500 m in yungas of La Paz and Cochabamba, Bol. (*melanotis*). Locally fairly common.

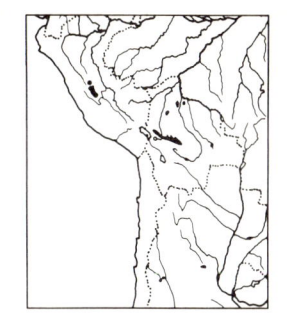

RUSTY-FACED PARROT *Hapalopsittaca amazonina* – Plates XXIV 3a-d and LXIV 6

23 cm, with short graduated tail. Green above, quite olivaceous in *theresae*, with **red wing-bends**, but with considerable difference between the sspp. Face ochraceous, with more or less brownish red to vermilion forehead and cheeks, and the slightly elongate ear-coverts with yellow streaks (not in *theresae*, which has tawny instead of red on face, or *fuertesi* and most **juv.** birds, which have red color at most on the lower forehead, and otherwise are greenish; in *fuertesi* and *pyrrhops* crown partly blue; unnamed ssp from C Andes of Col. has golden olive head extensively chestnut on face, save for a white lore). Breast quite rusty olive in *theresae*, belly partly red in *fuertesi*. The wing-bend and lesser coverts vary from signal-red (*amazonina, theresae*) to crimson or maroon (*fuertesi*), and rose-red (*pyrrhops*); greater coverts and secondaries more or less violet-blue (except in *pyrrhops*); primaries dull violet-blue; underside of wings blue with pale red lesser coverts. Tail violet-blue, red basally (green in *pyrrhops*, but c feathers all green).
Habits and **voice**: Alone, in pairs, or a few (max. 16 in Loja Ecu.) together, typically flying fast and high above the forest. Sometimes conspicuously in treetops, but may typically feed inside the canopy, and occ. low, on *Phytolacca* berries. In s Ecu. may be a *Podocarpus* feeder. Flies with shallower wingbeats than a *Pionus*. Calls undistinctive *chek-chek-chek…*.
Breeding: No information.
Habitat: Wet epiphyt-clad temp. forest up to dwarf and elfin forest, maybe esp. places with large trees and broken canopy, sometimes in places with small palms and tree ferns, or oaks. Occ. in small cloud forest patches and alders close to cultivation.
Range: At 2000-2700 m, in Caldas 3100-3600 m, and in s Ecu. at 2500-3200 (3500) m. Mérida mts and Boca de Monte in Táchira, nw Ven. (*theresae*). Paramo de Tamá, nw Ven., and from Norte de Santander and s along w slope of E Andes, Col. (*amazonina*) and at head of Magdalena valley (ssp?). W slope of C Andes, Col. near border of Quindío, Risaralda, and Tolima (*fuertesi*, on n slope of Nevado del Ruiz also an unnamed ssp related to *amazonina*: GG in press). Patchily along e Andean slope of Ecu. and in Cajas mts in Azuay, Chilla mts and Cajanuma in Loja, and in El Oro on the w slope, and s to Cerro Chin-

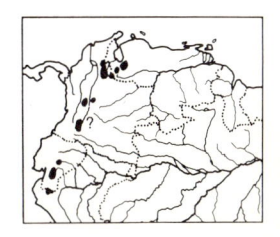

guela in Piura, n Peru (*pyrrhops*). Rare and local with few recent rec., *fuertesi* probably on the verge of extinction (no definite rec. since 1911, but a probable sighting 1980).

NOTE: These forms, lumped under one species name by Peters (1937) clearly represent 3 good species, *H. amazonina* (with *theresae*), RUSTY-FACED P.; *H. fuertesi*, **FUERTES' PARROT**, these 2 probably sympatric in the C Andes of Col. (GG); and *H. pyrrhops* (**RED-FACED PARROT**).

Parrots – Genera *Pionus* and *Amazona*

Altogether 35 species, mainly in trop. lowlands. All are stocky with square tails. Peculiar in flight, as the tail projects less far behind the wings than the head does in front: thus the birds appear to fly backwards. The voices are generally high-pitched in *Pionus*, squawking in *Amazona*.

RED-BILLED PARROT *Pionus sordidus* **Plate XXIV 6a-b** - of the subtrop. humid forest of n Ven. and Col. (several sspp) and through Ecu. and e Andean premontane zone of Peru to Cochabamba, Bol. (*corallinus*) – may reach lower limit of temp. zone (rec. to 2400 m in Col. and Bol.). 30 cm. Ssp *corallinus* dull green (other sspp generally more olive). **Feathers of head broadly edged or tipped with dark blue**, chin and throat purplish blue. Underside of wings green, under tail-coverts patterned red, blue, and green, tail-feathers green with some violet and red at base. **Bill red**, eye yellow surrounded by light gray skin. **Juv.** has pale green foreparts and blue feather-edges only on crown, vent only partly red. Flight call a loud inflected *keeank*.

PLUM-CROWNED PARROT *Pionus tumultuosus* – Plate XXIV 7a-b

29 cm. **Crown and forehead vinaceous pink** with faint scaled pattern, sides of head and throat purplish with white feather-bases shining through, neck reddish purple grading to purplish drab on breast. Rest of body green, lighter below. Vent red, tail green with violet-blue and some red basally. **Bill olive-yellow**, eyes dark, contrasting with white ocular skin. **Juv.** with forehead (-crown) reddish, throat rose-red, but cheeks, hind-crown, nape, breast, and vent lighter or darker green.
Habits: In pairs or small flocks, feeding quietly in treetops. All *Pionus* species fly with very deep wingbeats (from horizontal to right down; compare *Amazona*s). Nomadic, at least in n part of range.
Voice: Chattering and twittering, occ. with remarkable jumps in pitch.
Breeding: Breeds Nov-Dec.
Habitat: Mainly humid subtrop. forest, but locally in tall cloud forest with bamboo thickets.
Range: 1670-3300 m. Along e Andean slopes in c and s Peru, at least from Carpish mts (Huánuco) in the n to La Paz, Cochabamba to Florida in Sta Cruz, Bol. Not uncommon, but may be local.

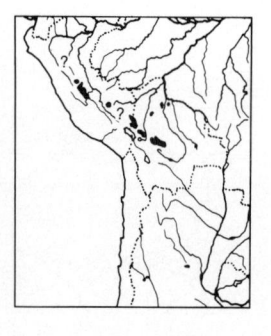

NOTE: A more detailed mapping of distribution and variation n of the Carpish mts is needed in order to state whether this taxon intergrades with the White-capped P., and should in fact include it as ssp. If lumped, the collective name **SPECKLE-FACED PARROT** applies.

WHITE-CAPPED PARROT *Pionus seniloides* – Plate XXIV 5

29 cm. **Forehead white, rest of head bluish to pinkish white with scaled pattern** of gray or dusky feather-edges. Back and upper and undersides of wings green. Underparts dark purplish gray with mainly green sides and thighs. Vent red, tail green with dull purplish red feather-bases. Bill olive-yellow, eyes dark contrasting with pale blue ocular skin. **Juv.** green with white bases of the head-feathers and mainly dull green underparts.
Habits: Singly or in small restless and noisy flocks. Feeds inconspicuously and quietly in treetops, bushes, or in corn. Often flies high over the forest with very deep wingbeats.
Voice: A mixture of smooth *chiank*, *keeweenk*, and *ah-eeek* calls with laughing quality, and soft high-pitched twittering.
Breeding: No information.
Habitat: Humid forest to cloud and elfin forests and wooded ravines in páramo habitat.
Range: Mainly around 3000 m, but seasonally down to 1400 m. Mérida and Táchira, nw Ven., and along E Andes and in C Andes of Col.; at 3300 m on mt Pichincha (JM, PG), on e Andean slopes of Ecu. to 2600 m, and in Piura, n Peru at 2050-2300 m (Chinguela); s and e of the Marañón at least as far as La Libertad and also w of the Marañón in Cajamarca, n Peru. Generally rare and highly nomadic, but common in Loja, s Ecu.

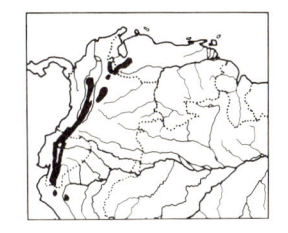

BRONZE-WINGED PARROT *Pionus chalcopterus* of wet and humid pre-montane forest of Ven. and Col. (*chalcopterus*), and Ecu. to nw Peru (*cyanescens*), occ. straggles to lower fringe of temp. zone (2400 m, once 2800 m near Bogotá). 28 cm. Head, neck, and underparts **very dark bronzy brown edged with violet-blue**, except for whitish area with narrow pink zone on throat. Mantle, wing-coverts, and tertials bronzy brown, rest of upperparts deep blue, as the flight-feathers. **Wing-linings cerulean blue.** Tail dark blue with red inner webs and under tail-coverts. **Bill yellow**, eyes dark contrasting with white ocular skin.

ALDER PARROT *Amazona tucumana* – Plate XXIV 4

31 cm, a stocky parrot with broad blunt wings. Green barred with blackish feather-margins on neck and body. **Forehead red**. Primary-coverts red, entire trailing edge of wings deep purplish blue. 'Thighs' orange-yellow in **ad.** Tail-coverts and tail-tip yellowish green, outer tail-feathers with partly concealed red spots near base. Bill horn, orange eyes surrounded by white skin.

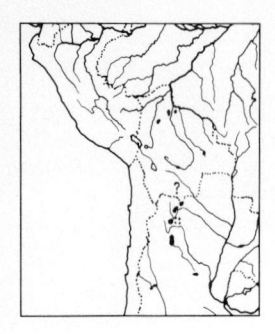

Habits: Usually shy. In flocks. Flight butterfly-like, with rapid and stiff, rather shallow wing-beats. In the non-breeding season, entire local populations may collect to roost in one or a few sites. Seeds of *Alnus acuminata* may be the principal food source.
Voice: Harsh squawking reported in flight.
Breeding: Eggs in Jan (Chuquisaca).
Habitat: Ravines with dense alders (*Alnus acuminata*) and probably also in areas with the conifer *Podocarpus* in hills dominated by deciduous forest. Sometimes descends to wooded foothills and mulberry plantations.
Range: Normally at 1500-2600 m, to lowlands in winter (but a rec. from Misiones is undoubtedly erroneous). Known from Andean foothills of Chuquisaca and Tarija, s Bol. to border Tucumán/Catamarca, nw Arg. Has disappeared from many areas, owing to deforestation in the n (with no certain rec. from Bol. since 1938); in other areas with good local remnant populations it suffers strongly from capturing for the cagebird market. Seems now to be threatened.

NOTE: May be a ssp of the Red-spectacled P. (*Amazona pretrei*) of s-most Brazil.

SCALY-NAPED PARROT *Amazona mercenaria* – Plate XXIV 1a-b

34 cm, a stocky parrot with broad blunt wings. Green, palest on face and below (these parts approaching olive-yellow in *canipalliata*). Crown, nape, upper mantle, and breast **barred** with more or less distinct blue-gray feather-margins. Flight-feathers purplish blue distally, outer 3 secondaries red except distally (*mercenaria*), or with concealed maroon spots basally (*canipalliata*). **Tail distinctive**, with yellowish green coverts and tip, and broad purplish subterminal band, except on c pair. Bill horn, ocular skin white.
Habits: Generally very shy. Single, in pairs, or sometimes parties. Sometimes in large flocks on their way to and from feeding grounds at dawn and dusk. Often flies high, usually with very shallow, stiff wing-beats (from behind resembles a small duck).
Voice: Loud and harsh, barked *krrea*, *rrrhee*, or *keoueeh koueeoeeh*, *kouee-oueeh*, alternating with small melodic whines with marked gliding fall in pitch.
Breeding: Probably Mar-May (Col.).
Habitat: In canopy of free-standing or large projecting trees, or in open forest, along high ridges, or at cloud forest edge, or in wooded ravines in páramo habitat.
Range: At 1600-3600 m. In Mérida and Perijá mts, nw Ven., Sta Marta mts, n Col., locally in all 3 ranges of Col., nw, sw, and e Ecu., and at 800-3200 m along e Andean slope of Peru and somewhat lower in Bol. to yungas of Sta Cruz. Northern *canipalliata* grades through middle of range into southern *mercenaria*. Always at low densities.

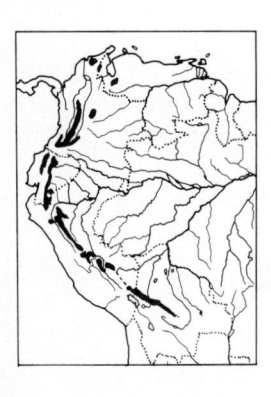

Cuculiformes

Cuckoos and anis – Family Cuculidae

Altogether 127 species, with almost worldwide distribution. In S Am. mainly tropical, but a few species reach the lower temp. zone (see *J. Orn.* 117(1976):75-99 for adaptations to highland conditions). Cuckoos are generally slim, with long graduated tails. The bill has more or less curved ridge. In the feet, 2 toes are forwards, 2 backwards. Most cuckoos are solitary and shy inhabitants of dense bushy vegetation. Many species move in the manner of passerines, with powerful hops, leaping from branch to branch with wings closed. The flight may be quite awkward. However, migratory species fly swiftly and fast. Cuckoos eat insects, and many are specialists in eating hairy and noxious caterpillars that other birds rarely take.

Cuckoos are widely known for their habits of laying eggs in nests of other birds, but this applies only to 1 of the species mentioned below. The rest raise their own young. However, in the social anis and Guira Cuckoo, young and subordinate birds contribute to the egg-laying and act as 'helpers' for a dominant breeding pair. The nests are coarse twig platforms. The young are nidifugous, S Am. species having sparse bristle-like down.

DWARF CUCKOO *Coccyzus pumilus* – patchily distributed in mainly drier lowland shrub in Col. and Ven. – occurs in small numbers at 2600 m in the Bogotá area. 20 cm, very **short-tailed** for a cuckoo. Brownish gray, **throat and chest rufous** in ad. (pale gray in juv.), rest of underparts whitish. Underside of tail gray, all feathers dark distally with narrow white tips. Sluggish and inconspicuous. Call a grating *trrr trrr....*

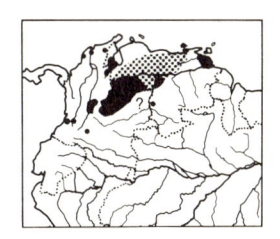

BLACK-BILLED CUCKOO *Coccyzus erythrophthalmus* – a migrant visitor from N Am. to trop. and subtrop. S Am. s to Peru and occ. to n Arg. – is rec. accidentally at 2700-2800 m near Quito, Ecu., and possibly reaches lower limit of temp. zone on e Andean slope of Peru. 27 cm. **Bill black**. Top of head gray, **eyering red**. Upperparts of body bronzy brown, all underparts off-white. Long graduated tail with indistinct blackish subterminal area, and **small white tip to each feather**. **Juv.** with some pale dorsal feather-edges.

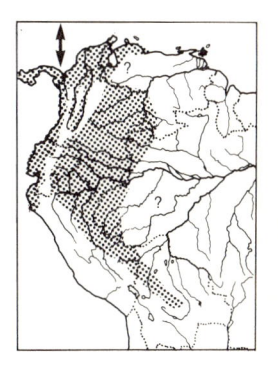

YELLOW-BILLED CUCKOO *Coccyzus americanus* – Plate XIV 13 a-b

28 cm. Slender curved bill has **yellow mandible**. All upperparts gray-brown, usually with some rufous visible in the closed wing. Ear-coverts dusky. Underparts pale grayish white from chin to vent. Tail long, graduated, the black underside with large white feather-tips. In **flight** shows **large bright rufous area in the wing**, visible below. Behaves

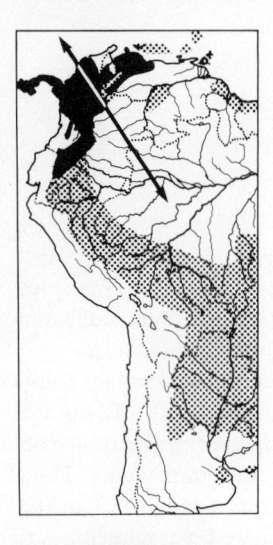

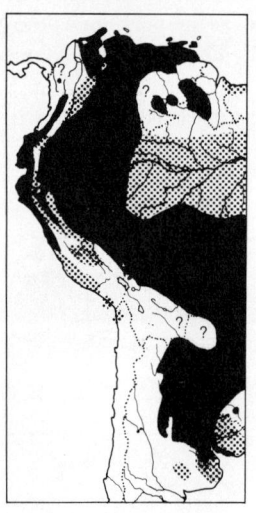

much as Dark-billed C., but usually silent on the wintering grounds.
   A N Am. migrant wintering Aug- early May in C and S Am. s to Buenos Aires and La Rioja, Arg., but absent w of Andes except in Col. Migrates at night. A few 'oversummer'. Mainly in woodland, scrub and bushy savanna in the lowlands, but rec. at 4200 m in Mérida, Ven. common in the migratory periods at 2500-3000 m at least in the E Andes of Col., and rec. at 2800 m in Mar and May in Quito, Ecu. (PG) (probably only ssp *americanus*).

DARK-BILLED CUCKOO *Coccyzus melacoryphus* – Plate XIV 12

22 cm. Slender curved **bill black**. Top of head gray with blackish ear-coverts and yellow eyering, upperparts of body grayish olive-brown. **Underparts warm buff with light gray sides of neck and breast. Tail long, graduated, the black lateral feathers with large white tips.** **Juv.** may show some rufous color in the closed wing, but wing-linings are buff, much paler than in Yellow-billed C.
Habits: Solitary or in pairs, concealed inside thick bushes (never in tall trees or on ground). Sits quietly rather low, watching for moving hairy caterpillars and other insects. Flies among the closely intervening branches with surprising ease.
Voice: Usually silent, but occ. gives guttural downward inflected coos, often in descending and accelerating series: *chow chow chow...chowchow-chochuc*, sometimes with some slower notes added. Also a dry rattle.
Breeding: Eggs in Oct (Col.) or Oct-Nov (n Arg.).
Habitat: Edge of deciduous forest, gallery forest, várzea borders and mangroves, bushy country, xerophytic areas, and second growth.
Range: Widespread through all warm parts of S Am. s to Río Negro, c Arg, e of Andes, and to extreme n Chile on the Pacific side. Mainly in lowlands, but ascends to 2400 m in Col., to 2800 m in semi-arid montane valleys in Ecu. and along the e Andean slopes of Peru, accidentally to 3600 m in Cuzco. In dry valleys of Cochabamba and Chuquisaca, Bol. regularly to c. 2500 m and occ. higher, and known to 1800 m in nw Arg.

GRAY-CAPPED CUCKOO *Coccyzus lansbergi* of the trop. zone in n Ven. and w Col. and Ecu. migrates to w Peru. A sight rec. from Bosque Ampay in Apurímac in c Peru (Venero 1988) suggests an accidental Andean crossing. Gray sides and top of head contrast rufous brown back and wings. Tail black, outer feathers tipped white. Underside tawny-rufous.

SQUIRREL CUCKOO *Piaya cayana* – Plate XIV 10

43 cm. Slender with **very long graduated tail**. Strong curved bill yellow, eyes red and orbital skin red (e of Andes) or yellow, or a mixture of the two. **All upperparts chestnut-rufous**, tawny in *mehleri*, lighter color of throat and breast grading to light gray on belly and slaty gray on vent. **Tail black with large white feather-tips**, the basal parts of

the feathers often more or less chestnut (esp. in *mehleri*, less clearly in *nigricissa*). Alone or in pairs in dense shrub, where they slip and glide through the branches in runs and powerful leaps. The call is a distinctive arresting **skwík**, *ahh* or **chí** *karah*, sharp *djit-^djit* or *djit-^jit-jit* at alarm, the song a prolonged *wyp-wyp-wyp...*. Also gives many non-birdlike grating and cackling sounds.

Widespread in gallery forest, edge of humid lowland and premontane forest, second growth, gardens, and plantations, with numerous sspp from Mexico s to the pampa region of c Arg., and w of the Andes locally s to nw Peru. Ascends to 2500 m in Zulia, n Ven. (*circe*) and further w in Ven. and to 2500-2700 m on e slope of Magdalena valley (*mehleri*), around head of Magdalena valley and e Andean slopes of Col. and Ecu. (*mesura*), around Cauca Valley, w Col. and maybe to similar elevation in w Ecu., nw Peru and to 2515 m in the e Andean zone of Peru (*nigricissa*). To 2350 m in yungas of Bol. (*boliviana*), but probably lower from Cochabamba, Bol., to Catamarca, Arg. (*mogenseni*).

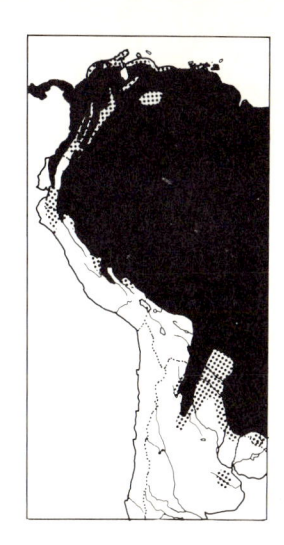

Anis – Genus *Crotophaga*

3 grotesque-looking **black cuckoos with very deep but narrow bills** (less striking in juv.). Apparently loosely put together, due to the untidy plumage, broad-tipped wings, and long narrow-based tail, incessantly cocked or twisted. Upon alighting, the tail is swung up and forward. The 2 smaller species feed mainly on grasshoppers, often near grazing cattle. Anis appear in separate species groups of up to 25 birds (Greater Ani sometimes in large flocks). When a flock moves, the birds always fly one at a time!

GREATER ANI *Crotophaga major* – Plate XIV 15

45 cm, slender with very large tail. Metallic blue-black with bronzy green feather-edges and conspicuous **white eye** (dark in **juv.**). Bill quite long, compressed, with the keel high proximally. Looks sleek and well-groomed compared to the smaller anis. The calls comprise a bubbling melodious *cu-curre*, and a monotonous, repeated *wau, wau, wau...*, and deep, mammal-like *oak*.

Inhabits swamps and forest thickets near rivers, often in vegetation overhanging water. Widespread through trop. and subtrop. parts of S Am. from Panama and Caribbean coasts to chaco swamps of n Arg., or to Ecu. w of Andes. Casually in savannas at 2600 m in E Andes, Col., and (accidental?) at Lake Alalay (2550 m) in Cochabamba, Bol.

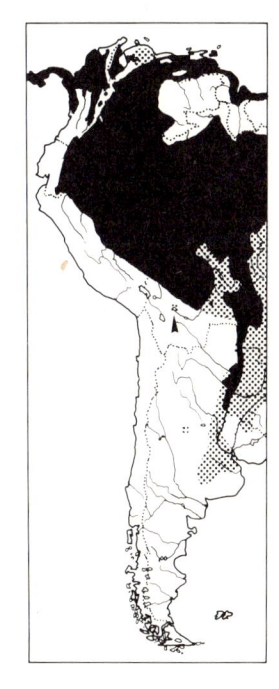

SMOOTH-BILLED ANI *Crotophaga ani* – Plate XIV 14a-b

C. 33 cm, highland birds largest. Bronzy black with paler pattern of bronzy feather-edges on much of the foreparts. Bill smooth, narrow, with the **keel highly arched, curving down to the forehead**. Juv. dull black. Larger and more thick-necked than Groove-billed A., and seems to hold head higher in flight. Best identified, however, by high-pitched **rising whine** *oueee^eenk*, often emitted in flight. Also gives various guttural calls.

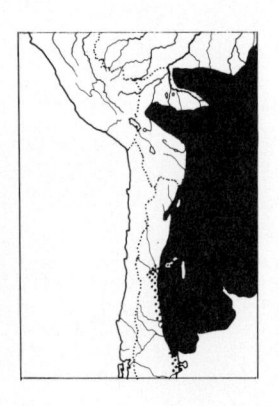

Inhabits clearings in humid forest and cloud forest, corn crops, sugarcane, and second growth, and occurs near water in scrubby and xerophytic areas. Widespread through trop. and subtrop. lowlands from C Am. to pampa region of c Arg., and w of the Andes to sw Ecu. Ascends to 2400 m n of Orinoco, Ven., and to 2000 and occ. 2700 m in Col., rec. to 2400 m in s Ecu., and breeds in temp. side-valleys of the Marañón valley, n Peru, and to 2800 m (once at 3200 m) in some Andean valleys of c Peru. Not high up in Arg.

GROOVE-BILLED ANI *Crotophaga sulcirostris*

30 cm. Dull black with paler pattern of bronzy feather-edges on much of foreparts. **Bill longitudinally furrowed, narrow, and deep, with the upper ridge continuous with the crown**. Juv. dull black. Most easily identified by a melodious, liquid, and downwards-slurred *kee-wuy* call frequently preceeded by throaty clucks, and often repeated 3 times. Also sharp, dry *hwik* notes.

Inhabits riparian thickets, tall *Gynerium sagittatum* canes, and marshy vegetation near cultivations and pastureland in desert-like or xerophytic areas. Warm arid parts of the Americas from Baja California (*pallidula*) and Texas s locally s to Tarapacá (Antofagasta), n Chile, (and reported from Salta, nw Arg., if right) (*sulcirostris*). Generally lowlands, but local populations exist to 2700 m in Andean valleys of n Peru and in Huánuco, and maybe some valleys of sc Peru. Accidentally near Huacarpay (3100 m) in Cuzco, and at 2550 m in Cochabamba, Bol. Rec. to 2700 m on Pacific slope of Peru and to 2300 m in the Chilean desert.

STRIPED CUCKOO *Tapera naevia* – Plate XIV 11 - inhabits trop. and subtrop. forest edge, swampy gallery forest, woodland, and second growth bordering open areas, and also in xerophytic areas, in C Am. (*excellens*), S Am. from Ven. and n Col. to sw Ecu., and n Peru (*naevia*) and from s of the Amazon to c Arg. (*chochi*), reaching lower edge of the temp. zone (2500 m) n of the Orinoco, Ven. 28 cm. Crown with bushy crest **rufous with black streaks, eyebrow white**. Rest of upperparts and tail brown with dusky streaks. Also cheeks somewhat streaked, throat and breast pale sandy grading to white on belly. **Juv.** with transverse buff spots on fuscous crown, and scapulars and wings extensively golden-spotted. Shy and skulking, usually recognized by very guttural, plaintive whistles given day and night: *suu-^see...suu-^see...* in sometimes interminable series. Brood parasite in nests of ovenbirds.

GUIRA CUCKOO *Guira guira* - Plate XIV 9 - is common and widespread in scrub, woodland savanna, and parks in trop. and subtrop. zones s of the Amazon forest to Chubut and Tucumán, c Arg., reaching transition to temp. zone in the latter province. - 37 cm. **Very shaggy-looking. Head and neck pale buff** with orange-buff crested crown, side of neck with thin fuscous streaks. Upperparts of body dark brown with pale buff feather-edges and large buff rump. **Tail fuscous with**

whitish lateral area at base and broad white tip. All underparts pale buff. Bill yellow. Often in small groups, flying like anis. Feeds on ground, but roosts in trees. Perches conspicuously with drooping wings, exposing the pale rump, and sometimes waves its tail. Very noisy, giving long series of 2-note phrases *pee-oop* gradually falling in pitch and musicality and ending very scratchily. Also moderately high, slow gargled trill and series of *creep*s at several pitches.

Strigiformes

A well-defined order of nocturnal raptors which apparently evolved from primitive roller-like (coraciiform) birds. Nightjars may represent a side branch to the owl line.

Barn Owls – Family Tytonidae

10 species, of which 1 is worldwide, the others local in the Eastern Hemisphere. Differs from the typical owls by a rather slender skull and a large heart-shaped facial disc. Feeding ecology described in *Bol. Mus. Nac. Hist. Nat. Chile* 38(1981):137-146; *Neotrópica* 30(1984): 250-2; and *Stud. Neotrop. Fauna Envir.* 22(1987):129-36.

BARN OWL *Tyto alba* – Plate XXVI 1a-b

36 cm. A large-headed owl with slender body, long legs, and a peculiar narrow profile when perched. Above light gray (*tuidara*) to dark slaty gray, distal parts of the feathers darkest (*contempta*), with fine black-and-white mark on most feather-tips, and some orange-buff areas. **Big heart-shaped facial disk** usually pale brown with brown outline, and diffuse dark area in front of black eyes (*contempta*), or lighter (*tuidara*). All underparts honey-colored with dense dusky mottling (*contempta*), or scattered dusky spots (*tuidara*). A lighter color phase (which dominates in s parts of range) is almost white on face and underparts. In **flight** looks indistinctly marked, light, or even totally white. Unlike in typical owls, plumage of ad. type directly replaces the white natal down.
Habits: Most active in open terrain evening and night. Flight buoyant. If disturbed by day, flies only short distance. When excited, violently rocks head and body from side to side.
Voice: A 1-2 s rasping *tsheeeerrr*, tremulous if given in flight. Also subdued screeches, purring, wailing, staccato squeaking, and (in nest defence) hissing and bill-snapping.
Breeding: In caves, hollow trees, lofts of houses, church towers etc. Eggs in Oct-Nov in the s of the continent. Eggs July (w Ecu.) and Sep (Bogotá, Col.), fledglings July (Ecu.).
Habitat: In villages and suburban areas, and a wide spectrum of open and lightly wooded, often man-disturbed landscapes. Presence of caves or crevices decisive.
Range: Cosmopolitan, with at least 34 sspp recognized. Present in all parts of S Am., but somewhat local, and not in heavily forested regions.

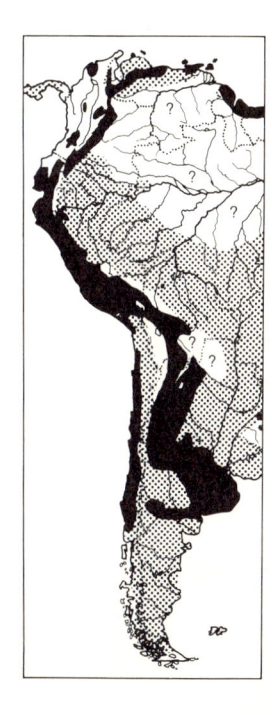

In Ven. only below 1500 m, but in other parts of the Andes c. up to the cultivation limit (viz. 3200 m in Col., above 4000 m in Peru and Bol.), with *contempta* from Andes of Col. and Ecu. to n Peru, *tuidara* further s all the way to Isla Grande. Probably not rare, but distribution poorly known.

Typical owls – Family Strigidae

124 species, with a worldwide total distribution. Owls are easily recognized by their compact shape, large heads with wide facial discs, and a very thick and soft plumage giving soundless flight, and with a cryptic coloration. Many species have brown and grayish color phases. The forward-directed eyes and the ability to blink with the upper eyelids give a semi-human expression. Owls have a wide repertoire of expressions: usually a relaxed owl has a round head, an alert bird raises its ear-tufts, and a scared owl freezes in a slim posture with slit-like eyes and narrow face. The vision is very sharp, and binocular in much of the field of vision. However, the big eyes are almost fixed in the orbits. To compensate for this, owls turn their heads 270°. A curious owl often twists its head to follow objects continuously, and often peers rapidly from side to side. The hearing is no less remarkable than the sight. The bristle-like feathers of the facial disc covers enormous parabola-shaped outer ear-tubes. Owls have good nocturnal vision, and can spot prey very precisely by means of their hearing. However, they also see well by day, and certain species hunt in full daylight. The ability to hunt in the dark rests in part on sedentary habits and a very precise knowledge of the territory. Other adaptations resemble those of diurnal raptors: strong talons, hooked bill with swollen cere, and a broad gape (however, with the bill partly concealed by bristles). Many species have feathered legs or toes, presumably a protection against possible bites from the prey. The wings are normally rounded, the tail short.

Owls usually eat ground-living prey such as rodents, frogs, and larger insects. A few often take birds. Unlike diurnal raptors, no owl has specialized in carrion, and there are no high-soaring species, and only one with falcon-like habits (the bat-eating Stygian Owl). Owls eject indigestible parts of the food (bones, hair, chitin) as pellets, which can be found at traditional roosts, and in nest-holes. These pellets are valuable sources of information about prey utilization.

Owing to their nocturnal habits, owls are rarely seen. However, loud mobbing by smaller birds sometimes reveal a day-time perch. Particularly in forest zones, listening by night (or playback with tape-recorder), mistnetting, and alertness to road-kills are almost the only ways to record owls. Most species have a wide repertoire of wide-ranging nocturnal calls varying from musical hoots to shrieks and hissing. The calls are usually diagnostic of species, but are incompletely known as far as Andean species are concerned.

Owls breed as solitary, monogamous pairs. They do not normally build nests, but lay their round white eggs directly in holes or abandoned raptor-nests. Owls can be very aggressive near nests, sometimes using their talons against human visitors. The clutch-size varies with the food supply. The nidicolous young are covered by white down, and are fed a long time in the nest. The natal down is replaced by a fluffy, cryptically colored down-feather plumage (mesoptiles), which is worn during the semi-flightless period immediately after the nest is left.

Further information about Andean forms are given in Burton, JA (1973) *Owls of the world*. London, and Hekstra, GP (1982) '*I don't give a hoot'. A Revision of the American Screech Owls*. Acad. Proefschrift. Thesis. Vrije Universiteit to Amsterdam. See also Clark, RJ, Smith, DS & Kelso, LH (1978) *Working bibliography of owls of the world*. Washington; *Auk* 94(1977):409-16, 98(1986):1-7; *Rev. Biol.*

Trop. 10(1962):45-59. Studies of feeding ecology are given in *Ann. Inst. Pat.* 9(1978):199-202; *Auk* 97(1980):895-6; *Bol. Mus. Nat. Chile* 38(1981):137-46; *Bol. Orn.* 9(1977):9-10; *Can. J. Zool.* 59(1981):2331-40; *Doñana Acta Vert.* 13(1986):180-2; *Raptor Res.* 20(1986): 113-6; *Rev. Chil. Hist Nat.* 60(1987):81-6, 93-9; and *Revista Peru Ent.* 22(1979):91-4. The classification of screech-owls follows Hekstra (see above), with some recent changes by JW Fitzpatrick and JP O'Neill (in *Wilson Bull.* 98(1986):1-14).

SAVANNA (or TROPICAL) SCREECH-OWL
Otus choliba – Plate XXV 5a-d

25 cm. Distinguished from other small Andean owls by combination of short ear-tufts, white spots on scapulars, pale gray to brownish (but never rufous) **facial discs distinctly outlined in black** (except in very rufous birds), and **yellow eyes** (not shiny). Eyebrows, throat, and incomplete nuchal collar whitish. The general coloration is gray or slightly fulvous, **intricately mottled**, spotted and streaked with black. Wings and tail dusky brown banded with cinnamon-buff. Underparts white to ochraceous with black 'herringbone' marks. Among Andean sspp, *montanus* is large and dark; *alticolus* and *luctisonus* light gray-brown, the former densely mottled below, the latter with only 3-4 herringbone marks per feather; *roboratus* rather brown, underparts with white blotches disrupting a mottled pattern of very dense 'herringbone marks'. Ssp *koepckei*, and to some extent *alilicuco*, have broad black shaft-streaks, and become almost black on crown; they are never fulvous and have only 4 bars per 'herringbone'. **Juv.** down-feather plumage is cinnamon-buff densely mottled with thin dusky bars, and with a pale 'mask' edged black.

Habits: In pairs. Very hard to trace by day. Flutters with silent wavering action quite unlike bounding undulating flight of pygmy-owls. When scared, freezes in an elongated posture with erect 'ears' (but 'ears' are not always evident).

Voice: Heard mainly after sunset and just before sunrise. Territorial song, often in duet between mates, or antiphonally between males: a 1-1.5 s bubbling series of 5-15 hoots ending suddenly with one or more accentuated strokes, as *ho-o-o-o-o-orr* **ook**, or short series of *churro* notes, often repeated at 3-5 s intervals. Also a shorter trill followed by 3-4 strokes.

Breeding: Breeds Jan-Feb (Col.); eggs rec. Apr (Ven.) and Oct (Bol.), pull/juv. Apr-July (n Ven.); juv. Aug (Cauca, Col.), July (Peru), Nov (Tucumán, Arg.).

Habitat: Open wood, scrub, savanna, and cultivated land from humid zones to semi-desert. In Andean valleys esp. in balsa woodlands, and in the highlands from broken cloud forest to *Polylepis* woods or dry scrub with occ. larger trees.

Range: Numerous sspp through S Am. lowlands and subtrop. Andean slopes from Ven. to c Arg., and w of Andes s to c Peru. Ascends locally to temp. zone: *alticolus* to 3000 m w of E Andes, Col., and in Ecu., *luctisonus* maybe to similar elevation in Atrato drainage and Dagua valley, w Col., *roboratus* to 2500 m in Marañón valley, n Peru; *koepckei* (Hekstra 1982, maybe actually 2 sspp involved) scatteredly or locally at

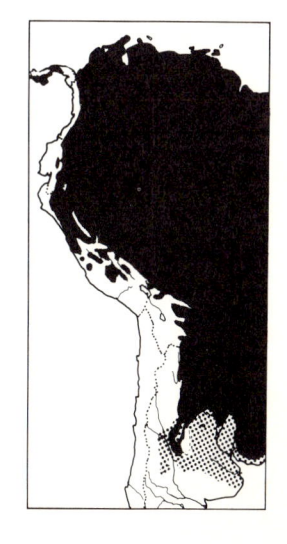

1500-3200 (locally to above 4000 m) through Peru, *alilicuco* (Hekstra 1982) from subtrop. to 2700 m on Andean slopes of Tucumán to Catamarca, nw Arg., birds from yungas and valleys of Bol. and Jujuy, nw Arg., variably intermediate between *koepckei*, *alilicuco*, and chaco forms *suturus* and *choliba*.

NOTE: Ssp *roboratus* was previously treated as a separate species **WEST PERUVIAN SCREECH-OWL**, possibly related to C Am. Cooper's and Balsas S-o. *Otus cooperi* and *O. seductus*. An apparent intergrade between *roboratus* and *O. choliba portoricensis* from the lower Marañón Valley of n Peru leads to inclusion under the Savanna S-o. (Hekstra 1982). However, the screech-owls of n Peru seem to deserve further study.

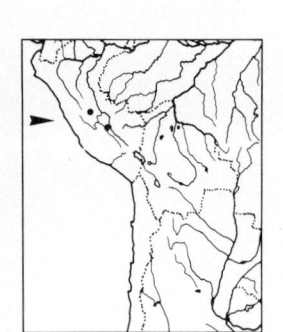

CLOUD-FOREST SCREECH-OWL *Otus marshalli* - **Plate XXV**
δ is rec. at 1920-2240 m in Cordillera Vilcabamba, Cuzco and 2050 m in Cordillera Yanachaga Pasco, Peru, and could be expected to reach the temp. zone. Inhabits wet cloud forest with impenetrable undergrowth, clinging bamboo, and some tall trees massively festooned with mosses, ferns, and orchids. 23 cm. A dark rufous owl distinguished from other small Andean owls by combination of short but distinct ear-tufts, richly **rufous face with broad, black outline, dark brown eyes**, mottled rufous-and-black upperparts with white spots across nape (distinctly bordered with black feather-tips), white scapular streaks, rufous breast with dusky bars, and increasing amount of white spots towards abdomen giving **ocellated appearance**. Behavior and voice unknown. Breeds July-Aug and possibly also at other seasons.

NOTE: Described by JS Weske & JW Terborgh in *Auk* 98(1981):1-7.

RUFESCENT SCREECH-OWL *Otus ingens* – **Plate XXV 2a-c**

26-28 cm. Distinguished from other small Andean owls by **rather uniform brown appearance**, buff (not white) underdown, and combination of buffy brown (not rufous) poorly outlined facial discs, vestigial ear-tufts, and **brown eyes**. Above tawny-olive to sandy or light gray-brown with fine dark vermiculations, and a few whitish spots, and 2 rows of white scapular streaks. Underparts finely barred light brown and white, with a few thin dark brown streaks. Face usually with pale (but not white) eyebrow and throat. A rufous morph is more uniform throughout. Averages deeper red-brown with more black at high elevations, and shows increasing amounts of white spots and less cinnamon-buff in n of range. **Juv.** down-feather plumage light buff with dense dusky mottling, and with indistinct 'mask'. Song *hu hu hu hu tutututu…*, typically starting slow and proceeding with c. 50 notes in 10 s, or in slower series.

Inhabits dense wet and humid premontane forest, locally to 2500 m on lower edge of temp. zone. Scattered rec. from coastal mts of Ven. (ssp?), Mérida, and Perijá and Táchira mts on Ven./Col. border (*venezuelensis*), and from Nariño, s Col., along e Andean zone of Ecu. and

Peru (*ingens*) and Cochabamba, Bol. (doubtful ssp *minimus*). Probably uncommon.

NOTE: 2 more forms of the *Otus ingens* superspecies may deserve species rank (see *Wilson Bull.* 98(1986):1-14): The allospecies **COLOMBIAN SCREECH-OWL** *O. colombianus* of the w Andean slope from Valle, w Col. to Nanegal in Pichincha, nw Ecu., and the synspecies **CINNAMON SCREECH-OWL** *O. petersoni*, known from the Cutucú mts and Cajanuma in Loja (PG), se Ecu., and Piura, Cajamarca, and Amazonas on both sides of the Marañón in n Peru (and a 'Bogotá' specimen). Both in the premontane zone, the Cinnamon S-o. to 2450 m, and may be expected to lower edge of temp. zone. Both resemble a dark buffy brown Rufescent S-o., with fine cinnamon freckles, few vermiculations, and no white on any body- or wing-feathers. Colombian S-o. is largest (wing chord 176.5-192 mm) with long legs with naked distal half of tarsus, and inconspicuous dark rim of the facial disc. Cinnamon S-o. is small (wing 153-165.5 mm) with short feathered tarsi. Song of Cinnamon S-o. consists of a 5 s series of 30-40 notes, first rising slightly in pitch and finally sliding back. Also gives series of 1-6 explosive whistles.

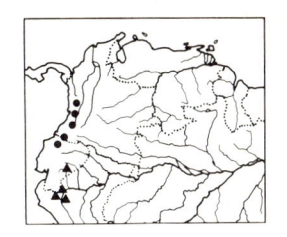

WHITE-THROATED SCREECH-OWL *Otus albogularis* – Plates XXV 4 and LXIV 7

28 cm, with large rounded head almost without ear-tufts, and with relatively longer tail than in other screech-owls. **Dark** fuscous brown with blackish mottles, and tiny rufous and white marks (but lacking white scapular streaks). Facial discs dark with whitish marks below yellow or orange eyes. **Throat extensively white**, appearing as 2 large moustaches in certain postures. Below lighter, with c and lower parts mainly tawny-buff with scattered dusky bars and streaks. Ssp *meridensis* (Plate LXIV) has whitish forehead and eyebrow and numerous pale freckles above suggesting nuchal collar; other sspp darker, *remotus* (Plate XXV) particularly dark with few white and rufous markings above, dark breast contrasting poorly marked lower underparts. **Juv.** down-feather plumage pale buffy gray with orbital discs and long bristles above bill forming black 'mask'.
Habits: Strictly nocturnal.
Voice: Gives 4-9 hoots per s, often both sexes together in more or less synchronized duet at somewhat different pitch. Primary song a descending trill of 10-14 hoots every 5-10 s; secondary song 7-30 gruff, barked notes *churrochurro-churro-chu chu chu chu*, and may sing uninterruptedly for nearly 1 min. Also gives single hoots.
Breeding: Ovary enlarged July (c Peru); pull. Oct; juv. Jan (Ecu.), Mar (c Peru), and Sep (Ven.).
Habitat: Rain and cloud forest, usually heavily loaded with epiphytes, and with bamboo thickets, possibly requiring glades or at least gaps in the canopy.
Range: Mainly at 2200-3000 m, sometimes higher, and outside breeding season lower. Andes of Mérida, Ven. (*meridensis*), from Perijá mts, Ven. (*obscurus*), through E Andes, Col. (*albogularis*), and C and W Andes of Col. and Ecu. s to Cerro Chinguela in Piura, n Peru (*macabrum*), and

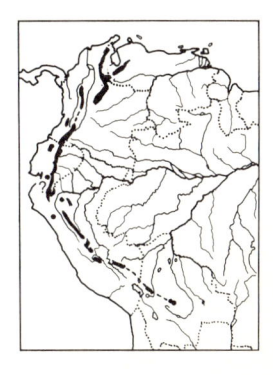

with scattered rec. along e Andean slope of Peru and in yungas of Cochabamba, Bol. (*remotus*). Probably not uncommon.

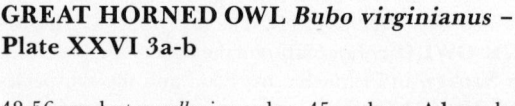

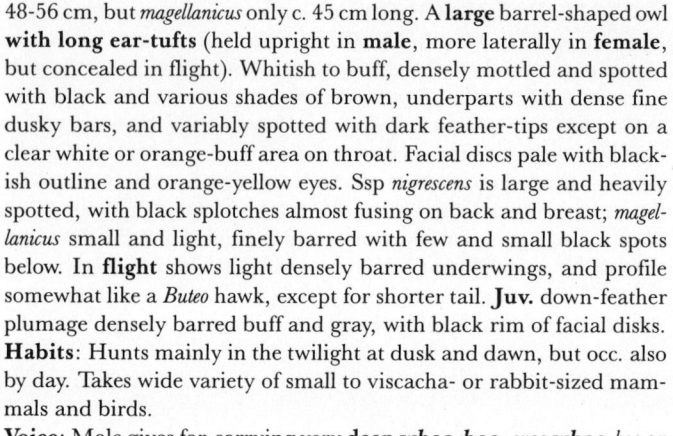

GREAT HORNED OWL *Bubo virginianus* – Plate XXVI 3a-b

48-56 cm, but *magellanicus* only c. 45 cm long. A **large** barrel-shaped owl **with long ear-tufts** (held upright in **male**, more laterally in **female**, but concealed in flight). Whitish to buff, densely mottled and spotted with black and various shades of brown, underparts with dense fine dusky bars, and variably spotted with dark feather-tips except on a clear white or orange-buff area on throat. Facial discs pale with blackish outline and orange-yellow eyes. Ssp *nigrescens* is large and heavily spotted, with black splotches almost fusing on back and breast; *magellanicus* small and light, finely barred with few and small black spots below. In **flight** shows light densely barred underwings, and profile somewhat like a *Buteo* hawk, except for shorter tail. **Juv.** down-feather plumage densely barred buff and gray, with black rim of facial disks.
Habits: Hunts mainly in the twilight at dusk and dawn, but occ. also by day. Takes wide variety of small to viscacha- or rabbit-sized mammals and birds.
Voice: Male gives far-carrying very **deep *whoo-hoo*** ,*woo **whoo**-hoo* or *guacouroutou-tou* followed by low tremulous *chuhuhurr*. Female usually gives long series of hoots on somewhat higher pitch. Sometimes mates call back to one another with strident screams.
Breeding: On ground, among rocks or in cave, or abandoned raptor nests in trees. Eggs in Oct-Nov, pull. Jan in the s and in c Peru.
Habitat: Open or wooded land of arid as well as humid climates. Prefers lonely hillsides with some rocks, but in some areas lives near villages.
Range: Throughout the Americas, except areas of dense humid lowland forest. In Andes, widespread mainly at 2500-4500 m through n Ven., s E and C Andes of Col., and through Ecu. (*nigrescens*); mainly above 3000 m in puna zone of Peru, w Bol. and nw Arg. and n Chile, continuing through s Andean zone (and in Córdoba mts) to Isla Grande (*magellanicus*), and locally in the lowlands to the e (*nacurutu*). Generally uncommon, but common in Patagonia and Magellanic Chile.

ANDEAN PYGMY-OWL *Glaucidium jardinii* – Plate XXV 10a-c

14 cm. A tiny owl with **small head with ill-defined mask**. Normally looks round-headed, but 'ears' are indicated in slim posture. Dull dark brown with white marks, or dark chestnut with buff marks: **crown with numerous small white dots** (normally not streaks, except on forehead, but some individuals have dots connected with thin streaks), nape with conspicuous black-and-white face-like pattern. Back and wings with large but scattered white spots. Tail with 3-4 white or buff bars. Underside boldly marked, the dark brown parts with some short pale spots or bars, and the c parts with coarse white or buff streaks. Short eyebrows, moustaches, and broad throat-patch white. **Eyes** yel-

low. **Juv.** lacks markings on crown and nape, or have some thin streaks here.
Habits: Often seen by day in top of small dead tree or other exposed perch (frequently mobbed by other birds), or in rapid bounding and undulating flight. Tail often twisted from side to side, or cocked as the bird alights. Takes more birds than other owls do, sometimes of nearly its own size, but also feeds on large insects.
Voice: **Series of whistled monotonous hoots**, often c. 2 introductory whistles followed by a rapid, slightly irregular series, *tyuk tyuk hy-hy-hy...* or *de hy-de hydehydehyde...* (Ferrugineous P-o. of adjacent lowlands gives long series, fast or slow, but always with regularly spaced hoots).
Breeding: In tree-holes.
Habitat: Semi-arid to humid open cloud, dwarf, and elfin forest, or wooded ravines with transition to swampy or grassy habitat.
Range: Mainly at 2000-3500 m, but locally to 4000 m. C Am. (*costaricensis*) and locally in all Andean chains from Perijá and Mérida, Ven., through Col. and Ecu., and along the slopes and inter-Andean valleys of c Peru and e Andean zone to Cochabamba, Bol., and recently rec. in Jujuy, nw Arg. (*jardinii*). Fairly common.

NOTE: Sometimes treated as a highland ssp of the widespread **FERRUGINOUS PYGMY-OWL (*Glaucidium brasilianum*)**, which has streaked rather than dotted crown, and is less contrasting, often very fulvous in color, sometimes with uniform rufous tail. The 2 taxa are clearly segregated altitudinally in Bol., and differ vocally. However, the distinction is not clear everywhere, as some Ferrugineous P-o. of w Peru approach *jardinii* morphologically. The Ferrugineous P-o. here lives mainly in the subtrop. zone, but has been reported as high as 2800 m in Moquegua (RH).

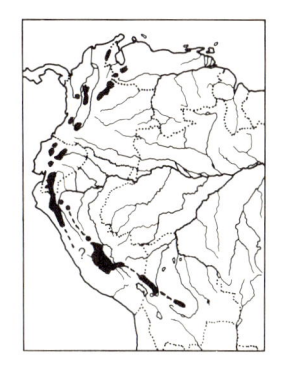

AUSTRAL PYGMY-OWL *Glaucidium nanum* –
Plate XXV 7a-b

17 cm. Resembles Andean P-o., but rarely strongly tinged rufous, crown minutely streaked (not dotted), and **tail with 7-10 distinct rufous or rarely white bars**. **Juv.** has uniform brown crown, nape, mantle, and chest.
Habits and **voice**: As in previous species, but long **series of whistled hoots**, sometimes slow, but always regularly spaced (as in Ferrugineous P-o.).
Breeding: In tree-holes, eggs in Sep-Nov.
Habitat: Forest, often rather open and stunted types with many dead trees on hill-sides or swampy ground. Also gardens and parks.
Range: To 2000 m. Wooded parts of Chile and s Arg. from Isla Grande to Neuquén. In winter migrates to Atacama, n Chile, and the lowlands of c and n Arg. Common.

NOTE: Often listed as a ssp of Ferrugineous P-o. (*Glaucidium brasilianum*). Marín *et al.* (1989) dispute the wisdom of accepting *nanum* as a separate taxon at all. It may simply represent a predominance of a darker rufous brown morph with several rufous tail-bars in the south.

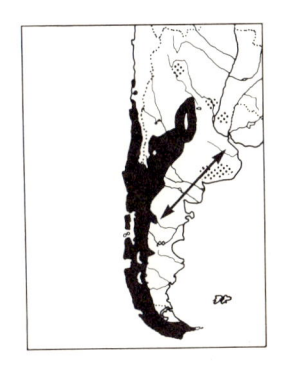

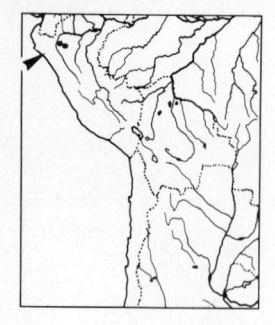

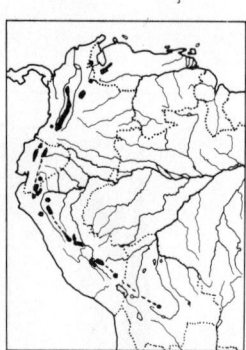

LONG-WHISKERED OWLET *Xenoglaux loweryi* - **Plate XXV 3**
is known from 1890 and 2350 m in a very wet (temp.) climate e of Abra Patricia in Rioja Amazonas/San Martín, and e of Bagua in Cordillera Colán in Amazonas, n Peru. On ridges with stunted wet forest heavily laden with epiphytes, and with bamboo thickets and scattered palms and tree ferns. 12 cm, possibly almost flightless. A **tiny**, almost tail-less large-headed owl with outer decomposed feathers of facial disc extending beyond main plumage of head like **loose whiskers**, and with bristles covering base of bill extending as vertical fan-like crest. Brown, minutely marked with dusky bars, and with white spots forming incomplete collar across nape, and irregularly on scapulars and wings, and as diffuse bars on belly. Eyebrows and throat white. Eyes orange. Voice of possibly this bird comprised series of short widely spaced mellow whistles, sometimes followed by series of faster, slightly higher-pitched notes.

NOTE: Described by JP O'Neill and GR Graves in *Auk* 98(1977): 409-16.

RUFOUS-BANDED OWL *Ciccaba albitarsus* – **Plate XXV 6**

35 cm. Compact, with large round head, and rather short tail. **Coarsely barred** tawny and black, below also marked with lighter orange, and below breast ocellated with white. Facial disc tawny, black centrally, and outlined in whitish towards bill. Eyes orange-yellow. Flight- and tail-feathers barred blackish and buffy. Thighs rufous. **Juv.** down-feather plumage very different: plain warm buff with blackish 'mask'. Eyes brown.
Habits: Active shortly after dark and before dawn.
Voice: **Rhythmic series of deep hoots** *rrho rrho rrho...* or *whoo whoo whoo...*, or of hoarse *whyh* calls, followed after a short pause by an accentuated *whooa*. Also single gruff hoots.
Breeding: Juv. Mar and Nov (Cauca), Apr (Caldas), June (Huila, Col.), Aug (Mérida, Ven.).
Habitat: Thick humid montane forest.
Range: At 1700-3000 m from Trujillo, Ven., spottily along most Andean slopes of Col. and Ecu. to n Peru (*albitarsus*), and with scattered rec. on e Andean slopes s to La Paz, Bol. (*tertia*). Common.

RUFOUS-LEGGED OWL *Strix rufipes* – **Plate XXV 12**

35 cm. Compact, with large round head. Sooty **black with white barring**, esp. on head and breast, and more or less spotted with orange-buff. Facial disc usually whitish with concentric dusky rings, outlined with black, or whitish towards bill. Eyes dark. Throat white or densely barred black. Legs and vent uniform orange-buff. In flight shows buff wing-linings and buff barring on dark flight- and tail-feathers. **Juv.** warm buff obscurely barred dusky brown, with many white freckles on head; face tawny.
Habits: Strictly nocturnal.

Voice: Irregular series of copious hoots, *hoo-hooo*, higher pitched than hoots of Great Horned Owl, and a high scream *crru crru*.
Breeding: In tree-holes; eggs probably in Oct; but one large pull. May (Chile).
Habitat: In the s Andean zone mainly in deep and dark mature forest, but extends to forest/steppe ecotone. Chaco woodland.
Range: From isls s of Isla Grande through lowlands and valleys in the Andean zone n to Bío-Bío (and casually to Santiago), c Chile, and in adjacent forest of sw Arg. n to Neuquén (*rufipes*); Chiloé isl, c Chile (*sanborni*). Chaco zone through lowlands of n Arg. to Parag. (*chacoensis*). Generally scarce.

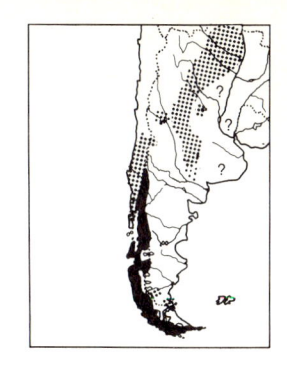

STYGIAN OWL *Asio stygius* – Plate XXV 1

42 cm, wingspan 90 cm. Slender, with rather small head with **long ear-tufts**. **Very dark**: upperparts sooty black with scattered buff and whitish mottling. Facial discs fuscous contrasting with lighter forehead and throat, eyes orange-yellow. Underparts ochraceous buff with coarse black streaks and 'herringbone' marks. Tail barred. In **flight** shows buff wing-linings; unlike Short-eared O., the black 'comma' on the wrist is inconspicuous because the flight-feathers are regularly barred from tip to base. The profile is somewhat like a small *Buteo* hawk, but shorter-tailed, and with broader wings than in the Short-eared O. **Juv.** down-feather plumage barred grayish buff with ear-tufts indicated; 'mask' and wings sooty black.
Habits: When relaxed, with round head ('ears' down). When alert raises 'ears', and when scared freezes in elongated posture with slim face and 'ears' right up. Crepuscular, hunting doves and bats near their roosts. Glides on level wings.
Voice: Loud, deep *oou* or *hoo-oou*, **hoots always well spaced**. Female gives catlike *miah*.
Breeding: Usually in abandoned stick-nests in trees. Possibly in May (Col.).
Habitat: Forest and combinations of dense wood (often small patches) and open country.
Range: Discontinuous in Carribean area, C and S Am. In the Andes at 2-3000 m from Perijá mts, Ven., rare or very local through all 3 ranges of Col., and along both slopes of Ecu. (*robustus*). In s Brazil (*stygius*), and chaco of Parag. and n Arg. to the humid forest at 1000 m (*barberoi*). Sighted by NK at 3900 m above Cochabamba town, c Bol. (ssp?).

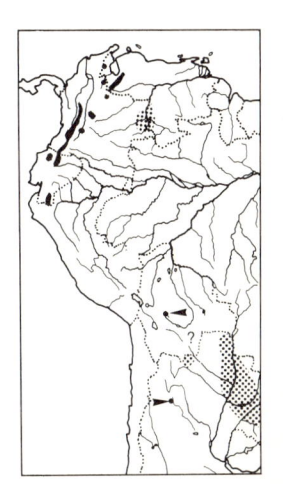

SHORT-EARED OWL *Asio flammeus* – Plate XXVI 2a-b

36 cm, wingspan 95 cm. Head small with large white X centered above bill, and black surroundings of yellow eyes. Upperparts densely spotted fuscous and tawny-buff, with some pale spots on scapulars and wing-coverts. Underparts warm buff with dark stripes, which are bold on breast and thin lower down, giving 2-toned effect. In **flight, long tapering wings have pale buff underside with black 'comma' on the wrist, and distinct dark-barred tip**; also above with an extensive buff area basally on the primaries. Fairly short tail regularly barred buff

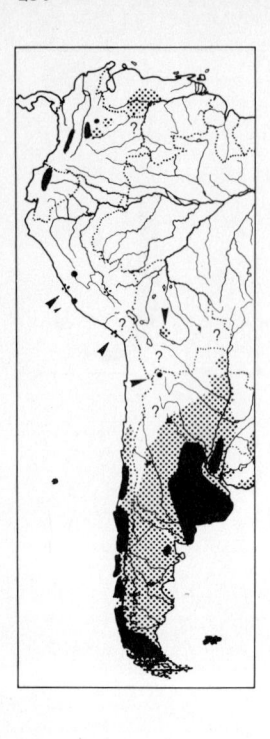

and black. Ssp *bogotensis* considerably darker buff than *suinda*. **Juv.** has very dark 'mask'.

Habits: By day, usually in tall grassy vegetation, but in breeding season often perched on poles etc. in less upright posture than other owls. Mainly hunts at dusk and dawn in wavering flight low over the terrain, with **deep flopping wingbeats and gliding and soaring with wings raised in open V** (but occ. glides on level wings as Stygian O.).

Voice: Sharp *chi-chi-chi*... and *wheechiz* as it circles over intruders on breeding territory. In display flight, male claps wings together and drops suddenly, giving deep hollow *boo-boo-boo*... (like distant slow puffing of old steam engine). Female replies from ground with harsh barked *ree-yow*.

Breeding: Directly on ground; season prolonged with eggs Oct-Dec (Jan) far s, Apr-May (Córdoba), and Sep (near Bogotá, Col.).

Habitat: Extensive reed-marshes, moorland, and areas of rushy pastureland.

Range: Widespread in Holarctic and Neotropical regions. Rather common in lowlands of c Arg. (to 2200 m in Córdoba) and from the Fuegian zone to c Chile, sparsely across the Patagonian steppe, and very patchily at 2500-4100 m through Andes of Arg., Bol. (one rec. Cochabamba 1920), Peru (Lake Junín area, Cuzco, a few coastal sites), and Ecu., and in E Andes of Col. Widespread southern *suinda* grades into northern *bogotensis* in Ecu. Llanos of Ven. and e Col. (*pallidicaudus*).

BURROWING OWL *Athene cunicularia* – Plate XXV 11a-c

20 cm (*nanodes* tiny, *juninensis* 24 cm). **Rather flat-headed and with stumpy tail and long legs**. Above with various shades of grayish to fulvous brown with round pale spots. **Broad white eyebrows** give angry expression. Gray-and-white facial discs small, spreading only out- and downwards from yellow eyes. Throat and foreneck mainly white with blackish semicollar, further down white more or less barred and spotted gray-brown. Several sspp overlap and intergrade: most are gray-brown with white markings, *punensis* and *juninensis* most fulvous, with the light dorsal marks mainly pale pinkish buff, and the ventral barring quite fawn-colored. The dorsal marks vary from small in *nanodes* and *partridgei* to very large in *punensis*. The underparts vary from densely barred dark in *cunicularia* and *nanodes* to fulvous white with just a few breast bars in *punensis*. In **flight** gray-brown with small black wingtips from above; below white heavily barred on primaries, but *partridgei* with very white underwings owing to incomplete barring. **Juv.** has unspotted crown, and much white on face. Down-feather plumage with all upperparts uniform gray-brown, underparts white or buff, and with small dark mask.

Habits May live in loose semi-colonies. Often seen by day in short undulating flight, or perched on poles, rocks, or mounds near entrance to its burrow, bobbing up and down when nervous. Mainly eats large beetles, but also takes other insects, scorpions, lizards, and rodents.

Voice: When disturbed, gives cackling chatter, screeches *zree tchichi chi chi*..., rasping calls, or short *chuck* notes. Nightly song a **long series of clear, well spaced musical coos** on constant pitch, occ. ending with short warble; female responds with rapid series of *eep* calls. Has several other calls.

Breeding: In burrows. Eggs Oct-Dec (in the s), Sep-Dec (Junín, Peru), Jan (Col.).
Habitat: In the Andes in desert, tola heaths, pastureland, and wide plains and slopes with very short vegetation and patches of bare earth.
Range: Widespread in the Americas, in arid and semi-arid zones and also in wet climates, wherever there are stretches of open land. Chile n of Valdivia, ascending to 2000 m (*cunicularia*). Throughout Arg., from plains and foothills of Patagonia (but now virtually extinct on Isla Grande) to the chaco and pre-puna of the nw (*partridgei*, described by C Olrog in *Neotrópica* 22(1976):107-8), grading in Jujuy into the populations living at 3500-4300 (4500) m across altiplanos of w Bol., ne Chile, and Peru n to Junín (*juninensis*). Coastal Peru (*intermedia, nanodes*), semi-arid valleys of n Peru to the lowlands of w Ecu. (*punensis*) and locally in temp. tablelands of Ecu. (*pichinchae*). Upper Magdalena valley, Col. (*tolimae*), in Boyacá E Andes, Col. (*carrikeri*), and lowlands of n Col. and Ven. (*brachyptera*). Locally common.

NOTE: Sometimes placed in a monotypic genus *Speotyto*.

BUFF-FRONTED OWL *Aegolius harrisii* – Plate XXV 9

19 cm. Unmistakable. Compact with short tail and large square head. **Above dark brown, frontal surface warm buff**. With amber-brown tone on the mantle, and scattered round white and buff spots, esp. across nape (forming large V), along humerals, and on upper tail-coverts (in *iheringi*). The buff underparts vary from pale to richly hued ochraceous buff. Odd visage (like Old World Masked Owls *Phodilus*): triangular buff forehead distinctly bordered by chocolate-brown (or blackish, in *iheringi*) bands slanting from lore to above eye, suggesting eartufts; facial disc spreading only out- and downwards from eyes has thin black outline. Black spot on chin. **Juv.** lacks buff spots above.
Habits: A hunting bird was seen hovering just before dawn over roadside shrub (PG).
Voice: A 3-15 s extremely fast, somewhat irregular trill (c. 9 notes per s, and high but sligthly fluctuating pitch). Possibly also rapid series of 5-7 whistled hoots every 6-7 s are given by this species.
Breeding: In tree holes.
Habitat: Rec. from fairly open humid forest, to the treeline, but also in drier zones, and in nw Arg. and Bol. mainly in ravines in dense *Podocarpus*, *Alnus*, and *Polylepis* wood.
Range: Lowlands of se Brazil to Parag. and ne Arg. (*iheringi*). At 375-2000 m from Tucumán to Jujuy, nw Arg. (*dabbenei*, described by C Olrog in *Acta Zool. Lilloana* 33(1979):5-7), and with highly scattered rec. in Bol. (w of Comarapa Sta Cruz and at 3900 m in Cochabamba), Yurinaqui Alto in Junín, Cushi in Pasco, Cajamarca and around Huancabamba deflection in Peru, Zambiza in ne and Pichincha in nw Ecu., Nariño, Cauca, and 'Bogotá', Col., and Mérida, Caracas (*harrisii*), and Neblina mt (ssp?) in Ven. Generally rare.

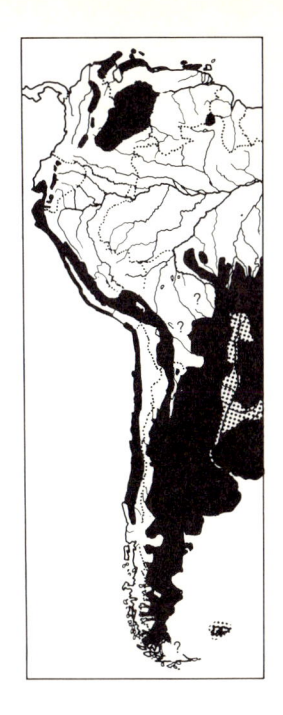

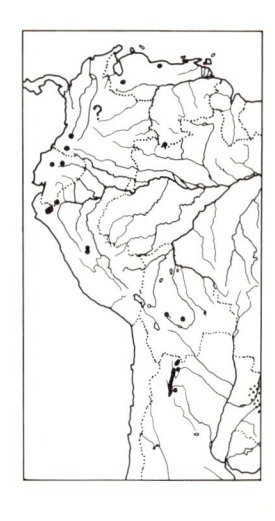

Caprimulgiformes

Nocturnal birds, which somewhat resemble owls, and evidently evolved from that group.

Oilbirds – Family Steatornithidae

A single species. Biology described in *Zoologia* 46(1961):27-47 and 47(1962):199-221.

OILBIRD *Steatornis caripensis*

45 cm. Nightjar-like, but much larger, and with strong hawk-like bill. Quite uniform rufescent brown, above dark with some fine dusky bars, below pinkish brown. Head, wing-coverts, outer flight- and tail-feathers with scattered white dots, the largest black-encircled. Tail long, stiff, tawny with thin black bars. Eyes reflect bright red. Nocturnal and gregarious. Very noisy when disturbed. Calls disagreeable *cree cree crrree...* or *crrau*. Feeds on fruits and palm nuts taken off the trees in flight. By day, many together in completely dark caves (but occ. roost in trees). Known for using high-pitched clicks for echo-location in the dark.

Local in premontane forest zones from Guyanas through Ven., Col., Ecu., and e Peru to Cochabamba, Bol. Oilbird caves have been reported as high as 2200 m (Cajamarca, n Peru) and in subtrop. Andean valley in Ecu. During seasonal wandering reaches temp. zone (to 3000 m) in all Andean ranges of Col. and also shows up on temp. tablelands of Ecu.

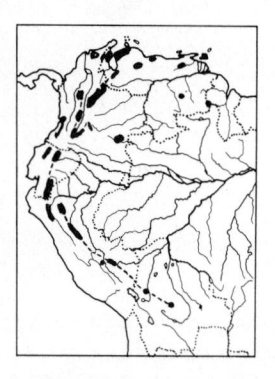

Potoos – Family Nyctibiidae

S Am. group of 5-6 species of nocturnal birds. Differ from the related nightjars by generally larger size, longer and broader tails, and an upright stance when perched in tree (never on ground). Resemble owls with large round heads, but when alarmed freeze in thin postures with the bill up, and with almost closed slit-like eyes (see plate). Then perfectly camouflaged as a dead end of the stumps on which it is perched. Bill extremely broad, triangular, and feathered, except on the bent tiny tip. Huge eyes light-reflecting.

ANDEAN POTOO *Nyctibius (leucopterus) maculosus* – Plate XXVI 7

38 cm. Blackish brown, freckled and barred with cinnamon, becoming **mostly black on crown**, lower back, and middle wing-coverts. **Humerals and some inner wing-coverts form large pale cinnamon to partly white band** with some black feather-tips. Belly heavily barred

and spotted buff and dusky brown, and not streaked as in Common P. (*Nyctibius griseus*) of the lowlands. Long tail dusky brown with many pale brown bars. Wing-linings blackish.

Habits and **voice**: By day quietly perched. Also by night mainly perched, hunting by hawking for passing larger insects. Normally detected by its song in moonlit nights, a loud drawn-out *ree-aa*, very different from the mournful descending series of hoots of better-known Common P.; more like hair-raising **waaa***uhm* of Great P. (*N. grandis*). Each bird has several song-posts within its territory.

Breeding: Probably as in other potoos, with egg placed directly on small depression on a branch, or on end of broken vertical stem.

Habitat: Humid forest with many clearings.

Range: 1800-2800 m. Rec. at Boca de Monte in Táchira, Ven., Norte de Santander, 'Bogotá', and Llorente in e Nariño, Col., Mindo in Pichincha (PG), w Pastaza and Tungurahua in e Ecu., Lucuma in Cajamarca and Cordillera Yanachaga in Pasco, Peru, Sacramento Alto and Chuspipata in La Paz, Bol. Known from very few specimens, but may be fairly common locally.

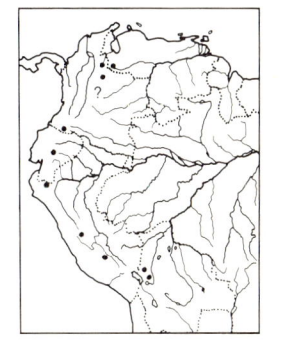

NOTE: It seems inconceivable, on basis of size and color pattern, that this bird should be conspecific with the White-winged P. *Nyctibius l. leucopterus* of Bahía, e Brazil (see *Gerfaut* 74(1984):61-62).

Nightjars – Family Caprimulgidae

A worldwide group of c. 72 species, with the majority in S Am. lowlands. 2 groups can be recognized: **nighthawks** are vespertine counterparts of swifts, recognized by sharp wings, short tails and high bounding flight; **true nightjars** have blunter wings, generally longer tails, soft plumage, and long bristles around the gape, and more nocturnal habits.

Being aerial insect-hunters, the nightjars share several characteristics with swifts: enormously broad triangular bills (with only the tip protruding from the plumage), sickle-shaped wings with very long primaries, a light body, and feeble feet. Being crepuscular or nocturnal, they also have owl-like traits: large light-reflecting eyes, and a finely mottled plumage that offers perfect camouflage by day. The birds also camouflage themselves by staying motionless among leaf-litter on the ground or along a dead branch. They are rarely discovered before they are flushed, sometimes just below one's foot. Then usually fly 20-100 m, with hesitant and irregular wingbeats, some tilting and inserted glides, to drop down into cover again. Nightjars are occ. mobbed by passerines by day.

Nightjars are only rarely glimpsed against the sky by dusk or night. However, all species have characteristic vocalizations. A twilight-singing bird can sometimes be approached closely.

The 2 cryptic eggs are placed among leaf-litter on the ground. If disturbed on the nest, the bird threatens with open gape, hissing, and sometimes raises one wing, or feigns injury. The young are downy, cryptically colored like nidifugous chicks, but scarcely mobile. They are fed for a long time.

No detailed studies have been made of Andean species.

SHORT-TAILED NIGHTHAWK *Lurocalis semitorquatus* – Plate XXVI 4a-b

21-25 cm, wingspan 60. When perched, the long dark wings reach well

beyond tip of very **short square tail**. Flight profile bat-like. **Fuscous-black** with conspicuous **white bar across throat**; above with tiny chestnut ocellations, and some cinnamon and pale gray mottling on scapulars. Wing-linings, belly, and vent tawny with dusky bars and mottles. Large ssp *rufiventris* has more gray and chestnut vermiculations above, paler gray scapulars with fine dark vermiculations and diamond-shaped spots along the feather-shafts, the blackish chest (sometimes with a whitish demarcation zone) contrasting with the almost **unmarked tawny belly**. **Juv.** is extensively whitish on the upperparts, with fine gray mottling, and smaller or larger dark spots near the tips, and also the breast quite light, mottled, and grading into rufous color of belly.
Habits: Single, in pairs, or small groups flying at dusk and dawn erratically with raised wings above the canopy, in clearings or over streams.
Voice: Distinctive call (Ven., nw Ecu.) in flight a *tor-ta quírrrt, tor quírrrt, tor,* or *quírrrt*. In short period of evening and early morning *rufiventris* gives series of mellow whistled hoots at falling pitch. The 2 lowland forms have different calls.
Breeding: Half-grown young June (Ven.).
Habitat: Near edges and glades of humid forest and second growth.
Range: Throughout trop. S Am. e of Andes. Northern *semitorquatus* and migratory southern *nattereri* in lowlands (and 2 *nattereri* specimens from 2550 m near Cochabamba, Bol., in LSUMZ). Ssp *rufiventris* (only shown on map) in Andes from Mérida mts, Ven., Perijá mts and C and E Andes of Col., and e Andean zone s to Bol., in pre-montane zone and ascending at least to 3000 m in Col. and 2550 m in c Peru and Bol. Uncommon and local, but certainly more widespread than the map suggests.

NOTE: Also called Semi-collared N. Ssp *rufiventris* may well be a distinct (semi)species, **RUFOUS-BELLIED NIGHTHAWK**, and species rank has also been proposed for *nattereri* (**Chestnut-bellied N.**).

LESSER NIGHTHAWK *Chordeiles acutipennis* - Plate XXVI 6 – a local resident in dry bushy country in trop. S and C Am., and migrant from w N Am. to nw Col., can be expected to reach temp. parts of the northern Andes (*texensis*). With slightly shorter tail and less pointed wings than Common N., less contrasting underparts, and the white (or buff, female) **wing-patch more distally on the primaries**. Usually flies low. Flight call a low soft *chuck*, or series of bleating calls.

COMMON NIGHTHAWK *Chordeiles minor* – Plate XXVI 5

22 cm, wingspan 55 cm. Wings pointed, tail medium long, notched. Drab gray with dark breast, and whitish belly and vent, more or less densely barred and vermiculated with dusky (the distinctness of the pattern varying between several sspp). In **flight** easily recognized by a **broad white patch across dark primaries midway between wing-bend and wingtip**. **Male** with white chin and patch across throat, and a white subterminal bar on tail.

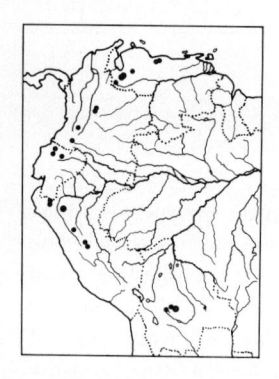

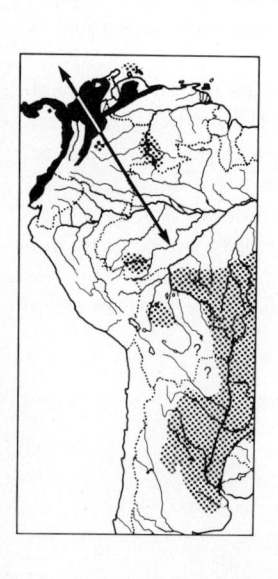

Habits: Active at dusk and dawn, but sometimes also on cloudy days. Usually flies high, with rapid and bounding very deep wingstrokes, swerving and changing speed abruptly. Migrates high.
Voice: In flight, gives nasal buzzy insect-like *beerp* accented at beginning and end of note, or occ. 2-3 short notes combined together.
Habitat: Open country of all kinds, often seen hunting insects over plazas and streets.
Range: Widespread in most warm parts of the Americas, but in much of S Am. only as a migrant visitor. Can be seen Aug-Nov and Mar-Apr in temp. parts of Col., e.g. commonly at least in Sep-Oct on Bogotá savanna 2600 m, and sometimes above 3000 m (possibly *henryi, howelli, minor*, as well as *sennetti*).

PAURAQUE *Nyctidromus albicollis* - a characteristic species throughout trop. and subtrop. woodlands of C and S Am. on both sides of Andes – possibly reaches lower fringe of temp. zone (rec. to 2300 m in Col.) (*albicollis*). Best recognized by its loud whistled *wuc wuc wheeeer* or *coo-wheeeer* calls. Resembles a large (28 cm), light Band-winged Nightjar, male with tail-feathers white to base (except on mottled c pair and black outer pair). Female has mottled tail with distinct white tip only on 2nd feather.

CHUCK-WILL'S-WIDOW *Caprimulgus carolinensis* – a sparse N Am. migrant to Col. and Ven. - is rec. to 2600 m in E Andes of Col. A chunky (28 cm) and heavy-headed nightjar with **no white in the wings, but the male showing white inner webs of outer 3 rectrices** as stripes in the large broad tail. Generally dark brown with fuscous mottling above and barring below, and with ochraceous bars and notches on flight- and tail-feathers. Buffy eye-brow and buffy-white bar across throat.

BAND-WINGED NIGHTJAR *Caprimulgus longirostris* – Plate XXVI 10a-d

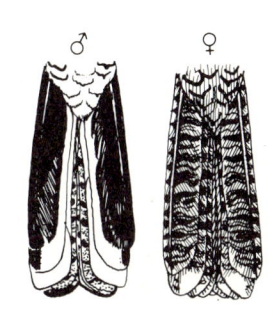

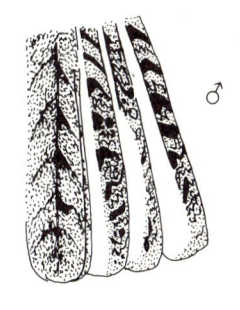

C. 22 cm (*decussatus* **very** small, *patagonicus* largest). Above blackish densely spotted and mottled black, chestnut, buff, and pale gray, with rufous nape-collar. Broad white or pale buff throat bar separates mottled face and breast; lower underparts being buffy with dense fuscous barring and mottling. Wings dark with conspicuous **buff (female)** or **white (male) bar across outer primaries** well outside middle. Tail rather long, c feathers grayish, the rest fuscous with buff bars; in **male with broad white terminal spots** on 3 outer feather-pairs, and a narrow white bar near base. **Juv.** similar, with the tail pattern fully developed, but white bar on primaries often restricted and occ. placed more distally, and with no rufous color on nape. Several in part intergrading sspp: wing-bar narrow, rather variable, often interrupted in *ruficervix* and *decussatus*. Ssp *ruficervix* is dark, the breast sooty brown with large pale cinnamon spots (resembling *Uropsalis* species), this condition being less clearly expressed in *atripunctatus*, while other forms have lighter and finely mottled breast, and more gray mottling above, *bifasciatus* and esp. *decussatus* being extensively sandy buff and pale gray. An ocellated pattern, with black centers of most buff backspots is esp. conspicuous in *atripunctatus*, which also has a particularly prominent rufous nape collar.

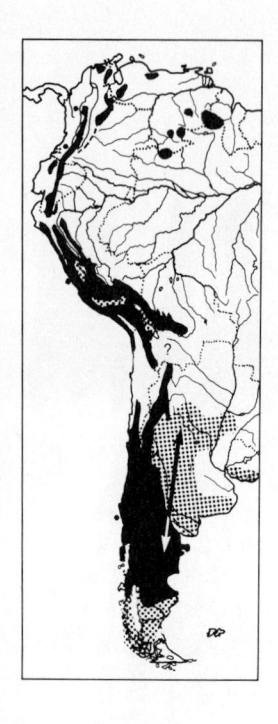

Habits: In flight blunter-winged and longer-tailed than nighthawks. More strictly nocturnal. Sallies short distances from ground or slightly elevated perch for passing insects. Occ. flutters low over ground with tail right down, sometimes flashing white spots (male). By day on ground, usually in a ravine, road cutting, gorge, or rocky or bushy slope.

Voice: When flushed, sometimes gives short, nasal *tchree-ee*. Sings shortly after dusk from ground or low perch, with loud, high-pitched, falling plaintive *chee-whit* calls, combined, or 1-3 s apart, in sometimes interminable series throughout the night. Also a thin sucking *zee-eorr* or *zueeeeert*, rising towards the end.

Breeding: Egg-laying in Nov. (Chile), Nov (c Arg.), July, Oct and Nov (Col. Andes), Mar-Apr (Sta Marta Col.) or July (Perijá mts), pull. Sep (Popayán), fledglings Mar-July (Mérida), Apr (Cauca Col.).

Habitat: From completely bare desert and stony semi-desert and steppe to puna grassland, shrubby páramo, and often high Andean woodlands and grassy slopes in humid to semi-arid elfin forest.

Range: Lowlands of southern S Am. and throughout Andes. In Chile from the s to Coquimbo and occ. sw Antofagasta and into adjacent w Arg., to 2500 m (*bifasciatus*, but unnamed ssp on Mocha Isl), and almost throughout Arg., migratory in the s (*patagonicus*). Semi-arid hills and bare desert from sea level to 3550 m along coast of n Chile and Peru (*decussatus*); at 2300-4200 m from Jujuy, nw Arg., and Arica, n Chile to yungas of Bol., and through the puna of w Bol., and Peru to the n Peru low (*atripunctatus*). At 1500-3600 m through Ecu. and Col. to Sta Marta and Mérida mts, and Aragua to Carabobo, n Ven. (*ruficervix*). Tepuis of s Ven. (*roraimae*). Common.

SWALLOW-TAILED NIGHTJAR *Uropsalis segmentata* – Plate XXVI 8a-b

22 cm. Very dark with long blunt **black wings, and long forked tail (in ad. male with scissorlike lateral streamers**, 40-50 cm or **twice body length**). Sooty brown with rich rufous bars and freckles throughout, rarely any gray mottles on tertials. Throat with light semi-collar, breast black with scattered tawny spots (larger in male), belly and vent predominately tawny to ochraceous with some barring. Underwings almost uniform black. Tail-feathers with narrow well spaced light bars. Streamers uniform fuscous with white shaft and buff notches along outer web, or pale-tipped with white outer web in the small *kalinowskii*.

Habits and **voice**: By day perched under bush or on vine. At dusk (and dawn?) hunts low over open grassy slopes, often along forest edge. Gives low *churrr* when startled. In courtship, several males may circle and chase females. Give Pauraque-like *puit-sweeet* calls. What may have been this form gave a 2-3 s vibrating *rrrrre^{eeee}err* with distinctly rising and finally falling pitch at the break of dawn.

Breeding: Downy young Feb and Aug (Col.).

Habitat: Humid cloud and elfin forest with glades with bamboo, shrub or coarse grass (hunting habitat).

Range: At 2300-3600 m. Patchily in C and E Andes of Col. and s through Ecu., where rec. in Imbabura (Apuela road, 2700 m: MECN), Pichincha? (road to Chones (*sic.*? = Chone?): BMNH), and e Chimborazo (Matus (3000 m)) (*segmentata*), and along e Andean slope of Peru to Cochabamba, Bol. (*kalinowskii*). Fairly common locally.

LYRE-TAILED NIGHTJAR *Uropsalis lyra* – Plate XXVI 9

Resembles previous species, but shows gray (not rufous) freckles on crown and tertials, conspicuous rufous nape-collar, and otherwise buffier mottles. Also slightly larger, with **considerably longer streamers** (**male; 75 cm, or 3 times body length** in *lyra*, slightly shorter in the 2 other sspp) curved into lyre-shape, broad, but with narrow tip, and with white edge of inner web. Calls at dusk a distinctive loud and rapid *tre-cuee, tre-cuee, trecuee...*, or accelerating series of *weeep* calls, given when perched and while flying at great speed low over forested slopes and landslides.

Local near grassy glades of wet and humid forest of Andean premontane zone from Mérida, Ven., through Andes of Col. and w and e Ecu. (*lyra*), and e Andean zone of Peru (*peruana*), and in Tarija, Bol., and 1 rec. in Jujuy, nw Arg. (*argentina*, described by C Olrog in *Neotropica* 21(1975):147-8). Although rec. at 2500-3000 m in Ven., 2500 m in Col., 2800 in nw Ecu. (PG), and 3150 m in Cuzco, it normally occurs below Swallow-tailed N.

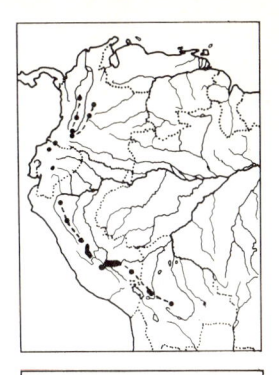

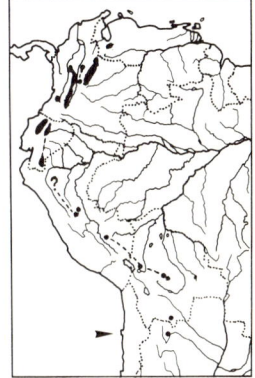

Apodiformes

Small birds with strongly specialized wings. Possibly related to caprimulgiform birds. It has been suggested that hummingbirds could be related to passerine birds, but recent biochemical evidence supports the view that swifts and hummingbirds belong together.

Swifts – Family Apodidae

A well defined nearly worldwide distributed group of 90 specialized fliers. They somewhat resemble swallows, but have narrower sickle-shaped wings adapted for high-speed sailing. As in hummingbirds, the wings have very long primaries and extremely short basal bones. The body is streamlined, neckless; head rounded with minute bill, but enormous gape. The feet are extremely small, uniquely adapted for grasping with toes opposing one another 2 and 2 (not all 4 toes forwards as stated in older textbooks). Many species have stiffened tail-feathers with projecting spiny shafts that (along with the strong claws) help the birds support themselves on vertical surfaces. All species are mainly sooty brown.

Swifts are the most aerial of all birds. They feed entirely in the air (often high up), on aerial plankton of tiny insects and 'ballooning' spiders, and drink and bathe by swooping down to the water surface. They can 'sleep' on the wing, although some species normally roost on vertical cliffs or tree-trunks. They never settle on the ground, in vegetation, or on wires, and have difficulty in taking off again if grounded. Swifts move around in sometimes large flocks, and often rise high up in early morning and before sunset to give communal concerts in the air. The voices are generally shrill, but vary from sharp chips in *Streptoprocne, Cypseloides*, and *Chaetura* to drawn-out, buzzy screeches in *Aeronautes*.

Both ad. and young can lose their temperature control and enter semitorpid condition to save energy by night or in periods of food shortage. Species of northern regions are known to travel 100s of km to avoid depressions with few insects. This seemingly applies also to Andean forms.

Breeding is normally semi-colonial, with tiny nests glued with saliva to tall, overhanging rock-faces. Most birds lay 4-6 elongate white eggs, and have long incubation and nestling periods. When feeding young, the birds assemble a large food bolus in the distended sublingual region of the mouth.

Some details about Andean taxa can be found in *Condor* 67(1965):449-56 and *Amer. Mus. Novit.* 1609(1953). Adaptations of swifts to high elevations are discussed by F Salomonsen in *Acta XVIII Congr. Int. Orn.* (1985): 541-58.

WHITE-COLLARED SWIFT *Streptoprocne zonaris* – Plate XLVIII 1a-b

20 cm, wingspan 45-55 cm (*albicincta* small, wing length 19-21 cm, *altissima* large, wing 21-24 cm). Unmistakable in the Andes. A **large black swift with broad white ring all around neck**. Ssp *altissima* is said to have pale forehead, but this is an exception rather than a rule. **Juv.** variable, some birds with somewhat mottled collar, and often white feather-tips on crown and lower back, and always with pale-margined flight-feathers.
Habits: Social, often in large flocks, and sometimes hundreds roost together. Often very far from roost or colony on their daily foraging, and make marked local population movements in connection with weather shifts. Usually seen high up in swirling flocks, and sometimes at enormous speed through deep canyons. The flight silhouette varies, as the tail may look narrow, forked, or fan-shaped, according to variations in flight-style. The wings are often somewhat truncate. Soars with wings characteristically bowed.
Voice: Nasal twitter *chee chee chee*, *whiss whiss*, and scratchy but not shrill *tseet*, *tchee*, and *chirrio* notes. Often many birds call in unison, like a distant parakeet flock.
Breeding: Big mossy nest with little saliva placed on damp rock surface in cave or behind waterfall. Pull. May-June (Mérida, Ven.), Apr (Tolima, Col.); sexually active Oct (Bol.).
Habitat: High over hilly and mountainous country of any kind, but mainly in semi-humid to semi-arid regions.
Range: Widespread in C and S Am. s to s Peru in the w, and Matto Grosso, Brazil in the e (*albicincta*, with *altissima* through highlands of Col. and Ecu., and apparently also this form in Andean valleys of Peru, although *albicincta* inhabits humid premontane parts). In s Brazil and e Bol. and along base of Andes to Mendoza in wc Arg. (*zonaris*). Mainly in premontane areas, but also in lowlands (e.g., May-Aug to coast of Peru) and to temp. zone in Ven., Col. and Ecu., where it breeds to 3600 m and sometimes in the highest parts (e.g., 4300 m at Papallacta, ne Ecu (PG), often seen above 4000 m in Cord. Blanca in Ancash, and

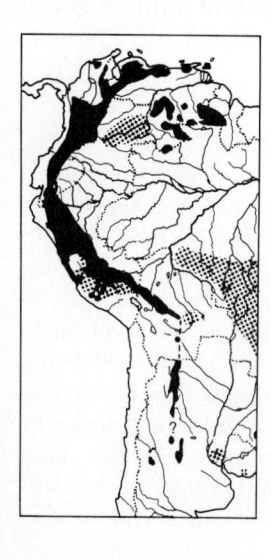

seen at 4350 m on Cuzco/Puno border, Peru). Common, at least locally.

NOTE: Possibly a separate ssp *caucensis* (with collar broad on nape, vs. narrow on nape in other sspp) is recognizable in Andes of Mérida through Col., below *altissima*, although rec. to 3000 m.

NOTE: Forms a superspecies with Biscutate S. (*S. biscutata*) of e and se Brazil.

CHESTNUT-COLLARED SWIFT *Cypseloides rutilus* – Plate XLVIII 5a-b

15 cm. Dark sooty brown and partially almost black. **Male with rufous throat, cheeks, forebreast, and narrow collar around neck. Female** often has little rufous, and **juv.** may totally lack this color. Tail scarcely notched, and with projecting spines. Longer-tailed than Chimney S., and wings much narrower (flight profile closer to White-tipped S.).
Habits: Single, or in small groups, and occ. hundreds migrating together during weather shifts.
Voice: High-pitched, hoarse, metallic, and buzzy chittering.
Breeding: On rock surfaces in damp shade near water.
Habitat: Bushy savanna and other open terrain to grassy ridges in montane forest zone. Roosts in trees.
Range: C and northern S Am. Mainly in upper trop. and subtrop. zones through Coastal Cord. of Ven. and Andes of Ven., Col., and Ecu., where rec. to 3400 m, and extending to the w slope. Locally on the e slope through Peru to Yungas of Cochabamba, Bol. In Col. to 2500 m, casually to 3300 m, in Ecu. said to be commonest on w temp. slope, and rec. to temp. zone also in Pasco and Ayacucho Peru, (*rutilus*).

CHIMNEY SWIFT *Chaetura pelagica* – **Plate XLVIII 4** – a N Am. migrant visiting coastal w Peru to n Chile (BW) Nov-Apr – has been seen up to temp. zone in s Ecu. (3200 m), Arequipa (rec. 3900 m), accidentally at 2500 m in Boyacá, Col. and at 3600 m in Cuzco town, e Peru. In open terrain, but often roosting in chimneys, churches, caves etc. 13 cm. Above **dark sooty** olive, slightly lighter on rump; below grayish brown with distinctly paler throat. In flight shows thick cigar-shaped body with short squared-off tail with projecting spines. Recognized by **short-tailed appearance** and its flight, with series of very rapid wingstrokes and much soaring. Sometimes gives rapid series of chips and ticks of hummingbird-like quality.

WHITE-TIPPED SWIFT *Aeronautes montivagus* – Plate XLVIII 2

13 cm. Dark sooty brown, becoming black on wing-linings and back. Throat white, becoming marbled dirty white and dusky on breast and along thin mottled stripe of c belly, and with **white area around feet contrasting with black vent**. Tertials and tail-feathers with white

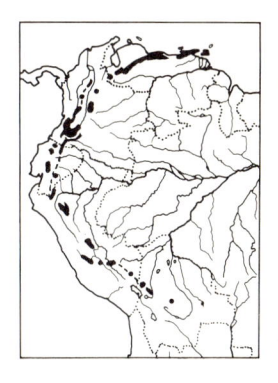

White-tipped Swift

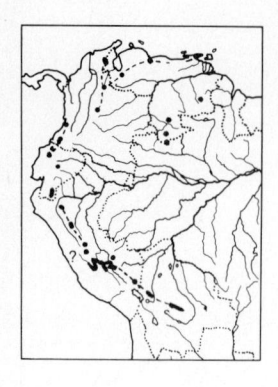

margins distally. In **flight** not unlike Chestnut-collared S., with sickle-shaped wings, cigar-shaped body, but with a slightly forked or **notched tail**.
Habits: In swirling flocks over high hills and ridges.
Voice: A rather low-pitched tickling or buzzing *shritte* noted.
Breeding: In ravines (sometimes behind waterfall) and in holes in buildings. Apr-July in n Ven.
Habitat: In humid to semi-humid zones, usually over forested hills and high, grassy and bushy slopes and ridges.
Range: In Tepuis of n Brazil/s, Ven. (*tatei*). Very locally, at 500-2600 m from Sucre to Carabobo and in Zúlia and sighted in Andes of n Ven., recently sighted in Col. (Sta Marta mts, w slope of E Andes in Santander and e slope in Meta, Pacific slope in Valle and Nariño). Locally on both Andean slopes of Ecu., and patchily in the e Andean zone through Peru to Sta Cruz, Bol. (*montivagus*). Locally common.

ANDEAN SWIFT *Aeronautes andecolus* – Plate XLVIII 3

C. 14 cm, *andecolus* largest. Generally quite uniform gray-brown, below lighter, mostly white in *andecolus*, and always white on vent, and with **white nape-collar** and broad **whitish bar across rump**. Sspp *peruvianus* and esp. *andecolus* are rather light with little contrast, and with interrupted nape-collar. However, ad. *parvulus* is more contrastingly colored, the dark parts becoming almost black near eyes, on mantle, and flanks, c underparts white from throat to vent contrasting with brown sides, and the nape-collar broad and complete. In **flight** looks longer- and broader-winged and longer-tailed than White-tipped S., the tail usually looking narrow and slightly forked, but fan-shaped when spread during maneuvers.
Habits and **voice**: Often come in 'waves' along the mountain slopes, filling the air with shrill *zeezeezeezeeer* or weaker *trritrrritrri...* for a few minutes until they move on to another area. Often fly high up along tremendous cliff-walls. Glide with wings bent down a little. Low *trp-rrie* given by night, in the air.
Breeding: In holes, usually under overhangs in tall road-cuttings, canyons, and sometimes tremendous cliff-walls. We have no data on breeding season.
Habitat: Semi-arid mountainous country, mainly over bushy slopes (Leguminaceae, Anacardiaceae), sometimes with some wood (*Podocarpus*, *Polylepis*), but also in rocky desert country with scattered cacti, *Jatropa* scrub etc.
Range: Andean slopes and montane valleys (occ. to the coast, but not on altiplanos). Mainly at 2500-3500 m (occ. 340-3900) in Peru-Bol., or 2000-2500 m furthest s. From Cajamarca, n Peru, to Tarapacá, n Chile, on w slope (*parvulus*), through valleys of Huancavelica to Cuzco, Peru (*peruvianus*), and in e Andean zone of Bol. and Arg. s to Río Negro (*andecolus*). Common.

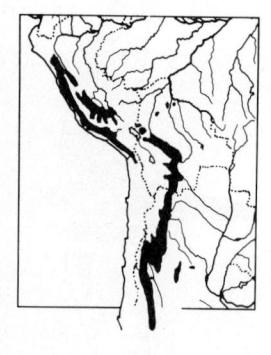

Hummingbirds – Family Trochilidae

Hummingbirds live nearly throughout the Americas, but the majority of the 320 species inhabiting humid forest near the Equator.

Hummingbirds include the smallest of all birds, with body weights down to 2 g. However, most species are larger, with 20 g as a maximum. Hummingbirds are adapted to extract nectar from flowers with their specialized tongues and usually long and thin bills. Most species feed hovering; feeding perched is seen mainly in larger species of hummingbird and at certain plants. The wings are highly specialized, with long primaries and greatly reduced proximal parts. Very flexible shoulder-joints permit hovering flight both for- and backwards. The wingbeat rates are 22-78 per s, which usually produce a pronounced humming sound. Hummers are also characterized by small feet and by brilliantly iridescent colors, which may be glittering or shining (metallic). The body plumage is usually metallic green, but females may become dull gray below due to strong wear of the feathers while breeding. The throat often has a strongly reflective gorget: gorgets are usually present only in males, but sometimes also in females, and then apparently function as a threat signal. The colors depend much on the inclination of the light. **In the species accounts a designation like 'golden/green' means shining golden seen with the light, green (usually darker) against it**. In many lights hummingbirds look black. Field identification must therefore consider size and shape of bills and tails, profile, the inclination of the bill, lowering of the wings, etc. As to vocalizations, metallic ticks and chips identify the group, but may seem insufficient for separating species. However, the calls are useful locally, as localities in the temp. zone rarely have more than 8-10 sympatric hummingbird species. Most species are poorly known biologically, and many are known mostly from 'trade skins' with minimal collecting data exported by millions in the late 19th century from Bogotá, Trinidad, and Bahia, and sold to collectors in London and Paris (see *L'Oiseau* 14(1944):126-155). As many species are very local, the unexplored parts of the Andes certainly still have undescribed forms.

Hummers are very aggressive, to secure an adequate supply of flowers. Many species hold large territories, and outside the main flowering season males and females may hold small feeding territories around blooming trees. Other species are 'trapline feeders' that roam widely in search of particular kinds of flowers. They do not defend territories, but yet are very pugnacious. Interspecific aggression is common, certain species dominating others. Hummingbird flowers, e.g. *Heliconia*, *Passiflora*, many bromeliads, gesneriads, and labiates, are chiefly red, yellow, pink, or white (rarely blue), usually with widely separated stamina and nectaridiae, and often long corollar tubes. The structure permits exploitation mainly by one of the local hummers, which ensures effective pollination. A long-billed species usually exploits a number of plants with long tubes, and supplements this with visits to smaller flowers. The smallest hummers often 'trapline' for small insect-pollinated flowers, and many species also feed through holes made by flowerpiercers (*Diglossa*) in the corollas. Certain species (e.g., *Chlorostilbon*) are 'parasites' piercing the bases of flowers. Most hummers also take insects, and a few species are specialized on an insect diet.

Most hummers are resident, but many show local or elevational movements in response to the flowering seasons, often with males higher than females. A few species are truly migratory. The high activity of hummingbirds requires a high metabolism and frequent intake of energy-rich food. In order to conserve energy, hummers can become torpid at night, and let their body temperatures fall, sometimes to near the ambient temperature, although below 18° C is lethal. This torpor is best known in the hillstars (*Oreotrochilus*) of the high Andes, but any hummer is capable of becoming torpid in stress situations.

The breeding seasons are linked with flowering seasons, and highland hummers mainly nest early or late in the rainy season. Hummingbirds never establish sustained pair-bonds, and most species show promiscuous polygamy. Males sing from exposed posts or at traditional leks (the singing assemblies of hermits). Some species have a high display-flight. Except perhaps in the Sparkling Violetear (*Colibri coruscans*), the female builds the nest, incubates, and rears the young unaided by the male. The nests are usually minute cups of plant fibers and cobweb. Most are placed on thin twigs, but some species weave their nests into moss hanging below branches, or have pendant nests, and a few highland forms have larger nests in caves and crevices. The normally 2 eggs are elongated and white. The chicks are nidicolous and nearly naked.

Information about Andean species can be found in the regional handbooks and in Behnke-Pedersen, M (1972) *Kolibrier,* Skibby (Denmark), Vol.1 (Vol.2 not published); Grant, KA & V (1968) *Hummingbirds and their flowers,* New York.; Greenewalt, CH 1960 *Hummingbirds,* Garden City N.Y. See further Todd (1942b), Bond (1954b), *Amer. Mus. Novit.* 1449, 1450, 1463, 1474, 1475 (1950), 1513, 1540 (1951), 1595 (1952), and 1604 (1953); *An. Mus. Hist. Nat. Valparaiso* 13(1980):299-300; *Ardeola* 21 (1975):933-43; *Auk* 99(1982):172-3; *Bol. Soc. Venez. Cienc. Nat.* 15(19?):153-62; *Bonn. Zool. Beitr.* 19(1968): 225-34, 24(1973): 24-47; *Breviora* 230(1965); *Bull. Brit. Orn. Club* 97(1977):121-5, 104(1984): 95-7, 105(1985):113-6; *Bull. Brit. Mus. Nat. Hist.* 38(1980):105-39; *Comp. Biochem. Physiol* 41A(1972):797-813, 73A(1983): 689-9; *Condor* 52(1950):145-52, 55(1953):17-20; *Hornero* 11(1985):265-70, 12(1986):298-300; *Ibis* 111(1969):17-22, 116(1974):278-97 and 347-59; *J. Orn.* 112(1971):205-15; *L'Oiseau* 26(1956):165-93, 32(1962):95-126, *Proc. Acad. Nat. Sci. Philadelphia* 106(1954):165-83; *Proc. Biol. Soc. Wash.* 99(1986):218-24; *Senckenbergiana Biol.* 58(1978):137-41, *Trochilus* 3(1982): 90, 4(1983):58-9, 91-2, 6(1985):2-8, 110-15, 7(1986):79-84 and *Univ. Calif. Publ. Zool.* No.106(1975).

Hermits – *Glaucis* and *Phaethornis*

Rather dull-colored with **long, arched bills**. Species of *Phaethornis* have a characteristic appearance caused by flat head and dark 'mask' through the eyes. Tails are **graduated and pointed**. Hermits are characteristic of understory of humid forest of lowland and premontane zones. They feed much on *Heliconia* and bromeliad flowers, and glean insects, hovering. They usually sit with raised bills and hanging wings. During group displays, males incessantly bob their tails and call, and sometimes rise slowly and hover over their individual perches. All Andean forms have pendant nests from tips of palm leaves.

RUFOUS-BREASTED HERMIT *Glaucis hirsuta* of humid trop. forest and várzea forest from Panama through n Col. and Ven. and across the Amazon area to the premontane zone to n Bol., has been rec. to 3500 m in Ven. (ssp *affinis*, KLS pers. comm.). 13 cm (incl. 3.3 cm arched bill). Above bronzy-green, **below cinnamon-rufous. Tail rounded, black with chestnut basal part and white feather-tips**.

PLANALTO HERMIT *Phaethornis pretrei* – **Plate XXX 2** – inhabits wooded savanna of uplands of se Brazil, and is rec. to 2500 m in e

Rufous-breasted Hermit

Andean zone of Sta Cruz and Tarija, s Bol., and Salta, nw Arg. 15 cm (incl. 3 cm bill and 6-7 cm tail). Above **bronzy green** with crown dull brown, **rump tawny**. **Tail with long and thin c feathers**, greenish, black subterminally, and **each feather with long white tip**. Black mask framed by pale buff superciliary and rictal stripes. Below cinnamon-buff. Song loud and sustained *ti-ti-tri, ti-ti-tri....* In courtship pursuit male calls *tri-tri-tri...*, female calls with a lower but very continuous *che-che-che....*

GREEN HERMIT *Phaethornis guy* – Plate XXX 3

16 cm (incl. **4 cm bill** – even longer in *emiliae*, and 6 cm tail). **Dark bronzy green**, bluish towards rump, **underparts mostly deep dull gray**, but male has distinctly green-glossed breast. Black mask framed by rusty superciliary and rictal stripes, stripe down c throat rusty brown. Also lower belly rusty. Tail black with the base green (*guy*) to blue (*emiliae, apicalis*), elongate c feathers tipped white. Calls loud for a so small bird, *tsweep tsweep tsweep...*, song a nasal metallic 2-note *heweet-heweet-heweet...* repeated c. once per s in sometimes interminable series.

Inhabits humid forest undergrowth, second growth, and dense *Heliconia* thickets mainly in the premontane zone, but in the n part of the Andes at least it goes higher up in the flowering season, casually to *Polylepis* woods above 3000 m. Costa Rica to Panamá and extreme nw Col. (*coruscans*), e coastal cordillera of Ven. and Trinidad (*guy*), around Magdalena and Cauca valleys, Col. (*emiliae*); from Perijá and Mérida mts, nw Ven., along e Andean slope of Col. and Ecu., and maybe continuously all the way to Puno, e Peru (*apicalis*).

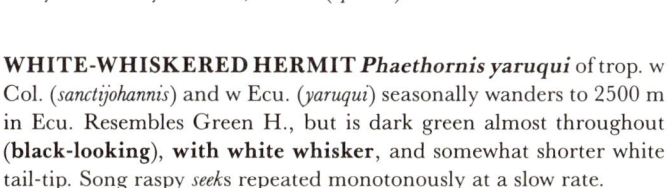

WHITE-WHISKERED HERMIT *Phaethornis yaruqui* of trop. w Col. (*sanctijohannis*) and w Ecu. (*yaruqui*) seasonally wanders to 2500 m in Ecu. Resembles Green H., but is dark green almost throughout (**black-looking**), **with white whisker**, and somewhat shorter white tail-tip. Song raspy *seek*s repeated monotonously at a slow rate.

SOOTY-CAPPED HERMIT *Phaethornis augusti* of premontane zone of Sta Marta mts (*curiosus*), from Zúlia and Táchira e to Sucre and Guiana (*augusti*), and locally s of the Orinoco (*incanensis*), shows seasonal vertical movements and is rec. to 3500 m in Ven. (KLS pers. comm.). - **Resembles Green and White-whiskered H.**s, but upperparts glossed more bronzy, supercilia and whiskers white, and **rump dull rufous**. Song a very rapid twitter monotonously repeated.

TAWNY-BELLIED HERMIT *Phaethornis syrmatophorus* – **Plate XXX 4** – inhabits humid premontane forest of the w slope of W Andes s of Choco Col. to n Ecu. (*syrmatophorus*), C Andes, around head of Magdalena valley, and e Andean slopes of Nariño and Ecu. to extreme n Peru (*columbianus*), and trop. San Martín, n Peru (*huallagae*). Acc. to KLS (pers. comm.) seasonally to 3100 m on w slope and to 2800 m on e slope. Resembles Planalto H., *viz.*, bronzy with tawny underside and rump, but has **very long bill**, and **broad tawny tips to the**

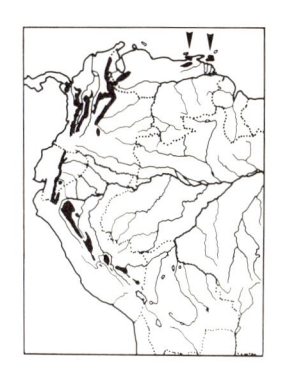

short lateral tail-feathers, and white tips only to the long c feathers. In *columbianus* rump and tail-feather tips are rich tawny, in *syrmatophorus* more ocher. Song a squeaky *tseep* repeated c. twice per s.

BUFF-TAILED SICKLEBILL *Eutoxeres condamini* – Plate XXX 1

16 cm (incl. 2.7 cm **exceptionally curved bill**). Stockily built. Above dark bronzy with ear-coverts and crown blackish (naked stripe along mid-crown). Blue-green to blue patches on neck sides, their size varying with sex and age. All **underparts cinnamon buff with heavy black streaks**. Tail 5 cm, bluntly graduated, all feathers tipped white, the 4 c feathers otherwise oil-green and black, **rest of tail buff** with black only basally on each feather. **Juv.** barred with tan feather-tips above. Retiring and hard to see. Behaves much like a hermit, e.g. with frequent tail-bobbing. Feeds in shady places, mainly clinging, from *Heliconia* and *Centropogon* flowers, but also gleans and hawks for insects. The nest is attached near the tip of a big leaf, on the underside.

Seems typically to inhabit hilly terrain with streams, forest edge, shrub, and *Heliconia* thickets, and is seemingly favored by forest clearance. Trop. zone (below White-tipped S.) in s Col. through e Ecu. s to the Marañón river, n Peru (*condamini*); premontane zone and ascending to temp. humid forest and sometimes cloud forest (rec. 3300 m) on e Andean slope of Peru from Amazonas to Cuzco and Madre de Dios (*gracilis*), and in La Paz, n Bol. (unnamed ssp).

WHITE-TIPPED SICKLEBILL *Eutoxeres aquila*

inhabits wet subtrop. forest from Panamá to n Peru, and is expected sometimes to reach lower limit of temp. zone in e Ecu. (*aquila*). Unlike Buff-tailed S. **all tail-feathers are greenish black tipped white**. Also lacks blue patches on nape-sides and naked mid-crown.

GREEN-FRONTED LANCEBILL *Doryfera ludoviciae*

13 cm (incl. **3.5 cm, straight, thin bill**). **Bronzy black**, coppery on nape, and with frontlet glittering green, upper tail-coverts glossy greenish blue, vent coppery brown. Tail steel-blue. Proposed sspp within S Am. doubtfully distinct. Larger than Blue-fronted L. (*D. johannae*) of lower elevations, and male **not** blue on forehead. Usually sits with **long bill strongly raised**. Flies fast and directly, sometimes very low. A typical 'trap-liner', visiting pendant flowers with long corollas. Feeds hovering without clinging. Also often seen flycatching. In display male shows pendular flight in front of perched female, singing at her, iridescent frontlet erected. Seemingly breeds any season, with hanging nest in rock-cave.

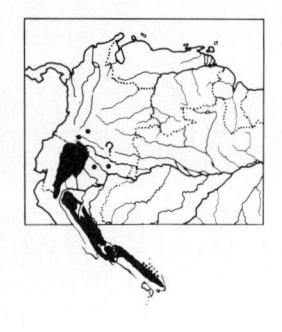

In understory in ravines in wet and humid mainly premontane forest, but seasonally moves to the lowest *Polylepis* woods, occ. at least to 2850 m (maybe not on the w Andean slope). From Mérida, Ven., through E Andes, Col., and along e Andean slopes of Ecu. and Peru and La Paz to Cochabamba, Bol., in C Andes to Cauca valley and w slope of W Andes, Col., and in w Ecu. (*ludoviciae*). W Panamá (*veraguensis*). Common locally.

LAZULINE SABREWING *Campylopterus falcatus* – Plate XXXI 1a-c

14 cm (incl. 2.5 cm **slightly arched heavy bill**). **Ad. male** above glittering green with blue crown, below glittering blue ranging from violet-blue on c throat to blue-green on belly. Vent and 5 cm long, square **tail chestnut** with bronzy green c feathers. **Imm. male** above glittering green, bronzy on crown, below gray except for glittering blue on throat, and bronzy-green spots on flanks. **Female** similar, but gorget incomplete, chin gray. **Juv.** like female, but lacks glittering throat-feathers and has broad, glittering greenish tips on all tail feathers (Plate XXXI 1a shows imm. male with its 1st tail-feather of ad. type). – Generally perched in shade or near cover. Feeds (hovering), chiefly from *Heliconia* flowers. Male shows pendular flight in front of female. Song reported to be an irregular series of sprutting calls. Cup-like nest on fairly high branch. Breeds Dec.

In the interior of humid montane forest, shrub, gardens, and coffee plantations, mainly in the shadiest parts, at 900-2500 (rarely 3000) m. From Miranda, n Ven., to Perijá mts and along the E Andes to ne Ecu., and in the n and s of the W Andes, Col. Common in n parts, rare in Ecu.

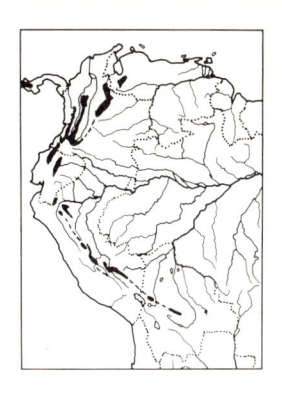

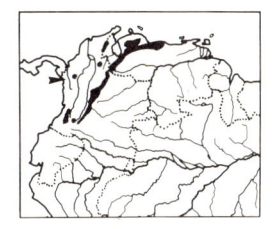

SANTA MARTA SABREWING *Campylopterus phainopeplus* – Plate LXIII 5ab

13 cm (incl. 2.5 cm **slightly arched bill**). **Ad. male** above metallic emerald-green. Face black, throat deep purplish blue, underparts dark with strong yellowish to blue-green reflection, which seen against the light turns deep purplish blue centrally. **Tail 5 cm, square, deep steel-blue** inclining to green on c feathers. **Female** above and on flanks grass-green, face dusky contrasting light gray throat and underparts of body, tail dark metallic green with gray outer corners.
Habits: Shy. Generally perched low, in shade, but sings from high exposed dead branch.
Voice: Marked 'bur' of the wings and a sharp double note. Song comprises plaintive notes. In flight and display a plaintive *twit-twit*.
Breeding: Probably in the rainy season June-Oct.
Habitat: Bushy páramo slopes, in winter in banana plantations.
Range: N Col. in Sta Marta mts. In June-Oct to 4800 m on s slope, in Feb-May mainly at 1300-2000 m.

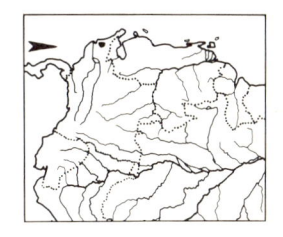

GREEN VIOLETEAR *Colibri thalassinus* – Plate XXXI 9a-b

11 cm (incl. 1.8 cm **slightly curved bill**). Blue-green with strong metallic sheen, glittering on throat, and with blue c belly. Cheek and elongate **erectile ear-feathers purple**. Broad 2-lobed tail metallic green

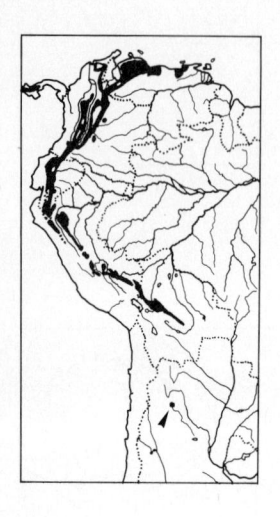

(light metallic blue below), with **dark subterminal bar**. Vent blue-green with buff feather-margins (*cyanotus*) to nearly uniform cinnamon-buff (*crissalis*). **Juv.** duller, with buff feather-margins above, rusty rictal stripe, and not glittering on throat, or blue below.

Habits: Often sings from exposed dead top branch, sometimes flapping wings and reversing position. In frontal display-flight fans the 'ears'. Also shows an undulating, low courtship flight, and male and female may fly side by side in undulating flight. However, there is **no high display flight** as in Sparkling V. Sometimes several males form loose 'leks', and can be seen chasing a female in a wild race.

Voice: Sings all day: short hard notes variably combined 2-3 together, e.g., *tusup-chip* or *tsip-chup*, sometimes admixed with *rrt* notes, or a loud *huitta huitta...*, these calls repeated over long periods. In sexual chasing a soft *zesesoorr* and ticking calls.

Breeding: Season apparently long or variable.

Habitat: **Humid forest** of generally quite open types, in windfall gaps and landslides with bamboo, at forest edge, clearing, and plantations, from mid-height to tree tops.

Range: C Am. (*thalassinus*). Mainly at 600-2800 m through all mountain ranges of n Ven., Col. (but hardly in W Andes) and Ecu., and in e Andes through Peru to Sta Cruz, Bol., and 1 rec. from Trances in Tucumán, nw Arg. Gradual transition in n Peru between northern *cyanotus* and southern *crissalis*. As abundant as Sparkling V., but in different habitat, and generally at lower elevations.

NOTE: the forms inhabiting S Am. are often treated as a separate species **MOUNTAIN VIOLETEAR** *C. cyanotus*.

SPARKLING VIOLETEAR *Colibri coruscans* – Plate XXXI 22a-b

14 cm (incl. 2.5 cm **slightly curved bill**). Blue-green with metallic sheen, and sparkling green breast, where the velvety black feather centers cause a scale-like pattern; belly purplish blue. **Purplish-blue color of the erectile ear-plumes continues across chin** (unlike in other violetears). Tail 6 cm, broadly 2-lobed, metallic green (light metallic blue below) with **dark steel-blue subterminal bar**. Melanistic form occurs. **Ad. male** does not have white spot behind eye. **Juv.** more bronzy green, not glittering, with buff feather-margins, and pale rictal stripe.

Habits: Often sings continuously for long periods from exposed bare twig in tree-top. In **display-flight** (seen commonly in breeding season) male ascends steeply c. 10 m from its perch, then returns twittering same way with widespread tail. Feeds from typical hummingbird flowers as *Centropogon, Elleanthus, Guzmania, Salvia, Siphocampylus*, and other flowers with curved corollar tubes, and also frequents *Clusias, Nicotiana*, and flowering *Eucalyptus* trees.

Voice: Song hard *tzirp* notes monotonously repeated, c. 2 per s, and sometimes admixed with *rrt* notes. In display flight a hurried *tzeezee-zirrr....* During aggression hard *trrrr*.

Breeding: There is an inconclusive rec. of a male participating in raising young, but this is not the case normally. Quite large cup-nest in

trees. Seems to breed much of the year, at least in the northern Andes.
Habitat: Adaptable and common in areas with **not too humid climates** with all kinds of rather open vegetation with scattered trees, e.g. bushy slopes with some trees, gardens, forest edge, *Eucalyptus*, *Puya* stands, and rarely *Polylepis* woods. The commonest hummer in gardens of, e.g., Bogotá, Quito, and Cuzco.
Range: Mainly at 2-3000 m in 'winter', but in Peru and Bol. higher up in the breeding season, locally to 4500 m. All mountain slopes (except densely forested humid parts) from n Ven. and Sta Marta mts, n Col., through Col. and Ecu., in w Peru s to Arequipa and Omate district in Moquegua, at Mamiña in n Chile (Marín *et al.* 1989), and in valleys of the C and E Andes through Peru and Bol. and along Andean slope to Catamarca (and further s?), wc Arg. (*coruscans*). Widespread, common and sometimes abundant. Tepuis of s Ven. (*rostratus*, *germanus*).

WHITE-VENTED VIOLETEAR *Colibri serrirostris* – Plate XXXI 10a-b

13 cm (incl. 1.8 cm **slightly curved bill**). Lighter green than Green and Sparkling V., with **ear-plumes more reddish purple**, and **vent white**. **Female** somewhat gray below, thus appearing rather pale. **Juv.** with white rictal stripe, more or less completely gray underparts (Plate XXXI 10b an extreme case), and with whitish outer corners of tail. Behaves much like other violetears. The song is an alternation of 2 or more call types repeated faster than in Sparkling V.: *zip zip zap, zip zip zap...* (zamba rhythm). In courtship pursuit a fast *chep-chep-chep....*

Inhabits semi-open terrain and bushy gorges across the s Braz. uplands and chaco, to lower temp. zone (normally below 1500 m, but accidentally to 3600 m) on e Andean slopes locally in La Paz, Cochabamba, Sta Cruz, and Chuquisaca, Bol., and Tucumán, nw Arg.

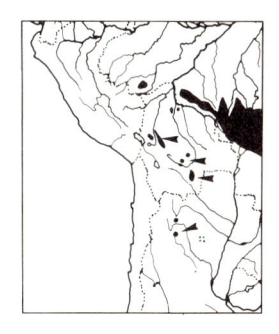

Emeralds – Genera *Chlorostilbon* and *Amazilia*

Small rather compact hummers with mainly green colors. Their classification is greatly in need of revision. Bills are rather long, thin and slightly curved; tails quite short, broad, and square. Usually slightly lower their wings below tail. The flight is fluttering or whirring, rather slow, the hum strong for a hummingbird. The nests are tiny cups camouflaged with leaves on branches. A tropical group, with a few species reaching the lower temp. zone.

BLUE-TAILED EMERALD *Chlorostilbon mellisugus* – Plate XXXI 5

7.7 cm (incl. 1.8 cm, thin bill), with rather short tail. **Male strongly glittering dark green**, above somewhat bronzy. Rump emerald-green **contrasting deep blue forked tail**. Gorget emerald-green, belly more golden. **Female** more bronzy, with **grayish white c underparts** from chin to vent, thin white line behind eye, **dark mask** through eye, and white tips to the 2 tail lobes. **Juv.** similar, but **imm. male** has glittering emerald-green discs along periphery of throat. Like other emeralds perches low, but feeds at various heights, often from typical bee-pollinated flowers, or taking nectar by piercing flowers. Voice loud and hard *chewp chewp* (*chirrrt*), and a short twittering song.

Widespread and common in savanna, clearings in 'terra firme' forest, xerophytic areas, gardens, and fields with scattered scrub in trop. and subtrop. S Am., virtually throughout from the Caribbean to c Bol., with several (unsatisfactorily defined) sspp. Ascends to upper subtrop. zone in Col. (to 2200 m, *pumilus*) and w Ecu., and is characteristic of Andean valleys of Ecu. to well above 2500 m, and once at 3050 m in Cajanuma in Loja (*melanorhynchus*). Rec. to 2200 m in Huánuco, c Peru (*phaeopygius*), and 2275 m in yungas of La Paz, Bol. (*peruanus*). Common.

GLITTERING-BELLIED EMERALD
Chlorostilbon aureoventris – Plate XXXI 8

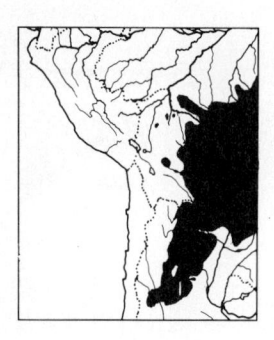

Resembles Blue-tailed E., but readily recognized by strongly **red basal part of bill**. Often quite coppery, belly strongly glittering golden (*aureoventris*), or distinctly coppery with golden-orange belly (*igneus*). **Female** light gray on c underparts from chin to vent. The song is a short rapid series of ticks.

Widespread in semi-arid to moderately humid bushy habitats in trop. and subtrop. zones s of the Amazon area to foot of Andes of Bol. (*aureoventris*) and s-wards to Mendoza and Córdoba, c Arg. (*igneus*), southern populations migrating n Mar-Sep. Probably not uncommon at c. 2500 m in Bol., with extreme rec. at 3070 and 3500 m (migrants?). To 2600 m in Tucumán, Arg.

COPPERY EMERALD *Chlorostilbon russatus* - a very local bird of subtrop. shrubby forest of Sta Marta mts and Perijá mts, Col./Ven., and lower Magdalena valley has been sighted once in temp. zone in Sta Marta, and rec. at 2600 m near Lake Fuquéne in E Andes, Col. Male differs from other *Chlorostilbon* by a coppery sheen, esp. more coppery underparts, and golden coppery tail.

SHORT-TAILED EMERALD *Chlorostilbon poortmanni*

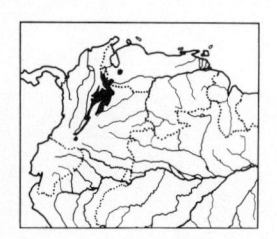

9 cm. **Tail very short and shining golden**, the lateral feathers not attenuated as in the Narrow-tailed E. When perched, tail exceeded slightly by wing-tips (reaches wing-tips in Narrow-tailed E.). Also differs by crown being more extensively glittering green (**male**), or bronzy brown (**female**). Visits non-pendant flowers (*Elleanthus, Guzmania, Phaseolus*).

Uncommon in light humid woodland, shrub, and second growth mainly in suptrop. zone to 2400 m, but locally to 2800 m in temp. zone. Mérida and Táchira, nw Ven., and along both slopes of E Andes, Col. (maybe with a local variant *euchloris*).

NARROW-TAILED EMERALD *Chlorostilbon stenura* – Plate XXXI 6a-c

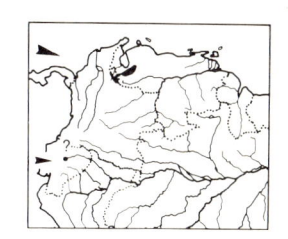

8 cm. Unlike in other emeralds, the **short tail shines golden green**, **outer tail-feathers of ad. male acute and very thin**. Outer tail-feathers of female grayish white, green on the middle, with blue subterminal demarcation.

Inhabits subtrop. gallery forest, second growth, scrub, and humid forest. Coastal mts of Ven. (*ignota*), at 1950-3000 m from Trujillo to Táchira, nw Ven. (*stenura*), and at 1000-2300 m along e base of n end of E Andes, Col., and at Baeza, ne Ecu. (doubtful ssp *acuticaudus*).

NOTE: Possibly a megassp of *C. poortmanni*.

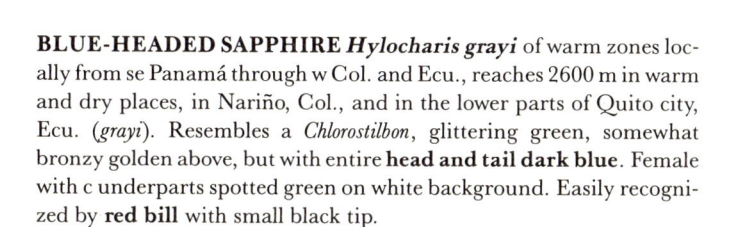

BLUE-HEADED SAPPHIRE *Hylocharis grayi* of warm zones locally from se Panamá through w Col. and Ecu., reaches 2600 m in warm and dry places, in Nariño, Col., and in the lower parts of Quito city, Ecu. (*grayi*). Resembles a *Chlorostilbon*, glittering green, somewhat bronzy golden above, but with entire **head and tail dark blue**. Female with c underparts spotted green on white background. Easily recognized by **red bill** with small black tip.

GREEN-AND-WHITE HUMMINGBIRD *Amazilia viridicauda* – Plate XXXI 3

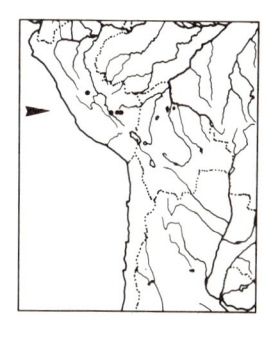

11 cm (incl. 2.7 cm slightly curved bill). Light metallic green or somewhat golden bronzy above, spotted on sides, and **white on c underparts from chin to vent**. Tail dull grayish green above and below, sometimes narrowly tipped white on outer feathers (imm.?), but **lacking any white at base of these feathers**. Perches on top of bushes in clearings and fields, and sings from such positions *tsi tzi tziu, twi*, sequences of 5-6 calls in c. 2 s, the pitch alternating, but not as regularly as in White-bellied E., and weaker, the quality like cork rubbed against glass. Nest-building July (Cushi), sitting on nest Jan. (Cuzco) (KR, TS).

Very poorly known, in landslides, cleared slopes, and second growth at forest edge at 1000-2500 m, probably replacing White-bellied E. in more humid zone in c Peru: near Cushi in Pasco, Machupicchu (where common), Huyro, and Hacienda Cadena near Marcapata in Cuzco.

NOTE: This and the following species were previously placed in genus *Leucippus*.

WHITE-BELLIED HUMMINGBIRD *Amazilia chionogaster* – Plate XXXI 2ab

11 cm (incl. 2.5 cm bill). Above rather light metallic green, **below white** (or pale gray in **juv.**) with green spots along sides. Tail dull bronzy green, hardly glossed below, and with more or less **white inner webs** from base to tip on outer feathers (often all white web in ad. *chiono-*

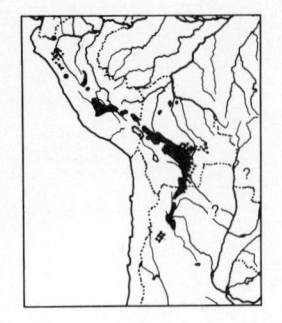

gaster, but sometimes a blackish subterminal patch in *hypoleuca*). Call sharp *zwit*. Song, given from high up in a tree (but rarely from the top) long sequences of *tzy* and *tji* notes on slihtly different pitches alternating fairly regularly, usually with c. 2 notes per s. Also an accelerating *tsee titititiiti*. In the morning shows territorial behavior, with much chasing high up in the air, and incessant *zeekeekee-kee...* and *tzrrrr* notes. Frequents *Nicotiana* flowers and agaves. Breeds in Jan.

Common in bushy terrain with cacti and agaves, gardens, *Alnus* groves, *Eucalyptus* rows, at forest edge and woodland at 800-2800 m in dry or not too humid valleys (but occ. in humid situations outside range of Green-and-white E.). In n and c Peru from Cajamarca to río Santa valley in Ancash on the w slope, and in the C Andes to Cuzco (*chionogaster*), in Sandia valley in Puno, through valleys penetrating E Andes of Bol., and Andean slopes from Taríja, Bol., to Catamarca and La Rioja, wc Arg. (*hypoleuca*). Migratory in Arg. Common locally.

STEELY-VENTED HUMMINGBIRD
Amazilia saucerottei – Plate XXXI 4

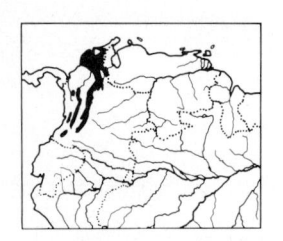

9 cm (incl. 1.8 cm bill). **Very dark**; **male** metallic greenish black, **female** and **juv.** more bronzy or slightly barred coppery, esp. on lower back. **Tail steel-blue**. Small leg-puffs white, vent steel-blue with whitish feather-edges. Base of mandible flesh. Gives chitting calls and a very thin song. Males are extremely territorial.

Inhabits forest clearings, second growth and cultivated areas (foraging low, in flowers of shade trees). In premontane zones, below the cloud forest, in C Am., in Sucio, Dagua, Patía, and Cauca valleys (*saucerottei*), Guáitara valley in Nariño, s Col. (*australis*), and across n Col. and in Magdalena valley, to 3000 m in Perijá mts (*warscewiczi*), and also high in Andes of Mérida and Trujillo, nw Ven. (*braccata*). Common.

RUFOUS-TAILED HUMMINGBIRD *Amazilia tzacatl* – widespread in dry to wet trop. and subtrop. zones of C Am. and from w Ven. through Col. and Ecu. – reaches 2500 m in lower parts of Quito, Ecu. (warm, dry valley), and may visit the lowest *Polylepis* woodlands in the flowering season (*jacunda*). Metallic bronzy green above, grass-green below, with gray belly, and easily recognized by **deep rufous tail** tipped deep bronzy, and by red base of bill.

AMAZILIA HUMMINGBIRD *Amazilia amazilia* of arid trop. and subtrop. zone of sw Ecu. and w Peru, has been sighted at 2850 m above Arequipa, at base of Volcan Misti, s Peru (BW pers. comm., but Zimmer – in *Amer. Mus. Novit.* 59(1950):26 – found it unlikely that an *Amazilia* could occur on Misti; *coeruleigularis?*). Inhabits gardens and shrubby and wooded terrain. Dull metallic green, **below chest and on tail chestnut.**

BLOSSOMCROWN *Anthocephala floriceps* is very local in Col., rec. in subtrop. zone of Magdalena Valley (*berlepschi*) and Sta Marta mts (*floriceps*). One was seen on a nest as high up as 2400 m in Sta Marta mts (O Frimer). 8.4 cm, incl. 1.3 cm bill. **Crown buffy-white**, more chestnut on rear crown (male) or brownish (female). Otherwise metallic green, rump bronzy, underparts grayish buff. Tail broadly tipped whitish.

An unnamed form of **Taphrospilus** (otherwise represented by the Many-spotted H. *T. hypostictus* in trop. e Ecu.-n Peru and e Bol.-n Arg.) is rec. at 2800-3400 m in the *Podocarpus* forest Ampay right above Abancay in Apurímac, c Peru (P Hocking, JF). Resembles an emerald, 12 cm (incl. 2.2 cm bill slightly curved distally), with upperparts green (crown dusky), **underparts buffy white thickly spotted with green**, except on c belly, tail blue-green below with **purplish blue subterminal bar** and gray tip. Frequently calls *dick-dick-dieck*.

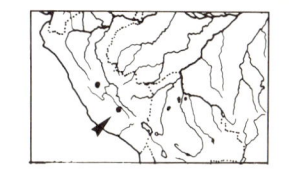

SPECKLED HUMMINGBIRD *Adelomyia melanogenys* – Plate XXXI 14ab

10 cm (incl. 1.4 cm bill). Characteristic round head with fuscous crown, broad **white supercilium** from right above eye slanting backwards and **partly encircling fuscous auriculars**. Above dark bronzy, below whitish buff (or more rufescent in *cervina* and *connectens*), with darker, somewhat bronze-spotted flanks. **Ad.** with **throat finely dotted** with dusky and sometimes with a little green (light blue discs in *inornata*). Tail slightly forked, but blunt-tipped, dark bronzy, outer feathers buffy gray basally, purplish black inside buffy tips (pale bases of tail-feathers large in *maculata*; light tips small in *melanogenys*, particularly wide and pale in *maculata*, wide and deeply colored in *inornata*).
Habits: Often sits with slighty lowered wings. Submissive. Mainly feeds low at shady borders from non-pendant flowers (*Aphelandra, Disterigma, Palicourea, Phaseolus*), and piercing flowers of acanthacads, ericads, and rubiads, composites (*Vernona*), labiates (*Beloperone*), the orchid *Elleanthus*, and the probably night-flowering *Posoqueria*.
Voice: Fine *zit*, *zi*, and *zidik* calls usually at 2 s intervals, but sometimes combined to brief twitter.
Breeding: Maybe most of year. Nest a hanging construction often beneath fern leaves.
Habitat: In wet to humid montane forest, mostly in lower story near clearings and in second growth. Partial to bamboo thickets.
Range: At 1200-2500 m in Col., but often towards 3000 m in Peru and Bol. On virtually all forested Andean slopes. Lara to Miranda, n Ven. (*aeneosticta*). W and C Andes (*cervina*) and around head of Magdalena Valley, Col. (doubtful ssp *connectens*), from Mérida, Ven., through E Andes of Col. and e slope of Ecu. to e side of W Andes in n Peru, along e Andean zone to Cuzco (*melanogenys*), in Cord. Vilcanota (hybrid form '*chlorospila*') and Puno, Peru, through Bol. to Salta and Jujuy, nw Arg. (*inornata*). W Ecu. to w side of n Peru low (*maculata*). Common.

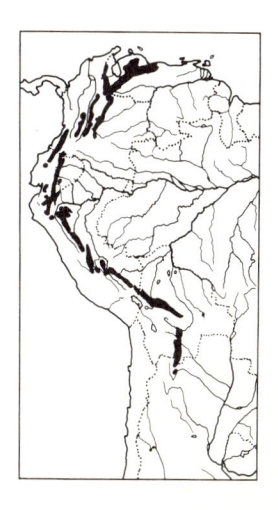

FAWN-BREASTED BRILLIANT *Heliodoxa rubinoides*

11 cm (incl. 2.5 cm **straight, rather thick bill**). Above bronzy green, wing-coverts coppery. Below with green chin, otherwise cinnamon with some green spots on breast and flanks, and small glittering violet throat-spot.

Inhabits wet and humid premontane forest in w Col. and Ecu. (*aequatorialis*), along C Andes, and locally on w slope of E Andes, Col. (*rubinoides*), and through e Andean slopes of Ecu. and Peru (*cervinigularis*). Locally and seasonally reaches 2600 m both in Col. and Peru.

Also **VIOLET-FRONTED BRILLIANT** *Heliodoxa leadbeateri* – common in humid premontane forest from Ven. through E and C Andes of Col. and s to n Bol. – may ascend to lower fringe of temp. zone in Col. (rec. 2400 m, *sagitta, parvula*). Size like Fawn-breasted B., bronzy green, male with glittering violet or blue forehead, and emerald-green gorget. Outer feathers of the deeply forked tail steel-blue. Female white below, throat and breast with green disks, belly tinged buff.

GREEN-CROWNED BRILLIANT *Heliodoxa jacula* of premontane forests of Col. and Ecu. has been rec. in the lower temp. zone of Cajanuma, Loja, Ecu. (Didre Platt). Male differs from Violet-fronted B. by glittering green crown, purple throat spot, and deeply forked tail uniform steel-blue. Female very similar to Violet-fronted B., but belly burely white.

WHITE-TAILED HILLSTAR *Urochroa bougueri* of humid premontane forest of W Andes of Col. to nw Ecu. (*bougueri*), and e slope of s Col. through Ecu. to n Peru (*leucura*), reaches lower fringe of temp. zone (rec. 2500 m). 14 cm (incl. 3 cm bill). Dark coppery green, throat and upper breast glittering blue. **Tail white**, patterned much as in Chimborazo and Andean Hillstars. Ssp *bougueri* has rufous malar streak.

Oreotrochilus hillstars

Typical of **very high parts of the Andes**. Relatively large, with slightly curved bills and fairly large tails usually with a conspicuous pattern. Males are adorned with a strongly shining hood or gorget (which can be strongly expanded or spread in certain displays), and dark midbelly. Hillstars are adapted to the harsh highland climate both by showing elevational movements, by avoiding low night temperatures in caves and crevices, and by metabolic adaptations, being torpid by night and during afternoon hailstorms. Large woolly nests in gullies and caves, or occ. in a *Polylepis* tree.

CHIMBORAZO HILLSTAR *Oreotrochilus estella*; *chimborazo* group – Plate XXVII 4a-e

14 cm (incl. 2 cm slightly curved bill). **Ad. male** above dark bronzy green, **hood covering entire head and throat shining purplish blue** with black demarcation towards breast (with green lower part of gorget in *chimborazo*, indicated in *soderstromi*). Underparts white with vent and thin midline on belly dusky. **Tail white** with dark c and lateral feathers and dark tips to the other feathers. **Female** dark gray-brown, bronzy above, and only slightly lighter below, and with dense dark speckles on throat, tail greenish black with white spots distally on the inner webs. **Juv.** similar, but juv. male recognizable by pale zone across upper breast, and **imm. male** by dark glaucous hood.
Habits: Very aggressive. Often perched conspicuously for long periods on top of a shrub, but uses low perches in windy weather (PG). Feeds almost exclusively from orange flowers of *Chuquiragua insignis* scrub. Also Malvaceae and *Puya* rec. as food plants.
Voice: Apparently as in Andean H.
Breeding: No information.
Habitat: On humid as well as rather dry shrubby or grassy slopes, esp. near secluded ravines (PG).
Range: Ecu., at 3500-5000 m. On páramos Pichincha, Iliniza, Cotopaxi, and just n of mt Antisana, and sighted at headwaters of Río Pantavi, Imbabura, and 70 km n of Pichincha (*jamesoni*); Quilotoa in w Cotopaxi (*soderstromi*); volcán Chimborazo area and on border Azuay/Morono-Santiago (*chimborazo*). Fairly common.

NOTE: Sometimes given rank as an (allo)species of the *chimborazo-estella-leucopleura* superspecies. Ssp *soderstromi* may be an intergrade between the other 2 sspp.

ANDEAN HILLSTAR *Oreotrochilus estella*; *estella* group - Plate XXVII 5a-h.

14 cm (incl. 2 cm slightly curved bill). **Ad. male** above drab brown (dark and distinctly bronzed in *stolzmanni*). Throat **gorget strongly shining emerald-green** (often more blue-green in *boliviana*) with black lower demarcation. Rest of underparts white, with prominent mid-line (black in *stolzmanni*; rufous in *estella*; broad, chestnut, somewhat spotted with black in *boliviana*). **Tail white** with bronzy black c feather-pair and outer part of outer feather (more extensively patterned in *boliviana*). **Female** gray-brown throughout (distinctly bronzed in *stolzmanni*), but pale below, tail greenish black, outer 3-4 feathers with **white inner webs at base and tip** (*estella* with extensive white color, usually with dark zone only on the middle, and the outer feather with entirely white inner web, but individual or age-related variation makes it difficult to define population differences on basis of the material presently available). **Juv.** more drab gray, with yellow base of mandible, and lacking throat discs; **juv. male** usually with rather dull dark glaucous gorget.
Habits: Aggressive, sometimes even seen attacking Giant Hummingbird. Often perched conspicuously on top of shrub for long periods. Flies fast and sometimes high over the terrain. Often flutters low over

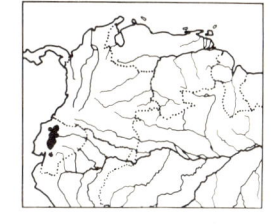

the grassland in search of concealed flowers, but normally feeds perched. Spreads tail when hovering. Sometimes hangs on cliffs using tail for support. In winter, feeds much on *Eucalyptus*, and otherwise depends on the orange flowers of the *Chuquiragua spinosa* scrub, but in the rainy season also frequents orange-red *Cajophora* flowers, *Barnadesia, Bomarea, Berberis, Buddleia, Centropogon, Ribes*, and *Puya* as well as prostrate plants like cushion-cacti, low *Castilleja*, and malvacaeans. In breeding season, females occupy territories in rocky terrain, while males wander, without territories, in the open grassland. Steep display-flight of male to 30-40 m, often in spiral with diving return to right above perched female. Sometimes also shows a low undulating display-flight.

Voice: Fine *tij* and *zirr* notes. In pursuit, sustained rather metallic *tjit-jitji...* or *tzitzi....* Song a fine passerine-like trill.

Breeding: Eggs in Sep-Dec throughout range.

Habitat: Puna grassland and *Puya* and *Polylepis* stands, with rocks and access to water and suitable food plants. Outside breeding season often descends to scrub and open woodlands of Andean valleys.

Range: Puna zone, normally nesting at 3500-4500 m, but can be seen to 5000 m, and descends in winter to 2400 m. N Peru from Cajamarca through Cord. Blanca, Ancash, to Lake Junín and Huánuco/Pasco border (*stolzmanni*). From w Ayacucho in c Peru through nw Bol. to Tarapacá, n Chile (*estella*), grading on Bol. altiplano to populations of s Bol. and Jujuy to Tucumán, nw Arg., and accidentally Antofagasta, Chile (*boliviana*). Common.

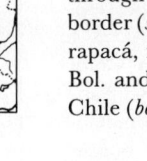

WHITE-SIDED HILLSTAR *Oreotrochilus leucopleurus* – Plate XXVII 8a-b

Very much like Andean H., but **male** has very broad blue-black median zone on lower breast and belly, and tail more rounded, with slightly shortened and **narrow outer feathers curving inwards**, and white only at base (narrow shape indicated in *boliviana* specimens of Andean H.). **Females** and **Juv.** as in Andean H., but generally with narrower outer tail-feathers with broad and complete dark transverse bar.

Habits and **voice**: Probably as in Andean H. Feeds on *Barnadesia, Berberis, Chuquiragua*, and *Puya*. On rock surfaces uses tail for support, like a woodpecker.

Breeding: Eggs in Nov-Dec.

Habitat: As for previous species.

Range: Known from 1200-4000 m (seasonal movements?) in Tarija, s Bol., and in Andes of Arg. from Jujuy to Mendoza, and in Chile from Antofagasta to Colchagua and Bío-Bío; 1 rec. near coast of s Antofagasta (Marín *et al.* 1989) suggests that Chilean birds may migrate. Common.

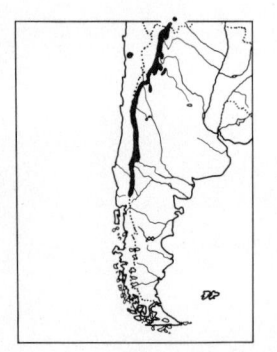

NOTE: Due to some indications of intergradation, and lacking evidence of sympatry, often regarded as a (mega)ssp of Andean H.

BLACK-BREASTED HILLSTAR
Oreotrochilus melanogaster – Plate XXVII 6a-b

13 cm (incl. 2 cm slightly curved bill). **Ad. male** above dark bronzy brown, gorget glittering emerald-green, **below black** with drab brown extreme sides. Tail square, blue-black. **Female** above bronzy or somewhat coppery, below light gray-brown. Upper tail-coverts green, tail steel-blue with white spots distally on all except c feathers. Very similar to sympatric Andean H. female, but darker, more metallic, and **not showing any white at base of tail-feathers**.

Habits: Seems to be submissive to Andean H., living more secludedly and lower in vegetation, wherever the 2 occur together. Frequents flowering *Chiquiragua spinosa* scrub, but also seen feeding from *Cajophora*, *Castilleja*, and occ. from *Cassia* bushes.

Voice: *Zit*, during pursuit *zee pitepitepit*....

Breeding: In rocks and sometimes under roofs of houses. Eggs in Feb-Mar.

Habitat: Grassy slopes with much *Chuquiragua* shrub and usually some sheltered places among rocks, with fertile soil and nitrophilous plants as *Cajophora*. Sometimes in villages, and in stands of *Puya raimondii*.

Range: C Peru. At 3500-4400 m in Junín and Huancavelica, and locally in adjacent Lima, Pasco, and in Huaron and Paron, Ancash. Also rec. near Tambo in Ayacucho. Fairly common.

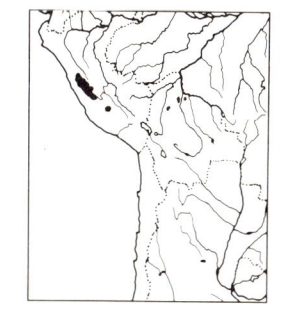

WEDGE-TAILED HILLSTAR *Oreotrochilus adela* – Plate XXVII 7a-b

12 cm (incl. 2.5 cm slightly curved bill). Tail wedge-shaped. **Ad. male** above gray-brown, very faintly bronzed, and with somewhat scaly effect. Gorget glittering vivid green, **below chestnut with bold black c zone**. Tail blue-black with cinnamon inner webs, except on the c feathers. **Female** with throat white spotted with fuscous, and with some green discs, **below rufous, pale centrally**. Tail-feathers dusky with rufous bases and white tips, c feathers blue-black. **Juv.** plumage female-like, juv. males without the partly developed gorget of the congeners.

Habits and **voice**: Seen feeding from *Barnadesia* flowers and occ. from labiates and columnar cacti. Apparently submissive to Andean H. Gives a melodious song from exposed perch, in slim upright pose and with strongly expanded gorget. Rattles wings if female comes near. Then repeatedly rises c. 1 m, singing. Copulation may follow (KS).

Breeding: Eggs Nov, juv. Feb and June.

Habitat: Semi-arid, esp. in *Polylepis* woodland with ravines with denser scrub, esp. *Barnadesia*. Also in disturbed habitats as *Dodonea* heaths, if scattered columnar cacti and taller bushes are present.

Range: Bol.; at 2600-4000 m in s La Paz, Cochabamba, Potosí, and Chuquisaca. Might migrate. Uncommon.

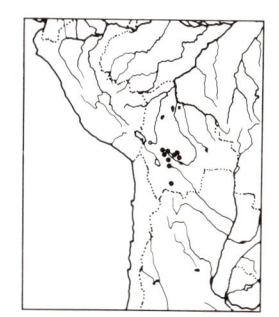

GIANT HUMMINGBIRD *Patagona gigas* – Plate XXVII 1a-c

23 cm (incl. 3.5 cm rather thick bill), wing-span 30 cm (can be confused only with female or juv. Greater Sapphirewing). Recognized by size, long and narrow wings (as in a swift), **flight style**, and slightly forked **tail with pale base and white bronze- or drab-tipped coverts**. Above rather dull bronzy, cheeks and underparts cinnamon-rufous (*peruviana*), or duller cinnamon, obscurely washed with gray-brown, esp. on cheeks and breast (*gigas*). **Females** can generally be recognized by dusky spots below. **Juv.** still grayer, but not necessarily spotted below, and finely scaled above, with white edges distally on the flight-feathers.

Habits: Generally territorial and aggressive, chasing violetears and hillstars. Flies with erratic wing-beats and inserted glides, a flight-style resembling a long-winged swallow rather than a hummer. Hovers with rather slow deep wingstrokes, and with tail spread, but often sits or climbs while feeding. Feeds esp. from puyas (both small forms and the gigantic *Puya raimondii*), but also visits *Cylindropuntia* and other tall cacti, *Buddleia*, *Passiflora*, and *Sittacanthus*, and may crowd around flowering *Agave* and thickets of flowering *Mutisia* or *Nicotiana*. Sometimes shows sustained hovering for swarming insects, bouncing up and down, often high up in the air.

Voice: A characteristic drawn-out plaintive *cueeet*. In aggression repeated *weeet* calls.

Breeding: Tiny nest on branch or cactus stem, or on cliff face. Eggs in Dec-Feb (Mar) (Ecu.), or Oct-Jan (c Chile).

Habitat: Arid, shrubby hillsides, in western Chile and Peru in varied shrub-steppe vegetation often with many columnar cacti, in the puna zone frequenting bushy slopes and *Puya* stands and open *Polylepis* and *Buddleia* woods.

Range: Mainly at 2000-3500 m in Andean valleys, locally to 4500 m in Peru, but lower in the dry season and in the s parts. Chile from Valdivia (occ. Aysén) to Atacama, breeding below 2000 m, and migrating across Andes to adjacent w Arg. (*gigas*). Through puna zone from Mendoza, c Arg., and c Chile through Bol. and Peru, and now established almost through Ecu., and probably in extreme s Nariño, Col. (*peruviana*). Páramo de Chingaza E Andes, Col.? (see *Wilson Bull.* 95(1983):661-2). Generally common.

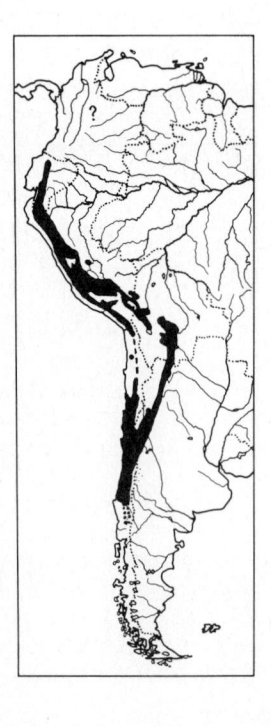

Sunbeams – Genus *Aglaeactis*

Fairly large hummers with long broad wings, thin and straight bills, and mainly dark brown colors, usually with tuft of pale feathers on c breast. Strongly iridescent colors occur only on rump and lower back, and are most conspicuous at a very sharp angle from behind (possibly functioning as a backward-directed signal in flight). Often raise feathers of fore-crown and c breast (unlike in the flat-headed *Coeligenas*). Small cup-shaped nest in a bush.

SHINING SUNBEAM *Aglaeactis cupripennis* – Plate XXVIII 18a-c

12-13 cm (bill c. 2 cm in the n, 1.5 cm in s). Above bronzy brown, **lower back glistening lilac and gold**, upper tail-coverts **shining** silvery green (occ. feathers purple or blue in *caumatonotus*). Face and **underparts cinnamon-rufous** (*cupripennis*) to dull brown with fuscous face and upper throat and dusky brown pectoral band with several pinkish buff feathers centrally (*caumatonotus*, its characteristics more or less clearly also in the more rufous forms *parvula*, *cajabambae*, and *ruficauda*). Wing-linings rufous. Tail rufous with bronzy distal part (the rufous being limited in *aequatorialis* and *parvula*, extensive in *ruficauda*). **Female** may have little iridescence on back. **Juv.** generally more uniform brown, *viz.*, with very vaguely expressed pattern in *caumatonotus*, and lacks glittering colors. Full development of a lilac lower back takes at least 1 year.

Habits: Usually on exposed perch in tree-top, and often seen in high sustained flight, and sometimes gliding downhill on spread wings. Interspecifically territorial. Often holds wings out in V or flaps them slowly after alighting. Feeds perched, often with raised wings, mainly from *Bomarea*, *Centropogon*, *Embothrium*, *Mutisia*, *Passiflora*, *Puya*, and *Tristerix* flowers, and also from smaller melastome flowers. Takes many insects, e.g., gleaning from foliage. Displaying male raises breast-feathers and shows some wingbeats (perched), and then flutters in front of female with the back exposed towards her. In 20-30 m display flight rises and descends at c. 45° angle.

Voice: Rapid thin twittering *tzee-tzee-zee-zee...* or *tzr-zr-tzrrr*. In flight fine *zit* notes, which in *Polylepis* woods can be confused with calls of Giant Conebill.

Breeding: Nests Mar and July (Col.), Feb (Ecu.), Apr (Lima), and Nov (Cuzco).

Habitat: Bushy subpáramo and low rather open and dry montane forest (humid to semi-arid cloud forest and in second growth), preferably with some larger trees with domed or flat canopies with dense stiff leaves (*Escallonia*, *Eugenia*), and seasonally ascends to *Polylepis* woods.

Range: At 2500-4300 m, but not so high where sympatric with White-tufted S. Northern E Andes of Col., southern C and W Andes of Col. through Ecu. (*cupripennis*). From extreme sw Ecu. s to Taulis and Chugur in s Cajamarca and La Lejia, Chachapoyas, and Leimebamba in Amazonas, n Peru (*parvula*). Cajabamba s Cajamarca, and Hda Motil (and possibly Huamachuco) in w La Libertad (*cajabambae*), Patás and Quebrada La Caldera (e of Tayabamba) on w slope of C Andes in e La

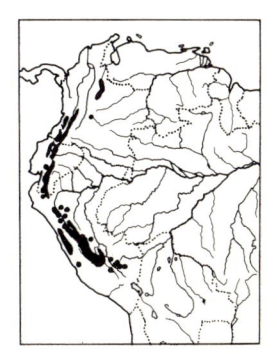

Libertad and rec. near Nuevas Flores in Huánuco (*ruficauda*), patchily on Pacific slope from La Libertad (Cochabamba) and Ancash (Yanac) to Lima, and locally from La Unión to Acomayo in Huánuco and se to Cord. Vilcanota in Cuzco (*caumatonotus*). Locally common.

NOTE: The Sunbeams form a superspecies, and the White-tufted S. may have arisen by isolation of the c part of a former cline, or maybe by isolation of a hybrid population of Shining and Black-hooded S. (KS). Presently, Shining and White-tufted S. are sympatric locally, and seem fully compatible.

WHITE-TUFTED SUNBEAM *Aglaeactis castelnaudii* – Plate XXVIII 19

12 cm (incl. 1.8 cm bill). **First impression fuscous with tawny tail**. Above fuscous black, somewhat bronzy, and with **magenta reflection on rump** and lower back. Below rich rufous brown (*regalis*) or more dull earthy brown (*castelnaudii*), with blackish upper throat and face and blackish pectoral band, sometimes entirely blackish below except for the tuft of **white feathers on c breast**. Tawny secondaries form wing-patch, wing-linings tawny, tail tawny with dark bronzy tips and c feathers (less bronzy in *regalis*). **Juv.** lacks glittering rump and is more uniform brown below.
Habits: Like Shining S., often perches conspicuously on dead uppermost twigs, and often very confident. However, where the 2 overlap, it is submissive and skulking, and usually perches on low side twigs of dense bushes. Sometimes flies high during pursuit. Holds wings raised some time after alighting. Seen feeding from *Berberis*, *Barnadesia*, *Brachyotum*, *Centropogon*, labiates, *Lupinus*, and *Siphocampylos*.
Voice: *Pypypypy...*, in pursuit *zreetzreetzreet...*, and very faint and thin *zeee*s.
Breeding: No data.
Habitat: Rather open shrub, and *Polylepis* and *Escallonia* wood, and in glades in semi-arid forest.
Range: Peru. At 2500-4300 m, above Shining S. where the 2 are sympatric. In semi-humid pockets around arid head of the Marañón in Huánuco and s to Aconcocha, Millpo, and Rumicruz in adjacent Pasco, on Nevado Ampay in Apurímac, and mainly semi-arid valleys in Cordillera Vilcanota, Cuzco (*castelnaudii*; individuals with *regalis* traits occur in many places from Huanuco to Cuzco, apparently mainly at those localities that lie most isolated from the humid Andean slopes, KLS). Common only at a few sites.

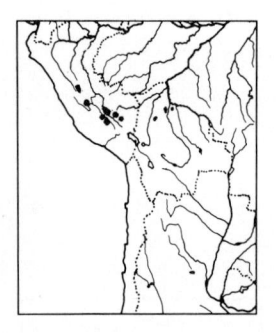

PURPLE-BACKED SUNBEAM *Aglaeactis aliciae* – Plate XXVIII 20

13 cm (incl. 1.5 cm bill). Dark earthy brown, slightly coppery on wing-coverts, lower back and rump glistening amethyst shading to golden green on upper tail-coverts. **White on lore, across upper breast, as spots on the dark c breast, and on vent**, and at base of bronzy tail (**female** slightly buff on these parts). Wing-linings buffy white with black

spot near wrist. Outer primary very narrow. **Juv.** lack glittering rump.
Habits: Seen visiting flowers of a (*Tristerix*), and a leguminaceous bush.
Breeding: Juv. taken Feb, Mar, and June.
Voice: Humming flight sound occ. with high overtone.
Habitat: Known from shrubby slopes with some alders *Alnus* and other trees.
Range: N Peru; at 3000-3200 m on e-slope of W Andes at Succha and nearby Soquián (45 km e of Huamachuco), and 20 km nw of Succha at Molino (10 km nw of Aricapampa) (TS) in La Libertad. Probably common locally, at least during visits by OT Baron (1895) and MA Carriker (1932), and some rec. in Oct and Nov 1979 (LSUMZ).

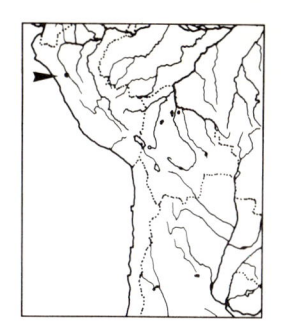

BLACK-HOODED SUNBEAM *Aglaeactis pamela* – Plate XXVIII 21a-b

12 cm (incl. 1.5 cm bill). Purplish **black** with tuft of **white feathers on c breast**, rump and sometimes entire **lower back glittering golden green/blue** (some birds from Unduavi with feather-tips bronzy, or golden on upper tail-coverts). Vent and **tail chestnut** with bronzy feather-tips. **Juv.** slightly browner, with often inconspicuous breast-patch, more olivaceous tail, and orange mandible (but apparently shows metallic colors on rump from an early age).
Habits: Feeds from flowers of lobeliads (*Siphocampylus*), labiates (*Centropogon*) and melastomes (*Brachyotum*), sometimes perched. Wings are raised a few s after alighting.
Voice: High-pitched *zeet-zeet-zeet* reported.
Breeding: Probably Sep-Mar, but no nest rec.
Habitat: Humid cloud forest to the timberline.
Range: Bol. Locally at 1800-3500 m in yungas of La Paz and Cochabamba. Probably uncommon.

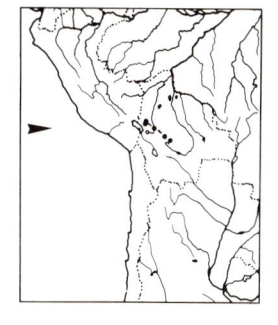

MOUNTAIN VELVETBREAST *Lafresnaya lafresnayi* – Plate XXXI 11a-c

11.5 cm (incl. **2.5-3 cm thin, curved bill**). With rather broad, notched **tail with distinct pattern: pale horn (*tamae*), creamy buff (*lafresnayi*) or white (other sspp)**, with dark bronzy tips and c feathers. **Ad. male** with glittering green upperparts, throat, and sides, c underparts velvety black, vent white with broad silvery-green feather-tips. **Female** with crown more dusky, underparts creamy buff densely dotted with glittering green discs, but belly clear white. **Juv.** female-like but more diffusely green on sides (not clear-cut discs). **Imm.** males develop fully black breast and belly while still showing white anterior underparts. Ssp *saul* has a more blue-green breast, a usually broad tail-bar (4-9 mm on inner webs of outer feathers), and a distinctly curved bill, but *orestes* has reduced dark tail-tips with only 2-3 mm dark border on inner web of outer feather, and *rectirostris* is greener below, with intermediate (variable) dark tail-tip, and a slightly curved bill.
Habits: Males often hold territories while females 'trap-line'. During fluttering foraging flight incessantly spreads and closes tail, the white

(or buff) color then making the bird conspicuous (like a fluttering white butterfly) at a distance. Feeds hovering in low vegetation, from typical hummingbird-flowers as *Salvia, Centropogon* and other labiates, *Castilleja, Pentanderia, Siphocampylus, Symbolanthus*, and from yellow leguminaceous flowers.

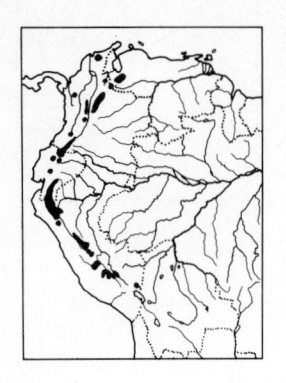

Voice: Frail *tzrr* notes and a clear whistle *zeee*; during pursuit *tititi...rrr*. In flight the wings give a very prominent hum.
Breeding: Nest in Jan (Col.), nests Jan-Feb (Papallacta ne Ecu.).
Habitat: From inside primary forest (at low levels) to fields and open slopes with half-meter tall herbaceous and bushy growth.
Range: Mainly at 1900-3200 m, with marked seasonal movements. Páramo de Tamá, nw Ven. (*tamae*, described by Phelps in *Bol. Soc. Venez. Cienc. Nat.* 41(1987):7-26) and Perijá mts, and E Andes Col. (*lafresnayi*). Sta Marta mts, n Col. (*liriope*). S Trujillo to Táchira, nw Ven. (*greenewalti*). W Andes of Col. throughout Andes of Ecu. s to the n Peru low (*saul*); Chachapoyas area in Amazonas, n Peru (*orestes*), and spottily from Huánuco to Apurímac, Cuzco, and in Sandia valley of Puno, Peru (*rectirostris*). Generally common, but somewhat local.

GREAT SAPPHIREWING *Pterophanes cyanopterus* – Plate XXVII 2a-b

19 cm (incl. 3 cm bill). With very long wings and slightly forked tail. Told from Giant Hummingbird by thin slightly upturned bill (usually held more upwards), **dark general coloration, and lack of white area near base of tail** (instead, **female has whitish lateral edge of tail**). **Ad. male** black with dark green sheen, dark blue wings (but **imm. male** lacks iridescent blue on remiges). **Female** dark metallic green with underparts cinnamon-rufous spotted with green, esp. on sides. **Juv.** is similar, but with few green lateral spots. Proposed sspp are doubtfully distinct.
Habits: Flies extremely fast, with sometimes erratic wingbeats and some gliding, but decidedly more like a typical hummer than the Giant H. Always makes a few deep wingstrokes when just perched. Feeds hovering or perched on the outer surface of thickets, usually rather low. Feeds from *Barnadesia, Bomarea, Inga, Mutisia, Passiflora*, and *Puyas*, and sometimes from larger herbs (*Centropogon, Loasa, Siphocampylus*). Occ. flycatches. Said sometimes to associate with mixed flocks of tanagers, flower-piercers, and warblers. 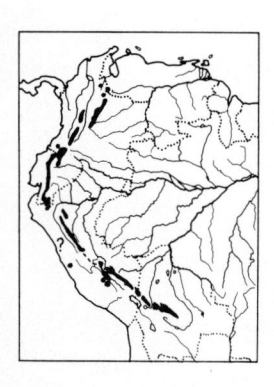
Voice: A drawn-out piercing high-pitched *zeeee* and agitated *tititirr*.
Breeding: Open cup suspended from roots. Eggs May (Pichincha, Ecu.).
Habitat: Not too dense cloud forest, elfin forest, treeline shrub, to bushes in the lower páramo, in wet to sometimes rather dry zones.
Range: At 2600-3700 m, in n half of E Andes and C Andes, Col., and through Ecu. (but not often in dry Andean valleys), in e Andean zone through Peru to yungas of Cochabamba, Bol., and occurs in some isolated woods in w Peru s to Lima. Locally common.

NOTE: Should probably be included in the genus *Coeligena*.

Incas and starfrontlets – Genus *Coeligena*

Rather large hummers with **long straight bills**, and pointed flat heads. Fairly large notched tails, and long broad wings. Generally more metallic (green) than sunbeams. Exceedingly fast fliers. Although usually foraging along borders of dense vegetation, and partly inside forest, they sometimes fly straight and high up, e.g. right across a valley. Thus, all the species will occ. occur in small or even quite isolated forest patches. Probably all species are wide-ranging 'trapline-feeders', which ecologically replace hermits at high elevations. Nest surrounded by loose camouflage sheath of moss, and suspended below an overhang.

BRONZY INCA *Coeligena coeligena*

14 cm (incl. 3.6 mm bill). Dark bronzy brown, greener on lower back, throat and chest dusky diffusely streaked with whitish feather-edges, vent with cinnamon feather-edges.

Inhabits edges of humid and wet premontane forest, and occ. rec. to 2600 m in Col. and Peru. In Falcón (*zuloagae*), n Lara, nw Ven. (*coeligena*), Perijá mts on Ven./Col. border (*zuliana*), and C and W Andes, Col. (*ferruginea*). From Lara through Andes of nw Ven. and e Andean slope through Col. (*columbiana*), and through Ecu. and Peru (*obscura*) to yungas of Bol. (*boliviana*).

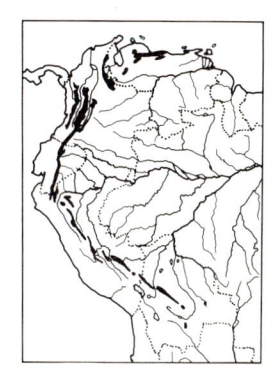

BLACK INCA *Coeligena prunellei* – Plate XXVIII 16

13.5 cm (incl. 2.6 cm bill). **Male** purplish **black with white half-collar** on each side of breast contrasting with dark metallic blue shoulders. Small glittering greenish-blue throat-patch, and some whitish feather-tips on vent. **Female** less blue on shoulders. **Juv.** also lacks gular spot. Feeds hovering in tree crowns and low at forest edges, from vines and climbers (*Aphelandra, Palicourea, Psammisia, Thibaudia*). Rare *ick* calls noted.

Col. In humid montane forest at 1400-2600 m in Santander, Boyacá, and Cundinamarca on w slope of E Andes. Thought to be threatened by deforestation, but recently re-discovered near Facatavitá and Cerro Carare ese of Toqui, Boyacá, near Virolin in Santander, and in several other places s to Pedropalo in Cundinamarca (G Andrade, C Rocha *fide* DW Snow).

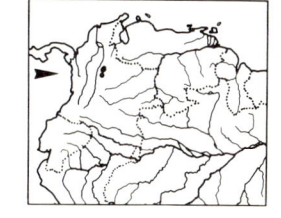

NOTE: Forms a superspecies with Brown I. *C. wilsoni* of the pre-montane Pacific slope of w Col. and nw Ecu. These 2 again form a species group with Collared I.

COLLARED INCA *Coeligena torquata* –
Plates XXVIII 15a-d and LXIV 5

14.5 cm (incl. 3 cm bill). **Ad. male** generally dark, adorned with **broad white triangular patch on breast (rufous patch in *inca* and *omissa*)**, and with **white tail** with greenish c feather-pair, and bronzy outer corners (very extensive white in *inca*, less in northern forms). The main color varies from black with very faint green sheen (*fulgidigula*) to strongly glistening emerald or golden green throughout (*conradi*), or a glittering emerald to golden green with velvety black head (*inca*; *omissa* and *eisenmanni* less contrasting, and the latter with coppery upper tail-coverts). Top of head black (except in *conradi*) with glistening blue crown-spot (*fulgidigula, torquata*), or green frontlet (*omissa, inca*) or both. **Female** similar, but crown green, throat and chin buff with dark dots (more rufescent and without dots in *eisenmanni*, and uniform buff in *conradi*), belly rather gray. **Juv.** normally as a female, but outer tail-feathers with mainly dark outer web (a clear-cut age difference in *inca*). In northern forms, juv. have dark throat, with white patches laterally (as in Black and Brown I.); *inca* juv. has fully developed rufous breast patch, but dull gray-brown throat. **Imm.** male retains some bronzy green gloss on crown.

Habits: In the exceedingly fast flight through the forest reveals itself by briefly 'flashing' the white of the tail. When perched does not raise its bill as much as other incas. Feeds low, mostly hovering, along edges of dense entangled vegetation with *Bomarea*, *Cavendishia*, *Fuchsia*, *Loasa*, and *Vriesea* flowers, but also visits short-crowned ones like malvaceans.

Voice: Generally silent, but occ. gives repeated *tsc* notes. Song consists of *tzee*, *zi-tee* or *tzi-zi-tee* notes several s apart. Rapid trill during pursuit.

Breeding: Nest-building rec. Nov and Mar (Col.), eggs Mar (Pasco, Peru).

Habitat: Usually in understory and lower canopy of dense heavily moss-clad and entangled cloud forest, mainly in very humid zones.

Range: At 1500-3000 m on humid Andean slopes. Mérida mts, nw Ven., and e slope of E Andes of Col. in Norte de Santander (*conradi*), Tamá in Táchira, nw Ven., through Andes of Col., and e slopes in Ecu. to Chaupe in Piura, n Peru (*torquata*). W Andean slope of Ecu. (*fulgidigula*). Chachapoyas area in Amazonas (*margaretae*), and most e Andean slopes of Peru from Cord. Colán s (continuously?) to Ayacucho (*insectivora*). Cord. Vilcabamba, Cuzco (*eisenmanni*, published by JS Weske, pp. 41-45 in Buckley *et al.* (1985)). From around Urubamba Valley in Cuzco through Puno, se Peru (*omissa*) and yungas of La Paz and Cochabamba, Bol. (*inca*). Generally common.

NOTE: Ssp *inca* (with *omissa*) was previously treated as a full species.

NOTE: *Coeligena traviesi*, known only from 'Bogotá' trade skins, has been considered a hybrid between Collared I. and Buff-winged Starfrontlet. Above mostly black with glittering blue-green frontlet and some blue-green on crown; rump coppery purple. Throat green with glittering violet patch, breast with white patch, lower underparts glittering dark green. Tail bronzy.

WHITE-TAILED STARFRONTLET *Coeligena phalerata*
– Plate LXIII 6a-b

14 cm (incl. 3 cm bill). **Male dark with white vent and tail**, often with bronzy tip, but becoming totally white by age. Above metallic green

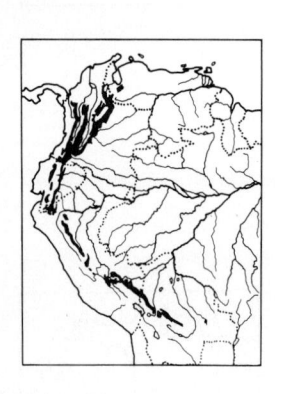

with strongly glittering turquoise forehead and crown, below glittering emerald-green with l cm wide blue gular patch. **Female** above shining green, crown mainly dusky blue-green, **below uniform cinnamon-rufous**, tail dark bronzy with small pale tips on outer feathers.
Habits and **voice**: Pugnacious and territorial, but at times may 'trap-line'. Rec. feeding from *Fuchsia*s. Gives a pretty twittering.
Breeding: Possibly Feb-Apr.
Habitat: Humid and wet montane forest and shrubby forest borders.
Range: N Col., at 1400-3200 m in Sta Marta mts.

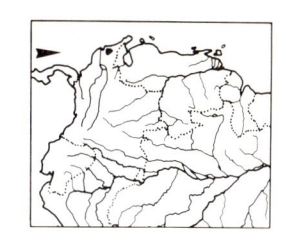

GOLDEN-BELLIED STARFRONTLET *Coeligena bonapartei* – Plates XXVIII 13a-c, 14; LXIV 12

14 cm (incl. 2.8 cm bill). **Male** sspp *bonapartei* and *consita* metallic emerald-green with velvety black hood with glittering emerald-green frontlet (conspicuous seen frontally), wings dark, or with rufous patch (*consita*), rump glittering golden (seen from above, but not conspicuous from behind). Belly fiery glittering yellowish green to gold. **In ssp *eos*** (Plate LXIV) **the secondaries are cinnamon-rufous** (forming conspicuous patch) with dark tips; rump and belly partly cinnamon, with orange-golden glittering, the vent uniform cinnamon with no bronze. The species has **tail and vent mainly orange-buff tipped golden bronze (tail totally golden bronze in *consita*)**. **Male** has 0.7 cm wide glittering violet-blue gular patch; **female** similar, with green crown, cinnamon-buff somewhat dotted chin and throat. **Ssp *orina* is very dark**, mainly black with weak lime-green sheen and no bronzy or cinnamon, gular spot small, spectrum blue, and belly glittering yellowish green. **Juv.** quite uniform, with white tips to outer tail-feathers. **Imm. male** retains some green gloss on crown.
Habits: Feeds on tubular flowers around outer edge of thickets.
Voice: High-pitched *twee-tzee* in flight.
Breeding: Nests in Jan-Mar.
Habitat: Patches of cloud and dwarf forest, incl. rather dry *Tillandsia*-clad woods. Mainly low to mid-height.
Range: At 1400-3200 m. Cerro Tetari in Perijá mts (*consita*), from Trujillo and Barinas through Mérida to Táchira, nw Ven. (*eos*), and locally on w slope of E Andes, Col. (*bonapartei*). At 3500 m on Páramo Frontino at n end of W Andes (*orina*). The latter form is known only from the imm. type specimen; *bonapartei* is rare or at least rarely seen, *eos* more numerous.

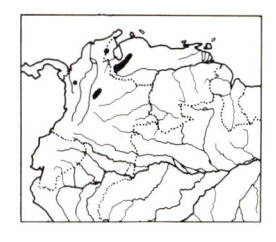

NOTE: Forms a superspecies with a leapfrog pattern of variation with Blue-fronted, Buff-winged (?), and Violet-throated S. Ssp *eos* is phenetically more similar to a Violet-throated S. than to other forms of Golden-bellied S. The very dark (melanistic) *orina*, based on 1 imm. specimen, was previously regarded a separate species **DUSKY STARFRONTLET** (see *Bull. Brit. Orn. Club* 108(1988):127-131).

BLUE-THROATED STARFRONTLET
Coeligena helianthea – Plate XXVIII 22a-b

14 cm (incl. 3 cm bill). Identified by **strong rosy glittering on belly**

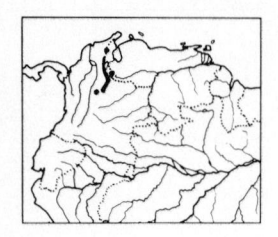

and vent. **Male very dark** with slight emerald-green gloss, hood velvety black, but glittering emerald-green frontlet conspicuous seen frontally. Lower back and rump blue with some violet (conspicuous seen from above, but not from behind), tail slightly bronzed black. Throat with 0.8 cm purplish blue gular patch. **Female** with stronger green gloss, throat and breast cinnamon-buff with green spots on lower parts, tail bronzy purple, glittering rump restricted. **Juv.** similar, with little rosy glitter on belly, but imm. male soon gets violet gular patch, but lacks black hood.
Habits: *Cavendishia, Palicourea, Passiflora, Rubus*, and *Symbolanthus* noted as food plants.
Voice: Hum of wings sometimes with marked overtone.
Habitat: Cloud and dwarf forest, flower gardens (e.g. in Bogotá), and (seasonally?) to open páramo habitat with some bushy growth.
Range: 1900-3300 (mostly 2400-2900) m. Páramo de Tamá on border of Col./Ven. (*tamae*), Perijá mts and both slopes of E Andes of Col. s to Bogotá (*helianthea*). Locally fairly common.

NOTE: A number of old trade skins in BMNH with no collecting data are intermediate between Blue-throated and Golden-bellied S., suggesting a local intergradation or (judging from the constancy of the characters) a distinctive local form.

BUFF-WINGED STARFRONTLET *Coeligena lutetiae* – Plate XXVIII 17a-b

14 cm (incl. 3.5 cm bill). **Dark**. Pale **buff patch on secondaries** appears transparent in flight; however darker cinnamon in Peruvian specimens. **Ad. male** above velvety black with glistening green forehead, below glittering dark green with blue gular patch. Tail bronzy black. **Female** and **juv.** have cinnamon-buff chin and throat, and generally a stronger green sheen, and more bronzy tail. **Imm. male** adopts blue gular patch, but lacks glittering frontlet.
Habits: Territorial. Often chasing around high up, sometimes several together. Feeds low at vegetation borders, hovering and clinging.
Voice: A characteristic nasal *szac* or *eernt* (like Grass-green Tanager).
Breeding: No information.
Habitat: From dense cloud forest and elfin woodlands mixed with bamboo to humid shrubbery in the páramo grassland.
Range: At 2600-3300 m. Col. from Caldas on w slope and Cauca on e slope of C Andes s through both slopes of Nariño to ne Ecu. (Oyacachi, Papallacta, upper Sumaco), and along w Andean slope in Imbabura (?), Pichincha (mts Pichincha, Atacazo, and Corazón), and Azuay (Cajas mts). In Morona-Santiago (Macas, Zapote-Najda mts) and Loja (PN Podocarpus) se slope of Ecu.; Cerro Chinguela in Piura, n Peru (unnamed ssp). Common locally.

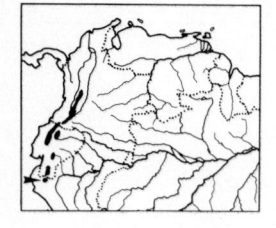

VIOLET-THROATED STARFRONTLET *Coeligena violifer* – Plate XXVIII 12a-d

14 cm (incl. 3 cm bill). Best recognized by **orange-buff tail (white in**

unnamed form from Ayacucho and Apurímac, Peru) with bronzy tip, 1.5-2 cm wide in northern forms to vestigial in southern *violifer*. Dark metallic green, above somewhat bronzy, turning coppery on wing-coverts, and with secondaries more or less rufous at base, and also lower underparts mainly cinnamon, vent uniform cinnamon. **Male** with glittering gular patch (lilaceous in *violifer* to more blue in *dichroura*) and glittering blue-green frontlet. Ssp *violifer* lacks frontlet and has extensively blackish crown, and whitish band indicated across breast. **Female** has partly buff throat (with indications of blue spot). **Juv.** has still more obscure throat, a distinct rictal stripe (Plate XXVIII 12c), and more extensive dark tail-tip.
Habits: In display, male shows butterfly-like pendular flight rapidly back and forth above perched partner. 'Traplines', visiting mainly flowers of *Fuchsia*, *Vriesea*, and esp. *Bomarea* at the periphery of thickets.
Voice: Series of *zwit* calls. In pursuit rapid twittering and *trrt* notes.
Breeding: Nov-Jan (Cochabamba, Bol.).
Habitat: Vine-tangled forest edges or in clearings and in rather open cloud forest (incl. semi-arid types) and second growth, at middle and high levels (often feeding inside canopy).
Range: In small relict forests at the tops of some valleys from Chachapoyas to Lima, w Peru, and commonly at 2800-3300 m (occ. 2000-3700 m) in C and E Andes from Abra Patrícia Amazonas/San Martín through C Andes of Peru, passing s side of Nevada Ampay (Apurímac) to nw end of Cord. Vilcanota (*dichroura*), where grading into populations that continue to yungas of Bol. (*violifer*). At Tambo, S Ayacucho, on n side of Nevado Ampay, and through Cord. Vilcabamba on Apurímac/Cuzco border (unnamed ssp, J Terborgh, JP O'Neill, and our own data). Generally common.

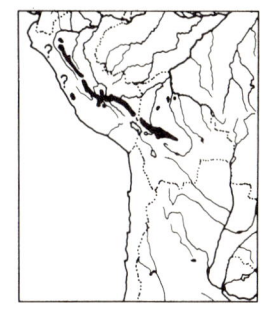

RAINBOW STARFRONTLET *Coeligena iris* – Plate XXVIII 11a-e

14 cm (incl. 3 cm bill). Recognized by extensively **brick-red to chestnut posterior parts**, incl. flight- and tail-feathers. In the most northern form *hesperis* and the southern *eva* the metallic green of the anterior parts extends over the entire back and much of the underparts. In *flagrans* this sheen changes towards copper, but in *fulgidiceps* and particularly *aurora* the iridescence is restricted to the chest, and contrasts with the rather light rufous posterior parts. In **male forehead and crown are fiercely shining and multicolored**, contrasted by black hindcrown and nape: *iris*, *hesperis*, *fulgidiceps*, *flagrans*, and *eva* have yellow/green forehead grading to orange or red on crown (most fiery in *fulgidiceps*), with blue c stripe (lilaceous in *fulgidiceps*). Ssp *aurora* has crown blue-green/deep purplish blue, sometimes golden far back, with concealed purple c spot. Males of *hesperis*, *fulgidiceps*, and *iris* have glittering violaceous gular spot, and occ. males of other sspp may show rather dull gular spots (except *aurora*, which has entire throat strongly glittering emerald). **Female** has much less metallic colors, and more uniform rufous and coppery colors, **juv.** dull gray-brown throat. **Imm. males** maintain a coppery or bronzy sheen on the hind-crown long after the glistening cap is acquired, and thus lack the constrasting black nape of ad.

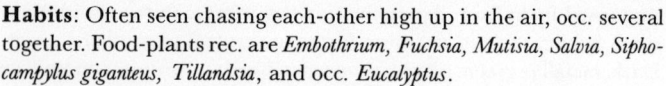

Habits: Often seen chasing each-other high up in the air, occ. several together. Food-plants rec. are *Embothrium, Fuchsia, Mutisia, Salvia, Siphocampylus giganteus, Tillandsia*, and occ. *Eucalyptus*.
Voice: In flight gives distinctive, sharp and strong *tzip*s.
Breeding: Nov-Jan. 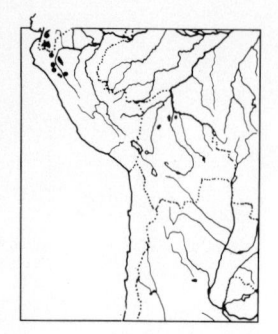
Habitat: From edge of humid or rather dry cloud and elfin forest to gardens and riparian scrub; generally below Buff-winged S.
Range: S Ecu. to n Peru, at 1700-3200 m. W of Cuenca in Azuay, sw Ecu (*hesperis*), and Loja region, s Ecu., to w slope to Huancabamba area of Piura, n Peru (*iris*). W side of W Andes in nw Cajamarca (*flagrans*), Cerros de Amachonga near Cutervo in Cajamarca (*aurora*), Utcubamba Valley in Amazonas (*fulgidiceps*), and Cajabamba and Succha on e slope of W Andes in s Cajamarca and on w slope of C Andes near Tayabamba, La Libertad (*eva*). One sighting (straggler?) of probably this form in Cord. Blanca in Ancash (O. Frimer). Common locally, but habitat much fragmented.

NOTE: Previously placed in genus *Diphlogaena*.

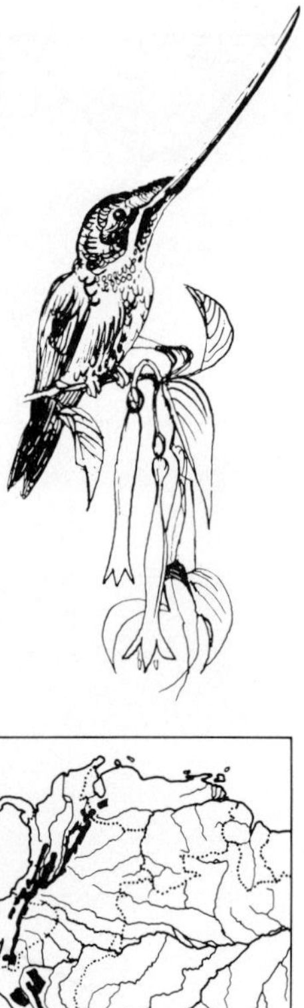

SWORD-BILLED HUMMINGBIRD *Ensifera ensifera* – Plate XXVII 3a-b

C. 22 cm (**incl. 6-11 cm straight bill**, thus with much individual variation). Generally dark green. **Ad. male** approaches copper on head and glittering emerald-green below the dull fuscous throat area, and has dark rather gray belly. Tail blackish blue, forked. In **female** and juv. underparts have buff main color with dense dusky streaks on throat and dense green discs or blurry spots (**juv.**) on the rest, the tail being rather short.
Habits: Like *Coeligena*s a wide-ranging 'trapline-feeder' often seen in long high direct flight between remote feeding sites. Usually feeds in mid or upper levels, hovering or perched right below pendant flowers with long corollar tubes, e.g. *Aethanthus, Brugmansia, Datura, Fuchsia, Passiflora, Salpichroa*, and *Tacsonia* (but also visits small flowers). Perches quietly for long periods, usually with bill sharply raised.
Voice: Humming wing-sound very apparent. Plaintive whistle and a low guttural *trrr*. 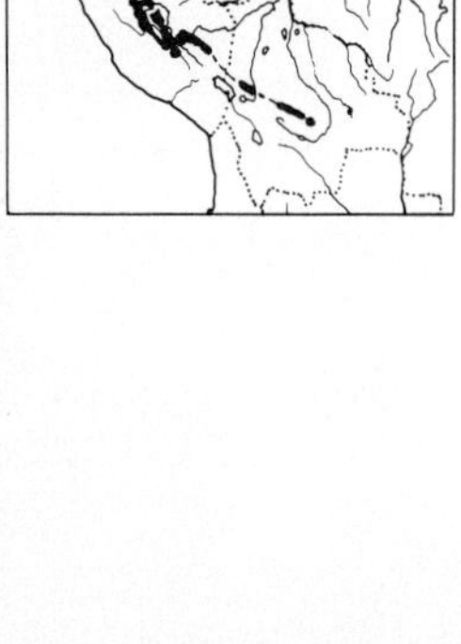
Breeding: No information.
Habitat: Semi-humid to wet forest, and in glades and at forest edge, mostly places with some large vine-tangled trees, but in some areas in low elfin forest, and visits patches of páramo shrub.
Range: At 1700-3500 (but mainly 2500-3000) m from Mérida, Ven. through all Andean ranges of Col., Ecu. and n Peru, and continuously in the e Andean zone s to yungas of Sta Cruz, Bol. Rather local, but not uncommon.

NOTE: Should possibly be included in *Coeligena*. The small bluebreasted ssp *caerulescens*, of unknown origin, may be a valid taxon (see *Hornero* 12(1986):301-2).

GREEN-BACKED FIRECROWN
Sephanoides sephaniodes – Plate XXVIII 5a-c

8 cm (incl. 1.5 cm bill). Compact with rather large head and short tail. Above bronzy green, below pale buff with dense bronzy spots fusing almost to uniform light bronze on sides. In **male, forehead and crown have strongly fiery red/yellow(/green) reflection**. **Juv.** has rusty feather-edges on head, and is quite cinnamon below.
Habits: Rises steeply in high aerial display. Feeds esp. from red *Abutillon*, *Embothrium*, and *Fuchsia* flowers. Sometimes found hanging inert in thick foliage.
Voice: Has a loud and melodious song and high-pitched *tsee-ee* rapidly repeated during interaction. In high vertical display flight long *sirrrrrrrr*.
Breeding: Tiny nest often overhanging water, eggs in (Sep) Oct-Nov.
Habitat: Forest edge and glades, thickets, and gardens.
Range: Chile s of Atacama, and in adjacent Arg. from Neuquén to Sta Cruz. Seen regularly in spring near the Beagle Channel, Isla Grande. Locally to above 2000 m. Migratory in the s, wintering in lowlands of Arg. Juan Fernandez isls. Quite common.

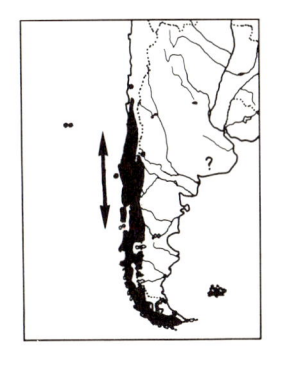

BUFF-TAILED CORONET
Boissonneaua flavescens – Plate XXXI 13a-b

12.5 cm (incl. 1.8 cm virtually straight bill). Metallic green, very glittering golden green on head, and with tawny wing-linings (conspicuous in flight) and small white leg-puffs. Vent and fairly short notched **tail creamy buff** except for dark tip and 2 c feathers, very pale buff and with narrow dark tip in *flavescens*. In **female** and **juv.** throat and rest of underparts are less solidly green with some buff shining through.
Habits: Highly territorial over concentrated nectar sources. Mainly feeds at mid-level and occ. in forest crowns, esp. on *Cavendischia, Disterigma, Palicouria*, and the hummer-adapted melastome *Huilaea*. Often hawks from high perches, and typically clings to flowers, holding wings up, and also holds wings up momentarily when alighting. Occ. clings like a woodpecker to tree trunks, gleaning insects.
Voice: Sometimes several singing from adjacent trees, with series of rapid, sharp *chip*s.
Breeding: May breed Apr-Aug (Col.).
Habitat: Humid and wet montane forest, forest edge and shrub.
Range: At 2-3000 m, sometimes lower. From sw Táchira, w Ven., and in E Andes, Col., and rare or local in e Ecu. (*flavescens*), and in W Andes of Col. and commonly along w Andean slope of Ecu. s to Illiniza (*tinochlora*).

CHESTNUT-BREASTED CORONET
Boissonneaua matthewsii – Plate XXXI 12

13 cm (incl. 1.8 cm bill). Above strongly metallic green. All **underparts, wing-linings, and tail chestnut-rufous** except for numerous yellowish green discs on throat, and bronzy 2 c feathers and tip of tail. In Peru, resembles Violet-throated Starfrontlet, but more strongly rufous below, and with shorter bill. Behavior apparently as in Buff-tailed C. Known mainly from canopy and interior of premontane forest

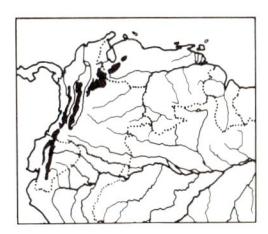

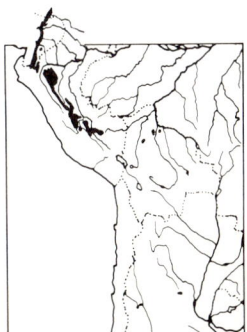

along e Andean slopes from s Col. through Ecu. and Peru to Cuzco, in the Peruvian part sometimes ascending to 3000 m. Fairly common.

Sunangels – Genus *Heliangelus*

Medium-sized, compactly built, with rather short straight bills, rather heavy rounded heads, broad and blunt wings, and variable (broadly rounded, square, or forked) tails. Males and many females normally have broad **sometimes fiercely shining throat gorgets**, and there is a widespread tendency for developing a distinct (buff, white, or metallic) **pectoral bar**. All species seem to be territorial and feed particularly from flowers of melastome shrub. Often feed perched. Sometimes take insects by sallying from favourite perches. Nest well hidden in long strand of moss hanging from tree.

ORANGE-THROATED SUNANGEL *Heliangelus mavors* – Plate XXIX 19

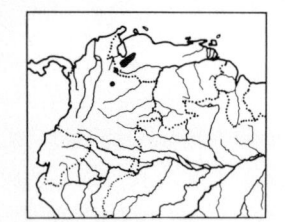

11 cm (incl. 1.5 cm bill). Metallic grass-green with cinnamon-**buff pectoral bar**, and some cinnamon mottling below and buff c belly. **Ad. male** has gold-glittering frontlet and large iridescent **orange/greenish-yellow gorget**. Tail square, bronzy black. **Female** has throat buff and speckled as rest of underparts. Tail square, bronzy black. **Juv.** similar, but males soon become dusky brown on the throat area.
Habits and **voice**: Apparently as in Amethyst-throated S. A red *Salvia* rec. as food plant.
Habitat: Edge of cloud and dwarf forest and open páramo with scattered trees and shrubs, often rather dry and disturbed habitats with scattered second growth only.
Range: At 2000-3200 m. Locally in the n of E Andes, Col., and in Táchira, Mérida and Trujillo to s Lara, Ven. Common locally.

MERIDA SUNANGEL *Heliangelus spencei* – Plate LXIV 4a-b

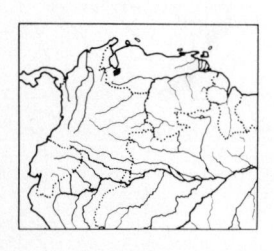

10 cm (incl. 1.3 cm bill). Above shining dark green with crown dusky, glittering **frontlet a peculiar silvery green**, underparts dusky buff with numerous green spots laterally, and with **white pectoral bar**. Tail square, dark bronzy green, lateral feathers with very little sheen. In **male, gorget violet**, often with a distinct coppery sheen, and not strongly glistening. In **female**, throat usually dull brown with some orange discs or bronzy feather-tips. **Juv.** closely resembles juv. Orange-throated S., but has white pectoral bar.
Habits, voice and **breeding**: Appears to be very similar to Amethyst-throated S.
Habitat: At low to mid heigths, mainly at borders of humid cloud and dwarf forest and páramo.
Range: Ven., at 2000-3600 m in c Mérida mts. Common.

GORGETED SUNANGEL *Heliangelus strophianus* – Plate XXIX 20a-b

10 cm (incl. 1.5 cm bill). Resembles Amethyst-throated S. ssp *laticlavius*, but with broader and more purely **white pectoral bar**, and the green-spotted **underparts have dark slaty gray basal color with no indication of buff**. Tail dark steel-blue and more forked. Male has more fiercely rosy gorget.

Inhabits humid and wet premontane forest and thickets, but may locally or casually reach temp. zone (rec. 2725 m). On w Andean slope in nw Ecu. from border to Col. s-wards, recently rec. as far s as Cajas area in Azuay. Common in Pichincha area, but may have declined much due to habitat destruction.

NOTE: The rather coppery, blue-throated *Heliangelus violicollis*, known from 2 specimens, may be a variant of this species.

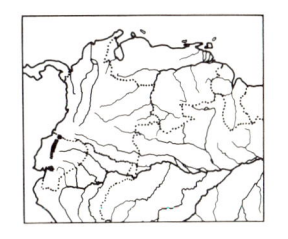

AMETHYST-THROATED SUNANGEL *Heliangelus amethysticollis* – Plate XXIX 17a-d

11 cm (incl. 1.8 cm bill). **Dark metallic grassgreen, with pectoral bar white (*clarissae*), buffy white, or all buff (*amethysticollis*)**, and with vent mostly white. Tail rather short, even in the male, broad and square, with broadly wedge-shaped feather-tips; bluish black, occ. with white corners, and with bronzy c feathers. **Male** has glittering blue-green frontlet (*violiceps* has purplish crown and capri-blue frontlet, *laticlavius* green frontlet). **Throat area varies with age, sex and season**, being cinnamon-orange, sometimes speckled with green in juv., otherwise usually sooty, more or less strewn with glittering discs, or in ad. male often developed as a shiny gorget varying from flashy rosy (*clarissae*) through reddish purple (*laticlavius*, *verdiscutatus*) to violet (*amethysticollis*) or deep violet (*decolor*). Below the pectoral bar, the breast is dark metallic green grading to mostly ochraceous buff on belly (*laticlavius* with rather grayish buff belly below thin buffy-white pectoral bar).
Habits: Holds feeding territories over concentrated nectar resources, but also hawks much for insects. Often perched while feeding on small *Fuchsias* and *Brachyotum*, *Disterigma*, *Guzmania*, and *Palicourea*. Upon alighting, and after each feed, often holds wings raised and quivering for a moment (threat display, seen less often in other sunangels, except Orange-throated and Mérida S.). In display flight rises 5-10 m, bounces up and down and then descends to the same perch.
Voice: A low short and dry *trrr*. During pursuit, gives rapid series of *tzt* notes, sometimes with longer *zhrrrr* inserted. In flight, wings rattle rather than hum.
Breeding: Breeding condition May-Aug (ne Col.); eggs in Mar (Pasco, Peru).
Habitat: At mid-heigth, near edge and often inside vegetation in mossy cloud, dwarf, and elfin forest.
Range: At 1800-3200 m. Col./Ven. border in Perijá mts (*violiceps*) and sw Táchira (*verdiscutatus*), and E Andes of Col. s to Bogotá (*clarissae*). Along e Andean slopes of s Ecu. and in Condor mts. n Cajamarca and Cerro Chinguela in Piura, n Peru (*laticlavius*). In e Andean zone of Peru

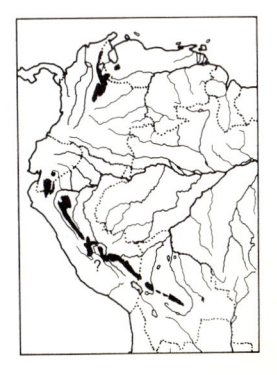

s of the Marañón, probably all the way to Apurímac (*decolor*), and from Cuzco e Peru to yungas of Cochabamba, Bol. (*amethysticollis*). Common and sometimes abundant.

NOTE: A megaspecies, with the northern forms sometimes considered a separate species, **LONGUEMARE'S SUNANGEL *H. clarissae*.**

NOTE: *H. claudia* and *dubius* from 'Bogota' collections may be aberrant specimens of *clarissae*.

TOURMALINE SUNANGEL *Heliangelus exortis* – Plate XXIX 18a-d

11 cm (incl. 1.5 cm bill). **Dark metallic green with glittering emerald-green pectoral bar**, and somewhat gray belly, **vent more or less purely white**. Tail 4 cm, broad and somewhat forked (esp. in *exortis*), steel-blue. **Ad. male** with glittering blue-green frontlet and purplish blue chin grading to a **fiercely reflective gorget** that is **rosy pink** (*exortis*) or **orange-yellow**/green (*micrastur*) to more reddish orange (*cutervensis*). **Ad. female** has throat white (but chin black), sometimes with shining green to dusky spots, or with glittering rosy feather-tips which sometimes give rise to a male-like gorget (great individual and geographic variation, with male-like females dominating in n of range). **Juv.** white-throated, occ. with some buff feathers, and black chin in female. **Imm. male** may have white chin some time after the development of a dark throat with more or less extensive glittering feather-tips.
Habits: Perches on outer side of vegetation, usually low, clinging to flowers, with wings held outstretched.
Voice: Ascending *tschirr-tschirr*. Guttural song of male audible only at close range.
Breeding: Probably Mar-Aug and one rec. May (Col.).
Habitat: At mid-height in interior of dense mossy cloud and elfin forest, often with some bamboo; females in shrubby pasture. 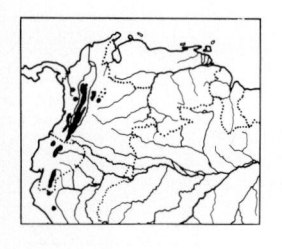
Range: At (1400-) 2300-3400 m. In all Andean ranges of Col., but widespread only in C Andes, and along e slope of Ecu. to w Pastaza and very locally on nw slope of Ecu. (*exortis*). Locally on e Andean slope from e of Cuenca in Morono-Santiago to Loja, se Ecu., and Cerro Chinguela in Piura, extreme n Peru (*micrastur*); Cerros de Amachongo near Cutervo in Cajamarca, nw Peru (*cutervensis*). Ssp *exortis* fairly common, the others uncommon, except locally in Loja.

NOTE: Ssp *micrastur* (with *cutervensis*) form a megassp., previously listed as a full species.

NOTE: Olive-throated Sunangel *Heliangelus squamigularis*, known from Antioquia, Col. and from a few 'Bogotá' trade skins, resembles Tourmaline S., and may be a variant of it, but has glaucous/steel-blue gorget. Other quite similar forms known from 'Bogotá' trade skins are Golden-throated S. ***H. speciosus***, with more golden green gorget; ***H. barrali***, with leaden green gorget; and Glistening S. ***H. luminosus***, with flashy golden green gorget. Rotschild's S. ***H. rotschildi*** is possibly a hybrid with the Purple-mantled Thornbill.

PURPLE-THROATED SUNANGEL *Heliangelus viola* – Plate XXIX 21a-b

11-12 cm, depending on tail length (incl. 1.4 cm bill). **Dark metallic green** with glittering blue-green forehead and pectoral bar, **vent buff spotted metallic green** (unlike in Tourmaline S.). **5-6 cm tail deeply forked**, esp. in male, green centrally, with black outer feathers. **Gorget deep violet**, strongly iridescent in some males, generally less shining in females, but with much individual and seasonal variation. **Juv.** has tawny throat more or less speckled with green.
Habits: Male often on exposed perch. 3 birds displaying in top of dead tree flew around one-another, and raising and quivering wings while perching. Rec. congregating around blooming *Eucalyptus* trees.
Voice: *Tjrrrr* or *tjrrr-tjrrr*, more rolling than voice of Tourmaline S.
Breeding: No information.
Habitat: Wide range, from cloud forest to riparian thickets and *Alnus* wood.
Range: Common at 2150-3000 m along w Andean slope of s Ecu. from Azuay s-wards (and one 1898 specimen labelled 'n side of Pichincha') to Cajamarca, in side-valleys to Marañón valley, n Peru. Fairly common locally.

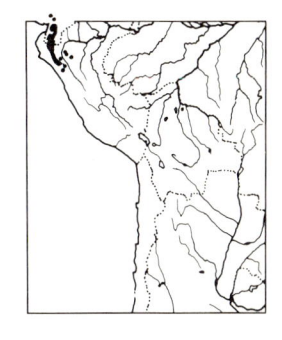

ROYAL SUNANGEL *Heliangelus regalis* – Plate XXIX 22a-b - of subtrop. n Peru (very locally in n Cajamarca and ne of Jerillo in San Martín) – ascends to near lower limit of temp. zone (2200 m) above San José de Lourdes in Condor mts on border to Ecu. Inhabits edge of stunted forest and wooded ravines in very rainy areas with bleak soils. 11-12 cm. **Tail deeply forked**, longest in **totally bluish-black male**. **Ad. female** resembles frail fine-billed and long-tailed female Amethyst-throated S., with buff pectoral band, spotted vent, and totally blue-black tail. **Juv.** similar, with mottled gray instead of green-speckled throat, molt to **Imm. male** plumage starting with appearance of dark blue gorget. Calls comprise high-pitched *tick* notes and a 3 s warble.

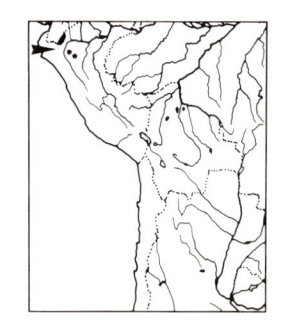

NOTE: Published by D Fitzpatrick, DE Willard & JW Terborgh in *Wilson Bull.* 91(1979):177-186.

Pufflegs, Genera *Eriocnemis* and *Haplophaedia*

Rather small to medium large montane hummers with thin medium long bills, more or less **forked tails, greenish black strongly glittering and sparkling plumage**, and large and conspicuous **white (or black) downy leg-puffs**. The puffs may have a signal function, but a thermoregulatory role has also been proposed. When perched, pufflegs show remarkably raised bills and steeply rounded foreheads compared with other similar-sized hummers. Pufflegs seem generally to be sedentary and territorial, although vertical movements occur. Nests often in caves or in vegetation on steep slopes.

BLACK-BREASTED PUFFLEG *Eriocnemis nigrivestris* – Plate XXVIII 7a-b

9 cm (incl. 1.5 cm bill), 3.5 cm tail moderately forked. **Dark**, with white puffs and purplish **blue gular patch** and vent, leg-puffs white. **Ad. male black with only faint green sheen and dark blue rump. Female (juv.?)** glittering dark bronzy green grading to blue-green on back and rump and golden green below, lacking the admixture of cinnamon-buff seen in Glowing P. Some females show strong golden glittering on neck, but unlike Turquoise-throated P. it has a blue gular spot.
Habits: Prefers flowers of the small rubiad tree *Palicourea huigrensis*, but also feeds (frequently while perched) from flowers of shrubs, herbs, and vines, e.g., *Thiboudia floribunda*, *Disterigma*, *Rubus*, *Tropaeolum*, and *Psychotria* (see *Wilson Bull*. 95(1983):656-61).
Voice: Song a monotonous, metallic *tzeet tzeet tzeet*....
Breeding: No information.
Habitat: Grassy slopes with patches of stunted montane forest with blackberry brambles, ericads etc. Males are high up, on ridges, outside breeding season.
Range: At 2440-3700 (4725?) m, possibly highest in Nov-Feb, males highest. Very locally in nw Ecu., on volcans Pichincha and Atacazo, possibly also in Imbabura (Intag). A 'Bogota' specimen is similar. Only recent rec. from ridge crests (Frutillas, Yanacocha, Cerro Alaspungo, and latest (1983) Cerro Pugsi) on n side of Volcán Pichincha. Apparently near extinction.

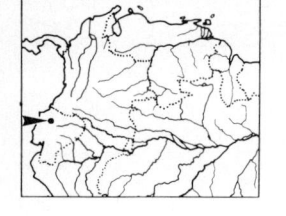

GLOWING PUFFLEG *Eriocnemis vestitus* – Plate XXVIII 9a-b

10 cm (incl. 1.8 cm bill), tail 3.5 cm, or 4.5 in s of range, with 1.5 cm fork. **Ad. male metallic green** (grass-green, to emerald-green below in *smaragdinipectus* and in n Peruvian birds), belly strongly glittering green (except in *paramillo*), and rump very strongly golden green (to greenish yellow in n Peru) when seen from above. **Gular patch** and vent shining purplish blue. Leg-puffs white. Tail blackish blue. **Female** is similar, but with buff malar stripe, and much **cinnamon-buff on throat and breast**, numerous small golden green discs, and incomplete coerulean blue patch. **Juv.** lacks blue gular patch, and juv. male does not have particularly black head.
Habits: Normally darts around, low down in fairly open vegetation, feeding mostly on flowers of low ericads and rubiads, hovering and perched, but also seen visiting *Tillandsia* flowers. May raise wings briefly on alighting.
Voice: Metallic *tzeet* notes.
Breeding: Dec (S Col.). Nest often in grass stands.
Habitat: Mainly in ecotone habitats in humid páramos with heather-like shrubs (e.g., *Hypericum* and *Pernettya*) and along borders of dwarf forest and stunted cloud forest, but occ. low in taller forest.
Range: Mainly at 2250-3850 m from Mérida and Táchira, nw Ven., through E Andes to Cundinamarca, Col. (*vestitus*), Paramillo and Frontino in n of W Andes, Col. (*paramillo*), Páramo de Sonson in Antio-

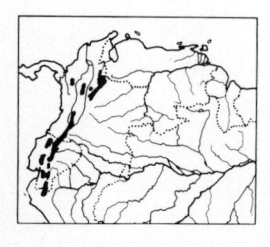

quia and s Huila in C Andes, through e Nariño, Col., and locally along e Andean slope and in w Azuay (río Mazan), Ecu. to Cerro Chinguela in Piura, extreme n Peru (*smaragdinipectus* and unnamed ssp in the s). Common.

TURQOUISE-THROATED PUFFLEG *Eriocnemis godini* – Plate XXVIII 4a-b – is known definitively only from hot ravines at 2100-2300 m in Guayllabamba, nw Ecu. However, old unconfirmed statements, a specimen labelled 'Tungaragua' (LSUMZ), and its representation in 'Bogota' collections suggest occurrence in temp. zone. Possibly s of Pasto in s Nariño. 11 cm (incl. 1.5 cm bill), 3.5 cm tail slightly forked. Resembles Glowing P. male, but **male** has prominent **golden glittering on throat and side of neck**, and not distinctly outlined blue gular spot. **Female** is bronzy green (coppery gold/deep emerald green) with strongly glittering golden belly. Mid-throat emerald green without blue spot or any cinnamon color.

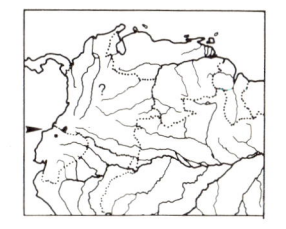

NOTE: Hard to separate from Glowing P., and not definitely an authentic species.

SAPPHIRE-VENTED PUFFLEG *Eriocnemis luciani* – Plate XXVIII 8a-c

12-14 cm (incl. 2.2 cm bill), **blackish blue tail varying from 5.5 cm with 3 cm fork in *luciani* to 4.5 cm with 2 cm fork in *marcapatae*)**, feather-tips blunt. Metallic grass-green, glittering golden green around neck, vent shining violet (*luciani*, blue (*catharina*), or purplish blue (*sapphiropygia*). Puffs white. Ssp *luciani* has blue forehead, *sapphiropygia* and to some extent *marcapatae* coppery back of head. Females of *catharina* and *sapphiropygia* have white stripe on c belly.
Habits: Darts around low at vegetation borders. Often feeds perched, from a wide variety of flowers from tiny terrestrial ones on grassy slopes to *Barnedesia*, *Embothrium*, mistletoes, and large *Bomarea* and *Siphocampylus* flowers at lush forest edge.
Voice: Sharp *tirr tirr* noted.
Breeding: Rec. in Feb (above Quito).
Habitat: Mainly low, at edges and in glades of tall wet montane forest to bushy and grassy páramo slopes.
Range: Mainly at 2000-3500 m, but even higher in Peru, and seasonally sometimes down in trop. zone. From Ipiales in Nariño, s Col., through Andes of Ecu. (*luciani*), Utcubamba valley in Amazonas, n Peru (*catharina*), Mashua in e La Libertad (intergrade), e Andean slope locally from Pasco and Junín to Apurímac/Cuzco border (*sapphiropygia*), grading into population of Cordillera Vilcanota to Valcón in Puno, e Peru (*marcapatae*). Locally very common.

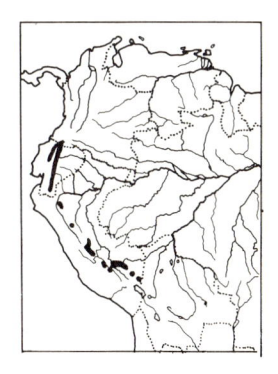

COPPERY-BELLIED PUFFLEG
Eriocnemis cupreoventris – Plate XXVIII 10

9.5 cm (incl. 1.8 cm bill), 3.5 cm **deeply forked tail with acute tips**.

Metallic green with glittering underparts varying from emerald-green on breast to **golden coppery on belly**. Leg-puffs white, vent purplish blue, tail blackish blue. **Juv.** dark, with green throat, and totally black breast and belly with no copper, and with red base of mandible. 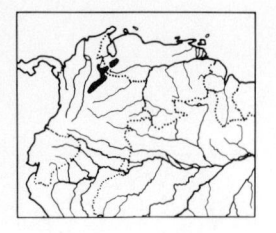
Habits: Territorial, around flowering *Cavendischia*, *Palicourea*, and *Pernettya* shrubs. Clings momentarily to flowers to feed.
Voice: High-pitched *tee teee* noted.
Breeding: Sep-Jan. Nest large, often in dense vegetation.
Habitat: Montane forest borders and páramo vegetation with scattered, low shrub.
Range: At 2-3000 m. From n Mérida mts, nw Ven., s to Cundinamarca in E Andes, Col.

NOTE: May form a superspecies with the following species. A possible melanism of this species has been described under the name *E. dyselius*. Another variant of possibly this species is more gray-brown.

GOLDEN-BREASTED PUFFLEG *Eriocnemis mosquera* – Plate XXVIII 3

13 cm (incl. 1.7 cm bill), **5-6 cm long, dark green tail with 3 cm fork, narrower** and with more pointed tips than in Sapphire-vented P. Metallic grass-green with neck and **breast glittering golden bronze**, leg-puffs white, **vent olivaceous to dull dark gray**.
Habits: Active and aggressive. Feeds hovering or clinging at low flowers on the edge of dense shrub.
Voice: *Trit* notes.
Breeding: No information.
Habitat: Elfin forest and bushy clearings in stunted montane woodland.
Range: Known mainly from 1600-3600 m, but seasonally down to the tropics. In C Andes and s part of W Andes, Col., and in nw Ecu. Common only in Pichincha.

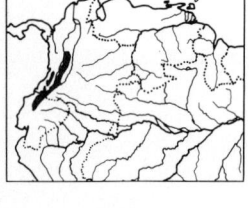

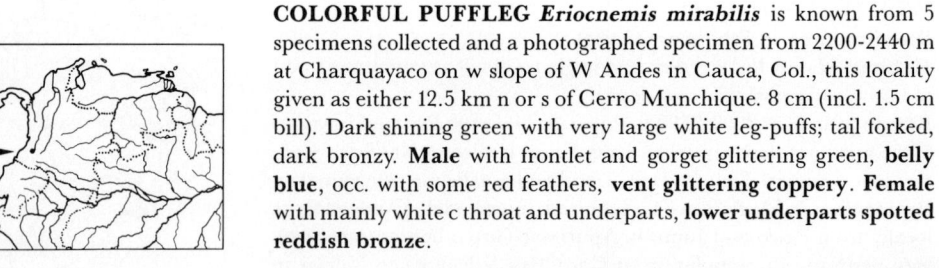

COLORFUL PUFFLEG *Eriocnemis mirabilis* is known from 5 specimens collected and a photographed specimen from 2200-2440 m at Charquayaco on w slope of W Andes in Cauca, Col., this locality given as either 12.5 km n or s of Cerro Munchique. 8 cm (incl. 1.5 cm bill). Dark shining green with very large white leg-puffs; tail forked, dark bronzy. **Male** with frontlet and gorget glittering green, **belly blue**, occ. with some red feathers, **vent glittering coppery**. **Female** with mainly white c throat and underparts, **lower underparts spotted reddish bronze**.

NOTE: This very distinctive species was published by R Meyer de Schauensee in *Notulae Naturae* 402(1967):1-2.

EMERALD-BELLIED PUFFLEG *Eriocnemis alinae*

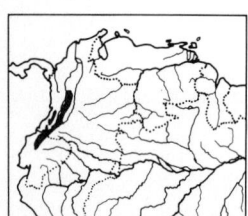

7.5 (*alinae*) to 9 cm (*dybowskii*). Strongly shining green, rump brighter, vivid emerald-green, frontlet (**male**) and all underparts to vent strongly glittering vivid emerald-green/blue-green, **breast with large white patch** centrally dotted with some green discs. **Short forked tail dull green**. White leg-puffs very large. Female lacks frontlet, and has bluish sheen below and short tail-fork.

Inhabits glades in cloud forest ridges (oak-dominated in Col.), mainly in upper premontane zone, but occ. to 2800 m both in the n and the s. Locally in E Andes and around head of Magdalena valley in Huila to Cauca, Col., and in se Ecu. (*alinae*), and along e Andean zone to Cuzco, Peru (*dybowskii*).

BLUE-CAPPED PUFFLEG *Eriocnemis glaucopoides* – Plate XXVIII 2a-b

9.5-11 cm (incl. 1.6 cm bill), 4.5 cm tail of male forked 1.5 cm. **Ad. male** black with green glittering and strongly shining **light blue forehead**. Leg-puffs white, vent purplish blue. **Imm. male** lacks blue of forehead. **Female with underparts brightly cinnamon-buff** except on sides and vent, which are strongly spotted with green. Only puffleg in its range.
Habits: Darts around low, at vegetation borders.
Voice: High-pitched *zee-zee*.
Breeding: Nest rec. from Nov, juv. males Dec and Feb (Cochabamba).
Habitat: Rather low in dense shrubbery and forest edge, esp. in cloud forest and deciduous 'barranca' forest with grazed glades, but requirements hardly known.
Range: At 1500-2900 m and probably also higher, from humid slope above Cochabamba city and in yungas and valles of Cochabamba, Bol., to Tucumán, nw Arg. Rare or local.

BLACK-THIGHED PUFFLEG *Eriocnemis derbyi* – Plate XXVIII 1a-b

11 cm (incl. c. 2 cm bill – longest in *longirostris*), 2.5 cm tail slightly forked. **Ad. male** metallic green with emerald to golden green glittering, upper tail-coverts and **vent strongly glistening malachite-green**, **puffs and tail black**, outer corners of tail very acute. **Female** and **juv.** more golden glittering, with slight white barring on throat, and with grizzled puffs.
Habits and **voice**: Probably as in Glowing Puffleg.
Breeding: Large gonads in Feb.
Habitat: Humid forest borders and bushy pasture and ravines.
Range: At (2500) 2900-3600 m in C Andes, Col. (*longirostris*), and in Nariño, s Col., and both slopes of páramo de Anhíl (Angel), Carchi extreme nw Ecu. (*derbyi*). Uncommon and local.

NOTE: Söderströms Puffleg *Eriocnemis soderstromi* from Pichincha is possibly an hybrid, or at least not a valid taxon. Also a number of other forms described on basis of 'Bogotá' specimens are probably not valid taxa. 3 specimens originally described as *E. isaacsonii* may be hybrids between *Eriocnemis* and *Coeligena* species.

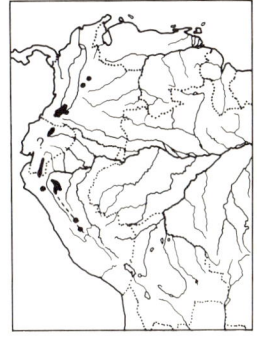

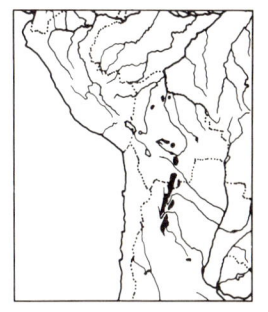

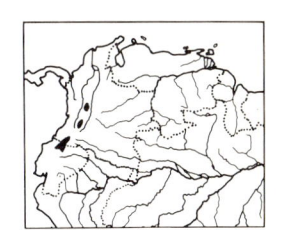

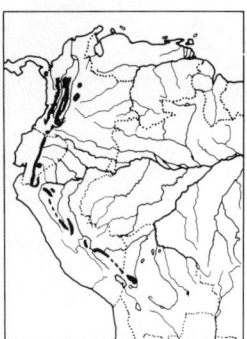

GREENISH PUFFLEG *Haplophaedia aureliae*

10.5 cm. Duller than *Eriocnemis*, and with smaller puffs (partly cinnamon in male). Green, but back of head and rump more or less coppery, tail slightly forked and blue-black, underparts dull green and usually inconspicuously scaled grayish white. Several sspp poorly differentiated and in need of revision; populations with very coppery upperparts and gray-scaled underparts occur disjunctly in Cauca area, w Col. and Cutucú mts and further s in e Ecu.

Inhabits understory of humid premontane forest, but wanders seasonally to lower temp. zone at least in Col. (rec. 3100 m). All Andean ranges of Col. through Ecu, and along e Andean zone of Peru to yungas of Bol. Not uncommon.

H. aureliae

HOARY PUFFLEG *Haplophaedia lugens* of humid and wet premontane forest of sw Col. and nw Ecu., mainly below the Greenish P., has been rec. occ. to 2500 m in w Nariño. Resembles a Greenish P. with somewhat coppery crown and rump, but **underparts dark gray, decidedly scaled grayish white on throat and breast**. Tail almost square, bluish black.

BOOTED RACKET-TAIL *Ocreatus underwoodii* – Plate XXX 14a-g

Tiny, but 8-13 cm depending on varying tail lengths (bill 1.3 cm). **Large leg-puffs tawny** (*peruvianus, annae, addae*) or **white** (other sspp). **Ad. male** dark metallic green with large glittering emerald-green gorget. Narrow tail with **2 outer feathers 5-8 cm long partly bare-shafted, ending in 1 cm-wide steel-blue rackets** (somewhat variable, fairly small disks in *ambiguus* and *melanantherus*, much smaller and with shafts crossing each-other in *annae* and *addae*). Ad. males of *addae* have white-spotted breast and belly (immature-looking). **Ad. female** has white underside densely speckled with green dots (c underparts unspotted white in *melanantherus*, and with few spots in *discifer* and *ambiguus*), vent buff. A thin white line separates dark cheek from dark **rictal stripe**. Tail only 3 cm long, narrow, and deeply forked, blue-black with white feather-tips. **Juv.** similar, with 2.5 cm tail, and recognizable in much of the range by lack of green spots on white underparts (female) or on throat (male).

Habits: Usually seen briefly, as it darts back and forth and suddenly disappears. Flight wavering and bee-like as in woodstars. Holds wings outstretched a few s after alighting. Often bobs tail when perched and opens and closes it when hovering, and male in aerial display swings tail 180° up and down (which produces strong sounds). Holds feeding territories. Perches low, but feeds in canopy on many kinds of tubular flowers, e.g., *Clusia, Inga* and other legumes, and from smaller flowers of *Buddleia* and *Palicourea*. Often clings to the flowers.

Voice: Frail twittering. Produces distinctive wing hum, like woodstars.

Breeding: Probably most of year, but seasonal in certain areas.

Habitat: In mid story of wet and humid montane forest, low in glades (incl. roads), and in damp shady ravines with tall trees.

Range: Mainly in premontane zone, but seasonally to lower temp. zone, with larger amplitude (to 3100 m) at least in W Andes, Col. Coastal and interior range of Miranda to Carabobo (*polystictus*) and in Falcón through Mérida and in Perijá mts to Táchira, nw Ven. and e Norte de Santander, Col. (*discifer*), through E Andes (*underwoodii*), and C and W Andes of Col. (*ambiguus*), w slope in Nariño, s Col., and both Andean slopes of Ecu. (*melanantherus*). From w Cutucú mts, se Ecu., spottily along e Andean slope to Huánuco, c Peru (*peruvianus*), locally from Pasco to Cuzco, Peru (*annae*), and patchily in yungas of La Paz to Sta Cruz and in Chuquisaca, Bol. (*addae*). Generally uncommon.

NOTE: Has sometimes been treated as 3 separate species, *O. underwoodi*, *O. peruvianus*, and *O. annae*, defined by color of leg-puffs and whether tail-feathers cross each-other or not.

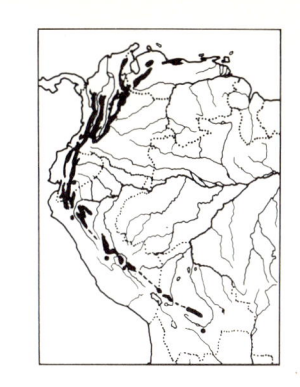

BLACK-TAILED TRAINBEARER *Lesbia victoriae* – Plate XXX 19a-d

Tiny, but 13-26 cm, depending on tail length (bill 1-1.5 cm, longest in the n, shortest in *juliae*). **Tail extremely forked with narrow lateral streamers** varying from 6 cm in juv. to 16.5-18 cm in ad. male of *victoriae*. **Ad. male** shining bronzy green, with glittering emerald-green **gorget with V-shaped lower outline**, slight leg puffs, and tawny c belly and vent, **tail-feathers black with slight bronzy feather-tips**. **Female** with underparts buffy white densely strewn with green dots and rows of tiny speckles on throat, streamers black with narrow white outer webs. **Juv.** similar, but with greener, more diffusely marked sides, and **juv. male** with larger spots on throat. Resembles Green-tailed T-b., but streamers generally longer, and tail-feathers have very little bronzy gloss near tip. **Bill slightly curved** and usually longer, but *berlepschi* has small bill almost identical to that of sympatric Green-tailed T-b. Position of lower point of gorget is then a useful character, even in imm. male (see Plate).
Habits: Aggressive and territorial. Streamers often spread, but in rapid flight curve back distally. In high display flight the streamers are widely spread, and the wings produce canvas-ripping sound. When perched, the tail is frequently bobbed. Feeds at mid and high levels, esp. from yellow flowers of leguminaceous bushes and bignoniads, but also visits puyas, and takes insects in composite flowers.
Voice: Fine *tick* notes, call a thin *zeeet*, song *ti ti tit tttrrrr tic tic tic*.
Breeding: Tiny nest often suspended from thin twig among rocks. Breeds in Sep-Oct (Bogotá), June-Aug (Nariño), and Oct-June (Quito).
Habitat: Not too humid bushy ravines and slopes with forest edge, bushes, gardens, *Polylepis* wood, or at least half-meter tall grassy, herbaceous, and bushy growth.
Range: In semi-arid to semi-humid Andean valleys, at 2600-4000 m. E Andes s to Bogotá, Col. (*victoriae*); Nariño, s Col., and tablelands of Ecu. (*aequatorialis*), n Peru from Amazonas s to n Lima and Huánuco (*julia*), and locally from Tarma in Junín to Cuzco, se Peru (*berlepschi*). Common and locally abundant, e.g., in outskirts of Bogotá and Quito cities.

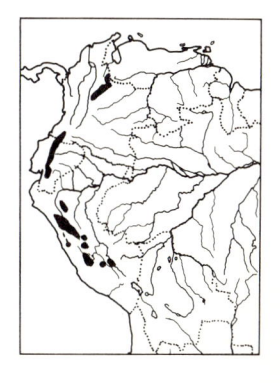

NOTE: Occ. intermediates between Black-tailed and Green-tailed T-b. may be hybrids, but normally the 2 are fully compatible.

NOTE: Probable hybrids of Black-tailed T-b. with Long-tailed Sylph and Purple-mantled Thornbill have been described from Col. and Ecu. as *Metallura purpureicauda, Zodalia glyceria, thaumasta*, and *ortoni*.

GREEN-TAILED TRAIN-BEARER *Lesbia nuna* – Plate XXX 18a-d

Tiny, but 11-16 cm, depending on tail length (bill 0.7-1.4 cm). **Tail long, extremely forked, with narrow lateral streamers** varying from 5.5 cm in **juv.** to 10 cm in **ad. male** of northern forms, 12 cm in *nuna*, 14 cm in *eucharis*. Differs from Black-tailed T-b. by less extreme tail and extensively glittering **emerald-green distal parts of all tail- and rump-feathers**, more round gorget not tapering towards mid-breast (a useful character in **imm. male**, see Plate XXX 18d and 19d), green-spotted vent, and usually **small straight bill**. However, the bill length increases s-wards, *nuna* almost matching sympatric Black-tailed T-b. by its 1.23-1.5 cm slightly curved bill and fairly long tail. Also pattern of tail varies geographically: seen from above, the closed tail is totally green, except on the exposed part of the outer feather (northern forms), or with some black visible on 2nd longest feathers (*eucharis* and esp. *nuna*); white lateral edge of tail extends beyond tip of 2nd longest feather, except in *nuna*.

Habits: Aggressive, but submissive to Black-tailed T-b. Feeds from *Buddleia, Castilleja, Cavendischia, Cuphaea, Rubus* etc., and occ. makes aerial sallies for insects. Aerial display differs from that of Black-tailed T-b. by a terminal zig-zag flight in front of perched partner.
Voice: A very buzzy *bzzzt* reported.
Breeding: Nests Nov-Apr (Quito area).
Habitat: Bushy slopes, esp. with composite shrub, bignoniads, and *Cassia* bushes. Also patches of second growth and locally in open *Polylepis* woodlands. Usually lower and in drier areas than Black-tailed T-b., but occ. in humid forest, where it lives in the treetops, or in bamboo-covered landslides.
Range: At 1700-3800 m, but in most areas below 2800 m. Mérida, nw Ven. (1 old rec.) and in Boyacá and Cundinamarca in E Andes of Col. (*gouldii*); C Andes in Cauca and e slope of Nariño, s Col., through Ecu., esp. in s parts, and quite low (*gracilis*); n Peru on both sides of Marañón valley and in n Lima on w slope (*pallidiventris*), near Huánuco, c Peru (*eucharis*); and Andean valleys from Huancavelica to Cuzco and in Mapiri canyon and (in humid habitat) at Charazani in La Paz, Bol. (*nuna*). Recently rec. in Cochabamba, Bol. (ssp?). Fairly common locally.

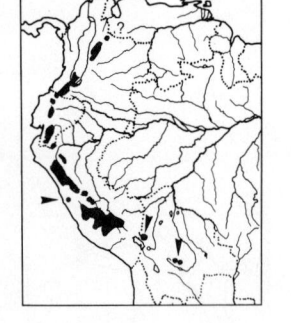

RED-TAILED COMET *Sappho sparganura* – Plate XXX 17a-c

12-19 cm, depending on tail length (incl. 1.7 cm slightly curved bill). Tail deeply forked, the **rather broad and blunt streamers** varying from 5 cm in juv. to 10 cm in ad. male. **Ad. male** shiny bronzy green with V-shaped glittering emerald-green gorget, much of back and **rump purple. Tail strongly glistening orange-red/golden** to green against light (*sapho*), **or purplish rosy**/golden (*sparganura*), with velvety black spots near the blunt feather-tips, and black underside with silky golden sheen. **Female** with pale buff underparts finely spotted with green; moderately long tail without black spots, and with white lateral edge. **Juv.** female-like, but with bronzy green back, only slightly coppery rump, and with white of outer tail-feathers invading inner web distally and along the shaft.

Habits: Submissive to Violetears and Hillstars. Hovers and sits in very upright position. Often bobs tail when perched or agitated. Feeds hovering, esp. from flowers of *Castilleja*, *Salvia*, and flowers with longer corollas, as *Nicotiana*, *Tripodanthus* mistletoes, and *Lamourouxia*.

Voice: When feeding, often gives unmelodic *tjrrrt*. From exposed perch gives rather harsh *tsha* or *zack* notes, which may be monotonously repeated at more than 1 s intervals (compare faster song of Sparkling Violetear, which often lives in the same places). Gives faint high-pitched warble.

Breeding: In cliffs and steep ravines. Eggs known from Apr, fledglings Apr, June (Bol.), or eggs Nov (Córdoba, Arg.).

Habitat: Rather dry slopes with scattered trees and some bushy or herbaceous cover and in gorges with dense tangled scrub or semi-arid deciduous forest.

Range: At 2-4000 m, but winters much lower in the s. Mapíri and La Paz canyons in La Paz, nw Bol. (*sparganura*). E Andean slopes from s of Tunari range in Cochabamba, Bol., to Mendoza and n Neuquen and on isolated mountains of Catamarca, San Luis, and Córdoba, c Arg., reaching Chilean territory in the Portillo pass (*sapho*). Generally fairly common.

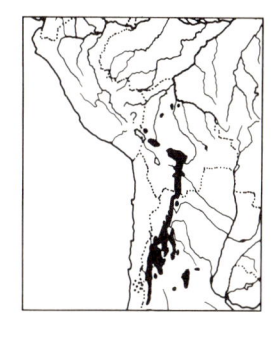

BRONZE-TAILED COMET *Polyonymus caroli* – Plate XXX 16a-b

11-13 cm, depending on tail length (incl. 2 cm slightly curved bill). **Dark bronzy green. Ad. male** with large, rosy violet gorget, 6 cm tail deeply forked, with rounded tips, steel-blue with white lateral edges. **Ad. female** with glittering orange gorget, somewhat gray belly, the slightly shorter and less forked tail with bronzy green c feathers. **Juv.** rather pale, mottled below with scattered golden discs and some green spots on throat, or more rosy discs in male; tail rather short, and only slightly forked, and with white tips on the outer feathers.

Habits: Apparently submissive to Violetears and Hillstars. Shy, lives rather inconspicuously and low in dense bushy growth. Seen feeding from flowers of *Phrygilanthus* mistletoes.

Voice: No rec.

Breeding: Nov-Dec.

Habitat: Semi-arid shrubby slopes with dense shrub, scrubs, cacti, or agaves, and near small montane woods.

Range: Peru; at 1500-3600 m on w Andean slopes and in Andean valleys from Cajamarca to w Arequipa on w slope, and in high side valleys to the Marañón valley and s to Huánuco, and at Yauli in Huancavelica. Locally common.

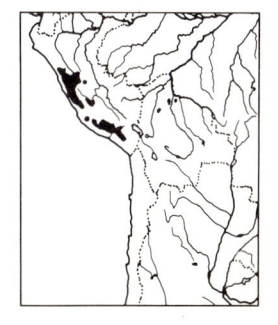

PURPLE-BACKED THORNBILL
Ramphomicron microrhynchum – Plate XXIX 14a-c

C. 8.5 cm (incl. **0.5 cm bill, smallest of any hummingbird**). Head pointed, with flat forehead. **Ad. male above deep metallic purple**, below bronzy green with glittering greenish yellow gorget of V-shaped outline. 4.5 cm tail with rather broad feathers with rounded tips, black. **Ad. female** above metallic green, some females with white on lower back, below white with dense green discs, except on belly; tail 4 cm, not very deeply forked, purplish black, outer feathers broadly tipped white. **Imm. male** with partly purple dorsal parts, but shows white-tipped tail (longer than in juv. and ad. female). Vent white (or rich buff in *microrhynchum*) with coppery feather-centers (small in *albiventer*), or all dark (*bolivianum*, which also has steel-blue tail). **Juv.** female-like, but more buffy below, and with tawny rump.
Habits: Flight bee-like and weaving. In the n Andes mainly feeds on insects in flowers, e.g., of *Espeletia*s, in canopies, by gleaning from foliage, and by hawking. In Peru always seen feeding near the tops of the highest (flowering) trees in stunted forest along ridges – higher up than other hummers in this habitat. Feeds hovering, on outer surface of vegetation. Sometimes many assemble near flowering trees.
Voice: Long *ti, ti, ti...* series. Song weak.
Breeding: Possibly May-Sep in Col., fledgling Dec. (Ecu.).
Habitat: In the n Andes at borders of (and occ. inside) humid montane forest of all kinds to open páramo grassland with espeletias. In Peru rec. along ridges with stunted cloud forest with many ecotones and some taller trees flowering.
Range: At 1700-3400 m, but above 2500 m most of year. Very local in Mérida (*andicolum*), and Táchira, nw Ven., and in all Andean ranges of Col. and Ecu. to the n Peru low in Piura (*microrhynchum*), and s of Porculla pass from Chota s through the C Andes in Cajamarca, La Libertad, and very locally in not too humid areas in Huánuco, Pasco, Junín, Apurímac, and Cuzco (*albiventer*). Rec. at Cocapata, Cochabamba, Bol. (*bolivianum*, published by KS in *Bull. Brit. Orn. Club* 104(1984):5-7). Generally uncommon.

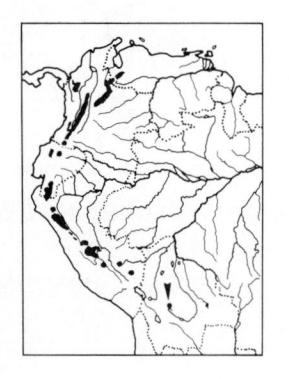

BLACK-BACKED THORNBILL
Ramphomicron dorsale – Plate LXIII 3ab

9-10 cm (incl. **0.5 cm bill**), with forked tail. **Ad. male above velvety black** with purplish red gloss on upper tail-coverts, below dark gray with green spots and glittering golden olive gorget with V-shaped outline. 5 cm tail bluish black, the long lateral feathers exceptionally broad towards tip. **Female** as in Purple-backed T., but whiter below, and with coppery upper tail-coverts (this latter character also seen in odd females of Purple-backed T.). **Juv.** similar, but **imm. male** has large black tail with white feather-tips.
Habits and **voice**: Much like Purple-backed T. Often exposed on top of canopies.
Breeding: No information.
Habitat: Borders of humid forest and elfin woodland to páramo grassland, and seasonally up to the snow-line.

Range: N Col. in Sta Marta mts, at 2000-4600 m, lowest in the dry season Jan-Mar. Uncommon.

Metaltails – Genus *Metallura*

Rather small hummers of high Andean shrub and forest. With rather short bills and very broad-feathered tails incessantly spread and folded in flight. Metallic colors of tail change according to angle of light. Glittering gorgets are usually quite narrow.

Feed on nectar and on insects, perch-gleaning and flying. During fluttering foraging, thoroughly search outer surfaces of canopies of trees and bushes, sometimes also interior parts of canopies. Otherwise usually fly low. Nest suspended from overhanging vegetation on escarpment, or in small cave.

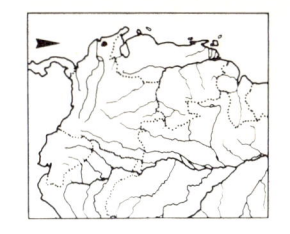

BLACK METALTAIL *Metallura phoebe* – Plate XXIX 4a-b

13 cm (incl. 1.5 cm bill). **Purplish black** with narrow glittering emerald-green gorget (incomplete in **female**, and lacking in **juv.**) and small whitish leg-puffs. White spot behind eye conspicuous. **Tail dark coppery purplish**/coppery golden, most strongly glistening below.
Habits: Usually open to view, and sometimes flies high. Foraging flight a butterfly-like fluttering, and often wriggles tail in flight, and flicks tail and wings while perched. Feeds hovering from a variety of bushes, e.g., flowering *Mutisia*, herbs, as *Cajophora* and *Loasa*, or from smaller flowers (occ. tiny *Ludwigia* flowers in matted grass), and sometimes in canopy of trees, e.g. *Eucalyptus*, and *Polylepis* with flowering *Tristerix* mistletoes.
Voice: Ticking, dry *th th th ...*, *drrt drrt...*, and *ti tyrrdi trr thr thr* noted, and fine *zeeep*s when chasing. Song a fine *zit* or *tzee* repeated every 1.5 s.
Breeding: Eggs probably July-Dec and Jan.
Habitat: Semi-arid open montane woodlands (e.g., *Polylepis*), vegetated canyons, and bushy and shrubby slopes.
Range: At 1800-4300 m. Along w Andean slope of Peru from Cajamarca to Moquegua, and along Ríos Santa and Marañón, upper Huallaga in Huánuco and Mantaro Valley (Concepción) in Junín, Peru. Possibly into extreme n Chile, and 2 19th century specimen labelled Bol. Fairly common.

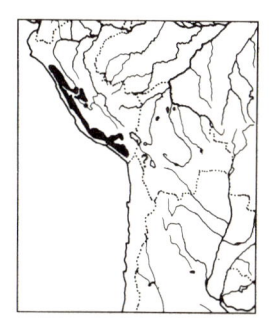

VIRIDIAN METALTAIL *Metallura williami* – Plate XXIX 5a-d

10 cm (incl. 1.4 cm bill). **Ad. male** dark metallic green, approaching bronze on side of neck, and with **throat glittering emerald or golden green** (or black in the middle; *atrigularis*). **Female** buff below, densely spotted with green, esp. on breast and sides. **Tail above dark purplish blue (*williami*) or green shot with blue**; below silky purplish blue (*williami, recisa*) or golden bronzy as in a Scaled M. (*primolinus*), in **female** and **juv.** with whitish tips on outer feathers.

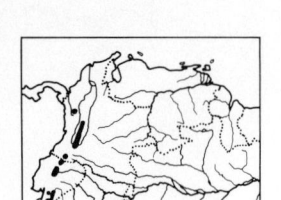

Habits: Perches on exposed twigs, both low and high. Feeds mainly from ericaceous shrub and melastomes.
Voice: Series of *zeeen* notes and fine, sharp *trrrt* calls. Song short, hurried series of *zee* notes.
Breeding: Possibly Feb-Mar (Col.), several nests found in loose colony over mountain torrent.
Habitat: Páramo grassland with espeletias or pine plantations and humid montane dwarf forest and elfin forest clad with mosses and lichens.
Range: Rather common at (2100-) 2900-4000 m, male high up outside breeding season. Isolated e of Popayán in W Andes, Col. *(recisa,* published by A Wetmore in *Proc. Biol. Soc. Wash.* 82(1970):767-776). Both slopes of C Andes from Tolima to Cauca, Col. *(williami)*. Nariño, s Col., and into ne Ecu. *(primolinus)*. Morona-Santiago, Loja, and Azuay (Sigsig, río Mazan), s Ecu. *(atrigularis)*.

NOTE: Together with Violet-throated, Neblina, Coppery, Fire-throated, and Scaled M. forms a superspecies with a checkerboard pattern of character variation, or a leap-frog pattern, as ssp *primolinus* of the Viridian M. resembles the southern Scaled M. closely. Recently, Viridian and Violet-throated M. have been found to be syntopic (segregated, with the latter species highest up) in Río Mazan in Azuay, Ecu.

VIOLET-THROATED METALTAIL *Metallura baroni* – Plate XXIX 6a-b

10 cm (incl. 1.3 cm bill). Very dark and **uniform bronzy green**, **male** with an often partly disintegrated **violet gorget**. Tail above shining olivaceous violet-purple, below silky green. Female more mottled below, with only centers of feathers bronzy green, the margins tan.
Habits: Seen feeding from *Durantha, Macleanaia, Tristerix* mistletoes, and from red *Castilleja* flowers near the ground.
Voice and **breeding**: No information.
Habitat: At edges of moist elfin forest, and in open *Polylepis* wood with boulders drapped with bromeliads, ericads, ferns and mosses.
Range: Sw Ecu. at 3150-3650 m. Known from Mihuir on w side and Río Mazan on e sides of Cajas plateau near Cuenca, Azuay. A reported sighting at 1900 m near Oña. May be fairly common in treeline habitats, but apparently very local.

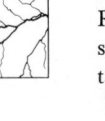

NEBLINA METALTAIL *Metallura odomae* – Plate XXIX 7a-b

9 cm (incl. 1.4 cm bill). **Ad. male** shining dark bronzy green with quite coppery wing-coverts, cheeks and vent, but lighter underparts than in other allospecies. **Gorget reddish purple**. Has slight or hardly visible scaled pattern caused by tan subterminal bars on the abdominal feathers and vent. Tail golden green, shot with blue above. **Female** similar but tan below, extensively spotted with bronzy, and with scarlet discs all over throat.
Habits and **voice**: Probably as in Viridian and Coppery M. Seen feeding at flowers of *Brachyotum*, a *Berberis*-like shrub, and white flowers of

dwarf ericaceous shrub. A rather loud *seet-seet-seet-ti-tttt* reported.
Habitat: Pajonal of humid tree-line type with forested ravines and patches of elfin forest bordering against grassy slopes.
Range: N Peru, at 2850-3350 m on Cerro Chinguela near Huancabamba deflection, n Piura. In small numbers.

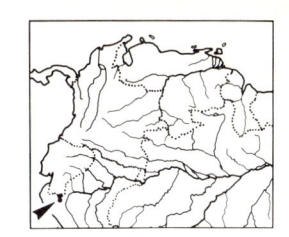

COPPERY METALTAIL *Metallura theresiae* – Plate XXIX 3a-b

10.5 cm (incl. 1.3 cm bill). Dark, **deep purplish coppery**, esp. on head (or with green crown, *parkeri*). Throat with narrow, glittering **golden-green gorget** (incomplete in female, lacking in **juv.**). Tail dark blue above, with underside golden green shot with purple (*parkeri*), or dark and dull bronzy shot with blue.
Habits: Usually perched in bush tops. Fluttering feeding, mostly from melastome flowers (*Brachyotum, Macleania*). In courtship, male repeatedly bounces 1 m up and down, or alternates rapidly between 2 perches in front of female.
Voice: Series of frail *zeee* calls, often alternating with *ttrrrt* notes.
Breeding: No information.
Habitat: Open glades with melastome shrub and ericads in humid dwarf and elfin forest clad with mosses and lichens, and shrub bordering boggy grassland.
Range: N Peru, at 2900-3700 m along e ridge of Andes. Cord. de Colán in Amazonas (*parkeri*, published by GR Graves in *Auk* 98(1987): 382), and s-wards to Carpish mts n of Río Huallaga, Huánuco (*theresiae*). Very common locally.

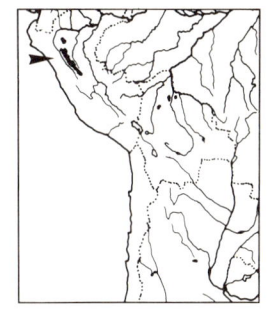

FIRE-THROATED METALTAIL *Metallura eupogon* – Plate XXIX 8a-b

10 cm (incl. 1.25 cm bill). Almost uniformly dark bronzy green, somewhat coppery above (and the available specimens from Ayacucho are coppery all over). **Gorget rather small and not very broad, orange-red (male)** or just some orange discs centrally on throat (**female**). Upper tail-coverts and tail blue-green or bronzy shot more or less with blue above, golden green below.
Habits, voice, and **breeding**: Probably as in Coppery, Scaled, and Viridian M.
Habitat: Humid moss-clad dwarf and elfin forest alternating with pajonal, with shrubs along the ecotones.
Range: C Peru, near the tree-line (3000-3700 m) from immediately s of Río Huallaga in Huánuco locally to río Apurímac (maybe 2 sspp). Locally common.

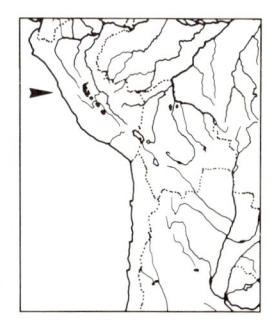

SCALED METALTAIL *Metallura aeneocauda* – Plate XXIX 9a-c

11 cm (incl. 1.5 cm bill). **Male olive-green with 'scaled' underparts**

Tyrian Metailtail

caused by tan feather margins, and with glittering yellowish green c throat. Tail bronzy green shot with blue above, and golden green below, La Paz birds often coppery towards the feather margins (*aeneocauda*), or more purplish blue above and distinctly purplish coppery below (*malagae*). **Female** similar, with slight indications of gorget along mid-throat, and with underparts predominatingly tan (not the rich orange-buff of female Tyrian M. or ad. female of Rufous-capped Thornbill), with dense olivaceous mottling, esp. on flanks. Tail with pale tips (also stated for some males from Cuzco).
Habits: Forages within 2 m of the ground, from *Berberis*, *Brachyotum*, *Centropogon*, *Gentiana*, *Ribes*, and other flowers with 2-4 cm corollas, to which they often cling.
Voice: A buzzy *zew-zew-zew, zwhizizi-zwhizizi*, with the first 3 notes forming a descending series.
Breeding: No information.
Habitat: Bushy glades in cloud forest, tree-line and sometimes humid rocky slopes with herbs, low shrubs, and occ. bushes even higher up.
Range: At 2800-3600 m, sometimes to 4000 m. E Andean slopes from Cord. Vilcabamba in Cuzco through Puno, Peru, to yungas of La Paz, n Bol. (*aeneocauda*); Incachaca in yungas of Cochabamba, Bol. (*malagae*). Fairly common locally.

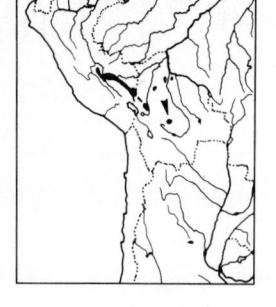

TYRIAN METALTAIL *Metallura tyrianthina* – Plates XXIX 2a-c, LXIV 2

9 cm (incl. 1.3 cm bill). **Ad. male bronzy green** grading to bottle green on rump, below more bronzy, somewhat **'scaled'** with cinnamon feather-margins, and with rather narrow emerald-**green gorget**. Ssp *oreopola* is very dark bronzy brown below, and apart from its distinctly green upperparts it resembles a miniature Perijá M. The lightest forms are *quitensis*, *septentrionalis*, and *tyrianthina*. **Female orange-cinnamon below**, normally with almost unspotted breast, green spots concentrated to the sides, and with occ. glittering green discs on throat (lacking in female *districta*, and in **juv.**). **Tail varies much**, from dark purplish or violet (*districta*) through purple shot with gold (*oreopola*), and to golden bronzy shot with purple (*quitensis* and *tyrianthina*), or purplish blue/coppery (*septentrionalis*) to deep blue (*smaragdinicollis*); females, juv. and males of some *smaragdinicollis* populations have whitish tips to the outer tail feathers. Females and juv. are hard to tell from ad. females of Rufous-capped Thornbill.

Habits: Rather aggressive hummingbird of low and mid height. Often spreads its tail widely in flight. Slowly and thoroughly searches for flowers and insects in the outer canopy of medium large trees, sometimes fluttering continuously for several minutes within one canopy, and occ. feeds clinging. Visits various open flowers, e.g. *Berberis, Escallonia, Eucalyptus, Eugenia, Hesperomelas, Palicourea*, and *Rubus*, some melastomes, solenads and ericads, some typical hummingbird flowers as *Salvia*, and occ. pierces bases of flowers with deep corollas.
Voice: Mainly short *shric* and hard *ttrr* notes. Song from exposed (but usually low) twig *tz tz tz* .... In pursuit hurried **zee** *tee* ..., *titititi-zweee, dzi-zi, dziri, dzi-zi...* etc.
Breeding: Probably Apr-Aug (Col.), eggs Oct-May (June) w Ecu., nest in a pendent mass of moss in a rock niche or among roots of overhanging bank.
Habitat: Interior of rather open humid forest, both second growth, cloud forest and moss-clad tree-line shrub, even in small patches of *Tillandsia*-clad wood in semi-arid valleys or in elfin forest patches (but locally displaced from tree-line habitats by members of the *Metallura aeneocauda* superspecies).
Range: Mostly at 1700-3800 m, with seasonal movements (casually down to 600 m). Sta Marta and Perijá mts, n Col. (*districta*), Andes of Mérida, Trujillo and Lara, Ven. (*oreopola*), and coastal mts of Ven. (*chloropogon*). From sw Táchira, Ven., through all Andean chains of Col., and through e and s Ecu., generally below Viridian M., and to w side of Andes in Piura, nw Peru (*tyrianthina*), and in nw Ecu. (*quitensis*). N Peru, mainly in the W Andes locally s to Lima on the w slope, to e slope of Cord. Blanca, and near Cajamarcilla in C Andes (*septentrionalis*), and from Río Utcubamba in Amazonas and widespread on the e Andean slopes, and in humid places in Andean valleys through Peru to yungas of Bol. to Cochabamba/StaCruz border (*smaragdinicollis*, incl. '*peruvianus*'). Common and sometimes abundant.

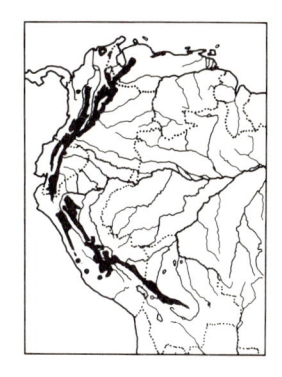

PERIJÁ - METALTAIL *Metallura iracunda* – Plate XXIX 1

10.6 cm (incl. 1.35 cm bill). Larger and darker than sympatric Tyrian M. **Male mainly black**, lustre of upper parts ranging from deep green on forecrown to coppery on wing-coverts and rump, below with slight bronzy sheen. Narrow gorget glittering emerald-green. Tail metallic coppery violet or carmine. **Female** pinkish cinnamon below with quite clear breast, but occ. bronzy spots on throat, the abdomen more heavily spotted bronzy; tail more golden copper than in male, outer tail-feathers broadly tipped buffish.
Habits, voice, and **breeding**: No information.
Habitat: Open bushy country near summits.
Range: At 1800-3100 m on Cerros Pintado and Tres Tetas in Perijá mts on Col./Ven. border.

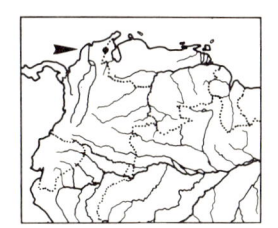

Thornbills – Genus *Chalcostigma*

Mainly insectivorous highland hummers, somewhat passerine-like due to tiny bills and rather long legs. Narrow rainbow-colored 'beard' and somewhat elongate crown-feathers sometimes give a shaggy appearance. The first species is easily mistaken for a metaltail (the female even in the hand); the others are larger, very **long-winged** and with **rather long 2-lobed tails.**

RUFOUS-CAPPED THORNBILL *Chalcostigma ruficeps* – Plate XXIX 13a-b

10 cm (incl. 1.1 cm bill). Metallic bronzy green, tail bronzy with coppery underside with silky gloss. **Ad. male** has dark face, **chestnut cap** and rather narrow, emerald-green gorget often with golden tip; underparts somewhat mottled or scaled with tan. **Females** lack rufous cap, has vestigial gorget, and an almost uniform orange-cinnamon breast with green spots concentrated to the sides. In this dress **closely resembles female Tyrian Metal-tailed (a coppery-tailed ssp)**, but can be distinguished by shorter bill with slightly upturned tip, and more dull bronzy or coppery, narrower (0.7 cm wide) tail-feathers. **Juv.** similar, but diffusely spotted bronzy all over the cinnamon underparts, which gives a quite uniform blurry impression.
Habits: Feeds from melastome flowers (*Miconia, Tibouchina*) and *Vallea stipularis*, and pierces the bases of *Fuchsias* and other flowers with deep corollas, or uses slits made by flowerpiercers (*Diglossa*). Also takes insects. Flight fluttering, as in a metaltail.
Voice: In flight fine *tzee* notes. Song a soft frail trill.
Breeding: Dec-Mar (Bol.).
Habitat: From humid second growth and riparian thickets to borders and glades in humid montane forest and moss-laden cloud forest.
Range: On Pan Azucar in C Andes, Col. (sighting), and along e Andean slope from se Ecu. probably continuously through Peru to yungas of Cochabamba, Bol. Generally rare or local around 2000-2700 m, occ. to 3600 m, but males of this species was one of the commonest birds at 2900 m above Incachaca in Cochabamba in Jan 1984 (seasonal crowding?).

OLIVACEOUS THORNBILL *Chalcostigma olivaceum* – Plate XXIX 11

12-14 cm (depending on tail length, incl. 1.2 cm bill). Very long-winged; tail 2-lobed. Dull **dark bronzy brown** (distinctly lighter, Wood Brown below, in *pallens*), tail bronzy green. **Ad. male** with 4 cm tail; dark sooty face and throat, and a long, glittering c 'beard' grading from green on chin through yellow and purple to blue at tip. **Female** with small partly disintegrated beard. **Juv.** has 3 cm tail, some tan feather-margins, esp. below, female with pale-margined throat-feathers, male with throat area uniform gray-brown.
Habits: Largely insectivorous. Often feeds walking over short grass, cushion plants, or thick mats of hair covering cushion cacti, and seen feeding from prostrate red flowers (*Castilleja?*) on boggy slope. Sometimes sallies for flying insects.
Voice: No rec.
Breeding: Jan in Bol., juv. in Feb-Apr (c Peru) and May (La Paz).
Habitat: Usually rather humid puna grassland and cushion plant

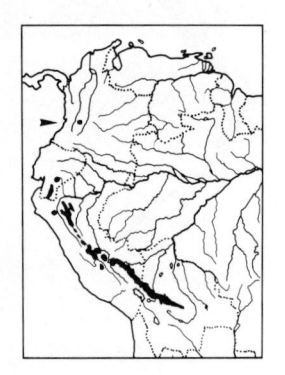

communities, sometimes in low composite brush, but also rec. at edge of dense *Polylepis-Gynoxys* woods (breeding habitat?).
Range: At 4000-4600 m, locally from nc Peru to LaPaz, n Bol. May be restricted to Cord. Blanca, Ancash, and Cord. Huayhuash to W Cord. above Lima (*pallens*) and E Andes from Cuzco s-wards (*olivaceum*). Uncommon.

BLUE-MANTLED THORNBILL *Chalcostigma stanleyi* – Plate XXIX 10a-d

12-13 cm (incl. 1.2 cm bill). First impression black with 2-lobed tail. **Very dark sooty brown**, crown and nape with slight bronzy gloss grading into purplish blue on mantle and steel-blue on tail. **Ad. male** with entire head and neck almost black, tail 4.5 cm.; long 'beard' grades from emerald-green on chin to rosy or purplish distally (*stanleyi*), or with small purplish tip (*versigularis*) or grayish violet-blue or glaucous tip (*vulcani*). **Female** has slightly shorter tail, and whitish throat with bronzy dots and incomplete (green) 'beard', or at least some green discs. **Juv.** is very uniform dark below, with unspotted throat, or mouse-gray (the latter possibly the initial plumage).
Habits: May feed walking on slopes and rocks, but mainly feeds gleaning while clinging, often with fluttering wings, to the foliage, esp. gleaning sugary secretions and some tiny cicadas and aphids from the undersides of *Gynoxys* leaves. Interacts aggressively with the Tit-like Dacnis, which exploits the same resources. Occ. hawks, or feeds from small flowers of *Berberis*, *Gauteria*, *Ribes*, and a tiny red *Gentiana*.
Voice: Mainly weak *dzr* notes.
Breeding: No information. Probably rainy season (juv. Mar in La Paz).
Habitat: Slopes with rather humid páramo and jalca vegetation, esp. steep and rocky places with some *Gynoxys-Polylepis* woods or scrub, at higher elevations than Rainbow-bearded T.
Range: At 3700-4400 m. In the large páramos of n Ecu. and probably Nariño, Col., and in Cajas mts in Azuay, s Ecu. (*stanleyi*). Very local in C Andes in Amazonas and La Libertad, and W Andes from Cajamarca to Junín and Pasco, and at Mamaccocha in Ayacucho, Peru, most common in Cord. Blanca (*versigularis*). Along E Andes from Ocobamba in Cuzco to Tunari range in Cochabamba, Bol. (*vulcani*). Common locally.

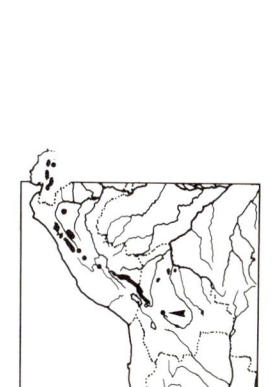

BRONZE-TAILED THORNBILL
Chalcostigma heteropogon – Plate XXIX 15a-c

10-14 cm, depending on tail length (incl. 1.3 cm bill). Shiny **bronzy**, glossed green on crown, more coppery on nape and rump, with fairly long, deeply **2-lobed golden bronzy tail**. Below mainly bronzy brown, more or less mottled with tan feather-margins. **Ad. male** has long glittering 'beard' grading from blue on chin through green to golden or rosy tip. **Female** has small, partly disintegrated beard without rosy point. **Juv.** male has a uniform throat with no glittering feathers, and juv. approaches Rainbow-bearded T. by having partly dark rufous crown.

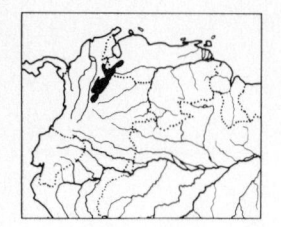

Habits: Often looks shaggy with raised crown-feathers and beard. Hovers with rather erratic wing-beats. Takes tiny insects from *Espeletia* flowers (often sitting below or clinging to them), by gleaning from foliage and by hawking. Also feeds from flowers of *Bartsia*, *Brachyotum*, *Castilleja*, *Hesperomeles*, *Rubus*, and ericads.
Voice: Male sings with monotonously repeated chips from exposed perch.
Breeding: Probably in Sep-Jan, but possibly also at other seasons.
Habitat: Semi-arid to humid páramo habitat, mainly steep rocky slopes with ferns, bromeliads, shrubs, espeletias, and at the edge of *Polylepis* and other stunted wood.
Range: At 2900-3450 m on Páramo de Tamá, w Ven., and E Andes s to Cundinamarca, Col. Common.

RAINBOW-BEARDED THORNBILL
Chalcostigma herrani – Plate XXIX 16a-c

12 cm (incl. 1.2 cm bill). Uniform dark **earth-brown** with some bronzy gloss, forehead and **crown deep rufous** centrally, rump bronzy. **5.5 cm long notched tail black with large white tips. Ad. male** with long glittering 'beard' grading from blue-green on chin through golden to fiery red on the distal part. **Female** like male, but mid-throat pale with some glittering discs. **Juv.** male has uniform dark throat.
Habits: Aggressive, also displacing other hummingbirds. Often spreads tail, showing white spots. Displays almost stationarily, singing and beating wings while perched. Feeds in *Brachyotum* and *Puya* and probably a number of other flowers, clinging to them. May walk on the ground.
Voice: Repeated low-pitched *cheet-dee-dee-cheeet* noted.
Breeding: Nest-building July, eggs Sep-Dec.
Habitat: On well-drained rocky slopes with ferns, bromeliads, shrubs, edge of brush, and sometimes in humid *Polylepis* woods.
Range: Local, at 2700-3600 m on Volcán Tolima in C Andes, Col. (*tolimae*); and from Munchique area in W Andes through Nariño into n Ecu., near Loja-Zamora road and Cajanuma in Loja, s Ecu., and near Huancabamba deflection in Piura, n Peru (*herrani*). Common in w Col.

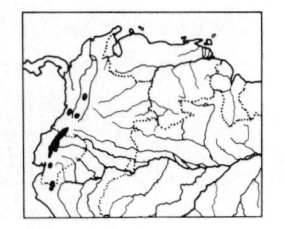

BEARDED HELMETCREST *Oxypogon guerinii* – Plates XXX 21a-e, LXIII 4 and LXIV 3

12 cm (incl. 0.8 cm bill). Generally drab brown, slightly bronzed, but lighter below. **Dark brown hood demarcated by whitish band slanting from nape to below throat**. 5 cm long tail blunt-tipped but deeply notched, coppery with **white shafts on outer tail-feathers** (thin stripes in *lindeni* to broad stripes in *guerinii*, extensively buffy white in *cyanolaemus*). **Ad. male** with nearly **black head with white forehead ending in long and thin white crest, and with long white 'beard'**, the latter with small green speckles centrally (*lindeni*), or with frontlet and c zone of beard light green (*guerinii*), or light purplish blue (*cyanolaemus*). In ssp *strubelii* all white is replaced with tan, the outer tail-feathers with 5 mm broad isabelline lateral stripe; 'beard' coppery blue. **Fe-**

male has white chin and c throat with many bronzy dots; **juv.** has incomplete white collar, and uniform gray-brown hood. **Imm. male** has partly developed beard and white-mottled c forehead.
Habits: Esp. males are very elusive. Singing and displaying males look shaggy with raised crest and beard. Mainly feeds by picking tiny insects from flowers of *Espeletia*, sitting below them or clinging to them, but also finds insects walking on matted grass, or by hovering jumps.
Voice: Song a squeaky *seep* (or trilled *ti-e-o*) at 3 s intervals.
Breeding: Large woolly nest on cliff faces, eggs Aug-Oct (Col.).
Habitat: Humid grassy or barren páramo slopes with espeletias, females often in gullies. Also on slopes with scattered elfin forest.
Range: At 3100-5200 m. Sta Marta mts, n Col. (*cyanolaemus*). Andes of Mérida and Trujillo, nw Ven. (*lindeni*). E Andes of Col. s to Cundinamarca (*guerinii*). Nevado del Ruiz in Tolima and Quindío in C Andes of Col. (*strubelii*). Sighting on Páramo de Angel in extreme nw Ecu. (C Mattheus). Common locally.

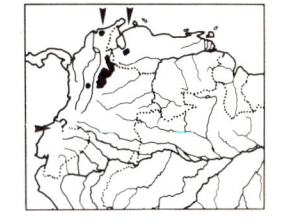

BEARDED MOUNTAINEER *Oreonympha nobilis* – Plate XXX 20a-c

14-17 cm, depending on tail length (incl. 2.2 cm bill). **Tail deeply forked, 8.5 cm in male to 6.5 cm in juv., bronzy with white outer feathers** only with dark margin on inner web. **Head, incl. throat bronzy black**, this dark hood demarcated by a whitish band slanting from nape to breast. Body bronzy with some chestnut spots above and on sides, to almost white on c underparts. **Ad. male** has particularly dark hood, and a long narrow 'beard' changing from emerald-green basally to purplish, with blue tip. Forehead and crown deep purplish blue, velvety black centrally (*nobilis*), or blue-green with white demarcation of deep chestnut c part (*albolimbata*). **Female** has small beard, disintegrated, and with white scaling basally (continuing as a white line across lore up to crown-side in *albolimbata*). **Juv.** has yellow mandible, and dull hood with green, somewhat scaled crown, dull earth-brown throat, and white-mottled rictal stripe and lore.
Habits: Nervous, and submissive to most other hummers. When perched often cocks tail to horizontal position. Feeds from flowers of columnar cacti, agaves, *Nicotiana*, and *Eucalyptus*, hovering with almost vertical body and incessantly closing and opening the tail (showing the white parts in glimpses), or clinging to the flowers while feeding.
Breeding: Probably nests in caves or crevices in rocks. Juv. Dec and Jan.
Voice: We have never heard this species vocalizing.
Habitat: Dry Andean valleys with scrubby slopes and open woodland, from mixed *Schinus*, *Tecoma*, and *Carica* woods with many columnar cacti to mixed *Polylepis/Escallonia* woods with dense thorny undergrowth, and sometimes around planted *Eucalyptus*. Often near rocky outcrops.
Range: C Peru. Locally at 2500-3800 m in Urubamba valley and around Cuzco city, and in part of Apurímac valley (*nobilis*); Cotaruse to Mutca in the s of the Apurímac drainage, and at Yauli, Acoria, and Lircay in Huancavelica (*albolimbata*). Common locally.

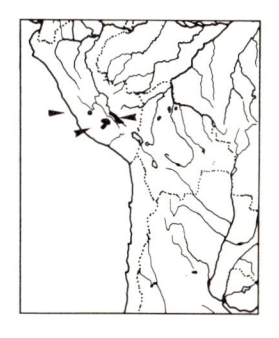

MOUNTAIN AVOCETBILL *Opisthoprora euryptera* – Plate XXIX 12

10 cm (incl. **1 cm bill**). Above bronzy green, below white with rufous on flanks and vent, and densely spotted with metallic green. Tail short, broad, notched, dark bronzy with small pale feather-tips. Distinguished from juv. or female metaltail by thicker and distally upcurved bill (looks short at a distance, the upcurving being difficult to see), a **streaked general effect** and more prominent white postocular spot. Clinal geographic variation, southern birds being quite grayish (GG).
Behavior: Single. Perches low. Generally somewhat sluggish. Feeds hovering as well as climbing, from a large variety of flowers.
Voice and **breeding**: No information.
Habitat: Clearings in open cloud forest, forest edge, humid shrubby hillsides, and elfin woodlands.
Range: At (1700) 2900-3600 m. Locally in C Andes of Col. and in ne Ecu., in Zapote-Najda mts Morona-Santiago, and below Sabanilla in Zamora-Chinchipe in s Ecu. In Cord. de Colán in Amazonas and in e La Libertad, n Peru. Well-known only in Puracé in Cauca, Col., and in Papallacta pass in n Ecu., elsewhere uncommon.

GRAY-BREASTED COMET *Taphrolesbia griseiventris* – Plate XXX 15a-b

14-17 cm, depending on tail length (incl. 2.2 cm rather thick bill). **Tail deeply forked, 8.5 cm in male to 6 cm in juv.**, blunt-tipped, bottle-green grading to orange or golden on the feather-tips. Above variably colored, deep chrome-green to more or less coppery orange, **below light gray**, throat with numerous metallic blue discs., esp. in male. **Juv.** with tan throat and breast with slightly dusky green speckles.
Habits: The species may live singly among rocky and inaccessible places in the canyons. Submissive to Giant Hummingbird in the competition for flowers of agaves and other amaryllidaceans.
Voice and **breeding**: No information.
Habitat: Semi-arid areas, such as barren partly cultivated hills with cacti, agaves, etc., and in small brushwood with some small trees or impenetrable thickets in the canyons.
Range: N Peru at c. 2600-3500 m. On the Pacific slope of Cajamarca (Paucal) and along the Marañón valley to its head in nw Huánuco (Cajamarca, Cajabamba, Cullcui, sighted near La Union). Apparently rare.

NOTE: The emendation of the name to *Tephrolesbia* has been accepted by some, e.g. Zimmer (1930) and Meyer de Schauensee (1966), but not by others, e.g. Peters (1945), Zimmer (1952), Bond (1954).

LONG-TAILED SYLPH *Aglaiocercus kingi* – Plate XXX 5a-c

10-19 cm, depending on tail length (1.4 cm bill). **Ad. male** metallic green, rather bronzy below, forehead and median crown-stripe

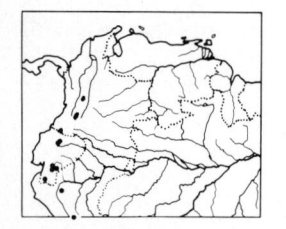

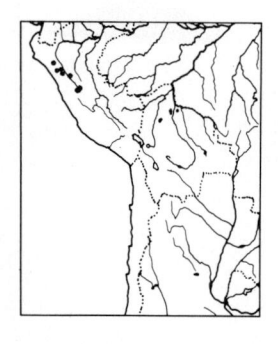

strongly shining emerald-green, throat with small glittering blue or violet patch (absent in *emmae* and *caudatus*). **Tail narrow, deeply forked, with fully 12 cm long lateral streamers**, emerald-green (*emmae*) or blue-green to dark purplish with peacock-blue outer half of each feather (*kingi, caudatus*), blackish below. **Female** has head and upperparts metallic green, short malar streak white, throat white dotted with green, rest of **underparts pale buff to cinnamon** with some green spots laterally (*kingi* females vary locally regarding extension of rufous). 4 cm tail narrow, forked, steel-blue with white tips. **Juv.** similar, but usually with weak tinge of buff below (sometimes almost covered with diffuse green spots), white patch concealed on lower back. **Imm. males** soon develop medium-long forked tail, and partly acquire the glittering cap.
Habits: Feeds on territories or 'trap-lines', and sometimes gathers with other hummers in flowering trees. Mainly feeds in the treetops, but low down in glades, hovering or clinging. May pierce the corollas of deep flowers. Bobs tail, but usually without spreading it. Male participates in ritualized nest-building.
Voice: Frail *tzit* and *trrt* notes. Also series of explosive *tzrrt* calls 0.7 s apart (*coelestis*).
Breeding: Possibly most of year. The female apparently sleeps in the bulky domed nest every night.
Habitat: In rather tall wet and humid forest, often in places with broken canopy and many vines, around clearings, and sometimes in shrub and coffee plantations, alder woods etc.
Range: Mainly in premontane zones, but at least seasonally ascends to 2500-2700 m, or locally 3000 m. In the coastal ranges and from Miranda to Falcón, n Ven. (*margaretae*), Lara through Mérida to Táchira, nw Ven. (*caudatus*) grading into populations of Perijá mts and E Andes of Col. (*kingi*). C and W Andes and w slope of Nariño, s Col., and nw Ecu. (*emmae*), and from head of Magdalena valley, Col., along e slope of Ecu. to Cajamarca, n Peru (*mocoa*), and e Andean slopes of Peru to Cochabamba/Sta Cruz border in Bol. (*smaragdinus*).

NOTE: Forms a superspecies with Violet-tailed S. (*A. coelestis*, incl. *pseudocoelestis*, *berlepschi*, and *aethereus*) of Pacific Col. and Ecu. (*berlepschi* of coastal Ven. often treated as a ssp of *A. kingi*), which is a solitary 'trap-liner' inhabiting understory of mossy premontane forest. Some authors treat *emmae* (incl. *mocoa* and maybe *caudatus*) as a distinct species, Green-tailed S.

WEDGE-BILLED HUMMINGBIRD *Schistes geoffroyi* – Plate XXVIII 6a-b

9 cm (incl. 1.5 cm rather thick but very acute bill). Metallic green with glittering lighter blue-green face, and a characteristic pattern of **purplish neck-side patches with white patches right below**, white postocular streak, and dusky cheek. In *albogularis*, at least some plumages (imm., females?) have white mid-throat and breast, and the purplish color of the neck is more or less replaced by blue-green, and ad. males may have slight white scaling on the blue-green throat. Tail *Colibri*-like: rather broadly rounded, blue-green with deep blue band across the middle. Reputedly a great singer.

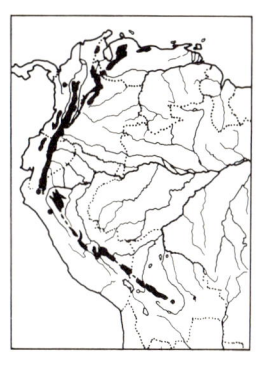

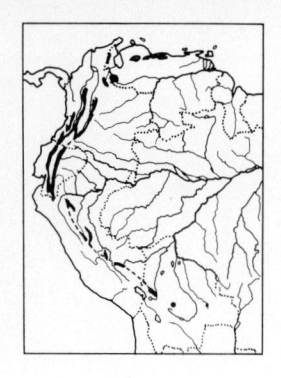

Associated with low thick understory inside humid premontane forest in n Ven., Col., and Ecu., and along e Andean slope of Peru and Bol. Reaches lower temp. zone (at least 2500 m) in W Andes and w slope of Nariño, Col., to w Ecu. (*albogularis*), and casually to temp. zone also in Cuzco, e Peru (rec. 2800 m, *geoffroyi*). Common locally.

MARVELLOUS SPATULETAIL *Loddigesia mirabilis* – Plate XXX 6a-c

Tiny, but 10-15 cm long because of the tail (1.3 cm bill). **Tail varies with sex and age, and reaches extreme development in ad. male: narrow, the 15 cm long lateral feathers strongly curved and crossing each-other, bare-shafted** and ending in large purplish black **rackets**; 2 c feathers (actually modified under tail-coverts) 8 cm long, attenuated and arched; remaining 2 rectrices rudimentary. In **female** and **juv.** the racket feathers are 5 cm long, straight, narrow, ending in gray (**female, juv.**) or blackish (**imm. male**) rackets. Above bronzy, below mainly white (unlike male Booted Racket-tail). **Ad. male** has blue crest-like cap and blue gorget continuing as bold black line down c breast and belly; normally cap and gorget are adopted while the male has the imm. tail, but occ. after ad. tail is acquired.

Habits: Seems to move constantly throughout the day, with great maneuvreability, always within dense thickets. Often chased and displaced by other hummers. Rec. feeding from *Bomarea formosissima* (which may be favored), *Rubus, Satureja sericea,* and solenads. Feeds perched. Female- and imm.-like birds greatly outnumber ad. males, even at the leks. In displaying male, rackets are lifted to protrude in front of the bill.

Voice: Modifications of the wings cause a peculiar humming and loud flapping sounds.

Breeding: Display observed May, Oct, and Nov. Enlarged gonads Dec-Feb.

Habitat: Edge of humid forest, second growth, and patches of impenetrable shrubbery (thorny *Rubus* admixed with a few *Alnus* trees) in open land.

Range: N Peru, at (1830?) 2100-2900 m (males showing elevational movements ?). Appears to be confined to c. 100 km stretch along the right bank of Río Utcubamba in s Amazonas. Within this small range uncommon, but apparently not immediately threatened.

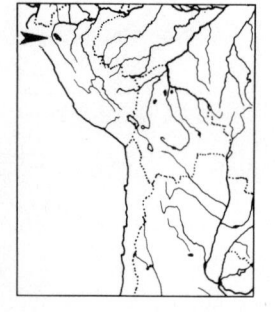

LONG-BILLED STARTHROAT *Heliomaster longirostris* is a fairly erratic inhabitant of premontane woodland and partly open terrain n and c S Am. Not uncommon to 2500 m on e slope of W Andes of Col. near Cali (KS). 11 cm, incl. 3.8 cm bill. Bronzy green with partly light gray underparts and white line down c belly, and white tips to outer tail-feathers. Male with crown glittering blue-green, gorget shining red demarcated with **prominent white malar stripe**.

Woodstars

Several poorly differentiated genera of tiny insect-like hummers of which some reach the lower temp. zone. Although they can fly fast during pursuit and on their way between remote feeding sites, their normal foraging flight is slow and wavering, as in a bumble-bee. Also the wing-hum is bumblebee-like, high-pitched and powerful. When hovering, the tail is incessanly bobbed. Occ. perches high on exposed twig or wire. Nest suspended low in a bush. Males of this group seasonally adopt an eclipse plumage of almost female-like colors.

OASIS HUMMINGBIRD *Rhodopis vesper* – Plate XXX 11a-b and XXXI 21

Tiny, but 11-13 cm owing to variations in length of bill and tail. **Bill arched, 1.9 cm in *koepckeae*, 2.2 cm in *atacamensis*, 3.3 cm in *vesper*.** Above gray-brown, back with light green to golden bronzy sheen, **rump cinnamon** (least in *koepckeae*), c underparts pale. **Ad. male** has gorget purple to blue laterally, contrasting white breast, and **slender black tail-feathers, the outer 4 cm long and curving inwards**. Male in eclipse plumage has scaled drab brown throat, but typical ad. tail. **Female** and **juv.** have underparts uniform pale gray, vent white, small 2-lobed, tail bronzy with black subterminal bar except on the c feathers (broad bar in *vesper* and *tertia*), and large **white tips** on the outer 3 pairs. **Imm. male** has mottled throat with occ. glittering discs, and medium-long tail patterned as in female.
Habits: Often on exposed perch, e.g., top twig or electric wire. Male shows a U-shaped courtship flight in front of female. Mainly feeds from flowers of leguminaceous trees, agaves, *Nicotiana*, and from cacti (flowers as well as ripe fruits partly eaten by ants or birds).
Voice: Melodious song *tzee-tzee-dee-dee*, first ascending then descending. Series or first fast, then slower *tick*s.
Breeding: Eggs at any season, but possibly mainly in Aug-Dec.
Habitat: Fog vegetation of arid loma zone, and in oases, riparian thickets, and gardens.
Range: Littoral Peru in Cerro Illescas in the Sechuran desert of Piura, nw Peru (*koepckeae*, described by J Berlioz in *l'Oiseau* 44(1974):281-90); along base of Andes from Piura s to La Libertad (*tertia*), and further s to Tarapacá (*vesper*), and in Atacama, n Chile (*atacamensis*). Mainly subtrop. zone, but in s parts of range to 3100 m. Common.

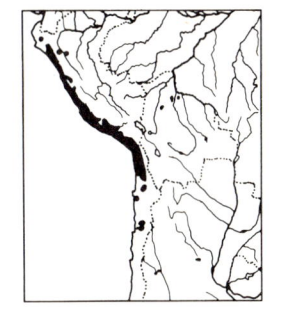

PERUVIAN SHEARTAIL *Thaumastura cora* – Plate XXX 9

Tiny, but 7-15 cm owing to variations in tail length (incl. 1.3 cm bill). Upperparts and sides glistening light green; below whitish. **Ad. male** with glittering rosy purple gorget; 2nd feather from mid of narrow black tail developed as slender **10 cm long streamers** with partly white inner webs, 3rd feathers medium long. Streamers break off easily, and are molted when the male has its **eclipse** plumage with mottled gray-brown throat. **Female** and **juv.** whitish cinnamon below, and differing from corresponding Purple-collared W. by smaller size and much **shorter bill**, and white-tipped 2 cm long tail notched centrally. **Imm. male** similar, but with mottled throat, white pectoral bar, and white base of slightly longer tail. Song and pursuit-calls a fine hurried insect-like twittering.

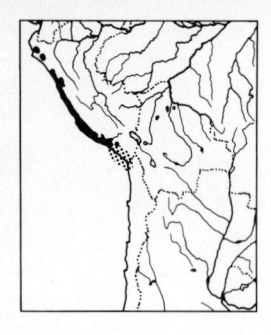

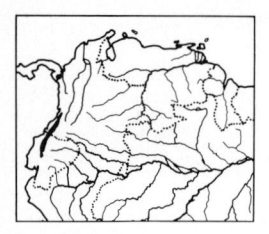

Inhabits oases, riparian thickets, gardens, and fog vegetation of arid littoral Peru and extreme n Chile, sometimes ascending to above 3000 m in Ancash and esp. in s parts of range. Common.

PURPLE-THROATED WOODSTAR *Philodice mitchellii* – **Plates XXX 8 and XXXI 18** - uncommon at shrubby forest borders and pastureland in the premontane zone of Cerro Pirre in e Panama and W Andes of Col. to Pichincha (Gualea), w Ecu., possibly reaches lower edge of temp. zone in s of range. 8 cm (incl. 1.5 cm bill). **Dark bronzy green above, flank with white patch** behind wing. **Ad. male** with gorget dark purple demarcated by **white pectoral band**; grayish bronzy on rest of underparts except for rufous lower belly. Deep purplish, 3 cm long tail narrow and forked. **Eclipse male** has mottled gray throat. **Ad. female** has sooty cheeks, underparts tawny with 2 **dusky bars, and pale c bar across breast**. **Tail tiny**, square, but notched centrally, cinnamon with rather narrow black subterminal bar, but bronzy c feathers. **Juv. male** similar, tail tipped cinnamon, but dark almost to base. **Juv.** has rusty-barred upperparts.

SLENDER-TAILED WOODSTAR
Microstilbon burmeisteri – **Plates XXX 13 and XXXI 17**

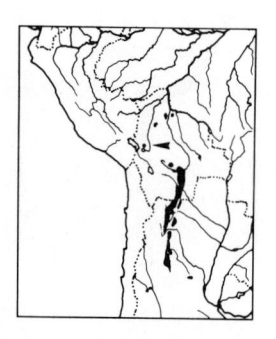

7-9 cm, depending on tail-length (incl. 1.5.cm, slightly curved bill). Bronzy green above. **Ad. male with gorget flared as long auricular plumes, reddish purple**; underparts light gray with bronzy green sides, vent cinnamon. Tail 2-lobed, with 2 outer pairs of feathers slender, 3 cm long. **Eclipse male** has pale throat. **Female** has dark cheeks, white line behind eye, uniform cinnamon-buff entire underparts, except for white patch on rump-side; **tail short and square**, cinnamon with rather narrow black bar, but dark c feathers. **Juv.** similar, but barred above, and male has 3rd tail-feather (from outside) black, slightly elongate. Frail *zhrzhrzhr...* and *tthrrr* notes rec.

Inhabits shrubby slopes and ravines with vine-tangled thickets and thorny scrub and patches of deciduous wood at 1600-2600 m from Tin-Tin to Cochabamba, Cochabamba Bol. (and 1 specimen in BMNH from Tilo-tilo in the yungas of La Paz) to Catamarca, wc Arg. Possibly not uncommon, but easily overlooked due to retiring habits.

PURPLE-COLLARED WOODSTAR *Myrtis fanny* – Plates XXX 12a-b and XXXI 19

8-10 cm, depending on tail length (incl. 2 cm slightly curved bill). Above shining light green, sometimes golden, unlike Oasis H. lacking cinnamon rump (except rusty feather-edges of juv.). **Ad. male** has **gorget glittering turquoise** with purple lower zone contrasting white breast; also c belly, vent, and rump-side white. Tail dark, forked, with narrow and slightly incurved feathers, 3.1-3.3 cm long in *fanny* to 3.2-3.5 cm in *megalura*. **Eclipse male** has mottled pale gray-brown throat. **Female uniform pale cinnamon-buff below**, tail small, with white tips on 3 lateral feathers, but unlike in Oasis H. black nearly to base, and tail broadly rounded without c notch.

Thaumastura cora, *juv. male*

Myrtis fanny, *female*

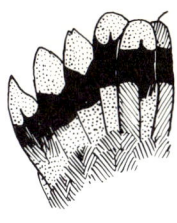

Acestrura bombus, *female*

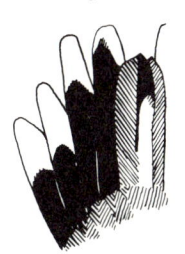

Thaumastura cora, *female*

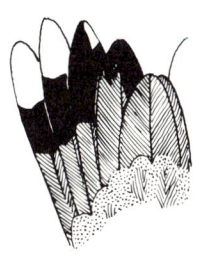

Rhodopis vesper, *female*

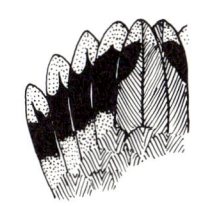

Acestrura mulsant, *female*

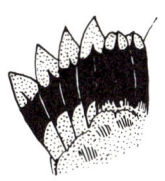

Acestrura h. heliodor, *female*

Chaetocercus jourdani, *juv. male*

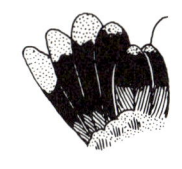

Philodice mitchellii, *juv. male*

Acestrura h. cleavesi, *female*

Chaetocercus jourdani, *female*

Philodice mitchellii, *female*

Acestrura astreans, *female*

Tails of female and juvenile male Woodstars

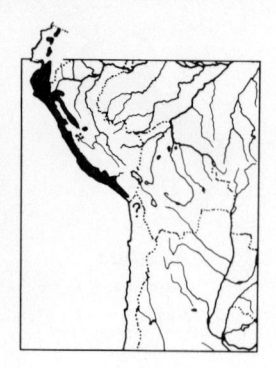

Habits: Aerial display up to 10 m. Male also shows pendular display in front of female.
Voice: Weak *shrih*s reported for courthip flight. A weak trumpet-like note also heard.
Breeding: At least in some areas seems to breed all year.
Habitat: Inhabits arid semi-open Andean slopes with light wood, agaves, and gardens.
Range: Mostly in trop. and subtrop. zone, locally (Arequipa) to 3550 m, on the Pacific slope of Ecu. and Peru almost to the Chilean border (*fanny*). At 2300-3200 m from Cajabamba region in w Andes of Cajamarca crossing to w side of C Andes in La Libertad, and to nw Huánuco, and accidentally above 4000 m on Junín altiplano (*megalura*). Common.

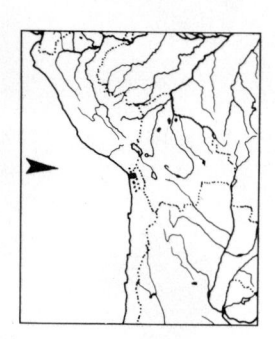

CHILEAN WOODSTAR *Eulidia yarrellii* seems to breed only in the heavily cultivated Lluta and Azapa valleys in the desert near Arica, n Chile, and straggles to adjacent s Peru and s to n Antofagasta. Mainly known from gardens in villages and suburbs near the coasts, but a rec. from Mamiña at 2600 m suggests that it may seasonally move to high elevations, where small patches of bush-steppe vegetation and scrub occurs in the canyons. The Lluta and Azapa valleys represent the only continuous and significant strips of vegetation from the coast to the temperate zone in the Atacama desert. Previously reported to be common, but may have declined. 7 cm (incl. 1.3 cm long, slightly curved bill). Above shining olive-green. **Male** with throat glittering violet-red, underparts white, sides tinged green; tail 2-lobed, short c feathers green, 2.5 cm long outer feathers blackish brown. **Female** buffy white below; outer tail feathers black, buff at base and broadly tipped white. Rec. feeding from *Lantana* and *Hibiscus*. May today mainly rely on garden flowers.

WHITE-BELLIED WOODSTAR *Acestrura mulsant* – Plate XXXI 15a-b

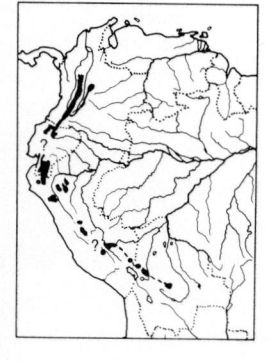

8.5 cm (incl. 1.5 cm thin, slightly curved bill). With tiny tail, but **male** has specialized bristle-like outer tail-feathers. **Ad. male** dark blue-green, white postocular line curves down side of neck encircling the rosy gorget, and joining the **broad white breast patch**. Belly, flank spot, and vent white. **Eclipse male** white-throated. **Female** more bronzy, fuscous sides of head separated by white postocular line; below tawny with paler cinnamon throat, and white c breast, belly, and flank patch, but vent tawny, the contrasting colors of the underparts being the best difference from female Slender-tailed W. Note also that the tail is rounded, not square. **Juv.** similar, but edged rusty above, throat speckled with dusky. **Juv.** and **imm. male** have decidedly narrow tail-feathers with rufous tips. Feeds much from *Lantana* and other smaller flowers.

Inhabits semihumid shrub, forest glades, and bushy pasture in premontane zone, and locally to at least 2800 m (and casually to 4000 m). W slope of E Andes and throughout C Andes, Col., Ecu. (mainly on w slope) and n Peru, and mainly in Andean valleys locally through C and E Andes of Peru to yungas of Cochabamba, Bol. Most common in c parts of range.

LITTLE WOODSTAR *Acestrura bombus*

6-7 cm (incl. 1.4 cm bill). **Ad. male dark bronzy blue-green on most of body**, although somewhat grayer below; buffy white postocular line curve down to encircle the rosy gorget; c tail-feathers very short, next pair narrow, 1.8 cm long, rest becoming shorter and more bristle-like outwards. **Eclipse** male has cheeks dark, throat cinnamon-rufous. **Female** more bronzy above, **all underparts cinnamon** to tawny on sides and vent (with only the thighs white); tiny rounded tail tawny with black subterminal bar.

Lives in canopies in the transitional zone between humid and semi-arid regions, maybe depending primarily on the much threatened deciduous forest habitat. Sea level to 2250 m, but rec. to 2700 and 3050 m in the s. Pacific slope from Esmeraldas, nw Ecu. s-wards to Piura and Lambayeque, nw Peru, and further known from a few sites in e Ecu., along the upper Marañón valley, and isolated in the upper Huallaga drainage in Huánuco, c Peru. Used to be common, but appears to have become rare, and not rec. in Peru in recent years.

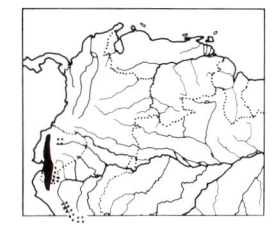

GORGETED WOODSTAR *Acestrura heliodor* – Plates XXX 10 and XXXI 16

C. 7 cm (incl. 1.3 cm bill). **Ad. male dark metallic blue-green on most of body, gorget with long auricular tufts pinkish purple** contrasting grayish white patch on upper breast; flanks just behind wings with white patch. Tail rudimentary, except for thin, somewhat elongated 3 outer feathers (1.85-2.3 cm and blunt-tipped in *heliodor*, c. 1.8 cm, very thin, in *cleavesi*). **Eclipse male** with cheeks black encircled by white line, throat pale. **Female** bronzy with lower **rump rufous**, tiny 2-lobed tail cinnamon with black bar all the way across (not interrupted by green c feathers of Purple-throated and Santa Marta W.). Below cinnamon-buff with white postocular line and prominent dark auricular band. **Juv.** barred rusty above, tail with very narrow bar in female, broad in male. Feeds from tiny flowers in thick vegetation. Fledglings Oct (Ecu.).

Inhabits edge and glades of humid forest and shady plantations, at (500-) 1200-3000 m. Mérida mts, nw Ven., and Norte de Santander and 'Bogotá' in E Andes, very locally in C Andes (and 1 rec. in W Andes), Col. (*heliodor*). N Ecu. along the e slope and at Pallatanga in the w (*cleavesi*).

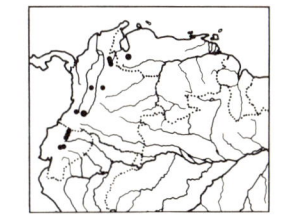

NOTE: Forms a superspecies with the following form (see *Proc. Biol. Soc. Washington* 99(1986):218-224.).

SANTA MARTA WOODSTAR *Acestrura astreans* – Plate LXIII 2

C. 7 cm. Like Gorgeted W., but male still more blue-green, with dark red rather than purple gorget, and 2-2.5 cm long, sharply pointed outer tail-feathers, and vestigial c feathers. Female lacks the rufous rump-

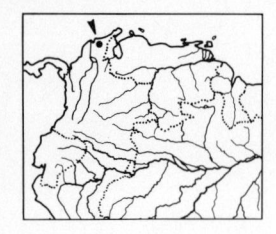

bar and has 2 c tail-feathers uniform green, and is paler below.
W and s slopes of Sta Marta mts, n Col.

NOTE: Previously classified as a ssp of Gorgeted W.

RUFOUS-SHAFTED WOODSTAR
Chaetocercus jourdanii – Plates XXX 7 and XXXI 20

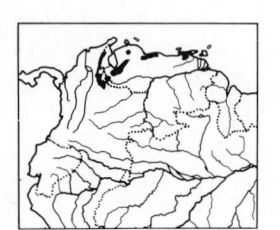

6-8 cm, depending on tail length (incl. 1 cm bill). **Ad. male** dark bottle-green, **gorget glittering scarlet** demarcated by broad buffy white pectoral bar; flank with white patch behind wing. Tail deeply forked, with outer feathers narrow, 2 cm long, dark with **orange shafts**. **Eclipse male** has cinnamon throat. **Female** with upperparts bronzy continuing as green c feathers of the 2-lobed tail, the outer feathers of which have cinnamon bases and tips. **Juv.** with rusty barring above, which may give the impression of a rufous rump, as in Gorgeted W.

Subtrop. shrub, plantations etc. Local. From nw Zulia to Distr. Federal, n Ven. (*rosae*), and Sucre and Monagas, ne Ven. (*jourdanii*). Locally to 3000 m in Andes from s Lara, n Ven., and in subtrop. n end of E Andes, Col. (*andinus*). Not well known.

NOTE: Probably best placed in the genus *Acestrura*.

Males of Crested and Golden-headed Quetzals

Coraciiformes

An old order of landbirds, apparently maintaining many anatomical traits that may also have been present in the common ancestors of the piciform and passeriform birds.

Trogons – Family Trogonidae

Previously placed in a separate order, but now regarded as a coraciiform family related to kingfishers. 35-40 species inhabit trop. Asia, Africa, and Am., with the highest species richness in the Amazon area.

Trogons are readily recognized by their strongly green and blue metallic colors, and often strikingly red or yellow bellies, and by an upright posture with the long and square-tipped tail hanging straight down. The tail tends to be closed so that the short outer feathers are positioned ventrally to the others. **The underside of the tail has distinct black-and-white patterns, important for species recognition**. Usually, upperside of tail is hidden by long metallic-colored coverts. The bill is short, broad, with serrated edges; the feet feeble with 2 toes forwards, 2 back.

Trogons are solitary forest birds that often sit motionless for long periods at mid-heights. They may suddenly hawk for a passing larger insect, and then pass on to another perch. Sometimes they feed on swarming termites, or take frogs etc. The main bulk of their diet is, however, fruits and berries, taken from the canopy in hovering flight. The normal flight is deeply undulating. Trogons are not often seen, but their whistled and rolling ventriloquial calls at dawn and sunset are readily noted. Trogons nest in holes in trees or termite-hives.

Detailed studies of Andean species are lacking. See Todd (1943) and *Amer. Mus. Novit.* 1380 (1948).

CRESTED QUETZAL *Pharomachrus antisianus*

32 cm. **Male** with short compressed crest over base of bill. Plumage **strongly metallic green**, except for **crimson-red belly and vent, and white underside of tail**. When perched, loose-barbed, attenuated, and plume-like metallic green wing- and tail-coverts hide the dark flight-feathers and the black c tail-feathers. **Female** similar, but lacks crest, and has bronzy brown head and underparts, except for red lower belly and vent; **tail black with white barring** distally and as notches along outer webs of most feathers. **Juv.** has loose black plumage with buff spots and tawny-buff belly, but metallic green feathers begin to appear before fledging. However, younger birds can be recognized a long time by pale webs on outer secondaries.
Habits: Perches quietly high up in dense foliage. Most often in fruiting trees, e.g., *Persea* and *Ocotea*.

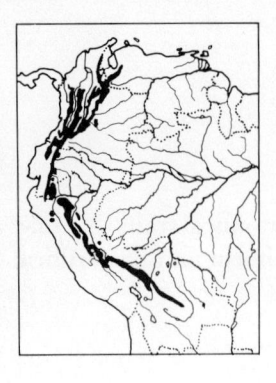

Voice: Gives loud rolling *way-way-wayo* and a muffled whistle **whee-eoo**, *chuk*. Responds to imitations of these calls. Alarm a short series of *ka* notes.
Breeding: Probably mainly Feb.-June (Col.).
Habitat: Wet and humid montane forest.
Range: Mainly premontane, at lower elevation than Golden-headed Q., but rec. to 3000 m in the n parts. From Perijá mts and Trujillo, nw Ven., through all 3 ranges of Col. (also in Macarena mts) and Ecu., and along e Andean zone to yungas of Cochabamba, nw Bol. Fairly common.

NOTE: Sometimes classified as a ssp of the C Am. Resplendent Q. *Pharomachrus mocinno*, but more probably a separate (semi)species.

WHITE-TIPPED QUETZAL *Pharomachrus fulgidus* inhabits humid forest, coffee plantations, and second growth at 1500-2500 m in Santa Marta mts, Col. (*festatus*), and at 900-1900 m from Yaraquy to Sucre in n Ven. (*fulgidus*). Only Q. in range. Much like Crested Q. Outer 3 tail-feathers of male with basal 2/3 black, those of female tipped white and barred white near tip.

GOLDEN-HEADED QUETZAL *Pharomachrus auriceps*

33 cm. Differs from other quetzals by **entirely black tail** (partly covered above by green plume-like coverts), and **more coppery golden head**. Except in *hargitti*, rather golden green all over, and with wing-coverts shorter, less attenuated and plume-like than in Crested Q. **Female** has pale outer webs of primaries, **imm.** pale edges also on secondaries, and some vague pale mottling at tips of outer (shortest) tail-feathers.
Habits and **breeding**: Probably much as in other trogons.
Voice: The most frequently heard call is a melodic sad *wi-dwyyi* (much like last 2 notes of song of Chestnut-crowned Antpitta *Grallaria ruficapilla*). Other calls include rapid whinnying **why**-*dy-dy-dy-dyyrrr*, *hoo-whooooy* (beginning of the long note highest pitched), and plaintive *ka-kaaaur* of guttural quality, sometimes repeated.
Breeding: Probably Apr-June (Col.).
Habitat: At mid elevation in wet and humid montane forest, forest edge, and tall second growth, and sometimes in temp. cloud and elfin forest.
Range: 2000-3100 m in Perijá mts. and Andes from Trujillo to Táchira, nw Ven. (*hargitti*), (600) 1400-2700 m in all 3 ranges of Col. and Ecu. to n Peru, and in the e Andean premontane zone (rarely above 2500 m) of Peru s to Cochabamba/Sta Cruz border, Bol. (*auriceps*). Fairly common.

NOTE: Ssp *hargitti* sometimes (probably erroneously) listed under Pavonine Q. *P. pavoninus* of Amazonian lowlands. The form *heliactin* of subtrop. w Ecu. is usually listed under *P. pavoninus*, but may belong under *P. auriceps* or be a separate species.

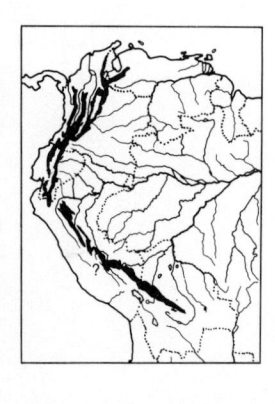

MASKED TROGON *Trogon personatus* – Plate XXXI 23a-b

25 cm. Black mask contrasts pale yellow bill and orange eye-ring. **Male** otherwise metallic brassy- to bluish green on anterior parts, back, and upper side of tail, wings minutely and densely mottled black-and-white. Green breast separated from scarlet-red belly and vent by white bar. **Underside of tail densely barred black and white with broad white distal bar on each feather**, in *temperatus* with bars of outer feathers narrow, indistinct, and often confined to the outer web; those of *assimilis* also faint, often broken, or almost obsolete. Other sspp seem doubtfully distinct. **Female** differs from male in being sandy brown instead of green (*temperatus* with whitish vermiculations above, vs. brown), and usually less red belly (but pale belly color of most museum specimens is to a considerable extent a result of fading). **Juv.** is buffy brown almost throughout, but with black mask and typical colors on wings and tail.

Habits: Sits quietly for long periods on mid-elevation perch, then suddenly flies 50-100 m to another perch. Hovers for fruit and sallies for insects. Sometimes active in bird-waves.

Voice: Male calls *zooorrh hr hr*, and gives short series of copious whistled *whyh* notes. Female responds with a descending *trrrrrr*.

Breeding: In hole carved into half-rotten wood. Probably breeds much of year (Col.).

Habitat: Interior of not too dense wet or humid montane forest and cloud forest.

Range: Tepuis of s Ven. and adjacent Brazil (many sspp). Subtrop. Sta Marta mts (*sanctaemartae*); at (700-) 1400-2900 m from Perijá mts and Andes of Trujillo, nw Ven., through E and locally in C Andes, Col., and e Andean zone s through Ecu. (incl. río Mazan w of Cuenca in Azuay) to Cuzco, se Peru (*personatus*); Puno, Peru, to Samaipata in Sta Cruz, Bol. (*submontanus*). Subtrop. (and once 3300 m) in W Andes of Col., and locally in w Ecu. (*assimilis*). At 2500-3500 m throughout E and C Andes of Col. and Ecu. (*temperatus*). Generally common.

NOTE: Ssp *temperatus* was by Meyer de Schauensee (1964) listed as a separate species, although intergradation towards *personatus* is indicated, and definite overlap is unproven.

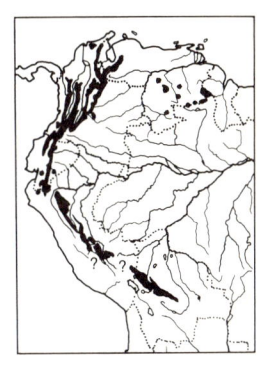

Kingfishers – Family Alcedinidae

Kingfishers comprise a homogenous group of altogether 86 species mainly in the tropics of the E Hemisphere. They are readily recognized by heavy heads with huge dagger-shaped bills, feeble feet, and usually rather small tails. The 6 Am. species mainly inhabit lowland habitats. They frequent streams, ponds, and lakes, and take fish by plunge-diving from a perch or from the air. These species may hover over water (the smallest species with hummingbird-like wing frequency), or otherwise fly straight and very fast. Kingfishers are generally solitary and very aggressive. The nests are in tunnels excavated in banks.

Amazon Kingfisher   *Green Kingfisher*   *Pygmy Kingfisher*

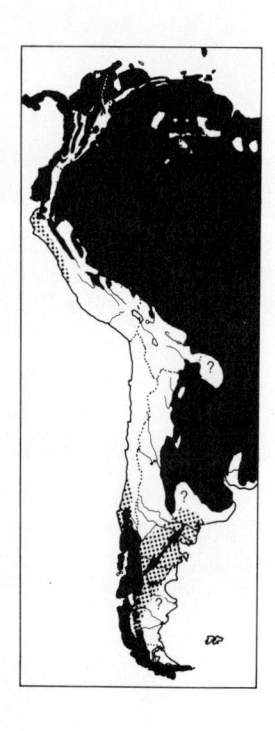

RINGED KINGFISHER *Ceryle torquata* – Plate XXXI 25a-b

40 cm; by far the **largest S Am. kingfisher**. Elongate feathers on forehead and hind-crown often raised as **crests**. Above light blue-gray (*torquata*) or darker and duller gray (*stellata*), with occ. black streaks. Loral spot, throat, and broad neck-collar white, underparts of body chestnut (**male** with white vent, **female** and **juv.** with dark gray and white bars across upper breast). Slate-colored flight- and tail-feathers have white spots. Although usually solitary, can sometimes form loose colonies in high banks. Sometimes seen in high, unsteady flight, even over land, but usually flies low over rivers. Dives with a great splash from high perch. Often returns to permanent perch after brief fishing excursion. Frequently bobs tail. The voice is a loud *ktick*, a loud, wild rattle in alarm, and in flight repeated *tchack* notes.

In general associated with large rivers, lagoons, and lakes, and can be seen along the fiords and channels of the Magellanic zone. Widespread in the lowlands of C and S Am., and casually to 2600 m on Bogotá savanna, Col. (*torquata*). Common in the lake districts in the Andes of Chile, mainly s of Bío Bío, and into adjacent sw Arg. (*stellata*).

AMAZON KINGFISHER *Chloroceryle amazona* lives along streams of fair size not arched over by vegetation, almost throughout trop. parts of the Americas, and casually to 2500 m in Ven., and probably also moving to this elevation in some valleys in Bol. **28 cm. Above shiny dark oil-green**. Throat and collar around neck white, underparts of body white with more or less complete chestnut (**male**) or green (**female**) **pectoral bar**, and green streaks on sides. Black flight- and tail-feathers minutely notched with white. Gives sharp *click* notes, a hard rattle, and long series of flute-like notes accelerating, rising, and falling rapidly towards end of strophe.

GREEN KINGFISHER *Chloroceryle americana*

19 cm. **A smaller version of Amazon K.** with **conspicuous white spots on flight- and outer tail-feathers**. Sides spotted rather than

streaked with green. **Male** with chestnut pectoral bar, **female** with light buff throat and breast with 2 green pectoral bars. Gives series of clicking *tick* calls.

In swamps and lagoons, along wood-fringed streams, and sometimes bouldry mountain streams. Almost throughout trop. and lower subtrop. zones of the Americas, and ascends to 2000 and casually to 3200 m in semiarid valleys of La Paz and Cochabamba, nw Bol.

PYGMY KINGFISHER *Chloroceryle aenea* – Plate XXXI 24a-b – inhabits swamps and várzea forest with small often shallow ponds and streams with overhanging trees in trop. parts of the Americas. Casual in Bogotá area (2600 m) in Col. 13 cm, **smallest Neotropic kingfisher**. **Above shiny dark oil-green**. Cinnamon stripe from bill to eye. Throat orange-buff usually sending thin collar around neck, **underparts of body chestnut** with white c belly and vent. **Female** has green pectoral bar: **juv.** similar bar with some white spots. Usually seen in swift flight following curves of streams. Quiet, but sometimes gives series of rather weak *dzit* notes.

Motmots – Family Momotidae

Altogether 8 species, all in Neotropical lowlands. See *Bull. Amer. Mus. Nat. Hist.* 48 (1923): 27-59.

BLUE-CROWNED MOTMOT *Momotus momota*

41 cm, the elongate (25 cm) c tail-feathers narrow subterminally, this portion becoming bare-shafted by wear, so that terminal **rackets** are clearly set off from rest of tail. Crown turquoise (plus extensive violet in lowland forms) with black c patch (smallest in *chlorolaemus*), **Mask through eye black** tapering backwards, and narrowly outlined in turquoise. Otherwise olive-green grading to cinnamon below, but the large highland forms *aequatorialis* and esp. *chlorolaemus* are grass-green to light green below. Alone or in pairs. Inconspicuous, as it usually sits motionless half-way up in a tree, but swings tail from side to side when uneasy. Gives low owl-like *hoo-doot* call. The nest is an unlined burrow. Fledglings Nov. (Cuzco).

Widespread in thickets and second growth, and at forest borders. With several sspp almost throughout humid trop. and lower subtrop. zones of C and S Am., but *aequatorialis* at 1500-2400 and rarely to 3100 m in all Andean chains of Col. and along e Andean slopes of Ecu. (crossing over to el Oro in the sw), and *chlorolaemus* at 1200-2200 m in e Andean premontane zone through Peru to La Paz, Bol. Uncommon.

NOTE: Ssp *aequatorialis* is almost certainly a high-elevation (semi)species **HIGHLAND MOTMOT** (incl. ssp *chlorolaemus*).

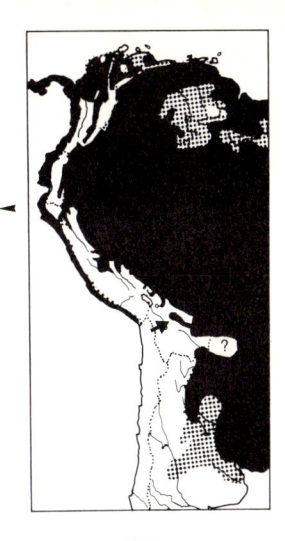

Piciformes

Traditionally regarded as related to passeriform birds, connecting this order with the older groups of landbirds. However, recent biochemical data raise doubts about this view. It has been suggested that puffbirds, together with jacamars (Galbulidae), are primitive Coraciiformes, but woodpeckers, barbets, and toucans are universally recognized as a close-knit group. Anatomical as well as molecular evidence clearly point out the Neotropical barbets as the closest relatives of toucans (see *J. Linn. Soc.* 92(1988):313-43). Peruvian forms dealt with by Bond (1954a).

Puffbirds – Family Bucconidae

Altogether 32 species, all in C and S Am. The only highland species is generally representative of the family.

WHITE-FACED PUFFBIRD *Hapaloptila castanea* – Plate XXXI 27

24 cm. Head large and square, gray with black demarcation on forehead towards **white area around base of long hooked black bill**. Otherwise dark olive-brown with rufous chestnut underparts. Eye red.
Habits: Confident. Often watches quietly, in hunched forward posture, from exposed mid-level branch. Occ. flies up to 15 m to take beetles and caterpillars from other branches, hovering, then returning to its post. Also dislodges beetles from decaying tree-stumps and logs. Several may gather in good feeding sites in clearings with older wind-felled or logged trees.
Voice: Weak insect-like trill *wuoooooooo*, down-slurred at the end, short to minute-long. Also weak, mournful, ascending whistle (PG).
Breeding: Possibly Apr (W Andes Col.).
Habitat: Humid montane forest, esp. rather open places with many decaying dead trunks. Possibly partial to ridges.
Range: Subtrop., locally to 2900 m in temp zone. Along w slope of W Andes and Nariño, s Col., nw and extreme ne Ecu., on e slope of Cerro Chinguela in Piura, and in La Libertad, n Peru. Rare and local.

NOTE: Usually called White-faced Nunbird, but does not belong in the nunbird group.

Barbets – Family Capitonidae

Altogether 78 species in trop. Asia, Africa, and Am. 1 species in the Andes.

TOUCAN BARBET *Semnornis ramphastinus* – Plate XXXI 26 – of wet subtrop. forest of Valle and Cauca in W Andes (*caucae*) and w Nariño, s Col., to nw Ecu. (*ramphastinus*), ascends to near lower fringe of temp. zone (rec. 2400 m). - 17 cm. Cylindrical, heavy-headed, and apparently neckless. Thin tail sometimes bobbed. **Bill short, thick, yellowish white with dusky band** and brown spot at base. **Head ashy gray with black mask, crown, and napeside, and a white eyebrow**. A thin crest is normally concealed. Back dusky brown, rump yellow, underparts yellowish gray with red breast and belly (yellow lower breast and belly in *caucae*). Along or in pairs, perched stolidly at midheights to canopy, but can also be seen moving about gathering fruit, sometimes in bird-waves. Gives sometimes endless series of **resonant, metallic coo's** (higher pitched than similar calls of Highland Motmot, the calls often given as duet) audible 1 km away. Also gives a scolding chuckle. A cooperative breeder, nesting Feb-May, in tree-hole.

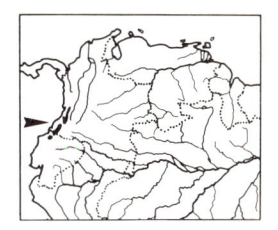

Toucans – Family Ramphastidae

Altogether 42 species, all in C and S Am. **Instantly recognized by enormous and colorful bills**. In spite of the size, the bills in no way seem to interfere with the birds' actions. They are filled with air, very light, but reinforced by a spongy web of thin bony struts. The wings are short and rounded, the tail rather long and graduated.

Andean toucans comprise rather small green toucanets (*Aulacorhynchus*) and large, blue-breasted, but otherwise multi-colored mountain-toucans (*Andigena*). The former are typical of humid premontane forests and second growth, the latter prefer cloud forest with heavily epiphyt-laden canopies.

Toucans usually appear in pairs or small parties. The birds often twist their heads and tails in various angular positions. Sometimes one bird is seen preening or feeding another. Asleep, toucans rest bill on back and folds tail forwards over the bill. When the group travels, one bird flies away and the others follow one by one. The flight of toucanets is direct and swift, but large species, as mountain-toucans, show a more labored flapping, sometimes with inserted glides. The diet comprises berries, small fruit, and *Cecropia* catkins, insects, frogs, eggs, and young of other birds etc. Mountain-toucans feed much by probing among the leaves of

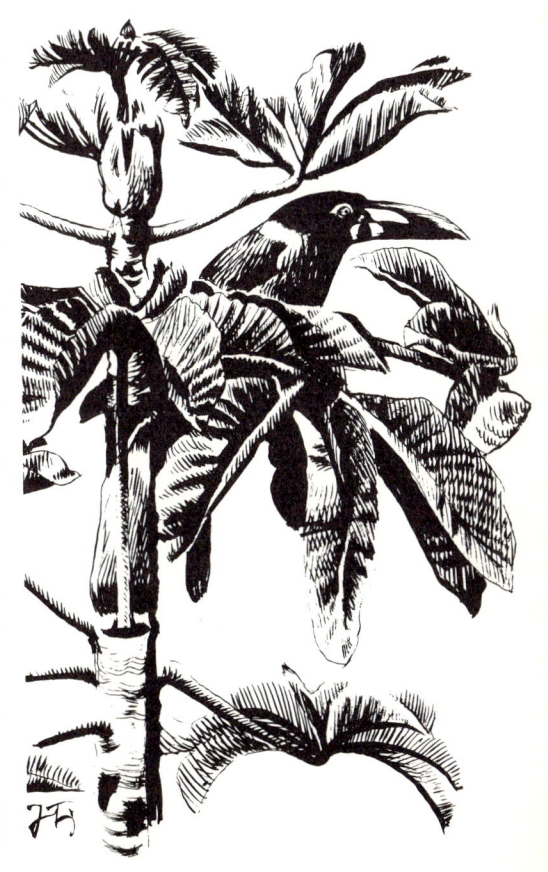

Gray-breasted Mountain-toucan in Cecropia tree

bromeliads. Seizing the food in the bill tip, a toucan tosses its head up to throw the food back into the throat.

Toucans nest in natural tree-holes or abandoned woodpecker holes, and have naked nidicolous young.

Detailed information about all the species is found in Haffer (1974). See also *Bol. Soc. Venez. Cienc. Nat.* 29(1968):459-76 and *Wilson Bull.* 56(1944):65-76 and 133-51.

EMERALD TOUCANET *Aulacorhynchus prasinus* – Plate XXXII 3a-d and LXIII 8

34 cm. **Ad.** has **bill black with greenish yellow upper ridge**, and a broad white demarcation basally (*albivitta* with red spot on lower mandible). Plumage **grass-green**, lightest below, **throat variably colored**: white (*albivitta*), light gray (*griseigularis, lautus*), blue-gray (*phaeolaemus*), deep blue (*cyanolaemus*), or black (*atrogularis, dimidiatus*). **Vent and tips of tail-feathers chestnut-tawny**. **Juv.** has pale orbital skin and rather small horn-colored bill.
Habits: Elusive, usually in small restless parties. Often in rather horizontal posture. When alert, frequently leans forwards, then pulls back. Nods energetically between sweeping lateral jerks of the head, while cocking the tail upright.
Voice: Varied, mostly unmelodic croaks, barks, and rattles. The song is a hurried series *churt churt chirt churt*.... Human watchers are scolded in tones that resemble the chatter of an angry squirrel.
Breeding: probably Jan-June (Col.).
Habitat: Inhabits misty and cool mossy montane forest, but wanders through adjoining clearings with second growth and scattered trees.
Range: C Am. and Andes s to Bol., mostly premontane, below other highland toucans, and *dimidiatus* only in trop. zone. Displaced to 2-3000 (once 3700) m in Ven. and Col., but mainly below 2200 m from Amazonas to Ayacucho in Peru. Sta Marts mts., n Col. (*lautus*); Andes from s Lara to Táchira, Ven., Perijá mts, and E Andes and e slope of C Andes of Col. to e slope of n Ecu. (*albivitta*); western C Andes and W Andes s to Patía canyon, w Col. (*griseigularis*), and s portion of W Andes of Col. (*phaeolaemus*). Se Ecu. to n Peru (*cyanolaemus*), grading into population of e Andean zone of Peru to yungas of Sta Cruz, Bol. (*atrogularis*). Lowlands of Loreto, e Peru, to Beni, nw Bol. (LSUMZ) (*dimidiatus*). Generally common.

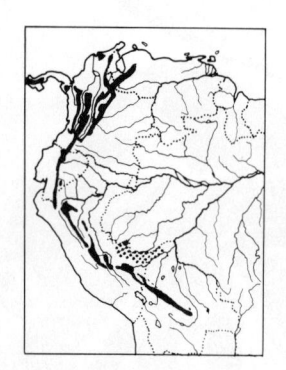

BLUE-BANDED TOUCANET
Aulacorhynchus coeruleicinctus – Plate XXXII 1

40 cm. **Bill gray with pale horn-colored tip** and cutting edge. **Eyes sulphur-yellow**. Dull green, palest below. **Throat white**, cheek and small eyebrow pale blue, and breast with indistinct light blue band. Rump with dark red patch, vent pale green, tail tipped chestnut (usually restricted to c feather pair in *coeruleicinctus*, or 2 c pairs in *borealis*.
Habits: As Emerald T., but usually more confident.
Voice: Song composed of a throaty rattle followed by one to several short notes, as ***kirrit it..kirrrik it ik...it...r***.
Breeding: No information.

Habitat: Humid montane forest and mossy cloud forest to adjacent forest edge and second growth.
Range: At 1700-2500 m (rarely to above 3000 m). Huánuco s of Huallaga river and Junín, c Peru (*borealis*); from Cuzco, e Peru, to yungas of Cochabamba, Bol. (*coeruleicinctus*). Fairly common.

NOTE: Forms a superspecies with Yellow-browed and Crimson-rumped T.s.

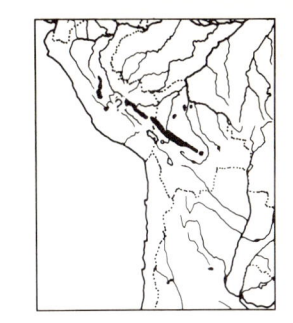

Several toucanets replace each other altitudinally, and those of the temp. cloud forest thus have near counterparts right below this zone.

YELLOW-BROWED TOUCANET *Aulacorhynchus huallagae* – **Plate XXXII 5** – is rec. only from (1830?) 2135-2450 m near Río Mishollo in La Libertad, n Peru, but might reach temp. zone elevation. Inhabits canopy of epiphyte-laden montane forest (esp. *Clusia* trees). The ongoing deforestation in the area takes place mainly below 2100 m. 38 cm. **Bill gray** with pale tip and cutting edge, and with **white basal demarcation**. Plumage as in Blue-banded T., but throat whitish gray (not purely white), eye-brow yellowish, and **vent strongly yellow**. In pairs and small groups. Harsh croaking calls reported.

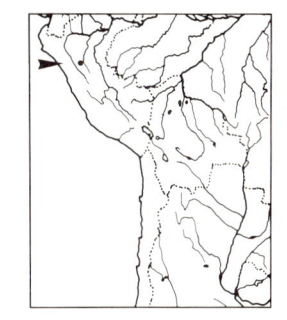

CRIMSON-RUMPED TOUCANET *Aulacorhynchus haematopygius* probably approaches lower temp. zone in Col. (*haematopygius*), e Ecu. (ssp?), and w Ecu. (rec. at 2750 m in Celica mts, HB, MKP, CR, JFR) (*sexnotatus*). This species is quite large – 40 cm – with **a peculiarly box-shaped black-and-red bill** with white demarcation at the base, and with dark red orbital skin. Dull green, lighter below, light blue below eye and across breast, **crimson on rump**, and with chestnut tail-tip. The song is a succession of barks and growls.

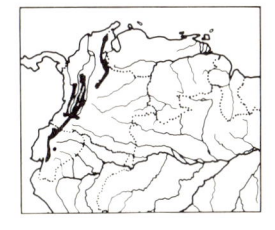

CHESTNUT-TIPPED TOUCANET *Aulacorhynchus derbianus* – **Plate XXXII 2** – of trop. and subtrop. areas in the Tepuis of s Ven./n Brazil and e of Andes in s Col., Ecu., Peru, and n Bol., almost reaches lower temp. zone in the s of its range (*derbianus*, which in this area lives between Emerald- and Blue-banded T.). Like an Emerald T. with grayish white throat, but with **longer black bill with red base and ridge** and white demarcation basally. Lacks chestnut vent. Song barking *guah hawk gahk*....

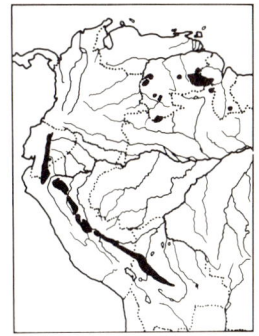

GROOVE-BILLED TOUCANET *Aulacorhynchus sulcatus* – **Plate XXXII 4** – inhabits subtrop. zone in n Ven. (*sulcatus*), and Lara to Táchira, along the Ven./Col. frontier in Perijá mts, and in Sta Marta mts, n Col. (*calorhynchus*), the latter ssp possibly meeting Emerald T. near lower limit of temp. zone. This ssp differs from Emerald T. by more extensive yellow keel and a lateral groove on bill, yellow eyes, and small light eyebrow. The blue-green **tail lacks chestnut feather-tips**. Song a rapid *coank coank coank coak*....

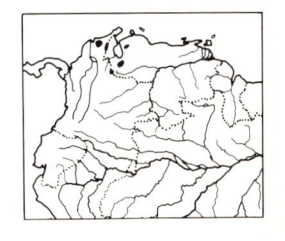

Plate-billed Mountain-toucan

PLATE-BILLED MOUNTAIN-TOUCAN
Andigena laminirostris – Plate XXXII 7

47 cm. **Bill black with reddish base and a square buffy white plate laterally** just inside the middle; orbital skin pale blue above, swollen and wax-yellow below eye. Head black-capped; back and wings olive-brown, rump pale yellow (prominent in flight), tail deep blue-gray with chestnut tip. Ear-coverts and all underparts gray-blue except for strongly **yellow lateral patch**. Thighs chocolate-brown, vent red. **Juv.** duller, lighter brown and red on thighs and vent, and the bill plate only indicated as a pale patch.
Habits and **voice**: Unlike Gray-breasted M.-t. inquisitive and very vocal. Male calls loudly, a metallic rising *tryyyyyk* repeated every 1-2 s. Bobs head and flicks tail for each call. Female replies with a very dry cracking *t't't't't'*.
Breeding: Fledglings May-June (Nariño).
Habitat: Humid and wet montane forest with much bromeliads and mosses.
Range: At (300) 1000-3000 m. Along w Andean slope from Nariño s of the Patía canyon, sw Col., to Río Chanchan, w Ecu. Common at least in Pichincha.

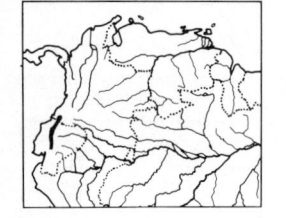

GRAY-BREASTED MOUNTAIN-TOUCAN
Andigena hypoglauca – Plate XXXII 6

48 cm. **Bill with red tip and ridge, and yellow base** separated by oblique black zone, and with a large black spot in the yellow part. Orbital skin blue and black, contrasting olive-yellow eyes. Top and side of head has black hood separated by **blue-gray collar** from olive-brown back; wings dull blue-green. Rump pale yellow (very prominent in flight), tail black with chestnut tip. Underparts grayish glaucous blue, ssp *lateralis* with a pale yellow-tinged patch suggested laterally; thighs dark chocolate-brown, vent red. **Juv.** has a more dull-colored plumage with lighter gray, brown, and red on lower underparts; eyes dark, maxilla and base of mandible lacks black marks.
Habitat: Singly, in pairs, or small groups. Feeds silently and well hid-

den in the canopy, but sometimes feeds low, on blackberries (*Rubus*). Like other mountain-toucans, generally more vertically perched than toucanets, but may hang head down when eating berries.
Voice: Silent much of the day, but has low-pitched *kek-kek-kek*... calls, and is quite vocal at dusk and dawn, giving loud, slowly rising, complaining, nasal, almost cat-like (*th*) *eaaaaaah*, or a complaining *weeeeep* (repeated). Also rattles bill.
Breeding: Possibly Jan-Feb (Col.), fledgling Nov (Huánuco, Peru).
Habitat: Humid montane forest, cloud forest, and wooded gullies, often in trees rising high above the others, and clad with large masses of mosses and bromeliads, but also frequents epiphyte-free cecropias.
Range: At 2500-3400 m, but sometimes down to 2000 m. Andes in Caldas and Cauca, and along e slope of Nariño, s Col., and continuing along e slope of Ecu. to extreme n Peru, also w of Cuenca (Río Mazan) in Azuay and in Cord. de Chilla in Loja, sw Ecu. (*hypoglauca*). Along e Andean slope from Cord. de Colán, Amazonas, to Ayacucho, and known also in Cord. Vilcabamba and Vilcanota, and Marcapata valley in Cuzco, se Peru (*lateralis*). Fairly common.

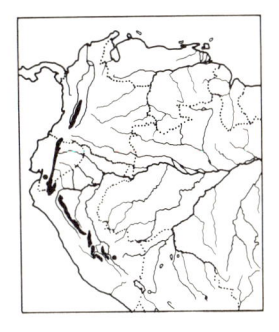

HOODED MOUNTAIN-TOUCAN
Andigena cucullata – Plate XXXII 8

50 cm. **Bill green** with black tip and a **black spot near yellow mandible-base. Appears very dark.** Head black with turquoise orbital skin. Hind-neck with light blue-gray collar, back dark reddish brown, wings dull blue-green, rump yellowish green. Tail black. Black of throat grades into dark grayish glaucous underparts. Thighs deep chocolate-brown, vent red. **Juv.** duller, base of shorter bill yellowish green lacking black spot.
Habits: Seemingly as in other mountain-toucans. Shy.
Voice: Rapid series of nasal *wuh* notes and irregular *ke-ke, ke-keke-ke*. Complaining 1.5 s whine *weeeei*.
Breeding: Juv. Feb and May (Cochabamba).
Habitat: Epiphyte-laden cloud forest.
Range: At 2500-3300 m, sometimes lower. Known from Puno, se Peru, to yungas of Cochabamba, Bol. (precise limit towards Gray-breasted M-t. not determined). Not common.

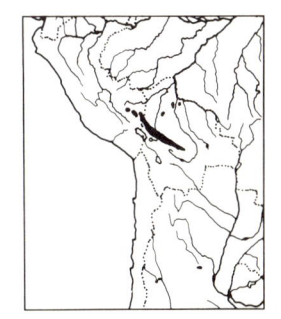

BLACK-BILLED MOUNTAIN-TOUCAN
Andigena nigrirostris – Plate XXXII 9a-c

50 cm. **Bill black**, entirely (*nigrirostris*), or with some red at base of culmen (*spilorhynchus*), or much of maxilla red (*occidentalis*). Orbital skin blue and yellow. Top of head capped glossy black, back and wings olive-brown, rump yellow (conspicuous in flight). Tail slaty gray with chestnut tip. **Throat and cheeks white, underparts of body light sky-blue**, thighs deep chocolate-brown, vent red. **Juv.** recognized by shorter bill and dull colors.
Habits: Apparently as in Gray-breasted M-t.
Voice: The song is a series of nasal yelping calls *cro-ak...co-ak...co-ak*. A hollow bill rattle is often heard.

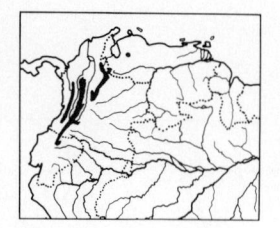

Breeding: Possibly in Mar (Huila); juv. July (Valle, Col.).
Habitat: Epiphyte-laden cloud forest, humid forest, or sometimes in small thickets in more scattered wood. Seems always to be near water.
Range: Mainly slightly lower than Gray-breasted M-t., at 1600-3200 m, but still lower on Pacific slope and e slope of E Andes. Andes of Trujillo and Táchira, nw Ven., and on both slopes of E Andes s to Cundinamarca, Col. (*nigrirostris*). W Caquetá (ssp?), and w slope of E Andes at head of Magdalena valley, both slopes of C Andes, and on e slope of Ecu. in Napo (upper Sumaco) (*spilorhynchus*). Both slopes of W Andes of Col. s to Patía canyon (*occidentalis*). Fairly common in areas with intact primary forest.

BLACK-MANDIBLED TOUCAN *Ramphastos ambiguus* of the Andean foothills of nw Ven. and Col. to Peru approaches temp. zone (rec. 2400 m) in Magdalena valley, Col. Large, black, with yellow keel of large bill, and yellow face and throat. Rump white, vent red.

Woodpeckers – Family Picidae

Woodpeckers inhabit, with 198 species, all continents except Australia. They range in size from very tiny to 60 cm, and occupy a wide variety of habitats with arboreal vegetation. A few species are adapted to grassland.

Woodpeckers are characterized by their unusual habit of excavating with their beaks, and by sometimes communicating by 'drumming' on resonant parts of tree-trunks. The bill is straight, chisel-tipped, the tongue extremely extensible, useful for extracting insects from crevices. Another aspect, the adaptations to climbing tree-trunks supported by the stiff tail, is lacking in a few species, and is shared with the woodcreeper family. All species have a zygodactyl toe arrangement, with 2 toes normally facing forwards, 2 back- or outwards.

The food, caterpillars, ants, and other insects, is searched for by excavating and tapping on decaying wood, flaking bark and moss, probing crevices, and gleaning. Some species (e.g., *Melanerpes*) eat seeds, fruit, and sap, and all species eat berries when available. Many areas have 2 sympatric species with quite similar plumage, which facilitates interspecific territoriality: one a specialized 'excavator', the other a generalized tree-surface forager. Most species are sedentary and territorial. They show complex color patterns (with sexual dimorphism) that are intricately involved in signalling social status or state of aggressiveness. Such patterns are often reduced in the more social terrestrial species, which instead have patches of bright 'flash colors'. Most woodpeckers show a characteristic flight in long undulating 'waves', but some large species with broadly rounded or square wing-tips show rather slow, vigorous, direct flight.

Woodpeckers breed and roost in holes which are normally excavated in trees, but some ground-living species excavate banks etc. The young are altricial, and hatch naked.

Details about Andean species are given in Short, L (1982) *Woodpeckers of the World*, Greenville Delaware (followed below regarding taxonomy). See also *Amer. Mus. Novit.* 1159(1942), 2349(1959), 2413(1970), and 2467(1971); *Bull. Amer. Mus. Nat. Hist.* 149(1972):5-109; *Condor* 90(1988):100-6; *Nat. Hist. Mag.* 80:66-74; *L'Oiseau* 26(1956):118-25; and *Wilson Bull.* 82(1970):115-29.

OLIVACEOUS PICULET *Picumnus olivaceus* – Plate XXXII 10

9 cm, resembling a small passerine with large head and stumpy tail with 3 white stripes (centrally and along sides, as in other *Picumnus* species). Face whitish with faint squamate pattern of dusky feather-tips, cap black with very fine (inconspicuous) dots, white, or yellow to red (**male**). Above olivaceous, wing-coverts with olive-buff edges; below pale yellow with **yellowish olive (*eisenmanni*) to olive-gray (*tachirensis*), or more brownish (*olivaceus*) all over breast and as streaks on belly**. Does not use tail for support, and usually perches like a songbird. Taps and excavates when feeding. Drums rapidly and loud for its small size. Vocalizations a laughing rattle and a sharp little *pss pss*. Breeding season protracted, with local variation.

Inhabits humid montane forest, usually in understory with many creepers, and near clearings. Mainly subtrop., but ascends locally to 2500 m or maybe even higher. With several sspp from C Am. to w Ecu. and nw Ven., highest in Táchira in Ven. and Norte de Santander in adjacent Col. (*tachirensis*) and Perijá mts on Col./Ven. border (*eisenmanni*), along w slope of E Andes, C Andes, and e slope n and s in the W Andes of Col. (*olivaceus*).

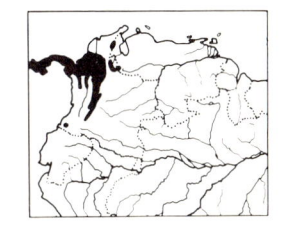

GRAYISH PICULET *Picumnus granadensis* of Cauca, Dagua, and Patía valleys in w Col. possibly approaches the lower temp. zone. It is gray-brown above, white with fine to almost imperceivable dusky streaks below. Tail 3-striped.

ACORN WOODPECKER *Melanerpes formicivorus* – Plate XXXII 12

22 cm. Unmistakable. Ssp *flavigula* bluish **black with white forehead connecting with yellowish throat**, and with white-streaked underparts and white rump with black streaks and bars, and a large white wing-patch clearly visible in flight. Most **males** have red on nape. In **ad.**, eyes gray to white, in **juv.** brown.
Habits: Often conspicuous on high dead stubs. Gregarious, living in groups of up to 15 birds which defend communal territories. Sometimes associate with quetzals and toucanets. In N Am. reputed as an acorn specialist storing acorn in prepared crevices. However, less specialized in S Am., not relying on acorn as a major food source, rarely storing, and feeding much on sap, insects hawked, or gleaned from trunks, and fruit.
Voice: Series of characteristic *ya-kup* and *karrit-cut* calls are given whenever birds meet, and a chattering *aak-a-ak-a-ak* and churring and rattling calls are given in groups.
Breeding: Apparently breeds much of the year, often with several 'helpers' at the nest. Breeding rec. from Mar (Col.).
Habitat: In open semi-humid terrain. In Col. may prefer pastureland with scattered trunks and adjoining patches of cloud forest dominated by tall oaks (*Quercus humboldti*), and with open understory, but also inhabits farmland with gardens and some shrub.
Range: W US and C Am.; ssp *flavigula* living isolated in Col., at

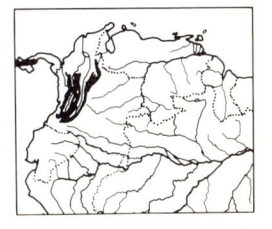

1400-2500 m and sometimes to 3500 m on both slopes of W and C Andes, and w slope of E Andes. Common near Popayán.

WHITE WOODPECKER *Melanerpes candidus* of semi-open trop. and subtrop. vegetation e of the Andes, possibly reaches lower limit of temp. zone in semi-arid valleys of Bol. 22 cm, **white**, with thin black line behind eye, and black mantle, wings, and tail. **Male** and **juv.** have yellow nape.

WHITE-FRONTED WOODPECKER *Melanerpes cactorum* of thorny scrub with large columnar cacti in the chaco, approaches the temp. zone (rec. 2500 m) in Chuquisaca and Sta Cruz, Bol. 16 cm. **Black head with white forecrown** connecting with white cheeks and throat with yellow patch. Black upperside with white line down midback, underparts of body pale smoky gray. Call nasal and drawn out.

STRIPED WOODPECKER *Picoides lignarius* – Plate XXXII 18

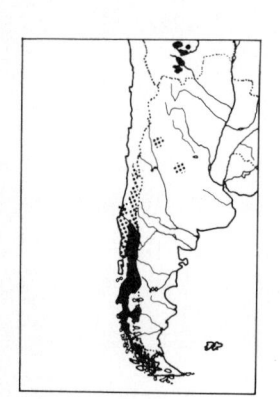

15 cm. **Black and white, densely barred and checkered above**, incl. on wings and tail. Striped below. Head appears whitish, the ear-coverts forming a dark brown band, and the malar region being streaked. Top of head black with red nape-patch (**male**), more extensive red (**juv. female**), or with no red (**female**).
Habits: Feeds by gleaning, probing, and prying, but also taps and excavates. Sometimes together with Thorn-tailed Rayadito.
Voice: Drumming loud. Vocalizations a loud *peek* and a rattle.
Breeding: Sep-Dec in Bol., Oct-Jan in s Chile.
Habitat: Scattered trees, edges of clearings, or open stunted Fuegian forest; in Bol. in cacti and *Acacia*-dominated xerophytic areas.
Range: At 1600-4000 m from s of Tunari range in Cochabamba and Sta Cruz to Potosi and Tarija, Bol. To 1800 m in Neuquén to Sta Cruz, sw Arg., and from Bío-Bío to Magallanes, s Chile, the southern populations migrating to La Rioja and Córdoba.

NOTE: Forms a superspecies with the Checkered W. *P. mixtus* of the chaco. Previously in genus *Dendrocopos*.

BAR-BELLIED WOODPECKER *Veniliornis nigriceps* – Plate XXXII 16a-c

19 cm. Small compact woodpecker with **upperparts of body uniform bronzy, and all underparts conspicuously barred pale olive-buff and blackish olive** (the darkest bars normally widest, but light and dark bars of equal width in *equifasciatus*). Cheeks and ear-coverts olive-brown, slightly streaked, with narrow white line above and below, the latter line continuing to the white lore. Top of head overlaid with red, and with red nape (**male**), or nape black (**juv. male**), only little red on crown (**juv. female**), or crown olive-black (**female**) or black (**female of nigriceps**).

Habits: Conducts a quiet life within dense vegetation, often on horizontal moss-clad branches, or in dense bamboo. Mainly feeds by probing, prying, and gleaning. Regularly follows mixed-species flocks.
Voice: Rather high-pitched but descending blurred *kzrrrrr*.
Breeding: Possibly Feb-Mar (Ecu.), Aug-Sep (Peru), or Apr-May (Bol.).
Habitat: Humid montane forest and cloud and elfin forest, mainly in glades with dense scrub or bamboo. Occ. in *Polylepis* woods.
Range: At 2000-4000 m, highest in Ecu. C Andes of Col., and through Ecu. (*equifasciatus*), and continuously along e Andean slopes of Peru (*pectoralis*) and yungas of La Paz to border Cochabamba/Sta Cruz, Bol. (*nigriceps*). Probably common.

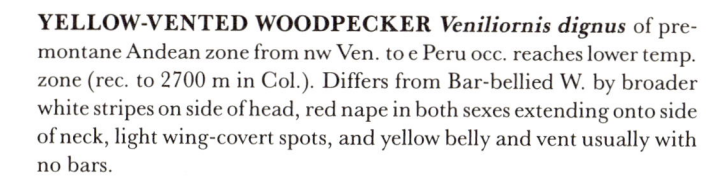

YELLOW-VENTED WOODPECKER *Veniliornis dignus* of premontane Andean zone from nw Ven. to e Peru occ. reaches lower temp. zone (rec. to 2700 m in Col.). Differs from Bar-bellied W. by broader white stripes on side of head, red nape in both sexes extending onto side of neck, light wing-covert spots, and yellow belly and vent usually with no bars.

SMOKY-BROWN WOODPECKER *Veniliornis fumigatus* – Plate XXXII 11a-b

16 cm. Because of its **small size** it resembles a barbtail (Furnariidae). Compact, **almost uniform olivaceous**, in all parts of the range with some individuals of a darker phase. Mantle golden-green to dull olive-brown, usually with orange tinge, wings browner barred only on the inner vanes; surroundings of eyes and ear-coverts usually lighter, throat sometimes lighter grayish, lower underparts yellowish brown to olive. Ssp *obscuratus* is darker, less golden, uniform umber-brown below, ear region uniform dusky, but some birds from the upper Marañón valley and those from Cord. Blanca have the entire headside brownish white (as in the Mexican *olaeginus*), contrasting the sooty brown throat and neck-side. Cap red (**male** and **juv.**) or dark olive-brown (**female**). **Juv. male** is very dark. In **flight** exposes white bars on inner webs of secondaries.
Habits: Pecks, taps, and gleans, often on larger branches, but also feeds on slender limbs and tangles, and on horizontal twigs. Often with mixed-species flocks.
Voice: A single *chuck*, a rattling *zur zurrrrr...*, and during aggression a wheezy sucking *whicker*.
Breeding: Eggs in Feb-Apr and maybe other seasons.
Habitat: Humid and wet montane forest, mainly in undergrowth near clearings, in shrub, and second growth. In Cord. Blanca ascending to *Polylepis* woods.
Range: C Am. and Andes at 600-3000 m (to well above 4000 m in Cord. Blanca, Peru). Sta Marta mts (*exsul*), and all Andean slopes of Col. and Ecu., and the e slopes through Peru and in the yungas, and locally in Tarija, Bol., and Cerro Calilegua in Jujuy, nw Arg. (*fumigatus*). Around Huancabamba and the Marañón bend in nw Peru (*obscuratus*), and locally around the upper Marañón valley and in Cord. Blanca and Negra, Ancash, Peru. Scarcely common.

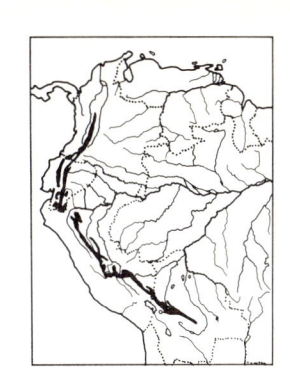

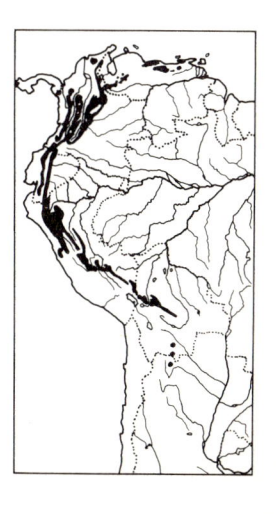

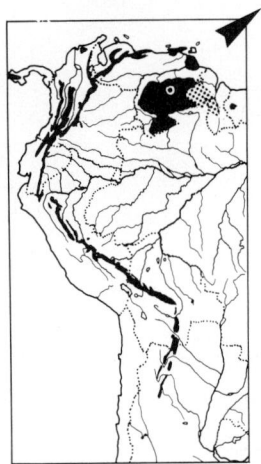

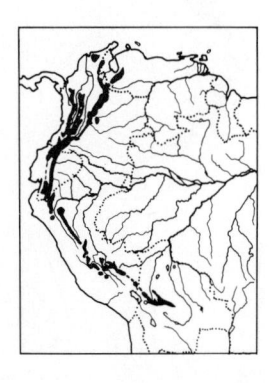

GOLDEN-OLIVE WOODPECKER *Piculus rubiginosus*

23 cm. Crown and nape crimson (or gray, only sides of crown and nape red in *alleni* and *michaelis*; gray with black forecrown and only nape red in female). **Sides of head whitish** bordered by red (male) or black (female) moustache. **Body generally olivaceous**: upperparts golden olive with yellow rump barred olive; throat black with white freckling, remaining underparts yellow with dense dusky olive barring (*tucumanus* much grayer). Behaves much like Crimson-mantled W. Calls loudly *geep* or *keer*, and gives a peculiar *utzie-deek*.

Inhabits broken forest and second growth. Widespread in premontane zones of C and S Am. s to nw Arg. Rarely to temp. zone (rec. 3100 m) at least in W Andes of Col. s to Cauca (*pacificus*), w Nariño, C Andes s to Cauca/Huila (*gularis*), Sta Marta mts (*alleni*), Perijá mts and Mérida, nw Ven. (*meridensis*), w slope of E Andes of Col. s to Santander (*palmitae*), e slope of E Andes (*buenavitae*), se Nariño (*michaelis*). At least to 2300 m in Peru (*coloratus, chrysogaster*) and yungas of Bol. (*canipileus*), but elevation poorly known in valles of s Bol. and nw Arg. (*tucumanus*).

CRIMSON-MANTLED WOODPECKER *Piculus rivolii* – Plate XXXII 17a-c

26 cm, bill fairly long except in s parts of range. Crown black (mostly red in male, except of *atriceps*), **side of head pale yellow** with dark red (**male**) or black (**female**) whisker, and black chin. **Nape and mantle crimson-red**, wings more bronzy, esp. in *atriceps*; rump and tail black. Throat and breast appear black at a distance, but at close range show densely scalloped pattern of black, red, and white, or just black and yellow (*atriceps*), lower underparts yellow, sometimes with black spots. The northern sspp differ little, *quindiana* being largest, *brevirostris* intermediate towards *atriceps*. **Juv.** duller, generally with little red, with some black on back, and scalloped breast lacking red.

Habits: Inconspicuous, as it moves slowly and silently in dense bushes or among dense epiphytes in trees, gleaning, probing, and tapping weakly. Sometimes at flowers of *Puya* and *Espeletia*. Accompanies mixed-species flocks.

Voice: Rather quiet. Drumming weak, slow, or fast as a brief cracking. Vocalizations *ky-ky-ky-ky...* and rolling *chrrr-r-r-*, ka-we**ep**, ka-we**ep**. Territorial 'song' a long series of *wik* notes.

Breeding: Possibly mainly Feb-Mar but also juv. Jan-Feb (Col.). Oct (Puno, Peru).

Habitat: Rather tall humid montane forest and sometimes elfin forest. In trees with loose bark and abundant epiphytes and mosses, and occ. in adjoining páramo habitat.

Range: At 1000-3200 (locally to 3700) m. Andes from Trujillo to Táchira, nw Ven. (*meridae*); extreme sw Táchira, Ven., and Perijá mts and E Andes, Col. (*rivolii*). C Andes (*quindiana*), and sw Col., through w and e Andes of Ecu. and along e Andean slope of Peru to Ayacucho (*brevirostris*), and Cuzco (from Marcapata) and Puno, se Peru to Cochabamba, Bol. (*atriceps*). Fairly common.

BLACK-NECKED FLICKER *Colaptes atricollis* – Plate XXXIII 3a-b

25 cm. **Head multicolored**: cap red (**juv.**), or gray with red (**male**) or black (**female**) lateral borders, becoming red towards nape; side of head off-white; throat area black (red whisker in **male**). **Entire body barred greenish and dusky, appearing uniform olivaceous at a distance**: back and wings bronzy green with narrow brown bars (*atricollis*), or buff to olivaceous with broad dusky bars (*peruvianus*), underparts pale yellow to whitish with black bars. In flight shows pale yellowish rump.
Habits: Feeds in trees by probing and gleaning, and also on the ground, often well hidden in dense cover of bushes.
Voice: A loud, rather stereotyped *wicwicwicwic...* and *chypp* calls often reveals its presence. Calls less copious and ringing than in Andean F., and usually in shorter series. Possibly drums.
Breeding: Eggs probably in July (*atricollis*) or Sep (*peruvianus*).
Habitat: Riparian thickets and orchards, woodlands and semi-arid cloud forest, and in bushy slopes nearby. Dry slopes with scattered scrubby vegetation and large columnar cacti.
Range: Peru. At 500-2800 m and occ. higher on w Andean slope from La Libertad and Ancash (Cord. Negra) to w Arequipa (*atricollis*), to 4000 m on w slope of Cord. Blanca in Ancash (ssp?), and at 1700-4300 m around Marañón valley from Piura to w Huánuco (*peruvianus*). Common at least locally.

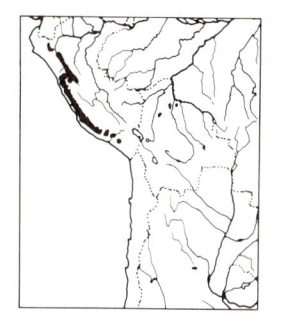

CHILEAN FLICKER *Colaptes pitius* – Plate XXXIII 1a-b

30 cm. Rather dark **gray-brown densely barred with white** above and below. Crown gray, face light cinnamon-buff with faint rictal stripe. In **flight** appears gray with yellowish white rump. **Juv.** darker, the crown blacker, somewhat barred with buff feather-tips, and eyes brown (yellow in **ad.**). Ssp *cachinnans* longer-tailed than *pitius*, but otherwise hardly different.
Habits: Feeds, mostly hopping, on the ground, on fallen logs, tree-stumps etc., but flies into trees when alarmed.
Voice: A characteristic loud *piteeu*, repeated. When alarmed *kwee-kwee-kwee....*
Breeding: In tree-holes or escarpments; eggs in Oct-Dec.
Habitat: Open *Nothofagus antarcticus* woods, and in scrub along streams, sometimes to almost open Patagonian grassland.
Range: Fuegian zone and to well above 1000 m along Andes from Magallanes to Chiloé, s Chile, and in adjacent w Sta Cruz to Neuquén, Arg. (*cachinnans*); Llanquihue to Atacama, c Chile (*pitius*). Common.

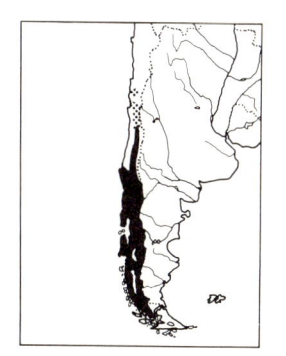

ANDEAN FLICKER *Colaptes rupicola* – Plate XXXIII 2a-b

30 cm, with very long, slightly curved bill. **Light cinnamon-buff with mantly and all wing-feathers densely barred dark gray-brown.**

Crown gray, dark rictal stripe conspicuous. In **flight** shows large **pale yellow rump**. Sspp *rupicola* and *puna* pale, with small dark spots on breast, mainly red rictal stripe (**male**; in *puna* also red nape patch). Ssp *cinereicapillus* larger, and deeper cinnamon, with distinct short bars on breast, and mainly gray rictal stripe. **Juv.** duller, more barred, with brown eyes (pale yellow in **ad.**), and often red nape.

Habits: Generally terrestrial, most often among rocks, but sometimes in trees if available, or out in open grassland. On the ground walks rather than hops. Mainly feeds by digging the ground with its long beak. Wary, but easy to observe. Small groups often assemble and display on open grassland, raising bills, and flicking wings (see fig.).

Voice: Very audible. Calls often in very long series. At alarm a whistled *tew-tew-tew...* or *kwee-kwee....* Also series of *peek* calls, a hoarser *tzy-tzy-tzy...* and sometimes rapid *kly-kly-kly....* In aggressive situation *kway-**ap***, *kyy-**ap***, or *kwa-kwa-kwa...* (*cinereicapillus*).

Breeding: Sometimes in loose colony. In holes between rocks, in banks, and in walls of usually abandoned adobe houses. Casually excavates nest-hole in *Polylepis* tree. Eggs in Sep-Oct.

Habitat: Puna grassland and fields, requiring mountain slopes with some banks, rocks, or abandoned houses to roost and breed, but may spend the mid-day far out on open grassland.

Range: At 2000-5000 m through the puna zone. N Peru from Cruz Blanca w of Huancabamba in Piura to Pasco (*cinereicapillus*); from a sharp hybrid zone in Pasco s to Peru/Bol. frontier (*puno*), through highlands of Bol. (occ. down to 850 m in Sta Cruz) to Catamarca, nw Arg., and in Tarapacá, n Chile (*rupicola*). Common.

CAMPO FLICKER *Colaptes campestris* is widespread in lowlands from Parag. to c Arg. Reaches the temp. zone in Río Negro (at 600 m at base of Somuncura plateau; *campestroides*). Cap black, broad white headside grades to yellow neckside and throat; whisker red, chin and upper throat black; body barred gray-brown much as in the Chilean F.; rump white.

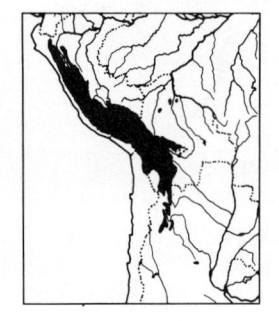

Andean Flickers in group display

POWERFUL WOODPECKER *Campephilus pollens* – Plate XXXII 15a-b

32 cm, large-headed. Mostly black. White zone around base of bill continues as **white malar stripe** and down side of neck, and into large white V of the mantle. **Male** has extensive cap red. Lower back and rump white (*pollens*) to cinnamon (*peruviana*), with cinnamon and black bars towards base of tail. **Pale rump and white wing-patches are conspicuous features in flight**. Underparts below black chest cinnamon with dense black barring. **Juv.** resembles ad., but is duller, with paler red on head.
Habits: Usually in pairs, working over the trunks of forest trees.
Voice: Uses a double drum-tap: an initial loud blow followed by a softer blow. Gives a loud reedy *pee-yaw, pee-yaw* call (occ. with 2nd note repeated).
Breeding: Possibly Apr-June in Col., fledgling Mar (Ven.).
Habitat: Steep hills with wet and humid montane forest with some large trees.
Range: At 1300-2250 m in sw Táchira, Ven., and to 3000 m and sometimes even higher throughout Andes of Col. to s Ecu. (*pollens*), continuing in lower temp. zone (and occ. to 3500 m) through e Andean zone s to Oxapampa in Pasco, c Peru (*peruviana*). Fairly common.

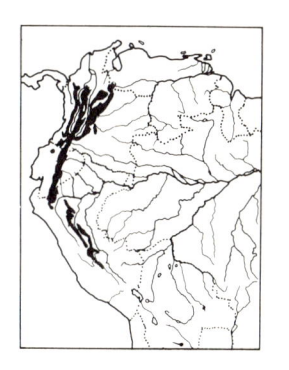

NOTE: This and the following species are often placed in the genus *Phloeoceastes*.

CRIMSON-CRESTED WOODPECKER *Campephilus melanoleucos* – Plate XXXII 14a-b

35 cm, large-headed, with sharply pointed crest, esp. in female. **Head red** with black chin and throat, and buffy white area around base of bill; **male** with black-and-white ear-spot, **female** with a broad white malar stripe more or less demarcated with black from red parts, and with black midline along crown and long crest. **Juv.** like female, but lighter red head with small crest. Rest of plumage mainly black, with white band down side of neck and merging on mantle to form large white V, and with dense cinnamon-buff barring below black chest (*malherbii*; other sspp with buffy white bars). Bill blackish (*malherbii*), or ivory. In flight shows large white wing-patch, but not white rump. Generally silent and inconspicuous. Chiefly forages by excavating rotten wood, delivering powerful blows. Has a typical double 'drum-tap', but also gives longer drumming. Calls comprise *cow*, ringing *kawarr* or *kwirr-ad*, and *ttt-he-he-he*. Eggs known from Nov-Mar (Col.).

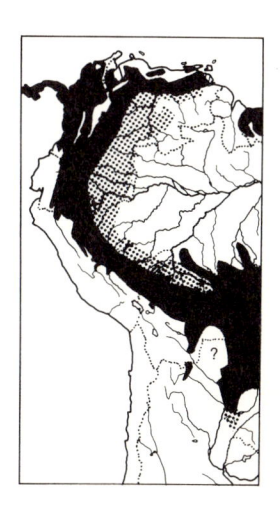

Inhabits densely forested hills, forest edge, second growth, and gallery forest in most trop. lowlands e of Andes. Ascends to 2500 and sometimes 3100 m almost throughout w of E Andes in Col. (*malherbii*), and seen to near border of temp. zone in the yungas of Bol. (*melanoleucos*). Fairly common.

NOTE: Forms a superspecies with Pale-billed and Guayaquil W.s *C. guatemalensis* and *guayaquilensis*.

RED-HEADED WOODPECKER *Campephilus rubricollis* inhabits dense lowland forest from Ven. and Col. across the Amazon basin to n Bol. Specimen rec. from 2400 m elevation in Sud Yungas of La Paz, Bol., suggests that it might reach lower temp. zone. Head red, crested, in female with large white malar zone edged black; the red color continues onto breast, and grades into tawny of belly and vent. In flight shows large tawny area on the primaries.

MAGELLANIC WOODPECKER
Campephilus magellanicus – Plate XXXII 13a-b

36 cm; the largest S Am. woodpecker, and **only big woodpecker in its range**. **Black** with white wing-patch (extensive wing-bar in flight) formed by inner vanes of most flight-feathers, and usually a little white on rump and flanks. Head entirely red (**male**), or deep red around base of bill (**female**), and with a curled crest (long and tapering in **female**). **Juv.** resembles female, but is browner, less glossy, with short crest and often some white barring below.
Habits: In pairs or family groups. Often raises crest. When feeding, very mobile, with frequent flights from tree to tree. Uses a variety of feeding methods: Pecks, excavates rotting logs, and flakes bark mostly by deliberate series of 3 or 4 laterally delivered pecks. Female often feeds clinging to thin branches.
Voice: The drumming is a single or double loud blow. Loud *kee-eah*, and more gargling *weerr-weeeerrr* calls, are given singly or in series. Feeding birds sporadically give *toot* or *toot-toot* calls. Often gives a distinctive flapping sound in flight.
Breeding: Nest-hole, usually of droplet-shape, in large partly dead tree. Egg-laying in Oct-Dec.
Habitat: Mature forest of large beech (*Nothofagus antarcticus* and *betuloides*, but also the more montane *N. pumilo*) and cypress (*Cupressus*), often with bamboo thickets.
Range: Forested zone to the timberline near 2000 m in s Chile from Curico to Isla Grande, and in the forest refuges of adjacent sw Arg.

Woodpecker chasing Woodcreeper

Passeriformes

The largest order of birds, comprising 58.5 % of the world's species. Normally believed to be related to piciform and coraciiform birds, the origin of the group is now disputed. Probably the first passeriform birds lived before the final break-up of the ancient southern continent in the late cretaceous. With the breaking up of this continent, the order was split into an assembly of primitive groups, the **suboscines (*Oligomyodi*)**, and the true **songbirds (*Oscines* or *Passeres*)**. The first group reached its highest diversity in S Am. (*Furnarii* = Woodcreepers, Ovenbirds, Antbirds, Tapaculos, and *Tyranni* = Tyrants, Tityras, and Cotingas, with Manakins, Plantcutters, and Sharpbill), but some small families also inhabit the Old World tropics and New Zealand. The songbirds mainly radiated in other parts of the world and more recently made several independent invasions of S Am. Thus the passerine fauna of S Am. falls into 2 large systematic groups. The basis for the traditional classification of these large groups has been outlined in *Bull. Peabody Mus. Nat. Hist.* 32(1970):1-131. A revised classification based on molecular evidence has been proposed in *Acta XVIII Cong. Int. Orn.* I(1985):83-121, and in *Neotropical Ornithology, Orn. Monogr.* 36(1985): 396-428. As numerous details should be corroborated by other evidence before they find wide acceptance, we have not adopted the new classification, but in some cases comment on the possible relationships in the introduction to families.

Woodcreepers – Family Dendrocolaptidae

44-50 species, closely related to the Furnariidae, and distributed from s Mexico to n Arg., mostly in the trop. zone. Medium-sized to fairly large passerines that hop up tree-stems in the manner of woodpeckers or creepers, using their stiff tails for support. A specialization to this habit can be seen in the tips of the tail-feathers, that are strongly stiffened bare shafts curved abruptly inwards. 2 species are partly terrestrial.

When alarmed woodcreepers will 'freeze' for long periods. They are monogamous, and both sexes are alike. Their plumages are brownish, wings and tails reddish brown to chestnut. Most species are streaked with whitish on throat, breast, crown, and upper back. Strong streaking may be correlated with rapid hitching along dark trunks; barred and unstreaked forms tend to wait more or use well-illuminated trunks. The bill, used for probing into crevices, under moss etc., is straight or slightly curved, in one genus thin, long, and sickle-shaped. The nest is placed in a natural cavity in a tree, or in an abandoned woodpecker-hole, rarely in an epiphyte, or in a hornero's nest. The young are fed by both parents, and after leaving the nest, they stay with the female for a considerable time, and are then only rarely attended by the male. For relationship with the furnariids, see: Feduccia, A (1973) *Evolutionary Trends in the Avian Families Furnariidae and Dendrocolaptidae*. Orn. Monogr. 13. See also *Amer. Mus. Novit.* 728, 753, and 757 (1934), Meyer de Schauensee (1945c), Esteban (1948), and Bond (1953a). Detailed studies of some tropical forms by E Willis.

TYRANNINE WOODCREEPER *Dendrocincla tyrannina* – Plate XXXIII 9

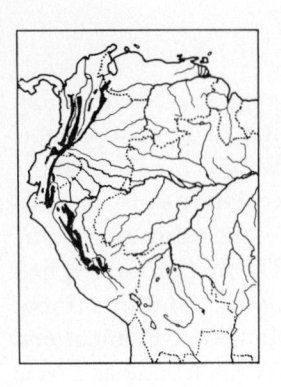

23 cm. **Bill** fairly stout and **straight** with a hooked tip. **Almost uniform olivaceous brown**, throat somewhat paler; wings and tail chestnut. Feathers of forecrown, earcoverts, chin, throat, and upper breast with indistinct pale shafts. **No dark moustache**. Ssp. *hellmayri* more olivaceous, less rufous, esp. above; feathers of fore-crown more distinctly dusky-edged. **Juv**. unknown.
Habits: Solitary or in pairs, sometimes in mixed-species flocks. Climbs trees, probing into moss and crevices.
Voice and **Breeding**: No data.
Habitat: Middle strata of tree-ferns and rather thin to medium-sized trunks in humid forest and cloud forest.
Range: 1800-2300 m in w Táchira, Ven. (*hellmayri*), and 1500-3000 m (rarely down to 900 m) in all 3 ranges of Col., nw and e Ecu. s to Cuzco, Peru (probably to Bol.) (*tyrannina*). Local and uncommon to rare.

NOTE: It is uncertain whether *D.t.macrorhyncha* represents aberrant individuals or a separate taxon. It is known from 2 specimens from the Pun region at the Colombian border in ne Ecu., presumably at 2200 m. They are paler, larger, with longer bills (bill 35-38 mm, wing 138-144, tail 120, vs. bill 31-32, wing 118-125, tail 110-118 in 24 *D.t.tyrannina*).

OLIVACEOUS WOODCREEPER *Sittasomus griseicapillus* – Plate XXXIII 8

widespread with many sspp in the trop. and lower subtrop. zones of C and S Am., ascends to 2300 m in La Paz, Cochabamba, and Sta Cruz, Bol. (*viridis*). **16 cm**. **Bill small** and thin. **Above dark olive** turning brown on lower back, ochraceous tawny on rump; inner remiges and tail rufous chestnut. Below buffy olive, vent pale cinnamon rufous. **Juv**. unknown. Usually alone, rarely in mixed-species flocks. In humid forest and second growth, actively climbing trunks, thick and thin branches in rather open parts below the canopy. Sometimes takes insects in the air and searches dead leaves. Call a woodpecker-like, very sharp fine rattling *cirrrrrr*. Song a prolonged, faintly rising or falling series of well spaced *weep*s, gradually accelerating. Also single *weep*s at 1 s intervals. Display call a harsh chatter on constant pitch, but fluctuating in intensity and speed; when prolonged usually reaches peak speed and amplitude near the end.

STRONG-BILLED WOODCREEPER *Xiphocolaptes promeropirhynchus* – Plates XXXIII 6a-b and LXIII 16

29 cm. **Bill heavy, culmen curved**. Brown, crown slightly darker; wings, rump, and tail chestnut. Crown and upper (and sometimes middle) back with distinct narrow pale streaks. Throat buffy white; sides of head and most of underside with buffy white black-bordered streaks; belly warm buff, more or less barred blackish. Birds from higher parts from the Marañón Peru to Bol. have narrower streaks, sometimes virtually lacking on crown (though still present on nape), dark

streaks on lower earcoverts and most of throat, and rump like back.
Juv. ?
Habits: Solitary or in family groups, only rarely in mixed-species flocks. Climbs trees, probing into crevices like most other woodcreepers, but also splinters wood and searches bromeliads, making much noise.
Voice: Song given at long intervals at dusk and dawn unmistakable: a whistle-like series of 4-8 descending double notes, 1st of each on same pitch or slightly lower than last of preceding series; often initiated with a single higher note. Call *gue what*.
Breeding: Fledgling Jan (Cochabamba); imm. Feb (Mérida).
Habitat: Mostly large trunks and branches at all levels inside, or at edge of humid forest.
Range: Subtrop. zone and to 3000 m. Sta Marta and Perijá mts (*sanctaemartae*), from w slope of C Andes to w bank of Magdalena river (*virgatus*), from Trujillo, Ven., through E Andes and s end of C Andes of Col. (*promeropirhynchus*), Andes of Ecu. (*ignotus*), n Peru in Cajamarca, Amazonas and San Martín (*compressirostris*), c Peru in Junín (*phaeopygus*), ec Peru in Junín and Huánuco (*solivagus*), and from Cuzco, se Peru, to Sta Cruz, Bol. (*lineatocephalus*). Other premontane sspp in C Am. and coastal mts and Tepuis of Ven., and several trop. ssp in n Col., upper Amazon area, and Bol. Arid premontane sw Ecu. and adjacent Peru (*crassirostris*). Generally rare to uncommon, but locally common.

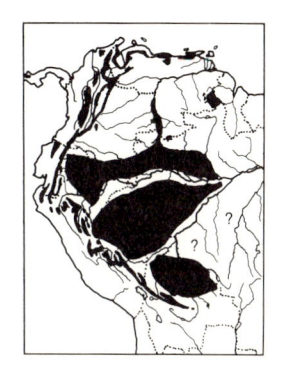

BLACK-BANDED WOODCREEPER
Dendrocolaptes picumnus – Plate XXXIII 7a-b

26 cm. **Bill smaller than in Strong-billed W., almost straight**. Brown, rump rufous; inner remiges and tail chestnut. Crown and sometimes mantle with thin buffy white streaks. Throat buffy white, sometimes faintly streaked; sides of head and neck, breast, and upper belly with broader buffy white streaks; belly more or less barred with dark brown, least so in *casaresi*, where barring sometimes absent, and where back and entire wing reddish brown. **Juv**. has buffy throat and supercilium. Alone or in pairs, usually in bird-waves. Territorial. Female, which is dominant, but shyer than male, is sleek-headed, male rough-headed. Follows army ants sometimes to tree-tops. Waits or moves slowly on big trunks, sallying out to peck at insects, spiders, or small amphibians and reptiles. Song mostly at dusk and dawn, a descending 2 s series of 15-20 simple chirps *kie-ie-ie-...ie-ie-ee-eu-eu-er*. After disputes a dissatisfied series of short chirps, 6-8 per s, fluctuating in volume, changing irregularly in quality, and sometimes grading into song *ee-ee-ee-ee-ie-ie-e-e-ie-ei....* During fights several loud *squeeh* notes. Alarm a growling noise followed by a squeal *chauhhh-eesk*. Competitors are warned with short faint grunts *uk-uk-uk*. Peepsong of fledgling at times rising and then falling *wh-e-e-e-e-e-e-e-e-up*. For further notes on vocalizations and behaviour see *Condor* 84(1982):272-85. Fledgling Nov (Cundinamarca); juv. Nov (W Andes Col.), Dec (Mérida).

At 300-2700 m in rain and cloud forest, woodland, and plantations. Sta Marta mts and coastal mts of Ven. (*seilerni*), all 3 ranges of Col., Táchira, Mérida, and Barinas, w Ven. (*multistrigatus*), La Paz, Cochabamba, and Sta Cruz, Bol. (*olivaceus*), Jujuy, Salta, and Tucumán, nw

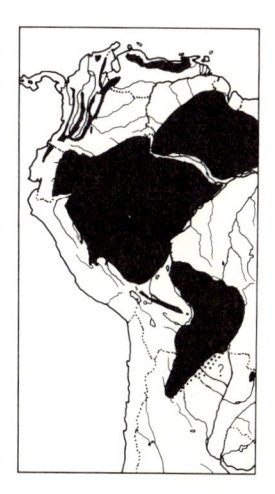

Arg. (*casaresi*). Other montane sspp in C Am., and several in the tropics e and w of Andes. Rare in the temp. zone.

OLIVE-BACKED WOODCREEPER
Xiphorhynchus triangularis – Plate XXXIII 10

19.5 cm. **Bill = head, slightly decurved.** Plumage **olivaceous**, except for chestnut lower rump and tail. Crown, nape, and occ. upper mantle lightly spotted to streaked, feathers of crown tipped blackish; wings grayish olive. Throat whitish, densely scaled by narrow dark feather-tips, **breast** and c belly with irregular **large whitish triangular spots** partly concealed by olive-gray tips and edges that widen downwards from the throat. Sspp *intermedius* and *bangsi* are browner above, back distinctly streaked buff, and bill ivory-white with dusky base and tip. **Juv**. unknown.
Habits: Solitary, in pairs or small groups, often in mixed-species flocks, and then regularly associated with Lineated Foliage-gleaner (Plate XXXVIII). Climbs trees probing into epiphytes and crevices.
Voice: Loud calls reported.
Breeding: Imm. Sep (Amazonas).
Habitat: Middle to upper strata. Medium stems, lianas, and moss-covered branches in tall as well as low humid forest and cloud forest, often at edge.
Range: 400-2700 m, mainly above 1500 m. W Ven., all 3 ranges of Col., e Ecu., and Peru n of the Marañón (*triangularis*); c Peru (*intermedius*), intergrading in se Peru into populations of La Paz and Sta Cruz, Bol. (*bangsi*). Another ssp in coastal Ven. (*hylodromus*). Fairly common.

NOTE: Forms a zoogeographical species with Spotted W. (*X. erythropygius*) of pacific Col. and Ecu.

SPOT-CROWNED WOODCREEPER
Lepidocolaptes affinis – Plate XXXIII 4a-c

19 cm. **Bill 1.5 x head, slender, distinctly curved, pale**. Crown, back, wing-coverts, and edge of outer primaries rufous olive; rump, tail, and rest of remiges chestnut; crown-feathers pointed, with buffy white spots and black tips, the spots becoming narrow indistinct streaks on nape and upper mantle. Headside streaked. Throat white or buffy white, rest of underparts olive-brown with very **distinct long white shaft-streaks outlined in black**. Sspp *lacrymiger* and *sneiderni* have the throat-feathers outlined in black, and the streaks below with rounded black-bordered tips, almost as spots in *sneiderni*. Ssp *sanctaemartae* has broader streaks below. Southern birds have palest bills and are most olivaceous. **Juv**. unknown.
Habits: Alone or in pairs, usually in mixed-species flocks, where often seen with Montane Foliage-gleaner (Plate XXXVIII). Aggressive towards Olive-backed W. Climbs trees, probing into mosses and other epiphytes, frequently on the underside of branches. See *Condor* 69(1967): 522-5.

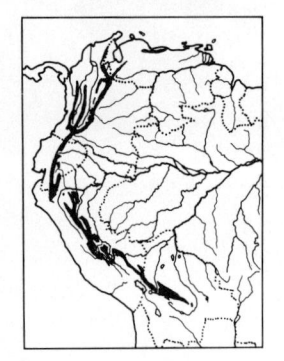

Voice: Its calls include a series of 3 reedy squeaks, a 2-note cry, and a little laugh with an introductory *ah*.
Breeding: No data.
Habitat: Stems and branches of rather thin moss-covered trees at all levels of open as well as dense humid forest and cloud forest, woodland, park-like pastures, and clearings.
Range: From upper trop. zone to 3000 m. Sta Marta mts (*sanctaemartae*); from Lara, Ven., to e slope of E Andes of Col. (*lacrymiger*), w slope of E Andes, C and W Andes of Col. s to Río Patía (*sneiderni*), sw Col. and both slopes of Ecu. (*aequatorialis*), n and c Peru (*warscewiczi*), Cuzco and Puno, se Peru (*carabayae*), La Paz to Sta Cruz, n Bol. (*bolivianus*). Other sspp in C Am. and coastal Ven. Fairly common.

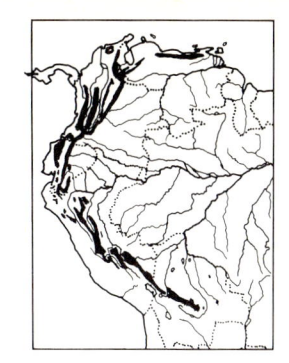

GREATER SCYTHEBILL *Campylorhamphus pucherani*
– Plate XXXIII 5

24 cm. **Bill long, slender, deeply curved**, pinkish gray to creamy brown. Plumage brown, crown dark brown, buff-spotted, headside blackish brown. Remiges, rump, and tail chestnut. Nape, upper mantle, breast, and upper belly with indistinct buffy gray streaks. **Sides of head with 2 distinct whitish lines** above and below eye. **Juv**. much like ad.
Habits: Solitary. Usually in mixed-species flocks. Climbs trees, probing into moss and crevices.
Voice: No rec.
Breeding: Fledglings Oct (e Ecu., Amazonas); small gonads July (Cuzco); 50 % oss. skull July (Ayacucho).
Habitat: 1-5 m up on moss-covered branches, tree ferns, and stems of tall thin trees in open cloud forest, elfin forest, and humid montane forest.
Range: 900-3250 m, mainly above 2000 m in Valle and Cauca W Andes and Huila, Col. (also 'Bogotá'), nw and e Ecu. s to Puno, Peru. Rare.

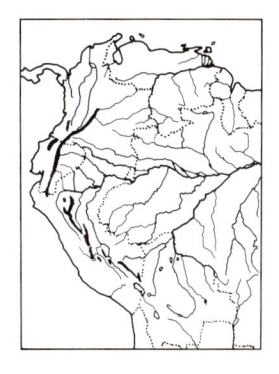

Ovenbirds – Family Furnariidae

The ovenbirds embrace a diverse assembly of some 215 exclusively Neotropical species. Short, curved, and rounded wings reflect the poor flying abilities of most species. The plumage is brownish or gray-brown (greenish in one species), and most forms have a rufous or whitish bar across base of flight-feathers. The primitive number of tail-feathers is 12, but in the subfamily Synallaxinae a reduction has taken place in several species. Thus, *Leptasthenura yanacensis*, *Schoeniophylax*, *Oreophylax*, *Hellmayrea*, *Certhiaxis mustelina*, *Poecilurus*, *Schizoeaca*, *Thripophaga cheriei*, and normally *Thripophaga macroura* have 10, members of the genus *Synallaxis* 10 or 8, and *Sylviorthorhynchus* only 6. Varying numbers of tail-feathers in closely related species is also seen in other families, e.g. wrens (genus *Cyphorhinus*) and antbirds (genera *Myrmotherula*, and *Grallaria*). Many furnariids have decomposed tail-feathers, sometimes ending in bare, more or less stiffened shafts, but only in the genus *Pygarrhichas* are the tips of the shafts strongly stiffened and curving abruptly inwards, as in woodcreepers.

The sexes are alike (female slightly smaller), and often they both sing the fairly monotonous trilled song. They are monogamous, usually keeping in close pairs, and both sexes building nest and tending young. The nest is closed, with a side entrance. The well-known mud-nests of the horneros have given name to the family. Many build enormous stick-nests, whilst others place their nest at the bottom of a tunnel, in a crevice under a rock, in a hole in a tree or a wall, etc.

The family is evidently closely related to the woodcreepers, and more distantly to the antbird lineage. The ovenbirds fall into 3 subfamilies: the stout, mainly terrestrial *Furnariinae*, with miners, earthcreepers, cinclodes, and horneros; the smaller *Synallaxinae* with long ragged tails, incl. canasteros, spinetails, and some aberrant forms; and the larger, mainly arboreal *Philydorinae*, which includes foliage-gleaners and tree-hunters. The group was revised by C Vaurie (1971) *Classification of the Ovenbirds (Furnariidae)*, London. The published information on the family was compiled by Vaurie, C (1980) in *Bull. Amer. Mus. Nat. Hist.* 166(1). However, we follow Peter's *Checklist of Birds of the World*, rather than Vaurie's classification. Comments on field identification of foliage-gleaners are given in *Continental Birds* 1(1979):32-7. See also Bond (1945), Meyer de Schauensee (1945c), and Esteban (1951).

Miners – Genus *Geositta*

10 species found in fairly dry habitats ranging from highlands and coastal deserts of Peru s through the Andes, most of lowland Arg., and the coast of Chile to Isla Grande. The Coastal Miner (*G. peruviana*) is confined to the coast of Peru, the remaining 9 species are found in the high Andes, 4 of them exclusively so. A closely related monotypic genus (*Geobates*) inhabits the campos of s Brazil. Miners somewhat resemble Old World larks in size, colors, vocalizations, and strictly terrestrial habits of running and walking, and digging the earth with their bill in search of seeds and insects. They nest in burrows (one species under a rock) on level ground or in low banks. Some species can dig their own burrows, whilst others depend on those provided by earthcreepers (*Upucerthia*) or rodents (usually *Ctenomys*). Most species, perhaps all, have courtship flights where they fly low (occ. high) over the ground with vibrating wings whilst singing. Many of them wag their short tails a little when excited. The tail and rump pattern is important for identification. See also *Amer. Mus. Novit.* 860 (1936).

GRAYISH MINER *Geositta maritima* – Plate XXXIV 9

15 cm. In **flight wings dark** without any rufous; **tail dark**, only outer web of outer feather buffy white. Bill fairly short and straight. Above gray, wings brownish gray, tips of greater and median coverts forming faint narrow bars; below whitish, **flanks pinkish**. **Juv**. buffier throughout. Wings with panel on tertials and edges on wing-coverts paler and buffier; ferrugineous area on inner flight-feathers very poorly defined and much restricted.

Habits: Usually in close pairs. Makes small hops and walks, digging the earth with its bill, picking up seeds and insects. Often wags hindbody or tail a little.

Voice: Song an unmusical *te-cherrrr* given on ground or during low flights, sometimes followed by a longer series of notes *ke ke ke ke* ... rising

and falling in pitch. Call (both on ground and in flight) *plyt-plyt* or *plyt*. Sometimes a softer *jyp*.
Breeding: Eggs Nov (Chile).
Habitat: Flat or sloping, sandy, gravely, or rocky, **extremely dry areas** with very sparse vegetation. Sometimes at edge of, but never within fields.
Range: Sea level to 2600 m (locally rarely to 3500 m) in arid littoral desert from Ancash, Peru to Tarapacá, n Chile. Fairly common.

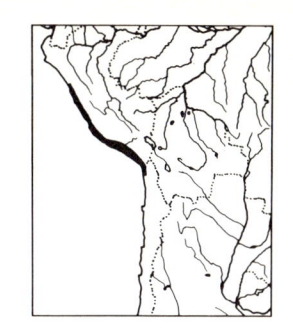

DARK-WINGED MINER *Geositta saxicolina* – Plate XXXIV 8

15 cm. **In flight dark wings; rump and basal half of tail buffy white**, terminal half blackish brown; outer rectrices mostly white. Bill fairly straight, blackish. Above rather dark gray-brown; below uniform, buffy to buffy white; **sides of head and neck**, and sometimes forehead and sides of crown **buffy cinnamon**. **Juv.** with a warm buffy tinge throughout, and dusky scales on breast.
Habits: Alone, in pairs, or small groups. Wags hindbody very often, and even cocks tail upright when excited and during what appears to be a communal display.
Voice: Shrill trill during fights. Flight-call *tirr tirr tirr*.
Breeding: Fledglings Dec.
Habitat: Gently **sloping** grassy or stony hills with banks and bare patches.
Range: 3700-4900 m in Junín to border of Lima, and into adjacent w Pasco and Huancavelica c Peru. Fairly common.

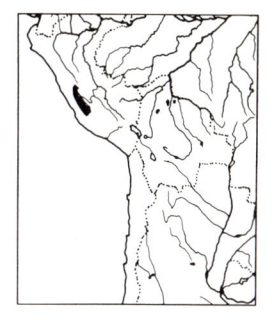

CREAMY-RUMPED MINER *Geositta isabellina* – Plate XXXIV 5

19 cm. **Large**. In **flight** remiges dull rufous with a hardly contrasting ill-defined broad dark brown subterminal bar; **rump whitish**, as are upper tail-coverts and base of tail; tail more buffy when spread, c rectrices and **broad terminal bar blackish**. Bill relatively long and thick, slightly curved, black. Above pale cinnamon brown; closed wing without, or with only small dull rufous patch. **Below unstreaked**, light grayish cream, upper throat and belly white. **Juv.** has the crown pale-spotted.
Habits and **voice**: Very active, running and making short flights, while continuously giving loud and piercing calls. Surveys and sings from rock-tops. Flight call *fit fit*. Song a loud strident series of 3-12 notes at 0.5 per s, the first 1-3 weaker than the rest. In flight the series is often preceeded by a jumbled rattle of 1-4 s (BW).
Breeding: Nov-Jan (Feb) (Chile, Arg.).
Habitat: Valleys with barren rocky slopes, at very high elevations.
Range: 3000-5000 m, down to 2000 m in winter. From Atacama to Talca, c Chile and nw Arg. in Jujuy, Catamarca, San Juan, and Mendoza.

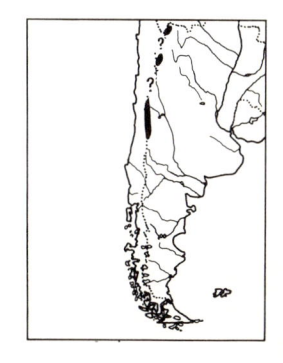

RUFOUS-BANDED MINER *Geositta rufipennis* – Plate XXXIV 6

17 cm, *fasciata* smaller. **Fairly large, upright. In flight conspicuous bright rufous wingbar outlined in black**, and rufous trailing edge, and **no white on rump or tail. Tail bright rufous** with fairly restricted blackish (sub-terminal) bar; **tip rufous**. **Bill** very **straight** and **fairly short**. **Underparts unstreaked**. Ssp *harrisoni* like *rufipennis* but smaller, with slenderer bill, whiter belly and no rufous on tips of longest primaries. Ssp *rufipennis*: above, incl. most of c tail-feathers buffy brown; closed wing black with small contrasting rufous patch at base of remiges; tips of all remiges incl. tertials rufous; throat white, rest of underparts warm buffy white; ssp *fragai* grayer above, rather cinnamon on breast; *hoyi* pure gray above and uniformly pinkish white below, with grayish tinge on breast. Sspp *fasciata* and '*hellmayri*' intermediate, '*hellmayri*' palest. Ssp *ottowi* differs from *rufipennis* by larger size, grayer coloration, narrower rufous wingbar more broadly bordered blackish, outer remiges uniform, and paler underside of wings. **Juv.** much like ad., but more buffy throughout, and with faint brownish tinge on breast; upperparts with pale feather-edges.

Habits: Often wags tail. Regularly moves in flocks, even in breeding season. Undulating flight like other miners, but flies more often. Feeds mainly on insects in summer. In winter partial to the seeds of *Coliguaya odorifera* (*Euphorbiaceae*).

Voice: Nuptial song persistent and shrill, with short low unmusical trills regularly interspersed *de rrrr dirdirdirdirde de rrr dirdir*.... Alarm a monotonous *keekeekeekeekeekeekee*. Appreciable differences between voices of *rufipennis* and *fasciata* have been noted (BW).

Breeding: Oct-Feb and Mar (Chile, Arg.).

Habitat: High flats or gentle slopes with matted short vegetation, patches of bare earth and scattered bushes, stones, and rocks. Banks and ridges in dry river beds.

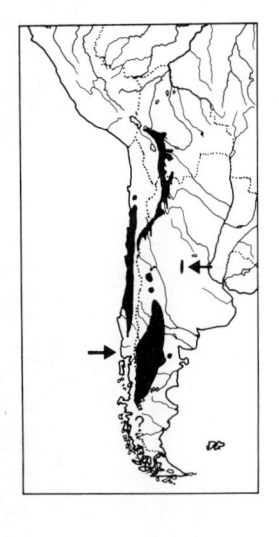

Range: Breeds at 3000-4400 m in La Paz, Cochabamba, Potosí, and Oruro, Bol., and from the coast to middle altitudes of Andes from s Antofagasta to s Cautín, Chile, where moving to lower altitudes in large flocks in winter (*fasciata*); Quebrada Paposa in sw Antofagasta (*harrisoni*, see Marín, Kiff & Peña in *Bull. Brit. Orn. Club* 109(1989):73); Atacama and Coquimbo, n Chile ('*hellmayri*'), nw Arg. from Jujuy to San Juan (*rufipennis*), Cerro Famatina in La Rioja (*fragai*, see M Nores in *Hornero* 12(1986):262-73), the Andes and pre-Andean foothills of Arg. in w San Juan and Mendoza (*giaii*, JR Contreras in *Physis B. Aires (C)* 35(1976):213-20), and from Mendoza to Sta Cruz, and Chile in s Aysén (*hoyi*, Contreras in *Hist. nat. Mendoza* 1(1980):137-48; the ranges of the 2 are difficult to separate from the rec. given by Navas & Bo, in *Rev. Mus. Arg. Cienc. Nat. 'Bernardino Rivadavia'* 14(1987):55-86). Above 2500 m (with little altitudinal movement) in Sierra de Córdoba in Córdoba, Arg. (*ottowi*, G Hoy in *J. Orn.* 109 (1968):228-9). Fairly common.

NOTE: The ssp name *hellmayri* is preoccupied by a form of *G. cunicularia*. A renaming has not been suggested, to our knowledge, presumably because the population is doubtfully distinct.

PUNA MINER *Geositta punensis* – Plate XXXIV 7a-b

13.5 cm, body quite round. **In flight** shows strongly rufous wingbar contrasting with broad black trailing edge. No white on rump or upper tail-coverts. **Tail fairly short, isabelline above** with fuscous terminal bar and whitish whole outer web of outermost feather; inner webs of rectrices rufous above the terminal bar. Bill slightly decurved at tip (shorter than in *cunicularia*), black with paler base of mandible. Above pale brown; **below uniform** whitish, sometimes with slight buffy tinge. Outer web of remiges (except 4 outer) rufous, forming conspicuous **rufous patch in closed wing**; greater coverts rufous with dark subterminal bar. **Juv**. has crown and back darker and grayer, lower back with dull rufous tips.
Habits: Alone or in pairs. Fairly often wags hindbody, esp. upon landing. Often bows head. Often perched very upright.
Voice: Song given in flight with vibrating wings low over ground or from bushtop, a shrill *tree tree* **trier-trie-trie**-*trie-trie*..., lasting 10-15 s, 3rd-5th notes accentuated and the rest descending towards the end. Flight-call a sharp *kveee*, when distressed *krrip*.
Breeding: (Oct) Nov-Jan (Feb) (Chile); fledgling Jan (Potosi, Jujuy).
Habitat: Depends on *Ctenomys*-burrows for nesting. Level parts of the puna zone. On dry pumice flats with scattered mats of grass and tussocks of bunchgrass. Places with patches of bare earth and scattered stones and rocks in areas with matted grass, but occ. on grassy slopes bordering streams. Fairly bare slopes in bushy watered canyons in arid regions.
Range: 3000-5000 m in Mocquegua, Tacna, and Puno, s Peru, La Paz, Oruro, and Potosi, Bol.; Jujuy, Salta, and Catamarca, nw Arg., and neighboring Chile from Tarapacá to Atacama. Fairly common.

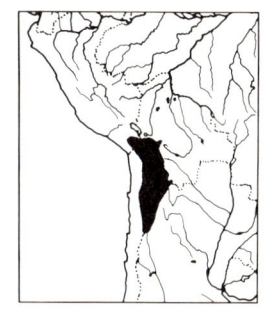

SHORT-BILLED MINER *Geositta antarctica* – Plate XXXIV 4

14.5 cm. In **flight** remiges **gray with only slight rufous tinge**. Upper tail-coverts whitish, **outer 3 rectrices mostly whitish, rest of tail blackish brown forming dark triangle**. Wings long, tips extending 2 cm beyond tertials. Above **grayish** brown, supercilium and upper tail-coverts whitish; wing-coverts with ill-defined narrow grayish **tips**, forming narrow bars. Below buffy white, breast mottled with pale brown. Bill **slightly** shorter, straighter, and slenderer than in Common M., base of mandible pale. **Juv**. has pale-tipped crown-feathers and less defined breastband.
Habits: Like a typical miner, walks picking up insects and seeds. Alone or in pairs, during migration in flocks up to 50, and then often in company with Common M.
Voice: Flightcall a shrill *tjeek, de trrrit, trritrritrritrritrrie*, or *tjiriidetjiherri-rri*, quite different from the nasal call of ssp *cunicularia* of Common M.
Breeding: Nov-Jan (Isla Grande).
Habitat: Breeds in sandy coastal areas and on the sheep-grazed grasslands. Possibly also on barren basaltic uplands, as well as in heavily grazed areas. On migration like Common M.

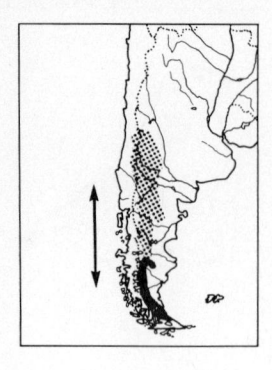

Range: Breeds on northern Isla Grande, and on isls in the Straits of Magellan, possibly sparsely at 500-1000 m on Andean foothill plateaus through Sta Cruz. As early as Feb migrates along the barren Andean foothills, wintering n to c Chile and Mendoza, Arg. Abundant on the unforested part of Isla Grande (BW) and fairly common as migrant in w Sta Cruz (Feb).

COMMON MINER *Geosutta cunicularia* – Plate XXXIV 3a-d

17 cm. **In flight remiges rufous**, primaries edged dark brown all around, secondaries with dark subterminal bar. **Upper tail-coverts**, except c, **creamy white** in *juninensis* (where most extensive) and *titicacae*, buffy gray in sw Peru (*frobeni, georgei, deserticolor*); c tail-feathers dark brown, rest dark brown with cinnamon-tawny base (base creamy in *juninensis* and *titicacae*); outer web of outer 2 feathers white. Bill fairly long, slightly curved. **Tertials approximately reach wingtip (unlike Short-billed M.)**. Above gray-brown to buffy brown; **wing-coverts with pale margins forming conspicuous irregular mottling** (unlike narrow **bars** in Short-billed M.); **long rufous panel in closed wing** (largest in se Peru, Bol., and Arg.). Below whitish, **breast streaked dark brown (unstreaked in *juninensis*)**. Juv. faintly pale-barred above; breast-streaks less well defined.

Habits: Alone, in pairs, or family groups; flocks during migration. Often wags hindbody a little, but very rarely cocks tail upright. When nervous wipes bill. During nuptial flight flies in long waves low over the ground, circling and swinging erratically, occ. rising to 10-15 m and then descending with vibrating wings.

Voice: Nuptial song a descending series of shrill notes slowing towards the end and sometimes initiated with a slightly lower note: *de dirr-rr-rrrrr* (or *ta whit-ta whit*... in *cunicularia*). Alarm a monotonous fast *kee-kee...kee*. Flight-call a highpitched *keep*, (or a nasal *dee-dijer* or *er?* in *cunicularia*) as well as trilled notes resembling song.

Breeding: Eggs Nov (Salta), Oct (San Luis), ult Aug-Feb (Chile, 2 broods); fledglings Oct-Dec (s Chile, s Arg.), Dec (La Paz), Oct (Puno), Oct, Nov (Junín); nestlings Nov (Jujuy).

Habitat: Level or very gently rolling ground, mostly on short grass, and much less on patches of bare sand or earth than Puna M. Sometimes in cultivated fields.

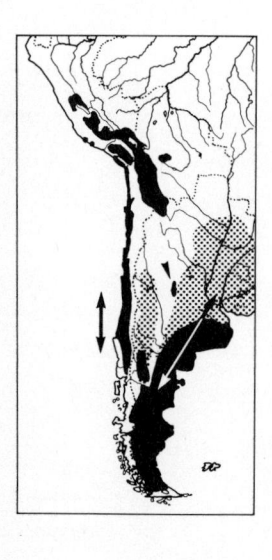

Range: 4000-5000 m in Junín and Huancavelica, c Peru (*juninensis*), 3000-4800 m in highlands of Ayacucho and Puno, Peru, Bol. on altiplano and in Cochabamba, and Arg. in Jujuy, and rec. also in La Rioja and Mendoza (*titicacae*), 4300-2300 m in highlands and on the upper pacific slope of s Peru and n Chile (*frobeni*), 'lomas' near coast in extreme s Ica and Arequipa (*georgei*, M Koepcke in *Beitr. neotrop. Fauna* 4(1965):154-7), and coast and foothills from Arequipa to n Chile (*deserticolor*), c Chile from s Atacama to Llanquihue (*fissirostris*), mts of Córdoba, Arg. (*contrerasi*, Nores et al. in *Hist. nat. Mendoza* 1(1980):169-72), from c Neuquén s to n Chubut, Arg., at lower elevations in winter, some migrating n (see also *Neotrópica* 25(1979): 167-71) (*hellmayri*), and in s Arg. and s Chile, scarce and local s of Strait of Magellan (*cunicularia*). Southern birds winter n to s-most Brazil, Urug., and Parag. Common s to Straits of Magellan, but very scarce and local on Isla Grande (BW).

NOTE: We suggest that the vocally aberrant *cunicularia* may be a distinct species. The remaining forms should then be called *G. fissirostris*. Revision of Arg. forms by JR Contreras in *Hist nat. Mendoza* 1(1980):33-40.

SLENDER-BILLED MINER *Geositta tenuirostris* – Plate XXXIV 1a-c

18 cm; **bill long** (29 mm), **slender**, slightly decurved. With rather slim jizz and long legs, but **tail short like in other miners** (long in Earth-creepers). In **flight wings extensively rufous**, dark subterminal band on remiges lacking or restricted to inner secondaries. **No white in rump or tail**; tail mostly rufous, c feathers and tips of 2-4 c pairs dark brown. Above gray-brown to brown, crown and back faintly mottled with dark feather-centers, supercilium buffy white. Distinct long rufous patch in closed wing. Below buffy white; breast mottled or faintly streaked dark brown (or distinctly streaked fuscous in Lima and Junín). **Juv**. has more distinct pale feather-edges on crown and back, and has the breast streaked dark brown.
Habits: Alone or in pairs. During nuptial flight circles at great height singing and vibrating wings, then gliding to the ground, still singing and vibrating, sometimes continuing song from ground or bushtop. Rarely wags hindbody.
Voice: Song during nuptial flight with a quality of other miners, a high-pitched, continuous *tji-tji-tji-tji*.... Alarm a complaining sharp, but not short *keek*.
Breeding: Nests in *Upucerthia*-burrows in banks. Large gonads Dec (Cuzco, Puno). Nuptial flight noted Aug (Lima), Oct (Arequipa), Oct, Dec (Puno).
Habitat: More often near water than other miners, though not always so. In ploughed fields, grassy flats, or on gentle slopes with bare earthy patches. Meadows or the vicinity of rivers, lagoons, and bogs. Small hills with scattered low scrub.
Range: 2500-4600 m in w Andes of Peru from Cajamarca to Arequipa, and from Huánuco s through highlands of Bol. to Tucumán, nw Arg. Fairly common.

THICK-BILLED MINER *Geositta crassirostris* – Plate XXXIV 2

18 cm. **In flight** conspicuous rufous wingbar outlined in blackish; **no white in rump or tail. Tail mostly rufous with indistinct dark terminal or subterminal bar. Bill** rather long and stout, slightly curved, esp. **heavy at base. Legs whitish**. Above dark grayish to brownish, mottled with dark brown, almost uniform in worn plumage, inner part of closed wing sometimes with rufous patch. Supercilium and eye-ring pale. Below whitish to grayish buff, breast more or less streaked and mottled with brownish or grayish. Ssp *fortis* more reddish brown above,

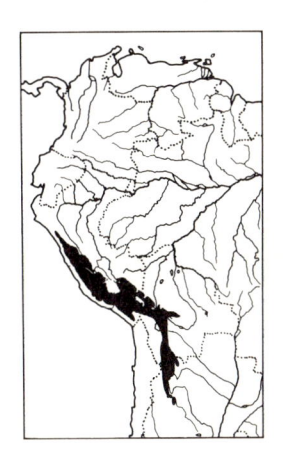

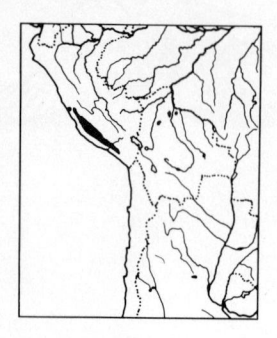

scaling of back and wings not so blackish, rufous wingbar broader, terminal black wingband narrower. **Juv**. undescribed.
Habits: Appears as a typical miner. Surveys and sings from ledges of cliffs, or the top of a rock or cactus. Alone or in pairs walks on the ground picking up seeds and insects.
Voice: Rather unlike other miners. In flight gives sparrow-like *chirp* (M Koepcke on ssp *crassirostris*). On the ground a fairly loud deep plaintive call. Other calls incl. a loud *keen keen*. The song is a series of c. 10 rather drawn-out, wheezy notes, rising through the 1st couple to a steady pitch, then falling in pitch and slowing noticeably through the last syllables, the whole lasting c. 5 s (BW on ssp *fortis*).
Breeding: No data.
Habitat: Ssp *fortis* in very dry valleys and slopes with scattered large rocks, bushes, and cacti, *crassirostris* in rocky hills with fog vegetation (lomas).
Range: W Peru at 1500-3000 m in Lima, w Ayacucho (where reaching 3850 m on Pampa de Nazca), and Arequipa (*fortis*), and at 600-800 m in coastal lomas of Lima (*crassirostris*). Generally uncommon.

Earthcreepers – Genus *Upucerthia*

9 species, 2 of them in the chaco and adjacent Andean slopes, the other 7 Andean, from c Peru s to Isla Grande. Earthcreepers are insectivorous and resemble the N Am. thrashers in size, by their long curved bills (straight in 1-2 species), and by their sneaky behaviour as they hop about mostly on the ground, and hide in or behind bushes at the slightest disturbance. The bill grows slowly, so younger birds have considerably shorter bills than ad. (see Plate XXXIV 11b). The fairly long tail is rufous or rufous brown, and is frequently cocked. The nest is normally placed in a burrow excavated in a bank by the bird, but in the 1-2 smallest species a fissure in a rock, cavity in a tree-trunk, or the used nest of a Canastero (*Asthenes*) or Thornbird (*Phacellodomus*) may be used (perhaps they can build their own stick-nest). For localities and biology of Arg. forms see Navas (1971b).

SCALE-THROATED EARTHCREEPER
Upucerthia dumetaria – Plate XXXIV 18a-c

19 cm. **Bill long, curved**. Ssp *dumetaria* dark gray-brown above, ear-coverts streaked; supercilium white. Tail dark with outer web of outer feather and tips of following 2 pale rufous. Wings with rather restricted dull rufous patch on base of inner remiges. Below buffy brown with distinct dark tips and edges giving **scaled** effect; throat and c belly white, sides gray-brown. Ssp *hypoleuca* slightly more rufous. Ssp *saturatior* darker with a shorter, blacker bill. Ssp *peruana* darker than *hypoleuca*, nearest *saturatior*, but bill longer, less abruptly curved; rufous wingbar less sharply defined distally. Ssp *hallinani* very pale, only lower throat and upper breast faintly scaled, underparts white. **Juv**. sometimes has the whole underside scaled and usually has narrow pale streaks on forehead and upper mantle.

Habits: Alone or in pairs, boldly hops on the ground sometimes cocking tail 30°. Surveys from rocks or bush-tops. When alarmed makes long low flights between the bushes or hops away at considerable speed. Sings with slightly raised tail from elevated post.
Voice: Song a rapid *chippy chippy chippy chip*. Call a wheezy, abrupt *keet*.
Breeding: Eggs Oct-Dec (Chile, Arg.)
Habitat: Rolling grassy hills or arid brush-covered slopes with earthy banks or rocky outcrops. Edge of rocky basaltic plateaus with sparse vegetation. Among densest growth of *Atriplex* and other scrub on seasonally inundated plains. In winter in tracts with dense thorny brush. Open woods.
Range: 2000-4000 m from Puno, se Peru (*peruana*, J Zimmer in *Proc. Biol. Soc. Wash.* 67(1954):189), through Oruro, Cochabamba, and Potosí, w Bol., and Jujuy to Mendoza and Córdoba, Arg.; also in C Cordilleras of Chile, in the n from Antofagasta to Coquimbo (where descends to C Valley and coastal hills in winter), and in c Chile from Aconcagua to Tarapacá (*hypoleuca*). C Valley and coastal hills of c Chile from Aconcagua to Valdivia, and in w Neuquén and w Chubut, Arg. (*saturatior*), and in Arg. from Neuquén s to Tierra del Fuego, in winter n to the pampas region (*dumetaria*). Birds from Jujuy, nw Arg., and Río Loa in Antofagasta to n Coquimbo have been separated (*hallinani*), but intergrade with *hypoleuca* in Atacama. Fairly common in s parts of range, uncommon in n, not reported in Peru in recent years.

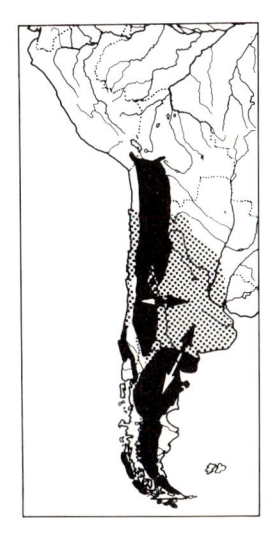

WHITE-THROATED EARTHCREEPER
Upucerthia albigula – Plate XXXIV 12

19.5 cm. **Bill** long, curved, **relatively thick at base**, tail shorter, and upperparts averaging browner than in Plain-breasted E. (with which it overlaps), crown clearly darker and grayer than back. **Chin and sometimes entire throat** and c lower belly **pure white**; broad supercilium and rest of underside pale buff; **lower throat and upper breast** finely **scaled** by dark feather-tips. **Wing-coverts and outer web of all but outer 2 primaries dark rufous; no sooty shade on margins of secondaries**; tail dull rufous. **Juv** has a distinctly darker crown than back, and scaling of upper breast and lower throat more pronounced.
Habits: Terrestrial. Fairly shy and sneaky, hopping quickly away to hide between bushes when alarmed. Usually alone. Often cocks tail to c. 40°.
Voice: Call a short sharp high-pitched **chit**. Song, delivered from a place with good view, is a chattering series of 4-10 sharp dry *chit* notes of scolding quality, c. 6 per s (but unlike in Plain-breasted E. slow enough to be counted) (BW).
Breeding: In banks. Eggs Nov (Chile).
Habitat: Arid zone, in ravines with streams and bushes, hedgerows in agricultural areas, and desert scrub. Wet meadows.
Range: 2300-3550 (3900) m, overlapping with Plain-breasted E. Only known from w slope of Andes in Ayacucho (slope below Pampa de Nazca at 2675 m), Arequipa, Tacna, and Tarapacá (near Putre) in s Peru and n Chile, but fairly common (see *Condor* 89(1987):654-8).

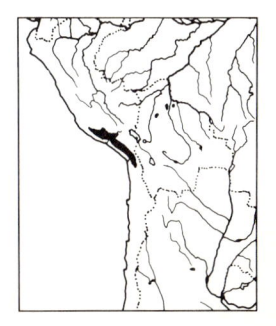

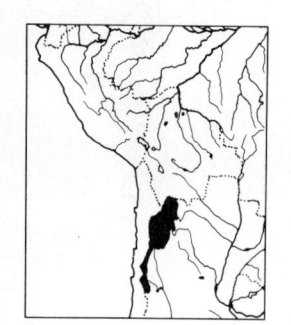

BUFF-BREASTED EARTHCREEPER
Upucerthia validirostris – Plate XXXIV 10

20 cm. **Bill long, deeply curved**. Above light gray-brown; supercilium and **entire underside dark buff**. C tail-feathers like back, rest dull dark rufous. Remiges with dull rufous edging in closed wing, in **flight** looking mostly rufous. Ssp *rufescens* is more rufous throughout. **Juv**. has the breast mottled with dark brown, and has faint pale bars above.

Habits: Alone, in pairs, or in family groups. Terrestrial. Tail often half-cocked. Moves in long powerful hops, or in short, undulating flight just above the bushes. Secretive, hiding between bushes. Surveys from rocks or bushtops. Sings from bushtop with hanging, vibrating wings and spread tail.

Voice: As a rule silent, but occ. a low *chwit* is given. Song, given from a ravine edge, bush top etc., is a shrill long trill of usually more than 10 *tyik* notes of different quality than in Plain-breasted E., given at c. 5 per s, the first couple of notes fainter than the rest (BW).

Breeding: Eggs Nov (Salta), nestlings Feb (Arequipa), Mar (Jujuy); fully grown imm. Mar (Mendoza).

Habitat: Along gravely or sandy stream-beds with bare earth between rather dense *Lepidophyllum* brush, or on slopes with scattered bushes.

Range: From 2500 m to well above 4000 m. Nw and w Arg. from Jujuy to Mendoza (type from Mendoza) (*validirostris*), and in Sierra de Famatina, La Rioja (*rufescens*, see M Nores in *Hornero* 12(1986):262-73). Fairly common.

NOTE: The paraspecies Plain-breasted E. is sometimes included in this species.

PLAIN-BREASTED EARTHCREEPER
Upucerthia jelskii – Plate XXXIV 11a-b

19 cm. **Bill long, curved**. Above dark brown, supercilium buff; tail rufous chestnut tinged dark brown (or plain dark brown in *saturata*); edges of remiges dark gray, sometimes with **small rufous area at base of inner primaries**. **Below** buff (grayish buff in *saturata*), **fairly uniform**, breast occ. with very faint mottling. In **flight** remiges show indistinct dusky chestnut base of inner webs. Southern *pallida* has more rufous

wings and tail, (approaching Buff-breasted E., but wing still paler and less rufous, and with sooty shade to edges of secondaries, which also distinguishes it from White-throated E. of generally lower elevations). **Juv**. has the breast mottled with dark brown.
Habits: Terrestrial. Prefers running or hopping to flying. Usually alone. Tail often half-cocked. Inquisitive, but when alarmed quickly hides between bushes or rocks. Surveys and sings from top of rock. Shakes wings and holds tail slightly lowered during song.
Voice: Call a metallic *click*. Song of *jelskii* a long, drawn-out, high-pitched, and rather weak trill *drrrrrrr...*, first rising, then falling in amplitude. Song of *pallida* (BW) too fast to count notes, and higher-pitched than in White-throated E., lower-pitched and slightly slower than in *jelskii*, a series of evenly pitched *tik* and *sik* notes usually preceded by a few, evenly spaced, harder notes, with no change in amplitude. 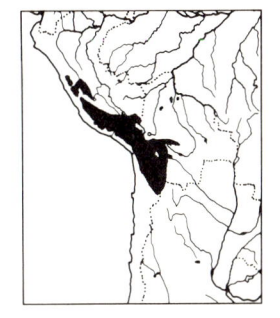
Breeding: Eggs Nov (Chile), Feb (Ancash); nestlings Feb (Arequipa); fledgling Jan (Lima); juv./imm. Aug (Arequipa).
Habitat: Dry rocky slopes with scattered shrub like *Chuquiragua* and *Ribes*, open grassland with rocks or banks.
Range: 3000-5000 m. W Andes in Ancash and nw Pasco, c Peru (*saturata*), Lima, Junín, and Huancavelica, c Peru (*jelskii*), Ayacucho, Cuzco, Puno, and Arequipa, s Peru, to Tarapacá, n Chile, and through highlands of Bol. to Salta, nw Arg. (*pallida*). Fairly common.

NOTE: Often regarded sspp of Buff-breasted E., but further research in nw Arg. is needed to settle this question.

STRIATED EARTHCREEPER *Upucerthia serrana* – Plate XXXIV 13

18 cm. **Bill** long **curved**. Above dark gray-brown; forecrown and mantle narrowly streaked buffy. **Below** gray-brown **heavily streaked buffy to whitish**; supercilium, throat, and c belly whitish. **Wings and tail strongly rufous**. Ssp *huancavelicae* said to be darker and with less brown-tinged flanks. **Juv**. scaled on breast and faintly barred on upper belly, lower underparts almost unstreaked.
Habits: Alone or in pairs, hops on the ground with tail half-cocked. Flicks litter like a thrush. Occ. climbs *Polylepis* trees, holding tail down, and probing bill into crevices. Inobtrusive, hiding in dense bushes.
Voice: Song a harsh slowing trill introduced by 3 notes, the 1st longest, *keep kip kip trrrrrrrrrr-r-r r*, lasting c. 4 s. Call a single, drawn-out, somewhat wheezy *weeeee*, sometimes over 1 s long and given regularly at 5 s intervals for long periods. 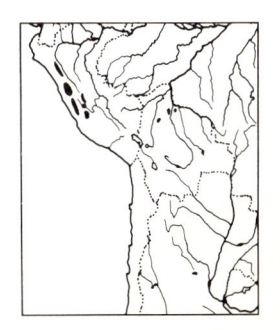
Breeding: Nests in adobe walls and banks. Fledglings Jan (Pasco); juv. Mar, Apr (Ancash), June (La Libertad).
Habitat: Bushy rocky slopes and open spaces with scree, rocks, and tall puna grass in *Polylepis* woodland.
Range: Peru. 3000-4200 m in Cajamarca, La Libertad, Ancash, and Lima, W Andes, crossing over to e-draining valleys in Junín and Huánuco (*serrana*), and in Huancavelica (*huancavelicae*). Fairly common.

ROCK EARTHCREEPER *Upucerthia andaecola* – Plate XXXIV 15a-b

16.5 cm. **Bill** long, **almost straight**. Above brown. **Tail uniform rufous** (occ. some rectrices have a little dusky at tip); closed wing with rufous patch on remiges. Supercilium buffy white (usually more buff than in Straight-billed E.), throat whitish; rest of underside buffy white to pale buff, feathers more or less with narrow, darker edges, esp. on sides and belly. Bill and **legs brown**. **In flight** remiges dusky with broad basal rufous bar. **Juv.** has the whole underside below white throat scaled; back with slightly indicated pale feather-edges.
Habits: Alone or in pairs. Climbs large rocks almost using tail for support. Also hops about between bushes in typical sneaky earthcreeper fashion with tail cocked 45° or more. Lowers tail when pecking on the ground.
Voice: Song, given from a shrub, or occ. from the ground, a piercing *veetveet-veeveeveeveeveeee-veet-viree-veeetveee-veee*, or occ. extended through 3 higher-pitched notes immediately followed by a rapid descending chatter (BW). Call wheeny and drawn-out (1 s) *weeee*, often repeated. Alarm *krik-krik*.
Breeding: Eggs Oct (Tarija), Nov (Salta); fledglings Nov (Salta, Sucre), Dec (Cochabamba); Imm. Nov (Cochabamba), Feb (Potosí).
Habitat: Stony slopes, steep banks of bushy canyons in morraine soil with *Prosopis* scrub forest, columnar cacti, and sometimes *Polylepis*. Also at edges of ploughed terraced fields.
Range: 3000-4500 m from La Paz and Cochabamba, Bol., to Catamarca, nw Arg., and e-most Antofagasta, n Chile. Fairly common.

STRAIGHT-BILLED EARTHCREEPER *Upucerthia ruficauda* – Plate XXXIV 14

16.5 cm. **Bill slender, only very slightly curved, or completely straight**. Ssp *ruficauda* pale brown above; supercilium, **throat and breast white, unstreaked**; rest of underparts white variably streaked with cinnamon-buff. **Tail-feathers rufous**, all except outer pair with mostly blackish inner webs, c pair more or less washed with dusky. Bill blackish, **legs black**. **In flight** remiges dusky with broad rufous basal bar. Ssp *montana* has a very white throat and breast, and a darker abdomen. Ssp *famatinae* has darker lower underparts contrasting with white breast; more southern birds have more cinnamon-tinged breast, and are less rufous above. **Juv.** much like ad., but tinged rufous throughout; feathers of crown and mantle pale-tipped.
Habits: Alone or in pairs. Secretive. **Tail normally partly raised and sometimes cocked 90°**, or even tilted slightly forwards. Restless, runs and hops on the ground and hides under or in bushes.
Voice: Contact call, a single sharp *queeee* or *wheeet*, is given frequently all year round. Unlike Rock E. sings from rocks and almost never from vegetation. The song is a series of *pu* notes, 5 per s, or faster, quickly rising into 2-3 loud, penetrating *pee* or *wheet* notes spaced farther apart, sometimes followed by extra *pu* notes. Also gives a discinctive series of 4-5 *tee-yeh* calls in rapid succession, maybe in territorial contexts (BW).

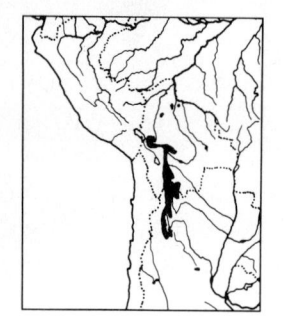

Breeding: Eggs late Nov-Dec (Arg., Chile); fledglings Dec (Catamarca, San Juan, Mendoza), Jan (Neuquén).
Habitat: Rocky slopes and outcrops, often in gullies with some scrub, bunchgrass, or arid *Polylepis* shrub.
Range: 2300-4500 m (generally higher than Rock E., but down to 1280 m in Neuquén). Arequipa, s Peru, to n Tarapacá, n Chile (*montana*), Oruro (intergrades) and from La Paz and Potosí, Bol., s through Andes of Arg. to Catamarca and in neighboring Chile (ssp ?); Cerro Famatina in La Rioja, Arg. (*famatinae*, see M Nores in *Hornero* 12(1986):262-73); Andes of San Juan to Chubut and adjacent Chile to Santiago (*ruficauda*); several populations distinct. Locally fairly common.

NOTE: Sometimes placed in the genus *Ochetorhynchus*.

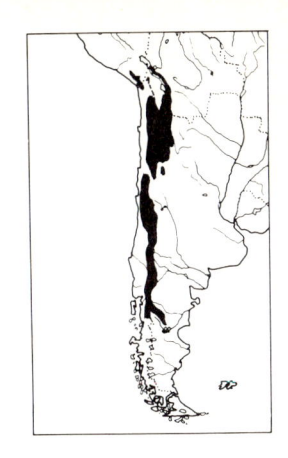

CHACO EARTHCREEPER *Upucerthia certhioides* – **Plate XXXIV 17a-b** - is included in this book because the available biological data are likely to apply also to the Bolivian E., which is Andean. Inhabits woodland and heavy brush and scrub fringing dry riverbeds below 1350 m in nw Arg. (but to 1800 m in Córdoba): e Formosa to Entre Ríos (*certhioides*), Jujuy and sw Chaco to Catamarca and Córdoba (*estebani*), and La Rioja and Mendoza (San Juan?), and sighted in La Pampa (BW) (*luscinia*). – **16 cm**. Bill hardly curved. Above rather light gray-brown tinged rufous, warmer on rump (*luscinia* larger and **grayer**); **forehead and supercilium dull dark rufous**. Throat white, rest of **underparts gray-brown** (darkest and most uniform in *luscinia*), paler than back; flanks tinged reddish brown (except in *estebani*). Wings like back or more rufous chestnut, **in flight** mostly dull rufous remiges without contrasting wingbar (*estebani*), or remiges edged rufous, and showing small rufous bar in flight (*luscinia*; *certhioides* being intermediate). Tail dull rufous, but c rectrices like back. **Juv.** tinged rufous throughout, but rufous of forehead less distinct. – Territorial. Cocks tail. Forages on the ground and low in bushes in such dense cover that it is difficult to observe, but its presence is often revealed by its loud calls. At any alarm gives very loud whistled calls *tio* and *tuit* (suggestive of N Am. Canyon Wren). Sings from high in tree or scrub, a 2-part series of 6-15 sharp penetrating notes, 4 per s, occ. trailing of with weaker notes (BW). Nestsite variable: builds sticknests in bushes or trees, or nests in a deep fissure or cavity among rocks, or in a hollow tree-trunk or hornero's nest. Eggs Oct-Dec (San Luis), Nov (Córdoba).

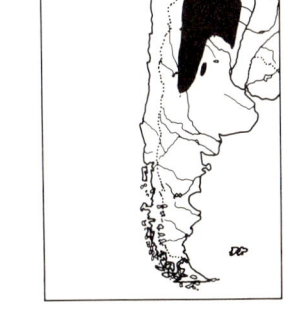

NOTE: Sometimes placed in the genus *Ochetorhynchus*.

BOLIVIAN EARTHCREEPER *Upucerthia harterti* – **Plate XXXIV 16**

16 cm. Bill slightly curved. Above gray-brown turning rufous brown on rump; supercilium pinkish buff; throat and upper breast white, rest of **underparts pale grayish buff**, flanks and vent pale brown; wings and tail dull rufous. **Juv.** tinged rufous throughout.

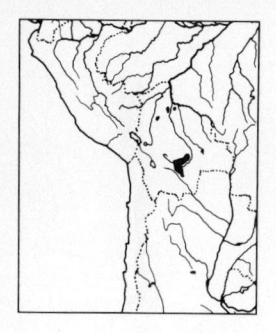

Habits, **Voice**, **Breeding**, and **Habitat** possibly like the closely related Chaco E.
Range: 1500-3050 m in Cochabamba, Sta Cruz, and Chuquisaca, Bol.

NOTE: Sometimes placed in the genus *Ochetorhynchus*. Closely related to the Chaco E., and by some considered a ssp of it.

Earthcreeper – Genus *Eremobius*

Monotypic genus. Jizz somewhat intermediate between *Upucerthia* and *Asthenes*. Bill straight. Tail rarely cocked. The large stick-nest is built in a low bush. For localities and biology see Navas (1971b).

BAND-TAILED EARTHCREEPER
Eremobius phoenicurus – Plate XXXIV 19

17.5 cm. **Bill fairly long and straight**. Tail hardly graduated. Above gray-brown to smoky gray; posterior supercilium whitish, **earcoverts dark rufous**. C tail-feathers dusky, bleaching to paler than the back, sometimes with some rufous chestnut at base; **rest of tail-feathers bright rufous chestnut at basal half** or more, blackish brown terminally, amount of blackish brown decreasing towards outermost. Throat silvery white, this color sometimes continuing as streaks down rest of underparts, that are pale gray-brown; c belly buffy white; vent white to ochraceous. Under-wing with pale rufous bar, not visible from above in flight. **Juv**. like ad., but forehead with faint pale tips, throat unstreaked, and streaks on rest of underparts vague, scarcely evident on breast.
Habits: Alone or in pairs. Mostly terrestrial, but sometimes perches on tip of side-branch of bush. Sings from bush-top. Probes into rock crevices, tussocks of grass, and low cushion-plants in search of insects and seeds. Tail usually carried low, occ. cocked 30°. Fairly secretive, hiding between bushes and rocks.
Voice: Calls include a trill *trrrrrrrrrt* much like Scale-throated Earthcreeper, a faint clicking *tick-tick-tick*, and upon landing, a very sharp mammal-like wheezy note *wheeet*.
Breeding: Eggs Nov (Río Negro), nestlings Jan (Chubut), juv. Dec (Rio Negro).
Habitat: Slopes with bare earth or sand and cushion-plants and tussocks of composite plants, and often some thorny scrub. Occ. between rocky outcrops.
Range: At least to 1200 m. W and s Arg. in La Rioja (occ.), Mendoza, Neuquén, Río Negro, Chubut, and n Sta Cruz. Fairly common in Neuquén and Río Negro, elsewhere sparse.

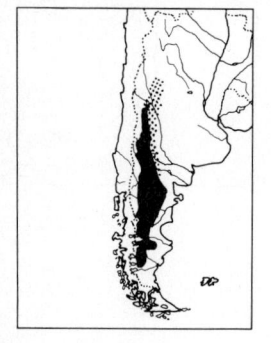

Chilia – Genus *Chilia*

Monotypic genus, only found in Chile. Much like an *Upucerthia*, but bill straight and most of tail black. Nests in a cavity under a rock.

CRAG CHILIA *Chilia melanura* – Plate XXXV 8a-b

19 cm. **Bill straight**. Crown gray-brown turning brown on back; rump rufous chestnut. **Tail** fairly long, **black** with rufous base, outer web of outer rectrix rufous chestnut. Supercilium, throat and breast whitish, not sharply set from gray-brown of underparts, grading to rufous chestnut on vent. Northern birds (*atacamensis*) paler. **Juv**. has faint dark bars on breast and belly, and pale tips on back-feathers.
Habits: Restless and elusive. Swiftly moves about cracks and crannies of rocky outcrops in search of insects and seeds, often cocking tail, and rarely flying. Suns itself early morning, upright on rock-top, exposing black rump to the sun (BW).
Voice: While foraging continuously utters a piercing metallic chatter. Sings from posts marked with considerable droppings on tops of rocky outcrops. The 1 s song is a staccato chatter with a short rapid jumble of notes at the beginning and end, and dominated by 4-7 loud piercing *teet*s (5 per s) in the middle (BW).
Breeding: Eggs Sep-Oct.
Habitat: Crags on semi-arid hillsides with scattered clumps of vegetation among rocky outcrops and boulders.
Range: Chile in the Andes up to 2500 m, foothills, and coast range in winter. Atacama and Coquimbo (*atacamae*), and from Aconcagua to Colchagua (*melanura*).

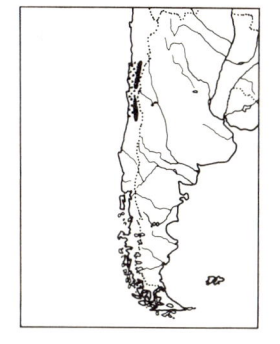

Cinclodes – Genus *Cinclodes*

12-13 species of medium-sized to large furnariids: 1 in mts of se Brazil, 3 only along coast of Peru, Chile, and Fuegian isls, the remaining 8 in the high Andes. All forage in the vicinity of water: along streams, at lakesides, on bogs, and coastal cliffs etc., sometimes wading into the water in the manner of a dipper (*Cinclus*). Upon landing or taking off, the tail is cocked, and during the display one or both wings are raised to expose wingbar on the underwing, and a simple trill is emitted. The nest is placed in a fissure under a rock, in an excavated burrow, or sometimes in the wall of a house.

STOUT-BILLED CINCLODES
Cinclodes excelsior – Plate XXXV 11a-b

20 cm. **Bill stout**, somewhat curved distally. Above dark brown, supercilium and throat whitish. Below gray-brown, irregularly mottled with buffy white on breast and belly, and with many light shaft-streaks on sides of breast and flanks. In **flight** shows **faintly outlined rufous wingbar** and dull dark rufous secondaries. Tail-feathers rather pointed, outer 3 variably tipped dull rufous, (sometimes only a trifle browner than rest of tail), this color extending up along outer web of outer feather as in Bar-winged C. Ssp *columbiana* slightly darker above, paler below, and dark bar on tawny secondaries more clearly defined. Ssp *aricomae* has considerably darker underparts, more buffy supercilium, distal half of secondaries darker, and **wingbar** somewhat narrower and **outlined in blackish**, making it prominent in flight. **Juv**. is warmer

hued and often more mottled on breast than ad., and can for several weeks be recognized by a smaller bill.

Habits: Usually in pairs or family groups. Walks probing the ground, tussocks of grass, moss, or animal droppings in search of insects and sometimes seeds and small frogs. When alarmed seeks shelter between bushes or in trees. Like Bar-winged C. (with which it often occurs) rarely wags tail, and only cocks it when excited. Ssp *aricomae* has been seen flaking off moss from a large rock near a stream, and poking at base of *Polylepis* trees at the edge of a wood.

Voice: During wing-flapping display a high-pitched trill *trrrrrrrieep* (rougher than in Bar-winged C.). In flight a fast *trrip-reep-reep* or *reep reep treep-reep* lasting c. 1 s. Call a nasal *kiu* or *kee*, quite unlike any call of Bar-winged C.

Breeding: Nests in self-dug burrows in roadcuts etc. (see *Condor* 90(1988):251-3). Nests Jan, Apr (Caldas); fledgling Dec (Pichincha); large gonads Dec (Cuzco); molt (post-breeding?) Mar (nw Ecu.).

Habitat: Like Bar-winged C. on short matted grass or bare soil, along streams, bogs and lakes, but never far from bushes or trees. Also in thick *Sphagnum* bogs among elfin forest. Ssp. *aricomae* may be typical of patches of semihumid *Polylepis* wood.

Range: 3200-5200 m. Páramos of C Andes of Col. (Páramo de Santa Maria, Nevado de Tolima) (*columbiana*); Nevado de Cumbal, sw Nariño, sw Col. and páramos of Ecu. (s to Loja), where fairly common, (*excelsior*). Se Peru in Cuzco (recent sightings and specimens from Abra Malaga) and Puno (1 specimen from Aricoma pass 1930), and Bol. in La Paz (1 specimen from Tilotilo 1876) (*aricomae*). The population of the latter form is unlikely to exceed 100-150 birds. It is threatened by continued logging of *Polylepis* for firewood.

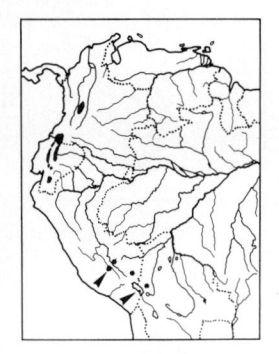

NOTE: Has been variously placed in the genera *Cinclodes*, *Geositta*, and *Upucerthia*; its behavior clearly places it in *Cinclodes*. Species status (and a separate vernacular name **ROYAL CINCLODES**) may be suggested for *aricomae* on basis of its habitat, coloration approaching Dark-bellied C., and emphasized wingbar.

Royal Cinclodes; see p. 847.

DARK-BELLIED CINCLODES *Cinclodes patagonicus* – Plate XXXV 10a-b

19 cm. Bill long, straight, and slender. Above very dark gray-brown, **below dark gray**; conspicuous supercilium and throat white; lower throat with faint barring; breast and upper belly with narrow white shaft-streaks. Tip of outer 3 tail-feathers and narrow wingbar pale dull rufous, forming small patch in closed wing; **axillaries brownish**. Ssp *chilensis* is more brownish below and has the ventral streaking less extensive posteriorly. **Juv**. has upper back with pale feather-tips and the streaking below broader, less defined, and confined to the breast.

Habits: Usually in pairs, frequents streams in dipper-like fashion, picking insects from rocks or shallow water. Also along earthy riverbanks, and sometimes on decomposing logs where it pecks the wood like a woodpecker, and makes small excursions to survey from a thin branch or tree-top. When alarmed raises crown-feathers a little. Only rarely cocks or wags tail.

Voice: Rather low-pitched. Call from ground a sharp *tjik*. In flight the same call, often repeated, and also *tili* and a sharp trill *tjrrrr*. Trills during wing-flapping display.
Breeding: Eggs Jan (Río Negro), Feb (Isla Grande), Sep (c Chile), Oct (s Chile), Nov-Dec (Sta Cruz, Isla Grande).
Habitat: Rocky or earthy banks of quiet streams and rivers, lake-sides and open swampy forest, and tussock-fringed rocky beaches (where it may be found alongside the commoner Blackish C. (*C. antarcticus*) in s Isla Grande, and Seaside C. (*C. nigrofumosus*) along c Chilean coast).
Range: To 2500 m. From Aconcagua to Aysén in c and s Chile and from interior Mendoza to Sta Cruz in w and sw Arg. (*chilensis*), and from Gulf of Penas, s Chile ,and w Sta Cruz, s Arg. s to Isla Grande and neighboring isls (*patagonicus*). Fairly common, very common on s Isla Grande.

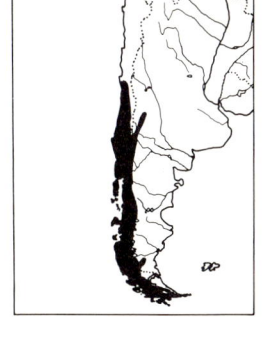

GRAY-FLANKED CINCLODES *Cinclodes oustaleti* – Plate XXXV 9a-b

17 cm. **Bill** fairly long and slender, **slightly curved**, dark. Above dusky gray-brown. Small supercilium and throat white; throat barred or spotted by dark feather-tips; rest of **underparts** except whitish belly **dusky** with ill-defined, narrow white spots on breast, becoming streaks on lower breast, upper belly, upper sides and vent; feathers of vent tipped whitish. **Axillaries white**, under wing-coverts mixed gray and white; rufous on secondaries hardly visible in closed wing. **In flight** shows buffy white, relatively restricted wingbar, tip of outer tail-feathers pale **rufous**, extending up on outer web of outer feather. Southern *hornensis* largest and grayest. Ssp *oustaleti* washed dark brown above, and usually much browner than *hornensis* below and esp. on flanks. Secondaries edged dark brown distally, outlining black basal patch. Bird with densely streaked breast, shown on the plate, is probably **juv**. of this species (after photo).
Habits: Much like Bar-winged C., but very frequently wags and cocks tail and sometimes clambers agilely over steep rock-surfaces. Very confident. Strong, undulating flights.
Voice: Call a sharp, emphatic note.
Breeding: Eggs Oct (Chile, Staten Isl.), nestlings Oct (Mendoza).
Habitat: Breeds at higher altitudes and is less dependent on water than Bar-winged and esp. Dark-bellied C. Occ. about houses and near torrents in mountain valleys. In winter much like other species of *Cinclodes*, along quiet channels and irrigation ditches, but also on steep rock faces.
Range: Highlands up to 4200 m from Antofagasta to Aysén, Chile, and in Mendoza, Arg., in winter commonest at 1500-2000 m (*oustaleti*), mts of Isla Grande and neighboring isls, only rec. on Isla Grande from Jan to Mar (migrants?) (*hornensis*). Juan Fernandez isls (*baeckstroemii*). Uncommon.

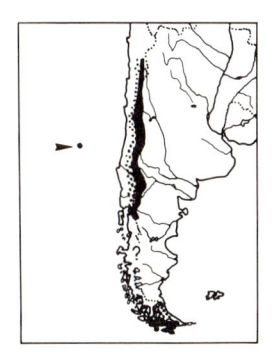

OLROG'S CINCLODES *Cinclodes olrogi* – Plate XXXV 12

15 cm. Bill relatively short. **Above dark gray-brown**, faintly tinged

chestnut, esp. on rump; crown darkest. **Extensive white wingbar** turning pale rufous on inner secondaries. Supercilium, throat, and breast white; breast-feathers with dark tips; rest of underparts gray-brown, white of breast continuing as flammulations down c belly. **Tips of** fairly narrow **outer rectrices dull pale rufous**. **Juv**. undescribed.
Habits: Alone or in pairs. Much like other species of *Cinclodes*, but wags and cocks tail very frequently.
Voice: No information.
Breeding: Eggs Nov-Dec.
Habitat: Sandy or stony edges of streams and lakes.
Range: Arg.; Above 1600 m in Sierras Grandes and Sierras de Comechingones, Córdoba, and in Sierras de San Luis, San Luis. In winter down to 900 m. Common.

NOTE: Described by Nores & Yzurieta, *Acad. Nac. Cienc. Córdoba Arg. misc.* 61(1979):4-8. Considered a ssp of Gray-flanked C. by Olrog (1979) and by Navas & Bo (in *Rev. Mus. Arg. Cienc. Nat. 'Bernardino Rivadavia'* 14(1987):55-86, while Nores (in *Hornero* 12(1986):262-73) regarded it a ssp of Bar-winged C., closely related to ssp *riojanus*.

SIERRAN CINCLODES *Cinclodes comechingonus* – Plate XXXV 14

17 cm. **In flight shows rather rounded wings, bright rufous wingbar**, patch on primary-coverts, and terminal part of outer rectrices. Above gray-brown. Supercilium and underparts buffy white; throat white; breast, lower throat and sides of throat scaled by dark feather-tips; flanks pale brown. **Basal half or more of mandible orange yellow**. Much like Bar-winged C., but smaller; tail longer, upperside grayer, flanks browner, wingbar brighter rufous, broader, more extensive (from 3rd primary inwards), and distinctly outlined in black; rufous of tail-tip more extensive than in passing migrant form of Bar-winged C. **Juv**. undescribed. 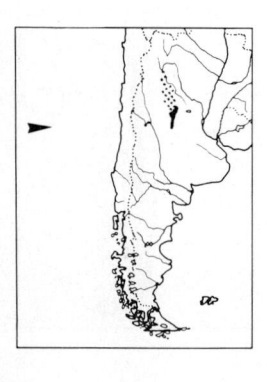
Habits: Alone or in pairs. Much like Bar-winged C., but wags and cocks tail much more frequently (though hardly more than 60°). More shy; hides in bushes and shrubbery, hopping from one branch to another.
Voice: No information.
Breeding: Eggs Nov-Dec.
Habitat: Valleys with bunchgrass, rocky outcrops and small patches of *Polylepis australis*. Usually near water.
Range: Arg., at 1600-2800 m in Sierras de Comechingones and Sierras Grandes of Córdoba. In winter down to 1000 m, and n to e Tucumán and n Santiago del Estero.

NOTE: By some authorities considered a well marked ssp of Bar-winged C. (see *Neotrópica* 18(1972): 54-6). As late as Oct *C. f. fuscus* occurs as a migrant on Pampa de Achala in Sierra Grandes, where also the 3 residents occur: Sierran, Olrog's, and White-winged C.s.

BAR-WINGED CINCLODES *Cinclodes fuscus* – Plates XXXV 13a-d and LXIV 24

16.5 cm. Bill shorter than in White-winged C., and much smaller than in Stout-billed C. Varies geographically: Birds from s Chile and s Arg. (*fuscus*) are gray or gray tinged with brown above, with a pure white supercilium; other populations dark brown above, usually with a darker crown, and with buff or buffy white supercilium, *riojanus* particularly chocolate-brown above and brown on breast; *yzurietae* very dark above, white on breast; *rufus* very rufescent. **Conspicuous wingbar whitish** in Peru, Bol., and n Arg., **or rufous** further n and s, palest rufous and most restricted in southern birds. Outer 3 rectrices vary from almost wholly rufous in Ven. to only tipped rufous (Col., Ecu.), or pale rufous gray (*fuscus*), but **this color always extends at least half-way up along outer web of outer rectrix**. Rectrices pointed and narrower than in White-winged C. Throat white, often with dark dots; below whitish to whitish buff, sides and esp. flanks buffy brown (or gray-brown in *fuscus*); breast often washed, mottled, or spotted brown or gray. Northern birds have base of mandible yellowish. **Juv**. more saturated than ad., and with most of underside marked with dark feather-tips.
Habits: Alone or in pairs. Territorial, birds often seen chasing one another. Walks searching for insects by probing bill into grass or turning cattle dung. Sometimes wades into shallow water. Often flies long distances low over the ground, thus exposing wingbar. Cocks tail upright on landing and take-off. During display from elevated ground, bush- or rocktop raises one wing for 1-2 s to expose pattern of underwing every 2-3 s. Song may begin in flight.
Voice: Song a drawn out fairly high-pitched trill *trrrrrrrrrt*. Flight-call a sharp whistle *tsip* or *pizzt*. During chase low over ground a hard *tee tee...tee* or *zeezeetezeetezeete*.
Breeding: In Chile and s to c Arg. 2 clutches: eggs Sep-Oct and Dec-Jan; eggs Nov (n Chile, Isla Grande), Apr (Cochabamba), Dec (Puno); fledglings Nov, Dec (Arequipa), Jan (Isla Grande, s Chile, Puno, s Ecu.), Mar (n Ecu., Ayacucho), Feb (Sta Marta); large gonads Nov (Pasco), Dec (Cuzco, Puno), Mar (Mérida).
Habitat: Mostly along streams or on wet bogs. Sloping ground with rocks and matted grass. May nest among rocks quite far from feeding grounds. Frequently nests in houses.
Range: 3500-5000 m on páramos of s Lara, Trujillo, and Mérida, w Ven. (*heterurus*); Sta Marta mts, E Andes in Boyacá, and C Andes in Cauca, Col. (*oreobates*); Nariño, sw Col., and immediate adjacent Ecu.(*paramo*), rest of Ecu. (*albidiventris*), 1500-5000 m from Cajamarca to Ancash and Huánuco, nw Peru (*longipennis*), c and s Peru from Junín to Arequipa and Puno (*rivularis*), from Tacna, s Peru s to Antofagasta, n Chile, and through highlands of Bol. (*albiventris*); above 2100 m in Salta and Jujuy to La Rioja (*tucumanus*, hardly distinct from *albiventris*); Cerro Famatina in La Rioja (*riojanus*, see M Nores in *Hornero* 12(1986):262-73); Campo de Arenal in Catamarca (*rufus*, see Nores *cit. op.*); Sierra del Manchao in Ambato se Catamarca (*yzurietae*, see M Nores *cit. op.*); and from Atacama Chile and Mendoza, Arg., s to Isla Grande, in winter n to Tucumán and s Brazil (*fuscus*). Common throughout range.

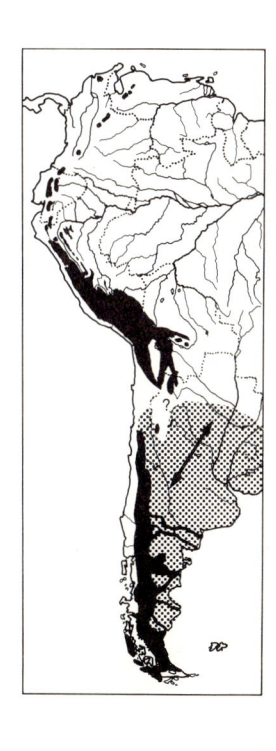

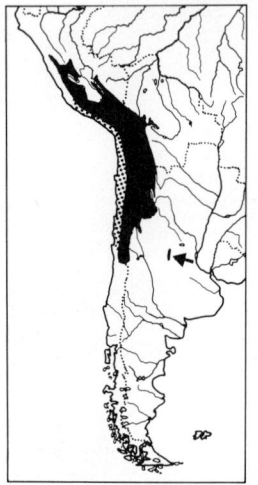

WHITE-WINGED CINCLODES *Cinclodes atacamensis* – Plate XXXV 15a-b

19.5 cm. Bill long and fairly straight. Above chestnut-brown becoming grayish on forehead; ill-defined supercilium whitish. Throat white, lower throat with dark feather-tips outlining throat from rest of underside, which is pale gray-brown; flanks and sides olivaceous brown; breast and belly with hardly discernible white shafts. **Shows broad white wingbar. Tail-feathers broad and rounded**, outer 3 with **well-defined white tips**, the **white not extending up along outer web of outer feather**. Ssp *schocolatinus* darker brown above and with underside below white throat (incl. c belly) rather dark gray; sides and flanks dark brown. **Juv**. has narrow pale feather-tips on lower back, and smudged lower breast.
Habits: Alone or in pairs, behaves much like Bar-winged C., but fairly shy.
Voice: Call a loud whistle, *queeet*, slightly harsher than in Bar-winged C. Song a trill. When flushed gives a *whuy tetededede*, 1st note highest.
Breeding: Eggs Nov (Chile, Salta, Córdoba), Dec (Salta); fledglings Oct (La Paz); molt (probably post-breeding) Jan (Cochabamba).
Habitat: Along streams running through bogs, bunchgrass, dry bushy country, *Alnus* or *Polylepis* wood, and bushy gulleys. Village ditches. Rarely found away from streams. Sometimes breeds in houses.
Range: 2500-5000 m in Peru and Bol. (2 rec. at sea level in Arequipa s Peru), lower further s. From Ancash, Pasco, and Junín, c Peru, s to n Santiago, Chile, and through w Bol. to Mendoza, Arg.(*atacamensis*); above 1600 m (900 m in winter) in Sierras Grandes, w Córdoba, Arg. (*schocolatinus*). Fairly common, but in much lower numbers than Bar-winged C.

WHITE-BELLIED CINCLODES *Cinclodes palliatus* – Plate XXXV 16

23 cm. Largest furnariid. Above chestnut (with narrow gray feather-tips in fresh plumage), crown gray tinged brown, wings and tail black; tail-feathers broad, rounded, outer 3 tipped rufous and white. **In flight** wings show very broad white wing-bar (also visible in closed wing). Lores blackish, earcoverts dark gray-brown. **Below white**, conspicuous at long distance. **Imm**. has chestnut earcoverts and crown paler, more tinged brown.
Habits: Alone, in pairs, or small groups of 3-4 birds. Runs and hops, **tail always cocked 60°**. Probes into the matted vegetation in search of insects and small frogs. Sometimes a small group will gather, cocking tails upright while vocalizing. Sleeps in cavity under rock.
Voice: 3 birds together gave a chattering 7-8 s long trill *rrrirrri*....
Breeding: No data.
Habitat: Seems to be ecologically tied to mineral-rich well-watered cushion-plant bogs (e.g. *Distichia* bogs) with rocky outcrops and stony slopes nearby. (Always below glaciers?).
Range: (4080) 4400-5000 m in Junín, Lima, and Huancavelica, c Peru. Rec. from San Martín undoubtedly erroneous. Rare.

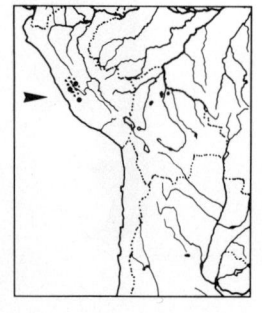

Horneros – Genus *Furnarius*

6 species of fairly large terrestrial furnariids, found in the lowlands of most of trop. and subtrop. S Am. with 2 species reaching highlands in Bol. These confident birds often live near human habitation, and are well known for their conspicuous domed clay-nests, placed on top of telegraph-poles, fence-posts, horizontal branches, or sometimes on the bare ground. A new nest may be constructed on top of an older one, and occ. 3-4 nests can be seen atop one another. See *Hornero* 3(1926): 409-11, *Verh. Orn. Ges. Bayern* 20(1933): 153-61, *Amer. Mus. Novit.* 860 (1936), *Abh. Verh. Naturwiss. Vereins.* 10(1965): 117-54, *Hornero* 11(1977): 384-6, and *Condor* 82(1980): 58-68.

RUFOUS HORNERO *Furnarius rufus* – Plate XXXIV 20

19 cm. **Stout. Dull rufous, tail rufous**. Throat white or whitish. **Juv**. like ad., but feathers of throat with faint, narrow, dark borders.
Habits: Largely terrestrial. Usually in pairs. Takes long strides, nodding its head. Picks up seeds, insects, earthworms, and other invertebrates. Confident. When alarmed takes to thick, horizontal branches.
Voice: A laughing, high-pitched, and often sustained chatter *kierkier-kierkier...*, given by both sexes in chorus.
Breeding: No data for highlands.
Habitat: Parks, gardens, fields, and woodland in fairly dry areas, often foraging out on open ground.
Range: Lowlands of se Brazil (*albogularis*), lowlands of most of Bol., and up to 3500 m in Cochabamba (*commersoni*), Tarija, s Bol. (*schuhmacheri*), Parag. and nw Arg. (*paraguayae*), s Brazil and ne Arg. s to Río Negro (*rufus*). Common.

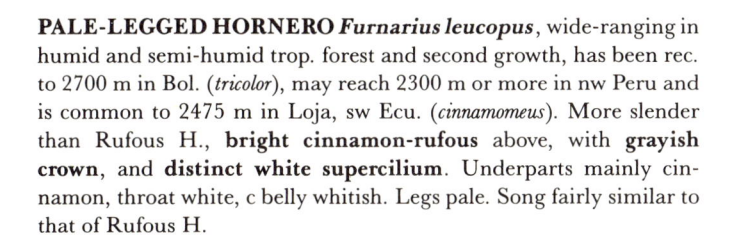

PALE-LEGGED HORNERO *Furnarius leucopus*, wide-ranging in humid and semi-humid trop. forest and second growth, has been rec. to 2700 m in Bol. (*tricolor*), may reach 2300 m or more in nw Peru and is common to 2475 m in Loja, sw Ecu. (*cinnamomeus*). More slender than Rufous H., **bright cinnamon-rufous** above, with **grayish crown**, and **distinct white supercilium**. Underparts mainly cinnamon, throat white, c belly whitish. Legs pale. Song fairly similar to that of Rufous H.

Wiretail – Genus *Sylviorthorhynchus*

Monotypic genus. One of the smallest furnariids, but with a remarkably long, decomposed tail. Restricted to damp thickets in c and s Chile and immediate adjacent Arg. Globular stick-nest with side entrance placed in low bush.

DES MURS' WIRETAIL *Sylviorthorhynchus desmursii* – Plate XXXVI 18a-b

7 cm + **16 cm tail**. Bill fairly long and straight. Forehead reddish

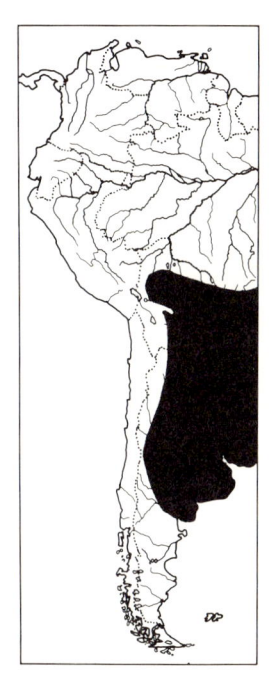

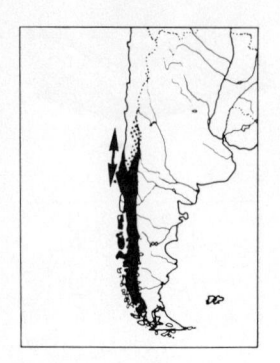

brown turning olivaceous brown on rest of upperparts; wings rufous; supercilium and headside mottled gray and whitish. Below buff, c belly white. **Tail very long**, graduated and composed of only 6 feathers, which are reduced almost to mere shafts. **Juv**. has no rufous on forehead, is darker above, and unevenly buffy and whitish below; breast-feathers with dark tips; tail shorter.
Habits: Forages in dense thickets. Difficult to observe. Feeds on insects.
Voice: Song composed of 2 notes repeated 4-5 times and lasting c. 1.5 s *de-djirr de-djirr...*, the whole song repeated every 4 s. Alarm is a loud descending trill ending with creaky sounds *de drrrrrrrrrrr di dr di dr drrrrrrrrrrr...*
Breeding: Eggs Nov (Chile, Sta Cruz); fledglings Jan-Feb (Chile).
Habitat: Damp bamboo thickets and among blackberry (*Rubus*) brambles, and other dense humid second growth.
Range: From Aconcagua to Magallanes (incl. Mocha and Guaitecas isls), Chile, and humid Andean valleys of adjacent Arg. from Neuquén to Sta Cruz.

Rayadito – Genus *Aphrastura*

2 species, 1 restricted to Patagonian *Nothofagus* forest and scrub, the other on Masafuera Isl in the Juan Fernández Archipelago off Chile. They are small inquisitive birds moving about in flocks, foraging much in the manner of titmice (*Paridae*), creepers (*Certhiidae*), or nuthatches (*Sittidae*) of other continents. The tail is rarely used for support, and is occ. held cocked over the back. The rectrices end in long almost bare shafts. The nest is usually placed in a crevice in a tree, and is lined with feathers.

THORN-TAILED RAYADITO *Aphrastura spinicauda* – Plate XXXVI 9a-b

14 cm. Wings, crown, and headside black; **very long broad supercilium buff**. Back dark olivaceous with faint darker scales, rump rufous. Tail chestnut with blackish markings, feathers ending in bare shafts. Wings with buff and white bars and panels. **Below white**, lower sides, flanks, and vent washed olivaceous. Feathers of lower throat sometimes with dark tips. **Juv**. has a pale-streaked crown, and for a short time pale feather-centers on back.
Habits: Agile and restless, curious and confident. In small flocks of up to 15 birds 0.5-6 m up. Searches bark, mosses, and lichens for insects. Sometimes half-cocks tail, sometimes use it for support. Occ. feeds on the ground, scratching among fallen leaves. Nuclear species in mixed-species flocks.
Voice: Very vocal. Song fast, composed of thin wheezy notes. Contact call a very high-pitched *zee zee*, a sharp *tick* and a metallic *ti ti....* Alarm a long trill *trrrrt-trrrrrrrr-trr...* lasting 5-10 s.
Breeding: Eggs Nov-Dec (Chile, Arg.); nestlings Nov-Dec (Chile), Jan (Neúquen), Oct-Nov (Río Negro); fledgling Dec, Jan (Isla Grande).

Habitat: Edge and interior of *Nothofagus* shrub and forest and nearby second growth. On isls with little or scarce vegetation also in bunch-grass and bushes like *Berberis* and *Ribes*.
Range: Sea level to 1800 m. From Coquimbo Chile and Río Negro Arg. s to Isla Grande and Staten Isl, accidental on Islas Malvinas (*spinicauda*). Other sspp on Mocha Isl, and on Melchor Isl in the Guaitecas isls and Chiloe Isl off coast of Chile. Abundant.

Rushbird – Genus *Phleocryptes*

Monotypic genus, found in reed- and rush-beds of the lowlands and high Andes from Peru and s Brazil to Isla Grande. The finely woven, globular or pear-shaped, closed nest with a small side entrance near the top, is built of grass, fibers etc., lined with feathers and hairs, and placed 0.5 to 1.2 m above water, attached to several reeds. See also *Hornero* 5(1933): 199-204, *Amer. Mus. Novit.* 860(1936), and *Hornero* 11(1977): 434-5.

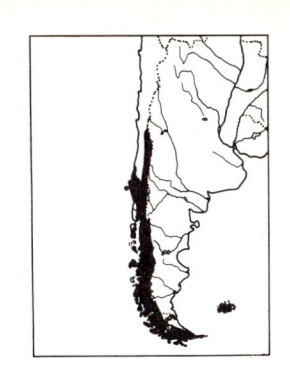

WREN-LIKE RUSHBIRD *Phleocryptes melanops* – Plate XXXVIII 19a-b

14 cm. Bill fairly long. **Above broadly streaked black, brown, and white**; wings black with rufous patches and bars. Supercilium whitish. Below whitish; breast buffy, sides and flanks olivaceous buff. Tail rather short, blackish, c pair of feathers dull dark rufous, rest tipped rufous gray. Ssp *schoenobaenus* larger, with longer bill, buffier below, but in worn plumage becomes almost white-breasted. **Juv**. with distinct rufous shafts on crown, and underside mottled with narrow, dark brown edges.
Habits: Alone or sometimes in pairs, crawling about in the rushes somewhat like an Old World *Acrocephalus* warbler or Marsh Wrens (*Cistothorus*). Only rarely picks insects from bases of stems and from low vegetation, or on the mud between the rushes, or on open mud-flats and floating vegetation, when undisturbed. At alarm several will gather to investigate, but usually stay within vegetation. During the tilting flight the tail is characteristically raised a little.
Voice: Song an endless series of monotonous notes sounding like two pebbles being struck together, or paper being flapped 4-5 times per s, and often initiated with a cicada-like sound *rrrr-clickclickclick*.... Contact-note a nasal *ehh*.
Breeding: In Chile 2 broods, 1st Oct, 2nd late Dec or Jan. Eggs Sep, Jan (coast of Arequipa), Sep-Jan (2 broods) (Lake Junín); nestlings Sep (Puno).
Habitat: **Tall rush and reed-beds** (esp. *Scirpus californicus tatora*) in marshes with fresh or brackish water.
Range: Widespread in lowlands of Chile and Arg. n to s Brazil, Parag., and (rarely) e Bol., and with local sspp in coastal wetlands in n Chile and Peru. From 4000 to 4300 m in Lake Junín area in Junín, wc Peru (*juninensis*), and highlands of se Peru and Bol. in Lake Titicaca region s through Cochabamba and Oruro, w Bol., to Jujuy, nw Arg. (*schoenobaenus*). Common in Lakes Junín and Titicaca. 1 rec. on Isla Grande (near Ushaia).

Tit-spinetails – Genus *Leptasthenura*

10 species inhabiting Andes from Ven. to Patagonia, and lowlands of Arg. and s Brazil. Not in very humid habitats. Small agile furnariids with very long, strongly graduated tails composed of narrow pointed rectrices, and appearing forked when closed. Tawny Tit-spinetail only has 10 tail-feathers, the others 12. Usually in pairs or small groups, foraging much in the manner of titmice of other continents, crawling about and often hanging upside down, searching bark and leaf-clusters for insects, and flowers for seeds. Song given by both sexes. They use the abandoned nests of canasteros, horneros, thornbirds, etc., or place their nest in holes in a bank, tree, cactus, stone wall or house, or in a crevice in a rock. The nest is lavishly lined with feathers.

ANDEAN TIT-SPINETAIL *Leptasthenura andicola* – Plates XXXV 2a-c, LXIII 13, LXIV 9

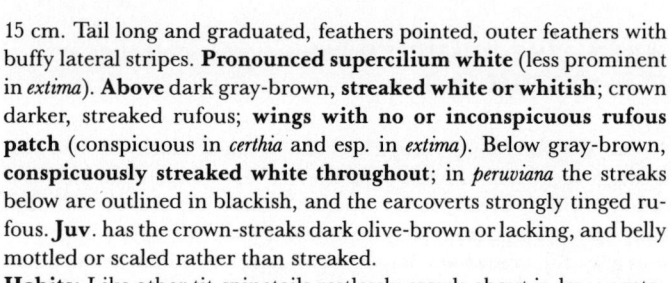

15 cm. Tail long and graduated, feathers pointed, outer feathers with buffy lateral stripes. **Pronounced supercilium white** (less prominent in *extima*). **Above** dark gray-brown, **streaked white or whitish**; crown darker, streaked rufous; **wings with no or inconspicuous rufous patch** (conspicuous in *certhia* and esp. in *extima*). Below gray-brown, **conspicuously streaked white throughout**; in *peruviana* the streaks below are outlined in blackish, and the earcoverts strongly tinged rufous. **Juv**. has the crown-streaks dark olive-brown or lacking, and belly mottled or scaled rather than streaked.

Habits: Like other tit-spinetails restlessly crawls about in low vegetation. In pairs or small groups, searches bark, foliage, and bases of bromeliads for insects. Fond of *Senecio* flowers. Sometimes perches on stone-fences. Tilting flight low over the ground. Hard to flush in regions with very low vegetation.

Voice: Call a weak tinkling *téz-dit* or *téz-dit-dit*. Contact note a loud fine *zik*. Alarm *tik-tik-trr-tik-trr-tik*.... Also a 2 s long monotonous trill and (song?) a descending series of notes starting and ending with a trill.

Breeding: Nestling Apr (Lima); fledglings Nov and May (Mérida), Sep (C Andes Col., Cundinamarca), Aug (Arequipa); juv. Apr (Junín); large gonads Sep (Boyacá); molt (probably post-breeding) Aug (Lima).

Habitat: In Ven., Col., and Ecu., where no other tit-spinetail occurs, in low bushes in humid grassland and *Espeletia* páramos, in *Lupinus* meadows, copses of *Stevia lucida* (Asteraceae), and *Polylepis sericea* woodland and shrub. In Peru and Bol. in *Polylepis*, *Puya raimondii*, and low montane shrub in the arid zone, at higher elevations than most congeners. At its lowest elevational levels it overlaps with Rusty-crowned T-s. in dry shrubbery in c Peru, where *andicola* may be the dominating t-s. in *Polylepis*, and with Streaked T-s. in *Polylepis*-shrub in sw Peru, where that species dominates and largely restricts the present form to low shrub at higher elevations. In the highlands of se Peru *andicola* only inhabits those *Polylepis* woods not occupied by White-browed T-s., being otherwise confined to low brush and shrub.

Range: 3000-4500 m. Trujillo and Mérida, w Ven. (*certhia*); Sta Marta mts (*extima*); E Andes of Col. in Boyacá and Cundinamarca (*exterior*);

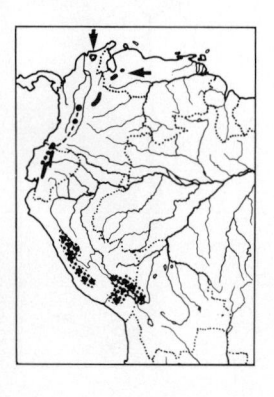

páramos of C Andes of Col., and Ecu. (*andicola*). From Ancash s through Lima, Junín, and Huancavelica to s Arequipa, Cuzco and Puno, Peru, and La Paz, Bol. (*peruviana*, apparently 2 sspp involved). Fairly common in n part of range, less so in Peru.

STREAKED TIT-SPINETAIL *Leptasthenura striata* – Plate XXXV 3a-c

14.5 cm. Tail long, graduated, feathers pointed, outer 3 pairs extensively buffy gray distally with black shafts. Ssp *striata*: Above gray with whitish streaks outlined in black; crown black, streaked rufous buff; faint supercilium white; **prominent pale rufous patch at base of secondaries and inner primaries**. Below grayish-white; **throat white, dotted blackish**, breast with **faint** white streaks occ. extending onto upper belly, flanks pale gray-brown. Ssp *superciliaris* similar, but white brow more pronounced, and belly unstreaked; *albigularis* has no supercilium, and has the throat white without spots; the pale gray breast and sides are faintly streaked with white. **Juv**. has crown gray-brown, almost unstreaked, breast-feathers edged dark, giving scalloped effect; wings extensively rufous.
Habits: In pairs or small groups, where the individuals usually feed several m apart. Climbs about in tops of bushes and low composite shrub (e.g., *Baccharis*), occ. descending to the ground.
Voice: Call a slightly wheezy *twet* and a *trit*. Song a long descending trill.
Breeding: Juv. June (Ancash); imm. Apr (Arequipa), June (Ayacucho); molt (post-breeding?) July (Arequipa).
Habitat: Arid zone. *Baccharis* and other composite shrubbery and scattered montane shrub, sometimes cacti and bromeliads, and in the s parts up into arid *Polylepis* woodland.
Range: Coast to 3500 (4000) m. Ancash and Lima, wc Peru (*superciliaris*), locally in Huancavelica, Ayacucho, and Apurímac, c Peru (*albigularis*), from s Ayacucho through Arequipa, s Peru, to Tarapacá, n Chile (*striata*). Fairly common.

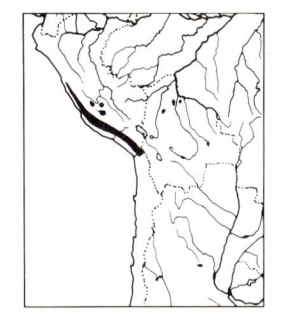

RUSTY-CROWNED TIT-SPINETAIL *Leptasthenura pileata* – Plate XXXV 5a-b

15 cm. Tail long, graduated, feathers pointed, outer 4 pairs extensively gray distally. Above gray with white streaks outlined in black; **crown rufous** (streaked with black in *cajabambae*); supercilium indistinct. **Narrow pale wing-panel at base of outer primaries, inconspicuous small rufous patch at base of inner primaries and outer secondaries. Throat well-marked**, white with black spots; rest of underparts gray, more or less streaked with narrow obscure whitish streaks. Ssp *latistriata* much like *pileata*, but differs by broader, more contrasting streaks on back, heavier streaks on breast (in *pileata* breast often spotted rather than streaked), and darker gray breast and belly. **Juv**. undescribed.
Habits: Usually in pairs, agilely climbs about in bushes, esp. in the thinner branches, and every now and then surveying from a top.

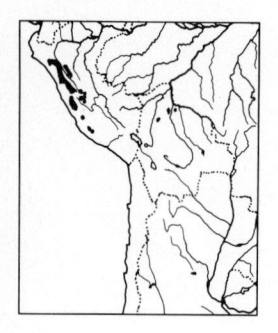

Voice: Contact call a simple *tsak*. Disturbance call a staccato *rick teek teek...* with varying pitch, speed, and amplitude. Song a c. 2 s long trill, slightly increasing in amplitude and speed, repeated at 3-6 s intervals; another bird (mate ?) may simultaneously give ticking notes. Also gives a slower, slightly descending trill.
Breeding: No data.
Habitat: Dry bush-covered slopes, riparian thickets, and edge of temperate woodlands in the arid zone (*pileata*). In alder thickets, *Polylepis* wood, and other fairly humid montane shrubbery (*cajabambae*).
Range: Peru. 2000-4000 m. Cajamarca, La Libertad (both sides of the Marañón), Ancash, Pasco, Huánuco and Junín (*cajabambae*) and n Lima (intermediate), and Chillon and Rimác valleys in Lima (*pileata*), w Huancavelica and Ayacucho (*latistriata*, M Koepcke in *Beitr. neotrop. Fauna* 4(1965):158-60). Fairly common.

WHITE-BROWED TIT-SPINETAIL
Leptasthenura xenothorax – Plate XXXV 4

15 cm, but more stocky with relatively shorter tail than Rusty-crowned T-s. Tail fairly long, graduated, feathers pointed, outer 3 pairs with whitish patch distally. Above gray-brown with whitish streaks outlined in black on lower back, nape unstreaked. Crown uniform rufous; supercilium white. Wings with 2 narrow whitish panels, but no rufous patch. **Throat white, coarsely checkered with black and sharply set from uniform drab gray underside. Juv**. undescribed.
Habits: In pairs or family parties. Works deliberately along thicker branches with mosses and lichens, unlike Rusty-crowned T-s. Aggressive, often seen chasing conspecifics.
Voice: Very vocal. Song a high-pitched trill *trrrrrrrrrie* lasting c. 2 s, but up to 5 s when excited, and sometimes with 1-4 introductory notes *tjit tjit tjit tjit trrrrrrr....* During chases and fights *rrrt tee teetee rrrt rrteetelit tit tit teeeeerr rrit tit trrrrrie*. Contact note a fine repeated *tjit*.
Breeding: Gonads small Dec and July; molt (probably post-breeding) Dec.
Habitat: Semi-humid *Polylepis* wood, occ. descending to *Escallonia* wood during snow-storms.
Range: Se Peru. 3800-4500 m, primarily in the upper parts. Restricted to the small *Polylepis* woodlands of Vilcanota mts above right bank of upper Urubamba Valley, probably also in southern Cord. Carabaya, Cuzco. Common within suitable habitat. Total population probably some 100s, but the species is very vulnerable to loss of its restricted habitat.

NOTE: We do not approve of Vaurie's synonymization of this bird with ssp *pileata* of the Rusty-crowned T-s.

PLAIN-MANTLED TIT-SPINETAIL
Leptasthenura aegithaloides – Plate XXXV 1a-c

15 cm. Tail long and graduated, feathers very pointed, outer 3-4 pairs with much buffy gray distally. Ssp *aegithaloides* olivaceous gray above;

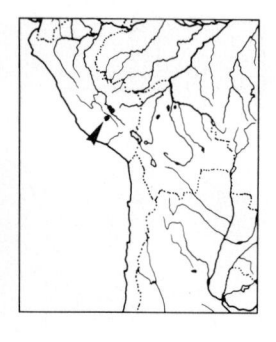

crown blackish streaked dull rufous, often somewhat raised; nape with a few white streaks that may extend faintly to upper mantle, but **back unstreaked. Wings with conspicuous rufous patch. 3 outer tail-feathers with large grayish white tips** (dull rufous in the very similar Tufted T.-s. (*L. platensis*) of lowland Arg. and s Brazil). Supercilium and throat white, throat sometimes with dark markings; below grayish, washed olivaceous on sides and flanks. The sspp show only slight differences in color hue: *berlepschi* has pale brown upperparts and dark rufous crown-streaks, buffy (vs. grayish white) edges to tertials, buffy gray outer tail-feathers with dusky markings, and pale buffy brown breast and belly; *pallida* is pallid smoky gray above with buffy white crown-streaks, much like *aegithaloides*, but primaries edged whitish (vs. rufous), and rufous patch in wing smaller; *grisescens* like *aegithaloides*, but darker below (breast and belly gray, belly slightly tinged buff). **Juv**. has indistinct crown-streaks and throat-markings. Back faintly spotted, tail shorter with rounded feather-tips.
Habits: Usually in pairs or small parties. Climbs about in bushes and herbs. Tilting flight low over the ground. Sings from bush-top.
Voice: During foraging occ. utters low complaining notes. Song lasts 2-3 s: *zi-zi-zi kre kie er ki kie er ki erkierkierrkierr*. Also an abrupt, fairly low-pitched monotonous trill lasting 0.5 s. During interactions may give rapid trills.
Breeding: Eggs Sep-Oct (Patagonia, c Chile), Oct-Nov (n Chile), Dec (Mendoza), Feb (Jujuy); fledglings Feb-Mar (Chile, nw Arg.).
Habitat: Arid zone. Rocky ravines or open slopes with scattered thorny bushes and cacti, thick brush on alkaline flats, riparian thorny thickets, dry open woods. Oases and gardens in winter.
Range: Coast-2500 m in (Puno?) and Arequipa, s Peru, to Atacama, n Chile (*grisescens*), 3500-4300 m from c Puno and e Tacna, s Peru through w Bol. and neighbouring Chile to Tucumán, nw Arg. (*berlepschi*), coast to Andean foothills from Coquimbo to Aysén, Chile (*aegithaloides*), and from La Rioja, San Juan and Mendoza e to Buenos Aires and s to Sta Cruz, Arg. (several observed s of Porvenir, Isla Grande by BW and RA Rowlett Dec 1987 presumably of this ssp) and Chile in Aysén and Andes of Bío Bío (see *Bol. Soc. Biol. Concepción* (Chile) 53(1982): 171-2) (*pallida*). Fairly common.

BROWN-CAPPED TIT-SPINETAIL
Leptasthenura fuliginiceps – Plate XXXV 6a-b

15 cm. Crest often partly raised. Tail long, graduated, feathers pointed. **Crown dark brown contrasting with remainder upperparts**, which is rather light gray-brown (Sayal Brown), **unstreaked** (1 otherwise normal specimen had distinctly light-streaked crown). Wings with large chestnut area, **tail rufous**. Below grayish tinged buff, sometimes wholly buffy; c throat and indistinct supercilium whitish. Ssp *paranensis* paler (Wood Brown) above. Birds from mt Illampu n La Paz are slightly darker on crown and back and grayer throughout than those from se La Paz s-wards. **Juv**. with faintly mottled breast and rounded tailfeather-tips.
Habits: In pairs or small groups. Forages in bushes and dense shrubbery near ground, occ. up to 3 m.

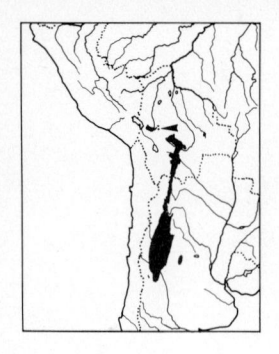

Voice: Call a very high-pitched faint insect-like *pree* given every 1-2 s.
Breeding: Eggs Feb (Salta); nestlings Jan (Córdoba, Salta), Feb (Salta); fledglings Feb and Apr (Tucumán), Mar (Mendoza); imm. July (La Paz, Tarija); large gonads Jan (Cochabamba).
Habitat: Bushy slopes in watered valleys and ravines in both arid and semi-humid zone.
Range: 1500-3250 m, in Cochabamba to 3900 m, down to lowlands in winter. Illampu massif nw La Paz (unnamed ssp), se La Paz and Cochabamba to Tarija, Bol.(*fuliginiceps*), and from Jujuy s to Mendoza and Córdoba, Arg.(*paranensis*). Fairly common.

TAWNY TIT-SPINETAIL *Leptasthenura yanacensis* – Plate XXXV 7a-b

16 cm. Tail very long, graduated, feathers strongly pointed; only 10 rectrices. **Above warm brown, below orange-buff**; forehead, wing-edging, and tail rufous chestnut; supercilium buff. Birds from Cochabamba, Bol., and s-wards are decidedly paler than northern birds.
Juv. has breast faintly barred, sometimes lacks rufous-chestnut of forehead, and has blunt tail-feather tips.
Habits: In pairs or small flocks. Rapidly works its way out to the tips of thin branches in search of insects. Sometimes climbs vertical trunks. Aggressive, often seen chasing conspecifics.
Voice: Very vocal. Song a trill, rougher than that of White-browed T-s., and always finished off with an extra note and a short trill *trrrrrrrrrrree-teeteerrr* lasting 3-5 s. Also gives various chatters *tjiketjitjitjitjitjtjt-t-t-t*, and short notes *t-t-tjk-tjketjketjketjketjk-tjk*, *rrrip rrrip tjktjktjke rrrhie tjktjke*, and a short snarling *eep*. Contact call a *tjip tjip*.
Breeding: Eggs Nov (Cuzco), Imm. Apr (La Paz).
Habitat: *Polylepis* woodland, feeding on the thinner branches. In Puno, se Peru, and La Paz and Cochabamba, Bol., also enters composite shrubbery and small patches of mixed *Gynoxys*, *Ribes*, *Miconia*, *Escallonia* and other bushes.
Range: 3200-4600 m, in the n parts in semi-humid situations and generally very high up, and remaining here during snow-storms. Along Cordillera Blanca, Ancash to n Lima, wc Peru, from Abra Malaga to above Urubamba town, Cuzco and Aricoma pass, Puno, se Peru, and La Paz (*yanacensis*), in more arid situations in Cochabamba, Potosi and Tarija, Bol. (unnamed ssp). Locally fairly common, mainly in the highest parts.

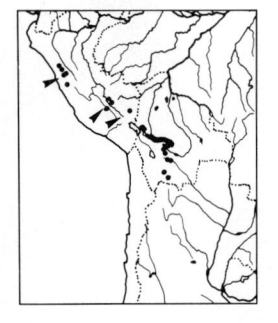

Spinetails – Genus *Synallaxis*

Small fairly apathetic birds that hop among twigs and vines within dense shrubbery. Rarely seen, but frequently repeating their simple song. They have fairly long graduated tails, that are usually held slightly cocked. The rectrices are loosely webbed at their tips and vary in number from 8 to 10, sometimes even between closely related forms. The nest is an oblong stick-structure with an entrance tube at the side. At least in some species both sexes sing. See *Amer. Mus. Novit.* 861(1936).

BUFF-BROWED SPINETAIL *Synallaxis superciliosa* – Plate XXXVI 17a-b

16 cm. Tail fairly long, graduated, with 8 rectrices. Forehead and back olivaceous brown; crown, wings, and tail rufous chestnut. Eye-ring whitish, continuing in a **distinct, buff postocular streak**. Below whitish, sides olivaceous brown. Ssp *samaipatae* less brown, the rufous cap less extended posteriorly, and feathers of forehead distinctly tipped rufous; side of head, neck, and chest grayer, underparts considerably whiter. **Juv**. pale brown above, white on throat, and below buffy distinctly mottled or barred with brown. – In pairs, foraging in dense vegetation. Eggs Oct-Nov (Tucumán); juv./imm. Oct (Tarija), Mar, Apr (Tucumán).

In thorny thickets and hedgerows, undergrowth in open deciduous forest at 1500-2500 (2900) m. Cochabamba to Tarija, Bol. (*samaipatae*), and Jujuy, Salta, and Tucumán, nw Arg. (*superciliosa*). Fairly common.

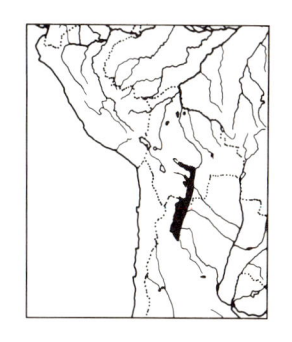

APURÍMAC SPINETAIL *Synallaxis courseni* – Plate XXXVI 13

18 cm. Much like Azara's S., but **tail considerably longer**, with 10 rectrices, and **with more dusky**, sometimes wholly dusky with only a little rufous along shafts near tip. Forehead, back, and flanks gray, only very faintly washed olivaceous; belly gray, not mottled with whitish. **Juv**. like juv. of Elegant and Azara's S.
Habits: Apparently like Azara's S.
Voice: Like Azara's S. a nasal *keet-weet?*, lasting c. 1 s and repeated every s or so for long periods.
Breeding: Large gonads Dec; imm. Mar.
Habitat: Understory and up to 3 m in vines, tangles, and bamboo in *Podocarpus hermsianus*-forest and nearby composite shrubbery, as well as semi-humid cloud forest and forest edge. Habitat somewhat drier than that of Azara's S.
Range: Sc Peru. 2450-3500 m. Restricted to the slopes of Nevada Ampay n of Abancay, Apurímac, sc Peru. Common.

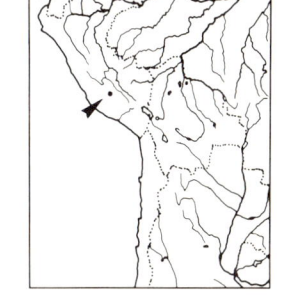

NOTE: Described by E Blake in *Auk* 88(1971):179. Possibly conspecific with Azara's S., responding to playback of its song.

ELEGANT SPINETAIL *Synallaxis elegantior* – Plate XXXVI 15a-c

17 cm. Like Azara's S., but **tail longer**, with 10 rectrices (8 in Loja sw Ecu.). Tail rufous, only washed with dusky in s part of range. **C belly white**. Ssp *ochracea* has a buff or ochraceous postocular streak. **Juv**. above uniformly brown; below strongly ochraceous to clay-colored, throat whitish.
Habits: Apparently like Azara's S. Very active in bird-waves.
Voice: As in Azara's S. *keet-weet?*, never with a 3rd note. When excited a series of notes like last note of song *weet-weet-weet-weet....* Alarm a short low complaining churr.

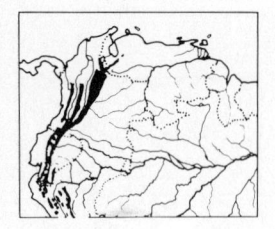

Breeding: Eggs Mar-Apr (*elegantior*), Feb-Apr (*media*); fledglings Sep (*media*), June (*ochracea*), May, July (*fruticicola*); juv./imm. Feb, Aug, Nov (*elegantior*), rec. all months (*media*), May, Aug, Sep (*ochracea*), June-Sep (*fruticicola*); large gonads Feb, Mar (*media*).
Habitat: Like Azara's S. Humid hedgerows, dense low shrubbery, bamboo thickets, forest undergrowth at edge of clearings etc.
Range: 1100-3800 m, rarely down to 900 m. From Trujillo, Ven. through E Andes of Col. (*elegantior*); W and C Andes of Col. and s through n Ecu. (*media*), s Ecu. and Piura, Lambayeque, and n Cajamarca, n Peru (*ochracea*), s Cajamarca, La Libertad, Amazonas, and San Martín, Peru (*fruticicola*). Common.

NOTE: Often considered conspecific with Azara's S., to whose song it responds. We have not examined specimens from San Martín.

AZARA'S SPINETAIL *Synallaxis azarae* – Plate XXXVI 14

15-16 cm. Tail fairly long, graduated, with 8 rectrices. **Forehead** and back **olivaceous**, **crown** and wings **rufous**. Tail rufous chestnut more or less washed with dusky at base and along inner webs. Throat-feathers blackish with whitish tips that sometimes almost cover the bases; rest of underparts gray more or less mottled with whitish in center; flanks olivaceous. Ssp *infumata* darkest and with the widest frontal band. Ssp *carabayae* long-tailed, long-winged and dark. **Juv**.: crown and back brown; throat-patch indistinct; underparts smudged with pale brownish ochre below throat, and somewhat mottled or indistinctly barred with dark feather-edges and tips; inner remiges brown.
Habits: In pairs or family groups foraging near the ground in the interior of dense thickets, where they hop along the branches picking insects from leaves and twigs, occ. searching dead leaves.
Voice: Heard continually throughout the day. The song, given by both sexes, consists of 2 notes lasting 1 s and repeated every 1-4 s. 2nd note rises in pitch *kheet-wee?*. Rarely adds a 3rd note resembling the 2nd or slightly higher *khee-wee-wee?*. Alarm a short low complaining churr.
Breeding: Juv./imm. Dec, Mar, May (*infumata*), Aug (*urubambae*), Nov, Dec, June (*carabayae*); imm. May (*azarae*).
Habitat: Always deep in tangles or other dense vegetation. Humid zone and in humid riparian thickets into the arid zone. Montane shrub, understory in cloud forest, bamboo thickets, and dense second growth. Generally in less wet habitat than Rufous S.
Range: 1250-3700 m. From San Martín to Junín c Peru (*infumata*), Urubamba and Occobamba valleys in Cuzco (*urubambae*), Puno, se Peru, and La Paz, Bol. (*carabayae*), Cochabamba and Sta Cruz, Bol. (*azarae*). Common.

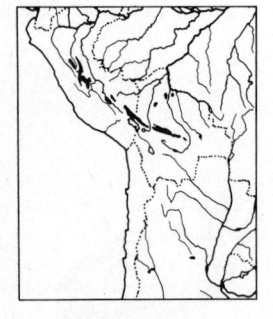

SOOTY-FRONTED SPINETAIL *Synallaxis frontalis* ascends to 2500 m in the drier parts of Cochabamba, Bol. (*fuscipennis*). Resembles Azara's S., but always has gray-brown inner remiges and darker rufous crown and wings, white c belly, less decomposed tail of 10 feathers, shorter bill with a slightly more curved maxilla, and longer wings. **Juv**.

without ochre or buff on belly and breast. Vocalizations with same pattern as Azara's S.

SLATY SPINETAIL *Synallaxis brachyura* of trop. and subtrop. zones has (erroneously?) been rec. as high as 3350 m in nw Ecu. (*chapmani*). 15.5 cm. Bill rather stout. Tail fairly long, graduated, dusky, feathers decomposed. Forehead, back, rump, inner remiges, sides of head, and underparts dusky; crown and most of wings rufous. Throat with thin white streaks, back and flanks faintly tinged brownish, c belly gray to light gray. **Juv**. like ad., but throat light gray faintly barred gray; crown like back. Song a fairly low-pitched *tjedededrrrr*, somewhat descending at end, lasting c. 0.3 s, and repeated every 1-2 s.

NOTE: Sometimes called Sooty S.

SILVERY-THROATED SPINETAIL *Synallaxis subpudica* – Plate XXXVI 16 a-b

18 cm. Tail fairly long, graduated, with 10 rectrices. Above, incl. **forehead** and headside, **brownish gray**. **Cap and wings rufous chestnut, tail dusky brown**. Throat-feathers blackish at base, whitish on tips; rest of underparts buffy gray, more buffy on sides and flanks, and whitish on lower belly. **Juv**. has crown like back.
Habits: In pairs, foraging low in dense vegetation.
Voice: Song *Cranioleuca*-like: a descending series of loud high-pitched notes answered similarly by another bird (mate?). Alarm a slow low trill (RR rec.).
Breeding: Juv./imm. May and Dec (Cundinamarca); large gonads June-Sep (E Andes Col.).
Habitat: Dense undergrowth at forest edge, second growth incl. rather low brush and introduced scots broom (*Genista*), overgrown clearings, hedgerows.
Range: Col. 1200-3200 m. Restricted to the plains and slopes of E Andes in Boyacá and Cundinamarca. An old rec. from e Ecu. is probably erroneous. Common.

RUSTY-HEADED SPINETAIL *Synallaxis fuscorufa* – Plate LXIII 17

16.5 cm. Back and tertials gray tinged olive, wings and tail chestnut. Tail with 10 rectrices. Lores gray; **crown rufous, headside and underparts pale rufous**, c belly palest; flanks and lower belly washed olive-gray. **Juv**. has crown like back, hindcrown margined with rufescent; below tawny-olive, palest posteriorly, and with faint barring.
Habits: In pairs. Forages low in dense vegetation. Confident.
Voice: Song consists of 3 notes, the first 2 identical, 3rd descending and lower pitched *ki ki kye*, lasting slightly less than 1 s and repeated every 1-4 s.
Breeding: Juv./imm. May, June, July; large gonads Jan-June.
Habitat: Humid zone. Bushes, tangled thickets, undergrowth at forest edge, open shrubbery.

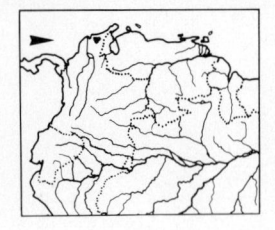

Range: Col. 760-3000 m, mainly above 2000 m in Sta Marta mts. Fairly common.

RUFOUS SPINETAIL *Synallaxis unirufa* – Plate XXXVI 11a-b

16.5 cm. Tail fairly long, graduated, with 10 rectrices. **Above chestnut, below rufous**, lower belly, flanks, and vent ochraceous. **Lores black**. Sspp *meridana*, *munoztebari*, and *ochrogaster* paler; *meridana* with a slightly longer tail, and the black bases of the throat-feathers more apparent; *munoztebari* with a faint, paler supercilium and forehead; *ochrogaster* with particularly pale underparts. **Juv.** brown above, sometimes tinged olivaceous on head and upper back; below paler, ochraceous-brown, darkest on breast.
Habits: In pairs, foraging very low in dense vegetation.
Voice: Song by both sexes, one of them slightly higher pitched than the other, a single (rarely 2 or 3) rising *wee*? repeated endlessly every 1.5-4 s. Sometimes a rapid *eeh-eeh-eeh*.... Contact call *tjyp*. Alarm a low-pitched churr. See also *Amer. Mus. Novit.* 2482(1972).
Breeding: Fledglings Apr (Cundinamarca), June (Huánuco); juv./imm. Feb (*meridana*), Apr (*unirufa*), rec. all months (*ochrogaster*); large gonads Apr (*munotzebari*), Apr-July (*meridana*).
Habitat: Dense shrubbery and bamboo thickets in very humid cloud forest.
Range: 1200-3700 m. From Trujillo, Ven., to Páramo de Tamá (*meridana*), Perijá mts (*munoztebari*, Phelps & Phelps in *Proc. Biol. Soc. Wash.* 66(1953):4-5), all 3 ranges of Col. s to nw and e Ecu. and probably Peru n of the Marañón (*unirufa*); s and e of the Marañón s to Vilcabamba mts, Cuzco (*ochrogaster*). Fairly common to common.

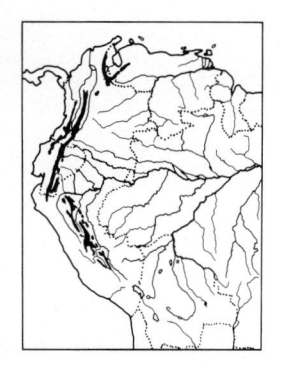

RUSSET-BELLIED SPINETAIL *Synallaxis zimmeri* – Plate XXXVI 12

17 cm. Tail fairly long, graduated, with 10 rectrices. Above gray, rump rufous; wing-coverts rufous chestnut; tail dark, each feather with rufous along basal part of outer web, amount of rufous increasing on outermost feathers. **Throat and sides of head gray**, lores blackish, eye-ring white, **throat with white streaks; rest of underparts russet**. **Juv**. undescribed.
Habits: In pairs, forages in dense vegetation, hopping along vines and twigs, probing dry clumps of moss. Rarely raises tail and then only slightly. Very confident. Territorial.
Voice: Song consists of 2 identical notes, a snarling *quick-quick* lasting c. 1 s and repeated every 1 or 2 s. When excited, the intervals become shorter *quick-quick-quick-quick*.... During disputes a chattering trill *trrrrrrrr*.
Breeding: No data.
Habitat: On dry mountain tops and slopes. Interior of impenetrable, thorny, 1-3 m tall, tangled bushes, with occ. small trees projecting.
Range: C Peru at 1800-2900 m. Confined to the w slope of the Andes in c and s Ancash (Casma and Huarmey valleys). Fairly common lo-

cally, but habitat restricted, and continuously opened and degraded by roaming cattle.

NOTE: Described by M Koepcke in *Publ. Mus. Hist. Nat. 'Javier Prado' ser.A (Zool.)* 18(1957):2.

Spinetail – Genus *Hellmayrea*

Monotypic genus. Small, agile spinetail with a fairly short tail of 10 pointed rectrices. Inhabits humid forest. Creeps about in vines, often hanging head down. The nest is undescribed. Biochemical and behavioral evidence have shown that merging with the genus *Synallaxis* is unwarranted (see pp. 333-346 *in* Buckley *et al.*, eds. (1985).

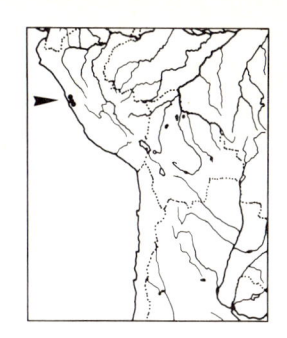

WHITE-BROWED SPINETAIL *Hellmayrea gularis* – Plates XXXVI 10a-b and LXIV 26

11 cm. **Tail rather short**, graduated, feathers pointed. Above rufous chestnut (or light brown in *brunneidorsalis*), tail chestnut. **Supercilium and throat white**; rest of underparts various shades of dull rufous to rufous clay (grayish in *brunneidorsalis* and *cinereiventris*). Birds from n Peru are paler rufous below than populations n and s of them. Some birds from Ecu. and the n end of E Andes Col. are considerably grayer than others, most evident in juv. **Juv**. has the entire underside mottled with dark feather-tips.
Habits: Usually alone, sometimes in mixed-species flocks. Often holds tail slightly raised and constantly makes little jerks. Very agile, climbing about in dense vegetation and tangles, often head down (sometimes crawling on underside of large *Gunnera* leaves) in search of insects.
Voice: Song an infrequently heard accelerating series of high-pitched notes *cheet teet-teet-ti-titititi*, last 2 notes slightly higher. Calls include a *chip* and a nasal descending trill.
Breeding: Juv./imm. Sep (Caldas), May, July, Oct (Cauca), Apr, July, Sep (nw Ecu.), June (Piura), Sep (Amazonas), June, July (Huánuco).
Habitat: From near ground to 4 m up in dense undergrowth, vines, and edges of clearings with second growth in cloud and elfin forest.
Range: 2300-3900 m. Trujillo, Ven., to Páramo de Tamá (*cinereiventris*), Perijá mts (*brunneidorsalis*, Phelps & Phelps in *Proc. Biol. Soc. Washington* 66(1953):6), locally in all 3 ranges of Col., Ecu., and Peru n of the Marañón (*gularis*), n Peru s of the Marañón bend from Amazonas to Junín (*rufiventris*). Fairly common in n part of range.

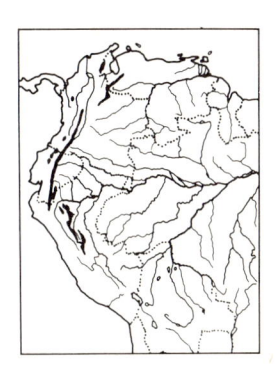

Spinetails – Genus *Cranioleuca*

A rather homogenous genus of spinetails with longer wings and shorter and stiffer tails than *Synallaxis*. The tail is graduated and the rectrices pointed in such a way that the closed tail looks forked. They sometimes climb trees like woodcreepers, and at other times acrobatically crawl about in the foliage. The large globular or ovoid nest is sometimes pensile, hung conspicuously far above the ground as in *Phacellodomus*, sometimes placed between 2 branches or against the trunk. It is constructed of green moss, twigs, vines, straw, and leaves. There is no tube on the side entrance. See *Amer. Mus. Novit.* 860(1936).

CRESTED SPINETAIL *Cranioleuca subcristata* – Plate LXIV 25
– inhabits humid premontane forest to near lower edge of temp. zone (700-2300 m) in n Ven. On s slope of the Andes in Barinas, in coastal mts, around lake Maracaibo and in interior mts in Guárico (*subcristata*), Andes of s Táchira and w Apure (*fuscivertex*, Phelps & Phelps in *Proc. Biol. Soc. Washington* 68(1955):50). 14 cm. Above light gray-brown becoming ochraceous on rump; **crown**-feathers pointed and slightly elongate, forming small crest, dull olive-brown or **buff, streaked dark brown**; indistinct supercilium whitish. Wings and tail bright russet. Below buffy gray to buffy white, chin and upper throat whitish. **Imm**. grayer, faintly mottled below, crown-feathers not pointed, streaks indistinct, sides of crown with some rufescent. Alone or in pairs. Climbs trees like a woodcreeper, but all the way to the thinner branches. Also hops in vines and tangles near edge of dense vegetation.

STREAK-CAPPED SPINETAIL *Cranioleuca hellmayri* – Plate LXIII 26

15 cm. Iris pale. Above olive-brown, wings and tail rufous; **crown rufous** lightly to heavily **streaked black** or blackish, densest on forehead. Supercilium white admixed black, headside streaked blackish and buff, throat whitish, rather sharply set from pale olive-gray underside. **Juv**. undescribed.
Habits: In pairs, families, or groups, often in mixed-species flocks. Actively hops among the branches, gleaning foliage, and probing cracks, frequently feeding among bromeliads. Also climbs.
Voice: Call, often monotonously repeated every few s, a weak high-pitched and squeaky trill *ti ti't't'tttt*.
Breeding: Eggs Sep-Nov; large gonads Feb-Oct.
Habitat: Mainly from 5 m to subcanopy. Humid forest, forest edge, and tall second growth.
Range: N Col., at 1520-3000 m in Sta Marta mts. Common.

ASH-BROWED SPINETAIL *Cranioleuca curtata* – Plate XXXVI 7a-b

14 cm. Eye pale. Above dark brown, remiges reddish brown; **crown, wing-coverts, and tail chestnut**; forehead variably admixed dark gray. **Supercilium olivaceous gray**. Below olivaceous gray with very faint darker shaft-streaks; chin whitish, sides of breast, flanks, and vent olivaceous brown. Sspp poorly differentiated. **Juv**. has **supercilium, sides of neck, and entire underparts ochraceous-orange**, except for whitish chin; spot below eye white; crown like back, but usually admixed with some chestnut feathers, rarely all chestnut. **Imm**. like ad., but crown like back. Alone or in pairs, often in mixed-species flocks. Actively hops and acrobatically crawls in dense tangles, outer branch-

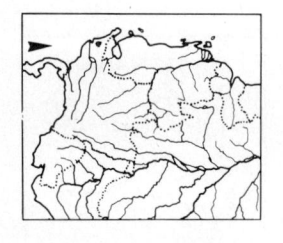

es, and epiphytes on larger branches, picking insects from leaves and twigs. Fledgling Mar (ne Ecu.); juv./imm. Sep (Huila), Aug (ne Ecu.), Apr (n Cajamarca) and Nov (Cuzco); large gonads Apr, May (Huila).

Understory and middle strata, among epiphytes on larger branches, in dense tangles and foliage, often in outstanding trees in humid montane forest. 1000-2500 m. E Andes of Col.(*curtata*), from head of Magdalena Valley in s end of C and E Andes of Col. s through e Ecu. to Amazonas and San Martín, n Peru (*cisandina*), from Divisoria mts in Huánuco, c Peru, s to Cochabamba, Bol. (*debilis*). Fairly common to uncommon.

NOTE: Closely related to, and considered a ssp group of the allopatric Red-faced S. **FORK-TAILED SPINETAIL** *C. furcata* appears to be based on a completely chestnut-crowned juv. of the present species, although in *Condor* 88 (1986):120-2 it has been interpreted as a molting juv. with ad. crown.

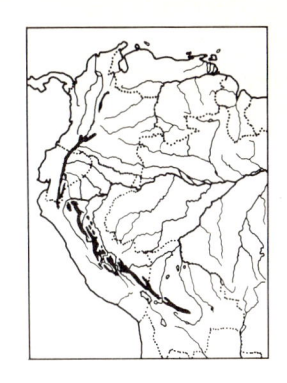

RED-FACED SPINETAIL *Cranioleuca erythrops* – Plate XXXVI 8

13 cm. Eye pale. Above olive-brown; **crown and headside rufous**, wings and tail rufous chestnut. Below olive-gray, chin whitish. Ssp *griseigularis* with grayer breast and sides of neck. **Juv.** usually like juv. of allopatric Ash-browed S., but crown apparently always like back. Alone or in pairs, frequently joins mixed-species flocks, often with Common Bush-tanager present. Hops across and along branches, and crawls among tangles and foliage, often reaching down to turn dead leaves, and probing into mosses and lichens in search of insects. Song a high-pitched fast chattering *seet-seet-seet-se'e'e'e'e'e'e'*. Fledgling Nov (nw Ecu.); juv./imm. Feb, Mar, May, and Sep (Narino), Sep (nw Ecu.).

In lower to middle strata, among leaves and twigs near tips of branches and in vines and tangles. Humid zone, at forest edge, clearings, and second growth. 500-2500 m. W Andes and w slope of C Andes of Col. (*griseigularis*), w Ecu. s to El Oro (*erythrops*); at somewhat higher elevations in Panamá and Costa Rica (*rufigenis*). Fairly common.

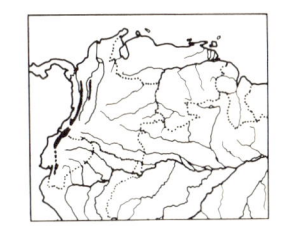

NOTE: Should possibly include Ash-browed S.

STRIPE-CROWNED SPINETAIL *Cranioleuca pyrrhophia* – Plate XXXVI 4a-c

13.5 cm. Above light gray-brown, forecrown streaked blackish and pinkish buff, feathers somewhat elongate and pointed; **white supercilium broad behind eye. Wing-coverts and tail rufous. Below whitish**, flanks and vent washed pale brown. Ssp *rufipennis* browner above and with dull rufous inner remiges and more extensively streaked crown; *striaticeps* intermediate, but with the entire crown streaked like *rufipennis*. **Juv.** has uniform crown, feathers not pointed. Underside darker, often stained dark, and sometimes faintly mottled, at least on breast and neck, sometimes on entire underside and upper back.

Habits: 1-4 birds together, territorial, but sometimes in mixed-species flocks. Active. Climbs trees like a woodcreeper, but also climbs head down, and acrobatically crawls about like a titmouse, swinging to underside of limb with the tail in an awkward position. Confident.
Voice: Song a fairly short descending trill with a few introductory notes, much like others of the genus, but fairly high-pitched. When excited a rapid sputtering explosive *spee-ee-ee-ee* or *tsee-ee-ee-a* accelerating into fast descending trill, sometimes continuing in rising and falling pattern. Calls include a rapid *tidrrit*, and a pleasant low trill.
Breeding: Eggs Nov (San Luis), Dec (Córdoba); nestlings Oct (Córdoba), Nov (San Luis), Dec (Córdoba), Feb (Salta); juv./imm. Dec, Jan (Chuquisaca), Jan (Tarija).
Habitat: 1-10 m up, on branches and in foliage in *Prosopis* and other arid thorny woodland and savanna scrub.
Range: 600-3100 m. La Paz (Tilotilo), Bol. (*rufipennis*); Cochabamba, Sta Cruz (Samaipata), and Chuquisaca, Bol. (*striaticeps*); s Sta Cruz and Tarija, Bol., and to 1500 m, occ. higher, in w Parag., Urug. and most of Arg. s to Río Negro (*pyrrhophia*). Locally common.

NOTE: May form a superspecies with Northern and Southern Line-cheeked S.s.

NORTHERN LINE-CHEEKED SPINETAIL
Cranioleuca antisiensis – Plate XXXVI 6

14.5-15 cm. Above olive-brown; **crown, wings, and tail dark rufous. Supercilium buffy white; earcoverts dusky, washed and slightly streaked rufous**. Throat whitish, turning dull **buffy brown on rest of underside**. Ssp *palamblae* has the back more grayish olive, and an almost pure white supercilium; it is slightly grayer below, has whitish instead of rufous streaks on the earcoverts, and paler under wing-coverts. **Juv**. is uniformly olive-brown above, forecrown more or less mixed with rufous feathers; below either buff, scaled with very narrow, faint feather-edges, or very diffusely mottled with buffy brown edges on c underparts, becoming uniform buffy brown on belly and flanks.
Habits: Alone, in family parties, or in mixed-species flocks. Probes into dead leaf-clusters, epiphytes, crannies, and crevices.
Voice: Song said to resemble that of Elegant Spinetail (MKP, CR).
Breeding: Nest Mar, fledglings Aug (Loja); juv./imm. May (Lambayeque), June (Piura).
Habitat: Mainly on middle strata of humid montane forest and montane shrub.
Range: 900-2800 m. Carchi (sight rec. O Læssøe) and Pichincha (sight rec. RR), nw Ecu.; Azuay, El Oro, and Loja, sw Ecu. (*antisiensis*); both slopes of W Andes in Piura and n Cajamarca s to n Lambayeque, n Peru (*palamblae*). Fairly common.

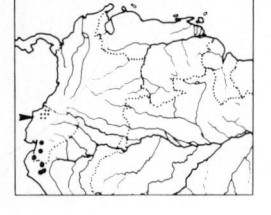

SOUTHERN LINE-CHEEKED SPINETAIL
Cranioleuca baroni – Plate XXXVI 5a-b

15-19 cm. Above gray with slight brownish olive tinge; **crown, wings,** and tail rufous to rufous chestnut. **Superficium white, earcoverts streaked black and white, usually admixed a little rufous; throat and upper breast white, slightly mottled and turning grayish on rest of underparts**, flanks with olive-brown wash. Ssp *capitalis* has a lighter crown (tawny) and somewhat lighter wings, distinct white shaft-spots on feathers on sides of neck and breast, and a darker gray abdomen; c tail-feathers often with more or less distinct dusky terminal spot. Ssp *zaratensis* is **much** smaller than most *baroni* and *capitalis* (however, there is a complex local size variation, birds from Cochabamba in La Libertad being almost as small), and has more rufous on the under wing-coverts. **Juv**. is olive-gray above, crown occ. admixed with rufous feathers, sometimes with feathers rufous basally. Supercilium grayish buff, underparts below whitish throat grayish buff distinctly scaled with diffuse, dark feather-margins.

Habits: Alone or in pairs climbs stems and branches, often in the manner of a woodcreeper, picking insects from crevices in the bark or leafclusters. Difficult to see as it usually feeds in the very dense c canopy, and, in *Polylepis*, esp. on the dense masses of dead twigs characteristic of the interior canopy. Often feeds high up, but flies low over the ground.

Voice: Song a high-pitched descending 1-3 s trill, initiated with a lower note than the beginning of trill *keek kéee kee-keekeerrrr*.... Sometimes the song continues, ascending again, repeating hysterically *rreep-rreep-rhe-eee-rrr-rr-r-r-rr-rrr-rree-rreep-rreep-rheeee*.... Also a growling *aaa-haah* (female?).

Breeding: Fledgling Apr (Chugur Cajamarca); juv./imm. July (Huánuco), Feb (Ancash), Aug (Cajamarca), June (Lima).

Habitat: *Polylepis* woodlands and semi-arid cloud forest; in some areas at edge of semi-humid montane forest and in second growth, esp. with admixed alders (*Alnus*).

Range: 1700-4500 m. S Amazonas, c and s Cajamarca (where also on Pacific slope), La Libertad, Ancash (where crosses to Pacific slope of Cord. Negra) and wc Huánuco (Cullcui) (*baroni*); Huallaga drainage in Huánuco (Carpish, Panao mts, Caina) and la Quinua in Pasco (*capitalis*); Lima (*zaratensis*, described by M Koepcke in *Publ. Mus. Hist. Nat. Javier Prado*, ser. A (*Zool.*) 20(1961):1-17). Locally fairly common.

nest

NOTE: Often considered sspp of Northern Line-cheeked S., and distribution still needs clarification. They may form a superspecies with Stripe-crowned S.

MARCAPATA SPINETAIL *Cranioleuca marcapatae* – Plate XXXVI 3a-c

14.5 cm. **Above rufous. Crown rufous** (*marcapatae*), **or white** (*weskei*), **outlined in black; supercilium whitish above eye**, widening and turning olive-gray posteriorly; thin line through eye blackish. Sides of head and neck olive-gray, throat whitish, underside of body gray, washed olivaceous on sides, flanks, and vent. In *weskei*, malar region washed with buff. **Juv**. like ad. above, below with ochraceous wash and heavily mottled with dark feather-edges.

Habits: Alone or in pairs, often in mixed-species flocks. Crawls about

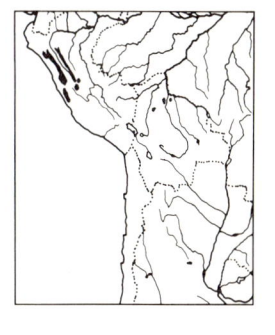

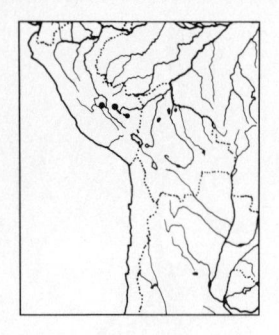

in epiphytes and thickets, hops along branches and creeps like a woodcreeper, probing into moss, bark etc.
Voice: No rec.
Breeding: Fledglings Dec, Apr and May; juv. Feb; juv./imm. Aug (Cuzco).
Habitat: Near the ground in dense bamboo thickets and undergrowth, and also higher in trees. Edge of humid forest, cloud forest, elfin forest, and second growth.
Range: Se Peru at 2400-3500 m in Vilcabamba mts, Cuzco (*weskei*, V Remsen in *Wilson Bull*. 96(1984):515-23); e slope of Vilcanota and Carabaya mts, Cuzco, se Peru (*marcapatae*). Fairly common.

NOTE: Closely related to, and possibly sspp of the allopatric Light-crowned S. For a right interpretation of the leap-frog variation in crown-color, a thorough study is desirable.

LIGHT-CROWNED SPINETAIL *Cranioleuca albiceps* – Plate XXXVI 2a-c

14 cm. **Back, wings and tail rufous chestnut. Crown outlined in black**. In *albiceps* the crown is white, or white more or less washed buff, the nape olive-gray; in *discolor* both **crown and nape are orange-buff**. Supercilium dark gray, palest above eye; headside mixed gray and black, neckside and underparts dark olive-gray with occ. buffy feathers; chin whitish or buffy white. **Juv**. undescribed. Buff-crowned specimens of ssp *albiceps* may be imm.
Habits: 1-3 birds together, often in mixed-species flocks. Crawls acrobatically through the interior of dense bushes, climbs trees like a woodcreeper, and hops along limbs to probe moss, bromeliads, dead leaf clusters, and bark, much like other *Cranioleuca*s. Rarely sallies.
Voice: A descending series of loud high-pitched notes, repeated 2-3 times when excited *weeweewee-weeweeweewee*.
Breeding: Nest-building July-Aug (La Paz). Large gonads Jan (Cochabamba).
Habitat: Lower and middle strata in open cloud forest, bamboo, and other second growth.
Range: 2200-3400 m. Se Puno (Valcón) se Peru to La Paz, Bol. (*albiceps*), Cochabamba and Sta Cruz, Bol. (*discolor*). Intergrades between the 2 sspp are known from prov. Ayopaya, Cochabamba, Bol. Fairly common in Bol., but so far only known from 1 locality in Peru.

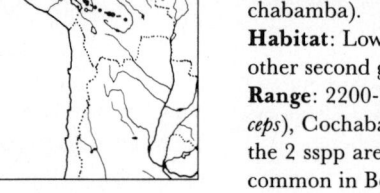

NOTE: Should possibly include the allopatric Marcapata S.

CREAMY-CRESTED SPINETAIL *Cranioleuca albicapilla* – Plate XXXVI 1a-b

17 cm. Above grayish olivaceous, paler and browner on rump; **crown creamy**, feathers elongate, forming small crest; supercilium white. Wing-coverts and tail rufous chestnut. Earcoverts somewhat streaked, throat white, gradually becoming darker on rest of underparts, sides of neck and breast ochraceous; sides, flanks, and vent olive-gray. Ssp *albi*-

gula has crown-feathers more cinnamon, back and underside more washed ochraceous, throat more extensively and purer white, and tail-feathers with some dusky shades on inner webs. **Juv**. pale yellow on forehead and throat, strongly sooty on nape and upper belly, and more or less washed ochraceous and scaled below.
Habits: In pairs. Climbs stems and thick branches like a woodcreeper, and hops along the branches, usually in horizontal posture, sometimes hanging down, and occ. cocking tail 30-45°. Often raises crown- and throat-feathers. Pecks and probes into moss, bark, lichens, epiphytes, and bases of twigs. Usually keeps well hidden.
Voice: Calls include a song-like high-pitched descending chattering 2-6 s trill, sometimes lingering on, and a sharp irregular *kjeep kjeep-kjeep-kjeepkjep*. Song lasts c. 1 s, and starts with 2 notes, the 2nd higher and stressed, followed by a descending trill.
Breeding: Fledgling May (Apurímac).
Habitat: 2-8 m up in **not very humid** montane forest, often feeding rather low in dense and tangled parts, dense tangled or moss-clad parts of e.g. *Podocarpus* forest, mixed *Eugenia/Escallonia* woodland, and sometimes into old *Polylepis* wood.
Range: C and s Peru. 2400-3650 m (locally to 4080 m) in Junín (Huacapistana) and Huancavelica (*albicapilla*), grading through Ayacucho and Apurímac (Cotaruse, Chapimarca, Bosque Paragay, Nevada Ampay) to the population of the Urubamba and Quillabamba valleys in Cuzco (*albigula*). Locally common.

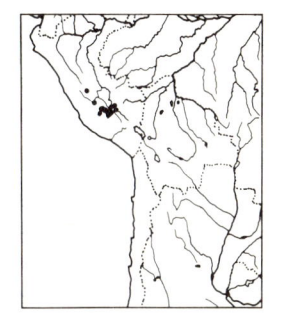

Spinetail – Genus *Siptornopsis*

Monotypic genus. Fairly large spinetail with no rufous in wing. Confined to a small area in nc Peru. Nest undescribed.

GREAT SPINETAIL *Siptornopsis hypochondriacus* – Plate XXXVII 21

19 cm. Tail fairly long and graduated, tips of feathers rounded. Above, incl. tail, dark brown to gray-brown, feathers of forehead tipped paler, and outer tail-feathers somewhat pale. Lesser and median wing-coverts rufous; **in flight no rufous band across remiges**. Lores black, earcoverts dusky streaked white on lower part, neckside with some black and white streaking; **supercilium and underparts white**, sides of body streaked dark gray, streaks almost meeting on middle of breast; flanks, vent, and abdomen washed umbraceous brown to buff. **Juv**. undescribed.
Habits: In pairs. Forages in dense shrubbery.
Voice: A loud chatter.
Breeding: Very worn ad., probably nesting, Apr (Ancash).
Habitat: Dense, humid, montane shrubbery, sometimes admixed with alders (*Alnus*).
Range: N Peru at 2450-2800 m in the Marañón drainage in s Cajamarca (Malca, Cajabamba) (and probably adjacent Amazonas), La Libertad (Patas, Cochabamba), and Ancash (Santa Clara). Apparently rare, and habitat under strong pressure.

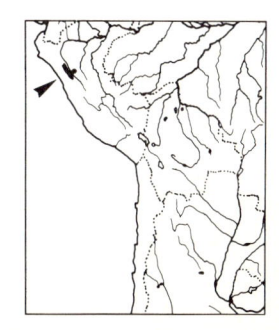

Thistletails – Genus *Schizoeaca* – Plates XXXVI 19a-k, LXIV 32

The thistletails are a complex of 12-13 allopatric forms of temp.-zone spinetails, disjunctly distributed at 2600-4000 m in the Andes from Ven. to Bol. They are medium-sized brownish to grayish birds with fairly long and thin, graduated, strongly decomposed tails. They forage in the undergrowth, esp. at border of dense humid shrubbery, and mainly differ from one another in general coloration, and in the lack or presence of a supercilium, eye-ring, and chin-patch. The call-notes rec. do not differ significantly between the forms, but there is a slight difference in song between several forms, and some evidence of negative response to playback of *S. f. plengei* to *S. helleri*. They have variously been treated as 1 or 8 species. Electrophoretic comparisons of enzymes from *griseomurina*, *fuliginosa*, and *helleri* suggest that at least these 3 forms deserve species status, as is probably also the case with the distinctly colored *coryi*. As more material becomes available, this may well be shown to be the case for additional forms. The morphology of all 13 forms are here dealt with separately.

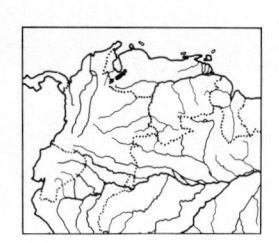

OCHRE-BROWED THISTLETAIL *S. coryi* – Plate LXIV 32

19.5 cm. Above earthy brown; long supercilium, forehead, and upper throat ochraceous, chin tawny. No eye-ring. Below grayish, whiter on c abdomen. C tail-feathers like back, outer ones dull rufous.
Range: W Ven. in Trujillo, Mérida, and Táchira.

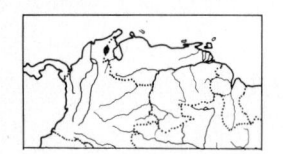

PERIJÁ THISTLETAIL *S. perijana* – Plate XXXVI 19a

20 cm. Above gray-brown to gray. Half-ring below eye whitish; ear-coverts dark gray-brown. Chin pale rufous. Below gray, paler, grayish-white on c belly.
Range: Perijá mts. Described by Phelps in *Bol. Soc. Venez. Cienc. Nat.* 33(1977):43-53.

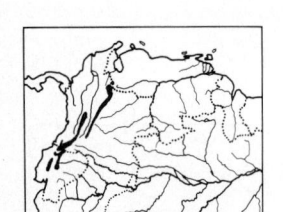

WHITE-CHINNED THISTLETAIL *S. f. fuliginosa* – Plate XXXVI 19b

20 cm. Above warm dark brown, tail sometimes brighter. Supercilium gray or olivaceous gray, whitish or at least palest over eye. Below dirty gray, chin white, sides and flanks washed olivaceous.
Range: Páramo de Tamá and El Cristo w Táchira, w Ven., E and C Andes and Nariño, Col., n and c Ecu. *S. f. fumigata* (JI Borrero in *Noved. Colomb.* 1(1960):238-42) from C Andes and w Nariño, Col., doubtfully distinct (it may have grayer supercilium and underparts, and a smaller chin-spot).

MOUSE-COLORED THISTLETAIL *S. griseomurina* – Plate XXXVI 19d

19.5 cm. Above olivaceous. Very faint supercilium grayish; distinct eye-ring white. Below gray, washed olivaceous on breast and sides;

flanks olivaceous brown, vent brown; no chin-patch, but extreme upper chin whitish.
Range: S Ecu. from Zapote Najda mts, Morona-Santiago s to nw Peru in Piura (Chinguela).

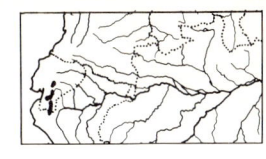

PERUVIAN THISTLETAIL *S. fuliginosa peruviana*

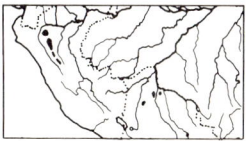

19.5 cm. Intermediate between *S. f.fuliginosa* and *S. f.plengei*.
Range: S Amazonas and San Martín, n Peru.

PLENGE'S THISTLETAIL *S. fuliginosa plengei* – Plate XXXVI 19c

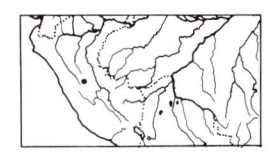

20 cm. Above reddish brown. Supercilium gray, pale gray over the eye. Chin rufous; rest of underparts gray, very faintly tinged brown on belly; sides mixed dark gray and brown, flanks dark brown, vent mixed gray and buff.
Range: Ec Peru in Carpish mts Huánuco. Described by O'Neill & Parker in *Bull. Brit. Orn. Club* 96(1976):136-41.

EYE-RINGED THISTLETAIL *S. palpebralis* – Plate XXXVI 19e

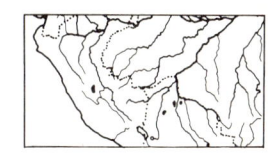

20 cm. Above rufous brown. Distinct eye-ring white. Chin rufous; below gray, sometimes mottled with whitish.
Range: C Peru in Junín.

VILCABAMBA THISTLETAIL *S. v. vilcabambae* – Plate XXXVI 19g

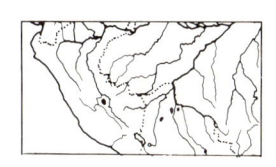

20 cm. Above olive-brown. Supercilium lacking or faint, grayish. No eye-ring. Upper chin rufous; below gray scaled white.
Range: Se Peru in n end of Vilcabamba mts Cuzco. This and the following form described by Vaurie et al. in *Bull. Brit. Orn. Club* 92(1972):142-4.

AYACUCHO THISTLETAIL *S. vilcabambae ayacuchensis* – Plate XXXVI 19f

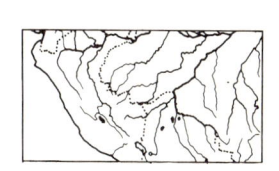

20 cm. Above olive-brown. No or very faint grayish supercilium. No or incomplete eye-ring. Chin and upper throat rufous, rest of underparts gray with faint white scaling.
Range: S Peru w of Apurímac river in n Ayacucho.

PUNA THISTLETAIL *S. h. helleri* – Plate XXXVI 19h

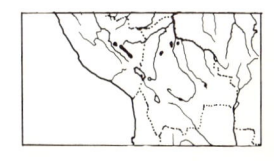

18.5 cm. Above brown, tail rufous. Faint supercilium gray, eye-ring indistinct. Chin with small rufous patch; feathers of throat with blackish gray bases showing through; below gray-brown, paler on c belly; sides olive-brown.
Range: Se Peru e of Urubamba Valley, Cuzco.

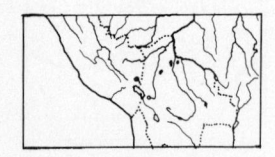

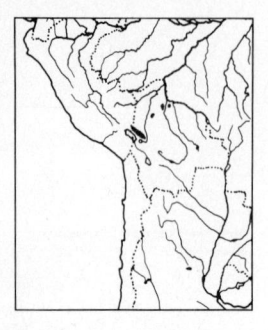

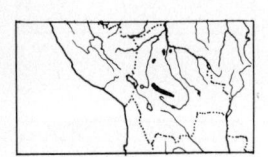

SANDIA THISTLETAIL *S. helleri ssp* – Plate XXXVI 19i

18 cm. Above olivaceous brown, tail brown. Supercilium and underparts gray, chin rufous.
Range: Region near Limbani and n and s of Sandia, Puno, se Peru (specimens in LSUMZ and ZMUC).

BLACK-THROATED THISTLETAIL *S. h. harterti* – Plate XXXVI 19j

17.5 cm. Above reddish brown, crown slightly redder. Supercilium gray turning grayish buff behind eye; narrow eye-ring whitish. Below grayish buff, throat blackish.
Range: Bol. in La Paz.

COCHABAMBA THISTLETAIL *S. harterti bejaranoi* – Plate XXXVI 19k

17.5 cm. Above olivaceous buff to brown. Eye-ring white; supercilium ochraceous behind eye. Below mottled dirty white and dark olivaceous buff, chin washed blackish.
Range: Bol. in Cochabamba and Sta Cruz. Described by V Remsen in *Proc. Biol. Soc. Washington* 94(1981):1068-75.

Juv.: Breast weakly barred; facial and throat-pattern indistinct.
Habits: Tail normally raised 45° during foraging. Often in close pairs. Rarely in mixed-species flocks. Agile. Hops and climbs deliberately along moss- and lichen-covered thin branches, acrobatically gleaning insects from leaves and twigs, often hanging or reaching up and down. Feeds on the ground if ground-cover not too dense. Occ. eats berries. Flicks wings and jerks tail when excited.
Voice: Song given by both sexes: a trill increasing in speed and amplitude, and slightly descending in pitch, lasting c. 2 s and repeated every 6 s *kee kee-keekeekrkrrr...rrrrrr*. Alarm a rapid, sharp *tzeek-tzeek-tzeektzeek* (irregular). Contact call a loud sharp *riek*, *tzeeek*, or *weeek* repeated every 2 s.
Breeding: Eggs Aug (Cundinamarca); fledglings Oct (Puno); imm. July (Perijá), May (Mérida); large gonads Jan (Cochabamba), Oct (n Ecu.), Nov (Huánuco), July (Perijá), Mar (Táchira, *fuliginosa*), Mar (Táchira, *coryi*).
Habitat: Humid zone. Bushes, coarse grass, and ferns in open páramos; within dense humid shrubbery; undergrowth in humid *Polylepis* forest; or 1-2 m up in elfin forest and edge of cloud forest. Fairly common.

Canasteros – Genus *Asthenes*

The canasteros range from lowland Chile and Arg. n through the high Andes to Ven., mostly in the arid zone. There are at least 2 examples of habitat displacement in the genus. Canasteros are small, restless, mostly terrestrial birds with a fairly long graduated tail that is occ.

cocked in certain species. The grassland species whizz about between the tussocks like mice. In many species the rectrices are pointed and decomposed. Most have rufous chin-patch and rufous bar across base of remiges, this bar usually being conspicuous in flight. The species can be very difficult to tell apart, but almost always differ in tail-pattern, which is best seen when the bird lands. The nest is a sometimes enormous stick construction with a side entrance, placed conspicuously in a tree, bush, cactus etc., but Cordilleran C. may also nest in a hole in a wall, when no sticks or bushes are available, though in some cases it will fly several km to obtain sticks, and place its nest between rocks in the bare desert. At least in some species both sexes sing. See *Amer. Mus. Novit.* 860(1936).

LESSER CANASTERO *Asthenes pyrrholeuca* – Plate XXXVII 4a-b

15 cm. **Tail fairly long**, graduated, feathers relatively broad and slightly pointed. Above gray with brownish cast (Patagonia), or dark reddish brown (Chile and wintering birds); crown with dark shafts; lesser wing-coverts dull rufous, secondaries edged reddish brown, esp. at base. **Tail with blackish 3 c pairs, and wholly rufous 3 outer pairs**; 3rd and 4th pair sometimes with a little blackish and rufous along base. **Distinct chin-patch pale rufous**; rest of underparts pale gray, more or less tinged brown; flanks and vent pale brown. Lores white, short postocular streak gray. **In flight** only shows small rufous wingbar. **Juv**. has white chin and dusky feather-margins on throat and breast.
Habits: Active, but clumsy. Climbs along the branches through c **bushes**, constantly wagging tail, and often cocking it 60-70°. Rarely on the ground under dense cover. Fairly rapid tilting flight low over the ground. Sometimes calls from top of bush.
Voice: Calls very high-pitched and fine *ziut, zit, zitzit* repeated every s, sometimes longer *zee-ee-ee-ee zee-ee-ee-ee*. Alarm a scolding chatter.
Breeding: Eggs Oct (Córdoba), Nov (Córdoba, Mendoza), Dec (Río Negro); fledglings Feb (Cautín); imm. Nov (Río Negro), Feb (Cautín, Sta Cruz Arg., Tucumán), Mar (Chubut, Tucumán).
Habitat: Dense, sometimes small patches of dry, dense, and thorny, 1-2 m tall bushes in sand-dunes, on slopes, or other sheltered places, usually near water. In winter also found in dense tall grass, and bushes in marshes (see *Hornero* 11(1971): 93-7.
Range: Sea level to 3000 m. From Aconcagua to Aysén, Chile, and w Arg. near Lago Nahuel Haupí (*sordida*); from Mendoza, n Córdoba, and s Buenos Aires s to Sta Cruz, Arg., in winter n to s Bol., Parag., and s Urug. (*pyrrholeuca*). Fairly common.

NOTE: The names *affinis* and *flavogularis* are synonyms of nominate *pyrrholeuca* (see Olrog, *Acta Zool. Lilloana* 18(1962):111-20).

CREAMY-BREASTED CANASTERO *Asthenes dorbignyi* – Plate XXXVII 6

NOTE: This 'species' is under revision (JF and TS). Apparently 3 bio-

Pale-tailed Canastero, ssp usheri

species are involved, judging from vocal differences, marked habitat shifts, and lack of intergradation. Sympatry has not been proven, but 2 forms come close together in an area lacking obvious dispersal barriers, and 2 may be altitudinally separated in c Peru. Although we will not formally introduce new scientific names here, we treat the forms separately, and with special vernacular names for the component species. Within these respective biospecies, the names *huancavelicae*, *arequipae*, and *dorbignyi* have priority.

The Creamy-breasted C. superspecies forms a zoogeographic species with Berlepsch's, and Chestnut C., and Short-billed C. (*Asthenes baeri*); actually Berlepsch C. is possibly best treated as a ssp under *arequipae*. All are unstreaked, with fairly broad tail-feathers, the c black, the outer more or less rufous. Most forms have rufous rump, flanks and vent.

PALE-TAILED CANASTERO

Unnamed ssp – Plate XXXVII 6d
16 cm. Resembles *Asthenes d. dorbignyi*, but with **tail extensively cinnamon**, outer webs pale, and only the c 2 pairs mainly black. Above umber-brown grading to orange-cinnamon on rump; closed wings almost uniform tawny, primaries more cinnamon. Sides of face lighter, pale eyebrow not prominent; below creamy with buffy gray breast with faint barring, flanks clay-colored, vent pinkish buff. Bill bluish horn; feet light blue.
Range: Nc Peru, at 1830-2700 m. Known from río Santa drainage in northern Ancash (below Huaylas, between Yuracmarca and Yanac; TP, P Hocking, NK). Possibly this form sighted at 4050-4200 m in *Polylepis* wood in quebrada Ulta in Cord. Blanca (O. Frimer). All populations must be very small.

Unnamed ssp

Only known from reports of nests in cacti above Santa María del Valle on the right bank of río Huallaga 12 km downstream of Huánuco town, Huánuco (P Hocking).

Asthenes dorbignyi huancavelicae

15 cm. Resembles the above ssp, but primaries fuscous without light outer webs, contrasting rufous secondaries. In general more richly hued, with pale loral patch, and short supercilium; flanks and vent pinkish buff; sides of tail light pinkish cinnamon.
Range: C Peru, at 2650-3700 m. Known from right bank of río Mantaro in Huancavelica (Huancavelica, Yauli) and Ayacucho (above Quinua, Huari, Quebrada del Agua Potable). Common at Huancavelica and Yauli.

Asthenes dorbignyi usheri – Plate XXXVII 6c

16 cm; very strong-billed. Very **dull-colored, and sides of tail appearing white** (outer 2 feather-pairs white, next 2 pale cinnamon, the 4th with black streaks, but only the c 2 pairs mostly black. Above hairbrown, crown more olive-brown, rump sepia-brown, with no rufous at all. Headside lighter, pale lore and supercilium conspicuous; wings quite uniform ochraceous, secondaries more rufous. Below light smoky gray, always faintly scaled on breast, and grading to white on chin (occ. with rusty gular patch indicated) and c underparts. Vent with only slight cinnamon tinge. Feet light blue. **Juv.** with quite distinct dark feathertips below.
Range: Sc Peru at 2135-3500 m. In Apurímac along río Pampas (above Ahuayro, Ninabamba), río Chalhuanca (near Mutca) and río Vilcabamba (Vilcabamba, Chuquibambilla, Piyai, Bosque Paragay). Rare.

Habits and voice: Forages on the ground with half-cocked tail, and low in trees and bushes. Sings from exposed perch in top of small tree. The song (*usheri*) is a monotonous, ringing, fairly high-pitched, and intense trill almost on even pitch – 2 s long (quite unlike song of Dark-winged C.), given at 1-4 s intervals.
Breeding: Apparently always in columnar cacti (*Pseudoespostoa* and esp. *Cylindropuntia*). The stick-nests are semispherical, c. 40 cm in diameter, *viz* much smaller than nests of Dark-winged C. Young raised in Feb (*usheri*, P Hocking)
Habitat: Arid zone, in valley bottoms and on slopes with columnar cacti and some trees and shrubs (*Schinus molle*, *Carica*, *Tecoma*, *Spartium*), on stony and rocky ground with only a few herbs or short grass. Never seen in the widespread disturbed habitats dominated by *Eupatorium* and *Dodonea* shrub. All forms may be endangered.

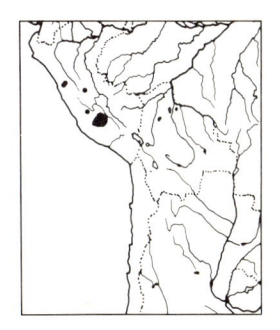

DARK-WINGED CANASTERO

Unnamed ssp – Plate XXXVII 6a

16 cm, and extremely dark; tail black with only part of outer web of outer rectrix rufous. Upperparts dark earthy brown, head particularly dark, ear-coverts almost black; rump normally not rufous, at most with a slight cinnamon tinge; wings black, the rufous wingbar narrow and hardly detectable in flight. Below pale smoky gray, gular patch mottled blackish, and flanks and vent barely tinged rufous. **Eyes pale**; bill and feet dark.

Range: C Peru, at 3700-4200 m. S Lima, near Hortigal (habitat widespread in this area); Ayacucho, 15.5 road-km w of Lucana, Pampa Galeras, and near Yauriviri. Wider occurrence suggested by rec. of nest possibly of this form in southern Apurímac, presence of patches of suitable habitat high above valleys of c Apurímac, and by rec. of intergrades towards ssp *arequipae* near Nevado Coropuno, Arequipa,.

Asthenes dorbignyi arequipae – Plate XXXVII 6b

15.5-16.5 cm. **Dark, with almost black wings and tail, but unlike the above ssp with rufous gular patch, tawny rump, cinnamon-rufous vent, and with 2 outer rectrices more or less extensively rufous**. Above generally umber-brown, distinctly grayish across nape, and with gray supercilium and **blackish ear-coverts**; wings with only narrow rufous margins on the coverts (except for chestnut lesser coverts), and rufous wingbar narrow, hardly visible in flight. Below creamy washed smoky gray on breast and sides. **Eyes pale** (but hue variable); bill and feet dark gray. **Juv.** lack gular spot, and has scaled breast.

Range: At 3500-4800 m, on Pacific slope occ. down to 2500 m, in Arequipa, Mocquegua, and Tacna, s Peru to valley right below Putre in Tarapacá, n Chile, and very locally in Puno, Peru (near Nuñoa, Lampa, Ayaviri, Putina) and n Bol. (Gral Perez at Río Mauri in w La Páz and Nevada Sajama in w Oruro). Common in its patchy habitat.

Habits: Alone or in pairs, but family groups (often pair and 2 imm.) roost together in the same nest much of the year, and often gather during disturbance, and give alarm calls, rising and lowering the closed tail. Eats seeds and insects, usually feeding on the ground and on low (horizontal) limbs of *Polylepis* trees, hopping with tail cocked upright, but sometimes working higher up along the branches, with tail less cocked. When disturbed, always moves into cover in bushes or trees.

Voice: Song much less intense than in Rusty-vented C. (see this species), a 3 s long chattering series of descending, rattling, quite ringing notes, often quavering or rising slightly again at the end. No introductory notes. Alarm a series of single descending notes, 3rd note emphasized *kee kee* **kie** *kee kee kee*.... At the nest, a dry *krrrt*.

Breeding: Build enormous nest in *Polylepis* trees and occ. in columnar cacti, *Puya raimondii*, or shrubs. Some nests are 3/4 m long. Often several nests close together, representing loose colonies or several successive nestings by the same pair or family. Eggs Nov (Arequipa, Tarapacá) and Dec (Puno).

Habitat: Usually in *Polylepis* woodland, which in some areas has character of bushy heathland or semi-desert. Also stands of *Puya raimondii*, and occ. in areas with scrub and columnar cacti.

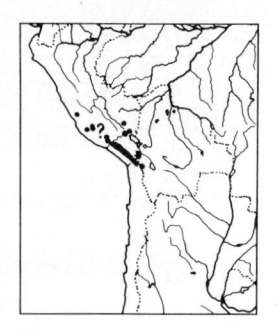

RUSTY-VENTED CANASTERO

Asthenes dorbignyi consobrina – Plate XXXVII 6e

15 cm. Resembles *A. d. dorbignyi*, but **lacking rufous webs on the primaries, which thus are dark and contrast with the rufous secondaries**. Unlike Dark-winged C., less dark-faced, wingcoverts more broadly edged rufous, **rufous wingbar** much broader (to c. 2 cm inside tip), only outer rectrix with rufous outer web, but next 1-2 feathers rufous **at base**.
Range: N Bol., down to 2500 m in the La Paz canyon, and scattered near 4000 m on the Altiplano w-wards into n Oruro (where contact with Dark-winged C. can be expected) and s to Potosí, at higher elevations than ssp *dorbignyi*, but apparently intergrading with it.

Asthenes dorbignyi dorbignyi – Plate XXXVII 6f

16 cm. Above snuff-brown becoming russety cinnamon on rump; wings extensively rufous, with only narrow dusky centers of tertials, cinnamon outer webs at base of primaries, and **broad rufous wingbar** conspicuous in flight. Face lighter without distinct pattern; underparts creamy, gular patch rufous, sides of breast washed gray, flanks and vent cinnamon. **Tail black with narrow rufous side** (compare Chestnut C.): normally only outer rectrix with rufous outer web, next 1-2 with rufous base, but in much of Cochabamba and Sta Cruz and in the s of the range also 2nd or even 3rd feather has more or less rufous outer web. **Juv**. lacks gular patch, and has dark-tipped feathers of breast and upper belly, a character sometimes partly retained in ad.
Range: At 2200-3600 m, and occ. down to 700 m in winter, through Cochabamba and Sta Cruz, Bol., and along the edge of the Altiplano from Potosí s to Mendoza, w Arg. Common over most of its range, but local in Arg. There are old specimens labelled 'Parana' and 'pampas Argentinas' (BMNH)

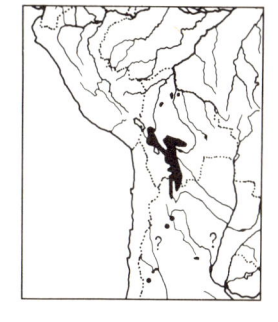

Habits: In open cactus country usually hops about on the ground under and around scattered bushes with the tail cocked upright, but moves up into the foliage when disturbed. Sometimes gathers to scold, like Dark-winged C. May esp. feed on ants.
Voice: Song normally 5-6 s long, of 10-12 intense introductory notes (2nd, 3rd, and 4th notes ofteh highest and longest) given at increasing speed and turning into a trill that fades and descends at the end *tri, tri tri tri tri – tri-tritritritititititrrrrrr*, sometimes longer, as the trill may be prolonged in 2-3 short bursts, or sometimes shorter (2-4 s). The introductory notes may be given alone, the first few notes often rising in pitch, amplitude and speed, the following slowing, descending and falling. Call a wheezy, earthcreeper-like *whee*.
Breeding: Globular stick-nest in scrubs, small trees, and columnar cacti, often several nests in the same tree. Eggs Nov (Salta, Mendoza), Feb (Jujuy); fledglings Jan (Cochabamba), Feb (Chuquisaca); imm. May (Jujuy), Jan, Feb (Tarija, Chuquisaca), Sep (Sta Cruz), Jan-Mar (Cochabamba), Jan, Mar, Apr (La Páz).
Habitat: Arid zone, mainly in fairly open country with scrubs, hedgerows, or columnar cacti, and sometimes in open *Polylepis* woodland. Rec. on *Lepidophyllum* heath on the Altiplano, in stands of *Puya raimondii* in Cochabamba, and in introduced poplars (*Populus*) in Arg.

BERLEPSCH'S CANASTERO *Asthenes berlepschi* – Plate XXXVII 5

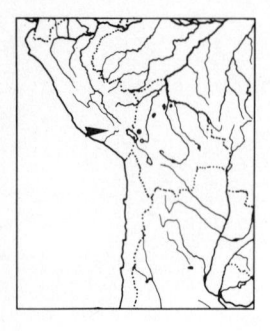

16.5 cm, quite strong-billed. Above snuff-brown, slightly grayer on nape, and rump dark rufous; **wings brown**, all coverts much more extensively webbed dull rufous than in Dark-winged C., the tertials with 2 mm wide webs. Supercilium and neckside gray, earcoverts dark. **Below cream-colored**, throat white with hardly any indications of rusty gular patch, breast always with some weak sooty scaling; **flanks and vent rufous-chestnut**. **Juv.** with well marked dark scales on breast. Much like Dark-winged C., but with **2 outer rextrices wholly rufous, 3rd rufous on outer half**. There are also several rather subtle plumage differences, and it is generally larger with bill longer and stouter. Clearly larger and differently patterned than *A. dorbignyi consobrina*.
Habits, **Voice** and **Habitat**: No information.
Breeding: Imm. July, Aug (mt Illampu).
Range: Bol. Only known from 2600-3700 m on mt Illampu, La Páz.

NOTE: This form shows a distinct combination of character states. However, any character is found in some population or another within the traditional 'Creamy-breasted C.' Ssp *arequipae* comes closest, and in fact many individuals from the local populations in Puno and w La Paz approach Berlepsch's C. in bill-size, tail-pattern, or other characters. Further data (biochemical and biological) are required to state whether species rank is justified for Berlepsch's C.

CHESTNUT CANASTERO *Asthenes steinbachi* – Plate XXXVII 7

15.5 cm. Above gray-brown to cinnamon-brown, grayest on nape; Wings and rump rufous chestnut. **2 outer rectrices rufous, rest blackish with rufous along outer web**, esp. at base, and even c rectrices with a little rufous. Chin and throat whitish, feathers narrowly tipped dusky, rest of **underparts pale gray, flanks and vent tawny. In flight** small ferruginous wingbar. **Juv.** undescribed, but presumably with barred breast and upper belly like Rusty-vented C.
Habits: Tail often raised, but never cocked upright.
Voice: No information.
Breeding: Eggs Jan (Salta).
Habitat: Bushy hillsides.
Range: Nw Arg. At 2-3000 m, and down to 800 m in winter, in Salta, Catamarca, extreme w Tucumán, La Rioja, San Luis (see *Hist. Nat.* (Mendoza) 1(1979): 9-12), and Mendoza (Las Heras) (*steinbachi*).

NOTE: Probably closely related to Rusty-vented or Pale-tailed C.s (see Olrog in *Acta Zool. Lilloana* 18(1962):111-20). Ssp *neiffi*, first described by JR Contreras as a form of *steinbachi*, was later, by Contreras, Nores et al., and Navas & Bo placed under Short-billed C. (*Asthenes baeri*), which it resembles in colors, measurements, and elevational distribution.

PATAGONIAN CANASTERO *Asthenes patagonica* – Plate XXXVII 11

15 cm. **Bill short and stout**. Above gray tinged brownish; lesser wing-coverts pale rufous gray. Gular patch slaty gray with white feathertips and occ. some rufous; breast pale gray turning ochraceous cinnamon on belly, flanks, and vent. **Tail relatively short, blackish**, outer web of outer feather rufous, narrow edge of rest cinnamon-gray. **In flight** a faint small dull tawny wingbar. **Upon landing tail looks blackish**. **Juv**. undescribed. 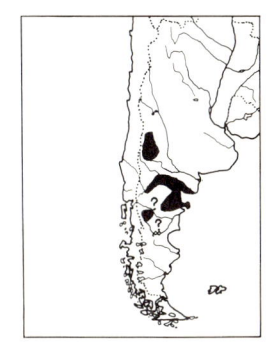
Habits: Singly hops about among twigs, or walks slowly on the ground with cocked tail, always under protection of a bush, and feeding in the foliage of the densest shrubs. Rapid tilting flight low over the ground. Not shy.
Voice: Song a musical even trill *tree-ee-ee-ee-ee*.
Breeding: Stick-nest with a long, tubular side entrance. Eggs Oct (Chubut), Nov-Dec (Río Negro, Mendoza).
Habitat: Semi-arid scrub. In dense, tall *Atriplex* stands, *Condalia microphylla*, and *Schinus* shrubs in river-bottom or on gravelly hillsides.
Range: Arg., in w Mendoza, in the Andean foothills, and from s Neuquen across Río Negro into n Chubut.

NOTE: For more information see JR Contreras in *Hist Nat. Mendoza* 1(1980):101-8.

DUSKY-TAILED CANASTERO *Asthenes humicola* – Plate XXXVII 10a-b

15 cm. **Rectrices fairly broad**. Above dark gray-brown, crown dark brown, rump rufous brown; lesser and median wing-coverts rufous chestnut. Sides of neck and breast gray; flanks and vent olivaceous brown (*polysticta*), tawny (*humicola*), or brighter cinnamon (*goodalli*). **Supercilium** and chin **white**; headside, **throat, breast, and c belly** narrowly **streaked** to spotted dark brown and white, streaking most conspicuous in *polysticta*, which also has distinct black spots on throat, and black lores. **Tail blackish, only slightly graduated**; outer 2 feathers more or less chestnut along shaft. **In flight** rufous wingbar small or lacking. **Imm**. like ad., but pattern below more diffuse. 
Habits: Slowly works its way along branches or hops underneath bushes, picking insects from twigs and leaves.
Voice: Song a loud, clear trill.
Breeding: Both sspp have 2 broods, *humicola* laying in Aug and again in Nov.
Habitat: Dense thickets of low brush on hillsides, esp. in semi-arid regions with thorny bushes like *Acacia cavenia*.
Range: Coast to 2200 m. Chile in Quebrada Paposa in sw Antofagasta (*goodalli*, see Marín, Lloyd & Peña in *Bull. Brit. Orn. Club* 109 (1989):74-5), from Atacama to Maule, and Arg. in n Mendoza (*humicola*), Chile in coastal zone of s Maule, Concepción, and Arauco, and in the Nahuelbuta massif in Malleco (*polysticta*).

CORDILLERAN CANASTERO *Asthenes modesta* – Plate XXXVII 9a-c

14.5 cm. Tail graduated, feathers pointed. Above gray-brown (darkest in *cordobae*, *rostrata*, and the slightly shorter-winged *proxima*, quite gray in *hilereti*, more rufescent in *serrana*), crown with dark shafts. **Rufous patch at base of secondaries visible in closed wing**; below pale gray-brown turning **buffy on belly** and flanks. Narrow supercilium buffy; gular patch pale rufous, bordered all around by obscure whitish streaks that continue over earcoverts, neckside, and breast (streaks most extensive in *rostrata* and *proxima*, slightest in *modesta*). **Each tail-feather with most of outer web rufous, and most of inner web blackish. In flight** a fairly broad rufous wingbar. Sspp *australis* and *navasi* long-billed, the latter very pale and gray, with insignificant buff tinge below, and with an inconspicuous rufous wing-patch. **Juv**. lacks gular patch and has dark gray tips to the breast-feathers.

Habits: Mostly terrestrial. Usually alone. Often cocks tail, sometimes straight up. Walks, runs like a mouse, and hops under or around bushes and rocks, picking up insects from the ground. When disturbed usually hides behind tussock or under low dense brush. Sings from top of rock, small bush, or grass tussock. Occ. perches atop *Polylepis* bushes.

Voice: Song a monotonous or slightly ascending sharp trill lasting 2 s. Also a descending series of notes initiated with 1-2 notes, sometimes followed by 3 even notes *tch teeteetee dju-dju-dju-*.... The descending single notes may be replaced by double notes, and the initiating note can be longer *trrtjeet tju-tju-tju-djuti-djuti-djuti-dju*, or double notes may be given on one pitch *tjí* djuti-djuti-djuti-djuti. Call a sharp *tjit*, at high excitement given in rapid succession. During disputes a chattering *keekeekeeke-ekee...*, descending towards the end, or a series of *tick*-notes ending with a descending trill.

Breeding: Eggs Oct (Salta), Nov, Jan (Chile), and Dec (Córdoba); nestlings Jan (Jujuy); fledglings Apr (La Páz); imm. Jan, Feb, May (Potosí), Apr (Arequipa), Feb (Sta Cruz Arg.).

Habitat: Dry bushy canyons; 'tola' heath; arid puna, rolling hills with low scattered brush, or even desert with no vegetation. Among rocks on grassy slopes, and locally in arid *Polylepis* shrub. In Arg. also in dry open woods.

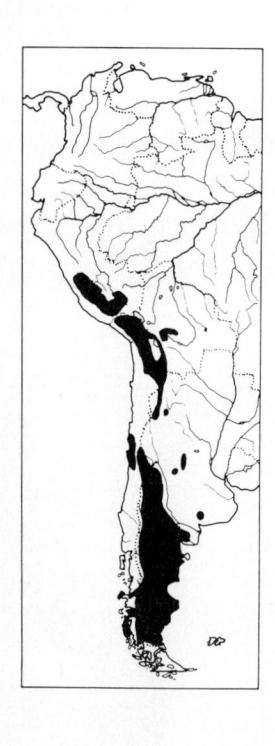

Range: In Peru at 3000-4600 m, in Patagonia down to sea level. From Junín to Cuzco, Peru (*proxima*), highlands of Cochabamba, Bol. (*rostrata*), at high elevations from Arequipa and Puno, s Peru, s to Antofagasta, Chile, and through extreme w Bol., in Catamarca, and maybe La Rioja, nw Arg. (*modesta*); Sierra de Aconquija in Tucumán (*hilereti* Oustalet, see *Hornero* 12(1986):267); Cerro Famatina in La Rioja (*serrana*, M Nores in *Hornero* 12(1986):262-73); Córdoba Arg. (*cordobae*, see Nores et al. in *Hist nat. Mendoza* 1(1980):169-72), c Mendoza and s Buenos Aires to Sta Cruz, Arg. and in Aysén and n Magellanes, s Chile (*navasi*, JR Contreras in *Hist. Nat. Mendoza* 1(1979):13-6; see also *idem* 1(1980): 49-68), C Andes of Chile from s Atacama to Colchagua (*australis*). Fairly common.

CACTUS CANASTERO *Asthenes cactorum* – Plate XXXVII 8

14.5 cm. **Bill long**. C tail-feathers not pointed, and no dark spots bordering throat as in Cordilleran C. Ssp *cactorum* palest; light ochre-sandy above; lesser and median wing-coverts rufous ochre, median with dark centers; base of all but outer remiges rufous; rest of remiges edged gray-brown. Throat pale ochre, **breast unstreaked**, **whitish** like belly; flanks buffy. **Outer 2 rectrices wholly rufous, next rufous with dark edge of inner web, next 2 blackish with narrow grayish ochre edge of outer web** and sometimes along shaft, c pair like back, very narrowly edged grayish ochre. Ssp *monticola* larger than *cactorum*, with longer bill, more brownish gray (less ochre) above, darker c tail-feathers, and less uniform underside. Belly, flanks, sides and vent reddish-brown contrasting with whitish breast and upper belly. Rufous of wings and throat darker. **Juv**. undescribed. – Alone or in pairs. Territory rather large. Climbs on rocks and cacti, picking up insects and cactus fruit, tail often cocked high. Only flies reluctantly, and with wriggling flight. Call a repeated, fairly low-pitched, very fast, dry, weak trill lasting 0.5-2 s, given at 1-3 s intervals. Song a similar but more prolonged trill (up to 6 s). Eggs Jan (*cactorum*), Oct-Jan (*monticola*), Apr, and June-Aug (*lachayensis*).

Stands of large cacti, sometimes with low *Jatropa* and other desert brush, in rocky, hilly country in w Peru. 250-300 m in Loma de Lachay, Lima (*lachayensis*); 2200-2400 m (down to 700 m in Lima) from Lima to Arequipa (*monticola*); 50-1700 m in Loma de Atico, n Arequipa (*cactorum*). Uncommon.

NOTE: Species described by M Koepcke in *Beitr. neotrop. Fauna* 1(1959):243-8, sspp by same M.K. in *Beitr. neotrop. Fauna* 4(1965):162-6. We do not follow Vaurie's synonymizing of these forms with Cordilleran C., although *cactorum* is approached by the pale, faintly streaked, and fairly long-billed *modesta* of n Chile.

CANYON CANASTERO *Asthenes pudibunda* – Plate XXXVII 1

17 cm. Rectrices fairly broad, tail shorter and bill stronger than in Rusty-fronted C. Above dark brown turning reddish brown on rump and **rufous chestnut on tail**; c 1-2 pairs of rectrices with slight dusky wash on outer half of inner web. **Remiges edged rufous**. Narrow supercilium pinkish buff. Below gray, c belly palest; flanks brownish to grayish olive; throat pale rufous, bordered all around, incl. on earcoverts, with whitish and dark gray indistinct streaks, that very faintly continue onto breast. **In flight** shows rather restricted rufous wingbar. Ssp *neglecta* browner and darker; *grisior* grayer below, with more pronounced whiter throat-streaks and longer slenderer bill. **Juv**. lacks rufous chin-patch, has dark-tipped breast-feathers, and is washed with brownish below.
Habits: Alone or in pairs. Climbs about within dense shrub and on loose-barked trees, occ. with head down, or up trunks like a *Cranioleuca*,

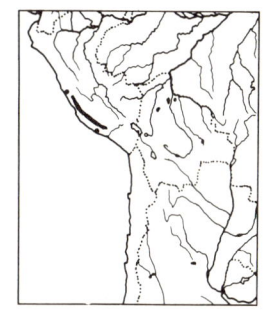

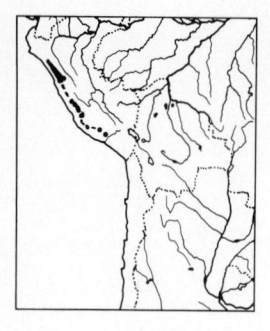

at times on the ground in shelter of rocks, trees, or bushes in search of insects. Regularly cocks tail, esp. upon landing.
Voice: Song a somewhat long accelerating loud trill, falling towards end, and with a few introductory notes *dee dee de trrrrrrt*, given all year, but mostly during the rainy season. Call a distinctive *keee-whit* repeated endlessly.
Breeding: Fledgling May (Lima); imm. May (La Libertad).
Habitat: Always in shade of large rocks or dense bushes. Rocky bushy slopes, sometimes with scattered cacti, *Polylepis* trees or in open semi-arid cloud forest.
Range: Peru. 2500-3700 m. Nw Peru in La Libertad and Ancash (*neglecta*), wc Peru in Lima (*pudibunda*), sw Peru in Huancavelica, Ayacucho, and Arequipa (*grisior*, M Koepcke in *Amer. Mus. Novit.* 2028(1961):13-5). Fairly common.

MAQUIS CANASTERO *Asthenes heterura* – Plate XXXVII 3

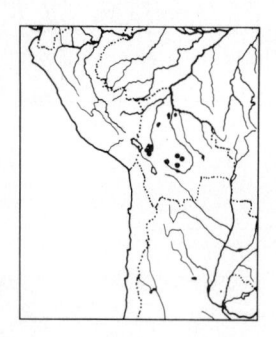

17 cm. **Tail fairly long, strongly graduated, feathers narrow**. Above gray-brown, remiges edged rufous; **outer 4 pairs of rectrices rufous**, c 2 pairs dusky rufous. Below buffy gray, sides brownish, flanks pale brown; chin orange-rufous, bordered below by a few, 1-2 mm long, indistinct streaks; eye-ring and fairly narrow supercilium buff. **In flight** shows rufous wingbar across both primaries and secondaries. **Juv**. undescribed.
Habits: Alone or in pairs, forages near the ground in low vegetation, occ. on the ground.
Voice: Presumably like Rusty-fronted C.
Breeding: Imm. June, July (Cochabamba).
Habitat: Arid zone. Thorny scrub and hedgerows, dry bushy slopes with scattered *Polylepis* trees.
Range: Bol. at 3000-4150 m in La Paz and Cochabamba. Fairly common.

NOTE: Sometimes called Iquico C. Should possibly include Rusty-fronted C. Previously treated as a ssp of Canyon C.

RUSTY-FRONTED CANASTERO *Asthenes ottonis* – Plate XXXVII 2

18.5 cm. **Tail long**, strongly graduated, **feathers rather narrow**. Told from Canyon C. by longer tail and smaller bill. Above brown, forecrown usually rufous chestnut; faint supercilium buff; **wings and tail rufous chestnut**, c rectrices sometimes with some dusky near tip (least extensively in birds of Cuzco). Gular patch orange-rufous, surrounded by faint, whitish streaks that may continue onto breast; rest of underparts gray-brown, flanks, and vent tawny olive. **In flight** rufous wingbar across primaries and secondaries. **Juv**. undescribed, but presumably without gular patch and rusty forehead.
Habits: Alone or in pairs, forages low in bushes and small trees, occ.

on rocks or ground. Usually climbs bushes when disturbed. Tail often raised.
Voice: Song a fine descending trill initiated with 3 single ascending notes *tee tee tee teetrrrrrrrrr*, lasting 2 s. Call a timid *veer* given at variable high pitch, and a nasal strained *djeeh*.
Breeding: No data.
Habitat: Valleys with clumps of dry shrub and at edge of *Polylepis* woodland, occ. in quite disturbed areas with only scattered low shrubs.
Range: Se Peru at 2750-3600 m in interandean valleys of Huancavelica, Ayacucho (La Quinua on rd to Tambo, LSUMZ), and Apurímac (unnamed ssp); and Cuzco (*ottonis*). Fairly common.

NOTE: May be sspp of Maquis C., and is certainly not close to Canyon C., as has been suggested.

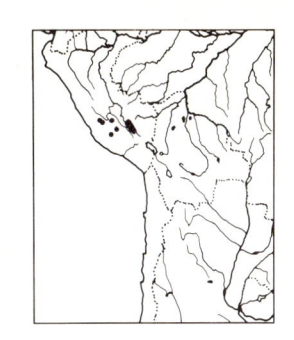

STREAK-BACKED CANASTERO *Asthenes wyatti* – Plates XXXVII 13a-b and LXIII 18

15.5 cm. Ssp *aequatorialis*: bill long and very straight (shorter in other sspp). **Above** rufous brown or gray-brown, **with broad blackish streaks** along shafts; **wings with conspicuous rufous chestnut patch** at base of secondaries and inner primaries. **Below buffy** with fine, dark brown shaft-streaks along sides, and hardly visible spots across breast; chin ferruginous, flanks tawny buff; narrow supercilium buff. Rectrices rounded, c ones dark brown, next 2 pairs with narrow, rufous edge of outer web, **outer 3 pairs wholly rufous. In flight** very conspicuous rufous wingbar. Ssp *sanctaemartae* small, very dark, strongly streaked on back, gular patch orange, underparts rather dark drab brown with slight pale streaking on breast, belly whitish; *wyatti* browner above, with dark centers of feathers smaller and slightly less black; tail less blackish, flanks and vent less bright; *azuay* more rufous in wings and tail; *graminicola* like *azuay*, but duller, underparts tawny with virtually no streaks. Differs from Puno C. mainly by tail-pattern. **Juv**. darker and less distinctly patterned above, and lacks gular patch; breast with sometimes quite distinct narrow dusky bars; rectrices pointed.
Habits: Alone or in pairs, runs swiftly between bushes or tussocks of grass like a mouse, tail half cocked. Regularly surveys from a rock. Probes into bases of tussocks, and hops to catch insects in the air. In Peru only locally (e.g. Limbani, Puno) seeks shelter in a bush, but in Ven. and Sta Marta mts, where no other canastero occurs, regularly ascends bushes and herbs. Song is given from top of rock, or in Ven. from an *Espeletia*.
Voice: Song of *graminicola* and *aequatorialis* a rather flat trill at 4-5 kHz, lasting c. 1 s, increasing in speed, amplitude, and sometimes pitch till just before the end. Also 3 short identical trills altogether lasting 1 s. Female (?) gives a wheezy complaint *wheedididi*. Contact call a sharp accelerating *tick tick*.... Alarm *chick* or *check*. Call a *tjit* much like Cordilleran C.
Breeding: Nest-building Mar (Mérida); eggs Jan (Puno); fledgling Apr (Sta Marta); juv. Apr (Sta Marta); imm. May (Chimborazo), Oct (Puno); large gonads Feb, Mar (Sta Marta), July (Perijá).
Habitat: High elevations. In Ven. in a dense mixture of Espeletias and

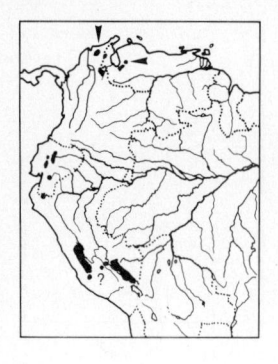

shrubs interspersed with more open grassy areas. In Ecu. and Peru in the puna zone, between tussocky bunchgrass on dry slopes with scattered rocks, and usually devoid of any bushy vegetation.
Range: 3000-5500 m, down to 2400 m in Sta Marta mts. Páramos of Ven. in Trujillo and Mérida (*mucuchiesi*); Perijá mts (*perijana*, Phelps in *Bol. Soc. venez. Cienc. nat.* 33(1977):43-53); Sta Marta mts (*sanctaemartae*); E Andes Col. in Santander (Páramo de Pamplona = SantUrbán) (*wyatti*); n Ecu. on Cotopaxi and Chimborazo mts, and in the hills between Illiniza and Tiupullo (*aequatorialis*); s Ecu. in Azuay (*azuay*); c and s Peru in Lima, Junín, Huancavelica, Apurímac (above Virundo), Cuzco, and Puno n of lake Titicaca to the Bol. border (*graminicola*). Fairly common.

NOTE: Regarded as forming a superspecies with Puno and Córdoba C.s. However, a great individual variation in the intermediate population (*A. p. punensis*) in Puno suggests that the differentiation has only reached megassp level.

PUNO CANASTERO *Asthenes punensis* – Plate XXXVII 14a-b

17.5 cm. **Above gray-brown with broad blackish streaks** along the shafts. Chin whitish with orangerufous patch, sides of chin mottled with black, rest of underparts warm buff to buffy white. **Tail** fairly long, strongly graduated and with rather broad feathers, dark gray-brown, **with large rufous tips to outer 3-4 feathers**. Wing-coverts edged rufous; **extensive rufous band across base of all remiges**, also **conspicuous in flight**. Ssp *punensis* like Streak-backed C. (ssp *graminicola*), except for tail pattern, warmer buff underparts, and only 1.5 cm wide rufous wingbar (2.5 cm wide in *graminicola*). Ssp *cuchacanchae* paler above, more heavily streaked blackish; rufous of the wing and color of underparts also paler. Ssp *lilloi* more rufescent and heavily streaked above. **Juv**. has breast mottled with dusky, and lacks gular patch.
Habits: Alone or in pairs rapidly runs between tussocks of grass, occ. jumping to snatch insects from the grass. Often cocks tail 45°. Escapes by running, only reluctantly taking to the wing.
Voice: The fledglings give a high-pitched trill to attract their parents. Song: 1 rec. of *cuchacanchae* was a 1 s trill like that of Streak-backed C., but slightly less rapid (BW).
Breeding: Eggs Jan (Salta).
Habitat: Open country covered by bunchgrass, and open *Polylepis* woodland with bunchgrass ground-cover.
Range: 2150-4000 m. Se Peru in Puno sw of lake Titicaca, and Bol. in La Páz (*punensis*), Bol. in Cochabamba (e in the Tunari range) and Potosí, and nw Arg. in Salta (*cuchacanchae*); nw Arg. in Catamarca, Tucumán, and La Rioja (*lilloi*).

NOTE: Forms a superspecies with Streak-backed and Córdoba (and perhaps Austral) C.s. Sometimes treated as sspp of Córdoba C.

CORDOBA CANASTERO *Asthenes sclateri* –
Plate XXXVII 15

17.5 cm. Rectrices rounded. Above pale gray-brown with blackish **streaks** bordered with dull rufous; supercilium buff; wing-coverts edged rufous; **remiges with very broad and distinct rufous basal band**. Throat white with indistinct, pale rufous patch; rest of underparts tawny-buff. C rectrices gray-brown, rest blackish with bright rufous tips, amount of rufous increasing outwards. **Juv**. darker, throat mottled with dark.
Habits: Presumably like Puno C.
Voice: No data.
Breeding: Eggs Nov, Jan (Córdoba).
Habitat: Rocky slopes with bunchgrass. Bushy pastures.
Range: Arg. at 2000-2900 m in the mts of Córdoba. Fairly common (see *J. Orn*. 106(1965): 204-7).

NOTE: Forms a superspecies with Streakbacked and Puno C.s, *wyatti* having priority, and possibly Austral C., in which case *anthoides* has priority. The name *sclateri* antedates *punensis* (see *Com. Mus. Arg. Cienc. Nat. 'Bernadino Rivadavia'* 4(1982): 85-93).

AUSTRAL CANASTERO *Asthenes anthoides* –
Plate XXXVII 16a-b

16.5 cm. Above buffy gray-brown, **distinctly streaked blackish; in flight conspicuous rufous bar across base of remiges outlined in blackish**; the bar becomes whitish on outer primaries, visible in closed wing. Supercilium whitish; gular patch orange-rufous, bordered by small dark streaks on a whitish background, sometimes continuing as small spots onto upper breast; breast warm buff with very indistinct shaft-streaks, rest of underparts buffy white, lower flanks streaked dark brown. **Tail dark brown to blackish, edged grayish**; rectrices fairly broad, very pointed, 3 outer ones with well-defined rufous to ochraceous spots on apical part of inner webs, and whitish outer web. **Juv**. has white throat and barred breast, sides and upper belly.
Habits: Forages mostly on the ground without cocking tail, picking food off ground and stems, occ. jumping up to pick from a higher stem. May crawl into clumps of grass, reemerging in a few s, or huddle by a tussock while picking intently at the base for several s. Sometimes forages quite out in the open, on sparse vegetation or even bare earth. Sings from an elevated perch, usually near top of a shrub (BW).
Voice: Call *tick*. Song a c. 1 s trill like Streak-backed C., but remaining constant in speed (BW rec.).
Breeding: Eggs Dec (Isla Grande); fledglings and imm. Feb (Cautín), Jan, Feb (Isla Grande); large gonads Sep (Isla Grande). Apparently feeding young Jan (Sta Cruz, BW).
Habitat: Open meadows, damp grassland, well-watered hillsides, dense thorny thickets in open terrain. Perhaps characteristic of the Patagonian long-grass prairie, until this habitat was destroyed by sheep-grazing. Recent (1987) observations (BW) have been in diverse hillside shrubbery (incl. *Berberis cuneata*) interspersed with native grass, and, at

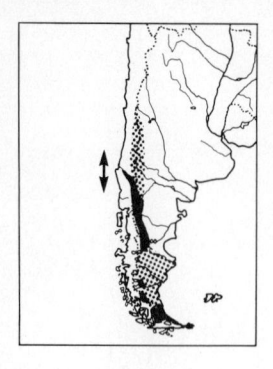

the penguin colony at Cabo Virgenes also in more homogeneous beach vegetation (BW).
Range: Sea level to 1500 m. From Concepción, Chile and from the e base of Andes in Neuquén, Arg. s to most of Isla Grande and Isla de Los Estados. Partly migratory, in winter n to Aconcagua, Chile. Now apparently vanishing, but still locally common near Cabo Virgenes, s Sta Cruz. Uncommon or rare on Isla Grande.

NOTE: Forms a superspecies with Streak-backed, Puno, and Cordoba C.s (see Olrog in *Acta Zool. Lilloana* 18(1962):111-20), *anthoides* having priority.

STREAK-THROATED CANASTERO *Asthenes humilis* – Plate XXXVII 12a-b

15 cm. Above dark gray-brown, faintly streaked with fuscous on crown and back. **Tail dark brown**, outer 2 pairs of rectrices with some pale rufous scribbling, rest with a narrow pale brown edge of outer web. **Headside, neckside, and throat streaked whitish and dark brown**, these parts appearing dark in contrast to the lighter underparts. Base of secondaries with inconspicuous rufous patch, often concealed by gray-brown scribbling. **In flight** only shows small rufous wingbar. Chin rufous; lower underparts light buff. Ssp *cajamarcae* slightly paler, less buffy below, and slightly grayer above, with better defined streaks; *robusta* darker. **Juv**. has less distinct streaks and lacks gular patch; nape mottled with rufous spots.
Habits: Alone, in pairs, or family groups, hops about on the ground between tussocks of grass, under scattered bushes, or on short, matted vegetation, constantly wagging tail, but not cocking it upright. Almost entirely terrestrial. Probes bill into grazed tussocks and matted vegetation in search of insects. Sings from top of a rock or isolated bush.
Voice: Song given at dusk and dawn, a soft double or triple trill *trrr-trrr, trrr trrr trrr* lasting c. 1 s. Also a longer soft trill lasting 2-5 s, ascending and descending in pitch once or twice through the trill *trrrrrrrrrrrrr*. Rarely a rising, soft trill *trrrrrr?*. Contact note fairly loud and given every 0.5 s *pit pit pit*....
Breeding: Fledglings Dec (Cajamarca), Mar (La Paz); imm. Dec, Apr (Junín), Mar (La Paz). 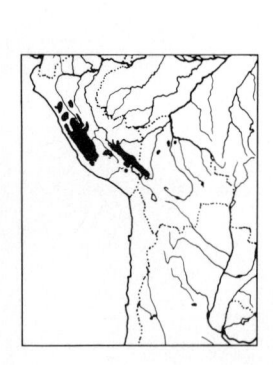
Habitat: Puna grasslands. On matted grass and rocky areas with bare soil between the larger bunchgrass tussocks, and occ. with scattered bushes. Most frequent near water.
Range: 2750-4800 m. Nw Peru in Cajamarca (*cajamarcae*); c Peru from La Libertad and Huánuco to Ayacucho (*humilis*); Cuzco and Puno, se Peru and La Paz, Bol. (*robusta*). Common.

SCRIBBLE-TAILED CANASTERO *Asthenes maculicauda* – Plate XXXVII 17a-b

16 cm. **Above blackish streaked buffy**; crown streaked dark tawny, densest on forehead. Supercilium, chin, throat, and c belly buffy white; sides of head, neck, and body buff, with some fuscous streaks on

sides and (except in southern birds) across breast. Wings vermiculated with blackish along the feather-edges, and with fairly inconspicuous rufous patch. **Tail-feathers** pointed, **not as narrow as in Line-fronted C., tawny vermiculated with blackish. Juv.** lacks rufous on forehead and is more mottled below. Birds from Valcón Puno seem to be dark with narrow tail-feathers.
Habits: Alone, in pairs, or family groups crawls about in low bushes and tussocks of bunchgrass, sometimes sallying 1-2 m straight up to catch insects in the air.
Voice: Like Many-striped C.: song a series of similar notes, the 2nd or 3rd being accentuated and the following ones descending. Call a rising whistle *tuuuiit*.
Breeding: Juv. Jan (Iquico Bol.).
Habitat: Fairly humid zone in páramo-vegetation with patches of bushes and shrubs interspersed with tussockgrass. Where the bushes have been cleared, it remains in tall bunchgrass.
Range: 2250-4300 m. Disjunct. From 8 km s of Limbani, Puno, se Peru to La Paz and Cochabamba, Bol. Isolated in the Aconquija mts on the Tucumán/Catamarca border, nw Arg. (unnamed ssp). Uncommon.

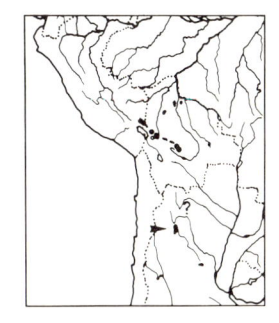

NOTE: Possibly sspp of Many-striped C.

JUNÍN CANASTERO *Asthenes virgata* – Plate XXXVII 18

17.5 cm. **Above** dark brown, rather narrowly **streaked blackish and whitish**; crown blackish narrowly streaked rufous buff. Chin and upper throat orange, c underparts whitish, breast, and sides broadly and indistinctly streaked olive-brown and whitish, flanks olive brown. **Tail dark brown, outer 3 rectrices and outer web of 4th rufous with black scribbling**. See Many-striped C. for distinguishing features. **Juv.** undescribed but presumably without gular patch.
Habits: Alone, in pairs, or family groups. Mainly terrestrial, rapidly runs between rocks, herbs, and tussocks of grass. Surveys and sings from top of a rock or bush, occ. sings in flight. When flushed from the grass, seeks shelter in or under bushes.
Voice: Like Many-striped C.: a descending series of notes initiated with 2-3 notes on even pitch *trree trree trree treeheeheeheeheehee*. Mewing call *tuuuiit*. Alarm a complaining *eek*.
Breeding: No data.
Habitat: Bunchgrass with scattered small bushes (*Lupinus*) adjacent to forest and *Polylepis* woodland. Sometimes found in rocky or hilly areas with no other vegetation than bunchgrass. Overlaps with Line-fronted C. in forest-edge shrubbery, and with Streak-throated C. in grassland. More frequently found in one of these habitats when one of those species is absent.
Range: Peru. Very locally at 3350-4300 m in Lima, Junín, Ayacucho (LSUMZ), Cuzco (Vilcanota mts), and Puno (Limbani).

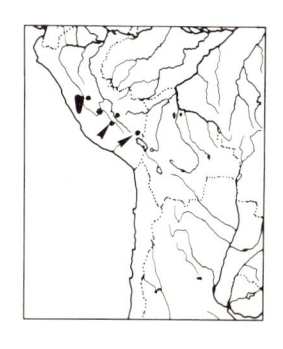

NOTE: Possibly a ssp of Many-striped C.

MANY-STRIPED CANASTERO *Asthenes flammulata* – Plate XXXVII 19a-c

16 cm. **Above** blackish brown with **prominent, white or buffy streaks** down centers of feathers of back, and tawny streaks on crown. Sides of head narrowly streaked, underside of body broadly so, least distinctly on c belly. **Wings conspicuously bright rufous at base and edge of remiges** (in *multostriata* wing-patch not always distinct); supercilium whitish (ochraceous in Col.). Rectrices narrow and pointed, dark brown; c pairs narrowly edged rufous, outer 3 pairs more or less vermiculated with rufous (mostly so in Peru). Chin and upper throat pale rufous. Peruvian birds approach Junín C. by their almost unstreaked whitish c underparts, but may be distinguished by whitish throat, more extensive rufous wing-patch, smaller size, and blackish rather than brownish streaking below. **Juv**. lacks gular patch and is less distinctly patterned than ad.
Habits: Alone or in pairs. Forages mostly on the ground with slightly raised tail, but also along branches of *Polylepis* trees, and in bushes, esp. when Line-fronted C. absent. Insectivorous. When alarmed seeks shelter in or under a bush. Sings from conspicuous post in the top of a tree or a bush.
Voice: Song a slightly descending series of buzzy notes initiated with 2-3 notes *tree tree tree treetreetreetreetree* lasting c. 3 s. Mewing call a distinctive *tuuuiit*. Alarm short *pytt* notes.
Breeding: Large gonads Feb (C Andes Col.), Sep (E Andes Col.); active nest Feb (Ancash).
Habitat: From ground to 2 m up. Edge of park-like cloud forest, humid bushy páramos with small patches of grass, bushes in *Espeletia* páramos, tall bunchgrass at edge of *Polylepis* forest. In W Andes of Peru in open puna.
Range: 2800-4500 m. E Andes of Col. from N.d.Santander to Cundinamarca (*multostriata*); from Caldas to Cauca, C Andes of Col. (*quindiana*); Col. in Nariño, and Ecu. (*flammulata*); nw Peru in w Cajamarca and w La Libertad (*pallida*); n and c Peru in s Amazonas, e La Libertad, Ancash, Huánuco, and Junín (*taczanowskii*). Most common in n part of range.

NOTE: Should possibly include Junín and Scribble-tailed C.s.

LINE-FRONTED CANASTERO *Asthenes urubambensis* – Plate XXXVII 20a-b

16.5 cm. **Tail dark brown, strongly graduated, rectrices very narrow and pointed**. Appears **dark**. Above dark brown, streaked on forecrown and nape to extreme upper back with tawny-buff; wings with inconspicuous dark chestnut-brown patch at base of inner primaries and outer secondaries. Supercilium white; headside, neckside, throat, breast, and sides streaked dark brown and whitish. Flanks brown, c belly whitish. Gular patch bright orange. Ssp *huallagae* is much darker and grayer, esp. below, and the entire underside, incl. c belly is streaked. **Juv**. undescribed.

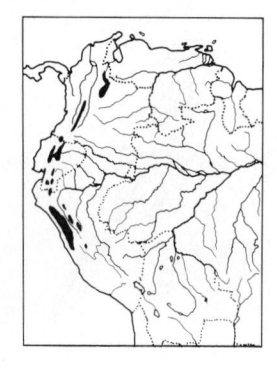

Habits: Though often feeding in humid moss below dense bushes, still the most arboreal of the streaked species of Canastero. Hops along thick branches with tail slightly raised. When on the ground walks like a pipit, but probes bill straight down.
Voice: Song an ascending trill.
Breeding: No data.
Habitat: Semi-humid *Polylepis* woodland, open elfin forest, and mossy slopes with low bushes (*Gynoxis*, melastoms, *Ribes* etc.) and mossy ferns and ericads.
Range: 3050-4300 m. In C Andes from s San Martín to Huánuco and Pasco (Millpo) (*huallagae*); se Peru in Cuzco and Puno, and Bol. in La Paz and Cochabamba (*urubambensis*). Very local.

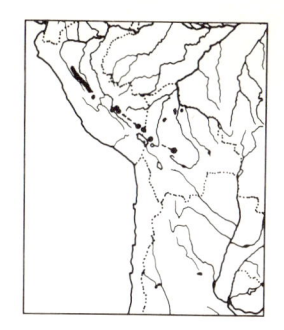

Softtails – Genus *Thripophaga*

The Softtails comprise 4 species of medium-sized arboreal furnariids with pale tawny graduated tails of 10-12 broadly rounded rectrices. 3 species are trop., 1 in Ven., 1 in se Brazil, 1 in the Amazon basin. The 4th species is restricted to the temp. zone of a small area in nc Peru. No nest of this genus has been described.

RUSSET-MANTLED SOFTTAIL *Thripophaga berlepschi* – Plate XXXVIII 4a-b

16.5 cm. Crown-feathers somewhat elongated, buffy olive to ochraceous brown, with buffy centers. Breast, neck, back, wings, and tail rufous chestnut, merging to clay-color on belly, rump, and forehead; chin whitish. **Juv.** very different: **Crown whitish** with darker feather-edges on forehead, hind-crown clay-colored; faint supercilium whitish; ear-coverts and nuchal collar ochraceous tawny, **neck scaled all around**; back pale brown tinged olivaceous, wings and tail as in ad.; below brownish clay, more buffy on breast, and whitish on throat; **scaled with dark brown feather-tips** everywhere below, except on flanks and lower belly.
Habits: Strictly arboreal.
Voice: No data.
Breeding: Juv. Sep (La Libertad), imm. July, Sep (Amazonas).
Habitat: Elfin forest.
Range: Nc Peru. Only found at 2450-3350 m from s Amazonas to e La Libertad (La Peca Nueva, Leimebamba, Atuén, Lluy, Mashua). Rare.

NOTE: Placed in the genus *Phacellodomus* in Vaurie's revision of the *Furnariidae* (1971).

Thornbirds – Genus *Phacellodomus*

7 species of medium-sized arboreal furnariids mainly of the arid trop. zone. They are noisy conspicuous birds usually seen in pairs or family groups. Characterized by a flat head profile, stiffened, pointed chestnut feathers of the forehead, and broad rounded rectrices. Their pre-

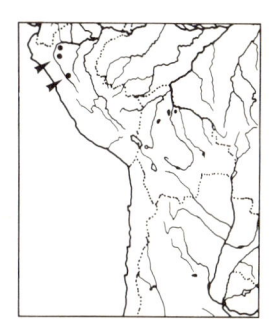

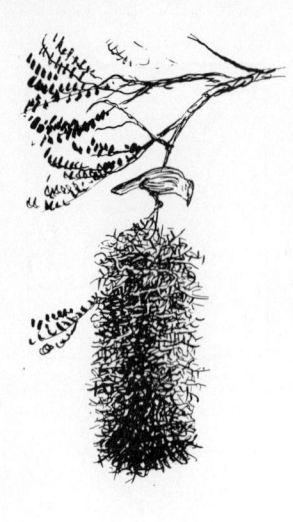

sence is usually revealed by their enormous triangular or oblong sticknests placed in a tree, often pending from the tip of a branch. Both sexes sing in duet.

RUFOUS-FRONTED THORNBIRD *Phacellodomus rufifrons* – **Plate XXXVIII 2** – common in arid thorny scrub and wood in lowlands, locally from Ven. to Arg, ascends to at least 2000 m on Andean slopes from Sta Cruz, Bol., to Tucumán, Arg. (*sincipitalis*). 15.5 cm. Ssp *sincipitalis*: **above olivaceous brown, incl. wings and tail; pointed feathers of forehead rufous chestnut**. Postocular streak and underside dull whitish, whitest on throat and belly; breast tinged grayish, flanks and vent ochraceous brown. Outer 3 rectrices paler and browner than rest, remiges sometimes edged rufous. **Juv.** distinctly mottled above and faintly below; forehead without rufous. In pairs. Slow but acrobatic. Climbs about in trees and bushes, occ. on the ground. Noisy. Sings all year, male somewhat louder than female. Song is an arhythmic series of shrill notes, falling and trembling at the end of the series. Eggs Oct-Dec, Mar (Tucumán).

STREAK-FRONTED THORNBIRD
Phacellodomus striaticeps – **Plate XXXVIII 3**

15 cm. More rufous in tail than Creamy-breasted C. Above graybrown; rump, **lesser wing-coverts, and broad zone on remiges rufous chestnut. In flight shows broad rufous wingbar. Forehead chestnut, feathers stiffened, with pointed gray tips, giving faintly streaked appearance**, supercilium pinkish buff. Below grayish white, flanks and vent tawny-buff. Tail graduated, rufous, center dusky, but outer 4 pairs rufous with decreasing amount of dusky on tips. Sides of neck with varying amounts of rufous. Ssp *griseipectus* is duller colored, earcoverts gray-brown. Birds from Cuzco and Apurímac washed grayish cinnamon below, and the rufous tail-side almost lacks dark distal bar. **Imm.** with poorly developed rufous forehead, and faint dark barring on throat and breast.
Habits: Usually in pairs or family groups. Sneaky. Forages in low bushes, on stone fences, and on the ground, sometimes on stony slopes some distance from the bushy breeding terrain, at mid-day regularly out in the completely open, often with half-cocked tail. Conspicuous while building the nest, that is sometimes placed on a telegraph-pole, but usually pending from the tip of a branch overhanging a cliff or ravine.
Voice: Call a loud trill, and a *Cranioleuca*-like cascade given in duet. Alarm an irregularly rising and falling series of *tsip*s that may accelerate and change to a long descending series of rapid notes. Song a loud series of slightly descending notes.
Breeding: Eggs Dec-Mar (Jujuy), Dec (Tucumán); fledgling Jan (Cochabamba); imm. Feb (Potosí).
Habitat: Semi-arid zone. *Polylepis* forest, sometimes with stands of *Puya raimondii*, hedgerows, gardens, *Eucalyptus*, thorny scrub, arid slopes with cacti. Often near human habitation.
Range: (2300-) 2800-5000 m, down to 1200 m in winter. Se Peru in in-

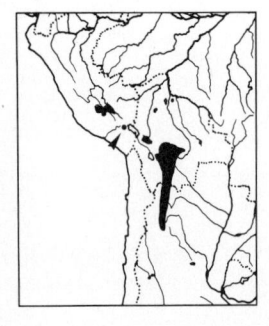

termontane valleys of Apurímac and Cuzco (unnamed ssp), Lampa in Puno (*griseipectus*), La Paz canyon and along the Andean slope from Cochabamba, Bol., s to Tucumán and Catamarca, nw Arg. (*striaticeps*). Common and widespread in s part of range, but very local in Peru.

CHESTNUT-BACKED THORNBIRD
Phacellodomus dorsalis – Plate XXXVIII 1

18.5 cm. **Back**, wings, and tail **rufous**; crown, nape and rump gray-brown; forehead with stiff and narrow dull rufous feathers edged whitish on extreme forehead. Below whitish, breast-feathers rufous, scalloped with whitish; lower belly, vent, and flanks more or less admixed rufous. **Imm**. has gray-brown forecrown without rufous.
Habits: Alone or in pairs. Forages in dense shrubbery, usually with vertical posture. May scold from bush-top.
Voice: No data.
Breeding: Extremely worn ad. (nesting?) May (s Cajamarca). The huge stick-nest is placed in a *Prosopis* tree.
Habitat: Dense and tangled arid to firly humid thorny shrubbery, and hedgerows on bushy slopes with scattered *Prosopis* trees.
Range: Peru. 2000-2800 m on e slope of W Andes in s Cajamarca (Malca, Cajabamba, Hacienda Limon) and La Libertad (Succha). Sighted 1988 at 3400 m at Río Rurichinchay on e slope of Cord. Blanca in Ancash (O Frimer). Uncommon.

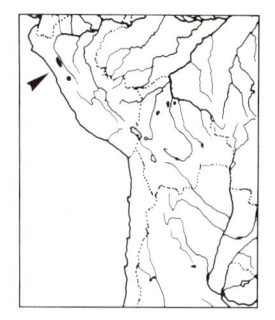

Prickletail – Genus *Siptornis*

Monotypic genus. Very small. Bill thin. Behaviour *Cranioleuca*-like. Nest not described in detail. Only in n part of Andes.

SPECTACLED PRICKLETAIL *Siptornis striaticollis* –
Plate XXXVIII 7 – is a rare inhabitant of humid premontane forest to at least 2200 m in Col. on w slope of E Andes in Cundinamarca and both slopes of Magdalena Valley in Huila (*striaticollis*), e Ecu. in Tungurahua and Cutucú mts Morona-Santiago, and extreme n Peru on Piura/Cajamarca border (Chinguela; see *Condor* 81(1979):319) (*nortoni*, Graves & Robbins in *Proc. Biol. Soc. Wash.* 100(1987): 121-4). Frequents lower to middle strata, along 2-5 cm thick mossy branches of larger forest trees, often at edge. **10.5 cm**. Tail rather short, moderately graduated, **rectrices pointed**. **Above unstreaked rufous**, crown and tail chestnut. **Conspicuous supercilium white** (but may almost lack), and occ. white feathers may occur over nostrils and on headside and crown. Below olive-gray, throat washed orange; **throat and breast with narrow, whitish shaft-streaks**; sides of head streaked rufous and white. **No white malar streak**. Juv. has crown dull rufous like back, and the whitish streaks below broader and reaching belly. Singly, usually in mixed-species flocks. Much like a *Xenops* or *Cranioleuca* spinetail. Works along branches and on twigs, occ. hanging upside down. Gives a high-pitched trill.

Treerunners – Genus *Margarornis*

3 arboreal species of humid montane forests of Costa Rica, Panamá, and the Andes from Ven. to Bol. Wings and tail are rufous, and the tail graduated, rectrices ending in bare shafts; breast or entire underparts conspicuously spotted. All are scansorial (climbing trees like woodcreepers). Pearled T. builds a closed nest of moss, placed under a limb or a rock. See *Amer. Mus. Novit.* 757(1934).

PEARLED TREERUNNER *Margarornis squamiger* – Plate XXXIII 13a-c

14 cm. Back, wings, and tail chestnut, tail ending in bare shafts; crown and nape olive-brown (crown chestnut like back in *squamiger*); below olive-brown; **supercilium, throat, and black-encircled drop-shaped marks over the entire underparts white** in *perlatus*, **or yellowish white** in *peruvianus* and *squamiger*. **In flight** pale cinnamon wingbar. **Juv**. like ad., but crown less uniform, and throat black-spotted.

Habits: Alone or in groups of up to 6, usually in mixed-species flocks. Actively climbs moss-covered stems like a woodcreeper, but also along thinner horizontal branches, often hanging at ends of fine twigs. Occ. searches dead leaves. Rarely in underbrush and low bushes.

Voice: Call a short, sharp, wet trill *trrrt-trrrt* or *trrrrt*.

Breeding: Fledglings June (Mérida), May (W Andes in Cauca), July (Huánuco), Mar, Dec (St Cruz); juv./imm. Sep (Risaralda), Mar, May, July, Sep (Cauca), Dec (w Narino), Nov (nw Ecu.), Feb, Sep (ne Ecu.), June (Piura), Mar (s Amazonas Peru), Dec (Junín), Jan (La Paz), Mar (Cochabamba); large gonads Oct (Boyacá).

Habitat: Mossy cloud and elfin forest, incl. places with second growth, humid *Polylepis* forest, and humid montane forest.

Range: 1350-4100 m, mainly in temp. zone. E Panamá (*bellulus*, sometimes considered a distinct species); Perijá mts and from Lara, Ven., s through all 3 ranges of Col., Ecu., and Peru n of the Marañón (*perlatus*), s of the Marañón to Cuzco (*peruvianus*), from Puno in se Peru to Sta Cruz, Bol. (*squamiger*). Common.

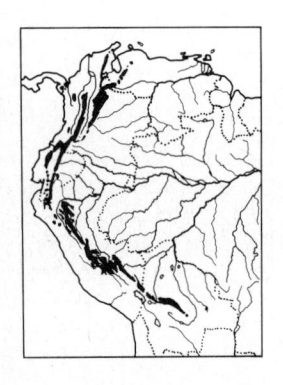

FULVOUS-DOTTED TREERUNNER *Margarornis stellatus* – Plate XXXIII 12

– of wet mossy premontane forest mainly above 1600 m may approach the temp. zone. Nw end of C Andes in Antioquia and W Andes from Risaralda s through w Nariño Col., not s of Imbabura in adjacent nw Equ. (a specimen from near Baeza, ne Ecu. probably mislabeled. 14.5 cm. Above, incl. wings and tail chestnut. Below rufous, throat white; **lower throat and breast with narrow, black-bordered, white streaks** sometimes continuing as rows of tiny dots down c underside. **In flight** cinnamon wingbar. **Juv**. undescribed, perhaps lacking spots below breast. Alone, in pairs, or small groups, usually in mixed-species flocks. Scansorial, sometimes using tail for support. Pecks and probes moss, twigs, and leaves. No information on breeding or vocalizations.

Barbtail – Genus *Premnornis*

Monotypic genus. Arboreal inhabitant of humid subtrop. to lower temp. forest from w Ven. to se Peru. Not scansorial. Nest undescribed. Placed in the genus *Margarornis* in Vaurie's revision of the *Furnariidae* (1971). See *Amer. Mus. Novit.* 757(1934).

RUSTY-WINGED BARBTAIL *Premnornis guttuligera* – Plate XXXIII 11a-b

13 cm. Crown dark olivaceous, faintly scaled by dark feather-edges; extreme forecrown and lores with buff spots; upper back dark brown turning reddish brown on lower back and chestnut on rump; upper back with buff streaks. **Tail chestnut**, rectrices stiff, but not ending in spines; **remiges with chestnut brown outer webs**. **In flight** obscure rufous wingbar. Throat buffy white; buff postocular streak and broad buff streaks on the gray-brown underside, streaks becoming obscure on lower belly; vent rufous. Ssp *venezuelana* has darker, more olivaceous, less brownish crown with blacker scaling; back more olivaceous with less rufous tint. **Juv.** has belly washed with rufous, and the whole back and wing-coverts with black-edged, rufous buff streaks, and rump narrowly barred dusky.
Habits: Usually in mixed-species flocks. Forages actively along branches and among (often dead) leaves in the manner of a small foliage-gleaner or *Cranioleuca* spinetail, but not scansorial, and does not use tail for support.
Voice: No data.
Breeding: Fledglings Jan (W Andes in Cauca), Sep (nw Ecu.), Jan, Feb, Sep (ne Ecu.), Nov (Loja); juv./imm. Aug (wc Amazonas Peru).
Habitat: Lower strata of cloud forest and humid montane forest.
Range: 1300-2750 m. Perijá mts in sw Táchira, Ven. (*venezuelana*, Phelps & Phelps in *Proc. biol. Soc. Wash.* 69(1956):160), all 3 ranges of Col. except w slope of W Andes, nw, e, and sw Ecu. s through C and E Andes of Peru to Puno; sight rec. from Cochabamba, Bol. (*guttuligera*). Uncommon, locally fairly common.

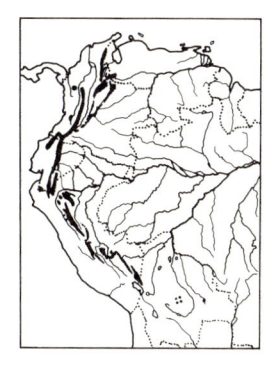

Barbtail – Genus *Premnoplex*

Monotypic genus. Much like *Margarornis*, but smaller, wings more rounded, and with weaker bones. Scansorial. The closed nest with a tubular entrance near the bottom, is built of mosses, lichens, and rootlets, sometimes also with a few twigs and leaves, and is placed between broken branches, under a flap of bark, in a depression in a tree, or the crotch of a fallen log. One nest only used for roosting was smaller and attached to the uneven rock-face of a vertical highway cutting. Placed in the genus *Margarornis* in Vaurie's revision of the *Furnariidae* (1971). See *Amer. Mus. Novit.* 757(1934).

SPOTTED BARBTAIL *Premnoplex brunnescens* – Plates XXXIII 14a-b and LXIII 13

13 cm. Above, incl. **wings, dark brown**, crown dark gray-brown; upperparts faintly scaled with darker feather-margins (upper back with pale shaft-streaks in *stictonotus*). **Tail blackish**, ending in soft spines. Throat orange-buff with a few narrow, dark edges; rest of underside dark gray-brown with drop-shaped buff spots, encircled blackish. Ssp *coloratus* has a browner tail and no rufous tinge above. **Juv**. has the back and wing-coverts spotted with buff; pattern below less regular than in ad.
Habits: Usually alone, sometimes in mixed-species flocks. Actively climbs trees and branches like a woodcreeper, sometimes creeping on the underside of branches. Occ. descends to the ground.
Voice: A squeaky, variable, toy-like note *tsip*, often repeated during foraging. Call a descending series of weakening *tsip*s. The infrequently heard song a high-pitched, thin *eep eep eep ti'ti'ti'titititi*, trilled at the end.
Breeding: Pull. Apr (e Huila); juv. Nov (Quindío).
Habitat: Lower strata. Mossy branches in the darkest parts of the understory in very humid montane forest and cloud forest.
Range: 600-2750 m. Sta Marta mts (*coloratus*); Perijá mts and from Mérida and Táchira, Ven., s through all 3 ranges of Col., e and w Ecu. to Cuzco, se Peru (*brunnescens*), from Puno, se Peru to La Paz and Cochabamba, Bol. (*stictonotus*). 3 sspp in Panamá and Costa Rica, 3 in coastal Ven., 1 in s Ven. Common through most of its range, but rare at high elevations.

Tuftedcheeks – Genus *Pseudocolaptes*

2 species of fairly large furnariids, found in humid montane forest in Costa Rica , Panamá, coastal Ven., and the Andes s to Bol. The only nest described was placed in an abandoned woodpecker-hole in a decaying tree-trunk. See *Amer. Mus. Novit.* 862(1936).

STREAKED TUFTEDCHEEK
Pseudocolaptes boissonneautii – Plate XXXIII 15a-c

20 cm. Head flat. **Ad**. above brown; crown black to blackish gray, and with pale rufous shaft-streaks; supercilium pale rufous; **back with** broad, pale rufous **streaks**; rump rufous, tail chestnut. **Conspicuous tuft of white feathers on neckside often erect**, throat and breast whitish, breast-feathers with brown borders, giving scalloped effect; belly dull rufous. The sspp differ mainly in color tone and amount of scaling below, some have the malar region and/or tuft yellowish white. Females have longer and straighter bills than males (least pronounced in ssp *auritus*). **Juv**. has black crown, rufous supercilium, and is brighter rufous and with scaling on underparts extending onto upper belly and throat. (Vaurie believes the true juv. to have streaked crown, unstreaked back, and bright, scalloped underside. We have not seen such a specimen. The bird here described as juv., he considered a succeeding plumage).
Habits: Solitary or sometimes in pairs, usually in mixed-species

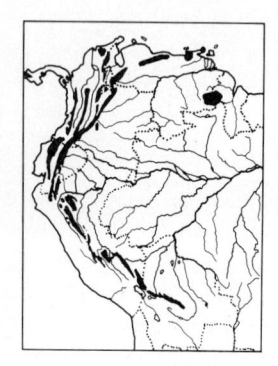

flocks, where almost invariably seen climbing about clumps of epiphytes, probing bill into mosses, lichens, and bromeliads in search of invertebrates, often using tail for support. Rather acrobatic, sometimes with head down; will also attain woodcreeper-postures. Occ. searches dead leaves.
Voice: Call a dry *chit*. Song 1-2 very sharp notes followed by a very fast trill that slows abruptly at the end.
Breeding: Juv Sep (Cauca); large gonads July, Aug (n Col.), Oct (E Andes Col.).
Habitat: Middle to upper strata of cloud and elfin forest, and fairly open humid montane forest. In vines, leafy clusters, and epiphytes.
Range: 1500-4400 m. Coastal Ven. (*striaticeps*), Perijá mts. and the Andes from Trujillo to Táchira, Ven. (*meridae*), in E, C, and W Andes of Col., and in w Ecu. except the s (*boissonneautii*), e and sw Ecu. (*orientalis*), extreme n Peru in Piura and adjacent Cajamarca (*intermedianus*), w slope of W Andes in Cajamarca s of Abra Porculla, nw Peru (*pallidus*); Cajamarca, Amazonas, and La Libertad, n Peru (*medianus*), c and se Peru in Huánuco, Cuzco, and n Puno (*auritus*), se Peru in s Puno and Bol. in La Paz, Cochabamba, and Sta Cruz (*carabayae*). Fairly common to common.

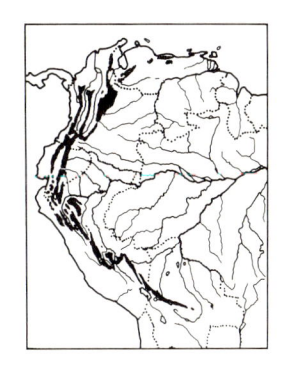

Foliage-gleaners – Genus *Syndactyla*

Foliage-gleaners are medium-sized to fairly large arboreal furnariids that actively forage along branches and in foliage. Present genus with 3 fairly slender, streaked, mainly montane species in C Am., Andes and lowlands of Parag., s Brazil, and n Arg. They forage in dense undergrowth within 5 m of the ground, and are difficult to observe. A nest of *S. guttulata* was placed in a cavity in a wall. The cavity was filled with twigs and there was an entrance at the bottom.

Perhaps the behaviorally and vocally similar Rufous-necked F-g. (*Automolus ruficollis*) and Recurvebills (*Simoxenops*) are better placed in this genus, which they also resemble by their streaked pattern. *Syndactyla* was synonymized with the genus *Philydor* in Vaurie's revision of the Furnariidae (1971). See *Amer. Mus. Novit.* 785(1935).

BUFF-BROWED FOLIAGE-GLEANER
Syndactyla rufosuperciliata – Plate XXXVIII 9a-c

17 cm. Bill rather slender and pointed; maxilla straight, **mandible slightly upturned**. Above olive-brown, nape sometimes with a few pale shaft-streaks. Rump reddish brown, tail chestnut, as long as wing. Wings brown. Supercilium buff, in *similis* rather indistinct and ochraceous; narrow, broken eye-ring buff. **Throat whitish, feathers with narrow dark tips**; rest of underparts olivaceous, with whitish or buffy white broad streaks, that become narrow and obscure on lower belly, and lack on flanks. **Juv**. with large spots rather than streaks below.
Habits: Solitary, sometimes in mixed-species flocks. Works under dense cover, therefore difficult to observe.
Voice: Alarm a loud sharp *set* or *setet*. Call *kssr*. Song, given at dusk and dawn, is a series of loud notes *kit-kit-kit...*, first faint and ascending, then louder, accelerating and descending.

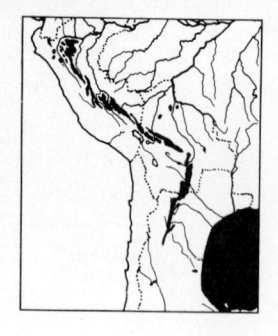

Breeding: Eggs Nov (Tucumán); fledgling Aug (Cuzco); juv./imm. Mar, Apr (Tucumán), Mar (Pasco), Feb, Apr (Cajamarca).
Habitat: Lower strata. Edge of humid montane forest and cloud forest, often with admixed bamboo and second growth. Sometimes in swamps.
Range: 1300-3000 m. N Peru from e Piura s to c Cajamarca, where also on w slope (Chugur) (*similis*), from Amazonas and s Cajamarca, Peru, s to La Paz and Cochabamba, Bol. (*cabanisi*), from Sta Cruz, Bol., to La Rioja, nw Arg. (*oleaginea*). 2 other sspp in n and ne Arg., Parag., Urug., and s Brazil. Common in Pasco, c Peru and nw Arg., but rare in n Peru.

LINEATED FOLIAGE-GLEANER *Syndactyla subalaris* – Plate XXXVIII 10a-b

17.5-18 cm. **Throat buff** to buffy white, often puffed; rump and tail chestnut. Birds from w part of range dark brown above, with upper mantle blackish; crown and lower mantle with narrow shaft-streaks, upper mantle with broader, buffy white streaks; **below** olive-brown **with narrow whitish streaks**; flanks brown, unstreaked. Birds from e part of range have blackish crown and most of back, both conspicuously streaked buffy white; they have a longer tail, wings and upper rump are olive-brown; entire underside below pale throat olivaceous with broader, buffy white streaks. Ssp *olivacea* most olivaceous below. **Juv**. has an ochraceous orange supercilium, neckside, throat, and breast; throat and breast unstreaked, and dorsal streaks ochraceous. Alone or in pairs, often in mixed-species flocks. Actively climbs about, picking insects from leaves and branches. Occ. searches dead leaves. Often in flocks with Spotted Woodcreeper. Voice much like a treehunter (*Thripadectes*): a dessicated *ki-ki-kikikiki-ki*, either monotonously or somewhat up and then down the scale, slightly speeded in the middle. Also a dry, nasal, or reedy *tcheck* or *tsuck*. Fledgling Dec (nw Ecu.); juv./imm. Jan (Cauca), Dec, Mar (w slope of W Andes in Valle), Jul (w Nariño), June (nw Ecu.), Jan (ne Ecu.), Feb, Mar (n Cajamarca).

Inhabits lower to middle strata of humid premontane forest, second growth, and edges, locally to border of temp. zone (rec. 2600 m in sw Ecu.). Sw Táchira (*olivacea*), Andes of w Ven. in s Lara and sw Barinas, E Andes at head of Magdalena Valley in Huila, Col. (*striolata*), from Antioquia W Andes and w slope of C Andes of Col. to w Ecu. (*subalaris*), e Ecu. (*mentalis*), n Peru in n Cajamarca (*colligata*), s of the Marañón to n Vilcabamba mts, Cuzco, Peru (*ruficrissa*). 2 sspp. in Panamá and Costa Rica.

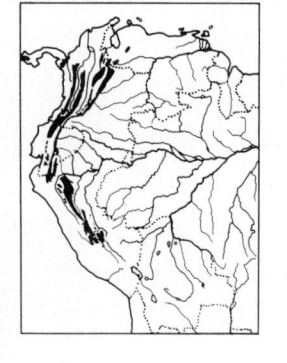

Foliage-gleaners – Genus *Anabacerthia*

3 species of fairly small montane foliage-gleaners with streaked underparts and distinctive head-patterns. Actively search bromeliads and vine tangles from mid-heights to sub-canopy, often in clear view. They are found from Mexico to Panamá, in coastal Ven., and the Andes s to Bol., 1 species in se Brazil. No nest of the genus has been described in detail, but a nest of Montane F-g. was located in a broken trunk of a thorn-studded *Bactris*-palm. See *Amer. Mus. Novit.* 785(1935).

SCALY-THROATED FOLIAGE-GLEANER *Anabacerthia variegaticeps* – **Plate XXXVIII 5a-b** – of humid premontane forest of C Am. (*variegaticeps*) and Pacific slope from Chocó, Col., to Loja, sw Ecu. (*temporalis*), may reach lower temp. zone (rec. at 2300 m). 16 cm. Crown dark grayish with faint, pale shafts; back and wings reddish brown, tail chestnut. **Eye-ring and conspicuous postocular streak ochraceous buff**; throat whitish, rest of underparts olive-brown, heavily **flammulated ochraceous buff on breast**. **Juv**. has less distinct general pattern than ad., and a more ochraceous wash on breast and abdomen. Alone or in pairs, usually in mixed-species flocks 5-10 m up. Actively crawls along twigs and moderately thick, often moss-covered branches and among foliage and epiphytes; frequently upside down. Easy to observe. Often spreads tail or holds it downwards and forewards at an angle. Call a peculiar scratchy rattle; also a lisping sharp squeak.

NOTE: Ssp *variegaticeps* has been treated as a ssp of Montane F-g., and ssp *temporalis* as a species.

MONTANE FOLIAGE-GLEANER
Anabacerthia striaticollis – Plate XXXVIII 6a-c

16 cm. **Mandible noticeably upturned**, esp. in *montana*. Ssp *montana*: Crown olive-gray turning pale olive-brown on back and tawny on rump; wings dull rufous, tail rufous chestnut. **Eye-ring and conspicuous postocular streak whitish**. Below pale olive-brown; throat whitish, continuing as streaks onto breast and sometimes upper belly. Ssp *striaticollis* somewhat paler and less rufescent above, crown olivaceous in decided contrast to back; breast streaks obsolete. Ssp *anxia* like *striaticollis*, but more olivaceous above, throat, and supercilium much brighter, breast and abdomen more yellowish; *yungae* like *montana*, but more rufescent throughout; *perijana* like *striaticollis*, but wings shorter, back lighter more yellowish olive (less rufous brown); below lighter, more yellowish olive (less buffy brown). **Juv**. has darker crown, more prominent and brighter buff eye-ring and postocular streak, and is more rufous throughout. Alone or in pairs, usually in mixed-species flocks. Hops along and crawls across moss-covered branches and twigs, often hanging head down to turn dead leaves. Also actively searches epiphytes and dead leaf-clusters. Dead leaf specialist. Easy to observe. Noisy. Calls a scratchy rattle and a sharp squeak.

Mainly mid-levels of humid forest and cloud forest at 950-2600 m in coastal Ven. (*venezuelana*), Sta Marta mts (*anxia*); Perijá mts (*perijana*, Phelps & Phelps in *Proc. biol. Soc. Wash.* 65(1952):89-91); Andes of Ven., all 3 ranges of Col., but only on w slope of W Andes in Antioquia (*striaticollis*), from e Ecu. s through e and c Andes of Peru to Junín (*montana*); from Cuzco and Puno in se Peru to Cochabamba, Bol. (*yungae*). Fairly common.

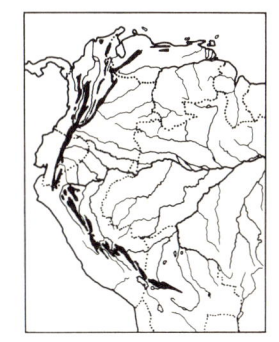

Foliage-gleaners – Genus *Philydor*

10 species found from Costa Rica to ne Arg. Most inhabit lowland rainforests. Slim birds, averaging 16.5 cm long. Arboreal, typically fre-

quenting mid-heights, but at least 2 are canopy-dwellers. Often seen gleaning live leaves or probing clumps of dead curled ones on slender limbs within crowns of middlestory trees. All are very active. Only the nest of Buff-fronted F-g. has been described: It is placed in a tree hole, in a tunnel under ground, or in a cavity in a wall. See *Amer. Mus. Novit.* 785(1935).

BUFF-FRONTED FOLIAGE-GLEANER *Philydor rufus* – **Plate XXXVIII 12a-b** – inhabits second growth in clearings in humid premontane forest and humid, low-lying terra firme forest, occ. higher, maybe to temp. zone. Costa Rica, Panamá, lower Cauca, and Magdalena valleys s to Cundinamarca in E and C Andes of Col. (*panerythrus*), from c Chocó W Andes of Col. s to nw Ecu. (*riveti*); from San Martín, n Peru, along Andean slopes to Sta Cruz and Chuquisaca, Bol. (*bolivianus*). Other disjunct populations in coastal Ven., s Ven., and in c and s Brazil to ne Arg. Common. 17.5 cm. Ssp *riveti*: Bill fairly long and slender, tail as long as wing. **Imm**.: above brown, dark gray-brown on crown; wings and tail (except c feathers) rufous chestnut. Sides of head, incl. long conspicuous supercilium ochraceous orange with thin dark line through eye; throat ochraceous orange, turning buffy brown on rest of underside. **Ad.**: less orange below, color of **forehead** more **buffy** and extensive. Alone, usually in mixed-species flocks. Active. Works half-way and further out the branches in canopy of middle-story trees, hidden in the foliage, acrobatically crawling about leaves and twigs, sweeping tail and twisting body. Calls comprise snorting or harsh croaking, almost frog-like churrs. Nestling June (Valle).

Foliage-gleaners – Genus *Automolus*

8 species found from Mexico to ne Arg. A 9th, Rufous-necked F-g. is probably best referred to *Syndactyla*. Birds of humid lowland forest. Several are easily confused with *Philydor* spp, but most are conspicuously larger, with proportionally larger thicker bills. All spend most of their time in undergrowth within 5 m of the ground. More vocal than *Philydor*, most make their presence known by loud 1-2-note calls. The nest is a broad shallow cup of plant fibers constructed at the end of a burrow dug into a bank. See *Amer. Mus. Novit.* 785(1935).

CRESTED FOLIAGE-GLEANER *Automolus dorsalis* – **Plate XXXVIII 11** – an uncommon inhabitant of trop. terra firme forest from Caqueta and Putumayo se Col. through e Ecu. to San Martín, Junín, and in Madre del Dios, se Peru, ascends occ. as high as 2200 m. 17 cm. No crest! Bill stout. Above incl. wings brown, reddish on nape; rump and tail rufous chestnut. Lores and eye-ring whitish, postocular streak buffy. **Throat and upper breast white, often tinged yellowish**; rest of underside gray, washed brownish on sides and vent, whitish down center. **Imm**. has crown and postocular stripe tinged rufous, and is washed pale ocher below white throat.

RUDDY FOLIAGE-GLEANER *Automolus rubiginosus* – **Plate LXIII 15** – of lower strata of dense humid terra firme forest and dark ravines with thick second growth, ascends into subtrop. zone, and occ. as high as 2450 m in Sta Marta mts. (*rufipectus*). 16 sspp in C Am., Col., Ecu., to n Peru w of Andes, and to Bol. e of Andes., and isolated in s Ven. and French Guiana. 19 cm. Ssp *rufipectus*: Above dark brown, rump chestnut, tail very dark chestnut. Below rufous, breast tinged chestnut, flanks and vent dark brown; **throat pinkish rufous**, palest in center. **Imm**. more rufous above, washed pale ocher below white throat; throat and upper breast slightly scaled by narrow brown feather-edges. Solitary. Not shy, but difficult to observe. Forages in thick undergrowth and on the forest floor. Call a distinctive whistled *ka-kweek*. Imm. Mar, Apr (Sta Marta).

RUFOUS-NECKED FOLIAGE-GLEANER
Automolus ruficollis – **Plate XXXVIII 8**

17 cm. Above incl. wings brown, crown dark brown. Headside streaked and mottled gray-brown and buffy to whitish; supercilium ochraceous, darkening behind eye, and merging with the **orange-rufous neckside** that extends to nape and lower throat. Throat warm buff, palest in center; feathers of lateral and lower throat, and upper breast with dark edges forming breastband; rest of underparts olive-brown with distinct buff streaks, narrowing and vanishing on belly; vent rufous, tail chestnut. Ssp *celicae* brightest, palest, most ochraceous, esp. above, and with ochraceous brow, less dusky ear-coverts and broader breast-streaks. **Imm**. has grayer crown and forehead tinged blackish; entire throat scaled, upper breast streaked by blackish feather-edges.
Habits: Solitary. Much like a *Syndactyla*. Probes arboreal bromeliads, ferns, and mosses, and hops through dense tangles in bamboo, where difficult to observe.
Voice: Song an accelerating series of notes *chi chi-chi-chi-chchchchchch*, strikingly similar in pattern and quality to Buff-browed F-g. Call an emphatic *check*.
Breeding: Juv./imm. Apr, June (Cajamarca); large gonas May (Cajamarca).
Habitat: Usually 2-3 m up, but occ. as high as 10 m. Dense to moderately open forest understory (esp. bamboo), and middle story of humid montane forest and in humid patches in fairly dry, deciduous forest.
Range: 400-2900 m. Coastal hills in El Oro, sw Ecu. (Las Lajas), Tumbes (El Caucho; Campo Verde), and Piura (El Angolo), nw Peru, and Celica mts. Loja, sw Ecu. (Cebollal; Guachanamá; Alamor; Celica) (*celicae*); Pacific slope of Andes in Loja, sw Ecu. (Gonzanamá), Piura (Canchaque-Cruz Blanca, incl. Palambla and El Tambo; Porcula) and Cajamarca (Chugur; Paucal; Seques; Taulis), nw Peru (*ruficollis*), birds between Celica mts and Cajamarca intermediate. Uncommon; habitat threatened by trampling of undergrowth and clearing.

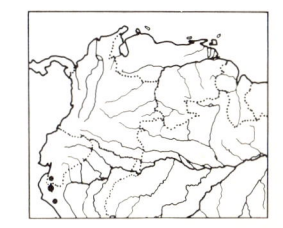

NOTE: May belong in the genus *Syndactyla*.

Treehunters – Genus *Thripadectes*

7 species, 1 in C Am., the rest in coastal mts of Ven. and Andes from Ven. to Bol. Fairly large stout arboreal montane furnariids with strong slightly hooked bills, foraging under dense cover in the undergrowth of humid forest and second growth. More phlegmatic than *Syndactyla*. Streaked, at least on lower throat and upper breast; rump, and upper tail-coverts rufous, the moderately graduated tail chestnut. Underwing bright pale rufous. Frogs and salamanders, which form an important part of their diets, are disclosed by tearing apart bases of bromeliads, and by searching in moss and other epiphytes. The nest is placed at the end of a burrow, dug in a bank by the bird itself. See *Amer. Mus. Novit.* 862(1936).

FLAMMULATED TREEHUNTER *Thripadectes flammulatus* – Plates XXXVIII 14 and LXIV 8

22 cm. Above and below blackish brown, belly and flanks olivaceous, vent blackish, **everywhere heavily streaked buff**. Rump dark rufous, tail chestnut. Wings dark reddish brown, wing-coverts with rufous buff streaks. Ssp *bricenoi* has rufous-olivaceous lower back and rump, less black-edged streaks on lower back; streaks below only black-edged on lower throat and upper breast, rest of underparts rufous tinged olive, with broad, buff streaks. **Juv**. has less distinct streaking than ad.
Habits: Solitary, occ. in mixed-species flocks. Clambers about in the undergrowth, usually under dense cover. Has a large territory.
Voice: Territorial song a c. 2 s long, loud series of grating machine-gun-like notes, accelerating and loudest towards the end, last note lower and weaker, repeated at 3-5 s intervals. Infrequently utters an emphatic *chek*.
Breeding: Large gonads June (Antioquia W Andes).
Habitat: Dense undergrowth of humid montane forest. On moss-covered trunks, tangles of dead bamboo and foliage within 2 m of the ground, rarely in higher branches.
Range: 800-3300 m, lowest in w Nariño. Andes of Mérida, Ven., on Páramos La Culata, Conejos, and Escorial (*bricenoi*), Ven. in sw Táchira, Sta Marta mts., all 3 ranges of Col., nw, sw, and e Ecu., and n Peru in Piura and Cajamarca n of the Marañón (*flammulatus*). Widespread but uncommon.

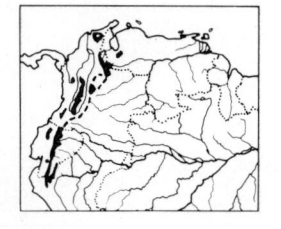

NOTE: Should possibly include Buff-throated T.

BUFF-THROATED TREEHUNTER *Thripadectes scrutator* – Plate XXXVIII 13

22 cm. Head and upper **back blackish, broadly streaked buffy**; back brown turning chestnut on rump, wings, and tail; lower back usually with a few inconspicuous pale shafts. Throat and breast blackish, broadly streaked buff; rest of **underparts** olivaceous to brown, **streaked buff**; flanks washed chestnut, unstreaked. **Juv**. undescribed.
Habits: Solitary. Slowly works its way up along bamboo or among

foliage. Spreads wings even at small hops, exposing chestnut of wing. Not shy, but sometimes difficult to observe.
Voice: Like Flammulated T.: territorial song a loud descending series of grating notes, accelerating towards end.
Breeding: Gonads small July and Aug (Cuzco).
Habitat: Within 2 m of the ground. Bamboo and other thick undergrowth in humid montane forest and second growth on clearings.
Range: 2100-3650 m. E Andes of Peru from Amazonas s and e of the Marañón to Cochabamba, Bol. Fairly widespread, but with a large territory.

NOTE: Perhaps best regarded a ssp of Flammulated T.

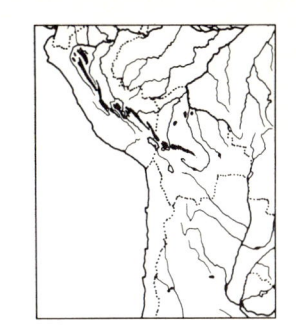

STRIPED TREEHUNTER *Thripadectes holostictus* – Plate XXXVIII 15

20 cm. Above dusky-brown narrowly streaked buffy. Wings dark brown, wing-coverts with buffy shaft-streaks; rump and tail chestnut. Throat buff, **feathers distinctly pointed and dark-edged**; rest of underparts olivaceous, streaked with buff or ochraceous buff on breast and faintly on upper belly, **only the streaks on upper breast with dark edges**. Ssp *moderatus* has fewer and narrower streaks below, an almost unmarked chin, and narrower and less defined streaks above. **Juv**. has belly unstreaked, but mottled by dark feather-tips; the streaks above narrower and less distinct.
Habits: Forages under dense cover, much like its congeners, usually alone, but sometimes in mixed-species flocks.
Voice: No data.
Breeding: Juv./imm. Sep (w slope of C Andes in Cauca), Feb (Pasco); large gonads Aug (e Santander).
Habitat: Within 2 m of the ground. Dense undergrowth in humid montane forest. Second growth along streams on clearings in humid forest.
Range: 800-3100 m. Generally at lower elevations than Flammulated and Buff-throated T.s, higher than Black-billed and Streak-capped T.s, but with considerable overlap. Cauca and Nariño, sw Col., and w Ecu. (*striatidorsus*), Ven. along Río Chiquito in sw Táchira, spottily in all 3 ranges of Col. (not n of Cauca in W Andes), e Ecu., and n Peru in Cajamarca (Condor mts) and Piura (Chinguela) (*holostictus*); Pasco and Junín, and Cuzco, Peru, and La Paz and Cochabamba, Bol. (probably also further n and in intervening provinces) (*moderatus*). Uncommon.

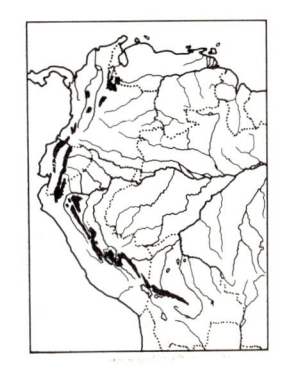

BLACK-BILLED TREEHUNTER *Thripadectes melanorhynchus* – Plate XXXVIII 17 – is included only for comparative purpose. Replaces Streak-capped and Striped T.s at lower elevations (900-1750 m) in Meta E Andes of Col. (*striaticeps*), and from e Ecu. s to Puno se Peru (*melanorhynchus*). 19.5 cm. **Above dark** dusky brown, crown darker; **crown and mantle with** narrow, well-marked, **pale streaks**. Throat ochraceous orange, **feathers distinctly pointed and dark-edged; below olivaceous**, breast sometimes with narrow streaks. **Juv**.

has more obscure throat-squamations and streaks above. Its habits are like those of congeners. Uncommon.

STREAK-CAPPED TREEHUNTER
Thripadectes virgaticeps – Plate XXXVIII 16

20.5 cm. Ssp *virgaticeps*: crown olivaceous, feathers edged blackish and with whitish shafts; **back unstreaked**, dark olive-brown; wings reddish brown; rump dark rufous, tail chestnut. Throat warm buff, feathers slightly pointed and dark-edged appearing scaled; rest of underparts dark ochraceous, upper breast with olivaceous wash and pale shafts. Geographical variation slight. Crown-streaks extend to nape in *sumaco*, throat most conspicuously scaled in *tachirensis*, least so in *magdalenae*. **Juv**. with scales pattern of throat absent or hardly suggested. - Like congeners solitary; hardly ever joins mixed-species flocks. Climbs about in the undergrowth, usually well hidden in the foliage. Territorial song a loud chatter. Alarm a raspy nasal *jwick*. Call a hard low-pitched fast *ju dut*. Juv. May (Valle).

Within 2 m of the ground in dense undergrowth at the edge of humid montane forest, and in dense second growth along streams. At 1000-2500 m, between Black-billed and Striped T.s. Coastal Ven. (*klagesi*); sw Táchira, Ven. (*tachirensis*, Phelps & Phelps in *Proc. Biol. Soc. Wash.* 71(1958):119-24); W Andes and w slope of C Andes from Valle and Antioquia s to Nariño, Col. (*sclateri*), nw Ecu. (*virgaticeps*); head of Magdalena Valley in Huila, Col. (*magdalenae*), ne Ecu. (*sumaco*). Uncommon, but the treehunter most often seen in Col.

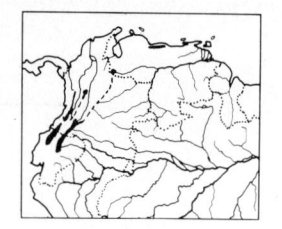

Xenops – Genus *Xenops*

4 species of very small furnariids with rounded rectrices. Mostly trop., they are found in humid forest from Mexico to n Arg. They climb along branches somewhat in the manner of woodpeckers, but do not use tail for support. Often hang upside down. See *Amer. Mus. Novit.* 862(1936).

STREAKED XENOPS *Xenops rutilans* – with 9 sspp found in rain and cloud forest as well as deciduous woodland in the lowlands and subtrop. zone from Costa Rica to n Arg., has been taken as high as 2800 m in Col. (*heterurus*). 11.5 cm. Mandible upturned. Above reddish brown, wings blackish with conspicuous rusty under-coverts and bar on remiges. Below olivaceous streaked whitish, distinct malar streak white. Tail rufous with diagonal dusky band.

Treerunner – Genus *Pygarrhichas*

Monotypic genus found in humid Andean forest of s Chile and s Arg. The nest is in a hole, excavated by the bird itself in decaying wood.

WHITE-THROATED TREERUNNER
Pygarrhichas albogularis – Plate XXXIII 16a-b

16 cm. **Bill long and straight**, mandible slightly upturned. **Short tail** ends in bare stiffened shafts which curve inwards abruptly. Crown and c mantle dark brown, sides of crown and mantle gray-brown; lower back, wings, rump, and tail rufous. **Below white**, feathers of belly with narrow dark edges, giving scaled effect; on the sides and flanks the edges are broader, brown, or rufous. Some birds (imm.?) much darker below. **Juv**. has crown and back broadly streaked with orange-buff or rufous, in some birds with the streaks merging on head to form a mottled orange-rufous cap; often somewhat scaled with black margins on white throat.
Habits: Usually alone, sometimes in mixed-species flocks. Scansorial, body held pressed against bark. Mostly creeps up the stems and out along the branches, but occ. will climb down a trunk head first; often along the underside of horizontal branches. Makes frequent jerky shifts of the body and changes in direction of movement; probes and pecks into crevices in the bark, and also spends much time in the smaller branches, working around the base of leaf-clusters. Flight undulating and erratic. Normally undisturbed by presence of man, although it may fly to a conspicuous perch while calling in distress.
Voice: On the wing often a clicking *tick*. Bill-tapping on the bark frequently heard. Call a loud *peet peet*. Contact note a very high-pitched *zeezeedeezee*....
Breeding: Eggs Nov (Isla Grande), Dec (Nuble); nestlings Dec (Río Negro); juv. Jan; juv./imm. Jan, Feb (Valdivia, Isla Grande), Mar (Sta Cruz).
Habitat: *Nothofagus* forest. Prefers tall forest, whether dense or open.
Range: From Santiago, Chile, and s Mendoza, Arg., to Isla Grande, where common.

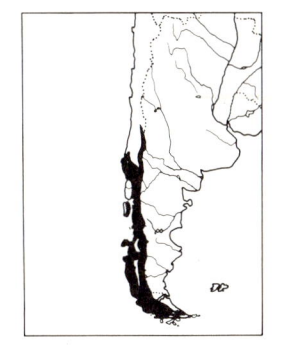

Leafscrapers – Genus *Sclerurus*

6 species of medium-sized dark short-tailed terrestrial furnariids of humid forest from Mexico to ne Arg., most in trop., but some also in subtrop. zone. They spend much of their time on the forest floor, flipping leaves in search of insects, hence they are often called Leaf-tossers. The nest is placed in a burrow, dug into a bank by the bird itself. See *Amer. Mus. Novit.* 757(1934).

GRAY-THROATED LEAFSCRAPER *Sclerurus albigularis* – Plates LXIII 31
– inhabits humid, dark, premontane forest to near lower border of temp. zone (2200 m) in Sta Marta and Perijá mts. Costa Rica, and Panamá (*canigularis*), Sta Marta mts (*propinquus*); Perijá mts (*kunanensis*, Aveledo & Gines in *Mem. Soc. Cien. Nat. LaSalle* 26(1950): 59-71); Andes and coastal mts of Ven., and e Col. in Meta (*albigularis*); from se Ecu. to Huánuco, c Peru (*zamorae*); Bol. in La Paz and Sta Cruz (*albicollis*). 16.5 cm. Jizz of a short-tailed thrush. Ssp *propinquus*: above dark brown, rump chestnut; **tail short, black**. Chin and upper throat whitish, becoming gray on lower throat. **Breast rufous** (very dark

brown in imm.), rest of underparts dark olive-brown. **Juv**. duller. Terrestrial, only rarely perching on root or low shrub. Alone, in pairs, or family-parties. Song given early morning and late afternoon, a five-syllable phrase persistently repeated: a rather soft note, then 2 loud higher notes, then 2 weaker but shorter and sharper notes. Alarm a sharp *chick* or *chick-chick*, repeated at intervals. Scolding a quick chatter, often given near the nest.

Streamcreeper – Genus *Lochmias*

Monotypic genus. Small dark terrestrial bird with spotted underparts, and a short tail ending in bare shafts. The nest is globular with a side entrance, and is placed in a tunnel in a bank.

SHARP-TAILED STREAMCREEPER
Lochmias nematura – Plate XXXVIII 18a-b

13 cm. Above rufous brown (*sororia*) or olive-brown (*obscurata*); tail black, fairly short. **Below** brown, densely patterned **with** large, black-edged **white spots** (*sororia*), or dark olivaceous, throat, flanks and belly blackish with small **white spots on throat**, c breast and upper belly (*obscurata*). Unnamed ssp in W Andes of Col. grayer below than *sororia*, and with smaller spots. **Imm**. has the white spots less regular and tinged with buff.
Habits: Alone or in pairs. Largely terrestrial. Silently hops on mossy stones, among limbs and roots on banks and low in bushes along dark streams overgrown with thick, mossy vegetation. Difficult to observe. Perhaps particularly active at dusk and very early dawn.
Voice: Alarm *tschett-tschett-tschett*. Call a loud *sea-sick* or *seesee-sick*. Song is softer, a rapid short trill *sisisisi…*, first ascending to a peak, then descending.
Breeding: No data for highlands, but see *Ibis* ser. 6, 6(1894): 484-94.
Habitat: Near or on the ground. Along streams in humid montane forest and cloud forest.
Range: 725-2780 m. W slope of W Andes of Col. in Cauca (*ssp*, also birds from n end and e slope of W Andes may belong here), coastal Ven., w slope of C Andes in Valle, both slopes of E Andes in Cundinamarca, Col., e Ecu., and n Peru in San Martín (*sororia*); from Huánuco in c Peru along e Andean slope to Jujuy, nw Arg. (*obscurata*). Other sspp in Panamá, s Ven., and se Brazil with adjacent Urug., Parag., and Misiones Arg. Uncommon, apparently local.

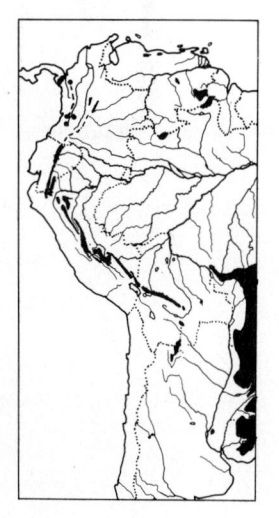

Antbirds – Family Formicariidae

249 species of small to large suboscine passerines of forest and dense shrub. Restricted to the Neotropical region, where found from c Mexico to n Arg. Ranging from wren-like *Myrmotherula* and warbler-like *Terenura*, to pitta-like *Grallaria* and shrike-like *Thamnophilus*, they occupy a wide range of niches, but generally live in shade, and the family finds its greatest diversity in the darkest parts of the Amazon forest, where many species follow army-ants and feed on the insects flushed by the ants.

The family should possibly be divided into true Antbirds (Thamnophilidae), and the more terrestrial Antthrushes and Antpittas (Formicariidae) which may be closer related to Tapaculos (Rhinocryptidae) and Gnateaters (Conopophagidae).

Antbirds keep together in close pairs, usually mated for life, and stay in incessant vocal contact. Both sexes sing the same stereotyped song, but are usually very different in plumage. Most males of 'true Antbirds' are clad in grays and black, often with white-tipped wing-coverts and tail, and often have a concealed white or buff interscapular patch that is exposed during display and threats. Females are usually brown and rufous with a distinct pattern, while males of the various species may be hard to tell apart. The juv. plumage is very loose and soon replaced by a female-like imm. plumage except in the antpittas. In at least one species this imm. plumage is immediately replaced by the ad. plumage, so that 3 feather-generations may be present simultaneously, but in most species the imm. plumage is carried for a longer period. Antthrushes and Antpittas do not show the sexual dimorphism outlined above.

Both sexes build the nest, brood, and tend the young. Some rely so much on their invisibility in the dark that they may sing from the nest, which is always an open cup placed above ground.

See Meyer de Schauensee (1945c) and Bond (1950).

Antshrikes – Genus *Thamnophilus*

21 species of C and S Am., mostly found in lowlands. Highly sexually dimorphic, medium-sized birds with hooked bills. They inhabit lower to medium strata of forest and shrub, where they perch-glean among foliage and branches. See *Amer. Mus. Novit.* 646, 647 (1933), and 917 (1937).

UNIFORM ANTSHRIKE *Thamnophilus unicolor* – Plate XXXIX 3a-b

14.5 cm. **Iris pale, body fairly stout, bill heavy. Male: plain deep gray**, outer tail-feathers minutely tipped whitish. **Female:** Crown rufous; back, wings, and tail rufous-brown. Below pale rufous-brown. **Supercilium, headside, and upper throat gray.** Outer tail-feathers minutely pale-tipped. Ssp *grandior* slightly larger, with a longer tail; outer tail-feathers more decidedly pale-tipped. Ssp *caudatus* even larger, with longer tail and inconspicuous pale tips and edges of greater secondary-coverts (white in male, fulvous in female). **Imm.** like female. - Alone or in pairs. Forages in dense shrubbery, therefore difficult to observe. Only rarely in mixed-species flocks. A female-plumaged bird uttered a low complaining *keee kee-kee-kee*, 1st note falling in pitch, the following 2-3 shorter and alike. Also mewing call. Song an accelerating series of *na* notes. Fledgling Aug (nw Ecu.); juv. Feb, July (Valle); large gonads May, June (Antioquia), Feb, Mar (Valle), Aug (Huánuco); imm. Mar (Cauca).

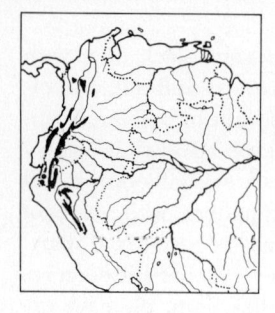

Within 2 m of the ground in dense undergrowth of humid montane forest, at 900-2700 m. W Ecu. (*unicolor*), Col. w of E Andes and in E Andes on w slope in Cundinamarca and e slope in Caquetá s through Nariño and e Ecu. to n San Martín, n Peru (*grandior*), and s San Martín to Huánuco c Peru (*caudatus*). Uncommon.

VARIABLE ANTSHRIKE *Thamnophilus caerulescens* – Plate XXXIX 1a-b

14.5 cm. Tail graduated. Varies geographically. **Ssp *subandinus*:** **Male** glossy-black, wing-coverts with narrow white edges; outer primaries narrowly edged white on middle portion; all except c tail-feathers tipped white; concealed interscapular patch white. **Female** much like *melanochrous*. **Ssp *melanochrous*: Male** has (unlike *subandinus*) white flanks and vent scaled with dark gray; **female** (unlike *aspersiventer*) lacks white wing-covert tips, and has more black crown, more grayish back, grayish of throat extending onto breast, belly much paler ochraceous. **Ssp *aspersiventer*: Male** has breast and belly barred or scaled with white and dark gray to black, throat and foreneck black; above like *melanochrous*. **Female** has crown mainly black, back brownish, underparts deep ochraceous; upper wing-coverts tipped white; throat grayish streaked paler, gray not extending onto breast, foreneck washed grayish olive. **Ssp *dinellii*:** eye-ring whitish. **Male** has whitish throat freckled gray; rest of underparts bright buff, c abdomen buff to white; supercilium and headside grayish. Crown black; back, wings, and tail gray with slight olive cast; c back admixed with black; concealed interscapular patch white; wing-coverts black tipped white; c tail-feathers gray, rest blackish tipped white, outer feather also with white spot on middle part of outer web. **Female** like male, but back olive-gray to olive-brown; throat buffy white, below ochraceous buff, crown rufescent olive. **Ssp *connectens*** is variably intermediate between *aspersiventer* and *dinellii*. **Juv.** fluffy; olivaceous buff, barred with buff on belly. Soon changes to succeeding plumage. **Imm.** female-like, but feathers of c back and wing-coverts tipped with small buff spots; belly buff with narrow, dark feather-tips. Almost immediately changes to ad. plumage, so that 3 feather-generations may occur simultaneously. - Alone or in pairs. Works its way up along vines to tips of branches, picking caterpillars and other arthropods from the leaves. Call a loud, clear, complaining *whar-whar-whar*, ascending, decelerating, and terminated with a *whaa*. Eggs Oct, Dec (Jujuy); fledgling Aug (Huánuco), Oct (La Paz), Nov (Puno); imm. July (Huánuco), Nov (Puno), Apr (Cochabamba).

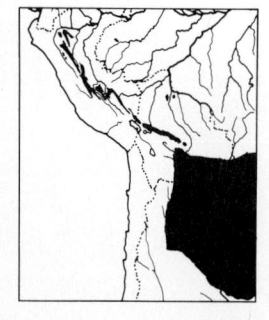

3-5 m up in thickets and undergrowth of humid forest. In less dense scrub than Marcapata A. (*T. marcapatae*; see under the following species). 1300-2750 m. E La Libertad, Amazonas, and San Martín (*subandinus*), Huánuco to n Puno (*melanochrous*), s Puno se Peru to Cochabamba Bol. (*aspersiventer*), subtrop. zone of Sta Cruz Bol. (*connectens*); 150-2300 m from se Cochabamba Bol. through the chaco of Arg. to Tucumán (*dinellii*). 7 other sspp from Parag. to e Brazil. Uncommon to fairly common.

RUFOUS-CAPPED ANTSHRIKE
Thamnophilus ruficapillus – Plate XXXIX 2a-e

15.5 cm. Rather slender, posture often horizontal. **Crown chestnut** extending to nape (except in *subfasciatus*), **wings rufous**. Tail graduated. Varies geographically. **Ssp *jaczewskii*: male**: above gray slightly tinged brown; tail black, c feathers unmarked, rest with 4 large white spots on each web. Sides of head gray, eye-ring whitish; below white barred black. **Female** similar but supercilium admixed with a little buff, lores and underparts buffy, palest on belly; flanks gray-brown. Tail dark rufous, outer feathers with suggestion of narrow, blackish bars. **Ssp *marcapatae*: Male** like *jaczewskii*, but more barred below, tail black with 4 tiny white spots on inner webs except on c feathers; throat obscurely barred; flanks and vent like back, occ. browner. **Female** above brown tinged gray; cap, wings, and tail chestnut. Headside gray, eye-ring buffy. Below ochraceous orange. **Ssp *subfasciatus*: Male** most heavily barred of the forms. Upper tail-coverts with white tips and black subterminal bar. Tail black, 3 outer feathers with white subterminations and narrow black tips, and outwards with 2, 3 and 4 (resp.) white bars on inner webs, not reaching shaft, and a tiny corresponding spot on outer web. **Female** said to resemble female of southern forms. **Ssp *cochabambae*: Male** gray-brown above; tail gray-brown, c feathers unmarked, rest narrowly tipped white and with 4-5 white bars on inner web, and minutely on outer web. Sides of head mixed gray and white; eye-ring white. Below whitish, barred black on breast, and washed with gray-brown on flanks. **Female** like male, but tail rufous with faint suggestions of dark bars, tips narrowly whitish except on c pair. Underside washed cream, breast unbarred or almost so. **Imm.** like female, but buff below. **Juv**. like female, but with hardly visible, dark tips to breast-feathers. - Alone or in pairs. Rather slow. Hops out along branches. Usually forages under heavy cover. Call a low whistle, also *kekeke*. Song by both sexes, a laughing *ker ker-ker-ker-ker-kekekekeker*, lasting c. 5 s; 2nd note accented and higher pitched than 1st, followed by several even notes which accelerate and fall to same pitch as 1st note. No breeding data.

Northern forms in humid forest undergrowth, occ. in canopy; southern forms in bushes in dry, secondary shrub, dense brush near water, at 2000-2600 m. N Peru in Cajamarca (Cutervo), Amazonas (Cocochó, Leymebamba) and San Martín (Abra Patricia, sighting) (*jaczewskii*). Se Peru in Cuzco (Marcapata) and Puno (Urohuasi) (*marcapatae*); La Paz (Sandillani) and Cochabamba (San Cristobal) Bol. (*subfasciatus*); lowlands to 2300 m from Cochabamba (Tujima) and Sta Cruz (Samaipata), Bol. s along Andean slope to Tucumán nw Arg. (*cochabambae*). Nominate ssp in lowlands of se Brazil and ne Arg. Rare in n of range, fairly common in s.

NOTE: The northern sspp are sometimes regarded as a distinct species **MARCAPATA ANTSHRIKE (*Thamnophilus marcapatae*)**.

Antbirds – Genus *Drymophila*

7 species, 5 in se Brazil and neighboring Arg. and Parag., 1 in Amazon-

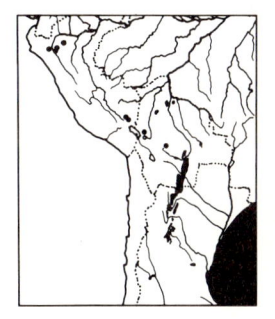

ian forest, 1 in the Andes. Fairly small antbirds with long, graduated tails. Females rarely differ strikingly from males. Throat often streaked, belly rufous; both sexes have a white interscapular patch. They inhabit bushes, tangles, and bamboo within 5 m of the ground. See *Amer. Mus. Novit.* 509(1931).

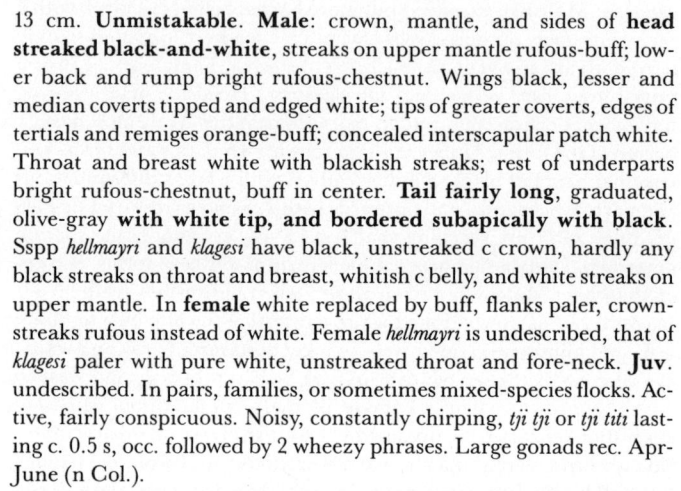

LONG-TAILED ANTBIRD *Drymophila caudata*

13 cm. **Unmistakable**. **Male**: crown, mantle, and sides of **head streaked black-and-white**, streaks on upper mantle rufous-buff; lower back and rump bright rufous-chestnut. Wings black, lesser and median coverts tipped and edged white; tips of greater coverts, edges of tertials and remiges orange-buff; concealed interscapular patch white. Throat and breast white with blackish streaks; rest of underparts bright rufous-chestnut, buff in center. **Tail fairly long**, graduated, olive-gray **with white tip, and bordered subapically with black**. Sspp *hellmayri* and *klagesi* have black, unstreaked c crown, hardly any black streaks on throat and breast, whitish c belly, and white streaks on upper mantle. In **female** white replaced by buff, flanks paler, crown-streaks rufous instead of white. Female *hellmayri* is undescribed, that of *klagesi* paler with pure white, unstreaked throat and fore-neck. **Juv**. undescribed. In pairs, families, or sometimes mixed-species flocks. Active, fairly conspicuous. Noisy, constantly chirping, *tji tji* or *tji titi* lasting c. 0.5 s, occ. followed by 2 wheezy phrases. Large gonads rec. Apr-June (n Col.).

Lower to middle strata, occ. in canopy, in dense as well as open cloud forest, second growth, abandoned plantations, mostly in vines, but also in tall bushes, bamboo thickets and braken-covered hills. At 500-2700 m. Perijá mts, coastal mts and Andes of Ven. (*klagesi*), Sta Marta mts and n end of E Andes of Col. (*hellmayri*), remainder of Col. Andes s through Ecu. and Peru to La Paz, Bol. (*caudata*). Fairly common in nw Ecu. and Peru.

Antwrens – Genus *Terenura*

5, mostly trop., little known species found from Panamá to Bol. and ne Arg. Very small, thin-billed, almost warbler-like antbirds that forage in the canopy. The sexes differ only slightly. See *Amer. Mus. Novit.* 584(1932).

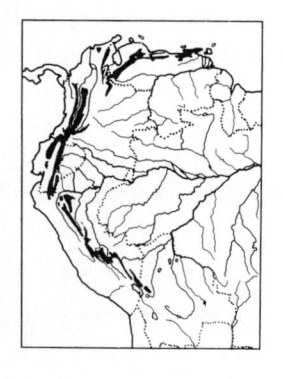

RUFOUS-RUMPED ANTWREN *Terenura callinota* inhabits dense, humid forest at 1000-2400 m in Perijá mts (*venezuelana*, Phelps & Phelps in *Proc. Biol. Soc. Wash.* 67(1954):107); w Panamá, W, C, and w slope of E Andes of Col., and e slope from Caquetá s through e and w Ecu. to n Cajamarca n Peru (*callinota*); San Martín and Junín, c Peru (*peruviana*). Another ssp in Guiana. 9 cm. **Male**: **crown black** bordered by white supercilium, nape gray, remaining upper-parts olivaceous, rump rufous. Wings dusky with yellow 'shoulders' and 2 wingbars; tail dusky with yellow feather-edges. Headside, throat, and upper breast pale gray, rest of underparts pale olive-yellow. **Female**: similar, but throat whitish, cap and 'shoulders' olive like back. Female of *venezuelana* (a distinct species?) is more brownish-olive above, with narrow, grayish rump-band, grayer on breast, and pale yellow on rest of underparts (male undescribed). Alone (or in pairs?), usually in mixed-species flocks. Active; **often exposing rufous rump** as it peers under branches or briefly hangs at tips of branches. Gives high-pitched chipping *ti-ti-ti'i'i'i'ti'-tzzs, tzzs, tzzs* (number of terminating notes variable).

NOTE: Forms a superspecies with Yellow-rumped A. (*T. sharpei*) of Puno se Peru and Cochabamba Bol.

Fire-eyes – Genus *Pyriglena*

3-4 poorly differentiated, medium-sized species, 2 mainly trop. from e Brazil to Arg., 1 also premontane. See *Amer. Mus. Novit.* 509(1931).

WHITE-BACKED FIRE-EYE *Pyriglena leuconota*, with 10 sspp in trop. and subtrop. zones from Col. to Bol. and in Brazil, ascends to 2700 m in Col., where found in clearings and at edge of humid forest within 3 m of the ground at head of Magdalena valley and on e slope of E Andes from w Caquetá southward (through e Ecu. to n Peru) (*castanoptera*). 18 cm. **Iris red**. Bill small. Tail fairly long. **Male** glossy-black with semi-concealed white interscapular patch. **Female** similar (also with white interscapular patch), but back and wings deep rufous-brown.

Antbirds – Genus *Myrmeciza*

18 species in C and S Am. Mainly trop., but 2 species premontane, 1 of them reaching lower temp. zone. See *Amer. Mus. Novit.* 545(1932).

GRAY-HEADED ANTBIRD *Myrmeciza griseiceps*

14 cm. Bill fairly long, slender, and straight. **Male** has crown and headside dark gray, back and rump reddish brown with semi-concealed interscapular patch white bordered black. Shoulders white, rest of wing-coverts black tipped white; remiges mostly reddish brown, basally blackish. Throat and breast black, sides and belly gray, flanks and vent tawny-olive. Tail dark gray tipped white. **Female** and **imm**. similar, but crown and headside paler gray and tinged brownish, eye-ring buffy white, interscapular patch less pronounced (completely lacking in a female with fully ossified skull from Abra Porculla); primary-coverts reddish brown like entire remiges. Lower headside, throat and upper breast mottled gray and whitish, feathers black basally, lower breast and belly buffy white, flanks and vent tawny-olive.

Habits: In pairs or family groups, not known to follow mixed-species flocks. Forages within dense undergrowth. Its exceptionally long, slender bill may be specialized for probing internodes and leaves of bamboo.

Voice: Nothing published. One was heard singing from within a clump of bamboo.

Breeding: Probably in the wet season, Jan-May. Juv. June; small gonads Sep-Dec.

Habitat: Low in densely tangled undergrowth, primarily patches of *Chusquea* bamboo, in humid forest and humid patches in semi-deciduous forest.

Range: At 600-2900 m. Pacific slope from El Oro, sw Ecu. s at least to Abra Porculla in sw Piura, nw Peru. Uncommon.

Antthrushes – Genus *Chamaeza*

4 species in humid forest of S Am. s to ne Arg., ranging from trop. to temp. zones. All are very local, with sometimes highly disjunct distributions. The sexes are alike, brownish with black and white spots, bars, or streaks below. The tail is fairly short, held cocked when the bird is excited, the legs long. Terrestrial, running with tail cocked, alone, in pairs, or family groups. In at least 1 species, they congregate in larger groups in thick, low bushes in the evening. Their presence is usually only revealed by loud calls, mostly given at dawn. See *Amer. Mus. Novit.* 584(1932).

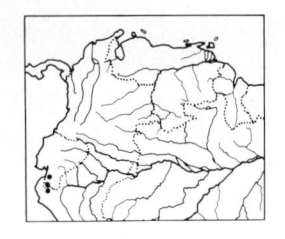

BARRED ANTTHRUSH *Chamaeza mollissima* – Plate XXXIX 4a-c

19 cm. Bill fairly short. Above chestnut to dark brown, incl. sides of head that are traversed by black-and-white barring above and below earcoverts. Entire underparts black to dark brown barred white to buff. Ssp *yungae* has pale-spotted rather than barred underparts. **Juv.** is coarsely streaked rather than barred below.
Habits: Terrestrial, running like a rail pumping cocked tail, lifting feet high for each stride. Very shy.
Voice: Song a very long trill of *cu* notes, lasting c. 20 s, and ending abruptly, each note increasingly emphasized and slightly louder.
Breeding: Juv./imm. Nov (Puno).
Habitat: Dense undergrowth with fallen trees covered in mosses and lichens, in cloud forest and humid montane forest. In the s rec. from same localities as Unadorned Flycatcher (*Myiophobus inornatus*).
Range: 1800-3100 m. E slope of W Andes and w slope of C Andes of Col. in Valle, Quindío, and Cauca, perhaps E Andes ('Bogotá'), e Ecu. s to n Cajamarca, Peru (*mollissima*); from Vilcabamba, mts Cuzco s to Cochabamba, Bol. (*yungae*). Locally fairly common.

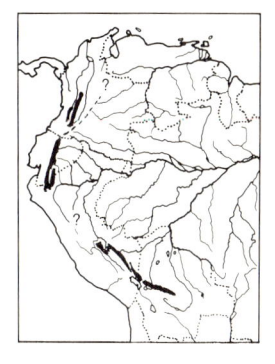

RUFOUS-TAILED ANTTHRUSH *Chamaeza ruficauda*

18.5 cm. Above dark rufescent-brown, brightest on crown, rump and tail; loral spot, eyering extending as long brow, streak below earcoverts, and throat white, rest of underparts whitish heavily streaked to scaled black. Song a loud trill of *cu* notes, lasting up to 50 s.
   Found at 1600-2600 m in humid mossy forest in Valle, w slope of C Andes, and Huila, head of Magdalena valley, Col. (*turdina*), in coastal Ven., and in se Brazil.

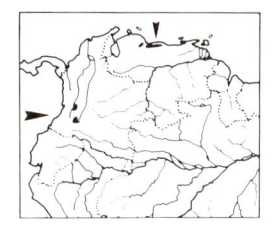

Antthrushes – Genus *Formicarius*

5 species found in humid forest from Costa Rica to Amazonian Brazil and n Bol., w of the Andes to Ecu. Thrush-sized, fairly long-legged, with short tails cocked. Terrestrial birds with distinctive whistled calls that are fairly easily imitated. The sexes are alike, the plumage rather uniform and dull. Primaries rufous at their bases, forming broad band on underside of wing, presumably shown in display. See *Amer. Mus. Novit.* 584(1932).

RUFOUS-BREASTED ANTTHRUSH *Formicarius rufipectus* – Plate XXXIX 5

18 cm. Blackish gray; crown, nape, neckside, rump and vent dark chestnut, breast rufous chestnut, c belly somewhat paler. Throat and sides of head black. Ssp *lasallei* has black fore-crown, *thoracicus* has entire crown and hindneck deep black. **Imm.** has the rufous-chestnut below duller and confined to the breast; black replaced by dusky; throat white barred dusky.

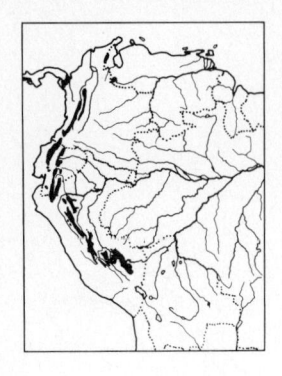

Habits: Alone (or in pairs?). Terrestrial. Responds to an imitation of its call, but keeps well hidden. Runs about at great speed, like a rail.
Voice: Song lasts c. 1 s, and consists of 2 clear, resonant, short whistles, the 2nd shorter and slightly higher-pitched, repeated every 4-5 s or more, esp. at dusk and dawn.
Breeding: Imm. Nov. (nw Ecu.); large gonads June (Antioquia).
Habitat: Humid forest with understory of saplings and fallen trees.
Range: 1000-2400 m, rarely to 3100 m. Perijá mts w Zulia, and sw Táchira, Ven. (*lasallei*, Aveledo & Gines in *Mem. Soc. Cienc. nat. LaSalle* 13(1952):203-206); W and C Andes of Col. s to w Ecu. (*carrikeri*); e Ecu. s to Cuzco, se Peru (*thoracicus*). Locally common. Nominate ssp in Panamá and Costa Rica.

ANTPITTAS – Genus *Grallaria*

27 species, 1 ranging from c Mexico to Bol., the rest restricted to S Am. They inhabit humid forest s to n Arg. Most are montane, and many have a fairly narrow elevational distribution.

They have very long legs (unlike in rails not trailing in flight), stubby tails (of 8-12, occ. 14 rectrices), a compact, apparently neck-less body and large eyes. They often stand motionless for long periods, bobbing and flicking wings and tail in sudden jerks between powerful hops and frozen postures. They are primarily terrestrial, and some species will come out onto open grassy or mossy spots to feed like a thrush at late dusk or early dawn.

Rarely seen more than 2 m above ground except when tending the nest, which is a loosely made open cup, and which may be placed as high as 3 m up. The sexes are alike. The plumage is brownish olive, rufous, white, and black. Juv. is usually barred dusky and buffy. Imm. resembles ad., but scattered feathers are barred as in juv. Both sexes sing. Songs and calls are often very loud and stereotyped, in many species composed of clear whistles that are easily imitated.

Due to their secretive habits most species are very poorly known. See *Amer. Mus. Novit.* 703(1934).

UNDULATED ANTPITTA *Grallaria squamigera* – Plate XXXIX 7a-b

20 cm. Above slaty gray tinged olivaceous except on crown and upper mantle. Throat, lores and area immediately behind eye buffy white (white in *canicauda*); moustache black; headside and rest of underparts orange-buff, barred or squamulated with blackish brown, only c lower belly unmarked, vent bright ochraceous orange. **Juv**. broadly barred slaty black and pale orange-buff; belly, flanks, and vent plain orange-buff.
Habits: Alone or in pairs. Forages on the ground. Takes to a low branch and freezes when alarmed.
Voice: Contact call by both sexes a soft *pt*, repeated every s. **Characteristic song** given at dusk, very early dawn, or during rain, a slow tremulous series of loud, low notes, slightly ascending in pitch, lasting 3-4 s, and repeated every 14-18 s. Sometimes distinctly rising at end.

Breeding: Fledglings Nov (nw Ecu.), Dec (La Paz), Mar (Apurímac); imm. Nov (nw Ecu.); large gonads Aug (Antioquia W Andes).
Habitat: Cloud and elfin forest with bamboo and mossy patches; humid, montane shrub.
Range: 1830-3800 m, mainly above 2600 m. Andes of Mérida, and Páramo de Tamá, sw Táchira, Ven., all 3 ranges of Col. s to e and w Ecu. (*squamigera*); from Cajamarca and Amazonas, n Peru s to Cochabamba, Bol. (*canicauda*). Uncommon.

GIANT ANTPITTA *Grallaria gigantea* – Plate XXXIX 6

22.5 cm. **Bill very heavy**. Ssp *hylodroma*: forehead rufous-chestnut, grading into slaty gray of crown and nape; rest of upperparts olivaceous tinged brownish. Headside and underparts rufous chestnut, narrowly barred with blackish; c belly and vent ochraceous, unbarred. Ssp *gigantea* more heavily barred below, esp. on flanks; forehead, loral region, edges of remiges, and underparts less bright; back browner, less olivaceous. Ssp *lehmanni* like *gigantea*, but c underparts even more heavily barred; color of back like *hylodroma*. **Juv**. barred rufous-chestnut and slaty-black throughout.
Habits: Possibly feeds mainly on tadpoles and frogs.
Voice: No data.
Breeding: Imm. Nov (Pichincha).
Habitat: Swampy places with shallow puddles of stagnant water in humid forest.
Range: 2300-3000 m on w slope of C Andes of Col. in Cauca (San Marcos and Tijeras, both in Moscopán) and Huila (*lehmanni*), (1373?) 2000-2400 m in Pichincha, nw Ecu. (Gualea; Pachijal; Cerro Castillo) (*hylodroma*), at 2900 m in extreme ne Ecu. in the Pun region (headwaters of Río Chingual), Napo, at 2200 m on ne slope of mt Tungurahua, Tungurahua, and on El Tambo immediately w of continental divide in c Loja (USNM) (*gigantea*). Rare and local, with no recent rec. of *lehmanni*.

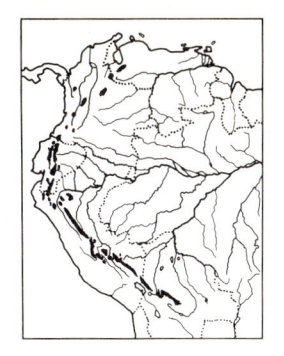

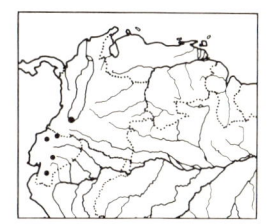

SCALED ANTPITTA *Grallaria guatimalensis* – Plate XXXIX 9a-b

15 cm. Ssp *regulus*: above dark brown with faint olivaceous cast; hindcrown and nape slaty gray, feathers of back indistinctly tipped blackish. Lores, **broad moustache, crescent on lower throat, and streaks on upper breast white**; rest of headside, throat, neckside, and breast dark reddish brown, remainder of underparts bright orange-rufous; white bases to feathers of c belly show through. **Under-wing bright rufous**. Ssp *carmelitae* has darker breast and duller wings. **Juv**. blackish above, crown and nape with fine, pale buff shaft-streaks, back with large buff spots; below blackish brown streaked pale buff and buff on throat and breast, belly buff faintly barred dusky. – Habits as in other antpittas. Song, mostly given at dusk and dawn, a slow trill of low notes lasting c. 4 s, increasing in amplitude, and repeated at 15-20 s intervals. Fledgling Apr (Valle), June (Mérida); juv./imm. Aug (nw Ecu.); large gonads Apr (Perijá).
Fairly open understory of humid, tall primary forest; humid second

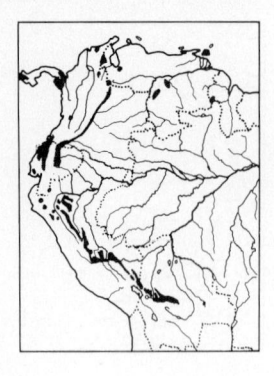

growth. Mostly upper trop. zone, but up to 2400 m in w Peru. N base of Sta Marta mts, Perijá mts and upper Río Negro Zulia Ven. (*carmelitae*, hardly distinct from *regulus*), Valle W Andes of Col. (*ssp*); c Mérida and extreme s Táchira, Ven., E Andes of Col., e and w Ecu. s through Peru (in Cajamarca also on Pacific slope) to Cochabamba, Bol. (*regulus*). Other sspp, not reaching temp. zone, in C Am., Trinidad, and lowlands and Tepuis 'of s Ven. Apparently local.

CHESTNUT-CROWNED ANTPITTA
Grallaria ruficapilla – Plate XL 14a-b

18 cm. **Crown, headside, moustache, nape, and upper mantle rufous**; rest of upperparts olivaceous. **Below white, streaked** olivaceous and blackish on sides and sometimes faintly on breast, feathers of breast with some ochraceous edges. Ssp *watkinsi* has pale shaft-streaks on crown and back, dusky earcoverts, and paler legs; *avilae* has a darker rufous forehead; *perijana* like *avilae*, but underparts with fewer and narrower streaks and more suffused with bright ochraceous orange, esp. on the chest, and extending to middle of abdomen and posterior flanks (where it is only faint in *avilae*); *albiloris* has white lores and moustache, and mostly white earcoverts; *interior* has a well-developed white eye-ring, and concealed whitish shafts on feathers of back and rump. **Juv**. is pale cinnamon-rufous to rufous, barred dark brown above, throat white, breast like back, belly buff, whitish in center.
Habits: As for genus.
Voice: Sings frequently. Song consists of 3 whistles, the middle note lower in pitch, and longer than the other 2 *hy huyy-huy*. Call a piercing single whistle *pyuu*.
Breeding: Fledglings June, July, Sep; juv./imm. Aug, Nov, Dec (all Mérida); fledglings Jan (Cauca), Apr (w Nariño), Feb, Sep (Pichincha); juv./imm. July (Boyacá, Mérida), Aug (El Oro); eggs Apr, May (n Cajamarca); juv./imm. Apr, June (Cajamarca), Aug (Lambayeque), Feb, Mar, Aug, Nov, Dec (Amazonas); large gonads Mar (Valle), Apr-Sep (n Col.).
Habitat: Both humid and dry montane scrub, second growth, and understory of humid forest.
Range: 600-3600 m. Coastal mts and Andes of Lara, Ven. (*avilae*), Perijá mts (*perijana*), Andes of Trujillo, Mérida, and Táchira, Ven. (*nigrolineata*), all 3 ranges of Col. and n Ecu. (*ruficapilla*), Chilla mts, ne El Oro/n Loja border and Celica mts, w Loja, s Ecu. (*connectens*), c and s Loja, s Ecu., Piura, Cajamarca, and Lambayeque n Peru (*albiloris*), w slope of c Andes of n Peru in Amazonas and San Martín (*interior*). Sighted at 3150 m near Chavin de Huantar in Ancash (O Frimer, ssp?). Fairly common to common.

NOTE: The morphologically and vocally distinct form *watkinsi*, living at 550-1400 m in extreme sw Loja and w El Oro, Ecu., and neighboring Tumbes, n Peru, is best threated as a distinct species **Watkins'** or **Scrub Antpitta (*Grallaria watkinsi*)** (Chapman 1919, Parker *et al*. MS).

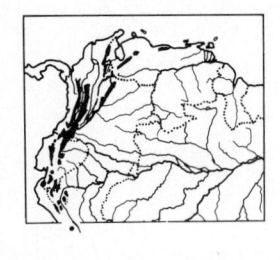

SANTA MARTA ANTPITTA *Grallaria bangsi* is confined to Sta Marta mts, Col., where it occurs between 1200 and 2450 m. Olivaceous above with brown wings; neckside gray, earcoverts and region below the eye brown with whitish shaft-streaks. **Throat rufous**; rest of **underside white**, breast and sides **streaked dusky**.

STRIPE-HEADED ANTPITTA *Grallaria andicola* – Plate XL 6a-c

16.5 cm. Above gray-brown, primaries edged dull olivaceous rufous; crown and back with blackish-bordered whitish streaks. Lores and eye-ring whitish; headside, neckside and underparts streaked and scaled whitish, buff and blackish; c throat whitish or buff, unstreaked. Ssp *punensis* has a blackish crown with orange-buff streaks, and an unstreaked back. **Juv.** is spotted to barred throughout. **Imm.**: like ad., but tips of wing-coverts and secondaries mottled, and tail-feathers pointed and pale-tipped.
Habits: Usually seen alone or in pairs. Forages on the ground, esp. in the early morning and late evening. Makes a few hops, flicks wings nervously, then freezes, often with tilted head, picks up a caterpillar with a quick movement, then makes another few hops and stands motionless again. When scared while feeding will flick wings frequently, flutter short distance (occ. making long flight down-hill), usually to a rock or tree, or makes long hops into cover. Spends the warm hours of the day 'frozen' in a tree.
Voice: In contrast to other Antpittas, has a weak and rarely heard song, which is a rolling series of wheezy notes, first slightly descending, then ascending and accelerating, last note sometimes drawn-out *ree ree...ree eeee*. In w Andes song may be confused with that of Andean Tapaculo. Call a single, somewhat wheezy, slightly descending note (*andicola*) or 1-2, usually mellower notes alike, somewhat resembling call of Great Thrush (*punensis*).
Breeding: Fledgling Feb (Huánuco, Lima), Mar (La Libertad); juv./imm. Feb, Dec (Junín), Jan (Pasco), Mar, Apr (La Libertad), May (Huánuco, Puno), Jan, Mar, May (Cajamarca).
Habitat: Semi-humid *Polylepis* forest and shrub, as well as adjacent rocky slopes, and in *Gynoxys* and *Buddleia* stands, but along the eastern Andean slopes also in dwarf forest and humid shrub bordering grassy slopes. Forages on open patches of short, moist grass or among rocks. May search for insects on vertical cliff faces or in the puna grassland, sometimes a considerable distance from trees.
Range: 3000-4300 m. S of the Marañón and Porculla pass in Cajamarca, La Libertad, Amazonas, Ancash, Huánuco, Pasco, Junín, Lima, Huancavelica, Ayacucho, and Apurímac (*andicola*), Cuzco and Puno, se Peru, and La Paz, Bol. (specimen from near Pongo AMNH, and heard near Zongo by JF) (*punensis*). Widespread, usually local and in fairly low numbers, but locally common.

BICOLORED ANTPITTA *Grallaria rufocinerea* – Plate XL 9

15.5 cm. Upperparts and throat rufous-brown, feathers of throat gray

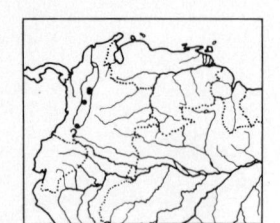

basally; rest of underparts gray, c belly mottled with whitish. Ssp *romeroana* has a solid rufous-brown chin and throat, and more extensive whitish scaling on belly. **Juv**. undescribed.
Habits: As for genus.
Voice: A high-pitched, clear, whistled *treeeee* or double-sounding *treeeee-aaaa* at 2.5-3 s intervals, last part slurred down.
Breeding: Large gonads June (se Antioquia).
Habitat! Dense, humid, montane forest and cloud forest near the treeline.
Range: Col. in C Andes at 2100-3300 m in Antioquia (e slope 8 km e of Medellín at Santa Elena and at Páramo Sonsón) and Quindío (w slope above Salento) (*rufocinerea*), and Huila (Puracé) (*romeroana*, J Hernández C & JV Rodriguez M. in *Caldasia* 12(1979):573-580); sight rec. from e Nariño. Rare and local.

CHESTNUT-NAPED ANTPITTA *Grallaria nuchalis* – Plate XXXIX 10a-c

17 cm. Bill black. Ssp *obsoleta*: upperparts olivaceous, slightly browner on crown and flight-feathers and with rufous-chestnut nape. Lores and eye-ring blackish, earcoverts chestnut. Underparts blackish-gray. Ssp *nuchalis* browner above with a dark chestnut crown, dusky brown below with a tinge of ashy on the abdomen; *ruficeps* has the whole crown, nape, and headside rufous-chestnut, and is gray below. **Juv**. plain-colored light brown, darkest on crown; light buffy brown below, darkest on breast; bare parts of face bright orange.
Habits: In pairs. Only rarely feeds in the open. Territorial.
Voice: Song loud, lasting 3-4 s, starts with a series of 4-5 metallic notes on same pitch, accelerates and terminates in a short trill rising hysterically at the end *tih--teh-teh-teh-teh-ti-ti-ti-tittttt*. Response to playback a strident *chee-chee-chee-chee-chee* (female?).
Breeding: Fledgling July (Antioquia); juv. Jan (Antioquia); juv./imm. Aug, Sep (w Ecu.); large gonads Jan (se Antioquia).
Habitat: More or less restricted to the darkest recesses of bamboo thickets and adjacent forest undergrowth on steep slopes and in stream ravines.
Range: 1900-3150 m, mainly 2600-2900 m. C Andes and w slope of E Andes of Col. in Antioquia, Cundinamarca, Caldas, Tolima, and Cauca (*ruficeps*), nw Ecu. (*obsoleta*), e Ecu. and Peru in Piura (Chinguela) (*nuchalis*). Locally fairly common.

PALE-BILLED ANTPITTA *Grallaria carrikeri* – Plate XXXIX 11a-b

20 cm. **Bill white**. Head blackish, back dark brown, wings and tail dark chestnut. Below gray, abdomen pale gray, flanks dark olivaceous. **Juv**. has head blackish gray with lores black, hindcrown, back, wingcoverts, and breast barred rufous and black, upper belly and sides barred buff and blackish, c lower belly white. Bill black with orange flecks at base.
Habits: Alone or in pairs. Found on or within 1 m of ground. Sings

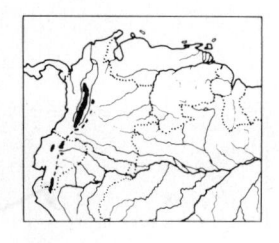

from a low perch. Territorial. Both sexes tend young. Does not sing near nest.
Voice: Song usually given at mid-morning, 6 staccato notes on the same pitch, lasting c. 3 s. There is a slight pause after the 1st and before the last note.
Breeding: Eggs Sep, Oct (La Libertad); juv. Aug, Oct (Amazonas).
Habitat: Dense undergrowth with large bamboo-thickets in up to 30 m tall cloud forest with broken canopy and lush epiphytic growth.
Range: Peru at 2350-2900 m in e Andes s of the Marañón, in Amazonas (Cord. Colán) and e La Libertad. Locally fairly common.

NOTE: Described by Schulenberg & Williams, *Wilson Bull.* 94(1982):105-113.

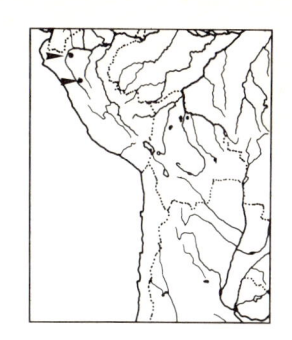

WHITE-BELLIED ANTPITTA *Grallaria hypoleuca* – Plate XL 11

13-15 cm. Ssp *hypoleuca*: above brown; nape, earcoverts, and sides of neck rufous-chestnut. Lores dark gray. Throat and c belly white, flanks and sides of breast and body rufous-chestnut, rest of underparts pale gray. Ssp *castanea* considerably smaller, and chestnut-brown above; *flavotincta* has chestnut-brown upperparts and sides of head and body (lores dark gray), and creamy white underparts, tinged pale gray on breast and sides of belly. **Juv.** undescribed.
Habits: Much like congeners, but less terrestrial.
Voice: Song given at dusk and dawn. Sspp *castanea* and *hypoleuca*: 3 loud whistles *tooo tew-tew*, the last 2 highest and alike, the 1st longer and (in *castanea* only?) somewhat slurred or double in quality. Also a soft *whee-whee-whee-whee* on one pitch (alarm?). Ssp *flavotincta*: song (2-)3 whistles; usually the 1st and last alike, middle note longer and slurred with a double quality *tew too tew*; in one variant both 2nd and 3rd notes slurred *tew too too*; in another only the 3rd *tew tew too*.
Breeding: Large gonads Mar-Sep (C and E Andes Col.), June (Antioquia).
Habitat: Landslides, slash-and-burn agriculture and other habitats with patches of second growth; dense understory at edge of humid montane forest.
Range: 1400-2350 m. On Pacific slope from n Antioquia, Col. to Carchi (Pichincha?), nw Ecu. (and w slope of C Andes, Col.?) (*flavotincta*), Antioquia, C Andes and w slope of E Andes of Col. from Santander s to Cundinamarca (*hypoleuca*), from headwaters of Río Magdalena in Huila, Col. s through e Ecu. to n Cajamarca and Piura, Peru (*castanea*). Locally common.

NOTE: Ssp *castanea* has been doubtfully rec. from nw Ecu. If this is confirmed, *flavotincta*, which also has 10 (not 12) tail-feathers and a tail shorter than tarsus (not longer), must be regarded a full species, **YELLOW-BREASTED ANTPITTA**.

NOTE: Rusty-tinged, Bay, and Red-and-white A. are sometimes treated as sspp of the present species.

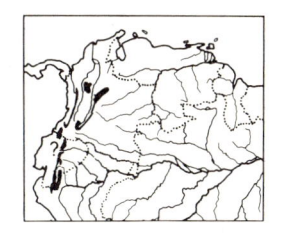

RUSTY-TINGED ANTPITTA *Grallaria przewalskii* – Plate XL 12a-b

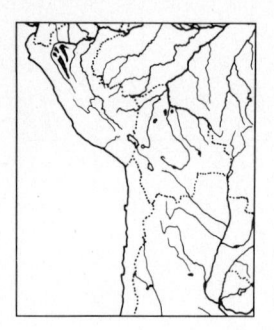

16.5 cm. Above dark rufous-brown; crown gray tinged blackish; forehead pale gray grading into color of crown. Broad eyering whitish; sides of neck rufous-chestnut. Chin whitish, throat rufous; breast, flanks, and vent rufous-brown; belly grayish with whitish center. Old males have buffy white throat, gray breast and sides, whitish belly, and often a few rufous-brown feathers on breast. **Juv.** has white throat and black breast; rest of plumage dark, finely barred with buff.
Habits: Presumably like the closely related White-bellied A.
Voice: Song consists of 3 notes, middle note lower than other 2.
Breeding: Fledgling July (Amazonas).
Habitat: Dense understory of cloud forest and humid montane forest, often with admixed bamboo. Second growth?
Range: Peru at 2200-2750 m and probably lower, s and e of the Marañón in Amazonas, e La Libertad, and San Martín. Uncommon?

NOTE: Possibly a ssp of White-bellied A.

BAY ANTPITTA *Grallaria capitalis* – Plate XL 10

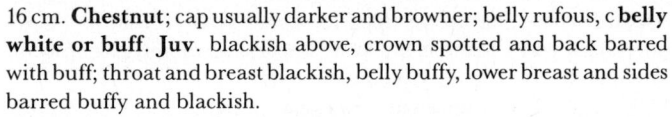

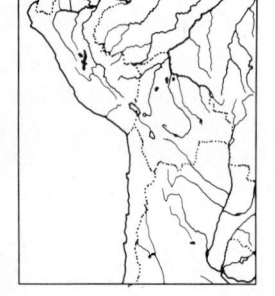

16 cm. **Chestnut**; cap usually darker and browner; belly rufous, c **belly white or buff**. **Juv.** blackish above, crown spotted and back barred with buff; throat and breast blackish, belly buffy, lower breast and sides barred buffy and blackish.
Habits: Alone or in pairs. Forages low in the vegetation, sometimes on the ground. Secretive.
Voice: Sings throughout day, 4 whistles *hy hyhyhy*, lasting c. 1 s and repeated every 5-10 s, 1st note highest, last 3 alike.
Breeding: Fledgling Dec (Huánuco).
Habitat: Bamboo-covered landslides and second growth at edge of cloud forest and coppice where cloud forest is cleared for slash-and-burn agriculture. Also found in sometimes fairly dry riparian extensions into the neighboring arid zone.
Range: C Peru at (1525) 2600-3000 m. Only known from Rumicruz and immediately se of Oxapampa, Pasco, and Divisoria and Carpish mts, Huánuco, where fairly common.

NOTE: Possibly a ssp of White-bellied A.

RED-AND-WHITE ANTPITTA *Grallaria erythroleuca* – Plate XL 13

17 cm. Bright dark rufous, back browner; throat, belly, c breast and spots on sides of breast white. **Juv.** Unknown.
Habits: Alone or in pairs, forages within 1.5 m of and often on the ground. Makes rapid jerks.
Voice: Song, given mainly at dusk and dawn, 3 loud whistles, the 1st followed by a brief pause, the last 2 slightly lower and alike *tew too-too*, lasting 1-2 s and repeated every 5-10 s. Sometimes the last note is

slightly downslurred *tew too-tooe*, and rarely a 4th note is added *tew too-too-too*. Call a single whistle *hye*.
Breeding: No data.
Habitat: Bamboo covered landslides and other secondary thickets in the humid zone.
Range: Se Peru at 2200-3000 m in Vilcabamba and Vilcanota mts, Cuzco. Locally fairly common.

NOTE: Possibly a ssp of White-bellied A.

GRAY-NAPED ANTPITTA *Grallaria griseonucha* – Plates XXXIX 8a-b and LXIV 27

15 cm. Crown dark brown; nape and sides of hindcrown slaty gray; back dark reddish brown. Sides of head and entire underparts rufous chestnut; c belly paler and tinged olivaceous. Underwing dark reddish brown. Ssp *tachirae* is brownish olive above, and has throat paler than rest of underside. **Juv**. has pale shaft-streaks on nape, back, and underparts.
Habits: Solitary and secretive, perhaps more arboreal than Chestnut-crowned A..
Voice: Song c. 2 s long, of 12-18 *hu* whistles, starting quietly and quickly getting loud and slightly higher in pitch. Given at 10-20 s intervals. Also 5 hollow whistles, alike, with a pause after 3rd *hu-hu-hu hu-hu*. Call a single or double, ringing *woik* at irregular intervals (BW, JP, JA, JC).
Breeding: Juv. Jan (Mérida).
Habitat: Understory of tall, dense, humid montane forest.
Range: Ven. at 2300-3000 m. Boca de Monte, ne Táchira (*tachirae*), Andes of c Mérida (*griseonucha*). Locally common, at least at 2600-2700 m (BW, JP, JA, JC).

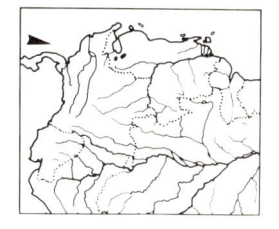

CHESTNUT ANTPITTA *Grallaria blakei* – Plate XL 7

15 cm. Very similar to nominate form of Rufous A., but differs from other Peruvian forms (*cajamarcae*, *obscura*, *occabambae*) by its darker reddish brown plumage, (indistinctly) barred lower belly, dark reddish brown head, lack of contrasting feather-tips on back and breast, dark brown under tail-coverts, thicker bill and tarsi, and lack of contrasting eye-ring. The Amazonas specimens have brightest underparts and least distinctive barring on belly. A single specimen from 2500 m, c. 12 km E Oxapampa, Pasco, lacks bars on belly and is particularly dark above. Another from ne Ayacucho may also belong here. **Juv**. a molting specimen has traces of streaks scattered on crown, nape, wing-coverts, throat and breast.
Habits and **Voice**: No data.
Breeding: Juv./imm. Aug, Nov (Huánuco).
Habitat: On or near the ground in dense cloud forest.
Range: Peru at 2135-2470 m in c Amazonas (Cord. Colán, Abra Patricia c. 30 rd km NE Florida) and Huánuco (Carpish mts), probably also in intervening region. A specimen from near Oxapampa, Pasco doubtfully referable to this form, as may also another from ne Ayacucho. Uncommon.

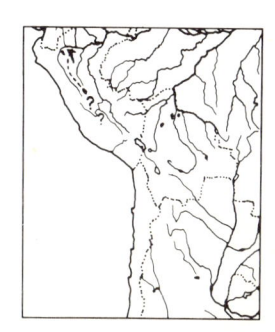

NOTE: Although formally described as a species, (G Graves, *Wilson Bull*. 99(1987):312-321), additional material (biochemical analysis, sound rec.) confirming the status is desirable.

RUFOUS ANTPITTA *Grallaria rufula* – Plate XL 8a-d

15 cm. Bill black. Ssp *rufula*: Above rufous brown; headside and underparts rufous, breast darkest; flanks dark gray-brown, c belly dull light rufous, feathers with paler tips. Abdomen buffy-white to white, dark bases more or less showing through, under tail-coverts varying from whitish through buffy-brown to dark brown. Ssp *spatiator* hardly differs; *saltuensis* and the southern sspp differ by being much duller, lighter, and more olivaceous, dullest in *saltuensis*, *obscura*, and *cochabambae*. Ssp *saltuensis* has a slender tarsus and contrasting clay-colored feathertips above. **Juv**. has back barred buff, crown and underparts streaked whitish.
Habits: Alone or in pairs. Comes out onto grassy or mossy spots to feed when undisturbed. Otherwise keeps well under cover, mostly off the ground.
Voice: Song (n Peru) a 3 s long monotonous trill of c. 20 emphasized notes, repeated at 3-5 s intervals. In ne and nw Ecu. the trill is fainter, slows towards end, and of somewhat ringing quality, repeated at 2-7 s intervals. Call (nw and se Ecu.) 0.5-1 s long, of 4 notes, 1st note apart and emphasized, last notes of a ringing double quality, repeated at 2.5-3.5 s intervals, sometimes given for several minutes. Call (Cuzco) lasting c. 0.5 s of 2 notes on same pitch, 1st slightly stronger, repeated at c. 4 s intervals, sometimes for several minutes. A 2-note call has also been rec. from e Ecu. and Col., and a 1-note call from c Peru in the region of overlap with Chestnut A.
Breeding: Fledgling Feb (ne Ecu.); juv./imm. Sep (Quindío), July (Cauca), Dec, May, Jun, Sep (nw Ecu.), Feb (ne Ecu.), Mar (Pasco); large gonads Mar-May (n Col.).
Habitat: Within 2 m of the ground in dense stands of bamboo and adjacent understory in humid montane forest, cloud forest, and elfin forest.
Range: 2300-3650 m. Sta Marta mts. (*spatiator*); Perijá mts (*saltuensis*); e and w Táchira, Ven., all 3 ranges of Col. (birds from w Col. perhaps separable) s through e and w Ecu. to Piura and n Cajamarca, Peru (*rufula*), s Cajamarca (*cajamarcae*), Huánuco, Pasco, and Junín (*obscura*); from Cuzco, se Peru to La Paz, Bol. (*occabambae*); Cochabamba, Bol. (*cochabambae*). Fairly common.

RUFOUS-FACED ANTPITTA *Grallaria erythrotis* – Plate XXXIX 12

17 cm. Above olivaceous; **sides of head and neck orange-rufous**. Throat and c abdomen white; breast orange-rufous obscurely streaked with white; sides olivaceous. **Juv**. above gray-brown, almost uniform on top of head, with cinnamon-buff spots and rather inconspicuous tawny barring on back to conspicuously barred on lower back; chin whitish, below very profusely barred dusky-brown and cinnamon-

buff, with some pale spots on breast and pale buffy belly. Mandible orange.
Habits: Solitary. Shy. When undisturbed will come out on open grassy patches or margins of ponds and puddles to feed in the manner of a thrush. Otherwise stays under cover within 2 m of the ground.
Voice: Song given mostly morning and evening: 3 loud whistles, very slightly ascending in pitch, esp. the last note; there is a slight pause after the 1st note. Call a single descending whistle *krie* or *hye* repeated every 3-4 s.
Breeding: Juv. Nov (Cochabamba).
Habitat: Elfin forest, cloud forest, and second growth, sometimes with admixed bamboo.
Range: Bol. at 2050-2900 m in La Paz, Cochabamba, and Sta Cruz. Common.

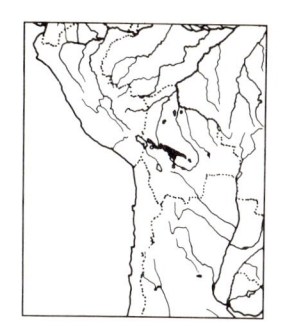

TAWNY ANTPITTA *Grallaria quitensis* – Plate XL 15a-c

17 cm. Above olivaceous, with a decidedly grayish wash on crown and back; rump dull rufous to clay-colored. Ocular region whitish, rest of headside mixed rufous olivaceous and blackish-brown. Underparts mixed rufous and white. Ssp *alticola* has a smaller bill; *atuensis* has smaller bill and legs, and has more pronounced whitish mottling below. **Juv**. is barred pale rufous and blackish brown almost throughout, belly mostly buffy-white.
Habits: Alone, rarely in pairs. Fairly confident. Perches on horizontal branches in the middle, occ. top of bushes. When singing, throws head back, body vertical, bill wide open, and throat-feathers fluffed. Flicks wings and tail, and bobs constantly. Searches for frogs and insects on moss-covered branches and on the ground. Sometimes will flap wings while clinging to a vertical stem and probing moss.
Voice: Song lasts c. 1.5 s, and consists of 3 (sometimes 2) loud, somewhat double notes, the last 2 lower-pitched and alike *kyr kerk-kerk*, or *kyr kerk*, repeated every 4-5 s for long periods. When excited gives a little complaining *eerrr*. Call a single, loud, sharp *tieu* repeated every 2-3 s.
Breeding: Fledglings Sep (Quindío), Mar (Huila); juv./imm. June (Cundinamarca), Sep (Cauca, Quindío), Mar (Huila), Jan (nw Ecu.), Jul, Oct, Nov (ne Ecu.), Jan (Azuay), Aug (Piura); large gonads Aug (Santander), Feb (Cauca).
Habitat: Humid montane shrub and mossy forest with adjacent grass. Pine plantations. *Espeletia* páramos with patches of elfin forest.
Range: 2200-4000 m. Páramos of Santander, Cundinamarca, and Boyacá, E Andes of Col. (*alticola*), from Caldas, C Andes of Col. s through páramos of Ecu. to n Cajamarca and Piura, Peru (*quitensis*); C Andes of Peru s of the Marañón in Amazonas and e La Libertad (*atuensis*). Very common in n Ecu.

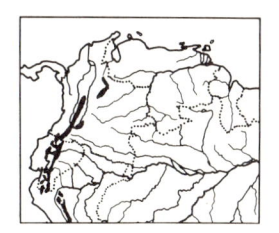

BROWN-BANDED ANTPITTA *Grallaria milleri* – Plate XL 5

14.5 cm. Upperparts and headside dark brown. Lores gray; long, very

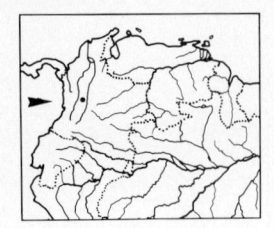

faint supercilium brown. Throat and belly gray, breast and flanks brown. **Juv.** has a streaked crown and a barred back.
Habits and **Voice**: No data.
Breeding: Juv./imm. Sep (Quindío).
Habitat: Forest.
Range: Col., at 2700-3140 m near Laguneta above Salento, Quindío, and at el Zancudo, Caldas, both on w slope of C Andes. Laguneta is now mostly deforested. Status indeterminate, but possibly extinct.

Antpittas – Genus *Grallaricula*

8 species found in the coastal mts of Ven. and the Andes s to Bol., 1 also on Tepuis of Ven., and 1 into C Am. Primarily subtrop., they are very locally distributed, and inhabit fairly steep slopes in humid premontane forest, where they live within 2 m of the ground, often in pairs.

They are small, elusive antpittas, less terrestrial and more active than *Grallaria*, but like them have large eyes, long legs, and short tails. All species have a more or less developed pale eye-ring and loral region, forming 2 'horns' on the forehead. Characteristic are also 2 buff spots on leading edge of wing. Only 2 species show marked sexual dimorphism. Most forms have a white crescent across lower throat, exposed during song, when the bird throws its head back and fluffs out the throat. In at least 1 species (all?), only the male sings. The fledgling is covered in uniform vinaceous pink fluffy feathers for the first 2 weeks, and then attains ad.-like plumage, except for rufous tips to the greater wing-coverts in most species. The nest of Rusty-breasted A. is a shallow cup of leaf-stalks on a concave platform of twigs 60-120 cm up in a bush or tangle.

Most species in the genus are very poorly known.

OCHRE-BREASTED ANTPITTA *Grallaricula flavirostris* – Plate XL 4a-b

10.5-11 cm. Ssp *boliviana*: above brown, crown tinged gray. **2 'horns' on the forehead and broad eye-ring buff.** Throat and breast buff, throat streaked blackish; buff mustache bordered on both sides with blackish, crescent across lower throat white, breast-feathers bordered blackish causing v-shaped scalloping. Belly white; flanks buffy brown. Small pale patch at base of primaries sometimes visible; outer web of alula buffy. Maxilla dark, mandible yellow. Ssp *similis* larger, upperside washed with olive, crown and nape darker than back; underside more heavily scalloped with black; maxilla with yellow tip in ad. male. The more northern sspp get increasingly darker, more ochraceous below with less extensive black markings, streaks rather than scallops. The white throat-crescent is presumably not acquired until 2nd year or later in these birds, and only very few old males have olive-brown flanks and an almost wholly yellow bill (flanks being orange like the breast in younger birds and females). **Juv.** uniform vinaceous pink; wing-coverts tipped rufous.
Habits: Solitary or in pairs. Quite sedentary, relatively inactive, and not particularly shy. Sits motionless for long periods. Sallies to foliage or briefly to the ground.

Voice: Its gargled call has a rattling quality, and is slightly descending. Song (Costa Rica) a short ascending series of 5-8 mellow whistles, faster at the end.
Breeding: Juv./imm. Dec (La Paz, Cochabamba).
Habitat: Within 1.5 m of, but not on the ground. Leafy shrubbery in denser portions of the understory in humid montane forest and second growth.
Range: 1150-2750 m, occ. down to 500 m. W Andes of Col. (*ochraceiventris*), nw Ecu. in Pichincha (*mindoensis*), e slope of E Andes of Col. s to e Ecu. (*flavirostris*), sw Ecu. in El Oro (*zarumae*); at Río Jelache and Afluente n and c San Martín, Divisoria mts Huánuco, and Pasco, c Peru (*similis*); Puno, se Peru, and La Paz and Cochabamba, Bol. (*boliviana*). Very local.

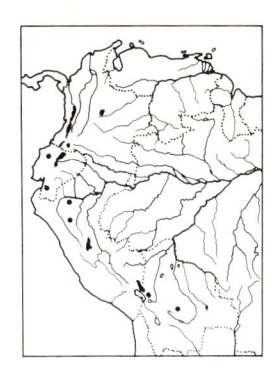

RUSTY-BREASTED ANTPITTA
Grallaricula ferrugineipectus – Plate XL 1a-b

12 cm. Headside and upperparts light to dark brown with slight olivaceous wash. **Spot behind eye white**; forehead with 2 faint white 'horns'; eye-ring buffy. Below bright rufous, crescent across lower throat and c belly white. Outer web of alula, and base of primaries buff. Bill black, base of mandible flesh-colored. Ssp *leymebambae* larger, more olivaceous above and darker below. Chin and sides of throat indistinctly streaked with blackish, breast slightly mottled with sooty olive. 'Horns' in the forehead buffy, half-ring over the eye blackish. Ssp *rara* has top and sides of head deep rufous-brown, back somewhat duller. Underparts brighter than in *ferrugineipectus*, and almost lacking white crescent. **Juv**. is covered by vinaceous pink fluffy down (least so on the head); wing-coverts tipped rufous.
Habits: Not shy. In close pairs, presumably mated for life. Sometimes active in bird-waves. Forages along moss-covered stems and branches a metre or so above the ground, and sallies to foliage. Occ. flycatches. Never spends more than a moment on the ground. Makes constant little jerks while flicking wings and tail. While singing from a perch throws head back exposing white crescent. When surprised, the fledgling 'plays possum'.
Voice: Song by males a rhythmic series composed of 3 parts of altogether 16-18 soft notes: 1st part 6-8 notes, 2nd 7-10 notes, the 3rd a single, or at times 2 notes. In one variant the 3 parts are distinct; the notes of part 1 alike or slightly ascending (and at times also accelerating) towards end; part 2 consists of higher notes more or less alike, louder than part 1 and slightly slower; part 3 is a brusque note at same pitch as part 1, or lower. In another variant part 1 and 2 are hard to tell apart. The pitch rises evenly during part 1, and often during most of part 2; the song slows unevenly during part 1, but is constant in the louder part 2. Part 3 is as in other variant. *Twa twa twa twa twe twe* **cwi...cwi** *cu tu*, *twu twa twa twe twi twi twe* **cwi cwi cuwi ...cuwi** *tu*. Call a loud *kew*, *cui*, or *kierk*. Alarm near the nest a double note.
Breeding: Eggs Oct (Sta Marta); fledglings Dec (Amazonas), May, June (Huánuco), Jan (La Paz).
Habitat: Lower strata of moderately mossy forest with sparse undergrowth, edge of bamboo-covered landslides in cloud forest and open parts of shady forest and woods.

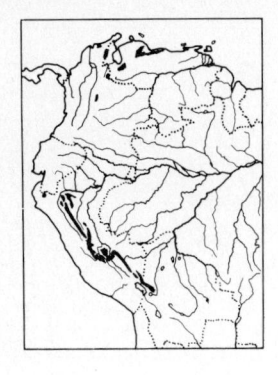

Range: 800-2200 m, rarely down to 250 m. Sta Marta mts, Col., Andes of c Mérida and w Lara, as well as coastal mts, Ven. (*ferrugineipectus*), Perijá mts and w slope of E Andes of Col. in Cundinamarca (*rara*). 1750-3350 m on Pacific slope in Piura and s and e of the Marañón from Amazonas, Peru to La Paz, Bol. (*leymebambae*). Local. Fairly common at some localities, territory c. 3500 sq m (coastal Ven.; see *Bol. Soc. Venez. Cienc. Nat.* 18(1957):42-62).

SLATE-CROWNED ANTPITTA *Grallaricula nana* – Plate XL 2

11 cm. Above dark brown, **crown dark gray**. 2 '**horns**' on the forehead, **eye-ring** and underside **bright rufous-chestnut**. Crescent across lower throat and c belly white; feathers of lower cheeks and underparts with varying amounts of narrow black edging. Ssp *occidentalis* and birds from n Peru are paler below. **Juv**. covered in vinaceous pink down; wing-coverts without rufous tips of juv. congeners.
Habits: In pairs. Actively hops and climbs about in bamboo thickets and on moss-covered trunks, searching each trunk for a half-minute, then flutters to next trunk; often makes short forwards sallies to foliage and stems, and occ. hawks insects from the air within a few cm of the perch. Frequently jerks and flicks wings. Sometimes on the ground, where it forages in the manner of a small thrush (*Catharus*). Not shy. Usually silent.
Voice: Song a soft, descending series of slightly buzzy ventriloquial whistles *bzree-zreee-zree-ree-ee-eeee* given during the 1st hour of the day (usually closer than it sounds). An emphatic *chep* given in response to playback of song.
Breeding: Fledglings Feb (ne Ecu.); juv./imm. Apr (ne Ecu.), Jun (Cajamarca); large gonads June (se Antioquia).
Habitat: Within 2 m of the ground on steep slopes in very dark parts of humid forest with moss-covered stems and sparse undergrowth.
Range: 700-2400 m. W Andes and w slope of C Andes of Col. (*occidentalis*), Andes of c Trujillo, Mérida, and Táchira, Ven., E Andes of Col. s to se Ecu. (*nana*). 2400-2930 m in Cajamarca (Chinguela), n Peru (*occidentalis?*). In Ven. 3 more sspp in the coastal mts and 1 in the Tepuis. Local.

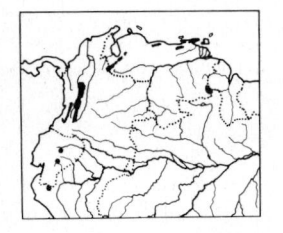

CRESCENT-FACED ANTPITTA *Grallaricula lineifrons* – Plate XL 3

11.5-12 cm. Cap slaty, **headside black. White frontal 'horns' continue down face to form a mustache**. Spot behind eye white. Back and wings olive-brown. Neckside from mustache buff, c throat and upper chin white, lower chin and sides of throat black, lower throat and breast buff, belly white, sides and flanks olive-brown, breast and belly streaked by black feather-edges, vent buffy. Outer webs of alula, and edge of outer primary whitish. **Juv**. Unknown.
Habits: Presumably much like others of the genus. Is attracted to strange sounds. Hops about within 1 or 2 m of the ground.
Voice: Alarm a series of hard notes *clip clip clip*.

Breeding: Gonads small Apr (Cauca).
Habitat: Thick underbrush in seepage areas in epiphyte-clad montane forest on fairly steep slopes, where landslides provide breaks in the canopy and permit smaller growths.
Range: Known from 2 localities: at 2500 m near Oyacachi ne Ecu., and at 3050 and 3220 m in Parque Nacional de Puracé, c. 7 km E Popayán Cauca C Andes Col. (see *Condor* 79(1977): 387-8). Rare.

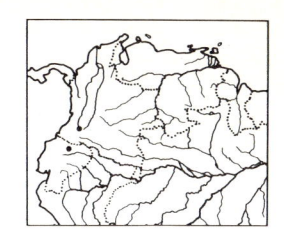

HOODED ANTPITTA *Grallaricula cucullata*

11 cm. **Bill orange. Head and throat bright orange-rufous**. Above olive-brown, below gray. Narrow white crescent across lower throat; c lower breast and c belly white. **Juv**. Unknown. - Usually alone. Said often to rock body from side to side without moving head and feet. No data on voice and breeding.

From near the ground to 1.5 m up in undergrowth of humid forest, at 1500-2700 m. Rio Chiquito, sw Táchira, Ven. (*venezuelana*, Phelps & Phelps in *Proc. Biol. Soc. Wash.* 69(1956):163); e slope of W Andes (once on w slope in Valle), e slope of C Andes in Antioquia, and at head of Magdalena valley in Huila, Col. (see *Condor* 79(1977):389) (*cucullata*). Common, but very local.

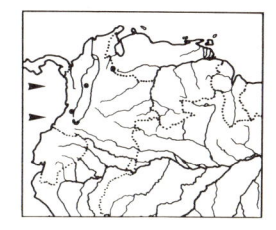

Gnateaters – Genus *Conopophaga*

8 species of small passerines found e of the Andes s to ne Arg., mainly in the trop. zone. They inhabit the undergrowth and forest floor, where they reveal their presence by whistles. They are rather compact, short-tailed, and long-legged, with an upright stance. Ad. males, and sometimes females, have a silky white postocular tuft which is gray or lacking in younger birds. The sexes usually differ. They are here placed in the antbirds for traditional reasons, but recent molecular data suggest that they are more closely related to tapaculos. See Bond (1953a).

SLATY GNATEATER *Conopophaga ardesiaca* – Plate XLI 12a-b

inhabits undergrowth of humid mossy forest at (800) 1000-1700 (2400) m, in e Cuzco, n Puno and s Madre de Dios, se Peru (*saturata*) and La Paz to Tarija Bol. (*ardesiaca*). 12 cm. **Male**: above dark olivaceous, feathers of back faintly margined with black; forehead slaty gray. **Long postocular tuft silvery white**. Sides of head and neck, and entire underparts gray to dark gray, sometimes with c abdomen white. **Imm**.?: browner above, wings brownish, greater wing-coverts narrowly tipped buff. **Female**: above reddish brown, forehead orange-rufous; below like male, but flanks and vent ochraceous brown. No postocular tuft. **Juv**. brown; crown with small, and back with large pale spots; below pale buff feather-centers give scaled appearance; c belly white. Sits quietly in the understory, making short sallies to snatch insects from the underside of leaves. Fledglings Nov (Puno).

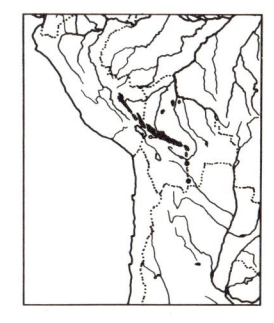

Tapaculos – Family Rhinocryptidae (Pteroptochidae)

Some 35 currently recognized species found in Costa Rica, Panamá and S Am., with the greatest diversity in the cooler parts of the Andes. Small to thrush-sized birds that inhabit dense thickets. Many are predominantly terrestrial, with strong legs and almost flightless. They live in pairs, presumably mated for life, and are almost entirely insectivorous. Most of them regularly cock the tail upright, but owing to their secretive habits they are more often heard than seen. The loud calls are stereotyped and frequently uttered. The nest is globular with a side entrance, and may be placed in a bush, tree, tussock of grass, or at the end of a tunnel.

Huet-huet and Turca Genus *Pteroptochos*

2 species. The nest is placed in a burrow.

HUET-HUET *Pteroptochos tarnii* – Plate XL 17a-b

22.5 cm. Appearance of a **large-bodied, short-tailed dark thrush**. Forecrown, supercilium, lower back and rump chestnut; rump with faint dark bars; wings dark brown; hindcrown, back, tail, throat, sides of head, neck, and breast slaty black. Rest of underparts rufous-chestnut, lower belly and flanks with blackish and pale rufous barring and squamations. Ssp *castaneus* has throat rufous-chestnut like breast. **Juv**. (*castaneus*) has the forecrown barred with black, headside mixed chestnut and black, greater wing-coverts tipped with black-bordered buff spots. Throat and breast chestnut, barred or mottled with blackish, remainder of underparts blackish brown, spotted and barred with whitish and pale cinnamon.
Habits: Mainly terrestrial, or walks on horizontal limbs and mats of bamboo 2-3 m above the ground. Skulks slowly, stopping often as it jumps and pecks at the ground or flips a piece of debris with its bill. Difficult to observe, usually managing to keep to the other side of a thicket when approached. See *Wildlife, London* 18(1976): 266-7.
Voice: Song, given from a perch up to 1 m above ground, a loud series of fairly low-pitched notes, given at 2 per s in irregular sequence of 1-4 at 1-2 s intervals: *huethuet-huethuethuet-huet-huethuethuethuet-...*
Breeding: Eggs (Oct) Nov-Jan.
Habitat: Dense undergrowth and bamboo-thickets in *Nothofagus* forest. Ecologically tolerant. Mature forest (both pastured and not pastured), second growth woodland, thickets in unforested areas, dense riparian brush amid narrow patches of low trees in shrub-steppe.
Range: Sea level-1600 m. From Colchagua s to Río Bío Bío, Concepcion, c Chile (*castaneus*), from Río Bío Bío s to Messier Channel, s Chile, and in extreme w Arg. from Neuquén to Sta Cruz (*tarnii*).

NOTE: Often divided into 2 species: **CHESTNUT-THROATED HUET-HUET** (*P.castaneus*), and **BLACK-THROATED HUET-HUET** (*P.tarnii*).

MOUSTACHED TURCA *Pteroptochos megapodius* – Plate XL 18

23 cm. Above gray-brown, forehead paler and browner, rump and tail dull chestnut, tips of rump-feathers barred blackish and whitish, feathers of crown with darker centers. Lores and earcoverts dark; **supercilium**, chin, upper throat, **and broad sides of throat white**. Lower throat and upper breast vinaceous cinnamon, remainder of underparts barred white, cinnamon, and blackish brown. Ssp *atacamae* is smaller and much paler, esp. below, where the rufous tinge is absent. Lower underparts mostly whitish with little barring. **Juv**. cinnamon to dusky above, without barring on rump; lower underparts vinaceous cinnamon like breast, flanks only very faintly barred with dark brown.
Habits: Alone or in pairs. Terrestrial. Usually cocks tail when running. Said to use feet for scratching debris in search of insects, but this needs confirmation, as no other tapaculo is known to use its feet in this manner. Skulking habits make it difficult to observe.
Voice: Song a gurgling crake-like *guerk-guerk-...* with 2-3 notes per s, accelerating and slightly descending in pitch towards the end. Call a series of 12 single notes, slowly descending in pitch in the beginning and the end, each note slightly rising, the notes 1 s apart *uur uur....* Alarm or contact note a single sharp whistle.
Breeding: Sep-Dec (Chile).
Habitat: Bush-covered hillsides and dense bamboo thickets and other undergrowth in temp. rain forest.
Range: Chile from sea level to 3050 m. Vallenar and Copiapó departments in Atacama, n Chile (*atacamae*); at somewhat lower elevations from Coquimbo to Concepción, c Chile (*megapodius*). Fairly common.

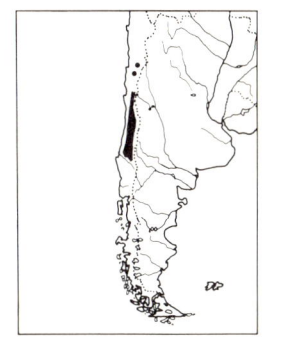

Tapaculos – Genus *Scelorchilus*.

2 species. The nest is placed in a burrow, often that of a rodent.

WHITE-THROATED TAPACULO *Scelorchilus albicollis* – Plate XLI 18

20 cm. Forehead vinaceous cinnamon, turning gray-brown on rest of upperparts; rump brownish, sometimes (younger birds?) slightly barred with blackish. Wings brown, wing-coverts tipped whitish and blackish. **Distinct supercilium white**, lores and earcoverts blackish. Below whitish, turning pale cinnamon on lower belly; **throat unmarked**, rest of **underside** and sides of neck **barred** dark brown, bars widest on flanks and vent. Ssp *atacamae* much paler, without brownish on upperparts; bill shorter. **Juv**. barred throughout.
Habits: Alone or in pairs. Terrestrial. Difficult to observe.
Voice: Song is a repeated phrase of 1-4, fairly low-pitched notes, the 1st note in each phrase higher than the others *uh-er uh-er-er uh-er uh-er-er uh-er-er uh-er-er-er*, the whole sequence lasting c. 6 s, and repeated at 4-5 s intervals. Call a short, harsh, pig-like grunt.
Breeding: Eggs Sep-Oct (both sspp Chile); fledgling Sep (Coquimbo).
Habitat: Bush-covered, rocky slopes.

Range: Chile from sea level to 1600 m in Andean foothills and parts of the coast range. Quebrada Paposa in sw Antofagasta and Atacama to n Coquimbo (*atacamae*), Coquimbo to Curicó (*albicollis*). Fairly common.

CHUCAO TAPACULO *Scelorchilus rubecula* – Plate XLI 17

19 cm. Above dark fuscous-brown; wings, rump and tail dark brown. Upper lores and supercilium rufous, lower lores to earcoverts fuscous, spot below eye white. **Throat and upper breast rufous**, rest of underparts dark gray barred white, flanks and vent rufous-brown to olive-brown. **Juv**. undescribed.
Habits: Alone or in pairs. Almost exclusively terrestrial. Runs with tail cocked. Difficult to observe.
Voice: The loud, resonant song is frequently given. It lasts 1-2 s, and is repeated every 4-5 s *cru-chuchuchuchuchu-ciu-r*, 2nd note accented, 1st and last 1-2 notes lower pitched. Also gives a higher pitched, neighing cry, and a low-pitched, croaking grunt.
Breeding: Eggs (Aug) Sep-Oct (Chile), Nov (Arg.).
Habitat: Seems confined to dense bamboo and other undergrowth of mature *Nothofagus* forest which has not been subject to heavy grazing. In such situations outnumbers the Huet-huet. However, also rec. in thickets in unforested areas.
Range: Sea level to 1500 m. From Bío-Bío to Aysén, Chile, occ. n along the Andean foothills to Colchagua. From Neuquén to Chubut, extreme sw Arg. (*rubecula*). A slightly larger ssp on Mocha, Isl Chile (*mochae*). Common in s part of range.

Gallito – Genus *Rhinocrypta*

Monotypic genus. The nest is globular with a side entrance. It is constructed of twigs, and placed 1 m above ground in a bush.

CRESTED GALLITO *Rhinocrypta lanceolata* – Plate XL 16

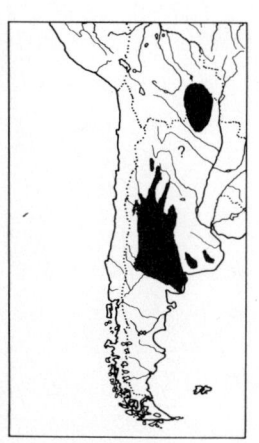

20 cm. Crown, headside and nape 'ground cinnamon' **streaked white**; feathers of crown elongated, forming a crest. Upperparts olive-gray. Throat pale gray, turning whitish on breast and c underparts; **sides of breast and body chestnut**; extreme sides of body, flanks, lower belly and vent olive-gray. **Juv**. undescribed.
Habits: Alone or in pairs. Runs with tail cocked. Does not fly. Forages on the ground and in low bushes. When disturbed runs, then hops onto a branch and clambers up the bush for lookout, giving alarm at 3-4 s intervals, which starts off other individuals nearby. Inquisitive, but much more shy outside breeding season.
Voice: A loud *chirrup*. Alarm a loud hollow *chirp*, and at intervals a violent scolding cry, repeated several times *prut prut prut*.
Breeding: Eggs Nov (Mendoza), Dec (Río Negro).

Habitat: Patches of rather heavy growth of open brush in arid regions. Second growth.
Range: Lowlands of Sta Cruz, Bol. and Parag. (*saturata*), and lowlands to 1800 m from Catamarca and Buenos Aires s to Río Negro, Arg. (*lanceolata*). Locally very common.

Gallito – Genus *Teledromas*

Monotypic genus. The nest is placed in a tunnel in a bank.

SANDY GALLITO *Teledromas fuscus* – Plate XLI 15

16 cm. Above pale cinnamon; tail blackish, c feathers like back. Supercilium white, underparts whitish. Crown and sides of breast washed with grayish, flanks and vent washed with pale cinnamon. **Juv**. undescribed.
Habits: Alone or in pairs. Terrestrial. Runs with long steps and cocked tail. Droops wings when curious.
Voice: Song, given at dawn and dusk from upper branches of a *Larrea* or other scrub, is a series of evenly spaced, far-carrying **tchowk** notes given at a rate of 4-5 per s (the first 1-2 notes less loud). The phrase counts 8-10 notes in Salta, 3-5 in Río Negro. An 'alternate' song (Río Negro) starts with 1-3 low, guttural croaks, then suddenly breaks into a loud, rising series of c. 8 ***quee*** notes slightly higher-pitched than the song. Also gives a series of 14 evenly spaced notes in 4 s, introduced by 2 low, trebled calls: *djiik, djiik, beer beer beer beert beert tchyeer tchueer tyeer teer teer teer tur tur tur* (descending in Río Negro; faster, first rising and then dropping sharply in Salta) (BW).
Breeding: Eggs Jan (Salta), Feb (Mendoza), Nov (Río Negro, Tucumán).
Habitat: Low open brush in arid regions, often far from water.
Range: Arg., Andean slopes to 3500 m from sw Salta to Río Negro.

Crescentchests – Genus *Melanopareia*

4 species inhabiting semi-arid regions in nw Peru and adjacent Ecu., e and c Brazil, Bol., and n Arg.

OLIVE-CROWNED CRESCENTCHEST
Melanopareia maximiliani – Plate XLI 14

15 cm. Tail fairly long. Above olivaceous. **Supercilium and throat** yellowish **white**; **sides of head**, connected to **band across breast black**. Rest of **underparts bright rufous-chestnut**. Ssps *argentina* and *pallida* generally paler, but the former has light cinnamon throat, against pinkish white in the latter. **Juv**. above heavily streaked black and cinnamon-buff, but back-feathers with broad olive-brown sides, grading to nearly uniform olive-brown. Facial pattern as in ad., but underparts white with dusky spots and bars across breast and on sides.

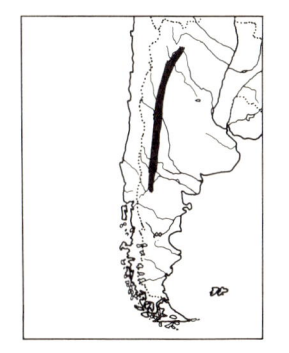

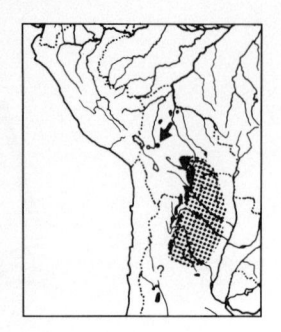

Habits: Insectivorous. Feeds on the ground or low in bushes, with horizontal body and tail. When disturbed will fly or climb into a bush, then rest, often giving a curious tilting jerk with the tail, sometimes cocking it upright.
Voice: Call a rapid *chit chuck*. Song a series of 3-4 (rarely 2 or 5-6) *chuck*s, lasting 0.5-1 s and given at 1-2 s intervals: *chuck-chuck-chuck, chuck-chuck-chuck, ...*
Breeding: Eggs Dec (Tucumán), nestlings Dec (Córdoba); juv. Apr (Cochabamba).
Habitat: Semi-arid regions, as well as landslides and second growth adjacent to humid forest. Heavy saw grass, brush, and low shrubs in woodland and savanna.
Range: 1700-2100 m in yungas of La Paz (*maximiliani*); somewhat lower elevations (but in Cochabamba up to 2950 m) from Cochabamba and Sta Cruz, Bol. and along base of Andes to Catamarca and w Córdoba (*argentina*); chaco of Parag. to Santiago del Estero and Córdoba, Arg. (*pallida*, Nores & Yzurieta in *Hist. nat. Mendoza* 1(1980):169-172). Uncommon.

Tapaculos – Genera *Eugralla*, *Scytalopus* and *Myornis*

Some 14 species found in dense humid thickets and undergrowth on elevated ground from Costa Rica to Isla Grande, in coastal mts of Ven., and in se Brazil and Misiones Arg. In spite of having extremely sedentary habits and being almost flightless, they have been found in Quarternary deposits in Cuba, and probably nested previously on the Malvinas Isls (see Woods 1988).

Wren-sized, partially terrestrial birds with large feet, and a tail that may be cocked. The species are extremely difficult to tell apart, even in the hand, and some appear to grade into one another. They are best distinguished by their voices, but often there is great local variation. It is not known whether this always represents dialects, or whether additional species are involved. (However, recent sequence data from DNA suggest a high number of full species; P Arctander). The plumage is light gray to blackish gray, usually with brown to rufous flanks that are barred or waved with blackish. Very old males of most forms may be identified by their size, relative length of tail, shade of gray, presence or absence of bars on flanks, white c forecrown, supercilium, and throat. However, most species take several years to attain this plumage. They may breed in imm. plumage, which is washed with brownish above and on the flanks, more or less barred or waved with dusky on wings, rump, tail, and flanks, and sometimes spotted or waved with buff on tips of inner remiges, most extensively on the tertials. The underparts are paler than in the ad., and there may be silvery feather-tips on the c lower breast and belly. Females resemble males, but are somewhat smaller and paler, and rarely attain full ad. plumage. Juv. brown to ochraceous, palest below, and either uniform, partially barred, or barred throughout with dusky or blackish. The succeeding imm. plumages show either a mixture of juv. and ad. feathers, or each feather may have alternating areas with ad. and juv. colors, producing a scaled or undulating pattern.

Alone or in pairs they forage in the darkest, dampest, and densest parts of the undergrowth, often on the ground, only reluctantly take to the wing and are extremely difficult to observe. Both sexes frequently give call and alarm notes, those of females slightly higher-pitched than males. The song is presumably only given by the male, but at least in some species there may be duetting. Both sexes incubate and tend the young. The globular nest may be fitted under a large tussock of grass, into a niche in a bank (dug by the bird?), a crevice in a tree, or placed in dense, tangled vegetation on or near the ground. The known nests of *Eugralla* were placed 1-2 (occ. 6) m up in a *Raphilthamnus cyanocarpus* or some other thick bush or small tree.

A thorough revision of the group was undertaken by JT Zimmer (1939) (*Amer. Mus. Novit.* 1044. See also Bond (1953a). On the basis of distributional sympatry, morphological and vocal differences we propose a slightly different classification, although we admit that the relationships of many forms are still obscure.

OCHRE-FLANKED TAPACULO *Eugralla paradoxa* –
Plate XLI 13a-b

14.5 cm (tail 58 mm). Maxilla elevated at base, in profile continuous with forehead. Above very dark gray, extreme lower rump cinnamon. Below gray, c lower belly pale gray, flanks cinnamon, abdomen and vent clay-colored. **Juv**. dark brown, everywhere barred with pale cinnamon to rufous, esp. on back, rump and tips of wing-coverts and inner remiges.
Habits: In pairs. Forages near the ground in dense thickets, following the same route every day with clock-like regularity.
Voice: Song a staccato series of hard *tek* notes at 5-6 per s, given at 3 s intervals, usually beginning with series of 2-3 notes, series gradually getting longer, 8 or more notes after a minute. Both sexes may sing simultaneously, female slightly higher-pitched. Alarm a rather low-pitched *eek* or *week* (BW). Contact call is a single sharp note, sometimes varied with a low clucking noise (given by female?) (Johnson 1967).
Breeding: 2 broods. Eggs Sep and Nov (Chile).
Habitat: Dense humid shrubbery and bamboo in forest or along streams.
Range: Coast and Andean slopes from s Santiago s to Chiloé Isl (incl. Mocha Isl), Chile, and at Lake Hess, nw Río Negro, Arg., and probably in adjacent Neuquén. Common in s part of range.

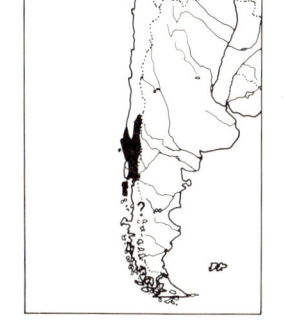

ASH-COLORED TAPACULO *Myornis senilis* –
Plate XLI 11a-c

14 cm. **Tail fairly long (7 cm)**. Above very dark gray; below gray, c belly light gray to pale gray; abdomen cinnamon. Female a trifle paler than male. **Juv**. virtually unbarred. Above brown (Brussels Brown), below ochraceous buff. Faint dark bars on rump, tail and tips of inner remiges; the flanks may be faintly barred and the loral region and c belly whitish. Alula edged ochraceous. No imm. plumage.
Habits: Alone or in pairs. Usually forages near (but not on) the ground, occ. 4 m up in thick foliage and tangles of bamboo. Tail often cocked.

Voice: Song up to 90 s long. It starts with a single very sharp note that is repeated at 5-10 s intervals; then these intervals become shorter, and the note is given in series of first 2, then 3-4, interspersed with single notes, finally turning into series of 3-5 hysterical, descending trills of 14-18 notes per s (sometimes starting as high as c. 7 kHz, falling to 5-3 kHz) that get successively shorter; sometimes 1st of these trills considerably longer than the others, and may only descend at the end (a similar trill up to 10 s long may be given alone). A single, descending trill may also be given without introductory notes, as may a variety of shorter trills and churrs at 3-4 kHz, some much like song of Andean T. (ssp. *affinis*), and best told by being descending or unevenly pitched. After playback of song may give a low-pitched (1.9-2.4 kHz) softer trill of c. 8 notes per s for 10-20 s. Alarm a sharp, 1 s *rikiki kirrr...* of 14-18 notes, falling at the end, much like the alarm of Narino T. (but lower). In response to playback a female gave a 1 s, dry, 22 per s trill, highest pitched in the middle, much like alarm of Andean T. (ssp *affinis*), but faster. (TP, TS, and NK).

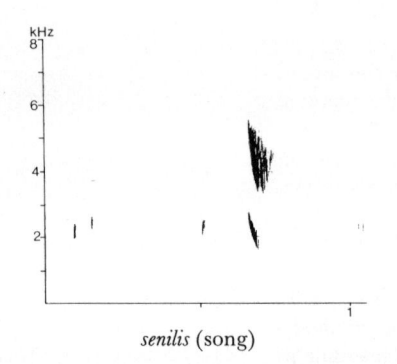

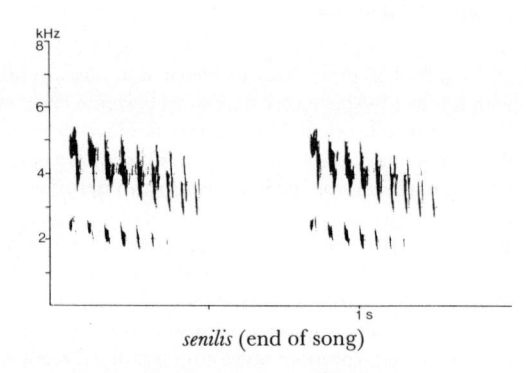

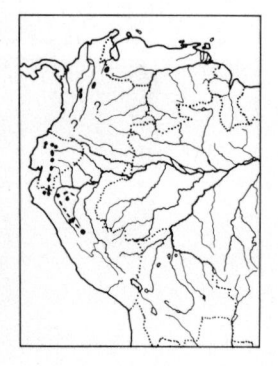

senilis (song)

senilis (end of song)

Breeding: Fledgling June (se Ecu.), Aug (La Libertad).
Habitat: Confined to dense bamboo thickets (*Chusquea*) and adjacent shrubbery on landslides in humid forest.
Range: 2300-3950 m. From Páramo de Tamá Ven. s through E and C Andes of Col., e and w Ecu., n Peru, and c and e Andes to Carpish mts Huánuco, perhaps to Cuzco. Common to fairly common.

NOTE: By some considered a long-tailed member of *Scytalopus*. However, the longer tail plus juv. plumage, bill-shape, and the large vocabulary, suggest a somewhat isolated position.

UNICOLORED TAPACULO *Scytalopus unicolor*

Ad. male generally uniform. At least 2 (partially sympatric) species involved. The allocation of the nominate form not yet definitively established.

latrans group – Plate XLI 3a-b

11 cm. Ssp *latrans*: **Uniform blackish gray**, female only very slightly paler. Birds from nw Ecu. and Antioquia Col. have bills as small as in Andean T., while the bill is sometimes fairly heavy in s Ecu. and n

Peru. No imm. plumage. Ssp *intermedius* very slightly paler, and with a fairly short, stout bill; female paler and often with silvery sheen on belly and with dark-barred, brown flanks. Ssp *subcinereus* has a long slender bill, male blackish gray, but female considerably paler and with dark-barred, brown flanks; imm. plumage present. **Imm.** (*subcinereus*) said to resemble juv./imm. *parvirostris*. **Juv**. *latrans* rather uniform; dark brown, slightly paler below, sometimes with obscure dusky waves on back and rump, and usually with bars on flanks. Juv. *subcinereus* light cinnamon-brown above with dark bars (bars sometimes absent from head and mantle), below dull ochraceous, in 2 males and a female with dark bars, in 4 females nearly without bars. Juv. *intermedius* said to have very contrasting barring throughout.

Habits: Relatively sluggish. Otherwise as for genus.

Voice: **Very low-pitched** (1-2 kHz). Call (song?) (all 3 sspp) given by both sexes, a soft, rising whistle *huy* or *huy-huy*, rarely *huy-huy-huy*, sounding like certain frogs, repeated every 3-10 s (in Ven. whistle sometimes slightly rough). Alarm a harder, repeated *uik*, during excitement given at increasing speed up to 8 times per s, sometimes in 0.5-1 s phrases at c. 0.5 s intervals. When faster than 2 per s, the quality changes to a harder, barking, somewhat thrush-like *chuck*. In the isolated Cutucú mts, se Ecu. alarm sharper, even more thrush-like than in the Andes, 1st note somewhat higher pitched, after repeated playback becoming a continuous series of 8 notes per s. The rec. alarm of *intermedius* a rather fast and rhythmic series of 2-4 notes. (BW, TS, and NK).

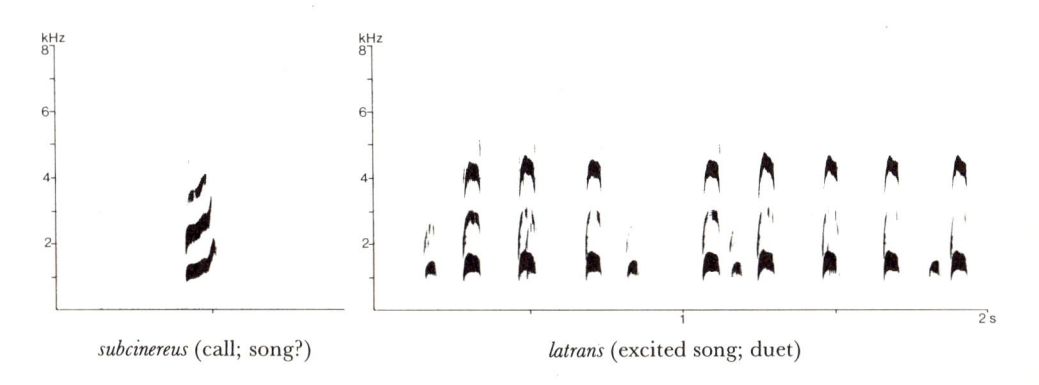

subcinereus (call; song?)   *latrans* (excited song; duet)

Breeding: Juv. Mar (Amazonas, Cauca), May (Piura), Mar, Apr, Sep, Oct (Mt Pichincha), May (Cotopaxi).

Habitat: Within 1 m of the ground in dense shrubbery and large dead bamboo-stalks in humid mossy forest and cloud forest. Sspp *latrans* and *intermedius* often in very swampy situations, *subcinereus* (at least sometimes) in fairly dry vegetation.

Range: 1500-3700 m. From Mérida, Ven. and all 3 ranges of Col. s through Ecu. (except sw) to Cajamarca (Chaupe; Lomo Santa; Chira ?), n Peru (*latrans*), from Azuay (Bestión), El Oro (Zaruma; Targuacocha; El Chiral) and Loja (Celica), sw Ecu. s in W Andes to s Cajamarca, n Peru, s of Porculla pass apparently only on w slope (*subcinereus*), and s of the Marañón (perhaps locally) in Amazonas (Cord. Colán: LSUMZ; La Lejía; Levanto; Chachapoyas; San Pedro; Leimebamba; Llui) and San Martín (Puerto del Monte: LSUMZ), possibly Huánuco (Bosque Unchog: LSUMZ) (*intermedius*). Fairly common in

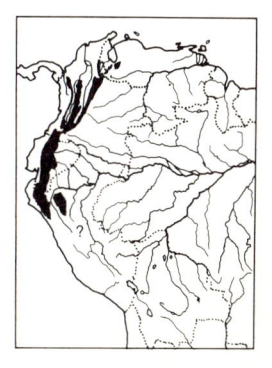

Col., very common in n Ecu. and wc Cajamarca, probably less common further s.

unicolor

11 cm. **Ad**. (both sexes) above gray to deep gray, below pale to light gray; rump, flanks, and vent with no or only a slight wash of light brownish. Most (younger, but some breeding) birds with silvery sheen of pallid gray on c underparts, esp. belly, and a brownish tinge on flanks; wings and tail with or without traces of barring. **Imm**. above brown, head-feathers inconspicuously tipped dusky, back scaled blackish and ochraceous, tail barred black and ochraceous subapically. Below dull ochraceous with grayish tinge, throat-feathers with fine dusky tips, breast and belly scaled with stronger dark tips and pre-subterminal spots or lunules, flanks barred ochraceous-brown and dusky.
Habits, **voice**, **breeding** and **habitat**: No information.
Range: 2000-3150 m. W Andes in s Cajamarca (Taulis, Chugur, Sunchubamba, Cajabamba) and La Libertad (Soquian, Succha, Huamachuco).

parvirostris – Plate XLI 4a-c

11 cm, bill fairly short and stubby. **Male** above deep gray; below slightly paler, gray to light gray (pale individuals often with throat palest), flanks with slight brownish and dusky barring. Younger individuals (?) frequently have silvery sheen of pallid gray feather-tips medially, esp. on belly, and a certain amount of brownish on wings and tail, with some barring. **Female** slightly paler than male, and usually

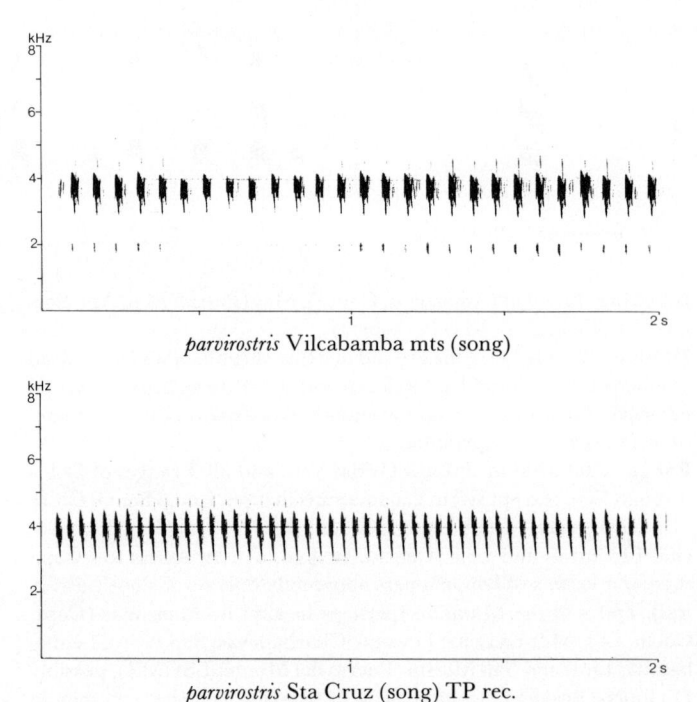

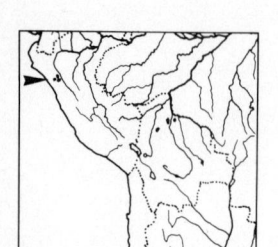

parvirostris Vilcabamba mts (song)

parvirostris Sta Cruz (song) TP rec.

washed with dark brown, esp. on wings, rump, tail and flanks, these parts being more or less barred with dusky. **Imm**. resembles female, or has remains of juv. plumage, either as scattered feathers, or as ochraceous spots, lunules, bars, or edges on the feathers, appearing scaled below.
Habits: As for genus.
Voice: Song (*parvirostris*) a 3-5 kHz trill of 12-16 notes (Cuzco) or 26-28 notes (Bol.) per s, either given once and lasting 10-15 s, or lasting 1-4 s and given in series at 2-9 s intervals. After playback the repeated trills get longer, 5-15 s long and occ. up to 2 minutes or more, a very long trill sometimes initiated with an irritated *kick*. Birds from e Cuzco (and Puno?) and Pasco also sing somewhat differently. (TP, TS, and NK).
Breeding: Juv./imm. July (Cuzco).
Habitat: Elfin forest and dense shrubbery and undergrowth in humid forest and cloud forest, at tree-line, on landslides or along streams.
Range: 1850-3200 m (in c Peru not found above 2500 m) s and e of the Marañón from Amazonas (Cord. Colán: LSUMZ) through c and se Peru to Sta Cruz, Bol. (*parvirostris*). Common.

NOTE: Local variation in pitch and speed may suggest that more taxa are involved.

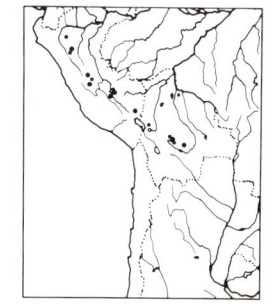

TAPACULO *Scytalopus* unnamed species – Plate XLI 5a-b

11 cm. Virtually indistinguishable from Unicolored T. (ssp *parvirostris*), but averages shorter-tailed (though with considerable overlap). The 4 males examined are darker and more uniform than *parvirostris*, and show no silvery sheen on belly. **Ad. male** dark gray above, gray below, flanks only with a slight wash of brownish and at the most with traces of barring. **Imm**. male can have dark-barred brown flanks and rump, being otherwise all **gray** (incl. **tail**). A juv./imm. probably belonging to this species resembles juv./imm. of *parvirostris*, but has a **browner tail** with more markings.
Habits: As for genus.
Voice: Call a monotonous series of 3-7 *keek*s at 2.5-3 kHz, lasting 1-2 s and repeated after 3-4 s pause. Song 2.5-3.5 kHz, an 0.5 s long, slightly fading trill of 6-9 notes with 1-2 louder, sometimes higher introductory notes *keek-krrrr* or *keek-keekrrr*, and repeated every 1-1.5 s, 4-10 times or more. After playback of song may prolong the trill to 1-1.5 s, sometimes

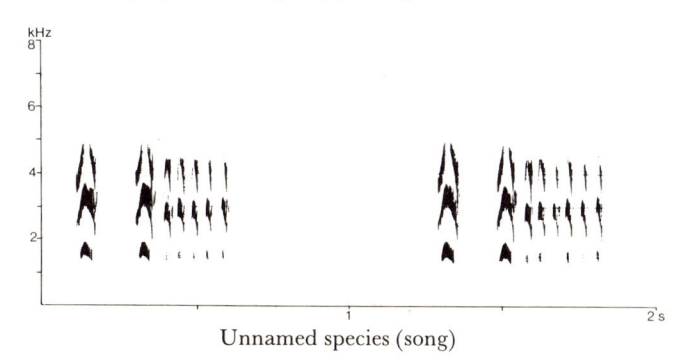

Unnamed species (song)

Unnamed species (song)

Unnamed species (call)

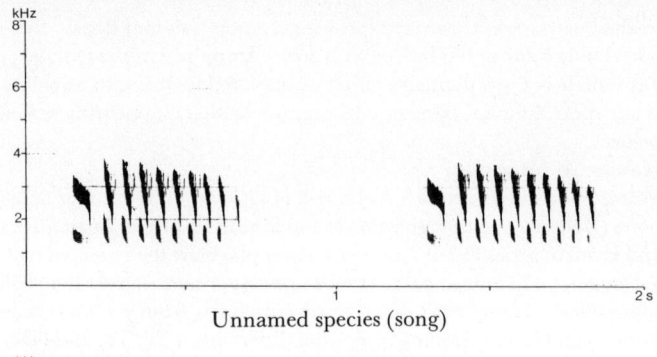

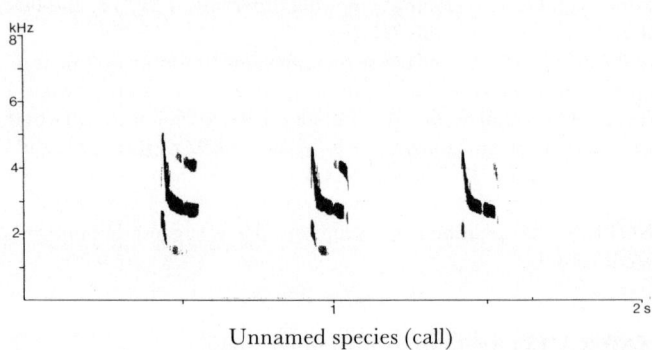

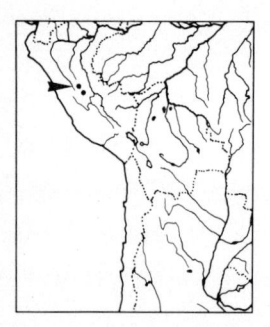

rising at first, and repeated after 1-2 s pause for up to 30 s or more. What may be a female call was a slowly descending and slowing series of notes (slightly higher pitched than in male) given in phrases of 1-4 notes. (TS and NK).
Breeding: Juv./imm. (species?) June (Huánuco).
Habitat: Dense shrubbery in cloud forest, only rarely in bamboo.
Range: 2675-3500 m. Presently only definitively known from Carpish mts Huánuco, and e of Panao in Pasco, c Peru. However, specimens without pale brow that have been treated under the name *acutirostris* have been taken in Junín (Maraynioc, Pariayacu). Fairly common.

NOTE: This puzzling species is being further investigated by NK and TS. The name *acutirostris*, hitherto used for c Peruvian populations of Andean T., could possibly represent the present form, but no material (with known vocalizations) comparable to the type (an imm. female) exists.

LARGE-FOOTED TAPACULO *Scytalopus macropus* – Plate XLI 10a-b

14 cm. Uniform blackish gray. Feet very large. Juv. everywhere irregularly barred, scaled and waved cinnamon-rufous and blackish; belly with pale buff rather than cinnamon-rufous.
Habits: As for genus. Perhaps primarily terrestrial. Shy.
Voice: Song given throughout most of the day, a monotonous series of low-pitched (1.5-2 kHz) notes given at 0.3 s intervals, sometimes lasting up to a minute or more, and usually finished off with a different (normally higher) note. (NK).

macropus (song)

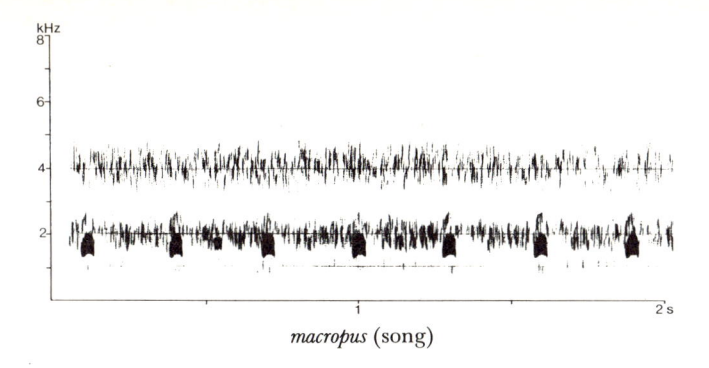

Breeding: Large gonads Aug (Huánuco).
Habitat: Dense thickets in mossy cloud forest and elfin forest, often along streams.
Range: Peru at 2400-3500 m s and e of the Marañón from Amazonas to Junín. Local and uncommon.

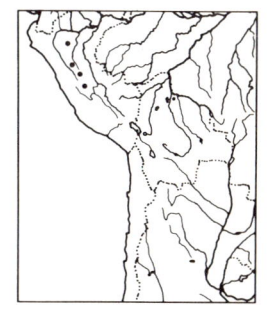

RUFOUS-VENTED TAPACULO *Scytalopus femoralis*

Fairly long-tailed and stout-billed. Mainly premontane. At least 2 (partially sympatric) species are involved, but all the 4 groups listed may deserve species rank, and perhaps even further division may be necessary. A systematic revision of these forms is being undertaken by TS.

femoralis group – Plate XLI 7a-b

14 cm. **Long-tailed, with rusty lunules on flanks**. Above dark neutral gray, below slightly paler (neutral gray). Flanks cinnamon-brown with fairly broad dusky lunules and wavy bars, c belly often with silvery or pale mouse-gray feather-tips. A single specimen (type !), apparently a very old bird, has brown of flanks rather restricted, and with only a few straight rather than wavy bars. **Juv**. brown above (dark Brussels Brown or Prout's Brown), light ochraceous buff below; flanks reddish cinnamon brown to ochraceous tawny; upperparts faintly and underparts (esp. flanks) distinctly barred with dusky. **Imm**. above darker brown than juv., below with feathers showing waves of alternating ad. and juv. color, giving scaled appearance; tail-feathers blackish with narrow pale subterminal bar. Most ad. have retained some dark brown barred dusky on tertials, rump and upper tail-coverts.
Habits: As for genus.
Voice: Song (*femoralis*) a single, resonant, 1.5-3 kHz note (actually composed of 2 notes) repeated endlessly at 0.3-0.7 s intervals *chuock*. Song (*micropterus*) starts similarly, then each note becomes double, occ. and irregularly triple, with the last 1 or 2 notes sharper and alike *chu-ock* or *chu-ock-ock*. (TP, BW, NK, and OF Jakobsen).
Breeding: Juv. (species?) Jan (ne Ecu.); juv./imm. July (Junín).
Habitat: Humid forest undergrowth and shrubbery at forest edge, along streams, clearings etc.
Range: 800-2400 m, very rarely to 2950 (3150?) m, primarily 1500-2000 m. In Col. only definitely known from Huila (at head of

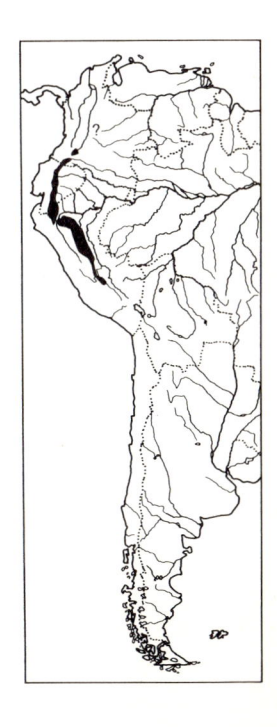

femoralis (song)   *micropterus* (song) TS rec.

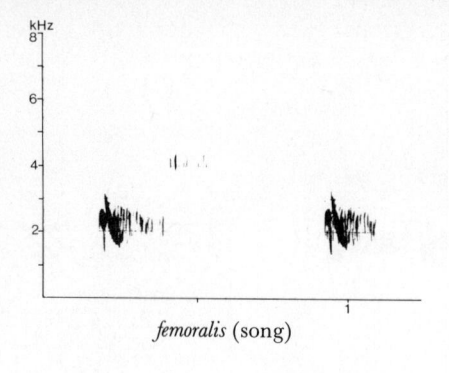

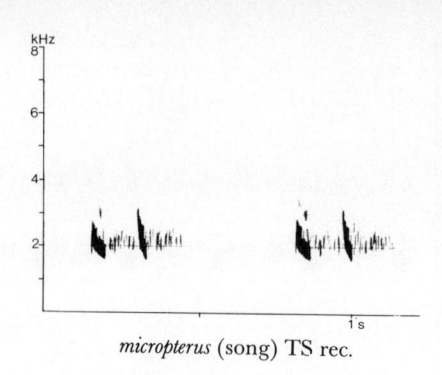

Magdalena valley), e Nariño and Putumayo, but old 'Bogotá' specimens may suggest occurrence in e Cundinamarca. Thence s through e Ecu. to n Peru n of the Marañón (*micropterus*), s and e of the Marañón from Amazonas to Ayacucho, s Peru (*femoralis*). Common.

atratus group – Plate XLI 9

12-13 cm incl. 5 cm tail. **Usually with white crown-patch**. Tail shorter than in *femoralis* group. Ssp *confusus* slaty-black above appearing blackest on forehead, patch on c crown white (white possibly sometimes absent); lower back tinged brownish; rump, flanks and vent reddish brown barred black. Below dark gray, belly usually with conspicuous whitish feather-tips. Most ad. have retained some dark brown barred dusky on tertials, rump and upper tail-coverts. Female somewhat paler than male, often tinged brownish above, and with a smaller and duller crown-patch. Sspp *atratus* (known from 2 males and 1 or 2 females) and *nigricans* (known from a male and a female) more blackish, with a larger crown-patch. Flanks with hardly any rufous barring. Ssp *atratus* with more slender bill, and sometimes with contrasting whitish (male) or grayish white (female) throat. In *nigricans* the male has throat-feathers whitish with dusky tips, the female grayish white chin and throat merging with gray of rest of underparts. **Juv**. heavily barred and scaled throughout, showing a pale c fore-crown.
Habits: As for genus.
Voice: Birds from Finca Merenberg in Huila, Col., Cutucu mts, se Ecu., and Jirillo in San Martín, Peru: a monotonous series of 20-30 sharp 2-3 kHz notes, 5-6 per s, at irregular intervals of 2-20 s. Cutucú

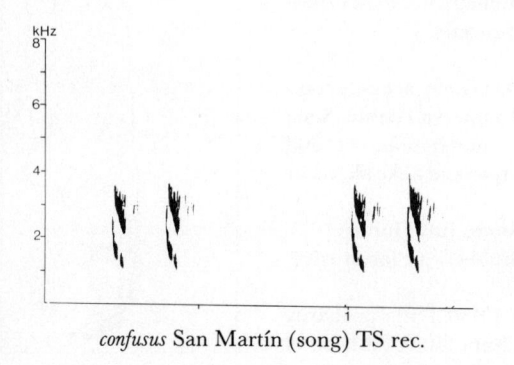

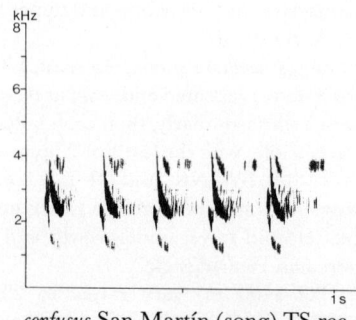

confusus San Martín (song) TS rec.   *confusus* San Martín (song) TS rec.

mts (TS): a similar quality, 0.5-0.6 s series of 4-5 notes at 2.4 kHz given every s. Jirillo, San Martín, Peru (TW): similar quality 2-3 kHz notes, usually at 5-6 per s. Very irregular series of 1-10 notes are given at 0.5 s intervals (sometimes 1-3 s), at peak of song up to 25 notes in a series, then the series get shorter, irregularly. After playback speed may increase to 8 per s.
Breeding: No data.
Habitat: Undergrowth of humid mature forest at forest edge, streams, clearings etc.
Range: 1150-1900 m in Perijá mts Zúlia (and sw Táchira?), Ven. (*nigricans*, Phelps & Phelps in *Proc. Biol. Soc. Washington* 66(1953):7), e slope of E Andes of Col. in w Casanare (altitude?) (*atratus*); 1050-2200 m (rarely to 2600 m?), primarily 1100-1800 m in Cauca and Magdalena valleys, and e slope of E Andes of Col. from Cundinamarca s through e Ecu. to Cuzco, se Peru (*confusus*). Fairly common to uncommon.

bolivianus – Plate XLI 8

11.5 cm. Like *atratus* group (ssp *confusus*), but slightly smaller, and with a better developed crown-patch; belly without silvery feather-tips. Some birds from s Peru perhaps intermediate.
Habits: As for genus.
Voice: Song a fairly low-pitched (1.2-2.5 kHz note repeated continuously about 5 times per s, at increased excitement up to 8 or 12 times per s (a trill) for up to 15 s. At alarm the trill may be given in series 3-5 s long at 3-5 s intervals, each phrase slightly accelerating. (RA Rowlett).

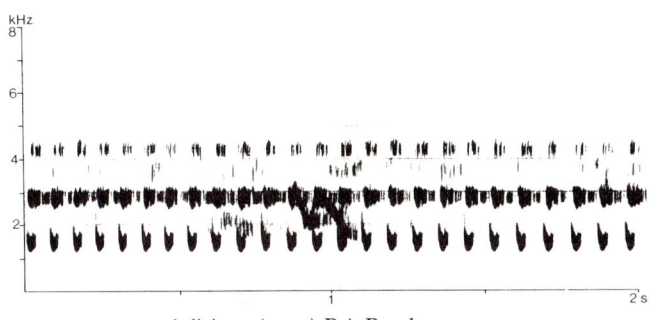

bolivianus (song) RA Rowlett rec.

Breeding: No data.
Habitat: Undergrowth of humid mature forest at forest edge, streams, clearings etc.
Range: 1200-2150 m. From n Puno, se Peru to Sta Cruz, Bol. Uncommon.

sanctaemartae – Plate LXIII 29

11.5 cm. Tail fairly short. Like *atratus* group (ssp *confusus*), but size smaller, crown-patch smaller, bill slightly slenderer, general coloration much lighter gray. Rump and flanks brighter and more extensively rufous with broad wavy dark bars. **Female** paler below, upperparts washed with brown, crown with only a trace of a patch. **Juv**. like juv. of

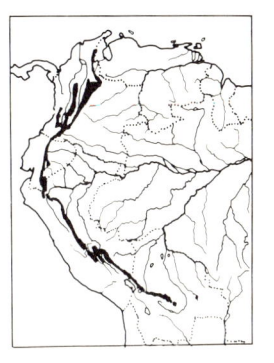

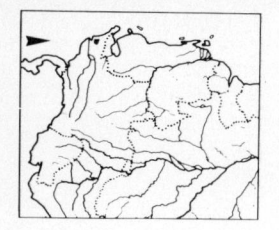

atratus group, but much less rusty below, esp. on throat and breast.
Habits: As for genus, (keeps on ground?).
Voice: 1 rec. gave a rapid trill lasting 7 s, but when excited up to 12 s long. (S Hilty).
Breeding: Juv. July.
Habitat: Dark tangled ravines and undergrowth in heavy forest.
Range: Col. at 1350-1700 m in Sta Marta mts. Fairly common.

NARIÑO TAPACULO *Scytalopus vicinior* – Plate XLI 6

12.5 cm. Type (ad. female) very dark reddish brown, brighter on the dark-barred rump; forehead and brow slightly tinged grayish; head-side, throat, and breast light gray in contrast to the dark brown belly; lower sides, flanks, and under tail-coverts deep Argus Brown barred blackish, c belly pinkish cinnamon. Birds from Ecu. (2000-2800 m): **male dark gray** above, rump and upper tail-coverts and sometimes nape dusky brown, lower rump and upper tail-coverts often slightly barred. Below slightly paler, belly usually with gray feather-tips, broadest and palest on lower belly; lower sides dark brown with dark lunules (at least posteriorly) turning reddish brown with rather narrow straight bars on flanks and under tail-coverts; **vent light orange-brown with a slight ochre hue, unbarred**. Female smaller, somewhat paler, and more extensively brown above.
Habits: As for genus.
Voice: Song at type locality (1800-1850 m; D Willis & F Lambert rec.) and P.N.Farallones on the Pacific slope in Valle at 1950 m (BW rec.): one song, given every 6 s for half a minute or more is a 2.5 s 3.1 kHz trill

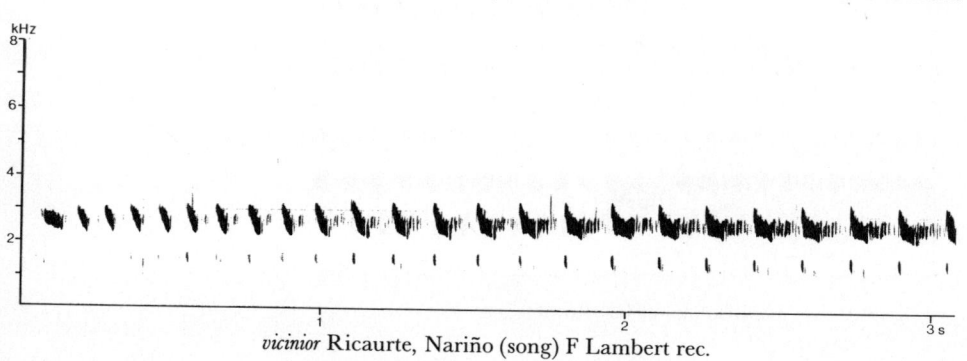

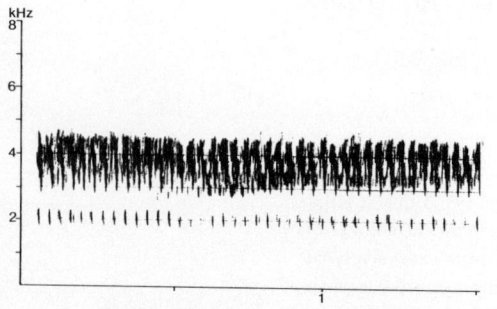

vicinior Ricaurte, Nariño (song) F Lambert rec.

"vicinior" Pichincha (song) TS rec.

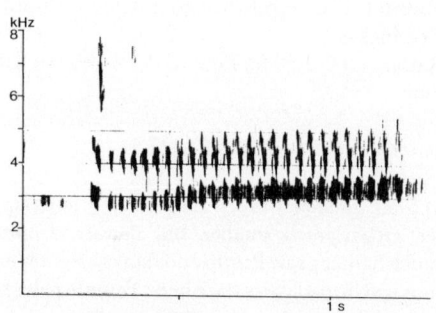

"vicinior" Pichincha (excited) TS rec.

(with a fairly loud 1.5-1.6 kHz fundamental), starting slightly higher (3.2 kHz) at 12-13 notes per s, and slowing gradually to 10-9 notes per s while increasing in amplitude, and ending abruptly. Another similar song is 15-20 s long, steady in amplitude and speed (10 notes per s) after c. 5 s, last s slightly slower and louder. Distress calls include a 3.5 kHz *ki* every 6-7 s, a *ke ki ki*, and a *kekikikikike*. Birds from Ecu. (TS and NK): alarm very similar to that of Ash-colored T., but higher pitched, a 1 s long trill at 3.5-4 kHz, rising at beginning and falling at end, and of 14-15 notes (*ke*) *cirrrrr*, given at 2-4 s intervals. Song a long, very fast trill (28 notes per s) at 4 kHz, lasting up to 20 s; another bird may simultaneously give a short, distinctly rising, 2.7-3.8 kHz trill of 26-30 notes per s, lasting 0.5-1 s, and repeated every 2 s; also gives a slower, lower 5 s long trill, or a similar but 2.5 s long trill repeated every 5 s. Distress call (of female ?) an explosive sharp *brzk*, sometimes in a descending series of 5-6. Call a 4 kHz note repeated every 1 s.
Breeding: No definite data.
Habitat: All 3 forms in shrubbery and undergrowth in humid forest, only rarely in bamboo.
Range: Not well known. The *vicinior* type was taken at 1500-1800 m at Ricaurte, w Nariño, Col. Birds rec. here vocalize similarly to birds at 1950 m at P.N.Farallones on the Pacific slope in Valle, Col.; a bird from 1525 m in w Valle (Las Lomitas) and 2 from Mayasquer at 1465 and 2380 m in w Nariño probably also belong here. Another form previously referred to *vicinior* is rec. at 1435-1525 m on mt Pirre in Panamá, at 650 and 1050 m on Pacific slope in Valle Col., and at 670 m in Esmeraldas nw Ecu. A bird taken at 350 m in w Nariño (La Guayacana) and one from 450 m on Pacific slope of Antioquia presumably also belongs here, while one from Ricaurte at 1200 m may be typical *vicinior*. A third form has been taken in ne Ecu. near the Col. border, at 2500 m, and in nw Ecu. at 2300-2800 m in Imbabura (a specimen from 3200 m may also belong here), and at 2000-2700 m in w Pichincha. Specimens from Ricaurte at 2000, 2200, and 2500 m may belong here or with *vicinior*. Other Col. specimens that have been referred to *vicinior*, but whose vocalizations are not known, are from W Andes in Antioquia at 3700 m (Paramillo), e slope of W Andes in Cauca at 1800-2000 m (Cocal, San Antonio), w slope of C Andes in Quindío at 1500-2150 m (Salento, Laguneta), e slope of C Andes in Tolima at 2225 and 2600 m (Toche, El Eden), and w slope of E Andes in Huila at 2300 m (Buena Vista).

NOTE: May include at least 3 species, differing only slightly morphologically, but with very different vocalizations. The relationships of these forms with *panamensis*, *confusus*, *meridanus*, and *spillmanni* need investigation.

BROWN-RUMPED TAPACULO *Scytalopus latebricola*

Flanks bright rufous with little or no barring. Probably several (allopatric) species involved, each ssp group dealt with separately below.

latebricola – Plate LXIII 30

11 cm. Bill elevated basally and compressed laterally. Above dark gray,

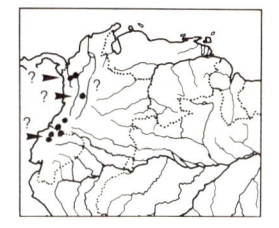

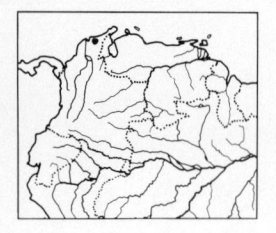

below gray. Rump, flanks and abdomen relatively bright, rufous-chestnut, either uniform or narrowly barred with blackish. **Juv**. brown, each feather black centrally, giving squamate or barred appearance.
Habits: As for genus.
Voice: No data.
Breeding: Juv./imm. Mar.
Habitat: No information.
Range: Col. at 2150-3650 m in Sta Marta mts. Fairly common.

meridanus – Plate XLI 2a

10 cm. Like *latebricola*, but slightly smaller, distinctly paler, often with a silvery sheen on belly, and with a much smaller bill that is not elevated basally. Upperparts washed with brownish even in ad. plumage. Both bill and plumage generally resemble *S. magellanicus griseicollis*, but *meridanus* has slightly darker rufous flanks with a few wavy bars, and a paler

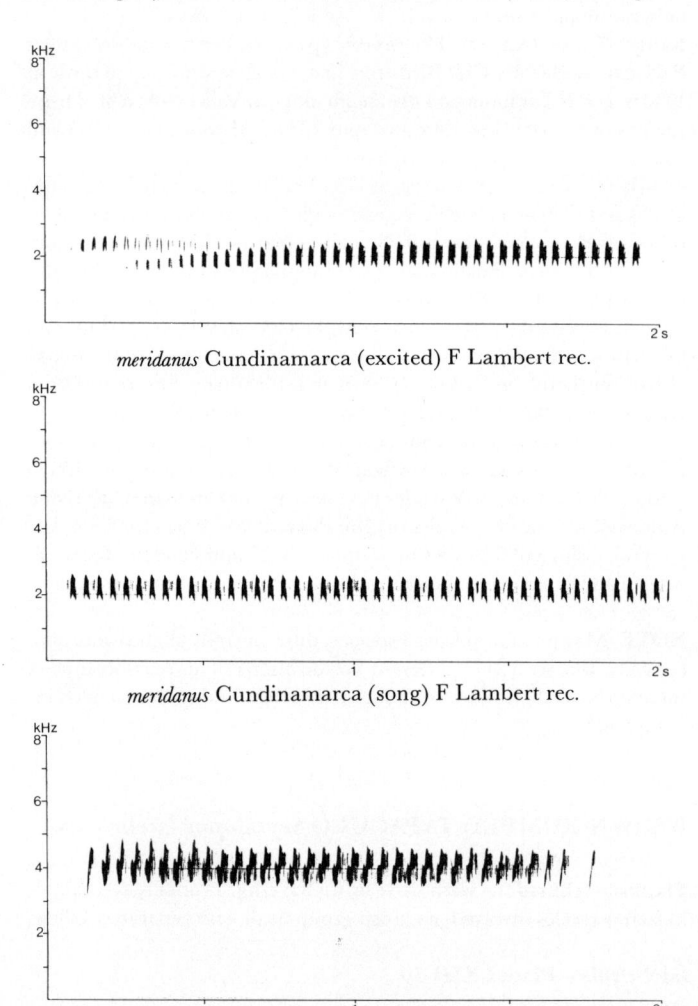

meridanus Cundinamarca (excited) F Lambert rec.

meridanus Cundinamarca (song) F Lambert rec.

meridanus Táchira (song; duet?) BW rec.

belly. Birds from Páramo de Tamá and C Andes of Col. are paler brown above and have larger bills, very much resembling *S. magellanicus fuscicauda*. **Juv**. Above brown, below light brown. Crown, back and throat with narrow dark feather-tips, rump and entire underparts with broader dark bars, appearing scaled on upper flanks.
Habits: As for genus.
Voice: In Ven. a 4 kHz, 1.5 s long trill of c. 20 notes per s, and a 4 kHz, 15 s long trill of c. 10 notes per s, but slowing at the end (S Hilty, BW). In Cundinamarca a low (2-2.5 kHz), rapid (24 per s) trill of c. 10 s (BW & F Lambert). At excitement may repeat a 2 s long slightly lower pitched but rising trill at 2 s intervals (F Lambert).
Breeding: Juv. June (Mérida).
Habitat: Bamboo thickets (Chingaza, Cundinamarca) and treeline shrubbery (not bamboo; Mérida).
Range: 2000-3650 m (4000 m?) from Lara to Táchira, on Páramo de Tamá, Ven., and in E Andes and e slope of C Andes of Col. A specimen taken at 3700 m at Paramillo in Antioquia, W Andes, may belong here (see under Nariño T.). Fairly common (at least locally).

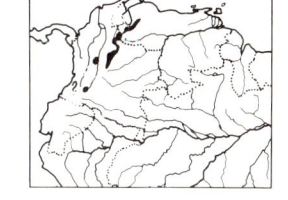

spillmanni group – Plate XLI 2b

12 cm. Bill not laterally compressed, but fairly broad at tip. Size larger and bill smaller than in *latebricola*. Also much like Narino T. (Ecu.), but slightly smaller, bill smaller, upperparts more extensively brown, and lower belly with a less ochre hue. Sexes alike, female barely averaging smaller. **Ad.**: Crown and mantle dark gray; nape, wings, and lower back dark brown; rump and upper tail-coverts reddish-brown with dark bars (in the grayest specimen seen, brown of wings is confined to tertials, and rump is unbarred). Tail dusky, feathers more or less edged with dark brown. Below gray to deep gray, belly with slight silvery sheen. Lower belly Ochraceous-Tawny, flanks and vent dull reddish brown with a slight olive hue, usually with dusky bars, but sometimes virtually unbarred. **Imm.**: A male in transitional plumage has remnants of bright dark reddish-brown and black scaling/barring on head-side, crown and nape, dark reddish-brown edges to remiges and wing-coverts, black vermiculations on wing-coverts and tertials, and buffy to pale buff bar near tips of inner remiges, secondary-coverts and alula. The tail shows traces of barring near tip, flanks are distinctly scaled, and belly with more contrasting and paler silvery feather-tips than in adult.

The above description is based on a series of an unnamed ssp from Zapote Najda mts, Morona Santiago, se Ecu. Another unnamed ssp from c Loja, s Ecu. and (s to) Chinguela mt, Piura, n Peru, has nape and most of wings gray, brown of wings confined to tertials, and the dark brown rump unbarred (upper tail-coverts barred as in other forms). The flanks are unbarred or only very faintly barred, and the lower belly more orange (between Ochraceous-Orange and Tawny) than in the Zapote Najda ssp. When viewed from above the bill tapers fairly evenly through most of its length, whereas it tapers fairly abruptly on the basal half and averages smaller in Zapote Najda birds. The type of *spillmanni* (a female) from mt Illiniza Pichincha/Cotopaxi, and a similar specimen (mt Pichincha?), resemble Loja-Piura birds in bill shape and barring, but have a warmer color of lower belly (Tawny), and are intermediate in amount of dark brown on nape and wings.

Habits: As for genus, perhaps spending less time on the ground.
Voice (both Zapote-Najda and Loja (-Piura) birds): Alarm (song?) a 3-4 kHz sharp trill of 10-12 notes per s, descending at first, lasting 1-2 s, and repeated at 1-8 s intervals. The trill may get faster (19 notes per s), lower (2.5-3.3 kHz), and repeated without intervals every 0.5 s for 10-20 s, or may be given for 5-60 s or more as one continuous trill (slower than song of Andean T., ssp *opacus*). During a duet male sustained a 19 per s, 2.2-3 kHz trill for over a minute, while female broke in with a 2 notes per s, descending series of 5-4 kHz notes (the first 2 or 3 very explosive and up to 7 kHz) and some alarm-like trills. (NK).

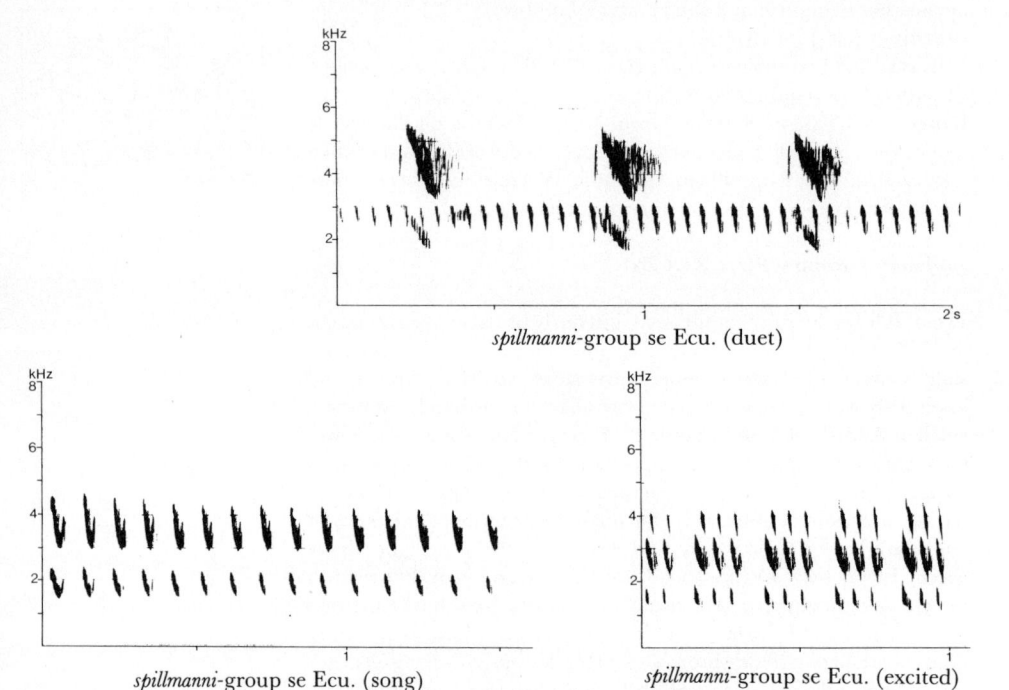

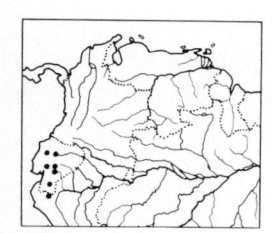

spillmanni-group se Ecu. (duet)

spillmanni-group se Ecu. (song)

spillmanni-group se Ecu. (excited)

Breeding: Imm. June (Morona-Santiago).
Habitat: In n Peru and se Ecu. found in bamboo, esp. tangles of dead stalks and foliage within 1 m of the ground, but occ. in branches and leaves as high as 2 m, and in adjacent dense shrubbery of other plants.
Range: 2200-3200 m. W Ecu. (on mt Pichincha ?), Pichincha/Cotopaxi (mt Illiniza), (and Chimborazo?) (*spillmanni*); e Ecu. in (Tungurahua?), Morona-Santiago (Zapote Najda mts) (ssp); c Loja (PN Podocarpus) se Ecu., and n Peru in Cajamarca (Chinguela) (ssp). Locally fairly common.

NOTE: The different habitat and vocalizations of birds from Mérida and Cundinamarca may suggest that more than one taxon is involved under the name *meridanus*. 2 specimens from ne Ayacucho s Peru (LSUMZ) may belong in this assembly. The subtrop. (1600-1900 m) and vocally very distinct ssp *caracae* in coastal mts of Ven. is undoubtedly a distinct species, as may also be the case with other sspp. The relationship between *meridanus* and the 2 sympatric sspp of Andean Tapaculo may deserve further investigation, as may also the relationships of *meridanus*, *vicinior*, and '*spillmanni*' from e Ecu.

ANDEAN TAPACULO *Scytalopus magellanicus* – Plates XLI 1a-m and LXIV 31

This assembly is composed of **the smallest and at the same time elevationally highest-ranging members of the genus**. There are some common tendencies in plumage texture, and in certain lights the lore appears blackish. Also most forms tend to have extensive, **straight and narrow bars on the flanks** (and usually rump and tail), and in juv. to have straight, narrow and fairly dense bars throughout.

Habits: As for genus. Largely terrestrial. Locally comes into open country, constantly hiding in rodent burrows, among tussocks of long grass and under rocks.

At least 2 (partly sympatric) species involved. The many allopatric sspp differ significantly from each other vocally and often morphologically, and are (at least in part) perhaps best treated as allospecies of a single superspecies. Awaiting biochemical analysis and further information on vocalizations and plumage variation, we have here treated each form separately. Recent DNA sequencing (P Arctander) suggests that the genetic variation in this 'species' is as high as in the rest of the genus.

magellanicus – Plate XLI 1l

10 cm. Dark gray, rarely blackish gray, usually with silvery white forecrown. Tail and flanks cinnamon-brown to grayish brown, narrowly barred dusky. Very old males have almost unbarred flanks, but tail usually still barred. **Juv.** cinnamon-brown to grayish brown, everywhere narrowly barred dusky.

Voice: Song *pa-trás pa-trás...* (Johnson 1967) or *pe-tjeh pe-tjeh...* (rec. by BW). In Isla Grande *cho-rín cho-rín...* (Clark 1986), and at higher excitement *chíuriuriuriú*. Alarm (Johnson) a harsh *chó-co chó-co...*, and a House Wren-like, dry, short trill.

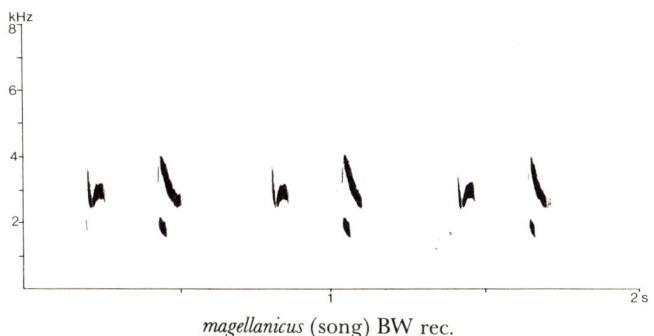

magellanicus (song) BW rec.

Breeding: Eggs Oct-Nov (Chile); juv. Nov (Río Negro), Mar (T.d.Fuego).

Habitat: In winter in dense, humid *Nothofagus* shrub, in summer mainly among rocks in humid grassland.

Range: From Isla Grande (previously also Islas Malvinas) n in the Andes at least to Neuquén Arg. and Santiago (Aconcagua?) Chile. In part of range sympatric with *fuscus*, judging from vocalizations.

fuscus – Plate XLI 1m

11 cm. **Blackish gray**, passing into silky blackish on c forehead and crown, flanks almost or wholly unbarred. Forehead white in some birds (both sexes). Bill and legs said to be stronger, and tail longer than in *magellanicus*. One specimen that is gray tinged brownish throughout may be **juv**, others are grayish brown and barred, resembling *magellanicus*.
Voice: Song an 0.3 s trill, rising almost one octave at 1-2 kHz (with a loud 1st harmonic at 2-4 kHz), repeated every 0.8 s *trrrrui trrrrui* ... (G Egli and BW). Alarm a House Wren-like, dry, short trill.

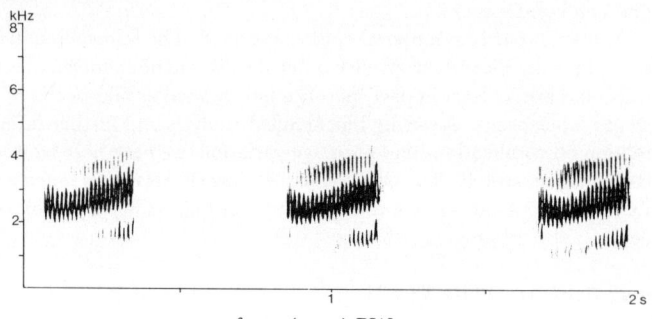

fuscus (song) BW rec.

Breeding: Eggs Oct-Nov.
Habitat: Dense undergrowth in dark ravines, generally in darker and moister situations than *magellanicus*.
Range: At least to 950 m, in mts of Atacama to 4000 m. From Río Bío Bío n to Atacama Chile and in Mendoza Arg. Sympatric with *magellanicus* at least from Bío Bío to Santiago (possibly to Aconcagua) (based on vocalizations). An additional taxon may be involved.

superciliaris – Plate XLI 1k

10.5 cm. **Ad: upperparts olivaceous-brown**, rump and tertials barred dusky, **supercilium and throat white**, headside and underparts gray, flanks and vent barred buffy and dark brown. **Juv**. Undescribed.
Voice: No data.
Breeding: Eggs Nov-Dec (Tucumán); nestlings Jan (Tucumán). Nest at end of 30 cm long, slightly turned tunnel.
Habitat: Undergrowth of alder forest (*Alnus acuminata*) in humid ravines.
Range: Nw Arg. at 1500-3350 m from (Catamarca?) Tucumán to Jujuy.

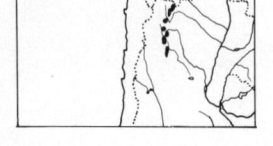

santabarbarae

Resembles *superciliaris*, but noticeably darker, blackish below.

Range: At 2000-2400 m in Sierra Santa Bárbara in extreme se Jujuy, nw Arg. (see M Nores in *Hornero* 12(1986):262-273).

zimmeri – Plate XLI 1j

10.5 cm. Ad?: upperparts brown tinged gray; tertials and tail barred with dusky; **supercilium pale gray**; **chin pale gray**, turning dark gray on rest of underparts; flanks cinnamon, barred blackish brown. **Juv**. undescribed.
Voice: No data.
Breeding: Nestlings Oct (Tarija).
Habitat: Alder forest (*Alnus acuminata*).
Range: Bol. in Tarija (3000 m) and Chuquisaca (2500 m).

simonsi – Plate XLI 1i

10 cm. **Male** above deep gray usually with a faint brownish wash. **Supercilium** above and behind eye **pallid gray to light gray**. Below gray to deep gray, sometimes lightest on chin (some specimens virtually indistinguishable from *zimmeri*). Flanks and rump dull cinnamon with dense, straight dark bars. Tail dark brown with irregular dark barring, rarely all dark. **Female** averages slightly lighter than the male. **Imm./juv.** lighter, more heavily washed with brown above, wing-coverts barred, tertials and tips of inner remiges with irregular light and dark markings, underparts barred with pallid gray feather-tips (at least on sides and upper belly), barring on flanks, rump and tail more extensive.
Voice: Song a repeated, 0.2 s churr of 6-8 notes at 2-4 kHz, usually with a louder, 3-4 kHz first note, churr evenly pitched in Bol., rising in Peru, *tírrr*, given twice a s. Alarm (female?) a rather high-pitched series of 4-5 notes, the first 2 or 3 highest (3.3-3.8 kHz) and alike, the last 1 or 2 lower (3-3.4 kHz) and alike *kikiki ker*, *kikiki kerker*, or *kikikerkr*, sometimes the notes become lower gradually. (TS and NK).

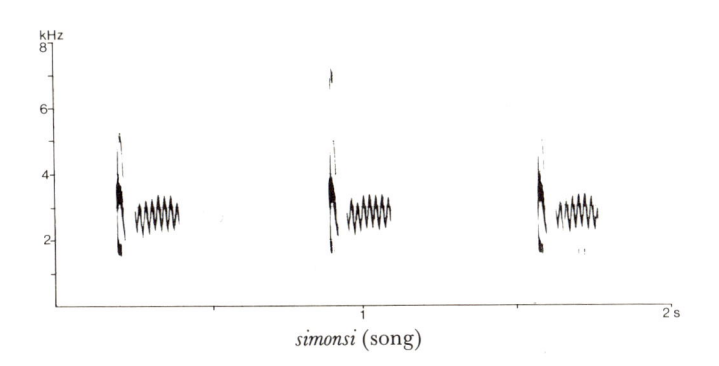

simonsi (song)

Breeding: Juv. Dec (La Paz).
Habitat: Dense undergrowth at treeline, in open terrain with low ericad-melastom brush, along streams in dense long-tufted bunch-grass, and in cracks and crannies of rocky ravines often in *Polylepis* woodland.

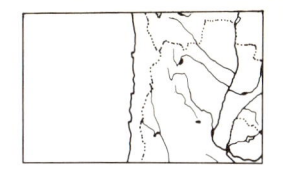

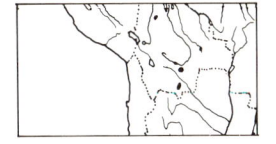

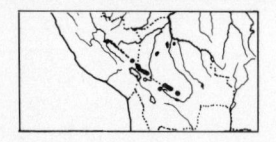

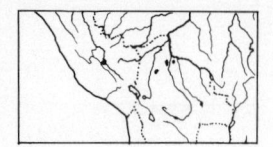

Range: 3000-4300. From s and e Cochabamba through La Paz, Bol. and e Andes in Puno to Vilcanota mts, Cuzco, se Peru. (Note that birds from Peruvian part of range previously were treated under the name *acutirostris*).

urubambae

10 cm. Known from a male and a female taken in 1915 at Cedrobamba (now Sayacmarca) (3650 m) 42 km sse of Machupicchu, Vilcabamba mts, Cuzco, se Peru. Differ from *simonsi* by their bright rufous unbarred flanks and lack of pale supercilium and forecrown.

Unnamed ssp – Plate XLI 1h

9.5-10 cm. Small. **Male** dark gray above, forecrown and **faint supercilium** silky-gray appearing pale gray in some lights. Below deep gray, only very slightly lighter on throat. Rump and tertials brown with dark bars, flanks light cinnamon with dense dark bars. Tail dark with irregular longitudinal and transverse light markings. **Female** and **juv.** unknown.
Voice: Song a single repeated 2-3.3 kHz note given at 2 per s for 90 s or more *chip chip....*, as well as a rougher churr given at 2 per s for up to 60 s *tras tras ....* Call (female?) a single 3.6-3.9 kHz note given in series of 3 at 3-6 s intervals up to a minute or more. (NK).

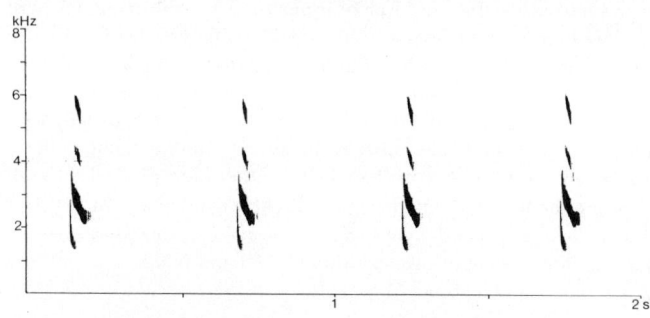

Unnamed ssp. Nev. Ampay, Apurímac (song)

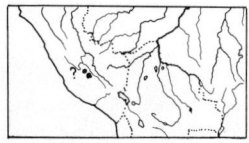

Breeding: No data.
Habitat: Dense undergrowth at humid treeline and in elfin forest.
Range: Only known from c Apurímac (Nevada Ampay, 3500 m: ZMUC), sc Peru.

Unnamed ssp

White-browed. Tail with longitudinal light markins.
Habits: No information.
Voice: Song (Millpo) a 1.6-3.3 kHz double or triple note twice a s for long periods *chirp chirp....* (TS).
Breeding and habitat: No information.
Range: Only known from Pasco (Millpo, 3650 m: LSUMZ; Chipa: AMNH), c Peru. Probably also in Junín.

NOTE: The specimen from Chipa has previously been treated under the name *acutirostris* (see unnamed species from c Peru).

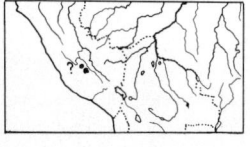

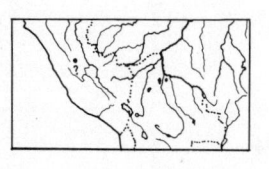

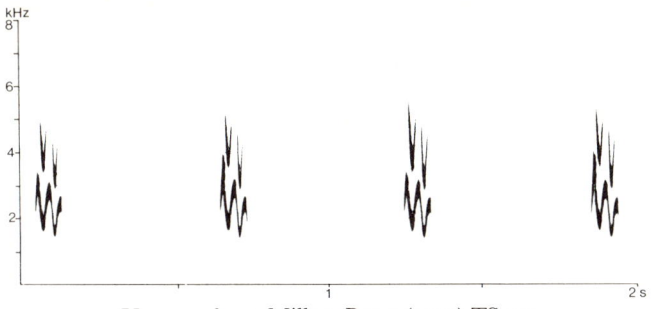

Unnamed ssp. Millpo, Pasco (song) TS rec.

altirostris – Plate XLI 1f-g

10 cm. Like the above-mentioned unnamed ssp, but slightly paler below, more grayish flanks with broader (but still dense and straight) bars; tail without longitudinal marks. An imm. female from within the range of *altirostris* (Cordillera Colán: LSUMZ) (Plate XLI 1f) is Brussels Brown above and pale mouse-gray below, c belly pallid mouse-gray; flanks and abdomen ochraceous tawny; there are dark tips to the feathers of the upperparts and flanks, but they are so minute that the bird appears unbarred; the tail is dark brown washed dusky.

Voice: Song a rough, 3-4 kHz, falling churr of c. 6 notes, lasting 0.12 s and given at 3-4 per s for up to 30 s or more.

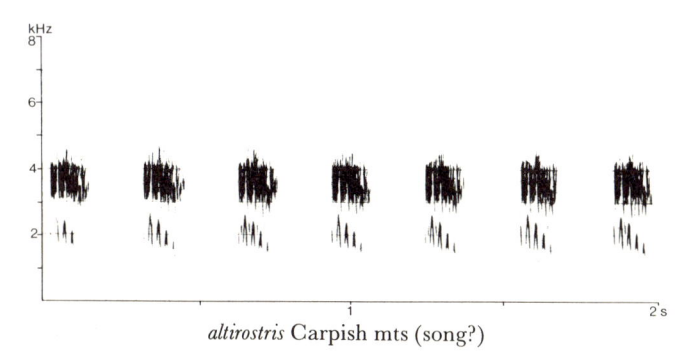

altirostris Carpish mts (song?)

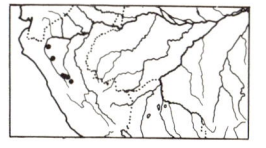

Breeding: Juv. July (Amazonas).
Habitat: Humid treeline and elfin forest.
Range: 2450-3300m from Huánuco (mts e of Huánuco: FMNH; Carpish mts: LSUMZ) n through e La Libertad (Patas) to s Amazonas (Atuen; Cord. Colán: LSUMZ), nc Peru.

NOTE: Two white-browed specimens from mts e of Huánuco (Pagancho) have previously been treated under the name *acutirostris* (see unnamed species from c Peru).

affinis – Plate XLI 1e

10 cm. **Palest form**. Above gray, lower back, wings, rump, and tail brownish, rump and tail with dark bars, irregular on tail. In some lights supercilium very faintly suggested behind eye. Below light gray,

flanks light cinnamon-rufous with fairly dense straight bars. **Juv**. gray-brown above, head densely and narrowly barred, wings with irregular dark markings. Below very light gray, with dark bars on sides and on the light brown flanks.

Voice: May give a short 3-4 kHz trill of 21 notes per s descending slightly towards the end, lasting 1 s, and repeated every 3-5 s (Piura, TP). Alarm near nest a House Wren-like dry trill of ca. 13 notes at 3-4 kHz, c. 0.7 s long and given at 3 or more s intervals (Ancash, NK).

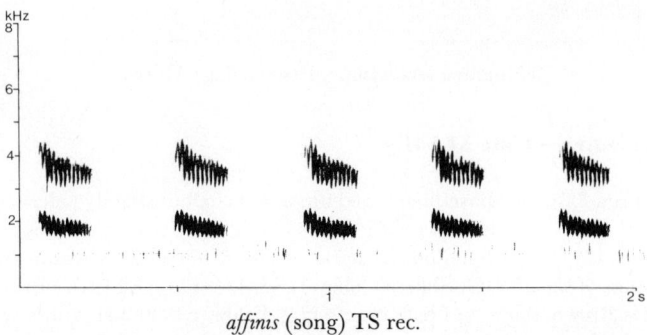

affinis (song) TS rec.

Breeding: Eggs Feb (Ancash); nestlings Feb (Ancash).
Habitat: Large tussocks of grass and rocky boulders, often near or in *Polylepis* or *Gynoxys* woodland.
Range: Nw Peru at 3050-4000 m in W Andes in Ancash (Cord. Blanca). Birds (LSUMZ) from Cajamarca (Chota, 2640-2670 m; Colmena, 2835 m) and possibly those from Piura (Chinguela, 2600-3500 m) may also belong here.

opacus – Plate XLI 1d

10 cm. **Male** above dark gray with little if any brown and dark bars on tertials and rump. Below deep gray. Flanks with suggestion of dense, straight, ochraceous tawny and dusky bars. **Female** lighter and with brown in the tail. **Juv**. undescribed.

Voice: Like Unicolored T. (ssp *parvirostris*) sings a fast (34 per s), dry, endless trill at c. 3.6 kHz, sometimes initiated with a lower note. Simultaneously another bird (female?) may break into a high-pitched, descending (5.7-4.2 kHz) series of 15-50 notes at 5-8 per s. Female may give a series of 5-9 *kee*s at 4-4.5 kHz, lasting c. 1 s and repeated at 2-6

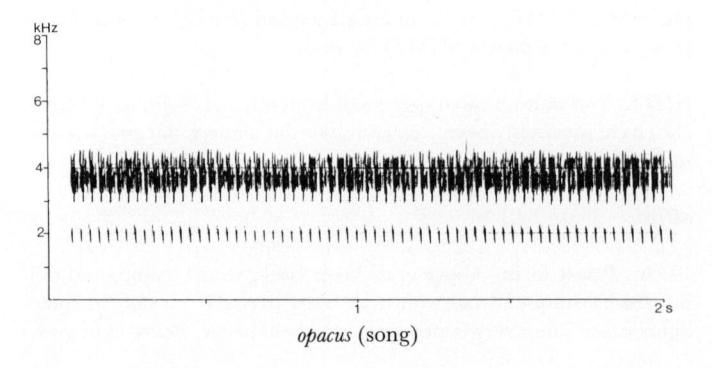

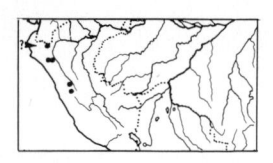

opacus (song)

s intervals, most *kee*s in each series and shortest intervals reached at highest excitement (after playback).
Breeding: Juv. Feb (e Pichincha), May, and Oct (mt Pichincha?).
Habitat: Upper edge of cloud forest, elfin forest, and ericad-melastom shrubbery above treeline. In more open and less swampy situations than Unicolored T. (ssp *latrans*).
Range: E Ecu. at (2450 m?) 3100-4000 m from Morona-Santiago n to Napo. Unconfirmed rec. (incl. possibly mislabeled specimens) from mt Pichincha.

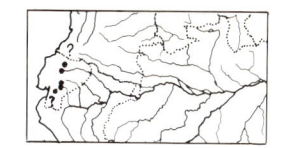

canus – Plate XLI 1a

10 cm. **Male** like *opacus* dark gray above and deep gray below, but without any brown or barring on flanks. **Female**: an imm. from Nariño has pale brown c belly, flanks with dense straight dark bars. **Juv**. undescribed.
Voice: In Nariño a measured series of double whistles *ty-ook, ty-ook, ...* has been noted (SL Hilty). In Huila (Puracé) like *opacus* (BW).
Breeding and **habitat**: No data.
Range: 2800-3800 m in n end of W Andes and Caldas/Tolima border C Andes of Col. Also rec. from e slope of C Andes in Huila (Puracé), w Nariño (Puerres) and e Nariño (e of Manizales), but systematic status of these birds not yet determined.

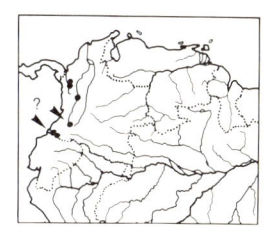

griseicollis – Plate XLI 1b-c

10 cm. Sex?: gray to dark gray, breast dark gray; back and wings fuscous-gray; **rump, flanks, and vent orange-brown, unbarred**; tail brown. **Juv**. cinnamon-brown, everywhere narrowly barred dusky.
Voice: No data.
Breeding: Juv. June (Cundinamarca).
Habitat: Rec. in dense elfin forest.
Range: 2500-3200 m in E Andes of Col. in s Boyacá and Cundinamarca.

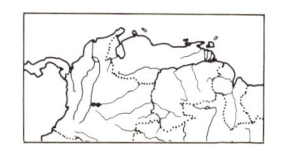

fuscicauda – Plate LXIV 31

10 cm. Like *griseicollis*, but with a darker tail (dark grayish- or dusky brown). Above grayish brown, below light mouse-gray. Rump and flanks plain cinnamon-rufous. **Juv**. with almost uniform dusky-brown tail.
Voice and **breeding**: No data.
Habitat: Cloud forest in ravines, often near streams.
Range: 2500-3200 m in Andes of Trujillo and Lara Ven.

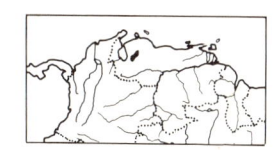

NOTE: The apparent sympatry of *magellanicus* and *fuscus* suggests species ranks. Sspp *superciliaris* and *zimmeri* have been treated as a distinct species, **WHITE-BROWED TAPACULO (*S. superciliaris*)**. The relationship of ssp *urubambae* is unclear, it may represent a different species occupying elevations between Unicolored and Andean T. The type of *acutirostris* may represent the same form as *S. unicolor parvirostris*, but, perhaps more likely, the 'unnamed' species from c Peru. Ssp *infasciatus* (known from 1 or 2 specimens) from e slope of Andes in Cundinamarca Col. is of uncertain affinities, but 1 (the type) is normally treated as an unusually dark *griseicollis*, the other as belonging to *S. latebricola meridanus*.

Tapaculo – Genus *Acropternix*

Monotypic genus with enormous feet and claws. The nest is undescribed.

OCELLATED TAPACULO *Acropternis orthonyx* – Plate XLI 16a-b

20 cm (tail 88-93 mm). **Black, spotted with white**; spots on crown buff. Forehead, supercilium, neckside, headside, and throat rufous; rump, flanks and vent chestnut; remiges narrowly edged reddish-brown. Feathers of back and underparts sometimes narrowly edged brown, esp. in ssp *orthonyx*. **Juv**. similar, but rufous of head replaced by blackish, underparts without spots, and feathers of breast and upper belly with whitish shafts.
Habits: Alone or in pairs. Insectivorous. Walks on the forest floor like a rail, and hops deliberately through dense vegetation, esp. bamboo, within 1 m of the ground. Occ. on mosscovered branches up to 2.5 m from ground. Inoffensive towards other birds.
Voice: Song a single, descending, clear whistle *piuew* or rarely *pew piuew*, repeated every 2-5 s, and answered by other birds. Alarm slightly lower-pitched and shorter, a single, slightly descending, somewhat raspy *keeek* repeated every 3-5 s.
Breeding: No data.
Habitat: Humid, mossy, foggy *Polylepis* forest with patches of bamboo.
Range: 2700-3900 m, rarely down to 2250 m. Andes of Mérida, Ven., and E and C Andes of Col. (*orthonyx*); Andes of Ecu., and n Peru in Piura (Chinguela) and immediately s of the Marañón in Amazonas (Cord. Colán) (*infuscata*). Locally common.

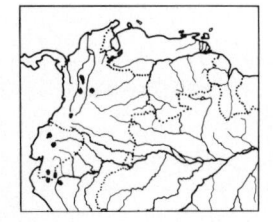

Cotingas – Family Cotingidae

59 very small to very large arboreal passerines found in humid forest of C and S Am. s to n Arg. They are closely related to manakins (*Pipridae*), and more distantly related to tyrant flycatchers, maybe as a specialized branch from within the tyrannid phylogeny, and with becards as a possible link (see pp. 396-428 in Buckley *et al.* (1985) and *Syst. Zool.* 34(1985):35-45). Many species are colorful, some have wattles or crests. They often spend most of the day motionless in the upper canopy, or several males may gather at leks, where they display plumage and/or vocalize loudly. The diet consists mainly of nutritious but inconspicuous berries, and an entire day's needs may be covered in half an hour. Only female incubates, but both sexes tend the young. Most species are trop., but 9 may be met with in the temp. zone of the Andes. For climate/molt relations see Snow (1976). A comprehensive compilation of life-history data is given by D Snow (1982) *The Cotingas*, Oxford. See also Sclater (1888), *Amer. Mus. Novit.* 893, 894 (1936), and Meyer de Schauensee (1945b, 1953).

Cotingas – Genus *Ampelion*

4 species found in the high Andes from Ven. to Bol. Thrush-sized, sluggish birds with a dull-colored plumage, and a rufous or chestnut crest which is spread during display and disputes (in White-cheeked C. the crest is only indicated). Usually alone or in pairs perched conspicuously in treetops, where they may remain motionless for long periods. Though predominantly frugivorous, insects are regularly taken by high vertical sallies from treetops. The sexes are alike in plumage and vocalizations, and both tend the nest, which is a shallow cup of mosses, lichens, and a few twigs, sometimes lined with rootlets, and placed 1-12 m up in a bush or tree, without firm attachment.

RED-CRESTED COTINGA *Ampelion rubrocristatus* – Plate XLII 4a-c

18 cm. Bill pale gray tipped blackish, appears white at a distance. **Dark gray**. Conspicuous, long, flat, dark **reddish-brown crest hanging from nape** often spread, but only erected during excitement. Wings, tail, and broad line through eye blackish. Feathers of rump and vent pointed and edged whitish, broadly so on vent; belly whitish, gradually turning gray towards breast. Inner webs of tail-feathers with large white median patch. Outer primary pointed. **Juv**. broadly streaked dusky and pale olive or warm brown above, faintly so on crown; rump, greater and median wing-coverts, and tertials edged whitish; below cinnamon to pale olive-buff streaked dusky. Bill dark.
Habits: Usually in pairs, sometimes alone or in family groups, and rarely up to 10 together in a fruiting tree. Perches upright, usually reaches up and down for berries (esp. mistletoes), but not uncommonly will shoot several m straight up to snatch an insect, plunging straight down to the same tree-top. During display and disputes in fruiting trees, 2 birds will face one another with raised, fanned crest, and half-cocked, fanned tail, flick wings, and bob head vigorously, while calling. During flight the wings make a rattling sound.
Voice: Short, guttural, frog-like *rrreh* or *babababrrr*, somewhat like a big alarm-clock being wound, sometimes prolonged and given at a higher pitch, but not distinctly patterned as in White-cheeked C. Also gives a series of softer notes *eh-eh-eh-eh-eh*, esp. during conflicts in feeding trees, and occ. gives a hoarse *puii* with neck stretched.
Breeding: Eggs Nov (Bol.), Jan (Huánuco), Oct (Huancavelica); fledglings July (nw Ecu.), Feb (ne Ecu.), Dec (La Paz); juv. June, Sep, Oct (nw Ecu.), Dec (ne Ecu.), June (Amazonas), Aug (Cochabamba); juv./imm. Aug (Piura), Feb (Amazonas), Sep (e La Libertad), June (Huánuco), Mar (Junín), Aug (Cuzco, Puno), June (La Paz), May, June (Cochabamba); nest-building Feb (Loja); display Feb, Mar (Loja). Molts, and possibly breeds throughout year between 10°n and 18°s (see Snow 1976).
Habitat: From edge of undisturbed cloud forest or fairly open, semi-humid temp. forest to groves, esp. of *Alnus* bordering cultivated land. In w Peru in relict patches of woodland such as *Polylepis*, *Escallonia*, and *Oreopanax*.
Range: 2500-3700 m. Sta Marta and Perijá mts, and Andes from Tru-

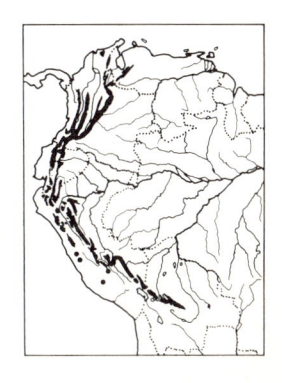

jillo, Ven., and all 3 ranges of Col. s through Ecu. and Peru to Lima, w Peru, and along e Andean slope to Cochabamba, Bol. Common.

CHESTNUT-CRESTED COTINGA *Ampelion rufaxilla* – Plate XLII 6

18.5 cm. Eye bright red. Forehead, supercilium, back, and wings gray. Very large crown-patch of elongated feathers chestnut, bordered on sides with black. Median and lesser wing-coverts brownish to chestnut, primaries and tail black. **Throat and headside rufous**, breast gray, **rest of underparts yellow with broad black streaks** except on c belly. Sspp hardly valid. **Juv.** unknown.
Habits: Much like Red-crested C., incl. display and flycatching, but usually perches in taller trees.
Voice: *reh* notes, and a long *eh-eh-eh-rrrreh*, much like Red-crested C.
Breeding: Eggs Dec (Puno); large gonads Mar, Apr (sw Huila); molt Jan-July (Col), Aug-Dec (Peru), but in the pair nesting in Puno in Dec both were molting.
Habitat: Tops of tall and slender trees in humid forest, stunted cloud forest and second growth at edges of clearings in cloud forest.
Range: 1860-2740 m. Scattered localities in sW and C Andes of Col., ne Ecu. (*antioquiae*), Loja in s Ecu. (s end of PN Podocarpus: D Platt) (ssp?), n, c, and se Peru, and La Paz Bol. (*rufaxilla*). Uncommon or rare, locally fairly common.

WHITE-CHEEKED COTINGA *Ampelion stresemanni* – Plate XLII 5

18 cm. Compact, often appearing very thick. Above fuscous streaked buffy, forecrown blackish, hindcrown gray-brown. **Lores, area below eye, and cheeks whitish**. Throat and breast gray-brown; rest of underside buff, broadly streaked blackish except on c belly. **Juv**. like ad., but throat paler (drab), and lores and cheeks only slightly paler.
Habits: Alone or in pairs, rarely in feeding aggregations of 4-10 birds. Much like Red-crested C. Will sit motionless and upright in tree-tops for long periods, but flies low over the ground, often with half-spread tail. When disturbed near nest, will assume horizontal posture with slender body, and bob head and flick wings and tail, sometimes stopping to turn head in jerks, while looking about with neck stretched. Seems to feed almost entirely on berries of mistletoes (*Tristerix* and *Ligaria*), and always regurgitates the seeds and wipes them off on a branch. Above 3000 m it may be the only disperser of these plants in many regions. Sticky mistletoe seeds not infrequently stick to the plumage of birds, which secures long-distance dispersal. During display, members of a pair will face one another, bob head and flick wings, but not give any vocalizations; after 30-60 s they both regurgitate. (See *Bull. Brit. Orn. Club* 101(1981):256-265).
Voice: Song given at dawn and late afternoon is a loud, low-pitched, nasal (from a distance frog-like) *reh-reh-reh-reh-rrrrrrrrrrr-ré-ré*, c. 4 s long, and repeated 3-6 times at 30-60 s intervals. Sometimes the last, higher notes are left out. Other birds will start singing simultaneously. Song

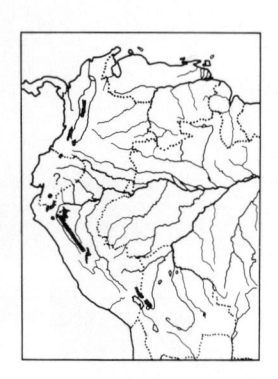

also given in feeding aggregations. Alarm at nest *raaaaaaaah*. Contact note shorter, but similar.
Breeding: Eggs May (Lima, Ancash), nestlings May (Ancash); juv. Aug (Lima). Large gonads Mar (Ancash, Lima, nest-building Mar (Lima). Probably no breeding in the dry season (Aug-Nov).
Habitat: *Polylepis* woodland with mistletoes present. In dry season (Aug-Oct) some birds descend to mixed woods with shrubs and, e.g., *Oreopanax, Escallonia*, also here with mistletoes present.
Range: 2700-4240 m. Only known from 10 localities in e La Libertad, Ancash, Lima, and nw Ayacucho, wc Peru. Generally uncommon, but locally common.

NOTE: On basis of its distinct skull structure, it was previously placed in its own genus *Zaratornis* (see also Bond (1956)).

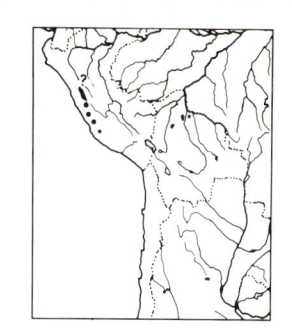

BAY-VENTED COTINGA *Ampelion sclateri* – Plate XLII 7a-b

20 cm. Cap shiny black with median, semi-concealed, reddish chestnut crest. Upperparts, headside, and throat dark gray; wings and tail blackish. Breast and belly dark drab-brown, vent rufous-chestnut. **Juv**. much like ad., but without crest, and crown gray like sides of head, and underparts with faint, paler and darker flammulations.
Habits: In pairs. Perches in tree-tops. Less active and vocal than others of the genus. Extremely sedentary, sometimes remaining in the same little patch of trees for weeks. Crawls about almost in parrot-like fashion while feeding, and often sits motionless for long periods. When excited bobs head. During display one bird will face another, fly straight up, to plunge right down, and repeat this exercise without pause for several minutes, not vocalizing. Besides berries, insects are included in the diet.
Voice: Call a wheezy *tjjjtetjjjtetjjjte*, varying in length and intensity, answered by others.
Breeding: Eggs Apr, May (Huánuco); juv./imm. July (Huánuco).
Habitat: Prefers rather low trees with very dense, flat canopies (esp. *Escallonia* and *Weinmannia*) in elfin forest.
Range: Peru at 2500-3500 m. Only known from e La Libertad, Huánuco, and Junín, c Peru. Birds sighted 1989 in Loja (PN Podocarpus: HB, MKP), s Ecu. undoubtedly represent a different ssp or sp. Local and uncommon.

NOTE: Previously placed in its own genus *Doliornis*.

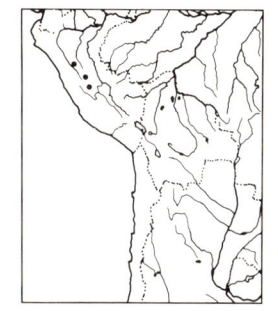

Fruiteaters – Genus *Pipreola*

8-11 species, found from the upper trop. to the temp. zone, one on the Tepuis of Guyana and Ven., the rest in coastal mts. of Ven., and in the Andes s to Bol. They are medium-sized cotingas inhabiting humid forest and cloud forest. They have red bills, are green above, usually with pale markings on the tertials; the underparts yellow, more or less barred, scaled, or streaked with green or blackish. The rump-feathers are very loosely attached. Sexually dimorphic, the male often has a black head. Lower elevation species are grass-green rather than olive-green above, and often have reddish on throat and upper breast. Found alone or in pairs, they sometimes follow mixed-species flocks, but usually sit motionless in shade in a fruiting tree, occ. swallowing a large quantity of berries in rapid succession. Their presence is often revealed by their very high-pitched, drawn-out call. The nest is a shallow cup of moss, lined with rootlets, and placed 1-2 m up in a bush or a vine. Both sexes tend the young.

GREEN-AND-BLACK FRUITEATER *Pipreola riefferii* – Plate XLII 9a-e

16.5-19 cm. **Male**: ssp *chachapoyas*: above green; remiges with narrow white tips, broadest on tertials. Head and upper breast black, throat with very faint greenish cast. Narrow yellow band bordering black of breast continuing round the neck, almost meeting at nape. Sides of lower breast green; rest of underparts yellow, with dark green, v-shaped centers of feathers, except on c belly. Ssp *tallmanorum* much smaller, head and upper breast shiny black; rest of underparts, except sides and vent, yellow; *confusa* hardly differs from *chachapoyas*, but head and upper breast more greenish, and lower breast and belly more strongly marked; *occidentalis* like *confusa*, but larger, paler above, less spotted below, inner greater wing-coverts very faintly tipped yellowish white when fresh; *melanolaema* like *occidentalis*, but head and upper breast glossy-black without greenish wash, pale tips of tertials and wing-coverts more conspicuous; *riefferii* like *melanolaema*, but larger, head and upper breast dark moss-green to blackish green; belly more extensively and darker yellow. **Female** like male, but black replaced by green. Upper throat with faint, yellowish barring. Eye-ring yellow. **Juv**. dark and dull olive-green above, crown and shoulders with yellow terminal streaks or spots. Below dull olive streaked pale greenish yellow. Bill dark, base of mandible pale. Molts to ad. plumage 3-12 weeks after fledging.

Habits: Alone, in pairs or loose groups of 4-6, sometimes in mixed-species flocks. Not shy. Sits silently in a fruiting tree eating berries (notably melastomes), occ. making a clumsy hover for them, or remains motionless for long periods. Also eats a few insects.

Voice: Song by both sexes, several soft, high-pitched notes followed by a very thin, high-pitched, drawn out *t-t-t-seeee*. During copulation *char-a-a-a-a*. In flight the wings make a rattling sound.

Breeding: Fledglings Mar (Mérida), Jan (Cauca), June (nw Ecu.). Molt suggests breeding all year, with peaks May-June (Mérida), May-July (n Col.), no peak (s Col.), all year except perhaps Aug-Sep (Ecu., n Peru).

Habitat: Lower to middle strata in thickest parts of humid forest and cloud forest. Occ. outside forest in fruiting trees.

Range: 1500-3050 m, rarely down to 900 m, below Band-tailed F. in zone of overlap. Coastal mts and Andes from sw Lara to c Táchira, Ven. (*melanolaema*); w Táchira, Ven., Perijá mts and C and E Andes of Col. (*riefferii*); from extreme s tip of C Andes and s end of W Andes of Col. s to El Oro, w Ecu. (*occidentalis*); e Ecu. s to Amazonas, n Peru, where intergrades with *chachapoyas* (*confusa*); from Amazonas s to San

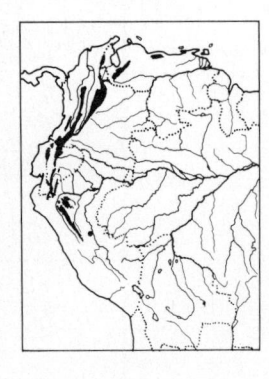

Martín and e La Libertad, n Peru (*chachapoyas*); in Sira and Carpish mts, Huánuco c Peru (*tallmanorum*, O'Neill & Parker, *Bull. Brit. Orn. Club* 101 (1981):294-296). Fairly common.

BAND-TAILED FRUITEATER *Pipreola intermedia* – Plate XLII 10a-c

18.5 cm. Ssp *signata*: **male**: above green; **inner remiges and tail-feathers black subterminally, and with narrow white tip**. Head, throat and upper breast glossy black, bordered all round (except on nape) by narrow, golden-yellow collar. Sides of lower breast green; rest of underparts bright yellow, with green bars and narrow, **rounded** (not v-shaped), blackish, subterminal markings on each feather except on c underparts. **Female** has head like back, eye-ring yellow, chin with faint yellowish barring; below the green breast, the feathers have rounded greenish-black markings at their bases, except on c underparts. Ssp *intermedia* similar, but c underparts marked like sides; female has less conspicuous eye-ring, and less yellow throat.
Habits: Like Green-and-black F.
Voice: Much like Green-and-black F.
Breeding: Molt suggests breeding all year, with peak Dec-Jan (Peru), and Nov-Dec (Bol.).
Habitat: Like Green-and-black F.
Range: 2000-3000 m. Above Green-and-black Fruiteater in zone of overlap. From e La Libertad and s San Martín s to Junín c Peru (*intermedia*); Cuzco se Peru to Cochabamba Bol (*signata*). Fairly common.

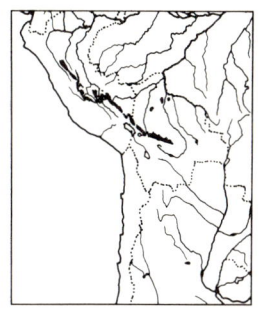

BARRED FRUITEATER *Pipreola arcuata* – Plate XLII 11a-c

22 cm. **Male**: bill crimson, legs scarlet, iris pale. Head and upper breast shiny black, upperside olive-green. **Underside** pale yellow, **barred blackish**. Inner tertials and greater secondary-coverts with whitish subterminal spot (often only on outer web), and black tip, extreme tips pale yellow in fresh plumage; secondaries and tertials with a tiny white dot at the extreme tip. Longest upper tail-coverts with black subterminal and pale yellow terminal bar; tail black, narrowly tipped white, and olive-green at base, amount of olive-green increasing towards c feathers, which are almost wholly green. Ssp *viridicauda* has olive-green of tail slightly more extensive. **Female** has olive-green crown and headside, throat and breast barred like rest of underparts, bill dark-tipped. **Juv**. like female, but crown, shoulders, lower back, and rump more or less barred; bill very dark red; tail-feathers pointed.
Habits: Alone or in pairs, sometimes in mixed-species flocks. Clumsily reaches out for berries, or sits motionless for long periods.
Voice: Often gives a drawn-out, high-pitched wheen, occ. a sharp shriek.
Breeding: Fledgling July (w Ecu.), Jan (Cochabamba); juv./imm. July (Ayacucho), Mar (Pasco); large gonads May (Perijá).
Habitat: Lower to middle strata. Humid mossy forest and cloud forest; also at forest edge.

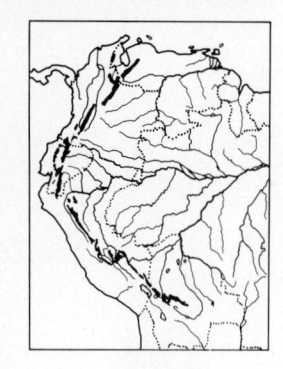

Range: 1800-3500 m. Perijá mts and Andes from sw Lara, Ven. s through E and s end of C and W Andes of Col., and both slopes of Ecu. to Peru n of the Marañón (*arcuata*); from Peru s of the Marañón to Cochabamba, Bol. (*viridicauda*). Fairly common to common.

SCALED FRUITEATER *Ampelioides tschudii*

20 cm. Eye pale. **Male: cap black; lores and stripe below eye whitish**, joining yellow nuchal collar. Upperparts blackish, conspicuously scaled olive-yellow; wings black, remiges tipped olive, greater coverts greenish yellow forming broad band. Throat white mottled dusky, rest of underparts yellowish white broadly scaled olive. Short tail olive, outer feathers black tipped buff. **Female** has olive-green cap, dark malar stripe, and more olivaceous underside. Alone, in pairs or groups of 3-4. often in mixed-species flocks. Actively hops along epiphyte-laden branches of middle and upper strata. Feeds on fruit. Song a loud, raptor-like whistle rising in pitch and amplitude, then fading, sometimes repeated at short intervals.

In wet forest at 650-2700 m. Occurs locally from Perijá mts and Táchira, w Ven., in the n and s of E Andes (and Macarena mts) and w slope of W Andes Col, nw Ecu, and locally along e Andean slope of Ecu. and Peru to n Bol.

Pihas – Genus *Lipaugus*

7-8 species found from trop. to temp. zones from Mexico to Bol. and se Brazil. Medium-sized to fairly large cotingas, thrush-like in proportions, but tail rather long in 2 species. Bill wide and somewhat hooked at tip. Plumage mainly gray-brown, olive, or rufous, 1 species sexually dimorphic. Food consists of large insects and fruit. At least some species have leks, where males give loud, piercing song, that of the Screaming Piha being a characteristic voice of the lowland rain forest. The only nest known is that of the Rufous Piha (*Lipaugus unirufus*), a very poorly made, shallow cup, constructed of a few twigs only, and placed high up in a fork or on parallel branches. The ad. has the peculiar habit of tearing apart the nest after breeding, and also if the nest has been robbed or breeding otherwise been unsuccessful.

DUSKY PIHA *Lipaugus fuscocinereus* – Plate XLII 8

31 cm. Above dark gray, wings and tail darkest. Throat light gray, rest of underparts gray faintly tinged brownish, vent cinnamon-gray. **Juv**. cinnamon to cinnamon-rufous throughout. **Imm**. like ad., but greater wing-coverts and some lesser coverts with rufous tips.
Habits: Solitary, sometimes in mixed-species flocks. One moved through the forest canopy with a group of Mountain Caciques. At leks a number of birds sing from treetops or high exposed branches, and frequently fly between the trees.
Voice: Song reminiscent of that of the Screaming P., loud and piercing *pee-a-weeee*, or *pee-a-weeee-a-weeee*, last syllable descending.

Red-ruffed Fruitcrow

Breeding: Molt suggests breeding all year (Ecu.); juv./imm. Nov, Jan (ne Ecu.); imm. Nov, Mar (ne Ecu.).
Habitat: Upper strata of humid forest.
Range: 1700-3000 m. All three ranges of Col. s through e Ecu. to n Cajamarca (Batán), n Peru. Rare to uncommon and local.

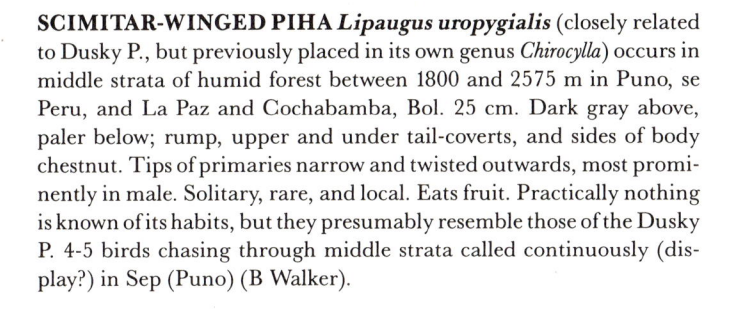

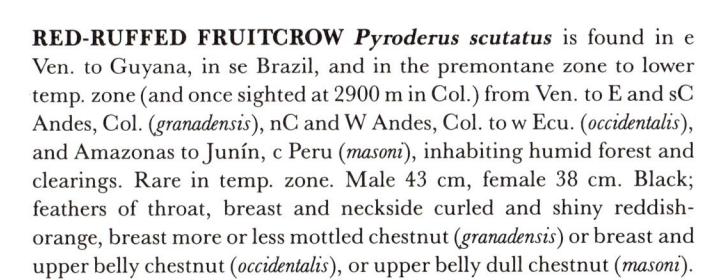

SCIMITAR-WINGED PIHA *Lipaugus uropygialis* (closely related to Dusky P., but previously placed in its own genus *Chirocylla*) occurs in middle strata of humid forest between 1800 and 2575 m in Puno, se Peru, and La Paz and Cochabamba, Bol. 25 cm. Dark gray above, paler below; rump, upper and under tail-coverts, and sides of body chestnut. Tips of primaries narrow and twisted outwards, most prominently in male. Solitary, rare, and local. Eats fruit. Practically nothing is known of its habits, but they presumably resemble those of the Dusky P. 4-5 birds chasing through middle strata called continuously (display?) in Sep (Puno) (B Walker).

RED-RUFFED FRUITCROW *Pyroderus scutatus* is found in e Ven. to Guyana, in se Brazil, and in the premontane zone to lower temp. zone (and once sighted at 2900 m in Col.) from Ven. to E and sC Andes, Col. (*granadensis*), nC and W Andes, Col. to w Ecu. (*occidentalis*), and Amazonas to Junín, c Peru (*masoni*), inhabiting humid forest and clearings. Rare in temp. zone. Male 43 cm, female 38 cm. Black; feathers of throat, breast and neckside curled and shiny reddish-orange, breast more or less mottled chestnut (*granadensis*) or breast and upper belly chestnut (*occidentalis*), or upper belly dull chestnut (*masoni*).

Plantcutters – Family Phytotomidae

3 species, 1 in coastal n Peru, 1 in Chile and s Arg., and 1 from Bol. to c Arg. Plumage, jizz, vocalizations, and nest very similar to cotingas of the genus *Ampelion* (but DNA sequence data do not support a particularly close relationship; P Arctander). Plantcutters are poorer fliers with shorter, more rounded wings, and also more rounded tails. They feed on grass, fruit, buds, and tender leaves, which they cut up, wastefully, with their finely serrated bills. Usually in pairs or small groups, but occ. congregate in large flocks that can cause severe damage to crops. Sexually dimorphic, male with rufous underparts or belly, female streaked.

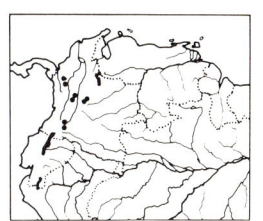

WHITE-TIPPED PLANTCUTTER *Phytotoma rutila* – Plate XLII 3a-d

18 cm. Bill short and stout, finch-like. Under wing-coverts whitish. Iris pale, conspicuous from a distance. **Male**: above dark gray, **forehead rufous**. 2 white wingbars, the upper broad, the lower sometimes lost through wear. **Tail** dusky, **conspicuously tipped white** except on c feathers. **Underparts orange-rufous**, extreme sides and flanks gray-brown. Ssp *angustirostris* larger with a stronger bill and broader white tip to tail. **Imm. male** obscurely streaked dark brown above; below buffy gray-brown, belly whitish, flanks (and occ. more of lower underparts) rufous-buff, sometimes faintly streaked dark brown on breast and upper belly. In fresh plumage with gray-brown feather-tips. **Female**: headside and upperparts blackish brown, streaked whitish on head and gray-brown on back, 2 white wingbars and edges of tertials. Tail like male, but outer web of outer feather edged white. Underparts whitish streaked blackish-brown. **Imm. female** similar, but with ochraceous vent. **Juv.** (both sexes): above gray-brown, feathers faintly tipped brownish; upper wingbar broadest, though not as broad as in ad. Tail like female. Underparts uniformly buffy-brown.
Habits: In pairs, families or small groups, occ. in flocks of 100s. Alights and often flies with tail spread, thus exposing white tip. Perches on telephone-wires, conspicuous side-branches and atop bushes or trees, but during the heat of the day or when disturbed hides in the foliage. Occ. forages on the ground. Regularly jerks tail. Never flies far.
Voice: Song is a mechanical *arrrrrrr*, sounding like the winding of an old-fashioned alarm-clock. Call similar, but shorter. Scratchy sounds during foraging.
Breeding: Eggs Jan (Córdoba, Jujuy), Jan-Feb (Salta), Oct-Nov (Tucumán); fledglings Jan (Tarija, Mendoza); juv. Feb, Mar, May (Cochabamba).
Habitat: Chaco, open woodland, brush, gardens, orchards, or agricultural land in semi-arid regions. At least in La Paz feeds mainly on *Prosopis* seeds.
Range: 600-3600 m in Andean highlands and slopes from La Paz, Cochabamba, and Sta Cruz s to Tarija, Bol., intergrading with *rutila* in nw Arg. (*angustirostris*), lowlands to 1800 m from Parag. chaco s to Río Negro, Arg., one rec. from Chubut (Estación Baltusa). Southern breeders winter n to Urug., Corrientes, and Entre Ríos (*rutila*). Fairly common, locally abundant seasonally. Fairly common in gardens of lower parts of La Paz city.

RUFOUS-TAILED PLANTCUTTER *Phytotoma rara* – Plate XLII 2a-c

18 cm. Under wing-coverts blackish, iris reddish. **Male**: crown rufous-chestnut, headside and upperparts light gray-brown, broadly streaked blackish brown. Wings blackish, broad band on median coverts and line across middle of outer webs of outer primaries white, greater wing-coverts and tertials edged buff. C rectrices blackish brown, rest of **tail rufous-chestnut with broad (sub-)terminal black band** on inner webs, outer webs blackish (with some rufous on outer feather), minute

tips rufous-gray. Underparts orange-rufous, belly palest. **Imm. male** like ad., but crown with some blackish-brown streaks; underparts buffy more or less admixed with rufous, and streaked blackish-brown; rufous feathers with buffy tips when fresh. **Female** has headside and upperparts light gray-brown broadly streaked dark brown. Wings with 2 narrow, buffy bars and edges of tertials. Underparts buffy, streaked dark brown on breast and sides. Tail as in male, but outer web of outer feather dark with narrow whitish edge. **Juv**. (both sexes) uniformly gray-brown, wingbar buffy and narrow.
Habits: In pairs in breeding season, small groups of 6-12 birds in winter; not rec. in large flocks. Like White-tipped P. perches conspicuously atop trees and bushes.
Voice: *ek-ek-eerrrrrr*, much like White-tipped P.
Breeding: Eggs Jan (Mendoza), Oct-Dec (Chile, Río Negro); nestlings Nov (Mendoza).
Habitat: Thorny scrub (mainly *Berberis*) in *Nothofagus* forest clearings; orchards, gardens, and cultivated fields.
Range: Coast to 2150 m. From Atacama to Magallanes (Torres del Paine; see *Anal. Inst. Patagonia* 8(1977): 317-8) Chile, and adjacent Arg. from Mendoza to Sta Cruz. Southern breeders wander n in winter (see also *Neotrópica* 11(1965): 38-40). Fairly common and widespread.

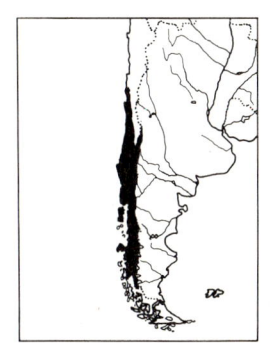

Tyrant Flycatchers – Family Tyrannidae

Some 375 species found in the Americas and on adjacent isls, northern and most southern breeders being migratory. A diverse group of birds ranging from 7.5 to 27 cm and living virtually throughout from bare desert to rain forest and from trop. to high alpine zone. Tyrant flycatchers make up for 18-23 % of the perching birds in any American bird community. As their name implies most feed on insects that may be captured by 'true flycatching' (aerial hawking), by warbler-like perch-gleaning and hover-gleaning or picked from the ground by perch-to-ground or run-and-pick methods. Most will also occ. eat fruit, some almost exclusively so. In most species the sexes are alike, but some are strikingly dimorph. The imm. plumage is carried for 3 months, in some a whole year. Many have a white, yellow, orange, rufous or red median, crown-patch more or less concealed, but exposed in excitement. Otherwise the plumage is usually dull, olive to brownish above and yellow to white below, but a few colorful species do exist. The vocalizations are poor and consist of simple repeated phrases. Song is given only at early dawn, simpler territorial phrases, contact notes and alarms throughout the day. Modification of the primaries, producing a whirring or buzzing sound when the wings are flapped, are seen in a large number of genera. The nest may be an open cup or domed, and placed in a hole in a tree, on a crotch of a branch or on the ground. In some species both sexes incubate (*Condor* 55(1053):218-19). More detailed information is given for the separate genera. For the genera *Serpophaga*, *Anairetes*, *Mecocerculus*, *Stigmatura* and *Tachuris* see *Condor* 73(1971): 259-86, where there are also references to papers on *Tyrannus*, *Sayornis*, *Contopus*, *Pitangus*, *Myiodynastes*, *Muscisaxicola* and *Pyrocephalus*. Data on *Ochthoeca*, *Myiotheretes*, *Xolmis*, *Neoxolmis*, *Agriornis* and *Muscisaxicola* can be found in *Bull. Mus. Comp. Zool.* 141(1971):178-268 and 148 (1977):129-84, *Fieldiana Zool.* 13(1982):1-22, and pp. 431-42 in Buckley *et al.*, eds. (1985). MA Traylor,Jr. & JW Fitzpatrick published a short synopsis of all the genera in *Living Bird* 19, 1980-1 (1982):7-50, and the adaptive radiation of the family is reviewed by Fitzpatrick on pp. 447-70 in Buckley *et al.* (1985). Phylogenies are discussed in *Amer. Mus. Novit.* 2797

(1984), 2846(1986), 2914, and 2915 (1988), and in *Syst. Zool.* 341(1985):35-45. See also Meyer de Schauensee (1945b) and Bond (1947).

Tyrannulets – Genus *Phyllomyias*

10 species distributed from Costa Rica to n Arg., inhabiting humid and semi-humid forest and forest edge, mostly in the subtrop. zone. Very small tyrants that hover-glean upwards, and also eat berries. The nest is an open cup. See *Amer. Mus. Novit.* 1109(1941).

SCLATER'S TYRANNULET *Phyllomyias sclateri* – Plate XLVII 19

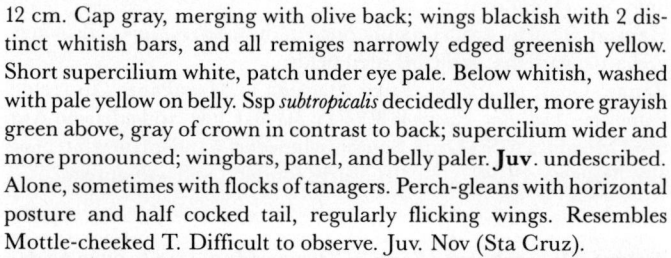

12 cm. Cap gray, merging with olive back; wings blackish with 2 distinct whitish bars, and all remiges narrowly edged greenish yellow. Short supercilium white, patch under eye pale. Below whitish, washed with pale yellow on belly. Ssp *subtropicalis* decidedly duller, more grayish green above, gray of crown in contrast to back; supercilium wider and more pronounced; wingbars, panel, and belly paler. **Juv**. undescribed. Alone, sometimes with flocks of tanagers. Perch-gleans with horizontal posture and half cocked tail, regularly flicking wings. Resembles Mottle-cheeked T. Difficult to observe. Juv. Nov (Sta Cruz).

In middle and mainly upper strata of humid forest at 1525-2400 m from Urubamba Valley Cuzco in se Peru to La Paz, Bol. (*subtropicalis*), and at 400-1500 m from Cochabamba and Sta Cruz, Bol., s to Tucumán, nw Arg. (*sclateri*). Rare, perhaps partly overlooked in n part of range, fairly common in s.

NOTE: Previously placed in the genus *Xanthomyias*. The type of **OLROG'S TYRANNULET** (*Tyranniscus australis*) proves to be a specimen of the present species.

BLACK-CAPPED TYRANNULET
Phyllomyias nigrocapillus – Plates XLVII 17 and LXIII 10

10.5 cm. Ssp *flavimentum*: cap dusky, rest of upperparts olive-green. **Wings with 2 rather broad whitish bands, black patch at base of secondaries**; rest of secondaries and primaries edged golden yellow, narrowest and palest on primaries; tertials edged whitish, inner one with whole outer web whitish. Narrow supercilium yellow, headside finely mottled yellow and blackish.
Below bright yellow, slightly darker on breast. Ssp *aureus* is more golden. Ssp *nigrocapillus* has **black cap, white supercilium** and an olive breast. **Juv**. unknown.
Habits: Solitary or in pairs, often in mixed-species flocks. Perches fairly horizontally. Forages inconspicuously in dense parts of the vegetation, making it difficult to observe. Active, warbler-like. Tail sometimes slightly cocked.
Voice: Call a clear high *peeeeep*, often repeated persistently.
Breeding: Large gonads Mar-Nov (n Col.).
Habitat: Upper strata of high dense cloud forest, old second growth,

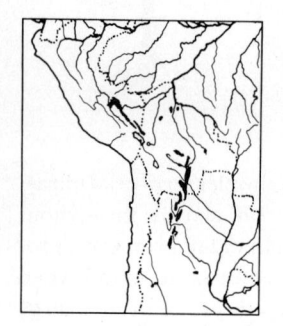

and more open woodland in the higher altitudes.
Range: 1600-3400 m. Sta Marta mts (*flavimentum*), from s Lara to n Táchira, Ven. (*aureus*), Perijá mts, Páramo de Tamá, all 3 ranges of Col. (but apparently not in Nariño), and from n Ecu. s to Vilcabamba mts, Cuzco, se Peru (*nigrocapillus*). Uncommon to rare, perhaps somewhat overlooked, locally fairly common.

NOTE: Previously placed in the genus *Tyranniscus*.

ASHY-HEADED TYRANNULET
***Phyllomyias cinereiceps* – Plate XLVII 16**

9 cm. Bill short. **Crown slaty gray**, back dark olive-green. **Wings with 2 yellowish white bars**, edges of tertials white; rest of remiges edged olive, except at the base of inner primaries and outer secondaries. Narrow supercilium and stippled eye-ring white; lores and area below eye grizzled with white, **earcoverts yellow with dusky patch**. Underparts sulphur-yellow, **flammulated with olive on breast**. **Imm**. similar, but without a dark patch at base of remiges. **Juv**. has darker cap, grayer back, whiter wingbars and panel, no dark patch at base of remiges, and yellowish white below. Throat and breast streaked with yellowish green (rather than olive-green).
Habits: Alone or in pairs, in mixed-species flocks. Upright or in slightly hunched posture. Catches flying insects among the outer branches of the middle canopy, occ. clinging briefly to leaves.
Voice: Distinctive song delivered from forest subcanopy, often while foraging in mixed-species flock: a high-pitched but far-carrying *psee* followed immediately by an equally high-pitched, very rapid descending trill, the whole lasting 1 s (BW).
Breeding: Fledglings Oct (Quindío), Nov (Cauca).
Habitat: Lower to medium strata (3-10 m up) in humid forest.
Range: 1150-2750 m. Sight rec. s of Ramon w Táchira, Ven. 1988 (BW, J Pierson, J Arvin, J Coons). Locally in all 3 ranges of Col. s through Ecu. and along e slope of Andes to Puno, se Peru. Uncommon to rare, locally common.

NOTE: Previously placed in the genus *Tyranniscus*.

TAWNY-RUMPED TYRANNULET
***Phyllomyias uropygialis* – Plate XLVII 18**

9.5 cm. Crown dusky, back dull brown, rump tawny. **Wings blackish with 2 broad buffy bars**, tertials edged white, rest of remiges edged buff, except at base of secondaries. Narrow supercilium and broken eye-ring whitish, area below eye finely mottled pale and dark gray. Throat light gray, breast washed olive, belly pale yellow. **Juv**. has whitish belly.
Habits: In loose small groups or pairs, often in mixed-species flocks. Warbler-like. Perch- and somtimes hover-gleans **across** the middle part of the thinner branches. Sometimes raises crown.
Voice: Call *pzit*. Song 2 wheezy notes, the 1st higher pitched.

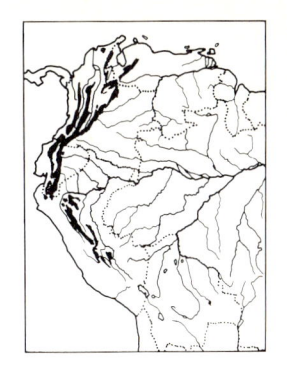

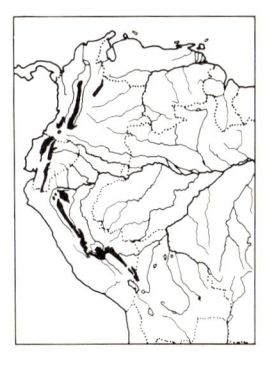

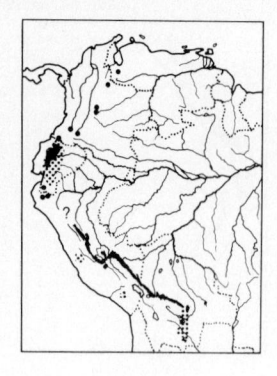

Breeding: Fledgling Dec (Sta Cruz).
Habitat: 0.5-5 m up in thickets, clearings, and edge of humid forest.
Range: 1500-3750 m. Locally from Mérida, Ven. s through all 3 ranges of Col., Ecu., and Peru to Tarija, Bol., 1 rec. from Arequipa. Fairly common in Bol., uncommon elsewhere.

NOTE: Previously placed in the genus *Tyranniscus*.

MOUSE-COLORED TYRANNULET *Phaeomyias murina* ascends at least to 2100 m in La Páz Bol. (*wagae*). It frequents fairly similar habitats as the previous species, but forages mainly by upward hover-gleaning, and also eats fruit. 12.5 cm. Tail relatively long. Above dark olivaceous; wings with 3 grayish to buffy white bars, the 2 lower bars broad; tertials edged whitish; rest of remiges very narrowly edged whitish, except at base of secondaries. Faint very short supercilium grayish, eye-ring whitish, lores and area below eye grizzled with whitish. Throat whitish, breast suffused with olive, belly whitish to pale yellow. For Peru see *Amer. Mus. Novit.* 1109(1941).

Tyrannulets – Genus *Zimmerius*

5-6 species distributed from s Mexico to Bol. and n Brazil, in humid forest and forest edge, mostly in the upper trop. and subtrop. zones. A fairly uniform group of small tyrants that flit about in the canopy. They seem to feed mainly on mistletoe berries. They are easily identified by narrow, conspicuous, pale edges to the wing-coverts. The ovoid closed nest with a side entrance is suspended from a branch. See *Amer. Mus. Novit.* 1109(1941).

PALTRY TYRANNULET *Zimmerius villissimus* – Plate XLVII 20

10.5 cm. Appears long-legged. Bill rather short and stout. Above olive-green, rump greenest; crown with some obscure, dark streaks. **Wing-coverts** and inner remiges **narrowly** but **conspicuously edged pale yellowish green**, except at base of secondaries (forming black patch). Short supercilium from bill whitish to pale yellowish white, broad in front of eye; lore blackish; broken eye-ring and area below eye whitish. Throat whitish, area below eye and moustachial region faintly mottled whitish and dark gray, rest of underparts pale yellow, breast washed with smoky gray. Ssp *tamae* has a darker crown, white supercilium, grayer breast, and a paler yellow belly. **Juv**. unknown.
Habits: Singly, but a number of individuals are often scattered about, sometimes in mixed-species flocks. Actively forages on outer foliage, hover-gleaning and making short sallies for insects and berries. The relatively long narrow tail is often half cocked.
Voice: A soft sadly whistled *peer*. Another call begins with a similar note, repeated several times, and is followed by 4 quicker notes, rising successively, the whole phrase repeated a number of times. Also a patterned *pee pee...*
Breeding: Large gonads Mar-Nov (n Col.).

Habitat: Humid zone. Medium to upper strata. In canopy of small to medium-sized well-foliaged trees, at edge of forest and cloud forest. Plantations, acricultural land, and glades in second growth woods.
Range: 800-3000 m. N.d.Santander, Col., and from Barinas to e Táchira, Ven. (*improbus*); Perijá mts, Páramo de Tamá, and Sta Marta mts (*tamae*, but birds from Sta Marta are variably intermediate). 2 sspp in C Am. to nw Col., 1 in coastal Ven. Fairly common.

NOTE: Previously placed in the genus *Tyranniscus*. 2 species may be involved.

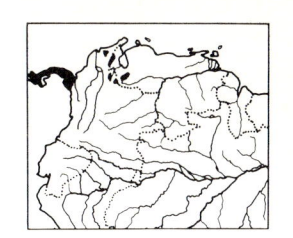

BOLIVIAN TYRANNULET
Zimmerius bolivianus – Plate XLVII 21a-b

11.5 cm. Iris greenish white. Maxilla black, mandible gray. Upperparts and sides of head dark olive, **greater wing-coverts** and inner remiges **edged golden-yellow**. Throat pale yellowish gray; breast grayish, somewhat flammulated, belly pale yellow. Ssp *viridissimus* averages lighter and greener above, and sometimes much yellower below. **Juv**. has brown iris.
Habits: Solitary, sometimes in mixed-species flocks. Feeds (entirely?) on mistletoe berries.
Voice: Has a wide repertoire of whistled notes. The commonest call is an imitable rapid *whee-whee-whee-**whee**oo*, with strongly accentuated last syllable, much like Golden-faced T. Also regularly gives a single, short, slightly rising clear whistle *wheeooo*.
Breeding: Fledgling Aug (Cochabamba).
Habitat: Clumps of mistletoes in canopy and edge of humid forest, occ. in isolated trees on forest clearings.
Range: 1100-2830 m. Cuzco, Madre del Dios, and n Puno, se Peru (birds from s Puno may be the following ssp) (*viridissimus*); La Páz and Cochabamba, Bol. (*bolivianus*). Fairly common.

NOTE: Previously placed in the genus *Tyranniscus*. Perhaps sspp of Golden-faced T.

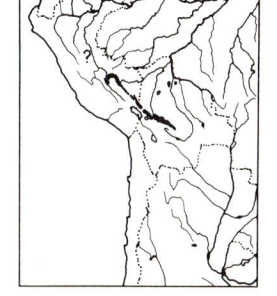

GOLDEN-FACED TYRANNULET *Zimmerius viridiflavus* – Plate XLVII 22a-b

7 cm. Ssp *albigularis*: above dark olive, **forehead yellow**. **Wing-coverts** and remiges **distinctly edged yellow**, except at base of secondaries. Region above and below the dusky lore mottled with yellow, earcoverts golden brown. Below pale gray, throat and belly whitish, sides tinged pale yellow, vent pale yellow. Ssp *chrysops* similar, but larger; yellow of face brighter, underparts more washed with yellow. Ssp *molestus* like *albigularis*, but larger; upperparts duller, less yellowish green. Ssp *flavidifrons* like *chrysops*, but upperparts darker, less greenish; forehead and underparts paler yellow. Ssp *viridiflavus* like *flavidifrons*, but crown grayish olive; underparts pale, flammulated with yellow. **Juv**. unknown.
Habits: Alone, in pairs or family-groups, usually in mixed-species flocks. Perches long-legged with a forward inclination and broad body,

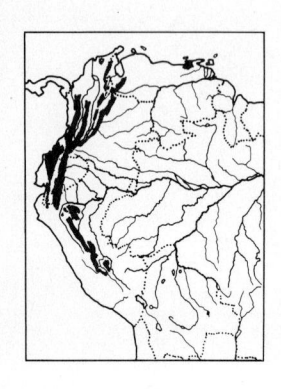

usually rather conspicuously. Forages with horizontal, more slender posture, and half cocked tail. Sometimes wags tail up and down, and occ. half spreads it just before take-off. Hops up along small twigs near the stem, picking berries, often reaching down for them. Rarely perch-gleans insects, occ. making short pursuits for flying prey. Males flick their wings during disputes.

Voice: '*hyy hy weeit*', lasting less than 1 s, last note distinctly rising and accented (c Peru). In Col. the call is often transcribed as *señor buenas días*, or *buenos días*.

Breeding: Nests Apr, June, Nov (W Andes Col.), Sep (Caquetá); large gonads Mar-Dec (Col.).

Habitat: Medium to upper strata, 6-10 m up in lower canopy, occ. the top of tall trees in humid forest, forest edge, second growth, plantations, and gardens. Usually in mistletoes.

Range: 300-2700 m. From Perijá mts to Táchira, Mérida, and w Barinas, Ven., and E and C Andes of Col. (not in Nariño), and from ne Ecu. s to San Martín, Peru (*chrysops*). W Andes of Col. n of Nariño (*molestus*, doubtfully distinct from *chrysops*), from Nariño, sw Col. to Guayas, w Ecu. (*albigularis*), from sw Guayas to w Loja, sw Ecu. (*flavidifrons*); from Huánuco (Carpish and Sira mts) through w Junín and Ayacucho to Vilcabamba mts in Cuzco, Peru (*viridiflavus*). 2 premontane sspp in Sta Marta mts and n Ven. Fairly common to common.

NOTE: Previously placed in the genus *Tyranniscus*. The nominate ssp may be a distinct species, possibly incl. the Bolivian Tyrannulet. The other sspp should then be called *Z. chrysops*.

Tyrannulets – Genus *Camptostoma*

2 species, 1 from s USA to n Costa Rica, the other from Costa Rica to n Arg., in arid and semi-arid country from trop. to lower temp. zones. The nest is globular with a side entrance (see *Hornero* 12(1983): 132-3). See also *Amer. Mus. Novit.* 1109(1941).

SOUTHERN BEARDLESS TYRANNULET
Camptostoma obsoletum – Plate XLVII 15a-b

8.5 cm. Ssp *sclateri*: above olive-gray, crown darker, brownish gray. **Wings with 2 buff bars**, tertials edged white, rest of remiges edged olive, except at the basal 3rd of secondaries. Rump tinged yellowish green, sometimes paler; tail with yellowish white edges and tips, feathers rounded. Supercilium, obscure narrow eye-ring and lunule below eye pale gray. **Underparts light gray**, c belly white, flanks and vent pale yellowish. Ssp *maranonica* slightly darker and grayer above, purer white below, without yellow; tail paler, esp. below, where it is creamy white with more buffy tips; rump creamy buff in contrast to sooty olive back; wingbars paler buff; forehead somewhat paler than crown, sometimes even whitish (more commonly than in *sclateri*). Ssp *bolivianum* like *sclateri*, but crown grayer, flanks tinged olive-buff rather than pale yellow. Ssp *bogotensis* brownish sooty olive above, dark crown grading into color of back. Wingbars yellowish white, tail brown,

feathers edged brownish olive and tipped yellowish white. Breast olive-buff, belly yellow tinged buff. Ssp *caucae* with contrasting sooty cap, whitish wingbars, and grayish throat and breast in strong contrast to sulphur-yellow belly. **Juv**. (*sclateri*) has no yellow on the belly.
Habits: Solitary. Characteristic posture: restlessly perch-gleans insects among branches, with horizontal body, often stopping to stretch neck, raise crest and tail, and spread wings in a jerk. Occ. hover-gleans and rarely makes short sallies to foliage or flying insects.
Voice: Vocal. Gives 5-7 explosive notes, either *tee teé tee...* whistled wearily down scale from the accented 1st or 2nd note, slightly accelerating at the end or equally valued, or *whee...* whistled in minor key. Also twitters and gives a loose or throaty chattering little rattle, and a *weéuh*.
Breeding: Juv. July (Cauca).
Habitat: 3-8 m up, inside canopy of trees, and in tall bushes. Arid zone. Riparian thickets, trees in gardens, scrubby forest, and second growth.
Range: Lowlands to 2600 m, in Bol. to 3125 m. C Col. on w slope of W Andes, Cauca Valley, and upper Magdalena Valley (*caucae*), 'Bogota' and nw Meta, Col. (*bogotensis*), w Ecu. and extreme nw Peru (*sclateri*), middle Marañón Valley from Amazonas to e Ancash (crossing over to w slope in e Piura?) (*maranonicum*); from La Paz and Cochabamba, Bol., s to Tucumán, nw Arg. (*bolivianum*). 10 other sspp in the lowlands from Costa Rica to Urug., absent only from areas with unbroken humid forest. Common.

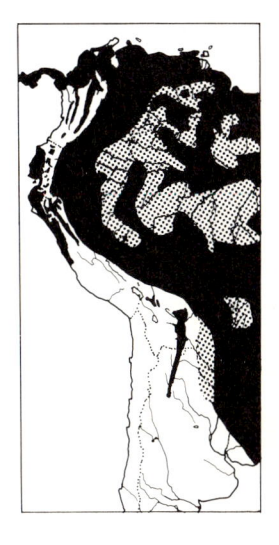

Elaenias – Genus *Elaenia*

17 species distributed in the Caribbean and from Mexico to n Arg., inhabiting both humid and arid regions from trop. to temp. zones. Medium-sized tyrants, usually with an upright posture, and often with a white median crown-patch exposed during excitement. Most species have dull olive upperparts and breast, pale yellow belly, 2 pale wingbars, a pale eye-ring, and a pale base of the mandible. They are extremely difficult to tell apart. The bill is fairly short, and berries form the main part of the diet. The nest is a neat open cup, placed in a bush or a tree. See *Amer. Mus. Novit.* 1108(1941).

WHITE-CRESTED ELAENIA *Elaenia albiceps* – Plate XLVII 13a-d

13 cm. Ssp *albiceps*: semi-concealed median whitish crown-patch. Above dark smoky gray, rump tinged brownish. Lores, eye-ring, and 2 wingbars whitish; remiges edged whitish, broadly on inner tertials; blackish patch at base of secondaries only very rarely absent. **Below smoky gray, throat somewhat paler, c belly white**, vent sometimes tinged yellowish. Ssp *chilensis* has a narrower bill, a pure white crown-patch, a very distinct eye-ring, narrow wingbars, and is more olivaceous above; 1st (outer) primary almost always longer than 6th (shorter in other ssp). Ssp *urubambae* is told from *albiceps* by paler brown upperparts, duller eye-ring and wingbars, yellowish belly, and heavier bill. Ssp *modesta* paler and duller, esp. on flanks, with hardly any eye-ring; wingbars weaker and grayish, the lower formed by narrow edges rather

than square spots; narrower, though distinct edging to tertials; secondaries often edged to base, leaving little or no blackish patch. Underparts ashy gray and white, only rarely with faint yellowish tinge; feathers of crown with more conspicuous dark centers, giving speckled appearance in fresh plumage. Ssp *diversa* like *urubambae*, but throat and belly whiter, and wingbars stronger; *griseogularis* similar, but wingbars even stronger, upperparts slightly paler, crown-patch usually duller and less extensive. **Juv**. lacks crown-patch, and has buffy white wingbars.

Habits: Alone or in pairs. Fairly shy. Active. Perches with a slight forward lean, protruding breast, and often with hanging wings, spread tail, and raised crest (forming 2 'horns'), thus **exposing white crown-patch**. Forages from within canopy of trees or bushes, making sallies up to 2 m for flying insects outside the tree, often changing site. Berries (e.g. *Schinus molle* and *Lanthana*) form an important part of the diet, and are picked by sally-gleaning, hover-gleaning, or reaching down. Sings from tree-top. During migration up to 100 birds has been seen gathering to feed in a flowering tree.

Voice: *chilensis*: call a distinct clear whistle *fío*, or *fío fío*. Song *trrie trrie er ee ee fío* or shorter phrases like *te río*, or *rire eéo*). Other sspp without the *fío*: calls *weee*, singly or repeated up to 4 times. Song *wee wee rríe*, or *rrrrie*, sometimes descending in middle or last part. Also shorter phrases *rrie, rrie rrie*, or a lower-pitched *rrear*.

Breeding: Eggs Nov-Jan (*chilensis*); song Dec-Feb (*modesta*); juv. June, Sep (La Paz).

Habitat: 1-10 m up. Edge and clearings in *Nothofagus* forest, riparian shrubbery, gardens, orchards. Fond of willows and peppertrees (*Schinus molle*). Not found in humid forest in n parts of range.

Range: From sea level to 2000 m in Andes from Isla Grande n to Atacama, Chile, and Chuquisaca, Bol., s of Córdoba also in lowlands of Arg.; winters in the Amazon basin (*chilensis*). At 1700-3300 m from Cochabamba, Bol., to n Puno, se Peru (*albiceps*), 2800-3300 m in Urubamba Valley Cuzco and ne Apurímac, se Peru (*urubambae*). Coast and lower slopes of W Andes from Atacama, n Chile n to La Libertad, Peru: winter-quarters unknown, but apparently in interior Peru as far as Amazonas, San Martín, and Huánuco (*modesta*). Subtrop. zone and up to 2750 m in C Andes and e slope of W Andes from Huánuco to s Cajamarca, Peru (*diversa*); 2000-3250 m (down to 350 m in Piura) from Lambayeque and c Cajamarca, n Peru through w and e Ecu. to both slopes of Nariño, Col. (1 rec. w of Bogotá in E Andes) (*griseogularis*). Fairly common.

NOTE: Ssp *modesta* is often considered a distinct species, **PERUVIAN ELAENIA**. Ssp *chilensis* apparently hybridizes with Small-billed E. along Andean slope of s Bol. Ssp *griseogularis* apparently hybridizes with Sierran E. in sw Ecu.

SMALL-BILLED ELAENIA *Elaenia parvirostris* – Plate XLVII
10 – breeds at least to 2000 m in s Bol., during non breeding season somewhat migratory, mainly to lowlands, but once rec. high on mt Illiniza, w Ecu. (type of *E. aenigma* (Stresemann)). With smaller bill than *E. albiceps chilensis*, but not safely told apart in the field.

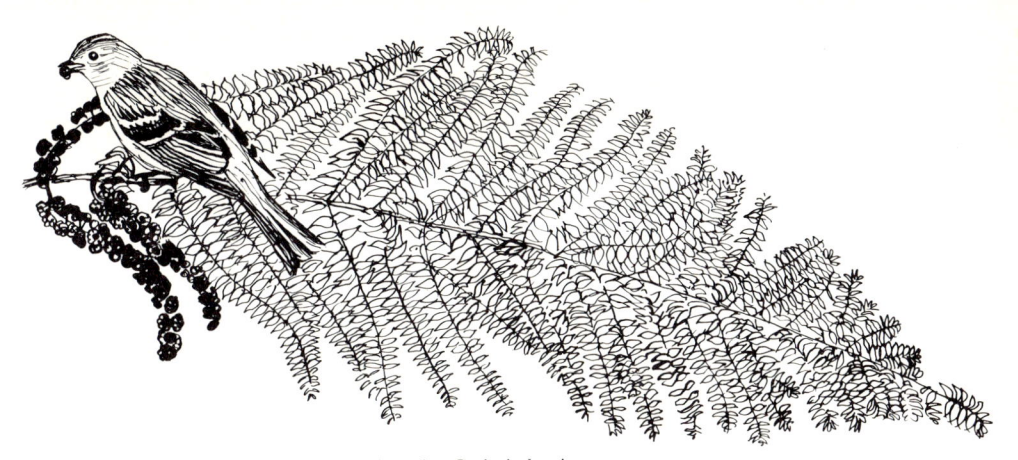

Small-billed Elaenia eating Coriaria berries

YELLOW-BELLIED ELAENIA *Elaenia flavogaster* occurs up to 2450 m in La Paz and Cochabamba, Bol. Similar to Small-billed E., but somewhat larger, elongate crown-feathers forming bushy crest with white center; upperparts olive; throat and sides of neck grayish white, deepening to pearl-gray on upper breast; rest of underparts pale yellow.

SLATY ELAENIA *Elaenia strepera* – **Plate XLVII 9a-b** - lives in bushes or trees along streams in wooded valleys at 1200-2000 m or maybe to the lower temp. zone, from Sta Cruz Bol. s to Tucumán nw Arg. In winter n to e Peru, e Col., and Ven. 15 cm. Short-billed. **Male dark gray**, throat gray, contrasting **belly whitish**. Eye-ring white; crown with somewhat elongated feathers and concealed white median patch. Wings with 2 faint gray bars and edges of remiges, except at base of secondaries; tail-feathers edged gray, narrowly tipped white when fresh. **Female** more greenish gray; c lower breast and belly yellowish white. Wing-coverts with 2 conspicuous bars, formed by ochraceous gray tips of outer webs, the lower bar grayest; eye-ring conspicuous. **Imm.** like female, but pale on belly more extensive, and pale yellow; wings with 3-4 ochraceous bars. **Juv.** unknown. Solitary. Posture very upright. Perches inside or openly in the top of a bush, making short sallies for insects, or hover-gleans for berries. Returns to same post. Regularly jerks tail. In winter sometimes follows mixed-species flocks. Song typically given from pre-dawn through the very early morning, a characteristic, insect-like, rising *trrrrrrit!*, often preceeded by dry *trrt*, the whole lasting 1.5 s. Also gives an emphatic unwavering buzz, like a fishing reel. (BW).

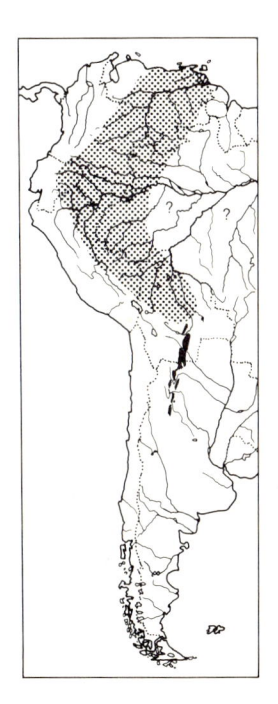

MOUNTAIN ELAENIA *Elaenia frantzii* – **Plate XLVII 14**

13.5 cm. **Neither crown-patch nor crest**. Above grayish olive, 2 wing-bars and broad edges to tertials conspicuously grayish white, secondaries edged yellowish white except at base. Below smoky gray with very faint yellowish green wash, throat slightly paler; c belly whitish, sometimes faintly streaked with pale yellow. Tail narrowly tipped whitish when fresh. Ssp *browni* smaller, and paler above (more greenish olive). **Juv**. duller, above dull brown, upper tail-coverts and tail tipped buffy. Below dull white, breast shaded with olive; a very faint yellowish median stripe on breast and abdomen.

Habits: Alone, in pairs, or family-groups. Aggressive. Flutters amidst branches and occ. picks prey from the ground. Also plucks berries (e.g *Coriaria* and *Rubus*). Seldom sallies for insects, and usually does not perch upright. Makes long fast flights.

Voice: Call a simple, loud *pseer* or *pseeur*, slightly falling at the end, and varied to an emphatic *pür*, sometimes rougher or softer. Song vireo-like, of short phrases, alternately burred and chirped, of 1-3 syllables, always accented on the rising last one. Several other vocalizations.

Breeding: Fledgling June (Huila), Aug (Valle); large gonads Mar-Nov (Col.).

Habitat: Inside low trees, shrubbery, and brambles. Forest clearings, edge of woodland, partly bare ridges, and cultivated areas.

Range: 900-3650 m, mostly 1800-2800 m. Coastal mts and Andes of Ven., all 3 ranges of Col. n of Nariño (*pudica*); Sta Marta mts (*browni*). Birds from Perijá mts variously referred to *pudica* and *browni*. 2 sspp from Guatemala to Panamá. Fairly common to common.

HIGHLAND ELAENIA *Elaenia obscura* – Plate XLVII 12

16 cm. Fairly short-billed, small-headed, broad-bellied, and long-tailed. Head rounded, without crest or median patch. Above dark grayish olive; eye-ring, 2 wingbars, and broad edge of inner tertial whitish; rest of tertials and inner secondaries narrowly edged whitish. **Most of underparts dull yellowish green**, throat somewhat paler; c belly pale yellow, faintly streaked with yellow-green. Tail tipped whitish when fresh. **Juv**. dark rufescent brown above, dirty grayish white below. Solitary. Perches on outer branches, 1.5 m below treetop, sallying to 1 m above top, dropping back to same place. Also eats berries, e.g. of *Coriaria*. Several calls: *trrrrié*, *rrrrhie*, snarling *rrrhit-rrhit rrrh*, *krit-ki rryt*, *krit-ki rret*, *kr kr rrhee*. When excited *rrhi rrhie rhi rrhie rhi rhitrreee*, or transscribed *de dée de dée de deedreee*. Eggs Oct (La Paz), Jan (Tucumán); fledgling Feb (Tarija); juv. Nov (La Paz).

1-4 m up in stunted cloud forest, second growth, and shrubbery at edge of humid forest at 1850-2500 m, occ. down to 1200 m, from Condor mts ne Cajamarca, n Peru, through C and E Andes of Peru, and Andean slope and foothills of Bol. s to Tucumán, nw Arg. (*obscura*). Another ssp in se Brazil and adjacent regions. Fairly common.

SIERRAN ELAENIA *Elaenia pallatangae* – Plate XLVII 11

14 cm. Ssp *intensa*: above grayish olive, crown with somewhat elongated feathers and concealed whitish to yellowish white median crown-patch. Eye-ring yellowish. **2 conspicuous wingbars yellowish white**, tertials broadly edged whitish, secondaries edged yellowish white ex-

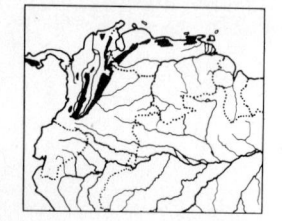

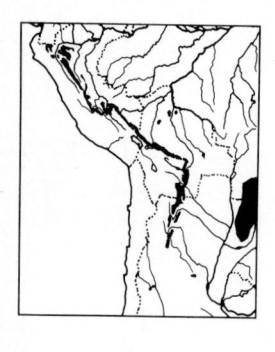

cept at base. **Most of underparts pale yellow**, throat and upper breast washed with greenish gray. Tail narrowly tipped whitish when fresh. Ssp *pallatangae* has crown darker than back, and paler yellow underparts. Ssp *exsul* like *intensa*; said to be brighter yellow, but this must be due to comparison with faded specimens. **Juv.**: no crown-patch; upperparts washed rufescent brown; below yellowish white, narrow band across breast buffy ochraceous.
Habits: Solitary. Sallies for insects and berries of *Coriaria* and *Rubus* from rather open perches.
Voice: Call (?) a sharp *whéee*. Dawn song a rapid *rrérit* repeated every 1.5-2 s. Another bird, probably female, may give a short lower-pitched *tre* immediately after, and after some time with mostly male singing they may synchronize to a perfect duet *reérit-tre, reérit-tre....*
Breeding: Fledglings Nov (w Ecu.), June, Sep (Huánuco), July (Cuzco); juv. Sep (Cajamarca); large gonads Jan-Aug (Cauca).
Habitat: Medium strata. 3-7 m up in humid second growth and epiphyte-laden trees at edge of humid forest, extending into riparian shrubbery of immediately adjacent arid zones.
Range: 1400-3650 m, mostly in temp. zone. From Valle, W Andes and Quindío, C Andes of Col. s to s Ecu. (*pallatangae*), E and C Andes of Peru s to Cuzco (*intensa*), from se Puno in se Peru to Cochabamba, Bol. ('*exsul*'). 1 ssp in s Ven. Fairly common, sometimes in dense populations.

NOTE: Apparently hybridizes with White-crested E. in Ecu. and perhaps in nw Peru (wc Cajamarca: ZMUC).

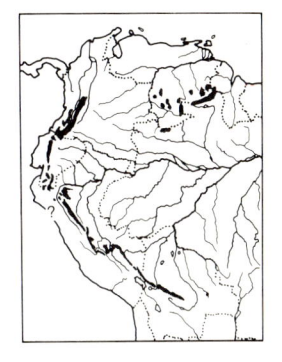

Tyrannulets – Genus *Mecocerculus*

6 species restricted to humid forest and shrubbery in the subtrop. and temp. zones of the Tepuis and coastal mts of Ven., and the Andes to nw Arg. Fairly small tyrants that actively and rather conspicuously forage at low to medium heights, regularly raising, fanning, or wriggling the tail. They all have 2 distinct wingbars, a very distinct blackish patch at base of secondaries, dusky lores, and the area below lores and eye finely mottled gray and white. Some sally and hover-glean, some perch-glean, and at least 1 species seems to feed mainly on berries. No nest of the genus has been described. See *Amer. Mus. Novit.* 1095(1940).

NOTE: Aparently polyphyletic, *minor* and *calopteryx* being related to *Phylloscartes*; *stictopterus*, *poecilocercus*, and *hellmayri* to the genera *Ornithion* and *Camptostoma*, which leaves only *leucophrys* in *Mecocerculus*.

WHITE-THROATED TYRANNULET
Mecocerculus leucophrys – Plate XLVII 5a-b

12.5 cm. **Relatively long tail; upright posture** when perched. Ssp *leucophrys*: above dark gray-brown with olivaceous cast. **Throat white**, contrasting with gray breast and pale yellow belly, belly fading to whitish with wear. Narrow white supercilium and broken eye-ring. 2 broad buffy wingbars; whitish to pale rufous wing-panel, widest on inner wing. The somewhat pointed tail-feathers are tipped pale buff when fresh. Ssp *brunneomarginatus* larger (13.5 cm), dark brown above, with buffier wingbar, and more distinct buffy panel, the latter whitish or buffy on tertials. The gray breast is tinged brown. Ssp *rufomarginatus* like *brunneomarginatus*, but darker above (sooty brown), with a darker crown, narrower and paler wingbars (tawny), and slightly paler underside. Ssp *notatus* like *rufomarginatus*, but paler below. Ssp *setophagoides* like *notatus*, but paler above, much like *leucophrys*, but more brownish above (less olive), and with buffier edges of secondaries. Ssp *montensis* much like *setophagoides*, but paler, less tinged with ochraceous. Ssp *gularis* has crown dusky drab, somewhat in contrast to smoky brown back; wingbars and edges to tertials whitish. Ssp *pallidior*, the largest and most distinct form, has whitish belly, pale gray-brown breast without any olive at sides, cinnamon (vs. pale yellow) carpal edge, bright cinnamon panel, and broad, dark hazel-brown wingbars. **Juv.** of the yellow-bellied forms with whitish belly.

Habits: Alone, in pairs, or flocks, often important in mixed-species flocks. Restless. Hover-gleans and sallies from branches near the stem inside the canopy, rarely perching the same place twice, taking most prey from under surface of leaves. Perches with upright posture, but is horizontal when perch-gleaning along the branches, wagging the tail from side to side. Also clings sideways on stems, looking up and down for insects.

Voice: Contact call a sharp twitter. Alarm a sharp *tjip-tjip-tjip...*, or *keepkeepkep*. Several other calls, some wheezy. Rarely heard song a soft, complex warble.

Breeding: Fledglings Mar (Cauca/Huila), Aug (nw Ecu), Sep (ne Ecu., Ancash); large gonads Jan-Aug (E and C Andes, Col.).

Habitat: 1-4 m up inside top canopy of trees or bushes. Humid *Polylepis* and *Alnus* forest, elfin forest, edge of cloud forest, humid montane shrub.

Range: 1850-4250 m in temp., rarely subtrop. zone. Perijá mts, and Andes of Ven. from s Lara to Táchira (*gularis*), Sta Marta mts (*montensis*), from N.de Santander to Cundinamarca n end of E Andes of Col. (birds from Perijá mts may belong here) (*setophagoides*), C and W Andes of Col. s to Cauca (*notatus*), from s Cauca and Nariño, s Col. through Ecu. to Piura and n Cajamarca, n Peru (*rufomarginatus*), from s Cajamarca through e and c Peru to Cuzco (*brunneomarginatus*), from se Puno, se Peru (intergrades) along Andean slope of Bol. to Salta and Tucuman, nw Arg. (*leucophrys*). Cord. Blanca Ancash, w Peru (*pallidior*). 5 sspp in Ven. Common.

NOTE: Species rank might be considered for ssp *pallidior*.

WHITE-TAILED TYRANNULET
Mecocerculus poecilocercus – Plate XLVII 3

10 cm. Crown and headside dark gray with faint narrow pale gray supercilium; back grayish olive; **rump pale yellow**; tail-feathers dusky, edged olive on outer webs, **inner webs mostly white**. Wings blackish with 2 broad, pale yellowish buff bars; primaries edged whitish, sec-

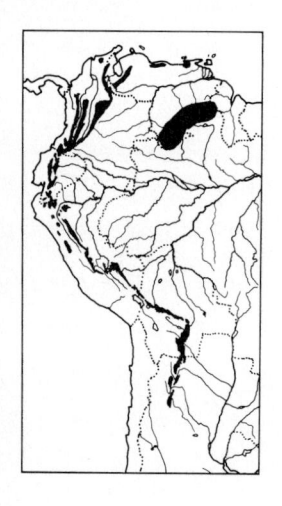

ondaries edged yellow-green, tertials broadly edged whitish. Throat and breast pale gray; rest of underparts, incl. extreme c breast yellowish white. **Juv**. unknown.
Habits: Alone, in pairs or up to 6 together, usually in mixed-species flocks. Restless. Hops along thinner branches, picking insects from underside of leaves. Fairly horizontal posture; regularly makes little jerks with fanning of tail (showing white), and sometimes also spreading wings (showing yellow rump). Tail not raised. Strict perch-gleaner (i.e. does not sally).
Voice: Call a wheeny *zeee*, and a distinctive high sibilant *tee-tee-tee-tee*.
Breeding: Large gonads Aug, Sep (Santander).
Habitat: Medium to upper strata of cloud forest and humid forest, sometimes in *Cecropias* trees.
Range: 1400-3050 m, usually at lower elevation than White-banded T. All 3 ranges of Col., e and w Ecu., n, e, and c Peru s to Cuzco. Fairly common in n Ecu. and c Peru, uncommon in Col.

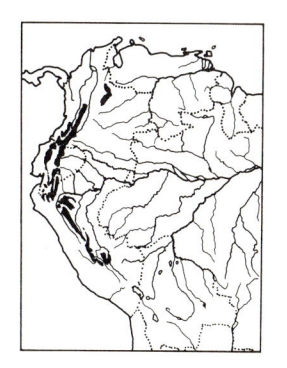

BUFF-BANDED TYRANNULET *Mecocerculus hellmayri* – Plate XLVII 2

10 cm. Cap gray, back olive-green, small rump-patch olive-yellow. Wings with 2 conspicuous yellowish bars, edges of remiges and tail yellow-green. Supercilium white, face light gray with dark line through eye; underparts pale yellow, clouded with olive-gray on breast. **Juv**. has buffy, rather broad wingbars; remiges narrowly edged olivaceous, tertials edged whitish; throat and breast mottled with olive-gray; feathers of upper belly faintly dark-tipped.
Habits: Alone in mixed-species flocks. A rather strict sallier, which makes short upward sallies to underside of leaves. Very active.
Voice: Song is a series of 4 clear whistled notes, each inflected upwards.
Breeding: Juv./imm. May (Cochabamba).
Habitat: Mid levels of humid montane forest, incl. *Podocarpus* forest.
Range: 1100-2700 m, occ. down to 800 m. From n Puno, se Peru, s along the Andean slope to Jujuy, nw Arg. Birds from Jujuy may be separable subspecifically. Uncommon.

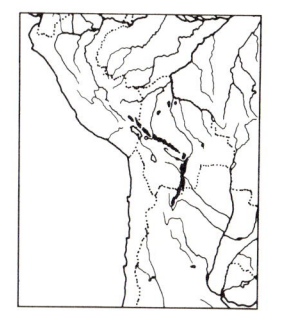

SULPHUR-BELLIED TYRANNULET *Mecocerculus minor* – Plate XLVII 1

11.5 cm. Crown dark gray, rest of upperparts dark olive-green. Wings dark with 2 buff bars and edges of secondaries (except at base); tertials edged paler, primaries only narrowly edged pale buff. Narrow supercilium and broken eye-ring white. **Below rather bright yellow**, chin white, breast flammulated with olive. **Juv**. unknown.
Habits: Usually in groups of 4-6, following mixed-species flocks. Rather actively perch-gleans outer branches with half cocked tail. Occ. hangs down, but resumes vertical posture when perched. Fruits form an important part of the diet.
Voice: No data.
Breeding: Juv. July (se Ecu.), Aug (Amazonas), Nov (e La Libertad).
Habitat: Middle strata. Mostly 2-4 m up in young trees and bushes at

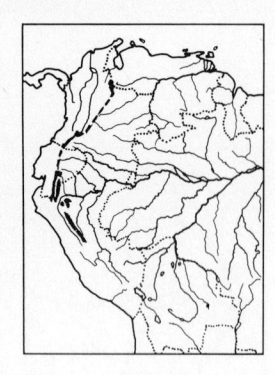

forest edge and clearings in humid forest and cloud forest.
Range: 1700-2600 m. Sw Táchira, Ven., and neighbouring ne Col., head of Magdalena Valley and e Nariño, s Col.; Morona-Santiago, se Ecu.; n Piura and in Amazonas, e La Libertad, and Huánuco (Carpish mts), Peru. Rare in c Peru, fairly common in se Ecu. and n Peru; perhaps sparsely, but continuously distributed further n.

WHITE-BANDED TYRANNULET
Mecocerculus stictopterus – Plate XLVII 4

11.5 cm. Ssp *stictopterus*: crown dark gray, rest of upperparts olive-brown. Dark wings with **2 broad white bars** and edges of tertials, secondaries edged buff except basally, primaries edged yellowish white. Long supercilium and broken eye-ring white, eye-line and earcoverts dusky. Below pale gray, belly white, flanks and vent with faint yellow tinge. Tail narrowly tipped whitish when fresh. Ssp *taeniopterus* similar, but back olive; all remiges except tertials edged pale olive, dark patch at base of secondaries small; vent pale yellow. Ssp *albocaudatus* has entire outer tail-feather and inner web of next 2 dull buffy white, whereas they are only slightly paler than rest of tail in the other sspp. **Juv.** undescribed.
Habits: In pairs or family groups, often in mixed-species flocks. Restlessly makes short sallies of 20-40 cm mainly to lower surfaces of foliage and flying insects, hover-gleans, and perch-gleans along the outer ends of thin branches, with a slightly raised tail and wriggling body. Also eats berries.
Voice: The frequently uttered call is a distinctive, slightly rising, high-pitched, snarling *wheeeit*, somewhat like the contact note of Rufous Wren. Song with a similar quality, but varied with short trills *weeeit trrrrt weeit*, *weeeit trrrrrt*, or *trrríerrrrt*.
Breeding: Fledgling Dec (Huila/Cauca); large gonads Jan, Aug (C Andes Col.).
Habitat: Medium to upper strata of humid forest and cloud forest, often at edges.
Range: 1800-3700 m. Andes of Ven. from Trujillo to Táchira (*albocaudatus*), C Andes, and locally in E and W Andes of Col., s through e and w Ecu. to e Piura, n Cajamarca, and Amazonas, n Peru (*stictopterus*), and C and E Andes from c Peru s to w Cochabamba, Bol. (*taeniopterus*). Common to fairly common.

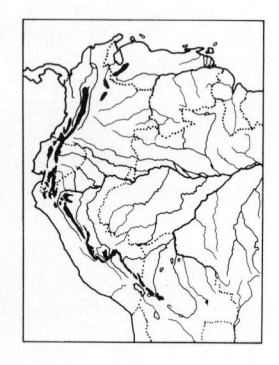

Tyrannulets – Genus *Serpophaga*

5 species found from Costa Rica to Río Negro Arg., practically throughout, from trop. to lower temp. zone. Small tyrants with black bill and feet, and usually with median white crown-patch and dark tail. Foraging involves both perch-gleaning, sally-gleaning, true flycatching and running on the ground. Most species are associated with water. The nest is an open cup.

TORRENT TYRANNULET *Serpophaga cinerea* – Plates X 7 and XLVII 6a-b

10.5 cm. Rather horizontal posture, crown partly raised. Gray, turning white on c belly; **head to below eyes, wings, and tail blackish**. Median crown-patch, 2 narrow wingbars, and edges of tertials white. **Juv**. similar, but cap dark gray without median patch; wing-markings buffy gray.
Habits: Alone or in pairs. Territorial in breeding season. **Perches on stones and rocks in streams**. Seems sometimes to hide between rocks when approached. Frequently flicks and sometimes cocks tail, makes sallies up to 5 m for flying insects; also runs and picks up prey on the ground or even from the water, and sally-gleans from low twigs or roots in banks.
Voice: Dawn song is usually given from a rock *chit-chitrrr-chitrrr...*, sometimes interspersed with with brief conflict chatters, that are also heard at other times. Also a repeated, single *chit*.
Breeding: Eggs July, Aug (Cundinamarca), fledglings May (Cundinamarca), Oct (Junín), juv./imm. Mar (Boyacá), May (Mérida), June, Aug (nw Ecu.), June (n Cajamarca), Oct (Tolima), territorial June (nw Ecu), not territorial Aug (nw Ecu); large gonads Sep (W and E Andes Col.).
Habitat: Rocky rapid streams in both humid and arid zone. At Lake Tota and on the Bogotá and Ubaté savannas also in reed marsh (*Typha, Scirpus*), and flooded alder forest (*Alnus*).
Range: 600-3500 m, mostly 1400-2300 m. Sta Marta mts, Perijá mts, and Andes from Trujillo, Ven., s to Cochabamba, Bol., on w slope s to Lima, Peru (*cinerea*). Another ssp in Costa Rica and Panamá. Common in the humid subtrop. zone, rare in w Peru.

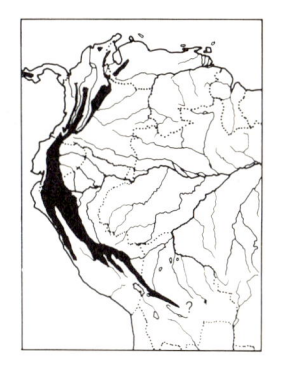

WHITE-BELLIED TYRANNULET *Serpophaga munda* – Plate XLVII 7

10 cm. Tail fairly long, loosely hinged. **Body lightly built. Above gray**, median crown-patch white bordered black. Dark wings with 2 distinct white bars and panel. Outer web of outer tail-feather white. Conspicuous supercilium and broken eye-ring white. **Below white**, with light gray sides of breast. **Imm**. has slight olive wash on lower back, and faint yellow tinge on lower abdomen. **Juv**. has buff wingbars. Alone, in pairs, or family groups. Restlessly perch-gleans and sally-gleans in the canopy among twigs and leaves, like an active warbler. When alarmed, may sit openly in a treetop for a while with half turned body and the long, slender legs apart, before taking off. Call a brief trill given with head thrown back *rrrtrtrtr* 0.5-2 s long, and may be given throughout the day. Dawn song similar, but regularly repeated with one burst every 4 s. Also irritated chatters 0.5-3 s long, at various times of day, and other calls *tseet*, *tjik-tjik*, and *trit*, *tjit tjit tjit*. Fledgling Feb (Cochabamba).

5-15 m up in the canopy of trees in arid and semi-arid regions. Found on slopes with low dry scrub and an occ. tree, as well as in dense thorny forest. Also in riparian trees, gardens, and hedgerows, at 50-2600 m. Lowland chaco and Andean slopes from Cochabamba and Sta Cruz,

Bol., s to Mendoza, Arg. Migratory in s, where present from Oct to Jan. Fairly common.

NOTE: Variously treated as a distinct species, and as a color phase or ssp of White-crested T. (*S. subcristata*) of e Brazil to Río Negro, Arg. See *Amer. Mus. Novit.* 1749(1955).

GREATER WAGTAIL-TYRANT *Stigmatura budytoides* – **Plate XLVII 8** – of thorny woodland from s Brazil to c Arg. ascends to 2600 m in the arid part of Cochabamba and Sta Cruz, Bol. (*budytoides*). **Unmistakable**. 16 cm. **Tail long, slightly graduated**, black, **broadly tipped white** except on 3 c pairs; outer web, and large median spot of inner web of outer feather white. Above dark grayish olive; **greater wing-coverts and tertials broadly edged white; median coverts black, broadly tipped white**. Short supercilium yellow. Underparts pale yellow, sometimes tinged buffy on breast. Normally in pairs. May perch upright, but usually perch-gleans actively, with fairly horizontal posture, somewhat hanging wings, and raised tail that is constantly wriggled. 1-7 m up in trees and bushes, mostly in the outer branches. Song loud, fast, and stereotyped, given in duet *tri ti treeowhit, tri ti treeowhit,....*

Tit-tyrants – Genus *Anairetes*

7 species found in the upper subtrop. and temp. zones of the Andes from Col. to Isla Grande and on Juan Fernandéz Isl. Chile. Small restless aggressive tyrants, with a crest that in some species is very long and usually raised, exposing white nape-patch. In 1 species the iris is pale yellow. The plumage is more or less streaked. The outer tail-feathers are edged white or whitish, and with a rare individual exception, all have a blackish patch at base of secondaries. Usually in pairs or family-groups, they make short forward sallies to foliage, occ. hover-gleaning, and also perch-glean twigs and small leaves, somewhat in the manner of a tit-mouse. In winter they also feed on seeds. The nest is a finely woven small compact open cup, placed in a bush. See *Amer. Mus. Novit.* 1095(1940).

NOTE: The genera *Yanacea* and *Uromyias* should probably be retained.

ASH-BREASTED TIT-TYRANT *Anairetes alpinus* – **Plate XLVI 2**

13 cm. Bill black. Above deep gray, very broadly streaked blackish on mantle. Top of head black, mixed gray on forehead, crown-feathers much elongated to form a long crest; nape white, at some angles appearing as a postocular streak. 2 prominent white wingbars and white panel. **Tail black, feathers broad with rounded tips, outer 2-3 feathers mostly white**. Below gray, c belly pale yellow (*alpinus*) or white (*boliviana*). **Juv**. duller, with slightly streaked throat and with short crest.

Habits: Alone or in pairs. Restlessly flitters about and climbs in the outer branches, perch-gleans, often reaching down for insects, and occ. sally-gleans and rarely descends to the ground. Does not wag tail, and only occ. raises crest.
Voice: Usually silent. Sometimes utters a complaining *eeeh*.
Breeding: Imm. Mar, July (Cuzco); small gonads May (Ancash), Mar (Cuzco).
Habitat: Outer branches, occ. on the ground. Restricted to semi-humid *Polylepis* woodland, mainly in high and steep rocky parts.
Range: 4000-4600 m. Cordillera Blanca, in Ancash, w Peru (*alpinus*). Nevada Veronica area above right bank of Urubamba Valley in Cuzco, se Peru, and 1 old rec. in La Paz, Bol. (*bolivianus*). Very rare and local, with usually only 1-2 pairs per occupied woodland. Considering the current cutting of *Polylepis* woods for firewood, probably endangered.

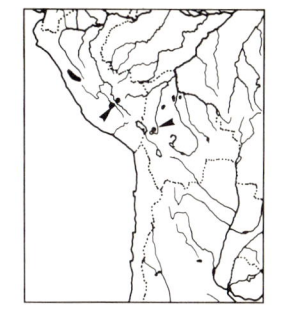

NOTE: Previously placed in its own genus *Yanacea*.

UNSTREAKED TIT-TYRANT *Anairetes agraphia* – Plate XLVI 6a-c

12 cm. Bill dark. Rarely raises crest. **Elongated crest black**; back gray-brown with very faint broad dark streaks. Wings fuscous with light brown to whitish panel. **Tail rather long** and graduated, outer web of outer feather whitish; tail-feathers pointed and tipped white in fresh plumage. Short supercilium and headside gray admixed white and with inconspicuous dark stripe through eye; sides of neck gray. Below mainly white; feathers of **throat and breast** with gray-brown centers, giving **scaled** appearance, flanks and lower belly tinged pale yellow. Sspp poorly differentiated; *squamigera* and *plengei* having quite distinctly scaled breast, *plengei* slightly darker and more olivaceous upperparts. **Imm**. like ad., but all underparts tinged yellow, supercilium slightly whiter; dark scales below less pointed, producing faint white pectoral band. **Juv**. has a prominent black stripe through eye; the supercilium widens behind eye, and is dirty white like side of head and faint pectoral band; wings with 2 narrow, but distinct, pale rufous bars and panel. Whitish in tail replaced by buffy. Scales below obsolete, breast and belly pale buffy yellow; back warmer brown, almost uniform.
Habits: In groups of 2-6 birds, sometimes in mixed-species flocks with hemispinguses and warblers. Restless; makes short upward sallies and hops to glean insects off the underside of bamboo-leaves and twigs, dropping back into cover. When horizontal, wags tail from side to side. During disputes, 2 birds will face one another with upright posture and raised crest.
Voice: Very vocal. Common call a high-pitched explosive *tzrrie*, sometimes *tzeeree*, or *tsit-trrrrrit*. A high-pitched varied *tsieweetsie* given at 1-2 s intervals may be the song. During disputes a wheezy *wheeet*, and *keke-ke*kekeek.
Breeding: Fledgling Aug (Huánuco).
Habitat: Usually within 0.5 m of the ground in low bamboo thickets, sometimes up to 3 m in neighbouring trees and undergrowth. Elfin forest and bamboo-covered landslides in cloud forest.
Range: Peru at 2700-3600 m. Cordillera Colán, c Amazonas (*plengei*,

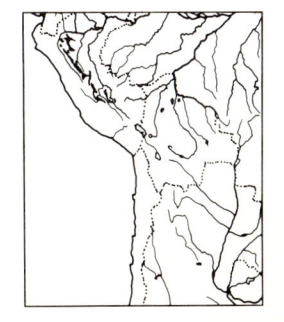

Schulenberg & Graham, *Bull. Brit. Orn. Club* 101 (1981):241-3), from e La Libertad to Huánuco (*squamigera*), Cordillera Vilcanota, Cuzco (*agraphia*). Uncommon, perhaps overlooked.

NOTE: Previously placed in the genus *Uromyias*, comprising this and the following form.

AGILE TIT-TYRANT *Anairetes agilis* – Plate XLVI 7a-b

13 cm. Base of mandible pale. Crown black ending in **fairly short median crest**, broadly bordered on the sides with white admixed gray-brown. Back coarsely streaked pale and dark brown; wings fuscous, tertials edged whitish. **Tail rather long**, graduated, outer feather with white outer web; the pointed tail-feathers tipped white when fresh. Earcoverts gray-brown; rest of headside, chin and upper throat streaked gray-brown and whitish; sides of neck and rest of underside lemon-yellow, more buffy on flanks, and *streaked* dusky except on belly. **Juv**. has 2 pale rufous wingbars and edges to tertials; rest of remiges edged brown. Lateral crown-stripe wider, buffy behind eye; whole forehead grizzled with whitish; dark streaks above and below obsolete.
Habits: In flocks of 4-5 birds. Swiftly hops in the outer branches, reaching out for insects. Occ. flycatches from treetops.
Voice: Vocal. Calls include explosive, fine trills *trirrrr*, *tttttrrrie*, and *trrrt tie trrie*.
Breeding: Fledglings Dec, July (nw Ecu.); large gonads Feb (Cauca/Huila).
Habitat: From near the ground to 7 m up in tops of trees. Humid forest, cloud forest, humid shrubbery, and second growth, always with admixed bamboo.
Range: 1800-3500 m, mainly above 2700 m. From Páramo Zumbador, n Táchira, Ven., in E Andes, s end of C Andes and Nariño in Col. s to Pichincha, nw Ecu. and Zapote Najda mts, se Ecu. Common.

NOTE: Previously placed in the genus *Uromyias*, comprising this and the former species.

BLACK-CRESTED TIT-TYRANT
Anairetes nigrocristatus – Plate XLVI 5a-c

12.5 cm. **Male**: **Bill orange tipped dusky**. **Very long recurved black crest**; prominent white median nape-patch. **Above black**, back narrowly **streaked white**; wings with 2 distinct pure white bars, and a whitish panel. **Tail broadly tipped white**, mostly on inner webs, 11-18 mm on outer feathers, decreasing gradually to 4 mm on c pair; outer web of outer feather white. Forehead, lores, and headside black, with scarcely any white streaks; throat, breast, and sides streaked black and white; belly yellowish white fading to whitish. **Female** duller and with shorter crest; maxilla dark, or only pale at sides of base. **Imm**. like female, but maxilla wholly dark. **Juv**. has dusky crown, whitish nape-patch, and back obscurely streaked blackish and pale olivaceous. Wings and tail like ad. Below dirty white, streaked dark brown across breast; bill dark, washed with flesh at base of mandible.

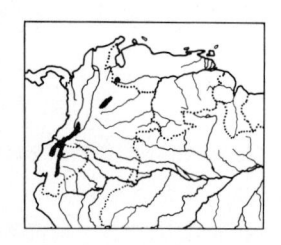

Habits: Alone, in pairs, or family groups. Makes short forward sallies to foliage, and also perch-gleans twigs and small leaves.
Voice: Calls include an explosive rapid series *wheeh-tritritrirrrrritri*, c. 3 s long, repeated at 40-60 s intervals. Also a shorter *wheek-trie-tritritritri*.
Breeding: Eggs June (Huánuco), juv. June, July (La Libertad).
Habitat: Fairly dry zones, in composite shrubbery, *Lupinus*, or *Berberis*, often adjacent to rivers and streams. Also enters *Polylepis* woodland.
Range: Peru at 2350-3600 (-4200) m, rarely down to 2000 m. W Andes from Piura s to n Ancash, where found e of Santa river, and at higher elevations than Pied-crested T-t. In La Libertad crosses to right bank of Marañón river. Also along upper Huallaga river in Huánuco and Pasco, where common. Otherwise uncommon.

NOTE: Possibly a ssp of Pied-crested T-t., although the 2 are perhaps sympatric in Cord. Blanca in nc Peru (see below).

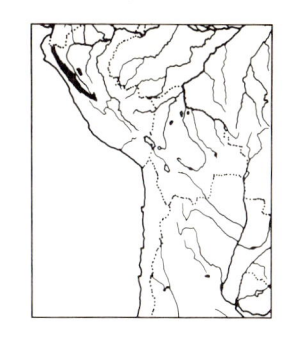

PIED-CRESTED TIT-TYRANT *Anairetes reguloides* – Plate XLVI 4a-c

11.5 cm. **Ssp *albiventris*: male** with mandible and base of maxilla pale; pale on maxilla more restricted than in Black-crested T.-t., and reddish flesh rather than orange. **Above black prominently streaked white**; crest black edged white, nape and distinct median crown-patch white. Wings with 2 whitish bars and panel. **Tail tipped white**, broadest (**4-5 mm**) on inner webs of outer feather, and decreasing inwards; outer feather with white outer web. **Forehead, lores and headside black**; throat black with faint white streaks; rest of underparts white, streaked black on breast and sides; belly with faint yellow tinge in fresh plumage. **Female** (breeding imm. female?) like male, but crest short, and black and white of back and wings replaced by fuscous and pale buffy gray. Forehead and headside with faint streaks, half-ring below eye white, throat and breast white streaked black, belly yellowish, flanks tinged buffy. Maxilla dark. Ssp *reguloides* has a shorter crest, more conspicuous white occipital patch, and narrower white tip of tail (1.5-4 mm on inner web of outer feather). **Juv**. has dark brown back streaked buff. Supralores buffy, almost broken right in front of eye, and turning to streaks posteriorly; lower headside and half-ring below eye buffy. Crest short, crown-patch tinged buff. Underparts light buff with narrow breastband and sides streaked fuscous, streaks sometimes virtually absent.
Habits: As for Black-crested T.-t. Sometimes follows mixed-species flocks, where may be seen in company with Yellow-billed T-t.
Voice: Call a loud descending chatter *trrrrr*. An *eek eeh rrrrr* may be the song. Often gives a loud descending series of whistles, somewhat resembling the song of Southern Beardless Tyrannulet.
Breeding: Juv. Nov (Lima); imm. Nov (Lima), Feb (Ancash), Apr, May (Arequipa); large gonads Mar (Apurímac).
Habitat: Riparian thickets, *Polylepis* shrub; composite shrubbery in the loma zone.
Range: Sea level to 2900 m, in s up to 3550 m (and a bird with *reguloides* tail-pattern seen at 4200 m in Quebrada Matara in Cord. Blanca; O Frimer). From w slope of W Andes in c Ancash s to Ica and w Ayacu-

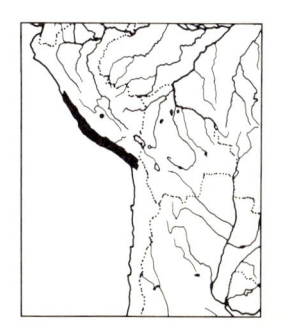

cho, and in nw Apurímac (Andahuaylas, 1 rec.) (*albiventris*); w slope of W Andes from s Ayacucho s to n Tarapacá (Arica), n Chile (*reguloides*). Uncommon.

YELLOW-BILLED TIT-TYRANT *Anairetes flavirostris* –
Plate XLVI 3a-b

10 cm. Base of mandible orange-yellow. Long crest. Sexes alike. Ssp *arequipae*: Above gray-brown, back narrowly and indistinctly streaked dusky; crown with well developed white median patch, and 2 long 'horns' of black feathers with narrow white edges; forehead dusky, more or less streaked white. Wings blackish with 2 well marked, cinnamon-buff bars and panel. Outer tail-feathers with pale outer web. White supercilium from bill broken at eye and turning to streaks posteriorly; white half-ring below eye; rest of headside, throat and breast streaked boldly blackish and white. **Belly light yellow, flanks with hardly any streaks**. Ssp *huancabambae* smaller; *cuzcoensis* large and dark, with broad streaks on breast, strong bill, and long crest; *flavirostris* has an almost unstreaked back. **Told from Pied-crested T-t. by tail-pattern, yellower belly, lighter foreparts, and rather uniform gray-brown upperparts**. Juv. like ad., but has shorter crest and duller, more buffy colors.

Habits: Alone, in pairs, or family groups of 3-4 birds, sometimes in mixed-species flocks with Conebills, Tit-spinetails, and also Black-crested T-t. Much like Tufted T-t. Hops about in the tops of low bushes, occ. darting up to secure insects in the air. Also eats grass-seeds. When excited, wags tail, occ. over the back. Often raises crest to vertical.

Voice: Contact call a short, often descending trill *tirrrr* or *tititirrrr*, lasting c. 0.7 s. Alarm a loud *keer*. Song a series of piercing notes, first slightly ascending, then descending, esp. in amplitude *kee kee* **kee** *kee kee kee kee ke*, or *seet zwee-ee seeta seeta seeta*, lasting about 3 s, and repeated every 10-15 s.

Breeding: Fledglings Feb (Chuquisaca); juv. Feb (Potosí).

Habitat: Fairly open low scrub and riparian bushes and trees in the semi-arid zone. Generally in more open habitat than Pied-crested and Black-crested T-t.s, and in more arid regions than Tufted T-t.

Range: 1000-3650 m, lower in s, rarely on coast in Peru. W Andes from Piura and Cajamarca s to Ancash, and in C Andes to w Pasco and Junín (Andamarca), c Peru (*huancabambae*); w slope of W Andes from Lima, w Peru, locally to Tarapacá, n Chile (*arequipae*); Huancavelica, Ayacucho, and nw Apurímac, sc Peru (*cuzcoensis?*), Cuzco, se Peru (*cuzcoensis*); w Bol., and from Buenos Aires, Córdoba, and Mendoza s to Sta Cruz, Arg.; in winter n to Jujuy and Entre Ríos (*flavirostris*). Fairly common.

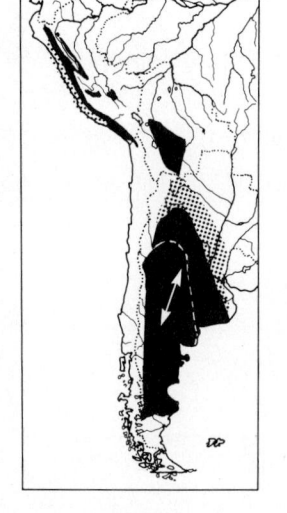

TUFTED TIT-TYRANT *Anairetes parulus* –
Plate XLVI 1a-c

9.5 cm. **Iris pale yellow**. Bill black. **Thin recurved crest**; crown dark gray, median white patch poorly developed or lacking; forehead more or less streaked white. Back mouse-gray; wings with 2 narrow faint white bars and panel. Outer web of outer tail-feather whitish; tail tipped white when fresh. **Supralores white, forming a distinct curve** above black loral patch; white half-ring below eye, and double white postocular streak. Below pale yellow fading to whitish, throat, breast, and sides streaked dark gray. Sspp poorly differentiated, *patagonicus* palest, with more distinct wingbars and broader streaks below. **Juv**. slightly duller than ad., crest shorter, crown more uniform without white; wingbars and panel washed buffy.

Habits: In pairs or family groups, sometimes in mixed-species flocks. Perch-gleans among twigs and foliage, rarely upside down. Also makes short sallies for insects on the vegetation or in the air, and hover-gleans leaves and flowers. Very restless; frequently pivotes on the perch, and occ. flicks tail. Aggressive, often chasing one another, making short undulating flights with pumping tail and bursts of rapid wing-strokes, producing whirrs of snapping, rattling quality.

Voice: Very vocal. Gives loud sharp high-pitched and rising trills, chatters, and irregular phrases and trills, often in duet. Chatters frequently drawn-out and descending towards end. Contact note a fine, irregular *pluit pluit*... **Breeding**: Probably usually double-brooded. Eggs (Sep) Nov-Jan (Isla Grande, s Chile), Aug-Nov (c Chile), Aug (c Arg.); fledglings Jan, May (nw Ecu.), Jan (sw Ecu.), Mar (Huila), Apr (s Cajamarca), June (w La Libertad); juv. Nov (nw Ecu., Jujuy), Dec (La Páz), Feb (Junín), July (Cuzco, Huánuco), Aug (La Libertad, Nariño); large gonads Jan/Feb (Cauca/Huila).

Habitat: Shrubs to forest edge. Has a broad ecological amplitude, but seems to prefer bushes with rather small leaves. Found both in dry thorny scrub and in humid second growth adjacent to cloud forest and humid montane forest. Penetrates denser vegetation than Yellow-billed T-t.

Range: 1830-4200 m, lowest in dry season, and down to sea level in s. From Cauca/Huila, s Col., through Ecu. and Peru to La Paz, Bol. (*aequatorialis*); from Cochabamba, Bol., to Jujuy, nw Arg. (unnamed ssp); from sw Antofagasta and Atacama to Chiloe and Wellington isls, Chile, and sw Arg. from extreme w Neuquén to Tierra del Fuego (*parulus*); from s Mendoza and s Buenos Aires to n Sta Cruz, Arg., e of previous ssp (*patagonicus*). Common.

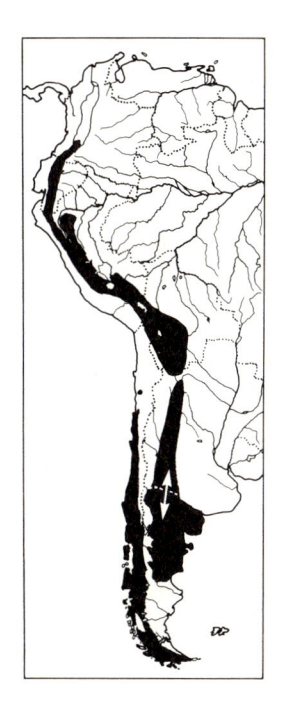

Rush-tyrant – Genus *Tachuris*

Monotypic genus. Legs fairly long, well adapted for perching on vertical reeds. The nest is a very delicate cone-shaped structure of dry reed-fibers, attached along one side to a single reed; it is smoothed on the outside with a strange gum, giving the appearance of having been cast in a mold. See *Amer. Mus. Novit.* 1095(1940).

MANY-COLORED RUSH-TYRANT *Tachuris rubrigastra* – Plate XLVI 26a-b

10 cm. **Unmistakable**. Crown black, bordered laterally by long greenish white line, and with crimson feathers centrallly. Sides of head dark blue; back dark greenish blue, with golden hue on hindneck; wings

black with very broad, buffy white bar, and white outer web of inner remiges. Tail black, outer feathers white. Chin and throat white, breast chrome-yellow, belly buffy yellow, vent red; broken black band on border of breast and belly. **Juv**. duller, without red on crown and vent; feathers of upperparts tipped buff; underparts, incl. throat creamy buff, with only a faint indication of transverse bar.

Habits: In pairs or family groups. Restless, agile. Normally droops wingtips and half cocks tail. Frequently flicks wings and tail. Perch-gleans low in tules and rush, picking or occ. flight-gleaning insects off the water surface, and hops on adjacent patches of short grass, mud, or floating *Azolla* etc., to half jump, half fly back to the nearest tules when alarmed. Occ. will feed high up in the tules when large swarms of midges are present. Curious; when disturbed, several will gather in the tules near the intruder, call continuously, but usually keep out of sight. Raises crest when excited. During display exposes white in tail and wings, and makes wing-whirring.

Voice: A characteristic rather faint nasal *piuh-piuh* 1-5 times, given throughout the day, and sometimes followed by a dry trill. No particular dawn song.

Breeding: Eggs Oct (Junín); nestlings Sep (Puno); fledglings Feb (Puno).

Habitat: Near the water surface in tall tule marshes (*Scirpus*, *Typha*). In large marshes scattered in small colonies.

Range: 3600-4100 m. From Lake Junín and adjacent Lima, c Peru, locally s through s Cuzco to Lake Titicaca and Lake Poopó, Bol., and somewhat lower in Jujuy and Tucumán, nw Arg. (*alticola*). Common. Also in lowlands, 2 sspp on coast of w Peru and n Chile, 1 widespread in Chile, c Arg., and se Brazil, southern populations migratory.

Tachuris – Genus *Polystictus*

2 species, one restricted to e Brazil, the other disjunct in northern and southern S Am. The nest is an open cup.

BEARDED TACHURI *Polystictus pectoralis* – Plate XLVI 23

8-9.5 cm (northern sspp smallest). Above brown, rump rufous; crown-feathers elongated, gray-brown streaked blackish; concealed crown-patch white. Wings black; lesser coverts, 2 broad wingbars, and broad edges of remiges rufous. Tail-feathers narrowly edged pale rufous. Supercilium buff, sides of head streaked buff and blackish; eye-ring pale buff; upper earcoverts dark brown. Chin and throat whitish, some feathers black basally; rest of underparts rufous buff. **Female** like male, but with less conspicuous or lacking markings on head (throat unmarked), and a browner cap. **Juv**. like female, but crown paler, wings browner; head without black markings.
Habits: Usually in pairs. Strict perch-gleaner. Difficult to see.
Voice: An infrequently given, weak *feee*.
Breeding: No data.
Habitat: Reed and rush-beds, and neighbouring grassland, thistles, bushes, and trees.
Range: Only known from Lake Suba, 2711 m on the Bogotá savanna, Cundinamarca (where not rec. in recent years), and Pavas above Dagua 1350 m on w slope of W Andes in Valle, Col. (*bogotensis*). Lowlands of e Col., s Ven., s Guyana, and s Surinam (*brevipennis*); from se Brazil and e Bol. to Buenos Aires and Córdoba, Arg. (*pectoralis*). Rare and local, its savanna habitat being destroyed everywhere.

NOTE: The very small, northern sspp have no distinct throat-markings and may be a different species (*P. brevipennis*).

Doraditos – Genus *Pseudocolopteryx*

4 species found mainly in Arg., migrating n, but one breeds locally in the Andes n to Col., and one has isolated breeding populations in Trinidad and Guyana (originally migrants?). See *Amer. Mus. Novit.* 1095(1940).

SUBTROPICAL DORADITO
Pseudocolopteryx acutipennis – Plate XLVI 21a-b

10.5 cm. **Above greenish gray; below bright yellow**. Inner primaries narrow and pointed in male. **Juv**. has upperparts very faintly tinged cinnamon, crown with faint dark shaft-streaks, 2 narrow, cinnamon-yellow wingbars, a narrow, buffy supercilium and eye-ring, pale mandible, and paler yellow underparts. Primaries normal.
Habits: Alone or in pairs. Forages near the ground within dense vegetation, peering about for a few s, then darting to snatch prey before landing on another stem. Difficult to see.
Voice: Calls include *tzit* while foraging, and rapid *tzididit-tzididit* during territorial chases. Display song from top c. 2 s long, of 3-6 notes, often followed by snore-like wing-whirring during very rapid horizontal turn-around or verticals dart up to 4 m *tzit-tzit-tzit t-konk* (BW).
Breeding: Eggs Nov (Jujuy), Jan (Tucumán).

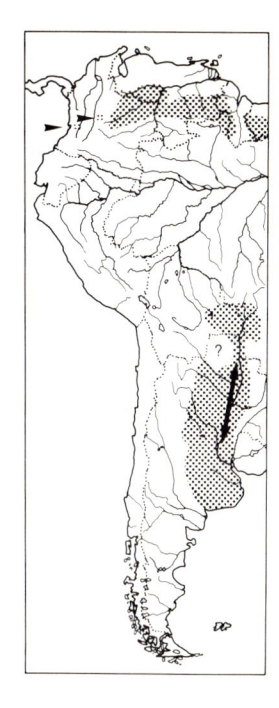

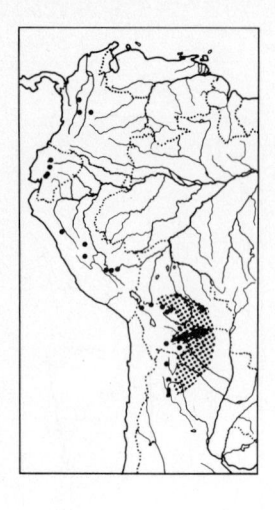

Habitat: Within rush-beds, dense shrubbery, corn-fields, bushes and trees at the borders of streams and lagoons in arid and semi-arid zone, often in quite disturbed areas.
Range: Breeds at 1500-3000 m. Found from C Andes of Col. s through interandean plateau of Ecu. to Tucumán, San Juan (see *Hist. Nat. Mendoza* 1(1980): 75-6), and Mendoza (see *Hist. Nat. Mendoza* 2(1981): 32), nw and w Arg. Birds from Bol. and Arg. winter in adjacent chaco of nw Arg., Parag., and Bol. Fairly common in s part of range, but rare and local in Col., Ecu., and Peru, the only presently known sign of breeding being a small resident population on the interandean plateau of Cotopaxi, c Ecu. (3000 m), and a sight rec. from Bogota in Feb (2600 m). Some, or all Peruvian rec. may be of wintering birds from the s.

WARBLING DORADITO *Pseudocolopteryx flaviventris*
Plate XLVI 22

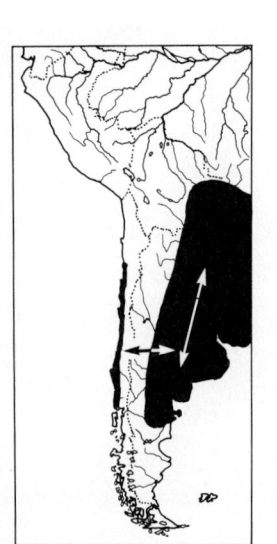

11 cm. **Above dull brown**; feathers of **crown** faintly edged **rufous brown**. **Cheeks darker**. **Below yellow**, brightest on throat. Bill black, base of mandible tinged buff in female. **Juv**. like ad. but duller; upperparts and edges of remiges and wing-coverts buffy brown (rather than mouse-brown). **Underparts bright buff**, throat buffy slate, vent and c breast and belly pale creamy yellow.
Habits: Alone, in pairs, or family-groups. In winter very retiring; works quietly through the inside of the vegetation. In breeding season more active, alert, and curious, sometimes perching in the tops of bushes, making short sallies for flying insects.
Voice: Call while foraging *tek*. Song of 4-6 single notes followed by 2 louder double-notes, the last often ending with slight vibrato *tek tek tek tek tek tick-it tick-idt*, head snapped sharply upwards with each double-note. (BW).
Breeding: Eggs Dec, Jan (Chile).
Habitat: Riparian thickets, marshes with saw-grass (*Cortaderia*), tall reeds, thistles, and low bushes over the water. During migration in swales, weed-patches, and corn-fields.
Range: Lowlands from Antofagasta to Valdivia Chile, and from e Bol. and s Brazil s to Chubut, Arg. Southern birds migrate n in winter. Chilean birds apparently cross the Andes to winter in Arg. Uncommon and local in Chile, more common in Arg.

Flycatchers – Genus *Mionectes*

5 species distributed from Costa Rica to se Brazil and Bol., inhabiting humid (rarely semi-arid) forest from the trop. to the lower temp. zone. The food consists largely of fruit, and most foraging is upward hover-gleaning. The long pendant nest is suspended from the tip of a branch, and the enclosed nest chamber has a side entrance. See *Amer. Mus. Novit.* 1126(1941).

STREAK-NECKED FLYCATCHER
Mionectes striaticollis – Plate XLVII 26a-c

12.5 cm. Fairly long and straight bill. Upright posture. Crown dark gray, sometimes faintly tinged olive (or crown dark olive in *viridiceps*). Back bright yellowish olive-green; *columbianus* and *viridiceps* have indistinct buffy tips to median wing-coverts, and edges to greater coverts, these markings suggested in *palamblae*, but absent in *striaticollis* and *poliocephalus*, where they are colored like the back. Headside like crown, but with whitish shafts; **white spot behind eye**. Throat somewhat paler than headside, medium gray (or grayish olive in *viridiceps*), **whitish shaft-streaks** more **pronounced** than on headside, and continuing onto olive breast. Sides and upper belly pale yellow (or yellow in *palamblae*) variously streaked gray. Ad male with 2nd outermost primary attenuated. **Juv**. duller, head more olive, no attenuated primary.
Habits: Alone or in family-groups, sometimes in mixed-species flocks. Perches upright within the vegetation. Makes short sallies within dead shrubbery, and hover-gleans insects and berries among foliage, but outside breeding season will spend most of its time picking berries in the canopy, sometimes reaching down for them. Occ. will make a quick sally for a flying insect up to 1 m outside foliage, to return to cover immediately.
Voice: Calls include a sharp fine *tzie*, and a repeated *tselee*.
Breeding: Fledglings Jan, Feb (ne Ecu.), May (Huila), June (e Narino), Apr (Junín, Cuzco), Mar (Sta Cruz); juv./imm. Feb (Pasco), Apr (e La Libertad), May (Ayacucho), Aug (San Martín), Dec (La Paz); large gonads Jan-Apr (Valle).
Habitat: Lower to upper strata, usually in trees, but also in bushes and dead shrubbery. Humid montane forest and cloud forest.
Range: 1300-3000 m, rarely down to 550 m. W slope of W Andes at La Selva (Risaralda/Valle) Col. (*selvae*), w Nariño, Col. and w Ecu. (*viridiceps*), E Andes on both slopes in Cundinamarca, and C Andes of Col., e Ecu, (intergrading with *viridiceps* in Loja s Ecu.) (*columbianus*), from se Piura (possibly Loja, Celica, and San Bartolo, sw Ecu.) to Huánuco (*palamblae*), Pasco, Junín and Ayacucho, c Peru (*poliocephalus*), from Cuzco, se Peru, to Sta Cruz, Bol. (*striaticollis*). Common in the subtrop. zone.

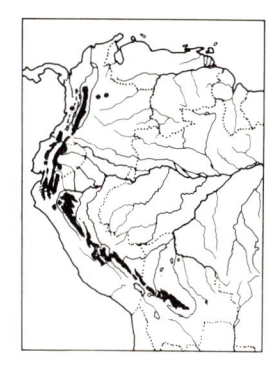

OLIVE-STRIPED FLYCATCHER *Mionectes olivaceus* – Plate
XLVII 25. A lower elevational counterpart of the previous species, occuring up to 3000 m in the Perijá mts and the Andes of Ven., a region where the Streak-necked F. does not occur. 11.5 cm. Crown olive, back olive-green, wing-coverts with obscure buffy markings. **Spot behind eye white**. Headside, **throat, breast**, and sides olive **with** narrow **yellow streaks**; c belly dull yellow. 2nd outermost primary is narrow at its c part, somewhat widening again towards tip. Habits as in Streak-necked F.

MCCONNELL'S FLYCATCHER *Mionectes macconnelli* ascends
to 2400 m in La Paz, Bol. (*peruana*). 12.5 cm. Smaller than previous species, but with a proportionally longer tail. **Unstreaked**. Above bright

olive-green, throat ochraceous olive with indistinct pale shafts, breast and belly ochraceous orange. It forages quietly and inconspicuously in the foliage for berries and insects from the undergrowth to medium heights, and is unique among tyrants in having leks with several males displaying, in spite of the sexes being alike.

Previously placed in the genus *Pipromorpha*.

Flycatchers – Genus *Leptopogon*

4 species distributed from s Mexico to n Arg., inhabiting humid forest of the trop. and subtrop., rarely lower temp. zones. Medium-sized tyrants with an upright posture like in the previous genus, and like them they upward hover-glean, and often feed on fruit. They have a peculiar habit of flicking one wing while perched. The long pensile nest is globular with a side entrance, and is suspended from the tip of a branch. See *Amer. Mus. Novit.* 1126(1941).

RUFOUS-BREASTED FLYCATCHER
Leptopogon rufipectus – Plate XLVII 24

13 cm. **Upright posture**. Above olive-green, cap dark gray; wings with 2 obscure, ochraceous bars (along lower edge of outer web); edges of remiges ochraceous on primaries, turning olive-yellow on secondaries, and white on whole outer web of tertials. Eye-ring, lores, ear-coverts, rest of headside, throat, and breast rufous; lores and headside admixed some gray; belly pale olive-yellow. Ssp *venezuelanus* has slightly darker crown and purer green back. **Told from Handsome Flycatcher by size, grayer crown, rufous facial area, and habit of flicking wing**. **Juv**. undescribed. - In pairs, often in mixed-species flocks. At lower levels associates with warblers and hemispinguses, and at mid height and lower canopy with tanagers, other tyrants, and furnariids. Sluggish. Primarily makes upward sallies to the underside of leaves, returning to same perch. Frequently gives an explosive *kweek*. Juv. Mar (Cundinamarca); large gonads Oct (e Boyacá).

Lower to medium strata of humid forest and cloud forest, at 1600-2700 m. Extreme sw Táchira on Páramo de Tamá and Río Chiquito, Ven. (*venezuelanus*), E and C Andes of Col. s through e Ecu. to Piura (Chinguela), n Peru (*rufipectus*). Generally rare, but locally fairly common.

INCA FLYCATCHER *Leptopogon taczanowskii* – Plate XLVII 23

12 cm. Replaces previous species s and e of the Marañón. Upright posture. Crown dark olive, back olive-green. 2 distinct ochraceous wingbars formed by triangular markings at tip of outer webs; remiges edged olivaceous, strongest on inner ones, white along inner tertial. Headside mottled blackish and whitish, forming long white bar across cheek; narrow eye-ring whitish. Chin and upper throat grayish, lower throat and breast olivaceous with faint orange sheen, belly pale yellow.

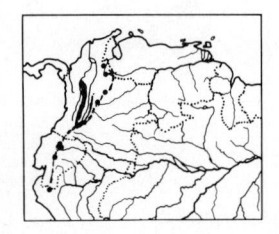

Imm. has broader, more rufous edges to remiges, also on inner tertial. Behavior as for genus (i.e. flicks one wing, esp. before take-off). Makes short sallies for flying insects, or to the underside of leaves within the undergrowth. Regularly repeats a sharp explosive *tziet* every 1-1.5 s. Also a *trrt-trrrt*.

Within 3 m of the ground in dark undergrowth in humid forest and forest edge. Peru, at 1350-2650 m from c Amazonas to Cuzco. Fairly common.

SLATY-CAPPED FLYCATCHER *Leptopogon superciliaris*, widespread in humid lowland forest, ascends to 2400 m in La Paz, Bol. (*albidiventer*). 14 cm. Tail fairly long. Above olive-green, crown dark gray. 2 whitish wingbars formed by spots on outer webs of coverts; wing-panel olive-green, tertials edged whitish. Headside mottled gray and whitish, posterior part of ear-coverts blackish. Throat and breast grayish olive, belly pale yellow, upper belly faintly flammulated with pale olive-gray. Tends to be more active, and forage more openly, and often higher than others of the genus.

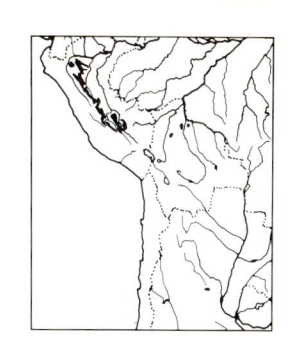

Tyrannulets – Genus *Phylloscartes*

19 species found from Nicaragua to n Arg. Most of them have a very limited distributions primarily in humid premontane forest. Some species (esp. those formerly seperated in *Pogonotriccus*) forage in an active gnatcatcher fashion, with tail almost horizontal and wings slightly drooped while the bird hops incessantly from perch to perch. Other species sit more quietly, with the tail almost right down, while they scan surrounding vegetation for prey, using mostly upward strike as foraging method. The nest of one species was a partly domed cup. See *Amer. Mus. Novit.* 1095(1940).

MOTTLE-CHEEKED TYRANNULET *Phylloscartes ventralis* – **Plate XLVI 12** – is found at 1125-2400 m, from San Martín C Andes of Peru s to Tarija, Bol. (*angustirostris*), from n Salta and Jujuy to Tucumán and Catamarca, nw Arg. (*tucumanus*); at lower elevations in se Brazil and adjacent regions (*ventralis*). 11.5 cm. **Tail fairly long**, feathers slightly pointed. Crown dark olive turning olive-green on back and edges of tail (compare Sclater's T.); wings black with 2 yellowish white bars, and bright olive-green panel, tertials black tipped white on outer web. Short supercilium, broken eye-ring, and grizzling below eye whitish; patch on ear-coverts greenish white. Throat yellowish white; breast washed pale yellow and olive, belly lemon-yellow. Ssp *tucumanus* large, pale and green above, with broader and yellower wingbars; below yellower, throat and breast more uniform, with little or no olive tinge. Alone, in pairs or 6-8 together, often in mixed-species flocks. Actively perch-gleans insects along twigs and thinner branches in lower canopy, body horizontal, **tail half cocked**. Loud calls, resembling furnarid, given irregularly at c. 0.5 s intervals *trip-trip-trip....* In Brazil: *spit, tirr, zi-ti tititi* (calls), *spit-spit-spit-tutututu-spit* (song). Eggs Oct (Tucumán); fledglings Jan (Tarija). Not common.

Pygmy-tyrants – Genus *Pseudotriccus*

3 species found in e Panamá and the Andes from Col. to Bol., inhabiting humid forest from upper trop. to temp. zone. Small dumpy tyrants that live in the undergrowth. They make short upwards strikes for insects on the undersides of leaves. No nest of the genus has been described. May be related to genus *Corythopis*. See *Amer. Mus. Novit.* 1066(1940).

HAZEL-FRONTED PYGMY-TYRANT *Pseudotriccus simplex* – Plate XLVI 14 – is rec. at 1670-2000 m from n Puno, e Peru, to Cochabamba, Bol., and is included here only for comparison with Rufous-headed P.-t. 9.5 cm. Forecrown and headside dull rufous brown, turning olive-brown on hindcrown and darker olive on back; wings and tail edged dull dark rufous. Throat olive-yellow, c belly pale yellow, breast and sides dull green. No eye-ring. **Juv**. has less rufous on forecrown and is yellower below. Solitary. Makes short upward sallies to the underside of leaves. The song is a long trill. Often clicks bill. Forages low in bushes at edge of humid forest. Uncommon.

NOTE: May be a ssp of Bronze-olive P-t. (*P. pelzelni*).

RUFOUS-HEADED PYGMY-TYRANT
Pseudotriccus ruficeps – Plate XLVI 13a-b

11 cm. **Head, wings, and tail rufous**, throat somewhat paler. Above dark dull green, below slightly paler, c belly pale yellow. Birds from c Peru have browner wings. **Juv**. only has face and cheeks brownish olive.
Habits: Solitary or in family-groups. Restless. Regularly jerks body as it flits around vines and tangles, and makes short sallies for insects.
Voice: Gives 2-3 s long, first slightly descending, then rising trills *trrrrrrrt* interspersed with very high-pitched, often drawn-out rising notes, and often clicks bill. Also gives a c. 0,3 s long rising abrupt trill.
Breeding: Fledglings June (Cauca, e Nariño), Nov (Amazonas); juv. June (n Cajamarca), July, Aug (Huánuco), Oct (e La Libertad), Mar (Pasco).
Habitat: 1-4 m up in vines and tangles in cloud forest, second growth, and edge of humid forest.
Range: 1400-3600 m, rarely down to 400 m. From w slope of E Andes, Antioquia in C Andes, and Valle W Andes of Col. s through e and w Nariño, e and w Ecu., and C Andes of Peru to Cochabamba, Bol. Fairly common to common in Peru and Ecu., rare in Cochabamba.

Tody-tyrants – Genus *Poecilotriccus*

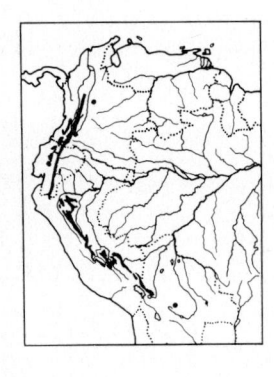

4-5 species patchily distributed in the northern half of the Andes and in the Amazon forest. 3 are trop. to upper trop., 1-2 mainly subtrop. Very small tyrants with a short flat bill, specialized for scooping insects from undersides of leaves. They all have a distinct color pattern, and the 3 trop. species are strongly sexually dimorph. No nest described. See *Amer. Mus. Novit.* 1066(1940).

RUFOUS-CROWNED TODY-TYRANT
Poecilotriccus ruficeps – Plate XLVI 10a-d

9 cm. **Head-pattern distinct**. Ssp *peruvianus*: Crown and headside rufous, supralores paler; forehead with a few black feathers; crown bordered laterally from eye round nape with blackish-gray. Half-collar on nape and sides of neck gray, back bright olive. Wings black with 2 narrow yellowish bars, broad yellowish white edges to tertials, and narrow olive edges of remaining remiges. Moustache black; chin, throat, and upper breast ochraceous white, fading to whitish, and separated from chrome-yellow lower breast and belly by indistinct, narrow black bar; sides of breast olive. Northern sspp poorly differentiated: *melanomystax* has black line from eye round nape bordered gray below, and cheeks paler; *rufigenis* has a paler, more uniform head, with more rufous on throat, and virtually no black moustache or border of crown; *ruficeps* has pale cheeks in strong contrast to the crown, like *melanomystax*, but has a distinct black moustache and border of crown. Unnamed form distinct: chin and lower throat whitish, in sharp contrast to rufous upper throat, crown and headside; half-collar round nape gray, but no black border of crown, and no moustache; breastband olive without black, belly warm yellow-buff. Remiges without olive-green borders, except faintly on innermost. **Juv**. has paler throat than ad.
Habits: Alone or in pairs, rarely in mixed-species flocks. Perches inconspicuously in the middle of a bush, making short upward sallies of up to 1 m, gleaning insects from undersides of leaves. More active than Black-throated T-t.
Voice: While foraging often utters a *tick-trrrrt*.
Breeding: Large gonads Mar-Sep (C and E Andes Col.).
Habitat: Within 1.5 m of the ground. Bamboo and second growth in humid forest and cloud forest, favoring bushes and small saplings.
Range: 1600-2750 m, rarely down to 1000 m. From w Nariño, s Col., to Chimborazo, Ecu. (*rufigenis*), n part of W and C Andes from Antioquia to Tolima, Col. (*melanomystax*), from Táchira, Ven., s through E Andes of Col. (birds from s end of C Andes intermediate with *melanomystax*) to Morona-Santiago (Zapote Najda mts), se Ecu. (*ruficeps*), Piura and n Cajamarca, n Peru, birds from PN Podocarpus, Loja, se Ecu. probably also referable here (*peruvianus*). S of the Marañón in Amazonas, Peru (unnamed form). Fairly common in nw Ecu., locally common in Col., otherwise uncommon.

NOTE: The unnamed form from Amazonas Peru (LSU specimens) may deserve species rank.

Tody-tyrants – Genus *Hemitriccus*

20 species found e of the Andes s to Córdoba, Arg., most of them endemic to very restricted areas in Brazil. Very small tyrants with a broad bill, specialized for scooping insects from the underside of leaves. The

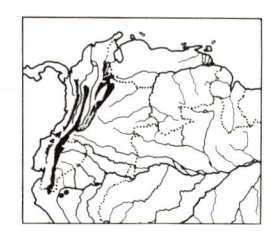

nest is closed with a side entrance, purse-shaped and suspended from a branch.

BLACK-THROATED TODY-TYRANT
Hemitriccus granadensis – Plate XLVI 9a-c

9 cm. **Iris straw**. Above olive, wing-panel brighter. **Lores and eye-ring** (and in some sspp forehead) **pale** ochraceous (white in *caesius*, *granadensis*, and *intensus*). **Chin and throat black**, lower c throat whitish, breast gray (sometimes washed brownish, esp. in *lehmanni*), and faintly streaked white; belly white, more or less tinged pale yellow; vent yellowish buff. **Juv**. unknown.
Habits: Alone or in pairs, rarely in mixed-species flocks. Makes short, usually upwards sallies, occ. up to 1 m, for insects on the underside of leaves. Stays in the same bush for some time, moves little, and only in short hops.
Voice: Dawn song ends with a hard sharp trill: *kee kee kee krrrrrrrt*. While patrolling gives a short trill *krrrt*. Alarm a *Synallaxis*-spinetail like *keep keep*....
Breeding: Large gonads Mar-July (Col.).
Habitat: 0.5-2 m up, occ. to 6 m, in the middle or low in the vegetation. Small openings in dense mossy cloud forest, and second growth at edge of humid montane forest.
Range: 1500-3300 m, mostly 2000-3000 m. Sta Marta mts (*lehmanni*); sw Táchira, Ven., and Perijá mts (*intensus*); E Andes of Col. s to w Santander (*andinus*, doubtfully distinct from *intensus*), rest of 3 ranges of Col., and Carchi, nw Ecu. (*granadensis*), from Morona-Santiago, se Ecu., s through C Andes to Cuzco, se Peru (*pyrrhops*), from n Puno in se Peru to Cochabamba, Bol. (*caesius*). Uncommon, locally fairly common.

NOTE: Previously placed in the genus *Idioptilon*.

YUNGAS TODY-TYRANT *Hemitriccus spodiops* - Plate XLVI 8 – uncommon in humid second growth at (800-)1250-1600 m in La Paz, Cochabamba, and Sta Cruz Bol., is rec. occ. up to 2450 m. 10 cm. **Iris pale straw**. Above olive, feathers of crown with darker centers. Faint pale brown loral spot. 2 indistinct wingbars olive; inner remiges edged bright yellow-green. C throat and upper breast dull olive, often with whitish streaks centrallly, belly pale yellow. **Juv**. has dark iris. Solitary, not in mixed-species flocks. 1-3 m up, making very short upward sallies to the underside of leaves, and sometimes hover-gleaning. Call a short harsh trill, lasting c. 1 s, and usually given 3-4 times in rapid succession.

NOTE: Previously placed in the genus *Idioptilon*.

Tody-flycatchers – Genus *Todirostrum*

14 species distributed from s Mexico to n Arg., mostly in the trop. zone.

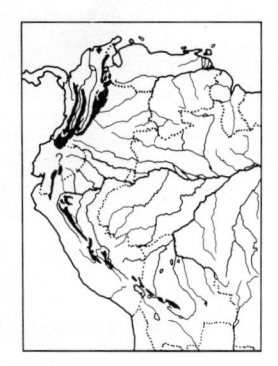

Very small tyrants with a broad bill, specialized for scooping insects from the underside of leaves. The closed nest, purse-shaped with a side entrance, is suspended from a branch. See *Amer. Mus. Novit.* 1066(1940).

OCHRE-FACED TODY-FLYCATCHER
Todirostrum plumbeiceps – Plate XLVI 11

9.5 cm. Iris brown, rarely straw. Ssp *viridiceps*: crown gray; back olive, palest towards rump. Wings with 2 distinct ochraceous bars and edges of tertials; rest of remiges edged yellow-green. Chin and sides of head and throat pale rufous, spot on earcoverts brown. Underparts white, flammulated with pale gray on breast; lower belly pale yellow. Ssp *obscurum* has darker spot on earcoverts, and stronger gray wash on breast; almost entire throat pale rufous. Solitary, making short sallies for insects. Call a series of some 5 fairly low-pitched churrs lasting about 2.5 s *drrr drrr drrr drrr drrr*. Juv. Mar (La Paz).

1-3 m up in vine-tangles and dense forest edge, and around tree-falls at 750-2600 m. From n Puno, se Peru, to Sta Cruz, Bol. (*obscurum*), from Chuquisaca, Bol., to Salta and Jujuy, Arg. (*viridiceps*). 2 other sspp in se Brazil. Fairly common in premontane zone.

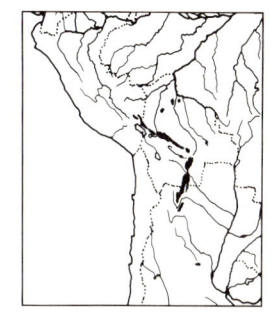

Flycatchers – Genus *Myiophobus*

Genus possibly polyphyletic. 9 species, 8 in humid forest in coastal mts and Tepuis of Ven. and Andes s to Bol., 1 in semi-arid country from Costa Rica to n Arg. Fairly small tyrants with a yellow to orange semi-concealed median crown-patch. Upright posture and protruding breast. Some species live in small groups, follow mixed-species flocks, and move quickly through the vegetation; others are more sluggish and solitary. The bill gets darker with age. Only nest of Bran-colored F. known: an open cup, placed in a bush or tree. See *Amer. Mus. Novit.* 1043(1939).

FLAVESCENT FLYCATCHER *Myiophobus flavicans* – Plate XLVI 17a-b

12 cm. **Upright posture and fairly broad bill, recalling an *Empidonax***. Maxilla black, mandible brownish or dusky flesh. Ssp *flavicans*: above yellowish olive, crown with semi-concealed orange to yellow patch, lacking in female. **Broken eye-ring yellowish**. 2-3 **wingbars cinnamon**, edge of remiges cinnamon-olive, tertials edged buffy-white. Tail dusky, edged buffy olive. **Below yellow, breast, sides** of throat and upper sides of body **flammulated with olive-green**. Ssp *venezuelanus* less flammulated below; mandible flesh; crown-patch smaller. Ssp *perijanus* like *venezuelanus*, but with black mandible. Ssp *superciliosus* with black mandible, olive-green upperparts, and deeper yellow lores and eye-ring than *flavicans*; only 1, narrow, brown wingbar (on greater coverts); edge of remiges duller and paler brownish. **Juv**. lacks crown-patch.

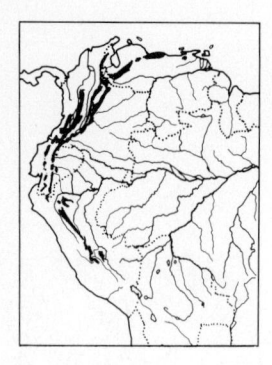

Habits: Alone, in pairs, or 2-3 birds together. Typically forages by sallying out from a low dark perch, picking insects from leaves, branches, or the ground.
Voice: No data.
Breeding: Nestling Feb (Huila); fledglings July (w Nariño); juv./imm. Sep (Cundinamarca), Jan (ne Ecu.), Aug (nw Ecu.); large gonads June (Perijá), Mar (sw Huila), Oct (Boyacá).
Habitat: Lower to medium strata, occ. high. At edge and inside humid forest and cloud forest, esp. in bamboo tangles.
Range: 900-2700 m, mainly above 1500 m. From coastal mts in Distrito Federal to Andes of Táchira, Ven. (*venezuelanus*), Perijá mts and e slope of Tamá massif (*perijanus*), all 3 ranges of Col. s through e and w Ecu. to Peru n of the Marañón (*flavicans*), C Andes of n and c Peru from s Amazonas s to n Vilcabamba mts in Cuzco, se Peru (*superciliosus*). 1 ssp in Ven. Fairly common.

UNADORNED FLYCATCHER *Myiophobus inornatus* – Plate XLVI 18

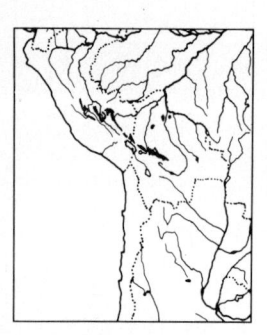

12 cm. **Bill rather small**. Above olive, median crown-patch yellow to rufous. **Broken eye-ring yellow**. 2 wingbars and edges to all remiges orange-brown; tertials edged whitish. Breast olive-yellow appearing faintly streaked; rest of underparts yellow. Birds from Cochabamba, Bol., are more yellowish olive above, and yellower on the abdomen. **Juv**. unknown. - Alone or in pairs. Much like Flavescent F. Call a high-pitched *zit* or *zib*. Eggs Oct (La Páz), Nov (Puno).
    Humid forest, at 1400-2600 m. Barred Antthrush occurs on some of the same localities. From Pasco, Ayacucho, and Vilcabamba mts in Cuzco, Peru, to Cochabamba, Bol. Uncommon.

ORANGE-BANDED FLYCATCHER *Myiophobus lintoni* – Plate XLVI 19

12.5 cm. **Tail rather long**. Base of mandible orange, rest of bill black. Above dark grayish olive, crown slightly grayer, and with median patch, orange to rufous in males, dull dark brown or absent in females. Ill-marked broken eye-ring pale yellow. 2 broad wingbars mixed ochraceous and white; remiges edged white, olivaceous, or ochraceous; inner remiges, tips of primaries, and base of secondaries unmarked. Below yellow, palest in female; sides of breast clouded with olive. **Juv**. lacks crown-patch, is slightly paler below, and has the wingbars ochraceous without admixture of whitish. Both juv. and imm. have base of maxilla and entire mandible yellow.
Habits: Small groups of 2-6 birds, in mixed-species flocks. Perches upright and conspicuously atop leaves. Makes mainly short forward sallies to glean insects off upper leaf-surfaces; less frequently upward sallies to foliage from within the canopy. Restless, calling almost constantly. Not shy.
Voice: Calls include *chip* and *chep* notes.
Breeding: Juv. June (Piura); imm. June (se Ecu.).
Habitat: Medium, rarely upper strata. Elfin forest, cloud forest ca-

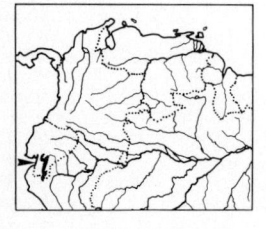

nopy; humid second growth at edge of landslides in forest.
Range: 2250-3050 m. Known from interandean plateau in Azuay (El Portete), e slope of E Andes on Morona-Santiago/Azuay border (Zapote Najda mts), ne Loja (mt Imbana), ec Loja (PN Podocarpus), s Ecu., and Piura (Chinguela), n Peru. Uncommon.

HANDSOME FLYCATCHER *Myiophobus pulcher* – Plate XLVI 15a-b

9.5-10 cm. **Tail rather short**. Ssp *pulcher*: above olive, **crown greenish gray** with median orange patch. Narrow supraloral streak white. Wings with 2 broad, pale ochraceous bars; outer web of inner tertial whitish, secondaries edged pale ochraceous, except at base (forming dark patch). **Throat and breast pale ochraceous yellow**, rest of underparts lemon-yellow. Females have less extensive and duller orange crown-patch. Ssp *bellus* grayish green above with darker (pale orange-buff) wingbars, slightly darker underparts, and a longer tail; *oblitus* larger than *pulcher*, with darker crown, and buffy panel on basal portion of primaries; upperparts brownish olive, underparts paler than in *pulcher*. Told from Rufous-breasted F. (*Leptopogon rufipectus*) by lack of rufous on face, smaller size and greener crown. In small groups of 2-4 birds, following mixed-species flocks. Gleans insects from foliage, or captures them in the air, making short sallies. Does not perch very openly. Large gonads Feb-Aug (Col.).

Middle to upper strata, occ. low, in humid forest, forest edge and second growth at 1400-2600 m, rarely down to 800 m. From w slope of W Andes, s Col., to Pichincha, nw Ecu. (*pulcher*), e slope of W Andes in Valle, C and E Andes of Col. s to ne Ecu. (*bellus*); 1600-3050 m from e Cuzco to n Puno, se Peru (*oblitus*). Fairly common in n part of range, rare in s.

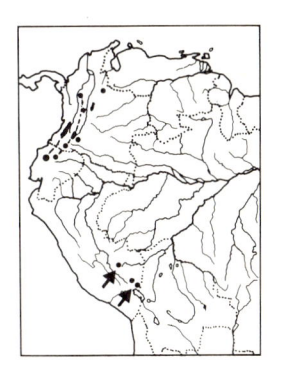

OCHRACEOUS-BREASTED FLYCATCHER *Myiophobus ochraceiventris* – Plate XLVI 16a-b

12 cm. Above olive; median crown-patch yellow to orange in males, dull orange in females. **Wings with** 2 distinct, buffy to whitish **bars**. **Lores dusky**, supra-loral line and eye-ring buff, throat and breast ochraceous yellow, belly yellow. **Juv**. lacks crown-patch.
Habits: Alone or in groups of 4-5 birds, often in mixed-species flocks. Rather conspicuous, and not shy. Perches on the tips of middle and upper branches, making short sallies and hovers for insects, and sometimes perch-gleans and hangs. Restless, moves through the vegetation rapidly.
Voice: No data.
Breeding: Juv. Mar (Pasco).
Habitat: Humid thickets, stunted cloud forest, and elfin forest.
Range: 2800-3500 m, rarely down to 2500 m. S and e of the Marañón bend from c Amazonas and e La Libertad, c Peru, to La Paz, Bol. Locally common.

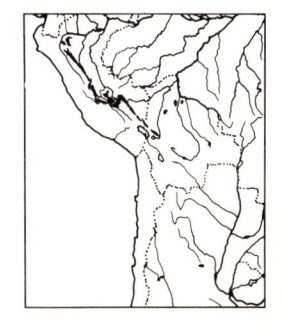

BRAN-COLORED FLYCATCHER *Myiophobus fasciatus*, common in lowlands from Costa Rica to n Arg., has been rec. to 2650 m in Bol.(*auriceps*), and to 2600 m in E Andes, Col. (*fasciatus*). A few rec. from Tarapacá (Araya Arica), n Chile. 12 cm, upright. Above dark brown, crown more or less reddish brown (entire upperparts reddish-brown in *fasciatus*), with median yellow to rufous patch. **Wings with 2** fairly **broad, buffy bars**, and narrow edge terminally on secondaries; tertials edged whitish. Tail relatively long, dusky brown. Below whitish (buffy white in *fasciatus*); throat lightly, **breast and upper belly distinctly streaked gray-brown**. Alone or in pairs flitting among low twigs, and sallying forth for insects. Call a low whistle *tjlup* or *tjulu*. Dawn song *djili-dju-djili-dju*.... Inhabits lower to medium strata of dry forest and edge, as well as riparian thickets in the arid zones.

Flycatcher – Genus *Pyrrhomyias*

Monotypic genus. A fairly small tyrant with upright posture and protruding breast. Makes short sallies for insects in the air or on the foliage, and also eats berries. The bill is fairly broad, almost like a *Contopus* or *Empidonax*. The 3 nests described were open cups, all placed 1.5 m up in a niche in a tree or a bank. See *Amer. Mus. Novit.* 1043(1939).

CINNAMON FLYCATCHER *Pyrrhomyias cinnamomea* – Plates XLVI 20a-b and LXIII 12

11 cm. Upright posture, crown somewhat raised. Ssp *pyrrhoptera*: above dark olive, crown darker and browner, with concealed, golden yellow median crest; rump cinnamon. Wings and tail black, 2 broad wingbars and conspicuous panel rufous. **In flight** shows broad rufous bar across inner remiges. **Sides of neck, and entire underparts rufous**. Ssp *cinnamomea* hardly distinguishable, slightly darker on crown and underparts; *assimilis* distinct, all rufous, primarycoverts and tips of primaries black, inner web of tertials dark brown, tail with 10 mm wide subterminal black band and 1 mm wide rufous tip. **Juv**. browner than ad., with no crown-patch; feathers of lower back and tail with ochraceous tips.
Habits: Alone or in pairs; joins, but does not follow mixed-species flocks. Sallies out to hawk insects in the air, or glean them of the foliage. Perches conspicuously, but not in tops. Not shy.
Voice: A characteristic 0.5-1 s trill *trrrrt*, given throughout day at 3-30 s intervals or more. Flight call *tjip tjip*.
Breeding: Eggs Jan-June (e Huila); fledglings Nov, Dec; juv. Mar, Apr, May (Sta Marta, probably all year with most breeding Aug-Feb); fledglings June (Valle), Nov, Feb (Santander), Nov (Quindío), Aug (El Oro), Mar (Amazonas), Feb (Tarija); juv./imm. Mar (Junín), May (e La Libertad, Cochabamba), June (San Martín), July (Cuzco), Nov, Feb, Mar (ne Ecu.).
Habitat: 1-6 m up on open or semi-open horizontal perches, not in tops. Clearings in humid forest, stunted cloud forest with admixed bamboo, and humid second growth.
Range: 670-3350 m. Sta Marta mts (*assimilis*); Andes of sw Táchira,

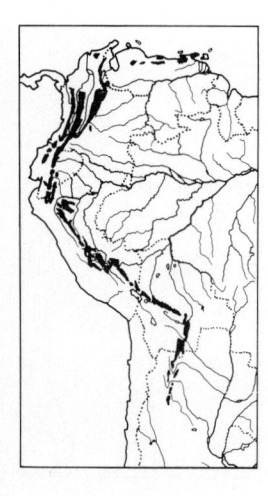

Ven., Perijá mts, and in all 3 ranges of Col. s through both slopes of Ecu. to e La Libertad and San Martín, Peru (*pyrrhoptera*), from San Martín to Huancavelica, Peru, and through Bol. to Tucumán, nw Arg. (*cinnamomea*). 3 other sspp in coastal Ven. Common in the subtrop. zone throughout.

Pewees – Genus *Contopus*

10 species found from northern N Am. s to n Arg., also on Caribbean isls. Medium-sized to fairly large tyrants, inhabiting both humid and semi-arid regions, found mostly at edges. They are aerial hawkers, with a broad bill and an upright posture. The sexes are similar, and the plumage grayish. The nest is a neat shallow cup, placed in a bush or a tree. See *Amer. Mus. Novit.* 1042, 1043 (1939).

OLIVE-SIDED FLYCATCHER *Contopus borealis* – Plate XLV 3

17 cm. Upright posture. Broad body. Relatively short tail. Above dark smoky gray; crown and headside (incl. lores) fuscous gray. Poorly marked eye-ring whitish. Wings and tail blackish; indistinct wingbar on tips of greater secondary-coverts, and panel on tertials and inner secondaries grayish white; wingbar and panels disappear with wear. Below dark smoky gray, feathers with dark shafts; **throat and c underparts white or yellowish white**; silky white tuft on flanks show on sides of rump, esp. in flight. Bill dark, somewhat paler at base of mandible.
Habits: Solitary. Perches conspicuously in the top of a bush or on an open branch. Hawks insects in the air, and returns to the same perch.
Voice: Song may rarely be heard in S Am.: 3 notes *quick three beers*. Disturbance call a slightly descending *pip-pip-pip*, recalling Greater P.
Habitat: Forest clearings and second growth in humid zone. 2-40 m up, preferring high dead branches or slender tops.
Range: Breeds in N Am. In winter at 400-3300 m in coastal mts of Ven., and the Andes s to Sta Cruz, Bol. Departs from Sta Marta mts Mar-Apr, rarely as late as mid May. Accidental on lower Amazonas, Brazil, and on coast of sw Peru. Uncommon, declining.

NOTE: Previously placed in the monotypic genus *Nuttalornis*.

GREATER PEWEE *Contopus fumigatus* – Plate XLV 10a-c

16-17 cm. Upright posture. Mandible pale. Ssp *zarumae*: Uniformly **dark** gray, somewhat paler below, **cap of elongated feathers** darker (an odd specimen has outer web of outer tail-feather whitish). Ssp *fumigatus* larger, with a less blackish cap and sooty gray upperparts; *ardosiacus* like *fumigatus*, but darker, more slaty gray, esp. below; *brachyrhynchus* like *fumigatus*, but paler throughout; above light smoky gray tinged olive, crown hardly darker; throat suffused with whitish; c belly exten-

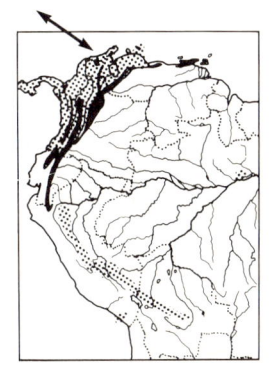

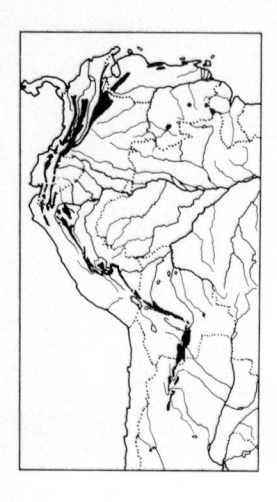

sively yellowish white. **Juv.** similar to ad., but overall tinged brownish, and with pale rufous feather-tips, esp. on nape. Wings with 3 or more narrow, ochraceous bars; belly and vent tinged ochraceous.
Habits: Solitary. Perches openly. Often changes site. Crown-feathers usually more or less raised. Shakes tail upon landing. Not shy.
Voice: Song rarely heard: a hoarse seasawing *zur zur zur zúr zur zur zur*. Calls include a clear whistle repeated every 2-3 s *peeew* (not given in w Ecu.?), a single *rrrrhye*, and a short *pjeek* or *pjeek pjeek*, sometimes varied with the whistle.
Breeding: Fledglings June (Cauca), Sep (w Nariño), Mar, Oct (ne Ecu.), Aug (Loja, w La Libertad), Mar (c Cajamarca), Apr (Junín); juv./imm. Feb (Cauca), Mar (Chimborazo), Sep (El Oro), Oct (Piura), Jan, Apr (Cochabamba); nest building June (Valle).
Habitat: 2-6 m up, on dead branches or fallen trees. Clearings in humid forest and cloud forest.
Range: 500-3000 m, locally down to 300 m. From w Nariño, sw Col. s through w Ecu. to sw Cajamarca, nw Peru (*zarumae*); Andes of Ven., Perijá mts, all 3 ranges of Col., s through e Ecu. and C and E Andes to Cuzco, se Peru (*ardosiacus*); from n Puno, se Peru, to Cochabamba, Bol. (*fumigatus*), from Sta Cruz, Bol., to Tucumán, nw Arg. (*brachyrhynchus*). Fairly common. 6 other sspp from s USA to Panamá, and in Tepuis and coastal mts of Ven.

NOTE: S. Am. forms often considered a distinct species, SMOKE-COLORED PEWEE.

WESTERN WOOD-PEWEE *Contopus sordidulus* - Plate XLV 8 – has been rec. up to 2600 m on the Bogotá savanna. 7 ssp breed in N and C Am., and winter in the premontane Andes s to Cochabamba, Bol. Habitat as in Eastern W-p, but also in woodland. - 13 cm. Upright posture. Mandible averages darker than in Eastern W.-p., and is sometimes all dark in ad. Above dark smoky gray, more or less tinged olivaceous; crown darker, feathers with contrasting dark centers when worn. Faint eye-ring and base of dusky loral feathers whitish; wings with 2 grayish white bars (the upper disappearing with wear), and white edges to tertials. Throat whitish gray, breast gray tinged brown, belly whitish with creamy cast. Not safely distinguishable from Eastern W-p. except on call, but averages grayer or browner above, and more extensively dark below (often without white on c breast). Birds in ad. plumage with an all-dark bill belongs to the present species. Behaves like Eastern W-p. Calls essentially one-syllabled: a rather sweet *peeír*, *peweér*, a level *péee*, or *peeep*, resembling Eastern W-p. **The distinguishing call is a burred, slightly descending *weeér*, or a throaty scratchy *prr'z*.** Dawn song a 3-note phrase *tswee tee teet*, mixed with the burred note.

NOTE: By some authorities considered sspp of Eastern W-p., but hybridization in zone of overlap not yet documented. For geographic variation see *Proc. Biol. Soc. Washington* 77 (1960):141-6.

EASTERN WOOD-PEWEE *Contopus virens* – Plate XLV 9 - has been rec. up to 2850 m in Col. on migration. Breeds in N Am. and winters in power premontane zone from Nicaragua s through Col. and w Ven. to se Peru. In low- to mid-strata, occ. higher, in clearings and cuts in humid forest. 15 cm. Upright posture. Wings relatively long, reaching over half-way down tail. Mandible pale. Headside and upperparts dark olive-gray to smoky gray, ill-defined eye-ring pale, feathers of lores with pale gray bases. Wings and tail dusky; indistinct grayish white wingbar on greater and median secondary-coverts; panel on inner remiges whitish. Chin, breast, upper sides of body and sides of throat pale gray; c throat white; belly and vent pallid yellowish white. **Imm**. has 2 buffy wingbars and tips of lesser wing-coverts, a yellowish tinge below, and a darker mandible. Solitary. Perches rather openly, and makes both short and long sallies for insects, then returns to same post. Continually moves head as it watches for flying insects. Feeds even during heavy rain. During migration, and sometimes in winter, sings a plaintive whistle *pee-weé* or *pee-a-wee*, the 2nd note lower, occ. as downslurred *pee-ur*. Calls include a sharp *chip*, and clear whistled rising *pweee* notes.

NOTE: Should perhaps include Western W-p., and may be also Tropical P.

TROPICAL PEWEE *Contopus cinereus* – Plate XLV 7

13 cm. Upright posture. **Wings reach less than half way down tail**. Ssp *pallescens*: Above gray with slight brownish cast, cap blackish brown. Eye-ring and **base of loral feathers white**. 2 narrow wingbars and panel on inner secondaries and tertials pale gray. Throat pale gray, breast gray, belly pale yellow. Ssp *punensis* similar, but grayish olive above, crown grayer. Throat often more extensively and purer white. Wing-panel and belly paler (yellowish white). 1st (outer) primary shorter than 5th. **Juv**. has 2 distinct, buffy white wingbars.
Habits: Solitary. Perches conspicuously in top of a bush or on an open branch. Makes short sallies for insects. Often changes post. Almost invariably shakes tail upon landing.
Voice: Song an emphatic *tzee-zée* repeated every 10-15 s. Call a chirpy *psee*, sharper than similar note of Eastern W.-p., and clearer than that of Western W.-p.
Breeding: Juv. Aug (Santander), Apr-May (Sta Marta).
Habitat: 1-2 m up, only rarely higher. Semi-open places in dry country. Plantations, agricultural land, pastures with trees, woodland borders, dry bushy slopes with scattered *Carica* and balsa trees. Often near rivers. Does not enter woodland.
Range: 0-2600 m. From Manabí, w Ecu., to Ica, w Peru, and in the Marañón drainage and C Andes s to Junín, c Peru (*punensis*); from La Paz, Bol., to Tucumán, nw Arg., and again from ne Parag. across sc Brazil to s Maranhao (*pallescens*); n and w Ven. and e Col. w to Sta Marta mts and s to middle Magdalena valley (*bogotensis*). 1 ssp in se Brazil, 1 in the Guianas and e Ven., 2 in C Am. n to Mexico, and 1 on Coiba Isl, Panamá. Fairly common.

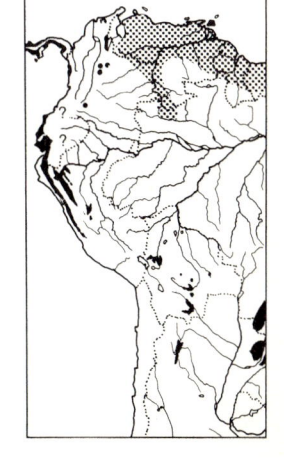

NOTE: Perhaps belongs to an Eastern W-p. superspecies.

Flycatchers – Genus *Empidonax*

16 species found from northern N Am. s to n Arg., also on Caribbean isls. Medium-sized tyrants with a broad bill and an upright posture. They hawk for insects from an enclosed perch, and are usually solitary. They have a pale mandible, dull plumage, pale eye-ring and 2 indistinct wingbars. The species are so similar that they often cannot be told apart in the hand, but must be identified on vocalizations and habitat. The nest is an open cup placed in a bush or a tree. See *Amer. Mus. Novit.* 1042(1939).

ACADIAN FLYCATCHER *Empidonax virescens* - Plate XLVI 24, breeds in N Am. and winters at 80-1200 m from Costa Rica s to w Ecu. and nw Ven. During migration occ. up to 2850 m on Bogotá savanna, Col. 13 cm. Upright. Wings do not reach half-way down tail. Mandible pale. Headside and **upperparts** grayish **olive**. Eye-ring, 2 wingbars and panel on inner remiges yellowish white. **C throat white**, occ. yellow; breast and sides of throat pale gray; sides of breast grayish olive, breast appearing very faintly streaked; rest of underparts yellowish white, fading to whitish when worn. **Imm**. has buffier wingbars. Not safely told from Alder F. in the field. In the hand, difference between longest primary and 5th outermost usually more than 6 mm (vs. less than 6 mm in Alder F.). Outermost primary usually longer than 6th (vs. usually shorter in Alder F.). Solitary. Hawks flying insects. Fairly active and regularly changing perch. Call a fairly loud emphatic unclear lisping, often squeaky *psip*, *pisp*, *peesp*, *pseeip*, *weece*, or *peesth*, often repeted. Song an explosive *wick up*. Usually within a few m off the ground, in the shade of thickets, woodland borders, and selectively logged forest; occ. in a densely foliated small tree in a shaded plantation.

ALDER FLYCATCHER *Empidonax alnorum* – Plate XLVI 25 – breeds in N Am. Winters from lowlands to 1100 m in w S Am., w of Andes to Limal Peru, e of them to nw Arg. During migration rarely to temp. zone in Cochabambal Bol. and on Bogotá savannal Col. 13.5 cm. Upright. Bill broad. Mandible pale. Wings do not reach half-way down tail. Above olivaceous gray, crown slightly darker. Eye-ring, supra-lores, 2 wingbars, and panel on inner remiges whitish. Throat whitish; breast like back, paler in center; rest of underparts yellowish white, sides washed olivaceous. **Imm**. browner with buffy wingbars; sometimes with buff wing-linings and thighs. For differences from Acadian F. see under that species. Solitary. Sallies for insects, flitting among the thin branches. Wags tail (which pewees do not). Call a loud piping *peep*, *whit*, or *kit*. Song distinctive, a falling buzzy *fee-beeo*. Found in lower strata, occ. high. Low bushes, often near water; second growth, thickets, edge, and clearings in humid forest.

NOTE: Previously regarded a ssp of Traill's Flycatcher (*Empidonax traillii*).

FUSCOUS FLYCATCHER *Cnemotriccus fuscatus* is mainly trop., but has been rec. to 2400 m in La Paz, Bol., in Sep and Oct (*bimaculatus*) (see *Amer. Mus. Novit.* 994(1938), and Gyldenstolpe (1945). 13.5 cm. **Tail fairly long. Bill black**. Above grayish olive; **wings** dusky **with 2 fairly broad, buff bars**, and narrow edges of inner secondaries; tertials edged whitish. **Line through eye dusky; relatively short supercilium yellowish white**. Throat yellowish white, faintly smudged with olive; breast pale grayish olive, somewhat contrasting with yellowish white belly. Its call is a **rough shrill whistle**, remarkably loud for the size of the bird. Song of jumbled clear notes *chip-weeti-weeti-weetiyee*, *chip-weety-weety-weety-teepip*, or *chewy-chewit-cheepeer*. Dawn song of clear high excited notes *p-pit-pit-peed*it. Lives inconspicuously from dense dark wet undergrowth in humid forest and tall second growth, to semi-arid areas with deciduous trees and scrub, foraging both high and low.

Phoebes – Genus *Sayornis*

3 species distributed from northern N Am. s to Panamá and the Andes s to nw Arg. Medium-sized tyrants with fairly long tails. They forage for insects near the ground, catching them in a variety of ways. The sexes are alike. The nest is a sheltered cup. See *Amer. Mus. Novit.* 1043(1939).

BLACK PHOEBE *Sayornis nigricans* – Plate XLV 17

14.5 cm. Thick neck, **tail fairly long. Black**; c belly, 2 wingbars, conspicuous wing-panel, and edge of outer tail-feather white. Ssp *latirostris* differs from *angustirostris* by being slightly larger, with more white on under wing-coverts and edges of tertials, inner primaries and basal part of tail-feathers.
Habits: Solitary. **Perches horizontally on rocks in streams**. Sallies for flying insects, returning to same rock. **Frequently wags tail**. Active. Not shy. Sometimes perches on low bushes, exposed branches, or telephone-lines.
Voice: Dawn song a repeated *peep-feebreew* (Panamá). For further notes on behavior and vocalizations see *Behavior* 37(1973):64-112.
Breeding: Eggs July (Cundinamarca), Oct (Tucumán); juv. Oct (Huancavelica), Nov (Puno), May (Sta Marta); large gonads Apr (Sta Marta).
Habitat: Rocky streams and ponds, often with overhanging vegetation.
Range: 100-2500 m, occ. to 3000 m. Coastal mts and Andes of Ven., Perijá mts, Sta Marta mts, e Panamá, and along all 3 ranges of Col. s through e and w Ecu., n, c and se Peru (along pacific slope to c Ancash) to Sta Cruz Bol.(*angustirostris*), and Andean slope from s Cochabamba, Bol., to Tucumán, nw Arg. (*latirostris*). Local in coastal Ven., common through most of remainder of range, esp. in humid regions. 4 other sspp from sw USA to Panamá.

NOTE: E Panamá and S. Am. birds are often considered a distinct species, **WHITE-WINGED PHOEBE (*S. latirostris*)**.

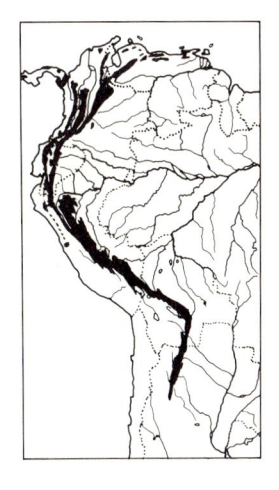

Flycatcher – Genus *Pyrocephalus*

Monotypic genus. The nest is a shallow cup. See *Amer. Mus. Novit.* 1126(1941), *Condor* 69(1967): 601-5, *Mitt. Inst. Colombo-Alemán Investig. Cient.* 6(1972): 113-33, *Neotrópica* 19(1973): 125, *Gefiederte Welt* 100(1976): 44-5, and *Hornero* 11(1977): 380-3.

VERMILION FLYCATCHER *Pyrocephalus rubinus* – Plate XLV 4a-c

13 cm. **Male**: **Cap and underside flame scarlet** to vermilion. Upperparts and headside fuscous; narrow tip of tail-feathers and outer web of outer feather whitish. In Lima, and to a lesser degree further s on the pacific coast, a dark phase occurs in which the male is dark or blackish throughout. Appears to have 2, complete annual molts. **Female**: Above gray-brown; throat and breast white, turning vermilion on belly, flanks, and vent; breast streaked gray-brown; sides of breast gray-brown. Tail like male. **Juv**. dark drab above, feathers with buffy white tips. Below white, breast and sides streaked drab; flanks and vent faintly tinged buffy. Changes to ad. plumage after 3 months; 1st ad. plumage of male sometimes with a few female-like feathers.

Habits: In pairs in breeding season, otherwise solitary. Perches conspicuously on open branches, bush-tops, and wires. Makes short flights to capture insects in the air or on the ground, then returns to post. Often flicks wings, raises crown, and wags tail when perched. Flight undulating. Territorial and aggressive, often chasing away larger birds. Confident. During display male flies 10-15 m up, levels off, and flutters forward in a curved or straight path, every m or so rising to a stall with head thrown back, crown and chest ruffled, wings held high, and tail raised to above horizontal, and giving RRV (see voice). Between stalls the wingbeats are rapid and shallow, wings kept above back. After several stalls he drops to a perch in a series of swooping glides, often with the tail cocked high.

Voice: Usually silent, but when very active both sexes may give sharp, usually loud *peent* at any time of day. Occ. the call has a double quality, and sometimes (during stress) is shorter *tp*. Clicks bill and makes a whirring sound with its wings when aggressive. Wing-whirring may also be part of the nuptial flight. At dusk, sometimes utters a mournful *churinche*. During nest-site presentation male gives a 1.5 s chattering repeated vocalization (RV) of c. 14 notes, sometimes accelerated and turned into a regularly repeated vocalization (RRV) *t-t-ti-ti-tee-teeree*, or *tit terrhiée*, last note accented. RRV also given when patrolling territory, esp. at dawn, and also at the presence of predators.

Breeding: Fledglings June (n Ecu., Piura). May breed throughout year on coast of Ecu. and w Peru.

Habitat: Arid and semi-arid open country. Open *Acacia* savannas; bushy thickets, gardens, hedgerows, and dry woods, or clearings in adjacent humid zone.

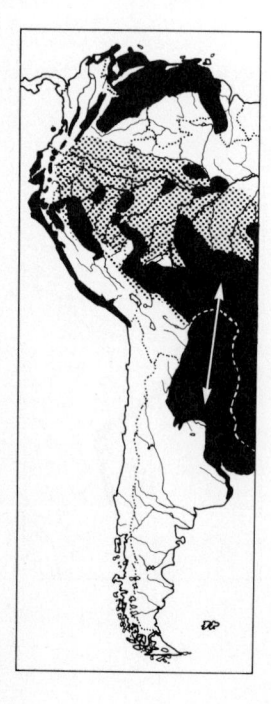

Range: 13 poorly differentiated sspp in lowlands from sw USA through C Am. and most of S Am., and in the Galapagos isls. W of the Andes s to Tarapacá, n Chile, e of the Andes in Ven. and from se Bol. to Urug. and Buenos Aires, Arg. Southern birds migrate n to winter in the drier parts of the Amazon basin n to Río Negro. Regular on the Bogotá savanna, Col., (2600 m) and in the vicinity of Quito, Ecu. In n Peru ascends to 3050 m, and in s Peru to 2500 m. Common.

CHAT-TYRANTS – Genus *Ochthoeca*

11 species restricted to subtrop. and temp. zones from coastal Ven. through the Andes to Isla Grande. Medium to small tyrants showing a certain similarity to Old World chats in size, shape, insect-catching from a perch, frequent wing- and tailflicking, and aggressive behavior. Prominent white, yellow, or cinnamon superciliaries are usually long, almost meeting on nape. Most species have forms with or without 2 rufous wingbars. Species differ mainly by color, width, and length of supercilium, color of underparts, habitat, size, habits, and elevation. The nest is an open cup placed in a cavity on a cliff, among roots projecting from a bank, in an *Espeletia*, or a bush. It has been suggested to place some species (*diadema*, *pulchella*, *frontalis*) in a separate genus *Silvicultrix*, but not all species have been investigated in that context. See *Amer. Mus. Novit.* 930(1937), 1203(1942), and 1263(1944).

SLATY-BACKED CHAT-TYRANT *Ochthoeca cinnamomeiventris* – Plates XLIV 9a-c and LXIV 10

13 cm. Rather hunch-backed. Ssp *nigrita*: **slaty gray**, with **conspicuous, short white supercilium** and under wing-coverts. Ssp *cinnamomeiventris* has chestnut breast and belly. In sspp *angustifasciata* and *thoracica* the supercilium does not reach bill, and only the breast is chestnut, narrowest in the former. **Juv.** of all dark or dark-bellied forms sometimes show varying degrees of a chestnut wash on belly, while all-dark juv. have been rec. once or twice in n Peru.
Habits: In close pairs. Sallies for flying insects or to foliage. Less conspicuous than Rufous-breasted, but slightly more so than Crowned C-t.
Voice: A very high-pitched drawn-out whine *eeeee*, recalling a fruiteater, or when excited a short fine *trieeee*.
Breeding: Eggs Feb (Huila); fledglings Mar (Junín); juv. Feb (E Andes Col.), June (Huánuco), Mar (Pasco), Aug (Cuzco); large gonads Oct (Boyacá).
Habitat: Lower to medium strata. 2-4 m up in very dark places in humid forest and second growth, almost invariably above or near streams.
Range: 900-3000 m, mostly 1800-2400 m. Andes of Mérida and Táchira, Ven. (*nigrita*); from Páramo de Tamá in sw Táchira, Ven., and all 3 ranges of Col. s to Ecu. (except sw.) and Piura (Chinguela), n Peru (*cinnamomeiventris*); e Cajamarca and s and e of the Marañón in Amazonas and San Martín, n Peru (*angustifasciata*), thence s through c and se Peru to Sta Cruz, Bol. (*thoracica*). Fairly common.

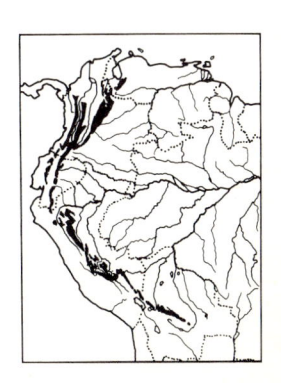

YELLOW-BELLIED CHAT-TYRANT
Ochthoeca diadema – Plate XLIV 15a-c

12 cm. **Small, rounded. Long supercilium yellow**, narrow behind eye. Back olivaceous brown to reddish brown, crown darker. Wingbars indistinct, olivaceous or dark brown (*meridana, jesupi, diadema*), or distinct, dull rufous (*rubellula, gratiosa, cajamarcae*). **Underparts yellow**, breast and sides olive (least so in Peru). **Juv**. has supercilium tinged ochraceous posteriorly, ochraceous vent and rufous wash on back.
Habits and **Voice**: Like Golden-browed C-t.
Breeding: Juv. Mar, June, July, Aug, Sep (nw Ecu.), Nov (Cauca), Dec (ne Ecu.).
Habitat: Like Golden-browed C-t. Dark undergrowth in interior of mossy forest, bamboo, and dense growths at edge of streams and roads.
Range: 1750-3100 m, rarely down to 800 m. Andes of Mérida, Ven. (*meridana*); Perijá mts (*rubellula*); Sta Marta mts (*jesupi*); extreme sw Táchira, Ven., and n E Andes of Col. (*diadema*), from C Andes and s of Valle in W Andes of Col. s through Carchi to Chimborazo, nw Ecu., and e Ecu. to Piura and n Cajamarca, n Peru (apparently not in Cord. Condor) (*gratiosa*), ec Cajamarca (Chira), n Peru (*cajamarcae*). 1 ssp in coastal Ven. Common, but rarely seen.

NOTE: Should probably include Golden-browed C-t.

GOLDEN-BROWED CHAT-TYRANT
Ochthoeca pulchella – Plate XLIV 13a-b

12 cm. **Small, round. Long supercilium yellow, narrow, and paler behind eye**. Above dark brown, crown darkest, and with olivaceous tinge. 2 wingbars rufous. **Below gray**, belly paler, flanks washed with brown; vent gray (*pulchella*), or ochraceous (*similis*). **Juv**. has brown crown, buffy posterior part of the supercilium, breast washed with olivaceous, belly and flanks ochraceous buff.
Habits: Little seen. Forages very close to the ground. Often flicks wings and tail. Sits silently, watching, then flutters from perch to strike small beetle or other insect from nearby leaf or mossy branch. Only flies short distances.
Voice: A descending whine *treeeee*. Also *trrr-rr-eeh-eh-errh-eehr-eeee?*, or *trrr-rr-rr rr rrrr eeee rr rr*, esp. given at dawn. Clicks bill.
Breeding: Juv. Mar (Junín).
Habitat: Usually 0.5-1 m from ground. Interior of dark, humid, extensive thickets, bamboo, and mossy forest. Sometimes (esp. juv.?) at forest edge.
Range: 1700-2750 m. E and C Andes of n and c Peru from Amazonas s (*similis*); from Vilcabamba mts, Cuzco in se Peru to w Sta Cruz, Bol. (*pulchella*). Fairly common to common, but little seen.

NOTE: Probably sspp of Yellow-bellied C-t. The form *jelskii* has previously been considered a ssp of the present species, or a distinct species. It is here treated under Crowned C-t.

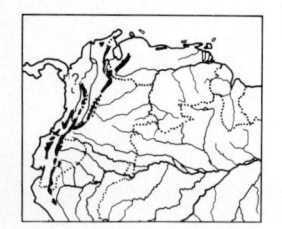

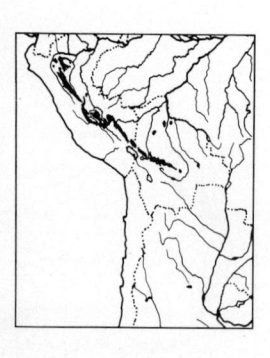

CROWNED CHAT-TYRANT *Ochthoeca frontalis* – Plate XLIV 12a-c and 14

13 cm. Posture rather slenderer and more upright, and with a more protruding breast than Golden-browed and Yellow-bellied C-t.s. **Long and broad supercilium yellow in front of eye, white posteriorly**. Ssp *frontalis* has blackish crown contrasting with dark brown back; 2 rufous wingbars faint or lacking; underparts gray, belly whitish, vent rufous. Ssp *albidiadema* like *frontalis*, but entire supercilium white; *boliviana* has gray crown and upper back, gradually washed with brown towards rump; wings with 2 rufous bars; vent white; *spodionota* like *boliviana*, but wingbars darker and narrower (absent in w Cuzco), and crown usually darker. Ssp *jelskii* like *boliviana*, but wash on lower back and rump brighter, more reddish brown; crown gray like *spodionota*. **Juv.** has brownish wash on flanks and belly, and is brighter and browner above; wingbars brighter and wider, supercilium with buffy tinge.
Habits: Alone or in pairs. Reaches down or sallies to foliage and branches for flying insects. Perches somewhat more conspicuously and higher than Golden-browed and Yellow-bellied C-t.s.
Voice: *te tirrrr*.
Breeding: Fledglings Sep (C Andes, Col.), Jan (Cochabamba). Juv. Nov (Huánuco), Apr (nw Ecu.).
Habitat: 0.5-2 m up. Dark places at edge of humid montane shrubbery, elfin forest, edge of mossy forest, and in dead bamboo on mossy ground. In rather more open places, and at higher elevations than Golden-browed and Yellow-bellied C-t.s. Not in the interior of forest (except *jelskii*, which may prefer closed forest). Often found in very small patches of shrubbery.
Range: 2100-3660 m, rarely down to 1300 m. Bogotá savanna and N.d.Santander, E Andes, Col. (*albidiadema*); at n end of W Andes and from C Andes, Col. s through n and e Ecu. and C Andes of Peru to La Libertad (*frontalis* incl. *orientalis*); W Andes from sw Ecu. s to Lima, Peru, crossing the Marañón in La Libertad (*jelskii*); c and sc Peru (*spodionota*) s to w Sta Cruz, Bol. (*boliviana*). Common.

NOTE: *O.f.jelskii* has previously been considered a ssp of Golden-browed C-t., or a distinct species.

RUFOUS-BREASTED CHAT-TYRANT *Ochthoeca rufipectoralis* – Plates XLIV 10a-b and LXIII 11

14 cm. Conspicuous supercilium white. Ssp *rufipectoralis*: upperparts, chin and throat gray, **lower throat and breast rufous**, belly white. N-wards the back gets progressively browner, wings get a single, broad, rufous bar, and the belly becomes washed with gray; *poliogastra* also has the crown brown, hardly differing from back. **Juv.** much like ad.
Habits: Alone or in pairs. Perches conspicuously, with upright slender posture; more horizontal and rounder when foraging. Makes short sallies and perch-gleans among the branches, taking most prey from upper leaf surfaces. Frequently jerks. In flight the tail is 'pumped', and the wings produce a whirring sound. Territorial.
Voice: A faint *chic-chic-chica* while perched. During territorial disputes

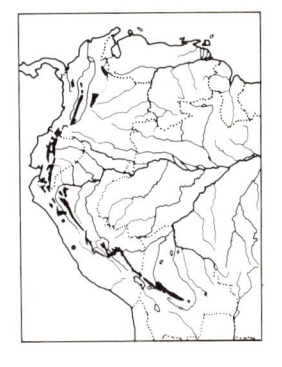

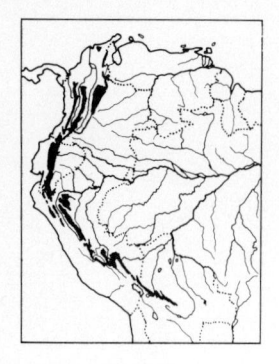

gives an irregular rapid series of dry short trills resembling churrs of Andean Tapaculo (ssp *altirostris*) *tjrt-tjrrrrt-tjt-trrrt*.
Breeding: Nest Oct (Puno).
Habitat: 1-3 m up, sometimes higher, 0.5 m below top of bush or small tree, often at forest edge. Open cloud forest, humid montane shrubbery, clearings, second growth; sometimes in riparian thickets and *Polylepis* wood in semi-arid zone.
Range: 2000-3600 m (occ. 4110 m). Sta Marta mts (*poliogastra*), Perijá mts (*rubicundula*), E Andes of Col. n of Bogotá (*rufopectus*), from W and C Andes of Col. s through Ecu. to Cajamarca and nw San Martín, n Peru (*obfuscata*), Peru from s La Libertad, Ancash, and Huánuco (*centralis*), Pasco to Cuzco w of E Andes (*tectricialis*), and on e Andean slope from Cuzco, Peru, to Cochabamba, Bol. (*rufipectoralis*). Common.

BROWN-BACKED CHAT-TYRANT *Ochthoeca fumicolor*
– Plates XLIV 7a-b and LXIV 11

16 cm. Long supercilium usually whitish, becoming ochraceous behind eye. Above various shades of dark brown; 2 wingbars rufous, the lower always conspicuous. **Below rufous**, washed with brown on breast and throat; chin washed grayish. Sspp *superciliosa* and *fumicolor* have the whole supercilium cinnamon, *brunneifrons* almost so. Northern birds have the outer tail-feather edged white; this disappears clinally s-wards. Ssp *berlepschi* has the whole head grayish, gradually merging with brown on back, and has a short narrow dirty white supercilium. However, birds from Cochabamba Bol., have brown-tinged crown.
Juv. is warmer brown, and lacks gray on chin.
Habits: Solitary. Territorial. Perches on thick branches near main stem (or clings to main stem), or conspicuously in bush-tops and fence-posts in the open, taking insects from the ground and in the air. Occ. hover-gleans. Often changes perch. Spreads wings and tail widely and for long periods upon landing. During territorial disputes makes short, wobbling flights, and wing-whirring.
Voice: Call a high-pitched *tsiu* from perch or in flight. When excited gives a sharp *keek-kee-keek*... with c. 2 notes per s, or *keek-tede-keek* lasting c. 1 s and repeated at 0,1-1 s intervals from perch, or sometimes in flight with accompanying wing-whirring. 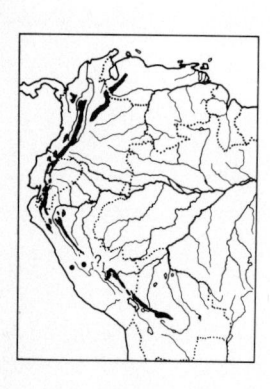
Breeding: Eggs Mar (Mérida); molt (probably post-breeding) Nov (Huánuco), Jan (Cochabamba); large gonads Dec (Boyacá), Feb-Sep (C and E Andes Col.).
Habitat: 0-2 m up, occ. higher. Borders of open páramo shrubbery and dwarf forest; *Espeletia* páramos; clearings in cloud forest and semi-humid *Polylepis* woodland.
Range: 2400-4400 m, rarely down to 1800 m. Andes from Trujillo to e Táchira, Ven. (*superciliosa*); w Táchira, Ven., and n E Andes of Col. (*fumicolor*), Antioquia, n end of C and W Andes, Col. (*ferruginea*), from s of Antioquia W and C Andes of Col. s through Ecu. to Lima and Junín, c Peru (*brunneifrons*), Vilcabamba mts on Ayacucho/Cuzco border (unnamed ssp), and from Cuzco to La Paz, Bol. (*berlepschi*), and Cochabamba, Bol. (unnamed ssp). Fairly common.

D'ORBIGNY'S CHAT-TYRANT *Ochthoeca oenanthoides* – Plate XLIV 6a-b

15-17 cm. **Very long broad supercilium pure white**. Ssp *polionota*: above sooty gray. Outer tail-feathers edged white. Throat and upper breast gray flammulated with whitish; rest of **underparts cinnamon rufous**, paler on vent and c belly. Ssp *oenanthoides* generally much paler, and usually with a weak cinnamon wingbar (2 wingbars as well as slightly rufous rump in extreme s). Told from Brown-backed C.-t. by grayer back, broader, pure white supercilium, paler, inconspicuous or lacking wingbar, and white edge of outer tail-feather. **Juv**. has the supercilium creamy-white.
Habits: Alone or in pairs. Perches upright and conspicuously on dead branches, bush-tops, or fence-posts, flying down to take insects from the ground. Often returns to same post.
Voice: Song a rhythmical loud *reek a teek a reek a teek a ...* lasting 1-3 s. Often gives song-like notes by excitement. Call a sharp *kvee*. Contact note *chic* or *chic-chic* repeated at 3-5 s intervals. Flicks wings while calling.
Breeding: Eggs Dec (La Paz); fledglings Dec (La Paz); juv. Jan (La Paz), Nov (Cochabamba), Feb (Chuquisaca).
Habitat: Temp. and puna zone, with *Polylepis* woodland, *Puya*, *Gynoxys*, or other bushes or low trees on slopes and in ravines. May stay a few km outside these habitats, on rocky slopes with bunchgrass and very scattered bushes or fence-posts, or in desert washes with a minimum of vegetation. In more arid regions than Brown-backed C-t., but both are found in semi-humid *Polylepis* woodland. Found at higher elevations than White-browed C-t. (overlapping between 2800 and 3600 m), but in fairly similar habitats.
Range: 2800-4500 m, down to 2000 m in Arg. From La Libertad and w Huánuco s to n Puno, Peru, and Tarapacá (Arica), n Chile (*polionota*); from Titicaca basin, Puno se Peru, to Tucumán, nw Arg. (*oenanthoides*; birds from extreme s Peru and n Chile intermediate). Fairly common.

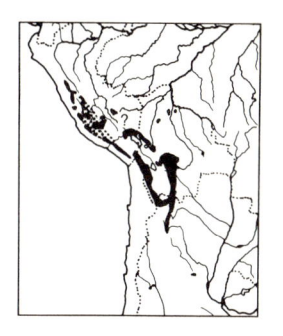

PATAGONIAN CHAT-TYRANT *Ochthoeca parvirostris* – Plate XLIV 16

12.5 cm. **Head gray with black median zone of crown and ear-patch**; back dark gray-brown, brownish on lower back and rump. Wings with 1-2 rufous bars, secondaries narrowly edged rufous, except at basal 2 mm. Below light gray, belly and flanks buffy white. **Juv**. much more rufous above without dark cap; below darker than ad., and with more ochraceous abdomen.
Habits: Solitary. Sallies for passing insects from low perches on weeds or bushes in winter. In breeding season searches for insects in the open amongst topmost foliage of the trees, flying only at long intervals.
Voice: Song a long high-pitched sibilant *seee*, recalling Slaty-backed C-t. Call a sad *peeoo*.
Breeding: Nov-Feb (Chile).
Habitat: Breeds in *Nothofagus* forest. In winter found in the forested Andean foothills, C Valley, and coastal region of Chile, where it also frequents gardens. Prefers the native trees *Quillaja saponaria* and *Maitenus boaria*.

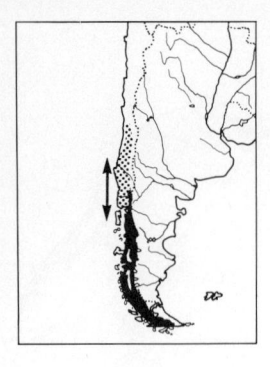

Range: Breeds from Valdivia, s Chile, and Neuquén, s Arg., s to Isla Grande, where it arrives in Oct and departs in Mar. In winter found as far n as Coquimbo, Chile. Uncommon.

NOTE: Previously placed in its own genus *Colorhamphus* and called Patagonian Tyrant.

WHITE-BROWED CHAT-TYRANT
Ochthoeca leucophrys – Plate XLIV 8a-c

14.5 cm. **Conspicuous long broad supercilium white**. Above gray, tinged brownish towards rump. Mottled whitish lunule below eye. **Underparts paler gray**, c belly and vent whitish. Outer tail-feather usually edged white. Peruvian sspp have white wing-panel, and 2 faint narrow buffy bars, that may be lost through wear; *leucophrys* has 2 rufous wingbars, and lower back browner; *tucumana* has broader (and rufous?) wing-panel, broader and deeper rufous wingbars, and is browner above. **Juv**. tinged brownish, and with more conspicuous wingbars than ad.
Habits: Solitary. Perches conspicuously, in upright posture near top of large herbs or small bushes. Catches insects on the ground or in the air, returning to same post. Often jerks wings and tail.
Voice: Call a piercing *keeu*, frequently uttered. Sometimes gives a *keeukeukeukeukeu* in flight. Alarm *tee teeti*.
Breeding: Probably Oct-Jan (Peru), and Nov-Feb (Arg.).
Habitat: 1-2 m up, in tops. Arid zone. Prefers bushes in ravines, gorges, or on slopes with grasses and herbs. Often found near water or at edge of irrigated farmland.
Range: 0-3600 m, and to 3980 m in the absence of D'Orbigny's C.-t. (e.g., sw Ayacucho). Loja, sw Ecu. (Celica mts: sightings) (ssp?), upper Marañon Valley in n Peru and on pacific slope in Cajamarca (Huacraruco) (*dissors*), Huánuco and Pasco, c Peru (*interior*), from Junín to Huancavelica, ne Ayacucho and Cuzco, Peru (*urubambae*), w slope of W Andes from La Libertad (Santiago) and Ancash, Peru, to Tarapacá, n Chile (*leucometopa*), e edge of highlands of Bol. intergrading with *tucumana* in se Potosi (*leucophrys*), highlands of nw Arg. from Salta to San Juan (*tucumana*). Common.

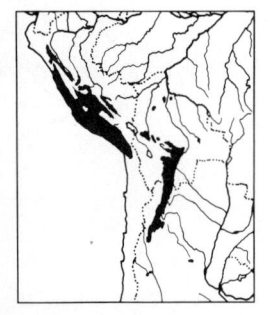

PIURA CHAT-TYRANT *Ochthoeca piurae* – Plate XLIV 11

12 cm. Conspicuous supercilium white. Above dark brown, tinged rufescent on lower back and rump; **2 broad rufous wingbars** and white edges of tertials and secondaries. Outer tail-feather edged white. Below gray, c belly and vent whitish. Entire outer web of outer primary white. Mottled whitish lunule below eye less conspicuous than in the considerably larger and slenderer White-browed C-t., which also lacks conspicuous wingbars in the sympatric form. **Juv**. undescribed.
Habits, **voice** and **breeding**: No data.
Habitat: Shrubby hillsides in arid regions.
Range: Only known from 5 localities on Pacific slope of Peru: Piura

(Palambla 2000 m), Lambayeque (Porculla 15-1800 m), La Libertad (Samne 1500 m), and Ancash in valleys of river Casma (Colcabamba 2800 m), and Huarmey (San Damián 1500 m). Uncommon.

Bush-tyrants – Genus *Myiotheretes* (sensu lato)

6 species in the temp., rarely subtrop. zone of the Andes s to nw Arg. Fairly large tyrants of both humid and arid, usually open country, forest edge, and clearings. They perch conspicuously in the top of shrubbery, trees, or on telephone-wires. Their posture is upright, and in flight all but one show rufous in the wings. They usually return to the same post. Their insect prey is hawked in the air, or taken by sally-gleaning, or from the ground. They are seen alone or in pairs. A nest of Red-rumped B-t. found was an open cup placed low in a bush, one of Rufous-webbed T. a simple open cup placed in a *Polylepis* tree. The first 2 spp are probably best placed in separate genera. See *Amer. Mus. Novit.* 930(1937).

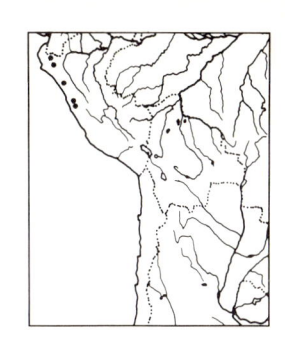

RED-RUMPED BUSH-TYRANT
Cnemarcus erythropygius – Plate XLIV 4

20 cm. **Forehead white** merging with hoary gray crown; back dark gray-brown; large **white patch on tertials**; **rump rufous**, tail dark rufous with blackish c feathers and terminal quarter. White wing-patch and extensive rufous color of tail very conspicuous in flight. Chin and streaks on throat white, breast dark gray-brown grading to rufous on belly and vent. **No rufous in wings**. Bill small. Sspp poorly differentiated. **Juv**. with little white on forehead.
Habits: Alone, rarely in pairs or family groups. Perches conspicuously and upright in tops of bushes, 1-3 m up, taking most of its prey from the ground. Occ. on dead branches higher in trees.
Voice: Generally silent. Sometimes utters faint scratching sounds *kerk-kekkek*, or gives a short slightly blurry whistle.
Breeding: Eggs Nov (Puno); large gonads Mar, Sep (Sta Marta).
Habitat: Humid and semi-humid zone at treeline. Park-like cloud forest, semihumid *Polylepis* wood and adjacent pajonal with scattered bushes, *Puya* slopes, shrubbery, and hedgerows.
Range: 3000-4300 m, occ. higher. Páramos of Sta Marta mts and n end and in s Cundinamarca in E Andes of Col. (*orinomys*), páramos from s Nariño, s Col., locally s through Ecu. and Peru to Cochabamba, Bol. (*erythropygius*). Rare in n Col., uncommon elsewhere.

NOTE: By some placed in the genus *Myiotheretes*.

RUFOUS-WEBBED TYRANT *Polioxolmis rufipennis* – Plate XLIII 15a-b

18 cm. **Above and below smoky gray**, wing-coverts sometimes with lighter edges at tips, c belly and vent whitish. Iris pale. Throat shows silvery white streaks in some lights. Underside of tail rufous with dark

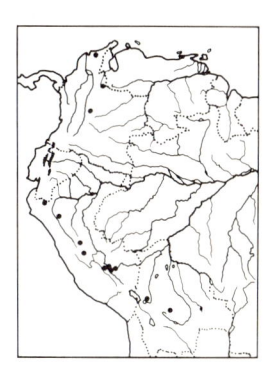

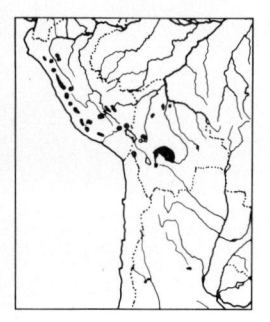

terminal quarter, outer tail-feather with white outer web. **In flight shows extensively rufous wings and tail**, but terminal dark band on secondaries broad (2 cm). Bol. birds smaller, duller gray, and with narrower tail-bar. **Juv**. like ad., but belly more decidedly tinged buff.

Habits: Solitary or in pairs. Perches conspicuously on cliffs or in top of a bush or tree, gliding down to take prey from the ground. Sometimes forages in open country, watching from hummocks and making short runs and sallies. Frequently hangs or glides on aerial updraughts or hovers like a kestrel with tail spread, thus eyeing prey on the ground (see plate). This behavior is probably derived from the aerial display seen in related genera (e.g. *Muscisaxicola* and *Knipolegus*). Aggressive.

Voice: Call a fairly high-pitched liquid vibrant *tree* given throughout day. Alarm similar, but shorter.

Breeding: Pair probably nesting late Dec (Puno); fledgling Feb (Ancash).

Habitat: Semi-arid grassland with occ. cacti and shrubs, often along tall rock walls and rocky ridges, but at least in breeding season narrowly associated with *Polylepis* groves. In C Andes of Peru and along the Tunari range of Bol. also sometimes associated with grassland adjoining cloud forest edge. Hunts in open country and along rock-walls and forest edge.

Range: 3050-4300 m, occ. to subtrop. zone in dry season. At scattered localities from Lambayeque, s Cajamarca, and s Amazonas s through Peru to Cuzco and Arequipa, and in w Oruro, Bol. (*rufipennis*), and from Puno, se Peru, to Tunari range in Cochabamba and Sta Cruz and in Potosí, Bol. (unnamed ssp). Uncommon, locally fairly common.

NOTE: Previously placed in the genus *Xolmis*, and by some in the genus *Myiotheretes*.

STREAK-THROATED BUSH-TYRANT
Myiotheretes striaticollis – Plate XLIV 2a-b

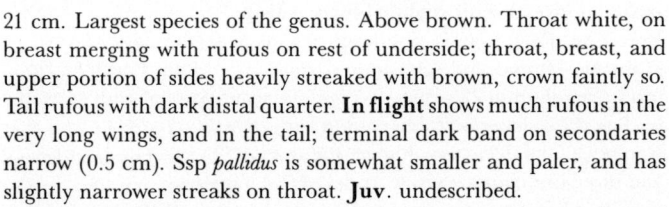

21 cm. Largest species of the genus. Above brown. Throat white, on breast merging with rufous on rest of underside; throat, breast, and upper portion of sides heavily streaked with brown, crown faintly so. Tail rufous with dark distal quarter. **In flight** shows much rufous in the very long wings, and in the tail; terminal dark band on secondaries narrow (0.5 cm). Ssp *pallidus* is somewhat smaller and paler, and has slightly narrower streaks on throat. **Juv.** undescribed.

Habits: Confident and conspicuous. Perches on telephone-poles and wires, and halfway up large trees (*Eucalyptus*). Frequently flies long distances in woodpecker-like long waves. May circle tail after landing, when excited leans forward wagging fanned tail. Cathes prey in the air and on the ground.

Voice: Both sexes often give a melancholic whistle while perched.

Breeding: Large gonads Jan-June (Perijá and Sta Marta mts and C Andes Col.).

Habitat: Often near human habitation. Prefers open country: extensive clearings in cloud forest, second growth, hedgerows, gardens, *Eucalyptus* groves, and dry open valleys and gorges with scattered vegetation.

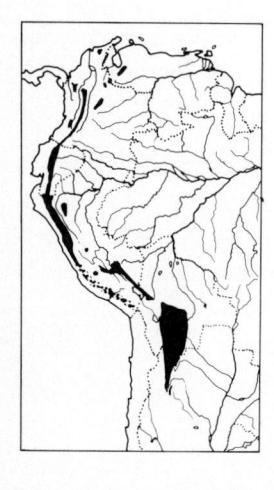

Range: 1700-3700 m, rarely down to 600 m. From Mérida, Ven., Perijá mts and Sta Marta mts, Col., s through the Andes to Apurímac, Arequipa, and Moquegua, s Peru (*striaticollis*), from Cuzco, se Peru s through Bol. to Tucumán, nw Arg. (*pallidus*). Widespread, but in low numbers.

SANTA MARTA BUSH-TYRANT *Myiotheretes pernix* – Plate LXIII 9

18 cm. Above dark brown; wings with 2 faint narrow pale rufous bars; outer web of outer tail-feathers ferruginous. Throat white streaked dusky, rest of underparts rufous, irregularly clouded with dusky on the breast, appearing slightly spotted. **Juv**. unknown. 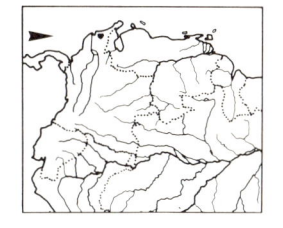
Habits: Usually alone perched exposed on top of bush or smaller tree. Feeds by long sweeping aerial sallies; rarely sallies within canopy. Occ. briefly with mixed-species flocks.
Voice and **breeding**: No data.
Habitat: Forest edge, in shrubbery and shrubby ravines.
Range: Col. at 2100-2900 m in Sta Marta mts. Uncommon.

SMOKY BUSH-TYRANT *Myiotheretes fumigatus* – Plate XLIV 3a-b

18 cm. **Wholly dark brownish gray** with dark line through eye (vent ochraceous in *lugubris*). Chin and upper throat faintly streaked whitish; throat tinged ochraceous; whitish eyebrow faint, rather short (lacking in *lugubris*, long in *olivaceus*). **No rufous in tail**, outer tail-feathers edged whitish. Wings with 1-2 faint, narrow, buffy gray bars. **In flight** shows **much rufous in wings**, but terminal dark band on secondaries broad (1.5 cm). **Juv**. has ochraceous vent.
Habits: Solitary or in pairs, sometimes in mixed-species flocks. Perches in middle strata, more or less conspicuously on free branches inside trees, or on the tips of open branches, occ. in treetops. Takes insects from the ground, or sally-gleans leaves, epiphytes, and moss-covered branches.
Voice: Dawn song a monotonous series of whistles *pew-pew-pew*... c. 3 notes per s, sometimes uttered at a different pitch. Call 2 notes 0.7 s apart. 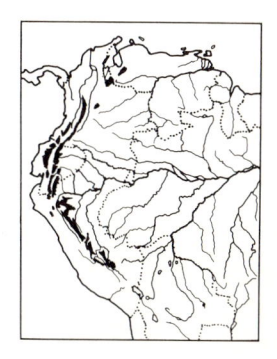
Breeding: Fledglings Jan (Huánuco), Feb (Cauca); large gonads Nov (Cundinamarca), July, Aug (Perijá mts, n E Andes Col.).
Habitat: Humid forest below treeline. Cloud forest, mainly on dead branches, both high and low, often at edge of clearings and on shrubby slopes with scattered trees. In s part of range, where it occurs with Rufous-bellied B-t., rare and apparently confined to unbroken cloud forest.
Range: 1800-3600 m. Mérida region, Ven. (*lugubris*), w Táchira, Ven., and Perijá mts (*olivaceus*), all 3 ranges of Col. and n Ecu. (*fumigatus*), intergrading with *cajamarcae* that is found from s Ecu. s to Cuzco, se Peru. Fairly common in Ecu., but rare in se Peru.

RUFOUS-BELLIED BUSH-TYRANT
Myiotheretes fuscorufus – Plate XLIV 5

18 cm. Above brown; wings blackish with 2 rufous bars and buff edges to secondaries. **Throat white**, merging with rest of underparts, which are rufous with an irregular mixture of brown and white. Underside of tail dark with rufous down center, probably not visible in flight; outer web of outer tail-feather rufous in some birds. **In flight** shows much rufous in wings, terminal dark band on secondaries 1 cm. **Juv**. undescribed.
Habits: Alone or in family groups of 2-4. Catches insects on the wing, and sally-gleans foliage from understory to canopy. Much like Smoky B-t., but although intraspecific aggression is frequent, aggression is rare interspecifically.
Voice: Dawn song a series of 2-4 monotonous whistles followed by 2 *pip pip pip pip pi-dóo*, lasting c. 1.4 s, notes shorter and higher pitched than in Smoky B-t. Clicks bill when angry.
Breeding: No data.
Habitat: All levels, but mainly upper strata. Edge of clearings in primary cloud forest, and in adjacent secondary woodland; in bamboo and other understory, as well as canopy of alders (*Alnus*), melastoms, and other trees up to 8 m tall.
Range: 2130-3550 m. Pasco, c Peru, s to Cochabamba, Bol. Uncommon to rare.

Monjitas – Genus *Xolmis*

6 species, 5 of which near swamps in semi-arid regions from Surinam to Río Negro Arg., 1 in the southern part of the Andes and adjacent lowlands.

FIRE-EYED DIUCON *Xolmis pyrope* – Plate XLIII 14

18.5 cm. Upperparts and headside dark gray. Throat and belly pale gray, breast light gray. Tail light gray, edged pale gray to grayish white. Primary-coverts and remiges black; inner remiges edged light gray. Iris red. **Juv**. undescribed.
Habits: Alone, in pairs, or family groups. Confident. Perches conspicuously on the branch of a tree, hawking for insects in the air by dashing out with sudden turns and whirls.
Voice: Call a low *tick tick* given infrequently. Also a weak plaintive whistle, sometimes followed by a musical double note.
Breeding: Eggs Oct, in s part of range in Dec or even Jan; juv. Jan (Río Negro).
Habitat: Mostly in bushes in openings between trees. Transitional zone of, or clearings in *Nothofagus* forest. In gardens, *Berberis* thickets, along hedgerows and small streams.
Range: Sea level to 3050 m. Andes from Atacama, Chile, and Neuquén, Arg., s to Isla Grande. Southern birds leave the breeding grounds to winter further n May-Nov (*pyrope*). Ssp *fortis* is resident on Chiloé Isl, Chile. Common.

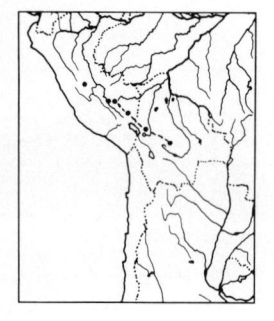

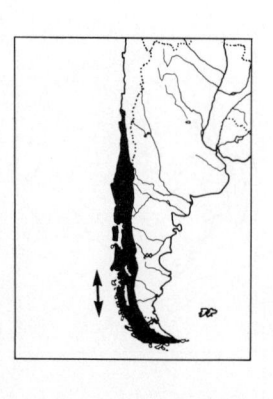

Tyrants – Genus *Neoxolmis*

2 species found in Arg. and extreme southern Chile. Large tyrants, in many ways resembling the Shrike-tyrants, but even more terrestrial.

RUSTY-BACKED MONJITA *Neoxolmis rubetra* – Plate XLIII 12

19 cm. **Above rufous brown**, sides of rump grayish white. **Long supercilium white**, upper earcoverts like back, rest of **headside and sides of breast streaked black and white**, almost meeting on breast. Rest of underside white, flanks rufous. Lesser wing-coverts like back, median coverts and tertials gray edged white, greater and primary-coverts black; **wings long**. C tail-feathers blackish brown, rest black, outer web of outer 3 edged white. Ssp *salinarum* much smaller (16 cm), markings below reduced to fine and inconspicuous streaks on sides of breast and neck, flanks with much less rufous, hindneck white with rufous tinge. White of wings and tail-coverts much more extensive. **Imm.** heavily washed buff on headside, throat, breast, and sides. **Juv.** undescribed.
Habits: Usually in small groups of up to 5 birds. Terrestrial, occ. perches on a wire or a low bush. Runs swiftly to pause with head erect, somewhat resembling a small thrush. During nuptial flight the male will fly up to rattle down (wings make a metallic rattle). Constantly opens and closes tail, and sometimes the wings.
Voice and **breeding**: No data.
Habitat: Open bushy country; bushy shores and margins of salt-lakes. In winter on the Argentine pampas and w chaco.
Range: Arg. Breeds from Mendoza to Chubut, in winter n to Tucumán (*rubetra*); Salinas Grandes and Ambargasta in Córdoba and Matará in Santiago del Estero (*salinarum*, Nores & Yzurieta in *Acad. Nac. Cienc. Córdoba Arg. misc.* 61(1979):7-8). Declining, being fairly rare now throughout its range.

NOTE: Previously placed in the genus *Xolmis*. Ssp *salinarum* may be a distinct species.

CHOCOLATE-VENTED TYRANT *Neoxolmis rufiventris* – Plate XLIII 13

22 cm. Resembles a large, long-legged, and long-winged thrush. Foreparts and back dark gray, lores blackish, ear-coverts brownish. Median wing-coverts mostly white, greater wing-coverts and tertials edged white, secondaries rufous, broadly tipped white, primaries black. Tail black, outer web of outer feather white. Belly and vent rufous chestnut. **In flight** wings look long and sharply pointed, **white and rufous zones of wings give a very characteristic pattern**. **Juv.** like ad. above, but with faint, dark, rather broad streaks. Earcoverts tinged rufous. Throat buff like belly; breast, sides, and upper belly very broadly streaked dark gray.
Habits: Terrestrial, feeding mainly on larger beetles and lizards. Takes

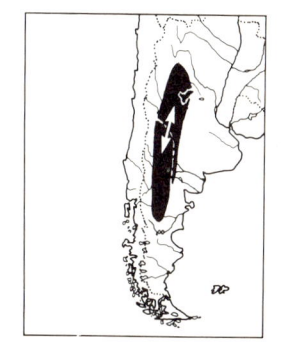

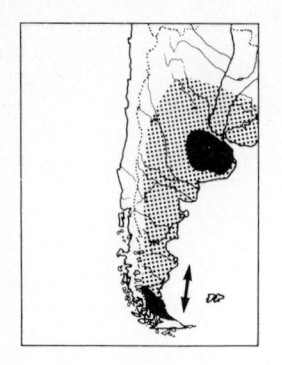

to the wing readily, flies low, skimming the ground, and perches on any elevation, threatening intruders by fluttering its wings. Runs along the ground in sharp rushes, stopping with its head erect. Alone or in pairs while breeding, but may congregate in large flocks during migration and in winter. May associate with plovers in winter.
Voice: A long, low, plaintive whistle.
Breeding: Eggs Dec (Sta Cruz) (*Auk* 86(1969):144-5).
Habitat: Flat or gently rolling open wind-swept tussocky grassland dotted with occ. *Berberis* or *Verbena* bushes; high black-looking peaty wind-swept moorland, where there are stretches of quaking bogs, and on the firmer ground with no other vegetation than hummocks of *Azorella*, scattered grasstufts, and crowberry (*Empetrum*). In winter on the pampas.
Range: Breeds from Sta Cruz to nw Tierra del Fuego, s Arg., and in Magallanes, s Chile, both in lowlands and on upland plateaus at least to 1200 m. Winters n to La Rioja, Córdoba, Sta Fé, and extreme se Brazil. Numbers fluctuate greatly. Some years abundant at the Strait of Magellan and plateaus of Sta Cruz, other years almost absent.

Shrike-tyrants – Genus *Agriornis*

5 species found in southern S Am. and in the Andes n to s Col. Mostly very large tyrants with an upright posture, a heavy hooked bill, and a long gliding flight with half spread tail. Preferring open habitats, they perch conspicuously on rocky outcrops or in bushtops, taking most of their prey from the ground, or by hawking in the air. Besides large insects, their diet consists of small mammals, lizards, frogs, and eggs or nestlings of other birds. They are aggressive and territorial, generally solitary or in pairs, and most species are relatively silent. The sexes are alike, brownish gray with a dark-streaked throat. Ad. males have attenuated 2 outer primaries that produce a low-pitched, almost inaudible whirring sound during the simple aerial display. The nest is a large loosely made open cup of dry sticks and grass lined with wool, placed in a rock-crevice, low in a bush, under the roof, or in the wall of a house. See *Amer. Mus. Novit.* 930(1937).

BLACK-BILLED SHRIKE-TYRANT *Agriornis montana* – Plate XLIII 19a-b

23 cm, bill of moderate size, black. Above dark gray-brown. Inner remiges edged and tipped whitish. Narrow and broken supercilium buffy white, throat whitish with obscure dark brown to blackish streaks, breast and sides gray-brown. Under wing-coverts cinnamon buff to buffy white. Belly and vent whitish. C tail-feathers like back, rest variable according to ssp: in *solitaria* all white, in the other sspp the feathers have an outwardly decreasing amount of gray brown at their base, least so in *insolens*, most in *maritima*. The latter form has whitish tips to c tail-feathers. **Juv.** has most of mandible from base yellowish, otherwise like ad., with breast somewhat darker.
Habits: Solitary, sometimes in pairs. Perches conspicuously on rocks, trees, or bushes with a vertical posture, but spends much time on the

ground. On search for prey, glides slowly from rock to rock, sometimes hovering briefly, and drops down on prey spotted in flight. Prey may also be searched out on the ground or in foliage. Sometimes quite bold, breeding in houses. Male will click bill during territorial disputes. May roost in caves.
Voice: Fairly vocal. Call a loud clear whistling, esp. at break of dawn *wheet hyou*, *huy-tchau*, or *pyuh*.
Breeding: Eggs Nov (n and c Chile); nestlings Oct (Arequipa); juv. Dec, Jan, Aug (Ecu.), Dec (Junín, Amazonas), Mar (Cuzco), May (w La Libertad), Feb, Apr (Cochabamba). See also *Physis C* (Buenos Aires) 35, 90 (1976): 205-9.
Habitat: Open arid country. Dry bushy hillsides, edge of *Polylepis* forest, puna grasslands with fenceposts or scattered rocks and boulders, rocky slopes and cliffs, villages, ploughed fields, low bushes near streams.
Range: 2400-3700 m in s-most Col. and in Ecu. (*solitaria*), 1830-4100 m through Peru (*insolens*), 3300-4300 m in Tarapacá (Arica), n Chile, and La Paz and Oruro, w Bol. (*intermedia*), 2000-3800 m from Cochabamba, Bol., to Tucumán and La Rioja, Arg. (*montana*), usually above 3300 m but rarely breeds down to sea level from Tarapacá to Coquimbo, Chile (*maritima*), above 2000 m in n, lower further s in the Andes of c Chile from Aconcagua to Cautín, and in sw Arg. from Mendoza to Sta Cruz, and away from the Andes in Córdoba, se Buenos Aires and Somuncura plateau of Río Negro (*leucura*). Fairly common, but has a large territory.

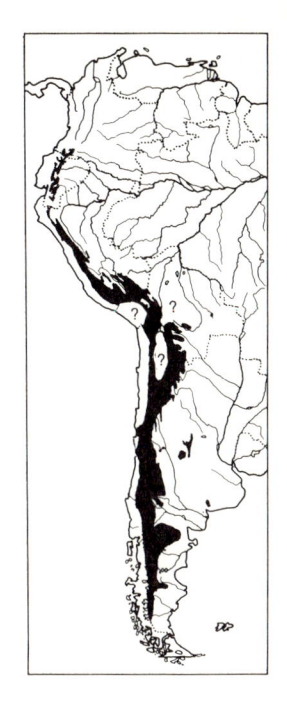

WHITE-TAILED SHRIKE-TYRANT *Agriornis andicola* – Plate XLIII 16

25 cm. Much like white-tailed forms of Black-billed S-t., but **larger; bill much heavier, more powerfully hooked**, and mandible pale horn-colored to yellowish; **tail longer; streaks on throat wider and blacker**; more white in tail; generally found at higher elevations. Above dark gray-brown. Narrow supercilium buff, headside streaked, throat white streaked blackish, upper breast and sides gray-brown; c lower breast, most of belly and vent whitish, irregularly tinged buff. C tail-feathers like back, rest white, with small, dark markings at their tips. Under wing-coverts cinnamon-buff. Sspp poorly differentiated. **Juv**. undescribed.
Habits: Much like Black-billed S-t., with which it sometimes occurs. During aerial display circles silently, alternately rising to partial stall, and dropping forward on closed wings.
Voice: No data.
Breeding: Large gonads June (nw Ecu.).
Habitat: Much like Black-billed S-t., but restricted to the puna zone. Open slopes and floors of high Andean valleys and areas having sparse and xeric vegetation of low shrubs, with *Puya*s, scattered rocks or boulders used as posts.
Range: Above 2450 m in Ecu. (*andicola*), 3660-4500 m from Peru s to Antofagasta, n Chile, and through w Bol. to Tucumán/Catamarca border, nw Arg. (*albicauda*). In Ecu. reported as 'common' in the 1800's, but there is only 1 recent report from the country (Nudo de Cajanuma,

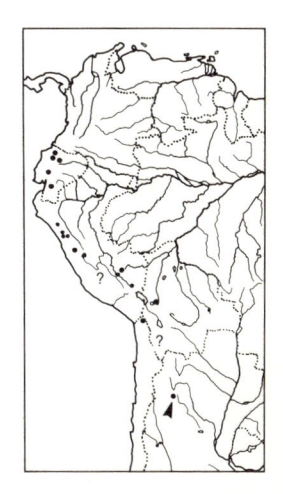

Loja, 1965). Generally local and rare, maybe declining, and outnumbered by Black-billed S-t. (in Bol. by 1:10). In n Chile said to be the commoner of the 2 (Johnson 1967), but this is not in agreement with our own observations.

NOTE: Sometimes called *A. albicauda*.

GREAT SHRIKE-TYRANT *Agriornis livida* – Plate XLIII 17a-b

26 cm, ssp *fortis* larger. **Thrush-sized**. Above gray-brown. Earcoverts tinged cinnamon, rest of headside and throat white streaked blackish, **no white supercilium** (but lores mottled with whitish); breast gray-brown, turning cinnamon-buff on lower belly and vent. Tail black, narrow tip and outer web of outer feather buffy white. Maxilla black, mandible pale horn becoming darker at tip. **Imm**. is browner and indistinctly streaked with dusky on head, back, and upper breast; entire belly cinnamon-buff.
Habits: Solitary, sometimes in pairs. Perches conspicuously in bushtop. When taking prey on the ground drops to pick it up, then rapidly runs a few steps and pauses abruptly with head thrown up and body erect.
Voice: *t-eek* or *t-eek-ek*, but usually silent.
Breeding: Oct (coastal Chile), Nov (interior Chile).
Habitat: Avoids both very xeric and very wet habitats, as well as wooded or heavily populated areas. Found in subarid countryside with large patches of bushes, thick scrub (such as *Chiliotrichium amelloideum*), cacti, and bromeliads; open flats; pastures dotted with bushes on surrounding slopes; trees in transitional *Nothofagus* forest.
Range: Sea level-1800 m. From Atacama to Valdivia, Chile (*livida*), s Chile in Aysén and Magallanes and in s Arg. from w Río Negro s to Tierra del Fuego (*fortis*). Though never common, the center of abundance is in c Chile from Aconcagua to Bío Bío, and again in sc Patagonia in Chubut and Aysén.

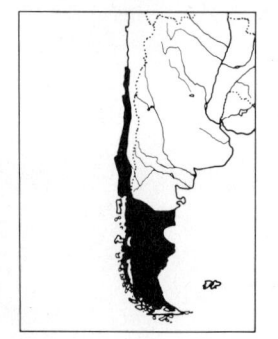

GRAY-BELLIED SHRIKE-TYRANT *Agriornis microptera* – Plate XLIII 20a-b

24 cm. Much like Great S-t., but **smaller and paler, belly washed grayish, only flanks and vent washed cinnamon, bill slightly smaller**. Above gray-brown, edge of wing-coverts paler and grayer. Tail black, narrow tip and outer web of outer feather white. **Supercilium whitish**, earcoverts tinged rufous, rest of headside and throat white streaked black (male) or dark brown (female); below gray-brown, c belly pale gray-brown to whitish, flanks washed with cinnamon, vent buffy white. Ssp *andecola* is larger, slightly darker, and has the vent tinged with tawny. **Juv**. browner; underparts pale cinnamon-brown, throat markings obsolete.
Habits: Usually alone. Perches on bushtops and runs on the ground. Wary. Flies long distances low over the ground.
Voice: Males pursuing one another give high-pitched petulant calls.

Breeding: Eggs Dec (Catamarca), Jan (Tarapacá).
Habitat: Patagonian birds breed in open scrubby steppes, birds from n Chile in brush steppes (*Baccharis*).
Range: 2900-5000 m. Resident from sw Puno s Peru s to Tarapacá, n Chile, and through w Bol. to Catamarca, nw Arg. (*andecola*); lowlands and Andean foothills from Río Negro and Neuquén s to Chubut, s Arg. Leaves southern breeding grounds in Feb., wintering from Tucumán, s Bol., Parag., and Urug., s to Córdoba and n Buenos Aires (*microptera*). Fairly plentiful in n Chile (Collacagua river valley), but scarce elsewhere.

MOUSE-BROWN SHRIKE-TYRANT *Agriornis murina* – Plate XLIII 18

17.5 cm. Above light smoky gray; tips of wing-coverts, edge of inner primaries, outer web of outer tail-feather and narrow tip of tail grayish white. Supercilium and eye-ring whitish, lores mixed blackish and white; throat white or whitish, streaked fuscous. Breast gray-brown to buffy gray-brown, turning buffy white on belly; flanks pinkish buff. **In flight** tail appears black. **Juv**. undescribed.
Habits: Runs rapidly on the ground or perches on bushes.
Voice: During pursuits gives sharp squeaky notes.
Breeding: Large gonads Dec (Tucumán).
Habitat: Dry scrub with scattered trees.
Range: 100-2000 m. Breeds at least from Tucumán s to Chubut, w Arg., but not well known. In winter migrates n as far as Cochabamba Bol. and w Parag. Uncommon.

NOTE: Previously placed in the genus *Xolmis*, and called Mouse-brown Monjita.

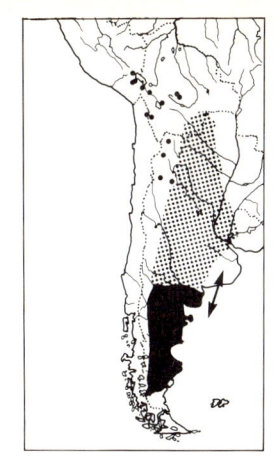

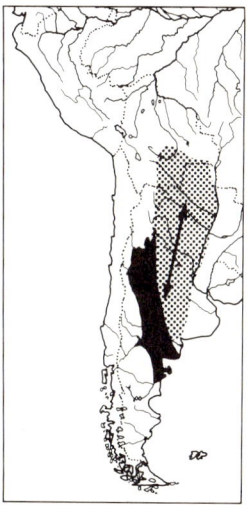

Ground-tyrants – Genus *Muscisaxicola*

12 species, 11 restricted to the Andes from Col. s-wards, most species breeding only in the southern third of S Am., wintering in the Andes n to Ecu., 1 found in the trop. and subtrop. zones e of the Andes. Medium-sized terrestrial tyrants with a striking resemblance to the Old World wheatears (*Oenanthe*) in shape and habits. They have long slender legs and an upright stance, open and close the conspicuous black tail constantly, and will make short pursuits and bouncy flights after insects spotted on the ground. During the aerial display, the male flies high up with dangling legs, comes to a stall, then drops to the ground. When defending territories most species will raise or shake their wings, and expose the hindcrown, which is usually distinctly colored in ad. Juv. and even ad. of some species are very difficult to tell apart. The nest is an open cup placed in a crack or cavity in a rock, under a stone, in a rodent burrow, or (at least in 1 species) directly on the ground. See *Amer. Mus. Novit.* 930(1937), and *Hornero* 11(1975): 242-54.

SPOT-BILLED GROUND-TYRANT
Muscisaxicola maculirostris – Plate XLIII 1a-b

15 cm. **The smallest highland Ground-tyrant**. Ssp *maculirostris*: above smoky-gray with **brownish** tinge; **wing-coverts and inner remiges edged buffy brown, giving mottled appearance**. Lower rump and tail black, outer web of outer tail-feather buffy white. Narrow, broken supercilium whitish, lore dusky. Throat whitish; breast pale smoky gray, turning pale buffy white on belly and vent. Under wing-coverts pale buff. Base of mandible yellow, but this is difficult to see, except from below. Ssp *niceforoi* is much like *maculirostris*; *rufescens* much browner, underside cinnamon-buff, under wing-coverts tawny-buff. **Juv**. has rufous edgings in the wing.
Habits and **voice**: Alone or in pairs, in winter sometimes 2-3 together. Does not fan tail as often as others of the genus. Regularly perches on bushtops, walls, and halfway up banks. During aerial display utters an accelerated *t - t-tk-tk-cleeoo* while fluttering up to stall where flicks and fans tail, and holds wings high while the *cleeoo* is uttered, repeats 1-15 times, and swoops to the the ground. Also utters a *tek*.
Breeding: Eggs Oct-Nov (Chile), fledgling Feb (Jujuy); nuptial flight Feb (Arequipa).
Habitat: Has a broad habitat preference, but is always close to vertical rocks, banks, or walls. Found from fairly arid country to desert with scattered bushes on sparsely vegetated hillsides, and on rocky slopes.
Range: 2000-3500 m in n part of range, mainly 2000-4000 m in s Peru, where apparently small numbers are resident in the coastal range and occ. descend to sea level (seen performing nuptial flight in Jan at 400 m in coastal Arequipa: RH), 3000-4000 m in Bol., in Patagonia below 1500 m. E Andes of Col. from c Boyacá to mts of Bogotá (*niceforoi*), Ecu. (*rufescens*), Peru (recs. from the low-lying Saposoa and Moyabamba, San Martín undoubtedly erroneous and referable to the following species), Bol., Chile, and w Arg. s to Straits of Magellan, and (isolated?) on Somuncura plateau in Río Negro; accidental at coast of Chubut, on Isla Grande and Isla Navarino (*maculirostris*). Generally fairly common, but uncommon and local in Col.

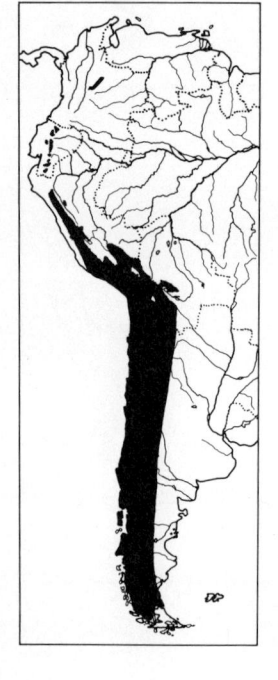

The very similar trop. and subtrop. **LITTLE GROUND-TYRANT** *Muscisaxicola fluviatilis* is a scarce straggler to the temp. zone of Cochabamba and the Titicaca basin from its limited breeding grounds in e Peru, n Bol., and adjacent Brazil. 13 cm. Resembles Spot-billed G-t., but tail shorter. Above gray-brown; tail black, outer web of outer feather white. Lores mottled whitish, eye-ring buffy (vs. blackish lores with more distinct whitish supraloral streak and whitish spot below eye); earcoverts uniform (vs. with whitish patch sometimes only formed by pale shafts). **Wings dusky**, inner remiges very narrowly edged pale grayish cinnamon, greater wing-coverts with narrow buffy brown tips forming **straight lower wing-bar** (vs. secondaries and tertials more broadly edged dull rufous, greater wing-coverts gray-brown edged and tipped buffy, giving mottled appearance). **Throat and breast buffy white**, lower breast washed pale gray-brown (vs. throat whitish, breast pale gray-brown, feathers broadly edged and tipped whitish giving mottled or flammulated appearance); belly whitish in contrast to breast (vs. whitish without contrast, often flammulated like breast, and often more buff). Less developed or no bill-spot. Almost invariably on sandbars and banks along rivers.

DARK-FACED GROUND-TYRANT
Muscisaxicola macloviana – Plate XLIII 8

15.5 cm. **Small and rounded**. Above dark smoky gray, **face blackish, turning dark umber-brown on crown**. Lower rump and tail black, outer web of outer feather and narrow edge of next white. Below pale smoky gray, c belly and vent whitish; throat tinged brownish. **Juv**. has streaked throat and buffy edges to wing-coverts.
Habits: Sometimes hovers for longer periods with spread tail, spotting prey on the ground. Usually catches insects on the ground or on ledges. Watches from a high point, hovers momentarily, then flutters and parachutes down, tail spread, to land and snap up insect almost simultaneously; then sometimes walks, runs, or hops after another insect, pursuing it with hops and bouncy flights. During aerial display male flutters up to 16 m up, to drop to a rock. While displaying on the ground, raises and lowers wings in front of female, making shiny underwing contrast the darker flank-patch. Sometimes stands near another bird (female?) with its wings raised high for 5-10 s, then moves c. 1 m and repeats. In winter often in loose flocks of up to 100. Occ. perches on trees and telegraph-poles.
Voice: Alarm or during pursuits a loud reedy *cheep*, uttered in a rapid series. Also a low hard *tu* that may be repeated rapidly and sometimes combined with the *cheep*-note *chee-**tu***. Territorial song(?): a pleasing series of warbling sounds.
Breeding: Large gonads Nov (Aysén).
Habitat: Breeds on the prairies and in the vicinity of rivers and marshy places in the valleys, as well as at great elevations in the mountains, where dwarf trees have been replaced by rocks and cushions of *Azorella*. During migration and in winter on floating sea-weed, flats near the sea-shore, and in irrigated fields in desert, sometimes in totally bare, rocky, and sandy deserts.
Range: Sea level to 1200 m. Along Andes from Llanquihué, s Chile, and Neuquén, sw Arg., s to Isla Grande, and (isolated?) on Somuncura plateau of Río Negro. Migrates n on the coast to Trujillo, Peru, and e of Andes to Entre Ríos and Urug. Arrives on the breeding grounds in Sep (occ. as late as Nov), and departs Feb-Apr. In winter fairly common on the coast of s Peru (*mentalis*). Resident on Islas Malvinas (*macloviana*).

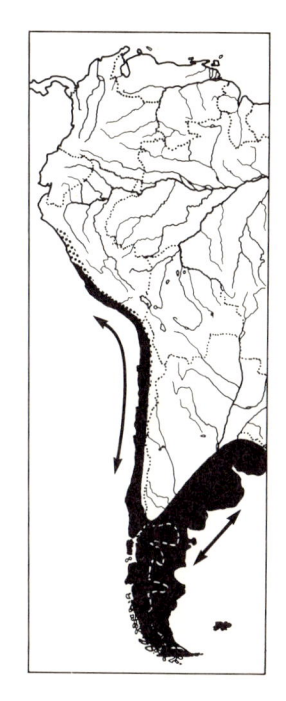

CINNAMON-BELLIED GROUND-TYRANT
Muscisaxicola capistrata – Plate XLIII 9a-b

16.5 cm. Above gray-brown, brownest on lower back. **Forehead and sides of forecrown black, rest of crown rufous chestnut**. Lower rump and tail black; outer web of outer tail-feather, and narrow edge of next 2 white. Throat, breast and upper belly pale buffy gray-brown; **lower sides, flanks, lower belly and vent cinnamon**. **Juv**. has paler belly (buff); rufous chestnut of crown replaced by fuscous or raw umber, feathers narrowly tipped blackish; breast faintly mottled or flammulated, belly pale buff. Wings broadly edged buff.

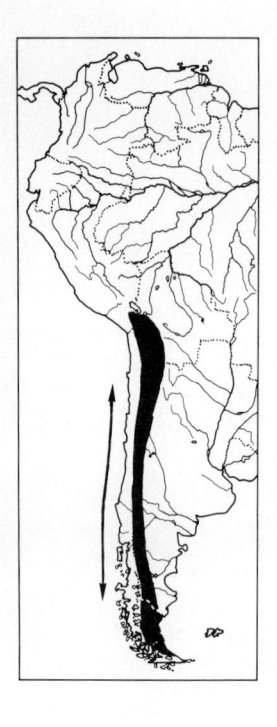

Habits: Forages much like congeners. During aerial display flies high up (higher even than Dark-faced G.-t.), halts, suddenly hovering like a kestrel with dangling legs and spread tail, the wings making a whirring sound, then drops to the ground. When the wind is strong these maneuvers are performed close to the ground. When aggressive exposes crown, ruffles chest and flanks, but apparently does not raise wings.

Voice: When nervous gives a high-pitched (pitch like call of Spotted Sandpiper) *wee tee*, *wee tee tee*, or *wee tee tee tee*. Upon landing after making short glides in the presence of mate(?) utters a *weetee weetee weetee wee wee wee* (*wee*-sounds rising, *tee*-sounds falling in pitch).

Breeding: Eggs Nov-Dec.

Habitat: Sometimes nests in rodent burrows. Rolling grassy hills with rocky outcrops; rocky canyons with grassy patches; overgrazed eroded soil with bunchgrass; pastures. Prefers flat, short moist grass. In winter on dry rolling hills and dry fields, on cushion bogs, and lakeshores. May roost in caves.

Range: Breeds in unforested n part of Isla Grande (see *Condor* 72(1970): 361-3) and neighboring Magallanes, Chile, and Sta Cruz and maybe further n along the Andes, and rec. on Somuncurá plateau of Río Negro, Arg.; present here from Sep to Feb (Mar). Winters in the Andes n to Puno and Arequipa, s Peru, at increasing elevation n-wards (4000 m in s Peru). Quite common on breeding grounds, uncommon or rare in s Peru in winter.

RUFOUS-NAPED GROUND-TYRANT
Muscisaxicola rufivertex – Plate XLIII 10 a-b

16.5 cm. **Very upright posture. Bill long and slender, downcurved at tip**. Ssp *occipitalis*: above **very pale and gray**, hindcrown with conspicuous chestnut patch. Outer 1-2 feathers of dusky tail very narrowly edged whitish. White supercilium ending just behind eye. Below pale gray, turning whitish on belly and vent. In fresh plumage very narrow white edges in wing. Ssp *pallidiceps* smaller (15.5 cm) with pale ashy gray upperparts. Supercilium narrower, underparts almost white, crown-patch pale (cinnamon), tail black instead of dusky. Ssp *rufivertex* like *pallidiceps*, but darker (ashy gray) above; crown-patch darker (rust). **Juv.**: crown-patch duller and browner, wing-coverts edged cinnamon.

Habits: Solitary. Stands erect on flat ground or a slightly elevated point, eyes an insect, then dashes or runs to snatch it, then nervously jerks while opening and closing the tail. During territorial disputes flies to the top of a rock, spreads wings and tail, exposes hindcrown and shakes wings; then flies to another rock and repeats. In the aerial display the male flies up to about 16 m up, then hangs vertically with wings stretched fully above back, at this point uttering a thin, high-pitched *twee-it*, the procedure being repeated a few times.

Voice: Calls include a fairly loud *tweet* and *tit*. See also under aerial display.

Breeding: Eggs Sep (c Chile), Jan (Córdoba); nestlings Sep (Cuzco); fledgling Dec (Puno).

Habitat: Flat, grassy, or ploughed fields, grassy plains with neighbouring rocky slopes; dry gravely slopes and sometimes dry desert; rocky ravines and small valleys with cliffs and rocky edges, and sometimes bushes or trees. In Chile sometimes occurs with (but usually above) Spot-billed, below White-fronted and Black-fronted, and together with Ochre-naped and Puna G-t.s. May roost in rocky caves.
Range: 3000-4520 m in Peru (except sw) and n Bol. in La Paz and Cochabamba (*occipitalis*), 1000-4000 m (regularly to below 300 m albeit without evidence of breeding: RH) from Arequipa, sw Peru s to Antofagasta, Chile, and in s Bol. and nw Arg. s to Córdoba, apparently migratory in the s (*pallidiceps*), medium altitudes, descending to coast in winter, from Atacama to Colchagua, Chile, and in Andes of Mendoza, Arg. (*rufivertex*). Common in s Peru.

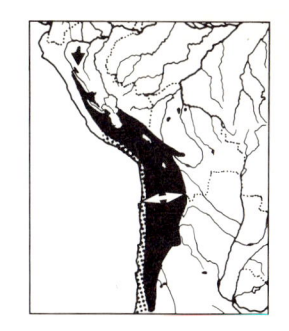

PUNA GROUND-TYRANT *Muscisaxicola juninensis* – Plate XLIII 6

16.5 cm. Above smoky gray, most of crown dull cinnamon. Narrow white supercilium stopping just behind eye. Below pale gray, throat and belly whitish; breast with very faint, whitish streaking; **underside with a faint buffy yellow tinge**. Also distinguishable from White-browed G-t. by **smaller bill, more conspicuous panel on inner remiges, and less white on lores**. Tail black, outer web of outer feather and narrow edge of following 2 whitish. **Juv**. has rufous gray edges to wing-coverts, and is tinged pinkish buff below.
Habits: Much like other ground-tyrants.
Voice: No rec.
Breeding: Eggs Oct (Chile), juv. Jan (Potosí).
Habitat: Bogs with matted vegetation and slightly drier grass, stones at the edge of bogs, and grassy steppes interrupted by rocky outcrops and small cliffs. Slopes with uneven soil (ploughed, or solifluction soil with turned tussocks etc.).
Range: 4000-5000 m, usually above 4200 m. From Lima and Junín, c Peru s to Tarapacá, n Chile, and through Bol. to Tucumán, nw Arg. Common to fairly common.

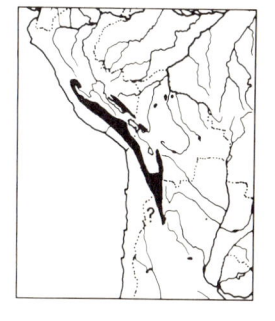

WHITE-BROWED GROUND-TYRANT
Muscisaxicola albilora – Plate XLIII 5

16.5 cm. Bill long. Forehead brown, **broad hindcrown tawny**, back gray-brown to brown. Below dull pale gray, vent whitish. Wings with only slight pale edges. **Supercilium white**, turning gray behind eye. **Juv**. has the whole crown brown; tertials, median and greater wing-coverts narrowly edged pale dull rufous.
Habits: Solitary, in loose flocks on migration. In the aerial display climbs slowly with dangling legs and spread tail, usually silent, then swoops to ground. During territorial disputes raises one or both wings, fans tail, and raises hindcrown; sometimes also bows. When very aggressive will even fly with raised, shaking wings and fanned tail.
Voice: *tseet*, *tseek*, *tut*, *tsk*, *tchk*, or *tk*, with increasing excitement. Flight-call *tseet*. Contact call *tut*. During aerial display rarely a *clee ip*.

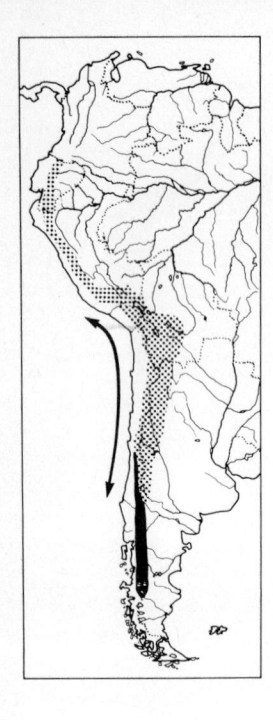

Breeding: Eggs Oct-Jan (Chile).
Habitat: Breeds on rocky barren slopes with almost no vegetation, overlooking watered valleys. In winter often found on marshes and in the vicinity of lakes.
Range: Breeds from 1500 to 2500 m from Aconcagua to Magallanes, Chile, and in neighboring Arg. from Neuquén to Sta Cruz and isolated on Somuncurá plateau in Río Negro. Winters at 3000-4000 m in Bol., Peru, Ecu., and probably Nariño, Col. Fairly common in Peru in winter.

PLAIN-CAPPED GROUND-TYRANT
Muscisaxicola alpina – Plate XLIII 3a-c

18 cm. Bill short. Sspp *alpina*, *columbiana*, and *quesadae*: above dark gray-brown, crown very dark brown, lower rump and tail black. Broad supercilium white, but stopping shortly after the eye. Chin silvery white; breast and sides pale gray-brown, turning whitish on rest of underparts, entire underparts with very faint, whitish streaks. Ssp *grisea* much paler and grayer, smoky gray above, crown only very slightly darker and browner; streaked effect below more conspicuous; supercilium narrower. **Told from Cinereous G-t. by larger size, darker and browner upperparts (Mouse-Gray or Mouse-Brown), whitish supercilium extending slightly beyond eye, gray breast, contrasting with the whitish throat, and white abdomen tinged with pale buff; wing-coverts not tinged buff. Juv.** has wing-coverts and inner remiges narrowly tipped and edged cinnamon; belly and vent tinged buff; whitish streaking below more pronounced; feathers of crown and back with faint, darker tips.
Habits: Usually solitary, but sometimes in flocks of up to 30. Much like others in the genus, but posture less upright. Will rarely alight on a dead bush. 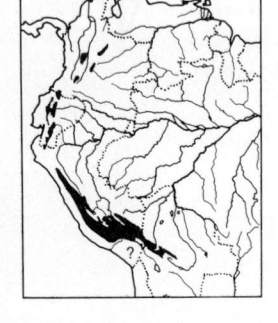
Voice: A fine *zit* uttered while foraging or on the alert.
Breeding: Juv. Dec (Puno/Cuzco), Jan (La Paz); large gonads Sep-Dec (Boyacá).
Habitat: **Level** or gently sloping grassy soil. Sometimes on dry, sparsely vegetated ground.
Range: 2700-4800 m. Páramos of E Andes of Col. in Boyacá and Cundinamarca (*quesadae*), locally from Páramo de Sta Isabel in Quindío to Cauca, C Andes of Col. (*columbiana*), Ecu. (*alpina*), Peru, and high Andean slopes of Bol. from La Páz to Cochabamba (*grisea*). Generally fairly common, but uncommon in Col.

NOTE: Should perhaps include *M. cinerea argentina*.

CINEREOUS GROUND-TYRANT *Muscisaxicola cinerea*
– Plate XLIII 2a-b

16 cm. **Much like Plain-capped G-t., but smaller, breast paler, hardly contrasting with throat**; supercilium narrower and shorter, in most specimens not extending beyond eye; above paler and grayer, but wing-coverts warmer tinged. Ssp *argentina* slightly larger than *cinerea*

(only very slightly smaller than Plain-capped G.-t.), upperparts and supercilium like Plain-capped G-t.; below also much like that species, but upper belly less pure white. Virtually indistinguishable.
Habits: Much like others of the genus.
Voice: No rec.
Breeding: Eggs Oct (Chile), fledglings Dec (Tucumán), Jan (Puno). See *Hist. Nat. Mendoza* 1(1980): 180.
Habitat: In Chile on rocky slopes close to snowline and overlooking streams or lakes, in Bol. on the altiplano. In winter on level bogs and lakesides, preferring short, matted vegetation.
Range: 2700-4000 m. Puno, s Peru, and the altiplano of Bol. (where it meets Plain-capped G-t. along the eastern mts), Andes of n Chile from Coquimbo to Talca (see *Bol. Soc. Biol. Concepción* (Chile) 53(1982): 171-2), and Arg. in Mendoza (*cinerea*). nw Arg. from Jujuy to Catamarca, and in Tucumán (*argentina*). In winter migrates n as far as Lima and Junín, c Peru (rec. on 12 and 25 Oct in Huancavelica, breeding?). Uncommon in winter in Peru.

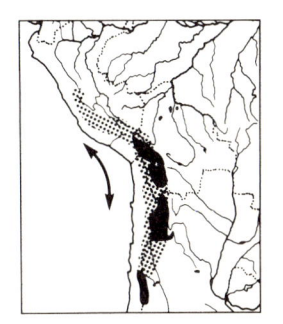

NOTE: Formerly regarded a ssp of Plain-capped G-t. Ssp *argentina* may belong in that species.

WHITE-FRONTED GROUND-TYRANT
Muscisaxicola albifrons – Plate XLIII 4

20 cm. **Largest of the genus. Very upright posture**. Wings long, almost reaching tail-tip. Above smoky gray, **secondaries and their coverts conspicuously edged silvery gray**. Extreme forehead and broad supercilium to eye white (smallest in female?); hindcrown dull brown. Tail blackish, outer 2-3 feathers pale-edged, outer web of outer wholly pale from below. Below pale smoky gray, turning whitish on c belly; vent white; underparts with very faint pale flammulations. **In flight pale wings. Juv**. undescribed.
Habits: Solitary. Much like others of the genus.
Voice: No rec.
Breeding: Eggs Nov and Jan (Chile).
Habitat: Apparently always forages on flat bogs with matted grass and cushion-plants, and on neighbouring gentle grassy slopes. Nests on slopes with sparse and overgrazed low shrubs and bunchgrass, interrupted by cliffs and boulders.
Range: 4000-5600 m. From Ancash, c Peru, s to Oruro, w Bol., and Tarapacá, n Chile. Fairly common, but usually outnumbered by Puna G-t.

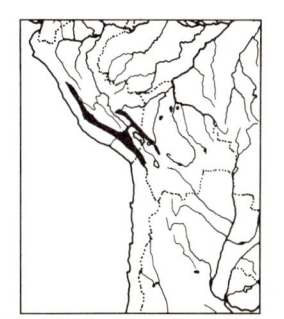

OCHRE-NAPED GROUND-TYRANT
Muscisaxicola flavinucha – Plate XLIII 11a-b

18.5 cm. **Large, wings long, bill long**. Above smoky gray. Conspicuous **patch on hindcrown yellow-ocher**. Lores white above and below a dusky patch; narrow line across forehead white, as the supercilium, that is broad in front of eye, but narrow behind. Throat and breast pale smoky gray, turning whitish on belly and vent. Tail blackish, outer web of outer feather, and narrow edge of next 2 white. In fresh plumage the remiges, esp. the inner, have broad white edges that are sharply defined, these also present on the greater wing-coverts and tips of remiges. White edges in wing disappear with wear. Ssp *brevirostris* is darker, and has slightly smaller bill and wings. **Juv**. has a faint or no crown-patch, and has rufous buff edgings in the wing.

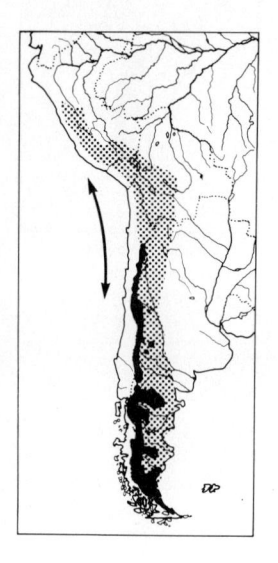

Habits: Solitary, during migration in flocks of up to 50. Not agile; flies even short distances, rather than hopping or running. Breast protrudes when bird comes to an erect stand, but mostly the body is held more horizontal than in other g-t.s. Spreads wings and tail simultaneously upon landing. Exposes crown-patch when aggressive.

Voice: When excited gives short, fairly high-pitched bursts *tsee tee tsee tseet*.

Breeding: Eggs Oct-Jan (c Chile). 2 broods?

Habitat: In winter on high elevation bogs and gentle slopes with short grass. Breeds on barren rocky slopes with small cliffs or boulders and very little herbaceous vegetation, overlooking well-watered valleys; often near lakes and marshes.

Range: Breeds at 2000-4300 m in c and n Chile, lower further s. Mts of Chile from Antofagasta to Colchagua and w Arg. from Mendoza s to Chubut (*flavinucha*), and on Somuncurá plateau in Río Negro (ssp?); from Aysén, s Chile (see *Bol. Soc. Biol. Concepción* (Chile) 53(1982): 171-2), and Sta Cruz, s Arg., s to Isla Grande (*brevirostris*). Winters at 3200-4700 m in the puna zone of Bol. and Peru n to La Libertad, returning to the breeding grounds on Isla Grande in Nov-Dec. Fairly common in Peru in winter. A few rec. from Oruro, Bol., and Ayacucho and Junín, Peru, during the austral winter may suggest breeding.

BLACK-FRONTED GROUND-TYRANT
Muscisaxicola frontalis – Plate XLIII 7

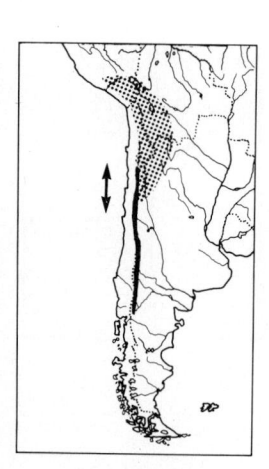

18 cm. **Bill long and distinctly curved at tip. Above** pale **gray** with slight brownish tinge; **forehead and c crown black**, supra-lores, broken eye-ring, and area below eye and lores white. Below whitish to grayish white, with slight creamy cast. **Juv**. with forehead sooty, wing-coverts edged pale buff.

Habits: Like others of the genus, but stance less upright than most. In winter solitary, sometimes roosting in caves with Rufous-naped G-t.

Voice: No rec.

Breeding: Pull Dec (Santiago).

Habitat: Stony or rocky hillsides.

Range: Breeds above 2900 m (higher than any sympatric congener) from Antofagasta to Santiago, Chile, and from Mendoza to Río Negro, w Arg. Winters above 3600 m in Jujuy, nw Arg., on the Bolivian altiplano, and in s Peru n to nw Arequipa. Rare in s Peru during the austral winter.

Negritos – Genus *Lessonia*

2 species found in Arg. and Chile and in the Andes n to c Peru. The nest is an open cup placed in a niche at the side of a tussock, or on the ground, hidden in the grass.

WHITE-WINGED NEGRITO *Lessonia oreas* –
Plate XLIV 17a-c

12.5 cm. **Male**: Black, back rufous. Underside of primaries and outer secondary silvery white. **Female**: Blackish brown, chin whitish, back and sides of breast dull dusky rufous. Underside of wing like male. **Juv**. much lighter.
Habits: Usually in pairs or family-groups. Terrestrial. Perches on tussocks. Makes flights to catch insects in low air or on the ground. Also makes short, quick runs on the ground in pursuit of prey. Often flashes wings when perched, thus exposing pale underwing. Sometimes fairly confident.
Voice: Contact call given almost constantly, esp. when in family groups, a very short, rather faint note repeated at 0.5 s intervals *tyt-tyt-*.... Alarm a very high-pitched *zi*.
Breeding: Eggs Oct-Nov (Junín), Dec, Jan (n Chile).
Habitat: Level bogs with very short vegetation, mudflats, and heavily grazed lakeshores, and seasonally inundated grassy plains.
Range: Breeds at 3000-4900 m in nw Arg. and n Chile, occ. down to 1000 m. In Peru only above 4000 m, but rarely down to coast in winter. From Junín, c Peru, s to n Coquimbo, n Chile, and through highlands of Bol. to Catamarca and Tucumán, nw Arg. Locally fairly common.

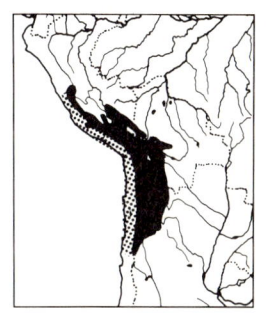

RUFOUS-BACKED NEGRITO *Lessonia rufa* –
Plate XLIV 18a-b

11.5 cm. **Male**: Black, back rufous chestnut. 2nd and 3rd outermost primary acuminate. No pale patch on underwing. **Female**: Crown dusky, feathers edged pale gray-brown; nuchal collar dark brown; back and scapulars dull rufous; wing-coverts bordered with reddish brown in fresh plumage. Rump and upperside of tail blackish brown, outer web of outer tail-feather whitish. Lores, supercilium and fore-cheeks whitish streaked gray-brown; below buffy white, broadly and indistinctly streaked gray-brown on breast and sides; flanks and c belly buff, vent whitish. Small patch on underwing cinnamon. **Juv**. like female, but back more rufescent. After c. 2 months, changes to imm. plumage, which is carried for 6 months.
Habits: Much like White-winged N. During migration in flocks of up to 50, all of the same sex. In winter in small, loose flocks. Fans tail when nervous. Regularly perches in bushtops, but spends most of its time on the ground like White-winged N. Sometimes follows people to feed on insects flushed from the grass. In warm weather frequents sea-shore, even down to watermark. During display the male performs a butterfly-like flight at 10-15 m height.
Voice: Alarm *tjit-tjit-*.... Contact note a short twitter.
Breeding: Eggs Sep (n Chile), Oct (coast of Chile), Nov (mts of Chile and s Chile), Dec (Sta Cruz), Oct-Jan (Isla Grande); fledgling Dec (Neuquén). Probably 2 broods.
Habitat: Like White-winged N. mostly on marshes and lakeshores, but also found on dry pastureland, and on Isla Grande in hilly as well as level, open country.

Range: Breeds from sea-level to 2000 m. From Tarapacá (Arica), n Chile, nw Arg., and Urug. s to Isla Grande. Migratory in s part of range. On Isla Grande males arrive mid Sep and depart Dec-Jan, while females arrive early Oct and depart with the young Feb (Mar). Common.

NOTE: Previously included the White-winged N.

Black-tyrants – Genus *Knipolegus*

10 species found e of Andes from Ven. to Río Negro, Arg., and in the Andes from Ven. to nw Arg. Medium-sized tyrants with an upright posture and rather thick-necked, head square. Often strongly sexually dimorphic: male black, female brownish or grayish, with streaked breast and rufous in the tail. The male attains a female-like imm. plumage after the post-juv. molt, and possibly breeds in this plumage the following year. Aerial hawkers. The nest is an open cup, placed in a tree or a bush. See *Amer. Mus. Novit.* 930, 962 (1937).

PLUMBEOUS TYRANT *Knipolegus signatus* – Plate XLV 12a-d

14.5 cm. **Male**: ssp *cabanisi* **slaty gray**, with light abdomen. Loral region, wings and tail blackish, **underside of wing with whitish patch** formed by inner webs of secondaries and inner primaries. Ssp *signatus* uniformly black; white patch on underwing as in *cabanisi*. **Female** (see also *Neotrópica* 8(1962): 99-100): ssp *cabanisi* **iris red**; above olive-brown, crown and mantle usually somewhat darker and with indistinct narrow streaks. 2 wingbars and wing-panel pale ochraceous, wingbars fading to white; tertials edged whitish. Upper tail-coverts and edges to tail-feathers rufous, broadly on inner webs of tail-feathers. Loral region black to gray. Below olive-brown, throat and c underparts faintly streaked olivaceous white; vent dull rufous; entire underparts with a brownish wash. Ssp *signatus* has darker and warmer olive-brown upperparts and flammulations below, and darker, duller and more restricted rufous edging of tail, hardly showing on outer webs. **Imm.** like female. **Juv**. like female, but iris brown, upperparts washed rusty, wingbars white, and panel pale yellow; the pale streaks below are more extensive and tinged pale yellow.
Habits: Solitary. Perches upright, with head tucked in between shoulders, breast-feathers puffed, and tail constantly vibrating from side to side. During display, the male repeatedly shoots 10 m up into the air, to drop straight down with folded wings, spreading them just before landing (see plate). Flies in long waves with wings close to body between the short series of wingflaps.
Voice: No rec.
Breeding: Eggs Nov-Dec (Tucumán), Jan (Jujuy).
Habitat: 1-4 m up in dark places of lower to middle strata, often on dead twigs. Humid montane forest. Along paths, edges, clearings in patchy woods, and in alder thickets and second growth on slopes along streams.

Range: 1900-3050 m from Condor mts on Ecu./Peru border s to Junín, c Peru (*signatus*); 1100-2500 m from e Cuzco, se Peru, along Andean slope to Tucumán, nw Arg. (*cabanisi*). Rare in c Peru, fairly common in nw Arg.

NOTE: Previously treated as 2 species: Jelski's Bush-tyrant (*Myiotheretes signatus*) and Plumbeous Tyrant (*Knipolegus cabanisi*).

RUFOUS-TAILED TYRANT *Knipolegus poecilurus* – Plate XLV 16

14.5 cm. Sexes almost alike. Slenderer than Plumbeous T. Iris red. Above dark gray; 2 faint wingbars grayish white to grayish buff, tertials fairly narrowly edged buffy gray to pale gray. Upper tail-coverts more or less washed with rufous; closed tail dark above, rufous below, outer tail-feather mostly rufous. C throat whitish, changing to pale rufous on c breast, and on belly and vent; sides and faint, broad streaks across breast gray. Female slightly browner than male. Sspp poorly differentiated. **Imm**. has brown iris; **juv**. more washed with rufous. Alone or in pairs. Shy. Perches on herbs and low bushes, sometimes with slightly cocked tail, making short sallies for flying insects or dropping to the ground, and returning to perch. Rarely seen. Breeds Mar (Piura); large gonads Mar-Sep (Col.).

Within 2 m of the ground. On humid grassy slopes with some tall herbs and scattered bushes; edge of forest and cloud forest. 900-2500 m, occ. to 3100 m. Distrito Federal coastal mts, and Mérida and Táchira, Andes of Ven. (*venezuelanus*), Perijá mts and all 3 ranges of Col. (*poecilurus*); se Ecu. s along C and E Andes of Peru from Cajamarca to Puno (*peruanus*). A single specimen from trop. c Bol. probably a straggler, perhaps from unknown breeding grounds on the low mts of e Bol., or Andes of Sta Cruz. 2 other sspp on Tepuis of Ven. Local. Generally uncommon, but fairly common in n Peru.

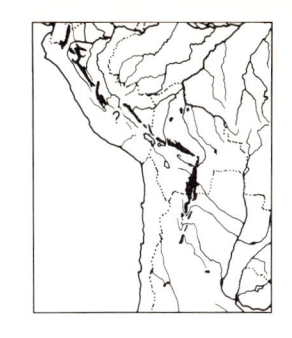

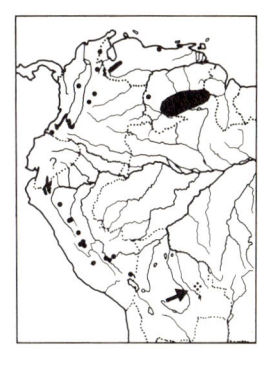

WHITE-WINGED BLACK-TYRANT
Knipolegus aterrimus – Plate XLV 13 a-f

16-17 cm. **Fairly upright. Head large and square. Iris dark brown. Male** *anthracinus* dull **black**, slightly lighter below. **Underwing with large white patch** on most of inner webs of primaries, **conspicuous in flight**. Bill blue-gray tipped black. Ssp *heterogyna* much like *anthracinus*; *aterrimus* larger and blacker, more uniform and glossy. **Female**: ssp *anthracinus* has dark ear-coverts and buffy white base to loral feathers, seen as 2 'horns' in the forehead at some angles. Upperparts dark gray-brown, darkest on c crown; wings with 2 buffy white bars, lower bar palest; whitish panel on tertials and secondaries, narrower and brown on primaries. **Lower rump and upper tail-coverts distinctly rufous**; closed tail dark above, rufous below, except for broad, dark tip and c feathers; outer tail-feather wholly rufous except for dark terminal spot. Upper throat whitish, turning pale rufous on rest of underparts; lower throat and breast heavily washed with faint brownish gray flammulations; c belly paler than flanks. Ssp *heterogyna* differs from *anthracinus* by darker upperparts, paler underside, and paler rufous in tail. Ssp *aterrimus* has blackish crown; c tailfeathers rufous basally, the rest edged rufous also on base of outer web; underparts darker rufous, breast and lower throat without flammulations. Outer web of outer primary rufous. **Imm**. like female *anthracinus*. **Juv**. more washed with rufous.

Habits: Solitary. Not shy. Flicks tail upon landing. Perches rather well hidden inside canopy, making short sallies into the open for flying insects, quickly returning to canopy. During aerial display the male performs a rapid vertical circular flight near the canopy (see plate). Ssp *aterrimus* perches low, and sometimes makes short runs on the ground.

Voice: A female-plumaged bird (Cuzco) uttered a *trrie rrhi rrhi – trrhie rrhie rrhie - trrhie rrhit*, while foraging near the ground. Migrants give a faint *tseet*.

Breeding: Dec-Feb in s. Eggs Oct-Nov (Cochabamba); fledgling Nov (Sucre).

Habitat: 0.5-3 m up, usually 0.5 m below top of small trees and bushes. Ssp *anthracinus* in second growth on forest clearings and landslides, humid open shrubby forest, and in outstanding trees in shrub. Ssp *aterrimus* in dense thickets of thorny trees, riparian willow thickets in the arid zone. During migration in open brush on gravely hillsides. Ssp *heterogyna* also in arid zone, and at higher elevations than *anthracinus*.

Range: 600-3250 m, occ. down to 250 m, and in Bol. regularly to 3700 m. Marañón Valley in Cajamarca, La Libertad, and Ancash n Peru (*heterogyna*); Junín (Andamarca), Huancavelica (Anco), and from n Ayacucho, s Peru, to La Paz, Bol. (*anthracinus*); from Cochabamba, Bol., s to Chubut, Arg., and e to Parag. chaco. Southern birds migrate n through breeding grounds in Mar and Apr (accidentally to Entre Ríos), and return in Nov (*aterrimus*). 1 ssp in ce Brazil. Fairly common.

Tyrant – Genus *Hymenops*

Monotypic genus. The nest is an open cup, attached to reeds.

SPECTACLED TYRANT *Hymenops perspicillata* – Plate XLV 15a-b

14.5 cm. **Conspicuous broad yellow bare eye-ring. Male: black; primaries** mostly **white**. Bill yellow, but appears white at a distance. **Female**: above dark brown with black streaks and pale brown edges. Remiges rufous with dark tip, 2 wingbars buffy. Below pale buffy, with sharp, dark brown streaks on breast; belly sometimes buff. Tail dark, outer feather edged rufous buff on outer web. Sspp poorly differentiated. **Juv**. much like female.

Habits: In close pairs on breeding grounds. Often perches on fenceposts and wires. Sallies upwards for flying insects, returning to same perch, or feeds on the ground like a pipit, or running on floating leaves. During display, the male shoots 3-4 m up and drops straight down with half closed wings, while the vibrating primaries produce a buzzing sound.

Voice: Usually silent, but occ. the male utters a squeaky series of thin notes.

Breeding: Nov-Jan (Chile).

Habitat: Reedbeds and marshes, nearby flats, and fields.
Range: Breeds in lowlands; during migration up to 3350 m. From e Bol. and s Brazil s to Chubut, s Arg., migratory in the s (*perspicillata*). From Atacama s to Valdivia, Chile, and adjacent w Río Negro to w Chubut, Arg., Chilean birds migrate across Andes to winter in lowlands of Arg. (*andina*). Accidental in puna zone of Cuzco, se Peru. Fairly common.

Tyrant – Genus *Satrapa*

Monotypic genus. The nest is cup-shaped.

YELLOW-BROWED TYRANT *Satrapa icterophrys* – Plate XLV 5a-b

15 cm. Upright posture. **Long, broad supercilium yellow**, contrasting with dusky stripe through eye and dark earcoverts. Above greenish gray; 2 **wingbars**, the upper **broad and pale gray**, the lower yellowish white; edges of tertials and outer tail-feather yellowish white. Below yellow, sides of breast like back. Female is less bright yellow below than male, and has the breast washed with olive. **Juv**. has a narrow upper wingbar, and elongated olive spots on the breast. Solitary. In winter and during migration may roost in townparks in flocks of 100s, together with 1000s of Eastern Kingbirds and many Crowned Slaty Flycatchers (*Empidonomus aurantioatrocristatus*). Perches openly at the side of a bush or a hedgerow, hawking insects in the air. No rec. of voice. Fledgling Jan (Jujuy).

At edge of vegetation. Patchily wooded ravines, hedgerows, open second growth, edge of fields, fairly dry scrub and shrub, swamps. Mostly trop., but up to 2600 m in Bol., and to 2000 m in nw Arg. From Cochabamba, Bol., Mato Grosso, Maranhao and Piaui Brazil s to Tucumán and Buenos Aires, n Arg. In late Apr southern breeders migrate to northern part of breeding range, and to e and c Peru, occ. as far as Ven. Returns to breeding grounds mid Sep. Common.

Flycatcher – Genus *Hirundinea*

Monotypic genus. The nest is an open cup, placed on a ledge of a cliff.

CLIFF FLYCATCHER *Hirundinea ferruginea* – Plate XLIV 1

15.5 cm. **Long and slender, perched horizontally**. Sexes alike. Ssp *pallidior* dark brown above, wings with large rufous area. Below rufous; chin and sides of head finely mottled gray and whitish. Tail rufous with broad terminal blackish bar and outer web of outer feather. **In flight** the long wings are mostly rufous. Other sspp have less or no rufous in the tail, and are darker above. **Juv**. undescribed.
Habits: Usually in pairs. Perches on outcrops or small branches on

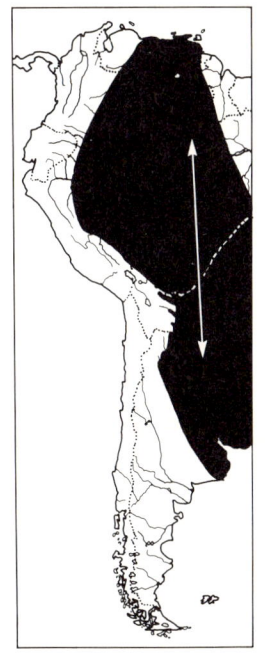

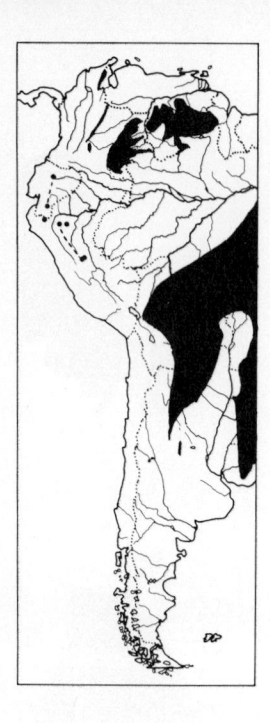

vertical rockfaces, occ. in trees. Sallies out, sometimes quite far, in swallow-like fashion, to catch insects in the air, then returns to the same post. Not shy.
Voice: An agitated *tee-trrr-rr* lasting c. 1 s, and repeated every 1-2 s.
Breeding: Juv. Dec (Cochabamba).
Habitat: Steep banks, cliffs, and bridges in forest edge and second growth, in dry valleys of Bol. on sparsely wooded slopes. Spreads rapidly with the building of roads on the humid Andean slopes.
Range: From trop. zone to 3600 m, in Cochabamba to 3900 m. From e Bol. and w Parag. s to Córdoba and La Rioja, nw Arg. (*pallidior*). 3 other sspp in trop. and subtrop. zone e of Andes from Col. to ne Arg. Fairly common and spreading.

Flycatchers – Genus *Myiarchus*

22 species found from s Canada to n Arg., also on Caribbean and Galapagos isls. Fairly large tyrants with upright posture, gray breast and pale yellow underparts. They catch insects by outward hovergleaning, and nest in holes. See *Proc. Biol. Soc. Washington* 35 (1922):181-218, *Amer. Mus. Novit.* 994(1938), and *Bull. Amer. Mus. Nat. Hist.* 161(1978): 427-628.

DUSKY-CAPPED FLYCATCHER *Myiarchus tuberculifer*
– Plate XLVI 29

17.5 cm. Bill dark. Crest half raised. Above greenish gray, **cap blackish**. Wings with only faint, narrow, pale olivaceous bar on greater coverts and edges of inner remiges; primaries with almost impercievable rufous edging; wings may become plain with wear. Tail appears uniformly dark, with outer web of outer feather only slightly paler. Throat and breast pale gray, rest of underparts rather bright lemon-yellow. Underwing with ochraceous edges of remiges. **Juv**. dark gray-brown above, without olive tinge. Cap duller black than in ad., feathers of rump edged reddish brown, wing-coverts and inner remiges edged cinnamon-rufous. Belly paler than in ad., very pale yellow, vent yellow-ochre.
Habits: Solitary. Hawks insects in the air, often flying up to 10 m from perch. Regularly changes site.
Voice: Call a distinctive, sad whistle repeated every 2-5 s. Also a soft, brief *whit*. Sometimes a rapid series of short whistles.
Breeding: Juv./imm. Sep (Cajamarca), June (Huánuco), July (La Paz), Mar (Cochabamba); large gonads Apr (Perijá mts), Apr-Aug (Cauca).
Habitat: Perches half way up in bushes or trees on open tangles or branches. Near edge of both dry and humid forest; and in shrubbery, second growth, hedgerows, gardens, and riparian thickets.
Range: 1100-3000 m, highest in s part of range, in Cochabamba to 3300 m. Andean slopes from s Ecu. through Peru and Bol. to Tucumán, nw Arg. On w slope of Andes at least to Huancavelica, Peru. Southern breeders are migratory (*atriceps*). 10 other sspp inhabit trop. and lower subtrop. zone from s USA to s Ecu. and n Ven. Nominate ssp

inhabits the Amazon basin and coastal Brazil, and ascends the Andean slopes along roads. Where secondary contact has occurred in Cochabamba, Bol., there seems to be a stepped cline in characters, whereas the cline in w Ecu. may be more gradual. Fairly common.

PALE-EDGED FLYCATCHER *Myiarchus cephalotes* – Plate XLVI 31

18 cm. Bill black. Above olive-gray, crown darker and browner, feathers with dusky centers. 3 **well-marked**, narrow **wingbars** and conspicuous edges of tertials **grayish white**; panel on inner secondaries yellowish. Tail-feathers dusky, sometimes narrowly edged whitish; outer web of **outer feather edged yellowish white**. Throat and breast pale gray, throat indistinctly flammulated white, rest of underside pale lemon-yellow. **Juv**. is dark gray-brown above, feathers of rump with slight rufous cast. Wingbars ochraceous, turning whitish towards body; inner primaries, and secondaries edged ochraceous. Tailfeathers narrowly edged rufous on outer web, outer feather like in ad. Alone, in pairs, or family groups. Perches in small trees and bushes. Hawks insects in the air, or gleans them from the vegetation or the ground. Also feeds on berries. Short sharp whistles *pyt pyt* much like Greater Pewee, and sometimes given in series. Also gives longer, plaintive whistles. Eggs Mar, Apr (Huila); juv Nov, Dec (e Narino), Nov (Huánuco); large gonads Feb-July (W and C Andes Col.).

2-5 m up. Perches in bushes or small, outstanding trees on clearings in humid forest and cloud forest at 1125-2750 m, mainly 1500-2400 m. All 3 ranges of Col. s through e Ecu. and e Peru to Sta Cruz, Bol. (*cephalotes*); coastal mts of Ven. (*caribbaeus*). Uncommon to locally common.

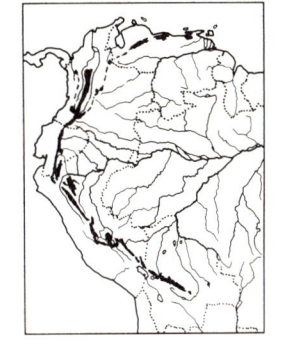

GREAT-CRESTED FLYCATCHER *Myiarchus crinitus* – Plate XLVI 30 – breeding in N Am., wintering mid-Oct to early May below 1200 m from Florida and e Mexico s to Cauca and Huila, s Col., and w Ven. During migration up to 2600 m on the Bogotá savanna. Accidental in n Peru. Usually found high in the canopy in tall, dense forest, open woodland, plantations, forest edge, dry forest and semi-arid regions. 20 cm. **Base of mandible buffy orange**. Crown-feathers usually somewhat raised. Above greenish gray; 2 blurred whitish wingbars, the upper widest; tertials broadly edged whitish; secondaries narrowly edged yellowish white in contrast to rufescent edges of primaries. Tail dark brown above (outer webs), rufous below (inner webs). Throat and breast gray, rest of underside pale lemon-yellow; remiges edged rufous on inner web. Solitary and aggressive. Alert, perching on high branches, dashing out to hawk insects in the air, or sally-glean them from bark-crevices, or the ground. Also eats berries. Its most distinctive call is a loud and harsh, slightly ascending whistle *wheeeep*, regularly followed by several *wick-wick-wick*s. Also gives a throaty, rolling *prrrrrreet* (rarely heard in S Am.).

Kiskadees – Genus *Pitangus*

2 species found from s USA to c Arg. Supreme generalists in foraging

behavior. The nest is globular or an open cup. See *Amer. Mus. Novit.* 963(1937).

GREAT KISKADEE *Pitangus sulphuratus* – Plate XLV 11

24.5 cm. Stocky, with stout bill. Ssp *bolivianus*: above drab brown, crown black, broadly encircled by white forehead, supercilium and nape; concealed c crown-patch golden yellow. Wings dark drab, borders of wing-coverts and edges of remiges rufous. Feathers of rump more or less bordered and tinged with rufous; tail dark drab, narrowly edged rufous on outer webs, broadly on inner webs, esp. at base. Headside black, throat white, rest of underside yellow. Ssp *argentinus* has less rufous in wings and tail, and is slightly paler above and brighter below. **Juv.** lacks crown-patch.
Habits: Solitary; in breeding season in pairs or family-groups. Perches fairly conspicuously in tree or on telephone line. Flies out to glean insects from leaves, branches and stems, but also picks small lizards and frogs from the ground, and eats nestlings of other birds. Regularly feeds on carcasses.
Voice: A sharp *kis ka-dee*, (also written as *bem tji vee*), the middle note lower, and the last drawn out, repeated at intervals, and given throughout the day. The most well-known call of any S Am. bird. During display the male clicks bill.
Breeding: Eggs Feb (Cochabamba); fledglings Feb, Mar (Cochabamba).
Habitat: 2-15 m up in trees or bushes. Clearings and edge of humid forest, second growth, open woodland and savanna in the arid zone, where most numerous around water. Gardens. In the highlands both in *Eucalyptus* and *Alnus*.
Range: 1500-3300 m from Cochabamba to Tarija, Bol. (*bolivianus*). Elsewhere below 2000 m, from Parag. and se Brazil s to Mendoza and Buenos Aires, Arg.; introduced and thriving in Río Negro and Chubut (see *Alauda* 47(1979): 116), s Arg., accidental in Tarapacá and Bío Bío (Los Angeles) (and Ñuble?), Chile (see *Bol. Mus. Nac. Hist. Nat. Chile* 29(1967): 121-4); southern birds migratory (*argentinus*). 7 other sspp distributed from Mexico to Brazil. Very common, and spreading with man.

Flycatchers – Genus *Myiodynastes*

5 species found from e Mexico s to n Arg. Fairly large tyrants with median crown-patch. They forage by upward striking and aerial hawking, and also eat much fruit. The nest is placed in a hole or a niche, that of the Golden-crowned F. fairly high up on a rock-ledge. See *Amer. Mus. Novit.* 963(1937).

GOLDEN-CROWNED FLYCATCHER
Myiodynastes chrysocephalus – Plate XLV 14a-b

19 cm. Upright posture. Ssp *minor*: crown and nape dark gray with semi-concealed, yellow median crest. **Headside blackish, with buffy**

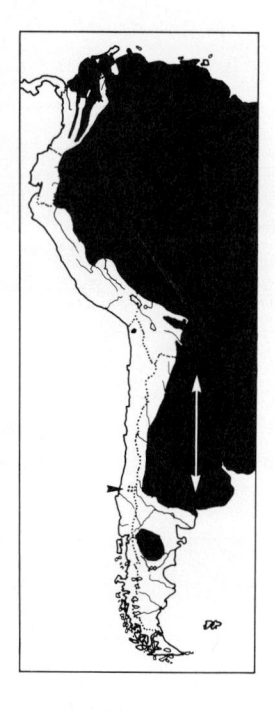

white supercilium and line below eye. Upperparts grayish olive, wings dusky, inner greater wing-coverts and tertials edged whitish, rest of wing-coverts and remiges edged rufous. Upper tail-coverts tipped rufous; tail dusky, feathers edged rufous. Chin whitish, throat and tinge on breast buffy, rest of underparts yellow; breast, sides, and upper belly broadly streaked grayish olive. Ssp *cinerascens* almost unstreaked below; *chrysocephalus* slightly larger than *minor*, and less tinged with buffy on fore-neck. **Juv**. lacks crown-patch, is grayer above and less streaked below. Supercilium and streak below eye more buffy. Solitary or in pairs. Hawks insects, and returns to same perch. Sits motionless for long periods. While singing, bows for every call. Song a short *tuie-wee, tuieweet*, or *pieuee*, repeated every 1-3 s. Alarm a sharp *kvuie*, or *kvuie-kvuie*. Eggs Aug (Valle); fledglings Aug (Sta Marta), June (E Andes Huila), Jan (ne Ecu.), June (El Oro), July (Chimborazo), Sep (Loja), May (Junín); juv. Dec, June (Sta Marta), May, Aug (Huila), Apr (w Nariño), June (nw Ecu.); large gonads May-July (Col.).

Middle to upper strata, sometimes down to 2 m. Clearings and riversides in humid forest and second growth. Perches on branches, occ. telephone-lines. At 600-2300 m, rarely to 2800 m. Sta Marta mts, Perijá mts, Andes and coastal mts of Ven. (*cinerascens*); extreme e Panamá, all 3 ranges of Col. s through e and w Ecu. to Loja, s Ecu. (*minor*); C and E Andes of Peru from San Martín s to Tarija, Bol. (*chrysocephalus*). Generally uncommon, but fairly common in nw Ecu., and increasing, spreading along roads.

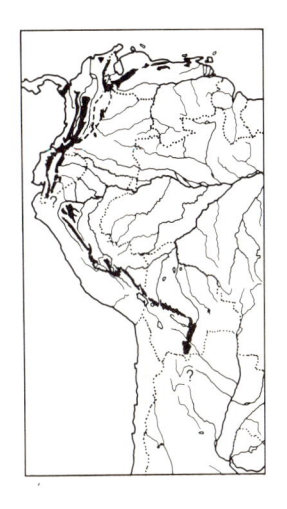

SULPHUR-BELLIED FLYCATCHER *Myiodynastes luteiventris* of open woodland and forest edge of C Am. winters in e Peru and Bol., and occurs up to 2600 m in Col. during migration. It resembles Golden-crowned F., but has a streaked back, more conspicuously streaked underparts and white wing-edgings.

VARIEGATED FLYCATCHER *Empidonomus varius* – widespread in woodlands and at forest edge in the trop. zone e of Andes – has been mistnetted once (Oct) at 3950 m in Quebrada Rurec in Ancash Peru (O Frimer; ssp *rufinus*). Resembles a small Golden-crowned F., but has much smaller bill, mottled back, and more densely olive-streaked underparts, and white wing-edgings.

Kingbirds – Genus *Tyrannus*

13 species found from Canada to Patagonia, also on the Caribbean isls. Large tyrants with upright posture, and usually holding the wing-bends somewhat out (look broad-shouldered). Aerial hawkers that perch conspicuously in open country. The Eastern K.'s aggressive habit on the breeding grounds, where it chases away almost any species of bird, has given rise to the name of the family. Sexes are alike. The nest is an open cup, placed in a tree or a bush. See *Nuttall Orn. Club Publ.* 6 (1966) on communication and relationships in the genus. See also *Amer. Mus. Novit.* 962(1937).

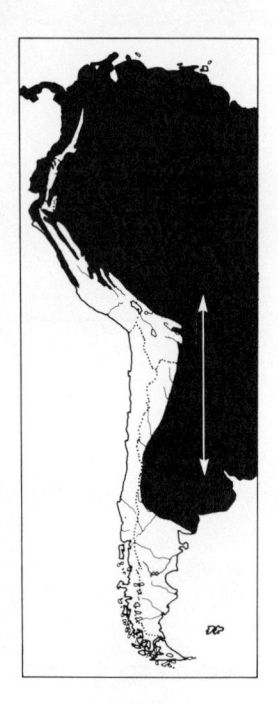

TROPICAL KINGBIRD *Tyrannus melancholicus* – Plate XLV 6

19 cm. Upright. Tail slightly forked. Above light gray to gray, back grayish olive, or more gray; crown with concealed median orange patch. Wings and tail fuscous, inner remiges and outer tail-feather edged yellowish white, wing-coverts edged olivaceous; outer 5 primaries attenuated, least so in female. Chin whitish, turning light gray on throat and extreme upper breast; breast yellowish olive, admixed gray on sides; rest of underparts bright yellow. Sspp poorly differentiated.
Juv. lacks crown-patch, and has rufous borders to the tail-feathers, feathers of rump and wing-coverts. Sometimes also narrow rufous edges of inner primaries; primaries not attenuated.
Habits: Solitary. Perches conspicuously on wires, fenceposts, and dead branches, usually above 3 m. Hawks large insects in the air, and returns to same perch. Not shy.
Voice: Song a thin trill *trrrr*, lasting c. 1 s.
Breeding: No data for highlands.
Habitat: Almost anywhere without regular frost, as long as there is at least an occ. tree present and an easy retreat to trop. areas without dense intervening forest. Not within forest, but everywhere along rivers, clearings, and forest edge, both humid and arid. In the arid zone mostly along rivers and streams. Particularly common in gardens and along roads, where its habit of perching on telephone-lines makes it the most conspicuous bird of S Am.
Range: All of trop. S Am. and immediately adjacent subtrop., and rarely temp. zone (though common to 3100 m in cleared parts of Cundinamarca Col.), s to Río Negro, Arg. W of the Andes s to Ica, w Peru. Migratory furthest s (*melancholicus*). Not rec. in Chile. Another ssp in ne Brazil, and one from se Arizona USA to n Ven. Ubiquitous. Spreading with the construction of roads in the jungle, and planting of gardens in the arid zone. See *J. Orn.* 117(1976): 75-99.

FORK-TAILED FLYCATCHER *Tyrannus savana* – Plate XLV 1a-b

38 cm (tail 29 cm). Crown and headside glossy black, concealed median crown-patch yellow, back light gray (darkest in *savana*). Wings dark gray-brown with light gray feather-edges. Lower rump blackish; tail black, outer web of outer feather white on basal half or more; **tail enormously long** and forked, outer feathers longest, curving inwards and slightly twisted at tip. Below white, neckside light gray. Ad. has attenuated outer 2-4 primaries, esp. in male; sspp differ mainly in details of the attenuation. **Juv**. has fuscous cap without median patch, smoky gray back and cinnamon rufous borders of wing-coverts. Primaries not acuminate, and tail not elongated.
Habits: Alone, in pairs, or family groups. In winter or during migration, many may roost together in trees, hedgerows, or rush-beds. Perches on wire-fences and tops of plants in the open. Darts out to hawk insects, and also eats berries. Opens and closes tail while twisting and turning in the air, making outer tail-feathers form a semi-circle. Attacks hawks and other large birds. During display, male flies in a

spiral while calling, then flies a short distance and spirals again.
Voice: While displaying a *tzig-tzig-zizizi...ag ag ag ag*. Call *tzig*. Alarm a little explosive *dzeep*.
Breeding: Eggs Feb (Valle), May (n Col.); fledglings Sep (Cundinamarca), Jan (Tucumán); juv. Feb (Cauca); large gonads Jan (Huila).
Habitat: Open pampas, esp. where there are scattered trees: savanna, cattle-fields, reed-beds, alders, or other tall vegetation near water.
Range: The lowlands of most of S Am. A regular, though generally scarce visitor to the high Andes, where rec. to 4100 m in c Peru, but common at 2600-3100 m in the Bogotá area, Col. From Mexico s to most of Col., Ven., Surinam, and Amapá, and nc Brazil, northern birds migrating s to Amapá and nc Brazil (*monachus*); n Col. and neighboring Ven. (*sanctaemartae*); from middle Río Tapajós to the mouth of the Amazon (*circumdatus*); from e Bol., Parag., Urug., se Brazil, and most of Arg. s to Patagonia, Arg. and Islas Malvinas; in winter n almost over all of S Am. e of Andes, and casually to Cuba, e USA, coastal Peru, and Chile (Tarapacá) (*savana*). Common.

NOTE: Formerly called *Muscivora tyrannus*.

EASTERN KINGBIRD *Tyrannus tyrannus* – **Plate XLV 2** is a migrant from N Am. wintering Sep-May in humid and dry forest in Panamá and trop. S Am. e of Andes, notably in e Bol., where roosting in large flocks in parks. It has rarely been found s to Tucumán, nw Arg., and Tarapacá, Antofagasta and Valparaiso, Chile. In s Peru it is an uncommon visitor w of the Andes, and is a rare straggler to the temp. zone. 19.5 cm. **Crown and headside glossy blackish**, with concealed, yellow to orange median patch. Back dark gray; wings fuscous, coverts and inner remiges edged whitish. Rump blackish, feathers with narrow white tips; **tail black**, conspicuously, though not very broadly **tipped white**; outer web of outer feather edged white. Below white, breast light gray. Birds with remains of **juv.** plumage, esp. dark gray on crown, can be seen in S Am. Hawks insects from a conspicuous perch, often high in the canopy.

Becards – Genus *Pachyramphus*

16 species found from s USA to n Arg., in humid to semi-arid forest and scrub from the trop. to the lower temp. zone. Medium-sized largeheaded tyrants with a relatively flat bill and somewhat sluggish actions. Sexually dimorphic. Both sexes tend the nest, constructed of twigs, globular with a side entrance, and placed in a tree. Becards have a rather isolated systematic position among the tyrants, and were formerly placed in the Cotinga family. See *Amer. Mus. Novit.* 894(1936).

BARRED BECARD *Pachyramphus versicolor* – Plate XLVI 27a-b

11.5 cm. Hunch-backed. **Male: above glossy black**, rump and tail gray; broad tips to humerals and median wing-coverts, and edges of greater wing-coverts and inner secondaries white. Lores, eye-ring, sides of head and neck pale greenish yellow, throat paler; rest of underside whitish with pale gray sides, sometimes with faint yellowish wash; **headside and underparts narrowly barred with dusky gray**. 2nd outermost primary short and pointed. **Female** and **juv**.: **crown dark glossy gray** with slight olive cast; headside, back and rump olivaceous, wing-feathers tipped or edged rufous, inner remiges edged whitish. Eye-ring pale yellow. **Throat, breast, and upper sides** pale olive-yellow, **barred** olive-gray; rest of underparts pale yellow. Ssp *meridionalis* is less barred below.

Habits: Alone, in pairs, or small groups, often in mixed-species flocks. Fairly active. Occ. perch-gleans with horizontal posture and bill pointing down, but usually perches upright, often on leaves and epiphytes, making sallies to glean insects from the underside of leaves.

Voice: Call a rapid series of insistent *pee*s. Also a somewhat similar soft even 'spinking'. Song by male frequent, a pretty whistled *treedididee*, rising, then falling.

Breeding: Fledgling May (nw Ecu.), May, Oct (Valle); juv. Dec (ne Ecu.); large gonads June (Antioquia).

Habitat: 1-5 m up in lower to medium strata, rarely high, in denser foliated parts of the vegetation in humid forest, cloud forest, and forest edge.

Range: 1500-2900 m, occ. to 3500 m, once down to 400 m. Perijá mts and Andes from s Lara, Ven., s through E, C and s end of W Andes of Col., Nariño, to n and c Ecu. (*versicolor*); from s Ecu. s through n, c, and e Peru to Cochabamba, Bol. (*meridionalis*). 1 ssp in Costa Rica and Panamá. Fairly common.

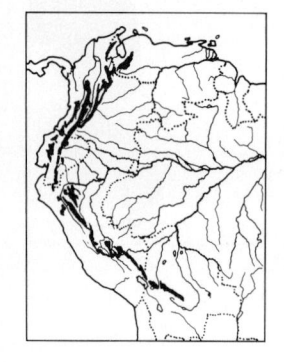

BLACK-AND-WHITE BECARD *Pachyramphus albogriseus* of trop. and subtrop. humid and dry forest from n Ven, Sta Marta mts, n Col. and Nicaragua s through e and w Ecu. to Piura, Cajamarca, and San Martín, n Peru, has been rec. to 3050 m on pacific slope of Piura nw Peru. 12.5 cm. In **male** crown glossy blue-black, lores and narrow forehead white, back gray, wings black, coverts broadly, and remiges narrowly edged white. Tail graduated, black broadly tipped white. Throat and breast pale gray, belly white. **Female** with ocher-brown crown bordered black at sides and behind, white narrow stripe from forehead to above eye; back olivaceous, wings dark, coverts as well as remiges broadly edged cinnamon-buff. Tail graduated, c feathers olivaceous with dark subterminal area and pale tip, rest black broadly tipped cinnamon-buff. Headside gray, throat white, rest of underparts yellow.

CRESTED BECARD *Pachyramphus validus* –
Plate XLVI 28a-b

17 cm. Crown-feathers often erect. Tail relatively short. **Male**: above blackish gray, headside very dark gray; below light gray with faint brownish tinge. Under wing-coverts washed with cinnamon, inner webs of remiges broadly edged white basally. **Female**: cap dark gray; back, wings, and tail dark rufous, back sometimes with occ. gray

feathers; outer remiges blackish. Underparts creamy buff, under-wing rufous. **Juv**. presumably like female. Solitary, occ. in mixed-species flocks. Slow. Often with horizontal posture and bill pointing forwards. Carefully watches for large insects for 3-4 s, then moves to new site. When undisturbed may perch conspicuously. Ssp *validus*: call a fine rising *tsri*; also *si-i-it*, and *tuit*. Song a distinctive low clear vibrating whistle, sometimes given in a slightly descending series of 6-8, 2nd note highest *dui dui dui dui*.... Eggs Jan (Jujuy).

Half-way out the middle branches inside rather open canopy. Wooded slopes in watered ravines in semi-arid and semi-humid regions. At 300-2200 m, locally (above Urubamba town Cuzco) at 3500 m. In Ayacucho and Cuzco, s Peru, and s along the Andes of Bol. to La Rioja and Córdoba, nw Arg. (*audax*). Generally fairly common, but rare in Peru. Nominate ssp in e Bol., Parag., ne Arg., and e Brazil.

NOTE: Formerly called *Platypsaris rufus*.

SONGBIRDS – suborder OSCINES or PASSERES

Larks – Family Alaudidae

Some 77 primarily terrestrial miner-like species, widespread in open country of the E Hemisphere. The Horned Lark is the only member of the family found in N Am. and with a small isolated population in Col. The nest is an open cup placed on the ground. This family possibly is an ancient offshoot from the large assembly of finch-like birds, which include pipits and the nine-primaried oscines.

HORNED LARK *Eremophila alpestris* – Plate XLVIII 18a-b

14.5 cm. Above pale brown streaked dark brown, wing-coverts pale rufous; tertials dark brown. Tail black, c pair dark brown, outer feather with white outer web. **Distinct facial pattern: forehead black, extending as 2 'horns' along side of crown; white supercilia** meeting over base of bill; black line from nostrils broadening below eye; cheeks grayish, throat to lower cheeks white tinged yellow. Breast-band black, lower breast mottled with grayish, sides of lower breast pale rufous, rest of underparts white, sides streaked with dark brown and dull rufous. **Juv**. lacks head-markings and is heavily dotted to scaled with whitish above; supercilium and underparts white; headside and breast mottled with brown.
Habits: Alone, in pairs, or small groups. Terrestrial. Walks on open ground, picking up seeds and insects. The young are mainly fed insects.
Voice: N Am. birds sing a tinkling, high-pitched, irregular, but often long sustained song from high in the air or on the ground. Call a clear *tsee-ee* or *tsee-titi*.
Breeding: Fledgling Feb (Cundinamarca).

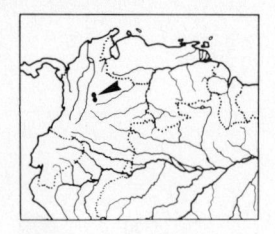

Habitat: Open country with sparse or low vegetation near lake-shores and in fields.
Range: 2500-3000 m (rarely down to 1320 m?). Confined to the Bogotá and Ubaté savannas in Cundinamarca and Boyacá, Col. (*peregrina*). Uncommon. Other sspp. in N Am. and n Eurasia.

NOTE: Called Shore Lark in Old World literature.

Swallows and Martins – Family Hirundinidae

82 species of worldwide distribution. They superficially resemble swifts by their habit of catching insects while constantly on the wings, but can be told by their slower, more maneuverable flight, broader and more bent wings, less shrill calls, and by their habit of perching on telephone-lines or exposed branches. Martins of the genus *Progne* are exceptionally large, with heavy heads and stiff triangular wings. Species with deeply forked tails are able to make quick maneuvers and actively pursue insects near the ground; species with slight fork mainly take 'aerial plankton' in gliding flight, often high up. The nest is often built of mud and placed under roofs of buildings or on cliffs, but some species may also nest in holes in trees or banks, some exclusively so. The family may have a rather isolated systematic position nearest to Old World warblers, tits, etc. See *Proc. Zool. Soc. London* 1872:605-9, Meyer de Schauensee (1946a), *Amer. Mus. Novit.* 1723(1955), Bond (1956), and *Hornero* 11(1969): 1-19.

PURPLE MARTIN *Progne subis* – Plate XLVIII 6a-c

20 cm; fork of tail 1.4-2.15 cm. **Male uniform dark glossy violet-blue** with a small **whitish patch on lower sides**. **Female** like male above, but forehead, nape, and sides of neck washed with dark gray-brown in contrast to dark cheeks. Throat, breast, and sides dark gray-brown, breast-feathers with whitish tips; belly and vent whitish, feathers with dark shafts, those of vent also with gray-brown centers. Under wing-coverts blackish narrowly edged whitish. **Imm. male** like ad. female. Plumage kept until 2nd year. **Imm. female** is glossy only on fore-crown and scapulars, rest of upperparts dark sooty or grayish brown, feathers of back, scapulars and rump with narrow, pale margins; feathers of anterior underparts more distinctly margined with whitish and remaining underparts more clearly and uniformly white.
Habits: Flies fairly low, often gliding on stretched triangular wings. Possibly roosts in large numbers (a flock of 5000 in Jujuy nw Arg. Dec, and 700 birds in sw Mato Grosso Brazil Dec probably of this species).
Voice: Song and calls a distinctive, low-pitched, liquid, rolling twitter.
Habitat: Open country, often near water.
Range: Breeds in N Am. Winters from s Ven., Guiana, and Surinam to se Brazil and irregularly in n Arg. From (late Aug) late Sep to early Oct, and again in late Apr regularly up to 3020 m in the E Andes of Col., and in Ven. to 4000 m, 1 rec. of 2 females from Andes of Ecu. (*subis*). Another ssp (with distinctly paler female) breeds on w coast of N Am. and winters s to Nicaragua (*hesperia*).

Southern Martin *Purple Martin*

SOUTHERN MARTIN *Progne modesta*

17.5-19 cm. Ssp *elegans*: **male** like Purple M., but tail deeper forked and no white patch on lower sides. Not safely told in the field. **Female** like male above; **below dark sooty brown**, feathers edged whitish; under tail-coverts white with dusky shafts. Ssp *murphyi*: **male** with shorter and less deeply forked tail; lower sides sometimes with a suggestion of a white patch. **Female** mouse-gray above, except for a broad shining steel-blue band across the mid-back involving scapulars and lesser wing-coverts, and similarly colored upper tail-coverts; feathers of crown and hind-neck with dusky centers. **Below mouse-gray**, somewhat paler than crown, under tail-coverts scarcely lighter than abdomen, apically edged whitish. **Juv**. brownish; probably as in Purple M. more uniform than imm. female, and with the faint gloss greener.
Habits and **voice**: As for Purple M. Large flocks roost in town parks.
Breeding: Present on southern breeding grounds Sep-Mar.
Habitat: Coastal cliffs, banks, fields, and buildings.
Range: Sea level to 2600 m. From Chubut Arg. n to Cochabamba and Sta Cruz, Bol., in winter n to e Peru (in Nov 1986 a flock of 30-40 at 3100 m in Cuzco (species ?)), n Brazil and possibly e Col. (accidental in Santiago, Chile and on Islas Malvinas) (*elegans*). Resident in coastal Peru from Piura to Ica, regularly in small numbers along coast of sw Peru, rarely inland to at least 1800 m in Arequipa (accidental in Arica n Chile and at 4080 m in Junín) (*murphyi*). Ssp *elegans* fairly common, *murphyi* rare. Galapagos Isls (*modesta*).

NOTE: The nominate ssp may be a distinct (allo)species, Galapagos Martin. The present species should then be called *Progne elegans*.

BROWN-CHESTED MARTIN *Phaeoprogne tapera* – Plate XLVIII 7

17.5 cm. Tail slightly forked. Ssp *fusca*: upperparts and headside dark gray-brown with slightly paler feather-margins; tertials tipped white in fresh plumage. Throat, sides of neck, belly, and vent white, white

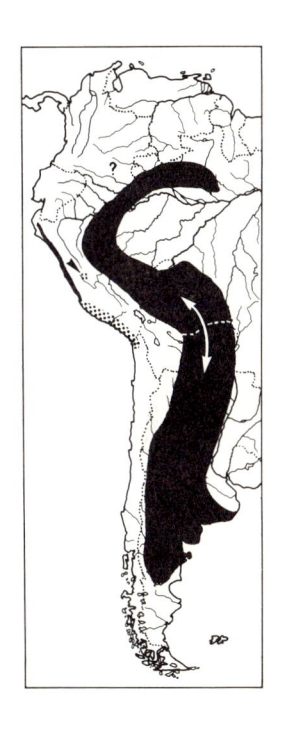

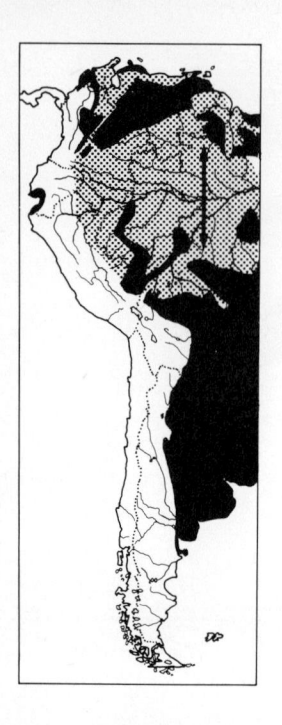

showing on sides of rump in flight; sides, **breast-band** of varying extent and shade **and spots down c belly dusky brown**; under wing-covert fuscous tipped white. **Juv**. has gray-brown sides of throat.
Habits: Alone, in pairs or groups, during migration in large flocks. Appears to be a weaker flier than other swallows, often resting. Flies with much gliding on lowered wings. Perches on wires and in reeds.
Voice: Call a rough, metallic *tschri* or *tch-tch*. Song *dju-it-dju* or *dchri-dchrie-dchrruid*.
Breeding: Breeds late (Nov-Feb or Mar), using holes in trees or hornero's nests.
Habitat: Towns, open woodland, fields. During migration from rivers and lakes in trop. rain forest to open highland habitat.
Range: Chiefly lowlands, but occ. during migration in Nov. a few straggle to 4000 m in La Paz, Bol., and Peru, and flocks of thousands occur on the savannas at 2600-3100 m in E Andes Col. (Sep-Oct). From Chubut and Buenos Aires, Arg., to La Rioja and n to s Brazil and Sta Cruz, Bol.; wintering to the Guianas, and nw Brazil, e and n Col. and Panamá; casual on coast of sw Peru in July-Sep (*fusca*). Resident in lowlands from Col. to nw Peru, Guiana, n Bol., and c Brazil (*tapera*).

NOTE: Sometimes placed in the genus *Progne*.

BROWN-BELLIED SWALLOW *Notiochelidon murina* – Plate XLVIII 9

12.5 cm. Lightly built, much like Blue-and-white S. Tail forked. Ssp *murina*: upperparts and sides of head black with blue-green gloss, wings and tail dusky. **Below, incl. under wing-coverts dark smoky-gray**; vent glossy-black. Ssp *cyanodorsalis* with more purplish violet gloss; *meridensis* with slightly bluer gloss than *murina*, underparts paler, throat whitish. **Juv**. has dusky upperparts; upper back with slight gloss, more greenish than in ad. Feathers of rump narrowly and faintly pale-tipped, tail less forked; throat dark brown, underparts except dusky vent grayish white.
Habits: In flocks of up to 50 or more birds. Forages 5-20 m up. Like Blue-and-white S. does not soar much. Nests in holes in banks and cliffs and in rock crevices.
Voice: Contact note during foraging *tjrip*, *tjrip-tjrip*, or *tjrrrp*, prolonged and more intensely at dusk *tiderrreh tide rrrheh rrreh rrrheh* ....
Breeding: Juv. July (nw Ecu.); juv./imm. June-july (La Paz), Aug (s Cajamarca); nests Sep-Oct (Cundinamarca); large gonads Jan-Aug (C and E Andes Col.).
Habitat: Forages over humid and semi-humid montane shrubbery, *Polylepis* woods, elfin forest, páramo, and sometimes arid puna grassland as well as outskirts of towns. Often along cliffs.
Range: 2100-4300 m. From Trujillo to Páramo de Tamá in Táchira, Ven. (*meridensis*), Sta Marta and Perijá mts, all 3 ranges of Col., Ecu., and Peru s to Arequipa and Cuzco (*murina*), La Paz and Cochabamba, Bol., birds from immediately adjacent Puno, Peru probably also referable here (*cyanodorsalis*). Common locally.

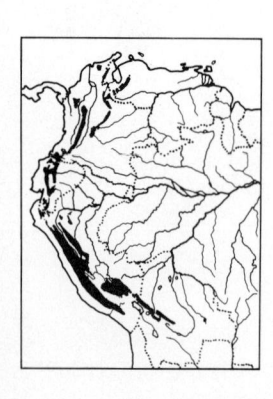

BLUE-AND-WHITE SWALLOW
Notiochelidon cyanoleuca – Plate XLVIII 8a-c

Lightly built. Ssp *patagonicus*: 12 cm, tail forked. Headside and upperparts black with violet-blue gloss, wings and tail dusky; **below white, lower vent glossy black**, extreme sides of body smoky gray; **under wing-coverts** smoky brown tipped silvery **white**. Ssp *cyanoleuca* **smaller** - 10.5 cm – tail only slightly forked; extreme sides of breast glossy black, **sides of body darker** and more extensively smoky gray; **entire vent glossy black**; **under wing-coverts darker** and with only few, narrow whitish tips. Ssp *peruviana* like *cyanoleuca*, but **underwing all black**; **sides of body blacker** and more extensive. **Juv.** has less forked tail than ad. Juv. *patagonicus* dusky above; feathers of rump with faint narrow pale tips, upper back with slight gloss; underparts white, **throat faintly tinged buffy**, obscure breastband pale smoky gray with buffy tinge; flanks tinged buffy; vent smoky gray, feathers broadly tipped whitish. Ssp *cyanoleuca* similar, but sides and underwing darker, vent dusky with hardly distinguishable paler narrow tips.
Habits: Usually in loose flocks weaving back and forth with frequent beats of the angular wings and no or little soaring. Rests in company with conspecifics and other swallows on wires and leafless branches. At least non-migratory populations pair for life. Territorial near nest, that is placed on cliffs, in a hole in a bank, tree, or under a roof. Sometimes lands on the ground. During migration in flocks of up to 1000. For further notes on habits see *Auk* 69(1952): 392-406, and *Noved. Colomb.* 1(1960): 256-76.
Voice: Song a long thin weak trill, often with introductory notes *tsi tsi tsi trrrrrrrrili* given in air throughout the year, though most frequently in the breeding season. Frequently twitters from perch in tree or on cliffs. In Patagonia the song is a shorter *trip*. Alarm and contact notes mono- or disyllabic *tsi tsi ...*, *tsritit* or a clear, somewhat descending *tsie*. Low harsh notes near the nest. Most vocal towards end of day, and when leaving roost-sites before sunrise.
Breeding: In most of Chile 2 broods with eggs Sep-Oct and Dec. In s and at higher elevations 1 brood with eggs Nov-Dec. Eggs Nov-Dec (Mendoza), Jan-Mar (Tucumán), Oct-Mar (Cochabamba), Sep-Oct (La Paz), Apr-May (n Ecu., sw Col.), Mar-Apr (Huila), Feb-July (Valle), Jan, Mar (Sta Marta), Feb-Apr (Ven.).
Habitat: Both humid and arid zones. Open country and forest clearings, often near habitation.
Range: Sea level to 4000 m. From Costa Rica to Guiana and s through Ecu., Peru, and Bol. to Tucumán, nw Arg., Parag., Urug., s and e Brazil (*cyanoleuca*); coast of Peru from La Libertad to Arequipa (*peruviana*). From c Chile and c Arg. s to Isla Grande, migrating n to n Chile and e of Andes to Ven. and Panamá, casually as far as s Mexico (*patagonica*). Very common throughout, but generally absent or rare in the puna zone except some villages on Bol. altiplano.

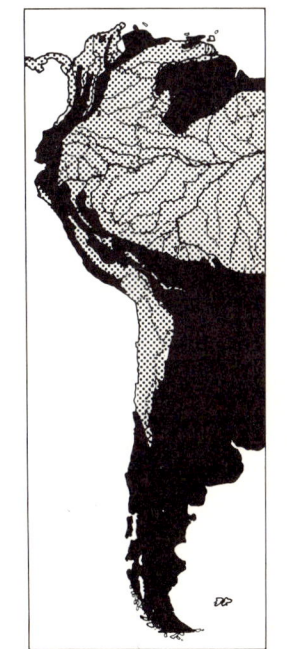

PALE-FOOTED SWALLOW *Notiochelidon flavipes* – Plate XLVIII 10

11 cm. Lightly built. Tail slightly deeper forked than in Blue-

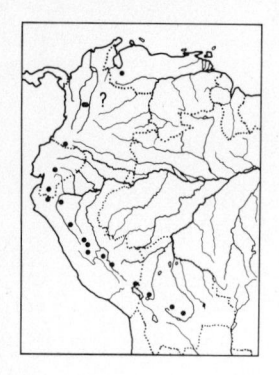

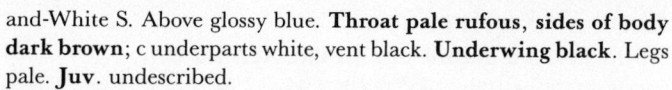

and-White S. Above glossy blue. **Throat pale rufous, sides of body dark brown**; c underparts white, vent black. **Underwing black**. Legs pale. **Juv**. undescribed.
Habits: In flocks of 10-15, rarely up to 50, and sometimes with Brown-bellied S. Much like Blue-and-white S., but with a faster, more direct flight and generally at higher elevations, and in different habitat.
Voice: Much like Blue-and-white S. (see *Auk* 97(1980):173).
Breeding: 5 birds in Vilcabamba mts Cuzco all had small gonads July-Aug.
Habitat: Glades in elfin forest and upper reaches of cloud forest and over adjoining pajonal slopes. In bad weather sometimes at lower elevations over man-made clearings and forest edge in association with Blue-and-white S.
Range: 2800-3600 m, rarely down to 2000 m. Locally from Mérida Ven. s in E and C Andes of Col., Ecu., C and E Andes of Peru to Sta Cruz, Bol. Fairly common, but very local.

NOTE: By some called Cloud-forest Swallow.

SOUTHERN ROUGH-WINGED SWALLOW *Stelgidopteryx ruficollis* of lowlands to 2200 m from Costa Rica to n Arg. straggles to 3600 m in Col. 13.5 cm. Tail notched. Above sandy brown, crown darker, **rump whitish to pale buffy gray**. **Throat cinnamon-buff**, breast and sides light gray-brown fading to yellowish white on c belly.

BARN SWALLOW *Hirundo rustica* – Plate XLVIII 13a-b

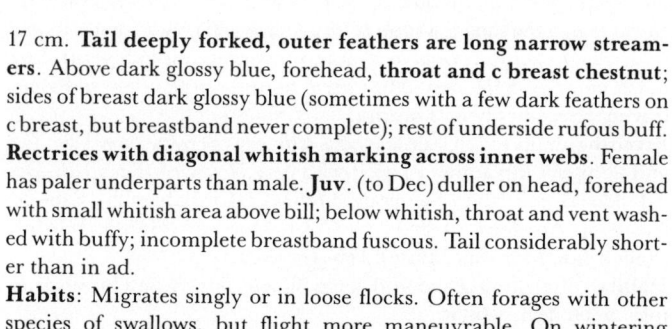

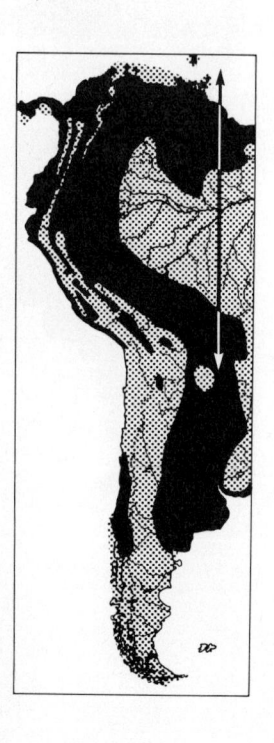

17 cm. **Tail deeply forked, outer feathers are long narrow streamers**. Above dark glossy blue, forehead, **throat and c breast chestnut**; sides of breast dark glossy blue (sometimes with a few dark feathers on c breast, but breastband never complete); rest of underside rufous buff. **Rectrices with diagonal whitish marking across inner webs**. Female has paler underparts than male. **Juv**. (to Dec) duller on head, forehead with small whitish area above bill; below whitish, throat and vent washed with buffy; incomplete breastband fuscous. Tail considerably shorter than in ad.
Habits: Migrates singly or in loose flocks. Often forages with other species of swallows, but flight more maneuvrable. On wintering grounds roosts in large flocks in reedbeds or on telephone-wires. Flies swiftly with angular wings and great maneuvreability, usually low.
Breeding: Eggs Nov-Jan (Buenos Aires!).
Voice: Calls include a *krik* and a metallic *whit*.
Habitat: Open country, esp. over meadows and marshes.
Range: Breeds in N Am. Winters throughout C Am., West Indies, and S Am. s to c Chile and n Arg., in small numbers to Isla Grande. In Mar-May and Sep-Nov flocks may be seen on migration through the high Andes, but Dec-Feb only scattered singles (rarely up to 7), are found at high altitudes; occ. birds seen at other seasons, and breeding near Mar Chiquita in Buenos Aires at least since 1980 (*erythrogaster*). Other sspp breed in subtrop. and temp. zones of the Old World, with northern populations wintering in s Asia and Africa.

BANK SWALLOW *Riparia riparia* – a migrant from N Am. – can occ. be met with anywhere in the high Andes, often in company with Barn S. and other swallows, but always in very small numbers. Passage peaks in Sep-Oct and Feb-Mar (May); singles may be seen throughout year. Rec. s to c Arg. (and sighted in StaCruz), and Antofagasta, Chile (*riparia*). 12 cm. Lightly built, tail slightly forked. Upperparts, head-side, **breastband**, extreme sides of body, and some feathers down center on lower breast **smoky brown**; underside apart from breastband white; under wing-coverts smoky brown narrowly tipped whitish.

NOTE: Called Sand Martin in Old World literature.

CLIFF SWALLOW *Petrochelidon pyrrhonota* – Plate XLVIII 12

13.5 cm. Rather stoutly built, tail short, hardly forked. Ssp *melanogaster*: **forecrown, throat, and headside** almost to nape **chestnut**; nape and neckside smoky gray; lores, eye-ring, back, and wing-coverts black with blue-green gloss, feathers narrowly bordered with whitish, **c back streaked whitish**; **rump rufous**; wings and tail dusky black with slight gloss, tertials and upper tail-coverts narrowly tipped whitish. Lower throat mixed with glossy black, breast and sides rufous buff tinged smoky gray, rarely dark shafts on lower breast and sides. Belly white, vent rufous; under wing-coverts pale smoky gray, conspicuously edged and tipped rufous chestnut. Ssp **pyrrhonota with buffy white forehead**; pale tips of back-feathers almost absent; streaks on c back conspicuous. Black patch on lower throat well-marked; breast and sides more washed with smoky brown, and more often with dark shafts; vent whitish or only tinged buff. Ssp *tachina* like *pyrrhonota* but smaller, forehead darker, underparts more rufescent. Ssp *minima* like *tachina* but forehead darker, much as *melanogaster*. **Juv.** with black replaced by dusky; breast and sides pale smoky gray, without or with only a trace of rufous buff; chestnut of head paler and mixed with smoky gray.
Habits: Alone or in flocks up to 35, often with other species of swallows. Flies with fairly stiff and relatively broad wings and regularly soars.
Voice: Calls include a metallic *et* or *ert*.
Habitat: Open country, sometimes over towns.
Range: Breeds in N Am., winters in s Brazil and n Arg., *pyrrhonota* mainly in ne Arg., *minima* and *melanogaster* mainly in nw Arg. Wintering grounds of *tachina* unknown. Rec. twice from Chile (Tarapaca) and a sight rec. from Isla Grande. Uncommon in the high Andes, primarily Oct-Nov, and less commonly Mar-Apr (May).

NOTE: Sometimes placed in the genus *Hirundo*.

ANDEAN SWALLOW *Petrochelidon andecola* – Plate XLVIII 11

14 cm. Rather **stoutly built with triangular wings, tail only very slightly forked**. Above black glossed blue (*oroyae*) or glossed greenish

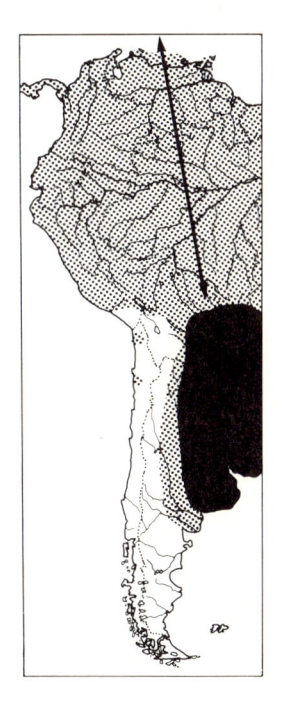

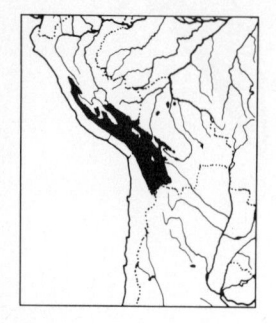

blue (*andecola*); **rump like back in older birds, but usually gray-brown like tail; throat brownish to grayish**, merging with dirty white of rest of underside; underwing and sides of body gray-brown. **Juv.** has rufous brown rump and pale rufous wash on lower belly and vent. Greater wing-coverts tipped pale clay-color; tertials edged and tipped pale.
Habits: In small flocks. Flies with fairly stiff wings and regularly soars.
Voice: Call a soft *trui*. Song a harsh short trill.
Breeding: Under roofs. Eggs and fledglings Jan (Titicaca); juv./imm. Jan (La Paz), Mar (Cochabamba); large gonads Jan (La Paz).
Habitat: Open country in the puna zone, incl. desert puna and tola heaths. Often over water.
Range: 3100-4600 m (in July occ. to below 2500 m in Arequipa). C Peru in Ancash, Lima, and Junín (*oroyae*), s to Arequipa, and through Huancavelica, Ayacucho, Apurímac, and s Cuzco s Peru to Tarapaca, n Chile, and Tarija, Bol., probably also in adjacent nw Arg. (*andecola*). Common locally.

CHILEAN SWALLOW *Tachycineta leucopyga*

12.5 cm. Relatively stoutly built; tail forked. Headside and upperparts black glossed greenish blue, occ. with a few white feathers over bill and lores. **Rump and tips of tertials white** (rump-feathers rarely with small subterminal glossy black spots). **Below white** with restricted black bar on side of breast; under wing-coverts, extreme sides of body and faint wash on breast pale smoky-gray, feathers of lower breast with dusky markings at base; longest under tail-coverts rarely with blackish subterminal spot and dark shaft. **Juv.** dusky above, only upper back and wing-coverts glossy; upper tail-coverts blackish tipped whitish instead of being all glossy black. Usually with a narrow line of white feathers above the lores.
Habits: Much like other swallows. Often forages low over water.
Voice: 3-4 high-pitched gurgling sounds followed by 2 lower, guttural gurgling sounds.
Breeding: 2-3 broods, eggs Sep-Nov, Dec and Feb. Nests in abandoned woodpecker-holes or under roofs.
Habitat: Often around human habitation. Lakes, streams, swampy lagoons, open *Nothofagus* forest, and occ. over dry slopes and steppes.
Range: Coast to Andean foothills. From Atacama in n Chile and Río Negro in Arg. s to Isla Grande. Southern populations migrate n as far as Bol., Parag., and Brazil. 1 Oct rec. from sw Peru.

TREE SWALLOW *Tachycineta bicolor* breeds in N Am. and winters in C Am., rarely into Col. and Ven., once to 2800 m in Nariño. 13 cm. Above steel-blue to greenish black, wings and slightly forked tail blue-black; **underparts white**. **Juv**. dark brown above, whitish below, sometimes with an incomplete smudgy brown breast-band. Usually in flocks.

NOTE: sometimes placed in the genus *Iridoprocne*.

Crows, Jays and allies – Family Corvidae

Some 104 species of large omnivorous passerines with a worldwide distribution. Sexes are alike, but younger birds of certain social species often have pale bills or other aberrant colors to signal lower rank. The nest is a large open cup constructed of twigs. Only jays are found in S Am.; they are colorful (vs. the black typical crows) and usually live in noisy family-groups moving through the vegetation, actively searching through all parts of the trees. New World jays are characterized by a peculiar adaptation of the lower mandible for chiseling, which may be important for stabbing nuts (*Auk* 104(1987):665-80). The family belongs to an ancient assembly of families originating in Australia, and its presence in S Am. is probably fairly recent. Data about S Am. jays are compiled in Goodwin, D (1976) *Crows of the World*. Ithaca N.Y. See also *Living Bird* 14(1975):5-43, Meyer de Schauensee (1946a), *Amer. Mus. Novit.* 1649(1953), Bond (1956), *Condor* 71(1969): 360-75, and *Contrib. Science, Los Angeles County Museum* 165: 1-16.

COLLARED JAY *Cyanolyca viridicyana* – Plates XIV 7a-b, LXIV 13

32 cm (tail 16 cm in Ven., 17-19 cm in Peru). **Bright dark blue** (*quindiuna* greenish blue, *meridana* purplish blue), crown paler and bluer (or like back, *meridana*). In northern forms narrow band on forehead, chin and headside curving as a narrow band below throat black. Southern birds are capri-blue with pale blue forecrown (*jolyaea*) or greenish blue with almost white forecrown (*cyanolaema* and *viridicyana*) contrasting with black of extreme forehead, and with white instead of black band encircling throat; *viridicyana* furthermore has the throat greenish black or deep glaucous green. Birds from the zone of overlap with Turqoise Jay ('*angelae*') are bluer than in C Andes of Col., and are told from Turqoise Jay mainly by lack of whitish blue on forehead, darker throat, and longer, more graduated tail. **Juv**. duller and without breast-band.
Habits: In groups of 3-7 birds, often in mixed-species flocks. Forages along the middle part of thinner branches and in dense undergrowth, bamboo thickets, and treeferns.
Voice: Very vocal with a large repertoire given in various combinations. Calls include chirking metallic sounds, a harsh *rrah* repeated at 2 s intervals, a single *warrah*, *craah-who-op*, a soft *crook*, a nasal ringing querulous *shree* or *reek*, and a descending clear whistle *weeee*. Also (young birds only?) a sharp, high-pitched *jip-jip*, *wetjirrrr* or *peep peep chee chee wee* with the last 3 notes highest pitched. See also *Condor* 69(1967): 513-21.
Breeding: Juv. Jan, Aug, Sep (Mérida), Jan, Sep (Cauca), Nov

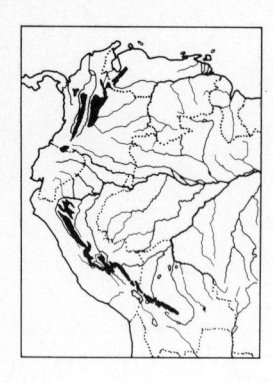

(Quindío), June (Amazonas), July (Cochabamba); large gonads Oct (Boyacá), June-Sep (W and E Andes Col.).
Habitat: Mainly middle strata. Cloud forest, humid epiphyt-laden forest, and humid second growth.
Range: 1600-3250 m. From Trujillo to n Táchira, Ven. (*meridana*), sw Táchira, Ven. and E Andes of Col., at least s to Cundinamarca (*armillata*), Antioquia W Andes and entire C Andes of Col.(*quindiuna*), extreme n E Andes of Ecu. and immediately adjacent region of Putumayo, Col., ('*angelae*', usually not considered distinct from *quindiuna*). From Amazonas to Junín, n and c Peru, probably intergrading with *cyanolaema* in sc Peru (*jolyaea*); Cuzco and Puno, se Peru (*cyanolaema*), La Paz and Cochabamba, Bol. (*viridicyana*). Generally fairly common, but uncommon to rare in most of Col.

NOTE: By some divided into 2 (allo)species, **BLACK-COLLARED JAY (*C. armillata*)** n of Ecu. and **WHITE-COLLARED JAY (*C. viridicyana*)** s of Ecu.

TURQUOISE JAY *Cyanolyca turcosa* – Plate XIV 8

27.5 cm (tail 15 cm). **Dark greenish blue to light cyanine blue**, on crown grading to whitish blue on upper forehead; forehead, upper chin, and headside black. Throat clear blue divided from darker blue underparts by black band. In n Ecu. told from Collared Jay by slightly shorter, much less graduated tail, **paler throat** and crown, and by the **contrasting pale forehead**. **Juv.** duller and without breast-band.
Habits: Like Collared Jay.
Voice: Calls include a clear descending whistle *hye* and a snarling *wharr* much like Collared Jay.
Breeding: Juv. Sep (nw Ecu.), Nov, Feb (ne Ecu.), June (se Ecu.), May (Piura).
Habitat: Humid second growth, forest, and cloud forest.
Range: 1850-3200 m. From e and w Nariño, s Col. (and 'Bogotá') along both slopes of Ecu. to n Cajamarca and Piura, n Peru. Fairly common.

NOTE: Formerly sometimes treated as a ssp of Collared Jay.

GREEN JAY *Cyanocorax yncas* – Plate XIV 6a-b

28 cm. Ssp *yncas*: erect feathers of fore-crown ultramarine-blue; supercilium, headside, and throat black, with ultramarine areas below eye, above rear of eye and on upper chin; crown and nape whitish separated from color of back by ill-defined blue band. **Back, wings, and c tail-feathers green**; rest of tail and **underparts yellow**. Ssp *galeatus* has larger frontal crest. Sspp *cyanodorsalis* and *andicolus* have most of crown bluish violet, narrow frontal band whitish, back and c tail-feathers tinged bluish. Ssp *longirostris* much like *yncas*, but slightly paler below, bluer above, and with longer slenderer bill. **Juv.** duller throughout, crown tinged blue even in otherwise white-crowned sspp.
Habits: In flocks. Actively forages in the vegetation, frequently twist-

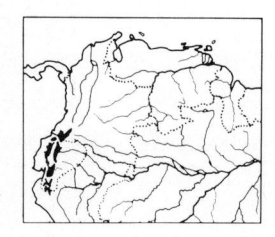

ing tail. Somewhat inquisitive. Breeds socially (see *Newsl. Memb. Lab. Orn. Cornell Univ.* 72(1974): 6-7, and *Living Bird* 14(1975): 5-43).
Voice: Very vocal. Calls include a sharp *kirr kirr kirr*, *keriie*, and a harsh *rrep rrep rrep*.
Breeding: Fledglings Jan (Mérida), Feb (Cundinamarca), Mar, Aug, Sep (Cauca), Sep (Nariño), Mar (ne Ecu.), Aug, Sep (Huánuco), Oct, Dec (Junín), Nov, Dec, Jan (Cochabamba).
Habitat: All levels of both dry and humid forest, cloud forest, forest edge, heavy second growth, plantations, and open country with scattered trees.
Range: 1000-2800 m, rarely to 3000 m, down to 300 m or less in the Marañón valley. W slope of E Andes, C Andes and W Andes s to Valle, Col. (*galeatus*), e slope of E Andes of Col. near Cucuta, N.d.Santander; Perijá mts and from s Lara to Táchira, Ven. (*andicolus*), e slope of E Andes of Col. s to Cundinamarca (*cyanodorsalis*), sw Col. in valleys of upper Cauca, Patía and San Miguel s through e Ecu. and Peru to Cochabamba, Bol. (*yncas*); upper Marañón valley, n Peru (*longirostris*). Locally fairly common. 5 other sspp from s USA to Honduras, 1 in coastal Ven.

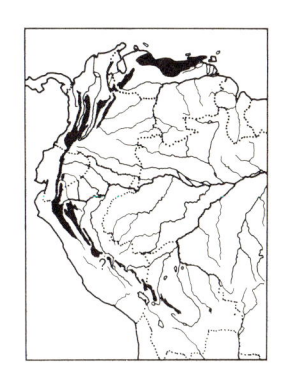

NOTE: Previously placed in the genus *Xanthoura*.

PLUSH-CRESTED JAY *Cyanocorax chrysops*, wide-spread in drier regions e of Andes, rarely ascends to at least 2100 m in nw Arg. (*tucumanus*). 37 cm (tail 17 cm). Above violet blue-black; **broad tip of tail white**. Head covered by erect black feathers; spot above and below eye, upper moustache and nape silvery blue. **Throat and breast black, rest of underparts white**. In flocks of 5-6 in forests and groves.

Dippers – Family Cinclidae

5 species found in temp. and cooler subtrop. parts of Eurasia and the Americas. Size like a small compact thrush, and with a short tail. Plumage mainly dark brown, gray, or blackish. Alone or in close pairs dippers forage for insects along water-courses, often preferring the wildest rapids and waterfalls where they may dive into the water to run on the bottom (however, the Andean species have never been reported to dive). When perched the 3 Holarctic species have a characteristic habit of bobbing the body up and down as if preparing for a take-off; in the 2 Neotropical species this behavior is replaced in part by wing-flicking. They are strongly territorial and place their nest in a crevice or on the ledge of a bank, regularly under a waterfall. The nest is a large closed construction of moss etc, with a side entrance. Often regarded as related to wrens, but recent molecular evidence suggests that it is a sister group to thrushes (*Turdidae*) and mockingbirds (*Mimidae*).

WHITE-CAPPED DIPPER *Cinclus leucocephalus* – Plates IL 19a-c, LXIII 27

15 cm. Ssp *leucocephalus*: black or blackish; **crown, throat, and upper breast white**, feathers of crown dark along the shaft, sometimes mostly dark. Inner webs of primaries white along their bases. Occ. upper back

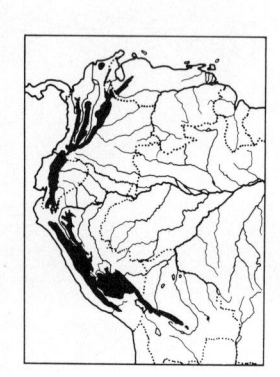

and lower breast streaked white, and *leuconotus* with a white patch on c back. Entire underparts except lower flanks and vent white, feathers of rump and vent white-tipped in fresh plumage. Ssp *rivularis* much like *leucocephalus*, but paler; white of throat spotted with gray and not extending to upper breast; belly sometimes mottled with whitish. **Juv**. much like ad; lower back, rump, and esp. belly with pale feather-tips; crown, throat, and breast narrowly dark-tipped, breast-feathers white at their bases and along their shafts, producing a somewhat streaked appearance.
Habits: In pairs. Unlike N Hemisphere dippers scarcely cocks tail, and only rarely bobs, but often flicks wings. Forages on wet mossy cliffs above streams and on the boulders in currents down to the waterline, but rarely swims, and never rec. diving. Flight straight and fast with whirring wingbeats, usually low over stream. Points bill up in threat.
Voice: Flight-call a sharp *tjik* which carries far despite the noise of the rushing water. Song undescribed.
Breeding: Eggs Mar (Mérida), Sep (e Ecu.), Nov (Cochabamba); fledglings Oct (Cauca), July (Puno), May (Cochabamba).
Habitat: Along clear, rocky, rapid streams from humid to semi-arid regions, often with wooded edges.
Range: (100-) 1500-3300 (-4200) m. Sta Marta mts (*rivularis*); from Lara, Ven., and Perijá mts s through all 3 ranges of Col. to Ecu. (*leuconotus*), Peru (see Bond 1956) (in W Andes s to Arequipa), and Bol. in La Paz, Cochabamba, and Sta Cruz (*leucocephalus*). Fairly common.

RUFOUS-THROATED DIPPER *Cinclus schulzi* – Plate IL 18

15 cm. Dark gray; chin light gray, throat rufous. White patch on inner webs of primaries more extensive than in White-capped D. and conspicuous in flight. **Juv**. unknown.
Habits: Wing-flicking and other habits as in White-capped D. Perhaps more often submerged. Unlike *Cinclodes* species of the same habitat holds tail down.
Voice: No data.
Breeding: Eggs Dec (Tucumán) (see Fraga and Narosky (1985) and *Gerfaut* 76(1986):63-6 for breeding data).
Habitat: Rapid, clear, permanent rocky streams fringed with alders (*Alnus*).
Range: At 800-3000 m, lowest in winter. Tarija (25 km nw of Entre Ríos, see *Condor* 85(1983):95-98), s Bol., Jujuy (Sierra de Zenta, ríos Yala and Quesera), and in Aconquija mts in Tucumán (Trancas, río de las Sosas drainage) and adjacent Catamarca (El Clavillo, río Charcas), nw Arg. Uncommon, total population a few 1000. Local threats include cattle encroachment, clearance, and water utilization.

NOTE: Sometimes considered a ssp of White-capped D.

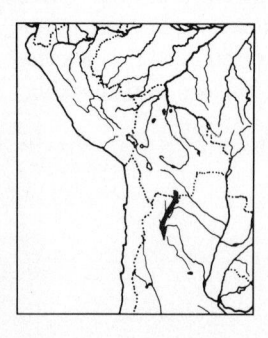

Wrens – Family Troglodytidae

Some 60 species, 1 found in temp. regions throughout the northern hemisphere, the remaining restricted to the Americas, from s Canada to Isla Grande, inhabiting both humid and arid regions. Small to medium-sized, mostly brown birds, reminiscent of small furnariids with similarly pointed, rather flat head, but more or less barred with dusky, esp. on wings and tail. The sexes are alike or differ only slightly. Wrens are found in pairs, and the majority of the species skulk in dense undergrowth and reveal their presence by their unusually loud melodic song, often given in duet. Insects form the main part of the diet. Only the female incubates, but both sexes tend the young. The globular nest is built by the male and may be placed in a cavity among roots etc., or in a bush or a tree. Some build several nests of which only one is used for breeding; one of the others may be used as a roost for the male, in some species as a communal 'dormitory'. Usually regarded as related to mockingbirds and dippers, although recent molecular evidence suggest relationship with the Old World assembly of warblers and tit-like birds. See Meyer de Schauensee (1946a) and Bond (1956).

GRAY-MANTLED WREN *Odontorchilus branickii* – Plate IL 1

- is uncommon (or overlooked) in premontane forest to at least 2400 m. On w slope of W Andes in s Col., and Valle to Imbabura nw Ecu. (*minor*); from head of Magdalena Valley Col. s through e Ecu. and Peru to La Paz, Bol. (*branickii*). 10.5 cm. Above gray, the fairly long tail barred black, rump spotted white. Crown brownish clay; sides of head streaked gray and white; narrow eye-ring white; **below white, underside of tail pale gray barred black**. Alone in mixed-species flocks, where usually in company with Ash-browed Spinetail, Buff-fronted Foliage-gleaner, and Streaked Xenops. Forages 15-30 m up, along more or less horizontal branches, in a distinctive manner, leaning over from side to side and peering at leaf-surfaces and the underside of the branch, moving constantly except for brief stops to probe clumps of moss or lichens. Rarely cocks tail. The infrequently heard song is a brief high-pitched monotonous trill. Call a thin weak *si-si-si-si*.

RUFOUS WREN *Cinnycerthia unirufa* – Plate IL 5a-b

15 cm. Tail fairly long; bill blackish gray. Ssp *unibrunnea* **uniformly rufous brown, flight- and tail-feathers with faint barring, lores dusky**. Inner webs of remiges narrowly edged dull pinkish cinnamon. Ssp *unirufa* brighter tawny, inner webs of primaries edged bright ochraceous tawny, some individuals with white feathers on forehead; *chakei* much like *unirufa*, but slightly darker and smaller, lores darker, tail longer, and wings with hardly any barring. **Juv**. has grayish cap and partly yellow mandible.
Habits: In family groups of 3-8 birds, slowly and acrobatically working their way through the undergrowth, probing moss and picking insects from foliage, occ. flicking aside leaves. Tail sometimes slightly raised and wagged from side to side. Keeps hidden. May climb vertical stems.
Voice: Calls very different from Sepia-brown W. Contact call, given constantly, is a wheezy *tsjip*. Also gives a somewhat croaking warble or chatter. Song bouts c. 10-15 s long, one bird repeating a melodious *tó-*

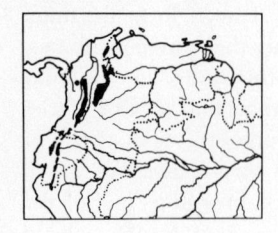

daly, *chu-woo*, or *chi-wee*, while another simultaneously gives a monotonous rapid trill.
Breeding: Juv. Mar (Cauca/Huila); large gonads June-Aug (Col.).
Habitat: Lower strata of cloud forest, often with admixed bamboo.
Range: 2200-3800 m, in Ven. down to 1800 m. Perijá mts (*chakei*); from páramo de Tamá and Río Chiquito, sw Táchira, Ven. to n end of E Andes of Col. (*unirufa*); from C Andes and n end of W Andes of Col. s through Ecu. to Piura, n Peru (*unibrunnea*). Fairly common to common.

SEPIA-BROWN WREN *Cinnycerthia peruana* – Plate IL 6a-c

15 cm (*fulva* 12.5 cm). Much like Rufous W., but **wings and tail more distinctly barred**. **Old birds have white face** (from se Ecu. s-wards). Ssp *olivascens* reddish brown above, somewhat paler below; crown and back with faint olivaceous cast. Lores dusky, faint postocular streak grayish, wings and tail with narrow black bars. Ssp *bogotensis* with much darker and deeper coloration; white-faced birds unknown. Ssp *peruana* slightly smaller than *olivascens* and without olivaceous cast, grayish postocular streak more conspicuous. Ssp *fulva* much smaller; coloration closest to *olivascens*, but underparts paler; broad supercilium buffy white. **Juv**. has grayish cap.
Habits: Much like Rufous W. In family groups. Tail at the most slightly raised.
Voice: Contact note a sharp, regularly spaced *zit* or *zee*, when excited a wheezy *tziee*. Alarm *trrek* or a chattering *trrr trrr tjk trrr* ... given by 2 or more birds. Song melodic, of loud and clear whistled phrases given by one or more birds; recorded variants (Ecu., c Peru) being *hy hehe hy he, hy hy hy hehe hy, whee whee whee ty de ly ly ly, tee dee hy de ly de ly ly, teedeweee hy hy hy hy a rry, weedeweedewee ty ly ty ly ty ly, rrrrhyy de de de de, lylylylylylyly, ahy-ahy-ahy-ahy-ahy, lydeee lydeee, dy ryy dyde ly, terrriidely dy lydyly* or *tjutjut-jutju*....
Breeding: Eggs Sep (Huánuco); large gonads June-Aug (Col.), Jan (La Paz, Cochabamba); fledglings Aug (Amazonas); juv. Feb, June, Nov (e La Libertad), Mar (Pasco), June (Huánuco).
Habitat: Within 2 m of the ground in tangles and undergrowth of humid forest and cloud forest, perhaps less frequently in bamboo and more in broad-leaved vegetation than Rufous W.
Range: (900) 1800-2400 m where range overlaps with Rufous W., up to 3370 m in Peru and Bol. W slope of E Andes of Col. (*bogotensis*), from W and C Andes of Col. s through e Ecu. to Piura, n Peru (*olivascens*); s and e of the Marañón from Amazonas to Junín, c Peru (*peruana*); Cuzco, se Peru, to Cochabamba, Bol. (*fulva*). Fairly common to common.

NOTE: Ssp *fulva* may represent a distinct species. Nest description and plumage variation in *Bull. Brit. Orn. Club* 99(1979): 45-7.

GRASS WREN *Cistothorus platensis* – Plates IL 9a-d, LXIV 28

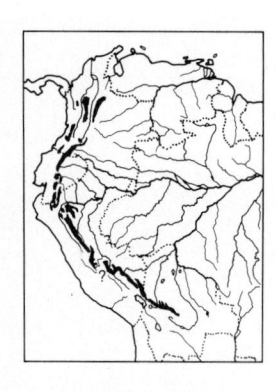

10-11 cm. **Streaked back** and usually with **distinct supercilium**. Generally gray-brown, some populations tinged rufescent, crown streaked or plain, the blackish mantle with buff or whitish streaks, rump streaked or plain; wings and tail barred blackish; underparts uniform buffy brown to whitish, palest in worn plumages. Andean forms of variable appearance: *aequatorialis* long-legged, with plain crown, lower back, and rump, tail barred throughout; *graminicola* similar, but with streaked crown s of c Peru, and with longer tail (tail/wing ratio .933-.937); *tucumanus* like birds of s Peru, but with shorter legs; *platensis* heavily and completely streaked above; *hornensis* similar, but longer-winged and shorter-tailed. Birds of the eastern group less variable, and recognized by solidly black inner webs of some rectrices: *polyglottus* with restricted dorsal streaking, as in *aequatorialis*, but with shorter wing and tarsus and longer tail, and with black instead of barred inner webs of rectrices 2-5; *alticola* shorter-tailed; *minimus* resembling a small *graminicola*. **Juv**. like ad., but streaks on back less white, bars on tail irregular.

Habits: Alone or in pairs, territorial, but may form loose colonies. Inconspicuously runs like a mouse or steals through bushes, grassy tussocks, and clumps of rushes. When flushed has tumbling flight with spread tail low over the vegetation, and soon drops back into cover. Tail usually cocked, during song sometimes even tilted forward to touch the nape. Perches rather conspicuously near the tip of a branch of a bush or in the top of a tussock while singing. Mainly feeds on insects, but takes some seeds in winter.

Voice: Song 2-4 s, given at 1-4 s intervals, consisting of warbling notes interspersed with trills *trtetiewehweh-zie-zie-zie*, or *veeveeveetrrrr-zie-zie-zie-zie*. Alarm a rasping *rrreh-rrreh*.

Breeding: Fledglings Jan, May (nw Ecu.), Mar (Huila), Apr (Cuzco/Puno); juv. Apr (Junín); song Mar (páramo de Tamá), Oct (ne Ecu.), Dec (Cuzco/Puno); large gonads Feb-Aug (Col.).

Habitat: Within 0.5 m of the ground in marshes and adjacent rushy pasture; mossy bogs in elfin forest to humid páramos with tall grass interspersed with low dense bushes. In the Bogotá area from partly flooded Alder forest (*Alnus*) to *Swallenochloa* bamboo bogs in páramo habitat, but not in tule marshes; in Ven. sometimes on open grassy savanna.

Range: At 2200-4000 m from Páramo de Tamá s to Cundinamarca in E Andes of Col., 1770-3875 m in páramo zone of n part of C Andes of Col. in Tolima and Caldas and 2240-4500 m from Cauca in s part of C and W Andes of Col. patchily to c Ecu. (birds from n Loja (San Lucas), s Ecu. may belong here) (*aequatorialis*). At 3500-4600 m from extreme n Peru along the C and E Andes to La Raya pass Cuzco/Puno, se Peru (*graminicola*). At 1675-4500 m in highlands of La Paz, Cochabamba and Sta Cruz, Bol., and 400-3200 m from Tarija, s Bol., s to Tucumán and Catamarca, nw Arg. (*tucumanus*). Pampas zone of Arg. w to Córdoba and n Mendoza (*platensis*), and from sea level to 2400 m in Coquimbo and sc Chile into adjoining Neuquén, Arg., and along the Patagonian Andes to Isla Grande (where resident!) (*hornensis*). Tepuis of se Bolívar Ven. and in Meta, e Col. (*alticola*); 500-3275 m in Sta Marta mts, Col., Perijá mts, coastal mts and Andes in Lara and Mérida, Ven., se Brazil, and isolated on lower Río Beni e Bol. (*polyglottus*); 2750 m at Oconeque in Puno, se Peru (*minimus*, known from 2 specimens). Fairly common, but very local; rare on Isla Grande. Several sspp in N and C Am., and one on Islas Malvinas.

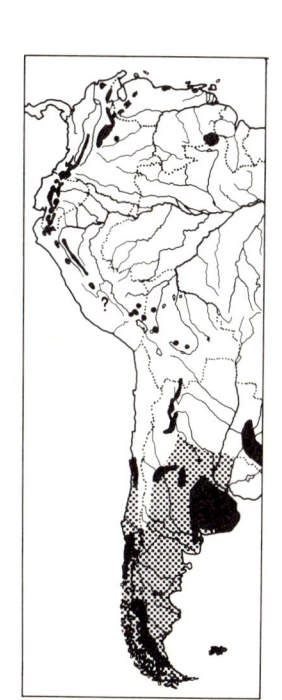

NOTE: Often called Marsh W. or Short-billed Marsh W. Sspp *alticola*, *minimus*, and *polyglottus*, which resemble each other closely, may represent a rather recent colonization from N Am. (MA Traylor, Jr in *Fieldiana Zool.* 1392(1988):1-35). They interbreed with ssp *platensis* of the Andean ssp group in the chaco zone. MA Traylor, Jr. has synonymized the former sspp *tamae* and *tolimae* (under *aequatorialis*) and *boliviae* (under *tucumanus*).

APOLINAR'S MARSH WREN *Cistothorus apolinari* – Plate X 12a-b

12.5 cm. Much **like a dark Grass Wren** (ssp *aequatorialis*), **but much larger**, bars on wings and tail straighter and often broader on wings and sometimes on tail. Pale streaks on back sometimes broader, scapulars dark warm buff with some dark brown blotches, greater wing-coverts with 4 bars. Pale bars on remiges (except inner remiges) paler than in rest of wing. Tail warm buff (light buff when worn) with 9-10 dark bars. Head dull buffy brown with **only slightly paler supercilium** and gray-brown eye-stripe; color of nape grades into the back. Throat and c underparts pale gray, sides, flanks and vent warm buff. **Juv**. distinct: **head** dark gray-brown **(bluish in the field) without supercilium**; lower nape buff; scapulars slightly paler than in ad., and without dark blotches, giving the back a less streaked look; greater wing-coverts concentrically marked light and dark. Legs gray-brown (paler in ad.).
Habits: In pairs or family-groups in loose colonies. Difficult to see, but frequently calls. Moves slowly through the cattails and tules, climbing from water-surface up 1 m, then drops to next stem. Gleans insects from leaves and stems.
Voice: Alarm a persistent, loud rasping churring. Song an energetic bubbly series of 1-3 or so rather low-pitched rythmic *toe-a-twée* phrases interrupted by short gravely churrs.
Breeding: Eggs July; juv. July, Aug, Oct; large gonads Mar, Aug.
Habitat: Confined to tall tules and cattails in marshes.
Range: Col. at 2000-3040 m on Bogotá savanna (e.g. in the city park La Florida), in páramo de Sumapaz, Lake Pedropalo on slope to Magdalena valley, and n to lake Tota and Guicán in Boyacá. Fairly common, but local (See *Lozania* 6(1953): 1-6, and Varty et al. 1986).

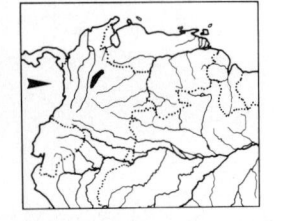

PARAMO WREN *Cistothorus meridae* – Plate LXIV 29

10 cm. Much **like Grass Wren** (ssp *alticola*), **but supercilium white and conspicuous** (faint or absent in *alticola*), and **flanks barred**. Crown brown, faintly streaked and barred with dark brown; back streaked black and whitish to rump; wings and tail pale brown barred blackish. Ear-coverts brown streaked dark brown. Lores and lower headside mottled whitish and blackish. Below whitish, faintly spotted brown on sides of breast; sides and flanks buffy, flanks barred dark brown. Iris yellow or blue; blue-eyed birds may be juv.
Habits, Voice and **Breeding**: No data.
Habitat: Bogs and moist spots with dense grass, moss, and low thicket cover; mossy, low open forest and bushy areas; wheat fields.

Range: Ven. at 3000-4100 m in Trujillo (Teta de Niquitao) and Mérida (La Culata, Mucuchies, La Negra, and Teleférico páramos).

NOTE: By some called Mérida Wren.

PLAIN-TAILED WREN *Thryothorus euophrys* – Plate IL 3a-d

16.5 cm. **Tail fairly long, faintly barred; generally rather uniform below**. Ssp *longipes* bright rufous above, crown browner and faintly bordered on the side with black. Headside grayish white streaked black, distinct supercilium pale gray, line through eye blackish; throat white with black moustache; below rufous, paler than back, c underparts distinctly paler and tinged gray, breast with some scattered blackish spots. Ssp *euophrys* similar, but crown gray, feathers faintly edged dark gray; hindcrown washed olivaceous; supercilium whiter, feathers of throat and breast tipped black; most of underparts gray, sides, lower belly, flanks, and vent pale olive-brown. Ssp *atriceps* smaller than *longipes*; crown gray, blackish at sides; headside unstreaked, throat whitish, breast and belly grayish tinged buffy. Ssp *schulenbergi* largest and with the crown dark gray or mixed with gray-brown; back and wings duller, less reddish brown; supercilium pale gray, headside unstreaked; throat whitish turning pale gray on breast and belly, flanks and lower belly brownish. **Juv**. has the crown more tinged with olive-gray, underparts below white throat rufous without gray tinge or blackish spots.
Habits: In pairs, not in mixed-species flocks. Climbs and hops through the densest shrubbery, vines, and bamboo, picking insects from the underside of leaves. Holds body horizontal and tail slightly raised. Sings in upright posture.
Voice: Duetting song a continuously repeated, loud and happy whistled *tuie huit huihui*, *hu hui huhe hui*, or *tu hehuhui-hu hehu*. Alarm a hard *tje tjekrrtekrr te tekrrr....* Calls include an often repeated sharp 2-note *tju tju*, 2nd note lowest (sometimes alike or higher), and a wheezy *whiee*.
Breeding: Juv. July (nw Ecu.), Nov, Jan (ne Ecu.).
Habitat: 0.5-1.5 m up in thickets of *Chusquea* bamboo and adjacent dense humid forest undergrowth and shrubbery.
Range: 1850-3500 m. From extreme sw Nariño, Col. s to Cañar (and probably Azuay), w Ecu. (*euophrys*); e Ecu. at least s to Zapote Najda mts (*longipes*); ec Loja (PN Podocarpus: ZMUC), s Ecu., and n Cajamarca and Piura, n Peru (*atriceps*); s of the Marañón in Amazonas and San Martín (*schulenbergi*, Parker *et al.*, p. 12 in Buckley *et al.* (1985)). Common.

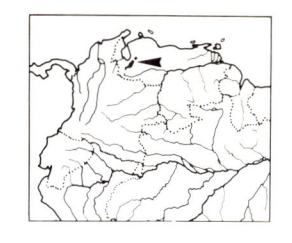

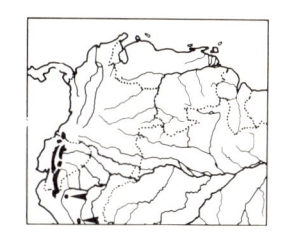

INCA WREN *Thryothorus eisenmanni* – Plate IL 2

15 cm. Tail fairly long. Crown dark gray; back, wings and tail cinnamon-rufous; tail very faintly barred blackish. Supercilium white, headside streaked black and white, throat white with black moustache, **breast and c belly white streaked black**; sides, flanks, lower belly, and vent olivaceous cinnamon-brown. **Juv**. has the crown and entire underparts uniform clay-brown.

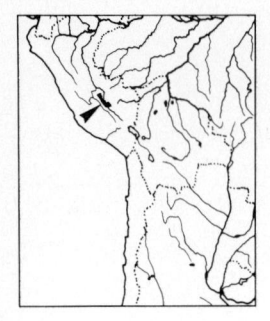

Habits: In pairs or groups of 5-6, not in mixed-species flocks. Much like Plain-tailed and Moustached W. Probes clusters of stems at the internodes of bamboo stalks, also searches dense foliage and curled dead leaves trapped in the tangled crowns of thickets.
Voice: Like most others in the genus sings a well-synchronized antiphonal duet; it is faster, less rhythmic, and with more overlapping notes, some of which are much higher pitched than in the duet of Plain-tailed W. Often sings in chorus of 5-6 birds. Sings through most of the day. Repeats a theme, e.g. *tui weee weet wee eh* over and over, then shifts to a new theme, e.g. *tui wee eeee weet wee eh* or *wheet ee eee*, and repeats that.
Breeding: Fledgling May (Cuzco).
Habitat: As for Plain-tailed W.
Range: 1830-3350 m. Confined to Vilcanota and Vilcabamba mts on both sides of Urubamba valley in Cuzco, se Peru. Common.

NOTE: Described by Parker *et al.* in Buckley *et al.* (1985):9-15.

MOUSTACHED WREN *Thryothorus genibarbis* – Plate IL 4

15.5 cm. Tail fairly long. Ssp *consobrinus* has dark brownish olive crown and upper back, and chestnut wings, lower back, rump and tail; **tail barred blackish**. Supercilium white, headside streaked white and black, throat white with black moustache, breast to c upper belly buffy white, sides of breast gray; sides of body, flanks and lower belly brown, vent rufous-brown. Ssp *saltuensis* has upper back rufous chestnut like lower back, and the entire breast gray without buffy. **Juv**. *saltuensis* like ad., but back duller, crown gray-brown grading to brownish on nape without contrast to back. Black and white of head replaced by grayish buff and dark gray, underparts washed brownish. Juv. *consobrina* has crown colored like back, and has buff breast and belly. Behavior like Plain-tailed and Inca W., but said occ. to join mixed-species flocks. Sings loud duets with very well synchronized phrases, that are even shorter and more spaced than in Plain-tailed W., and rarely with overlapping notes *tirru-okluk*, *dwuo-wio*, *tschrr-bo-wio*, *pai-avo*, *didu-didi-dudo*. Call *wu-tu-tu*. Fledglings June, Aug (Mérida), June (Cauca).

0.5-1.5 m up in undergrowth of humid and cloud forest, plantations, second growth, and clearings, but not in bamboo. Mainly premontane zone, up to 2400 m in Andes of Ven. and to 2800 m in W Andes of Col. Coastal mts of Ven. (*ruficaudatus*); Perijá mts and Andes from Lara to Mérida, Ven. (*consobrinus*); páramo de Tamá (*tachirensis*, doubtfully distinct from *consobrinus*), e slope of E Andes of Col. from N.d.Santander (where also on w slope) to Meta (*amaurogaster*), both sides of Magdalena valley (*macrurus*), w slope of W Andes from Chocó to Cauca, Col. (*saltuensis*), valley of Río Guaitara, interior Nariño, s Col. (*yananchae*), w Ecu. (*mystacalis*); La Paz, Cochabamba, and Sta Cruz, Bol. (*bolivianus*); n and nc Brazil s of the Amazon (*genibarbis*), c Brazil (*intercedens*), w Brazil and ne Bol., probably same form in e Ecu. and e Peru (*juruanus*). Locally fairly common.

HOUSE WREN *Troglodytes aedon* – Plate IL 10a-c

11-12 cm (*puna* largest). Ssp *chilensis*: above drab, rump brown, wings and tail dark reddish brown barred black; **no supercilium**; **conspicuous eye-ring whitish**. Below pale pinkish buff, whitish on c belly; lower sides, flanks, and vent washed rufous. In *tecellatus* the bill is much longer; upperparts smoky gray, rump and tail cinnamon; entirely barred above, least so on hindcrown and upper back; **supercilium and eye-ring buff**, ear-coverts dark brown streaked buff; below buffy white, palest on throat and c underparts; vent barred blackish. Ssp *puna* olivaceous brown above, crown sometimes very dark, **rump and tail rufous** cinnamon, wings and tail barred blackish, rest of upperparts indistinctly barred; supercilium and eye-ring buff, headside buff washed dusky; **below warm buff, sides, flanks, and vent rufous**; throat and c underparts sometimes paler, buffy white. Other sspp variously intermediate (*columbae* vinaceous buff below), in some the c underparts may wear to nearly white. **Juv.** has the feathers above and below narrowly tipped dark brown.

Habits: Alone or in pairs. Tail at the most slightly raised. Actively searches for insects in the bushes or on the ground. Confident and curious, often near human habitation.

Voice: Alarm a distinct, rasping, drawn-out, repeated *wheep* or *eeeeh*. Song a varied warble lasting 2-4 s and repeated at 1-2 s intervals.

Breeding: In Chile starts egg-laying in Aug in warmer parts, a month or so later further s (Nov-Dec on Isla Grande), raising 2, or 3 broods a year. Fledglings Jan, Feb (Tarija), Dec, Apr (Cochabamba), Oct, Feb, Mar, June, July, Aug, Sep (probably all year) (La Páz), Oct, Apr (Cuzco), Nov, Jan (Junín), Apr (Ancash), Nov (Amazonas), May (Cajamarca), throughout year (Cauca valley Col., coast off Peru).

Habitat: Within 3 m of the ground. Not in closed forest, but found in almost any other habitat in both humid and arid regions. *Polylepis* woodland and adjacent puna grassland with rocks and banks, shrubbery, second growth, brush, scrub, bushes, orchards, gardens, forest edge etc. On Patagonian basaltic plateaus among rocks bordering sheltered crater-lakes with no other vegetation than short grass.

Range: Sea level to 3400 m, *puna* at 2600-4500 m. 30 mostly poorly differentiated sspp found almost throughout from USA to Isla Grande, incl. Caribbean isls and Islas Malvinas. Andes of Ven. from Trujillo to Táchira, and C and W Andes of Col. (*striatulus*); E Andes of Col. (*columbae*); from Ven. (except range of *striatulus*), Trinidad, Nariño s Col., w Ecu., Piura n Peru, and Huánuco c Peru e to Guianas, and Brazil to nw Mato Grosso, Amazonas, and Maranhao (*albicans*); highlands from Cajamarca and Amazonas Peru s to w La Paz, Bol. (*puna*); e slope of E Andes of Peru from se Amazonas to Puno, grading into *puna* (*carabayae*); river valleys of coast from Arequipa, s Peru to n Tarapacá, n Chile (*tecellatus*); from e La Paz, w Beni, and w Sta Cruz, c Bol., s through Parag. and Arg. to San Juan, San Luis, Córdoba, and Chaco, Arg. (*rex*); from Coquimbo river, Chile, and Mendoza, Arg., s to Isla Grande, wintering n to Entre Ríos and Sta Fé, and to Atacama in Chile (*chilensis*). See pp. 297-304 in Ridgway (1924). Ubiquitous.

MOUNTAIN WREN *Troglodytes solstitialis* – Plates IL 11a-e and LXIII 28

9-10 cm. **Distinct eye-brow buff, vent conspicuously barred**. Ssp *monticola* large, **male** dark brown above, lower back, wing-coverts and tertials barred black, remiges blackish barred whitish. Distinct eye-brow buff, eye-ring buffy white, ear-coverts blackish brown; throat and breast buffy brown, belly and flanks buffy white barred dark brown; vent white barred blackish. **Female** paler; above brown, earcoverts dark brown. Ssp *solitarius* like *solstitialis*, but upperparts darker, throat and foreneck paler, vent tinged ochraceous. Ssp *solstitialis* cinnamon-brown above, outer primaries somewhat paler, tail grayish; wings and tail narrowly barred blackish, but back only with occ. faint bars. Conspicuous eye-brow warm buff passing around dark brown ear-coverts to lower headside, throat and breast buff turning white on belly, sides and flanks faintly washed buff, vent grayish white barred black. Ssp *macrourus* like *solstitialis*, but more reddish brown above, tail washed cinnamon and slightly longer; birds from Urubamba Valley white-browed. Ssp *frater* like *solstitialis*, but supercilium and eye-ring white, tail even longer than in *macrourus*; some (all?) birds from La Paz more buffy below, recalling *solstitialis*. Ssp *auricularis* like *frater*, but slightly less rufescent, and with shorter tail. **Juv**. has feathers dark-tipped above and below.
Habits: Usually in pairs or family groups, sometimes in mixed-species flocks. Perch-gleans insects along tangles and in dense growths. Tail not raised. Not shy.
Voice: Song distinctive, a weaving, very timid and high-pitched *trrii-tii-trii-trriie-trrie*, repeated. Alarm a repeated *zri di di* or *zri dr dre drrr*.
Breeding: Eggs Aug, Nov (Cauca/Huila), Nov (Tucumán); fledglings Apr (Tolima), Mar, July (Caldas), Sep, Oct (Quindío), Oct (Cundinamarca), Jan-Apr, Aug (Cauca), July (Nariño), Feb, May (Huila), June, Nov (e Ecu.), Apr (nw Ecu.), Feb (Cajamarca), Jan (Amazonas), Feb (Huánuco), Dec (Cuzco), Jan (La Paz), Nov, Feb (Cochabamba), Mar (Tucumán); nest-building Mar (Huila); large gonads Mar-Sep (Col.).
Habitat: 1-12 m up in elfin forest, cloud forest, and humid forest, sometimes with admixed bamboo. Occ. in felled trees on forst clearings.
Range: 1700-3500 m, down to 1200 m in nw Arg. Sta Marta mts (to 4600 m?) (*monticola*); Andes of Ven. from Lara to Táchira, Perijá mts and all 3 ranges of Col. (*solitarius*), in s Cauca and Huila (intergrades) and in w Ecu. (rec. in Carchi, Chimborazo and El Oro) and through e Ecu. s to Piura and n Cajamarca, n Peru (*solstitialis*). Peru from Amazonas to Cuzco, birds from Urubamba Valley intergrading (*macrourus*), from n Puno, se Peru through w Bol. (*frater*), and Jujuy, Salta, and Tucumán, nw Arg. (*auricularis*, doubtfully distinct from *frater*). 7 other sspp from Mexico to Panamá. 5 sspp of the closely related, perhaps conspecific Tepui Wren (*T. rufulus*) on Tepuis of s Ven. Fairly common, locally common.

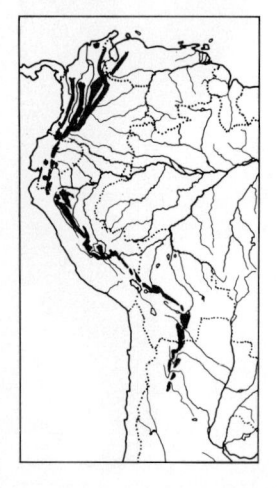

NOTE: Ssp *monticola* sometimes considered a distinct species, **SANTA MARTA WREN**.

GRAY-BREASTED WOOD-WREN
Henicorhina leucophrys – Plates IL 7 and LXIV 30

11 cm. **Tail short**. Ssp *leucophrys*: crown dark brown bordered laterally with black, crown-feathers with blackish bases; rest of upperparts cinnamon-brown to chestnut-brown. **Long conspicuous supercilium white, broad eyeline** dark, lower headside white with **black streaks**, and mottled moustache demarcating white throat. Breast and upper belly gray; sides, flanks, lower belly, and vent cinnamon-brown. Ssp *meridana* similar, but crown like back, feathers of throat with black edges, flanks and vent more rufous; *brunneiceps* has crown only slightly darker than back; *boliviana* has throat-feathers more or less edged blackish, belly faintly barred whitish; *anachoreta* has short bill, upperparts washed with russet, more or less black-streaked throat, and extensively brown flanks; *manastarae* like *meridana*, but lower breast paler and belly whitish rather than gray; *tamae* intermediate between *leucophrys* and *meridana*, though closest to the latter; *hilaris* like *leucophrys*, but foreneck and breast much paler, nearly grayish white, flanks more extensively cinnamon brown. **Juv.** has crown like back and throat indistinctly marked.

Habits: In pairs. Forages for insects in tangles and among moss-covered branches and stems. Territorial.

Voice: Song, often duet, given frequently throughout day, consists of clear melodious phrases of (1) 3-6 notes lasting 1-2 s and given 3-5 times at 1 s intervals (shorter intervals in excitement, e.g. in response to another bird). Usually the theme is constant through a series, but varied between bouts *whee tseea whyi*, *whee tzi whea why*, *whya tse whe weehe*, *teetelyi*, *trrlye tuiheeheehee*, *tui hui* or *wee tzie*. Alarm (both sexes) a rapid, hard ticking or a softer *trriut-trriut-trriut...* at irregular intervals, at times somewhat intermediate between the 2 or shifting between them.

Breeding: Fledglings May (Mérida), Jan, May, Sep (páramo de Tamá), Mar (Cundinamarca), Sep (Quindío), May (Huila), May, Dec (Nariño), Aug, Nov, Dec-Feb (ne Ecu.), Apr, May (nw Ecu.), Nov (Amazonas).

Habitat: Within 2 m of the ground in dark and dense, usually mossy undergrowth in humid forest, often along streams.

Range: Mainly premontane, but to 3600 m in Sta Marta mts (*anachoreta*) (ssp *bangsi* at 600-2100 m in the same mts!); to 3000 m in Andes from Trujillo to n Táchira, Ven. (*meridana*), páramo de Tamá and e slope of E Andes, Col. incl. Macarena mts (*tamae*), to 2300 m in Perijá mts (*manastarae*), to 2900 m on w slope of E Andes, C Andes and e slope of W Andes and e slope of Nariño, Col., Ecu. (except nw and sw) and C and E Andes of Peru s to Puno (to 3100 m in Peru) (*leucophrys*). W slope of W Andes from San Juan river Chocó/Valle, Col., s to Imbabura, nw Ecu. (not to temp. zone?) (*brunneiceps*), sw Ecu., intergrading with *leucophrys* in wc Ecu. (possibly not to temp. zone) (*hilaris*). To 2300 m from La Paz to Sta Cruz, Bol. (*boliviana*). 8 other sspp. in coastal Ven. and C Am. from Mexico to Panamá. Common.

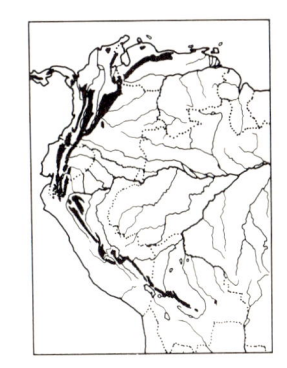

BAR-WINGED WOOD-WREN *Henicorhina leucoptera* (described by Fitzpatrick *et al.*, *Auk* 94(1977):195-201) is common at 1950-2450 m in Condor mts, n Cajamarca, at Abra Patricia at the border of Ama-

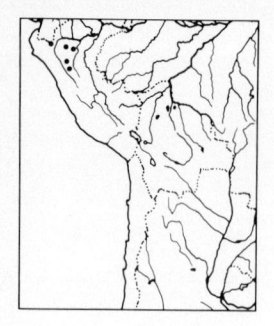

zonas and n San Martín, e La Libertad, and as low as 1350 m in E Andes se of Moyabamba in San Martín, n Peru. At the first localities it generally occurs at higher elevation than Gray-breasted W-w., but with much overlap. Told from Gray-breasted W-w. by its nearly **black wings with 2 conspicuous white bars**, narrow white edges of outer 3 primaries, paler underparts and a slightly different song: higher pitched, faster, with a more ringing quality and more frequent trills.

CHESTNUT-BREASTED WREN *Cyphorhinus thoracicus* – Plate IL 8

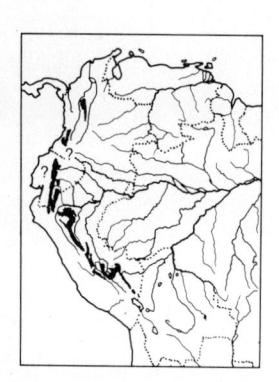

13.5-14 cm. Bill with high ridge at base (continuous with top of head). Above dark brown (*thoracicus*) or fuscous black (*dichrous*), tail blackish. Sides of head and neck, throat, breast, and c belly rufous to rufous chestnut; lower belly, flanks and vent dark brown (*thoracicus*) or fuscous (*dichrous*). **Juv**. has the dark brown parts lighter. Alone, in pairs or family groups, sometimes active beneath mixed-species flocks. Forages for insects low in the vegetation and is difficult to see. Pecks and pokes under leaves and debris. Responds to an imitation of its call. Raspy churr call usually repeated several times. Song of 2-4 pure melodious flute-like whistles *hy* or *hy hy*, 2nd note lowest, repeated at 1 s intervals for several minutes, ascending, descending or alternating in pitch in half-tone steps. After playback song may jump an octave higher. Fledglings Nov (Cuzco); juv. Apr (Valle); large gonads June (E Antioquia), Mar (Valle); nest-building Sep (Valle).

Within 1 m of the ground in dense very humid forest undergrowth, at 750-2650 m. From C and W Andes of Col. s through e (and w ?) Ecu. and E and C Andes of Peru to San Martín and e La Libertad (*dichrous*), Huánuco to Puno, c-se Peru (*thoracicus*). Uncommon.

NOTE: Closely related to, and perhaps conspecific with the widespread Musician Wren (*C. arada*) of the trop. lowlands, but with somewhat more compressed bill, more developed facial bristles, and with 10 instead of 12 rectrices.

Mockingbirds – Family Mimidae

27 species of the W Hemisphere, esp. in semi-arid regions. Fairly large passerines with a long tail. The family includes tremblers, thrashers, Mockingthrush, and Catbird, but only *Mimus* mockingbirds reach the temp. zone of the Andes. Of the 8 *Mimus* species only the N Am. and Caribbean *M.polyglottus* and *M.gundlachi* do not come into the scope of this book. Sexes are alike. Upperparts gray or brownish, underparts whitish, juv. with a spotted breast. All have a slender bill and a fairly long, graduated tail, which is conspicuously tipped white. Living in pairs they inhabit open country, and generally forage low in the vegetation or on the ground, running or moving in powerful hops. The food comprises insects and some fruit and seeds. The birds are aggressive. Their varied songs often include imitations of other birds and sounds, hence the name of the family. The flight alternates between series of rapid wingbeats and gliding with spread tail. The tail is often cocked upright. The open nest is a large loosely built construction of often thorny twigs, and it is frequently based on a smaller nest of another bird, placed in a tree or cactus, and visible at

a long distance. Traditionally regarded as near relatives of wrens, but recent biochemical evidence suggest thrushes (*Turdidae*) to be nearest relatives. The habitat segregation of Patagonian species is described in *Ibis* 120(1978):61-5.

CHALK-BROWED MOCKINGBIRD *Mimus saturninus* – Plate IL 15

23.5 cm. Ssp *modulator*: above gray-brown indistinctly streaked dark brown; wings fuscous with 2 white bars, inner remiges tipped whitish, tertials edged pale gray-brown; outer primaries narrowly edged white along their middle portion. Rump tinged cinnamon, **tail blackish, all but c feathers broadly tipped white**, outer web of outer feather edged white. Broad eye-brow whitish, line through eye blackish brown, lower headside and entire underparts whitish, neckside and flanks faintly tinged buff, sides of body rarely with a few faint dark shaft-streaks. Iris sometimes yellow. **Juv**. has mantle, rump, and edges of secondaries and wing-coverts brown, throat as in ad., rest of underparts buffy white, spotted and streaked with blackish on breast. - Alone or in pairs, frequently foraging on the ground, often jerking and cocking tail. Confident. Jumps into the air while singing. Calls include a *scha-scha-scha* and *krrra*. Alarm *check-check*. Song inferior to White-banded M., and mostly consists of mimicry. Breeds Sep-Jan (Tucumán).

In open country with scattered groves of trees, secondary growth, gardens. Mostly lowlands, but up to 2500 m in Tucumán, nw Arg. Much of Brazil and n Bol. (*saturninus*, *arenaceus*, *frater*), from Chuquisaca and Tarija, se Bol., Parag., and extreme se Brazil s to Catamarca, Tucumán, and Buenos Aires (Sta Cruz?), Arg. (*modulator*). Common.

TROPICAL MOCKINGBIRD *Mimus gilvus*, widespread, with 10 sspp, in lowlands from s Mexico through n and e S Am. to Rio de Janeiro Brazil, straggles to 2600 m on the Bogotá savanna, and to 2100 m in Tolima on e slope of C Andes, Col. (*tolimensis*). Recognized by its light gray upperparts and white underparts; the eye-brow white, wings and tail black, the tail much graduated and broadly tipped white. Inhabits semi-arid open country, gardens, and bushy pastures.

NOTE: Sometimes considered conspecific with Northern M. (*M. polyglottus*).

PATAGONIAN MOCKINGBIRD *Mimus patagonicus* – Plate IL 14

23 cm. **Bill fairly short**, tail slightly shorter than in Chalk-browed M., and less broadly white-tipped. Above **uniformly smoky gray**. Wings with 2 white bars, primary-coverts black tipped white, primaries blackish, outer ones edged white along the middle portion; inner remiges tipped whitish and edged buffy brown. Rump faintly tinged buff; tail blackish, all but c feathers broadly tipped white, outer web of outer feather narrowly edged white. **Short** broad **eye-brow** starting above eye whitish, thin line through eye blackish brown, rest of head-

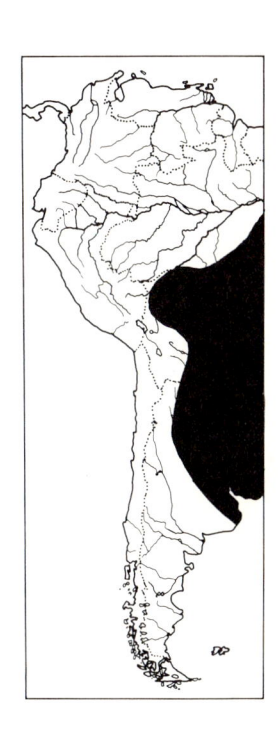

side pale gray-brown, ear-coverts tinged brownish. **Below pale drab gray**, upper throat and c belly whitish, sides and flanks washed cinnamon and sometimes with faint dark shafts, vent whitish or buff. **Juv.** has the breast spotted with blackish.

Habits: In pairs or family-groups. Watches alertly from bush-top, taking insects from the ground, where it may also hop around, but generally less terrestrial than Chalk-browed M. Sometimes cocks and wags tail. When alarmed takes to the inside of thorny bushes. Sings from top of bush or tree.

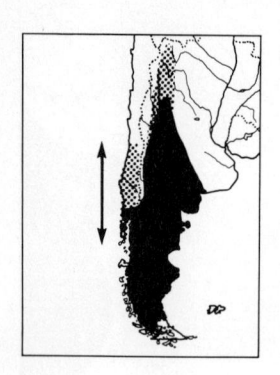

Voice: Song consists of weak, broken, fairly soft and level, warbling phrases with frequent imitations. Sings throughout year.

Breeding: Eggs Oct-Dec (s Chile), Jan (Tucumán); fledglings Feb (Sta Cruz), Nov (Neuquén). See also *Neotrópica* 22(1976): 103-4, and *Ibis* 120(1978): 61-5.

Habitat: Desert scrub, predominantly creosote bushes (*Larrea*); 'monte' woodlands of chanar (*Geoffroyia decorticans*), and mesquite-type trees (*Prosopsis*). Flood-plain stands of *Atriplex*.

Range: Lowlands to 1800 m. Breeds from Valdivia and Aysén, Chile, and n San Juan (Jujuy?), Córdoba, and s Buenos Aires, Arg., s to Isla Grande. In winter straggles n to Santiago and Jujuy. Fairly common to common.

WHITE-BANDED MOCKINGBIRD *Mimus triurus* – Plate IL 13

22 cm. **Unmistakable**. Above smoky gray. Wings black, most of **secondaries and their coverts white**, forming conspicuous band traversing the wing; median wing-coverts black tipped white; tertials and their covert edged cinnamon-gray. Rump cinnamon, 2 c tail-feathers black, rest white on inner web, and with decreasing amount of black on outer web towards outer feather, which is mostly white. **In flight large wing-patch and most of tail white**; short, broad eye-brow whitish. Underparts dirty white, throat and belly whitish, flanks washed cinnamon-brown, vent white to buffy. Eyes yellow to orange. **Juv.** presumably with spotted breast like others of the genus.

Habits: Alone or in pairs. Sings from tree-tops. During display the male runs a few steps on the ground while singing, then opens wings and tail and jumps up a foot or so, bringing the body nearly horizontal, then drops and runs a few steps further, and continues for a minute or longer.

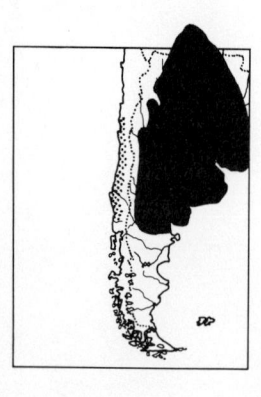

Voice: Song long, of loud resonant melodic phrases intermingled with mimicry of other species. Sings throughout year.

Breeding: In lowlands mainly in Dec.

Habitat: Chaco and 'monte' woodland, but not in desert scrub.

Range: Primarily lowlands, but nests up to 2600 m in Mendoza. From e Bol., extreme sw Mato Grosso and the extreme se Brazil, and Parag., s to n Chubut (see *Hornero* 11(1977): 436), Arg. Andean crossings can be inferred from 5 winter rec. in Chile (Vallenar Atacama, Zapallar Valparaiso, Santiago, n Pucón Cautín, and Valdivia; breeds?). Uncommon to fairly common.

BROWN-BACKED MOCKINGBIRD *Mimus dorsalis* – Plate IL 16a-b

24 cm. Above grayish rufous brown to grayish chestnut-brown; wings with 2 rather narrow white bars and a **large white patch formed by primary-coverts and base of primaries**; tertials edged with color of back and tipped white like secondaries. Rump pale rufous; **tail blackish, outer 4 feathers white**, the inner of these feathers with fuscous outer edge. Long eye-brow buffy white, lores and cheeks blackish, underparts whitish. Worn birds are gray-brown above with obscure dark streaks, and lack white wingbars and tips of remiges; underparts dirty white. **Juv**. has breast spotted with dark brown.
Habits: Alone, in pairs or family groups. Much like others of the genus. Feeds on the ground, often with cocked tail. During display flies up to spread wings and tail.
Voice: Sings from primo Dec. Song varied, mostly of harsh notes *terett terett tett tett....*
Breeding: Eggs Nov (Tucumán); fledglings Dec, Jan, Mar (Cochabamba), Feb (Chuquisaca); very worn ad. (presumably nesting) Jan (La Paz).
Habitat: Semi-arid regions. Bushy slopes, cactus country with scattered agaves, bushy slopes, agricultural areas with hedgerows. Often near villages.
Range: 1700-4200 m. From La Paz and Cochabamba, Bol., s to Tucumán, nw Arg. Sightings in Putre in Arica, nw Chile Nov 1986 (DS, BW). Common in Cochabamba and s Bol.

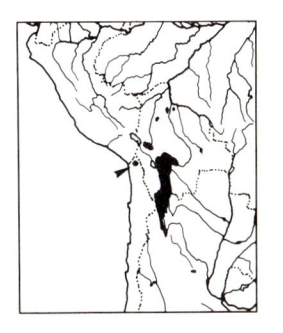

CHILEAN MOCKINGBIRD *Mimus thenca* – Plate IL 12

26 cm. Above gray-brown, rump browner (in worn plumage streaked dark brown above). Wings with 2 narrow white bars and edges of middle portion of outer primaries; primary-coverts black tipped white. Tail blackish, all but c feathers broadly tipped white, outermost feather edged white on outer web. **Long, broad supercilium whitish**. Line through eye, mottles on headside, and broad moustache blackish brown. Throat whitish, breast pale drab, sides and flanks tinged buff and streaked fuscous; belly and vent whitish with slight buffy tinge. **Juv**. presumably has spotted breast.
Habits: Alone or in pairs. Perches atop bushes, trees, and columnar cacti, taking most prey from the ground. When alarmed hides inside the vegetation.
Voice: Sings throughout year. Song includes much mimicry.
Breeding: Eggs prim Nov.
Habitat: Open scrub and heath-like vegetation covering large areas of rolling hills. Seldom in cultivated fields and orchards.
Range: Chile from sea level to middle elevations from Copiapó valley, Atacama s to Valdivia. Common.

LONG-TAILED MOCKINGBIRD *Mimus longicaudatus* – Plate IL 17

26 cm. Above gray-brown, streaked to spotted sooty black; nape with

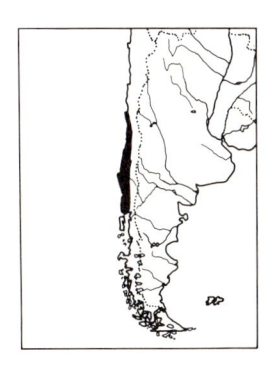

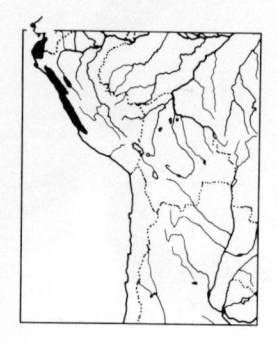

pale patch, esp. in males. Primary-coverts, 2 wingbars, edge of middle portion of primaries and tips of inner remiges whitish. **Very long tail** blackish, all but c feathers **broadly tipped white**, narrow edge of outermost white. Long eye-brow pure white, line through eye black, rest of headside mixed black, brown and whitish. Throat white, moustache black; breast gray-brown, feathers pale-tipped; lower underparts whitish, sides with a few fuscous streaks. **Juv**. has black-spotted breast. Behaves like other Mockingbirds.

Inhabits arid slopes with scattered thorny scrubs and trees, or hedgerows, gardens, and riparian thickets. Mainly lowlands, but ascends at least to 2600 m in many valleys in Peru. Plata isl. of w Ecu. (*platensis*), sw Ecu. to Piura (see Bond 1956), n Peru (*albogriseus*), along Pacific slope s to Ica, Peru (*longicaudatus*), and in the upper Marañón valley in La Libertad (*maranonicus*).

NOTE: May represent a ssp group (megassp) of Chilean M.

Thrushes – Family Turdidae

A cosmopolitan group of c. 307 species, sometimes treated as a subfamily of a larger assembly *Muscicapidae*, which includes Old World flycatchers, log-runners, babblers, parrotbills, rockfowl, and gnatcatchers. Recent molecular evidence suggest that many Old-World representatives in fact constitute a separate flycatcher family, while others belong with the crow-like passerines, and thrushes form a separate family related to mockingbirds.

All S Am. thrushes are well-sized birds with fairly long tails. A juv. plumage of usually spotted appearance is usually carried only a few weeks (3 months in Orange-billed Nightingale-thrush). Most species have the habit of flicking wings and tail. The solitaires are arboreal, but the true thrushes are partly terrestrial birds that move in strong hops. The diet consists of arthropods, worms, snails, berries, and fruit. They live in pairs. Usually only the male sings, and only during the breeding season. Some species are among the best singers of birds. The nest is an open cup, placed in a bush, tree, or on the ground. See Bond (1956).

WHITE-EARED SOLITAIRE *Entomodestes leucotis* – Plate L 8

22 cm. Long, slender, relatively small-headed, and with a long slightly graduated tail. Bill fairly short and broad at base. **Male**: **black** with **broad band below eye white**; back, rump, tertials, and flanks rufous chestnut; crown, esp. hindcrown, more or less tinged with brown. Outer tail-feathers mostly white, next 2 pairs with decreasing amount, 3rd with only the tip white. Pectoral tufts (probably exposed during display) and large patch on underside of primaries white. Mandible orange. **Female** with duller rufous colors and dark gray crown. **Juv**. undescribed.

Habits: Solitary and territorial. Sluggish. Quietly eats berries and keeps well hidden, hence difficult to observe. Shy. Only flies short distances to hide among foliage or behind mossy branches, where it may stay motionless for long periods.

Voice: Song a very high-pitched, ringing (almost telephone-like) wheen *zeeeeee*, repeated at 7 s or longer intervals.

Breeding: Fledgling Nov (Puno), Oct (Cochabamba).
Habitat: Middle strata of humid forest and cloud forest.
Range: 900-3350 m. From e La Libertad, Peru, s to Cochabamba, Bol. Locally fairly common.

ANDEAN SOLITAIRE *Myadestes ralloides* – Plate L 9a-b

16.5 cm. Bill short and broad at base. Ssp *ralloides*: mandible dusky with pale suffusions, sometimes almost wholly pale. Forehead dark gray turning bright cinnamon on rest of upperparts; wings with dark patch at base of secondaries. C tail colored like back, but **outwards the tail-feathers become increasingly white-tipped**, outer feather with terminal half mostly white. **In flight** shows white band across base of inner webs of remiges. Headside and **underparts dark gray**, c belly grayish white, flanks washed cinnamon. Ssp *venezuelensis* similar, but tawny above, mandible wholly yellow; *plumbeiceps* tawny above, whole crown gray, base of mandible yellow. Birds from Ecu. have hindcrown tinged brownish; *candelae* has dark tawny upperparts incl. crown, and darker gray breast, in greater contrast to belly. **Juv**. blackish brown densely spotted buff, c belly paler and grayer.
Habits: Solitary. Inconspicuous. For the most part sally-gleans or perch-gleans fruits and berries, but also flycatches and foliage-gleans insects. Very rarely on the ground. Regularly follows antswarms through the undergrowth.
Voice: Sings from lower canopy of tall tree, mainly at dusk and dawn, all year, but least outside breeding season. Song consists of 2 phrases of high-pitched, 'rusty' notes, 1st phrase of 3 notes, last note rising, after 5 s follows the 2nd phrase, of 2 notes, the last lowest. The cycle is repeated after 2-5 s. Female sings too, but not as well as male. Call a throaty clang *rraou*.
Breeding: Eggs Mar, July (Valle), Mar, Apr (Cauca), Mar (e Huila); juv. Nov (Mérida), Mar, Sep (Cundinamarca), June, Aug (Cauca), June (w Nariño), Jan (nw Ecu.), June, July (n Cajamarca), Aug (Huila, ne Ecu.), Oct (e La Libertad), Dec (Puno); juv./ad. Mar (La Paz, Pasco), June, July (n Cajamarca), July (w Nariño, Cauca), July, Aug, Sep (Huila), Nov (Huánuco).
Habitat: Lower to middle strata, 2-15 m up. Densely foliated parts of humid forest and cloud forest.
Range: 650-2900 m. From C and W Andes of Col. s to w Ecu. (*plumbeiceps*), head of Magdalena valley, Col. (*candelae*), Perijá mts, coastal mts and Andes of Ven s through E Andes of Col and e Ecu. to Peru n of the Marañón (*venezuelensis*), Peru s of the Marañón from Amazonas, San Martín, and e La Libertad s to Cochabamba, Bol. (*ralloides*). 2 sspp (sometimes considered 2 species) in Panamá with immediately adjacent Col., and Costa Rica. Fairly common.

SPOTTED NIGHTINGALE-THRUSH *Catharus dryas* – Plate L

2 – of humid premontane forest possibly reaches lower temp. zone locally. C Am. and w Ecu. (*equadoreanus*), E Andes of Col. along e Andean slope to Bol. (*maculatus*), and Tarija, Bol., to Jujuy, Arg. (*blakei*, Olrog in *Acta Zool. Lilloana* 30(1973):10). 14.5 cm. **Bill, legs and eyering orange**. Ssp *ecuadoreanus*: **head black**, rest of upperparts dark gray-

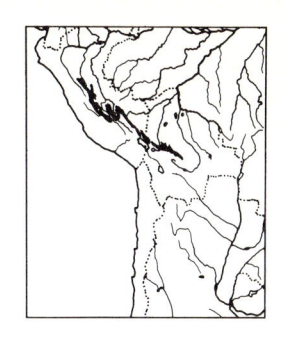

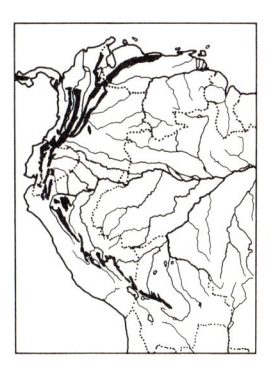

brown, c throat and rest of **underparts** below throat **apricot-yellow** (fades to whitish) spotted with blackish gray, sides of body dark gray, vent white. Ssp *maculatus* darker and browner above and heavier spotted below; entire throat yellowish and spotted; *blakei* has the c throat ochraceous. **Imm**.(?) has supercilium, lores and ear-coverts ochraceous streaked black. **Juv**. undescribed. Secretive and shy, keeps to the dark forest floor and interior of shrubs and undergrowth. Song repeated slightly guttural phrases, first 2 notes double and pleasant. Fledglings Feb (Tarija), Feb, Sep (Cajamarca); juv./imm. Apr (ne Ecu.).

SLATY-BACKED NIGHTINGALE-THRUSH
Catharus fuscater – Plate L 1a-b

18 cm. Crown flat. Ssp *mentalis*: **pale blue eyes with orange eye-ring, bill orange**. Mainly dark slaty gray, ranging from almost black on face to smoky gray below, throat smudged with buff, c belly whitish. Male averages darker and more richly colored than female. Other sspp larger and without brown tinge on throat, breast and upperparts, *sanctaemartae* darkest. **Juv**. uniformly brown, unspotted.
Habits: Elusive. Usually near ground in densest part of forest. At dawn may come out onto adjacent fields, trails, or other open places. Regularly half-cocks tail, and flicks wings upon alighting.
Voice: Song ventriloqual and melodic, 3 notes on a major scale followed by 2 higher notes (possibly given by another individual). Call a cat-like mew and a fuzzy, *Elaenia*-like *wheeety weer*. Alarm a low-pitched *khroum-khroum*.
Breeding: Eggs Jan (n Cajamarca), Dec (Puno); fledglings Feb (Puno), Apr (La Paz), July (nw Ecu.); juv. Apr-July (s Cajamarca), May (Cuzco).
Habitat: Near the ground in dense mossy forest, often near streams.
Range: 600-3250 m. Sta Marta mts (*sanctaemartae*); e Panamá, Perijá mts, Andes of Ven. from Trujillo to Táchira and w Barinas and from E Andes of Col. s to ne and w Ecu. (*fuscater*); W Andes of Col. at Herredura river, Frontino, Antioquia (*opertaneus*); n and c Peru from Piura and n Cajamarca w to Rio Zaña Lambayeque and s to Junín (Cuzco?) (*caniceps*), and from Puno, se Peru, to Cochabamba Bol. (*mentalis*). 2 sspp in Costa Rica and Panamá. Uncommon.

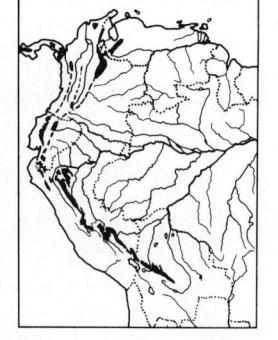

ORANGE-BILLED NIGHTINGALE-THRUSH
Catharus aurantiirostris – Plate L 5

15.5 cm. Bill and legs bright orange, narrow eye-ring orange. Ssp *aurantiirostris*: above olivaceous, wings light brown; headside like back, lores pale. **Below whitish, breast and sides pale gray**, flanks olivaceous, lower belly slightly tinged pale yellow. Ssp *barbaritoi* has upperparts olive-brown without rufescent shade; *phaeopleurus* with gray headside and crown grading into dark olive-brown back, breast and sides darker gray; *inornatus* with dark upperparts as in *phaeopleurus*, and pale underparts as in *aurantiirostris*; *insignis* with dull upperparts as in *aurantiirostris* and dark underparts as in *phaeopleurus*. **Juv**. has dark bill and the breast-feathers margined with dusky, giving spotted appearance.

Habits: Elusive. Forages for insects and worms and sometimes berries. Works through the dense undergrowth near the ground, but rarely on the ground. Often sits motionless for long periods. Sings well hidden low in a thicket, often while hopping about, occ. from an exposed perch. See *Publ. Michigan State Univ. Mus. Biol. ser.* 3(1964): 1-47.
Voice: Sings most of the year, but more consistently in the breeding season, and usually not right after. The song (varying geographically) is an abrupt, fairly high-pitched, explosive warble of 3-5 syllables, lasting 2 s or less, and repeated at 5-7 s intervals. Does not sing at dusk and dawn. Calls include a short, high- or low-pitched, sharp or soft *chirp*, a soft throaty *chirr-rr*, and a wren-like scolding chatter.
Breeding: Eggs Mar-May (Cauca); fledglings Apr, Aug (Cauca); juv. June, Oct (Cauca); large gonads Apr-July, Nov (Col.). Does not normally breed in the dry season Dec-Feb (Cauca).
Habitat: Thicket-studded pastures and fencerows, often near houses, bamboo thickets, plantations, dense understory of low rain forest, and cloud forest.
Range: 600-2900 m. From Cauca and upper Patía valleys s to the upper Guaitara valley w Nariño, Col. (*phaeopleurus*), Sta Marta mts (*sierrae*), e slope of E Andes in N.d.Santander Col., Andes of Táchira, Mérida (K Malling Olsen), and s Lara and coastal mts Ven. (*aurantiirostris*); Perijá mts (*barbaritoi*); w slope of E Andes of Col. in Sogamoso valley, Santander (*inornatus*), upper Magdalena valley and w slope of E Andes in Boyacá and Cundinamarca, Col. (*insignis*). 1 ssp in coastal Ven. and Trinidad, 7 sspp in C Am. Locally common.

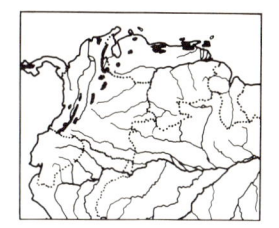

NOTE: Forms of C Am. and ssp *phaeopleurus* were previously considered a distinct species Gray-headed Nightingale-thrush (*C. griseiceps*).

SWAINSON'S THRUSH *Catharus ustulatus* – Plate L 3

16 cm. Bill fairly short. Ssp *swainsoni*: above olive-brown, **lores and eyering buffy white**. Throat and breast buffy, sides and flanks olive-brown, rest of underparts white; moustache, triangular spots on lower throat and breast dark brown becoming paler on flanks and upper belly. Ssp *almae* with more grayish upperparts, sides, and flanks. - Alone, or in flocks of up to 50 during migration and occ. in winter. Actively hovers and flutters for berries, e.g. of *Lantana* (and insects?), usually well inside canopy, occ. low or on the ground. May follow army ants. Flight swift and dashing; sometimes stiffly twitches tail. Call a quick, soft, watery, ventriloqual *pyt*, *pwink*, *pwk*, *pwet*, or *whit*, resembling Alder Flycatcher rather than a thrush, and usually accompanied by a flick of the wings. Sings from late April. Song delivered rapidly, rising in 2 or 3 stages, and terminated in a high-pitched, weak, warbling trill.

All levels, but mostly at mid-heights. Humid forest, forest edge, clearings, riparian thickets, plantations, gardens, second growth, and woodland. Breeds in N Am. Winters Oct-Apr from Mexico s to w Peru (occ. to sw Peru) and nw Arg., mostly in lowlands, but up to 2400 m in Bol., and during migration up to 3000 m in Col., 2900 m in Ecu. and 3300 m in Peru (*swainsoni*). Fairly common in premontane zones. 3 other sspp winter in C Am. (1 rec. of ssp *almae* from Lima Peru).

VEERY *Catharus fuscescens*, another N Am. migrant wintering in S Am. might turn up in the temp. zone of the Andes (known to 2300 m in Col.). Resembles Swainson's T., but is browner above and has less distinct eye-ring, and more faint spots below.

GRAY-CHEEKED THRUSH *Catharus minimus* – Plate L 4

16 cm. Ssp *minimus*: above olive-gray; below white, sides smoky gray, breast faintly tinged buff; moustache and triangular spots on breast dark brown, becoming pale on flanks and belly. Best told from Swainson's T. by lack of pale lore and eye-ring, grayer upperparts, and by a more distinct spotting below, on whiter background. - Very shy. Solitary. Generally silent. Call a flycatcher-like weak *wheesp*, or a downslurred *weeah*. Song thin and nasal, often rising sharply at end.

Usually near the ground, but sometimes at middle strata. Thickets along small shallow streams or densely tangled swales; brush-covered, log-strewn, relatively open parts of woodland, or in shrubs in clearings with scattered trees and breast-high ground-cover. Breeds in N Am. Winters Oct-early May from Nicaragua to Guiana, nw Brazil, and n Peru, mainly in trop. and subtrop. zones, but during migration rec. to 3000 m in Ven. and in the Bogotá area (*minimus*). Another ssp winters on Hispaniola. Uncommon.

YELLOW-LEGGED THRUSH *Platycichla flavipes* – Plate L 7a-b inhabits mid- to top strata of humid premontane forests, plantations, and gardens to 2500 m. Sta Marta mts, e slope of E Andes in N.d.Santander, Col., n and w Ven. to Gran Sabana and Alto Paragua (*venezuelensis*). 4 other sspp in Tepuis, coastal Ven., Trinidad, Tobago, and se Brazil. Partially migratory? 19 cm. **Bill fairly short**. Iris brown, bill and eye-ring orange-yellow, legs yellow. Ssp *venezuelensis*: **male** is mainly black, with gray back and posterior underparts; chin whitish, under tail-coverts tipped white, scapulars tipped black. **Female** olive-brown, rump grayer olivaceous; below drab gray heavily washed with clay-color, except on the belly; under tail-coverts marked with whitish. **Juv. male** pinkish slate above, feathers tipped black; feathers of supercilium and c back with buff shaft-streaks; wings black, coverts tipped buff. Below buff, c belly whitish, entire underparts barred with blackish brown. Alone or in pairs. Shy. Call *tsrip*. Sings from treetops, with loud, liquid, clear, and varied phrases at regular intervals. Often mimics other birds.

PALE-EYED THRUSH *Platycichla leucops* is mainly premontane, although rec. to 2900 m in Col. It ranges through the Tepuis of Ven. and adjacent Guiana and Brazil, and Andes from Lara, Ven. through Col. (both slopes of W Andes in Valle, upper Magdalena valley in Huila, e slope of E Andes in Macarena mts and Meta), e and w Ecu. and Peru to Sta Cruz, Bol., in humid forest and second growth. 18 cm. **Male**: **black**, bill and legs yellow, **iris bluish white**. **Female**: bill and eyes dark, narrow eye-ring yellow. Above dark brown, earcoverts with paler shafts; chin grayish white turning buffy on the virtually unstreaked throat; faint submalar streak brown; breast and sides brown, sides washed rufescent; c lower breast and belly dirty light gray, vent buffy gray. Secretively eats berries in lower to middle strata. Sings from treetops.

GREAT THRUSH *Turdus fuscater* – Plates L 16a-g, LXIII 20

28-30 cm (*gigas* largest). In zone of overlap best told from Chiguanco T. by **longer tail**, different habitat, and (except for Bol.) by darker plumage and eye-ring in ad. male. **Male**: Above blackish (*ockendeni*), dark gray-brown (*fuscater, gigantodes, quindio*), olive-brown (*gigas, cacozelus*), or light olive-brown (*clarus*), crown darkest in *fuscater*. Below somewhat lighter (c belly pale gray to gray in *fuscater*); throat pale with dark streaks (except in *ockendeni* and sometimes *gigantodes*). Eye-ring yellow, bill orange-red, legs orange. **Female** similar, or slightly lighter and browner with dark-streaked throat in *gigantodes*; female *fuscater* without darker crown and with uniform gray-brown underparts. Eye-ring dark (rarely yellow in older females), bill orange (orange-red in *ockendeni*), legs brownish yellow (orange in *ockendeni*). **Imm**. lighter and browner than ad., streaks of throat more or less continuing onto breast and belly; eye-ring dark, bill dull yellow, legs yellow (male) or brownish (female). **Juv**. with wing-coverts narrowly tipped buff; underparts buffy-white (*gigantodes*) or throat buffy-white in sharp contrast to smoky brown lower underparts (*quindio*), feathers with dusky tips giving a barred effect, but sometimes appearing spotted on upper breast. Eye-ring dark, bill yellow washed horn, legs brownish with slight yellow wash. Under wing-coverts dark in ad. *fuscater, ockendeni, gigantodes*, and *quindio*, but more or less cinnamon in lighter-plumaged birds (imm., juv., and northern sspp).
Habits: Alone or in pairs in breeding season, and then territorial; in non-breeding season often in small flocks, sometimes up to 40. Plucks berries at all levels and regularly feeds on the ground. May roost in flocks in trees, sometimes in tules in marshes. Sings from a branch at mid-levels. Regularly flicks tail.
Voice: The relatively loud song is only given in the breeding season, and mainly at dusk and dawn. It consists of less varied phrases than in Chiguanco T., and often has many clear whistles. Locally the repertoire may be limited and each phrase repeated many times. Alarm repeated *kee* or *tjuck* notes. Call a loud, liquid *sleeu*.
Breeding: Eggs Mar, Apr (Mérida), Apr and to a lesser degree Nov (Cauca), June (Cajamarca), Feb (Huánuco); fledglings Jan (Boyacá), Feb (Huánuco), Sep (Cundinamarca); juv. Jan, Feb, July (Sta Marta), Sep (Norte de Santander), May, Oct (Santander), Mar, May (Boyacá), Feb, Apr, June, Oct, Dec (Cundinamarca), Aug (Antioquia), Oct (Quindío), Apr, June (Cauca), July (w Nariño), May (n Ecu.), Sep (sw Ecu.), Aug (e La Libertad), Mar (Huánuco), July (Cuzco), Feb, July (La Paz), May (Cochabamba).
Habitat: Generally in forest, cloud forest, and at forest edge in humid zone, but where Chiguanco T. is absent it also moves well into páramos and other open country, and sometimes in garden trees in semi-arid

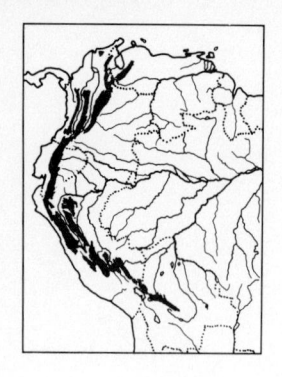

regions. Common in gardens of Quito. Overlaps with Chiguanco T. in *Polylepis* woodland, and both occur in gardens of La Paz city.
Range: 1500-4250 m. Sta Marta mts (*cacozelus*); Perijá mts (*clarus*); from Lara, Ven. s to E Andes of Col. (*gigas*); W and C Andes of Col. s to n Ecu. (*quindio*), from s Ecu. s in W Andes to n Lima, and through C Andes to Junín, c Peru, birds from both slopes of the Mantaro in Huancavelica and from Apurímac (and Vilcabamba mts?) probably also referable here (*gigantodes*); Cuzco and Puno, se Peru (*ockendeni*); La Paz and Cochabamba, Bol. (*fuscater*). Common.

NOTE: see NOTE under Chiguanco T.

CHIGUANCO THRUSH *Turdus chiguanco* – Plate L 15a-c

26 cm. **Best told from Great T. by shorter tail and different habitat**. Ssp *chiguanco* and the doubtfully distinct *conradi*: sexes alike; above olivaceous gray, below smoky gray, palest on throat and belly; throat faintly streaked fuscous, but streaks narrower than in Great T. Iris reddish, **eye-ring black**, bill and legs yellow. Under wing-coverts mostly cinnamon. Ssp *anthracinus*: **male blackish**; bill and legs orange-yellow, **eye-ring yellow**. In fresh plumage feathers of throat and breast with slightly lighter edges, appearing faintly scaled; under wing-coverts dark. **Female** paler and duller, throat more or less streaked with dusky; under wing-coverts dark; bill washed with dark at base. **Imm**. like female, but duller, more olivaceous; primaries edged paler; under wing-coverts tinged cinnamon. Plumage of female and imm. almost indistinguishable from Great T. **Juv**. spotted and barred with buffy throughout, wing-coverts with pale buff tips along shafts. Birds from s part of range may have longer bills.
Habits: Alone or in loose pairs. For the most part terrestrial; characteristic posture with protruding breast, somewhat lowered wings, and slightly raised tail. Regularly flicks tail. Sings from branch or telephone-line.
Voice: Only sings in the breeding season, and mainly at dusk and dawn. Song consists of pleasant, melodious, whistled phrases of 3-6 notes, each phrase repeated 2-3 times. Alarm in flight a loud, rapid *tsi tsi tsi*.... Call *duck-duck*. Alarm an up to 2 s long, high-pitched wheen. Juv. may give a short dry House Wren-like trill.
Breeding: Eggs Nov (Tucumán), Jan (Atacama), Mar, Apr (Arica); fledglings Mar, Apr (Tucumán), Jan (Tarija), Jan, Feb, Apr (Cochabamba), Jan, May (Puno), June, July (Cuzco), June (Arequipa), May (Lima), Apr, June, July (Junín), June (Huánuco), May (Amazonas), Apr, July (Cajamarca), Apr (Chimborazo); large gonads Dec (Cuzco).
Habitat: Open country with an occ. bush, or in bushy gorges in the puna zone, where it forages on short grass; in more arid regions mainly along streams, in irrigated farmland, and deciduous woods with cacti. Also in *Polylepis* woodland, montane shrub, town parks, and gardens, most arboreal in the far s. Overlaps with Great T. in *Polylepis* woodland and at forest edge.
Range: 2000-4300 m, locally down to coast in wc and sw Peru. From Chimborazo, Ecu., s to c Peru ('*conradi*'), coastal c Peru and from sc Peru s to Tarapacá, n Chile, and n La Paz, Bol. (*chiguanco*). 1000-3500

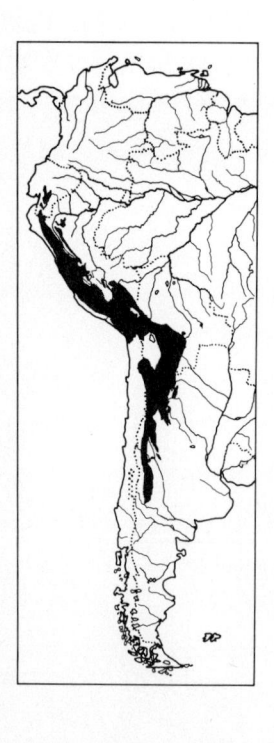

m from La Paz, Cochabamba (in the valleys, and only scarce along e edge of Bol. Altiplano), and Sta Cruz, Bol., s to Mendoza, San Luis, and Córdoba, Arg.; also breeds in Antofagasta and at Salar de Atacama, and straggles to Santiago, Talca, and Bío Bío, Chile (*anthracinus*). Common.

NOTE: Chiguanco T. (sspp *chiguanco* and *anthracinus*) and Great T. (ssp *fuscater*) seem to grade into one-another in Bol. (see Fjeldså & Krabbe 1989). 1 bird from Sandinilla in La Paz and 1 from Putre in Tacna n Chile are said to be intermediate between *chiguanco* and *anthracinus*. 4 *chiguanco* from Chicani on n slope of La Paz have larger feet, longer tail, browner hue, and less ochraceous underwings than typical birds. Birds from the Tunari range in Cochabamba, Bol., are very variable and appear intermediate between *anthracinus* and Great T., and also individuals with *chiguanco* traits as partly rufous under wing-coverts. An alternative interpretation is that a large, long-tailed and streak-throated ssp of Chiguanco T. inhabits the upper Andean slopes of La Paz and Cochabamba. Further study is needed to disclose the relationships of these forms, and to state whether they in fact represent 'ring species', connected in Bol. on the ecotone between humid and drier zones (although they appear as full sympatric species in Ecu. and Peru).

GLOSSY-BLACK THRUSH *Turdus serranus* – Plate L 6a-c

23 cm. Ssp *atrosericeus*: **male glossy black**. Iris dark brown, bill and eye-ring orange, legs orange-yellow. **Female** dull olive-brown above with little if any rufescent tinge on forehead and with dark olive-brown or fuscous (not rufescent) tail. **Below brownish gray, slightly washed with olivaceous or dull brownish on fore-neck, chest, and flanks**; under wing-coverts orange ochraceous. **Imm. female** above strongly rufescent, brightest on forehead and wings. Below, except for grayish median portion likewise rufescent, paler than upperparts. Ssp *fuscobrunneus*: **male** like *atrosericeus*, but bill smaller; **female** intermediate towards *serranus*; darker and more rufescent than *atrosericus*; darker, less rufescent, more olivaceous clove-brown, and slightly duller than *serranus*; **below raw umber brown tinged cinnamon**, throat somewhat paler and grayer, with faint dark streaks; under wing-coverts dark cinnamon. Bill blackish brown with yellow wash, eye-ring brownish, legs straw. Ssp *serranus* like *fuscobrunneus*, but wings and tail longer, female even brighter and more rufescent. **Juv.** dark olive-brown above, tail dusky; crown and back with pale buff shafts, densest on forehead; wing-coverts tipped buff; below buff, feathers tipped dark brown.
Habits: Alone or in pairs. Feeds on berries in the canopy and sometimes hops on the forest floor. Male sings hidden in an epiphyte in treetop.
Voice: Song given for one hour at dawn and again at dusk: short, fairly high-pitched interrupted phrases, each strophe only slightly different from the preceding. Alarm at dusk a rasping *rrrrrt-rrrrrt*. Flight call *tjick tjip-tjip-tjip-tjip-tjip* or *tjick tjick tji-tji-tji*.
Breeding: Eggs Apr, May, July, Oct, Nov (Valle); fledglings June (Huila), Feb (ne Ecu.), June (se Ecu.), July, Aug (n Cajamarca, Huánuco), Sep (Cochabamba); juv. July (Cundinamarca), Sep (nw Ecu.),

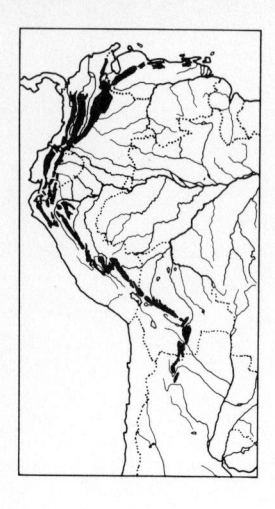

Sep, Oct (ne Ecu.), Aug (Cajamarca, Amazonas), Oct (e La Libertad), Mar (Pasco), Dec (La Paz), Jan (Cochabamba); juv./imm. Mar, July, Aug (Cauca), July (ne Ecu.), June (Cajamarca), July (Huánuco); large gonads Mar-Aug (C Andes Col., Perijá).
Habitat: Humid primary forest and tall cloud forest. In Ven. also in second growth, open fields, gardens, and clearings. In nw Peru also in dry forest.
Range: 900-3500 m, mainly below 3000 m. Perijá mts, Andes of Ven. from Táchira to Lara and coastal mts to Caracas (*atrosericeus*), páramo de Tamá, all 3 ranges of Col., and both slopes of Ecu. (*fuscobrunneus*), from n Peru (incl. w slope of W Andes to Cajamarca) s along Andean slope of Bol. to Salta and Jujuy, nw Arg. (*serranus* and doubtful *unicolor*, see Olrog in *Acta Zool. Lilloana* 27(1970):255-66). 1 ssp from Mexico to Honduras, 1 in coastal Ven. Fairly common.

RUFOUS-BELLIED THRUSH *Turdus rufiventris* – **Plate L 13** lives mainly in lowlands, but reaches 2600 m in Cochabamba, Bol. From Córdoba and Buenos Aires Arg. n to e Bol. and s Bahia e Brazil (*rufiventris*) and ne Brazil (*juensis*), in gardens, parks, open wood, agricultural land with tree groves or patches of bushes. 23 cm. Eye-ring yellow-ocher, bill pale olive-green, legs gray to purplish brown. Above dark olivaceous; throat white streaked dark brown; breast buffy gray-brown, **belly orange-rufous**; wing-linings orange-rufous. **Juv.** like ad., but wing-coverts with buff shafts and distal spots; feathers of underparts tipped dark brown; headside mottled with whitish; bill paler than in ad., legs more ashy. Often on the ground, feeding, but hides in thickets or in the foliage of trees. Sings from leafy branches in tops of low trees: loud, continuous, varied, locally more simple and monotone. Call *pup-pup* or *dru-uip*, and 'laughing' calls at dusk. Eggs Oct-Feb (Arg.); fledglings Jan, Feb (Tarija); juv./imm. Feb, May (Sta Cruz).

CHESTNUT-BELLIED THRUSH *Turdus fulviventris* – **Plate L 14**

23 cm. Eye-ring orange-yellow, bill yellow-orange. **Male**: entire **head incl. throat black**, upperparts dusky-gray, breast smoky gray; rest of **underparts ochraceous orange**; under tail-coverts dark gray with buff streaks and tips. Legs yellowish. **Female** similar, but legs fleshy brown; head darker and browner than back; throat dusky brown, feathers narrowly edged buffy. **Juv.** ? - Alone or in pairs. Usually feeds on berries in low vegetation. Song a notably varied series of choppy phrases with many trills and buzzes *che'e'e-chert chee-rt-ee e'r'r' chu-wurt titi t't't eet....* Call a wooden *peent*. Fledgling Feb (Cajamarca); juv. July (Perijá), June (Putumayo), Feb (ne Ecu.), Mar (Mérida); juv./imm. July (ne Cajamarca); large gonads Apr-Aug (Perijá), June (Putumayo).

Stunted cloud forest with many ericads and shrubby disturbed areas, at 1300-2700 m. Perijá mts and from n Trujillo, Ven. s along E Andes to Piura and n Cajamarca, n Peru. Generally uncommon, but fairly common in Condor mts in ne Cajamarca, n Peru, and w Putumayo, Col.

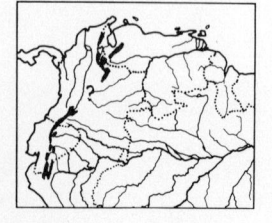

AUSTRAL THRUSH *Turdus falcklandii* – Plate L 11a-b

23 cm. Sexes alike. Bill and legs yellow. Above gray-brown, grayest on rump; crown and headside blackish. Throat white streaked blackish, breast and sides buffy gray-brown, belly buff to pale buff, under tail-coverts whitish; under wing-coverts pale buff. Plumage considerably paler with wear. **Juv.** has fore-crown tinged dark brown and back with pale shaft-streaks and blackish feather-tips. Some birds grayish above with almost whitish markings and whitish below, feathers tipped blackish; others are brownish above with buffy markings and deep ochraceous below with blackish feather-tips forming heavy spots.
Habits: Alone or in pairs. For the most part forages for worms, snails, and arthropods on soft, preferably bare soil in shady places, but also eats fruit. Sings from treetop well into the night. Not shy.
Voice: Song weak and ventriloqual, consisting of muted whistles, notes in varying number of order, incl. mimicry, and given continuously for many minutes at dawn. Calls include a low-pitched *huit*, a harsh *wreet* and a *trrrt trrrt*.
Breeding: 2, sometimes 3 broods. In c Chile first eggs Sep, last as late as Feb; on Isla Grande 1st in Oct-Nov, last Dec-Jan (Feb).
Habitat: Open understory of dark *Nothofagus* forest to open forest, plantations, second growth, gardens, parks, open woodland, riparian wilows and brushy country.
Range: From coast to 2150 m. Juan Fernandez isls and from Atacama, Chile and s Buenos Aires, Neuquén, and Río Negro, Arg. s to Isla Grande (*magellanicus*, incl. '*pembertoni*'). Nominate ssp on Islas Malvinas, another ssp on Mocha isl, Chile. Common.

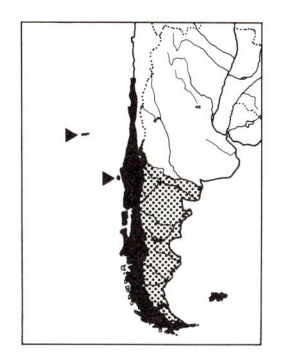

CREAMY-BELLIED THRUSH *Turdus amaurochalinus* – Plate L 12a-b

22.5 cm. Eye-ring dark; bill yellow in male, blackish (in older birds with some yellow) in female. Above gray-brown, grayer on rump, lore blackish brown. Below white tinged gray-brown on breast and sides; throat streaked dark brown, extreme lower throat white, unstreaked; under wing-coverts pale fulvous. **Juv.** browner; upperparts with buffy shafts and with buff spots on wing-coverts; underparts buffy white with buff breast and spotted with dark brown throughout; bill dusky. - Alone or in pairs, sometimes in small, loose flocks. At times runs about jauntily jerking wings and tail, but generally less terrestrial than most others of the genus. Song dry and fairly monotonous, composed of short simple phrases given at regular intervals. Calls include *bock*, *back*, *psib*, and *pchuo*. Fledglings Feb, Mar (Cochabamba), Mar (Sta Cruz, Tucumán).

Inhabits thorny thickets, open forest, and gardens in semi-humid to semi-arid regions. Primarily lowlands, but occurs at least to 2600 m in Cochabamba and Chuquisaca Bol., Brazil, se Peru, Bol., Parag., n and c Arg. s to Río Negro and Buenos Aires. 1 rec. from Atacama (Vallenar), Chile.

BLACK-BILLED THRUSH *Turdus ignobilis* of the premontane

zone from Col. to n Bol. has been rec. to 2800 m in E Andes of Col. It is found in clearings, parks, gardens, and lighter woodland, occ. humid forest or forest edge. 24 cm. Dingy with black bill. Uniform dark dull brown above; throat white streaked dusky, gradually becoming pale olive-brown on breast; c belly and vent white; under wing-coverts pale buff. Ssp *debilis* of e slope of E Andes Col. with more sharply streaked throat, and a narrow whitish band across lower throat; breast paler and grayer.

WHITE-NECKED THRUSH *Turdus albicollis* – widespread in humid forest and second growth in lowlands of most of S Am., ascends to 2400 m in Yungas of La Paz, Bol. (*contemptus*). 22.5 cm. Dark olive-brown above, tail blackish; **throat white streaked dark brown, lower throat white, unstreaked**; breast and sides pale gray-brown, flanks tinged fulvous, c belly and vent white; under wing-coverts buff. Best told from Creamy-bellied T. by much more conspicuously streaked throat, with more extensive white lower part.

SLATY THRUSH *Turdus nigriceps* – Plate L 10a-c

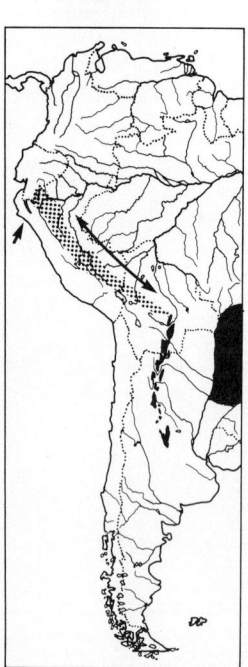

19 cm. **Male**: **crown, headside and tail blackish, remaining upperparts plumbeous** (crown sometimes dark plumbeous). Throat white streaked black, rest of underparts gray, c belly and most of under tail-coverts white. Iris dark brown, eye-ring yellowish green, bill yellow, legs pale yellow. **Female**: similar, but more brown, olive-brown above, slightly rufescent on forehead. Iris dark brown, eye-ring dark yellowish brown, bill dusky brown suffused with dark yellow, legs buffy. **Juv. male**: above dark olivaceous, slightly browner on head, hindcrown and back with a few pale shafts, wing-coverts tipped buff; below buffy to white on c belly, feathers tipped blackish brown; vent as in ad., but faintly tinged buffy. Under wing-coverts cinnamon. Bill dusky, legs dark brown. - Alone or in pairs. Stays inside dense vegetation, hence difficult to observe. Call *tsok*. Sings from well-hidden, fairly low perch. Song 2 alternating, only slightly different, unmusical notes repeated endlessly *tji tjihe tji tjihe...* or *tjie tjit tjie tjit....* Eggs Nov, Dec (Tucumán); fledglings Jan (Jujuy), Jan, Feb (Tarija); juv./imm. Apr (Sta Cruz).

2-4 m up in shady parts of the vegetation. Along streams in dense humid shrubbery and wooded ravines, sometimes alders (*Alnus*). Premontane zone, but up to 2550 m in Cochabamba, Bol., and to 2000 m in Tucumán, nw Arg. From se Ecu. and n Peru (where apparently breeds on w slope in Piura and Cajamarca) s along e slope of Andes to Córdoba, Arg. Southern breeders migrate n and return to breeding grounds in Oct (Tucumán). E Peruvian birds probably migrants only (see *Bull. Brit. Orn. Club* 107 (1987): 184-189) (*nigriceps*). Another ssp in se Brazil, Parag., and Misiones, Arg. Fairly common to uncommon in s part of range, rare in Peru and Ecu.

NOTE: Sometimes considered 2 (allo)species, Andean S-t. (*T. nigriceps*) and Eastern S-t. (*T. subalaris*).

Waxwings and allies – Family Bombycillidae

7 species, the waxwings circumpolar in the northern hemisphere. They are apparently related to thrush-like birds.

CEDAR WAXWING *Bombycilla cedrorum* – Plate XLII 1 of N Am. has straggled to Chocó (Feb), Valle (Jan) Col., and Perijá mts (at 1650 m). 15.5 cm, with pointed crest. Generally soft grayish cinnamon, rump light gray; tail gray turning blackish towards yellow tip; wings dark gray, inner remiges with shafts ending in red, varnish-like long drops. Narrow mask over the bill and around eye black narrowly outlined in whitish, chin blackish. Belly pale yellow, vent white. **Imm.** lacks red in wings and has narrower and paler yellow tailband. Usually occurs in small flocks feeding on berries at edge of humid forest and in second growth.

Wagtails, Pipits and allies – Family Motacillidae

52 species of which only the pipits are found in S Am. The 34 species of pipits have a worldwide distribution and inhabit open country. They are sparrow-sized birds with a light build and streaked plumage, and many species are extremely difficult to tell apart. The sexes are alike. Primarily terrestrial, they actively walk about in search of insects and have a habit of wagging the body and tail, though not as striking as in the longer-tailed wagtails. Most species have nuptial flights in which the male flies high up and then descends with vibrating raised wings and tail while singing. Both sexes tend the nest, which is an open cup placed on the ground among tufts of grass. Recent biochemical evidence places them near the old world weaverbirds (Ploceidae). For Peru see *Amer. Mus. Novit.* 1649(1953) and Bond (1956). See also *Hornero* 2(1921): 180-93.

SHORT-BILLED PIPIT *Anthus furcatus* – Plate XLVIII 15

14 cm. Ssp *brevirostris*: bill rather short and thick at base, head round. Crown blackish brown with small buffy streaks, eye-ring and **supercilium continuiung around ear-coverts whitish**, giving **pale-headed appearance**. Above blackish brown, nape with dense short buffy streaks; back-feathers with buffy edges and narrow tips, **appearing somewhat scaled** (esp. in fresh plumage); wings dusky, primaries narrowly edged white, wing-coverts, secondaries, and tertials edged buffy. Below white, lower throat, breast, and sides faintly tinged buffy, breast and sides streaked dusky, the breast-band relatively narrow, but densely streaked, **sharply demarcated from white belly**; long spots or streaks on sides often concealed by wings. Tail dusky, outer 2 feathers mostly white. Under wing-coverts white with faint buffy tinge; inner webs of primaries broadly edged whitish at their bases. Legs pale pinkish buff, hind-claw 10 mm, somewhat curved. **Juv.** undescribed.

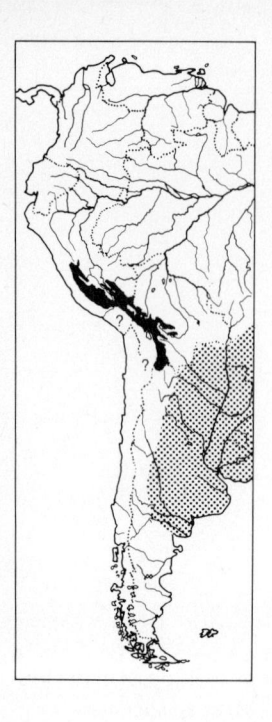

Habits: Alone or in pairs, outside breeding season in scattered flocks. Unlike Correndera P. (with which it often associates) does not wag tail. Otherwise forages much in the manner of other pipits. During nuptial flight flies high for 10-30 minutes, diving during each song-bout. Only rarely alights on twigs. Confident.
Voice: Flight call a fairly loud *tlit* or *tjip*. Song during nuptial flight (Brazil) **tzri**-*ze ze ze si si si si* or **tri**-*chi chi chi chi chi* repeated at 5 s intervals.
Breeding: No data.
Habitat: Puna zone. Lake-sides with short grass, meadows, flats, or rolling hills with bunchgrass interspersed with short grass. Fields. Generally in more arid country than other Andean pipits.
Range: 3500-4250 m, down to 2300 m in s Bol. From Lima (and probably Ancash), Junín, and Huánuco in c Peru s through w Bol. to Jujuy, nw Arg., but not yet rec. in Chile (*brevirostris*). Nominate ssp in lowlands from Parag. and s Brazil s to Mendoza and Río Negro, Arg. Fairly common.

HELLMAYR'S PIPIT *Anthus hellmayri* – Plate XLVIII 16a-b

14 cm. **Much like Paramo P.**, but slightly less buffy, streaks above paler; usually tips of wing-coverts are paler than the patch on secondaries. **Breast and sides with more extensive, broader and darker streaks**; throat almost pure white. Paler and less streaked and also with paler legs than Correndera P. Ssp *hellmayri*: above dusky streaked buff and buffy white; supercilium and eye-ring whitish streaked dusky; wings dusky with buff edges to coverts, appearing as **2 fairly distinct wingbars**; primaries narrowly edged whitish, secondaries edged buff appearing as a patch, tertials edged whitish. Tail dusky, outer feather mainly buffy white, next with a tiny white dot on tip (lost with wear); below buffy white, breast pale buff (or in worn plumage whitish, breast tinged buffy); breast and sides streaked dusky, streaks on sides conspicuous. Under wing-coverts buffy white, inner webs of remiges faintly edged dull rufous gray. Legs pinkish flesh, hind-claw 11 mm, somewhat curved. Ssp *dabbenei* similar, but penultimate tail-feather with some white at tip, even in worn plumage. **Juv**. undescribed.
Habits: Much like other pipits. During nuptial flight (Brazil) the bird sings while ascending almost vertically.
Voice: In Brazil, call *sclip* or *srip*; song varied, e.g. *zilid zilid* ..., *zidel-zi zi* **arrr**.
Breeding: Nov-Jan (Bol.).
Habitat: Open grassland. Peruvian specimens have been collected in fairly humid places, the Brazilian ssp in drier country than most other pipits (fields with short grass, arid land).
Range: 1650-3400 m. E Andean slopes from Puno, se Peru, to Tucumán, nw Arg. (*hellmayri*); highlands of w Neuquén, w Río Negro and w Chubut, Arg. and adjacent Chile (1 rec. from Lonquimay, Malleco); in winter migrates n to Sta Fé, La Rioja and Tucumán, Arg. (*dabbenei*); from Rio de Janeiro se Brazil to Buenos Aires, Arg. (*brasilianus*). Fairly common in Tucumán.

CORRENDERA PIPIT *Anthus correndera* – Plate XLVIII 14a-c

14 cm. Heavily streaked. **Legs fairly dark** (fleshy brown to vinaceous drab). Ssp *calcaratus*: bill long, head flat; **above** dusky **streaked cinnamon-buff**, long streaks on scapulars buff; wings dusky with buffy white edges to coverts (appearing as 2 wingbars) and inner remiges, primaries narrowly edged white. Eye-ring and supercilium whitish. Throat white with black moustache; **breast, upper belly, and sides buffy, heavily and broadly streaked** with blackish (streaks on sides rarely covered by wings); rest of belly and vent white with very faint, buffy tinge. Tail dusky, **outer 2 feathers mainly white**. Inner webs of remiges faintly edged pinkish gray to grayish white. Ssp *catamarcae* has somewhat shorter bill and less cinnamon-tinged upperparts, long streaks on scapulars whitish, breast only faintly tinged with buff. Ssp *chilensis* like *calcaratus*, but less white in tail, only outer feather mainly white, the penultimate just white along shaft; long streaks on scapulars pale buff; breast slightly less streaked, but a few narrow spots extend to upper belly. Ssp *correndera* much paler; long streaks on scapulars whitish; outer tail-feather mainly white, penultimate white along shaft, 3rd sometimes with white along tip of shaft. Hind-claw 16 mm, nearly straight. **Juv**. have primaries edged pale yellowish, wingbars white, feathers of upperparts edged whitish instead of cinnamon-buff, supercilium tinged yellowish.
Habits: Alone or in loose pairs. For the most part on the ground, but may perch atop a bush, a fence-post, or a tussock. During nuptial flight flies 30 m up, then circles a number of times while singing, and making short, downward glides, and finally a long glide to the ground with raised wings and tail while vocalizing. Not shy.
Voice: Flight call a distinct harsh *trrrit*, rarely *rrit rrie*; in Patagonia *srit* or *tsrrr*. During nuptial flight *tzi ti trrrrr* or *tzi ti tie tzi ti trrrrrr*, the dry trill at the end given during glides; the trill is prolonged during the long, final glide. Sometimes sings from the top of a tussock, giving abrupt, varied phrases *rrit ti triti*, *rrit ti tri ti rrrrrr*, *rrit rrit rrie rriii*, *ti trrie ti* or *ti trrie ti ti*, each phrase repeated 2-5 times.
Breeding: Eggs June, (Sep), Oct, perhaps all year (Junín), Oct (c Chile), Oct-Dec (Isla Grande); large gonads Dec (Cuzco).
Habitat: Bogs, lake-shores with short grass and scattered clumps of rush, as well as in adjacent puna grassland and salt marshes.
Range: 3350-4450 m. Lima, Junín, Cuzco, and Puno Peru s through La Paz, Cochabamba, and Oruro, Bol., to Jujuy, nw Arg. (*calcaratus*), Antofagasta, n Chile, and Salta, Tucumán, and Catamarca, nw Arg. (*catamarcae*); coast to 1250 m from Copiapó valley, n Atacama, n Chile s to Isla Grande, and in s and w Sta Cruz, Arg.; a few southern breeders may overwinter, but most migrate n in Apr and return in Sep (*chilensis*); from s Brazil s to Córdoba, Mendoza, Neuquén, and Chubut river, Arg. (*correndera*). Another ssp on Islas Malvinas. Common.

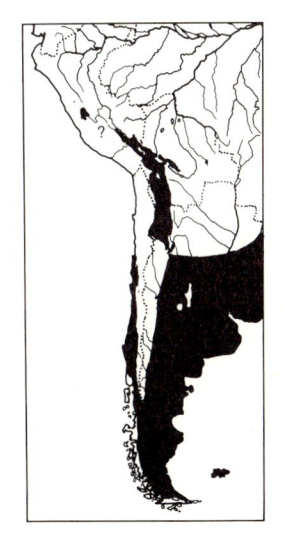

PARAMO PIPIT *Anthus bogotensis* – Plates XLVIII 17a-b, LXIV 23

15 cm. **Sparsely streaked below**. Bill relatively long and stout. Above

dusky streaked buffy (or streaked buff in *meridae*; wings dusky with 2 buffy bars in *meridae*, bars narrower in *bogotensis*; tertials edged buffy white, primaries and secondaries buffy brown, esp. at base, forming brownish patch in wing. Supercilium faint, buff streaked with blackish. Tail dusky, outer feather with whitish outer web and buffy inner web; penultimate with whitish tip (or buffy tip in *meridae*, which has pale of tail less sharply defined than *bogotensis*). **Below buff**, paler on throat and belly; breast with a narrow band of a few small streaks; sides and flanks inconspicuously streaked. Under wing-coverts buff; inner webs of remiges edged dull rufous at base, most conspicuous and sharply demarcated in *bogotensis* and *immaculatus*. Sspp *immaculatus* and *shiptoni* have shorter bill, wingbars almost lacking, and crown-streaks, supercilium, and underparts buffy white (sometimes entirely unstreaked on sides), whitest in *shiptoni*. Ssp *immaculatus* much like Hellmayr's P., but tips of wing-coverts are not in sharp contrast to patch on secondaries. Legs pale flesh, hind-claw 10-11 mm, somewhat curved. **Imm**. not tinged buffy above and below, the upperside being streaked whitish; outer tail-feathers buffy white on outer web and on tip of inner web, penultimate unmarked.

Habits: Alone or in loose pairs searches for insects in the grass much as other pipits. Sings from bush-top or other elevated point. Sometimes glides to song-post with vibrating wings without singing, and may also shake wings and tail between bouts of song.

Voice: Song (Ecu.) given from perch composed of broken phrases of thin thrush-like quality. In Cundinamarca, Col., nuptial flight noted, flying up and then gliding down while singing *sweet-sweet-sweez-twe e'e'e'e'e'e'e'e sr'r'r'r, tsee, tseez-tseez*. Flight call *tjirp* (Ecu.) or *pit-sit* (Col.). During territorial disputes a chattering *trieee chi chi chi* or *trieee chi chi* from the ground or in flight (Ecu.).

Breeding: Eggs Dec (Puno); imm. Sep (Santander), May (Ancash).

Habitat: Bogs and flats or slopes with short grass and bunchgrass in the puna and páramo zones.

Range: 2200-4500 m, in Cajamarca down to 1950 m. Mérida and Trujillo, Ven. (*meridae*); from páramo de Tamá, sw Táchira, Ven., s through E Andes to s Cundinamarca, n and s end of C Andes, and mts of Nariño, Col., through Ecu. to Cajamarca, n Peru (*bogotensis*); from Amazonas s to Huánuco, and in Junín, Lima and Huancavelica, c Peru, and in Cuzco and Puno, Peru, and La Paz and Cochabamba, Bol. (*immaculatus*); high arid plateaus of Cochabamba, Bol. (?) and Catamarca and Tucumán, nw Arg. (*shiptoni*). Locally fairly common.

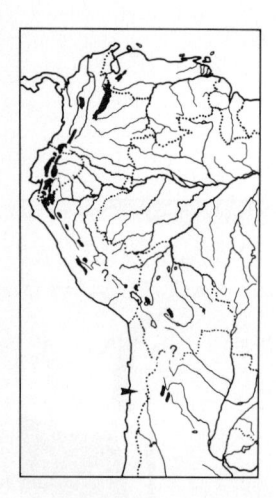

**Gnatcatchers and allies –
Family Polioptilidae**

C. 12 species restricted to the warmer parts of the Americas. Often treated either as a subfamily of the family *Muscicapidae* or the family *Sylviidae*. According to recent biochemical studies, they are aberrant wrens, which explains their American distribution.

MASKED GNATCATCHER *Polioptila dumicola* is found at 1500-2500 m in Sta Cruz, Cochabamba, and Chuquisaca, Bol. (*saturata*), other sspp from C Brazil and E Bol. s to Córdoba and Buenos Aires. 13.5 cm. **Male**: above plumbeous, **long tail black, outer**

feathers white, next 2 with decreasing amounts on tip. Headside black, tertials black edged white. Below blue-gray turning whitish on belly. **Female** similar, but sides of head gray like underparts, lores white. Alone or in pairs, perch-gleans along the branches, wriggles its cocked tail and frequently gives nasal wheezy calls. Inhabits open wood and scrub in semi-arid regions.

Vireos and allies – Family Vireonidae

Some 44 species restricted to trop., subtrop., and warmer temp. zones of the New World. Traditionally lumped with the following families in the 'nine-primaried oscines', but according to recent biochemical evidence this is due to parallellism, as the family may be derived from crow-like birds (*Wilson Bull.* 94(1982):114-28, *Condor* 90(1988):428-45).

All vireos are characterized by slower movements than warblers. Peppershrikes and shrike-vireos are fairly large, with stout shrike-like bills, vireos and greenlets warbler-like slender birds with slightly hooked (vireos) or pointed (greenlets) bills. Shrike-vireos are brightly colored, the others fairly dull, mainly olive, yellow, gray, and white. The sexes are alike or differ only slightly. Inhabiting dense canopy and shrubbery they are more often heard than seen; the stereotyped song is a loud, pleasant warble in vireos and peppershrikes, more high-pitched in greenlets. They are mainly insectivorous, but also eat some fruit. The finely woven nest is a deep cup, suspended by the rim from the crotch of a branch. Peruvian forms dealt with in *Amer. Mus. Novit.* 1127 (*Vireo*) (1941), and 1160 (*Cyclarhis*) (1942), and Bond (1953b).

RUFOUS-BROWED PEPPERSHRIKE
Cyclarhis gujanensis – Plate LII 4a-b

16 cm. **Bill heavy, slightly hooked**. Ssp *dorsalis*: above grayish olive, crown grayer; **broad eye-brow rufous**. Headside and half-collar around nape light gray, throat grayish white, lower throat and upper breast pale yellow, lower breast and belly pale buff; occ. the yellow is brighter and more extensive. Ssp *contrerasi* has most of crown chestnut, median crown-patch olive-green varying greatly in extent; yellow of throat washed olive; flanks and lower belly with hardly any ochraceous wash; *saturatus* much like *contrerasi*, but yellow much brighter and extending over the breast; sides more strongly suffused with ochraceous; *pax* has greenish olive back, gray crown, light gray upper throat, and bright greenish olive lower throat and breast-sides, breast grayish, belly whitish. **Juv**. has pale cinnamon (instead of chestnut) brow, pale buff forehead, and has crown-feathers pale ashy tipped buff, interscapulars and lesser wing-coverts tipped cinnamon-buff, lower throat irregularly washed with bright yellow. **Imm**. has brownish tinge to crown and nape, buffy brown tinge on flanks, and duller breast than ad.
Habits: Alone or in pairs, usually not in mixed-species flocks. Little seen as it forages inside the canopy with slow movements. Often sings incessantly for long periods.
Voice: Song loud and distinct, lasting c. 1 s, repeated at regular intervals. Composed of warbling notes and pure whistles; each phrase repeated for 5-10 minutes, then replaced by another phrase; most phrases end with a falling whistle *wee too wee tooe*, *wee too wee too wee tooe*, but sometimes rising *tee tuie te tui*.

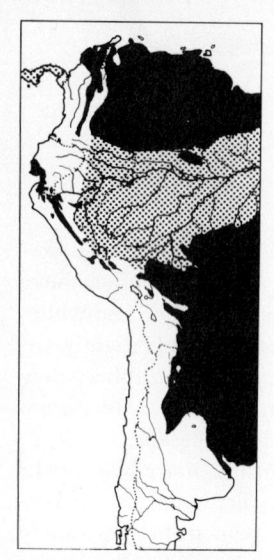

Breeding: No data for highlands.
Habitat: Middle strata in rather open canopies in humid forest clearings, open cloud forest, humid shrubbery, alder woods (*Alnus*), plantations, and second growth. In Bol. also in dry areas.
Range: 1350-3050 m mainly e of W Andes of n Peru from Marañón valley s to La Libertad and San Martín, on pacific slope in Piura and Cajamarca, intergrading with birds of c Peru in s Cajamarca (*contrerasi*), c Peru in upper Marañón Valley, Huánuco (*saturatus*); Yungas of La Paz, Bol. (*pax*); 1050-3200 m in Cochabamba, Sta Cruz, and Chuquisaca, Bol. (*dorsalis*). 17 other sspp in lowlands, mainly in the arid zone from Mexico to San Luis and ne Buenos Aires, c Arg. Fairly common. Spreads with the clearing of forest.

BLACK-BILLED PEPPERSHRIKE *Cyclarhis nigrirostris* is found in humid forest and forest edge at 1600-2400 m, on pacific slope down to 650 m and rarely up to 2700 m in Col., e and w Ecu. Resembles Rufous-browed P., but has a black bill, olive-green cheeks, crown and back, and grayer and darker underparts.

RED-EYED VIREO *Vireo olivaceus* – Plate LII 3

12 cm. Warbler-like. Iris brown. Ssp *chivi*: crown gray, **supercilium whitish, narrowly bordered above and below with blackish**. Most of upperparts grayish olive, rump greener, headside whitish tinged olive-yellow; **below white, sides, flanks and vent olive-yellow**. N Am. migrants (*olivaceus*) duller and grayer above; sides of head, neck, and breast pale buffy, sides of body mainly white, vent only tinged pale yellow, tail-feathers narrowly edged pale on inner web. **Juv**. above gray-brown tinged olive, below whitish to grayish, without greenish yellow; supercilium faint.
Habits: Alone or in pairs (migrants in flocks), sometimes in mixed-species flocks. Perch-gleans slowly, but continuously with horizontal posture on inner to middle branches of top canopy; sometimes fly-catches, sallies, or hangs. Territorial. Confident.
Voice: The fairly loud song is composed of 2 alternating simple phrases repeated incessantly, and varying locally; one of the phrases is double-noted with 1st note highest *trrrr tji-di trrrr tji-di...* or *zie trrri-dee zie trrri-dee....* Call a House Sparrow-like chirp *tijrr*.
Breeding: Eggs Nov (Tucumán); fledgling Jan (Jujuy).
Habitat: All levels, but mainly middle and upper strata of all woodland and forest (except dense and unbroken), groves and thickets, both humid and arid, primary and secondary.
Range: 450-1700 m, up to 3000 m in the drier parts of Cochabamba, Bol. From c Peru, Bol. (except altiplano), and wc Brazil s through w Parag. and to Buenos Aires, San Luis, and Córdoba, Arg.; 1 rec. from Atacama (Vallenar), Chile, and in winter n to e Col., s Ven., and nc Brazil (*chivi*). Fairly common to uncommon. 4 sspp breed in N and C Am. and winter in S Am. s to Bol., passing through Col. (up to 3600 m) in Sep and again in Apr-May. 9 other sspp resident in the lowlands and up to 1500 m in Tobago and S Am., w of the Andes to nw Peru, e of them to ne Arg., almost throughout.

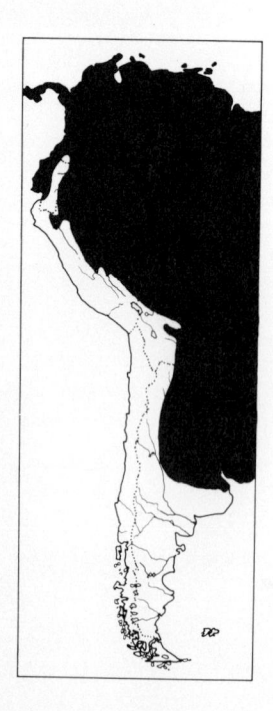

NOTE: S. Am. breeders are sometimes treated as a distinct species, *V. chivi*, and the populations of C Am. (*flavoviridis*) obviously deserve species rank (see *Wilson Bull. 97(1985):421-35*.

WARBLING VIREO *Vireo gilvus* – Plate LII 2

11 cm. Ssp *josephae*: **crown dark gray-brown, distinct supercilium white**. Back, wings, and tail grayish citrine. Headside pale gray-brown, throat and breast white more or less streaked with pale yellow, rest of **underparts pale yellow**, faintly streaked white on c upper belly. Tail-feathers narrowly edged pale on inner web. Other S Am. sspp slightly paler, esp. on crown. **Juv.** of *leucophrys* group undescribed. Alone, in pairs, or small groups, often in mixed-species flocks. Like a typical vireo has horizontal posture and moves slowly across thinner branches within the canopy, perch-gleaning insects and berries. Song consists of 6-8 clear whistles at 2 different pitches according to specific patterns, e.g. 2 high-pitched alternating with 1 lower-pitched or 2 high-pitched, alternating with 2 lower-pitched *hu he he hu hu he he hu, hu he he hu hu he* or *he hu he he hu he*. Alarm 3-noted, 2nd note lowest *werr a werr*.

In plantations, gardens, second growth, edge of humid forest, and cloud forest. Also in small humid patches in the arid zone. 700-2800 m. Sta Marta mts, Perijá mts, coastal mts, and Andes of Ven. (*mirandae*), e slope of n end of W and C Andes of Col. (also 'Bogota') (*disjunctus*), c part of C Andes and from cW Andes s to w Nariño, Col. (*dissors*), from Nariño to w Ecu. (*josephae*), both slopes of W Andes of n Peru, but not crossing to right bank of the Marañón (*maranonicus*), head of Magdalena valley and from E Andes of Col. s through e Ecu. to c Peru and probably further (*leucophrys*), from Urubamba valley in Cuzco, se Peru, to Cochabamba, Bol. (*laetissimus*). Fairly common. 10 other sspp from Canada to Panamá.

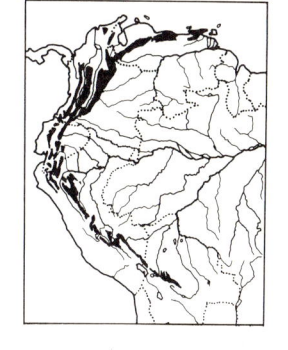

NOTE: The S Am. and most C Am. sspp are often considered a distinct species **BROWN-CAPPED VIREO** (*V. leucophrys*).

OLIVACEOUS GREENLET *Hylophilus olivaceus* – Plate LII 1, fairly common in middle strata of humid montane forest at 1000-1700 m, locally ascends to 2500 m. Found from e Ecu. s to Junín, c Peru. 12 cm. Above olive, below yellowish olive grading to yellow on belly. Iris pale yellow, bill pinkish, legs pale flesh. **Juv.** light gray-brown above, esp. on crown; wings and tail olive with brighter green edges than in ad.; underparts white. Iris straw. Alone, in pairs, or small groups, often in mixed-species flocks, foraging inconspicuously in the canopy. The closely related (conspecific?) Scrub G. has 2 songs, one of 2 differently pitched notes given alternately and repeated 2-3 times *chee cheer ...*, sometimes altered to a few, inconsecutive *chirees*, the other is a single, repeated note *weer weer....* Also burred, buzzy, scratching, or mewy sounds. Juv. Jan (Junín), Sep (San Martín).

NOTE: Sometimes treated as a ssp of Scrub Greenlet (*H. flavipes*).

Icterids – Family Icteridae

Some 95 species restricted to the Americas, as part of the '9-primaried oscines'. Mainly inhabiting lowlands, they occupy a variety of habitats from marshes, grassland, and semi-arid scrub to rain and cloud forest. In size they range from the sparrow-like Bobolink to the crow-sized oropendolas. They all have strongly built body and feet. Straight, conical, and sharply pointed bills are used for probing, and strong muscles for depressing the mandible enable them to force open small crevices, tangles etc. The diet consists mainly of seeds, fruit, and insects, but some scavenging and robbing of other birds' nests occur. Males are larger than females, in many species strikingly so. C. 1/3 of the species are sexually dimorphic in color, female duller than male. The majority of species are all black or contrasting black and yellow. Duetting and singing in chorus is normal, and many have elaborate courtships involving bowing, flicking of wings, cocking of tail, hanging, and aggressive interaction. Most cowbirds are brood-parasites, some laying their eggs in the nest of other icterids (sometimes to the advantage of the host as the young cowbird removes parasites from the other young), others using a variety of hosts. The nest may be an open cup placed on the ground or woven to a reed or suspended between branches, or it may be domed with a side entrance. The large penduline nests of the colonial oropendolas and Yellow-rumped Cacique are a well-known sight in the trop. zone, where they are often found near villages. See Bond (1953b).

SHINY COWBIRD *Molothrus bonariensis* – Plate LI 10a-d

17 cm. Bill shorter than in Bolivian and Scrub Blackbirds; bill and legs black. **Male** black with purple gloss, remiges with greenish gloss. **Female** ssp *bonariensis*: dark phase varying from uniformly smoky gray, upperparts with indistinct dark feather-shafts and slight gloss, throat paler, belly with an occ. black feather, to fuscous black throughout, wings browner and without gloss, throat paler and faintly streaked gray. Light phase: brownish above, feathers dusky along shafts, wings dusky, coverts and remiges edged buff, postocular streak whitish; throat whitish faintly tinged yellow, breast and sides buffy gray-brown, darkest and brownest on flanks, belly light grayish streaked buffy brown. **Female** ssp *aequatorialis* only with smoky gray dark phase. **Juv**. gray-brown above faintly streaked darker; below whitish with yellowish tinge streaked gray-brown. Bill and legs dark olive-green. Usually some juv. feathers are retained in imm. plumage.
Habits: Mainly in flocks, during migration sometimes up to 5000 together. Perches in trees and bushes, forages on the ground, often around cattle. During display the male bows head, though not as extreme as in the N Am. Brown-headed C., and more often on the ground than in a tree. More or less monogamous. A brood parasite, laying eggs in the nests of over 80 different species. See *Rev. Chil. Hist. Nat.* 17 (1913):172-9, *Rev. Bras. Biol.* 18 (1958):417-31, *Auk* 90 (1973):19-34, *Physis C* (Buenos Aires) 36 (1977):345-6, *Wilson Bull.* 90 (1978):271-84, *Living Bird* 17 (1978):41-50, *Amer. Nat.* 113 (1979):855-70, *Hornero* 12 (1979):69-71, *Hist. Nat. Mendoza* 1 (1980):151-2, *Hornero* No. Extrao (1983): 245-55, and *Hist. Nat. Mendoza* 3 (1983):149-58.
Voice: Song slightly buzzier than in Brown-headed C. *purr purr purr petsss tseeeee*, first 3 notes a low guttural bubbling given with quivering

body, the last 3 high-pitched. Call a piercing *tji-tji-tji-tji* and *tji-tjidi tjidi tjidi tji*.
Breeding: Eggs June (Cajamarca); fledglings Nov (Sucre), Dec (Santiago); juv. Feb (Cauca, Salta).
Habitat: Hedgerows, fields, groves of trees near farms, gardens, plantations, riparian thickets, often around cattle, and mainly in semi-arid regions.
Range: Mostly trop. zone, but up to 2500 m in Quito, Pichincha, Ecu., 2600 m in Cochabamba, and 3350 m in Potosí, Bol. Sw Col. s of Rio Patiá and w Ecu. s to Guayas (*aequatorialis*); from e Brazil s through e Bol., Parag. and Urug. to Chubut, Arg., and (1986) in Tarapacá, n Chile (see *Bull. Brit. Orn. Club* 109(1989):66-82) . First rec. in c Chile 1968 (introduced?), and has now spread and occurs locally from Atacama s to Aysén, one rec. from the Magellanic region (see *Anal. Inst. Patagonia* 13 (1982):183-7) (*bonariensis*). 5 other sspp in lowlands n to Panamá. Common.

BAY-WINGED COWBIRD *Molothrus badius* – Plate LI 11

17.5 cm. Ssp *badius*: smoky gray, throat paler; bill and lores to eye black, **wings mostly rufous**, lesser coverts like back, median and greater coverts and tertials dusky edged rufous, remaining remiges rufous tipped dusky; tail dusky. **In flight** wings short and rounded, rufous with dusky tip. Ssp *bolivianus* slightly larger and darker. **Juv.** similar, but obscurely spotted with dusky on crown and back, tertials conspicuously edged brown, lores and orbital region bare, distinct patch on ear-coverts fuscous; underparts faintly streaked whitish.
Habits: In pairs in breeding season, otherwise in flocks of 4-6 birds, occ. up to 50. Perches in trees, hedgerows, and on walls, and forages on the ground, usually near cattle. Not parasitic. Breeds in old nests of furnariids (e.g. horneros, Firewood-gatherer (*Anumbius anumbi*) or thornbird), but may also build its own nest; sometimes breeds semi-colonially, 2 or more females tending the same nest (see *Hornero* 3 (1926):416-7, and *Auk* 89 (1972):447-9). Often parazitised by Screaming C.
Voice: Several males sing in chorus, sometimes 1-2 females join in. Song pleasant, mellow, fairly monotonous and formless, given in a weary sing-song manner and frequently interspersed with a guttural *chuck* note. Calls include weak *kik errt*, *kik*, *riüe*, *kik krrr*, *krrrr* and *tji drrrrrrt*, as well as a sharp clear *tuit*.
Breeding: A late breeder in Arg. (mostly Jan-Feb, rarely before Dec). Eggs Jan (Tucumán); fledgling Apr (Cochabamba).
Habitat: Wooded patches near meadows and rangeland with cattle, dung-heaps, town-parks, sometimes in very open terrain.
Range: From lowland Bol. (intergrading with *bolivianus* in Tarija) and w Mato Grosso, Brazil, s through n Arg., Parag., extreme s Brazil and Urug. to Chubut (see *Neotrópica* 16, 49 (1970):11-6), Arg., accidentally crossing Andes to Chile (Curicó, and several sight rec. in c provinces) (*badius*); 1100-3350 m in La Paz, Cochabamba, Sta Cruz and Chuquisaca, c Bol. (*bolivianus*). Fairly common to common. Another ssp in campo region of ne Brazil.

SCREAMING COWBIRD *Molothrus rufoaxillaris* of lowlands from s Bol. and s Brazil to n Arg. is a brood parasite on Bay-winged C., and can be expected to follow this species to the temp. zone. 18 cm. Ad. glossy black, bill conspicuously shorter than in Shiny C.; axillaries occ. with chestnut spot. **Imm. female** blackish brown. **Juv**. resembles Bay-winged C., but has a darker subterminal area on the bill and is slightly larger. See *Wilson Bull*. 91 (1979):151-4, *J. Orn*. 124 (1983):187-93, and *Biotrópica* 16 (1984):223-6.

DUSKY-GREEN OROPENDOLA *Psarocolius atrovirens* – **Plate LI 3** is locally common at 1200-2400 m (occ. down to 850 m) in E Andes from Junín, c Peru, to Cochabamba, Bol. It inhabits ravines with primary humid forest, usually near streams. Calls include a subdued *jog* and a loud *tschuik*. Eggs Oct (La Paz); juv. Dec (Junín). Forages for fruit and berries in the canopy. Nests in small colonies. Male 38 cm, female 30 cm. Dark olive-green, rump and vent chestnut; c 2 pairs and **outer tail-feathers dark olive**-green, rest yellow with olive tips, penultimate with whole outer web olive. **Male** has yellow forehead. Iris blue in male, brown in female, legs black in male, dark gray in female, **bill** yellowish **pea-green** in male, pale pea-green in female. **Juv**. duller. Spreads tail before landing.

CRESTED OROPENDOLA *Psarocolius decumanus* – **Plate LI 1**, widespread in humid trop. forest and second growth e of Andes, ascends to the subtrop. zone in Yungas of Bol. and occ. to 2600 in Cochabamba (*maculosus*), in Col. rarely to 2600 m (*melanterus*). Male 45 cm, female 30 cm. Tail graduated. Glossy brownish **black** incl. long, thin crest (in ssp *maculosus* belly often admixed chestnut and body with scattered pale yellow feathers); lower back, rump, upper tail-coverts, and vent chestnut; c tail-feathers blackish, rest bright yellow. **Bill ivory-white**, elevated and somewhat swollen at base. Eyes bluish. **Juv**. like ad., but duller. Nesting behavior in *Zoologica* 42 (1957):87-98.

RUSSET-BACKED OROPENDOLA *Psarocolius angustifrons* – **Plate LI 2** is very common in humid forest and second growth at the base of coastal mts and Andes from Ven. to Bol. and w Ecu. and the upper Amazon, and occ. reaches lower edge of temp. zone in Ven. (*neglectus*), w Col. (*salmoni*), and in Yungas of La Paz, Bol. (*alfredi*), and breeds to 2700 m at Sozoranga, Loja, s Ecu. (*atrocastaneus*). Male 45 cm, female 35 cm. Tail graduated. Ssp *alfredi*: head incl. lengthened crest gray-brown tinged yellow; otherwise oily cinnamon-olive, rump and vent cinnamon-rufous. C tail-feathers blackish, next blackish on inner web, bright yellow on outer web, rest of tail bright yellow, feathers with dusky terminal half of outer web and tip of inner web. Bill yellow. Sspp *neglectus* and *salmoni* more rufescent and with some yellow on forehead, rarely continued as a short supercilium.

MOUNTAIN CACIQUE *Cacicus leucorhamphus* – Plate LI 9a-b

Male 27.5-29 cm, female 23-25 cm. Tail graduated, iris pale blue. Ssp *leucorhamphus*: **black; inner wing-coverts, lower back and rump golden yellow**, concealed collar white. Bill long, pale blue-gray with yellow-green tip. Ssp *peruvianus* similar, but bill heavier, blue-gray base more restricted and less abruptly defined; concealed white collar less extensive. Ssp *chrysonotus* largest, **wing without yellow** (occ. showing traces, esp. in Peru), and no concealed white collar; bill more arched at base and very slightly curved. **Juv.** duller black than ad.; bill dark horn with light tip, and not as strongly conical.
Habits: Alone, in pairs, or in flocks of 6-7 birds, sometimes in mixed-species flocks or with flocks of Hooded Mt-tanagers. Climbs about rather high in canopy eating berries, insects, and nectar. Also in smaller trees and bushes, and often probes the internodes of sprouting bamboo. Roosts communally, but apparently breeds solitarily.
Voice: Noisy. Calls include a single clear whistle *tjiue*, a quavering *wee-eueeueeu*, and varied phrases *arr* **tjie tjie** *tiuee arr arr tiue kik...kik tjiue*; sometimes one bird gives a *arrhee arrhee...* and another bird answers with a long, falling, hawk-like *weeeeeeee*. Contact note harsh.
Breeding: Fledglings Jan, Mar (nw Ecu.), Oct (ne Ecu.), Dec (Puno), Oct (Cochabamba); juv. Aug (Amazonas), May (Puno), Jan (Cochabamba); large gonads Feb-July (Col.).
Habitat: Humid forest and forest edge. Often in thickets of large bamboo.
Range: 1500-3300 m, highest in Bol. From Río Chiquito, sw Táchira, Ven. s spottily through all 3 ranges of Col. (w?) and e Ecu. to Piura, n Peru (*leucorhamphus*), s and e of the Marañón from Amazonas to Junín, c Peru (*peruvianus*), from s border of Junín, c Peru, to Cochabamba, Bol. (*chrysonotus*). Generally uncommon, but locally fairly common to common, esp. in s part of range.

NOTE: Ssp *chrysonotus* is sometimes considered a distinct species.

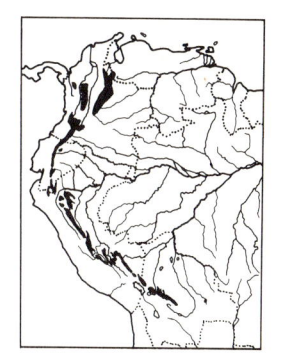

YELLOW-BILLED CACIQUE *Cacicus holosericeus* – Plate LI 8

Male 21.5 cm, female 21 cm. Tail graduated. Uniformly **dull black, iris yellowish white**; bill pale greenish yellow, base of mandible grayish at sides. **Juv.** duller.
Habits: Usually in pairs, sometimes singly or in family groups, often in mixed-species flocks. Scratches, tears and rummages among trapped dead leaves, mosses, tangles, and dry stems. Sometimes peels bark off slender limbs in search of insects, using beak as a plough.
Voice: While foraging both sexes regularly utter harsh *waak* notes.
Breeding: Fledglings Apr (Cundinamarca), Nov (ne Ecu.), Apr, Dec (Huánuco); juv. Mar (Pasco); large gonads June (Perijá).
Habitat: Lower strata. Thickets and undergrowth at edge of humid forest and cloud forest, often mixed with bamboo.
Range: 1500-3500 m. Sta Marta and Perijá mts, coastal mts of Ven., and Andes from Mérida, Ven. s spottily through E and C Andes of

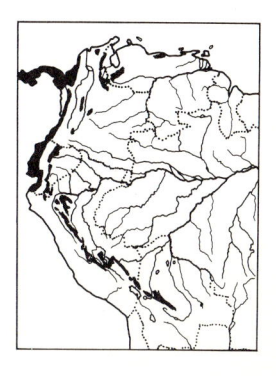

Col., and continuing through e Ecu. and Peru to Cochabamba, Bol. (*australis*). Fairly common. 2 other sspp mainly in trop., sometimes arid zone from se Mexico to nw Peru.

NOTE: Sometimes placed in the genus *Amblycercus*.

AUSTRAL BLACKBIRD *Curaeus curaeus* – Plate LI 7a-b

Male 24 cm, female 22 cm (*reynoldsi* larger). **Glossy black, bill long and black**, iris brown. Feathers of head narrow, stiff, and pointed. Tail square, fairly long. Female very slightly browner than male. **Juv**. deep slate, belly grayish fuscous; faint streaks on mantle and nape, only forehead with a few stiff feathers. Bill shorter and fuscous with pale tip.
Habits: In pairs or flocks of 6-10 following a leader. Forages for insects in dense vegetation, sometimes steals eggs and young from other birds. Perches openly on bushes. Curious.
Voice: When one starts singing others join in. Usual call a high-pitched *whee whee* followed by a low *chuck alah*.
Breeding: Eggs Oct-Dec, rarely in Jan or Feb (Chile); fledglings Feb (Isla Grande).
Habitat: Open woods, brushy, often well-watered slopes, heathlands with scattered bushes, or hedgerows in agricultural lands.
Range: Coast to 1500 m. Resident from Coquimbo, Chile and Neuquén, Arg. s to straits of Magellan (*curaeus*), Isla Grande (*reynoldsi*). Common.

SCRUB BLACKBIRD *Dives warszewiczi* – Plate LI 6

Male (*kalinowskii*) 27 cm, female only slightly smaller. **Black** with greenish gloss; **fairly long tail** only slightly graduated; **bill long**, black, iris dark gray. **Juv**. duller and grayer.
Habits: Alone, in pairs, or loose flocks. Forages in the canopies of trees, hedgerows, corn-fields (where it may be a pests), often perches atop bushes or agaves. During song the head is thrown back and throat-feathers puffed; sometimes 2 singing birds oppose one another, twist their bodies or throw back their heads simultaneously. Rather shy.
Voice: Song melodious, of loud clear whistles *teeweet*.
Breeding: Juv./imm. Apr (La Libertad).
Habitat: Riparian trees and thickets in arid regions, hedgerows, groves and orchards in irrigated farmland.
Range: Mainly coast, but locally (e.g. Lima) ascending to 3200 m in Andean valleys. Trop. zone from w Ecu. s of Chone river to Piura, nw Peru, intergrading with *kalinowskii* near Palambla in Piura (*warszewiczi*), w of Andes from La Libertad to Ica, w Peru, and a local population at 2500 m in Huancavelica (Anco) (*kalinowskii*). Fairly common.

NOTE: Sometimes regarded sspp of Melodious B. (*D. dives*) found from s Mexico to Nicaragua.

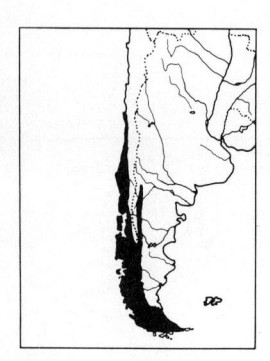

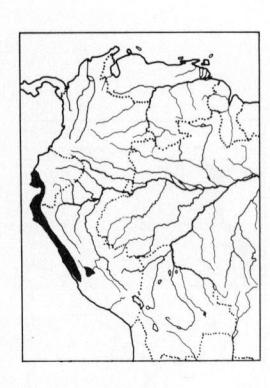

BOLIVIAN BLACKBIRD *Oreopsar bolivianus* – Plate LI 5

Male 22.5 cm, female 20.5 cm. **Black** with slight gloss, **primaries dark brown**; feathers of forehead and chin velvet-like. Bill slightly larger than in Shiny Cowbird, and maxilla somewhat decurved. **Juv**. has faint wash of fuscous below, and brown primary-coverts.

Habits: Usually in flocks of 3-8 birds. Flight grackle-like with frequent glides exposing the brown primaries. Perches openly in trees, bushes, or on cliffs. Forages on the ground for grass seeds and arthropods, gleans insects from leaves and branches of small shrubs (esp. *Dodonea*), and probes bromeliads. Also eats cactus fruit (*Cereus*). Nests socially, the incubating female being fed by several birds, although mostly by one male. During courtship male may point bill and closed tail upwards while flipping wings (much resembling female's begging posture), droop wings, or perform labored flights with deep wing-beats as if landing, thereby exposing brown primaries. See *Condor* 79(1977):250-6.

Voice: Fligh call upon take-off of a flock (single birds take off silently) a loud clear whistle *chu-pee*, 2nd note high-pitched, or a loud *cheep-cheep*; in flight sometimes 3-4 rapid croaking *churr*s or a *chu-pee-tit*. Song a rapid sequence of sharp *chip* notes given while tail is cocked and interspersed with *chu-pit* notes given while wings are flicked. Long songs may include harsh churring sounds while wings are being rapidly flicked. Alarm a short sharp *chip* (much as in other icterids) accompanied by a fast tail-flick. At presence of hawk gives a very high-pitched wheen.

Breeding: Eggs Apr (Cochabamba); fledglings June (Cochabamba).

Habitat: Requires cliffs for nesting. Dry inter-montane valleys with sparse ground cover, shrubbery, cacti, and scattered trees.

Range: C Bol. at 2400-3200 m, rarely lower. Cochabamba, ne Potosí and Chuquisaca. Fairly common.

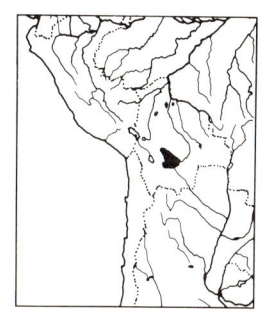

RED-BELLIED GRACKLE *Hypopyrrhus pyrohypogaster*

Male 30 cm, female 27 cm. Bill conical and pointed, black; eyes yellowish white. Black with **bright red belly and vent**, thighs black. Feathers of head narrow and stiff with shiny shafts. Usually in noisy groups of 6-8, sometimes with mixed-species flocks or oropendolas. Climbs about in the outer foliage. Nestlings May.

Inhabits humid forest canopy and edge, at 1200-2700 m. Col., in n half of W and C Andes, head of Magdalena valley and on e slope of E Andes in w Caquetá. Rare and local.

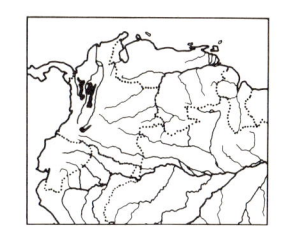

MOUNTAIN GRACKLE *Macroagelaius subalaris* – Plate LI 4

Male 28 cm, female 26 cm. **Tail long**, slightly graduated. Legs and fairly long conical bill black, iris brown. **Glossy black**, semi-concealed lengthened axillaries and under wing-coverts chestnut, feathers of rump with gray bases. **Juv**. undescribed.

Habits and **voice**: No data.

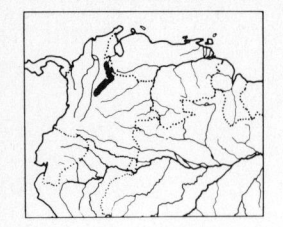

Breeding: Large gonads Sep (N.d.Santander).
Habitat: Middle to upper strata of humid forest.
Range: Col. at 1950-3100 m on both slopes of E Andes in N.d.Santander and from Santander to Cundinamarca. Uncommon.

NOTE: By some called Colombian G. The Golden-tufted (or Tepui) G. (*M. imthurni*) of the Tepuis of Guiana, se Ven. and immediately adjacent Brazil is sometimes regarded a (mega)ssp of the present species. It wanders in small flocks and calls *kur-a-leek* with a rusty sound.

YELLOW-WINGED BLACKBIRD *Agelaius thilius* – Plate LI 12a-c

Male 17.5-19 cm, female 16.5-18 cm. Bill long and slender, black in male, dusky in female. **Male** glossy **black, lesser upper wing-coverts**, under wing-coverts and axillaries **bright yellow**. In fresh plumage feathers of upperparts are tipped brown or pale gray along c crown-streak and supercilium; underparts have pale gray feather-tips, which normally wear off except on vent. Sspp poorly differentiated, *petersii* smallest, *alticola* largest and without pale supercilium. **Female** dusky streaked gray-brown above, c crown-stripe and supercilium whitish; underparts pale buff to pale gray-brown, streaked dusky; moustache dusky, rest of throat unstreaked. Underwing with yellow like male, lesser upper wing-coverts olive-green more or less mixed with yellow. Ssp *alticola* darkest and most heavily streaked below. **Juv**. like female, but washed reddish brown.
Habits: In small bands during breeding season, otherwise in large flocks. Forages on floating vegetation, muddy shores and cultivated fields nearby, taking mainly insects. Roost communally and breed semi-colonially in tule marshes, placing nest 10-60 cm above water.
Voice: Song *tjk-tjk-tjk-tjk-tjk-tjk-teyeey*, *tjk…tjk-dzeey* or *tetete zee-leeu zee-leeu zee-leeu*. Flight call a nasal *ch-eh*. Alarm a drawn-out *d-rrrrr*.
Breeding: Eggs Oct-Dec (Chile), Nov (Tucumán), Sep, Nov (Puno); fledglings Oct, Nov (Puno).

Habitat: Reed and tule marshes and surrounding open country.
Range: 3700-4300 m in the vicinities of lakes Titicaca and Poopó and in adjacent Cuzco, Peru, and Cochabamba, Bol. (*alticola*); lowlands from Copiapó valley Atacama to Llanquihue, Chile and in Chubut and Sta Cruz, sw Arg., also in lake Buenos Aires in Aysén, Chile, and possibly in Magallanes (P.N. 'Torres de Paine') (*thilius*); lowlands from extreme se Brazil and Urug. throughout Arg. s to Chubut (*petersii*). Common.

YELLOW-HOODED BLACKBIRD
Agelaius icterocephalus – Plate X 6a-c

Male 18.5 cm, female 16 cm. **Male glossy black, head yellow**, chin and lores to almost around eyes black; in fresh plumage with gray feather-edges on body and head except throat. **Female** dusky olive-gray, crown somewhat greener, upperparts indistinctly streaked dusky; faint supercilium olive, throat buffy yellow faintly tinged olive. **Juv**. more fulvous than female, and with buffier crown. In family groups or flocks. Probes mud and dense, floating mats of *Azolla*. When alarmed flies to top of tule. Roosts communally. Labored unmusical song (like rusty hinge) *took*, *tooweeeez*, 1st note faint, 2nd loud and rasping, or sometimes followed by a short musical *te-tiddle-de-do-dee*, down then up. Juv. Jan (Cundinamarca).

Tule marshes interspersed with flats of open mud and *Azolla*. At 2600 m on the Bogotá and Ubaté savannas in the E Andes of Col., now only at a few localities (*bogotensis*). Lowlands of n and e Col., n Ven., Trinidad, Guianas, and n Brazil from lower Amazon w to ne Peru (*icterocephalus*).

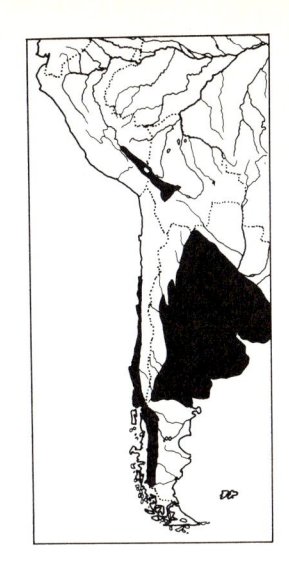

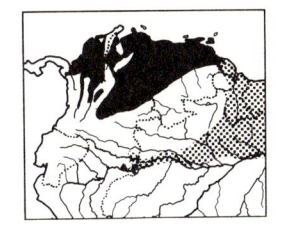

YELLOW-BACKED ORIOLE *Icterus chrysater* – Plate LI 13

21.5 cm. **Golden yellow**; face, throat, upper breast, wings, and tail black; lesser upper wing-coverts and under wing-coverts golden yellow. Bill black, base of mandible blue-gray; iris dark. Inhabits second growth, thorny scrub, open woodland, plantations, and edge of rain and cloud forest. Alone or in pairs forages in middle strata of trees. Frequently gives high-pitched, rich musical whistles, *wheer-hee who-hee who-hee ha-heet*, *wita-wita-wita*, or as more detached jerky series.

In humid forest edge, shrubby slopes and coffee plantations, mainly in the premontane zone, but straggles to the Bogotá savanna Col., and rec. to 2900 m in Ven. (*giraudii*). C Am., Col., in the n, w and Andean slopes, and nw Ven. mainly in the Andes and coastal cord.

NORTHERN ORIOLE *Icterus galbula* – Plate LI 14a-c, of N Am. winters s to n Col. and nw Ven. (*galbula*). 17.5 cm. **Male**: head, throat, back, wings, and c tail-feathers black; the rest orange-red to golden; remiges and distal portion of greater wing-coverts edged white. **Female**: upperparts and c tail-feathers gray-brown washed dull orange, crown dark-streaked, back grayish olive, feathers with dark centers;

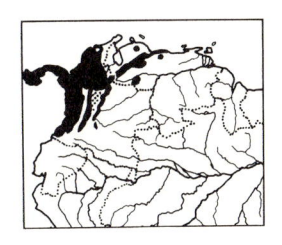

wings dusky with 2 white bars, that of the middle coverts broadest, remiges narrowly edged white. Below dull orange, palest on throat and c belly; flanks tinged gray. **Imm**. like female, but without dark markings on crown and back; below paler. Alone or in pairs forages for insects, fruit, and berries, often searching dead leaves, at mid- to upper strata of open woodland and scattered thickets in cultivated areas.

WHITE-BROWED BLACKBIRD *Leistes superciliaris* – Plate LI 15a-b

17 cm. **Less bulky and with bill and tail shorter than in meadowlarks**. Most of mandible blue-gray **in male, horn in female. Under wing-coverts black in male, dark gray in female. Male** black; throat, breast, and anterior belly ruby-red; supercilium **from above eye** white; c tail-feathers with faint gray barring. Fresh black feathers are edged brownish. **Imm. male** has broader edging, esp. on crown, c crown-stripe dirty white; supercilium tinged buffy gray, all tail-feathers barred with gray. **Female** and **juv**. above dusky streaked pale gray-brown, crown with pale c stripe, and whitish supercilium; tail dark gray barred dusky. Below buffy, more or less washed with red, or washed gray and streaked dusky on posterior underparts. In pairs or groups up to 15, territories clustered and shared with Long-tailed Meadowlark. In flocks in winter. On the ground, rarely on bush-tops. Frequently opens tail. Crouches to hide red breast when alarmed. Call is a low *chuck*.

Inhabits wet grassland, preferring grazed parts. E Brazil to the chaco and pampas, ascending Andean slopes to 2600 m in Cochabamba and 2300 m in Tarija, Bol. 1 rec. from Coquimbo (Baños de Toro 3050 m), c Chile.

NOTE: Often considered a (mega)ssp of Red-breasted B. (*L. militaris*) of lowlands from Panamá to Pando, n Bol., Guianas, and Amazonian Brazil.

LONG-TAILED MEADOWLARK *Sturnella loyca* – Plate LI 17a-b

25 cm. Bill and tail fairly long, **under wing-coverts white**. Ssp *loyca*: above dusky to black streaked gray-brown, c crown with palest streaks, shoulder red. Tail and upper tail-coverts grayish barred dusky. Long **supercilium red in front of eye**, white behind; underparts red, feathers tipped whitish in fresh plumage; sides, flanks, lower belly and vent black edged gray. Bill black, mandible and wash on maxilla pale slate. Ssp *catamarcanus* has considerably broader pale edges above and is less black ventrally. Ssp *obscura* like *loyca*, but darker; red less extensive. **Female** like male, but browner, red below less extensive; in fresh plumage breast has narrow, dusky streaks edged gray-brown. **Juv**. much browner; whitish markings replaced by buff; tail brown with narrower and straighter dusky bars. Throat whitish, rest of underparts buffy brown with broad dusky streaks, upper belly tinged pink; bill pale brown.

Habits: In pairs or family groups, with clustered territories. In winter sometimes in large flocks. Actively forages on the ground while nodding and bobbing, mainly feeding on beetles. Perches on rocks, bush tops, and fence poles. Flight direct. Crouches to hide red breast when alarmed. Food often gathered outside territory, which may be shared with White-browed Blackbird (see *Diss. Abstr. Int. (B)* 37(1976):592). For co-evolution with Shiny Cowbird see *Amer. Nat.* 113(1979):855-870. During song from elevated perch throat-feathers are puffed. See also *Condor* 80 (1978):251-2, and *Behavioral Ecol. Sociobiol.* 5 (1979):159-70.

Voice: Sings throughout year, mostly towards evening. In winter male sits a little away from family while singing, then female joins, starting and finishing after him. Song of 2 types, most commonly with a varying amount of introductory notes terminated with a long, harsh sound *chee chee rrrie chee chee rrrie*, *cha whit chee rrrie chee chee chee rrrie*, *cha cha chee rrrie chee chee chee rrrie*, *tui tui tui chee rrrie* or *cha whit zwee chee zwee rrrie*, but sometimes with harsh introductory notes. Also weaves and chatters without harsh notes. Contact call an almost thrush-like repeated *tjy*. Other calls include a short *peet*.

Breeding: Eggs Oct-Nov (Isla Grande), Sep-Jan (sometimes double-brooded) (Chile); fledglings Feb (Chubut, Chile).

Habitat: Requires wires, fence posts, or bushes for singing. Grazed pastures with bunchgrass, generally in slightly lower vegetation than White-browed Blackbird (see *Amer. Mus. Novit.* 2349(1968), scrub-covered hills and valleys, edge of *Nothofagus* forest, gravelly hills with scanty vegetation, partly vegetated sand-dunes. Often near cattle.

Range: Coast to 2450 m. From Salta s to Mendoza, nw Arg. (*catamarcanus*); Sierras Grandes, Sierras de Comechingones, and Sierras de San Luis in Córdoba and adjacent San Luis, Arg., breeding above 1600 m (*obscura*, Nores & Yzurieta in *Acad. Nac. Cienc. Cordoba, Arg. misc.* 61(1979):6-7). From Atacama, Chile and s Buenos Aires and Neuquén, Arg. s to Isla Grande, in winter n to Salta, Córdoba, and Buenos Aires (*loyca*). 1 ssp on Islas Malvinas. Common.

NOTE: Previously called *Pezites militaris*. Should possibly include Peruvian Red-breasted M. as a megassp.

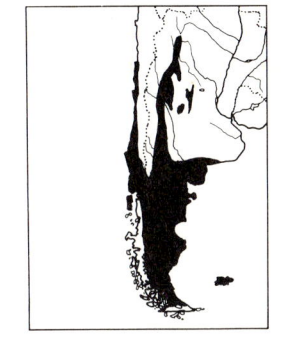

PERUVIAN MEADOWLARK *Sturnella bellicosa* – Plate LI 16a-b

21 cm (*albipes* smaller). Bill fairly long. **Male**: above pale gray-brown to brown streaked dusky, c crown whitish; shoulders red. Upper tail-coverts and c tail-feathers grayish barred dusky, rest of tail dusky with grayish bars only at tip; supercilium red in front of eye, white behind; sides of head and neck black. Underparts bright red; sides, flanks, lower belly and vent black, feathers edged grayish white; under wing-coverts and thighs white. **Female** like male, but smaller, crown streaked, feathers of flanks edged gray. **Juv**. has shorter bill, lacks red below, and is more streked on the breast.

Habits: In pairs or family groups. Much like Long-tailed M., but performs low, gliding nuptial flights.

Voice: Call *tzp*. Song includes harsh notes *piu-piu-chrrrrr*.

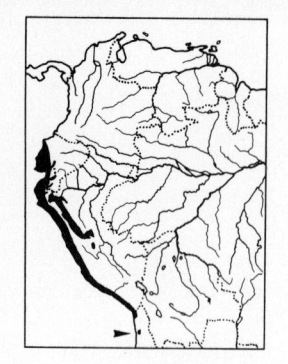

Breeding: On coast eggs Oct (Chile), Oct-Feb (Arequipa); fledgling May (Arequipa).
Habitat: Fields, hedgerows, open montane and desert scrub.
Range: Coast to 2600 m. From w Ecu. s to Lima, Peru, from n end of C Andes s in Maranon drainage to Huánuco and in Junín, c Peru (*bellicosa*), w slope of W Andes from Ica, Peru s to Tarapacá (Arica) and Antofagasta (Quillagua), n Chile (*albipes*). Common.

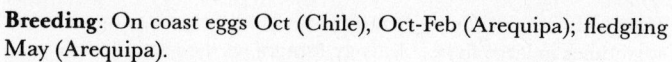

NOTE: Called Peruvian Red-breasted M. by Meyer de Schauensee. Previously considered sspp of Long-tailed M.

EASTERN MEADOWLARK *Sturnella magna* – Plate LI 18

22 cm. Bill long, tail short. Crown blackish, feathers edged brown, c stripe whitish; rest of upperparts brown streaked buffy white and barred blackish; bend of wing yellow. **Outer 3 tail-feathers mostly white**, conspicuous in flight. Long supercilium yellow in front of eye, whitish behind; **throat, breast, and c belly bright yellow**, **bordered by black U on breast**; sides, flanks, and vent buffy white with broad fuscous spots or streaks. **Juv**. much like ad., but buffy yellow below, breastband composed of spots.
Habits: In pairs or family groups, often polygamous. Forages on the ground, perches on ridges, bushes, fence posts, and wires, rarely low in trees. Flight with alternating flaps and glides. May sing while gliding low over the ground, but usually sings while perched.
Voice: Sings all year. Song a melodious series of whistled notes.
Breeding: Fledglings Mar (Mérida), Feb (Boyacá); large gonads Jan-Nov (n Col.), Oct (Cundinamarca), Jan (Boyacá), June (Huila).
Habitat: Grazed shore meadows and grassland at lower edge of páramo zone, man-made clearings, and agricultural areas.
Range: 1700-3500 m, from Trujillo to Táchira, Ven., and from N.d. Santander locally s through E Andes of Col. to head of Magdalena valley (*meridionalis*). Locally fairly common, spreading with deforestation. 13 other sspp in Cuba, N and C Am., and from lowland Col. to Surinam.

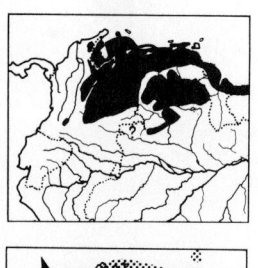

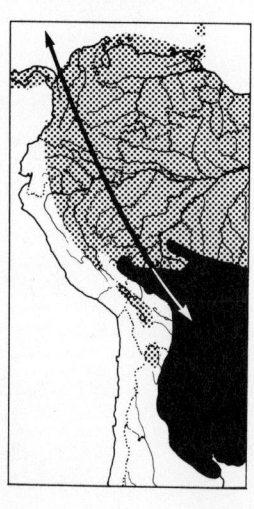

BOBOLINK *Dolichonyx oryzivorus* – Plate LI 19a-d

15 cm. **Sparrow-like**; **tail-feathers pointed**; bill short and pale. **Male winter**: crown dusky, feathers edged brown, c stripe pale grayish buff; nape pale grayish buff streaked dusky, rest of upperparts dusky streaked pale grayish brown; wings blackish, coverts tipped buff. Below white, breast, sides, and vent tinged buffy and streaked blackish; under wing-coverts white. Already in Jan with scattered black buff-tipped feathers. **Breeding male** black with buffy nape and mantle, white shoulder patch and rump, buffy streaks on c back, and black under wing-coverts. **Female** like winter male, but tinged yellow below. **1st year** birds have more pointed c tail-feathers. Usually a few birds together, sometimes up to 40. Forages in fields, roosts communally in reed-beds. Flight-call a characteristic *pink*.

Reed-beds, grassy fields, riparian thickets, shrubbery. Breeds in N Am., winters in S Am. s to Bol., Parag., and n Arg., w of the Andes rarely as far s as Arequipa, sw Peru, Tarapacá (1 rec. in Lluta valley) and Antofagasta (1 rec. Pedro de Valdivia), n Chile. Chiefly in trop. and subtrop. zones, but occ. to above 4000 m in the Andes (e.g., flock of 75 in lake Titicaca in Jan). Erratic.

Wood-warblers – Family Parulidae

Some 110 species restricted to the Americas s to C Arg. (1 also on Galapagos isls). Belongs to the '9-primaried oscines'. Northern breeders are migratory, wintering mainly in C Am. and nw S Am. Small and active, primarily insectivorous birds with thin pointed bills, 9 primaries, and inhabiting forest and shrubbery. Many are brightly colored; the sexes alike or different. The generally weak territorial song is given by the male; a few also sing duets. Some species are important in mixed-species flocks in undergrowth (*Basileuterus*) and canopy (*Myioborus*). Nest-building and incubation is by the female alone, natal care by both sexes. The nest is usually cup-shaped and placed in a tree or shrub, but may also be domed and then placed on the ground or in a tree cavity. Although traditionally called redstarts, members of the genus *Myioborus* are here called whitestarts. Further notes on the family in Todd (1929), *Amer. Mus. Novit.* 1428(1949), Bond (1953b), *Univ. Calif. Publ. 116: 1-107*, in Keast & Morton, eds. (1980), and in *Living Bird* 15(1977): 119-41.

BLACK-AND-WHITE WARBLER *Mniotilta varia* – Plate LII 15a-b – breeds in N Am., winters from sea level to 2500 m from n Mexico to w Ecu., c Col., and n Ven. 2 rec. from Peru (2115 and 2200 m). Uncommon. 12 cm. **Male** streaked black-and-white, crown with white mid-stripe, cheeks and throat black; 2 outer tail-feathers with large white terminal area on inner web. **Male winter** has white cheeks and throat, sometimes with black spots. **Female** similar, but underparts white with buffy tinge on flanks, sides of breast and extreme sides of body obscurely streaked dark gray; under tail-coverts with blackish centers. **Imm.** much like female, but vent buffy. Alone, often in mixed-species flocks. Actively creeps up and down trunks and heavy limbs, frequently dropping like a leaf. The rarely heard call is a weak chip. Middle strata of humid forest edge, second growth, and plantations.

GOLDEN-WINGED WARBLER *Vermivora chrysoptera* – Plate LII 14 – breeds in N Am., winters at 1000-3000 m from Yucatan to c Col., w Ecu., and n Ven. Uncommon. 11 cm. **Male**: forehead and crown golden-yellow grading to olive-green posteriorly. Back gray; median and greater **wing-coverts yellow**, remiges dusky edged grayish olive. Tail dusky edged gray, outer 2-3 pairs with mostly white inner web. Headside black, encircled by white supercilium and moustache; chin and **throat black**, rest of underparts whitish, sides tinged gray. Bill black. **Male winter** similar, but yellow of crown and gray of back obscured by olive feather-tips, bill brownish with pale mandible. **Female** similar, but yellow of crown less distinct, supercilium grayish on posterior portion, headside dusky, throat light gray; **female winter**

tinged olive-yellow. **Imm. male** like winter male, but black throat-feathers narrowly margined white. Alone, sometimes in mixed-species flocks. Actively forages in the foliage, at times clinging to dead leaves. Regularly fans tail and may halt on a twig and nervously twist body and esp. tail from side to side. Call a short *chup*. Usually low in trees and shrubbery, sometimes high up in humid forest, and rarely on the ground at forest edge.

TENNESSEE WARBLER *Vermivora peregrina* breeds in N Am. and winters s to n Ven., Col. and rarely n Ecu., to 2840 m, mainly in light woodland. 11.5 cm. Breeding plumage: crown gray in contrast to bright greenish olive upperparts. Supercilium and underparts white with dusky streak through eye. Female tinged olive on crown. Winter plumage (1st year?) greenish above with distinct yellowish brow and sometimes a faint whitish wingbar; below dingy yellowish, vent white.

TROPICAL PARULA *Parula pitiayumi* – **Plate LII 19** is common at lower elevations, with 14 sspp from Mexico to c Arg. Occ. reaches 2600 m from e Ecu. s to Huánuco, c Peru (*alarum*), sw Nariño in sw Col. through w Ecu. to Cajamarca, nw Peru (*pacifica*), Junín, c Peru, to Cochabamba, Bol. (*melanogenys*), Cochabamba and Sta Cruz, Bol., to e Brazil and s to La Rioja, Córdoba, and Buenos Aires, Arg. (*pitiayumi*). Around streams in dry forest, or in humid forest and second growth. 10.5 cm. Ssp *pitiayumi*: headside and upperparts plumbeous gray, **c back yellowish olive green**; **wings with 2 white bars**, outer 2 tail-feathers with subterminal white area. Lores blackish (dark gray-brown in female); underparts bright yellow (throat and breast ochraceous in male), lower belly and vent white. Sspp *pacifica* and *melanogenys* have plumbeous parts brighter blue, underparts brightest, outer tail-feathers with most white in *melanogenys*. Ssp *alarum* has only faint upper wingbar, at most. **Juv.** gray-brown above, whitish below, wings and tail like ad. Alone or in pairs, often in mixed-species flocks, flitters through the foliage, perch-gleaning undersides of leaves. Often on very thin twigs and tangles, mostly in canopy, but may prefer the understory when migrant warblers are present. Song *see see see see see dtrrrrrrr* accelerating in last half; also *swois swois swois see-ee-ee zee-ee-ee-ee-up* or *strsssss tsee-tsee-tsee*. At times give *chit* calls, single or in series.

YELLOW WARBLER *Dendroica petechia*. 5 sspp breeding in N Am. winter s to Bol., mainly in the trop. and subtrop. zones, but on migration regular to 2600 m on Bogotá savanna, Col., and occ. to 2900 m in Ecu. 29 lowland sspp are resident in the Caribbean, Galapagos isls, and C Am. and along Pacific slope s to Lima, Peru. 12 cm. **Male**: above bright pale yellowish green, outer greater wing-coverts edged yellow; below bright yellow, breast and sides streaked chestnut. Tail-feathers with yellow inner webs. **Imm. male** lacks chestnut streaks and has the sides faintly streaked green. **Female** like imm. male, but duller. Singly flits around in plantations, overgrown clearings, shrubbery, hedges, and gardens. Call a rather loud *chep*, a weaker *sit*, or still weaker *tseut*.

CERULEAN WARBLER *Dendroica cerulea* – Plate LII 13a-b
breeds in N Am., winters from Col. and Ven. s to Peru and Bol., mainly in trop. and lower subtrop. zones, but during migration rarely to lower temp. zone. 11 cm. **Male** (all seasons): above blue-gray, back streaked black, **wings with 2 broad white bars**; tail blackish edged blue gray, and with white subterminal areas. Headside blue-black, below white, narrow breast-band and streaks on sides slaty. **Female** (all seasons): above light blue-gray to grayish olive; wings and tail as in male, but edging light greenish. More or less distinct supercilium whitish, ear-coverts grayish or grayish olive, darker along upper margin; below dull white usually suffused with pale yellow; supercilium and underparts entirely yellow in fresh plumage. Yellower birds are possibly imm. Alone or in pairs, during migration in small flocks. May establish territory in winter quarters. Actively flitters and hovers amongst the foliage at all levels in semi-open country like shaded plantations and open woodland. Call a hissing *tsip*.

BLACKBURNIAN WARBLER *Dendroica fusca* –
Plate LII 11a-b

12 cm. **Male**: crown and headside black, long golden orange supercilium continues around cheek to golden orange throat and breast; spot below eye golden yellow. Above black streaked whitish, median and greater secondary-coverts white; outer tail-feathers mostly white. Streaks on sides of neck, breast, and body black, belly whitish tinged yellow. **1st winter male** has black feathers edged gray-brown, 2 fairly broad white wingbars, median crown-stripe only suggested, and the orange pale and restricted. **Female** like winter male, but black replaced by dusky and yellow even paler and more restricted. **1st winter female** duller, dark streaks obscure, hardly discernible on underside except at sides of breast. **Juv. male** like spring female, but without yellowish spot on crown, yellow of throat and breast much less orange. **Juv. female** like ad. winter female, but browner above, markings much less distinct, wingbars narrower, tail with less white, below pale buff-yellow, deepest on chest, sides and flanks indistinctly streaked gray-brown.
Habits: Alone or a few in mixed-species flocks, on migration in small loose flocks. Active. Horizontal, looking up and down, taking much prey from leaf-tops with a characteristic stretching head forward and down while perched. Sometimes searches dead leaves and may flutter in pursuit of insects. See pp. 309-17 in Keast & Morton, eds. (1980).
Voice: Call *ss-tsip* or a *tzik* repeated at 1 s intervals. In migrating flocks a sibilant *tsu*, like the dangling of a small chain.
Habitat: All levels, but mainly in canopy. Primarily in the humid zone, but sometimes in *Eucalyptus* and other trees in the arid zone. Clearings with scattered trees, park-like pastures, and along edges and breaks in and beside woodland.
Range: Breeds in N Am. Winters Sep-early Apr at 600-3600 m from Guatemala s to Ven., n Brazil, Col.(where abundant in C Andes in Jan), Ecu., Peru, and La Paz Bol. Winters fairly commonly at high elevations.

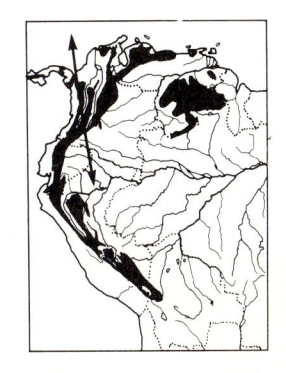

BAY-BREASTED WARBLER *Dendroica castanea* - **Plate LII 10a-c** breeds in N Am., winters from Panamá to n Col. and Ven., mainly in the trop. zone, but on migration to 3100 m in Ven. 13 cm. **Male spring**: chin, headside and forehead black, crown chestnut; pale **buff spot on neckside** continues as a narrow half-collar. Above gray-brown streaked blackish, wings with 2 white bars, narrow edges to tertials and tips of secondaries; outer 2 tail-feathers with large white subterminal patch on inner web. Throat, upper breast, sides of breast, and narrow sides of body chestnut, rest of underparts white. **Female** similar, but black of head mostly replaced by gray-brown, broken eye-ring and mottling on lores whitish, chestnut of underparts less extensive, paler, and mixed with buff. **Male winter**: crown and back light olive green broadly streaked black; no chestnut below except on flanks; breast buffy. **Juv.** much like juv. Blackpoll W., but brighter green above, more buffy (less yellowish) below and legs dark rather than pale. Often in small flocks, sometimes in mixed-species flocks. Actively perch-gleans insects, esp. from upper leaf surfaces in middle to upper strata. From dry, thorny vegetation to humid forest edge and shaded river banks.

BLACKPOLL WARBLER *Dendroica striata* – **Plate LII 12a-b** breeds in N Am., winters from Panamá s to Peru and e Arg. (1 rec. in Valdivia, Chile), primarily in the trop. zone, but on migration occ. to 3000 m. 12.5 cm. **Male spring: cap black, headside white** except for large black-streaked moustache to side of breast; narrow half-collar on nape streaked, but back gray-brown, dark-streaked; wings with 2 fairly broad white bars, tertials edged white, rest of remiges narrowly edged greenish. Outer 2 tail-feather pairs white subterminally. Below white. **Male winter**: above dull olive grading to dull gray on upper tail-coverts and narrowly streaked black; wingbars usually tinged yellow. Narrow and indistinct supercilium pale olive-yellow, upper eye-lid whitish; ear-coverts and sides of neck like upperparts. Below olive or straw-yellow to white on vent, sides and flanks indistinctly streaked dusky. **Female spring**: above olive to gray streaked black, narrow nape paler and grayer; wingbars narrow, whitish (tinged yellow in greenish birds), tertials edged white, rest of remiges narrowly edged olive. Eye-ring whitish, headside mottled dusky and whitish. Below olive-yellow to white, moustache and sides of breast and body streaked blackish. **Juv.** much like winter male, but more extensively yellow below (under tail-coverts white), upper tail-coverts dull olive like back, vs. grayish; back usually less streaked. Alone or in flocks, sometimes in mixed-species flocks. Forages from the ground to the canopy at edge of humid forest, in second growth, and in scattered trees in open country.

NORTHERN WATERTHRUSH *Seiurus novaeboracensis* breeds in N Am. and winters s to ne Peru, mainly in lowlands, but on migration up to 3050 m in E Andes of Col. 13 cm. Dull olive-brown above with prominent creamy eyebrow and dark eyeline. Below yellowish white to almost white, heavily streaked dark brown, **incl. on throat**. Mainly terrestrial, foraging along water, frequently bobbing hind part.

PROTHONOTARY WARBLER *Protonotaria citrea* breeds in N Am., winters in trop. zone s to Ven. and Ecu. Rec. as high as 3300 m in Sta Marta mts, Col. 13 cm. **Male**: entire head and underparts yellow with c belly and vent white. Back olive, wings uniform blue-gray; tail blue-gray with some white. **Female** duller, less orange.

CONNECTICUT WARBLER *Geothlypis agilis* – Plate LII 17 – breeds in N Am. and winters in lowlands from Nicaragua to se Peru. During migration to 4200 m in Ven. 13 cm. **Male spring**: **head, throat, and upper breast slaty, conspicuous eye-ring white**; above olive, below yellow, sides faintly olive. **Male winter** has forehead and crown tinged with brown and feathers of throat and breast indistinctly tipped paler gray. **Female** grayish olive instead of slate on head, chin and throat pallid brownish buff, breast darker. **Imm**. like ad. female, but crown always brownish olive, breast darker olivaceous, in female quite brownish buff. Solitary and shy, forages on or near the ground in humid forest and cloud forest. The closely related **MOURNING WARBLER** (*G. philadelphia*) has been rec. as high as 3000 m in E Andes of Col. It lacks eye-ring in male, but female and imm. are only safely told in the hand. Wing minus 0tail 19 mm or more in Connecticut W., 2-10 mm in Mourning. Hypothetical visitor Macgillivray's W. (*G. tolmiei*) very similar, but has a broken eye-ring in male, wing minus tail 10-18 mm (see *Bird Banding* 38 (1967):187-94).

NOTE: Previously placed in the genus *Oporornis*.

MASKED YELLOWTHROAT *Geothlypis aequinoctialis* – Plate LII 18a-b widespread in S Am. lowlands, with altogether 4 sspp, reaches 2400 m in Bol., and also edge of temp. zone in Arg. s to San Luis (*velata*). Within 0,2-1.5 m of the ground in dense thickets, esp. hardhack (*Spiraea*)-like vegetation, sometimes near marshes. 12.5 cm. **Male**: crown plumbeous, very narrow forehead and **anterior headside black**; above olive sometimes with gray cast, below bright yellow, sides washed olive. **Female** differs by having crown like back or slightly more grayish or brownish, headside gray-brown admixed yellow on lower parts, and eye-ring and lore whitish. **Imm. male** like female, but black mask suggested. **Juv**. undescribed. In pairs. Forages restlessly near the ground in dense vegetation, hopping up and down and between thin branches, circling tail and often cocking it almost upright. Male moves up 1-3 m to sing hidden in a bush top. Sometimes a pair will sing a harsh duet *trrep reh tri rreh trrep tri*, the *tri* sound given by the female. Territorial song of male a musical *tjip tjip tjip rreh dideledie diei*. Fledgling Mar (La Paz).

COMMON YELLOWTHROAT *Geothlypis trichas* breeds in N Am., and occ. winters s to n Col. Rec. once at 4500 m in Sta Marta mts. Resembles Masked Y., but male has white or pale gray border above black mask, and a whitish belly. Female has mainly dull whitish underparts, strongly tinged brown on the flanks.

CANADA WARBLER *Wilsonia canadensis* breeds in N Am., winters at 700-2600 m from Honduras s to c Peru and Amazonian Brazil; fairly common at lower elevations. 11.5 cm. **Male**: above gray, feathers of crown with black centers. **Eye-ring and supraloral streak yellow**, lore and cheek **black** continuing as **spots across breast**; underparts yellow, vent white. In **female** black is replaced by dusky olive, and upperparts are tinged olive. **Imm. male** much like ad. female. **Imm. female** is more strongly tinged olive above, markings on breast less distinct, or obsolete. Alone, usually in mixed-species flocks. Fans and often half-cocks tail, and jerks from side to side when excited. Sometimes follows army ants. Calls vary from *chick* or *check* to a distinctive, sharp, pebbly *sprit*. Found at all levels in shrubbery, second growth, and at edge of humid forest.

AMERICAN REDSTART *Setophaga ruticilla* – Plate LII 16a-c breeds in N Am., winters s to Peru and n Brazil, mainly in the trop. zone, but occ. to 3000 m. 12.5 cm. **Male**: black, base of remiges, base of tail (except c pair), pectoral tuft, and faint streaks on upper belly and flanks orange-red, belly white. Bill black in spring. **Imm. male**: crown and headside dark gray, back gray-brown. Red of tail replaced with yellow, like the small patch at base of inner remiges. Eye-ring and underparts whitish, pectoral tuft orange-yellow. In spring body with scattered black feathers. **Female** like imm. male, but back usually less gray, patch on breast-side yellow (to orange-yellow) and yellow patch in wing more extensive. **Imm. female** similar, but head in less contrast to back, yellow patch in wing very restricted, occ. concealed; throat and esp. breast tinged brownish buff. Forages singly in lower to medium strata in woodland, second growth, montane shrub, and gardens in both humid and (near water) in semi-arid regions. Acrobatically hops among branches and foliage, momentarily clinging to trunks, and also flycatching. Often fans tail. Call *chip*, *tsip*, or *stchit*.

SLATE-THROATED WHITESTART *Myioborus miniatus* – Plate LII 8

12 cm. **Upperparts and sides of head and neck dark gray** to blackish gray on anterior headside, throat blackish gray, c. crown rufous. Underparts below throat bright yellow, sometimes tinged ochraceous on breast, vent whitish. Tail blackish, outer 2-3 feathers mostly white; under wing-coverts whitish, axillaries blackish. S Am. sspp poorly differentiated, varying in intensity of yellow below, amount of blackish on forehead and sides of crown, and amount of white in tail. **Juv**. uniform brownish gray, palest and brownest below; wings and tail as in ad.
Habits: In pairs or small groups, usually very conspicuous in mixed-species flocks. Actively flycatches and perch-gleans along the branches wagging body from side to side, regularly flashing white in tail, and often dropping like a leaf with spread tail in pursuit of an insect. Frequently searches dead leaf clusters. Not shy. Aggressive towards Spectacled W. See *Living Bird* 15 (1976):119-41.
Voice: A *tick* is given during foraging. Song is an accelerated series of

chirking notes finished off by 1-2 single notes *t-t-trrrrrt t t* lasting 1-3 s, somewhat falling during the accelerated part.
Breeding: Eggs Dec-July (Valle), Apr (Huila); fledglings Apr (Huila), Mar, June (Cauca), Nov (Sucre); juv. Feb (Chimborazo), Mar (nw Ecu.), June, July (Cauca), Aug (Mérida), Dec (Junín); large gonads Mar-Nov (n Col.).
Habitat: Lower to middle strata using upper strata more when migrant warblers are gone. Humid forest and cloud forest, mostly at edge; second growth.
Range: 500-3000 m, mainly subtrop. zone. Sta Marta mts (*sanctaemartae*), Darién in Panamá, Perijá mts and Lara, Ven., s through all 3 ranges of Col. to n Ecu. (*ballux*), El Oro, sw Ecu. s to Piura, nw Peru (*subsimilis*), from Morona-Santiago, se Ecu, to Cochabamba and in Chuquisaca, Bol. Also on Tepuis of Ven. and adjacent Guiana and Brazil (*verticalis*). Common in subtrop. zone. 7 sspp from Mexico to Panamá, 1 in coastal Ven.

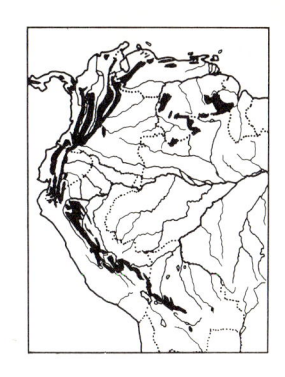

GOLDEN-FRONTED WHITESTART *Myioborus ornatus* – Plate LII 6a-b

13 cm. Above dark smoky gray, hindcheek, hindcrown, nape, neckside, and wings blackish, earcoverts mixed with a few white feathers. **Most of crown yellow, mask incl. narrow forehead and chin to around eye, white**; below bright yellow, sides of breast dark smoky gray, vent whitish. Tail blackish, outer 2 feathers mostly white. Ssp *chrysops* similar, but pale crown only reaches to above eye and is orange rather than yellow; **patch on ear-coverts pale yellow**, mask and underparts warm yellow. **Juv**. undescribed.
Habits: In pairs or family groups, by themselves, or nuclear in mixed-species flocks. Actively flits and sallies from low to high, keeping near the periphery of trees and bushes, and often very confiding.
Voice: Song (*ornatus*) a high-pitched rambling *pit it,t'chit, tswit tsweet, pits-whew! sits sweet iit…* for up to 15 s. Call *tsip*.
Breeding: Fledgling May, July (Cauca); juv. Mar (Huila), Apr, Nov (Cundinamarca); large gonads July (Santander).
Habitat: Middle to upper strata of trees in dense cloud and dwarf forest patches.
Range: 1800-3400 m, mainly above 2400 m. From Páramo de Tamá in sw Táchira, Ven., s to Cundinamarca, Col. (*ornatus*); W and C Andes of Col. from Antioquia s to Cauca and Huila and E Andes in w Caquetá (*chrysops*); n Nariño and immediately adjacent Napo, ne Ecu.? (sightings). Fairly common.

NOTE: Forms a superspecies with Spectacled, White-fronted, and Santa Marta W.

SPECTACLED WHITESTART
Myioborus melanocephalus – Plate LII 7a-d

13 cm. Ssp *ruficoronatus*: above dark gray, wings blackish. C. crown rufous; forehead, sides of crown, lores, headsides and usually nape black.

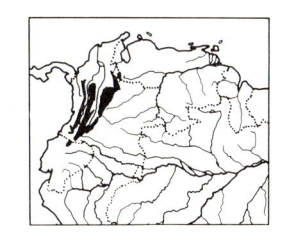

Supraloral streak, extreme forehead, **eye-ring**, and **underparts bright yellow**, sometimes tinged ochraceous, vent whitish. Tail blackish, outer feathers mostly white, next 1-2 with varying amounts; under wing-coverts blackish, axillaries whitish. Some birds from ne part of range (s to Cuyuja) have less black on head, and more extensive yellow forehead; *griseonuchus* similar, but nape with hardly any black, eye-ring broken in front, and side of head less deep black. Ssp *melanocephalus* has solid black crown; black feathers below eye basally yellow (in *malaris* wholly black); underparts slightly paler yellow, tail on average with more white. Ssp *bolivianus* with palest underparts, and with most white in tail (3rd outermost with considerable amount), black feathers below eye basally yellow, sometimes wholly yellow (always so in Urubamba Valley). **Juv.** gray-brown above, throat buffy, belly pale yellowish. 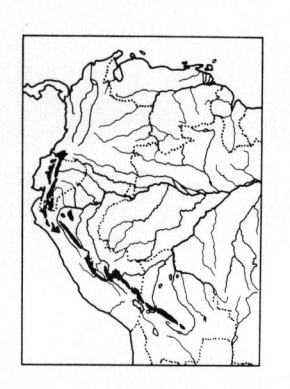
Habits: In pairs or groups of 4-5, conspicuous in mixed-species flocks. Sallies and perch-gleans among leaves and along branches and vines, sometimes head down, twisting body from side to side, and often fanning its slightly raised tail. Frequently drops like a leaf with spread tail. Aggressive towards Slate-throated W.
Voice: Male sings *vee vee vée vee te ttttt-t-t-t* or *tetelet-t-t-t-t* and female answers *tk-tk-tk-tk-*.... Contact call *tsit*.
Breeding: Fledglings Sep, Mar (nw Ecu.), June (Huánuco), Dec (La Paz); juv. Apr, July (nw Ecu.), Feb, Mar (Amazonas), June (Huánuco), Dec (Junín), Jan (La Paz).
Habitat: 2-10 m up in humid montane shrubbery, forest, and cloud forest.
Range: 2000-3950 m. From both slopes of Nariño, s Col. s to Loja, s Ecu. (*ruficoronatus*), W Andes in Piura and Cajamarca, n Peru (*griseonuchus*); Utcubamba drainage, Amazonas, n Peru (*malaris*), from s Amazonas s to Ayacucho, c Peru (*melanocephalus*), and from Cuzco, se Peru to Sta Cruz, Bol. (*bolivianus*). Common.

NOTE: The 2 northern sspp are sometimes considered a distinct species: **CHESTNUT-CROWNED WHITESTART (*M. ruficoronatus*)**.

WHITE-FRONTED WHITESTART *Myioborus albifrons*
– Plates LII 5 and LXIV 19

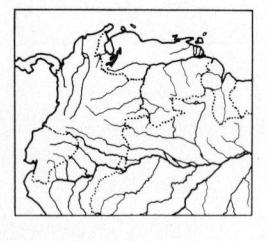

13 cm. Crown black with rufous center, rest of upperparts dark gray tinged brownish on back. **Lore and eye-ring white**, rest of headside and sides of neck dark gray. Below bright yellow, chin white, vent whitish. Tail blackish, outer 2 feathers mostly white. **Juv.** has crown and headside brownish gray like back, underparts buffy brown.
Habits: As in Slate-throated W.
Voice: No data.
Breeding: Juv./ad. June (Mérida).
Habitat: Cloud and dwarf forest.
Range: Nw Ven. at 2200-4000 m from Trujillo to Táchira.

SANTA MARTA WHITESTART *Myioborus flavivertex* – Plate LXIII 25

12.5 cm. **Crown yellow bordered all around with black**; back and wing-coverts dark olive, remiges slaty. Supra-lore and area above eye buff, chin and sides of head and neck black. **Below bright yellow**, vent sometimes almost pure white. Tail black, outer feather mostly white (only base of inner web black), next with less white, 3rd with white at tip only. **Juv.** has dull brown head, the yellow crest duller and restricted.
Habits: In pairs, usually in mixed-species flocks. Flits restlessly from branch to branch and tree to tree, bobbing tail, only rarely flycatching or spreading wings and tail like Slate-throated W.
Voice: Call a sharp *chip*. Song a weak sibilant series of phrases recalling Slate-throated W.
Breeding: Eggs May; juv./ad. July.
Habitat: Canopy of humid forest, cloud forest, forest edge, and light woodland.
Range: N Col. At 1500-3050 m in Sta Marta mts, mainly above 2000 m. Common.

NOTE: Called Yellow-crowned Redstart by M.d.Schauensee.

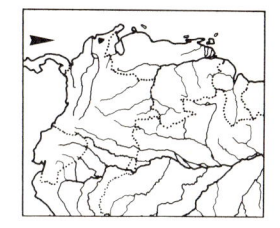

BROWN-CAPPED WHITESTART *Myioborus brunniceps* – Plate LII 9

13 cm. Crown rufous bordered narrowly on forehead with blackish and on the sides with dark gray; rest of upperparts dark gray, except for olive-green back. Faint supraloral streak and **broken eye-ring white**, loral region blackish, rest of **headside and side of neck dark gray**. Below bright yellow, vent whitish. Tail blackish, outer 2 feathers mostly white, 3rd white along terminal part of shaft. **Juv.** brownish gray above, throat buffy-brown, breast brownish buff with or without dark brown spots, belly pale yellow.
Habits: Solitary or in pairs. Silently and inconspicuously moves across the branches through the inside of canopy of small trees. **Does not fan tail**.
Voice: Song fine and wheezy, increasing in speed and amplitude and lasting 3-4 s: *tzeseeseeseeseeseesee* or *tzeseeseeseeseeseesee-see-see-see*.
Breeding: Fledgling Jan (Tarija); juv./ad. Jan, Feb (Tarija).
Habitat: Dense canopies of humid or semi-humid wooded slopes and ravines.
Range: 1400-3200 m, in warmer parts of Cochabamba up to 3800 m, in s part of range down to 400 m in fall and winter. Andean slopes in La Paz, and from Cochabamba, s to San Luis, Arg. (*brunniceps*). Fairly common in Arg. 3 other sspp on Tepuis of Ven. and adjacent Guiana and Brazil (sometimes regarded a distinct species Tepui Whitestart *M. castaneocapillus*).

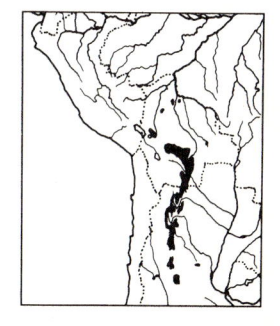

BLACK-CRESTED WARBLER
Basileuterus nigrocristatus – Plate LII 20

12.5 cm. Above citrine, forehead and **c crown black**, broad supercilium yellow, vanishing behind eye; line from bill to just through eye black, broken eye-ring yellow, rest of headside citrine. Below yellow, sides and sometimes faint clouding on breast citrine. **Juv.** dusky olive, darkest on head, belly olive.
Habits: In pairs or family-groups, often in mixed-species flocks. Restlessly forages for insects in the middle of tangles and bushes. Sings from near tip of middle strata branch.
Voice: Song an accelerated series of notes, falling at the end of the series, sometimes the accelerated last part repeated immediately after an introductory note *ti ti ti ti ti tierrrr* or *ti ti ti titierr ti tirrrr*. Alarm a repeated *tick*.
Breeding: Fledglings Jan, Oct (Cauca), Aug (Caldas); juv. Mar, July (w Nariño), May (nw Ecu.); large gonads May-July (Perijá, C Andes Col.).
Habitat: Near the ground in tangles and other dense parts of humid montane shrubbery, elfin forest, riparian thickets and edge of humid forest and cloud forest.
Range: 1300-3950 m, mainly above 2600 m, generally at higher elevations than Citrine W. From Perijá mts and coastal mts of Ven. s through all 3 ranges of Col., Ecu. and W and C Andes of n Peru to La Libertad, and recently mistnetted on e slope of Cord. Blanca in Ancash (O Frimer). Common.

NOTE: Previously included the form *euophrys*, now treated under Citrine W.

CITRINE WARBLER *Basileuterus luteoviridis* – Plate LII 22a-d

13.5 cm. **No eye-ring**. Ssp *luteoviridis*: above citrine, supercilium yellow, palest anteriorly, and disappearing behind eye; line through eye blackish; lower headside citrine, area below eye mottled with whitish. Underparts yellow, sides citrine. Legs usually buffy brown, occ. orange-brown, or even as pale as in Pale-legged W. (younger birds?). Ssp *richardsoni* paler, above olive-gray, wings olive; supercilium yellowish white to cream, sometimes whitish anteriorly; below dull olive-yellow, yellowest on c throat and belly; *quindianus* intermediate between *luteoviridis* and *richardsoni*; *striaticeps* like *luteoviridis*, but larger, sides of crown faintly washed with blackish, supercilium broader, longer, and much brighter; below brighter yellow, breast more clouded with citrine. Ssp *euophrys* has extensive black forehead and sides of crown, very broad and bright supercilium, blackish ear-coverts, and bright yellow underparts without citrine clouding on breast. **Juv.** dark olive-brown, faint supercilium ocher-citrine, c belly and vent yellow-ocher, legs pale. **Juv**. *euophrys* has longer, broader, and bright ocher supercilium, and faint dark forehead and line through eye.
Habits: In pairs or family groups, often in mixed-species flocks; *euophrys* usually in groups up to 12. Actively perch-gleans and sometimes

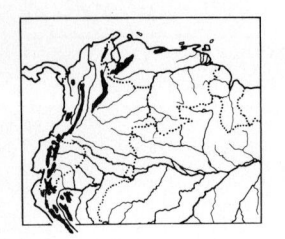

sallies along the branches, picking insects from upper and esp. lower surfaces of green (rarely dead) foliage.

Voice: Song and chatter much as in Pale-legged W. The chatter of *euophrys* is a 3-4 s long quavering trill starting with a few single notes. Contact note *tzik* or *tick-tick-tick*.

Breeding: Fledglings Feb (Cauca), Sep (Cundinamarca), Aug (Huánuco); juv. July (e Nariño), Jan (La Paz).

Habitat: As for Pale-legged W., but generally higher (2-5 m up) in the vegetation (*euophrys* low).

Range: 2300-3400 m, rarely down to 1700 m, usually higher than Pale-legged, and lower than Black-crested W. Antioquia and Cauca, W Andes of Col. (*richardsoni*), C Andes of Col. (*quindianus*), Mérida, Ven. s through E Andes of Col. and e Ecu. (sight rec from mt Pichincha, nw Ecu.) to Piura and n Cajamarca, n Peru (*luteoviridis*); s and e of the Marañón from Amazonas to Cuzco, Peru (*striaticeps*). Puno, se Peru, to Cochabamba, Bol. (*euophrys*). Common.

NOTE: (Mega)ssp *euophrys* was formerly treated as a ssp of Black-crested W.

PALE-LEGGED WARBLER *Basileuterus signatus* – Plate LII 21a-b

13 cm. Ssp *signatus*: headside and upperparts citrine, underparts yellow faintly clouded with olive on breast and sides outlining **yellow throat**. Yellow supercilium becomes faint behind eye, **line through eye dusky,**

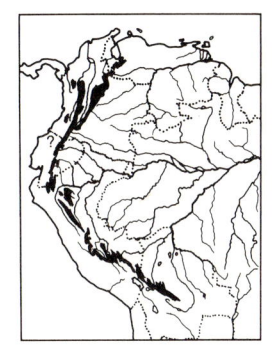

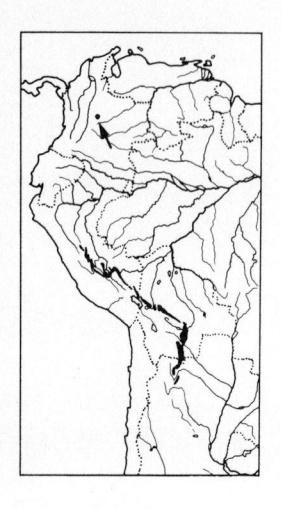

breaking yellow eye-ring; some yellow mottling below eye. Legs pale. Ssp *flavovirens* has a faint, narrow blackish forehead and line above supercilium. **Juv.** dark brown without olive tinge, c belly pale buff.
Habits: In pairs or family groups up to 6, often in mixed-species flocks. Actively perch-gleans and sallies in the undergrowth, picking insects from the upper surface and esp. by reaching up to under surface of green foliage. Frequently moves tail in circles.
Voice: Vocal. Song fast and continuous, of 2-3 notes given in varying order *tjidewitcheedewitchechiffewitcheedee...*; often another bird (mate ?) simultaneously gives a fast chatter *tjketjketjkeriketjekerike...*, or both chatter. During foraging a fine, tanager-like *tsit*. Contact note a slightly snarling *chiff*, different from that of Citrine W. Alarm a hard *tscheck*, sometimes prolonged to a chatter.
Breeding: Fledgling Feb (Huánuco, Tarija); juv Dec (Junín); large gonads June (Cundinamarca).
Habitat: 0.5-2.5 m up in undergrowth of humid forest and cloud forest and adjacent second growth, often with admixed bamboo.
Range: 1800-3050 m, rarely to 3300 m, in Arg. down to 1600 m; generally lower than Citrine W. From Huánuco, c Peru, s to Urubamba Valley, Cuzco, se Peru (*signatus*), e Cuzco in se Peru along Andean slope to Jujuy, nw Arg. (*flavovirens*). 1 specimen from Páramo de Guasca, 3400 m, Cundinamarca, Col. (ssp ?). Fairly common to common.

SANTA MARTA WARBLER *Basileuterus basilicus* – Plate LXIII 24

14.5 cm. **Head mostly black**, c crown-stripe, supercilium encroaching earcoverts, grizzling below eye, chin, and c throat white. Back, wings, and tail olive-green, underparts bright yellow, sides and flanks tinged olive. Female duller than male. **Imm. male** has the black and white areas of head and neck duller and less sharply defined, the black more brownish and the white tinged buffy.
Habits: In pairs or groups of 3-5, often in mixed-species flocks. Perchgleans like a hemispingus. Difficult to observe.
Voice: No data.
Breeding: Imm. July.
Habitat: 1-4 m up in thick shrubby edge of humid forest, esp. with bamboo, on slopes and ridges and along streams in humid forest.
Range: N Col., at 2100-3000 m, mainly above 2300 m in Sta Marta mts. Uncommon to fairly common.

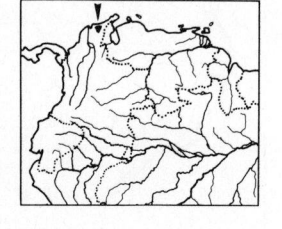

THREE-STRIPED WARBLER *Basileuterus tristriatus* – Plates LII 23a-b and LXIV 20

12-13 cm (*tristriatus* largest). Above olive to olive-brown or grayish olive (grayest in *canens*). Crown and headside black to blackish (headside buffy in *meridanus*), **median crown-stripe ocher yellow** (northern sspp) or **clay white** (southern sspp) becoming duller and grayer posteriorly; **long supercilium and patch below eye whitish to pale buffy clay**. Below yellow (*tristriatus*) to pale yellowish buff, throat palest; obscure moustache dusky, sides yellowish olive (*tristriatus*) to dull olive,

breast more or less clouded or spotted with olive. Younger birds have duller black on the head, feathers of forecrown with grayish buff tips.
Juv. very uniform olive-brown with slight indications of pale facial pattern, belly pale buff.
Habits: In pairs or groups, sometimes up to 30, often in mixed-species flocks. Forages along rather thick horizontal branches, bamboo and dense undergrowth, frequently making small tail jerks, and twisting its body from side to side as it moves. Rarely on the ground.
Voice: Song of 2 parts, first a single very long whistle, then a short weak (siskin-like) twitter ending in a chatter of 2-3 notes. Contact call *tsik*. Call a high-pitched husky *chee-wéep*.
Breeding: Eggs May-June (Valle), Apr (Huila); fledgling Jan-July, Sep, Oct (Valle); juv. June (Valle), July (w Nariño), Nov (Puno); large gonads May-July (Col.), Aug, Oct (La Paz).
Habitat: 1-3 m up at edge of cloud forest (sometimes very mossy), humid forest, and in adjacent second growth, often admixed with bamboo.
Range: 300-2700 m, *punctipectus* rarely to 3000 m. W Andes and from Caldas on w slope of C Andes, Col., s to Chimbo river, w Ecu. (*daedalus*), c Lara to ne Táchira, Ven. (*meridanus*), Macuira mts Guajira peninsula, Perijá mts, and from sw Táchira, Ven., s through E Andes and e slope of C Andes to Huila and Nariño, Col. (*auricularis*), e Ecu. s to n Morona-Santiago, se Ecu. (*baezae*), Morona-Santiago se Ecu. s to Cuzco, se Peru (*tristriatus*), Puno in se Peru to La Paz, Bol., intergrading with *punctipectus* in e La Paz (*inconspicuus*), Cochabamba (*punctipectus*) and Sta Cruz, Bol. (*canens*). Fairly common to common. 2 sspp in coastal Ven., 2 in Panama and immediately adjacent Col.

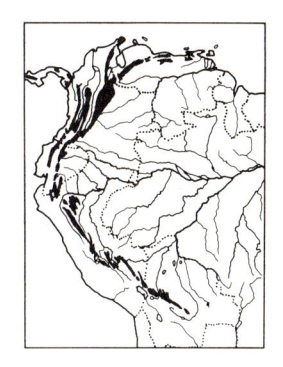

THREE-BANDED WARBLER *Basileuterus trifasciatus*

12 cm. **Crown gray bordered laterally by black bands** that continue to nape, c crown-stripe suffused with lemon-yellow anteriorly, olive-gray posteriorly (yellow most extensive in *nitidior*). Back, wings, and tail dull grayish olive, brighter and greener on rump and upper tail-coverts. Headside gray with black stripe through eye and pale spot on ear-coverts; throat and upper breast grayish white, rest of underparts bright yellow. **Juv**. undescribed.
Habits: Alone, in pairs, or small groups of 5-6, often in mixed-species flocks. Forages in the undergrowth and lower branches of trees. Inquisitive.
Voice and **Breeding**: No data.
Habitat: Usually low, but in feeding flocks up to 10 m up in lower canopy. Riparian thickets, forest edge, second growth, undergrowth in humid forest, and near streams in dry forest.
Range: 1500-3050 m. From El Oro and Loja, sw Ecu., to Tumbes, nw Peru (*nitidior*), Piura, Cajamarca, Lambayeque, and La Libertad, nw Peru (*trifasciatus*). Common.

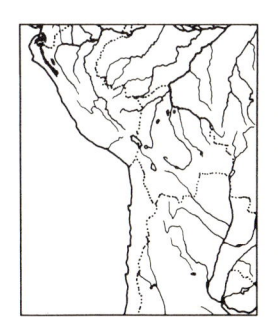

RUSSET-CROWNED WARBLER *Basileuterus coronatus* – Plate LII 24a-b

13.5 cm. Ssp *coronatus*: **head gray**, throat paler, **c crown-stripe orange-**

rufous bordered laterally by black lines to nape, and also with dark eye-stripe. Body citrine grading to bright yellow on belly. Sspp *elatus* and *inaequalis* have the c crown-stripe reaching bill, *inaequalis* with buffy or rufescent rather than orange-rufous forehead. Ssp *notius* is somewhat smaller and duller, and more shaded with olivaceous below; *regulus* has paler c crown-stripe. **Deviating c population** (sspp *orientalis, castaneiceps* and *chapmani*) have bronzy citrine upperparts, pale gray throat, light **gray breast and sides**, white belly (sometimes tinged yellow), bronzy citrine flanks, and buff-yellow vent. **Imm**. has feathers of c crown-stripe tipped olive. **Juv**. uniform grayish olive, with belly the same color as in ad.

Habits: In pairs or family groups, often in mixed-species flocks. Actively moves through the undergrowth near thick horizontal branches, often circling tail.

Voice: Alarm a nasal *rrr rrr-rrr* and a *tack-tack*. Call a wheezy *wheeee*. Song wheezy and melodic, distinctly rising at the end *tee tee tee teehe weede wéeh*, lasting 1.5-5 s, and repeated at 0.5-5 s intervals.

Breeding: Fledgling May, June, Sep, Oct (Valle), June (Mérida), Feb (Pasco); juv. Aug (Huánuco), Feb (Pasco); large gonads Feb-July (W and C Andes Col.), Oct (Boyacá).

Habitat: 1-6 m up in undergrowth of humid forest and cloud forest and adjacent second growth.

Range: 1300-3100 m. From s Lara, Ven., s through all 3 ranges of Col. to Cauca and Huila, on e slope of E Andes in (s/n to ?) Boyacá (*regulus*), w Nariño, s Col., s to Chimbo river, w Ecu. (*elatus*), e Ecu. s to Morona-Santiago (*orientalis*), Azuay in sw Ecu. s to w slope of W Andes in Piura, nw Peru (*castaneiceps*), e slope of W Andes in Cajamarca, nw Peru, possibly same form also on Pacific slope in c Cajamarca (Llama-Huambos) (*chapmani*), s and e of the Marañón in Amazonas and San Martín, Peru (*inaequalis*), from Huánuco to Puno, c-se Peru, intergrading with *notius* in La Paz, Bol. (*coronatus*), La Paz and Cochabamba, Bol. (*notius*). Fairly common to common.

NOTE: WHITE-LORED WARBLER (*B. conspicillatus*) of Santa Marta mts Col. (750-2200 m) often treated as a ssp of Russet-crowned W. (see also *Bull. Brit. Orn. Club* 95(1975): 173-75.

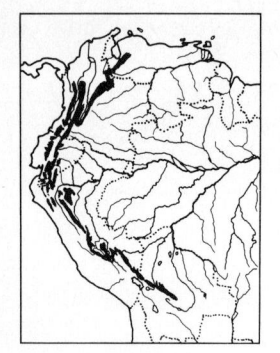

Honeycreepers – Family Coerebidae

Some 40 species of '9-primaried oscines'. The systematic unity of this group is highly disputed. It has been suggested that some genera evolved from American warblers (*Parulidae*), others from tanagers (*Thraupidae*). Recent biochemical evidence support inclusion of most (or all) genera in the tanagers. As the exact relationships of all the forms have not been fully established, the 'family' is here treated traditionally for convenience. They are small active birds; Andean forms mainly inhabit shrub and undergrowth. Sexes alike or different, plumage often brightly colored with blue dominating. The conebills have thin and straight, pointed bills, and feed on insects and berries. Most other members of the 'family' also eat nectar. Flowerpiercers have slightly upturned mandibles and hooked maxillas, and often pierce the bases of flowers without pollinating them; several taxa can be grouped into 'superspecies', each with a number of allospecies that may hybridize insignificantly in zones of secondary contact (see *Amer. Mus. Novit.* 2381(1969), and *Condor* 84(1982):1-14).

The flowerpiercer bill type apparently evolved twice (see Buckley *et al.* (1985):319-32), and it has therefore been suggested to place the forms *caerulescens, cyanea, indigotica*, and *glauca* in a separate genus *Diglossopis*. Biological data about most honeycreepers are compiled by Isler, ML & PR (1987) *The Tanagers*. Washington D.C.: Smithsonian Institution Press. See also *Amer. Mus. Novit.* 1193 a,d 1203 (1942), and Bond (1955a).

BANANAQUIT *Coereba flaveola* - **Plate LXIV 21** is widespread, with 41 sspp in the tropics of C Am., the Caribbean, and S Am., where also in premontane zone, and to 2600 m near Bogotá (*columbiana*), 1200-1900 m and probably higher in Andes of Mérida and Táchira, Ven. (*montana*), 1000-2400 m in upper Marañón valley, n Peru (*magnirostris*), to 2400 m in Huánuco and Junín, c Peru, and from La Paz to Sta Cruz, Bol. (*chloropyga*). Fairly common in subtrop. zone. 9 cm. **Bill** slender and pointed, **slightly curved**. Sspp *magnirostris* and *columbiana*, **male**: above dark gray-brown, crown and headside blackish, **supercilium white**. Primaries with white basal patch, rump yellow; outer 2-3 tail-feathers tipped white. **Below yellow** becoming more olive on sides and white on vent. **Female**: similar, but crown and upper headside scarcely darker than back, rump olive-yellow, throat whitish. Ssp *montana* darker gray and brighter yellow; *chloropyga* lacks wing-patch, and has olive-yellow rump. **Juv**.: above dull olive, supercilium faint and pale yellow, throat yellowish, rest of underparts dull yellow. Alone, in pairs, or family-groups. Actively forages among twigs for nectar and fruit. Sings from high branch. Song an insect-like, high-pitched trill. Eggs and fledglings Sep-Oct (La Paz). All strata. From humid rain and cloud forest to dry, thorny woodland and scrub, second growth, plantations, and gardens.

CINEREOUS CONEBILL *Conirostrum cinereum* – Plate LIII 5a-b

10.5-11.5 cm (*fraseri* largest). Ssp *cinereum*: above light slate-gray, crown dusky; wingbar, patch at base of primaries, edge of tertials, narrow forehead, and supercilium white. Below pale gray, vent buff. Ssp *littorale* similar, but supercilium shorter and underparts tinged buff; *fraseri* much darker, back brownish olive, fairly long and broad supercilium warm buff, underparts brownish clay to warm buff. **Imm**. has buffy wash to wingbar and edges of tertials. **Juv**. has feathers of back and breast faintly dark-tipped, supercilium and underparts faintly tinged yellowish.
Habits: Alone, in pairs or groups. Restlessly flitters about for insects and berries, often hanging, and spending most time along middle branches inside the vegetation.
Voice: Contact call a fine *zee* or *zee zeet*. Other calls include a fine twitter.
Breeding: Eggs Nov (Chile), Apr (La Paz, Cochabamba); fledglings June (Huánuco, w Ecu.), Aug (La Paz, Cochabamba); juv./imm. May (Puno), July (e Nariño); large gonads Feb (Cauca/Huila); food-carrying (green caterpillar) May (Moquegua).
Habitat: 0.5-2 m up in a variety of habitats. Arid to fairly humid scrub and shrubbery, forest edge, riparian thickets, gardens, *Polylepis* woodland, hedgerows etc.

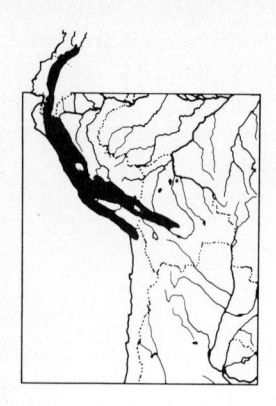

Range: Coast to 4500 m. From Cauca at s end of C Andes and Nariño, Col. to Loja, s Ecu. (*fraseri*), from w slope of W Andes and upper Utcubamba and Marañón valleys, n Peru, s through W Andes to Tarapacá, n Chile (*littorale*), and from Huánuco, c Peru, to Cochabamba, Bol. (*cinereum*). Uncommon in Col., otherwise common.

TAMARUGO CONEBILL *Conirostrum tamarugense* – Plate LIII 6a-b

11.5 cm. Above **gray**, wings with white patch at base of primaries, 2 rufous wingbars and buffy brown edges to tertials. Lores gray, short supercilium, **area below eye, throat, and upper breast rufous brown**; rest of underparts light gray, to buffy white on c belly, rufous on vent. **Imm. female** has narrower and pale buff wingbars, and has less, paler, or no rufous on throat. **Juv**. unknown.
Habits: In pairs or flocks up to 20, sometimes with Cinereous C. Moving flocks agile and restless, only perch occ., then move on through the interior of the bushes, constantly calling.
Voice: Contact call a fine *zie zizie* like Cinereous C., perhaps a trifle sharper.
Breeding: Presumably late Dec to early Mar. Unossified skull (2 males and 2 females) Aug, Sep (Arequipa); 50% ossified skull (2 males) June, Aug (Arequipa); imm.-plumaged July (Arequipa, some sighted: TP), Dec (Tarapacá, 1 female).
Habitat: *Polylepis* and *Gynoxys* shrub, *Prosopis tamarugo* and *Citrus* plantations, gardens and riparian scrub.
Range: No rec. from late Dec to early Mar, the presumed breeding season (a single was seen in Arequipa in Feb in a climatically abnormal year). Rec. at 3400-3950 m in Arequipa (Chachani, Pichupichu), at 4050 m in Tacna, sw Peru, and from sea level to 2600 (2950 ?) m in Tarapacá (Azapa, Vitor and Camarones valleys in Arica, Mamiña, Pica, and at Salar de Pintados at Panamerican highway at ca. 20°30' S). Common locally and temporarily. The breeding terrain not definitively known, but may possibly be the high elevation zone.

NOTE: Described by Johnson & Millie (1972). See also *Wilson Bull*. 90(1978):445-6, and *Condor* 89(1987):654-8.

WHITE-BROWED CONEBILL
Conirostrum ferrugineiventre – Plate LIII 4

12.5 cm. Crown slaty, remaining upperparts blue-gray, edge of primaries whitish in some lights. **Supercilium white**, headside blackish, occ. with paler streaks, **underparts rufous**, palest on chin; moustache connected with neckside slaty. Birds from Huánuco have slightly blacker crown, bluer back, and buff-tinged supercilium. **Imm**. similar, but tertials distinctly edged whitish, moustache and upperparts more gray, crown rather uniform fuscous black; supercilium buffy white, headside gray and more streaked with pale shafts on ear-coverts. **Juv**. similar, but with dusky feather-tips on mantle and breast.
Habits: Alone, in pairs, or family groups, sometimes in mixed-species

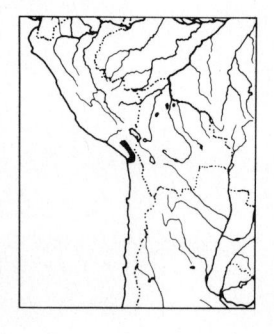

flocks, or associated with a pair of Giant C. Crawls along branches inside canopy and often hangs in thin outer twigs, but usually hidden in the vegetation. Flowerpiercer-like.
Voice: Birds seen by us were silent.
Breeding: Juv. Sep, Nov (Cochabamba); imm. Dec, Jan (La Paz).
Habitat: Low to high. Bushy slopes at edge of cloud and elfin forest and in páramo shrubbery, often admixed with some *Polylepis (multijuga)*. Sometimes ascends to genuine semi-humid *Polylepis* woods.
Range: 2000-3700 m (occ. 4100 m). In Huánuco and Junín, c Peru (unnamed ssp); from Cuzco, se Peru, to Sta Cruz, Bol. (*ferrugineiventre*). Fairly common.

NOTE: A hybrid with Giant C. is known from Puno se Peru.

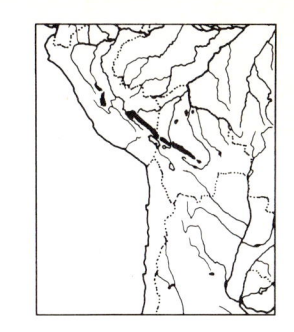

RUFOUS-BROWED CONEBILL *Conirostrum rufum* – Plate LIII 3

12 cm. Above slaty-blue, tertials edged white. **C forehead, supercilium, headside, and entire underparts rufous. Juv**. duller, the underside cinnamon fawn-colored.
Habits: In pairs, rarely small groups, sometimes in mixed-species flocks. Hops about among the tips of the branches. Relatively sluggish and confident.
Voice: A complicated series of high-pitched squeaky notes given very fast, resembling song of Blue-backed C.
Breeding: Fledglings July, Aug, Dec (Cundinamarca), Oct (Santander); large gonads Feb-Sep (Sta Marta, E Andes Col.).
Habitat: Low stunted trees and shrubbery at edge of cloud forest and adjacent second growth; sometimes in semi-arid land with cultivation, shrub, and groves.
Range: Col. at 2700-3400 m in Sta Marta mts and E Andes s at least to Cundinamarca. Uncommon to fairly common.

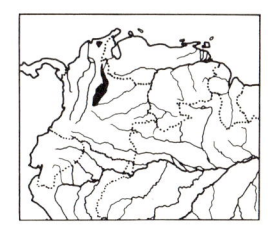

BLUE-BACKED CONEBILL *Conirostrum sitticolor* – Plate LIII 2a-b

12 cm. Ssp *sitticolor*: head, **throat and upper breast glossy black**; back and rump glossy blue; wings and tail black, feathers edged blue-black. **Underparts below upper breast rufous.** Ssp *intermedium* similar, but with blue supercilium from above eye. Ssp *cyaneum* has glossy purplish blue supercilium connecting with back, and dull dark blue instead of black throat and upper breast. **Imm**. duller, blue-gray on back and throat area. **Juv**. undescribed.
Habits: Alone, in pairs, or groups of up to 8, often in mixed-species flocks. Agile. Forages for insects and seeds among leaves at tip of otherwise naked branches and in bush tops, frequently hanging, occ. probing moss and flowers, and making short sallies.
Voice: Contact note a fine *zit*, often repeated. Song a complex chattering series of similar fine notes.
Breeding: Fledglings Jan (Cauca), Nov (Cundinamarca), July (Huánuco); juv./imm. Oct (Santander), Apr, Sep (Caldas), Feb (Cundina-

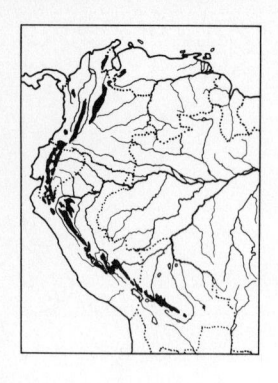

marca, ne Ecu.), Aug (Loja), July, Nov (Piura), Oct (Amazonas), Aug (Huánuco), June (Cuzco), Sep, Oct (Cochabamba); large gonads Feb, July (C Andes Col.).
Habitat: 1-14 m up in elfin, cloud, and humid *Polylepis* forest and humid open slopes with scattered low trees or shrubs.
Range: 2450-3800 m. Andes of Ven. in Mérida and n Táchira (*intermedium*), Páramo de Tamá, Perijá mts, and all 3 ranges of Col. s through Ecu. to Piura, Cajamarca, and s Amazonas, n Peru (*sitticolor*, birds from Carpish mts, Huánuco, intermediate with *cyaneum*), E Andes of Peru except range of *sitticolor*, and from La Paz to Sta Cruz, Bol. (*cyaneum*). Fairly common to common.

CAPPED CONEBILL *Conirostrum albifrons* – Plate LIII 7a-d

12 cm. **Male blue-black with blue cap (or white in n of range)**. Ssp *atrocyaneum* dull black with faint, dark blue tinge; crown, lower mantle to rump, and shoulders dark blue. Ssp *lugens* and the scarcely different *sordidum* similar, but back mainly dull black with only faint purple sheen and blue shoulder patch restricted to lesser coverts. Sspp *albifrons* and *centralandium* black, crown white (sometimes mixed with blue), back blue-black, wing-coverts, lower back, rump, and tail bluer (ultramarine). **Imm**. and **female**: crown blue, headside, neckside, and back glaucous gray; wings, rump, and tail greenish yellow; underside gray paler than back, except for greenish yellow lower sides, flanks, and vent (and sometimes tinge on lower breast and belly). In transitional plumage male first attains blue shoulders and back.
Habits: Alone or in pairs, often in mixed-species flocks. Forages at tips of branches, often working deliberately at a clump of leaves or berries for some time. Not shy.
Voice: Calls fine, resembling those of tanagers.
Breeding: Large gonads Mar-Sep (Col.)
Habitat: Humid forest canopy and adjacent second growth.
Range: 1500-3000 m, but mainly in the subtrop. zone. From Páramo de Tamá s to Cundinamarca, w slope of E Andes of Col. (*albifrons*); Antioquia to Cauca C Andes, w Huila and e Nariño, Col. (*centralandium*) from Antioquia, W Andes s through w Nariño, s Col., and e and w Ecu. to Piura and Cajamarca, n Peru (*atrocyaneum*); from Amazonas and e La Libertad Peru to La Paz, Bol. (*sordidum*); Cochabamba and Sta Cruz, Bol. (*lugens*). 1 ssp in coastal Ven. (*cyanonotum*). Fairly common.

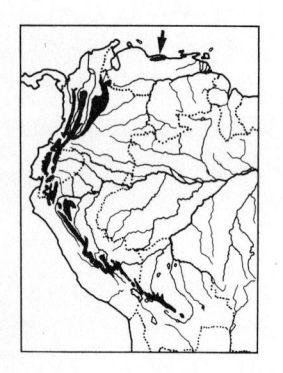

GIANT CONEBILL *Oreomanes fraseri* – Plate LIII 1a-d

14 cm; **large cone-shaped bill** with ridge continuous with crown. **Above plumbeous**; **supercilium and underparts** varying from russet to deep **chestnut**; forehead often mottled with grayish white and dusky, lores and sometimes mottling on chin and upper throat black, **lower headside white**. Sspp doubtfully valid, but in general the bill length increases s-wards, birds from Ecu. have very dark crown, birds from Bol. average lightest; as several characters vary independently, slight differences can be found between many local populations. **Imm**. has

crown mostly dusky; supercilium, lower cheek and throat white, throat mottled with dusky, and underparts slightly lighter russet than in ad. **Juv.** like imm., but browner above, many body-feathers narrowly tipped dusky; throat whiter. 
Habits: In pairs or family groups of 4-5. Slowly work through the canopies, as sometimes the whole group may stay several minutes within one small bush. Quiet, often detected only by the sound of bark being peeled off the tree. Opens bill under the bark to separate layers to retrieve beetles, small larvae, and spiders. Occ. clings upside down or climbs with legs spread out. Sometimes gleans aphids and sugary secretions from under *Gynoxys* leaves, and may probe flowers of mistletoes and *Puya*s.
Voice: Song a loud, irregular, or sometimes rhythmically repeated, siskin-like twitter *twee-de tzi-tzi-twee*, *tweezi twizidwee*, *tyi twi-twee*, or *zweete quidu*. When group moves, frequently utters high-pitched calls as *zit* or *whee*. 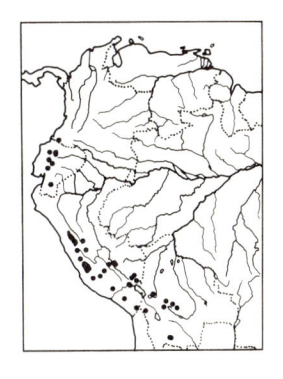
Breeding: Fledglings Nov (Puno), Dec (Cuzco), Aug (Lima).
Habitat: Mainly on medium thick branches just inside canopy, less often on thick stems or thin twigs. Both very dry and moss-covered humid *Polylepis* shrub, woodland and forest. May visit flowers of *Puya raimondii* when adjacent to *Polylepis*, but in mixed forest with several tree species keeps strictly to *Polylepis*.
Range: 2700-4850 m. Very local, in sw Nariño, s Col., and Ecu. (*fraseri*); from Ancash to Tacna, Cuzco, and Puno, Peru ('*binghami*'); La Paz, Potosí and Cochabamba, and w Oruro, Bol. (and presumably adjacent nw Chile) ('*sturninus*'). Fairly common, but distribution confined to the patches of *Polylepis*.

NOTE: A hybrid with White-browed C. is known from Puno, se Peru.

BLUISH FLOWERPIERCER *Diglossopis caerulescens* – Plate LIII 18a-c

12.5 cm. Iris red, bill rather long, slender, and hooked, mandible very slightly upturned. **Dull dark blue to blue-gray**, belly and vent paler and grayish. Ill-defined narrow forehead, lores, and chin blackish. Sspp vary as to tone of color and extent of pale gray below. **Juv.** slightly grayer than ad. and with faint streaks on upper belly; base of mandible yellow (black in ad.).
Habits: Alone or in pairs, at least seasonally in mixed-species flocks with tanagers and other flowerpiercers (esp. Masked F.). Relatively slow and inconspicuous. Seems to enter flowers straight on. Also gleans insects from leaves and eats fruit (e.g., *Rubus* and melastomes).
Voice: A loud moderately pitched chirk and distinctive flute-like note in flight. Song 2.5 s long with sweet notes running into a squeaky twitter or chatter, song regularly repeated.
Breeding: Fledgling Feb (Amazonas); juv./imm. Dec (Santander), June (Cajamarca), Aug (Cuzco); large gonads June-Aug (Col.).
Habitat: Middle or upper strata of low trees at edge of very humid forest and cloud forest, on windswept ridges and areas of poor soil. May favor *Clusia* trees.
Range: 1350-3200 m, mainly in upper subtrop. zone, once to 250 m in

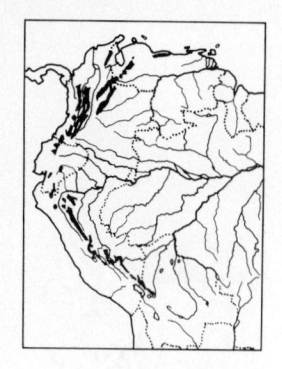

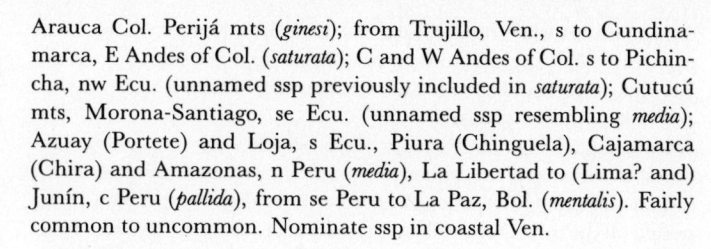

Arauca Col. Perijá mts (*ginesi*); from Trujillo, Ven., s to Cundinamarca, E Andes of Col. (*saturata*); C and W Andes of Col. s to Pichincha, nw Ecu. (unnamed ssp previously included in *saturata*); Cutucú mts, Morona-Santiago, se Ecu. (unnamed ssp resembling *media*); Azuay (Portete) and Loja, s Ecu., Piura (Chinguela), Cajamarca (Chira) and Amazonas, n Peru (*media*), La Libertad to (Lima? and) Junín, c Peru (*pallida*), from se Peru to La Paz, Bol. (*mentalis*). Fairly common to uncommon. Nominate ssp in coastal Ven.

MASKED FLOWERPIERCER *Diglossopis cyanea* – Plate LIII 17a-b

13.5 cm; bill long and black, mandible only slightly upturned; body long and slender. **Iris bright red. Dark blue** (sspp vary as to shade). **Conspicuous black mask** comprises broad forehead, sides of forecrown to over eye, headside and chin. Female slightly duller than male. **Imm.** and **Juv**. dark gray below.

Habits: Alone, in pairs, or pure flocks, sometimes in mixed-species flocks along with other flowerpiercers. Agile and fast, forages along the middle part of thin horizontal branches and vines, often well hidden in the c of bush-top, in thickets, or nearby isolated shrubs. Examines bark-crevices and reaches out and often hangs for berries (incl. blackberries (*Rubus*) and melastoms) and insects, and frequently wags slightly raised tail from side to side. Only rarely visits flowers, and then may probe them straight on. Sings from well-hidden perch about 30 cm below top of bush or small tree. Aggressive towards Black-throated F.

Voice: Song very **high-pitched**, either *tzi tzi tzi tzi tzideweedeleedeeleede* starting with fine notes and ending in a cascade, or *trr te tzíie tee tee tee*, (3rd note accented, last part descending), or *tzirr tr tr tr tee tee tee* (1st note accented). In Ven. just a 2-3 s accelerated jumble of twitters without the drawn-out end notes. Call a fine *zit* or *zik*.

Breeding: May breed twice a year, sometimes while in active molt. Fledglings Mar (Mérida), Dec (Cundinamarca), Apr (Cauca), July (Huánuco), Nov, May (Cochabamba), Mar (Sta Cruz); juv./imm. Nov (Cauca), Sep (Amazonas), Feb, Mar (Pasco); imm. Apr (Mérida), Nov (Antioquia), Apr (Cundinamarca), Mar, July, Sep (Cauca), Nov, Feb (nw Ecu.), Feb (ne Ecu.), Aug (sw Ecu.), Sep (Cajamarca), June, July (Amazonas), Sep, Oct (La Libertad), Mar (Pasco), Feb, Mar (Junín), June (Cuzco), June, Nov, Dec (La Paz); large gonads June-Sep (Col.), Apr (se Ecu.), Jan (Cochabamba).

Habitat: 0.5-7 m up in humid forest, cloud forest, and elfin forest without too many epiphytes, and immediate adjacent shrubbery and bushy slopes.

Range: 1500-3700 m, mainly above 2000 m. Coastal Ven. (*tovarensis*); Perijá mts (*obscura*), from Trujillo, Ven., s through all 3 ranges of Col. to Ecu. except sw (*cyanea*), sw Ecu. and nw Peru (*dispar*), C and E Andes of Peru s to Sta Cruz, Bol. (*melanopis*). Common.

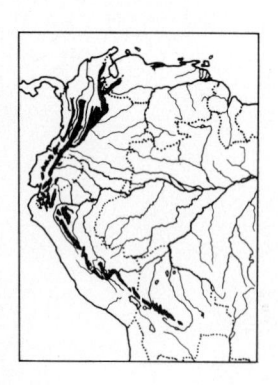

DEEP-BLUE FLOWERPIERCER *Diglossopis glauca* of the subtrop. zone from w Caquetá, Col. s along E Andes to c Ecu. (*tyrianthina*), and s and e of the Marañón to Cochabamba, Bol., is rec. to 2600 m (once 2800 m) in Peru. It frequents mossy humid forest and edge. 11.5 cm. Bill short and strongly hooked. **Iris bright golden yellow**. Dark dull indigo-blue to purplish blue; forehead and lores narrowly black (black difficult to see).

RUSTY FLOWERPIERCER *Diglossa sittoides* – Plate LIII 15a-d

11-11.5 cm (*sittoides* largest). Mandible distinctly upturned. **Male**: **above plumbeous**, tertials usually broadly edged buff; headside dusky, **underparts rufous buff**. Sspp vary in color tone, *dorbignyi* having crown darker than back, and narrow forehead and headside blackish. **Female**: above olivaceous, tertials pale-edged; below dirty yellowish white more or less obscurely flammulated on breast and sides, this most conspicuous in imm. and worn plumage. **Imm. male** like female, but vent buff, belly usually whiter, rest of underparts more or less suffused with buff to rufous-buff (may breed in this plumage). **Juv**. suffused and slightly streaked with olive-buff, appearing rather green.
Habits: Alone or in pairs. Forages by actively reaching down, picking, probing flowers, and sallying in small open trees, bushes and agaves, rarely on the ground. Feeds on exposed flowers, but into dense foliage when disturbed by other species. Sings from conspicuous perch.
Voice: Song of 2 types, one a thin monotonous trill lasting c. 2 s *trrrtrrrtrr'trrrr*, the other mostly twittering. Call a sharp *cheek*.
Breeding: Fledglings Sep-Nov (La Paz), Dec (Cochabamba), Feb (Tarija); juv./imm. Mar, May, Sep, Nov, Jan (Cauca), Sep (nw Ecu.); large gonads Apr (Lima).
Habitat: Open country with scattered bushes and trees, hedgerows, scrubby thickets, agaves, and gardens in the arid zone; also (in n part of range) at edge of humid forest and cloud forest, but hardly inside forest.
Range: 600-3500 m, occ. on coast of c Peru, in n part of range mainly in subtrop. zone. Sta Marta mts and coastal mts of Ven. (*hyperythra*); Perijá mts (*coelestis*); from mts of w Lara and Andes of Ven. s through all 3 ranges of Col. except w slope of W Andes (*dorbignyi*), Ecu. and Peru, in W Andes s to Lima (*decorata*, intergrading with *sittoides* in s Peru), from w Bol. s to Tucumán, nw Arg. (*sittoides*). Uncommon, except locally. 1 ssp in ne Ven. (*mandeli*).

NOTE: Often considered sspp of *D. baritula* of C Am., a group of 5 forms now treated as 2 species: Slaty F. (*D. plumbea*) and Cinnamon F. (*D. baritula*).

CHESTNUT-BELLIED FLOWERPIERCER
Diglossa gloriossissima – Plate LIII 10

14 cm. Bill black, mandible distinctly upturned. Glossy black, shoulders slaty blue, rump dark slate, lower breast and belly rufous chestnut. **Juv**. undescribed.
Habits: Presumably as for Moustached F.
Voice, **breeding**, and **Habitat**: Presumably as for Moustached F.

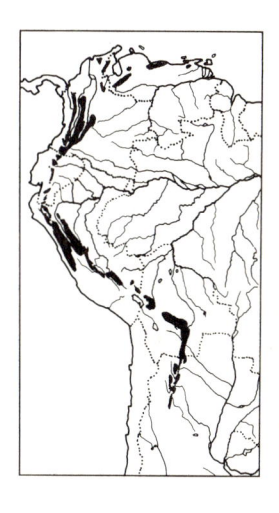

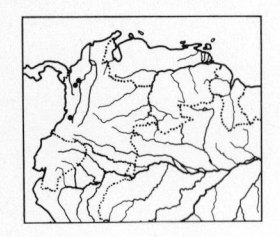

Range: Col. at 3150-3800 m at n and s end of W Andes in Antioquia and Cauca. Local and apparently scarce (?).

NOTE: Often considered a (mega-)ssp of Glossy F.

GLOSSY FLOWERPIERCER *Diglossa lafresnayi* – Plate LIII 11a-b

14 cm. Bill black, mandible distinctly upturned. Glossy black, shoulders slaty blue. Told **from Black F. by larger size, more glossy plumage, black bill, and generally more humid habitat. Juv**. like ad., but duller and grayer, mandible pale-based.
Habits: As for Moustached F.
Voice: Song apparently by both sexes, a fine continuous chirping that may go on for minutes *chiff chiff chee chiff chiff chee...* Calls include a *chick-chick* and a fine *zi zi*.
Breeding: Fledgling Mar (Cundinamarca); juv./imm. July (Cauca); large gonads Feb-July (C Andes Col.).
Habitat: 3-4 m up, sometimes lower. In dense patches of shrubs and trees at edge of cloud forest, humid forest, and elfin forest, and in humid páramo shrubbery, second growth, and riparian thickets.
Range: 2000-3750 m, mainly above 2700 m. From Trujillo, Ven., s locally in E and C Andes of Col. and from s Col. s through e and w Ecu. to Piura and n Cajamarca, n Peru. Common.

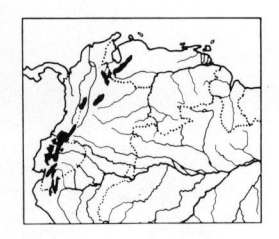

NOTE: Often Includes Moustached F.

MOUSTACHED FLOWERPIERCER *Diglossa mystacalis* – Plate LIII 9a-d

12.5-13 cm. Glossy black, rump tinged blue-gray. Ssp *unicincta* has **white moustache** and rufous breast-band, abdomen, and vent. Ssp *pectoralis* similar, but lower half of breast-band white. Ssp *albilinea* similar, but without breast-band and with blue-gray shoulders. Ssp *mystacalis* like *albilinea*, but moustache rufous. **Imm**. has buff belly and white-streaked breast. **Juv**. undescribed.
Habits: Alone, in pairs when breeding. Picks and acrobatically reaches down near tips of branches from base of tree to top, then sallies to base of next tree. May probe flowers of ericaceans straight on. Often feeds low, gleaning underside of leaves. Sings from bush-top or conspicuous perch. Territorial, also chasing other species. 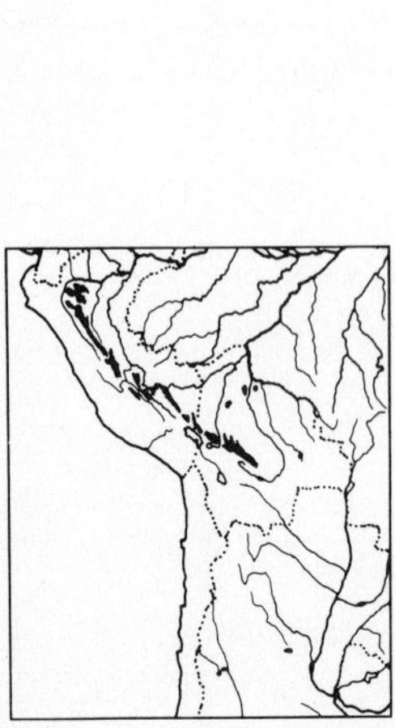
Voice: Song a fairly loud and continuous twittering with no fixed theme *tweeteedeterrteeteedeeretwetee...*, resembling Citrine Warbler, but amplitude constant from beginning to end.
Breeding: Fledgling Aug (Amazonas); juv./imm. Aug, Dec (Amazonas), Aug (San Martín), Aug-Oct (La Libertad), June-Aug (Huánuco), Mar (Pasco), Feb, Mar (Junín), July (Ayacucho), Oct (Puno), May, June (La Paz), Feb, May (Cochabamba).
Habitat: Elfin forest and humid páramo shrubbery, edge of cloud forest, mainly near treeline.
Range: 2500-3600 m. S of the Marañón in C Andes from Amazonas

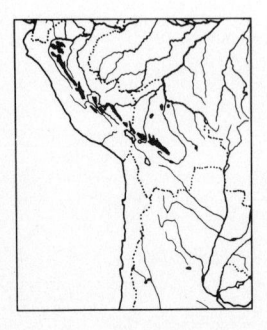

to La Libertad, Peru, intergrading with *pectoralis* in Carpish mts, Huánuco (*unicincta*), from Panao mts, Huánuco to Junín, c Peru (*pectoralis*); Ayacucho, Cuzco, and Puno, se Peru (*albilinea*), La Paz and Cochabamba, Bol. (*mystacalis*). Common.

NOTE: Often considered sspp of Glossy F.

BLACK-THROATED FLOWERPIERCER
Diglossa brunneiventris – Plate LIII 13a-b

12-12.5 cm (*brunneiventris* largest). Upperparts and headside black, shoulders and rump blue-gray. Throat black, moustache and remainder of underparts rufous chestnut, **sides pale gray, flanks light gray**, thighs blackish. **Imm**. dusky olivaceous above very faintly streaked dusky; wings blackish with 2 narrow grayish white bars and narrow edges to tertials; underparts buffy streaked dusky on throat, breast, and c belly. **1st ad**. plumage with the black feathers tipped olivaceous and the rufous feathers narrowly tipped buffy white. **Juv**. much like imm., but upperparts without brownish tinge; tips of wing-coverts and edges of tertials cinnamon-brown; underparts more uniform and tinged olivaceous. Bill pale.
Habits: Alone, in pairs, or family groups, usually not in mixed-species flocks. Actively searches tangles and bushes, spending most time within the vegetation reaching up and down and often hanging. Feeds esp. by piercing melastome flowers, but occ. on the ground. Territorial, often chasing conspecifics and sometimes Masked F. Makes courtship flights.
Voice: Song, by both sexes, a fine warble lasting 1.5-4 s *tee tree rri tee rri tee...*, rarely given in flight. Calls include a fine hummingbird-like *zit*, and a series of wheezy notes.
Breeding: Fledglings June (Huánuco), Dec, Mar (Junín), Dec (Huancavelica, Cuzco); juv./imm. Feb, Sep (Cajamarca), Nov (Ancash), Sep, Oct (Amazonas), Aug-Oct (La Libertad), June, July (Huánuco), Dec-Apr, Aug (Junín), May, July (Cuzco), Aug (Puno), May, June (La Paz); large gonads July, Aug (Col.). 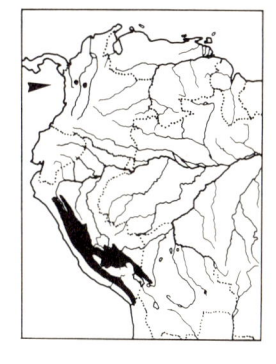
Habitat: Humid to rather dry zones, generally in drier areas or on poorer soils than Moustached F. In lower strata, rarely on the ground or high up, in dense, tangled, low montane scrub, slopes with *Brachyotum* shrubs, *Gynoxys* bushes, second growth, gardens, occ. *Eucalyptus*.
Range: (1450) 2000-4000 m. Seasonal altitudinal movements noted. Antioquia from Paramillo mts to Páramo Frontino in W Andes and at Angelopolis, Santa Elena, and Hacienda Zulaiba in C Andes of Col. (*vuilleumieri*, GG in *Bull. Brit. Orn. Club* 100(1980): 230-2); from c Cajamarca, n Peru s to Tarapacá (Arica), n Chile and La Paz, Bol. (*brunneiventris*). Very common.

NOTE: Previously considered sspp of *D. carbonaria*. It is disputed whether this and the 3 following forms should be classified as full (para-)species or as (mega-)sspp of *D. carbonaria* (see *Amer. Mus. Novit.* 2381(1969), *Bull. Brit. Orn. Club* 100(1980): 230-2, and *Condor* 84(1982):1-14). Suture zone with Black F. (with no hybridization proven, but mixed pairs seen) in c Cajamarca, n Peru, and a narrow hybrid zone with Gray-bellied F. just ne of La Paz city, Bol.

GRAY-BELLIED FLOWERPIERCER *Diglossa carbonaria* – Plate LIII 12

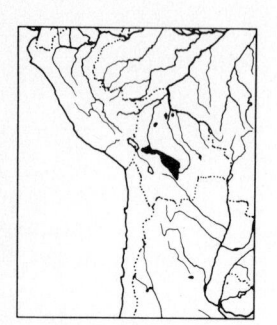

12.5 cm. **Male** black, shoulders light gray, rump more or less extensively dark gray, breast and belly broadly streaked light gray, vent rufous. **Female** has black replaced by blackish gray. **Imm.** above fuscous faintly streaked dark olivaceous; 2 faint wingbars and edge of tertials buffy brown; below gray, more buffy on belly, and streaked dusky on throat, breast, and c belly. **Juv.** dark gray-brown above, throat and breast dark gray-brown faintly streaked pale gray-brown to buffy, c belly pale buff, sides gray-brown faintly dark-streaked, vent rufous buff.
Habits, **voice**, and **habitat**: As for Black-throated F.
Breeding: Fledglings Dec (La Paz), May, July (Cochabamba); juv./imm. June (La Paz), Mar-July, Dec (Cochabamba); large gonads Jan (Cochabamba).
Range: Bol. at 2100-4300 m, highest in the rainy season. La Paz, Cochabamba, Sta Cruz, and Chuquisaca. Very Common.

NOTE: Insignificant hybridization occurs with Black-throated F. in La Paz. Forms a superspecies (Carbonated F.) with Black-throated, Black, and Mérida F.

BLACK FLOWERPIERCER *Diglossa humeralis* – Plates LIII 14a-c and LXIII 22

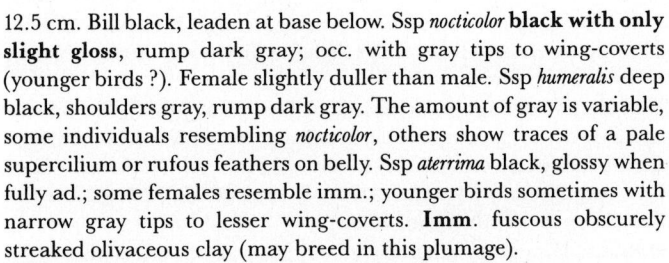

12.5 cm. Bill black, leaden at base below. Ssp *nocticolor* **black with only slight gloss**, rump dark gray; occ. with gray tips to wing-coverts (younger birds ?). Female slightly duller than male. Ssp *humeralis* deep black, shoulders gray, rump dark gray. The amount of gray is variable, some individuals resembling *nocticolor*, others show traces of a pale supercilium or rufous feathers on belly. Ssp *aterrima* black, glossy when fully ad.; some females resemble imm.; younger birds sometimes with narrow gray tips to lesser wing-coverts. **Imm.** fuscous obscurely streaked olivaceous clay (may breed in this plumage).
Behavior, **voice**, and **habitat**: As for Black-throated F. 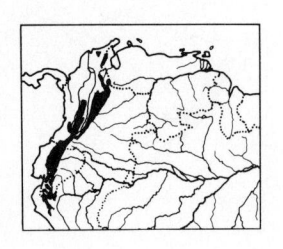
Breeding: Eggs Nov (Ecu.); juv./imm. Sep (Risaralda), Mar (Cauca), Aug, Sep (Imbabura), Nov (Pichincha), May, July (Chimborazo), July, Aug (El Oro), Sep (Loja), July, Aug, Oct (Piura), Sep (Cajamarca); large gonads Feb-Sep (Col.).
Range: 1500-4000 m. Sta Marta mts and Perijá mts (*nocticolor*); páramo de Tamá and E Andes of Col. (*humeralis*), W Andes in region of Cerro Munchique in Cauca, C Andes from Quindío and mts of Nariño, Col. s through Ecu. to Piura and Cajamarca, nw Peru (*aterrima*). Very common.

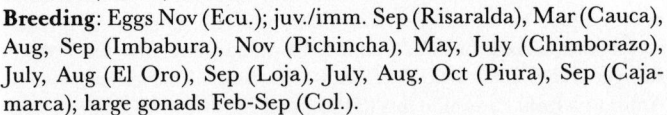

MERIDA FLOWERPIERCER *Diglossa gloriosa* – Plate LXIV 22

12.5 cm. **Black, belly chestnut**; rump, shoulders, short supercilium

starting just in front of eye, lower sides and flanks blue-gray. Many individuals have rufous malar spots suggesting a hybrid origin. **Imm.**: Above dark olivaceous faintly streaked dusky, tips of wing-coverts and edge of tertials dark dull cinnamon. Throat dark olivaceous, moustache and sometimes supercilium suggested by whitish mottling; rest of underparts buff flammulated with dusky on breast and sides.
Habits: Usually alone. Flits nervously. Forages actively inside the foliage.
Voice: Call a long, thin trill. Performs song flight.
Breeding: Imm. Mar, July, Aug, Nov, Dec (Mérida).
Habitat: Edge of dwarf forest, low, isolated trees and open stretches with bushes in páramos.
Range: Ven. at 2500-4150 m from Trujillo to ne Táchira. Fairly common.

NOTE: By some considered a ssp of *D. carbonaria*.

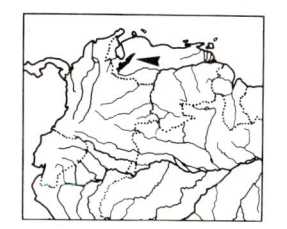

WHITE-SIDED FLOWERPIERCER *Diglossa albilatera* – Plate LIII 16a-d

11 cm. Bill black, distinctly hooked. **Male** blackish with faint dark blue cast (*affinis*), blackish slate-gray (*albilatera*) or bluish slate-gray (*schistacea*), flanks grayer. **Underwing and elongated pectoral tuft silky white**. **Female** olive-brown above, rufous brown below, flanks darker and more olivaceous (sometimes without brown), c belly yellowish buff, vent rufescent; **underwing and pectoral tuft as in male**. Much like buff-breasted individuals of Rusty F., but **tertials without whitish edges, bill black and pectoral tuft white**. **Imm. male** like female, but much duller; feathers of upperparts narrowly edged olivaceous, feathers of underparts broadly tipped buff, breast obscurely dark-streaked. **Juv**. dark gray-brown, feathers faintly dark-tipped; extreme center of belly buffy.
Habits: In pairs or small parties, often in mixed-species flocks. Forages much in the manner of other flowerpiercers, quickly and furtively. Often deep within foliage (to avoid attacks from hummingbirds). During frequent display shakes wings and spreads white tufts.
Voice: Both sexes frequently utter a monotonous, fairly loud flat trill resembling that of Cinnamon Flycatcher, but longer and usually preceded by 2 faint notes. Chirps during display.
Breeding: Eggs Mar (nw Ecu.); fledglings Nov (Antioquia), Feb (Cauca/Huila), Apr, Oct (nw Ecu.), July (ne Cajamarca); imm. July (Cauca), Dec (e Nariño), July, Sep (Cajamarca); large gonads Apr-Aug (Col.).
Habitat: 0.5-4 m up in stunted cloud forest, humid forest clearings with bamboo and shrubbery, bushy hillsides, and suburban areas.
Range: (1300) 1600-3300 m. Sta Marta and Perijá mts and Andes from Trujillo, Ven., s through all 3 ranges of Col. to Ecu. (except sw) and probably Piura (Chinguela) and Cajamarca (Cord. Condor), n Peru (*albilatera*), extreme sw Ecu. and nw Peru n and w of the Marañón (*schistacea*); from Utcubamba drainage Amazonas s to Huánuco c Peru, and locally in Ayacucho to extreme nw Cuzco (*affinis*). Fairly common to common. 1 ssp in coastal Ven.

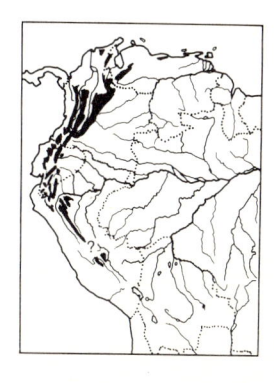

GOLDEN-COLLARED HONEYCREEPER *Iridophanes pulcherrima* is erroneously listed for the temp. zone in w Ecu. by Butler (1979) and was therefore included in **Plate LIV 11a-b** at an early stage. It is found uncommonly and locally in humid forest, forest clearings, and forest edge at 950-2150 m in sw Col. and nw Ecu. (*aureinucha*), and from Huila at head of Magdalena valley and Caquetá e slope of E Andes of Col. s through e Ecu. and e Peru to Cuzco, se Peru (*pulcherrima*). With slender, slightly curved bill. Male with hood, scapulars, and upper mantle black; wings and tail blue, otherwise golden olive with golden nape-collar. Female rather uniform olive, lighter below, with blue-green wings.

NOTE: Recently placed in the genus *Tangara*.

TIT-LIKE DACNIS *Xenodacnis parina* – Plate LIII 8a-e

12-13 cm (*parina* smallest, *petersi* variable, males from e slope of c Cord. Blanca, Ancash very large, but birds from Huánuco and Arequipa fairly small). Rounded, **bill stubby**. Ssp *petersi* (including birds from sw Peru) **male deep blue streaked** with shining light ultramarine, belly blue-gray. **Female** brown tinged blue-gray above, wing-coverts, rump, and edges of tail blue. Forehead fairly narrowly blue to above eye; cheeks, sides, and flanks buffy gray grading to dull pale rufous on throat and breast, belly buffy white. Ssp *bella* doubtfully distinct, female slightly duller and with a little less blue on forecrown. Ssp *parina* **male** blue, head brightest, belly blue-gray. **Female** like *petersi*, but entire crown blue, cheeks and sides pale rufous like breast. **Juv**. like female, but crown and wings like back, wing-coverts with clay spots at tip, underparts buffy clay with very faint, broad clay streaks.
Habits: Alone, in pairs, or small groups, sometimes in mixed-species flocks with flowerpiercers. In some places abundant, apparently with group territories, but pairs or families stay together within the flock. They are constantly disputing while calling. Actively moves about, often with horizontal body, mainly gleaning aphids, small cikadas, and sugary secretions (crystal and droplets) from underside of *Gynoxys* leaves. May probe mistletoe flowers. During nuptial flight makes small, often circular flights with shallow wingbeats and spread wings and tail while singing.
Voice: Call loud *tchiú tchiú tchiú* and *huit huit huit*. Contact note a fine *zit* and a snarling *wheit*. Song rambling and varied between loud calls and trilled *ziets: tiu tiu tiu ziet ziet zitzit*. Scolds predators with hissing notes. Vocalizations of ssp *parina* considerably weaker than in other sspp. In duet, male introduces the song, female then joins in with simultaneous notes, and then female ends the duet with a series of raspy notes.
Breeding: Tiny cup nest made of pale 'wool' from *Gynoxys*, placed in tree. Possibly multi-brooded when conditions suitable. Pull. and juv. at all molt stages Feb (Ancash); fledgling May (Ancash), July (Ancash/Huánuco/Lima border, Arequipa); song Feb, Aug (Ancash).
Habitat: Chiefly in *Gynoxys* bushes and trees, but also in other bushes and *Polylepis*, as well as lower to middle strata (0.5-5 m up) of cloud forest edge.
Range: 3000-4600 m. Distribution reflects patchy habitat. Páramo de

Angel in Carchi, nw Ecu. ? (possible sighting: O Læssøe); Cajas mts west of Cuenca, Azuay, s Ecu. (newly discovered population resembling *petersi*; see *Amer. Birds* 34(1980): 242-8); C Andes of Peru from s Amazonas to e La Libertad (*bella*), W Andes of Peru in Ancash, Huánuco, Lima, and (isolated ?) Arequipa and Moquegua (RH) (*petersi*); s Peru in Junín (Maraynioc), Ayacucho, Apurímac, and Cuzco, where also on e Andean slope (*parina*). Generally uncommon, but locally extremely abundant.

Tanagers – Family Thraupidae

The tanagers comprise some 215 species of '9-primaried oscines' restricted to the Americas, northern birds migratory. Most are brightly colored with a finch-like or sometimes fairly slender bill. Some members of the genus *Piranga* have a winter plumage. As far as known, postjuv. molt never involves rectrices. The majority of tanagers inhabit humid forest, where they feed on berries and insects. A few warbler-like species inhabit semi-arid shrub. Tanagers form the main part of the mixed-species flocks so well known in the neotropics, and sometimes a number of closely related species may be seen together feeding in the same manner in a fruiting tree. Only during the breeding season when largely feeding on insects do they seem to differ in foraging behavior. Hemispinguses and members of the genus *Thlypopsis* primarily forage in the undergrowth, where they flock with *Basileuterus* warblers, brush-finches etc.; most others are found in medium to upper strata. The majority have very similar high-pitched *zit* calls, like many other members of the mixed-species flocks. Songs vary from unmusical twitters in most species, to clear whistled phrases in some, and is given by the male, rarely both sexes in duet or a flock in chorus.

Most tanagers live in permanent or seasonal pairs, although a few are polygamic. Only the female incubates, but both sexes tend the young. The small, round, short-tailed chlorophonias and euphonias are unique among birds by lacking a gizzard, and differ from other tanagers by regurgitating food for the nestlings (sometimes with 2nd years' birds as 'helpers'), and by building closed nests with a side entrance; they are often placed in a separate subfamily Euphoniinae. In other species the nest is an open cup placed in a tree or a bush, rarely on the ground.

Tanagers are closely related to emberizine finches, and it is difficult to draw a line between them. Recent DNA comparisons suggest that tanagers constitute a major Neotropical radiation, which, beside typical tanagers should include Plush-capped Finch (*Catamblyrhynchus*), Swallow Tanager (*Tersina*), Neotropical honeycreepers ('*Coerebidae*') and several species traditionally included among the 'emberizine' finches (*Oryzoborus, Volatinia, Haplospiza,* and openland forms as *Sicalis, Diuca* and maybe others): see Sibley,CG & Ahlquist,JE in *Acta XVIII Congr. Int. Orn.* (1985): 83-121 and AH Bledsoe in *Auk* 105(1988):504-15. However, with no complete revision we maintain the traditional classification. Peruvian Andean forms have been dealt with in *Amer. Mus. Novit.* 1225 (*Chlorophonia, Euphonia, Chlorochrysa, Pipraeidea*), 1245, 1246 (*Tangara*) (1943), 1262 (*Iridosornis, Delothraupis, Anisognathus, Buthraupis, Compsocoma, Dubusia, Thraupis*) (1944), 1304 (*Piranga*) (1945), 1345 (*Creurgops, Thlypopsis, Piranga*), 1367 (*Sericossypha, Chlorospingus, Cnemoscopus, Hemispingus, Chlorornis, Schistochlamys*) (1947), and 1428 (*Catamblyrhynchus*) (1949). A recent compilation of data with abundant source references about the entire family is given by Isler, ML & PR (1987) *The Tanagers*, Washington D. C.: Smithsonian Inst. Press.

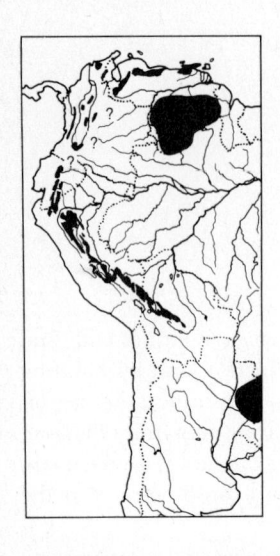

BLUE-NAPED CHLOROPHONIA *Chlorophonia cyanea* – **Plates LVI 3a-c, LXIII 21** inhabits humid forest at 600-2100 m, but is rec. to 2500 m in Ven. Sta Marta mts n Col. (*psittacina*), W Andes of Col. (*intensa*), and Andes from Ven. to Bol (*longipennis*), 3 additional sspp in Ven., 1 in Brazil. 10.5 cm, very short-tailed. Ssp *longipennis*: **head, upperparts, and upper breast bright glossy green** to turquoise on orbital ring, nape, and rump. **Underparts below upper breast yellow**. **Imm. male** duller with less turquoise, underparts yellow-green with yellow median line. **Female** even duller, with bright green rump (blue in *psittacina*) and olive-yellow lower underparts without median line. Sspp *psittacina* has yellow across forehead (narrow and ill-defined in *intensa*). **Juv.** green tinged yellow below. Apparently may breed in imm. plumage ('helpers only ?). In pairs or small groups, often in mixed-species flocks. Actively picks melastome and mistletoe berries, other fruit and insects, typically within canopy. Calls include a nasal *peent, ek, erk*, a plaintive down-slurred *teeeu*, and a short rattle *didle-itle-itle*. Eggs May (Sta Marta); fledglings May, July (Valle), Apr (Cuzco); large gonads Apr, May (Col.); breeding Jan, May, June (Col.).

CHESTNUT-BREASTED CHLOROPHONIA
Chlorophonia pyrrhophrys – **Plate LVI 2a-b**

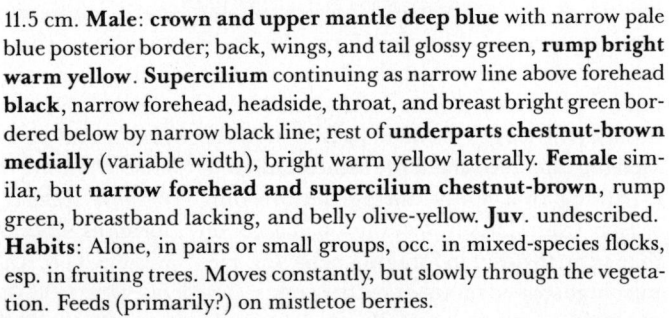

11.5 cm. **Male**: **crown and upper mantle deep blue** with narrow pale blue posterior border; back, wings, and tail glossy green, **rump bright warm yellow**. **Supercilium** continuing as narrow line above forehead **black**, narrow forehead, headside, throat, and breast bright green bordered below by narrow black line; rest of **underparts chestnut-brown medially** (variable width), bright warm yellow laterally. **Female** similar, but **narrow forehead and supercilium chestnut-brown**, rump green, breastband lacking, and belly olive-yellow. **Juv**. undescribed.
Habits: Alone, in pairs or small groups, occ. in mixed-species flocks, esp. in fruiting trees. Moves constantly, but slowly through the vegetation. Feeds (primarily?) on mistletoe berries.
Voice: Song siskin-like *tut tut tut too-dée too-dée...* or *na-deár, na-deár,...to-d'leép* with variations; nasal *near* notes often predominating.**Breeding**: Nest in hollow. Imm. May (Perijá); large gonads Feb, Apr (Valle), May (Perijá).
Habitat: Low in open humid terrain with scattered tall trees and shrubs at edge and clearings of cloud forest and from mid level to canopy in humid forest.
Range: 1400-3600 m. Perijá mts and from Trujillo, Ven., very locally s through all 3 ranges of Col., nw and e Ecu. to Pasco, c Peru (see *Amer. Birds* 28 (1974):960,962, and *Gerfaut* 74 (1984):57-70). Uncommon to rare.

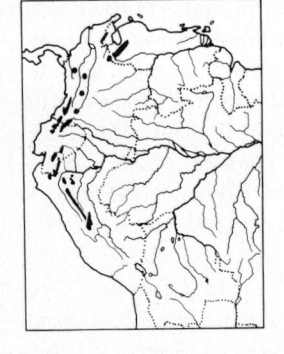

BLUE-HOODED EUPHONIA *Euphonia musica* –
Plate LVI 1a-d

10.5 cm. Tail short. Sspp *pelzelni* and *aureata*: **male**: **crown and nape blue**; rest of head incl. throat, wings, most of **back, and tail purplish**

black; lower back, rump, and belly bright warm yellow. Ssp *insignis* has golden yellow instead of black forehead. **Female**: with forehead orange, crown and nape light blue, eye-line black, otherwise greenish gray above grading to pale olive-yellow below. **Juv**.: above grayish olive tinged yellow on forehead and rump, crown and upper back sometimes gray with hardly any olive; below yellowish olive, yellowest on throat and c belly.
Habits: Usually in pairs or a few together, and occ. in mixed-species flocks. Active. Primarily feeds on mistletoe berries. Sometimes flies long distances in direct flight.
Voice: Call (Brazil) a characteristic slow series of 3-4 whistles. Song a fast varied warble in Brazil, in Ven. a conebill-like rapid twitter and trills lasting up to 30 s or more.
Breeding: Apr-May (Col.).
Habitat: From low thickets to treetops. Xerophytic growth, open deciduous forest, semi-humid evergreen forest, humid forest, cloud forest, second growth, plantations, parks, and gardens.
Range: Lowlands to 2700 m, chiefly subtrop. zone. Coastal mts, Tepuis and Andes of Ven., Col. (except extreme sw), e Ecu., nw and e Peru, e Bol., Parag., and Tucumán, nw Arg.; Guianas and Brazil n of lower Amazon and from Bahía s to Urug. and Corrientes n Arg (*aureata*); 1200-3250 m from extreme sw Col. s to Chimborazo, w Ecu. (*pelzelni*); at Gimá 3000 m, Azuay, s Ecu., only known from the type locality (*insignis*). 3 sspp from Mexico to Panamá, 3 in the West Indies. Uncommon to common.

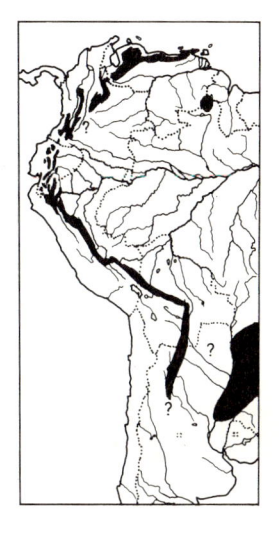

FAWN-BREASTED TANAGER *Pipraeidea melanonota* – Plate LIV 14

12.5 cm. **Male**: narrow forehead and sides of head black, crown and rump light blue; back, wings, and tail dark blue; below pale buff to buff. **Female**: similar, but mantle gray-brown with dull bluish sheen, crown and rump paler and duller. Iris dark red. **Juv**. like female, but feathers of throat and breast with faint dark tips, upperparts more brownish gray, less blue. **Habits**: Usually alone, rarely in mixed-species flocks, but often joins feeding aggregations at fruiting trees. Appears to travel over wide area in search of food. Picks insects and berries among leaves at tips of branches, frequently hanging, occ. sallying.
Voice: Infrequently heard song a 2-3 s long unmusical series of 4-11 squeaky high-pitched *see* notes, repeated at 2-3 s intervals.
Breeding: Fledgling July (Mérida); juv./imm. Aug (Junín), July (Cuzco); large gonads Mar, July (Col.).
Habitat: Mainly middle strata at edge of humid forest and woodland, clearings and overgrown pastures.
Range: 700-3000 m, primarily in subtrop. zone. Tepuis, locally in coastal mts and Andes of Ven. from w Lara s through all 3 ranges of Col., e and w Ecu., Peru (in Tumbes, and in Lima, maybe with a small population at 4000 m near Oyón, according to M Kessler), and e Andean slope through Bol. to Tucumán, nw Arg. (*venezuelensis*); at lower elevations from se Brazil to ne Arg. (*melanonota*, approached by Bol. population). Generally uncommon in the Andes.

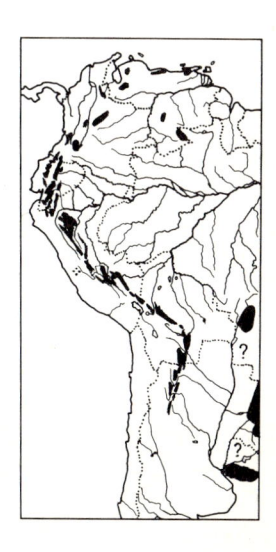

GLISTENING-GREEN TANAGER *Chlorochrysa phoenicotis*, fairly common in humid premontane forest of n tip of C Andes and in sw Col. and w Ecu., is rec. to 2400 m in Col. 13.5 cm. Glistening **green**, lesser and median wing-coverts opalescent; spots below and behind eye opalescent, spots behind earcoverts orange. Female slightly duller than male.

ORANGE-EARED TANAGER *Chlorochrysa calliparaea* could be expected to reach lower temp. zone. Fairly common in humid premontane forest from w slope of E Andes in Cundinamarca s to upper Magdalena valley and in Caquetá and Putumayo, Col. s through e Ecu. to n Ayacucho, c Peru, and from Cuzco, se Peru to Cochabamba, Bol. 13.5 cm. Glossy **green**, feathers with blackish bases; headside, lower back, and c underparts blue-green, patch on c crown, patch on headside, and lower rump orange, **throat black** (violet-blue in s part of range; gray in female). **Juv**. duller without orange or blue, eye-ring yellowish, feathers of throat buff tipped green, bill pale basally.

SAFFRON-CROWNED TANAGER
Tangara xanthocephala – Plate LV 1a-c

12.5 cm. Ssp *lamprotis*: above black streaked tuquoise-green, sides and back of neck black, rump turquoise-green. **Crown orange, headside yellow, triangular mask and upper throat black**. Below turquoise-green, c belly buff, upper belly indistinctly streaked dark gray. Ssp *venusta* similar, but less greenish turquoise and **crown yellow**; *xanthocephala* intermediate. **Juv**. similar, but duller and greener; yellow of head replaced by shining yellowish green, feathers of underparts edged buff. In pairs or up to 25 together, often in mixed-species flocks with other tanagers such as Flame-faced T. Restlessly forages in the outer branches, often hanging or peering under branches, mainly eating small fruit. Call a crisp, thin *tsit tsit* like other tanagers. Breeding Nov (Puno); fledglings Nov (Cuzco, Huánuco); juv. Apr (Mérida), Mar, Apr (Cundinamarca), Sep (Huila), Aug (San Martín), Feb (Pasco), Apr (Junín), Dec (La Paz), Nov (Cochabamba); large gonads July (Perijá), June (w Antioquia), Mar-May (Valle), Apr (Cuzco).

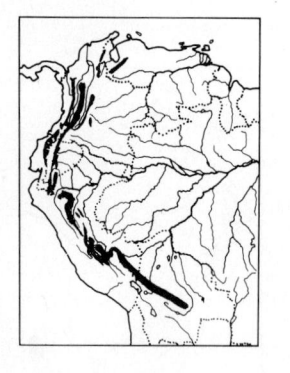

Middle to upper strata, rarely low in humid forest (mainly edge) and adjacent second growth. 1100-2600 m, once at 3150 m. Perijá mts and Andes from s Lara to Táchira, Ven., locally in E Andes, but widespread in C and W Andes of Col., both slopes of Ecu., and along E Andes to Huánuco, c Peru (*venusta*); Pasco and Junín, c Peru (*xanthocephala*), and Ayacucho, se Peru to Sta Cruz, Bol. (*lamprotis*). Common.

FLAME-FACED TANAGER *Tangara parzudakii* – Plate LV 4a-c

14.5 cm. Ssp *parzudakii*: **head orange-yellow** to red on forehead and lower headside; lores, ocular region to posterior earcoverts and throat black. Back, wings, and tail black; lower back, rump, lesser and median wing-coverts opalescent pale blue-green, greater wing-coverts edged turquoise. Below buff, breast and sides with opalescent turquoise gloss. Ssp *urubambae* with yellow of head more saturated. Ssp *lunigera* has the red of forehead replaced by orange to light cadmium, and red of lower headside replaced by lemon yellow; black of back extends to nape; the opalescent parts above and below much greener, secondaries edged turquoise. **Juv.** has crown and back golden-green instead of orange and black, underparts duller without greenish, throat whitish. In pairs, usually in mixed-species flocks with other tanagers. Actively searches open mossy branches and sometimes at *Cecropia* catkins, often hanging or reaching down. Call *zit* like other tanagers. Fledglings Mar (Huila), June (Nariño), Oct (La Libertad); juv. July (Cundinamarca), Sep (w Ecu.), May (Ayacucho); large gonads Mar (w Huila, Cundinamarca); nest-building June (Nariño).

Middle to upper strata, occ. low in humid forest and adjacent second growth. 1000-2625 m, rarely down to 700 m. Río Chiquito sw Táchira Ven., w slope of E Andes of Col. in Cundinamarca and at head of Magdalena valley to e slope of C Andes, and on e slope of E Andes in Caquetá (probably entire e slope), thence s through e Ecu. to Junín, c Peru (*parzudakii*), and through Ayacucho to Urubamba valley, Cuzco, se Peru (*urubambae*); from Cerro Tatamá, Río San Juan headwaters in w Risaralda, Col., along pacific slope to El Oro, sw Ecu. (not known to reach temp. zone) (*lunigera*). Fairly common to common.

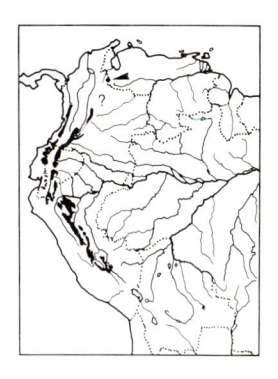

BLUE-BROWED TANAGER *Tangara cyanotis* –
Plate LV 6a-b

11-12 cm (*lutleyi* largest). Ssp *cyanotis*: crown and headside black; band over forehead ultramarine continuing as **silvery green brow**, patch below eye ultramarine. Back dark blue-green, rump silvery green; wings black, greater coverts tipped and narrowly edged turquoise, remiges narrowly and lesser wing-coverts broadly edged turquoise. Tail blackish edged blue. Below silvery green, c belly whitish buff turning warm buff on vent. Ssp *lutleyi* differs by having black back, crown, headside and chin without ultramarine, and by greener wing-feather edges. Alone or in pairs, rarely in mixed-species flocks with other tanagers. Active and fast, forages among tangles and leaves in the outer branches, occ. hanging. Call as in other tanagers. Juv./ad. Nov (Junín).

5-7 m up in canopy of humid forest and wooded ravined. At 1100-2600 m, but mainly below 2200 m, from head of Magdalena valley and Putumayo (also 'Bogota'), Col. s through e Ecu. to Cuzco, se Peru (*lutleyi*); yungas of La Paz and Cochabamba, Bol. (*cyanotis*). Generally uncommon, but fairly common in se Huila, Col.

METALLIC-GREEN TANAGER *Tangara labradorides* –
Plate LV 10

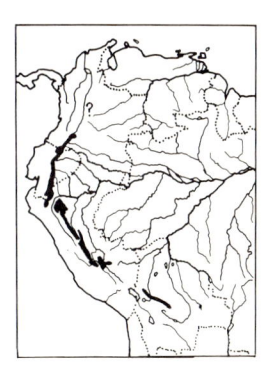

11.5 cm. **Mostly faded blue-green** with golden opalescent sheen. **Narrow forehead, lores, ocular region, and chin black**; also black patch from hindcrown to c upper mantle, wings (except wrist and edge of remiges) and tail. C belly and vent buff. Ssp *chaupensis* generally greener, forehead with little or no golden sheen, primaries edged golden green instead of blue, c belly paler (whitish) and much more extensive. **Juv**. like ad., but duller, c underparts yellowish white tinged gray. 1-3 pairs, usually in mixed-species flocks with other tanagers. Forages almost at tips of branches, mainly taking insects, but occ. melastome fruit and *Cecropia* catkins. Only infrequently peers under branches. Call a high coarse *jitt*. Also gives a squeaky *eeek* and a ticking twitter in flight. Fledglings June, Aug, Nov (Valle); juv./ad. July (Cundinamarca, Huila), Aug, Oct (Cauca); nest July (Valle); large gonads Mar-July (W Andes and n end of C Andes, Col.).

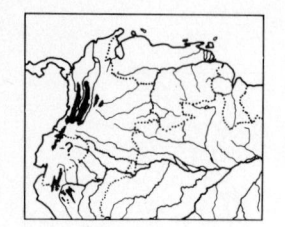

All strata of humid forest, forest edge and adjacent second growth. 1200-2750 m, rarely down to 500 m. All 3 ranges of Col. (but very locally in E Andes) through Nariño to nw Ecu. (*labradorides*); from Loja, se Ecu., to Piura (Chinguela), Cajamarca (Condor mts), and Utcubamba drainage in Amazonas and adjacent San Martín, n Peru (*chaupensis*). Uncommon, locally fairly common on w slope in Col. and Ecu.

GOLDEN-NAPED TANAGER *Tangara ruficervix* – Plate LV 3a-c

11.5-12.5 cm (*taylori* largest). Ssp **ruficervix: mostly blue-green**, area around base of bill and top of head black, eye-brow to nuchal half-collar ultramarine, **patch on hindcrown golden**, posterior earcoverts bluish white with golden tinge and with a black spot below. C belly buffy white, vent buff. Ssp *leucotis* with larger bill and shorter wings and tail; *taylori* with golden spot on earcoverts; *amabilis* like *taylori*, but forecrown lighter, and nape deeper orange; **inca** and **fulvicervix** like *amabilis*, but **general color bright ultramarine-blue** without black anterior margin of the rufous hindcrown; c belly white; *inca* with largest bill and palest flanks and vent. **Imm**: above turquoise, crown darkest, mantle grayish washed dusky, below grayish, c belly and vent as in ad. **Juv**. with gray on crown and nape, light gray below.
Habits: 1 to several pairs, often in mixed-species flocks with other tanagers, but often not. Rapidly moves out along the branches to the tip, searching all clusters. Frequently peers under branches, but also sallies for insects. Mainly feeds on fruit.
Voice: Call resembles other tanagers.
Breeding: Fledglings Apr (Cundinamarca), Aug (Valle), July, Sep (Nariño); juv. Sep (Valle); juv./ad. Aug (Valle), Apr, May (Nariño); nest-building Apr (Valle); large gonads Mar-June (C Andes Col.), Feb (Valle).
Habitat: Upper strata and near openings in humid forest, second growth, trees, and bushes in clearings.
Range: 1100-2700 m. All 3 ranges of Col. except e slope of E Andes (*ruficervix*), w Ecu. (*leucotis*), e Nariño, se Col., and e Ecu. (*taylori*); Peru s to Huánuco (*amabilis*), from Pasco to Cuzco (Puno?), c-se Peru (*inca*), La Paz and Cochabamba, Bol. (*fulvicervix*). Fairly common in Bol., Col., and w Ecu., uncommon to rare elsewhere.

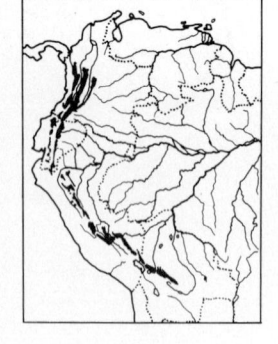

SCRUB TANAGER *Tangara vitriolina* – Plate LV 8

14 cm. **Male: crown rufous, headside blackish.** Above opalescent

pale blue-green, more silvery on rump; wings and tail edged turquoise; below opalescent buffy to pale silvery-blue, c belly buffy white, vent buff. **Female** duller and with blue cast replaced by green, vent paler buff. **Juv**. has indistinct headpattern.
Habits: Usually in pairs or groups, mainly feeding on insects, but also joins feeding aggregations at fruiting trees. Peers about somewhat like a vireo.
Voice: Call a shrill buzzy *ziit* note.
Breeding: Fledglings May (Cundinamarca), Feb (Cauca); juv./ad. Mar (Cauca); large gonads June-Aug (n end W and C Andes Col.), Jan (Huila), Mar (Valle), Nov (sw Cauca).
Habitat: All levels, frequently on the ground. Prefers fairly xeric or scrubby places, but also colonizes clearings and shrubby growth in adjacent humid areas. Often in town parks and shade trees of plantations.
Range: 1000-2600 m, rarely down to 500 m and up to 3000 m. C and s Col. from W Andes to w slope of E Andes and s to region of Quito, nw Ecu. Spreads with man-made clearings. Common.

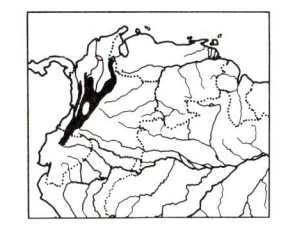

GREEN-CAPPED TANAGER *Tangara meyerdeschauenseei* – Plate LV 7 is only known from the vicinity of Sandia and w side of Abra de Maruncunca in Puno, se Peru. Fairly common in semi-arid riparian scrub, garden trees, and forest edge at 2000-2450 m. 14 cm. Like Scrub T., but back paler, opalescent in center, esp. towards rump. Crown opalescent golden bordered in front and on sides with greenish blue; cheeks bluish, lores and ocular region blackish. Underparts buffy tinged bluish, esp. on throat and breast. Female has blue of upperparts, cheeks, and underparts replaced by green. Alone or in pairs.

NOTE: Described by Schulenberg & Binford in *Wilson Bull*. 97(1985): 413-20.

BERYL-SPANGLED TANAGER *Tangara nigroviridis* – Plate LV 11

12 cm. Forehead, chin, upper headside and back black, rump turquoise; feathers of rest of crown, lower headside, neckside, and rest of underparts **black with broad turquoise tips**, tips on belly and vent broader and buffy white. Wing-coverts black edged turquoise, remiges and rectrices black narrowly edged blue. Female slightly duller than male, **juv**. duller. Sspp poorly differentiated.
Habits: In pairs, sometimes in groups up to 15, often in mixed-species flocks. Moves rapidly, usually on tips of branches picking berries (often *Miconia*) or insects from the underside of leaves. Often hangs and peers under branches. Sometimes near ground in herbs, ferns, and vines.
Voice: Call resembles other tanagers.
Breeding: Fledglings Apr (Quindío, Cundinamarca), May (Boyacá), Nov (Valle), June (Nariño), Sep (ne Ecu.), Feb (Pasco), July (Cuzco); juv./ad. Dec, Jan (Caldas), Aug (Huila), Apr, July (Nariño), July (Cajamarca), Feb (Pasco); large gonads May-July (n Col.), Mar-July (Valle).

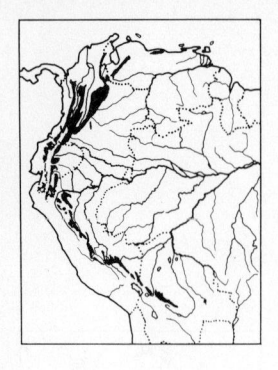

Habitat: Upper canopy of humid forest. Also at forest openings, edge, and in second growth.
Range: 900-3000 m, mainly 1500-2400 m. Perijá mts, coastal mts and Andes of Ven. from s Lara s through w slope of E Andes in N.d.Santander, C and W Andes of Col., and w Ecu. s to Peruvian border (*cyanescens*, birds from Col. and Ecu. sometimes separated as '*consobrina*'), from Boyacá, e slope of E Andes of Col. s to e Ecu. (*nigroviridis*), from Piura, Cajamarca and Amazonas, Peru, s along e Andean slope to Cochabamba, Bol. (*berlepschi*). Common.

BLUE-AND BLACK TANAGER *Tangara vassorii* – Plate LV 9a-d

12.5 cm. Ssp *vassorii*: **blue**, lores, chin, and eye-ring black; flight- and tail-feathers black very narrowly edged dark blue, greater wing-coverts black tipped blue. Female slightly duller than male. Ssp *branickii* similar, but top and sides of head dull bluish green, paler on hindcrown. Ssp *atrocoerulea* has crown, nape, upper mantle, and headside blue-green, **nape with** opalescent **yellowish patch** (rarely white); back black, shoulders with black feather-centers; underparts, esp. throat and breast paler blue, feathers below with black centers, c lower belly pale blue-green tinged gray. **Juv**. has wings and tail like ad., *vassorii* smoky gray, c belly pale gray, *atrocoerulea* grayish black above, crown dark gray, nape mottled pale gray, underparts streaked gray and whitish.
Habits: Up to 25 birds in mixed-species flocks with other tanagers and in feeding aggregations at fruiting trees. Rapidly forages across the thinner branches, occ. peering under branches a few times, and soon moving to another tree.
Voice: Call *tzit-tzit* much like other tanagers, sometimes in rapid ticking series. In flight a distinct *tzrr tzit*. Rarely heard song a 2-3 s series of high-pitched notes starting slowly *zieeu- zie-zie-zizizizeee*, given every 15 s.
Breeding: Fledglings May (Cauca/Huila), Jan (Valle); juv. Mar (Loja), Apr, Nov (Quindío), Apr (Cundinamarca), Jan (w Pichincha), Nov (La Libertad); juv./ad. Oct, Nov (Santander), May (Boyacá), June, July, Oct, Nov (Cauca), Sep (Cajamarca), June (Amazonas, Huánuco), Jan, Mar, June, Dec (Cochabamba), Mar (Sta Cruz); breeding Nov (Puno); large gonads Feb-Aug (Col.); nest June (Valle).
Habitat: Both high and low, but mainly 5-8 m up in humid shrubbery and middle strata of forest and cloud forest. Particularly fond of fruiting melastome trees (e.g. *Miconia*).
Range: (1300) 1650-3350 m, mainly above 2400 m. Andes from Trujillo, Ven., s through all 3 ranges of Col. (on pacific slope only in Antioquia and Cauca), e and w Ecu. to both slopes in Piura (apparently not in Condor mts Cajamarca), nw Peru (*vassorii*); C Andes of Peru s of the Marañón s to La Libertad and perhaps Huánuco n of Huallaga river (*branickii*); from Huánuco in c Peru to Cochabamba, Bol. (*atrocoerulea*); Sta Cruz, Bol. (ssp). Common.

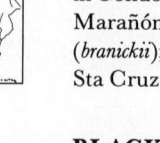

BLACK-CAPPED TANAGER *Tangara heinei* – Plate LV 5a-b

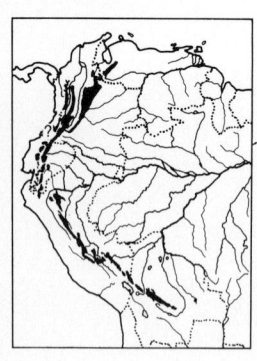

12.5 cm. **Male: cap black**, back, rump, and wing-coverts opalescent gray-blue, wings and tail dark dull blue; headside, throat, and breast greenish turquoise, feathers somewhat pointed and based black, rest of underparts dull dark blue with slight opalescent sheen, under tail-coverts edged white. **Female**: cap dark dull green with slight gloss; back, wing-coverts, and rump bright green with slight golden sheen, wings and tail darker green, primaries edged blue; headside, throat, and breast as in male, but slightly duller, sides of body like back, c underparts blue-gray more or less mixed with yellowish buff, under tail-coverts mixed with yellowish white. **Juv**. like female, but even duller. In pairs or rarely small groups in mixed-species flocks with other tanagers. Regularly peers under branches. Call like other tanagers. Eggs Aug (N.d.Santander); fledglings Aug, Dec (Sta Marta), Apr (Valle), Apr, May, Aug (Cauca), Apr, May (Huila); juv./ad. Jan (Nariño); nest-building Jan (e Huila); large gonads Jan-Sep (n Col.).

Medium to upper strata of cloud forest, humid forest, open woodland, clearings, second growth, and isolated trees. 700-2700 m, mainly 1500-2200 m. Sta Marta and Perijá mts, coastal mts and Andes of Ven. from Lara s through all 3 ranges of Col. (only locally in E Andes?) to nw and ne Ecu. Generally uncommon, but locally fairly common.

NOTE: A recently discovered, isolated population from Sira mts., Huánuco, c Peru (1300-1570 m), has been described as a full species, **SIRA TANAGER** (*Tangara phillipsi*) by Graves & Weske in *Wilson Bull*. 99(1987):1-6. It differs by being darker, esp. the male, which has lower breast and c belly heavily washed with black.

SILVERY TANAGER *Tangara viridicollis* – Plate LV 2a-b

12.5 cm. **Male**: cap and chin black. Lower headside and throat golden brown (*fulvigula*) or dull gold (*viridicollis*). Back, lesser wing-coverts, rump, sides and flanks silvery blue (lightest and most silvery in *fulvigula*, darker and bluer in *viridicollis*), wings and tail blue-black (lightest in *fulvigula*). Most of underparts black. **Female**: cap dark golden brown, lower headside and throat pale golden brown with opalescent green sheen (*fulvigula*), or head paler and duller (*viridicollis*). Back bright green with slight golden gloss; wings and tail green, some primaries edge blue. Breast-feathers dull fuscous tipped green, sides green, belly blue-gray, feathers of vent edged yellowish. Keeps body horizontal. In pairs or flocks up to 20, sometimes with a few Yellow-throated T.s, often in mixed-species flocks. Call a fine *tziu* and a lower *pew*. Juv./ad. Apr (Huánuco); breeding Nov (Puno).

1-15 m up (almost to top of trees) in humid forest, dwarf cloud forest edge and humid shrubbery as well as deciduous forest and riparian trees in the arid zone. 500-2750 m (once 3050 m) from both slopes of s Ecu. s to Lambayeque and Cajamarca w and n of the Marañon, n Peru (*fulvigula*); from s and e of the Marañon through c Peru and (perhaps only locally) Junín and Cuzco to Puno, se Peru (*viridicollis*). Fairly common (at least seasonally) in cord. Condor in n Cajamarca, n Peru.

NOTE: Sometimes called Silver-backed T.

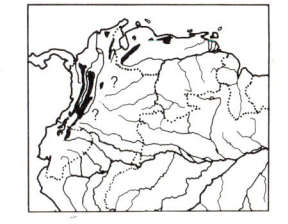

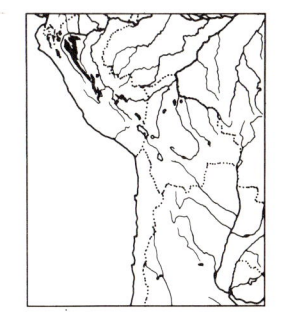

The closely related **GREEN-THROATED** (or **STRAW-BACKED**)

TANAGER *Tangara argyrofenges* inhabits the subtrop. (occ. lower temp.) zone of E Andes of Peru, with a relict distribution: Amazonas and San Martín, Conchapen mt. in Junín (*caeruleigularis*); and Andean slope of Bol. from La Paz to Sta Cruz (*argyrofenges*). **Male** has black crown, neckside, wings, tail, and most of underparts; silvery straw-colored back, rump, and flanks (less yellow in *caeruleigularis*); blue-green headside, throat, and upper breast (bluest in *caeruleigularis*). **Female** has dull blue-green crown and neckside, greenish yellow back, rump, and sides; greenish wings and tail; blue-green headside, throat, and upper breast, and blue-gray remainder underparts.

YELLOW-THROATED TANAGER *Iridosornis analis* – Plate LIV 1

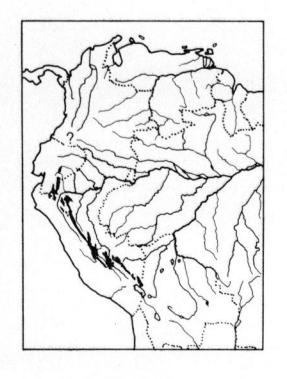

15 cm. Iris dark red. Headside and chin black, crown dusky faintly tinged purplish blue grading to dull turquoise of back, wings, and tail. **Throat bright yellow**; rest of underparts pale buff to buff, vent chestnut; breast and sides more or less washed with dull greenish blue. Female slightly greener and duller above than male. **Juv**. undescribed. In pairs, often in mixed-species flocks with other tanagers, e.g. Silvery and Grass-green T.s, and occ. in feeding aggregations at fruiting trees. Keeps body horizontal when active. Forages low in bushes. Probable song a squeezed, down-slurred *tseeeer*, repeated at 2-13 s intervals.

Stunted cloud forest and edge of wet mossy forest and shrub. 1130-2600 m, but mainly upper subtrop. zone. Along e Andean slope from Napo, e Ecu. to Puno, se Peru (and presumably adjacent Bol.). A sight rec. from w Putumayo se Col. is referable either to this species or Purplish-mantled T. Fairly common in n part of range, otherwise uncommon.

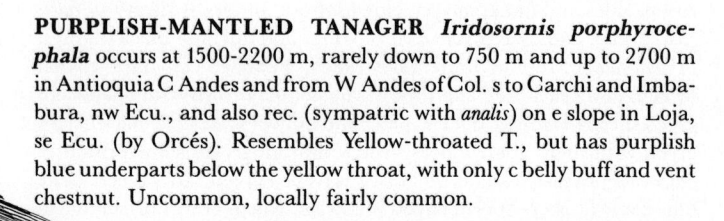

PURPLISH-MANTLED TANAGER *Iridosornis porphyrocephala*

occurs at 1500-2200 m, rarely down to 750 m and up to 2700 m in Antioquia C Andes and from W Andes of Col. s to Carchi and Imbabura, nw Ecu., and also rec. (sympatric with *analis*) on e slope in Loja, se Ecu. (by Orcés). Resembles Yellow-throated T., but has purplish blue underparts below the yellow throat, with only c belly buff and vent chestnut. Uncommon, locally fairly common.

GOLDEN-COLLARED TANAGER *Iridosornis jelskii* – Plate LIV 2

14 cm. Iris reddish brown. **Crown, nape, and posterior headside yellow**, feathers of forehead with black centers; supercilium, anterior headside and throat black; throat narrowly bordered below with yellow and gray-brown, nape narrowly bordered black. Rest of upperparts dark blue, rest of underparts rufous chestnut. Ssp *boliviana* slightly smaller, forehead more suffused with black. **Juv**. has rufous belly.
Habits: Alone, in pairs, or a few together, typically in mixed-species flocks. Perches upright. Forages from ground to mid-height, inside low bushes or trees, rather slowly searching for insects along slender, moss-covered branches. Occ. hovers to pick small fruit.

Voice: Very soft, high-pitched *seep*s and *cheep*s, and a rapid *ti-ti-ti* (2-5 notes).
Breeding: Juv./ad. June, July (Huánuco); large gonads Nov (Huánuco).
Habitat: In low trees, shrubs and bamboo in elfin forest and near treeline of humid forest.
Range: 3100-3650 m, rarely down to 2900 m (rec. down to 2150 m in Bol. probably erroneous). Known from e Andean slope in La Libertad and San Martín, Carpish mts in Huánuco, and Maraynioc in Junín, c Peru (*jelskii*); Cedrobamba (=Sayacmarca) in Vilcabamba mts in Cuzco, Limbani in Carabaya mts in Puno, se Peru, and La Paz, Bol. (*boliviana*). Uncommon.

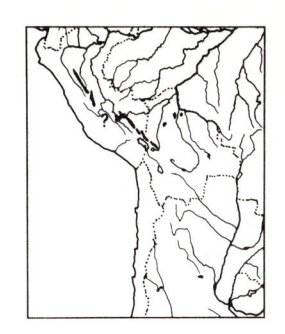

GOLDEN-CROWNED TANAGER *Iridosornis rufivertex* – Plate LIV 4a-b

16.5 cm. Unmistakable. Iris red. Sspp *rufivertex* and '*subsimilis*': head and upper mantle black, **c crown golden**. Rest of body dark purplish blue, wings and tail brighter blue, lower belly and vent chestnut; under wing-coverts dusky tipped tawny. Ssp '*ignicapillus*' has slightly more orange crown; *caeruleoventris* similar, but under tail-coverts dark blue (occ. with a faint wash of chestnut), under wing-coverts dusky. **Juv**. black, wings and tail as in ad.
Habits: In close pairs, often in mixed-species flocks. Restlessly forages for berries and gleans insects from leaves, body horizontal.
Voice: Calls include a *tsit* like other tanagers and a finer *tee tee*.
Breeding: Fledglings May (Cundinamarca, Cauca), Mar, July (nw Ecu.); juv./ad. May (Nariño), Aug (n Ecu.), June (n Cajamarca), Aug, Oct (Amazonas), Sep (La Libertad); large gonads Feb-Aug (E Andes Col.).
Habitat: Lower to medium strata. Humid tangled páramo shrubbery and edge of cloud forest and humid forest.
Range: (2000) 2300-3800 m. At Paramillo n end of W Andes, and n part of C Andes, Col. (*caeruleoventris*); from Páramo de Tamá and locally in E Andes of Col. s through e Ecu. to Piura (Chinguela) and Cajamarca (Condor mts), n Peru (*rufivertex*); s end of C and W Andes and Nariño (except e slope), Col. ('*ignicapillus*') and w Ecu. ('*subsimilis*'), the latter 2 sspp doubtfully distinct from *rufivertex*. Fairly common.

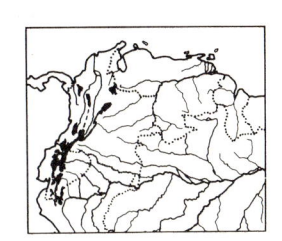

YELLOW-SCARFED TANAGER *Iridosornis reinhardti* – Plate LIV 3

16.5 cm. Unmistakable. Head and upper mantle black, **broad nuchal collar yellow**, rest of plumage purplish blue, wings brighter blue. **Juv**. black, wings and tail like ad., vent sometimes admixed a little chestnut.
Habits: 2-5 birds, usually in mixed-species flocks. Actively crawls about inside canopy, rarely staying the same place for long, but moves on through the middle branches of the next bushes and trees, body horizontal, jumping up from branch to branch. Probes moss and eats small berries.
Voice: Call a sharp *tzit*.

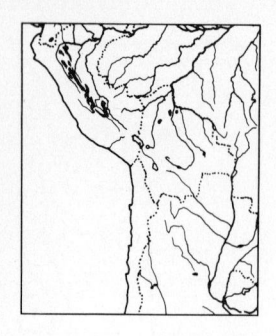

Breeding: Juv./ad. June (Huánuco).
Habitat: 1-5 m up inside low trees in humid montane shrubbery (particularly *Clusia*?), cloud forest, and humid forest. Occ. on moss on the ground.
Range: (1800) 2050-3400 m. S Ecu. in ec Loja (1989 sightings in P.N. Podocarpus: MKP, CR) and s and e of the Marañón in Peru from Amazonas to n Vilcabamba mts in Cuzco. Fairly common, locally abundant.

NOTE: Often regarded a (mega)ssp of Golden-crowned T. However, the two have recently been found sympatrically in s Ecu.

SCARLET-BELLIED MOUNTAIN-TANAGER
Anisognathus igniventris – Plate LIV 7a-b

17.5 cm. Unmistakable. Ssp *erythrotus*: above black, rump and shoulders blue, spot behind earcoverts scarlet. Throat, breast, and vent black, **remaining underparts scarlet**. Ssp *lunulatus* has vent mixed with red, *ignicrissus* almost wholly red; *igniventris* has blue-black back, narrow blue edges of greater wing-coverts and remiges, and red vent.
Juv. with black duller and scarlet replaced by orange-rufous feathers more or less conspicuously dark-tipped.
Habits: In pairs or small groups, occ. up to 15 or more, alone, or in mixed-species flocks with other tanagers. Forages half-way out along branches in c canopy and picks berries by reaching out or occ. making sallies or hanging among leaves at tips of branches of trees and low bushes. Mainly perches on or near thick, lichen- or bromeliad-covered branches. Usually sings from concealment, except at dawn.
Voice: Calls include a *tit*, as in other tanagers, and a characteristic *trrt tit*, sometimes prolonged *trrr-tit-trrr-trrr-tit tit trrr trrr trrr*. Song a rasping *tkd rrie-rrie-rrie-rrie-rreeeh* c. 3 s long. **Breeding**: Fledglings Feb, Apr, June (Cundinamarca), June, Sep (nw Ecu.), Jan (Azuay), Aug (El Oro), Dec (Puno), Mar (La Paz); juv./ad. Feb (Boyacá), Mar, May, Oct, Nov (Cundinamarca), Oct (Cauca), Sep (Cajamarca, La Libertad), Dec (Huánuco), Aug (Junín), July (Cuzco, La Paz), Jan, Aug, Sep (Cochabamba), Dec (Sta Cruz); song primo May (nw Ecu.); large gonads July (Santander), Dec (Boyacá), Nov (Cundinamarca), Feb (Cauca).
Habitat: 4-8 m up in elfin forest, humid páramo shrubbery, second growth, cloud forest, and edge of humid forest, in the n mainly in thickets and shrubs (disturbed areas), in the s often inside forest.
Range: (2250) 2600-3950 m. Páramo de Tamá and E Andes of Col. s to Cundinamarca (*lunulatus*); C Andes in Caldas, Cauca, and Huila, mts of Nariño, s Col., and Ecu. (both slopes and interandean shrubbery) (*erythrotus*), in Cajamarca and on e Andean slope from Amazonas to Junín Peru (*ignicrissus*); (probably Ayacucho), Apurímac and Cuzco, se Peru, to Sta Cruz, Bol. (*igniventris*). Common.

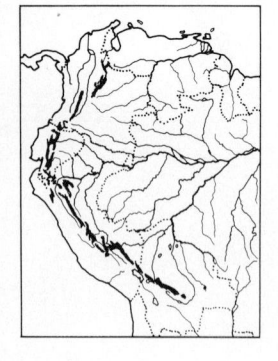

SANTA-MARTA MOUNTAIN-TANAGER
Anisognathus melanogenys – Plate LXIII 23

18 cm. Above glaucous blue, **crown** and shoulders **ultramarine**, head-

side and side of throat **black**, **spot below eye** and entire underparts **yellow**, flanks like back. **Juv**. undescribed.
Habits: In pairs or small groups, often in mixed-species flocks. Rapidly runs and hops along branches and in foliage.
Voice: Call a weak chirping note.
Breeding: Nest-building Oct; large gonads Jan-June, Sep.
Habitat: From 2 m to canopy, mainly subcanopy or lower in mossy forest edge, second growth woodland, overgrown pastures, fruiting trees, and shrubs, less frequently inside forest. Broader amplitude of habitats than Lacrimose Mt-t.
Range: Col. at 1500-3200 m, lowest during the wet season (June-Sep), in Sta Marta mts. Common.

NOTE: Called Black-cheeked Mt-t. by Meyer de Schauensee. Sometimes considered a (mega)ssp of Lacrimose Mt-t.

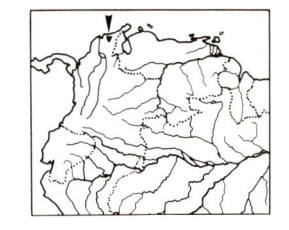

LACRIMOSE MOUNTAIN-TANAGER
Anisognathus lacrymosus – Plate LIV 5a-c

16-18 cm (*pallididorsalis* largest). **Yellow spot below eye** (except normally in *pallididorsalis*). Ssp *pallididorsalis*: above dull blue-gray, rump and shoulders blue; forehead and sides of head dusky green; below dark yellow. Ssp *palpebrosus* bluish slate above, rump and shoulders purplish blue, flight- and tail-feathers edged blue, headside and extreme side of throat greenish slate, yellow spot behind earcoverts, underparts bright warm yellow. Sspp *melanops*, *tamae*, and *intensus* like *palpebrosus*, but crown darker, headside slightly darker and underparts paler; *olivaceiceps* like *palpebrosus*, but forecrown, supercilium and sides of head tinged olivaceous, hindneck and back paler and browner, underparts paler; *lacrymosus* like *palpebrosus*, but headside blue-black and no spot behind earcoverts (spot occ. suggested); *caerulescens* intermediate between *palpebrosus* and *lacrymosus*, usually with yellow spot behind earcoverts. **Juv**. duller.
Habits: Alone, in pairs, or small groups, often in mixed-species flocks with other tanagers. Appears to move about widely. Often starts insect-foraging near base of shrubs, moving upwards to the subcanopy where it eats berries (e.g. *Miconia* and *Schefflera*). Agile, often difficult to see.
Voice: Call *zit* like other tanagers, and a longer *tziit*. Song a series of 2-6 high-pitched, sputtering, and excited phrases *ee-chut-chut-ee, ee-chut-chut-ee...*, much like song of Blue-capped T. Other variants *chuck-zit-it*, *swiik-id-dee-it-it*, and *suick-id-dit*.
Breeding: Fledglings Feb, July (e Cauca), Apr (w Cauca), Aug (El Oro); juv./ad. June, July (Huánuco); large gonads Feb-Aug (Col.), Aug (Huánuco); nest May (e Cauca).
Habitat: Lower to middle strata, rarely on the ground. Elfin forest, cloud forest, mossy humid forest, and occ. shrubby slopes near forest.
Range: 1800-3800 m, mostly above 2600 m. Perijá mts (*pallididorsalis*); Trujillo to Táchira, Ven. (*melanops*), Páramo de Tamá to N.d.Santander and Boyacá, Col. (*tamae*); n part of W Andes and C Andes s to Quindío, Col. (*olivaceiceps*), e slope of W Andes in Valle and both slopes in Cauca, sw Col. (*intensus*), s end of C Andes and both slopes of Nariño, Col., and e Ecu. s to n Morona-Santiago (*palpebrosus*), Loja, s Ecu., to Piura and

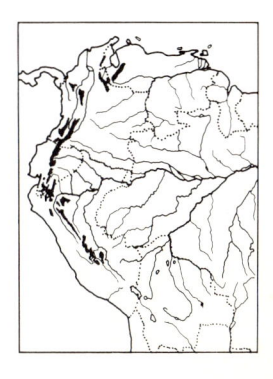

Cajamarca, n Peru (*caerulescens*), Amazonas to Ayacucho and extreme n Cuzco, se Peru (*lacrymosus*). Common to fairly common.

BLUE-WINGED MOUNTAIN-TANAGER
Anisognathus flavinuchus – Plate LIV 6a-b

17 cm. Ssp *somptuosus*: upperparts and sides of head black, **c crown and nape** and entire underparts **bright yellow**, shoulders purplish blue, **edge of flight- and tail-feathers turquoise**, rump and sometimes a few feathers on the back dull dark olive-green. Ssp *flavinuchus* similar, but with blue rump and less yellow on nape; *alamoris* intermediate between *somptuosus* and *cyanopterus*; *cyanopterus* differs from *somptuosus* by more extensively yellow nape, blue edges in wings and tail, rump sometimes darker, almost blackish olive; *antioquiae* like *cyanoptera*, but interscapulars tinged greenish, rump green, edges of wings and tail lighter blue; *baezae* like *antioquiae*, but back greener, edges of wings and tail lighter; *victorini* like *somptuosus*, but back moss-green, rump dark yellow-green. **Juv**. like ad., but yellow slightly duller, yellow feathers of nape narrowly dark-tipped. In pairs or groups, rarely up to 25, usually very conspicuous in mixed-species flocks. Restlessly forages for insects and *Miconia* berries among the foliage, frequently making short flights while calling. Also eats *Cecropia* catkins, and in Peru has been reported to search moss and bark. Surprisingly fast for its size. Calls include a long harsh *veeeeee*, an undistinctive *seet*, and faint, slightly buzzy ticking, chipping, and trilled notes. Song consists of a ticking series of notes, but in Bol. a more elaborate song is rec.: an explosive series of whistled notes *too-too tyoo-towoo-towoo too-wit too-wit too-wit*, 3-4 s long, also given in display and as duet, then preceded by a crescendo of *tsic* notes and subsiding to high-pitched *tic*s. Eggs Oct (La Paz); nestlings June (Valle); fledglings Nov (Valle); juv. Mar, Sep (Cauca), Oct, Dec (e Nariño), Jan (Azuay), Mar (Junín), Apr (Cochabamba); large gonads May-Sep (n end of W and C Andes Col.), Aug (La Paz).

Inside humid forest and in second growth, mainly at mid-heights, occ. high or low. At 900-2750 m, mainly in subtrop. zone. N part of W and C Andes Col., on e slope of C Andes s to Tolima (*antioquiae*); Río Chiquito, Táchira, Ven., and s along w slope of E Andes to head of Magdalena valley, Col., in Cundinamarca also on e slope (*victorini*), w slope of C Andes from Quindío s, s part of W Andes, w Nariño, and s to Río Chimbo, w Ecu. (*cyanopterus*), e Nariño, Col., and e Ecu. s to n Morona-Santiago (*baezae*), from Azuay to Loja, sw Ecu. (*alamoris*), from Morona-Santiago, se Ecu. to Ayacucho and Machu-Picchu in Cuzco, Peru (see *Bull. Brit. Orn. Club* 100(1980):149-50) (*somptuosus*), Puno in se Peru to Sta Cruz, Bol. (*flavinuchus*). 2 sspp in coastal Ven. Common.

NOTE: Sometimes placed in the genus *Compsocoma*.

HOODED MOUNTAIN-TANAGER *Buthraupis montana* – Plate LIV 9a-b

Unmistakable. Ssp *montana*: 21 cm. **Head black with glowing red**

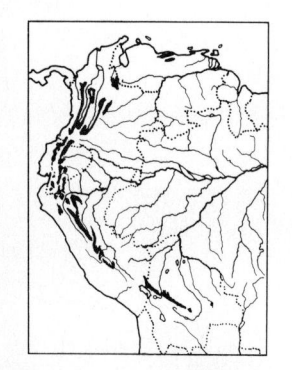

eyes, extreme sides of breast black; nuchal half-collar pale blue, rest of upperparts dark blue, rump and shoulder bright blue; below yellow, thighs black mixed with a little blue. Other sspp (23 cm) lack nuchal half-collar and have enamel-like feathers above varying from shining violaceous blue (*saturata*) to a duller purplish blue (*gigas*). **Juv**. like ad., but black and yellow duller.

Habits: In pairs or flocks up to 10 or more. Only rarely in mixed-species flocks. Roams up and down slopes, often flying long distances, frequently chasing one another while vocalizing. Rapidly moves about in the canopy of tall moss-clad trees, searching dense foliage near ends of branches, and occ. slowly and acrobatically hanging while eating berries. Flock is headed to new tree as one bird lands with tail slightly raised, then calling while wagging tail up. The others then follow one by one.

Voice: Flight call a characteristic loud *vee vee vee*.... Alarm a hoarse nasal *zhhhi*. Also gives weaker *zit-zit* notes like other tanagers. In the early morning individuals may join in a loud chorus. Possible dawn song by a single bird a **weeck** or *toot-weeck* every second.

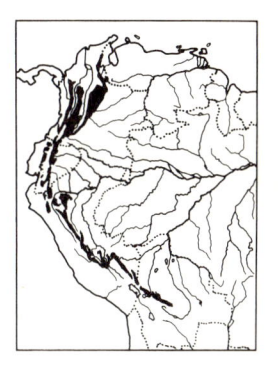

Breeding: Fledglings June (nw Ecu.), Nov (ne Ecu.), Apr (Cuzco), Nov, Jan (Cochabamba); juv./ad. Dec (e Nariño), July, Aug, Oct (Huánuco); large gonads Jan-Sep (Col.).

Habitat: Canopy of tall epiphyte-laden trees in humid forest and cloud forest, and on nearby shrubby hillsides.

Range: 1700-3500 m. Perijá mts, Páramo de Tamá, and through E Andes of Col. (*gigas*), W and C Andes and Nariño, Col., e and w Ecu. to Piura and n Cajamarca, n Peru (*cucullata*); Amazonas to Junín, c Peru (*cyanonota*); Ayacucho to Puno, se Peru (*saturata*); La Paz and Cochabamba, Bol. (*montana*). Common.

MASKED MOUNTAIN-TANAGER *Buthraupis wetmorei* – Plate LIV 13a-b

21.5 cm. Forehead, sides of crown, rump, and most of underparts bright yellow; lores, **supercilium, headside**, neckside, chin, sides of throat, wings, tail, and spots on sides of breast **black**; shoulders blue, greater wing-coverts black tipped blue; **crown, back**, flanks, and vent **dull golden olive**. Female is slightly duller than male and sometimes (younger birds?) has olive lores.

Habits: Alone, in pairs, or 3-4 birds, in mixed-species flocks with other tanagers, e.g. Black-chested Mt-t. Rather slow and silent. Forages along thick, moss-covered branches, and eats berries at tips of branches in treetops. May sally clumsily for a flying insect. Flies quickly across openings.

Voice: No rec.

Breeding: Large gonads Feb (Cauca).

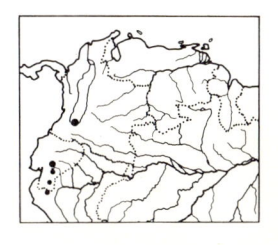

Habitat: Isolated epiphyt-clad trees in humid páramos, elfin forest and upper reaches of cloud forest.

Range: 2900-3650 m. Only known from San Rafael in Puracé Cauca w slope of s end of C Andes, Col., e slope of Ecu. at mt Sangay and Zapote Najda mts in nw Morona-Santiago, P.N.Podocarpus in e Loja, and Cerro Chinguela in Piura, n Peru. Apparently fairly common on mt Sangay, elsewhere rare.

GOLDEN-BACKED MOUNTAIN-TANAGER
Buthraupis aureodorsalis – Plate LIV 12

23 cm. Unmistakable. Crown dark blue, mantle, sides of head and neck, throat, breast, wings, and tail black; **back, rump, and belly golden yellow**, feathers chestnut-brown subterminally becoming more conspicuous with wear; lesser wing-coverts dark blue, vent chestnut. **Imm.** has little blue on lesser wing-coverts.
Habits: In pairs or family groups (sometimes up to 7), only occ. with mixed-species flocks. Slow. Watches for berries (e.g. *Miconia*) from middle branch, then hops up along the branches to flat tree-top (*Escallonia*), where it may spend hours in the same place, occ. eating berries. Raises crest when calling. May take insects from moss, or rarely leaves or grass.
Voice: Generally silent. Usually one bird begins calling prior to flight, sharp squeaky *chit*, *weet*, and *veet* notes, surprisingly weak for so large a bird. Flight call a sharp *steet-steet*. Song long and fairly elaborate, a rapid broken warble of both whistled and snarling notes given in varied phrases of 3-5 notes, each phrase repeated 2-4 times *tooit-zeee-tr-er tr-er tr-er tooit-tr-er....*
Breeding: No data.
Habitat: Elfin forest (esp. *Clusia* and *Escallonia*), surrounded by jalca grassland, and sometimes in small trees scattered over grassy slopes.
Range: C Peru at 3000-3500 m on e Andean slope in San Martín and La Libertad and in upper parts of Carpish mts, Huánuco. Uncommon.

NOTE: Described by ER Blake and P Hocking in *Wilson Bull.* 86(1974):323-4.

BLACK-CHESTED MOUNTAIN-TANAGER
Buthraupis eximia – Plate LIV 8a-b

21 cm. Bill stubby. Sspp *cyanocalyptra* and '*chloronota*': crown dark purplish blue, shoulders blue; **back, wings, and rump dark moss green**; primaries black, primary-coverts narrowly edged blue, tail black. Headside dull blue-black; throat and breast black, belly bright yellow, vent tinged orange. Semi-concealed patch on c throat white. Rump with some blue in *zimmeri*, wholly blue in *eximia*. **Juv.** presumably much like ad.
Habits: Alone, in pairs, or groups of 3-5, sometimes in mixed-species flocks, where possibly a nuclear species. Moves in jerks, flicking tail a little while searching epiphytes or foliage, or silently and slowly feeds on berries or takes pieces of *Schefflera* fruits in treetops.
Voice: Usually silent, but occ. gives a soft *zit* or *chip* like other tanagers. Song loud and fairly elaborate, lasting up to 30 s *tititi-turry-ti-titi-tee-ter-turry....*
Breeding: No data.
Habitat: All strata, but esp. middle and outer branches of dense trees and bushes in elfin forest, low mossy cloud forest, humid shrubbery, and humid *Polylepis* forest.
Range: 2750-3800 m, rarely down to 2000 m. Páramo de Tamá and in

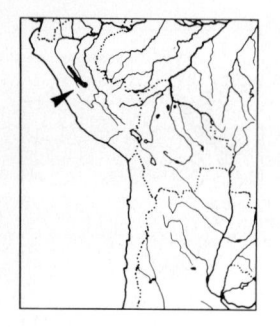

N.d.Santander and Cundinamarca in E Andes of Col. (*eximia*), n end of W Andes and C Andes in Quindío and in Cauca, Col. (*zimmeri*), e Nariño, Col., to n Ecu. (*chloronota*), e Ecu. in Morona-Santiago to Piura and n Cajamarca, n Peru ('*cyanocalyptra*', hardly distinct from *chloronota*). Fairly common in Ecu., uncommon and apparently local in Col.

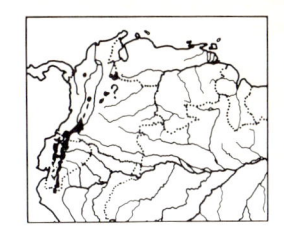

CHESTNUT-BELLIED MOUNTAIN-TANAGER
Dubusia castaneoventris – Plate LIV 15

15.5 cm. Above dull dark blue, below rufous brown; faint supercilium pale blue behind eye, headside and **moustache black**; *peruviana* similar, but bluer above, feathers of forehead and entire supercilium tipped pale blue. **Juv**. undescribed.
Habits: Alone or in pairs, sometimes in mixed-species flocks. Fairly restless, mostly with horizontal body along branches, sometimes searching moss and leaves for insects, occ. climbing down thick, vertical branches, and also reaches out and down for berries.
Voice: Call a short *zee zee* like other tanagers. Song a long, sharp high-pitched but slightly descending *tzeeeee*. Also rec. a sweet song of 2-3 notes *pee pay-oay* or *pee pee pay* (1st note highest), often repeated every 5-6 s for long periods.
Breeding: No data.
Habitat: 3-10 m up inside and just below rather dense canopy in elfin forest, open cloud forest, and adjacent second growth.
Range: 2150-3600 m. From e La Libertad to Puno, se Peru (*peruviana*); La Páz to Sta Cruz, Bol. (*castaneoventris*). Uncommon.

NOTE: Often placed in its own genus *Delothraupis*.

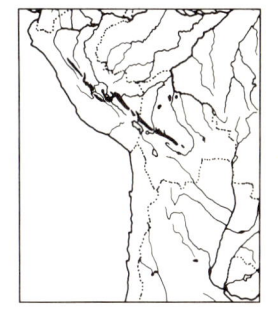

BUFF-BREASTED MOUNTAIN-TANAGER
Dubusia taeniata – Plate LIV 16a-c

18 cm. Ssp *taeniata*: head to upper mangle and upper breast black, feathers of forehead and very **long supercilium** to neckside with **pale blue** tips; back, wings, and tail dark blue sometimes mixed with black, shoulders light blue; band across breast and vent buff, belly yellow. Ssp *stictocephala* similar, but feathers of entire crown and nape tipped light blue, mantle dark blue like back, only narrow breastband buff, belly and vent yellow; *carrikeri* like *taeniata*, but crown dark blue, **mid-throat buff streaked black**, and breast cinnamon-buff. **Juv**. has dull olive-green back and is ochraceous instead of buff on breast and vent.
Habits: Alone, often in mixed-species flocks. Forages with horizontal body inside dense trees and shrubbery, often quite low, occ. searching moss on steep banks. Sings somewhat hidden in top of tall tree.
Voice: Song loud and distinct of 2 (nw Ecu.) to 3 (n Peru) clear whistles, the 1st highest and the last lowest pitched, the 1st 1 or 2 somewhat quavering *teeee tyyyy* or *teeee teeee tyyyy*, lasting 2-3 s and repeated at 5 s intervals.
Breeding: Juv./imm. Jan, Sep (nw Ecu.); song Oct (nw Ecu.), Aug (Piura); large gonads Feb-Sep, Dec (Col.).

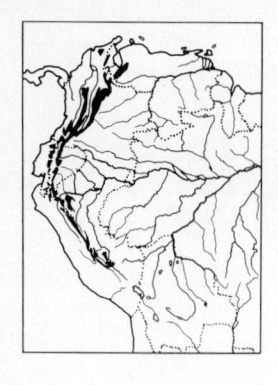

Habitat: 1-5 m up in dense humid montane shrubbery, elfin and cloud forest, incl. in mossy bamboo.
Range: 2000-3650 m. Sta Marta mts (*carrikeri*); Perijá mts and from c Trujillo, Ven., s through all 3 ranges (only s part of W Andes?) of Col., e and w Ecu. to Piura and n Cajamarca, n Peru (*taeniata*); s and e of the Marañón from Amazonas s to Cuzco, se Peru (*stictocephala*). Uncommon to fairly common.

BLUE-CAPPED TANAGER *Thraupis cyanocephala* – Plate LV 13

16 cm. Pot-bellied. Ssp *cyanocephala*: **crown and nape blue**, rest of upperparts olive-green; lores and ocular region black, rest of **headside dark blue**. Below blue-gray, sides olive-green, flanks and vent olive-yellow, thighs and **underwing bright yellow** (conspicuous in flight). Sspp *auricrissa* and *annectens* slightly darker below; *margaritae* smaller with dull blue tinge on throat and foreneck, *hypophaea* has the underparts strongly suffused with blue. **Juv**.: underparts gray without blue cast, blue of crown duller and restricted to tips of feathers.
Habits: Alone, in pairs, or small groups, occ. in mixed-species flocks. Forages along thick branches in top center of canopy, sometimes taking berries hovering, and flycatching upwards. May sit motionless for some time, but generally restless, and often conspicuous.
Voice: Calls include a high-pitched *tsit*, and several others. In flight *quivivivivi*. Song much like song of Lacrimose Mt-t., but usually more flowing and less heavily accented: 2-5 phrases of sharp squeaky notes *swick ik*, *sick-it-cheie-y* or *swichity-chee*, lasting 2-5 s. May also give 3 *tsee* notes followed by 10-12 *cha* notes at a slightly lower pitch, sounding like the rhythmic shaking of a gourd rattle.
Breeding: Juv./ad. June (Huánuco); large gonads Aug (La Paz), Oct (Cochabamba).
Habitat: Humid broken forest, cloud forest edge, frequently in *Cecropia*, second growth and alders (*Alnus*), sometimes coffee plantations.
Range: 350-3000 m, in Andes mainly above 2000 m. Páramo de las Rosas in Lara, Ven. (*hypophaea*, known from 3 specimens, maybe intergrades between *auricrissa* and the blue-bellied *olivicyanea* of coastal Ven.), Perijá mts and from Trujillo, Ven., s through E Andes to Meta, Col. (*auricrissa*); Sta Marta mts (*margaritae*), W and C Andes, both slopes of Nariño and w Putumayo, Col. (*annectens*), e and w slopes of Ecu., nw and e Andean slopes of Peru s to Cochabamba, Bol. (*cyanocephala*). 3 sspp in coastal Ven. Fairly common to common.

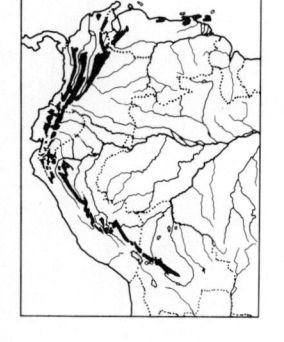

BLUE-AND-YELLOW TANAGER *Thraupis bonariensis* – Plate LV 14a-d

16.5 cm. Slender. Ssp *darwinii*: **male** head blue, narrow forehead, lores, ocular region, and chin black; back olivaceous, **rump bright golden-yellow**, feathers of wings and tail blackish edged blue, extreme sides of breast and body like back, rest of underparts golden yellow. **Female**: above gray-brown to grayish olive with dull blue tinge on crown, supercilium, and shoulders; rump pale olive-brown, underparts pale dirty

yellowish buff, throat gray-brown, breast and sides washed grayish olive, vent pale buff. **Imm. male** like female, but whole head tinged blue, rump with some golden yellow, wings and tail with dull blue edges, underparts golden olive-yellow, paler and yellow on c belly. Ssp *schulzei* similar, but smaller; male with back black, forehead without black, sides of neck and narrow broken breastband black, rump and underparts (esp. breast) bright orange, longest under tail-coverts white; female as in *darwinii*, but lacking blue on head. Ssp *composita* much like *schulzei*, but larger, and female with faint bluish wash on head. **Juv**. like female, but rump duller and less extensive, no blue tinge on head, breast and sides more buff. 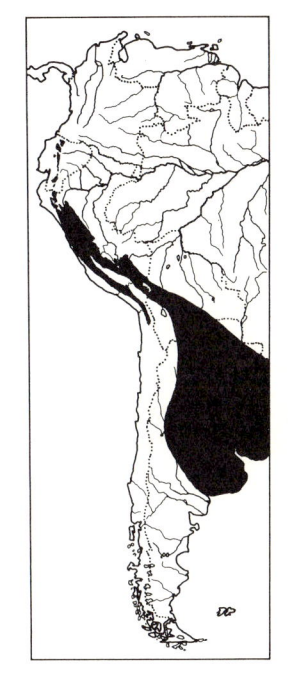
Habits: Alone, in pairs, or small groups up to 5, and sometimes in feeding aggregations, though aggressive towards conspecifics and other birds. Restlessly forages from low in the vegetation to treetops for insects, buds, *Schinus* berries, and fruits of cacti and other plants. Flies in swift undulating flight between groups of trees.
Voice: Song 4-6 sharp bisyllabic notes *sweet-sur, sweet-sur, ...*, that may end or start with a short trill. Incessantly delivers high-pitched *sweet* and *tick* notes, occ. in a jumbled series.
Breeding: Nest high in tree or cactus. Eggs late Mar to early Apr (Tarapacá), Dec-Jan (Bol.,Tucumán); fledglings Apr (Cajamarca); juv./imm. Apr (Chimborazo), June (La Libertad).
Habitat: Semi-arid areas with open trees and scattered cacti, second growth, small forest patches, open deciduous forest and shrubbery, parks, *Eucalyptus* stands, *Polylepis* woodland, and montane scrub bordering farmland.
Range: Sea level to 3600 m, in n and w part of range mainly above 2000 m, locally to 4000 m. From Ibarra in Imbabura, n Ecu., s to Tarapacá, n Chile, and La Paz, Bol. (*darwinii*); Cochabamba and Sta Cruz to Tarija, Bol. (*composita*, intergrading with *schulzei* in Tarija), and Tarija to Mendoza, w Arg., and Parag., partly migratory furthest s (*schulzei*); s-most Brazil, Urug., and ec Arg. (*bonariensis*). Common.

SAYACA TANAGER *Thraupis sayaca* – Plate LV 12

16 cm. **Male**: above blue-gray, rump, wings, and tail dull turquoise; below light blue-gray, whitish on c belly and vent. **Female** similar, but greenish gray instead of blue-gray. **Juv**. like female, but paler below, breast very faintly tinged brownish.
Habits: In pairs or groups of up to 12 working quickly through treetops and often flying long distances. Rather omnivorous.
Voice: Calls include a long *sssst*, *tzit*, and *chip* notes. 2-3 s long song quite variable.
Breeding: Fledglings Sep, Oct (La Paz), Feb (Cochabamba), Jan (Sta Cruz). 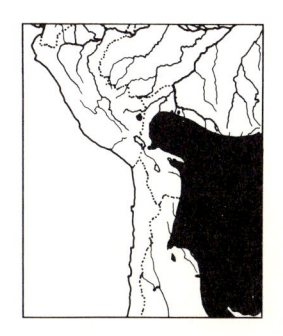
Habitat: Canopy of small trees and bushes in both dry and rather humid open country. Forest, woodland, and desert scrub, parks, gardens, *Eucalyptus* stands, orange groves, farmland. Often near human habitation.
Range: Lowlands to 3200 m, occ. to 3600 m in Bol. (e.g., in La Paz city). Andean slopes and adjacent lowlands in Bol. and w Arg. s to Tucumán, Córdoba, and Sta Fé (*obscura*), from e Brazil to Parag. and n Arg. (*sayaca*). Common up to 3000 m, less common higher.

BLUE-GRAY TANAGER *Thraupis episcopus*

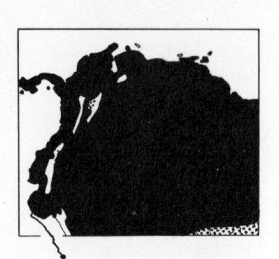

16 cm. Quite similar to Sayaca T., but *quaesita* has a paler head and azure-blue shoulder, *leucoptera* a darker head, bluish white shoulder, narrow whitish wingbar, and pale blue rump. In close pairs or small flocks roaming through the canopy, and often gathering in fruiting trees. Sometimes active in bird-waves. Call notes squeaky *seeee* or *che-eup*. Both sexes sing, though female weakest, a series of squeaky or drawn-out notes, often 1-2 high-pitched and 1-2 lower-pitched notes alternating.

In semiopen areas with second growth, forest edges, cultivated land, and often near human habitation, from arid to humid zones. From s Mexico to nw Peru and se Peru and c Brazil, mainly in trop. and subtrop. zones, but regularly ascendes to 2700 m in the Bogotá area of Col. (*leucoptera*), and in the vicinity of Quito, Ecu. (*quaesita*).

PALM TANAGER *Thraupis palmarum*

widespread in lowlands and premontane zones from Nicaragua to cw Ecu., c Bol., and s Brazil, is occ. rec. to 2600 m in Col. In semiopen savanna habitats, second growth, areas of human habitation, and at forest edge. - Grayish olive tinged violet except on crown; pale wing-patch formed by greater coverts, and in flight shows pale bar basally on the remiges. Calls include a lisping rising *seeeee?* or *wheerst?*, and a falling *see-you*.

HEPATIC TANAGER *Piranga flava* – Plate LV 17a-b

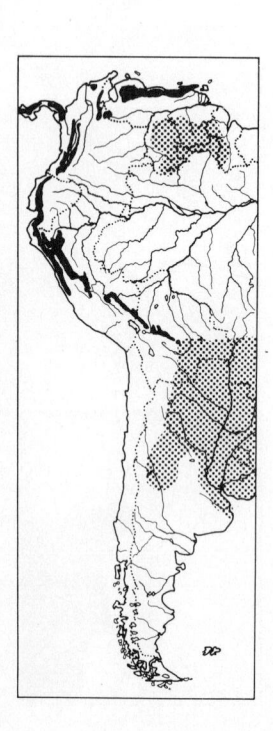

17.5 cm. **Large bill dusky** grayish yellow **with 1-2 'teeth'**. Outer primary shorter than next 2. Ssp *lutea*: **male: above dark red, below red**, earcoverts with faint white shafts. **Female** and **imm. male** to over 1 year old: above olive-yellow, **eye-ring yellow**, below yellow, sides and faint wash on breast olive-yellow. Ssp *desidiosa* barely differs; male deeper red and without white shafts on darker earcoverts; female with darker, more olivaceous green back; *faceta* male much paler, more orange-red. **Juv.** like female, but streaked with dark brown above and below. Alone or in pairs, rarely in small groups or mixed-species flocks. Active, frequently flying between trees in long woodpecker-like waves. Picks berries and looks for insects on underside of branches well inside canopy. May sit motionless for long periods. Wags tail slowly when excited. Alarm a repeated *tjik*. Call *yuhtitdit*, sometimes with an extra *yuh* at the end, given in flight or from perch. At other times *chuktik* or *chuktiti*. The jerky melodic song, buzzy in part, is a series of connected vireo-like phrases going up and down. Juv. rec. Mar (Ancash).

Tall trees in dry forest, edge of humid forest, woodland, parks, gardens, savanna, montane scrub, and scattered groves of trees in grassland. Sea level through premontane zone and rarely to 3150 m. Sta Marta mts, Perijá mts, Táchira to coastal mts of Ven. (*faceta*); middle and upper Cauca valley and w slope of W Andes from upper Dagua valley to upper Patía valley Cauca/Nariño sw Col. (*desidiosa*), from w Nariño, sw Col., through w Ecu. and Peru (w of Andes s to Lima), and along e slope s of the Marañón to Sta Cruz, Bol. (*lutea*). 12 other sspp from USA to Arg. Uncommon, locally fairly common.

NOTE: By some divided into 2 species, **HIGHLAND HEPATIC-T. (*P. hepatica*)** incl. Andean forms, and Lowland Hepatic-t. (*P. flava*).

SUMMER TANAGER *Piranga rubra* – Plate LV 16a-b

16.5 cm. **Large bill pale without conspicuous 'teeth'**. Outer primary equals 2nd. **Male** rosy red, palest on head and underparts. **Female**: above golden gray-brown, below pale yellow to dull gold; often the entire plumage is mixed with red feathers. **Imm. male** like female, but alwways with patches of red. Ad. plumage attained when over 1 year old. **Habits**: Solitary, sometimes in pairs with short-term territories. Often perches inactively within the foliage, then moves usually in a dashing manner, fluttering to leaves and hovering for insects and sometimes berries. Also flycatches and follows army ants.
Voice: Call *chikatuk* given alone or repeated, sometimes followed by a *chikik*.
Habitat: Variable, but generally more humid than Hepatic T. Lower to medium strata of open woodland, patches of woods, edge of humid forest and cloud forest, forest swamps, shaded plantations.
Range: Breeds e USA, winters Oct-Apr (mainly in upper trop. zone but on migration regularly to 3000 m, occ. higher) from Mexico to sc Peru, c Bol., w Brazil, se Ven. and Surinam, accidental in Antofagasta, n Chile (*rubra*). 1 ssp winters in Mexico. Uncommon.

SCARLET TANAGER *Piranga olivacea* of e N Am. winters from Panamá and w S Am. s to nw Bol., is rec. to 3000 m on migration through Col. 16 cm. **Male breeding plumage** bright scarlet with black wings and tail. **Female** and **imm**. much like Summer and Hepatic T.s, but wings and tail darker, and underparts dingier.

RED-HOODED TANAGER *Piranga rubriceps* – Plate LV 15a-b

17.5 cm. Bill black, slightly toothed. **Male: head and breast red**, lores and ocular region mixed with black; back golden olive-yellow, rump and lesser wing-coverts dark yellow; wings black, inner remiges and inner wing-coverts edged golden olive; **below dark yellow**. **Female** similar, but red only reaching upper breast. **Imm** with orange head grading to olive-yellow on throat and breast, greater wing-coverts narrowly tipped yellow, forming wing-bar. **Juv**.: upperparts incl. wings yellowish olive, 2 narrow wingbars buffy, broken eye-ring yellow; underparts pale yellow, breast and sides faintly streaked dusky, vent washed ochraceous; bill horn.
Habits: Alone or in pairs, but usually in small flocks, occ. in mixed-species flocks. Forages slowly for insects and fruit along limbs and among dense foliage, occ. surveying from conspicuous perch. Sings from open perch.
Voice: Rapid thin *ti-ti-ti-ti-ti-ti* alternate with sweet, evenly spaced, moderately pitched notes, e.g. *da-dee-dee*. Also repeats a rapid *ti-t-t-t-dee*.
Breeding: Fledgling Feb (La Libertad); juv. Nov (e Nariño); large gonads Sep (w Antioquia).

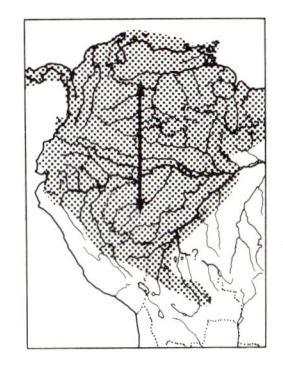

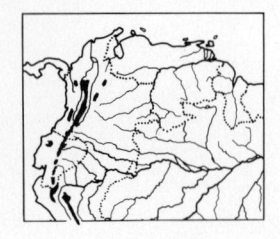

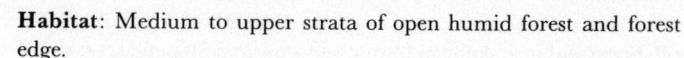

Habitat: Medium to upper strata of open humid forest and forest edge.
Range: 1700-3200 m. Patchily in all 3 ranges of Col. s to Pichincha, nw Ecu., and along e slope of Ecu. s to Huánuco, c Peru. Uncommon and local.

RUFOUS-CRESTED TANAGER *Creurgops verticalis* – Plate LVI 8a-b

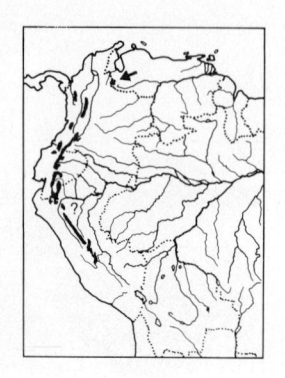

15 cm. Bill with 1 'tooth'. Headside and upperparts dark gray, underparts rufous. Male with rufous buff c crown-patch (absent in **female** and **juv**.). Solitary, usually in mixed-species flocks. Perches atop small trees, perch-gleans and occ. flycatches among thick, sometimes slender branches in lower canopy. Rarely at fruiting trees. Fast and agile. Juv. Aug (Huila), Dec (e Nariño).

Inhabits shrubby mossy humid forest and forest edge at 1150-2700 m. Río Chiquito, sw Táchira, Ven.; in C Andes; very locally in W Andes; at head of Magdalena valley and both slopes of Nariño, Col. to Imbabura, nw Ecu. and through e Ecu. to Junín, c Peru. Generally uncommon, but fairly common at head of Magdalena valley.

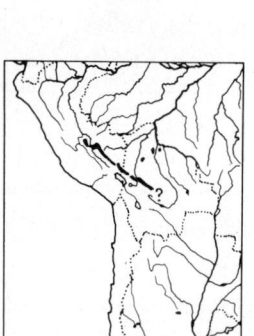

SLATY TANAGER *Creurgops dentata* – Plate LVI 9a-b, a generally uncommon bird of shrubby premontane forest at 1450-2150 m from Vilcabamba mts of Cuzco, se Peru, to Cochabamba, Bol., possibly ascends to the lower temp. zone. 13.5 cm. **Male** dark gray, forehead and brow blackish, crown chestnut; some birds faintly mottled with pale gray below. **Female** dark gray above, darkest on crown; supercilium from eye white mottled with dark gray; **headside, throat, breast, sides, and flanks rufous**, chin and moustache mottled whitish, belly whitish. **Imm. male** like female. Recently found to be fairly common in se Puno, where several have been rec. at mid levels in mixed-species flocks.

RUFOUS-CHESTED TANAGER *Thlypopsis ornata* – Plate LVI 5

13 cm. Ssp *ornata*: **head and underparts orange rufous**, palest on throat, **c belly white**; above grayish olive; *media* and *macropteryx* larger, the latter with darker crown, deeper color on throat and breast and more extensively white c belly. **Juv**. with the rufous replaced by yellow-ocher to almost yellow on throat, clay-color (occ. more gray-brown) on flanks, and buffy olive of upperparts continuing to crown.
Habits: Alone or in groups of 6-8, often in mixed-species flocks. Hops sideways up vertical branches with horizontal body and slightly raised tail, or wriggles body along bamboo etc. with slightly lowered wings, and restlessly flutters about, often in full view on the outside of tangles. Often searches dead leaves in trees.
Voice: Call *seep*. Song a warbling *tzee-wee-dee-zi*.
Breeding: Juv. Mar (nw Ecu.).
Habitat: Mainly 1-4 m up in tangles in humid montane shrubbery,

second growth, and edge of cloud forest. Also semi-humid *Polylepis* woodland, and in Lima in fairly dry situations.
Range: 1800-3500 m, occ. lower in Ecu., and to 3800 m in Ancash, Peru. Cauca w slope of C Andes of Col., w and ne (Pun), Ecu. (*ornata*), from Loja in extreme s Ecu. s to Lima and upper Marañón valley (*media*); from Huánuco to Junín, c Peru, and in n Cuzco, se Peru (*macropteryx*). Locally fairly common, elsewhere uncommon.

BROWN-FLANKED TANAGER *Thlypopsis pectoralis* – Plate LVI 6a-b

13 cm. Head, throat, and breast rufous, darkest on crown; back, wings and tail grayish olive; **sides gray-brown** turning buff along the white c belly; vent buff. Breast sharply demarcated from brownish sides and white belly. **Imm.** (or female?) has nape tinged olive and less bright throat. **Juv.** undescribed.
Habits: Alone or in pairs, occ. joins mixed-species flocks with Rufous-chested T. Hops about just inside outer foliage, working its way up from 1 to 3 m above ground. Reaches out and down for insects, often investigating curled dead leaves. Not as agile as Rufous-chested T.
Voice: Soft high-pitched *tsit* and twittering *seet seet seet* rec.
Breeding: Juv. Dec, Feb (Huánuco), Mar (Junín).
Habitat: Dense bushes on dry slopes and riparian bushes, alder thickets, and tangled trees adjacent to fields and pasture. Occ. at edge of cloud forest adjacent to arid zone.
Range: C Peru at 2000-3100 m in Huánuco, Pasco, and Junín. Uncommon.

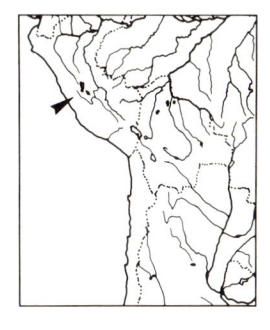

RUST-AND-YELLOW TANAGER *Thlypopsis ruficeps* – Plate LVI 7a-b

13 cm. **Crown and headside bright rufous** to rufous chestnut; back, wings, tail and sides olive-green, rest of underparts bright yellow. **Juv.** has crown and headside like back and lores dark, supralores and orbital region more or less washed with fulvous.
Habits: Occ. alone, but often in flocks of 5-6, sometimes in mixed-species flocks. Rapidly moves from one tree to the next, and actively perch-gleans and reaches out for insects mainly from the undersides of leaves, along outer branches in the canopy of small trees and bushes, occ. searching dead leaves. Usually hidden inside dense vegetation.
Voice: Song a 1.5 s long burst of chittering, given in duet.
Breeding: Fledglings Jan (Jujuy); juv. June (Cuzco), May, July (Puno), Feb (La Paz), May (Cochabamba).
Habitat: 1-9 m up in humid second growth, esp. with bamboo, shrubbery, and edge of cloud forest and humid forest, as well as in riparian thickets in more arid regions, but occ. up to *Polylepis* woodland.
Range: (1300-) 1850-3700 m. 1 rec. from Huánuco, c Peru. From Ayacucho, Apurímac, and Cuzco, se Peru, along e Andean slope of Bol. to Tucumán, nw Arg. Southern birds migratory. Fairly common, locally common.

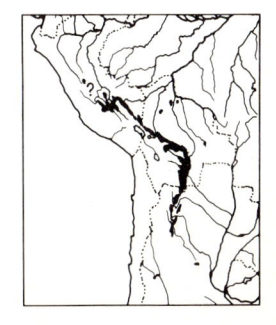

WHITE-CAPPED TANAGER *Sericossypha albocristata* – Plate LVI 12a-b

24 cm. Black, wings, and tail glossy; **cap down to lores white**; throat bright red (male), dark red (female), or black (juv.).
Habits: In close flocks of 5-20 (usually with one ad. male only) noisily sailing large distances through the air. Forages for fruit and insects actively and jay-like in tops of mainly large trees. Perches horizontally, bobs, leans forward, looks down stretching neck down. Not in mixed-species flocks.
Voice: Very vocal. Calls far-carrying, some recalling the whistles of Collared Jay *tamia-pischkoo*; often repeats a very loud *peeeaap*, sometimes followed by 1 or 2 sharp shrieking *keep* notes.
Breeding: Fledglings Oct (Amazonas); juv./ad. May (Cauca), Oct (La Libertad), Mar (Pasco).
Habitat: Humid forest and cloud forest.
Range: 1600-3200 m. Páramo de Tamá and Río Chiquito s Táchira, Ven., very locally on e slope of E Andes and e slope of C Andes near Caldas/Tolima border, at head of Magdalena valley in Huila, w slope of C Andes in Cauca, and locally through e Nariño, Col., and e Ecu. to Junín, c Peru. Uncommon and very local, but may wander seasonally (see *Novel. Colomb.* 1 (1960):256-76.

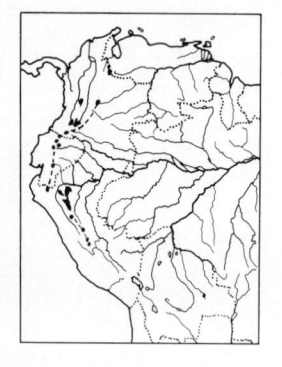

COMMON BUSH-TANAGER *Chlorospingus ophthalmicus* – Plates LVI 10a-d and LXIV 14

14 cm. **Dull-colored; usually with white postocular spot, but this is lacking in c parts of range**. Ssp *venezuelanus*: crown and headside dusky brown; **spot behind eye white**; **throat** buffy white **speckled blackish**; back, wings and tail olive-green, **breast ochraceous yellow**; sides, flanks, and vent olive-yellow, c belly grayish white. Ssp *jacqueti* similar, but crown and headside paler brown, throat paler and less spotted, breastband yellower; birds from Lara and n Trujillo with darker, less grayish crowns; ssp *ponsi* like *venezuelanus*, but crown dusky olive and throat darker buff; *eminens* like *venezuelanus*, but throat nearly white and breastband less ochraceous. Ssp *flavopectus* lacks white postocular spot; crown gray, throat white nearly without spots; *trudis* like *flavopectus*, but crown and back lighter and earcoverts brownish; *macarenae* much like *flavopectus*, but belly white; *nigriceps* without postocular spot, crown blackish, throat and sometimes part of breastband heavily speckled blackish; back rather dark and suffused with blackish; *exitelus* like *nigriceps*, but crown grayish, throat with fewer and paler specks, back more yellowish green; *phaeocephalus* like *flavopectus*, but throat pale buff, slightly speckled and breastband olive-yellow; birds from se Ecu. with dark gray crown and sometimes a suggestion of postocular spot. Ssp *hiaticolus* like *phaeocephalus* but brighter, throat slightly paler and less speckled, breastband slightly yellower; *peruvianus* like *hiaticolus*, but slightly smaller, breastband yellower, earpatch slightly darker; *bolivianus* much like *jacqueti*, but belly whiter, throat less speckled, sides of body brighter; crown and headside dusky gray, postocular spot white, posterior part of earcoverts tinged buffy brown; back, wings and tail bright olive-green; throat pale buff without or with indistinct specks, breast ochra-

ceous; sides, flanks, and vent olive-yellow, c belly grayish white; iris buffy white as in all S. Am. sspp further n; *fulvigularis* like *bolivianus*, but crown and headside paler (blackish sepia), throat darker buff, breastband darker (deep ochraceous), both light- and dark-eyed individuals (2 sspp?); *argentinus* like *fulvigularis*, but crown and headside lighter (gray-brown), throat buffy white with faint specks, breastband much paler and yellower (almost as yellow as in *bolivianus*); large white postocular spot, dark iris. **Juv**. has crown like back and throat olive-yellow like breast.

Habits: In pairs or more commonly in small flocks, conspicuous in mixed-species flocks. Moves rather slowly but constantly among 3 cm thick horizontal branches through c canopy, keeping body horizontal. Flicks unfanned tail (usually only laterally) and wings, and sometimes stretches neck and/or makes bowing movements and/or points bill up before take-off or when in conflict. See *Auk* 79 (1962):310-44.

Voice: A fine *tsit* is given infrequently. Song at early dawn and late dusk a long series of single notes finished off with a descending trill *tsi-tsi-...tsi-tse-tsederrrr*, the final trill slower in ssp *cinereocephalus*, sometimes (always?) omitted in *jacqueti* and *venezuelanus*.

Breeding: Fledglings Jan, Feb (Tarija); juv./imm. Aug, Sep (Huila), Nov (La Libertad, Huánuco), Dec (Puno); large gonads Oct (La Paz).

Habitat: Lower to middle strata, occ. in bushy understory, sometimes to treetops. Humid forest and edge, open mossy cloud forest, deciduous wooded valleys, and humid second growth.

Range: 900-3000 m, mainly in subtrop. zone. Coastal mts and Andes of Trujillo and extreme e Mérida, Ven.; w slope of E Andes of Col. in Cesar, N.d.Santander and n Santander (*jacqueti*); Lara and n Trujillo, Ven. (ssp); portions of Lara and from Mérida to Táchira, Ven. (*venezuelanus*); Perijá mts (*ponsi*); e slope of E Andes from s N.d.Santander to Boyacá, Col. (*eminens*); w slope of E Andes from Santander to Cundinamarca, Col. (*flavopectus*); La Pica, Santander, Col. (*trudis*, SL Olson in *Proc. Biol. Soc. Wash.* 96(1983):107); Macarena mts Meta, Col. (*macarenae*); w slope of s end of E Andes and both slopes of s end of C Andes, Col. (*exitelus*, SL Olson in *Proc. Biol. Soc. Wash.* 96(1983):103-9); both slopes of Ecu. (from Pichincha in the w) s to Piura and n Cajamarca, n Peru (*phaeocephalus*); s and e of the Marañón from Amazonas to Huallaga river in Huánuco (*hiaticolus*, O'Neill & Parker in *Bull. Brit. Orn. Club* 101(1981):296-9); from Sira mts in Huánuco to Ayacucho (birds from c Pasco intermediate with *hiaticolus*) (*cinereocephalus*); from Cuzco to Puno, se Peru (*peruvianus*); La Paz and n part of mts of Cochabamba, Bol. (*bolivianus*); ne and s part of mts of Cochabamba, Bol. (*fulvigularis*), from s bank of Mizque river, s Cochabamba, Bol. s to Tucumán, nw Arg. (*argentinus*). 8 sspp from Mexico to Panamá, 1 in coastal Ven. Fairly common to common.

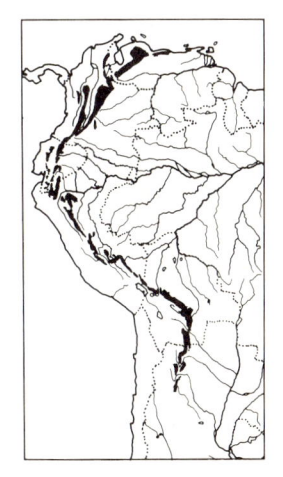

SHORT-BILLED BUSH-TANAGER
Chlorospingus parvirostris

14 cm. Iris gray to white. Above dull olive-green, below gray-brown, paler on c throat; **sides of throat deep yellow**, feathers often raised and extending tuft-like backward to below earcoverts. In pairs and small groups, typically with mixed-species flocks, foraging in dense

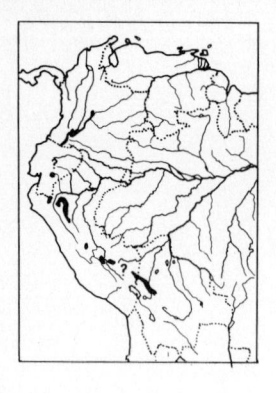

canopy. Call an incessant *tsip* (Col.), or a higher-pitched vibrating *seeep* sometimes preceded by an abrupt thin *tsip*.

Inside mossy and shrubby forest edge, esp. bordering mountain streams. In subtrop. zone, but occ. to 2500 m in Col., and to 2625 m in Peru. From Cundinamarca along e slope of E Andes to head of Magdalena valley and in e Nariño, in Morona-Santiago e Ecu., and Cajamarca and Amazonas s to La Libertad and San Martín, n Peru (*huallagae*); in Junín and Cuzco (*medianus*), and Puno in e Peru and La Paz, Bol. (*parvirostris*).

NOTE: Called Yellow-whiskered B-t. by Isler & Isler (1987). Previously considered sspp of *C. flavigularis* of lower elevations of the Trop. Andes.

DUSKY-BELLIED BUSH-TANAGER
Chlorospingus semifuscus – LVI 11

14.5 cm. Iris pale. Crown and headside dark gray-brown, rest of upperparts greenish olive. Throat and breast smoky brown to gray-brown, throat very faintly freckled, breast faintly washed olive-green, sides of breast and body, flanks and vent olive-green, belly smoky gray. Ssp *livingstoni* has much grayer crown and nape, gray throat and a more poorly defined and greener breastband; sometimes with a white spot behind eye. **Juv**. *semifuscus* has crown and headside tinged olive, otherwise like ad. **Juv**. *livingstoni* olive below faintly streaked grayish on belly, throat grayish, spot behind eye white. Small groups in mixed-species flocks. Agile. Hops along thick moss-covered branches and stems and inside bushes, body parallel to branch, or along thick vines inside canopy. Song much like Common B-t. *tzit-tzit…tzit-tzitit-tzititit-tzitititit-tzititititi-titrrrrrr*, given at late dusk and early dawn. Juv. July (w Nariño).

Inhabits all strata of humid forest, forest clearings and second growth, mainly in 5-7 m tall young trees, at 1200-2750 m, mainly in subtrop. zone, on pacific slope from Tatamá mt, w Risaralda s to w Cauca, Col. (*livingstoni*), and from w Nariño, Col. to Pichincha, nw Ecu. (*semifuscus*). Common.

NOTE: Called Dusky B-t. by Isler & Isler (1987).

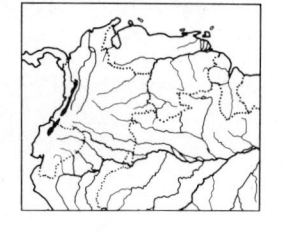

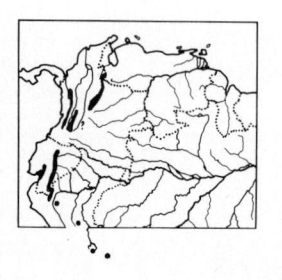

ASHY-THROATED BUSH-TANAGER *Chlorospingus canigularis* of premontane zones locally in Col., Ecu., and E Andean slope of Peru, has been rec. casually to 2600 m in Col. (*canigularis, conspicillatus*) and Ecu. (*signatus*). Inhabits tall mossy forest canopy and vine-tangled edges. 14 cm. Head gray, ear-coverts darkest, rest of upperparts bright olive-green. Below grayish white (**throat unspotted**), sides yellowish olive, **breastband light yellow**, under tail-coverts greenish yellow. **Iris dark**. In small groups, usually a core-species in mixed-species flocks. Active, forages from ground to canopy. Calls of chips and thin twitters much like congeners. Large gonads Sep, Mar-May (Col.)
NOTE: Called Ash-throated B-t. by Isler & Isler (1987).

BLACK-BACKED BUSH-TANAGER
Urothraupis stolzmanni – Plate LVI 13

16 cm. **Headside and upperparts black, underparts white** speckled or barred with dark gray, least so on throat; flanks and vent olivaceous gray, feathers of vent tipped whitish. Tail slightly graduated. **Juv.** undescribed.
Habits: Usually in flocks of 3-9, sometimes in mixed-species flocks. Actively perch-gleans insects and berries inside dense bushes, sometimes acrobatically hanging from tips of lower branches of trees and frequently stretching to peer above and below leaves. Also pecks at internodes of bamboo. Often tame and staying long at each stop.
Voice: Contact calls fine *zie, tsit* or *tzi*.
Breeding: Large gonads Feb (e Cauca).
Habitat: 1-3 m up in elfin forest and humid páramo shrubbery mixed with bamboo, and in adjacent isolated bushes.
Range: 2750-3950 m. In Caldas (Leonera), Quindío (Santa Isabel), and Cauca, C Andes of Col., and e Ecu (Papallacta, upper Sumaco, mt Tungurahua, Zapote-Najda mts of the e Azuay/n Morona-Santiago border). Uncommon, locally fairly common.

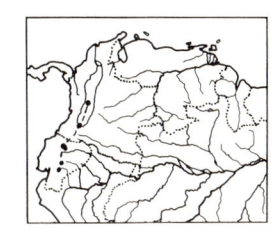

GRAY-HOODED BUSH-TANAGER
Cnemoscopus rubrirostris – Plate LVI 4a-b

13.5-15 (*chrysogaster*) or 17 cm (*rubrirostris*). **Bill pink** (northern ssp *rubrirostris*) **or black** (*chrysogaster*). Crown and headside smoky gray, **throat and breast light smoky gray**; back, wings, and tail greenish olive, belly yellow, sides faintly washed olive; *chrysogaster* has dark gray crown and headside, and pale gray to grayish white throat and breast. **Juv.** duller.
Habits: In pairs or small groups up to 8, usually in mixed-species flocks. Restless. May halt a few moments to look in various directions, then perch-gleans undersides of leaves along moss covered, rather thin branches, usually not all the way to the tip, or among leaves at tips where it may perch atop leaves, or sally for insects, returning to same perch. Keeps body horizontal when active. During song wriggles rear body and particularly tail.
Voice: Calls *tsit-tsit...*, *zit*, *pist*, and a sharp *tseeet* much like other tanagers. Song a broken unmusical jumble *tchee-tchee wee zt-z-tchee...*, sometimes speeded up to a twittering, and at intervals flows into smacking *chip*s.
Breeding: Fledgling Sep (Cauca); large gonads July-Aug (n end W and C Andes Col.).**Habitat**: 2-10 m up, often in c canopy in humid forest, cloud forest and forest clearings.
Range: 1900-3350 m. From Río Chiquito and Páramo de Tamá in sw Táchira, Ven., s locally in all 3 ranges of Col., (nw?) and locally in e Ecu. to Piura/Cajamarca border, n Peru (*rubrirostris*); s and e of the Marañón from Amazonas to Vilcabamba mts in Cuzco, se Peru (*chrysogaster*). Fairly common to common.

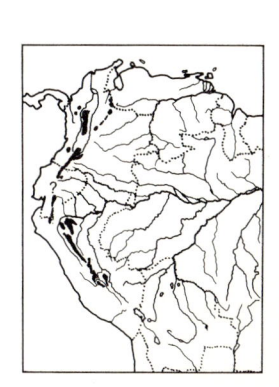

RUFOUS-BROWED HEMISPINGUS
Hemispingus rufosuperciliaris – Plate LVI 22

15 cm. Fairly heavy-bodied. Headside and c crown black; back, wings and tail blackish gray. Long very broad **supercilium** and most of **underparts bright rufous chestnut**, lower belly buffy gray-brown, flanks and vent dusky brown. **Juv**. undescribed.

Habits: Alone, in pairs or 3-4 together, sometimes in lower story mixed-species flocks with other hemispinguses, *Basileuterus* warblers, Sepia-brown Wren, and others. Perch-gleans insects and berries in the undergrowth, often hopping in thick moss on branches or on the ground, moving slowly, much as an *Atlapetes* Brush-finch. Flicks tail when calling.

Voice: Call a sharp, abrupt, sometimes rapidly repeated *tzit* much like other tanagers. Song consists of short squeaky nasal notes mixed with long slurred squeals, with a 'jazzy' rhythm, continued for minutes.

Breeding: No data.

Habitat: Within 1 m of the ground, rarely 2 m up. Thick understory of elfin and cloud forest, occ. in adjacent hedgerows and other second growth, but may prefer extensive thickets of bamboo and dense moss-laden shrub adjoining forest.

Range: Peru at 2550-3500 m from Cord. de Colán in Amazonas s to Carpish mts in Huánuco. Uncommon and local.

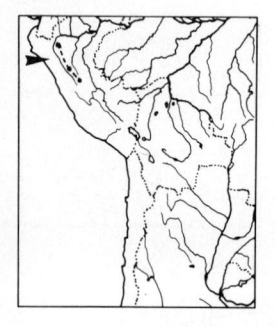

BLACK-CAPPED HEMISPINGUS
Hemispingus atropileus – Plate LVI 24a-b

15 cm. Above olive-citrine; crown and most of headside black, **long narrow supercilium** from bill **buffy white**. Lower earcoverts and chin olive-yellow, feathers narrowly tipped dusky; below olive-yellow, throat and upper breast brightest and tinged ochraceous. Birds from W (and C?) Andes of Col. duller above and more saturated below, esp. on c belly. Ssp *auricularis* has almost pure white supercilium not reaching bill; entire headside black, chin buffy white, throat and upper breast ochraceous yellow, belly bright yellow, sides of breast and body, flanks and vent olive. **Juv**. has pale maxilla, fuscous crown and headside, and buffy tinge to the supercilium.

Habits: Often 6-12 together, as nuclear species in mixed-species flocks, or sometimes in pure flocks. Typically feeds low in the vegetation, starting near the ground. Picks berries and perch-gleans insects from outermost crotches and undersides of leaves by reaching out and down, often bowing head 180°, and probes internodes of bamboo.

Voice: Call like other tanagers. Song snarling and twittering, reminiscent of Gray-hooded Bush-tanager. The light and lively flock songs consists of phrases of rapidly delivered, high-pitched notes, e.g. *d-d-d-d-d-dit* or *z-z-z-z-zeet*, quality varying from squeaky to chattering, occ. a rapid trill rolling up and down scale. Dawn song (*auricularis*) of single notes given slowly, evenly and distinctly, series of rich *chew* notes alternating with series of *zeet* notes, given continuously for long periods of time.

Breeding: Juv. Mar (nw Ecu.).

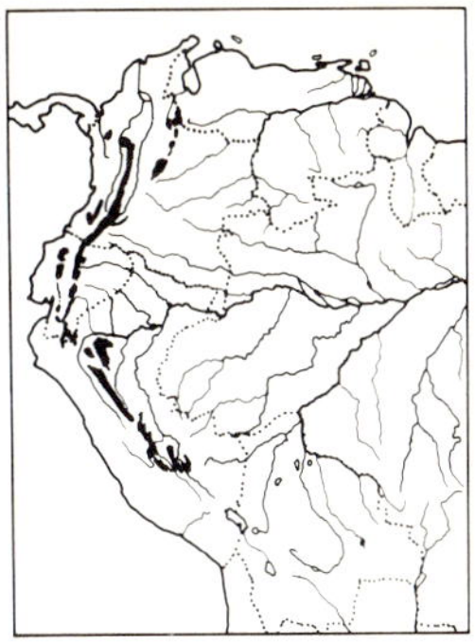

Habitat: 0.5-3 m up, rarely to 7 m. Undergrowth of very humid and dense forest, cloud, and elfin forest, preferring places with bamboo thickets or contiguous tracts of dense bushes or second growth.
Range: (1800) 2300-3600 m. Páramo de Tamá and Río Chiquito in sw Táchira, Ven., and locally in all 3 ranges of Col., e and w Ecu. to Peru n of the Marañón (*atropileus*); e Peru s and e of the Marañón from Amazonas to Cuzco (*auricularis*). See also following species. Common.

ORANGE-BROWED HEMISPINGUS
Hemispingus calophrys – Plate LVI 20

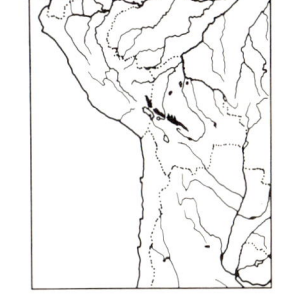

14 cm. Crown and upper half of headside black; back, wings and tail olive-citrine; long, fairly broad **supercilium, spot on earcoverts**, lower headside, **throat and upper breast ochraceous**, rest of underparts olive-yellow, c belly yellow. **Juv**. or **Imm**. with olive-tinged head.
Habits: As in previous species. In pairs or small flocks, often in mixed-species flocks, e.g. with *Basileuterus* warblers, *Cranioleuca* spinetails, and Scarlet-bellied Mountain-tanager. Agile; perch-gleans and often hangs from tips of branches and bamboo stalks, occ. searching dead leaves.
Voice and **breeding**: No data.
Habitat: Understory of humid forest and cloud forest, often in dense second growth bushes and bamboo inside forest or at edges.
Range: 2300-3500 m in Puno in se Peru and La Paz and Cochabamba, Bol. Fairly common to common.

NOTE: Often considered a ssp of Black-capped H.

PARODI'S HEMISPINGUS *Hemispingus parodii* – Plate LVI 19

15 cm. Maxilla black, **mandible gray, legs pale gray**. Above golden olive-green; **cap blackish, feathers faintly edged with color of back**. Long supercilium bright warm yellow; headside dark olive, lower headside mottled with yellow, below bright yellow; sides, flanks and vent olive-yellow. **Much like Citrine Warbler, but crown dusky olive** and bill stouter. **Juv**. undescribed.
Habits: In pairs or groups up to 9, often in mixed-species flocks that include the similar-looking Citrine Warbler. Gleans leaves, stretching out or down, and probes internodes. 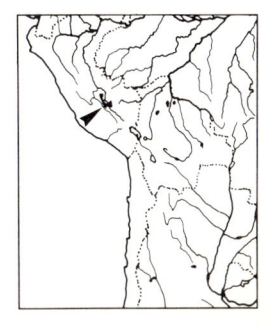
Voice: Song a string of labored, 1-2 s long rapid chittering phrases *p-p-p-psit-sit-sit, z-z-z-zit-dit-dit,* etc., often giving one phrase at an even pitch and then abruptly shifting to another pitch and quality. Dawn song alternating between 2 differently pitched notes, faster than 1 note per s.
Breeding: Large gonads July, Aug.
Habitat: Feeds for the most part low, sometimes 4-5 m up in elfin forest, mainly in dense bamboo.
Range: se Peru at 2750-3500 m in Vilcabamba and Vilcanota mts in Cuzco. Uncommon.

GRAY-CAPPED HEMISPINGUS *Hemispingus reyi* – Plate LXIV 18

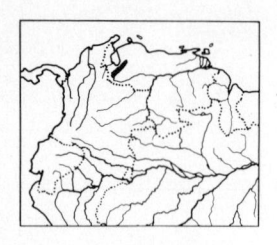

14 cm. Above olive, primaries edged olive-green, **crown gray**; broad line through eye olive; underparts yellow, flanks and vent olive. **Juv.** undescribed.
Habits: In small groups, often nuclear in mixed-species flocks.
Voice: Repeats a sharp insistent *tee-chew chew*, followed by several single *chew*s.
Breeding: No data.
Habitat: Esp. in cloud and dwarf forest, and also shrubby forest edge, bamboo, clearings, open woodland and páramos with bushes.
Range: Ven., at 1900-3200 m from c Trujillo to w Táchira.

SUPERCILIARIED HEMISPINGUS *Hemispingus superciliaris* – Plates LVI 18a-e and LXIV 16

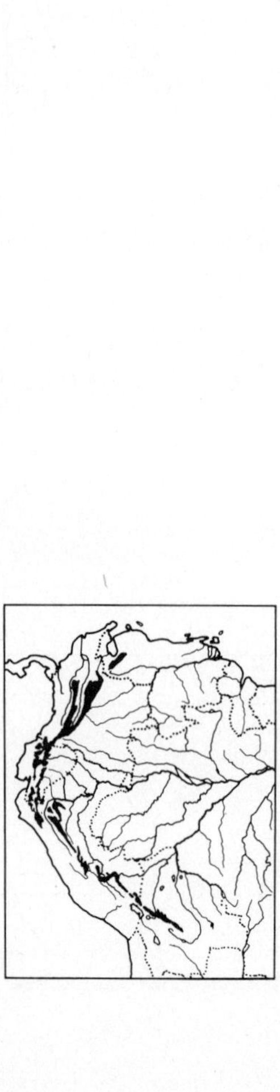

13.5 cm. **Greenish and yellow in n and s of range, gray in the c parts**. Ssp *chrysophrys*: above greenish olive, **below yellow; supercilium and half-ring below eye yellow**, headside mixed olive and yellow, line through eye dusky; sides of breast and body tinged olive. Sspp *superciliaris* and *maculifrons* have grayish forehead, **white supercilium**, dark olive earcoverts with shafts yellowish basally, area below eye finely mottled white and dark gray, throat and upper breast faintly mottled with olive. Ssp *nigrifrons* like *superciliaris*, but forehead usually mixed with blackish, line through eye blackish, feathers of side of throat with faint, narrow dark tips; *urubambae* like *superciliaris*, but entire forecrown and most of headside black, contrasting white supercilium and mark below eye. C populations: ssp *leucogaster* dark **gray above** and **grayish white below**, vent tinged buff; faint supercilium whitish, throat and upper breast mottled with gray; *insignis* still lighter, white breast barely tinged with gray. **Juv.** with less distinct supercilium, and with breast and sides more washed with olive.
Habits: In pairs or a few together, usually in mixed-species flocks (e.g. with Drab H.). Restlessly perch-gleans insects from leaves, occ. reaches out and down, or hangs, and hops along the branches of c canopy almost to treetops, but also in low bushes.
Voice: Frequent sharp *tsit* notes. Song a distinctive, 3-4 s long, rapid burst of harsh notes that become louder while accelerating. Mainly sings just after dawn, sometimes many birds together.
Breeding: Large gonads Feb-July (Col.).
Habitat: 1-5 m up in humid montane shrubbery incl. *Polylepis*, hedgerows, and edge of humid forest and cloud forest, preferring young trees and bushes.
Range: 1900-3600 m. **Citrine forms**: Trujillo to Táchira, Ven. (*chrysophrys*); Santander to Cundinamarca, E Andes of Col. (*superciliaris*); C Andes and both slopes of Nariño, Col., and Ecu. except sw (*nigrifrons*), sw Ecu. and Peru n and w of the Marañón (*maculifrons*). From Cuzco in se Peru to La Páz, Bol. (*urubambae*). **Gray forms**: in Peru, from Utcubamba drainage in Amazonas to Carpish mts in Huánuco (*insignis*) and Huánuco s of Huallaga river to Junín (*leucogaster*). Fairly common to common.

OLEAGINOUS HEMISPINGUS *Hemispingus frontalis* – Plates LVI 17a-b and LXIV 15

13 cm. Bill gray. Ssp *frontalis*: above dark olive, wings and tail brown, edges of primaries light olive-brown. **Very faint supercilium olive**, headside dark olive with a faint trace of light mottling below eye; underparts **ochraceous**, sides and flanks washed olive-brown. Ssp *ignobilis* similar, but with **ochraceous supercilium**, narrowing and turning greenish yellow behind eye; mottling below eye more conspicuous and whitish; throat pale ochraceous turning yellowish olive-brown on rest of underparts; vent rufous buff. Ssp *flavidorsalis* yellowish olive above; supercilium, throat, and breast greenish yellow, darker and tinged ochraceous on lower breast and belly, sides olivaceous. **Juv.** undescribed. **Habits**: Alone, in pairs, or 3-4 birds, regularly in mixed-species flocks. Frequently twitches tail. Deliberately searches the foliage, gleans leaves, and hangs on leaves at tips of branches, often probing dead leaves.
Voice: Song of 2 alternating squeaky notes or a squeaky rattle (duet). Calls include harsh *chip*s.
Breeding: Juv. Apr (Junín); large gonads June-Nov (Col.).
Habitat: From understory to mid heights of trees, esp. in patches of densely entangled vegetation, in humid forest and cloud forest.
Range: 1300-2900 m, mainly subtrop. zone. Perijá mts (*flavidorsalis*); from s Lara to Táchira, Ven. (*ignobilis*); E and C Andes and locally in W Andes of Col., Imbabura, nw Ecu. (specimen in ZMUC), and through e Ecu. and e Peru to Vilcanota mts in Cuzco, se Peru (*frontalis*). 2 sspp in coastal Ven. Uncommon in temp. zone.

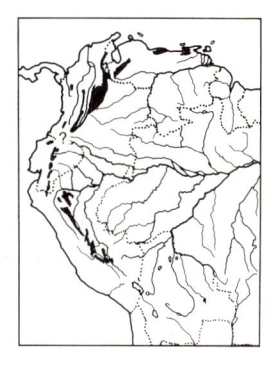

BLACK-EARED HEMISPINGUS *Hemispingus melanotis* – Plate LVI 23

14 cm. Ssp *castaneicollis*: above dark gray-brown, narrow, indistinct **supercilium white**, headside and most of throat **black**, extreme lower **throat and breast rufous chestnut**, belly and vent buff, sides and flanks buffy brown. **Imm. male** has the throat mottled with buffy white. Ssp *melanotis* has indistinct eye-brow, and black of throat confined to chin. Ssp *piurae* has crown, headside, and chin black, conspicuous long supercilium and spot below eye white, postauriculars and nape broadly gray; underparts below chin almost uniform ochraceous orange, slightly paler on c abdomen; *macrophrys* like *piurae*, but supercilium and nuchal band even broader. Alone, in pairs, or 3-4 together in mixed-species flocks. Forages for insects at mid-heights in trees, often hanging from the tips of branches. Call *tje*. Song rhythmic and accentuated *didadidadidadida*, lasting 2-3 s. Fledglings Nov (Cochabamba).

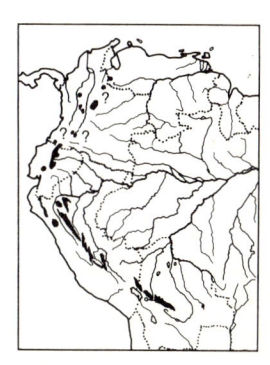

Edge of humid forest and cloud forest, plantations, and second growth, esp. in bamboo thickets. Mainly subtrop. zone, but to 2900 m in Col., 3050 m in nw Peru, and to 2450 m in La Paz, Bol. Táchira, Ven., and locally through E and C Andes of Col. and e Ecu. to Tungurahua (*melanotis*); Peru on both sides of W Andes in Piura and n Cajamarca s to Nancho in rather dry forest (*piurae*), upper Chicama valley, sw Cajamarca, n Peru (*macrophrys*); from Puno, se Peru to Cochabamba, Bol. (*castaneicollis*). Only subtrop. zone on Pacific slope of Nariño, Col., and Chimborazo, Ecu. (*ochraceus*) and C Andes of Peru from Amazonas to Ayacucho and n Cuzco (*berlepschi*). Fairly common.

SLATY-BACKED HEMISPINGUS *Hemispingus goeringi* – Plate LXIV 17

13 cm. Above dark blue-gray, crown and headside blackish; long supercilium starting in front of eye white; **below rufous**, flanks olivaceous chestnut to dark brown. **Juv**. undescribed.
Habits: In small flocks up to 5, sometimes in mixed-species flocks. Forages for insects and berries, near or perhaps also on the ground, usually under dense cover.
Voice: Song a continuous stream of moderately pitched harsh notes, and a more pleasant high-pitched stream of tinkling notes, the 2 types apparently given in duet.
Breeding: No data.
Habitat: Elfin forest and scattered trees in humid páramos and upper edge of cloud forest.
Range: Ven. at 2600-3200 m in Mérida and n Táchira.

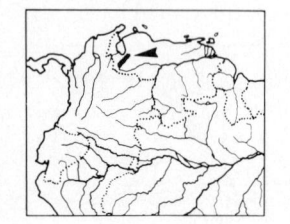

BLACK-HEADED HEMISPINGUS *Hemispingus verticalis* – Plate LVI 15

14 cm. Unmistakable. **Iris whitish**. **Above dark gray, head incl. throat black** with **broad buffy clay-colored c crown-stripe**; **below gray**, pale gray just below throat, c belly whitish. **Imm**. has white throat. **Juv**. undescribed.
Habits: Alone, in pairs, or small groups (once 12 or more) in mixed-species flocks, where they may play an important role. Crawls among outer leaves, sometimes walking atop leaves. Often stops and looks for berries and insects on the uppersides of leaves with characteristic stretching of neck and head-bobbing. Territorial.
Voice: Song by both sexes a fairly high-pitched, weaving and continuous rapid twitter without a fixed theme, lasting 5-15 s, often 2 or more birds singing simultaneously, and repeated at short intervals often for many minutes at a time, and recalling Black-capped H., Gray-hooded Bush-tanager, Plush-capped Finch, Moustached Flowerpiercer, or Citrine Warbler. During pauses, gives frequent high-pitched notes.
Breeding: Imm. Sep (w Antioquia); large gonads Sep (w Antioquia).
Habitat: Middle strata or top of small trees in elfin forest, cloud forest, and humid páramo shrubbery, regularly frequenting alders (*Alnus*).
Range: 2350-3600 m. On Páramo de Tamá, in Cundinamarca in E Andes, Antioquia in W Andes, and through C Andes of Col. to Napo, interandean plateau of Azuay (río Mazan w of Cuenca), on Loja/Zamora-Chinchipe border se Ecu., and on Cajamarca/Piura border (Chinguela) n Peru. Widespread, usually at low densities, but locally common in s Ecu. (PN Podocarpus).

NOTE: Sometimes placed in the genus *Pseudospingus*.

DRAB HEMISPINGUS *Hemispingus xanthophthalmus* – Plate LVI 16

13 cm. **Iris whitish**. **Above** dark **gray**-brown, grading to **whitish below**, faintly tinged buff on vent. **Juv**. undescribed.
Habits: As for Black-headed H. Frequently atop flat *Weinmannia* and other flat, dense canopies, deliberately walking on the leaves. Often bobs or stretches out and flutters to new location. Regularly seen with White-banded Tyrannulet, Spectacled Whitestart, Blue-backed Conebill, and Scarlet-bellied, Hooded and Chestnut-bellied Mountain-tanagers.
Voice: As for Black-headed H., a patternless, rapid sputtering series of staccato squeaky or spitting notes, often given by 2-3 birds together.
Breeding: No data.
Habitat: Elfin and cloud forest and humid shrubbery. Often on top of *Escallonia*, *Weinmannia*, and other flat-canopied trees with dense stiff leaves, but also large leaves of *Clusia* trees.
Range: 2375-3525 m. S and e of the Marañón from Amazonas, Peru, to La Paz, Bol. Fairly common, but always in small numbers.

NOTE: Forms a superspecies with Black-headed H.

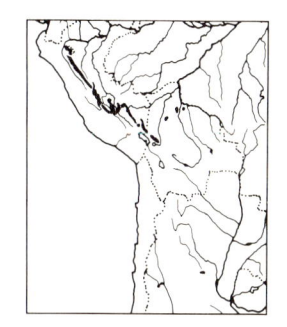

THREE-STRIPED HEMISPINGUS
Hemispingus trifasciatus – Plate LVI 21

13.5 cm. Above dark olive, **crown black with median dark olive-gray stripe, very long supercilium pale yellow-buff**, headside black; below orange-rufous, palest on belly; wings with ochraceous bar on outer median coverts and a faint bar on greater coverts. Plumage varies slightly, palest in n part, most ochraceous in s part of range. **Juv**. has supercilium tinged buff-yellow, crown and cheeks tinged brown.
Habits: In pairs or flocks up to 6 or rarely more, often in mixed-species flocks, where apparently nuclear. Restlessly forages along thick, horizontal, moss-covered branches within the canopy, gleans leaves, rarely hanging, reaches out, or makes short sallies.
Voice: Call a spitting, thin *swit*, sometimes given rapidly and often in chorus. Song of repeated *tzit* notes.
Breeding: Juv./ad. Dec (Huánuco).
Habitat: Canopy of elfin forest and cloud forest, mainly in small-leaved trees, more rarely in undergrowth or bamboo.
Range: (2800) 3000-3350 (4300) m. From Carpish mts, Huánuco, c Peru, to Cochabamba, Bol. Uncommon.

NOTE: Sometimes placed in its own genus *Microspingus*.

GRASS-GREEN TANAGER *Chlorornis rieferii* – Plate LIV 10a-b

18.5 cm. Unmistakable. **Bill and legs red, plumage strongly grass-green**. Ssp *boliviana* has forehead, headside, throat, and vent chestnut. Sspp *celata* and *elegans* with chestnut of forehead narrowly edged pale blue above, *elegans* with extreme lower throat green; *diluta* with blue on

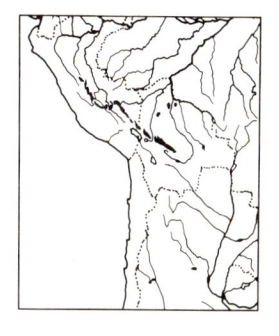

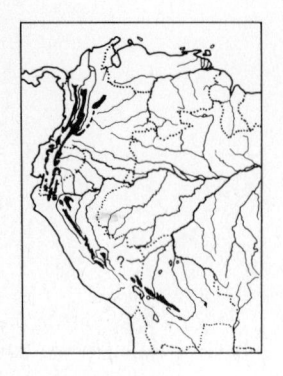

forehead, but brown of face and vent paler; *riefferii* without blue and below with only upper throat chestnut. **Juv**. like ad., but duller, bill dark.
Habits: Alone, in pairs or small groups of 3-6 or more, occ. in mixed-species flocks. Climbs about for berries near tips of branches, and inspects moss, epiphytes and leaves for insects. Conspicuous.
Voice: Noisy. Song a characteristic snarling *de-dede-detúitúi-detúitúi-...*, last part repeated 3-5 times. Call of a similar snarling quality.
Breeding: Juv. July (Cundinamarca), Mar, Aug (Cauca); large gonads Feb-Sep (W and C Andes Col.), Oct (e Boyacá).
Habitat: 1.5-15 m up in lower and middle strata of cloud forest with thick undergrowth mixed with bamboo.
Range: 1500-3350 m, but mainly 2000-2700 m. All 3 ranges of Col. (in E Andes n to Boyacá, e and w Ecu., and n Peru n of the Marañon (*riefferii*); Utcubamba drainage in Amazonas, n Peru (*diluta*), Huánuco to Cuzco, c Peru (*elegans*); Puno, se Peru (*celata*), La Paz and Cochabamba, Bol. (*boliviana*). Fairly common.

PARDUSCO *Nephelornis oneilli* – Plate LVI 14

14 cm. Above very dark brown, below pale brown, throat paler, sides and flanks brown, faint wingbar pale brown. Tail slightly graduated, feathers somewhat pointed. **Juv**. has underparts as dark as back.
Habits: In flocks of 5-15, sometimes in mixed-species flocks. Actively searches for insects in moss or under leaves while moving along the branches inside dense shrubbery, often from center upwards and out onto limbs. Frequently jerks tail from side to side and flutters wings. Often perches briefly atop a bush and peers from side to side before flying rapidly and directly to next tree or bush.
Voice: Call a repeated, high-pitched *zee* and a soft *chip*.
Breeding: Fledglings Aug (San Martín), Sep (Huánuco).
Habitat: Low trees, bamboo, bushes, and shrubs, rarely on the ground. Edge of elfin forest, seldom down to upper reaches of cloud forest.
Range: C Peru at 3000-3800 m from s San Martín to Carpish mts, Huánuco, and immediately s of the Huallaga bend in Panao mts Huánuco (LSUMZ). Common.

NOTE: Described by Lowery & Tallman, *Auk* 93(1976):415-28.

BLACK-FACED TANAGER *Schistochlamys melanopis*, widespread in the premontane zone, ascends to 2400 m in La Paz, Bol. (*olivina*). 18 cm. Bill blue-gray tipped black. Tail rounded, feathers rather broad. **Ad**. above dark gray, below gray to light gray, c abdomen whitish; **anterior crown, headside, throat, and breast black, outer 3 tail-feathers narrowly tipped white**. **Imm**. above olive, below yellow-green, palest and yellowest on c belly; outer tail-feathers tipped white when fresh (may breed in this plumage). **Juv**. like imm., but tail-feathers somewhat pointed and without white tip. Alone or in pairs forages low for fruits and seeds in open woodland, second growth, savanna with scattered bushes, scrub, low open vegetation, and cultivated fields.

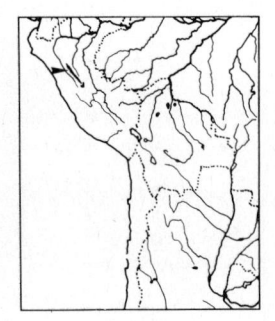

PLUSH-CAPPED FINCH *Catamblyrhynchus diadema* – Plate LIX 1a-b

13-14 cm. Unmistakable. Bill black, stubby, and rounded. Tail rounded, relatively long, feathers slightly pointed. **Forehead and forecrown with stiff plush-like golden-yellow feathers**, hindcrown and nape black; back, wings, and tail plumbeous; lores and eye-ring black, narrow supercilium and headside dark chestnut, underparts chestnut. Ssp *citrinifrons* has a yellow forehead, feathers not so stiff and plush-like, bordered behind by narrow chestnut band; color of headside and underparts lighter chestnut. **Imm**. has forehead and forecrown dark gray (*diadema*) or lighter gray (*citrinifrons*), feathers somewhat plush-like with yellow bases; rest of upperparts dark grayish olive (*diadema*) or brownish olive (*citrinifrons*); below dark gray-brown, feathers more or less edged rufous (*diadema*), or heavily washed ochraceous tawny (*citrinifrons*). Brief **juv**. plumage like imm., but grayer above with less contrast between fore- and hindcrown, no yellow at base of feathers of forecrown, underparts dark gray-brown (*diadema*) or brownish olive (*citrinifrons*) with no rufous or tawny.

Habits: In pairs or family groups, sometimes in mixed-species flocks. Actively feeds inside the vegetation or hangs from tip of branch or bamboo stalk. Eats insects and berries, perhaps also bamboo. Mainly forages in bamboo, pecking and tugging the dense axiles of leaf whorls, or running the bill along the bamboo stems with a series of tiny biting motions (see *Wilson Bull.* 91(1979):145-8, and *Bull. Brit. Orn. Club* 106(1986):161-3).

Voice: Song an unmusical, continuous, weaving twitter, reminiscent of Gray-hooded Bush-tanager.

Breeding: Fledglings Oct (Quindío), June (se Ecu.), June-Aug (Huánuco), Aug (Cuzco), July (La Paz); juv./imm. July (Piura), Aug-Oct (Amazonas), June (Huánuco), Aug (La Paz); imm. May (w Nariño), Dec (Amazonas), Sep, Nov (La Libertad), June, July (Huánuco), Aug (Cuzco); large gonads June, July (Perijá), June (se Antioquia).

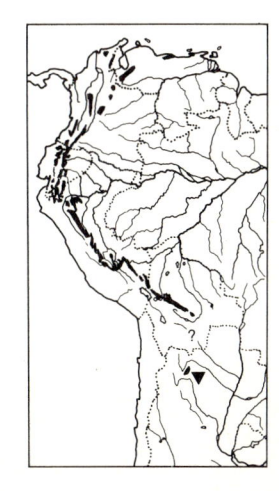

Habitat: 0.5-5 m up in lower strata. Bamboo in cloud and elfin forest, humid forest, humid shrubbery, and riparian thickets in semi-humid regions.
Range: 1750-3500 m. Sta Marta and Perijá mts, and from n Trujillo, Ven., s through all 3 ranges of Col. (in W Andes from Valle s), e and w Ecu. to Loja, s Ecu. (*diadema*), from Piura and n Cajamarca in n Peru to Sta Cruz, Bol., and on Calilegua mt in w Jujuy, nw Arg. (*citrinifrons*). Coastal mts of Ven. (*federalis*). Fairly common.

NOTE: Previously placed in its own family Catamblyrhynchidae.

Finches – Family Fringillidae

Finches comprise arboreal as well as ground-living passerine birds with conical bills specialized for eating seeds. However, many also eat insects, esp. in the breeding season. They are often separated into 2 families: the worldwide 10-primaried *Fringillidae* with the subfamilies *Fringillinae* (3 species) and *Carduelinae* (119 species, with only the siskins in S Am.), and the mainly American *Emberizidae* which belongs to the '9-primaried oscines', and have the subfamilies *Cardinalinae* (37 species), *Emberizinae* (276 species), and the tanagers (*Thraupinae*). The relationships of many forms yet need to be shown biochemically; hence the traditional classification is maintained here.

The arboreal saltators and grosbeaks are members of the entirely American *Cardinalinae*. They are large, some colorful, live in pairs, and have melodious whistled calls. *Sporophila* seedeaters are small, males mainly gray or black, females of the 28 S Am. species brownish and in most cases impossible to tell apart. Outside the breeding season several species gather in flocks, which are commonly seen around habitations. *Catamenia* seedeaters are grayer and inhabit the high Andes. Yellow-finches comprise 12 sparrow-sized species (that may in fact be tanagers). Primarily terrestrial and high Andean they frequent open country, often in large flocks. Several species carry an imm. plumage for over a year, and are difficult to tell apart. Sierra-finches are open-country birds of harsh climates, some occurring to over 5000 m. Brush-finches have rounded wings and fairly long graduated tails, and live in close pairs in shrubby undergrowth of humid forest to dry, bushy slopes (reviews by RA Paynter, Jr in *Bull. Mus. Comp. Zool.* 143(1972):297-320, and 148(1978): 323-60, and by KC Parkes in *Condor* 56(1954):129-38). Inca-finches are found in dry scrub of the W Andes and the Marañón valley in n Peru, and resemble the juncos of N and C Am. by morphology, general habits, and very high-pitched calls. Warbling-finches are often slaty and rufous, and usually have a white-tipped tail; they inhabit arid to semi-humid scrub, mainly in Arg. and Bol. Siskins are small birds with pointed beaks, forked tails and a yellow pattern. They often acrobatically hang from the tips of branches or herbs. Usually occurring in flocks, they amaze the observer with their continuous twitter. The Am. species were previously placed in the genus *Spinus*, which is dealt with in *Ann. Carnegie Mus.* 17 (1926):11-82.

Juv. of most finches have somewhat pointed tail-feathers. Only in the Grasshopper Sparrow is the postjuvenal molt known to include the rectrices. As ad., many have a dull non-breeding plumage washed with gray-brown or olive tips to the body-feathers, these tips wearing off at a predestined point, leaving the bird in the more contrasting breeding plumage. Finches build cup-shaped nests placed in the vegetation (a few species have domed nests, and yellow-finches and sierra-finches nest in holes). Both parents tend the young. See also Bond (1951), and *Proc. Acad. Nat. Sci. Philadelphia* 104 (1952):153-96.

GOLDEN-BILLED SALTATOR *Saltator aurantiirostris* – Plate LVII 4a-e

20 cm. **Heavy bill orange** in male, more or less suffused with dusky in female and imm., esp. in the s. Ssp *aurantiirostris*: above gray to dark gray, with olive wash except on crown. **Black face** formed by more or less black forehead, and black headside and sides of throat (blackish gray in female, and feathers tipped dark gray when fresh); **broad whitish supercilium from above eye** curve into a buff demarcation behind the earcoverts; chin and throat buffy, palest on chin, with 2-3 mm wide black demarcation towards breast. Rest of underparts buff, breast tinged grayish, vent warm buff; in worn plumage underparts paler and breast grayer. Outer tail-feathers often white-tipped. Ssp *nasica* similar, but bill larger, and tail never with any white; *tilcarae* combines large bill and presence of **conspicuocus white tips on outer tail-feathers**; *hellmayri* like *aurantiirostris*, but dark gray above with very faint olive cast, feathers of crown tipped black when fresh, headside blacker in female, supercilium pure white, earcoverts bordered gray posteriorly, throat almost pure white, breast and belly paler and grayer, tail with more white; *albociliaris* and *iteratus* like *hellmayri*, but pure dark gray above; forehead, headside, throat and breast black with only c chin and throat whitish, white rarely concealed and averaging smallest in *iteratus*. Ssp **nigriceps** like *iteratus*, but **entire head black** with only small, semi-concealed white spot on lower throat and sometimes suggestion of a white postocular streak; belly much grayer, vent buff. **Juv**. like female, but tinged olive above and yellow below.
Habits: Alone, in pairs, or family groups. Forages for insects, seeds, and berries on or near the ground. Sings well hidden high in a tree, but never in treetops. Ssp *nigriceps* somewhat more arboreal, frequently climbing about in middle canopy.
Voice: Characteristic song of 3-5 syllables, the last an explosive clear whistle *dee dee te teuie*, *tri te tuit trt te trrrt tiuiet*, or just *tiuit*, occ. given in winter. Alarm *rrrr*. Call a loud, harsh *zak*.
Breeding: Nov-Dec (Tucumán); fledglings June (Huánuco), Mar (Sta Cruz), Feb (Tarija); juv. Mar (Cochabamba), Feb (Jujuy).
Habitat: Both humid and dry montane scrub and shrubbery, second growth, *Eucalyptus* stands, hedgerows, and gardens, rarely *Polylepis* woodland; *nigriceps* also along streams in xeric forest.
Range: At 1350-2600 m on w slope of Andes from Loja, s Ecu., to Lambayeque, nw Peru (*nigriceps*). At 1500-3700 (4000) m, along upper Utcubamba and Marañón valleys in s Cajamarca, s Amazonas and La Libertad, and in n Ancash where also found on w slope at Yanac (*iteratus*, doubtfully distinct from *albociliaris*), Ancash and Huánuco, c Peru, s in the w to Tarapacá, n Chile, in the e to n La Paz, Bol. (*albociliaris*, intergrading with *hellmayri* above Yungas of La Paz), La Paz and Cochabamba s to se Potosí and n Tarija, Bol. (*hellmayri*), Andean slope of Jujuy to Tucumán, nw Arg. (*tilcarae*); lowlands to 2000 m from s Tarija, Bol., e Parag. and s Brazil s to Catamarca, Córdoba, ne San Luis, and s Buenos Aires, Arg. (*aurantiirostris*); pre-Andean zone of La Rioja through San Juan and w San Luis to Mendoza and w Córdoba and w La Pampa (*nasica*). Common.

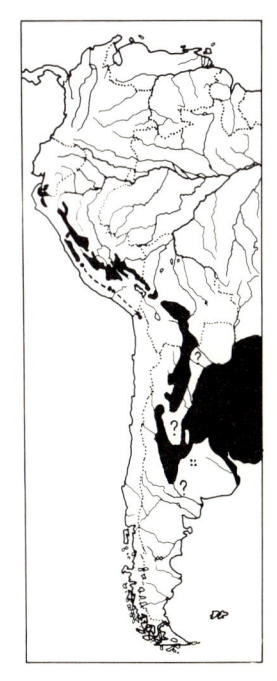

NOTE: Ssp *nigriceps* is often treated as a full species, **BLACK-**

COWLED SALTATOR. Variation and distribution is discussed in *Amer. Mus. Novit.* 261(1927), *Hist. Nat. Mendoza* 1(17, 1980):113-20, and *Resúmen 1° Reunión Argent. Ecol.* (1980): 82.

STREAKED SALTATOR *Saltator albicollis* of lowlands to 2000 m from Costa Rica to n Ven. and w Peru, rarely ascends to 2700 m in Col. (*perstriatus, flavidicollis, striatipectus*). 19 cm. Above olive-green or gray tinged olive on upper back, rump and tail grayer; short supercilium white; **below** white broadly but obscurely **streaked** dark olive except on mid-throat and belly. It inhabits dry shrubbery, gardens, and agricultural areas. Spreads with deforestation.

RUFOUS-BELLIED SALTATOR *Saltator rufiventris* – Plate LVII 3

20 cm. Brownish gray (rarely blue-gray), belly and vent rufous; long narrow supercilium whitish to white. **Juv.** like ad., but paler below and tinged olive above.
Habits: In pairs, rarely in small groups, in trees, bushes, and seldom on the ground. Forages for berries (esp. mistletoes). Also said to eat seeds and insects. Elusive.
Voice: Song of warbled notes on constant pitch. Calls include a sharp note and unmusical irregular sounds *rrhe-rrhe*.... Alarm strident.
Breeding: No data.
Habitat: Semi-arid. Trees, bushes, and scrub, in steep bushy valleys with scattered trees and steep slopes, riparian thickets and agricultural areas, seldom far from *Polylepis* trees.
Range: 2570-3800 m. La Paz (Inquisivi) and Cochabamba (a rec. from Ayupayo in Cochabamba has erroneously been referred to Chuquisaca), Bol., and Calilegua mt in Jujuy, nw Arg. Uncommon to rare, possibly somewhat overlooked.

MASKED SALTATOR *Saltator cinctus* – Plate LVII 2

21 cm, tail fairly long, graduated. Bill thick, dusky more or less washed with red. Iris orange. Upperparts, sides, and flanks dark slate; head-sides (almost meeting above bill) and **band across lower throat and breast black, upper throat and belly white; tail conspicuously tipped white**. **Juv.** somewhat duller, rectrices pointed, bill and iris dark.
Habits: Solitary, sometimes in mixed-species flocks. Usually reported to be shy, foraging within dense vegetation, but twice seen feeding confidently and openly at edges of clearings in s Ecu. (PN Podocarpus: PG; HB, MKP, CR, JFR). May sit motionless for some time, but usually restless.
Voice: Call *tzip*.
Breeding: Presumably Nov-Apr in Peru.
Habitat: Within 3 m of the ground in bamboo thickets (*Chusquea*) and adjacent forest undergrowth. One was seen feeding on *Podocarpus* cones.
Range: 1700-3000 m. At present known from 6 localities: Morona-

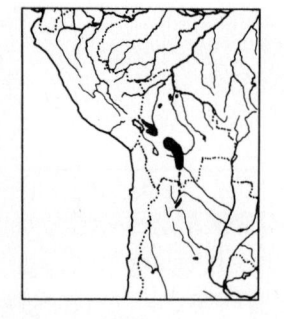

Santiago (Cord. de Cutucú), Loja, and Zamora-Chinchipe, (both PN Podocarpus), s Ecu., Piura (Cerro Chinguela), Amazonas (Cord. de Colán), and Huánuco (Cord. Carpish), Peru. Apparently rare, but possibly somewhat overlooked due to its retiring habits and impenetrable habitat. See *Auk* 96(1979):610-3.

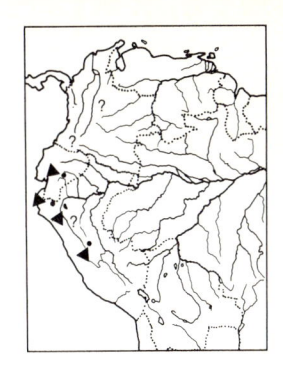

YELLOW GROSBEAK *Pheucticus chrysopeplus* – Plates LVII 5a-c and LXIV 34

20 cm. **Male**: *chrysogaster*: **head**, upper mantle **and underparts bright yellow**, back and tail black, feathers of back basally yellow, feathers of rump yellow with black bases and narrow black tip that wear off. Wings black with broad white band across base of primaries (conspicuous in flight), greater and median coverts and secondaries with large white terminal spots; longest upper tail-coverts black tipped white, outer 3-4 tail-feathers broadly tipped white on inner web. Vent white. Ssp *laubmanni* with bright yellow lateral edges of the black interscapular feathers (instead of black with semi-concealed subbasal yellow spots), scapulars narrowly tipped olive-yellow and outer 4-5 tail-feathers tipped white. **Female** has black replaced by brownish, narrower bar on primaries, bright yellow, dusky-streaked crown, dusky-streaked back, dull olive-edged dusky scapular feathers, and very narrow white tip of tail; *laubmanni* with less streaked crown. **Imm. male** much like female, but wings and tail browner. **Imm. female** with headside, breast, and sides faintly dark-streaked, rump olive-yellow, feathers with dusky centers; virtually no white in tail. **Juv.** buff above, browner on rump, and broadly streaked dusky; wings and tail like imm. female, underparts yellowish buff.
Habits: Alone or in pairs. Sedately forages for insects, seeds and berries, often open to view. Often in ripe corn. Sings from conspicuous perch high in a tree.
Voice: Call a sharp *kick* like Black-backed G. Song of gentle, abrupt whistles up and down, more or less alternating between 2 pitches *huit hy huiu hye huiii*, *huit hy huiu huit* or *hye huit hy hye*. 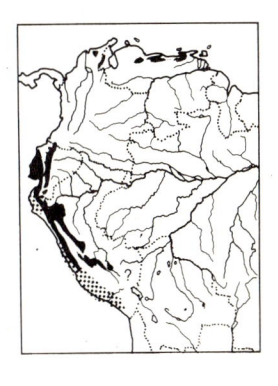
Breeding: Fledglings May (n Ecu.), June (Huánuco); large gonads Apr-July (Sta Marta and Perijá); juv. Jan (Lomas de Inari, Arequipa).
Habitat: Usually rather high in trees, but also low and on ground. Semi-arid regions. Lomas, riparian trees, bushy slopes, gardens, *Eucalyptus* groves, hedgerows, and fields.
Range: 900-3500 m, in w Ecu. and w Peru down to sea level. Sta Marta and Perijá mts and coastal mts of Ven. (*laubmanni*). From Nariño, s Col., through Ecu. to Arequipa and Cuzco (Puno?), s Peru (*chrysogaster*). 4 sspp in Mexico and C Am. Fairly common to common in Ecu. and n Peru, uncommon to rare further s.

NOTE: S Am. sspp by some considered a distinct species **GOLDEN-BELLIED** (or **SOUTHERN YELLOW GROSBEAK** (*P. chrysogaster*).

BLACK-BACKED GROSBEAK *Pheucticus aureoventris* – Plates LVII 10a-f and LXIV 35

20 cm. **Male**: *aureoventris*: **head, breast, and upperparts black**, rump admixed yellow, lesser wing-coverts yellow, median and greater coverts with large subterminal white spots, primaries with broad white basal band. **Belly and flanks yellow**, sides in some individuals with large black spots, vent whitish, feathers narrowly tipped black. Outer 2-3 tail-feathers with large white subterminal area; under wing-coverts yellow. Ssp *terminalis* similar, but upper tail-coverts barred to spotted white and tertials with small white spots near tip, vent yellow. Ssp ***crissalis* has yellow center of throat and most of breast, and is best told from Yellow G. by darker crown and headside, and by having black spots on sides and streaks on flanks**. It has large white spots on tertials; yellow black-barred rump, and semi-concealed yellow bases of back-feathers. Ssp *uropygialis* like *crissalis*, but rump showing less yellow; back, throat, and breast black, yellow feather-bases only rarely showing. Ssp *meridensis* with upperparts incl. upper tail-coverts black, back with some concealed yellow feathers, rump golden with a few black spots, throat and breast black, belly yellow always without spots. **Female**: above like respective male, but browner, below yellow more or less spotted black; wings with smaller white marks than male. **Imm**. like female, but streaked olive above. **Juv**. has crown and headside dusky streaked brown, back blackish brown streaked pale brown, wings and tail as in female, throat yellowish white tinged brown, belly white tinged yellow; mandible paler than in ad.
Habits: Alone, in pairs, or family groups, sometimes in mixed-species flocks. Mainly forages for insects and berries 1-2 m from treetops. Sings well hidden among foliage in center or top of a tree.
Voice: Call a sharp *kick* like Yellow G. Song melodious and varied whistles.
Breeding: Eggs Dec-Jan (Tucumán); juv. Nov (Nariño); large gonads Jan (Cauca).
Habitat: Cloud forest and humid forest and adjacent semi-arid zone in gardens, orchards, and corn fields. Alders (*Alnus*) bordering streams in bushy and patchily wooded ravines, and often in wild cherries (*Prunus capuli*). In Ecu. only collected in rather dry habitats.
Range: 1750-3700 m, rarely to trop. zone, in s at 600-2000 m. Mérida, Ven. (*meridensis*); N.d.Santander to Cundinamarca in E Andes and Cauca and Huila in C Andes, Col. (*uropygialis*), Nariño, s Col., and Ecu. (*crissalis*); Amazonas to Cuzco, e Peru (*terminalis*), from Puno in se Peru along Andean slope of Bol. to San Luis, nw Arg. and e through Bol. to w Mato Grosso and n Parag. (*aureoventris*). Uncommon.

NOTE: Birds from Ecu. should be studied for possible hybridization with Yellow G.

ROSE-BREASTED GROSBEAK *Pheucticus ludovicianus* – Plate LVII 6a-b

18 cm. **Ad. male breeding plumage**: head and upperparts black, rump and broad tips of upper tail-coverts white; **white band across**

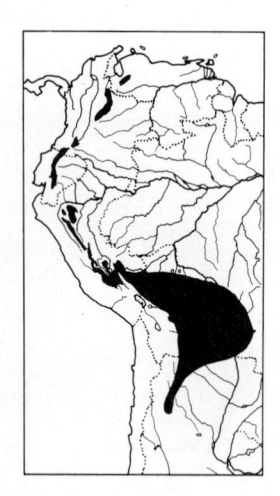

base of **primaries** (very broad on underside), and **white terminal spots on wingcoverts** (large on median and smaller on inner greater coverts and secondaries). Under wing-coverts and **c breast and upper belly rose-red**, rest of underparts white. **Male winter** has black feathers tipped and edged brownish, rump barred and spotted black; supercilium from above eye buff; throat, breast, and sides buffy brown with dark spots or streaks, feathers of c throat, breast, and upper belly more or less rosy basally, rest of underparts white. **1st fall male**: above brown streaked dusky, wings browner than in ad., white on wing-coverts tinged buff, rump barred; conspicuous supercilium buffy white, faint c crown-stripe pale. Below like winter male, but rosy color only weakly suggested, except on underwing; tail much more narrowly white-tipped. **1st spring male** much like winter male, but lacks supercilium, and is browner above, rump white. **Female**: above gray-brown streaked dusky, supercilium and feather-edges of nuchal band and c crown whitish; spots in wings whitish, primaries without or with a faint suggestion of a basal band. Below whitish streaked dusky, least on c belly; wing-linings yellowish buff; tail without white tip. In spring tinged buffy throughout.
Habits: Solitary, during migration and sometimes in winter in loose flocks up to 5. Actively searches for insects and berries and frequently flies to new site.
Voice: Call a metallic *chick* or *chink* and a gentle mew. Male rarely sings in S Am.
Habitat: All levels, but mainly middle strata. Edge of humid forest and adjacent second growth, orchards, and corn fields.
Range: Breeds in N Am., winters Oct-Apr at (600-) 1000-2000 m, rarely to lower temp. zone, and to 3800 m in Col. during migration. Winters through C Am. and Col. to Ven., and through Ecu. to ce Peru. Generally uncommon or rare, but during migration fairly common in ne and nw Ecu.

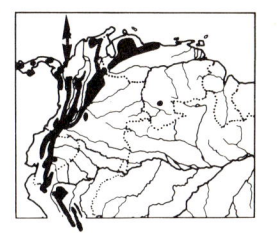

BLACK-AND WHITE SEEDEATER *Sporophila luctuosa* – Plate LIX 3

10.5 cm. **Male: in fresh plumage** above and on throat, breast, and sides gray-brown (tips of feathers), belly whitish; remiges black with **prominent white speculum**; extreme base of tail white (not visible in the field); **bill** blue-gray, **thick, maxilla curved**; in **worn plumage** gray-brown parts turn black. **Female**: above dull brownish olive, below buffy olive, c belly pale yellow; bill dark horn. **Juv**. undescribed.
Habits: In pairs or small flocks, often a few in flocks of Yellow-bellied S. Sings from conspicuous perch, often many males simultaneously in colony-like clusters.
Voice: Song by male *kiourrr-tititi-ti-tiou-tiou-tiou*, female answers with a *zizizi*, male finishes with a *tirrrr*, and may move to female. Duet lasts c. 2 s and may be repeated at 2-5 s intervals a number of times. *Tirrrr* of male sometimes repeated without 1st part or female answer.
Breeding: Juv Apr (Antioquia); large gonads Sep (Santander, Boyacá).
Habitat: On or near the ground in *Chusquea* bamboo (perhaps depends partly on flowering bamboo), other kinds of second growth at edge of

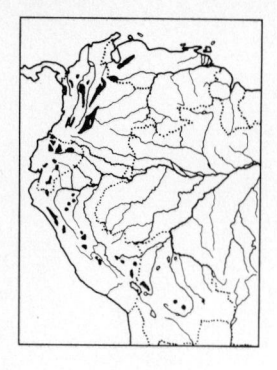

humid forest, in riparian shrubbery in semi-arid regions, and in bushy patches on grassy slopes.

Range: 900-3600 m, rarely down to 100 m, mainly subtrop. and lower temp. zone. S slope of Sta Marta mts and Andes from Lara, Ven., s through all 3 ranges of Col., e and w Ecu., and Peru (w of the Andes to n Ancash) to Sta Cruz, Bol. Lowland rec. e of Andes possibly of nonbreeding birds only. Generally rare, but locally common (e.g. wc Cajamarca).

YELLOW-BELLIED SEEDEATER *Sporophila nigricollis* – **Plate LIX 2a-b** is locally common at 1700-2400 m on w slope of Andes from nw Peru s to Lambayeque, e of W Andes from Cajamarca s to Cuzco, se Peru (*inconspicua*), and from lowlands to 2400 m in Nariño, Col., and w Ecu. (*vivida*); also in lowlands from Costa Rica to Surinam, e Ecu., and from se Peru to se Brazil (*nigricollis*). From ground to mid-heights from humid to semi-arid regions, in woodland, forest edge, second growth, clearings, and grassland, spreading with man. - **11 cm. Ssp *inconspicua* male**: bill blue-gray; above olive-gray turning **blackish on face**, throat and breast, belly pale yellow; base of primaries with small almost concealed whitish speculum. Ssp *vivida* with more extensive black on head and breast, and richer yellow below. **Female**: above dull olive, below pale olive-brown, c belly yellowish white; bill dull slate. Song includes *cheep*s and trill; in W Andes Col. a short musical *tsu tsu tsu chew-seesee-héet, tsu tsu tsu tswidle wés hére*, etc, last note typically higher. Call a sparrow-like chirp and a short musical *eeh?*.

DOUBLE-COLLARED SEEDEATER *Sporophila caerulescens* – **Plate LIX 4a-c** is widespread from e Brazil (*hellmayri*) to La Pampa and Chubut (see *Neotrópica* 15 (1969):63), Arg. and (winter; see *Bull. Brit. Orn. Club* 99 (1979):24-6) the Amazon lowlands, and breeds to at least 2100 m from Jujuy to Tucumán, nw Arg. (*caerulescens*), and in La Paz, Cochabamba, and Beni, Bol. (*yungae*, doubtfully distinct). In open country with bushes and grass, scrubby hills, dry forest, fields, and gardens. 11 cm. **Male**: bill stubby, greenish gray; upperparts and headside gray, lores black; small area below eye, **moustache and throat white, chin and bar on transition to breast blackish**, rest of underparts white. **Imm. male** tinged brown above, blackish feathers of throat tipped gray-brown, belly faintly buff, sides washed pale gray-brown. **Female** grayish olive above, grayish buff below, c throat and c belly whitish. **Juv. male** like female, but markings on throat suggested. **Juv. female** more brownish above, buffy below; bill horn. Usually in pairs, during migration in large flocks with other seedeaters. Sings from conspicuous perch, sometimes high in a tree; song c. 2.5 s long given at 5-10 s intervals, an explosive varied warble with a few introductory notes, e.g. *di di tzrídiriditzrídiri*. Eggs Dec, Jan (Tucumán).

BAND-TAILED SEEDEATER *Catamenia analis* – **Plate LIX 5a-g**

11-11.5 cm (*schistaceifrons* smallest). **Male**: bill stubby, yellow in breeding season, otherwise duller and dark-tipped. Above plumbeous; base of primaries with conspicuous white speculum in *soderstromi* of Ecu. and the southern *analis*, speculum faint or lacking in other sspp; primaries edged white in *soderstromi* and *griseiventris*; feathers around base of bill more or less black or blackish; below gray to light gray, c abdomen white in *analis*, gray in *insignis*, intermediate in remaining sspp; vent chestnut. **Tail-feathers with a large white area** of variable extent on middle of inner webs, forming conspicuous white band in flight. In fresh plumage back-feathers are tipped and tertials edged brown, feathers of underparts narrowly brown-tipped. **Ad. female** like male, but without black on face, and plumage slightly duller and paler, esp. below; upperparts with dark shaft-streaks; the brown tips and edges of fresh plumage broader; vent rufous. Some females have narrow streaks on breast and sides. **Imm.** buffy brown streaked dusky above, eye-ring whitish, underparts grayish white more or less tinged buff and streaked dusky except on c abdomen; vent rufous or buff. **Juv**. pale gray to pale buff above and streaked dusky, eye-ring whitish, below whitish streaked dusky, c belly white, tail as in ad.

Habits: In pairs or small flocks, often in mixed-species flocks with other finches. Forages for seeds in composite bushes or on the ground. See also *Hornero* 10 (1967):389-402.

Voice: Call *st*. Also gives a weak soft twittering. Song a faint note (often omitted), followed by a flat buzzy trill *tic bzzzzz* lasting c. 1 s and repeated at 7-8 s intervals.

Breeding: Fledglings Sep (Boyacá), May (Moquegua), June (Amazonas); juv. Feb, Mar (Sta Marta), Aug (La Paz); song Aug, Sep (Cundinamarca).

Habitat: Both dry and humid bushy country, lawns, hedgerows, and fields.

Range: 2500-3700 m, up to 4650 m in Sta Marta mts, and 4200 m in Puno, Peru, down to sea level on Peruvian coast (Lima) and in Arg. Sta Marta mts (*alpica*); E and C Andes of Col. (*schistaceifrons*), from Imbabura to Chimborazo, nc Ecu. (*soderstromi*); e slope of Andes from Cajamarca to Pasco, Peru (*insignis*, doubtfully distinct from *analoides*), w slope of Andes from Piura to Arequipa, Peru (*analoides*), birds from Apurímac intermediate with birds of Cuzco, se Peru (*griseiventris*); n La Paz, Bol. (*subinsignis*); from Tarapacá, n Chile and Puno, se Peru, s through w Bol. except altiplano to Mendoza, San Luis, w Córdoba, and se Buenos Aires, Arg.; in winter e to Buenos Aires, sight rec. from Aconcagua and Santiago, Chile (*analis*). Fairly common or locally very common.

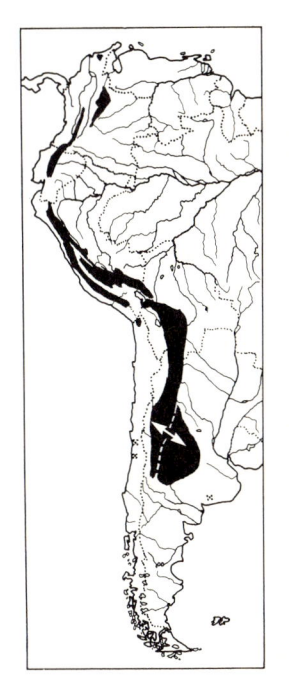

PLAIN-COLORED SEEDEATER *Catamenia inornata* – Plate LIX 7a-c

13-13.5 cm (*minor* smallest). **Male**: **bill stubby**, bright salmon-**pink** in breeding season, otherwise more brownish. Tail-feathers not pointed. Sspp *inornata* and *minor*: above dark gray, back with dusky shaft-streaks, below gray, paler and faintly tinged buff on c abdomen, flanks buff, vent chestnut; *mucuchiesi* darker above and on breast, buff wash on belly paler. **Female**: above buffy brown streaked dusky, below pale buffy brown, palest on c belly, vent buff; breast sometimes narrowly streaked; bill dark. **Imm. male** like ad., but tinged brownish, esp. on back;

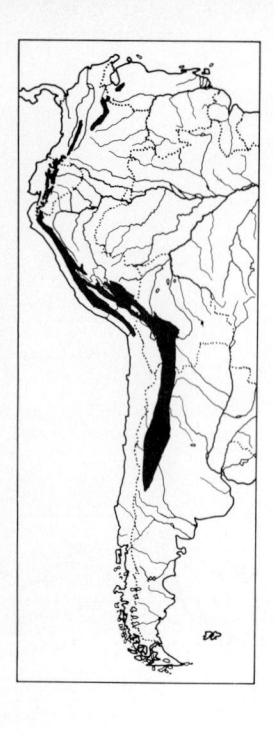

entire upperparts streaked. **Juv**. like female, but streaked dark brown on throat, breast, sides, and vent.
Habits: Alone, in pairs, or small flocks, often with other finches. Forages for seeds on the open ground and perches inside bushes or on wires.
Voice: Call a fine *zee*. Song (Cundinamarca, Nariño, Ayacucho) 2-5 musical notes, then 2-6 slow buzzy trills at different pitches and speeds *chit-tita zree, bzzz, bree*, lasting 3-6 s and repeated every 5-10 s.
Breeding: Eggs and fledglings Feb (Jujuy); juv. Oct (Cundinamarca), Oct, Apr (Cauca), Aug (Nariño), June (Huánuco); large gonads Jan, Feb (e Cauca), July, Aug (Santander), Sep (Boyacá), Aug (Cundinamarca); song Aug, Sep (Cundinamarca).
Habitat: Dry shrubby slopes, puna grassland, *Polylepis* shrub, *Espeletia* páramos, and humid shrub with grassy areas.
Range: 2900-4400 m, rarely down to 2000 m in Ecu. and on Pacific slope in Peru. Mucuchiés and San Antonio páramos in Mérida, Ven. (*mucuchiesi*); from Páramo de Tamá s through E Andes, and from Caldas C Andes of Col. s through Nariño and Ecu. to Junín, Peru (*minor*), and from Apurímac and Cuzco in se Peru through c Bol. to Mendoza and w Córdoba, nw Arg. (*inornata*). Birds from Arequipa, s Peru and Tarapacá, n Chile, not identified subspecifically. Uncommon in Mérida, common in Col., fairly common elsewhere.

PARAMO SEEDEATER *Catamenia homochroa* – Plates LIX 6a-e and LXIII 19

12 cm. Bill longer and more conical than in other seedeaters, bright pale yellow (appearing white at a distance) in breeding male, otherwise duller. **Tail-feathers pointed**. For plumage variation see *Auk* 103(1986):227-30. **Ad. male** uniform dark slate, vent chestnut; forehead and lores blackish, primaries edged whitish, flanks tinged dark olivaceous. **Juv**.: above brown, below pale brown, everywhere heavily streaked blackish. **1st** and **2nd year male** with progressively grayer and less streaked plumage, brown parts more olivaceous than in Plain-colored S. **Female** like 2nd year male. 1st and 2nd year female browner and more streaked. Ssp *oreophila* has a shorter, but still conical bill, and less olivaceous wash to brown colors, tail-feathers somewhat longer and broader.
Habits: In pairs or small flocks, sometimes in mixed-species flocks. Shy. Restlessly moves through dense understory and on the ground.
Voice: Flight call a high-pitched *tsit-tsit*.
Breeding: Juv. July (Cuzco); large gonads July (Perijá, e Antioquia).
Habitat: Lower strata and ground in humid páramo shrubbery, elfin forest, and edge of cloud forest. May enter grassland at forest edge. Presumably a bamboo specialist, thriving only when *Chusquea* and *Swallenochloa* are flowering.
Range: 2300-3800 m. Sta Marta mts (*oreophila*); Perijá mts, Mérida, Ven., all 3 ranges of Col., Ecu., and Peru to Cochabamba, Bol. (*homochroa*); Tepuis of Ven. and adjacent Brazil (*duncani*). Uncommon to rare, possibly abundant when bamboo flowers.

NOTE: Ssp *oreophila* previously considered a distinct species, **SANTA MARTA SEEDEATER** (see *Lozania* 23(1977):1-7).

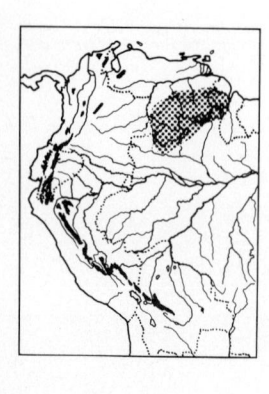

STRIPE-TAILED YELLOW-FINCH *Sicalis citrina* – Plate LX 2a-b

11 cm. Smallest Y.-f. **Bill distinctly smaller and more pointed than in other yellow-finches**. **Ad. male**: above greenish gray, forecrown (when worn) and rump bright golden olive; **back, but not crown streaked dusky**; all wing-feathers edged olive-green (greater wing-coverts and inner remiges edged, but not tipped buff when fresh). **Outer 2 tail-feathers with 2 cm long white patch at tip of inner web** (more extensive than in Grassland Y.-f.). Headside like crown, eye-ring yellow; below bright yellow, breast and flanks tinged olive. **Female**: above brownish gray streaked dark brown, rump olive-green; below pale buffy yellow turning yellow on belly, breast and sides streaked dark brown. Tail like male. **Imm**. like female, but underparts ocher-yellow washed cinnamon on breast, flanks and vent; tail only with white on outermost feather. Sspp poorly differentiated if at all valid.
Habits: In pairs or small flocks. Forages for seeds on or near the ground, in breeding season also takes insects. Flies far and high, always landing on the ground. Sings from perch or in fluttering nuptial flight.
Voice: Call a low chirp. Song (n Ven.) a musical *chu'u'u'u'u'u'u'u', zew-tew-tew-you*, the chattering notes or trill at the end on different pitch.
Breeding: Juv. Aug (Cundinamarca); brood-patch June (Puno); large gonads June (n end of E Andes Col., Puno); copulation July (Cundinamarca).
Habitat: Grassland, savanna, and fields in semi-arid regions, in Col. also reported (Hilty) to frequent wet swales and edges of marshes.
Range: Highly disjunct, at 400-2800 (-3700) m. In n Sta Marta mts (600-2200 m) and Perijá mts, Cauca (El Tambo 1700 m) e slope of W Andes, Antioquia (e slope Barro Blanco e of Medellín 2200 m) and Huila (Belén 2300 m) in C Andes, and in N.d.Santander (Guamilito 950 m) and Cundinamarca (Sabana de Bogotá 2500-2800 m) E Andes of Col.; coastal mts and Tepuis of Ven., Brazil, Guyana and Surinam (*browni*); e Brazil (*citrina*); 2100-3700 m in Cuzco (?) and Puno (Oconeque 2150 m, Sandia 2450 m) in se Peru, and n La Paz (Chilcani, Laripata), Bol., and in Tucumán (Cuesta de Macamala), nw Arg. (*occidentalis*). Local, uncommon or rare in the s, common locally in Col.

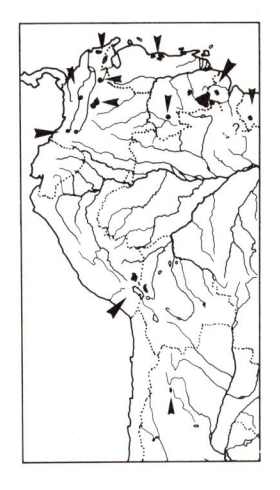

PUNA YELLOW-FINCH *Sicalis lutea* – Plate LX 1a-d

14 cm. Bill stout, even larger and with more arched proximal part of ridge than in Greenish Y-f., from which it mainly differs by being brighter yellow throughout. Sexes alike. **Ad. fresh**: **above bright golden green**, rump yellower, tail-feathers broadly edged bright yellow; eye-ring yellow, **underparts bright yellow**. **Ad. worn**: above dark gray-brown finely mottled with olive-green, rump yellowish olive-green; throat, breast, and sides yellowish olive, belly bright yellow. **Imm**. faintly streaked above, below yellow tinged greenish, fresh feathers of breast and sides with gray tips. Mandible orange-yellow (blackish in ad.). **Juv**. is obscurely streaked above, brownish olive crown turning browner on back; inner remiges very broadly edged warm pinkish brown; below buffy.

Habits: In family groups or pure flocks up to 20. Forages for seeds on the ground. Breeds and roosts in holes in banks. 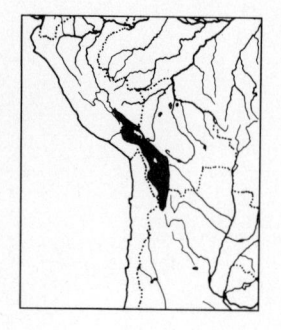
Voice: Contact call *tjyp* repeated at 1-2 s intervals, sometimes *tjyp-tjyp* given on ground or in flight. In flight also a questioning, snarling *eh ehr*, and a high-pitched, short *zit*.
Breeding: Nest-building and eggs Feb (Jujuy); juv. Apr (La Paz).
Habitat: Rather arid puna grassland and adjacent open brush-covered slopes and fields in open country. Rarely in outskirts of, but never **in** villages or towns.
Range: 3350-4500 m. From Cuzco and ne Arequipa se Peru through Bol. altiplano to Salta, nw Arg. Rec. from Mendoza (Potrerillos) Arg. probably erroneous. Fairly common, esp. in Titicaca basin and on Bol. altiplano.

BRIGHT-RUMPED YELLOW-FINCH *Sicalis uropygialis* – Plate LX 7a-e

13 cm. **Tail fairly short, primaries relatively long; in all plumages the yellowish lower rump contrasts the duller color of the back and upper rump.** Bill smaller and maxilla less arched than in Greenish Y-f. Ssp *uropygialis* and *connectens* **ad. male**: crown and headside dark golden olive-yellow (almost orange in some males), **posterior part of ear-coverts**, lores and half-ring below eye **gray**, eye-ring whitish. **Back, wings and tail gray**, shafts faintly darker, lower rump, upper tail-coverts and narrow edge of most tail-feathers olive-yellow; primaries narrowly edged white, smallest lesser wing-coverts golden olive-yellow. Below bright yellow, sides of breast washed olive-yellow, extreme sides of body gray, lower flanks yellowish white and gray. **Imm. male** similar, but less bright; gray replaced by gray-brown, crown faintly streaked, streaks on back broader, greater wing-coverts and inner remiges edged buffy white. **Ad. female**: crown gray-brown with dusky shafts, feathers edged olive-yellow; back gray-brown, broadly and obscurely streaked dusky, lower rump and upper tail-coverts olive-yellow, the latter with dark shafts; primaries narrowly edged whitish, no olive on lesser wing-coverts; headside mostly gray-brown, but yellowish on lower part, eye-ring whitish, neckside yellowish tinged gray-brown; below yellow, brightest and darkest on throat, lower throat and upper breast very faintly streaked dusky along shafts, breast-feathers tipped pale gray-brown, narrowly on c breast, broadly on sides of breast; sides of body pale brown, feathers tipped buffy white and faintly streaked along shafts. **Imm. female** like ad. female, but breast more conspicuously streaked and yellow of underparts paler. **Ssp *sharpei*** similar, but ad. male without gray on sides of body and flanks, bill considerably shorter, yellow-olive of rump more extensive, many feathers on back edged olive-yellow. **Juv**.: above warm brown broadly streaked dark brown, tertials with warm brown edges, lower rump golden brown, headside brown, throat pallid brown, breast and sides yellow-brown streaked dark brown, lower breast, belly and vent pale yellow, vent with brown shafts.
Habits: Sometimes in pairs, but usually in flocks even in breeding season, sometimes up to 500 outside breeding season, esp. in periods with much snow. Forages for seeds on the ground. Not shy. Breeds and roosts in cliffs, holes in banks, and occ. under roofs.

Voice: Songs, given from rocks or rooftops, include a rapid loud twitter 2-4 s long, and an irregularly descending series of twittering notes *de de de de dit chittit*, the last 1 or 2 notes stressed. Flight call with a similar quality *wee-weet* or *wee-weetju weetju*. Flocks give a twitter in flight *zedetzee-etdst*....
Breeding: Eggs May, June (Junín); nestlings April (Puno), May (Moquegua); fledglings Aug (Junín), May, July (Cuzco), June (Puno), Apr (Cochabamba).
Habitat: Breeds on rocky slopes with scattered shrubs, in stone fences, burrows in the ground or under roofs; forages in fields, short grassland, grazed shore meadows, and on rocky slopes.
Range: 3200-4800 m, in Chile at 2450-3950 m and rarely straggling down to 1200 m. Cajamarca to Junín, Peru (*sharpei*); upper Urubamba valley of Cuzco, se Peru (*connectens*, doubtfully distinct from *uropygialis*), from Apurímac and s Ayacucho through Puno and Moquegua, s Peru, to Antofagasta, n Chile, e to Cochabamba and s through Bol. altiplano to Catamarca and Tucumán, nw Arg. (*uropygialis*). Very common. The commonest yellow-finch through most of its range.

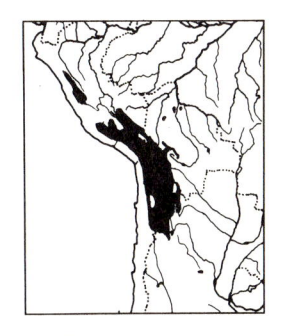

CITRON-HEADED YELLOW-FINCH
Sicalis luteocephala – Plate LX 6a-b

13 cm. Maxilla distinctly curved, **wing-tips short**. Male: **above gray to gray-brown**, most of head and wing-coverts yellowish citrine, **primaries** and outer tail-feathers **edged yellow**, throat and most of **narrow c underparts yellow**, c lower belly white, **sides extensively gray**. Female slightly browner above, and has even more extensive gray on sides and whitish on flanks. Apparently no distinct imm. plumage except for mainly orange mandible. **Juv.**: above brownish almost unstreaked, rump bright cinnamon, broad edges of tertials and tips and edges of wing-coverts cinnamon, flight- and tail-feathers narrowly edged greenish yellow; eye-ring pale yellow; below incl. chin ochraceous yellow washed cinnamon on breast, sides, and flanks, throat pale buffy brown.
Habits: In small flocks up to 20. Forages for seeds on the ground and in low herbs and bushes, in breeding season also takes insects.
Voice: Song of warbling notes. Calls include snarling notes *tee te te...*, *treh*, a clear *huit* (in flight only), and thin *tit* notes. Alarm call a harsh *tret*.
Breeding: Colonies in holes in banks. Fledgling May (Cochabamba).
Habitat: Arid, in places of short vegetation in dry quebradas in open country with scattered bushes and taller herbs, and in agricultural areas.
Range: Bol. at 2550-3800 m in Cochabamba (9 known loc.), Sta Cruz (Valle Grande), Chuquisaca (Sucre) and Potosí (Oploca), doubtfully rec. from La Paz. Locally common near Cochabamba city (e.g. between Arani and Vacas).

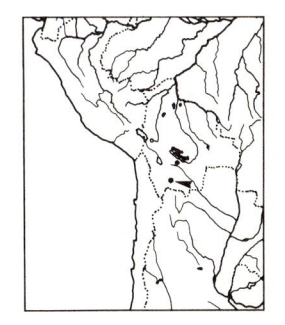

GREATER YELLOW-FINCH *Sicalis auriventris* –
Plate LX 3a-b

14.5 cm. Bill fairly long and conical. **Wing-tips long** (c. 3 cm beyond

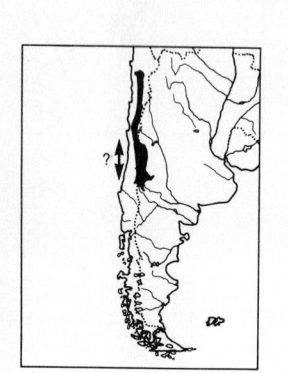

tertials). **Male golden olive-yellow, yellower on belly** and vent; wings grayish, lesser coverts like back, tail gray-brown, most feathers edged olive-yellow. Eye-ring yellowish white, lores grayish. In fresh plumage feathers of nape, back, rump, headside, breast, sides, and vent tipped pale gray-brown. **Female**: above gray-brown finely streaked dark brown, rump tinged olive-yellow; eye-ring white; below light buffy brown faintly streaked dark brown on flanks and sides, throat tinged yellow, c belly yellow. **Imm. female**: above gray-brown streaked brown, below very pale brownish to white with faint brown shafts, c belly mixed with a few pale yellow feathers; outer 4 tail-feathers edged golden at base. **Juv**. much like imm. female, but with 2 narrow wing-bars, no yellow below, outer tail-feathers edged pale olive and without golden at base, eye-ring almost interrupted (BW).
Habits: In small flocks in winter. Forages on the ground.
Voice: Song a rapid, explosive, rather unmusical warble, *trididetridet-ride*, repeated at 2-4 s intervals, and varying in length between bouts. Calls include a *trui*. (BW).
Breeding: Eggs rarely before Jan (Chile); fledgling Feb (Aconcagua).
Habitat: Open country, frequently around human habitation, breeding in stone walls and houses.
Range: Breeds above 1800 m, in Chile down to approaches of C Valley in winter. Antofagasta to lake Maule in Talca, Chile, and in adjacent Mendoza and Neuquén, Arg. Shares territory with Greenish Y-f. from Antofagasta to Coquimbo, and is replaced by Bright-rumped Y-f. to the n (see *Neotrópica* 10 (1964):36-9). Common in c Chile, but less so in winter, suggesting that some n-wards movements take place.

GREENISH YELLOW-FINCH *Sicalis olivascens* – Plate LX 10a-d

13.5 cm. Maxilla somewhat arched, small area below eye yellowish white or whitish. **Wing-tips rather short** (1.5-2 cm beyond tertials). Ssp *chloris*: **Male olive**-yellow **tinged gray**, c belly and vent yellow; upperparts tinged gray-brown grading to olive-yellow on rump. **Remiges edged olive**; most of tail-feathers narrowly edged olive-yellow; under tail-coverts with dark shafts. **Female**: above fairly uniform gray-brown, feathers very faintly edged olive-yellow, more conspicuous on rump. Remiges for the most part edged pale gray-brown, but edging olive at base of inner primaries and outer secondaries, lesser wing-coverts tinged olive-yellow. Headside gray-brown tinged olive-yellow, eye-ring whitish; underparts rather pale yellow, feathers tipped pale gray-brown except on c belly; under tail-coverts with dark shafts. **Imm. male** much like female, feathertips of underparts tinged buff, lesser wing-coverts hardly tinged yellow, remiges edged olive. **Juv.**: above gray-brown faintly streaked dark brown on crown and back. Tips and edges of wing-coverts and edges of inner remiges buffy; tail dusky brown edged yellowish. Eye-ring whitish; below pale grayish buff, buffier and paler on belly. Ssp *salvini* like *chloris*, but smaller, with shorter and blunter bill; *olivascens* like *chloris*, but tail slightly longer, bill weaker, and underparts yellower; *mendozae* like *olivascens*, but smaller, somewhat duller, more olivaceous on throat, breast, and sides, female

much like Patagonian Y-f., but darker, less grayish above, without white wing-panel, and with shorter wingtips.
Habits: In pairs or small groups, outside breeding season in flocks up to 30 or more. Forages on the ground.
Voice: Flight calls include a *trrui*, a snarling note, a clear *tjuit-tjuit...tjuit ee*, and *tu trui*. Song, often given from rocks or rooftops, a loud series of 10-15 chirps, finished off with a stressed higher chirp followed by 1-4 notes *tee-che-che...che-chéer-ti-ti*, lasting 2-3 s and repeated at 8-10 intervals.
Breeding: Eggs Dec (Cuzco), Jan (Chile), Mar (Jujuy); fledglings Aug (Ayacucho), Apr (Tacna), June (Cochabamba), Feb (Coquimbo), Aug Tucumán); song Dec (Cochabamba).
Habitat: Slopes with open montane scrub, cultivated and semi-cultivated land, often with some rocky parts like ravines, landslides, etc. At mid day spreads from bushy to open terrain. Also in towns, breeding under roofs and in stone-walls. Not in the puna zone.
Range: 1500-3600 m, rarely to 4250 m. From (Cajamarca, Amazonas?) La Libertad to Huánuco and possibly (ssp?) Junín and Ayacucho, Peru (*salvini*); w slope of Andes from Ancash, Peru, to Coquimbo, Chile (*chloris*), Apurímac (ssp?) and Cuzco, se Peru through w Bol. to n La Rioja, Arg. (*olivascens*), San Juan and s La Rioja to Mendoza and San Luis, in winter n to Salta, Arg. (*mendozae*). See also following species. Locally common, e.g. in La Paz city.

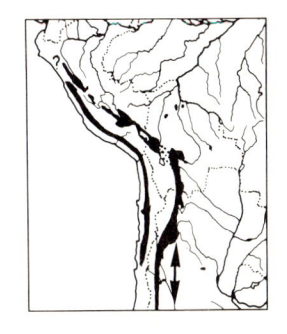

PATAGONIAN YELLOW-FINCH *Sicalis lebruni* – Plate LX 9a-c

14 cm. Bill relatively small, **wingtips long**. **Male**: upperparts and headside pale gray with olive-green showing through to varying degree; **primaries edged white**, outer 1-2 edged yellowish at base; **area below eye and lores light gray**, eye-ring whitish; below yellow, throat and breast with slight olive cast, breast admixed with some gray esp. at sides, **flanks pale gray**, **tail-feathers edged white**, pale yellow at base. **Female**: above gray-brown almost unstreaked and admixed with a little olive-yellow; headside and median and lesser wing-coverts washed olive-green, eye-ring white; below gray-brown, c belly bright yellow, throat admixed with some yellow, vent white; 2nd and 3rd outermost tail-feathers faintly edged olive-yellow at their bases. **Juv.**: above gray-brown very faintly streaked, back with slight buffy brown wash, rump buffy brown. Lesser wing-coverts tinged yellow-olive, greater coverts and tertials edged buffy, primaries narrowly edged buffy white. Most of tail-feathers edged yellow. Headside gray-brown, eye-ring whitish, throat grayish white, sides of chin tinged yellow, rest of underparts pale yellowish buff, yellowest on c belly.
Habits: In pairs Sep-Jan, otherwise in small groups. Forages for seeds and insects on the ground; perches on top of cliffs and low bushes.
Voice: Song elaborate and musical. Calls include a somewhat whistled, loud, often repeated *whit*, and a fine twitter, usually given in flight, and a peculiar *whi wheu*. Also gives weak monosyllabic notes.
Breeding: Eggs Nov-Jan (Isla Grande); fledgling Feb (Sta Cruz).
Habitat: Arid. Areas of short vegetation on grassland with earthy banks or rocky gorges, where it breeds and roosts in holes.

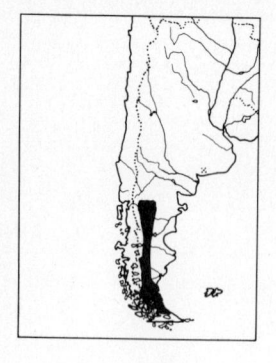

Range: Sea level to 1200 m. From Río Negro, Arg. (and occ. Buenos Aires), and Magallanes, Chile (Ultimo Esperanza), to n Isla Grande. Resident? Uncommon.

NOTE: Sometimes treated as a ssp of Greenish Y.-f.

SAFFRON FINCH *Sicalis flaveola* – Plate LX 4a-f

12 cm. **Wing-panel olive-yellow**, wing-tip reaches tip of under tail-coverts and exceeds tertials by 1-10 mm. Ssp *pelzelni*: **male**: above olive-yellow with dark shaft-streaks, feathers tipped pale gray-brown, rump and upper tail-coverts olive-yellow, feathers tipped gray; edge of flight- and tail-feathers and wing-coverts olive yellow. Forehead and lores tinged orange; **headside olive-yellow**, feathers tipped gray, eye-ring yellow; below yellow, brightest on throat, feathers tipped grayish white (except on throat), esp. on breast and sides; sides faintly dark-streaked. Gray feather-tips wear off leaving **bright yellow underparts and orange forehead**. **Female** much like Grassland Y-f., but wings edged olive instead of brownish. Above buffy brown streaked dusky, rarely with slight olive tinge on lower back; inner greater wing-coverts and tertials edged buffy white, tail-feathers narrowly so. Eye-ring white, earcoverts dark gray-brown; **below whitish**, moustache dark brown, **breast, upper belly, and sides** faintly tinged buffy brown and conspicuously **streaked dark brown**; underparts, esp. breastband and flanks occ. with scattered faint yellow edges. **Juv**. like female, but brown colors warmer, olive in wings and tail less distinct, sometimes absent in tail; inner greater wing-coverts and tertials edged buffy-brown. Birds from Cochabamba Bol. approach ssp *brasiliensis* esp. by brighter, more extensive orange forehead and greener crown and nape in male, and slightly longer, more laterally compressed bill. They may agree with ssp *koenigi*, which, however, was not available to us for direct comparison.
Habits: In pairs or small flocks up to 15 even in breeding season. Forages on or near the ground. Sings from a conspicuous dry branch in a tree or a rooftop. 1 male may attend 2 females.
Voice: Song *tri te tierit*, *tri te rrie rri rri rrhit* or *tee tie tr te ree te rii*, incl. varied, abrupt, fine notes.
Breeding: Eggs Dec-Feb (Tucumán); fledgling May (Cochabamba).
Habitat: Dry open wood, savanna, gardens. Usually breeds in hornero nest, stick nest, or under a roof.
Range: Lowlands to 1800 m, but up to 2100 m in La Paz, to 3200 m in the drier parts of Cochabamba, Bol., and at 2400-4000 m (lower in winter) in Salta and Jujuy, nw Arg. In n and e Col., n Ven. and n Guyana (*flaveola*); from sw Ecu. s to Ancash and Marañón drainage nw Peru (*valida*); ne and e Brazil (*brasiliensis*), se Brazil and e Bol. s to Mendoza, La Pampa, and Buenos Aires, Arg. (*pelzelni*), and mts of Salta and Jujuy (*koenigi*, G Hoy in *Stuttgarter Beitr. Naturk. Ser. A* 305(1970):4pp, birds inhabiting mts n to Cochabamba, Bol., seem also to be this form). Common in Cochabamba city, very common in lowlands. Introduced in Panamá and Jamaica.

NOTE: Ssp *pelzelni* previously considered a distinct species.

GRASSLAND YELLOW-FINCH *Sicalis luteola* –
Plates X 10a-b, LX 5a-e

11.5 cm. From Stripe-tailed Y-f. by stubbier bill and streaked crown, from Saffron Finch by **brown instead of olive wing-panel** in female and imm., from Raimondi's Y-f. by no or less decided gray on flanks, from all by **different habitat**. Ssp *bogotensis*: **male**: distinct facial pattern: narrow supercilium dark yellow lighter posteriorly, lores and area below eye yellowish, eye-ring yellowish white, moustache yellow, narrow submalar streak dark; rest of headside yellowish olive tinged gray, earcoverts mostly gray-brown. Above buffy brown to gray-brown, broadly streaked dusky, feathers edged olive-yellow basally, rump olive-yellow. Outer primaries narrowly edged whitish, inner primaries and secondaries narrowly edged olive, lesser wing-coverts olive-yellow. Below yellow, brightest on throat, sides washed olive, breast faintly so, breast and sides with white feathertips when fresh; tail dark gray-brown, feathers narrowly edged whitish, outer feathers with varying amounts of white near tip of inner web, but never as extensive as in Stripe-tailed Y-f. **Female**: above like male, but browner, olive on rump less extensive, basal edging of feathers of crown, back, and lesser wing-coverts golden-brown (vs. olive-yellow), secondaries edged warm buffy-brown (vs. olive), headmarkings duller and browner. Below paler yellow, throat tinged ochraceous buff, feathers of breast and sides with broader pale tips when fresh, sides of breast faintly streaked; tail as in male. **Juv**.: above pale buffy brown streaked dusky, below whitish, c belly pale yellow, breast and sides with dark shaft-streaks. Ssp *luteiventris* similar, but male has breast more decidedly washed olive, earcoverts dark olive-gray, tail-feathers rarely with any white on inner web.
Habits: In small flocks even in breeding season. Forages on the ground or in seeding rushes. During nuptial flight male flies 4-5 m up, then glides down with raised wings while vocalizing, sometimes continuing after landing. Not shy. Before mating, male hops around female with high-cocked tail, head back and rapidly shaking wings while singing.
Voice: Song *trr trr trr…*, sometimes with long thin trills *trrrr de de de de trrrr trrrr*. Flight call *tsi tsi tsidi*, *tsiri*, or *titeli*.
Breeding: In Chile 2-3 broods, first in Sep (rarely Aug); eggs June (Cajamarca); fledglings Sep (Nariño), Feb (Valdivia, Cautín); juv. Jan, large gonads May (Cundinamarca).
Habitat: Rush beds, esp. *Scirpus californicus* and tall *Juncus*, and adjacent rushy pasture and cultivated fields, sometimes in fields far from water. Nests in small colonies on or near the ground in tall grass or rush.
Range: Locally at 2200-3700 m in Mérida Ven., E Andes of Col., Ecu., and Peru s to Arequipa, also locally on coast of Lima, Ica, and Arequipa (*bogotensis*); locally at 3000-3700 m in Cuzco, Puno (and Moquegua?), s Peru, Cochabamba, Bol., and widespread from lowlands to 1500 m from Atacama to Aysén (incl. Islas Guaiatecas), Chile and from s Brazil and Sta Cruz, e Bol., to Río Negro, Arg; southern breeders migrate n presumably to s Peru and Brazil; a group of 5 rec. on Isla Grande in Feb (*luteiventris*). 6 sspp in lowlands from Mexico to Brazil (in Cauca valley up to 2900 m; *luteola*). Common at a few highland lakes.

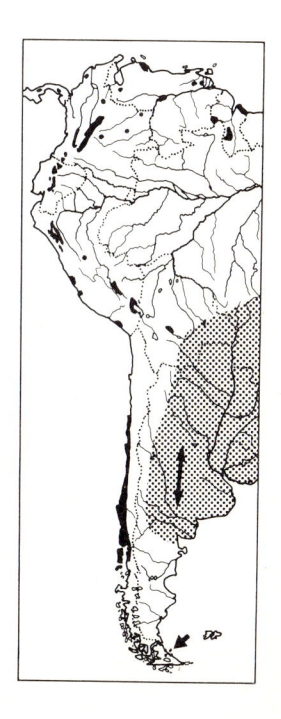

RAIMONDI'S YELLOW-FINCH *Sicalis raimondii* – Plate LX 8a-d

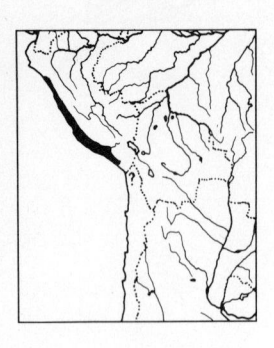

11.5 cm. Told from Grassland Y-f. by lack of conspicuous facial pattern and eye-ring, and by grayer plumage incl. flanks, and generally less conspicuously streaked back. Breeds in small colonies, outside breeding season sometimes occurs in compact flocks of 100s or 1000s (Oct-). No nuptial flight; song lower pitched and more full and rolling than that of Grassland Y-f. Call a somewhat snarling *che che chet* ...

Inhabits rocky places in lomas with fog vegetation and scattered stands of trees and slopes with cacti and steppic vegetation; hardly ever in arable land. Found locally from sea level to 2000 m, but occ. to 3450 m, on the Pacific slope of Peru from Cajamarca to Moquegua (n Chile?).

WHITE-WINGED DIUCA-FINCH *Diuca speculifera* – Plate LIX 14a-b

17.5 cm. Iris orange-brown, bill black. Wings very long, giving the whole bird an elongated shape. **Gray**; throat, c belly, and vent white; lores black, area below eye mottled with white; remiges blackish, **primaries with large white patch** on outer webs; tail blackish, outer feather with white outer web; under wing-coverts white. Slightly longer-billed ssp *magnirostris* doubtfully valid. **Imm.** is washed brownish above and buffy on throat.
Habits: In pairs or family groups of 3-4. Not shy. Walks slowly in search of seeds, sometimes giving a fast jerk. Surveys from rock.
Voice: Call during foraging or in flight a characteristic, snarling *wheit* or *wheit-wheitwheit*. In flight rarely a fine *tsit* or *suit*. 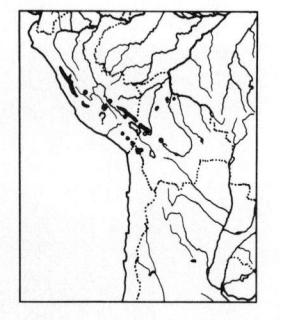
Breeding: Eggs Apr (n Chile); imm. July, Aug (La Paz), Aug (Cochabamba).
Habitat: Level and more rarely sloping bogs with densely matted cushion plants (*Distichlis*, *Plantago rigida*). Sleeps in adjacent rocky or bushy slopes or even in glacier cracks.
Range: 4500-5350 m, rarely down to 4000 m. From Ancash and Lima to s Junín and Huancavelica, wc Peru (*magnirostris*), presumably in intervening provinces and from Arequipa, s Peru to Tarapacá (Arica), n Chile, and adjacent nw Oruro (Sajama), Bol., and in e cordilleras from Cuzco and Puno Peru to La Paz and in Cochabamba, Bol. (*speculifera*). Fairly common.

COMMON DIUCA-FINCH *Diuca diuca* – Plate LIX 13a-b

16-16.5 cm (*minor* smallest). Above smoky **gray** (female brownest), remiges narrowly edged grayish white; **tail broadly tipped white**, decreasing and becoming restricted to inner webs towards the c feathers, which are unmarked. Lower eye-lid, **throat, and c belly white, flanks rufous**, under tail-coverts white edged rufous; **Juv.** similar, but duller; gray parts heavily washed with brown, white parts tinged buff, greater wing-coverts and shortest tertials broadly edged reddish brown, edge of longest tertials and wingbar on median coverts buffy white; outer 2

tail-feathers with large white terminal spot. Ssp *minor* much like *diuca*; *crassirostris* with larger bill and partially white outer web of outer tail-feather.
Habits: Alone or in pairs, in flocks in winter. Feeds on the ground in the early morning, perches in bushes during the midday heat. Sings from break of dawn.
Voice: Song 2-2.5 s long, of loud pleasant warbled notes given in series of 4-5 at 2 s intervals.
Breeding: In Chile 2-3 broods with eggs Sep-Feb; juv. May (Tucumán).
Habitat: Gravelly slopes with dry grass and *Acacia* scrub, steppes, brush-covered flats of river plains, open forest, farmland, parks, and gardens.
Range: Temp. regions from Antofagasta to n Coquimbo, Chile (see *Bol. Soc. Biol. Concepción* (Chile) 53 (1982):171-2), and from sw Bol. to La Rioja and in Córdoba, Arg. (*crassirostris*); lowlands to 1500 m from s Coquimbo to Magallanes (Ultimo Esperanza) and Isla Nueva (s of Beagle channel) Chile, and in adjacent Arg. from Mendoza to Sta Cruz (*diuca*), from lowlands of s Buenos Aires and Río Negro, Sta Cruz, Arg., migrating Mar-Sep to San Luis, Tucumán, and Entre Ríos, and (formerly) Urug. and immediate adjacent Brazil (*minor*). 1 ssp on Chiloe Isl., Chile. Common, but decreasing with the spreading of House Sparrow.

SHORT-TAILED FINCH *Idiopsar brachyurus* – Plate LIX 12a-b

17.5 cm. **Bill long**, gray, **tail short**. Above **dark gray**, below gray; area below eye mottled pale and dark gray; primaries narrowly edged white. Female slightly tinged with brown. Feet buffy. **Juv**. brownish gray, paler below and faintly streaked with dark brown on the flanks.
Habits: In pairs or small flocks. Forages on the ground and frequently perches on rocks with bill slightly raised. Elusive.
Voice: Call a sharp *ziht*, said to resemble the call of White-winged Diuca-finch. Song (according to Dabbene) a monotonous series of notes.
Breeding: Large gonads Jan (La Paz).
Habitat: 'Talus' slopes with long piles of rocks and with matted grass among rocky outcrops. Always among rocks, sometimes adjacent to bogs.
Range: 3950-4600 m, rarely down to 3500 m, in the E Andes in Puno (Huancasalani, Laguna Sallaku), se Peru, La Paz (Zongo rd, Yungas railroad, Illimani) and Cochabamba (Colomi, Tiraque), Bol; and in Jujuy (Sierra de Zenta), Salta, Catamarca, and Tucumán (Sierra Aconquija), nw Arg. Local and uncommon.

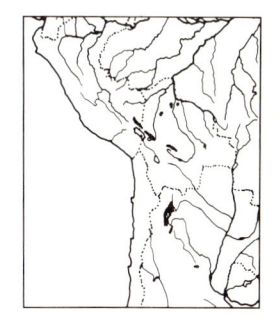

PATAGONIAN SIERRA-FINCH *Phrygilus patagonicus* – Plate LVIII 3a-c

14.5 cm. Bill more conical than in Gray-hooded S-f. **Male**: head dark to light plumbeous, darkest around bill; upper mantle olive, **back golden olive-brown to golden chestnut, contrasting with golden olive or gold rump; wings dark gray with **narrow** pale edges of primaries; lower breast and belly incl. **flanks bright yellow** (vs. gray in Gray-hooded S-f.) to golden yellow, c abdomen white, under tail-coverts gray edged white. Some (younger?) males have paler gray headside and throat, and olive tips to back-feathers in fresh plumage, or may even have dark olive back, olive rump, and rather pale yellow underparts. **Female**: crown dark gray faintly streaked dusky (no streaks in female Gray-hooded S-f.), back and rump dark olive-green (darker than male Gray-hooded S-f.), rump greenest; wings and tail gray-brown, tail-feathers edged gray, wings darker, remiges browner and pale primary-edges narrower than in Gray-hooded S-f.; headside and throat gray to light gray, breast and belly incl. **flanks olive-yellow**, c belly yellow, c abdomen whitish, under tail-coverts gray tipped and edged whitish. **Imm**. like female, but with faint whitish supercilium from above eye turning gray posteriorly, faint dusky moustache and short streaks on lower throat and upper breast, and duller underparts. **Juv**. like imm., but with olive-brown back, and yellow, more or less buff-tinged underparts and c throat.

Habits: In pairs or family groups. Forages on the ground, but spends much time inside open bushes and trees. Sings from tree. Territorial.

Voice: Song (Santiago, BW and G Egli) given at 3-5 s intervals, a 4-9 s series of 5-8 double-notes irregularly interspersed with 2-6 slightly higher-pitched single notes, 2nd part of each double-note stressed, e.g. *chi, cheerchi-cheerchi, chi, chi, cheerchi, chi, cheerchi*. Song in Sta Cruz (NK) similar, but series shorter, single notes not higher pitched, and double-notes with 1st part stressed, e.g. *cheerchi-cheerchi, cheerchi, chi, cheerchi, chi, cheerchi*.... At alarm gives a repeated, sharp *teck*, sometimes a trifle sharper than, but usually indistinguishable from that of Gray-hooded S-f.

Breeding: Nov-Feb.

Habitat: Often near water. Sometimes deep in beech (*Nothofagus*) forest, but more common in transitional zone towards grassland and in second growth where the forest has been burned; also in thickly vegetated river gullies, esp. where these have seeding bushes and rocky outcrops covered by vines and tangles.

Range: Coast to 1850 m from Curicó, Chile, and s Neuquén, Arg., to Tierra del Fuego; some migrate n to Aconcagua, Chile. Fairly common.

GRAY-HOODED SIERRA-FINCH *Phrygilus gayi* – Plate LVIII 2a-c

14.5-15.5 cm (*minor* smallest). **Male**: head and throat dark gray, area around bill darker; **back** dark **olive-green** to golden-green (in fresh plumage with gray feather-tips), **never with any chestnut**, rump more golden (lower rump gray like upper tail-coverts in *caniceps*); wings light gray, **primaries with silvery edges, broader than in Patagonian S-f**. Breast and upper belly golden yellow, **flanks pale gray**, lower belly **whitish**. **Female** similar, but back duller, head paler and sometimes washed brownish, lower headside and throat whitish; dusky crown-streaks, moustache and narrow band of streaks across lower

throat faint, becoming clearer with wear, breast ochraceous mixed with yellowish. **Imm. male** has head like ad. female, but often without streaks across lower throat, breast like ad. male. **Imm. female** like ad. female, but throat buff, breast ochraceous without yellow. **Juv**. like imm. female, but streaks on head more conspicuous, and belly buff.
Habits: In pairs, in winter in small flocks. Forages for seeds on the ground, perches on rocks, in bushes, or trees.
Voice: Song (Santiago, BW) 1-1.5 s long, of 3 equally stressed chirps, 1st lowest, and falling, 2nd level, 3rd highest, and rising; given at 0.5 s intervals, usually in bouts of 2-5 songs, bouts at 2-5 s intervals. In Sta Cruz (BW) song a 4-6 s long series of 7-9 virtually identical chirps at 1-3 s intervals. At alarm a repeated sharp *tack*, much like Patagonian S-f.
Breeding: Eggs Nov-Jan (Chile, Arg.).
Habitat: Grassland with rocky outcrops.
Range: (1300)1500-3650(4000) m, lower in s, in winter rarely down to coast in n Chile. C Chile from Santiago to Coquimbo, and in winter n to sw Antofagasta (and rec. in Nov. Potrerillos in Mendoza, Arg., see *Physis Buenos Aires (C)* 36(1977):349-50) (*gayi*); from Aysén, Chile, and Salta, nw Arg., to n Isla Grande, southern breeders migrating, occ. to Buenos Aires (*caniceps*, the southern populations possibly representing a distinct ssp, as may the population of the Somuncura plateau of Río Negro); lowlands from n Atacama (Chanaral) to Santiago (San Antonio), Chile (*minor*). Fairly common to uncommon.

NOTE: Breeds sympatrically with Black-hooded S-f. at Baños del Toro in Coquimbo, Chile, but may hybridize where one form is scarce (see *Bull. Brit. Orn. Club* 109(1989):66-82).

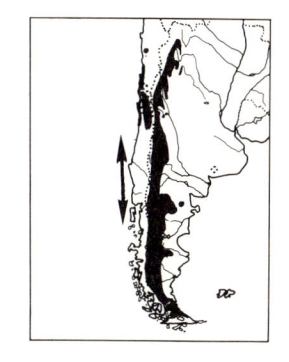

BLACK-HOODED SIERRA-FINCH *Phrygilus atriceps* – Plate LVIII 4a-c

15.5 cm. Male: head black or blackish, **back** golden **chestnut**, rump golden yellow, wings and tail blackish feathers edged gray; **below bright** golden **yellow**, breast and sides washed golden-brown, lower flanks and abdomen white, vent white with hardly visible blackish feather-centers. Males from nw Bol. altiplano (e.g., Sajama) have almost black wings. **Female** similar, but head dark gray palest on nape; almost indistinguishable from male Peruvian S-f., but with brighter yellow c underparts (and often accompanied by male). **Imm. male**: above dark warm brown, feathers with blackish centers along shafts; head blackish brown, c throat and line above dark moustache buffy; below orange-buff, palest on belly; wings with 2 vinaceous buff bars. **Juv**. like imm., but head gray-brown and c throat yellowish or whitish; narrow band of streaks across lower throat.
Habits: As for Peruvian S-f.
Voice: Song as in Peruvian S-f. *teer-zlip-teer-zlip* ..., the *zlip* notes perhaps slightly less shrill. Call a repeated *tep*.
Breeding: Eggs Dec (Oruro), Nov-Jan (Chile); fledglings June (Cochabamba), Jan, Apr (Oruro); juv. May (Oruro).
Habitat: In Peru and Bol. in open, rather dry *Polylepis* woodland, but may feed in open terrain part of the day, in Chile possibly also breeding in same habitat as Gray-hooded S-f.

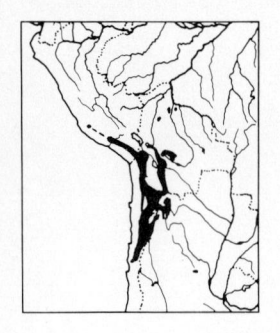

Range: 2400-4000 m. Pacific slope around Colca valley in Arequipa s to Coquimbo (Baños del Toro), n Chile, and in La Páz, Oruro, Cochabamba, and Potosí, Bol., and Jujuy, Salta, and nw Catamarca, nw Arg. Uncommon straggler (breeder?) at Pampa Galeras w Ayacucho. Common.

PERUVIAN SIERRA-FINCH *Phrygilus punensis* – Plate LVIII 1a-d

15.5 cm. Bill may shine pale in the sun. Ssp *chloronotus*: **male: head and upper breast dark plumbeous** bordered all around with dark olive; back and rump olive faintly tinged gold; wings, upper tail-coverts, and tail gray; below golden-yellow faintly tinged chestnut, lower flanks and abdomen white, under tail-coverts dark gray edged and tipped white. **Female** similar, but gray of head, wings and tail paler, forecrown usually with faint streaks. Ssp *punensis* similar, but back tinged brownish, less greenish in both sexes; some males more saturated with orange. **Imm.** like female, but crown and headside browner and faintly streaked, pale supercilium suggested, throat buffy with dark moustache, underparts duller (dullest in female), belly and abdomen washed buff, greater wing-coverts tipped pale buff forming narrow wingbar. **Juv**. has gray-brown head with buff supercilium, brownish back, 2 buff wingbars, underparts orange-buff, throat more yellowish, moustache and conspicuous streaks across upper breast dark gray.
Habits: In pairs, in winter in flocks up to 15 or more, often with other finches. Forages for seeds on the ground, perches on rocks and in trees. May feed from the flowers of *Puya raimondii*.
Voice: Flight call a sharp, fine, often repeated *tsit* or *tit ti*. Song of 2 (sometimes irregularly) alternating notes repeated 3-15 times or more *teer-zlip-teer-zlip*.... During mating a buzzy *prźeeer*.
Breeding: Eggs May (Puno); fledgling Apr (La Paz), June (Puno); mating Jan (Junín).
Habitat: Open country with scattered shrubbery on sloping or level ground, short grass with rocky outcrops, hedgerows, and ploughed fields. Often near farms or in villages, where frequently perches in *Eucalyptus* trees; also at edge of *Polylepis* shrub and in stands of *Puya raimondii*. May roost in puyas.
Range: 2900-4800 m, lowest in dry season. From c Cajamarca to Colca valley, Arequipa (perhaps sparsely to s Arequipa) and Cuzco, Peru (*chloronotus*), Puno in se Peru and La Paz, Bol. (*punensis*). Common.

NOTE: Variously placed with Gray-hooded and Black-hooded S-f.s. Occurs and possibly breeds sympatrically with the latter in w Ayacucho (Pampa Galeras) and n Arequipa (near Nevado Corunpuno). Some birds from La Paz and Oruro, Bol. (Comanche, El Pongo, Tanapaca, Carangas) approach *atriceps*, and may be hybrids, but parental types dominate in this area.

MOURNING SIERRA-FINCH *Phrygilus fruticeti* – Plate LVIII 9a-d

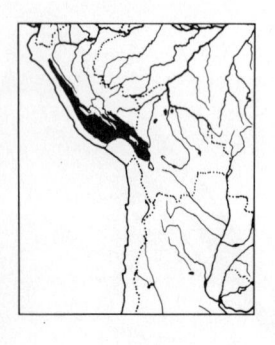

17 cm. Bill yellow in breeding male, otherwise brownish. **Male**: above gray streaked black on crown and back; uniform gray rump conspicuous in flight; wings black, feathers edged gray, median and lesser coverts black tipped white, forming 2 broad wingbars. Lores black, area below eye grizzled with white; **throat, breast, and upper belly black**; posterior headside to side of breast and flanks gray; lower belly, lower flanks, and vent white. **Fresh plumage** more uniform drab-colored, owing to broad pale brown to gray feather-margins, that wear of. **1st ad. male** similar, but brown feather-tips of head and back more pronounced, eye-ring buffy, moustache white, and vent tinged buff. **Female**: above pale brown streaked dusky, nape and rump smoky gray and streaked, **wings** brownish **with 2 white bars**; ill-defined supercilium whitish streaked dusky, broken eye-ring buffy-white, **ear-coverts reddish brown** admixed a little dusky, white whisker separated by dusky malar zone from white, slightly streaked throat, sides and flanks brownish, breast grayish white turning white on belly, flanks and vent pale buff, underparts streaked dusky except on belly. **Juv**. much like female, but sides pale buff instead of grayish, with thin streaks. No imm. plumage. Ssp *peruvianus* poorly differentiated (shorter wings, larger bill and more heavily streaked upperparts). In ssp *coracinus* the male is almost black throughout, with insignificant gray streaks on back and rump, narrow white wingbars, and hardly any white on lower belly or vent; female looks like male of other populations, except for paler lore.

Habits: In pairs. Forages on the ground, perches in bushes or on rocks. During nuptial flight male flies 2-3 m up from bushtop, then glides down with vibrating wings, all the while vocalizing. Breeds semi-colonially.

Voice: Song a dry *wreeee iu wreeui*, first syllable drawn out, last short, often followed by a *wrrrr* from the female.

Breeding: Eggs Oct (c Chile), somewhat later (s Chile), Jan (Tucumán); fledgling Dec (La Paz).

Habitat: Bushy slopes or rocky slopes with some brush in arid and semi-arid regions; *coracinus* probably mainly on pumice slopes with *Polylepis* shrub at very high elevations.

Range: 1500-3850 m in the n, *coracinus* around 4000 m, down to sea level in the s. S Cajamarca to Arequipa, Peru, and La Páz, Cochabamba, and Chuquisaca, Bol. (*peruvianus*); altiplano of Bol. w from lake Poopo, Oruro and Uyuni, Potosí to the Chilean border (Sajama and Pampa Olliaga in Oruro, Sacaya in Tarapacá) (*coracinus*); Arica (Putre) in Tarapacá, n Chile, s to Llanquihue, and from Jujuy to Sta Cruz, Arg., 2 rec. from Isla Grande (*fruticeti*). Common, but somewhat local.

NOTE: Hellmayr (1932) synonymized *coracinus*, regarding it as an individual variant. An examination of a larger material (JF) shows that this is a very well marked ssp, that hybridizes with other populations only in a narrow zone on the upper Andean slope of n Chile, and at the n corner of Lake Poopó.

PLUMBEOUS SIERRA-FINCH *Phrygilus unicolor* – Plate LVIII 6a-f

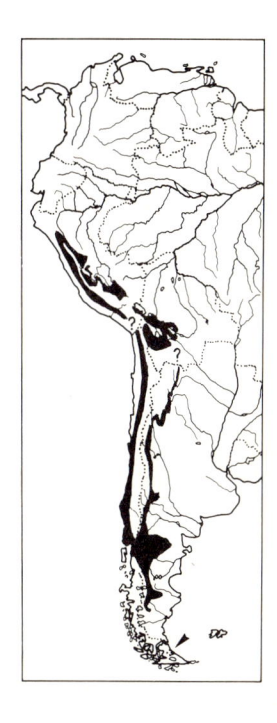

14.5 cm. Bill fairly long. Ssp *geospizopsis*: **male**: above dark **gray**, below light gray, abdomen and edge of primaries pale gray: **Female**: crown and back gray-brown streaked dusky, crown and lower back brownest, rump faintly streaked dusky; wings brownish with 2 faint, very narrow whitish bars. Obsolete supercilium buffy white streaked dusky, eye-ring buffy white, **underparts whitish broadly streaked dusky**. **Juv**. like female, but rump brownish. Sspp generally poorly differentiated, *montosus* and *nivarius* slightly smaller than others, but in *inca* **sexes alike**, gray, female only slightly duller and with traces of dark shafts on crown and sometimes back. **Imm**. like juv., but grayer. **Juv**. like female *geospizopsis*, female slightly browner than male.

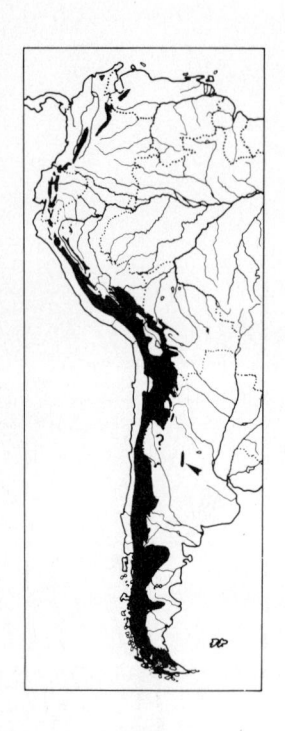

Habits: In pairs or small flocks. Forages for seeds, mostly on the ground, occ. from puddles, or the top of an *Espeletia*. Mainly perches atop rocks or well inside trees, sometimes in bushtops. May roost in rush beds, rocks, and glacier cracks. In some regions shy.

Voice: Song a short snarling *wheeze*, slightly vibrating and liquid, and somewhat accented towards the end. Contact note a faint *zee*. Flight call a timid *chip*.

Breeding: Eggs Nov-Jan (Chile), Jan-Mar (Tucumán); fledglings Sep (Quindío), Mar (Cundinamarca, Puno), Feb (Nariño); juv. Aug (Amazonas), June, July (Huánuco); large gonads July-Oct (E Andes Col.), Mar (C Andes Col.).

Habitat: In n in humid páramos with tall herbs and grassland, often near edge of cloud or elfin forest and on *Espeletia* páramos. In the s more in rocky puna grassland, highland bogs with cushion plants, short matted grass below rocky outcrops, and in open *Polylepis* woodland.

Range: 2700-5300 m, down to sea level in southern Sta Marta mts (*nivarius*); Trujillo and Mérida, Ven. (*montosus*); Páramo de Tamá, n part of E Andes, and from Caldas/Tolima C Andes, Col. s through Ecu. to n Peru (*geospizopsis*, birds from Amazonas intermediate towards *inca*), Peru and n La Paz, Bol. (*inca*), s La Paz and Cochabamba, Bol., to La Rioja, nw Arg. (*tucumanus*); w Córdoba, w Arg. (*cyaneus*); Tarapacá (Arica) to Magallanes (Isla Grande), Chile, and from Mendoza to Sta Cruz in adjacent Arg. (*unicolor*), Tierra del Fuego, Arg. (*ultimus*). Common.

RED-BACKED SIERRA-FINCH *Phrygilus dorsalis* – Plate LVIII 7a-b

14 cm. **Sexes alike**. **Back reddish brown**; crown, headside, wing-coverts and rump gray; flight- and tail-feathers dusky, primaries edged whitish, secondaries edged pale brown, outer rectrices narrowly tipped white. Area below eye grizzled with white, throat white, breast gray, belly whitish. **Juv**. similar, but duller, with warm tinge throughout, esp. on rump. Back with faint dusky streaks, throat white.

Habits: In pairs, small flocks in winter. Forages on the ground. During display male rises to a halt, then descends with vibrating wings. Tame.

Voice: Calls as in White-throated S-f.

Breeding: Eggs Feb (Potosí); fledglings June (Tarapaca); juv. Apr (Antofagasta), May (Tarapacá).

Habitat: 'Tola' heath; 0.5-1 m tall shrubs admixed with columnar cacti; bogs and lakesides in rocky country.

Range: 3350-4500 m. Chuquisaca (Nor Cinti) and Potosí (Karikari mts), Bol., Tarapaca (P. N. Llauca) and Antofagasta, Chile, and Jujuy,

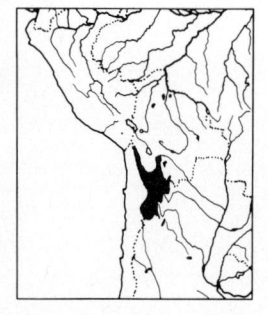

Salta, Catamarca, and Tucumán, nw Arg. Uncommon in n part of range.

NOTE: Owing to apparent interbreeding with White-throated S-f. in Llauca National Park (DS and our own obs.) perhaps best treated as a color phase of that species. The description in Meyer de Schauensee (1970) was based on a misidentified series of juv. Common Diuca-finch.

WHITE-THROATED SIERRA-FINCH
Phrygilus erythronotus – Plate LVIII 8a-b

14 cm (slightly larger than Red-backed S-f.). Sexes alike. Above dark **gray**; wing- and tail-feathers dusky, primaries narrowly edged whitish, secondaries pale brown (disappearing with wear), tail-feathers narrowly white-tipped. In worn plumage shows distinctly brownish remiges. **Throat white**, breast gray (but gray not extending down sides as in Common Diuca-finch), **belly** and vent **whitish**; lores and cheeks grizzled and streaked with white. **Juv.** more or less clearly tinged brown above and on breast, and with obscure dark streaks above, inner remiges edged warm brown, flanks and vent tinged buff.
Habits: In pairs or small groups up to 8. Forages on the ground. During aggressive behavior adopts horizontal posture. Seems to be rather shy.
Voice: Calls include a sharp *wheit* and a shorter *tsip*.
Breeding: Presumably Mar-May.
Habitat: Among rocks, bunchgrass slopes with dwarf shrubs, on lake shores and cushion bogs in barren country.
Range: 3650-4750 m. From s Arequipa and s Puno, s Peru, s to Tarapacá (P. N. Llauca), n Chile, and Oruro (Callipampa) and in Potosí (Kari Kari mts, Pulacayo), w Bol. Uncommon.

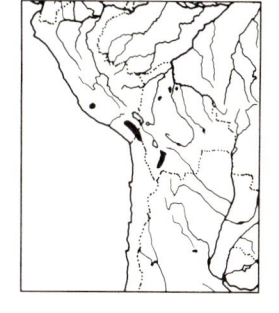

ASH-BREASTED SIERRA-FINCH *Phrygilus plebejus* – Plate LVIII 5a-b

11-11.5 cm (*ocularis* smallest). Sexes alike. Above gray-brown obscurely streaked dusky; ill-defined supercilium gray, eye-ring white, headside dark smoky gray; **below pallid gray**, breast and sides washed smoky gray. Ssp *ocularis* grayer and only faintly streaked above. **Imm.**: above gray-brown streaked dusky, below whitish streaked dark brown on throat, breast, and sides. **Juv.**: above very pale brown streaked blackish, below yellowish white narrowly streaked on breast and sides.
Habits: Usually in (wandering?) flocks, sometimes large, frequently with other finches. Forages on the ground, perches on rocks, cacti, or in bushes.
Voice: Song a fine trill usually followed by 2-3 notes *treedele tzie tzie*, or *treeee tzie tzie tzi*, lasting c. 1 s, or just a series of trills, each c. 0.3 s long and repeated at 1-3 s intervals, intervals sometimes left out between 2 trills. Call from ground or in flight a very fine *tsi*. During chases cracking sounds.

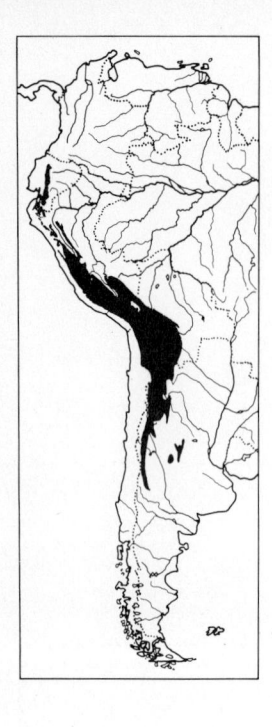

Breeding: Eggs Mar, Apr (Cuzco), July (Arequipa, Cuzco/Puno), June (Puno), Oct (Bol.), Oct, Mar (Arica).
Habitat: Found in a variety of non-forested habitats such as bushy rocky places and cactus country, but most common in sparsely vegetated areas, often areas strongly degraded by grazing and human activity. Roosts in bushes or open *Polylepis* woods and flies to open country during the day.
Range: 2400-4900 m. Ecu. (where also on coast in Huaquillas, El Oro) and extreme n Peru (Huancabamba) (*ocularis*), rest of Peru to Antofagasta, n Chile, and through w Bol. to Mendoza, Arg. (*plebejus*); mts of w Córdoba and San Luis (Sierras de San Luis), Arg. (*naroskyi*, Nores & Yzurieta in *Hornero* N° extra (1979):97). Very common.

BAND-TAILED SIERRA-FINCH *Phrygilus alaudinus* – Plate LVIII 10a-f

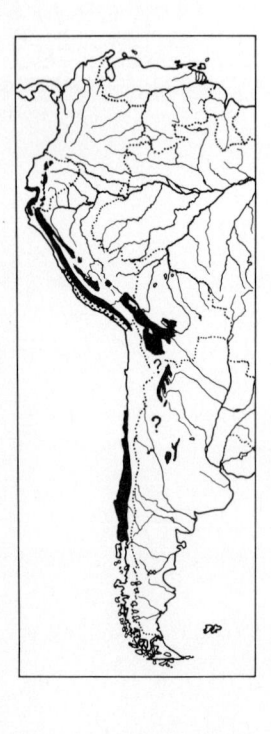

13.5-14.5 cm (*humboldti* smallest, *excelsus* largest). **Bill much longer and more conical than in the smaller Band-tailed Seedeater**. Breeding male with orange-yellow bill and legs, non-breeding with blackish ridge of bill. **Ad. male**: back earthy brown wearing to gray, and streaked black; head, breast, and sides dark gray; belly and vent white. Tail dusky, all but c pair with **white** median area appearing as a **band in spread tail**. Younger male (presumably 2nd year) has gray (instead of dark gray) headside and pale gray breast and sides, feathers of back only basally gray, wearing to a mixture of gray and brown. **Female**: above gray-brown streaked dusky; lesser wing-coverts gray, median and greater coverts narrowly tipped whitish forming 2 small wingbars, the lower smallest; lores and eye-ring whitish, headside gray-brown, underparts whitish streaked dark brown on breast and sides; tailband smaller than in male. Very old females grayer. **Imm. male** like ad. female, but with fewer and narrower streaks on breast and sides, these parts being grayish. **Juv**. like ad. female, but slightly browner above. Sspp poorly differentiated, most hardly valid and probably based on different age classes.
Habits: In pairs, sometimes with other finches. Forages on the ground, perches on rocks, stone fences, or bushes. During nuptial flight male flies in large circles 20-30 m up rising and falling, giving its trill while falling; occ. sings from the ground, bushtop, or telephone-pole.
Voice: Song *chit-chitdedrrrrr-tju tju tju tju*. Call a sharp, abrupt *chip*.
Breeding: Eggs Oct-Dec (Chile); fledglings May (Puno), June (La Paz), Mar (Tucumán); nuptial flight mainly Sep-Nov in years of breeding (coastal lomas of sw Peru).
Habitat: Ssp *excelsus* on level ground with bunchgrass, *Lepidophyllum* or other dense microphyllic bushes and herbs, and much bare soil, sometimes in fields. Other sspp also on slopes with sparse (irregularly abundant) vegetation of herbs, and in deserts with short fog vegetation.
Range: Coastal semi-desert from sw Ecu. to nw Peru (*humboldti*); 2500-3800 m from Ecu. through W Andes of Peru to mts of Tarapacá, n Chile, in c and s Peru almost to sea level (breeding in the 'lomas' in years with well developed vegetation) (*bipartitus*); coast to 1850 m from Atacama to Valdivia, Chile (*alaudinus*); 3000-4100 m or rarely higher or

lower from c Arequipa through Puno, se Peru, and in highlands of Bol., and recently rec. in Cuzco (S Englund & L Brøndal, Walker & Ricalde 1988) (*excelsus*); at somewhat lower elevations from Jujuy to Catamarca and Tucumán, and in San Luis and Córdoba, nw Arg. (*venturii*). Widespread, but common only locally.

CARBONATED SIERRA-FINCH *Phrygilus carbonarius* – Plate LVIII 11a-c

13.5 cm. Bill yellow in breeding male. **Male**: above gray streaked black, forehead unstreaked; lores, headside, and underparts blackish, sides of breast slaty, sides of body gray. In fresh plumage the entire underside is gray, the gray tips wearing off. **Female**: above pale brown streaked fuscous, below whitish narrowly streaked dark brown on breast and faintly on sides and flanks; earcoverts streaked; outer median wing-coverts with narrow, distinct pale edges at tip; outer tail-feathers narrowly edged white. **Juv**: above pale buffy white broadly streaked dark brown; 2 narrow wingbars buffy white; lores and eye-ring whitish, earcoverts streaked whitish and dark brown; below whitish streaked dark brown on breast and faintly on throat; outer tail-feathers narrowly edged white on outer web and on tip.
Habits: In pairs or family groups, during migration in small flocks often with other finches. In nuptial flight holds wings raised. Forages on the ground and in low bushes. Shy.
Voice: No data.
Breeding: No data. Present on the wintering grounds May-Aug.
Habitat: Open country with bushes and herbaceous vegetation; old pastures with tall bunchgrass.
Range: Arg., primarily in lowlands. Breeds in nw Córdoba and from Mendoza, La Pampa, and s Buenos Aires to Chubut. Southern breeders migrate n to Tucumán and Santiago del Estero. Uncommon.

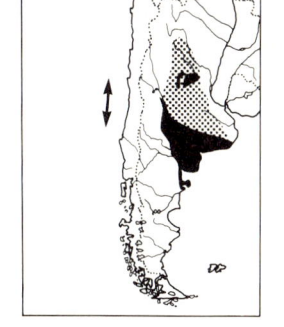

BLACK-THROATED FINCH *Melanodera melanodera* – Plate LVIII 13a-b

14.5 cm. **Male**: crown, headside, back, and sides gray, back and sides turning olive-green with wear; **wings mostly yellow** with dark trailing edge, inner remiges greenish gray; c tail-feathers brownish green, rest yellowish white with dark tips, dark decreasing towards outermost, which is unmarked. Lores and ocular region and **throat black outlined in white** that also continues as a supercilium; breast and belly yellow, flanks and lower belly white. Younger male has greenish gray wings with only edge of outer primaries yellow. **Female**: above pale brown streaked blackish, shoulders golden olive, **outer primaries narrowly edged yellow**, median coverts yellow with dark shafts; 2 outer tail-feathers mostly yellowish white with dark brown tip; throat whitish sometimes with narrow streaks, breast and sides buffy brown narrowly streaked, **belly** whitish admixed with a little pale yellow, **unstreaked**. **Juv**. like female, but streaks coarser, outer tail-feathers only edged yellow, shoulders and belly without olive or yellow.
Habits: In flocks of up to 1000 in winter, otherwise 6-25, except in

early breeding season, when territorial around nest. Forages inconspicuously in grass, picking seeds from ground or pulling down stems of grass, occ. climbing the tussocks. Sings from low rocks.
Voice: Song (*princetoniana*) a series of 4-25 or more, pleasant whistles lasting 4-17 s, and repeated at 2-8 s intervals. The notes alternate irregularly betweeen *wee* and *tuuee*, and either may be slightly rising or falling, also irregularly alternating (BW). In flight a very high-pitched *dzit* and rattling wings (BW). During frequent territorial chases low over the ground gives loud, explosive or buzzing calls (R Woods).
Breeding: Oct-Jan, sometimes double-brooded.
Habitat: Open grassland and pasture, often near settlements, but not attached to man. Seems to prefer a mosaic of short grass and tussocks (R Woods), at Cabo Virgines typically among tussocks with 20-40% of stems left by grazing sheep (BW). Winters at sea coasts.
Range: N Isla Grande (where now extinct?) and Sta Cruz, Arg., vanishing due to overgrazing by sheep, although recently rec. as fairly common at Cabo Virgines (BW) (*princetoniana*). Common on Islas Malvinas (*melanodera*).

YELLOW-BRIDLED FINCH *Melanodera xanthogramma* – Plate LVIII 12a-b

15.5 cm. **Male**: above slate gray wearing to olive, occ. (1st ad. plumage?) also tips olive, esp. on lower back; wings olive, edges of inner remiges and tips of greater coverts gray; outer 3 tail-feathers mostly yellowish white. **Forehead golden-olive**; moustache and short supercilium yellow, area just below eye whitish, lores and throat black; breast and sides like back, c belly yellow, vent whitish. Ssp *barrosi* larger with stouter, longer bill, more extensively yellow on breast, tail markings white. **Female** much like female Black-throated F., but larger, bill thicker at base, shoulders olive-green instead of golden olive, median wing-coverts brownish, outer remiges more narrowly edged yellow, and **entire underparts streaked**. **Juv**. undescribed.
Habits: Alone, in pairs, or family groups. Forages on the ground. Not shy and generally silent, hence difficult to find.
Voice: Song a long shrill note followed after a distinct pause by a shorter one of lower pitch *tweet wheu*. Alarm a short note.
Breeding: Oct-Mar, probably double-brooded.
Habitat: Rocky ridges above tree-line, and high plateaus with sparse high alpine vegetation and patches of snow. Rarely along beaches in winter.
Range: Above 2750 m, in the s down to 1500 m. From Aconcagua, Chile and Neuquén, Arg. to Straits of Magellan (*barrosi*); mountainous s part of Isla Grande along Beagle Channel and outlying isls; in winter most migrate n to n Isla Grande; also on Islas Malvinas, where now possibly extinct due to overgrazing on the poorer higher pastures (R Woods) (*xanthogramma*). Uncommon.

SLATY FINCH *Haplospiza rustica* – Plate LVII 1a-b

12 cm. **Bill rather long and conical**, tail slightly forked. **Male** uniform

slaty-gray. **Female**: above brownish olive, below yellowish gray tinged olive and obscurely streaked dusky, most conspicuously on throat and breast; flanks washed brownish olive, c abdomen whitish. **Imm. male** like female. **Juv**. like female, but browner, c abdomen tinged buff.

Habits: Alone, in pairs, or small groups up to 10. Forages for seeds on the ground under dense cover, perching within dense vegetation. Flocks slowly move on, the last individual currently flying to front of flock (see *Condor* 74(1972):99-101).

Voice: Call a sibilant high-pitched *seep* or *sssp*, louder during aggression. Song faster and higher pitched than in a conebill, a fat complicated burst of high chips, buzzes and trills, sometimes ending in a descending buzzy trill.

Breeding: No data.

Habitat: Within 2 m of the ground. Dense undergrowth of humid forest and cloud forest, notably *Chusquea* bamboo and in adjacent second growth and tall grass. May flourish only in areas with flowering bamboo, and presumably undertakes extensive migrations in search of such habitat.

Range: 1200-3300 m, rarely to 3500 m. Coastal mts of Ven.; Perijá mts; all 3 ranges of Col. s through e and w Ecu. and Peru (on Pacific slope to Cascabamba in Cajamarca) to Sta Cruz, Bol. (*rustica*). 2 sspp from Mexico to Panamá, 1 on Tepuis of Ven. Widespread but rare to uncommon. Possibly erratic throughout, becoming abundant in areas with flowering bamboo.

NOTE: Sometimes placed in the genus *Spodiornis*.

GRAY-CRESTED FINCH *Lophospingus griseocristatus* – Plate LVII 11a-b

14.5 cm. Above gray to **dark gray, crest** of slightly recurved narrow feathers darkest; below slightly paler gray, c lower belly and vent white. Tail blackish, **outer 3 pairs with white terminal half**. **Juv**.: above gray-brown, crest only suggested, wings with 2 buffy bars; supercilium and underparts whitish, throat, breast, sides, and lower belly washed brownish, breast streaked dark brown. White in tail as in ad.

Habits: In pairs or small flocks. Forages for seeds and insects on the ground, perches openly in bushes and trees. Territorial.

Voice: Flight call *kit-kit*. Song a melodious series of whistled notes and trills. Contact call a high-pitched *zit*.

Breeding: Juv./imm. Apr (Sta Cruz); imm. Dec, June (Cochabamba).

Habitat: Arid regions mainly with bare earth below thorny scrub, *Prosopis* trees, and cacti. Sometimes near habitation.

Range: 1500-3100 m. Locally in s La Paz (Tilotilo), Cochabamba s of Tunari range, w Sta Cruz, Chuquisaca, and Tarija, Bol., and Salta, nw Arg. Common.

TANAGER-FINCH *Oreothraupis arremonops* – Plate LXI 14, fairly common locally in mossy undergrowth of humid forest at 1700-2300 m from w Col. to Pichincha, nw Ecu., has been sighted at 2500 m in Col. and at 2400-2500 m in nw Ecu. 18 cm. Chestnut-brown, breast ru-

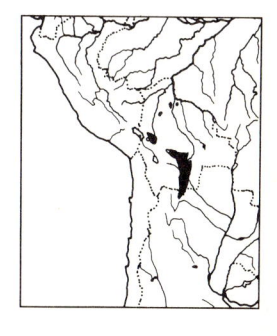

fous chestnut, wing- and tail-feathers blackish; head incl. chin and upper throat black; broad c crown-stripe and supercilium from above eye whitish blue-gray; c belly gray. Bill black. **Juv**. duller. Close pairs, or groups of 3-6, rummage slowly on or near the ground. Shy and inconspicuous. May be attracted to an imitation of its call, which is a soft whistle repeated at 2-3 s intervals.

MOUSTACHED BRUSH-FINCH *Atlapetes albofrenatus* – Plate LXI 1

15 cm. Ssp *meridae*: **crown and nape orange-rufous**, very narrow forehead and headside black, back, wings and tail citrine. **Moustache white**, submalar streak blackish, underparts yellow, sides and flanks washed citrine. Ssp *albofrenatus* with faint or no submalar streak, c throat white, crown darker and forehead more extensively black. **Juv**. above brownish olive, headside dusky, throat and belly partly yellow, breast washed olive and faintly streaked. - Alone, in pairs, or family groups, sometimes in mixed-species flocks. Forages in dense vegetation, occ. perching exposed. Very fast. Not shy. Call (Col.) a thin high-pitched *eeespe*, occ. trebled. Song (Ven.) *czeet, czeet, czeet, czeet, tsu-tsu-tsu-tsu-tsu*, last notes form a rattling series. Fledglings May, June (Mérida), Aug (Perijá); imm. Jan (Boyacá).

Mainly low, occ. 10 m up. Cloud forest and humid forest undergrowth with ferns and bamboo; moist oak woodland; second growth; dry, thorny scrub. 2100-2500 m from c Mérida to e Táchira, Ven. (*meridae*); 1600-2500 m, rarely down to 1000 m from N.d.Santander to Cundinamarca, E Andes of Col. (*albofrenatus*). Fairly common in Col.

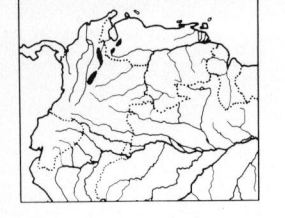

PALE-NAPED BRUSH-FINCH *Atlapetes pallidinucha* – Plate LXI 2a-b

17.5 cm. Above dark gray, **forehead orange-yellow turning whitish on broad c crown-stripe**, rest of crown and headside black; below yellow to olive-yellow with faint blackish submalar streak, sides and vent olive. Ssp *papallactae* somewhat darker. **Juv**. (*papallactae*) fuscous above with dark forehead and buffy clay-colored crown-stripe, underparts streaked olive and dark brown, throat yellowish, upper throat unstreaked.
Habits: In pairs or family-groups, often in mixed-species flocks, where sometimes with Rufous-naped, Slaty, Stripe-headed, and White-rimmed B-f.s. Forages for insects and berries within the shrubbery, sings from near top of bush. Occ. perches in the open. Not shy.
Voice: Contact note a fine *tsit*. Territorial song mostly given at dawn *tsie...tsie weu, tsi...tsie weu tsie weu, ti...ti wee tsits we weee* or *tsie...tsie tsieu*, 1st notes alike or descending, the *tsiu* sharply rising. Given in duet. Advertising song *wheet-tew-tew-tew*.
Breeding: Fledglings June (Cundinamarca), Mar (e Cauca), Nov (w Pichincha); juv./ad. June (Cundinamarca), Jan, May (Cauca), Apr (Chimborazo), June (se Ecu.); large gonads Feb-Sep (C and E Andes Col.).
Habitat: 0.5-1.5 m up in humid montane shrub at outer edge of stun-

ted cloud forest; also in interior forest with preference for bamboo thickets.
Range: 1700-3800 m, mainly 2000-3100 m. Páramo de Tamá and from N.d.Santander to Cundinamarca, E Andes of Col. (*pallidinucha*); from se Antioquia, C Andes of Col. s through e Ecu. to Piura (Chinguela), n Peru, and at Matos on interandean plateau in Chimborazo, Ecu. (rec. from below Nono, nw Ecu., probably erroneous (*papallactae*). Locally common.

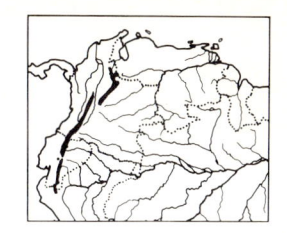

RUFOUS-NAPED BRUSH-FINCH *Atlapetes rufinucha* – Plate LXI 6a-g

15.5-17 cm (*carrikeri* smallest). **Above dark olive to black, crown rufous to chestnut, headsides black, below yellow with or without yellow moustache and black submalar streak; wings with or without white patch**. The geographic variation is complex: Crown chestnut (most sspp), dark chestnut (*simplex, caucae*), light chestnut paler on nape (*chugurensis*), rufous (*comptus*), or rufous very pale on nape (*baroni*); forehead like crown (most sspp), broadly black (*phelpsi*), or narrowly black (*melanolaemus, rufinucha, carrikeri*). Back, wings, and tail black (*melanolaemus, rufinucha*), gray to very dark gray (most sspp), moderately tinged olivaceous (*simplex, carrikeri*), slightly so (*phelpsi*), or strongly so (*elaeoprorus*). White speculum on primaries large (*elaeoprorus, caucae*), medium (*latinuchus*), small to absent (*simplex*), or always absent (other sspp). Yellow supraloral spot large (*comptus, baroni, rufinucha*), small (*elaeoprorus, caucae*), small to very small (*simplex*), vestigial or absent (*latinuchus*), or always absent (other sspp). Black submalar streak broad (*phelpsi*), distinct (*comptus*), moderate (*spodionotus, baroni, rufinucha, carrikeri*), faint (*elaeoprorus, simplex, caucae*), or very faint (*latinuchus, chugurensis*); in *melanolaemus* the black headside is continuous with the submalar area, and sometimes the entire throat is black; *phelpsi* has black chin. Below bright yellow, lower flanks and vent olive (olive most extensive in *melanolaemus, rufinucha*, and *carrikeri*). **Juv.** reminiscent of resp. ad., but brownish above and yellowish to brownish buff below, darkest on breast, underparts streaked dark brown at least in *chugurensis, latinucha, comptus, spodionotus*, and sometimes in *caucae, baroni, rufinucha*, and *carrikeri*.
Habits: In pairs or groups of 5-6, often in mixed-species flocks e.g. with Pale-naped, Slaty, Stripe-headed, and White-rimmed B-f.s. Restlessly perch-gleans and rarely makes short sallies, occ. searching dead leaves for seeds and insects within the vegetation. Sometimes in the open along vines to middle strata of trees; in Celica mts, s Ecu. often hops up vertical trunks of large trees, almost to canopy.
Voice: Contact note *tick tick*. Calls include a liquid, slightly descending *trrrr* and a long sharp *zeet*. Song of fine notes followed by whistles, the last falling *tsi-titititi tee teer tjiee*. Also a whistled *weee te weee* or *weee te weee weee*.
Breeding: Fledglings Apr, May (Cauca), May (Carchi), Aug (Loja), Apr (Cajamarca), Feb (Amazonas), June (La Libertad), Dec (Cuzco, Puno, La Paz), Feb (Cochabamba), Dec (Sta Cruz); large gonads Jan-June (Perijá and C Andes Col.), Apr (Cajamarca).
Habitat: 0.5-2, occ. 6 m up (in Celica mts, sw Ecu. up to canopy of tall trees) within thick undergrowth of cloud forest, dense humid *Polylepis*

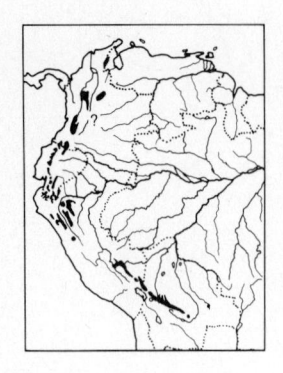

forest, humid forest clearings, second growth, humid montane shrubbery, sometimes mixed with bamboo. Often near streams, occ. in semi-humid shrub.
Range: (600) 1100-3800 m, mainly 2000-3000 m, highest in Ecu., se Peru, and Bol. Perijá mts (*phelpsi*); Antioquia in C Andes of Col. (*elaeoprorus*); E Andes of Col. in Boyacá (*simplex*); around Cauca valley in Valle and Cauca Col. (*caucae*); Nariño, s Col., and n and w Ecu. (*spodionotus*), intergrading in wc Ecu., and continuing s from Cañar to Piura, nw Peru (*comptus*); e slope from nw Morona-Santiago, se Ecu., to Amazonas, ne Peru (*latinuchus*); Pacific slope of c Cajamarca (Llamas-Huambos, Chugur), nw Peru (*chugurensis*); upper Marañón valley in s Cajamarca (and on Pacific slope at Sunchubamba and Huacraruco)), La Libertad, and Rurichinchay on e slope of Cord. Blanca in Ancash, Peru (*baroni*). Cuzco in se Peru to extreme n La Paz, Bol. (*melanolaemus*); La Paz and Cochabamba, Bol. (*rufinucha*); Samaipata in Sta Cruz, Bol. (*carrikeri*). Common.

NOTE: Forms a species group (zoogeographic species) with Santa Marta, Dusky-headed, and Tricolored B-f.s.

SANTA MARTA BRUSH-FINCH
Atlapetes melanocephalus – Plate LXIII 33

17 cm. **Head incl. upper throat black**, earcoverts silvery gray. Back, wings, and tail dark gray, underparts bright yellow, flanks and vent olivaceous.
Habits: In pairs or groups of 3-7, nuclear and usually the most numerous in mixed-species flocks. Very active and conspicuous. Forages in thickets.
Voice: Song loud and cascading, of chipping and twittering notes.
Breeding: Fledglings Nov-June; large gonads Jan-Apr.
Habitat: 1-10 m up in humid shrubby forest borders and overgrown bushy pastures. Not inside mature forest.
Range: N Col. at (600) 1500-3200 m in Sta Marta mts. Very common.

NOTE: Closely related to Rufous-naped B-f.

TRICOLORED BRUSH-FINCH *Atlapetes tricolor* – Plate LXI 5

16.5 cm. Much like Rufous-naped B-f., but larger, and **crown rich tawny gold**. Ssp *crassus*: **bill long and stout**; crown rich tawny gold; back, wings, and tail dusky olive, headside blackish, underparts yellow, sides and slight wash on breast olive. Ssp *tricolor* similar, but with a smaller bill, paler back (olive-citrine to dark olive), yellow to ochraceous orange crown, and much less extensively olive underparts. **Juv**.: crown dull dark rufous, back dark brown, underparts uniform brown (sometimes streaked?).
Habits: Alone or in pairs, often independent of mixed-species flocks. Forages within dense shrubbery, occ. surveying from bushtop. Shy except when breeding.

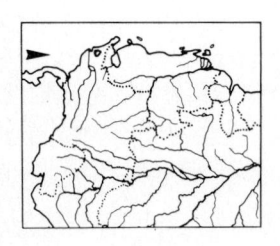

Voice: Song a long series of slowly accelerating notes finally terminated with a dry trill *tzit-tzit-tzit-tju-tju-tju-tju-tjrrrr*, sometimes initiated with a falling note. Other vocalizations include a squeak followed by 2 chirps and a *sweet churr* or *eeeu-tsit-tsit-tsueet*.
Breeding: Eggs Apr (Valle); fledglings June (Cauca), June, July (nw Ecu.), July (Huánuco); nest-building Nov, Feb, May (Valle).
Habitat: 0-2 m, rarely to 10 m up, in humid forest undergrowth and adjacent second growth.
Range: At 300-2000 m (below Rufous-naped B-f.) on w slope in Caldas, Valle, and Nariño (and e slope of W Andes in Valle and Cauca?), Col., and in Esmeraldas, Pichincha, and El Oro (La Chonta), w Ecu. (*crassus*). At 1525-2650 m from e La Libertad to Cuzco w of Urubamba valley, Peru (*tricolor*). Uncommon, locally fairly common.

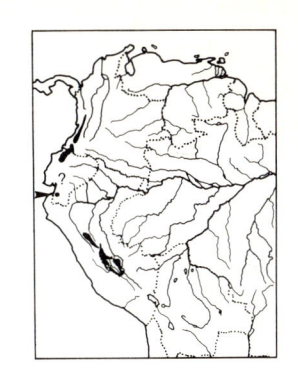

DUSKY-HEADED BRUSH-FINCH *Atlapetes fuscoolivaceus* is restricted to 1600-2400 m at head of Magdalena valley Col., where common in bushy overgrown pastures, woodland, second growth, and shrubby forest borders; less common within forest, where usually near tangled openings. 17 cm. Crown, headside, and indistinct submalar streak fuscous, rest of upperparts dark dull olive, wings and tail darker; below bright yellow, flanks washed olive.

OCHRE-BREASTED BRUSH-FINCH
Atlapetes semirufus – Plate LXI 7a-b

15 cm. Ssp *semirufus*: **head, throat, breast, and sides of upper belly rufous**; back, wings, and tail dark olive, c belly yellow, lower sides and flanks olive. Ssp *majusculus* slightly larger and much paler, throat tinged with yellowish; *zimmeri* like *semirufus*, but dark olive-gray above; *albigula* like *zimmeri*, but chin and c throat white; unlike all other forms has an inconspicuous dusky eye-ring, broken posteriorly; *benedettii* like *semirufus*, but back brighter and greener. **Juv.**: headside and upperparts dusky olive, throat ochraceous buff, submalar streak, breast, and sides dusky olive, c belly pale yellow.
Habits: In close pairs or groups up to 8, only rarely in mixed-species flocks. Feeds almost exclusively on the forest floor, knocking leaves and debris aside with its beak in search of seeds and insects.
Voice: Song monotonous, composed of some metallic notes, and sometimes terminated with a short trill. Call sharp and high-pitched.
Breeding: Juv. Sep (Ven.), Jan (e Boyacá); juv./ad. Oct (Cundinamarca); large gonads Nov (N.d.Santander).
Habitat: 0-1 m in dense underbrush of the better lighted parts of humid forest and cloud forest.
Range: 1200-3050 m. Coastal Ven. at 600-2100 m (*denisei*); San Luis mts, Falcón and in Lara and n Trujillo, Ven. (*benedettii*); Seboruco at 1300 m, n Táchira, Ven. (*albigula*); Trujillo and from s Táchira, Ven., s along e slope of E Andes to extreme ne Boyacá, Col. (*zimmeri*), w slope of E Andes in n Boyacá (*majusculus*), 3000-3500 m on e slope of E Andes of Col. in s Boyacá and Cundinamarca (*semirufus*). Previously common, now rare in Cundinamarca, where a small population was found at Une in 1967.

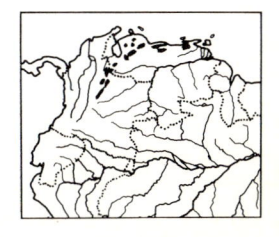

NOTE: May form a species group (zoogeographic species) with Fulvous-headed and Yellow-striped B-f.s.

FULVOUS-HEADED BRUSH-FINCH *Atlapetes fulviceps* – Plate LXI 11a-b

15 cm. Tail hardly graduated. **Headside, crown, and nape rufous**; back, wings, and tail dark olive. Supraloral spot and moustache yellow, submalar streak rufous (female) or blackish (male), below yellow, breast and sides washed with olive. **Juv**. undescribed.
Habits: In close pairs. Forages for seeds and insects within dense shrubbery, along moss-covered stems and thick branches, sometimes climbing head down, rarely on the ground; occ. flycatches. Jerks tail when alarmed.
Voice: Alarm *pi ziue*. Contact a repeated, thin *zit*.
Breeding: Eggs Jan (Jujuy).
Habitat: 0-3 m up. Semi-humid regions. Open spaces with second growth and clumps of trees in wooded valleys, often near streams, sometimes ascending to *Polylepis* woodlands.
Range: 1850-3150 m, in winter down to 400 m. La Paz, Cochabamba, W Sta Cruz, Chuquisaca, and Tarija, Bol., and Jujuy, Salta, and Tucumán, nw Arg. Locally fairly common.

YELLOW-STRIPED BRUSH-FINCH *Atlapetes citrinellus* – Plate LXI 10

16 cm. Above dark citrine, **broad supercilium yellow**, broken eyering yellowish white, most of headside black, earcoverts with yellow spot and shafts, moustache yellow, submalar streak black, below yellow, breast and sides citrine. **Juv**. undescribed.
Habits: In pairs or family-groups, sometimes in mixed-species flocks with Fulvous-headed B-f. Not shy.
Voice: Call *tsip*.
Breeding: Eggs Nov (Tucumán).
Habitat: Hedgerows, dense scrub, and forest undergrowth, coniferous woods, burnt forest, dense ferns in shaded humid ravines.
Range: nw Arg. at 1000-3100 m in Jujuy, Salta, and Tucumán (presumably also in adjacent Catamarca); doubtfully rec. from nw Parag. Fairly common in Tucumán.

RUFOUS-EARED BRUSH-FINCH *Atlapetes rufigenis* – Plate LXI 12a-b

17.5 cm. Crown, nape, and **headside russet**, supraloral 'horn' and moustache white. Back, wings, and tail dark smoky-gray. Submalar streak blackish, throat white, rest of underparts whitish obscurely flammulated gray on breast and sides, flanks brownish gray. Ssp *forbesi* purer gray above, and with black lower forehead, lores, and orbital region. **Juv**. *rufigenis* with gray-brown crown and all the gray parts washed brown; lores, supraloral spot, and throat as in ad., breast, and sides

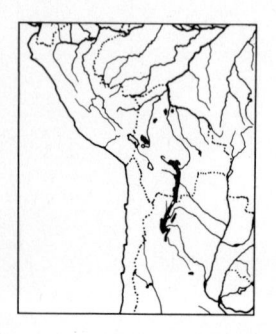

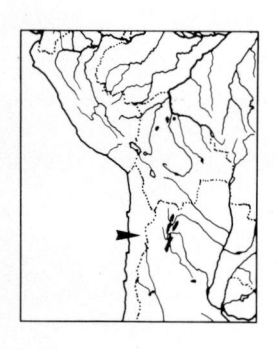

dark brown, c underparts below breast whitish; juv. *forbesi* with crown deep sooty brown, headside black, reminiscent of head of Rusty-bellied B-f.
Habits: In pairs, not in mixed-species flocks. Forages within dense shrubbery, along moss-covered branches both high and low, sometimes on the ground.
Voice: Contact *tzi*, not esp. faint. Song a variation on 4 different notes given every 4-6 s in different themes of 2-3 notes, each phrase 1-2 s long and repeated 1-5 times, *chit tet teeuit, chit teeuit, chi tui, chi tui teeuit*, etc.
Breeding: Fledglings May (Apurímac); juv./ad. June (La Libertad), Feb, Mar (Ancash).
Habitat: 0.5-1.5 m, rarely 5 m up. Dense bushy dry slopes and tangled undergrowth in dry and semi-humid *Podocarpus* woodland, in dense parts of *Polylepis/Gynoxys* woodland (where occ. forages along more open, thick, moss-covered branches), as well as dense, varied, fairly humid treeline shrubbery.
Range: Peru at 2750-4100 m. In cord. Blanca in Ancash and tributaries of upper Marañón from s Cajamarca to e Ancash and Huánuco (Cullcui) (*rufigenis*); Huancavelica (Pillala), Apurímac (Pomayaco, Bosque Ampay), Cuzco (Peñas, above Urubamba) and Puno (sight rec. at Limbani) (*forbesi*). Locally fairly common.

NOTE: It has been suggested that this species forms a species group (zoogeographic species) with White-winged, Slaty, and Rusty-bellied B-f.s. There is an apparent hybrid with Rusty-bellied B-f. in BMNH.

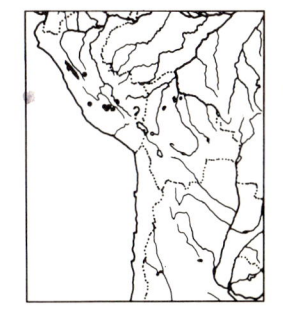

WHITE-WINGED BRUSH-FINCH *Atlapetes leucopterus*
– Plate LXI 8a-c

14 cm. Ssp *dresseri*: above drab-gray, **wings with white speculum**; crown rufous; forehead, sides of crown, neck, and thin and short submalar streak black, earcoverts dark brown, broad 'horns' in forehead, broad eye-ring, lores and area below eye white; below dirty white, flanks gray-brown, belly and vent tinged buff. **Frequently shows partial albinism on the head**, that rarely may be almost wholly white. Ssp *leucopterus* similar, but bill larger, forehead less extensively black, back darker, eye-ring absent, supraloral 'horns' smaller, earcoverts blacker, submalar streak less defined, belly and vent without buff tinge. Ssp *paynteri* like *dresseri*, but crown-patch nearly white, eye-ring lacking; back, wings, tail, and flanks pure gray. **Juv**. with a few faint brown spots or streaks on breast and anterior sides.
Habits: In pairs or small flocks. Restlessly forages low in bushes and undergrowth.
Voice: No data.
Breeding: Juv. June (sw Loja).
Habitat: Very dry, bush-covered hillsides with scattered *Bombax* trees, in nw Ecu. also in gardens and into shrubbery at edge of humid forest and cloud forest; in Condor mts in brushy edges of wet cloud forest.
Range: 600-2900 m. Interandean plateau of Ecu. s to Chimbo valley, and w slope in Imbabura (Apuela) and Pichincha (Mindo rd) (*leucopterus*); 700-2550 m on w slope from El Oro and Loja, sw Ecu., through Tumbes and Lambayeque to Cajamarca, nw Peru (*dresseri*); 1700-2200

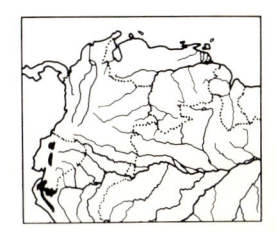

m in Condor mts and Huancabamba region Piura/n Cajamarca, n Peru (*paynteri*, Fitzpatrick in *Auk* 97(1980):883-7). Uncommon to fairly common.

SLATY BRUSH-FINCH *Atlapetes schistaceus* – Plates LXI 13a-c, LXIV 33

16 cm. **Very uniform dark gray with rufous crown, black headside with white on lore, and pale moustache bordered below by black submalar streak**. Ssp *castaneifrons*: above dark gray, crown rufous, headside, and submalar streak black, supraloral 'horn' and moustache white, throat white grading to gray on rest of underparts, extreme c belly pale gray; *tamae* similar, but crown darker, frontal edge slightly mottled with blackish, back blacker, supraloral spot obsolete, moustache less purely white or even grayish, throat shaded or mottled with gray, underparts darker gray; *fumidus* like *castaneifrons*, but slightly paler above and darker below; c belly hardly paler than rest of underparts; *schistaceus* like *castaneifrons*, but wing-speculum and area behind earcoverts white, supraloral spot larger, extreme forehead sometimes mottled with blackish. Ssp *taczanowskii* like *schistaceus*, but without wing-speculum and with black forehead; *canigenis* much like *tamae*, but darker, incl. on throat, and without white moustache; remiges edged grayish below (white in all other forms). **Juv**. washed brown, crown paler than in ad., underparts below throat obscurely streaked.
Habits: Alone or in pairs, usually in mixed-species flocks, sometimes with Pale-naped, Rufous-naped, Stripe-headed, and White-rimmed B-f.s. Rapidly forages along vines and twigs from undergrowth to top of small trees. Restless and shy.
Voice: Contact a fine *tzit-tzit-tzit*. Alarm a fine *krryt krryt*. Song an energetic series of high-pitched squeaks and trills, usually ending in a distinctive *chewy-chewy-chewy* or *t'chew, t'chew, t'chew*.
Breeding: Fledglings Aug (Ven), Apr (Cundinamarca), Apr, June, Aug (Cauca); juv./ad. July (Mérida), Mar (Caldas), June, July (Huánuco); large gonads Nov (Mérida), Apr-Sep (n Col.), Nov (Cundinamarca).
Habitat: 0-5 m up in undergrowth of humid forest and cloud forest and adjacent second growth; *Espeletia* páramos with elfin forest patches. In Ven. also in dense humid *Polylepis* forest.
Range: 1800-3800 m, mainly 2500-3000 m. Perijá mts (*fumidus*); Trujillo to ne Táchira, Ven. (*castaneifrons*), Páramo de Tamá (*tamae*); all 3 ranges of Col., nw and e Ecu. (in e s to Morona-Santiago) (*schistaceus*). From Huánuco to Junín, c Peru (*taczanowskii*); Cuzco, se Peru (*canigenis*). Common in Col., uncommon in Ecu., fairly common in Peru.

RUSTY-BELLIED BRUSH-FINCH *Atlapetes nationi* – Plate LXI 9a-d

16 cm. **Ssp *seebohmi***: above dark brownish gray, crown dark chestnut; forehead, sides of crown and neck, and short submalar streak black; earcoverts blackish gray, supraloral spot whitish; throat and belly white, breast and sides gray. Ssp *simonsi* similar, but crown somewhat

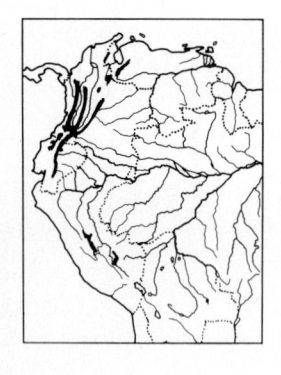

lighter and without black on forehead or gray on breast; *ceciliae* like *simonsi*, but crown even lighter (rufous), breast gray, no supraloral spot. **Ssp *nationi*** like *seebohmi*, but crown dark brown grading to blackish on forehead, crown often with scattered white feathers, no supraloral spot, and short black malars fusing on chin; lower **belly and vent orange buff**. Ssp *brunneiceps* like *nationi*, but much paler and browner above, crown lighter and more reddish (in n Arequipa head often with many white feathers), throat tinged buff. **Juv.** like ad., but washed brown above, throat and c belly buff, breast and sides brownish streaked buff.
Habits: In close pairs, sometimes small flocks. Relatively slow and deliberate. Forages within bushes and trees and rarely on the ground, occ. perching on rocks.
Voice: Sometimes noisy. Alarm *dee dee tee tee*.... Call a very high-pitched, repeated *zee*. Song *tschia-tschia-ziay-ziay-ziayziay*, sometimes terminated with a trill.
Breeding: Juv./ad. June (Lima).
Habitat: (0) 1-2 m up. Dry bushy slopes and *Polylepis* woodland and scrub.
Range: 1850-4300 m. Celica, w Loja, s Ecu. (*celicae*, based on a possibly aberrant *simonsi* specimen), e and c Loja, sw Ecu., and Porculla valley, Lambayeque, nw Peru (*simonsi*); Pacific slope from La Libertad to Ancash, Peru (*seebohmi*); Lima, Peru (*nationi*), Pacific slope from Ica to s Arequipa, Peru (*brunneiceps*). Fairly common in Lima and n Arequipa, otherwise uncommon, rare in s Arequipa.

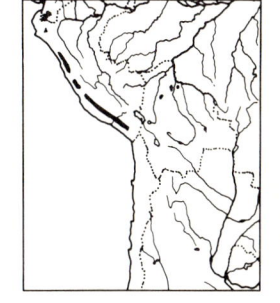

NOTE: Sspp *celicae*, *simonsi* and *seebohmi* are often treated as a separate species **BAY-CROWNED BRUSH-FINCH (*Atlapetes seebohmi*)**.

CHESTNUT-CAPPED BRUSH-FINCH
Atlapetes brunneinucha – Plate LXI 3a-b

18.5 cm. Ssp *frontalis*: forehead and headside black, **forehead with 3 white spots; crown rufous chestnut with narrow golden sides**; back, wings, and tail dark olive green. Throat white, **breastband black**; sides, flanks, and vent dark gray admixed with dark olive, belly white. Ssp *inornatus* smaller, breastband absent or only suggested laterally.
Juv.: heavily washed olive-brown throughout, top of head Walnut brown to more yellowish brown laterally, but without white spots in forehead; sides of head black; throat and c underparts usually somewhat streaked with dull yellow.
Habits: In close pairs or family-groups, sometimes joins (but never follows) mixed-species flocks. Slowly rummages among debris on or close to the ground. Frequently puffs white throat and sometimes nervously raises and lowers tail. Not shy.
Voice: Call a very high-pitched *spee* or *speee-tst*. Song 3-4 thin squeaky note (2nd often lowest) followed by a thin trill *tseee, tep, wee-teeeeeeee*.
Breeding: Probably throughout year in Col., Apr-June in Ven.; fledglings May, June (Ven.), Oct, Nov (Santander), Dec (Cundinamarca), Sep (Quindío), Apr (Cauca), June (Valle, w Nariño), Sep, Jan (ne Ecu.), Mar (Amazonas); juv./ad. Dec (Cauca); large gonads Mar-Aug (n Col.), Apr (se Ecu.); nest Mar, July, Aug (Valle).

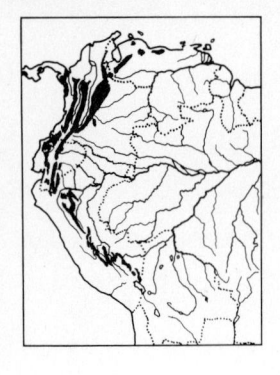

Habitat: On the ground and up to 1 m in the darkest, moist part of the undergrowth, preferring tangles of decomposing bamboo and other vegetation in tall humid forest.
Range: 800-3400 m, but mainly below Stripe-headed B-f., at 1000-2000 m, except in coastal Ven., where the opposite is the case (Chestnut-capped at 900-2400 m and Stripe-headed at 700-900 m). Mts of extreme e Panamá, Perijá mts, coastal mts and Andes from s Lara, Ven., s through all 3 ranges of Col., e and w Ecu. (except range of *inornatus*) to Cuzco se Peru (according to GG to Bol. border in Puno) (*frontalis*); vicinity of Chimbo and Chanchan rivers, cw Ecu. (*inornatus*). 6 sspp from Mexico to Panamá, 1 in Ven. Uncommon in temp. zone, fairly common at lower elevations.

STRIPE-HEADED BRUSH-FINCH *Atlapetes torquatus* – Plates LXI 15a-e and LXIII 32

19 cm. Sspp *assimilis* and *nigrifrons*: **crown black, supercilium and median crown-stripe white above bill and turning gray posteriorly**; back, wings and rump golden olive-green, headside and tail black. Below white, sides of breast dark gray, sides of body, flanks, and vent brownish olive. Sspp *basilicus*, *poliophrys*, *perijanus*, and *larensis* similar, but with black breastband and white supercilium not reaching bill; *fimbriatus* like *torquatus*, but supercilium almost reaches bill, tail-feathers edged olive, feathers of breastband margined with whitish; *borellii* like *fimbriatus*, but without breastband. **Juv**. very dark olive-brown, c underparts sometimes streaked yellowish white, head-pattern like ad., but duller and washed brown.
Habits: As for Chestnut-capped B-f. Sometimes in mixed-species flocks with Pale-naped, Rufous-naped, Slaty, or White-rimmed B-f.s.
Voice: Song high-pitched *tziiie-tzie tz ieeee wee tzee tzee wee tzeee tziiu trrrrrr tziee wee tzi...*, song of female weaker than that of male. Call a thin penetrating *tsit* or *tsint*. Also a vibrating *chirrr*.
Breeding: Fledglings Mar, July (Sta Marta), June (Cauca), Aug (nw Ecu.), July (El Oro), Sep (Loja), Aug (Amazonas), June (Huánuco), Nov (Puno), Feb (Tarija); large gonads Feb-Aug (Col.).
Habitat: As for Chestnut-capped B-f., but more often in bamboo and also in montane shrub.
Range: 1650-3300 m, generally above Chestnut-capped B-f., locally lower, esp. where the other species is absent. 600-2800 m in Sta Marta mts (*basilicus*); 700-1800 m in Perijá mts and e slope of E Andes in N.d.Santander (*perijanus*); w Lara to Táchira, Ven. (*larensis*), 1500-3600 m, mainly in temp. zone of all 3 ranges of Col., nw and e Ecu. to Cajamarca and Utcubamba drainage, Amazonas, Peru (*assimilis*); El Oro and Loja, sw Ecu, and Piura to La Libertad, nw Peru (*nigrifrons*); 1800-3650 m from Huánuco to Cuzco, Peru (*poliophrys*); 1650-3100 m in La Paz and w Cochabamba, Bol. (*torquatus*); 700-3050 m in e Cochabamba, w Sta Cruz, and Chuquisaca, Bol. (*fimbriatus*), 400-1200 m from Chuquisaca, Bol., to Jujuy and Salta, nw Arg. (*borellii*). 1 ssp in Costa Rica and Panamá, 2 in coastal Ven. Fairly common.

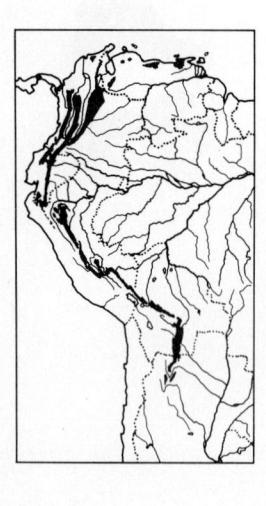

WHITE-RIMMED BRUSH-FINCH *Atlapetes leucopis* – Plate LXI 4

19 cm. Headside, sides of crown, back, wings and tail dull black, crown chestnut, **eye-ring white continuing as short line behind eye**; throat gray mottled with blackish, chin and upper throat sometimes dark fulvous; rest of underparts olive, breast more or less mottled with blackish. **Juv**. browner throughout.
Habits: Alone or in pairs, sometimes in mixed-species flocks where it may occur with Pale-naped, Rufous-naped, Slaty, and Stripe-headed B-f. Shy. Rapidly moves through the vegetation near the ground.
Voice: Song a soft chipping warble; begins with *twoo-twoo...* and ends with 4-5 musical chips.
Breeding: Juv./ad. Dec (e Nariño).
Habitat: 0.5-2 m up in dense understory admixed with some bamboo in cloud forest on steep slopes.
Range: 2350-3200 m. Known from C Andes in Huila at head of Magdalena valley, w Putumayo and se Nariño, Col., w Imbabura, nw Ecu., nw Morona-Santiago and adjacent Azuay, se Ecu. Rare and local.

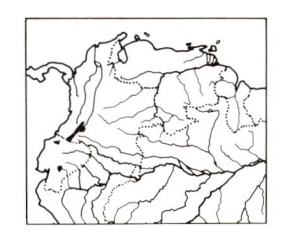

SAFFRON-BILLED SPARROW *Arremon flavirostris* – drawn adjacent to this text, inhabits interior e and s Brazil to e Bol. and nw Arg. s to Corrientes and n Catamarca, nesting at least to 2000 m in Jujuy and Tucumán (*dorbignii*). 14.5 cm. **Bill yellow with black ridge**. Crown and headside black, **long supercilium white**, median crown-stripe gray, nape dark gray, back, wings, and rump bright olive-yellow, lesser wing-coverts and bend of wing bright yellow, tail dusky. Below white, breastband black, sides of breast and body gray in male, pale brownish in female. **Juv**.: crown dark gray-brown with dull olive median stripe, back, wings and tail dull olive-green, shoulders brighter; throat and c belly whitish, rest of underparts dark olive gray-brown. Alone or in pairs forages on or near the ground in semi-humid shrubbery. Voice very high-pitched.

GRASSHOPPER SPARROW *Ammodramus savannarum* – Plate LVII 7

10 cm. Tail fairly short, feathers narrow and pointed, bill fairly large and conical, in profile continuous with flat crown. Crown blackish with whitish c stripe, **supercilium** dark **yellow above lore**, whitish posteriorly. Above grayish streaked dusky, bend of wing and wash on lesser wing-coverts yellow, patch at base of secondaries brownish; below pale buff, c belly whitish, flanks buffy white. **Juv**. has hindcrown and nape whitish streaked dusky, browner wings and dusky-streaked throat and breast. Postjuvenal molt when 4 weeks old includes rectrices. Some juv. males have a dull yellow superciliary spot, this spot always present in 1st winter of both sexes. Usually alone. Walks and runs in search of seeds, frequently flicking wings and tail. When flushed flies with dipping flight low over the ground, soon diving into cover in a tussock or behind a rock. Sings from stone, boulder, or exposed branch, first stretching neck, then raising head to sing. Between songs holds body hori-

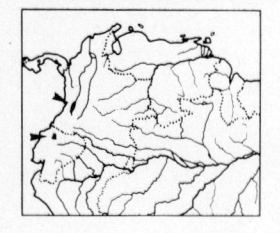

zontal and raises tail, then looks about in characteristic manner while flapping wings. Changes song-perch with fluttering flight. Song a buzzy, insect-like, protracted *tszzzzzz*, even or slightly quavering, sometimes twittering and usually preceded by a *tsick* or *tup tip*.

Cultivated land, esp. fields with long grass. 1000 m in Valle in the Cauca valley, Col., and 2850 m near Cayambe (fledgling) and 'Quito' (ad.), n Ecu. (*caucae*). 10 sspp in N and C Am. and the West Indies. Uncommon or rare.

SLENDER-BILLED FINCH *Xenospingus concolor* inhabits riparian thickets on the Pacific slope from Lima, Peru, to n Antofagasta, Chile, mainly low, but in Arequipa possibly to 1980 m, in Antofagasta to 2285 m, and a sighting has been reported at 3500 m in Putre in Tarapacá, n Chile. 17 cm. **Slender bill and legs yellow. Blue gray**, face blackish, paler below, c belly whitish. **Imm.** and **juv.** are streaked throughout. Alone or in pairs. Actively forages for insects and rarely seeds within dense thickets, only rarely perching in the open. Shy. Often cocks tail.

GRAY-WINGED INCA-FINCH *Incaspiza ortizi* – Plate LIX 11a-b

15 cm. Long bill orange, legs yellow or yellow-orange. Face black, slightly less extensive in female. **Above gray** tinged fuscous and faintly streaked dark brown, wings grayer; below gray (faintly tinged brown in female) grading to white (or whitish in female) on belly, vent tinged buff. Outer tail-feathers white with dark brown along terminal half of outer web and extreme base of inner web, next with less white, 3rd only with white spot at tip of inner web. **Imm**. like ad., but back dark brown streaked dusky, broad superciliary region and headside gray streaked blackish, underparts grayish white tinged buffy on lower sides and flanks and narrowly streaked dark brown on breast. **Juv**.: above gray-brown streaked blackish, below yellowish white streaked dark brown, c belly whitish, flanks and vent buffy. In pairs. More shy and hidden than other Inca-f.s. Silently and slowly hops about on or near the ground within dense vegetation, only rarely coming into the open and then never far from cover. Male sings from conspicuous perch in top of bush or small tree. At least in the breeding season feeds mainly on insects. Occ. gives a *tzik* like other I-f.s. Song of 3 very high-pitched insect-like notes. Nestling June (Piura); juv June, July (Piura), Aug (s Cajamarca).

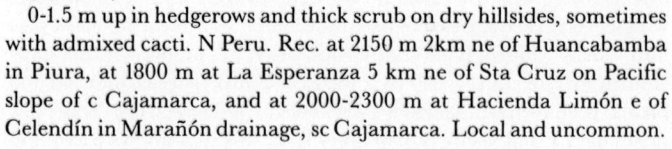

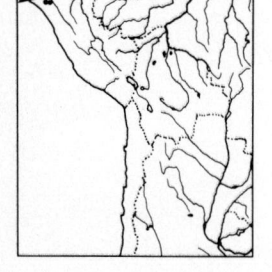

0-1.5 m up in hedgerows and thick scrub on dry hillsides, sometimes with admixed cacti. N Peru. Rec. at 2150 m 2km ne of Huancabamba in Piura, at 1800 m at La Esperanza 5 km ne of Sta Cruz on Pacific slope of c Cajamarca, and at 2000-2300 m at Hacienda Limón e of Celendín in Marañón drainage, sc Cajamarca. Local and uncommon.

GREAT INCA-FINCH *Incaspiza pulchra* – Plate LIX 9a-b

15.5 cm. Bill orange-yellow, legs yellow. **Lores from nostrils to around**

eye and throat black. Crown and back gray-brown, rump browner; shoulders, broad edge of greater secondary-coverts, and inner remiges rufous, rest of wing-coverts and edge of most remiges gray; superciliary region, sides of head and neck and breast gray, belly white, vent tinged buff. Outer tail-feather mostly white, next 2 with decreasing amount. Female slightly duller than male. **Juv.** lacks black on head and is streaked on crown, back, breast, and sides; throat grayish, rest of underparts buffy.

Habits: In close pairs. Outside breeding season in groups of 5-7. Feeds on cactus fruit (esp. *Melocactus*?). Hops about on the ground and low in bushes, often cocking tail.

Voice: Song a vibrating, very high-pitched, short note repeated every 5 s.

Breeding: Fledglings June (Ancash), May (Lima); juv. June (Ancash), May (Lima).

Habitat: Hot dry mountain steppes with xerophytic bushes and cacti, esp. *Melocactus*.

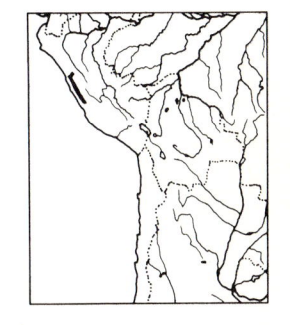

Range: W Peru at 1000-2750 m, below 2500 m where range overlaps that of Rufous-backed I-f. Pacific slope in Ancash and Lima. Known from n Ancash (Yuracmarca-Yanac 1920 and 2135 m, sight rec. to 2500 m), pacific slope of w Ancash in Casma valley (Pariacoto 1500 m, above Yaután 1675-1825 m) and Huarmey valley (Huambo-San Damián 1000-1030 m, 64 rd km E Huarmey 2000 m) and c and s Lima in Rímac valley (Santa Eulalia valley E Huinco at 1525, 1830 and 2135 m, Chosica mts 1675 m, Matucana 2440 m, Zarate 2440-2740 m) and Cañete valley (Cachui 2500 m). Generally uncommon or rare, but according to M Koepcke common in w Ancash.

RUFOUS-BACKED INCA-FINCH *Incaspiza personata* – Plate LIX 10a-b

16 cm. Bill and legs yellow. Face (**forehead to around eye and upper throat**) **black**; rest of head to upper mantle gray, back, inner greater wing-coverts, and edge of inner remiges rufous, rest of wing-coverts and edge of outer remiges pale gray when fresh; gray of headside continues to neckside and breast, flanks buffy gray, belly whitish tinged buff. Outer tail-feathers mostly white, next 2 with decreasing amount. **Imm**. similar, but only border of black forehead gray, rest of crown and nape brown faintly streaked dark brown, back and inner wing-coverts gray-brown, only scapulars rufous. **Juv**.: crown dark gray-brown washed black, back dark brown streaked blackish, throat brownish gray, breast and sides buffy faintly streaked dark gray, c belly whitish.

Habits and **voice**: No data.

Breeding: Fledgling May (Ancash).

Habitat: Dry montane scrub. Sometimes in *Puya raimondii*.

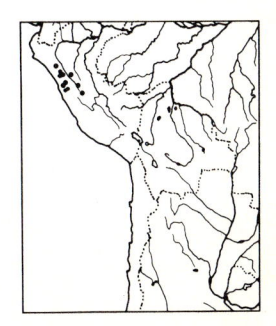

Range: W Peru at 2700-4000 m. Known from s Cajamarca (Cajamarca 3050 m, Cajabamba 3050 and 3200 m), e (Tayabamba 3325 m), and w La Libertad (Soquian, Succha 3050 m, mts near Otuzco), Ancash (Yanac 2700 m, Huaylas 2750 and 3500 m, Yungay, Cajamarquilla 4000 m), Huánuco (Cullcui 3170 m), and Pasco in upper Huallaga valley (sighting 4 km n of Huariaca on c highway c. 2900 m). Uncommon.

BUFF-BRIDLED INCA-FINCH *Incaspiza laeta* –
Plate LIX 8a-b

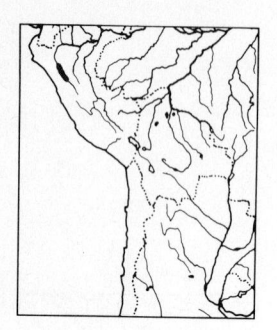

13 cm. Bill orange-yellow, legs yellow. Forehead to ocular region black, separated from black throat by **distinct pale buff moustache**; crown, headside, and nape gray, back rufous chestnut, rump deep gray, wings gray, tertials edged buff; breast gray, belly pale buff. Outer tail-feather mostly white, next 2 with decreasing amount. **Juv**. lacks black head-markings; crown grayish streaked brown, back rufous-brown to dull brown faintly streaked dark brown, throat dark gray, moustache and breast dirty white, breast streaked gray-brown, belly pale buff. Alone or in pairs. Forages on the ground or in bushes or trees, frequently open to view. Not very shy. Call *tzik* frequently uttered.

Bombax woodland admixed with cacti and with open understory of xerophytic thorny scrub. Peru at 1500-2750 m in Marañón valley from sc Cajamarca and extreme s Amazonas to La Libertad and n Ancash, possibly to w Huánuco. Fairly common.

RUFOUS-COLLARED SPARROW *Zonotrichia capensis* –
Plate LVII 13a-d

14.5 cm. Unmistakable. Slightly crested. Ssp *pulacayensis*: crown black with median gray stripe, supercilium gray, incomplete eye-ring whitish, headside blackish, earcoverts gray; **half-collar around nape rufous, meeting black patch on side of fore-neck**. Back and wings brown streaked dusky, rump gray-brown, wings with 2 narrow white bars. Throat white, rest of underparts whitish washed with smoky gray on breast, and with pale brown on sides, flanks and vent. Sspp generally poorly differentiated, varying in bill-size, length of crest, tail, and wings (wings longest in southern migratory forms), presence of white tip and edge of outer tail-feather, general coloration and amount of streaks on the back, width of c crown-stripe, extent of cheek-markings, tone, extent and contrast of rufous collar, extent of black neck-spot, presence of yellow bend of wing, and bars on under tail-coverts. The northern '*carabayae*' group (incl. *peruviana*) comprise fairly ochraceous and small forms that are quite buffy above. The southern '*pulacayensis*' group comprise large forms with a rich rufous hue above, notably on the edges of the tertials. Ssp *australis* has gray crown without black, chestnut collar, and grayer less streaked upperparts. **Juv**. like ad., but without rufous collar, crown dark brown streaked black, supercilium whitish streaked dusky; breast, upper belly, and sides streaked or spotted with dusky; wingbars smaller and tinged buffy. Juv. *australis* paler and with hardly any brown, wingbars white as in ad., sides and flanks almost unspotted. Molts to ad. plumage almost immediately in the n, considerably later in s.

Habits: In pairs or small groups, often with other finches. Forages for seeds and insects on the ground, perching in bushes and trees. Not shy. Often sings at night.

Voice: Song varies geographically, but always includes a drawn out whistle *zeeeeu* and/or a trill, e.g. *zeelit-zeeeeu trrrrt*. For further details see *Condor* 71 (1969):299-315, *Zeits. Tierpsychol.* 30 (1972):344-73, *Evolution* 30 (1976):802-17, and *Behavioral Ecol. Sociobiol.* 8 (1981):203-6.

Breeding: The breeding biology is described in *Caldasia* 10(1968):83-154. In Valle, Col. throughout year, each bird twice a year, peaking mid-Jan and mid-June at beginning of drier periods. Eggs Dec (Sta Marta), Nov (*peruviensis* Chile), Nov-Dec (*antofagastae* Chile), Sep-Feb (*chilensis* Chile), Jan-Feb (*sanborni* and *australis* Chile); fledglings Aug (Piura), Jan (La Libertad), May, June, Oct (Huánuco), Apr (Junín), May (Moquegua), Mar, Apr (Puno), July (La Paz), Jan (Cochabamba), Dec, Mar (Sta Cruz Bol.), Jan (Tarija), Dec, Feb (Sta Cruz Arg.), Dec (Tierra del Fuego); large gonads May, June (Perijá), Jan (Oruro, Cochabamba). Breeds at all seasons in coastal sw Peru (see *Condor* 73 (1971):127-46). See also *Auk* 97 (1980):400-4.
Habitat: Around human habitation and in a variety of habitats from bunchgrass with scattered bushes to arid scrub, or humid shrubby areas, but in dense humid forest only along roads. May nest in *Puya raimondii*.
Range: Sea level to 5300 m, n of Peru primarily montane. Absent from unbroken humid forest, otherwise found virtually throughout from Mexico to Isla Grande and in Dominican Republic. The 26 sspp recognized fall into 3 groups. The southern '*pulacayensis*' group passes n across the altiplano to meet the '*carabayae*' group around Ollantaitambo in Urubamba valley Cuzco Peru (for details see *Can. J. Zool.*, 63 (1985): 2383-8). Ssp *australis* breeds from Aysén, Chile and s Neuquén, Arg. to Cape Horn and migrates n to Bol. Ubiquitous, but generally decreasing with the spreading of House Sparrow. See also *Neotrópica* 18 (1972):95-102, *Hornero* 11 (1971):85-92, *Condor* 74 (1972):5-16, *J. Zool. London* 170 (1973):163-88, and *Ibis* 116 (1974):74-83.

WEDGE-TAILED GRASS-FINCH *Emberizoides herbicola* – **LVII 8** is found from lowlands to 2000 m, or 2300 m in Sta Cruz, Bol. (*herbicola*). Inhabits tall thick grass of open savannas and in adjacent tangled shrubbery, in premontane habitats in n Col. and Ven., Guianas, and n Brazil (*sphenurus*); se Peru through e Bol. to e and s Brazil and ne Arg. (*herbicola*). 3 sspp in Costa Rica and Panama, 1 on llanos of ne Col. and w Ven., 1 on Duida mt Ven. 20 cm. **Bill yellow** with black ridge, legs pale flesh. **Tail long**, double-pronged, **graduated, feathers narrow and rather pointed**. Above gray-brown streaked dusky, most of wing-feathers edged olive-yellow; **tail gray-brown**. Headside grayish, supraloral streak and **eye-ring whitish**, breast and sides pale gray-brown. **Juv.** warmer brown above; headside yellowish brown, supercilium and eye-ring pale yellow, throat pale yellow becoming pale buff on c underparts, c belly sometimes tinged yellow; buffy brown flanks and vent faintly streaked dusky. In pairs or small groups foraging low in tangled vegetation, and difficult to flush except when the vegetation is wet. Occ. perches rather erectly with head held high on low bushes or stems. Often flicks wings and tail and twists tail. Song an unmusical decelerating ticking rattle. Also may repeat a weak *tick* and give a warbler-like *stchip* and a swallow-like soft *tschrip*.

BOLIVIAN WARBLING-FINCH *Poospiza boliviana* – **Plate LXII 11**

15 cm. Above gray-brown, median wing-coverts with paler bar, greater

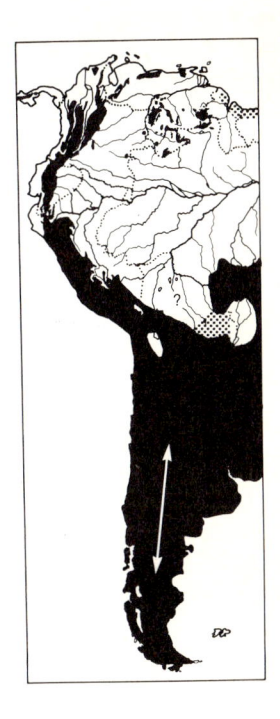

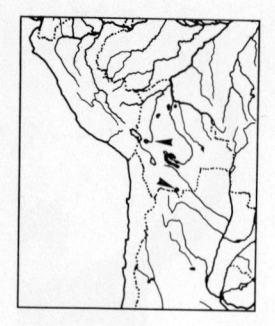

wing-coverts with narrow pale gray bar, inner greater coverts and inner remiges edged pale brown; outer tail-feather with outer half and entire outer web white, decreasing inwards to only 1 cm on tip of 3rd. **Headside gray**, long, narrow supercilium, throat, and c belly to vent white, **breast and sides broadly brick-red**. **Juv.** paler and duller, forepart of supercilium and throat tinged yellowish, lower throat and upper breast faintly streaked.

Habits: Alone or in pairs, sometimes with flocks of Rufous-sided W-f. Forages in dense shrubbery and on the ground. May perch atop bushes or cacti.

Voice: No data.

Breeding: Juv. Apr (La Paz).

Habitat: Dry shrubby hillsides and thorny thickets along stream beds.

Range: Bol. at 1700-3100 m in La Paz, Cochabamba, Chuquisaca, and Tarija. Uncommon.

PLAIN-TAILED WARBLING-FINCH *Poospiza alticola* – Plate LXII 15

15 cm. Above gray-brown, greater wing-coverts with paler tips forming indistinct bar, remiges and the fairly long tail edged gray. **Long broad supercilium** and moustache **white** bordering slaty headside, submalar streak gray-brown (sometimes mottled). Below whitish, **sides** of lower throat and breast **rufous**, flanks more dull cinnamon. **Juv.** dusky above and on throat and breast, darkest on head; distinct supercilium and moustache whitish; belly, flanks and vent buff, palest centrally.

Habits: Alone, in pairs, or small family groups. Forages in dense shrubbery. Mainly feeds on the sugary secretions (and associated insects) on the underside of *Gynoxys* leaves, thus probably subject to competition from Tit-like Dacnis and Blue-mantled Thornbill, but also gleans insects from leaves and stems of other shrubs (O Frimer, TP).

Voice: No data.

Breeding: Juv. Feb (Ancash), Mar (La Libertad); large gonads Sep (Cajamarca).

Habitat: 0.5-2 m up in humid montane shrubbery admixed with *Gynoxys*, and in mixed *Polylepis/Gynoxys* woodland.

Range: Nw Peru at 3500-4300 m, rarely down to 2900 m. Only known from s Cajamarca (Sendamal, 10 km e and 30 km sw of Celendin), e (Cajamarquilla, Quebrada La Caldera ne of Tayabamba), and w La Libertad (Huamachuco), and e Ancash (Quebradas Tutapac, Morococha, Llanganuco, Pucavado, Rurec, Rurichinchay, and Huanzala). Rare, except on e slope of Cordillera Blanca, where uncommon.

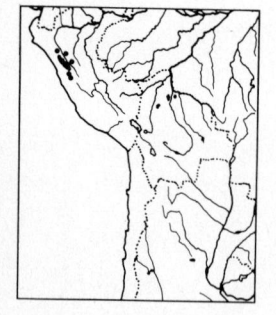

RUFOUS-SIDED WARBLING-FINCH *Poospiza hypochondria* Plate LXII 10a-b

13.5 cm. Tail relatively long. Above brownish gray, tail dusky, outer feather with white terminal half, next 2 with decreasing amounts. Headside dark gray, supercilium, spot below eye, and throat white, submalar streak dusky. Breast gray, **sides and flanks orange-rufous**,

rest of underparts whitish. Ssp *affinis* similar, but white tailtip on average less extensive, rump contrasting grayish, and sides of head gray.
Juv. like ad., but breast buffy gray, supercilium tinged yellowish, sides and flanks buff with slight suggestion of streaks, belly tinged buff.
Habits: Alone or in pairs, sometimes in mixed-species flocks incl. other warbling-finches. Restlessly forages for insects on the ground or in bushes, and deliberately probes for insects in leaf-clusters at tips of branches. Constantly moves tail a little up and down or from side to side, but never cocks it upright. Sings from rather exposed branch 0.5-5 m up in a bush or tree.
Voice: Song lasts 3-4 s *dyp dyp tee dee dy dy dyp*, first 2 notes alike, following series descending, starting higher, and ending lower than first 2. Also gives a short trill ending with 1-2 single notes.
Breeding: Juv. Mar, Apr (Cochabamba), May, June (Potosí).
Habitat: On or near the ground, occ. 15 m up in a tree. Bushy, watered valleys, corn fields, hedgerows, *Eucalyptus* groves, arid scrub, gardens.
Range: 2550-4200 m, (rarely to 4500 m?), down to 1000 m in s part of range. La Paz, Cochabamba, and Sta Cruz to Tarija, Bol. (*hypochondria*), Jujuy to Mendoza, Arg. (*affinis*). Common in Bol.

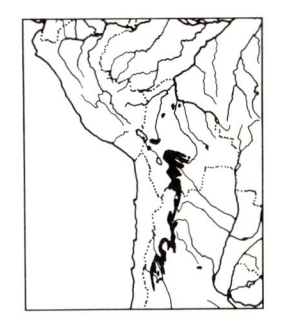

RINGED WARBLING-FINCH *Poospiza torquata* – Plate LXII 9a-c

12.5 cm. Above dark smoky gray, **wings with 2 white patches**, 1 formed by tips and edges of greater secondary-coverts, the other by the basal edge of some primaries; tail black, outer feather with outer web and terminal half of inner web white, next 2 with decreasing amount of white. Supercilium and a few feathers below eye white, headside black, underparts white with **black breastband** and rufous chestnut vent. Ssp *pectoralis* with broader pectoral band, more white in tail, and conspicuous whitish edges of tertials. **Juv**. like ad., but breastband brown.
Habits: Alone, in pairs or small flocks, often in mixed-species flocks, sometimes with Rufous-sided W-f. and Cochabamba Mt-f. Works among small branches in heavy brush.
Voice: No data.
Breeding: Eggs Dec (Tucumán); juv. Mar (Tucumán); large gonads Jan (Cochabamba).
Habitat: Near ground in dense thorny scrub, brushland, and grassland with scattered bushes (e.g. disturbed land with *Dodonea* shrub).
Range: 1500-3800 m in La Paz, Cochabamba, Sta Cruz, and Chuquisaca, Bol. (*torquata*), lowlands to 2500 m from Tarija, s Bol., and w Parag. s to Mendoza, La Pampa, and Buenos Aires and e to Entre Ríos, Arg. (*pectoralis*). Uncommon, locally fairly common.

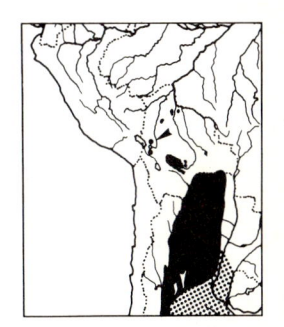

GRAY-AND-WHITE WARBLING-FINCH *Poospiza cinerea* – Plate LXII 14

12 cm. Ssp *melanoleuca*: **crown and headside blackish** turning gray on rest of upperparts; below white or whitish, flanks washed pale gray; tail blackish, outer feather with white terminal half, next 2 with decreasing

amounts. Female often has gray cap, and very old males may have cinnamon flanks. **Juv**. like ad., but duller, crown gray, earcoverts darker, underparts white.

Habits: In pairs or flocks up to 12 forages on the ground or within dense thickets, probing leaf-clusters of composite bushes. Very shy.

Voice: Alarm an excited chipping of alternating *dzhwi* and *dzhic* notes.

Breeding: Dec-Feb (Tucumán); eggs Nov (Tucumán); fledglings Mar (Sta Cruz), Jan (Tarija).

Habitat: On ground or low, rarely in tree canopies. Dense bushy thickets and thorny leguminaceous trees at edge of open country.

Range: Lowlands to 2500 m. From Cochabamba and Sta Cruz, Bol., and Parag. and s Mato Grosso, Brazil, s to La Rioja, San Luis, and n Buenos Aires, Arg., and Urug. (*melanoleuca*); c Brazil (*cinerea*). Fairly common.

NOTE: Often considered 2 species: **BLACK-CAPPED WARBLING-FINCH** (*Poospiza melanoleuca*), and **Cinereous Warbling-finch** (*Poospiza cinerea*).

BLACK-AND-RUFOUS WARBLING-FINCH
Poospiza nigrorufa – Plate LXII 13

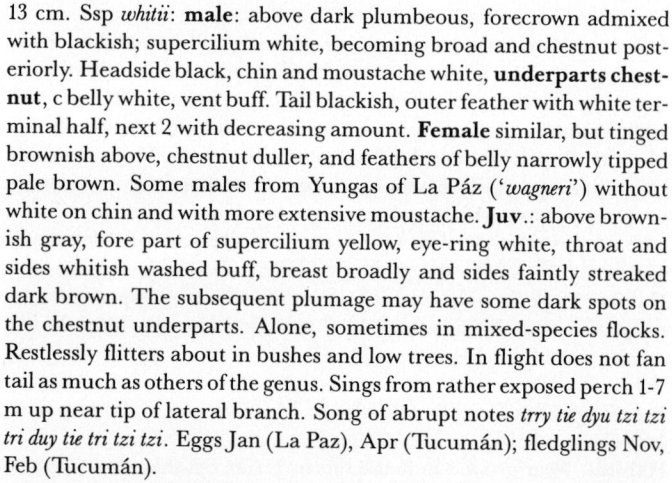

13 cm. Ssp *whitii*: **male**: above dark plumbeous, forecrown admixed with blackish; supercilium white, becoming broad and chestnut posteriorly. Headside black, chin and moustache white, **underparts chestnut**, c belly white, vent buff. Tail blackish, outer feather with white terminal half, next 2 with decreasing amount. **Female** similar, but tinged brownish above, chestnut duller, and feathers of belly narrowly tipped pale brown. Some males from Yungas of La Páz ('*wagneri*') without white on chin and with more extensive moustache. **Juv**.: above brownish gray, fore part of supercilium yellow, eye-ring white, throat and sides whitish washed buff, breast broadly and sides faintly streaked dark brown. The subsequent plumage may have some dark spots on the chestnut underparts. Alone, sometimes in mixed-species flocks. Restlessly flitters about in bushes and low trees. In flight does not fan tail as much as others of the genus. Sings from rather exposed perch 1-7 m up near tip of lateral branch. Song of abrupt notes *trry tie dyu tzi tzi tri duy tie tri tzi tzi*. Eggs Jan (La Paz), Apr (Tucumán); fledglings Nov, Feb (Tucumán).

1-3 m up, rarely high or on the ground. Patchy secondary shrubbery and clumps of trees along roads, fields, and in gardens, undergrowth of woodland bordering grassland. At 1200-2300 m from La Paz, Cochabamba, and Sta Cruz Bol., s along Andes to Mendoza, San Luis, and w Córdoba, Arg. (*whitii*). Fairly common. Lowlands from s Brazil and e Parag. to e Córdoba and Río Negro, Arg. (*nigrorufa*).

NOTE: Ssp *whitii* sometimes considered a distinct species.

RUSTY-BROWED WARBLING-FINCH
Poospiza erythrophrys – Plate LXII 12

13.5 cm. Slender. Crown and headside gray, back dark brownish olive; **outer greater secondary-coverts broadly tipped and outer primaries edged white**, outer tail-feather with white terminal half, decreasing to 1 cm along shaft of 3rd. **Short supercilium from bill rufous chestnut**, bordered above by thin black line; **below rufous**, palest in female, c belly ochraceous white. Ssp *cochabambae* has less contrast between crown and back (mantle grayer), primaries more extensively edged white and greater secondary-coverts with entire outer web white. **Juv.** duller, and very pale rufous below, wing-panel buff.
Habits: Alone or in pairs, sometimes in mixed-species flocks. Spreads tail in flight, but does not cock it, or only slightly so, while foraging. Restlessly works along and sometimes across branches inside tops of bushes or small trees. Shy.
Voice: Call a fine *zit zit zit*.
Breeding: Nov-Jan (Tucumán); eggs Nov (Tucumán); fledglings Jan, Feb (Tarija), Dec (Jujuy).
Habitat: 3-4 m up in semi-humid second growth and shrubbery in patchily wooded watered ravines.
Range: 1800-3150 m. Cochabamba (Tujma), Sta Cruz, and Chuquisaca, Bol. (*cochabambae*), Tarija, s Bol., to Tucumán and Catamarca, nw Arg. (*erythrophrys*). Uncommon.

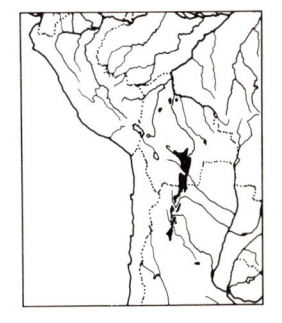

RUFOUS-BREASTED WARBLING-FINCH
Poospiza rubecula – Plate LXII 17a-b

14 cm. **Ad. male** (4 specimens): above bluish slate; headside, lower forehead, and chin black bordered above by **rufous supercilium and forecrown**; **underparts rufous**, extreme c belly white. **Ad.** (breeding condition) **female** (1 specimen) like imm., but without rufous below.
Imm. (2 males and 1 female): head and upperparts gray-brown, below whitish with many admixed rufous feathers, breast and sides streaked dark gray (incl. some of the rufous feathers).
Habits: Alone (probably also in pairs or family groups), sometimes in mixed-species flocks. Forages for seeds within the vegetation.
Voice: Song a prolonged twittering warble.
Breeding: Imm. Jan (Cajamarca), May (Ancash), Oct (Lima); large gonads Oct (Lima); song late Mar (Lima).
Habitat: Composite shrub (*Eupatorium*) on slopes and in woodland (and 'thick paramo sage'). Recently rec. in dry scrub-forest and at edge of *Polylepis* woodland with other bushes admixed; this may be its typical habitat.
Range: W Peru at 2350-3700 m on both slopes of W Andes. Known from extreme s Cajamarca (Cajabamba), La Libertad (Huamachuco), Ancash (Quebrada Rurichinchay on e-slope of Cord. Blanca, at Quitacocha above Huaylas on e-slope of Cord. Negra, Bosque San Damián in upper río Huarmey on w-slope of Cord. Negra), and Lima in Chillón valley (Obrajillo), and Rimac valley (Santa Eulalia valley, Surco, Bosque Zárate). Possible sight rec. in upper Pisco valley, Ica/Huancavelica (Koepcke). Apparently very rare, maybe vanishing.

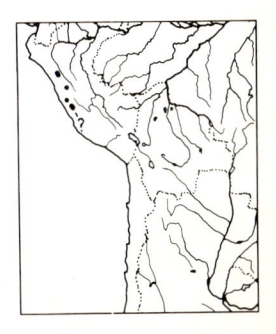

CHESTNUT-BREASTED MOUNTAIN-FINCH
Poospiza caesar – Plate LXII 16

17 cm. Above dark gray, headside blackish; thin **supercilium** from just in front of eye, **throat**, and c belly **white**; **breast** and vent **tawny** (fresh breast-feathers with broad, pale buff tips), sides and flanks gray-brown. Some birds (imm.?) somewhat duller and faintly tinged brownish above. **Juv.** with gray and black parts tinged brownish and with faint buffy tinge on throat. 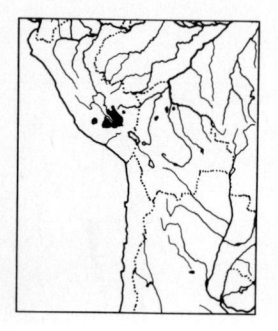
Habits: Alone, occ. in mixed-species flocks with other finches. Silently feeds on the ground, never far from cover. Seeks shelter within the vegetation. Song given from perch, usually near the ground, but during territorial clashes sometimes high in trees.
Voice: Song is a chirping warble lasting 2-3 s and repeated at some 10 s intervals.
Breeding: Song Mar (Cuzco).
Habitat: Semi-arid zone. Dense low shrubbery mixed with columnar cacti and hedgerows, maybe preferring mosaics of shrubbery, small tree groups or woods, and small fields.
Range: Se Peru at 2900-3900 m in Apurímac (Cotaruse, Curahuasi) and Cuzco in upper Urubamba, Paucartambo (and Marcapata? (Huaynapata)) valleys. Uncommon, locally fairly common.

NOTE: Often placed in its own genus *Poospizopsis*.

TUCUMAN MOUNTAIN-FINCH
Poospiza baeri – Plate LXII 19a-b

16.5-17 cm. Forecrown, supercilium, area below eye, throat, upper breast, and vent orange-rufous; rest of plumage gray, palest on c belly, flanks washed olivaceous, wing-lining occ. with some orange-rufous. **Juv.**: above gray-brown, crown indistinctly streaked, below light brown with narrow and obscure dusky streaks.
Habits: Alone or in loose groups. Forages in bushes and on the ground. Perches atop bushes. 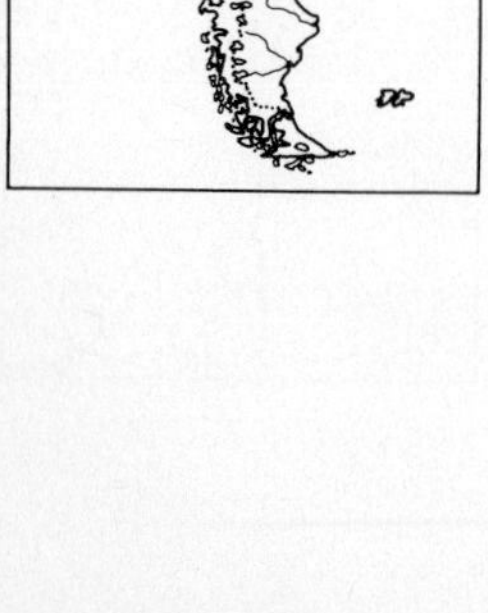
Voice: No data.
Breeding: Fledglings Mar, Apr; large gonads Mar; slightly enlarged gonads Dec.
Habitat: Semi-humid bushy canyons and shrubbery mixed with patches of grass and trees such as *Polylepis* and *Alnus*.
Range: Nw Arg. at 2000-3000 (3400?) m on e slope of Sierra de Medina and Sierra del Aconquija in Tucumán and immediately adjacent Catamarca (Sierra de Ambato), and recently rec. from Salta (above Chicoana: M Rumboll & RR) and Jujuy (near Thermas de Reyes (río de La Quesera)). Locally common, but found at less than 10 localities. Total population in Sierra del Aconquija probably only 2-300 birds (M Nores).

NOTE: Sometimes placed in the genus *Compsospiza*. May best be treated as a ssp of the following species.

COCHABAMBA MOUNTAIN-FINCH *Poospiza garleppi* – Plate LXII 18a-b

17 cm. Tail fairly long. Forecrown, supercilium, area below eye, and most of underparts orange rufous, rest of headside and upperparts dark gray, sides olivaceous, c belly pale buff; tail (esp. outer feathers) pale-tipped when fresh. **Imm**. gray tinged brownish above, headmarkings indistinct, below very pale rufous, darkest on breast, and vent, and obscurely streaked to spotted on throat, breast and sides. **Juv**.: upperparts and headside dusky, throat and breast dusky streaked buff, belly buffy narrowly streaked dusky; tail slightly graduated, pale-tipped when fresh.
Habits: Alone or in pairs, forages 1-3 m up within dense shrubbery, probably also on the ground. When disturbed near the nest may warn from treetop. Often twists tail. Secretive and shy.
Voice: Calls include hummingbird-like *tzit* and *zeeep*.
Breeding: Fledglings Apr, May; juv. July, Aug.
Habitat: Semi-arid zone. In watered ravines with dense bushes and scattered trees such as *Polylepis* and alders (*Alnus*), and a variety of dense thorny bushes.
Range: Bol. at 3000-3800 m in Cochabamba on the slopes surrounding Cochabamba town (Quehuiñapampa (LSUMZ), Pocona-Vacas, Tiraque, Faldas del Monte del Abra, Huacanqui mt (and Colomi), Liriuni, Toncoma, Cheñwa Sandra (ZMUC)). Uncommon, locally fairly common (e.g. Huacanqui mt).

NOTE: Sometimes placed in the genus *Compsospiza*. May include Tucumán M-f.

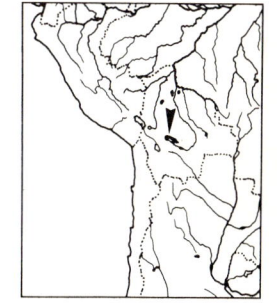

GREAT PAMPA-FINCH *Embernagra platensis* – Plate LVII 9a-b

20.5 cm. **Bill orange** with dark dorsal stripe or spot, legs flesh. Tail relatively long, slightly graduated, **feathers broad** and fairly rounded. Above dull olive-gray, rump contrasting grayish, wings olive-yellow, forehead darker and grayer and with faint dark shafts. Headside dark gray, throat and breast smoky gray; flanks and vent buffy gray-brown, belly white. Ssp *catamarcensis* paler gray below, and grayer on head, more yellowish green on wings. **Juv**.: above pale buff streaked dusky; wing-coverts and tertials dusky edged pale buff, rest of wing olive-yellow; **tail dark olive**, feathers slightly pointed and pale-tipped. Forehead and streaked **supercilium tinged olive-yellow**, eye-ring pale yellow; below buffy white, breast and sides streaked with small dark brown spots.
Habits: Usually in pairs, sometimes alone or in groups of 5-6. Forages on the ground, usually not far from cover, perches on low bushes or grass. Flight tilting with dangling legs and often half-cocked tail. Often quite confident.
Voice: Call *tzit*. Song sharp, but not high-pitched *tri te trierit*.
Breeding: Eggs Nov-Dec (Tucumán); fledglings Jan, Feb (Tarija), Jan (Tucumán); juv./ad. Mar (Cochabamba).

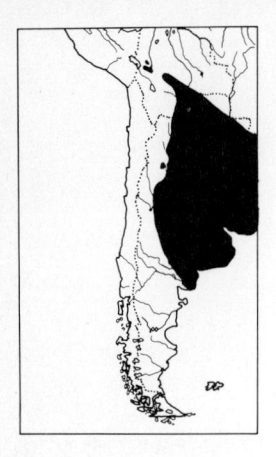

Habitat: Bushes, hedgerows, fields (mainly corn fields), farmland, dry meadows with thistles. and saw-grass swamps.
Range: Primarily lowlands, but up to 3500 m in the drier parts of Cochabamba, Bol., and to 2000 m in Tucumán and Catamarca, Arg. From La Paz, Cochabamba, and Sta Cruz, Bol., and w Parag. s to La Rioja, San Luis, and Córdoba, Arg. (*olivascens*); Campo del Arenal in Catamarca (*catamarcanus*, see M Nores in *Hornero* 12(1986):270); Mendoza (*gossei*, doubtfully distinct from *olivascens*; see *Bol. Mus. Cienc. Nat. Antr. Juan C. Moyano'* 1 (1980):23-4), se Brazil to Río Negro, Arg. (*platensis*). Fairly common.

ANDEAN SISKIN *Carduelis spinescens* – Plate LXII 6a-d

11 cm. Ssp *nigricauda* **male**: **cap black**, back and wings blackish, primaries with bright yellow bar across base, rump olive; upper tail-coverts blackish, feathers with olive edges that wear off; tail black. All underparts olive-yellow. **Female** much like male, but cap dusky-olive. Ssp *capitaneus* similar, but with yellow at base of tail, back paler with less pronounced dark centres, wings with more yellow, under tail-coverts unstreaked. Wings and tail slightly shorter. Female like female *spinescens* but grayer; *spinescens* like *nigricauda*, but back dark dull olive, upper tail-coverts olive-yellow, under tail-coverts bright yellow as is base of tail. Female lacks black cap and has dull grayish underparts, sides dull olive-yellow, **under tail-coverts** yellowish **white**.
Habits: Alone, in pairs or flocks up to 20. Forages on *Espeletia* flowers, in open fields and in thick shrubbery, actively moving from perch to perch, often near lichens.
Voice: Flight call like that of Hooded S. Song much like other siskins.
Breeding: Fledglings Mar (e Cauca); nest-building June (Valle); large gonads Aug (w Antioquia, Cundinamarca).
Habitat: *Espeletia* páramos, elfin forest, and farmland.
Range: 1500-4100 m. Sta Marta mts (*capitanea*); Perijá mts, locally in coastal mts and from Trujillo, Ven., s through E Andes, Cauca, and Putumayo at s end of C Andes, Valle e slope of W Andes, and w Nariño, Col., and in adjacent w Carchi, nw Ecu. (*spinescens*); n end of C and W Andes, Col. (*nigricauda*). Erratic.

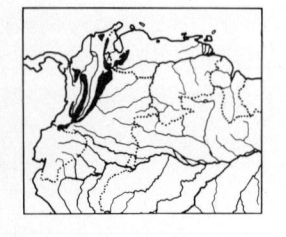

THICK-BILLED SISKIN *Carduelis crassirostris* – Plate LXII 8a-f

12.5 cm. **Bill very heavy**, *crassirostris* with strongly swollen base of maxilla (esp. when seen from above). Ssp *amadoni* **male**: **head and extreme upper breast black** contrasting with **yellow patch at side of neck**; back, rump, and wing-coverts olive, feathers with ill-defined darker centres. Wings blackish with bright yellow bar across base of primaries; tail blackish narrowly edged yellow at base. Below olive-yellow, flanks, lower abdomen, and vent whitish. **Female** gray, mousegray above and on head, pale on neck-side and c belly; hardly any yellow except in wings and a slight olive tinge to rump; back, breast, sides, and vent sometimes faintly streaked. Ssp *crassirostris*: **male** much brighter, upperparts olive-yellow, underparts yellowish, under tail-

coverts yellow ochre, base of tail yellow. **Female** colored like female Hooded S., but vent unstreaked. **Juv**. like female, but more washed with buffy brown or citrine. Unnamed ssp from wc Peru much like *crassirostris*, but with less swollen base of maxilla, and approaching *amadoni* by olive-tinged breast.
Habits: In pairs or flocks up to 30, sometimes with Black S. Feeds on buds and seeds of *Polylepis*.
Voice: Song often given in chorus, a twitter much like other siskins, but lower pitched and perhaps a trifle faster. Call a very drawn-out *wheeep* generally hoarser than in other siskins.
Breeding: Fresh eggs late Feb (Jujuy); large gonads Mar (Tacna); nest building Mar (Lima).
Habitat: *Polylepis* shrub and woodland, rarely in adjacent *Gynoxys* shrub, and after breeding season juv. apparently move down to bushy slopes below the elevation of *Polylepis*.
Range: 3400-4800 m and probably higher. Ancash, Pasco, Lima, and Ayacucho (unnamed ssp), intergrading in Ayacucho and Apurímac with birds from Cuzco, Puno, Arequipa, Moquegua, and Tacna, s Peru, and nw Oruro, Bol. (*amadoni*); Potosí, Bol., Jujuy, Salta, Catamarca, La Rioja (Cerro Famatina), and Mendoza, Arg., and Aconcagua and Santiago, Chile; in winter at lower elevations, southern birds migrating n (*crassirostris*). Locally common.

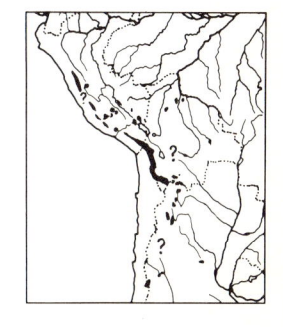

HOODED SISKIN *Carduelis magellanica* – Plate LXII 7a-d

11 cm. Ssp *capitalis*: **male: head incl. throat glossy black** (but odd individuals have yellowish throat or even cheeks); above olive, feathers usually with faint dark centres, rump olive-yellow to yellow. Wings blackish with bright yellow bar on greater coverts, and broad band across base of remiges, tertials edged white; tail blackish, all but c pair yellow basally. Below yellow, flanks and c abdomen whitish. **Female**: above gray-brown, feathers edged olive, rump olive-yellow; yellow in wings and tail less extensive than in male, below pale gray-brown faintly tinged yellowish olive. **Juv**. light buffy olive above, rump slightly paler, wings like ad., but with 2 buffy bars on median and greater coverts; below buffy gray-brown strongly washed yellow. Tail-feathers more pointed than in ad. Ssp *paula* similar, but male with back, rump and underparts yellower, back less variegated with dusky, yellow of wings and tail more extensive, no white on flanks. Female grayer above with yellower rump; *peruana* much like *paula*, but male greener above, rump duller yellow, female usually more tinged with olive-yellow, less grayish. Ssp *urubambensis* like *peruana*, but larger; *boliviana* variable, but black throat of male usually extending well onto breast, back rather black-streaked, underparts often washed with citrine, upper tail-coverts mainly black, and tail with much yellow; *hoyi* has black throat of male rather sharply demarcated from breast, less yellow in the tail than *boliviana*, and no white on flanks; *tucumana* much like *hoyi*, but male paler yellow below, and with white flanks.
Habits: In pairs or flocks, often with other siskins. Forages for seeds in bushes, trees, herbs, and on the ground. Not shy.
Voice: Song a continuous twitter like other siskins, often given in cho-

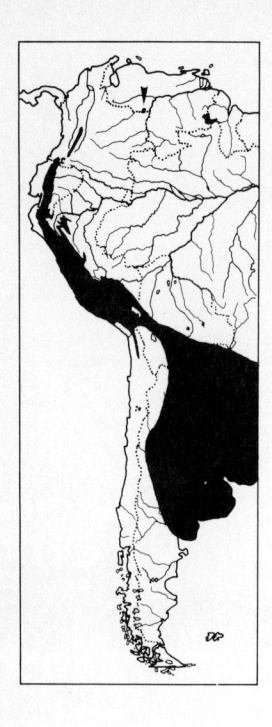

rus. Call a long trill *trrrrrr* also much like other siskins.
Breeding: Eggs June (Cajamarca), Mar (Tacna); fledglings July (Tacna), Feb (Tucumán); nest-building Oct, Nov, Jan (coastal sw Peru).
Habitat: Both humid and semi-arid shrubbery, farmland, gardens, and at forest edge.
Range: Lowlands to 5000 m. From Caldas/Quindío, C Andes s through Nariño, Col. (in Col. at 2300-3300 m), through Ecu. to La Libertad, nw Peru (*capitalis*), trop. and subtrop. zones from s Ecu. (Milagros) and up to 3500 m further s along W Andes to Arequipa, s Peru (*paula*); Huánuco and Lima (where only at high elevations) s to Ayacucho and Cuzco, Peru (*peruana*), 2000-4000 m from Cuzco s Peru to Tarapacá (1 rec. from Salar de Pedernales, Atacama), n Chile (*urubambensis*), highlands of c and s Bol. (*boliviana*), Salta, nw Arg. (*hoyi*, C König, *Stuttgarter Beitr. Naturk., ser. A* 350(1981):1-10), from Jujuy, Santiago del Estero and Sta Fé s to Mendoza, Arg., in winter casually to Buenos Aires (*tucumana*). 5 other sspp on Tepuis of Ven. (and immediately adjacent Col.) and Guyana, and from e Bol. and s Brazil to Río Negro, Arg. Generally very common, but some places erratic (e.g. C Andes of Col.).

NOTE: Perhaps hybridizes with the subtrop. (1100-2500 m) Olivaceous S. (*C. olivacea*) at edge of humid forest in e Peru, and apparently with Yellow-rumped S. in n Arequipa s Peru. Ssp *boliviana* resembles some of the hybrids from Arequipa, and could thus have a hybrid origin. However, the 2 normally seem to maintain their integrity where they coexist.

YELLOW-BELLIED SISKIN *Carduelis xanthogastra* – Plate LXII 2a-b

10 cm. **Male**: **above incl. rump black**, wings with broad yellow bar across base of remiges, tail yellow at base; throat and **upper breast black**, belly yellow, thighs dusky tipped yellowish or whitish. **Female** similar, but black replaced by dark olive. **Juv**.: above dark olive faintly streaked, below pale buffy yellow-olive, belly dull yellow, wings with 2 pale buff bars on coverts, and without yellow at base of remiges, extreme base of tail yellow. Ssp *stejnegeri* slightly larger with a longer bill; male has yellow thighs, female has throat clouded with dusky.
Habits: In pairs or small flocks, often with Black S. In open land forages in composite shrubbery (such as the tall herbaceous composite *Viguera*), and on the ground, in trees from mid-heights to top.
Voice: Song much like other siskins, but rather clear and varied.
Breeding: Fledgling Sep (Cauca); juv. Nov (Cauca); large gonads Sep (w Antioquia, Valle); nest-building Mar (Valle). 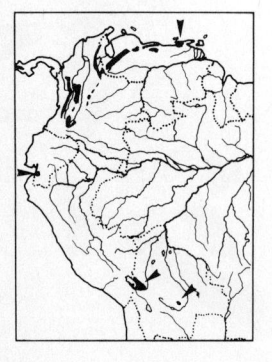
Habitat: Park-like pastures bordered by woodland; dry bushy semicultivated hillsides with scattered rocks.
Range: 800-3700 m. Locally in Costa Rica, w Panamá, Andes of Col. (except Nariño), Perijá mts, and Andes and coastal mts of Ven., 1 rec. from El Oro and irregular sightings in Pichincha (Tinalandia), w Ecu. (*xanthogastra*); Puno in se Peru to La Paz and in Sta Cruz, Bol. (*stejnegeri*). Uncommon, locally fairly common, but erratic.

BLACK SISKIN *Carduelis atrata* – Plate LXII 3a-b

12 cm. **Male glossy black**, extensive wingbar across base of remiges, basal 2/3 of tail, lower belly, vent, and some under wing-coverts yellow. **Female** like male but dull **black** (vs. glossy), feathers edged olive-brown when fresh. **Juv.** like female, but with ochre-yellow tips of median wing-coverts.
Habits: In pairs or flocks, often with other siskins. Sometimes performs nuptial flights, but also sings from bushtops with hanging wings. Feeds in bushes and composite shrub, and often on the ground. May perch on telephone lines.
Voice: Song and call much like other siskins.
Breeding: Eggs Nov (Arica); fledglings Nov, Dec (Puno); juv. Apr (Junín), June (Oruro).
Habitat: Puna zone. Watered rocky gullies with occ. bushes or trees such as *Ribes* and smaller *Polylepis* trees. When not breeding found in open country with other siskins, e.g. dry bushy hillsides, hedgerows, gardens, farms, villages, and plantations, and unlike Thick-billed S. not dependent on *Polylepis*.
Range: 3500-4800 m in Peru, n Chile, and nw Bol., 1800-3600 m further s. Huánuco and s Ancash, c Peru s to Cochabamba, Bol. and through Bol. altiplano and n Chile to Mendoza, Arg. and Antofagasta, rarely Santiago, Chile. Uncommon, locally common.

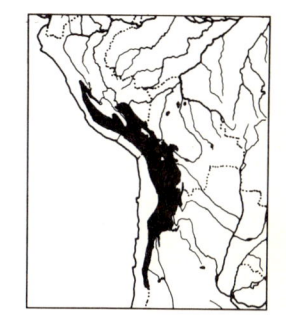

YELLOW-RUMPED SISKIN *Carduelis uropygialis* – Plate LXII 4a-b

12.5 cm. Head and **entire neckside and breast black**, feathers narrowly tipped olive when fresh, wing-coverts and **back black, feathers tipped and edged olive**, very worn birds may become black above; **rump yellow**. Wings blackish with yellow wingbar and a broad yellow band across base of remiges, tertials edged white; belly bright yellow; tail black, basally yellow. **Juv.** differs from juv. Hooded S. by size, more buffy gray appearance, golden wash on face, and conspicuously streaked back and flanks.
Habits: In pairs, when not breeding then in flocks with other siskins.
Voice: Song and calls much like other siskins.
Breeding: Nest with young late Feb (Santiago). Juv. Apr (Arequipa), May (La Paz, Potosí).
Habitat: At least in Chile apparently breeds in desert with scrub and vertical cliff-faces. Otherwise with other siskins in a variety of habitats such as tola heath, bushy hillsides, *Polylepis* woodland, and small fields.
Range: Breeds at 2500-3500 m, but when not breeding covers large distances, and may occ. be found down to 300 m and up to 4000 m. Locally in Ancash, Lima, Huancavelica, (Ayacucho?), Puno, and Arequipa, Peru, La Páz and Potosí, Bol., thence s to Mendoza, Arg. and Bío Bío, Chile. Erratic and uncommon in Peru, Bol. and n Chile, common in c Chile from Atacama s.

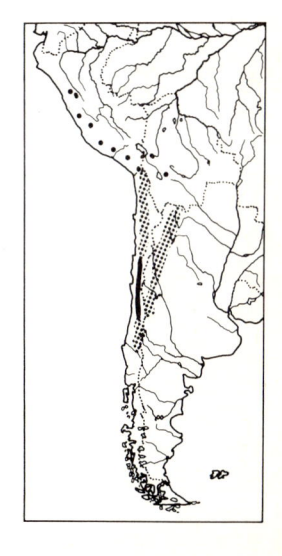

NOTE: Apparently hybridizes extensively with Hooded S. at Chuquibamba, n Arequipa. Males are intermediate, but always differ from Yellow-rumped S. in having yellowish neckside. One intermediate male (CNHM) from Ancash has neckside black with olive patch. Birds resembling Yellow-rumped S., but with the yellow parts of the body strongly washed with citrine have been imported to W Germany from Col. in recent years (H Heintzel).

BLACK-CHINNED SISKIN *Carduelis barbata* –
Plate LXII 5a-b

12 cm. **Male**: **crown black**, rest of upperparts olive, feathers with faint dark centers and narrow grayish tips, rump yellowish; **wings blackish with 2 yellowish bars** and a broad yellow band across base of remiges, tertials edged whitish. Headside yellowish mixed with olive, **chin and c throat black**, feathers tipped yellowish when fresh; rest of underparts yellow, sides washed olive, c belly whitish, vent pale yellow streaked dusky; tail-feathers broadly edged yellow. Occ. blue-gray birds occur. **Female**: above olive-gray smudged with gray-brown, rump olive-yellow; wings as in male, but browner, and with duller and more restricted yellow. Supercilium and neckside yellowish, throat and breast pale olive-yellow turning whitish on belly, flanks washed grayish, vent white streaked dusky, tail only edged yellow at base. **Juv**. like female, but wingbars on median and greater coverts more conspicuous, upperparts duller and grayer, entire underparts pale yellow washed grayish olive on breast and sides.
Habits: In pairs when breeding, otherwise in flocks up to 100. Forages for insects and seeds (and sometimes fungi!) from the ground to treetops. Sings from conspicuous perch in treetop. Generally more arboreal than other siskins (except Lesser Goldfinch).
Voice: Much like other siskins.
Breeding: In c Chile already from Sep (rarely Aug or even July); juv. Feb (Sta Cruz).
Habitat: *Nothofagus* shrub and transitional zone of forest, gardens, parks, plantations, and farmland.
Range: Sea level to 1550 m. From Copiapó valley in Atacama, Chile, and Neuquén, Arg., s to Isla Grande; accidental on Islas Malvinas and in La Pampa. Common.

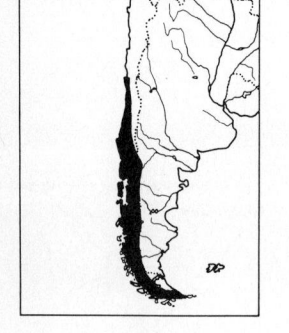

LESSER GOLDFINCH *Carduelis psaltria* –
Plate LXII 1a-c

10 cm. **Male**: **headside and upperparts glossy black**, underparts bright yellow; band across base of primaries (conspicuous in flight), **tips of tertials** and extreme base of outer tail-feathers **white**; feathers of rump white basally; in fresh plumage feathers of back and rump tipped olive. **Female**: above light olive, lesser and median wing-coverts with 2 whitish bars, tertials edged whitish; below light yellow; no white in tail. **Juv**. like female, but washed brown above and tinged pale brown below, breast and sides faintly streaked.
Habits: In pairs or small flocks, rarely up to 50. Restless and sometimes shy, flying far. Forages from the ground to treetops. Sings from conspicuous perch.
Voice: Call a sad, sweet, musical, lingering *pseee* or *psee-ee* from perch

or in flight. In flight also *chek-ek*. Song a rambling musical series of twittering notes that rise and fall; not always well connected.
Breeding: Eggs May, July, Dec, Jan (Valle); fledglings Mar, May (Valle), Aug (N.d.Santander); juv. Apr, May (Cauca), Sep (Nariño).
Habitat: Open country in arid as well as humid zone. Forest clearings and plains with shrubbery and scattered trees, farmland, parks, gardens, rush beds (*Scirpus*), dry forest, and woodland with thorny scrub.
Range: Sea level to 2500 m, up to 2850 m in nw Peru and to 3100 m in the E Andes of Col. Mexico through n Col. (incl. Sta Marta mts and E Andes s to Cundinamarca) to Andes, coastal mts, and lowlands of n Ven., and through Col. (except lowlands e of Andes) and w Ecu., and in W Andes of n Peru s to La Libertad (*columbiana*). 4 sspp in USA and Mexico. Local and uncommon in Peru, fairly common at lower elevations in nw Ecu. and Col., but uncommon or rare in the temp. zone, except in the marshes of the E Andes of Col.

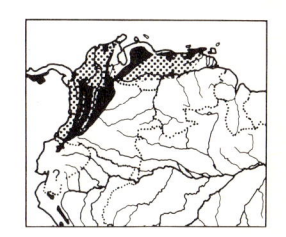

NOTE: Sometimes called Dark-backed G.

Weaverbirds – Family Ploceidae

Old World. 1 species introduced (see Summers-Smith, JD (1986) The sparrows. A study of the genus *Passer*. Calton, U.K.: T & AD Poyser).

HOUSE SPARROW *Passer domesticus* – Plate LVII 12a-b

14.5 cm. **Male**: crown gray, sides of neck and sides of hindcrown chestnut, back reddish brown broadly streaked blackish, rump and tail grayish. Lesser wing-coverts rufous chestnut, median coverts broadly tipped white, greater coverts dusky edged rufous. Lores and ocular region black, rest of headside whitish to gray, throat and breast black, feathers gray-tipped when fresh; sides of breast and rest of underparts grayish white, often with faint dark shafts. **Female**: crown and headside brown, broad postocular streak buff, median wing-coverts with buffy white bar; rump and tail gray-brown. Below pale gray-brown, palest (dirty white) on throat and c belly; belly and vent often with dark shafts. **Juv**. much like female, most males with a faint dusky throat-patch.
Habits: Alone, in pairs, or flocks. Forages on the ground, usually perches within vegetation. Breeds clustered when suitable nest-sites present, and a large flock may sing in chorus within a bush or a hedge.
Voice: Main call of the male a usually disyllabic *chirrup* (occ. *chirp*), used as a nest ownership proclamation, as mating invitation, or (away from the nest) as a vasic communication call of both sexes in a variety of social situations. In the breeding season this chirping becomes higher pitched and is speeded up into a rhythmic sequence (song). Many birds may assemble in a bush and call together (social singing). An aggressive churr *chur-r-r-it-it-it-it* is esp. given by the female, but also by males against intruders near the nest. Appeasement call a long, soft *deee*, alarm a nasal *quer*, *quer-quer-quer*, *quer-it*, *ki-quer*, or *ki-quer-kit*.

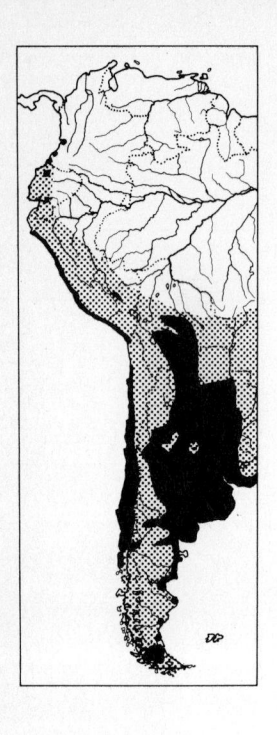

Breeding: Presumably throughout year. On coast of sw Peru mainly Sep-Dec, but occ. as early as late June and as late as Feb.
Habitat: Almost exclusively around human habitation in cities, villages, and on farms, nesting under roofs, in vines on walls of houses, or in palms.
Range: Mainly lowlands, but locally to 4100 m in the Andes. First introduced from Europe to Buenos Aires and Santiago de Chile in the 1870's, Aconcagua, Chile 1918, and has expanded rapidly since, mainly in lowlands. First rec. on n Isla Grande 1932, reached Ushuaia in the s Isla Grande 1957; first rec. in Tacna, s Peru 1951, and presumably introduced to Lima c. 1953. Now common throughout Chile, Arg., Urug., Parag., s and e Brazil, e Bol., coastal Peru, and locally common in highlands of Bol. and Peru (e.g. Ondores near lake Junín (4100 m), and Sicuani (3500 m), Cuzco, where a small and apparently isolated population established since the early 1970s), and also found in some Amazonian towns in e Peru (e.g. Tingo María). Also spreads rapidly into lowlands and highlands of Ecu. (since 1969) and pacific sw Col., (first rec. 1979), but not yet well established in these regions, in 1984 only found in a few towns in Ecu., but a large flock seen in Cauca, Col., 1983 (*domesticus*). Generally spreads at the cost of Rufous-collared S. and in the s also Common Diuca-finch. 10 other sspp in the Old World. Now introduced or found virtually throughout the World.

House-sparrow mobbing Dicua-finch

General Ornithological Literature for the Andean Zone

Recent papers relating to particular taxonomic groups or species are referred to in the introductory remarks to each family, or occ. under the species in question. For brevity, these latter references contain only the absolutely necessary data for retrieving the paper through a library (viz. journal, volume, year, and pages). In general, the references do not include all the older systematic and faunistic literature. For more comprehensive lists of references for Venezuela see Paynter (1982); for Colombia see Chapman (1917), Paynter & Traylor (1981), and Hilty & Brown (1986); for Ecuador see Chapman (1926a) and Paynter & Traylor (1977); for Peru see Stephens and Traylor (1983); for Chile see Hellmayr (1932); for Bolivia see Paynter *et al.* (1975); and for Argentina see Paynter (1985). References to type-descriptions of each form (some holding considerable additional information) can be found in Mayr & Cottrell (1979), Mayr & Greenway (1960, 1962), Mayr & Paynter (1964), Paynter (1968, 1970, 1987), Peters (1934, 1937, 1940, 1945, 1948, 1951), and Traylor (1979) *Check-list of birds of the world*, 16 vols.; references to forms described after publication of the respective volume are given under 'Range'. An almost complete list of synonyms and references to each ssp. is given in Cory (1918, 1919), Cory & Hellmayr (1924, 1925, 1927), Hellmayr (1929, 1934, 1935, 1936, 1937, 1938), and Hellmayr & Conover (1942, 1948a,b, 1949) *Catalogue of birds of the Americas*, 15 vols. References to all ornithological literature is given in the *Zoological Record*.

Adger Smyth, I (1971) Observaciones ornitológicas en la región del Lago Titicaca, Perú-Bolivia. *Rev. Univ. Nac. Técnico del Altiplano* III 4:76-79.

Albert, F (1898-1901) Contribuciones al Estudio de las Aves Chilenas. *Anal. Univ. Chile* 100:301-325, 593-627, 863-895; 101:23-54, 229-264, 497-520, 655-679, 909-941; 103:209-255, 445-461; 579-591, 677-690, 829-847; 104:95-134, 267-283, 987-1008; 106:579-591; 108:193-237, 243-306, 547-564.

Allen, JA (1889a) Notes on a collection of birds from Quito, Ecuador. *Bull. Amer. Mus. Nat. Hist.* 2:69-76.

Allen, JA (1889b) List of birds collected in Bolivia by Dr. H. H. Rusby, with field notes by the collector. *Bull. Amer. Mus. Nat. Hist.* 2:77-112.

Allen, JA (1900a) List of birds collected in the district of Santa Marta, Colombia, by Mr. Herbert H. Smith. *Bull. Amer. Mus. Nat. Hist.* 13:117-184.

Allen, JA (1900b) North American birds collected at Santa Marta, Colombia. *Auk* 17:363-367.

Allen, JA (1905) Supplementary notes on birds collected in the Santa Marta District, Colombia, by Herbert H. Smith, with descriptions of nests and eggs. *Bull. Amer. Mus. Nat. Hist.* 21:275-295.

Allen, WR (1921) The birds of Lake Poopó, Bolivia. *Auk* 38:340-344.

Alvarez, HL (1987) *Introducción a las aves de Colombia*. 2nd ed., Bogotá.

Aparicio P, M (1957) *Aves del Titicaca*. Tesis doctoral., Univ. Nac. Cuzco.

Araya Modinger, B, G Millie Holman & M Bernal Morales (1986) *Guía de campo de las aves de Chile*. Santiago: Editorial Universitaria.

Arias C, S (1986) Clasificación y distribución de las aves de la Laguna Alalay. Unpublished.

Aveledo, HR & H Gines (1952) Quatro Aves Nuevas y dos Extensiones de Distribución para Venezuela, de Perijá. *Noved. cient. contr. occ., Mus. Hist. Nat. La Salle, Caracas, Zool.* 6:1-15.

Aveledo H, R & AR Pons (1952) Aves nuevas y extensiones de distribución a Venezuela. *Noved. Cient. Mus. Hist. Nat. La Salle, Ser. Zool.* 7:3-21.

Baer, W (1904) Sur une collection d'oiseaux du Tucumán. *Ornis* 12:209-234.

Bangs, O (1898a) On some birds from Santa Marta, Colombia. *Proc. Biol. Soc. Washington* 12:131-144.

Bangs, O (1898b) On some birds from Pueblo Viejo, Colombia. *Proc. Biol. Soc. Washington* 12:157-160.

Bangs, O (1898c) On some birds from the Sierra Nevada de Santa Marta, Colombia. *Proc. Biol. Soc. Washington* 12:171-182.

Bangs, O (1899a) On some new or rare birds from the Sierra Nevada de Santa Marta, Colombia. *Proc. Biol. Soc. Washington* 13:91-108.

Bangs, O (1899b) On a small collection of birds

from San Sebastian, Colombia. *Proc. New England Zool. Club* 1:75-80.

Bangs, O (1908) Notes on birds from western Colombia. *Proc. Biol. Soc. Washington* 21:163-164.

Bangs, O (1910) New or rare birds from western Colombia. *Proc. Biol. Soc. Washington* 23:71-75.

Bangs, O & GK Noble (1916) List of birds collected on the Harvard Peruvian expedition of 1916. *Auk* 35:442-462.

Baron, OT (1897) Notes on the localities visited by O.T. Baron in northern Peru and on the Trochilidae found there. *Novit. Zool.* 4:1-10.

Barros, O (1954) Aves de Tarapacá. *Invest. Zool. Chilenas* 2:35-64.

Barros, R (1919) Aves del Valle de Nilahue, Curicó. *Rev. Chil. Hist. Nat.* 23:12-17.

Barros, R (1921) Aves de la Cordillera de Aconcagua. *Rev. Chil. Hist. Nat.* 25:167-192.

Barros, R (1924) Notas sobre algunas pájaros Chilenos. *Rev. Chil. Hist. Nat.* 28:31-35.

Barros, R (1926) Notas ornitológicas. *Rev. Chil. Hist. Nat.* 30:137-143.

Barros, R (1927) Apuntes para el estudio de la alimentación de las aves de Chile. *Rev. Chil. Hist. Nat.* 31:262-265.

Barros, R (1928) Segundas notas ornitológicas. *Rev. Chil. Hist. Nat.* 32:36-42.

Barros, R (1929) Nuevas observaciones sobre aves de la Cordillera de Aconcagua. *Rev. Chil. Hist. Nat.* 33:355-364.

Barros, R (1930) Sobre algunas aves de la Alta Cordillera de Mendoza. *Rev. Chil. Hist. Nat.* 34:312-330.

Behn, F (1944) Notas ornitológicas de un viaje a Laguna de Maule. *Bol. Soc. Biol. Concepción* 18:105-114.

Belcher, CF (1936) In north-west Patagonia. *Oologists' Record* 16:49-53.

Bennet, RS, ed. (1983) *Field guide to the birds of North America*. Washington, D.C.: National Geographic Society.

Berlepsch, H von (1901) Sur quelque espèces nouvelles ou peu connues recueillis dans le département de Cuzco (Pérou central), par M. Otto Garlepp. *Ornis* 11:197-198.

Berlepsch, H von & J Stolzmann (1892) Résultats des recherches ornithologiques faites au Pérou par M. Jean Kalinowski. *Proc. Zool. Soc. Lond.* 1892:371-411.

Berlepsch, H von & J Stolzmann (1896) On the ornithological researches of M. Jean Kalinowski in central Peru. *Proc. Zool. Soc. London* 1896: 322-388.

Berlepsch, H von & J Stolzmann (1902) On the ornithological researches of M. Jean Kalinowski in central Peru. *Proc. Zool. Soc. Lond.* 1902: 18-60.

Berlepsch, H von & J Stolzmann (1906) Rapport sur les nouvelles collections ornithologiques faites au Pérou par M. Jean Kalinowski. *Ornis* 13:63-133.

Berlepsch, H von & L Taczanowski (1883) Liste des oiseaux recueillis par MM. Stolzmann et Siemiradski dans l'Ecuadeur occidental. *Proc. Zool. Soc. London* 1883: 536-577.

Berlepsch, H & L Taczanowski (1884) Deuxième liste des oiseaux recueillis dans l'Ecuadeur occidental par MM. Stolzmann et Siemiradski. *Proc. Zool. Soc. London* 1884: 281-313.

Bettinelli, MD & JC Chebez (1986) Notes on the birds of Meseta de Somuncura, Río Negro, Argentina. *Hornero* 12:230-284.

Blake, ER (1962) Birds of the Sierra Macarena, eastern Colombia. *Fieldiana, Zool.* 44:69-112.

Blake, ER (1977) *Manual of Neotropical Birds*. Vol. l. Chicago: University of Chicago Press.

Bó, NA (1958) Nota sobre una colección de aves del este de Chubut. *Rev. Mus. La Plata* (Zool.) 7:35-50.

Bó, NA (1965) Notas preliminares sobre la avifauna del nordeste de San Luis. *Hornero* 10:251-268.

Bond, J (1945) Notes on Peruvian Furnariidae. *Proc. Acad. Nat. Sci. Philadelphia* 97:17-39.

Bond, J (1947) Notes on Peruvian Tyrannidae. *Proc. Acad. Nat. Sci. Philadelphia* 99:127-154.

Bond, J (1950) Notes on Peruvian Formicariidae. *Proc. Acad. Nat. Sci. Philadelphia* 102:1-26.

Bond, J (1951) Notes on Peruvian Fringillidae. *Proc. Acad. Nat. Sci. Philadelphia* 103:65-84.

Bond, J (1953a) Notes on Peruvian Dendrocolaptidae, Conopophagidae and Rhinocryptidae. *Notulae Naturae* 248.

Bond, J (1953b) Notes on Peruvian Icteridae, Vireonidae and Parulidae. *Notulae Naturae* 255:1-15.

Bond, J (1954a) Notes on Peruvian Piciformes. *Proc. Acad. Nat. Sci. Philadelphia* 106:45-61.

Bond, J. (1954b) Notes on Peruvian Trochilidae. *Proc. Acad. Nat. Sci. Philadelphia* 106:165-183.

Bond, J (1955a) Notes on Peruvian Coerebidae and Thraupidae. *Proc. Acad. Nat. Sci. Philadelphia* 107:35-55.

Bond, J (1955b) Additional notes on Peruvian birds. Part I. *Proc. Acad. Nat. Sci. Philadelphia* 107: 207-244.

Bond, J (1956) Additional notes on Peruvian birds. Part II. *Proc. Acad. Nat. Sci. Philadelphia*

108:227-247.

Bond, J (1974) *Birds of the West Indies*. Revised edition. London: Collins.

Bond, J & R Meyer de Schauensee (1940) On some birds from southern Colombia. *Proc. Acad. Nat. Sci. Philadelphia* 92:153-169.

Bond, J & R Meyer de Schauensee (1941, 1943) The Birds of Bolivia. Parts I-II. *Proc. Acad. Nat. Sci. Philadelphia* 94:307-391; 95(1942): 167-221.

Borrero H, JI (1945a) Aves migratorias en la sabana de Bogotá. *Caldasia* 3:407-414.

Borrero H, JI (1945b) Aves migratorias en los parques y jardines de Bogotá. *Caldasia* 3:415-418.

Borrero H, JI (1946) Aves ocasionales en la sabana de Bogotá. *Caldasia* 4:169-173.

Borrero H, JI (1947) Aves ocasionales en la sabana de Bogotá y las lagunas de Fuquene y de Tota. *Caldasia* 4:495-498.

Borrero H, JI (1952a) Algunas aves raras en la sabana de Bogotá. *Lozania* 1:7-12.

Borrero H, JI (1952b) Apuntes sobre aves Colombianas. *Lozania* 3:1-12.

Borrero H, JI (1955) Apuntes sobre aves Colombianas, 2. *Lozania* 9:1-15.

Borrero H, JI (1958) Aves de caza Colombianas. *Rev. Univers. Nac. Colombia* 23:111-168.

Borrero H, JI (1961) Notas sobre aves Colombianas. *Noved. Colombianas* 1:427-429.

Borrero H, JI (1968a) Notas ecologicas sobre aves del Valle del Cauca. *Bol. Depto. Biol., Universidad del Valle* 1:5-11.

Borrero H, JI (1968b) Notas sobre aves del departamento del Valle del Cauca. *Bol. Depto. Biol., Universidad del Valle* 1:26-34.

Borrero H, JI (1968c) Notas sobre aves del Pacifico Colombiano. *Bol. Depto. Biol., Universidad del Valle* 1:35-45.

Borrero H, JI (1972) *Aves de caza Colombianas*. Universidad del Valle, Depto. Biol.

Borrero H, JI & J Hernández C (1957) Informe preliminár sobre aves y mamíferos de Santander, Colombia. *Anal. Soc. Biol. Bogotá* 7:197-230.

Borrero H, JI & C Hernández, J (1958) Apuntes sobre aves Colombianas. *Caldasia* 8:253-294.

Borrero H, JI & A Olivares (1955) Avifauna de la región de Soatá, Departamento de Boyacá, Colombia. *Caldasia* 7:51-81.

Borrero H, JI, A Olivares & J Hernández C (1962) Notas sobre aves de Colombia. *Caldasia* 8:585-601.

Brabourne, L & C Chubb (1913) *The birds of South America*. Vol. I. London: R. H. Porter.

Brack E, A (1969-1972) Catálogo de las Aves del Perú. *Biota (Lima)* 8:1-68, 69-112, 141-176, 177-224, 241-272, 273-320; 9:1-56, 57-112, 113-160, 161-216.

Buckley, PA, MS Foster, ES Morton, RS Ridgely, & FG Buckley, eds. (1985) *Neotropical ornithology*. Washington, D.C.: American Ornithologists' Union (Orn. Monogr. 36).

Budin, E (1931) Lista y notas sobre aves del N.O. Argentino (Prov. de Jujuy). *Hornero* 4:401-411.

Budin, OA (1976) Contribución al conocimiento de las aves del Parque Biológico. *Parq. Biol. Reserv. Natural 'Sierra de San Javier', Escuela Naturaleza (Univ. Nac. Tucumán)*, Publ. 2.

Bullock, DS (1923) Sobre algunos nidos de Aves Chilenas. *Hornero* 3(1):90-94.

Bullock, DS (1929a) Aves de los pinares de Nahuelbuta. *Rev. Chil. Hist. Nat.* 33:121-127.

Bullock, DS (1929b) Aves observadas en los alredores de Angol. *Rev. Chil. Hist Nat.* 33:171-211.

Bullock, DS (1935) Las Aves de la isla Mocha. *Rev. Chil. Hist. Nat.* 39:232-233.

Butler, TY (1979) *The birds of Ecuador and the Galapagos Archipelago*. Portsmouth, New Hampshire: The Ramphastos Agency.

Buechner, HK & JH Buechner, eds. (1970) The avifauna of Northern Latin America. *Smithsonian Contrib. Zool.* 26.

Burmeister, CV (1888) Relación de un viaje a la Gobernación del Chubut. *Anal. Mus. Nac. Buenos Aires* 3:175-251.

Burmeister, CV (1890) Expedición a Patagonia por encargo del Museo Nacional. *Anal. Mus. Nac. Buenos Aires* 3:253-326.

Burmeister, H (1858) Zur Fauna von Süd-America; Briefliches aus Mendoza. *J. Orn.* 1858: 152-162.

Cabanis, J (1878) Ueber eine Sammlung von Vögeln der Argentinischen Republik. *J. Orn.* 1878: 194-199.

Cabot, J & P Serrano (1986) Data on the distribution of some species of raptors in Bolivia. *Bull. Brit. Orn. Club* 106:170-173.

Cabot, J & P Serrano (1988) Distributional data on some non-passerine species in Bolivia. *Bull. Brit. Orn. Club* 108:187-193.

Cabot N, J (1988) *Dinamica anual de la avifauna en cinco habitats del Altiplano Norte de Bolivia*. Thesis. Estacion Biol. de Doñana. 317 pp.

Cardiff, SW & JV Remsen, Jr (1981) Three bird species new to Bolivia. *Bull. Brit. Orn. Club* 101(2):304-305.

Carriker, MA, Jr (1930) Descriptions of new birds from Peru and Ecuador. *Proc. Acad. Nat.*

Sci. Philadelphia 82:367-376.

Carriker, MA, Jr (1931) Descriptions of new birds from Peru and Bolivia. *Proc. Acad. Nat. Sci. Philadelphia* 83:455-467.

Carriker, MA, Jr (1932) Additional new birds from Peru with a synopsis of the races of *Hylophila naevia*. *Proc. Acad. Nat. Sci. Philadelphia* 84:1-7.

Carriker, MA, Jr (1933) Descriptions of new birds from Peru, with notes on other little-known species. *Proc. Acad. Nat. Sci. Philadelphia* 85:1-38.

Carriker, MA, Jr (1934) Descriptions of new birds from Peru, with notes on the nomenclature and status of other little known species. *Proc. Acad. Nat. Sci. Philadelphia* 86:317-334.

Carriker, MA, Jr (1935a) Descriptions of new birds from Bolivia, with notes on other little known species. *Proc. Acad. Nat. Sci. Philadelphia* 87:313-341.

Carriker, MA, Jr (1935b) Descriptions of new birds from Peru and Ecuador, with critical notes on other little-known species. *Proc. Acad. Nat. Sci. Philadelphia* 87:343-359.

Carriker, MA, Jr (1954) Additions to the avifauna of Colombia. *Noved. Colombianas* 1:14-19.

Carriker, MA, Jr (1955) Notes on the occurrence and distribution of certain species of Colombian birds. *Noved. Colombianas* 2:48-64.

Carriker, MA, Jr (1959a) New records of rare birds from Narino and Cauca and notes on others. *Noved. Colombianas* 1:196-199.

Carriker, MA, Jr (1959b, 1960) Itinerario del autor durante sus recolecciones en la región de Santa Marta, Colombia de junio de 1911 a octubre de 1918. *Noved. Colombianas* 1:214-222, 330-335.

Carriker, MA, Jr (1959c) New records of rare birds from Nariño and Cauca and notes on others. *Noved. Colomb.* 1:196-199.

Castellanos, A (1935, 1937) Observaciones de algunas aves de Tierra del Fuego e Isla de Los Estados, 1-2. *Hornero* 6:22-37, 382-94.

Chapman, FM (1917) The distribution of birdlife in Colombia. *Bull. Am. Mus. Nat. Hist.* 36.

Chapman, FM (1921) The distribution of bird life in the Urubamba Valley of Peru. *Bull. U.S. Natn. Mus.* 117.

Chapman, FM (1925) Remarks on the life zones of northeastern Venezuela with descriptions of new species of birds. *Amer. Mus. Novit.* 191.

Chapman, FM (1926a) The distribution of birdlife in Ecuador. *Bull. Amer. Mus. Nat. Hist.* 55.

Chapman, FM (1926b) An ornithological reconnaissance in Southern Chile. *Bull. Brit. Orn. Club* 46:119-120.

Chavez C, J (1957) Apuntes para una ornitología cuzqueña: Aves de Taray. *Rev. Universitarias Univ. Cuzco* 108:157-74, 112:227-34.

Chebez, JC & S Heinonen Fortabat (1987) Novedades ornitogeográficas Argentinas, 2. *Nótulas Faunísticas* 3:1-2.

Chubb, C (1919) Notes on collections of birds in the British Museum, from Ecuador, Peru, Bolivia, and Argentina. *Ibis* 1919: 1-55, 256-290.

Clark, R (1986) *Aves de Tierra del Fuego y Cabo de Hornos*. Guía de Campo. Buenos Aires.

Collar, NJ & P Andrew (1988) *Birds to watch*. The ICBP World Checklist of Threatened Birds. Cambridge, U.K.: ICBP (Techn. Publ. 8).

Contreras, JR (1978a) Ecología de la avifauna de la región Puerto Lobos, Provincias de Río Negro y del Chubut. *Ecosur* 5:169-181.

Contreras, JR (1978b) La avifauna de la Isla Victoria, Parque Nacional Nahuel Haupí: ecología y composición. *Resumenes Comun. VII Congr. Argent. Biol. (Mendoza)* p. 11.

Contreras, JR (1979) Catálogo sistemático y descriptivo de la avifauna mendocina. I. No Passeriformes. *Resúmenes VII Reunión Argent. Ecol. (Mendoza)* 21.

Contreras, JR & A Fernández (1980) Ecología de la avifauna de la región del Vibrón, Departamento Maipú, Provincia de Mendoza. *Rev. Mus. Hist. Nat. San Rafael* (Mendoza) 8:3-14.

Contreras, JR & AG Giai (1978) Lista faunistica y comentarios ecológicos acerca de la avifauna del Parque Nacional de Nahuel Huapí (*sic.*) y regiones adyacentes. *Resúmenes Comun. VII Congr. Argent. Biol.* (Mendoza), p. 10.

Contreras, JR & VG Roig (1977) Biota centroandina. VI. Tres especies de aves nuevas para la Provincia de Mendoza. *Neotrópica* 36:349-350.

Cook, RE (1971) Origin of the highland avifauna of southern Venezuela. *Syst. Zool.* 23:257-265.

Cory, CB (1918) Catalogue of birds of the Americas. Part 2, no. 1. *Field Mus. Nat. Hist. Publ.* 197.

Cory, CB (1919) Catalogue of birds of the Americas. Part 2, no. 2. *Field Mus. Nat. Hist. Publ.* 203.

Cory, CB & CE Hellmayr (1924) Catalogue of birds of the Americas. Part 3. *Field Mus. Nat. Hist. Publ.* 223.

Cory, CB & CE Hellmayr (1925) Catalogue of birds of the Americas. Part 4. *Field Mus. Nat. Hist. Publ.* 234.

Cory, CB & CE Hellmayr (1927) Catalogue of birds of the Americas. Part 5. *Field Mus. Nat.*

Hist. Publ. 242.

Cracraft, J (1985) Historical biogeography and patterns of differentiation within the South American avifauna: areas of endemism. Pp. 49-84 *in* PA Buckley *et al.*

Crawshay, R (1907) *The Birds of Tierra del Fuego.* London: Bernard Quaritch.

Crespo, JA (1939) Estudio sobre la ecología de aves y mamíferos en la puna de la Provincia de Jujuy. *Publ. Facult. Cienc. Exactas y Nat., Univ. Buenos Aires.*

Dabbene, R (1902) Fauna magellanica: mamíferos y aves de la Tierra del Fuego é islas adyacentes. *Anal. Mus. Nac. Buenos Aires* 8 (ser. 3, tomo 1): 341-409.

Dabbene, R (1910) Ornitología argentina. Catálogo. Sistemático y descriptivo de las aves de la República Argentina, de las regiones limítrofes inmediates del Brasil, Paraguay, Bolivia, Chile y de los archipiélagos é islas al sur y sureste del continente americano hasta el Círculo Polar Antárctico. *Anal. Mus. Nac. Buenos Aires* 11:1-513.

Dabbene, R (1919) Especies de aves poco comunes o nuevas para la República Argentina. *Hornero* 1:259-266.

Dabbene, R (1926) Aves nuevas y otras poco comunes para la Argentina. *Hornero* 3:390-396.

Darlington, PJ, Jr (1931) Notes on the birds of Río Frio (near Santa Marta), Magdalena, Colombia. *Bull. Mus. Comp. Zool. Harvard* 71:349-421.

Darwin, C (1838-1841) *The Zoology of the Voyage of the Beagle.* Part III. London.

Davis, LI (1972) *A Field Guide to the Birds of Mexico and Central America.* Austin, Texas: University of Texas Press.

Davis, TJ (1986) Distribution and natural history of some birds from the departments of San Martín and Amazonas, northern Peru. *Condor* 88:50-56.

de la Peña, MR (1982) *Las Aves Argentinas y sus Ambientes.* Univ. Nac. Litoral, Fac. Agron. Veterin., Santa Fe.

de la Peña, MR (1983) *Reproducción de las aves Argentinas.* Univ. Nac. Litoral, Fac. Agron. Veterin., Santa Fe.

de Vries, T, J Black, C Solis & C Hernandéz (1983) *Historia natural del Curiquinque.* Quito.

Dinelli, L (1918) Notas biológicas sobre las aves del noroeste de la República Argentina. *Hornero* 1:57-68, 140-147.

Dinelli, L (1922) Notas biológicas sobre aves de Tucumán. *Hornero* 2:312-313.

Dinelli, L (1924) Notas biológicas sobre aves del noroeste de la Argentina. *Hornero* 3:253-258.

Dinelli, L (1929) Notas biológicas sobre aves del noroeste argentino. *Hornero* 4:272-277.

d'Orbigny, AD see Orbigny, AD, d'.

Dorst, J (1956a) Étude d'une collection d'oiseaux rapportée des hauts plateaux andins du Pérou méridional. *Bull. Mus. Natn. Hist. Nat., Paris* (2), 28:435-445.

Dorst, J (1956b) Recherches ecologiques sur les oiseaux des plateaux peruviens. *Trav. Inst. Franc. Etudes Andines* 5: 83-140.

Dorst, J (1957) Contribution a l'etude écologique des oiseaux du haut Marañón (Pérou septentrional). *Oiseaux* 27:235-269.

Dorst, J (1967) Historical factors influencing the richness and diversity of the South American avifauna. *Proc. 16th Int. orn. Congr.*:17-35.

Dorst, J (1967) Considérations zoogéographiques et écologiques sur les oiseaux des hautes Andes. *In* CD Deboutteville & E Rapoport, eds. *Biologie de l'Amerique Australe*, III:471-504. Paris.

Dott, HEM (1984) Range extensions, one new record, and notes on winter breeding of birds in Bolivia. *Bull. Brit. Orn. Club* 104:104-109.

Dott, HEM (1985) North American migrants in Bolivia. *Condor* 87:343-345.

Dugand, A (1945-48) Notas ornitologicas Colombianas, I. *Caldasia* 3:337-341, 4:157-199.

Dunajewski, A (1938) Über einige interessanten Vögel aus Peru (non Passeriformes). *Acta Orn. Zool. Polonici* 2:319-325.

Dunajewski, A (1939) Über einige interessanten Vögel aus Peru (Passeriformes). *Acta Orn. Zool., Polonici* 3:7-15.

Dunning, JS (1981) *South American Land Birds, a Photographic Guide to Identification.* Newton Square, Penn.: Harrowood.

Durnford, H (1877) Notes on some birds observed in the Chuput Valley, Patagonia, and in the neighbouring district. *Ibis*: 27-46.

Durnford, H (1878) Notes on the birds of central Patagonia. *Ibis*: 389-406.

Durnford, H (1880) Harry Durnford's last expedition to Tucumán and Salta. *Ibis*: 410-429.

Ernst, A (1870) Apuntes para la fauna ornitológica de Venezuela. Estracto de las publicaciones de P. L. Sclater y O. Salvin sobre las collecciones de pájaros Venezolanos hechas por Antonio Goering. *Vargasia, Soc. Cienc. Fisicas Nat.* 7:195-198.

Ernst, A (1887) Catálogo de las aves en el Museo Nacional de Caracas. *Rev. Cienc. Univ. Central Venezuela* 1:25-44.

Esteban, JG (1948) Contribución al conocimiento de los Dendrocolaptidos argentinos. *Acta Zool. Lilloana* 5:325-436.

Esteban, JG (1951) 'Furnariinae' de la República Argentina. *Acta Zool. Lilloana* 12:377-441.

Esteban, JG (1953) Nuevas localidades para aves argentinas. *Acta Zool. Lilloana* 13:349-362.

french, R (1973) *A Guide to the Birds of Trinidad and Tobago*. Wynnewood, Penn.: Livingston Publishing Company.

Fitkau, EJ, I Illies, GH Schwabe & H Sioli, eds. (1968) *Biogeography and ecology in South America*. The Hague: W Junk.

Fitzpatrick, JW (1980). Foraging Behavior of Neotropical Tyrant Flycatchers. *Condor* 82:43-57.

Fitzpatrick, JW & DE Willard (1982) Twenty-one bird species new or little known from the Republic of Colombia. *Bull. Brit. Orn. Club* 102: 153-158.

Fjeldså, J (1981) A comparison of bird communities in the temperate and subarctic wetlands in northern Europe and the Andes. *Proc. 2nd Nord. Congr. Orn.*:101-108.

Fjeldså, J (1983) Vertebrates of the Junín area, Central Peru. *Steenstrupia* 8:285-298.

Fjeldså, J (1985) Origin, evolution, and status, of the avifauna of Andean wetlands. Pp. 85-112 *in* PA Buckley et al.

Fjeldså, J (1987) *Birds of relict forests in the high Andes of Peru and Bolivia. Technical report from the Polylepis forest expedition of the Zoological Museum, 1987, with some preliminary suggestions for habitat preservation*. Copenhagen.

Fjeldså, J (1988) Status of birds of steppe habitats of the Andean zone and Patagonia. Pp. 81-95 *in* PD Goriup, ed. *Ecology and conservation of grassland birds*. Cambridge, U.K.: ICBP (Techn. Publ. 7).

Fjeldså, J (1988) Aves de la Laguna Lagunillas, en los Andes del sur del Peru. *Boletin de Lima* 58:61-68.

Fjeldså, J & N Krabbe (1986) Some range extensions and other unusual records of Andean birds. *Bull. Brit. Orn. Club* 106: 115-124.

Fjeldså, J & N Krabbe (1989) An unpublished major collection of birds from the Bolivian Andes. *Zool. Scripta. (in press)*.

Fraga, R & S Narosky (1985) *Nidificación de las aves Argentinas (Formicariidae a Cinclidae)*. Buenos Aires: Asoc. Orn. del Plata.

Fraser, L (1843) (On the Collection of Birds brought to England by Mr. Bridges). *Proc. Zool. Soc. London* 11:108-121.

Fraser, L (1845) (On Birds from Chile, and description of Leptopus Mitchellii). *Proc. Zool. Soc. London* 12: 157.

Fraser, L (1859) Letter: an expedition from Riobamba to Chimborazo. *Ibis*: 208-209.

Frauenfeld, GR von (1860) Ueber den Aufenthalt in Valparaiso und Ausflüge daselbst, während der Weltfahrt der k. k. Fregatte Novara. *Verhandl. Zool. Bot. Ges. Wien* 10: 635-640.

Frenzel, J (1891) Uebersicht über die in der Provinz Córdoba (Argentinien) vorkommenden Vögel. *J. Orn.*: 113-126.

Friedmann, H (1927) Notes on some Argentina birds. *Bull. Mus. Comp. Zool. Harvard* 68:137-236.

Friedmann, H (1947) Colombian birds collected by Brother Nicéforo. *Caldasia* 4:471-494.

Friedmann, H & HG Deignan (1942) Notes on Tschudi's types of Peruvian birds. *Zoologica* 27:49-53.

Frimer, O & SM Nielsen (1989) The status of *Polylepis* forests and their avifauna in Cordillera Blanca, Peru – Technical report from an inventory in 1988, with suggestions for conservation management. Copenhagen.

Gazari, R (1967) Notas sobre algunas aves no señaladeas o poco conocidas al sur del Río Colorado. *Hornero* 10:451-454.

Gadow, H (1883) *Catalogue of the birds in the British Museum*, 8. Passeriformes: Paridae and Laniidae. London.

Germain, F (1860) Notes upon the Mode and Place of Nidification of some of the Birds of Chili. *Proc. Boston Soc. Nat. Hist.* 7:308-316.

Giacomelli, E (1907) Catálogo sistematico de la avifauna riojana. *Annal. Soc. Cient. Argent.* 63: 280-301.

Giai, AG (1951) Notas sobre la avifauna de Salta y Misiones. *Hornero* 9:247-276.

Gigoux, EE (1924) Aves que nos visitan. *Rev. Chil. Hist. Nat.* 28:83-87.

Gigoux, EE (1930) Contribución Ornithológica. Aves Chilenas de las Familias Psittacidae, Picidae, Alcedinidae, Caprimulgidae, Trochilidae, Pteroptochidae, Bubonidae, Tytonidae y Cuculidae, y especias que hay en la provincia de Atacama. *Bol. Mus. Nac. Santiago de Chile* 13:37-49.

Gilliard, ET (1959) Notes on some birds of northern Venezuela. *Amer. Mus. Novit.* 1927.

Gines, H & R Aveledo (1958) *Aves de caza de Venezuela*. Caracas.

Continued on p. 833

Color Plates I to LXIV

Introduction

THE PLAN OF THE PLATES

For practical reasons **we often abandon the strict systematic sequence of the species** on the plates. It is, e.g., convenient to have waterbirds more or less together, and we therefore maintain the tradition (prior to the 2nd edition of Vol. 1 of Peters' *Check-list of Birds of the World*) of placing waterfowl after wading birds, and before diurnal birds of prey. Families too small for one full plate are often placed together with another small family: e.g., 3 families of long-tailed arboreal birds, guans, jays and cuckoos, on Plate XIV. Since the diurnal birds of prey would fill slightly more than 3 plates, the harriers are placed with gulls (with which they often share habitat). Doves and pigeons would fill slightly more than Plate XXII, and 1 species is therefore shown in line-drawing adjacent to the plate, and 1 – the Maroon-chested Ground-dove – near the Barred Parakeet (Plate XXIV). Both inhabit bamboo. In some cases the place of a species is determined by possibilities of confusion rather than by taxonomic position. The Great Grebe is placed with cormorants (Plate VIII) instead of with other grebes (Plate II), since it can be confused only with a cormorant. The hummingbirds are grouped mainly according to some conspicuous features, the separate plates showing large species and hillstars, species with long bills, species with small bills, species with long or otherwise remarkable tails, and species with curved bills.

There are 3 plates for specific geographic areas, all rather peripheral in the area covered. The wetlands of the Bogota area Col. (**Plate X**) are situated in the lower temp. zone, and most of the endemics are just sspp of lowland birds. However, we find it important to emphasize this fauna, because of the needs for conservation oriented research here. The many endemic forms of the Santa Marta mts near the Carribean coast of Col. and of the Mérida mts of nw Ven. are assembled on **Plates LXIII** and **LXIV**. The areas are small and discrete and placed at the extreme n end of the area of the book. Ornithologists visiting these areas may find it useful to have all local specialties together.

Other peripheral areas (Sierra Perijá on the Col./Ven. border, the Córdoba mts in c Arg. and the Fuegian grass- and boglands or Patagonian semi-deserts) have too few endemics to fill separate plates. The forest on the s Andes has a too diverse endemic fauna for one plate.

THE TEXT PAGES: The page to the left of each plate has an explanatory text. The bold figures refer to the numbers on the plate, and there is a reference also to the main text. For each form, a concise diagnosis points out the most significant identification clues. A short remark about the occurrence within the range of the book will tell whether a first tentative field identification is plausible. The text specifies which plumage categories and sspp are shown. Where no ssp name is given, the species is monotypic or represented only by one ssp in the area covered by this book.

SCALES USED: The linear scale is indicated in the introductory text to each plate. Because of the size range from condors to woodstar hummingbirds it is impossible to show all birds to the same scale. Even within one bird group there may be problems. Here, lines on the plates usually separate birds painted in different scales. Hummingbirds and Passerines were all painted 1:1 on the original plates, and are thus 30 percent of the original size on the printed plates (the only exceptions are jays and oropendolas).

PLATE I. Tinamous (compare with quail Plate XIX)

Tinamous are chicken-like birds with round, almost tail-less bodies. Usually seen startled into sudden, short flight with whirring, noisy wingbeats, often accompanied by loud whistles.

1 RED-WINGED TINAMOU *Rhynchotus rufescens*. Large. Only t. with strongly rufous wings in its range. Flies with great noise. **1a** ad. *maculicollis*, **b** pull. Ascends to lower limit of temp. zone in dry parts of Bol. and nw Arg. Text p. 57

2 DARWIN'S NOTHURA *Nothura darwini*. Ocher-brown conspicuously spotted and streaked throughout. Usually gives single scream when flushed. Ascends to bushy or grassy uplands of Bol. and Puno and Cuzco, Peru. Text p. 61

3 SPOTTED NOTHURA *Nothura maculosa*. More long-legged than Darwin's N. (tarsus 3-4 cm, against 2.8-3.4 cm) and with denser barring and mottling above, esp. in the rather drab-colored forms of the brush-steppes of the uplands of c and s Arg. Ssp *pallida* shown. Text p. 62

4 ELEGANT CRESTED-TINAMOU *Eudromias elegans*. Large. Minutely patterned, thus appearing rather uniform drab-gray at a distance. With white-striped face and thin crest, and unlike Patagonian T. lacking red in wings. **4a** ad. ssp *patagonica*, **b** 2-weeks chick. To 2500 m in semiarid parts of Arg. Text p. 63

5 TAWNY-BREASTED TINAMOU *Nothocercus julius*. Large, rich chestnut with rufous crown. White throat and cinnamon-rufous c underparts diagnostic. **5a** ad. **b** pull. Local in temp. forest of Col. to Peru. Text p. 56

6 HIGHLAND TINAMOU *Nothocercus bonapartei*. Large, deep rufous brown, not white on throat. **6a** *intercedens* of W Andes, Col., **b** *bonapartei*. In humid montane forest from Ven. to n Peru. Text p. 56

7 BROWN TINAMOU *Crypturellus obsoletus*. Small, uniform dark rufous with dark gray head with orange eyes. Ssp *punensis* shown. Shrill bursting *whee dwydwy* distinctive. Local in lowlands, and ascending to humid temp. forest in Peru and Bol. Text p. 57

8 HOODED TINAMOU *Nothocercus nigrocapillus*. Dark chestnut. White-throated, but differs from no. 5 by very dark 'hood' and conspicuous buff wingbars. **8a** northern *cadwaladeri*, **b** southern *nigrocapillus*. Very local in humid montane forest in Peru and Bol. Text p. 56

9 CURVE-BILLED TINAMOU *Nothoprocta curvirostris*. Rich tawny-buff, densely vermiculated and white-streaked above, and wings rufous-barred. **9a** ad. Ssp *curvirostris*, **b** pull. Temp. grassland and forest edge in Ecu. and n Peru. Text p. 61

10 ORNATE TINAMOU *Nothoprocta ornata*. Speckled grayish ocher. Head whitish densely dotted with black, breast gray. When flushed gives series of very high-pitched screams. **10a** ad. *branickii*, **b** head of juv. Puna grasslands, mainly above 3500 m, from Ancash, Peru to c Arg./Chile. Text p.59

11 ANDEAN TINAMOU *Nothoprocta pentlandii*. Rather drab gray (on e Andean slopes, e.g., ssp *pentlandii* of **11a**), or more ochraceous (w slopes, e.g., *ambigua* **11b**), back of body with gray and mottled brown stripes separated by white lines, and gray breast spotted white. **11c** juv. (more spotted), **d** pull. When flushed gives long descending series of calls *pyuc-pyuc-pyuc*.... Mainly at 1500-3500 m, on bushy Andean slopes from s Ecu. to n Chile and c Arg. Text p. 60

12 CHILEAN TINAMOU *Nothoprocta perdicaria*. Grayish (depicted ssp *perdicaria*) or more ochraceous. Resembles Andean T., but not white-spotted on breast, and wing-coverts and secondaries barred cinnamon. Chile. Text p. 60

13 KALINOWSKI'S TINAMOU *Nothoprocta kalinowskii*. A dark version of Ornate T., with rufous wings with the coverts densely mottled gray. Peru, with only 2 definite rec. Text p. 59

14 TACZANOWSKI'S TINAMOU *Nothoprocta taczanowskii*. Dark gray brown with pale streaks and spots, and wing-feathers barred. **14a** ad., **b** pull. Rare in temperate woodland ecotones in c and e Peru. Text p. 58

15 PUNA TINAMOU *Tinamotis pentlandii*. Finely mottled with olive-gray. Head whitish with dusky stripes. Vent rufous. Flute-like song *kewla-kewla*... distinctive. **15a** ad., **b** pull. Very high parts in w Peru to n Arg./Chile. Text p. 63

16 PATAGONIAN TINAMOU *Tinamotis ingoufi*. Large. With rufous flight-feathers. **16a** ad., **b** pull. S Arg/Chile. Text p. 64

PLATE I

Tinamous

PLATE II. Grebes (see also Plates VIII and X).

Expert divers with pointed bills, vestigial tails and lobed feet placed far back. Smallest species often look fluffy and round, but become slim when scared. 1, 2 and 3 may be very skulking. Rarely fly (2 species flightless). Identification problems mainly with young birds ('stripeheads').

1 LEAST GREBE *Tachybaptus dominicus*. Tiny and dumpy, gray with pale eyes. **1a** ad. breeding dress with black bib. **1b** off-season bird with white throat. **1c** 2-weeks old chick. Often revealed by mellow trills. In temp. zone only very locally. Text p. 67

2 PIED-BILLED GREBE *Podilymbus podiceps*. Stocky, with thick chicken-bill. **2a** ad. nuptial plumage with parti-colored bill and black throat bib. **2b** off-season with buffy brown neck. **2c** 2-weeks old chick. Vibrant chatter characteristic. Local in the temp. zone. Text p. 67

3 WHITE-TUFTED GREBE *Rollandia rolland*. Small, often high-sterned. **3a** ad. in nuptial plumage black with large white-streaked ear-patch and brown underparts of body (ssp *chilensis* shown). **3b** off-season bird browner, with buffy neck (large-billed *morrisoni* shown). **3c** juv. and pull. (on back of **3a**) with warm colors, head buff striped black. Widespread from Andes of Peru to southern lowlands. Text p. 65

4 TITICACA FLIGHTLESS GREBE *Rollandia microptera*. Large and fairly slim. Ragged cap dark, nape chestnut, cheeks streaked, grading to white throat. **4a** ad. ruddy below, **b** 2nd year lighter with white on breast. **4c** juv., and pull. on back of 4a light drab-gray with rufous-striped head. Very vocal. Lake Titicaca area of Peru/Bol. Text p. 66

5 HOODED GREBE *Podiceps gallardoi*. Looks shiny white at a distance. **5a** ad. black-hooded whith white forehead and rufous crest. Juv. has black cap contrasting white throat and nape. **5b** pull. Local in Patagonia. Text p. 70

6 SILVERY GREBE *Podiceps occipitalis*. Looks silvery gray, high stern whitish. Nape black, eyes glowing red. **6a** highland ssp *juninensis* with white throat and drab brown ear-plumes, **6b** southern *occipitalis* with gray throat and golden ear-plumes. **6c** juv. with less black nape and no plumes, **d** pull. Calls short whistles. Widespread in the Andes and southern lowlands. Text p. 69

7 JUNIN FLIGHTLESS GREBE *Podiceps taczanowskii*. When calm with fluffy plumage resembles Silvery G., but more slim and long-necked when active. Bill 3 cm long and lighter gray, flanks paler. **7a** ad., **b** juv., **c** pull. Lake Junín c Peru. Text p. 70

Flying grebes are recognized by trailing feet and often somewhat declining necks. The wing pattern is important for species recognition. Only juv. birds shown.

25 per cent of nat. size

PLATE II

Grebes

PLATE III. Herons (see also Plate X).

Medium-sized to large, lanky wading birds with angled necks, strong pointed bills and coarse plumages. Back-folded neck in sustained flight makes the chest look bulky. Wing-beats slow, rhythmic, the arched wings broad. Inhabits marshes, streams and lake shores. Most species are widespread, but breed mainly in lowlands, and only the Night Heron is well established in the high Andes.

1 CATTLE EGRET *Bubulcus ibis*. 50 cm, fairly stocky and heavy-headed. White, stained buff in breeding season. In flight, legs do not project much beyond tail. Widespread, mostly as migrants, but resident locally in the Andes (breeding at 4080 m in Peru). Text p. 76

2 SNOWY EGRET *Egretta thula*. 60 cm. Slim and very elegant. White. Slender bill black. Widespread in Neotropical lowlands, and visits highland marshes in small numbers. Text p. 77

3 GREAT WHITE EGRET *Casmerodius albus*. 110 cm, standing high above other herons. White. Bill yellow. Widespread in Neotropical lowlands, and not unusual in highland marshes. Text p. 75

4 LITTLE BLUE HERON *Egretta caerulea*. 60 cm. Slim. **4a** ad. deep gray, **b** juv. white with dark wing-tips. Imm. spotted. Bill parti-colored. Mainly in trop. coastal swamps. Casual in highlands. Text p. 77

5 WHITE-NECKED HERON *Ardea cocoi*. Dark cap reaches below eye; thighs white (Great Blue H. *A. herodias*, which may visit the n Andes, has small cap and reddish thighs). Widespread in lowlands, but reaches lower edge of temp. zone in Bol. Text p. 81

6 BOAT-BILLED HERON *Cochlearius cochlearius*. Stocky, with big slipper-like bill. Ad. with large black crest, gray wings and back, and black breast-sides. Juv. and depicted imm. mainly brown. Trop. swamps, but straggles to temp. zone in Col. and Ven. Text p. 79

7 PINNATED BITTERN *Botaurus pinnatus*. Ochraceous, streaked and mottled with fuscous (not barred as juv. tiger-herons). Note white-striped throat. In lowland savannas and in large reed-swamps in Andes of Col. Text p. 80

8 FASCIATED TIGER-HERON *Tigrisoma fasciatum*. Ad. black-capped and otherwise densely barred slate and buff. Depicted juv. boldly barred black and tawny, differing from lower-altitude Rufescent T-h. *Tigrisoma lineatum* by cinnamon (not white) primary-tips. Streams in hilly country in premontane zone from Ven. to nw Arg. In valleys of c Peru esp. juv. birds ascend to temp. zone. Text p. 79

9 NIGHT HERON *Nycticorax nycticorax*. Stocky, apparently neck-less, and short-legged. **9a** sooty phase, which dominates in the southern ssp *obscurus*, **9b** light phase *hoactli*. **9c** juv. is striped and spotted. Ad. plumage adopted 3rd year. Throughout S Am., in all kinds of wetlands, low and high. Text p. 78

10 STRIATED HERON *Ardeola striata*. Small, dark-winged. **10a** ad., above dark gray with neckside gray (chestnut in the rarely visiting Green H. *A. virescens*). **10b** juv. striped and spotted. Widespread in lowland swamps, and wanders to highland marshes. Text p. 77

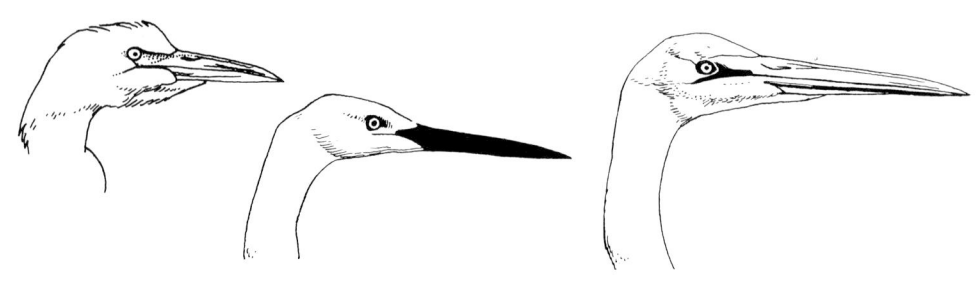

Heads of, from the left, Cattle, Snowy and Great White Egrets.
Note the heavy jowl of the former, and the dark line below the eye in the latter species.

Flying birds in 7 per cent of nat. size; rest in 12 per cent of nat. size

PLATE III

Herons

PLATE IV. Flamingos and Ibises

Flamingos are large, very lanky wading birds that stain minute invertebrates or algae from the water. Typical of shallow, weakly alkaline to hypersaline lakes with open surroundings. Sometimes in very large flocks. Ibises are medium-sized grassland and marsh birds with thin curved bills.

1 ANDEAN FLAMINGO *Phoenicoparrus andinus*. C. 110 cm. Recognized by black tertials, which form large triangular patch in standing birds. Ad. (**1a**) pink with magenta tinge, esp. around base of neck, and on wing-coverts, and with red triangular patch on the axillaries. Bill with pale yellow base demarcated with violet line on lore. Legs pale yellow. Juv. (**c**) drab-colored, with distinct dark stripes above, and with very thin stripes on flanks. **1b** imm., **d** pull. In the arid puna from s Peru to n Arg. and Chile, but mainly at moderate elevations in the Chilean desert.
Text p. 86

2 PUNA FLAMINGO *Phoenicoparrus jamesi*. 90 cm, more moderately long-legged than other flamingos. Ad. **a** pinkish white streaked purplish rosy around base of neck, and with elongated rosy humerals almost hiding black secondaries and primaries. In flight shows red triangular patch on axillaries. In breeding season quite strongly pink with distinct magenta hue. Short swollen bill mainly deep yellow demarcated with red lore. Feet buff to red. Juv. **b** very faintly striped, appearing uniform fawn-colored. **2c** pull. In the arid puna zone, breeding well above 4000 m in Bol. and nw Arg.
Text p. 86

3 CHILEAN FLAMINGO *Phoenicopterus chilensis*. C. 100 cm, more long-legged than the other species. Ad. (**a**) pinkish white with red upper and under wing-coverts, and black primaries and secondaries. Pink of bill not clearly demarcated from face. Legs greenish with red 'knee' and toes. Juv. (**b**) light fawn-colored, distinctly striped above; bill and feet gray. In subsequent plumages white and then progressively more pink (**c**). **3d** pull. Southern lowlands and through the Andes to nc Peru.
Text p. 85

4 BLACK-FACED IBIS *Theristicus melanopis*. Stocky. Foreparts buff, with gray breast-bar; mantle and upper wing-coverts mottled gray. Below black, incl. on underside of very broad wings and short tail. **4a** ssp *melanopis*, **b** juv., **c** puna form *branickii*, **d** pull. Southern lowlands and locally along coast of Peru, and above 4000 m mainly in Peru and Ecu.
Text p. 83

5 WHITE-FACED IBIS *Plegadis chihi*. Rather uniform bronzy chestnut, and more lanky than Puna I. Widespread in lowland marshes from se Brazil to c Chile, and casual in temp. zone of Arg. and Bol.
Text p. 82

6 PUNA IBIS *Plegadis ridgwayi*. Purplish fuscous, generally black below. **6a** breeding dress with dark red bill and chestnut head and neck. **6b** off-season (juv. similar with broader white streaks), **c** pull. Puna zone of Peru to Jujuy, nw Arg.
Text p. 83

Lowland storks accidentally visiting the Andes. From the left: Maguari Stork, Wood Stork, and Jabiru.

Flamingos in 7-9 per cent of nat. size;
Ibises in 12 per cent of nat. size

PLATE IV

Flamingos and Ibises

PLATE V. Geese and Steamer-Ducks (see also Plate VII)

Geese are conspicuous inhabitants of the Patagonian and Andean grassland habitats, esp. along rivers and lakes and on upland bogs. Males give low whistles, females grunt.

1 RUDDY-HEADED GOOSE *Chloephaga rubidiceps.* Tiny goose with rufous head contrasting gray neck. Rufous vent conspicuous in swimming birds. **1a** flying, **b** walking. Islas Malvinas and Isla Grande, visiting valleys of the s Andes on migration. Text p. 120

2 ASHY-HEADED GOOSE *Chloephaga poliocephala.* Small. Tricolored, with gray head; rufous chest and densely barred sides contrasting white lower underparts. **2a** flying male, **b** flying juv. (note dull colors and broad dark wingstripe). **2c** walking female, **d** pull. Mainly wood-fringed bogs in the Andean zone of s Arg. and Chile. Text p. 119

3 MAGELLAN GOOSE *Chloephaga picta.* Female (**3a** and **c**) with buffy brown head and neck grading into densely barred brown body (not the sharp change seen in Ruddy-headed G.). Male white with gray-brown back, those of Isla Grande and the wooded Andean zone densely barred on most of body (*dispar* type **3b**); those of open Patagonian grasslands barred only on the sides (*picta* type **3d**). **3e** juv. male of *dispar*, **f-g** light and dark pull. Very common in s Arg. and Chile. Text p. 118

4 ANDEAN GOOSE *Chloephaga melanoptera.* Stocky. White with streaked mantle grading to glossy black humerals. **4a** flying bird, **b** standing male, **4c** large pull. High parts of Andes of Peru to s Chile. Text p. 118

5 FLYING STEAMER-DUCK *Tachyeres patachonicus.* A heavy grayish duck with pointed tail often raised, exposing white vent. Colors vary seasonally and between sexes and age groups, from dark brown to light blue-gray with scaled pattern, and sometimes white head (see main text). Bill more or less yellow. **5a** typical spring male, **b** typical female, **c** juv., **d** and **e** pull. Southern Arg. and Chile from coast to highlands.
Text p. 120

Flying Andean Geese

8-9 per cent of nat. size

PLATE V

Geese and Steamer-ducks

PLATE VI. Dabbling Ducks (also Plates VII and X, and flying birds on Plate IX)

Dabbling ducks feed in surface position ('dabbling' or up-ending), but do not dive under normal circumstances.

1 CRESTED DUCK *Anas specularioides*. A large, mottled buffy gray-brown duck, dark around eyes, and with dark pointed tail. Esp. ad. males show hanging crest. **1a** female of southern *specularioides*, **1b** male *alticola*, **c** juv., **d** pull. Open lakes and tarns from the Patagonian lowlands through the high Andes to n Peru. Text p. 125

2 SPECTACLED DUCK *Anas specularis*. Dusky brown with sooty rear, and with black and white harlequin pattern on head. **2a** ad., **b** pull. Mainly wood-fringed rivers in the Andean zone of s Arg. and Chile. Text p. 126

3 NORTHERN PINTAIL *Anas acuta*. Female and juv. resemble a slim Yellow-billed P. with gray bill. Drake Plate IX 2b. Migrant visitor to Col. Text p. 126

4 YELLOW-BILLED PINTAIL *Anas georgica*. Slim and fairly long-necked. Mottled buffy brown with rather light unpatterned face. Bill yellow. **4a** ad., **b** juv., **c** pull. Widespread in southern lowlands and most of the Andes to s Col. Text p. 126

5 SPECKLED TEAL *Anas flavirostris*. Small and chunky. With dark gray head, the yellow-billed forms with pale rear. **5a** pale-bodied puna form *oxyptera*, **b** southern *flavirostris*, **c** *andium*, **d** pull. Most widespread Andean duck, with yellow-billed forms (called Yellow-billed T.) in the southern lowlands and the Andean highlands n to n Peru, and gray-billed forms (called Andean T.) from Ecu. to Ven. In all kinds of wetlands. Text p. 124

6 WHITE-CHEEKED PINTAIL *Anas bahamensis*. Head dark-capped with white lower face. Brown with pale cinnamon rear. Trop. lowlands, and casual in highlands. Text p. 127

7 RED SHOVELER *Anas platalea*. Large bill usually held near water surface. **7a** male ruddy, profusely dotted, and with pinkish head. **7b** female anonymously mottled brown. **7c** pull. Southern lowlands, abundant in plankton-rich lakes on Patagonian uplands, and with a small population in Puno and Cuzco, Peru. Text p. 130

8 SILVER TEAL *Anas versicolor*. A southern version of Puna T., small, more buffy, and coarsely marked. Bill usually with yellow spot basally. Lowlands and low Andean passes in southern S Am. Text p. 128

9 PUNA TEAL *Anas puna*. Blue bill fairly long. Black-capped with white lower face. Otherwise gray-brown with pale ashy rear. **9a** female, **b** male, **c** juv., **d** pull. Open weedy lakes in highlands of Peru to nw Arg. Text p. 128

10 CHILOE WIGEON *Anas sibilatrix*. Triangular head with characteristic white face. Also forewing (of ad.) and tail region mainly white. Male gives loud whistled calls. **10a** ad., **b** juv., **c** pull. Widespread in the s, with large numbers staging on Patagonian uplands. Text p. 123

Chiloe Wigeon

25 per cent of nat. size　　　　　　　　　　　　　　　　　　　　　PLATE VI

Dabbling Ducks

PLATE VII. Dabbling and Diving Ducks (see also Plates VI and X, and flying birds on Plates VIII and IX

1 TORRENT DUCK *Merganetta armata*. A small, agile duck of river torrents in the premontane and montane zones along the entire Andes. Female gray with rufous underparts, **1a** northern *colombiana*, **g** *berlepschi*. Male unmistakable, with white head with black lines, but very variably streaked bodies: **1b** *colombiana* of Ven.-Ecu., **d** *leucogenis*, **e** *berlepschi* (which intergrade with *turneri* and *garleppi* through Peru to c Arg./Chile), **f** southern *armata*. **1c** juv., **h** pull. Text p. 121

2 CINNAMON TEAL *Anas cyanoptera*. Forewing blue. **2a** male ssp *cyanoptera* (sometimes more black-spotted), **b** female, **c** eclipse male, **d** pull. Resembles Blue-winged T., but ruddier and slightly longer-billed (approaching Red Shoveler), the bill set off by a steeper forehead, and face less distinctly marked. Local in lowlands as well as in the high Andes. Text p. 129

3 BLUE-WINGED TEAL *Anas discors*. Small. Male (**a**) unmistakable. Female (**b**) and juv. like a faded Cinnamon T., with distincly pale loral spot, and ridge of bill and crown forming a smooth curve. Northern migrant to mainly the northern Andes. Text p. 128

4 SOUTHERN POCHARD *Netta erythrophthalma*. Male (**a**) deep purplish brown. Female (**b**) gray-brown with partially white face, and unlike Rosybill female with black rear. **4c** pull. In weedy lakes from coast to highlands, but extremely rare and local. Vanishing. Text p. 131

5 ROSY-BILLED POCHARD *Netta peposaca*. Male (**a**) unmistakable. Female (**b**) like a light female Southern P., but with white vent, and with cheeks vaguely patterned or uniform buffy. Lowlands in southern S Am., visiting Patagonian uplands in spring. Text p. 131

6 MASKED DUCK *Oxyura dominica*. Tiny and skulking, often i rich floating vegetation. Barred ferrugineous, male (**a**) with black mask, female (**b**) with striped head (incl. pale supercilium). In lowland swamps, and to temp. zone in Col. and Ecu. Text p. 133

7 RUDDY DUCK *Oxyura jamaicensis*. Stockily built with large, rather angular head, and short thick neck. Males often cock stiff tail. Male (**7a**) with blue bill, in breeding season chestnut with black head, in eclipse more female-like. White-cheeked form see Plate X 5. Female (**b**) quite uniform dark brown, juv. (**c**) with more conspicuous stripe on cheek. Pull. **d**. In lakes with rich submergent vegetation in the entire Andes.
Text p. 133

8 LAKE DUCK *Oxyura vittata*. Resembles Ruddy D., but smaller, with more round head and parallel-sided bill (not distally expanded), and longer, more graduated tail. Note the extension of black on neck of male (**a**), and barred pattern of female (**b**). Lowlands of southern S Am. and some lakes in the S Andes. Text p. 134

White-faced Whistling Duck

25 per cent of nat. size

PLATE VII

Dabbling and Diving Ducks

PLATE VIII. Flying Swans and Ducks (see also Plate V and IX)
Great Grebe and Cormorants

1 COSCOROBA SWAN *Coscoroba coscoroba.* Huge (100 cm) white 'duckswan' with rosy bill and big feet, and black wingtips. Characteristic bugling call. **1a** ad., **b** juv. Southern S Am., in lowlands, and visiting Patagonian upland lakes. Text p. 117

2 BLACK-NECKED SWAN *Cygnus melanocoryphus.* 125 cm. Unmistakable. **2a** ad., **b** juv. Southern S Am., in lowlands as well as Patagonian uplands and lakes of the S Andes. Text p. 117

3 SPECTACLED DUCK *Anas specularis.* Dark gray-brown duck with blackish wings with white axillaries and trailing stripe. White spots on face. Mainly wood-fringed streams in the S Andes. Text p. 126

4 CRESTED DUCK *Anas specularioides.* Large buffy gray-brown duck with pointed dark tail. In flight broad white secondary tips are conspicuous. Mainly barren lakes in Patagonia and in the high Andes n to n Peru. Text p. 125

5 FLYING STEAMER-DUCK *Tachyeres patachonicus.* Heavy dark gray duck with big white wing-patch. S Arg. and Chile from coast to highlands. Text p. 120

6 LESSER SCAUP *Aythya affinis.* White wing-linings and white wing-stripe on secondaries. **6a** male, **b** female. Northern migrant to marshes in the northern Andes. Text p. 132

7 SOUTHERN POCHARD *Netta erythrophthalma.* Almost black, except for broad, but not very long white wing-stripe. **7a** male, **b** female with white on face. In weedy lakes from coast to highlands, but extremely rare and local. Vanishing. Text p. 131

8 COMB DUCK *Sarkidiornis melanotos.* Wings all dark. Head white dotted black. **8a** male (75 cm, with frontal comb), **b** female. **8c** juv. obscurely mottled buffy-brown. Scattered in lowland savannas, and casual in the Andes. Text p. 121

9 ROSY-BILLED POCHARD *Netta peposaca.* Extensive white zone on wings, and with white vent. Male. Female on Plate VII 5b. Lowlands of southern S Am., visiting Patagonian uplands in spring. Text p. 131

10 FULVOUS WHISTLING-DUCK *Dendrocygna bicolor.* Lanky with big feet trailing, and all dark wings. Otherwise tawny with white base of tail. Casual in the Andes. Text p. 116

11 TORRENT DUCK *Merganetta armata.* Heavy flight rarely seen. Wings dark gray, with scarcely visible white line. **11a** male, **b** female ssp *colombiana.* Andean torrents from n to s. Text p. 121

12 MASKED DUCK *Oxyura dominica.* Tiny. In rare buzzy flight shows white patch on both wing-sides. Female; male on Plate VII 6a. In lowland swamps, and to temp. zone in Col. and Ecu. Text p. 133

13 RUDDY DUCK *Oxyura jamaicensis.* In rare buzzy flight looks compact and all dark. **13a** male, **b** female. In lakes with rich submergente vegetation in the entire Andes. Text p. 133

14 GREAT GREBE *Podiceps major.* A large diving bird with long bill and long neck often S-curved. **14a** breeding dress of ssp *navasi* of the southern Andes; **b** ssp *major*, in rather light-throated non-breeding dress. **c** juv. Lowlands of southern S Am., and the lake districts in the southern Andes. Text p. 68

15 NEOTROPIC CORMORANT *Phalacrocorax olivaceus.* Large dark diving bird. On water and in gooselike flight usually lifts bill. **15a** ad. breeding dress, **b** non-breeder, **c** juv. Widespread. Text p. 72

16 IMPERIAL SHAG *Phalacrocorax atriceps.* White below incl. cheeks and neck-sides. Southern seacoasts, and Andean lakes in wc Arg. Text p. 73

1 and 2 in 5 per cent of nat. size
3-13 and 15c in 6.5 per cent of nat. size
14-16 in 13 per cent of nat. size

PLATE VIII

Flying Swans and Ducks. Great Grebe and Cormorants

PLATE IX. Flying Dabbling Ducks (see also Plates VI, VII and VIII)

Dabbling ducks rise steeply without need for running start used by most other waterbirds. Esp. females and the buffy brown species look quite similar. Usually, however, patterns of wings and tail provide reliable differences. The **speculum** is the dark metal-colored patch on the secondaries.

1 YELLOW-BILLED PINTAIL *Anas georgica*. Mottled buffy brown with gray underwings. Fairly long-necked. **1a** ad., with speculum distinctly bordered with buff stripes; **b** juv. with narrowly demarcated speculum. Widespread in southern lowlands and most of the Andes.
Text p. 126

2 NORTHERN PINTAIL *Anas acuta*. Slim and elegant. **2a** male with dark head and needle-pointed tail; **b** female, like Yellow-billed P., but bill dark. Rare migrant to Andes of Col.
Text p. 126

3 WHITE-CHEEKED PINTAIL *Anas bahamensis*. Capped, white on lower face. Otherwise brown with green speculum with broad cinnamon framing, and with pale rear. Trop. lowlands, and casual in the Andes. Text p. 127

4 SILVER TEAL *Anas versicolor*. Resembles a small buffy Puna T. Lowlands and low Andean passes of southern S Am. Text p. 128

5 PUNA TEAL *Anas puna*. Capped, with white lower face. Otherwise grayish, with white band under wing, thinly framed speculum and pale rear. Common in open, weedy lakes in the puna zone. Text p. 128

6 SPECKLED TEAL *Anas flavirostris*, dark-billed ssp *altipetens* – yellow-billed forms with similar flight-marks. Grayish with restricted white band under wing, and with buff stripes demarcating dark speculum. Widespread in all kinds of wetlands in the Andes and in southern lowlands.
Text p. 124

7 GREEN-WINGED TEAL *Anas crecca*. Small with broad cinnamon stripe in front of, and white stripe behind green speculum. **7a** male dark-headed with yellow patch on vent; **b** female. Accidental in Andes of Col. Text p. 124

8 BLUE-WINGED TEAL *Anas discors*. Small duck with blue forewings and white band under wings. **8a** male in nuptial plumage, dark with white patch on face and near the rear, and with white-edged speculum. **8b** female, resembling Cinnamon T., with ill-defined speculum, and generally grayer. Northern migrant mainly to the northern Andes. Text p. 128

9 CINNAMON TEAL *Anas cyanoptera*. Small duck with blue forewing and white band under wing. **9a** nuptial male chestnut, broadly white in front of mirror. **9b** female uniform mottled brown, the bronzy speculum not clearly demarcated. Local, both in lowlands and in the Andes.
Text p. 129

10 CHILOE WIGEON *Anas sibilatrix*. With extensive white areas. Forewings gray in juv. Lowlands as well as uplands of southern S Am.
Text p. 123

11 AMERICAN WIGEON *Anas americana*. Grayish duck with square white belly. Male (**11a**) with white forehead, forewing, and bar in front of black tail. Female (**b**) and juv. grayish, with white band under wing. Rare migrant in Col.
Text p. 123

12 RED SHOVELER *Anas platalea*. Because of its large bill, wings seem set far back. Blue forewings and white stripe in front of speculum. **12a** ruddy male, **b** female. Mainly southern lowlands, but with huge molt assemblies on Patagonian uplands, and resident in Puno and Cuzco Peru. Text p. 130

13 NORTHERN SHOVELER *Anas clypeata*. Like previous species, but male (**13a**) with another, unmistakable pattern. **13b** female. Casual migrant in Andes of Col. Text p. 130

9 per cent of nat. size

PLATE IX

Flying Dabbling Ducks

PLATE X. Birds of the Bogotá wetlands.

The wetlands of the E Andes of Col., esp. on the Bogotá and Ubaté savannas at 2600 m, and in Lake Tota (3020 m), once had an outstanding fauna of waterbirds, both endemic species and highland sspp of lowland forms. This fauna now vanishes because of habitat destruction and inadequate protective measures.

1 NICEFORO'S PINTAIL *Anas georgica niceforoi*. A dark, short-tailed version of Yellow-billed P. (Plate VIa-c). Probably extinct.
Text p. 126

2 AMERICAN COOT *Fulica americana columbiana*. By behavior, voice, and bill-tip patches resembles N. Am. C. rather than Andean C. (*Fulica ardesiaca*, Plate XVI 6a-e). Range may extend to Andes of s Col. and n Ecu. Text p. 154

3 CINNAMON TEAL *Anas cyanoptera borreroi*. Male very dark cinnamon, below nearly black, and somewhat spotted. For separation of other plumages from Blue-winged T. see Plate VII. Previously the range extended to n Ecu. Now vanishing. Text p. 129

4 COLOMBIAN GREBE *Podiceps andinus*. Dark head with high crown and chestnut ear-plumes. Neck and sides partly white outside breeding season. Probably extinct. Text p. 69

5 COLOMBIAN RUDDY DUCK *Oxyura jamaicensis andina*. Very variable, as many drakes resemble southern black-headed *ferruginea* (Plate VII 7d) except for some white feathers on lore and ear, others northern white-cheeked *jamaicensis a&b* ad males, except for some black cheek spots. Local. Text p. 133

6 YELLOW-HOODED BLACKBIRD *Agelaius icterocephalus bogotensis*. **6a** ad. male unmistakable, black with yellow head; **6b** ad. female browner with buffy throat; **c** juv. with more extensively buff head. Reed-marsh. Very local.
Text p. 577

7 TORRENT TYRANNULET *Serpophaga cinerea*. Also shown on Plate XLVII 6ab. In most of Andes along montane streams, but in reed-marsh in the Bogotá area. Text p. 467

8 SPOT-FLANKED GALLINULE *Gallinula melanops bogotensis*. Slaty with brown back and white flank-spots. Note green bill. **7a** ad., **b** juv. Often swims among floating vegetation. Rather flat-backed compared with Common G. (Plate XVI 7). The commonest waterbird of the area. Text p. 148

9 LEAST BITTERN *Ixobrychus exilis bogotensis*. Tiny heron buffy-brown with cap and back black (male **9a**) or dark brown (female). Streaked juv. on **9b**. Note rich buff forewing. Rare and local in reedmarsh. Text p. 80

10 GRASSLAND YELLOW-FINCH *Sicalis luteola bogotensis*. **10a** male, **b** female. Also shown on Plate LX 5a-e. The commonest passerine in the marsh habitats of the Bogotá area.
Text p. 657

11 BOGOTÁ RAIL *Rallus semiplumbeus*. A long-billed rail with gray foreparts and chestnut wings. **11b** juv. Fairly common in marshes, and also inhabits páramo fens. Text p. 145

12 APOLINARS MARSH-WREN *Cistothorus apolinari*. Larger and darker, more uniform than Grass-wren (Plate IL 9a-d), but conspicuously streaked on back. Most readily recognized by its persistent rasping churring when disturbed. Local, in tall reeds. Text p. 542

The Bogotá area also has the only highland population of Pinnated Bittern, and is exceptional among localities of such elevation for the number of casual and accidental visitors among lowland marsh-birds (Anhinga; Great Blue, Agami, and Boat-billed Herons; Snail and Plumbeous Kites; Muscovy Duck; Azure Gallinule; Sungrebe), and occurrence of rare migrants as Northern Harrier, American Wigeon, Green-winged Teal, and Northern Pintail.

About 25 per cent of nat. size

PLATE X

Birds of the Bogotá wetlands

PLATE XI. Caracaras, Buzzard-eagle and Vultures

Caracaras have long rectangular wings with free 'fingers'. They fly with vigorous but stiff and shallow wingbeats, some gliding, but very little soaring. They are often seen walking or running in grassland, sometimes several together. Calls harsh, often rasping. New World Vultures are large to huge scavengers with deeply 'fingered' wings.

1 CRESTED CARACARA *Polyborus plancus*. Wingspan 120 cm. Unmistakable with deep bill and flat, dark-capped head. Dark, barred (ad., **a**) or streaked (juv., **b**), except for whitish lower part of head, primary-bases, and basis of tail. Widespread in rangeland and savanna in the lowlands, and local in the temp. zone. Text p. 109

2 MOUNTAIN CARACARA *Phalcoboenus megalopterus*. Wingspan 100 cm. Ad. (**a**) pied, black chest contrasting white lower underparts. Juv. (**b**) dark brown with whitish base of primaries and tail. Puna zone of Peru to c Arg. and Chile. Text p. 108

3 WHITE-THROATED CARACARA *Phalcoboenus albogularis*. Ad. (**a**) differs from Mountain C. by white c underparts from chin to vent. Juv. (**b**) scarcely differs. Replaces Mountain C. s of Neuquen in c Arg. Text p. 108

4 CHIMANGO CARACARA *Milvago chimango*. Resembles a miniature, light *Phalcoboenus* juv., with more extensive pale areas in wings and at base of tail. Frequent voice a petulant chatter. Ssp *chimango* shown. Southern lowlands into Andes of Arg. and Chile. Text p. 110

5 CARUNCULATED CARACARA *Phalcoboenus carunculatus*. Differs from Mountain C. by heavily black-streaked belly of ad. (**a**); juv. normally more rufous and spotted, and showing narrowly barred base of tail (**c**). Imm. (**b**) quite variegated. Páramos of Ecu. to s Col. Text p. 107

6 BLACK-CHESTED BUZZARD-EAGLE *Geranoaetus melanoleucos*. Wingspan 175-200 cm. Flight profile triangular due to extremely broad-based wings and very short tail. Flight-feathers appear uniform gray. **6ab** ad. gray with rather pale upper wing-coverts and blackish chest contrasting whitish underparts of body. **6c** juv. blackish brown with many white streaks on nape, and mainly orange-buff breast and dark vest; tail mottled and only moderately short. Imm. with mainly black chest grading towards lighter belly. For plumage variation see drawing in the main text. Widespread in wooded and open, rugged habitats in the southern lowlands and in the entire Andean zone. Ssp *australis* shown. Text p. 98

7 BLACK VULTURE *Coragyps atratus*. Wingspan 150 cm. Broad wings and short tails recall Black-chested Buzzard-eagle, but wings more rectangular and 'fingered'. Flies with vigorous flaps and glides. Black with pale base of primaries (*brasiliensis* **b**; *foetens* **a** with restricted pale patch). Widespread in S Am., esp. near villages, but to the temp. zone only in the northern Andes. Text p. 88

8 TURKEY VULTURE *Cathartes aura*. Wingspan 180 cm. Glides and soars on slightly raised wings, with much tilting. Blackish, with 2-toned wings. Fairly long, rounded tail. **8a** ad. *falklandicus*, **b** juv. Throughout S Am., but rare and local in highlands. Text p. 89

9 ANDEAN CONDOR *Vultur gryphus*. Wingspan over 300 cm. Long rectangular wings. Glides and soars with little tilting. Ad. with white collar and whitish panel on upperside of wings, **9a** casqued male, **b** female. **9c** dusky brown juv. In páramos from Ven. to Ecu., and from coast to highlands from Peru s-wards. Text p. 90

To plate XII:

11 HOOK-BILLED KITE *Chondrohierax uncinatus*. Rather small hawk-like bird with rectangular wings distinctly narrowed near body. Tail always conspicuosly barred. **11a** rare dark phase, **b** male (normally gray), **c** female (normally browner with rufous collar), **d** juv. black above, with pale collar, and mainly white below. Large parrot-bill and odd visage distinctive at short range. Forest edges in all warm parts of the Americas, ascending in the Andes to lower temp. zone. Text p. 92

3-4 per cent of nat. size

PLATE XI

Caracaras, Buzzard-eagle, and Vultures

PLATE XII. Larger, broad-winged Hawks

Hawks with large broad wings, which are distinctly (but not very deeply) fingered, and with medium long tail often spread. Many species are extremely variably colored, and safe identification requires training and familiarity with the species' jizz.

1 WHITE-RUMPED HAWK *Buteo leucorrhous*. Small dark hawk with broadly barred tail, white rump and vent, and light wing-linings contrasting dark flight-feathers. **1a** ad., **b** imm., **c** juv. Wooded hills in the premontane zone (occ. to 3650 m). Text p. 101

2 WHITE-THROATED HAWK *Buteo albigula*. Rather small and light hawk with flight- and tail-feathers inconspicuously barred and gradually darker towards rear edge. Underparts light, with heavy streaking on sides of breast. **2a** ad., **b** juv. Rare, but quite widespread, mainly in montane woodlands at 2100-3500 m from Ven. to Bol., and in Andes of c Arg. and Chile. Text p. 102

3 SHORT-TAILED HAWK *Buteo brachyurus*. Small, decidedly short-tailed. Flight- and tail-feathers weakly barred and gradually darker towards rear edge. Underside and wing-linings white or black, acc. to phase. Always white near base of bill. **3a** dark, **b** pale phase. Soars over wood-fringed soggy meadows and lake shore in warm lowlands, and occ. to 2500 m in Col. Text p. 102

4 BROAD-WINGED HAWK *Buteo platypterus*. A small and chunky hawk with broad wings and fairly large and distinctly barred tail. Underside generally light, with distinct dark trailing edge of wings and tail. **4a** ad., **b** juv. Northern migrant to forested hills (mainly 500-3000 m) s to Bol. Text p. 101

5 OSPREY *Pandion haliaetus*. Quite large and slender raptor which hold wings as an open M, both when seen from below and head-on. Mostly white below, with dark wrist-spots. Migrant visitor along coasts and lowland rivers of S Am., and casual in much of the Andes. Text p. 91

6 RUFOUS-TAILED HAWK *Buteo ventralis*. Resembles a large Red-backed H. with somewhat square wing-tips. Does not normally hover. Note dark patch of lesser under wing-coverts. Ad. in light phase has tail rufous, white-tipped with 8-10 dark bars; dark phase with weak tail-bars. Juv. (illustrated) have denser tail-bars. Forests of s Chile and adjacent Arg. Rare. Text p. 106

7 RED-BACKED HAWK *Buteo polyosoma*. A medium-sized hawk of extremely variable colors often seen hovering heavily over bushy slopes. Lighter built, looking less broad-winged but with relatively larger tail than Puna H. 3rd primary (from outside) longer than 5th. Ad. with pale, finely barred flight- and tail-feathers, tail with broad black terminal bar. Colors of undersides (incl. wing-linings) usually white (**a**), but sometimes almost black, occ. with some rufous (**b**). Juv. (**c**) with thin tail-bars only, and otherwise extremely variable. Widespread in the Andes, mainly at moderate elevations, and in Patagonian lowlands. Text p. 103

8 PUNA HAWK *Buteo poecilochrous*. Unlike Red-backed H. with 3rd primary shorter than 5th, and wings larger and more rounded. Similar color variation, but with a predominance of dark-phase birds. **8a** dark ad. male, **b** light imm. female, **c** medium dark juv. female. See also line-drawings in text. Mainly in rugged puna terrain at 3000-5000 m from Peru to nw Arg., and in páramos of Ecu. into sw Col. Text p. 104

9 BLACK-AND-CHESTNUT EAGLE *Oroaetes isidori*. A large, powerful eagle of humid montane forest. Ad. (**a**) chestnut and black (looks all dark at a distance) except for light primaries and tail-feathers, the tail with dark distal bar. Juv. (**b**) with head and underside light, slightly streaked, and 4 tail-bars. This tail is kept with the partly chestnut imm. plumage. In undisturbed premontane forest (locally to 3500 m) from Ven. to nw Arg. Text p. 98

10 SWAINSON'S HAWK *Buteo swainsoni*. With characteristic long wing-tips and teetering flight. Flight- and tail-feathers always rather dark and inconspicuously barred. Common light phase with brown band on chest shown. Also dark and rufous phases exist. Juv. has streaked breast. Migrant from N Am. to plains of Arg. passing the n Andes, usually high up. Text p. 103

Text for 11 on previous page

4.5 per cent of nat. size

PLATE XII

Larger, broad-winged Hawks

PLATE XIII. Kites, *Accipiter* Hawks, and Falcons

1 SWALLOW-TAILED KITE *Elanoides forficatus*. Unmistakable, graceful kite, almost constantly in the air. Casually crosses the Andes of Col., Ecu., and Peru. Text p. 93

2 WHITE-TAILED KITE *Elanus leucurus*. Quite gull-like, pale gray, with large black shoulders and white tail. Often hovers. Casually crosses the Andes. Text p. 93

3 SNAIL KITE *Rostrhamus sociabilis*. Dark with very broad, fingered, but somewhat narrow-based wings, and pale base of large square tail. **3a** female, **b** ad. male. Casual in temp. zone.
Text p. 93

4 ROADSIDE HAWK *Buteo magnirostris*. May resemble an *Accipiter*, except for the more rectangular wings and heavier head. 4-banded tail. Ad. has extensively brick-red wings. **4a** ad. of the dark-headed southern ssp *saturatus*, **b** juv. A long gliding whine, sometimes wheezy, is characteristic. Forest edge in all warm parts of S Am., and sometimes to lower edge of temp. zone.
Text p. 100

Accipiter hawks have long, distinctly barred tails, and rather broad-based wings not conspicuously fingered. Usually concealed in vegetation, and mostly fly low, weaving through the forest, with series of choppy wingbeats alternating with glides.

5 COOPERS HAWK *Accipiter cooperi*. Fairly large (c. 40 cm), with tail-tip wedge-shaped. Juv. depicted. Accidental migrant to the n Andes.
Text p. 97

6 BICOLORED HAWK *Accipiter ventralis*. Usually 2-toned gray with chestnut thighs. **6a** ad. *chilensis* of the Andean zone of s Arg. and Chile, with spotted underside and rusty wing-linings. **6b** ad. of the chaco form *guttifer*, with rusty underside and wing-linings, **c** ad. *bicolor*, with pale gray underparts (juv. plain buff below). **6d** juv. *chilensis*. Text p. 96

7 SHARP-SHINNED HAWK *Accipiter striatus*. Tiny, with square or notched tail. **7a** juv., **g** ad. of the rare dark phase of ssp *ventralis* of the n and middle Andes; **b** and **f** ad. males, **c** imm. of paler *ventralis*. **7d** ad., **e** imm. of the chaco form *erythronemius*. Widespread in woodlands and forest of the premontane zone, locally to well up in the temp. zone. Text p. 96

Typical falcons have slim, pointed wings, and usually rather long tails. Most take avian prey, in the air.

8 MERLIN *Falco columbarius*. Small falcon of somewhat *Accipiter*-like habits. Indistinct facial pattern. **8a** female, **b** small male. Migrant visitor to lowlands and highlands of Col. and Ecu.
Text p. 112

9 PEREGRINE *Falco peregrinus*. Powerful falcon with rather short tail (shorter than wing width in flight). **9a** juv. female of the northern migrant *tundrius*, **b** ad. male *tundrius*, **c** juv. female *tundrius*; **d** juv. of the rare pale phase of the southern ssp *cassini* (Kleinschmidt's or Pallid F., previously named *Falco kreyenborgi*), **e** ad. normal *cassini*, with large dark hood. In most of western S Am., but generally scarce in the Andes. Text p. 113

10 APLOMADO FALCON *Falco femoralis*. An elegant falcon with long wings and tail, easily recognized by light eyebrow, thin but distinct moustache, and dark 'vest'. **10a** perched ad. *pichinchae*, **b** ad. *femoralis*, **c** juv. with streaked underside. Ssp *pichincha* widespread in open or ligthly wooded habitats in the Andes, *femoralis* in open lowland habitats, and entering valleys of the S Andes. Text p. 112

11 AMERICAN KESTREL *Falco sparverius*. Small ruddy falcon often seen hovering. Characteristic pattern on head. **11a, f, g,** and **h** females, with rufous, barred upperparts; **a** and **h** *aequatorialis*, the others *cinnamominus*. **11b-e** males, with gray wings, and conspicuous distal bar on tail; **b** and **d** the deeply colored ssp *aequatorialis*, **c** the pale-breasted *cinnamominus*, **11e** the chaco form *caerea*. **11f** *cinnamominus* juv. female. Widespread in all kinds of open country in lowlands and highlands. Text p. 111

Flying birds 5 per cent of nat. size
Perched birds 15 per cent of nat. size

PLATE XIII

Kites, *Accipiter* Hawks, and Falcons

PLATE XIV. Larger, long-tailed arboreal birds

1 SICKLE-WINGED GUAN *Chamaepetes goudotii*. Lacking crest and dew-lap, but with naked light blue face. Very dark, except for rufous underparts of body. Ssp *fagani* depicted. Steep forested hills at 1100-3000 m in Andes from Col. to Bol. Text p. 138

2 BEARDED GUAN *Penelope barbata*. With bushy crest. Red dew-lap partly feathered. Short but distinct white streaks on foreparts, esp. on head-sides. Tail with cinnamon tip. **2a** perched, **b** flying. At 1700-3000 m in montane forests of sw Ecu. and nw Peru. Vanishing. Text p. 138

3 BAND-TAILED GUAN *Penelope argyrotis*. As Bearded G., but throat naked, and white streaking more extensive. Ssp *argyrotis* shown here; *colombiana* on Plate LXIII. At 800-3050 m in wet virgin forest in Ven. and nw Col. Text p. 137

4 ANDEAN GUAN *Penelope montagnii*. With bushy crest and orange-red dew-lap of throat partly feathered (except in *sclateri*). Dark with silvery gray feather-margins on anterior parts. Tail uniform blackish, unlike in Band-tailed G. **4a** ssp *brookei*, **b** southern ssp *sclateri* (with whitish superciliary), **c** pull. In humid forest at 1800-3500 m from n Ven. to nw Arg. Text p. 136

5 RED-FACED GUAN *Penelope dabbenei*. Silvery gray crest contrasts black forehead and naked red face and lappet. Quite rich brown with some white streaks. In large forest tracts at 1800-2500 m in s Bol. and nw Arg. Text p. 136

6 GREEN JAY *Cyanocorax yncas*. Unmistakable, green above, yellow below and on tail-sides. Head black and dark blue, with pale yellow crown except in Ven. and E Andes of Col. **6a** *cyanodorsalis*, **b** *yncas*. Humid premontane forest (to 2800 m) from Ven. to Bol. Text p. 536

7 COLLARED JAY *Cyanolyca viridicyana*. Blue, with silvery or pale blue forehead and black mask. Breast-band white in Peru-Bol., black in Ven to n Ecu. **7a** ad. *viridicyana* (Bol.), **b** juv. *meridana* (ad. *meridana* on Plate LXIV), **c** ad. *jolyaea* (Peru). Montane forest at 1600-3250 m.
Text p. 535

8 TURQUOISE JAY *Cyanolyca turcosa*. S Col. to extreme n Peru at 1850-3200 m. 2 specimens show range of color hues. In the n told from Collared J. by lighter turquoise crown and throat and whitish forehead. **a** ad *se* Equ., **b** ad. n Equ.
Text p. 536

9 GUIRA CUCKOO *Guira guira*. Very shaggy-looking, with characteristic tail pattern. Gives long gradually falling series of *pee-oop* shrieks. Lowlands s of the Amazon forest reaching lower temp. zone in nw Arg. Text p. 220

10 SQUIRREL CUCKOO *Piaya cayana*. Rufous with gray underparts of body. Large tail mainly black with large white feather-tips. In shrub and forest in all warmer lowlands of S Am. to lower edge of temp. zone on adjacent Andean slopes. Text p. 218

11 STRIPED CUCKOO *Tapera naevia*. Streaked buffy brown with brown bushy crest and white eyebrow. Most easily recognized by its repeated plaintive guttural whistle *suu-see*. Widespread in warm lowlands, and to edge of temp. zone in Ven. Text p. 220

12 DARK-BILLED CUCKOO *Coccyzus melacoryphus*. Gray-brown upperparts and warm buff underparts separated by gray sides of neck and breast. White feather-tips in tail. No rufous visible in closed wing. Widespread in warm lowlands and sometimes to temp. zone. Text p. 218

13 YELLOW-BILLED CUCKOO *Coccyzus americanus*. Gray-brown above, white below and with white feather-tips in tail. Rufous under wings and some rufous also visible in closed wing. **13a** perched, **b** flying, **c** flying Long-tailed Mockingbird for comparison. Northern migrant visiting highlands of Andes. Text p. 217

14 SMOOTH-BILLED ANI *Crotophaga ani*. Grotesque-looking, black. Differs from Groove-billed A. *C. sulcirostris* of arid zones by more highly arched bill lacking longitudinal furrows, and by its rising whine *oueeeenk* instead of down-slurred *kee-wuy* of Groove-billed A. Both species occur locally in temp. valleys, esp. in Peru. Text p. 219

15 GREATER ANI *Crotophaga major*. 45 cm, very large-tailed, and ad. with white eye. Lowland swamps, and accidental in the temp. zone.
Text p. 219

15 per cent of nat. size

PLATE XIV

Larger, long-tailed, arboreal birds

PLATE XV. Jacana, Rails, and Crakes (see also Plates X and XVIII)

Birds of marshes and fens. Except for the Jacana, these species are heard much more than seen.

1 WATTLED JACANA *Jacana jacana*. Unmistakable, with its extremely long toes and greenish yellow wings. **1a** ad. ssp *intermedia*, **b** juv. Casual in highlands. Text p. 155

2 PLUMBEOUS RAIL *Rallus sanguinolentus*. Slate-colored with long green bill with red and blue base. Usually revealed by series of rolling whines. **2a** ad. of highland form *tschudii*; **b** huge ssp *luridus* of the Fuegian zone; **c** juv.; **d** pull. Reedmarsh in lowlands as well as the Andes from Isla Grande to n Peru. Text p. 143

3 BLACK CRAKE *Laterallus jamaicensis*. Tiny, blackish, runs like a mouse through the vegetation. May be confused with chicks of other rails. **3a** ssp *tuerosi* of Lake Junín, c Peru with white-barred back, **b** russet-mantled *salinasi* of Chile into adjacent wc Arg., **c** juv., **d** pull.
Text p. 146

4 LESSER RAIL *Rallus limicola*. Long bill dark with red base. **4a** ad., with underparts fawn-colored, wings rufous. **4b** blackish juv., **c** pull. In fens and marshes at 2000-3120 m in s Col. and Ecu. Text p. 144

5 *Rallus peruvianus*. Painted on basis of Taczanowski's original description. Peru? Text p. 145

6 AUSTRAL RAIL *Rallus limicola antarcticus*. Recognized by small size, reddish bill, gray foreparts, buff-streaked back, rufous wings, and boldly barred flanks. **6a** ad., **b** large chick. Along the Patagonian Andes. Vanishing. Text p. 145

7 SORA or CAROLINA RAIL *Porzana carolina*. Plump with small pointed bill yellow. Mainly olivaceous with small white dots. Vent white. Migrant visitor to lowlands and highlands s to Peru. Text p. 146

8 PAINT-BILLED CRAKE *Neocrex erythrops*. Small bill yellow and red. Coloration uniform, but vent barred. Local in lowland swamps, and casual visitor in the Andes. Text p. 147

9 COLOMBIAN CRAKE *Neocrex colombianus*. Small greenish bill with sometimes red base. Unlike in Paint-billed C. with buffy vent. Possibly to temp. zone in Ecu. Text p. 147

10 SPOT-FLANKED GALLINULE *Gallinula melanops*. A dull gray and brown gallinule with green bill and white-dotted flanks. **10a** ad. ssp *crassirostris*, **b** pull. Juv. on Plate X 8b. Often swims among floating plants. In the s part of the continent, and E Andes of Col. Text p. 148

11 PURPLE GALLINULE *Porphyrula martinica*. A very lanky gallinule with all white under tail-coverts. **11a** imm., **b** the unmistakable purplish-blue ad. Casual in highland marshes.
Text p. 147

Plumbeous Rail

Jacana, Rails, and Crakes

PLATE XV

25 per cent of nat. size

PLATE XVI. Coots and Common Gallinule

Coots are mainly chicken-sized plump waterbirds which, unlike rails, often appear open to view on grazed shores, along reed-borders, or out on open water. Ad. birds are deep gray with black heads. Part of the food is fetched diving. Often in large flocks, but pugnacious and territorial in breeding season.

1 GIANT COOT *Fulica gigantea*. Goose-sized, very heavy-bodied, but imm. may not appear much larger than Andean C. Ad. (**a**) with knob above each eye, bill red-tipped with white upperside, huge feet red. **1b** juv., **c** pull. Open high Andean lakes (mainly at 4000-4500 m) of Peru to extreme nw Arg. Text p. 151

2 HORNED COOT *Fulica cornuta*. Resembles Giant C. Bill yellowish with black erectile 'proboscis'. Feet olive. **2a** ad., **b** pull. Barren lakes of the desert puna near junction of territories of Arg./Bol./Chile. Text p. 152

3 RED-FRONTED COOT *Fulica rufifrons*. Gallinule-like with large white patch below raised rail, and with characteristic straight profile. **3a** ad., **b** pull. Lowland marshes of southern S Am. to base of Andes. Text p. 150

4 WHITE-WINGED COOT *Fulica leucoptera*. 2 white stripes below tail. Conspicuous white rear edge of wings visible in flight. Bill and forehead yellow or orange. **4a** ad., **b** juv., **c** pull. Lowland marshes and weedy montane lakes in Arg. and Chile. Text p. 153

5 RED-GARTERED COOT *Fulica armillata*. 2 white stripes under tail. Bill and forehead pale yellow apparently disconnected due to dark red mark. **5a** ad., **b** juv., **c** pull. Lowlands of southern S Am., ascending to Andean foothills of Patagonia. Text p. 150

6 ANDEAN COOT *Fulica ardesiaca*. Previously treated as 2 species, or as ssp of American C. (Plate X 2). Large, but distinctly smaller than ad. Giant C. 2 white stripes under tail except in northern ssp *atrura*. Most often with bill yellow, forehead maroon, feet green (**c**). In parts of range with bill and forehead often white, feet gray (**a**), or sometimes forehead yellow (**b**). **6d** juv., **e** pull. Highlands lakes from s Col. to n Arg. and Chile, in Peru also along coast. Text p. 152

7 COMMON GALLINULE *Gallinula chloropus*. More in marsh vegetation than the coots, and swims with strong nodding, and with the rear distinctly raised. Often gives sustained cackling. Recognized by 2 broad white stripes under tail, and by thin white side-stripe. **7a** *pauxilla*, **b** *hypomelana*, **c** juv., **d** pull., **e** large highland form *garmani*. In marshes almost all over S Am. Text p. 149

Coots need a running start to become airborne, and show the big feet trailing (like a thin tail) in flight.

18 per cent of nat. size

PLATE XVI

Coots and Common Gallinule

PLATE XVII Quail and aberrant Plovers
(see also Plate XX, and compare Plates I, XVIII and XIX)

Quail are rotund gallinaceous birds of mainly premontane forest. They live under the cover of vegetation, but are often revealed by far-carrying whistled calls.

1 CRESTED BOBWHITE *Colinus cristatus*. With pointed sand-colored crest, and profusely white-spotted body. Call a whistled *quoit, bobwhite*. **1a** ad., **b** fledgling, **c** pull. Northern S Am., mainly in arid parts. Ssp *bogotensis* at 2600-3200 m in E Andes, Col. Text p. 140

2 STRIPE-FACED WOOD-QUAIL *Odontophorus balliviani*. Chestnut. Rufous crest, headside with 2 long rusty stripes. Call a bubbling, rapidly repeated *whydly-i, whydly-i...* Humid forest at 2000-3000 m from Cuzco, Peru to yungas of Bol. Text p. 142

3 GORGETED WOOD-QUAIL *Odontophorus strophium*. Chestnut, head black-and-white. Rare in premontane oak woods n of Bogotá, Col. Text p. 141

Seedsnipes inhabit Andean grasslands, desert, and alpine habitats. They have a somewhat chicken-like shape, but are shorebird-like in flight.

4 WHITE-BELLIED SEEDSNIPE *Attagis malouinus*. 28 cm. With densely scalloped pattern and white belly. Wing-linings white (vs. black in the smaller Gray-breasted S.). When flushed calls continually *tu-whit tu-whit tu-whit...* **4a** ad., **b** pull. Very bleak habitats in s Arg. and Chile. Text p. 181

5 GRAY-BREASTED SEEDSNIPE *Thinocorus orbignyianus*. 23 cm. Male (**a** and **b** showing color variation) with gray foreparts without black median stripe. Female (**c**) is cryptically colored with a scalloped pattern above. **5d** pull. Often revealed by the cooing song *poo**cuy**-poo**cuy**-poo**cuy**...* Wing-linings black. Grassland from Patagonia through the Andean highlands to nc Per. Text p. 182

6 LEAST SEEDSNIPE *Thinocorus rumicivorus*. 17 cm. Male (**6b**) with gray foreneck with broad black midline. Female (**a**, *rumicivorus*) differs from Gray-breasted S. by small size, heavier black demarcation of white chin, and spotty rather than scalloped pattern (not in *bolivianus*). May be recognized in flight by more wedge-shaped tail. **6c** pull. Song resembles Gray-breasted S., but somewhat variable. Quite social. Semideserts of Patagonia, desert puna of Bol., and along coast of Peru. Text p. 183

7 RUFOUS-BELLIED SEEDSNIPE *Attagis gayi*. 29 cm. Rufous, intricately mottled, **a** the pale southern *gayi*, **b** dark northern *latreilli*. **7c** half-grown pull. When flushed calls continuously *gly-gly-gly...*. Barren high alpine habitats in Ecu., and from nc Peru s-wards. Text p. 181

Plovers comprise rather compact shorebirds with normally short bills (see Plates XVIII and XIX). 3 aberrant species are shown here.

8 TAWNY-THROATED DOTTEREL *Oreopholus ruficollis*. 27 cm, incl. long and thin bill. Cryptically colored, with striped back. Belly with black patch. **8a** ad., **b** juv., **c** pull. Semi-desert and rangelands of Patagonia, and through the Andes to Peru, where also near the coast. Text p. 164

9 MAGELLANIC PLOVER *Pluvianellus socialis*. Resembles a small dove or a uniform gray version of a *Thinocorus* seedsnipe. **9a** ad., **b** juv., **c** pull. By Fuegian and Patagonian claypan lakes. Rare. Text p. 166

10 DIADEMED SANDPIPER-PLOVER *Phegornis mitchellii*. Tiny plover with long bill. Dark head with white 'diadem'. **10a** ad., **b** juv., **c** pull. In bogs at very high elevations from nc Peru to c Arg. and Chile. Text p. 165

PLATE XVII

Quail and aberrant Plovers

PLATE XVIII Plovers, Stilts, and Oystercatchers

1 ANDEAN AVOCET *Recurvirostra andina*. Unmistakable by colors and strongly upturned bill. Alkaline puna lakes mainly around 4000 m in c Peru to nw Arg. Text p. 158

2 BLACK-NECKED STILT *Himantopus mexicanus*. Slim, with extremely long, spindly legs. Pointed wings black above and below. **2a** southern ssp *melanurus*, **2b** and **c** male *mexicanus* (juv. Plate XX 15). Widespread in lowland savannas of S Am., but very local in the high Andes.
Text p. 157

3 SOUTHERN LAPWING *Vanellus chilensis*. Large plover with black forehead and midthroat contrasting pale gray of rest of head. Chest black. Broad and blunt wings with white diagonal above (see drawing p.xx). Scolds intruders with series of loud, harsh notes, or (southern populations) with parakeet-like shrieks. **3a** ad. *fretensis*, **b** pull. Widespread in savannas and grasslands of S Am. lowlands, ascending Patagonian foothills, and entering valleys of the S Andes. Occ. in highlands of Col., Ecu., and Bol. Text p. 159

4 ANDEAN LAPWING *Vanellus resplendens*. Large plover with pale gray foreparts and broad white diagonal on upperside of wings (see drawing p. 159). **4a** ad., **b** pull., **c** juv., **d** recently landed bird standing with raised wings (unlike Southern L.). Mobs with incessantly repeated shrill notes. Páramo and puna terrain from s Col. to nw Arg. Text p. 160

5 TWO-BANDED PLOVER *Charadrius falklandicus*. White face and throat demarcated with black, with double bar across breast. Male. Patagonia. Text p. 163

6 PUNA PLOVER *Charadrius alticola*. Above light gray-brown with black demarcation on forehead against white face. Indications of breastbars. **6a** ad., **b** pull., **c** juv., **d** ad. of pale form from Catamarca. Near lakes of arid puna from c Peru to nw Arg. Text p. 162

7 RUFOUS-CHESTED DOTTEREL *Charadrius modestus*. Dark foreparts contrast white belly. Light supercilium, which in the breeding dress (**7a**) forms white 'diadem'. **7b** pull, **c** juv. **d** ad. winter plumage. Patagonia, from coast to highlands. Text p. 163

8 SEMI-PALMATED PLOVER *Charadrius semipalmatus*. With white collar around nape, and dark breast-bar. Juv. Call a plaintive rising *cheweeet*. Migrant, accidental in the high Andes.
Text p. 162

9 MAGELLANIC OYSTER-CATCHER *Haematopus leucopodus*. Unlike in other oystercatchers entirely white secondaries form a large triangular wingpatch in flight. Gives plaintive, very high-pitched pipes. **9a** ad., **b** pull. Patagonia.
Text p. 156

10 AMERICAN GOLDEN PLOVER *Pluvialis dominica*. Medium-sized plover finely mottled gray and golden olive. Underparts spotted black in Mar-Apr. Gives 2-note whistle. Northern migrant crossing the Andes esp. in Mar-Apr.
Text p. 161

11 GRAY PLOVER *Pluvialis squatarola*. Casually with migrating Golden P. in the high Andes. Stronger-billed and paler gray, and has a drawn-out, plaintive 3-note call. Text p. 161

Flying Magellanic Oystercatcher and American Oystercatcher

20 per cent of natural size, but 1, 2c, and 4d not to scale

PLATE XVIII

Plovers, Stilts, and Oystercatchers

PLATE XIX. Snipes and Sandpipers (see also flying birds on Plate XX)

1 NORTH AMERICAN SNIPE *Gallinago gallinago delicata*. Smaller than Noble S., and rufous color of tail not extending to rump. White trailing stripe on wings. Zig-zag flight when flushed. Migrant to Andes of Col. and Ven.
Text p. 178

2 MAGELLANIC SNIPE *Gallinago magellanica*. A larger, rather richly hued version of N Am. and Puna S. Aerial 'bleating' a sustained deep and pulsating rumble. **2a** ad., **b** pull. Widespread in fens and marshes in lowlands and in the Andes of s Arg. and Chile.
Text p. 179

3 PUNA SNIPE *Gallinago andina*. Small. When flushed, flies with small jumps. Short yellow feet hardly project beyond tail-tip. Bleating a hoarse wheezy *shushushu....* **3a** ad., **b** pull. Bogs and marshes in the puna zone of Peru to n Arg. and Chile.
Text p. 179

4 NOBLE SNIPE *Gallinago nobilis*. Quite heavy, with long 2-toned bill. Rufous color of tail extends to rump. In flight shows rather dark underwings and lacks white trailing edge of wing. Flight direct, not twisting. **4a** ad., **b** pull. Marshes in the temp. zone of Col. and Ecu.
P. 177

5 CORDILLERAN SNIPE *Gallinago (s.) stricklandii*. Heavy. Lacks buff stripes and rufous or white in tail. Almost unmarked c underparts. **5a** ad., **b** pull. Crepuscular. In swampy bogs in s Chile.
Text p. 175

6 ANDEAN SNIPE *Gallinago (stricklandii) jamesoni*. Heavy, densely speckled above and below. **6a** ad., **b** pull. Right after sunset gives long series of piercing *djyc* calls, with rising and falling pitch, and aerial displays with loud *wityeeu* calls sometimes interrupted by a wheezy tone resembling a distant jet plane. Boggy treeline and paramó habitats from Ven. to Bol.
Text p. 175

7 BANDED SNIPE *Gallinago imperialis*. Heavy. Dark rufous densely barred. In crepuscular flight gives long series of rough staccato notes that increase and then decrease in volume, with double and triple notes in climactic section. Very rare and local in boggy cloud forest.
Text p. 176

8 GREATER YELLOWLEGS *Tringa melanoleuca*. Told from Lesser Y. by heavier, slightly upturned bill and by giving series of 3-5 loud ringing notes when flushed. Common migrant visitor in the Andes.
Text p. 172

9 LESSER YELLOWLEGS *Tringa flavipes*. Told from Greater Y. by thin straight bill and by the flight call of soft *tew* notes given 2-3 together. Common migrant visitor in the Andes.
Text p. 172

10 SOLITARY SANDPIPER *Tringa solitaria*. Smaller than Lesser Yellowlegs, with white eyering, olive-gray legs, and dark rump. Call a thin *peet* or *peet-weetweet*. Migrant visitor in mainly the northern Andes.
Text p. 173

11 SPOTTED SANDPIPER *Tringa macularia*. Note gray-brown breastside, but best recognized by incessant bobbing action and teetering flight (plate XX 31). Ad. have black spots below in the migratory period. Visits esp. the northern Andes.
Text p. 174

12 BAIRD'S SANDPIPER *Calidris bairdii*. Drab brown with buffier brown foreparts. Dark mid-rump. Calls low, raspy. Winters commonly in the high Andes and Patagonia, often feeding on grassland far from water.
Text p. 169

13 WHITE-RUMPED SANDPIPER *Calidris fuscicollis*. Dull grayish, recognized by rather thick bill, white rump (see Plate XX 24), and fine mouse-like calls. Migrant to the Patagonian steppe, and visits the high Andes.
Text p. 168

14 PECTORAL SANDPIPER *Calidris melanotos*. Richly hued brown, streaked breast contrasting white belly. Juv. shown. Call raspy. Migrant to boggy habitats in the high Andes and Patagonia.
Text p. 168

15 WILSON'S PHALAROPE *Phalaropus tricolor*. Slim and elegant, with needle-thin bill. Swims buoyantly. In winter plumage (**15b**) pale gray above (juv. eventually with some black and buff streaks). From Mar adopts breeding dress with chestnut band on neck-side (**a**). Migrant, appearing in enormous flocks in the flamingo lakes of the altiplanos of Peru and Bol., and in Córdoba, Arg., and also common in Patagonia.
Text p. 171

25 per cent of nat. size

PLATE XIX

Snipes and Sandpipers

PLATE XX Flying shorebirds (see also Plates XVII, XVIII and XIX)

1 RUDDY TURNSTONE *Arenaria interpres.* Unmistakable pattern of back and rump. Accidental in the high Andes. Text p. 167

2 COLLARED PLOVER *Charadrius collaris.* Black loral stripe and breast-bar. No wingbar. To lower temp. zone in Bol. Text p. 163

3 KILLDEER *Charadrius vociferus.* Double breast-bar. Conspicuous white wingbar and rufous base of long tail. Calls loud *kill-deeah* and *dee-ee*. Casual in the n Andes. Text p. 162

4 TAWNY-THROATED DOTTEREL *Oreopholus ruficollis.* Resembles American Golden Plover, but striped above, and with white under-wings. Patagonia to nw Peru. Text p. 164

5 RUFOUS-CHESTED DOTTEREL *Charadrius modestus.* Lacks wingstripe, and has white tail-sides. Patagonia. Text p. 163

6 GRAY PLOVER *Pluvialis squatarola.* Note broad white wingstripe and rump, and black axillaries. Plaintive 3-note whistle. Casual in the Andes, with American Golden P. Text p. 161

7 AMERICAN GOLDEN PLOVER *Pluvialis dominica.* Golden grayish (more or less black below in Apr). Loud whistled 2-note call. Migrant visitor. Text p. 161

8 SEMIPALMATED PLOVER *Charadrius semipalmatus.* White collar, wing-stripe, and tail-sides. Accidental. Text p. 162

9 TWO-BANDED PLOVER *Charadrius falklandicus.* Breast-bars outline white throat. Partly white sides of tail, but lack conspicuous wing-stripe. Patagonia. Text p. 163

10 PUNA PLOVER *Charadrius alticola.* Below white with faint breast-bars. Partly white tail-sides, but lack conspicuous wingstripe. Arid puna zone. Text p. 162

11 MAGELLANIC PLOVER *Pluvianellus socialis.* Note white tail-sides. Fuegian and Patagonian claypans. Text p. 166

12 GREATER YELLOWLEGS *Tringa melanoleuca.* Dark upper wing-side and whitish base of tail. Told from Lesser Y. by loud ringing calls 3-5 together. Common migrant. Text p. 172

13 HUDSONIAN GODWIT *Limosa haemastica.* Has partly rufous underparts when on n-wards migration. Winters in Patagonia, accidental in highlands. Text p. 175

14 WHIMBREL *Numenius phaeopus.* Large streaked buffy brown shorebird with curved bill. Accidental in the Andes. Text p. 175

15 BLACK-NECKED STILT *Himantopus mexicanus.* Juv. May resemble Greater Yellowlegs, but note dark under-wings. Local in the high Andes. Text p. 157

16 LESSER YELLOWLEGS *Tringa flavipes.* Dark upper wing-sides and white base of barred tail. Unlike Greater Y. gives 2-3 soft *tew* notes when flushed. Common migrant. Text p. 172

17 PUNA SNIPE *Gallinago andina.* Snipes are brownish, striped above. Long bills point partly down. For species identification see Plate XIX. Text p. 179

18 UPLAND SANDPIPER *Bartramia longicauda.* Buffy brown densely marked. Long tail for a shorebird. Rare in the Andes. Text p. 174

19 WILSON'S PHALAROPE *Phalaropus tricolor.* Migrant visitors, in enormous numbers in some lakes in the arid puna. Text p. 171

20 DIADEMED SANDPIPER-PLOVER *Phegornis mitchellii.* Flight undulating and weak, passerine-like. High Andean bogs. P. 165

21 WANDERING TATTLER *Tringa incana.* Looks uniform gray. Yellow legs short. Accidental in the Andes. Text p. 173

22 SOLITARY SANDPIPER *Tringa solitaria.* In flight shows dark undersides of wings and dark rump and tail. Migrant visitor in mainly the northern Andes. Text p. 173

23 LEAST SEEDSNIPE *Thinocorus rumicivorus.* The 2 small seedsnipes resemble sandpipers in flight. Remark black wing-linings. Patagonia to w Peru. Text p. 183

24 WHITE-RUMPED SANDPIPER *Calidris fuscicollis.* White rump distinctive. Migrant to Patagonia, sparse in the Andes. Text p. 168

25 SANDERLING *Calidris alba.* Note rather thick bill and large white stripe on dark wings. Casual in the Andes. Text p. 170

Text for 26 to 31 on next page.

15 per cent of nat. size   PLATE XX

Flying Shorebirds

PLATE XXI Harriers and Gulls

Harriers are raptors of grassland, rushy shore meadows, and marshes. With their light gray colors, males may resemble gulls at a distance. They are, however, readily distinguished by conspicuously fingered wings. The wings are raised in an open V as the bird soars buoyantly, with much tilting and quartering, back and forth low over the terrain. Gulls and terns are long-winged whitish birds of seacoasts, lakes, and marshy plains. Gulls often glide and soar. Terns have very pointed wings and forked tails, and fly with deliberate elastic wingbeats.

1 CINEREOUS HARRIER *Circus cinereus*. White rump. Ad. male (**1a**) gray with rufous-barred belly. Female (**b**) chestnut to gray-brown with dense white dots and bars. **1c** imm. male, **d** juv., with streaked underparts. Through the southern lowlands and locally in the high Andes. Text p. 95

2 NORTHERN HARRIER *Circus cyaneus*. As Cinereous H., but male (**2a**) whiter below, tail faintly barred. Female brown streaked below, juv. (**b**) very fulvous. Fewer dark bars on flight- and tail-feathers than Cinereous H. Occ. in Col. Text p. 94

3 FRANKLIN'S GULL *Larus pipixcan*. 36 cm. With rather deep ashy gray back and upperside of wings, and dark semi-hood while in S Am. Ad. (**3a**) with white (transparent-looking) zone inside black bar near wingtip. **3b** 1st winter, with all-dark primaries and black tail-bar. Northern migrant to coast of Peru and Chile, uncommon in Andean puna zone. Text p. 184

4 KELP GULL *Larus dominicanus*. 62 cm. Ad. (**4a**) white with slaty black mantle and upperside of wings (unlike Band-tailed Gulls *L. atlanticus* and *belcheri* of adjacent seacoasts lacking tailbar). Juv. (**b**) obscurely mottled dusky brown. Ad. plumage acquired after 3 years. Seacoasts, and inland in s Arg. and Chile. Text p. 186

5 BROWN-HOODED GULL *Larus maculipennis*. 40 cm, very pale. **5a** ad. of the unnamed Patagonian form, with extensive white along leading edge of hand. Dark brown hood in breeding season. **5b** juv. mottled with some brown, and with restricted tail-bar; inner primaries gray except at tip (darker in juv. Andean G.). **5c** pull. Mainly lowlands of southern S Am. Text p. 185

6 LARGE-BILLED TERN *Phaetusa simplex*. Note heavy bill and distinctive wing-pattern. Non-breeding plumage (black cap in breeding season). Casual in the northern Andes. Text p. 187

7 SOUTH AMERICAN TERN *Sterna hirundinacea*. Very pale and rather heavy-billed compared to related terns of southern S Am. **7a** ad., **b** 2nd year. Inland in s Arg. Text p. 187

8 ANDEAN GULL *Larus serranus*. The only typical high Andean gull. Always with white 'mirrors' in black wing-tip. **8a** ad. breeding dress, **b** winter, **c** 2nd year bird with tail-bar retained, **d** juv., **e** pull. Text p. 186

To plate XX:

26 STILT SANDPIPER *Micropalama himantopus*. Resembles Lesser Yellowlegs, but gives unmelodic calls. A rare visitor to some puna lakes on passage. Text p. 170

27 PECTORAL SANDPIPER *Calidris melanotos*. Note long white patches on sides of black midrump. Migrant to Patagonia and the puna. Text p. 168

28 BAIRD'S SANDPIPER *Calidris bairdii*. Rather uniform with black mid-zone of rump. Common migrant on short-grass terrain in the Andes and Patagonia. Text p. 169

29 LEAST SANDPIPER *Calidris minutilla*. Tiny with needle-thin bill. Thin white wingstripe. Accidental in the Andes. Text p. 169

30 BUFF-BREASTED SANDPIPER *Tryngites subruficollis*. Fairly uniform buff with pale face. Accidental in the Andes. Text p. 170

31 SPOTTED SANDPIPER *Tringa macularia*. Teetering flight with sloping wings characteristic. Migrant esp. to the n Andes. Text p. 174

Harriers in 5-8 per cent of nat. size
Gulls in 8 per cent of nat. size

PLATE XXI

Harriers and Gulls

PLATE XXII Doves and Pigeons (see also Plate XXIV)

1 EARED DOVE *Zenaida auriculata*. Pinkish drab brown, the graduated tail with broad white tips of the outer feathers (or tawny tips in northern S Am.). **1a** ssp *penthera* of nw Ven. and Col., **b** southern ssp *auriculata*, **c** *hypoleuca* of the Andes from Ecu. to n Bol., **d** juv. Widespread in savanna and cultivated areas all over S Am., ascending above 3000 m in many parts. Text p. 191

2 SPOT-WINGED PIGEON *Columba maculosa*. Resembles a Feral P., with white-spotted back, and the depicted highland form *albilinea* with white diagonal in the wing. Throughout the chaco, and in woodlands at 2000-4200 m from nw Arg. to Peru. Text p. 191

3 BAND-TAILED PIGEON *Columba fasciata*. Heavy gray p. with light gray tail-tip. **3a** ad., **b** juv. In humid forest mainly at 2000-3000 m from Ven. to nw Arg. Text p. 189

4 CHILEAN PIGEON *Columba araucana*. More vinaceous purple than previous species. **4a** ad., **b** juv. Forests of Chile into adjacent sw Arg. Text p. 190

5 LARGE-TAILED DOVE *Leptotila megalura*. Like White-tipped D., but more saturated brown, with purplish but no green gloss on hindneck. Ssp *megalura* shown. To 2800 m in the 'valles' of Bol., and in nw Arg. Text p. 198

6 WHITE-TIPPED DOVE *Leptotila verreauxi*. Gray-brown, hindneck with green-glossed feathertips in addition to purplish sheen. In flight shows larger, more fan-shaped tail than Eared D., with white outer corners and rufous underwings. Ssp *decolor* shown. Widespread in Neotropical woodlands, reaching 3000 m locally. Text p. 197

7 BLACK-WINGED GROUND-DOVE *Metriopelia melanoptera*. Gray-brown with white wrist-spot above and below. Medium long square tail black. Bushy slopes (feeding on grassland) in high parts of the Andes from sw Col. to Isla Grande. Text p. 196

8 GOLDEN-SPOTTED GROUND-DOVE *Metriopelia aymara*. Pale gray-brown. In flight only outer corners of the short black tail visible. Chestnut wing-patch and golden wing-covert spots not easily visible. At 3500-5000 m through the desert puna of c Peru to c Arg. Text p. 196

9 BARE-FACED GROUND-DOVE *Metriopelia ceciliae*. Earthy brown with light-spotted back; large white corners of tail. **9a** *gymnops*, **b** and **c** *ceciliae*. Mainly at 2500-4500 m on semi-arid, stony, and bushy slopes of Peru to nw Arg. Text p. 195

10 MORENO'S BARE-FACED GROUND-DOVE *Metriopelia morenoi*. Uniform gray-brown, with small white tail-corners. Vent brown. Pre-puna above 2000 m in nw Arg. Text p. 195

11 COMMON GROUND-DOVE *Columbina passerina*. Sparrow-sized. Grayish with scalloped breast, rufous flight-feathers and white corners of tail. Arid range-land of northern S Am., ascending to arid temp. zone in Col. and Ecu. Text p. 192

12 CROAKING GROUND-DOVE *Columbina cruziana*. Sparrow-sized, with black wing-feathers and yellow bill. Croaking call heard incessantly. Pacific slope from s Ecu. to n Chile. Text p. 194

13 RUDDY QUAIL-DOVE *Geotrygon montana*. Glossy purplish brown (male, **a**) or olive-brown (female, **b**), with light stripe on cheek. Call a low-pitched cow-like hoot. Floor of humid forest in the tropics, but to lower temp. zone in Col. Text p. 199

14 WHITE-THROATED QUAIL-DOVE *Geotrygon frenata*. Large brown dove with pink head with gray nape and black bar on cheek. **14b** juv. (molting). Call a deep *hoohooo*.. Humid forest at 1500-2500 m, or higher, from Col. to Bol. Text p. 199

PLATE XXII

25 per cent of nat. size, but
flying birds (except 7) smaller

Doves and Pigeons

PLATE XXIII Conures
(see also plates XXIV, LXIII and LXIV)

Conures are mainly medium-sized psittacines with long, pointed tails. Almost always in flocks, screeching incessantly in flight. However, they may be silent, and difficult to see, when feeding.

1 MITRED CONURE *Aratinga mitrata*. Green with more or less red on face and sometimes elsewhere (but not red wing-linings of the larger Red-fronted Macaw). Red color of forehead grades to pink on crown (contrasting the green color), and often reaches the cheek. Calls somewhat harsher than in Scarlet-fronted C. **1a** ad. ssp *mitrata* of forests and woodlands to above 3000 m from c Peru to nw Arg.; **b** ssp *alticola* of high parts in Cuzco Peru; **c** juv. Text p. 203

2 SCARLET-FRONTED CONURE *Aratinga wagleri*. Green with more or less scarlet at least on forehead. This color does not cause similar contrast against green crown as in Mitred C., and it never reaches cheek. **2a** ssp *wagleri* of Ven. and Col., **b** small *minor* of c Peru, **c** moss-colored ssp *frontata* of w slope of Ecu. and Peru, **d** juv. Humid to semiarid woodlands, mainly in premontane zone. Text p. 202

3 GOLDEN-PLUMED CONURE *Leptosittaca branickii*. Green with rusty forehead, yellow ear-plume, and dull red underside of pointed tail. Rare or local at 1300-3500 m from Col. to c Peru. Text p. 203

4 YELLOW-EARED CONURE *Ognorhynchus icterotis*. Very strong-billed. Green with yellow face and ears. Note also contrast between dark green upperparts, more yellowish underparts, dark flight-feathers and rufous underside of tail. At 2000-3400 m in Col. and n Ecu. Rare. Text p. 204

5 BLUE-CROWNED CONURE *Aratinga acuticaudata*. Grass-green with partly azure-blue head, and red color basally on golden underside of tail. **5a** ad., **b** juv. In woodland savannas n and s of the Amazon area, ascending to 2650 m in Bol. Text p. 201

6 MAROON-TAILED CONURE *Pyrrhura melanura*. Like other *Pyrrhura*'s rather small (25 cm), long-tailed, with barred or scaled gray-brown breast. Differs from other highland forms by red patch near wing-bend, and maroon-colored tail changing to green near base. **6a** ssp *chapmani*, **6b** *souancei*, **6c** the rusty-fronted western form *pacifica*. Locally from lowlands to 3200 m from Col. to c Peru. Text p. 206

7 FLAME-WINGED CONURE *Pyrrhura calliptera*. Differs from other highland *Pyrrhura*'s by striking yellow wing-bend and dark rufous brown ears and breast. In humid forest at 1850-3400 m in E Andes of Col. Text p. 207

8 MONK PARAKEET *Myiopsitta monachus*. Green with pale gray face and breast (ssp *luchsi*), and dark blue flight-feathers and tail (appear black). Voice very harsh. Widespread in savanna woodlands s of the Amazon forest, ascending to 2700 m in Cochabamba, Bol. Text p. 208

9 GREEN-CHEEKED CONURE *Pyrrhura molinae*. Differs from other highland *Pyrrhura*'s by pale gray ear-spot, deep blue primaries, more red tail, but no red in wing. Across the chaco, and ascending to the humid montane forest in adjacent Bol. and Arg. Text p. 205

10 SLENDER-BILLED CONURE *Enicognathus leptorhynchus*. Olive-green, faintly barred above, and with pointed red tail. Unlike in the smaller Austral C. quite extensively red on forehead and lores, and maxilla 2.7-3.7 cm long. Flight calls less trilled than in Austral C. Forest of s Chile and adjacent Arg. Text p. 208

11 AUSTRAL CONURE *Enicognathus ferrugineus*. Olive-green faintly barred above and below, with pointed red tail. Shrieks not unlike those of Southern Lapwing. Forests of s and c Chile, and adjacent Arg. Text p. 207

12 BURROWING PARROT *Cyanoliseus patagonus*. A mainly dark olive-gray conure with yellow or olive-yellow rump, unmistakable within its range in the semiarid parts of Arg. and Chile. **12a** ssp *patagonus* of Arg., **12b** juv. (light bill), **12c** *byroni* of c Chile, **12d** *alticola* from Andean foothills of c Arg. Text p. 204

18 per cent of nat. size

PLATE XXIII

Conures

PLATE XXIV Parrots, Parakeets, and Ground-dove
(see also Plates XXIII and LXIV)

1 SCALY-NAPED PARROT *Amazona mercenaria*. Stocky green parrot barred with dark feather-margins. Tail with purplish subterminal zone except on c feathers. **1a** northern *canipalliata*, **1b** *mercenaria*, with conspicuous red mark on outer secondaries. Humid forest at 1600-3600 m from Ven. to Bol. Text p. 216

2 BLACK-WINGED PARROT *Hapalopsittaca melanotis*. Green with mostly black uppersides of wings. **2a** black-eared *melanotis* of yungas of Bol., **2b** buff-eared *peruviana* of c Peru. Local in humid forest at 2800-3450 m. Text p. 212

3 RUSTY-FACED PARROT *Hapalopsittaca amazonina*. Quite olivaceous green with usually rusty face and red or rosy wingbends. **3a** rosy-wristed *pyrrhops* of Ecu. to n Peru. **3b** juv. **3c** *amazonina* of E Andes, Col. **3d** green-faced and red-bellied *fuertesi* of C Andes of Col. Ssp *theresae* of Ven. on Plate LXIV. Rare and local in humid forest at 2500-3500 m. Text p. 213

4 ALDER PARROT *Amazona tucumana*. Stocky green parrot with barred pattern, red forehead and outer secondaries, and yellowish-green tail-tip. Local and vanishing on Andean slopes of s Bol. and nw Arg. Text p. 215

5 WHITE-CAPPED PARROT *Pionus seniloides*. In ad. head whitish with dark scales. Below mostly purplish gray. Otherwise green with red vent and base of tail. Around 3000 m in humid forest from Ven. to n Peru, where possibly intergrades with Plum-crowned P. Text p. 215

6 RED-BILLED PARROT *Pionus sordidus*. Orange-red bill. Olive-green with head scaled blue. Vent red. **6a** ad. *corallinus*. **6b** juv. Premontane forest from Ven. to Bol. Text p. 214

7 PLUM-CROWNED PARROT *Pionus tumultuosus*. Unlike White-capped P. with almost uniform vinous crown, and unlike Red-billed P. with olive-yellow bill. Otherwise green with purplish breast and red vent and base of tail. **7a** ad., **b** juv. At 1670-3300 m in humid forest in Peru and Bol. Text p. 214

8 MOUNTAIN PARAKEET *Bolborhynchus aurifrons*. Fairly light green with 8-9 cm slender tail. Purplish blue wing panel formed by outer webs of primaries (less conspicuous blue-green panel in Andean P.). Best recognized by rather rolling twittering calls. **8a** male *margaritae*, **b** male of the dull green *rubrirostris*, **c** female *aurifrons*, **d** male *aurifrons* (yellow face also in male *robertsi*), **e** juv. Widespread in open terrain at 1000-4500 m in Peru to c Arg. and Chile. Text p. 209

9 ANDEAN PARAKEET *Bolborhynchus orbygnesius*. Rich green, with 6 cm long, rather broad but pointed tail. Closed wing with only hint of a blue panel. Calls richer than in Mountain P. Bushy and wooded valleys at 3000-4000 m in Peru and n Bol. Text p. 210

10 GRAY-HOODED PARAKEET *Bolborhynchus aymara*. With long thin tail. Dark gray cap contrasts pale gray lower face and breast; otherwise green, light below. In flight gives series of c. 3 agitated *tjic* calls. Arid bushy slopes at 2000-4200 m in Bol. and n Arg. Text p. 209

11 RUFOUS-FRONTED PARAKEET *Bolborhynchus ferrugineifrons*. Rich dark grass-green, rusty around base of bill. 6-7 cm rather blunt tail. Very locally at the treeline in C Andes of Col. Text p. 211

12 BARRED PARAKEET *Bolborhynchus lineola*. Grass-green, partly olive in Andean ssp *tigrina*, with dense dark barring developing into distinct black tips on tail-coverts and on the 5-6 cm long tail (but other parakeets can get dark smudges resembling bars by wear of the plumage). Associated with bamboo at 100-3300 m on humid slopes from Ven. to e Peru. Text p. 211

13 MAROON-CHESTED GROUND-DOVE *Claravis mondetoura*. With 3 bars of purplish spots across wings. Male **13a**, female **1b**. Outer tail-feathers with large (male) or small (female) white tips. In seeding bamboo at 1300-3000 m from Ven. to Bol. Text p. 194

25 per cent of nat. size

PLATE XXIV

Parrots, Parakeets, and Ground-dove

PLATE XXV Owls (see also Plates XXVI and LXIV)

1 STYGIAN OWL *Asio stygius*. Dark with rather small head with eartufts. Ssp *robustus* shown. Deep hoots always well spaced. Rare and local in the northern Andes (and sighted at treeline in Bol.). Text p. 229

2 RUFESCENT SCREECH-OWL *Otus ingens*. Rather uniform brown, buffy brown face not distinctly outlined. Gives long series of whistled hoots, typically starting slowly. **2a** gray-brown type, **2b** *colombianus*, now regarded as a distinct species, **2c** rufous type. Wet premontane forest from Ven. to Bol. Text p. 224

3 LONG-WHISKERED OWLET *Xenoglaux loweryi*. Tiny, almost tail-less with loose whiskers. Very local in wet forest in n Peru. Text p. 228

4 WHITE-THROATED SCREECH-OWL *Otus albogularis*. Very dark with white moustache. Ssp *remotus* shown. Gives gruff barked calls often in long series, and sometimes as duet. Humid mossy forest at 2200-3000 m from Ven. to Bol. Text p. 225

5 SAVANNA SCREECH-OWL *Otus choliba*. Small gray-brown owl with black rim of facial disk, yellow eyes, white scapular spots and mainly white underparts with dense black 'herringbone marks'. Short series of whistles ends abruptly with 1 or more accentuated strokes. **5a** ssp *roboratus*, **b** *koepckei*, **c** *luctisonus*, **d** *alilicuco*. Widespread in lowland savanna and woodlands, and reaches temp. zone locally in the Andes. Text p. 223

6 RUFOUS-BANDED OWL *Ciccaba albitarsus*. Compact and large-headed, coarsely barred tawny. Deep gruff hoots given single or in short rhythmic series. At 1700-3000 m in humid forest from Col. to Bol. Text p. 228

7 AUSTRAL PYGMY-OWL *Glaucidium nanum*. Fairly small-headed little gray-brown owl with white spots and streaks. Frequently cocked tail with 7-9 light bars. Gives series of regularly spaced whistled hoots. **7a** ad., **b** juv. In fairly open forest in southern Arg. and Chile. P. 227

8 CLOUD-FOREST SCREECH-OWL *Otus marshalli*. Small rufous brown owl with rufous facial disk with black rim, dark eyes, and white spots below. Wet premontane forest in c Peru. Text p. 224

9 BUFF-FRONTED OWL *Aegolius harrisii*. Unmistakable. Song an extremely fast trill at high but slightly fluctuating pitch. Se Brazil to n Arg., and very scattered in Andean woodlands n to Ven. Rare and local. Text p. 231

10 ANDEAN PYGMY-OWL *Glaucidium jardinii*. Fairly small-headed little owl with normally round white dots on crown. Tail with 5-7 light bars. Gives long series of whistled hoots at somewhat irregular speed. **10a** ad., **b** rufous phase, younger bird with streaked forehead, **10c** juv. Broken forest habitat at 2000-3500 m from Ven. to Bol. Text p. 226

11 BURROWING OWL *Athene cunicularia*. Characteristic, rather flat-headed, with stumpy tail, and long legs. Thick white eyebrows. Has a variety of screeching and cackling calls and melodic hoots. **11a** ssp *juninensis*, **b** *partridgei*, **c** juv. with unicolored crown. Widespread, but somewhat local in savannas, rangeland, and desert, high and low. Text p. 230

12 RUFOUS-LEGGED OWL *Strix rufipes*. Compact and heavy-headed. Densely barred. Gives deep hoots and high screams. Chaco woodlands and mature forest of s Chile and adjacent Arg. Text p. 228

Young Burrowing Owls in down-feather plumage.

25 per cent of nat. size

PLATE XXV

Owls

PLATE XXVI Owls and Nightjars (see also Plate XXV)

1 BARN OWL *Tyto alba*. With large heart-shaped facial disk. **1a** white phase, **b** dark phase *contempta*. Strictly nocturnal. Gives long wheezy scream. Widespread in lowlands and in the Andes, often in villages. Text p. 221

2 SHORT-EARED OWL *Asio flammeus*. A long-winged buffy brown owl sometimes seen in daylight over wide meadows and marshes. Note the dark 'comma' on the wrist, and large clear basal part of the primaries separating the dark wingtip. Widespread in grassland zones of S Am., but rare in the Andes. **2a** *bogotensis*, **b** *suinda*. Text p. 229

3 GREAT HORNED OWL *Bubo virginianus*. Long eartufts are not evident in flight. Resembles a short-tailed *Buteo* hawk in flight. **3a** the small southern *magellanicus*, **b** the northern *nigrescens*. Most often revealed by low, deep calls *whoo-hoo*, *woo **whoo**-hoo* or *guacouroutou-tou*. Widespread in S Am., low and high. Text p. 226

Nighthawks are vespertine counterparts of swifts. They have short tails.

4 RUFOUS-BELLIED NIGHTHAWK *Lurocalis (semitorquatus) rufiventris*. Bat-like, with long dark wings and short tail. Note white-mottled humerals (**a**) and white throat and tawny belly (**b**). In humid premontane zone, sometimes to 3000 m, from Ven. to Bol. Text p. 233

5 COMMON NIGHTHAWK *Chordeiles minor*. Mottled grayish with white wing-bar midway between wrist and wing-tip. Usually flies high, at dusk, with rapid and bounding wingstrokes. Call buzzy, insect-like *beerp*. Ssp *chapmani* shown. Visits temp. zone of Col. in the migratory periods. Text p. 234

6 LESSER NIGHTHAWK *Chordeiles acutipennis*. Mottled, with less contrasting pattern than Common N., and white bar more distally on hand. Expected in temp. zone of the northern Andes. Text p. 234

7 ANDEAN POTOO *Nyctibius (leucopterus) maculosus*. Owl-like when calm, but freezes in slim pose when alarmed. Spotted, with more or less extensive white shoulderpatch. Call a loud *ree-aa*. Very local on humid Andean slopes from Ven. to Bol. Text p. 232

Nightjars are nocturnal birds with longer tails and blunter tipped wings than nighthawks.

8 SWALLOW-TAILED NIGHTJAR *Uropsalis segmentata*. Blackish with tawny mottling, wings all dark below. Tail forked, in male with scissor-like streamers 2 x body length. Not rufous nape-collar. **8a** male, **b** female ssp *segmentata*. Gives pauraque-like *puit-sweet* calls. Grassy glades at 2300-3600 m in humid forest. Text p. 236

9 LYRE-TAILED NIGHTJAR *Uropsalis lyra*. Outer tail-feather of male curved, 3x body length. Replaces Scissor-tailed N. mainly in the premontane zone, and differs by having rufous nape-collar and otherwise more gray-mottled plumage, much as in Band-winged N. Calls at dusk series of loud *tre-cuee* calls or accelerating *weeep*s. Text p. 237

10 BAND-WINGED NIGHTJAR *Caprimulgus longirostris*. With pale bar across primaries, and male with broad white tail-tip (**a** male, **b** female *atripunctatus*). Mottled and spotted dusky buff or gray, ad. with rufous collar, the w Peruvian *decussatus* (**d**) and Chilean *bifasciatus* (**c**) grayest, the northern *ruficervix* very dark brown. May give plaintive *che-whit* calls in sometimes interminable series throughout the night. Also a thin sucking *zee-eorr*. Widespread throughout the Andean highlands, and to the coast in Peru and southern S Am. Text p. 235

Owls and Nightjars

PLATE XXVI

15 per cent of nat. size;
1a, 2b, and 3 in 10 per cent of nat. size,
and 2a, 4a, and 6 even smaller

PLATE XXVII Hummingbirds; Hillstars and very large species (see also plates XXVIII-XXXI)

1 GIANT HUMMINGBIRD *Patagona gigas*. Gigantic hummer with whitish base of tail. Flies with erratic wing-beats and inserted glides, somewhat like a long-winged swallow. **1a** ad. *peruviana*; **b** female, **c** male *gigas*. Call a characteristic, plaintive drawn-out *cueeet*. Widespread on arid Andean slopes from Ecu. to c Arg. and Chile.
Text p. 255

2 GREAT SAPPHIREWING *Pterophanes cyanopterus*. Very large, but with thinner and straighter bill than Giant Hummingbird, and the flight more typical hummingbird-like, although with some erratic wing-beats and glides. Male **2a** very dark, female **b** with cinnamon **c** underparts and white lateral edge of tail. Call a piercing, high-pitched drawn-out *zeeeee*. Widespread in cloud forest.
Text p. 260

3 SWORD-BILLED HUMMINGBIRD *Ensifera ensifera*. Dark greenish with enormous bill and forked tail. Gives a plaintive drawn-out whistle. **3a** male, **b** female. Widespread in humid montane forest.
Text p. 266

4 CHIMBORAZO HILLSTAR *Oreotrochilus estella; chimborazo* group. With slightly curved bills. Males with hood glossy dark blue and tail white, dark on center and sides; **a** *chimborazo*, **b** *jamesoni*, **e** imm. *chimborazo*. Female (**c**) and juv. (**d**, male) dark bronzy brown with small white tips of the tail-feathers. Open páramo habitat in Ecu.
Text p. 253

5 ANDEAN HILLSTAR *Oreotrochilus estella; estella* group. Puna habitat and woodland in high parts of Peru and Bol. Males with shining green gorget and mainly white tail, with dark center and sides, **a** and **d** *estella*, with rufous midline on belly; **c** northern *stolzmanni*, with black midline; **g** imm. male of *boliviana*. Female gray brown with white spots at tips and bases of tail-feathers, **b** female of the rather bronzy *stolzmanni*, **e** of *estella*, **f** juv. *estella* female.
Text p. 253

6 BLACK-BREASTED HILLSTAR *Oreotrochilus melanogaster*. Male **6a** with shining green gorget, and black underside and tail. Female **b** differs from Andean H. by lacking white bases of tail-feathers. High parts of Andes in c Peru.
Text p. 254

7 WEDGE-TAILED HILLSTAR *Oreotrochilus adela*. Tail wedge-shaped with partly rufous inner webs of feathers. Male **7a** with green gorget, below chestnut with black median zone, female **b** with white **c** underparts and tips of tail-feathers. Dry, scrubby Andean slopes in c Bol.
Text p. 255

8 WHITE-SIDED HILLSTAR *Oreotrochilus leucopleurus*. Like Andean H., but male **a** with broad blue-black midline below, and outer tail-feathers curved and rather short and narrow. Female **b** also has somewhat narrow outer tail-feather, with complete dark transverse bar (unlike *O. estella boliviana*). Andes from s Bol. to c Arg. and Chile.
Text p. 254

To plate XXVIII:

18 SHINING SUNBEAM *Aglaeactis cupripennis*. Earthy brown to more or less rufous, esp. in n and c parts of range. Back of ad. glistening lilac to silvery green on upper tail-coverts. **18a** ad. *cupripennis*, **b** *ruficauda*, **c** southern *caumatonotus*. Woodlands from Col. to e Peru.
Text p. 257

19 WHITE-TUFTED SUNBEAM *Aglaeactis castelnaudii*. Fuscous with tawny tail; rump violet. White tuft on breast. **19a** *castelnaudii*, **b** *regalis*. In semiarid woodlands in Andes of Peru.
Text p. 258

20 PURPLE-BACKED SUNBEAM *Aglaeactis aliciae*. Dark earthy brown with white on face, breast, vent and wing-linings. Local in La Libertad, n Peru.
Text p. 258

21 BLACK-HOODED SUNBEAM *Aglaeactis pamela*. Black with white breast-patch and chestnut tail. Back golden blue-green. **21a** male, **b** female. Yungas of La Páz and Cochabamba, Bol.
Text p. 259

22 BLUE-THROATED STARFRONTLET *Coeligena helianthea*. Dark tail and glittering rosy belly. Male (**a**) black-hooded. Female **b**. E Andes, Col., to border to Ven.
Text p. 263

30 per cent of nat. size

PLATE XXVII

Hummingbirds; Hillstars and very large species

PLATE XXVIII Hummingbirds; species with fairly long, straight bills (see also plates XXVII, XXIX-XXXI, LXIII and LXIV)

1 BLACK-THIGHED PUFFLEG *Eriocnemis derbyi*. Uniform dark with malachite-green vent. Male **a** with acute tips of tail-feathers; **b** female. C Andes of Col. to nw Ecu. Text p. 275

2 BLUE-CAPPED PUFFLEG *Eriocnemis glaucopoides*. Ad. male **a** with blue forehead and vent; female **b** brightly cinnamon below. Humid Andean slopes from c Bol. to nw Arg. Text p. 275

3 GOLDEN-BREASTED PUFFLEG *Eriocnemis mosquera*. With long, narrow and deeply forked tail. Breast glittering golden bronzy, but vent dull. W Andes, Col., and nw Ecu. Text p. 274

4 TURQOUISE-THROATED PUFFLEG *Eriocnemis godini*. Male **a** and female **b** dark with golden glittering neck. Unlike Glowing P., the female lacks blue throat-patch and tawny color on breast. Known definitely only from Guaillabamba valley, nw Ecu. Text p. 273

5 GREEN-BACKED FIRECROWN *Sephanoides sephaniodes*. Bronzy, spotted below, male **a** with fiery orange cap, female **b**, juv. **c**. C and s Chile and adjacent sw Arg. Text p. 266

6 WEDGE-BILLED HUMMINGBIRD *Schistes geoffroyi*. White streak behind eye. With purplish neck-side patches and white patches below (ad. male **a**) or more extensively white throat (**b**). Ssp *albogularis* of w Andean slope in Col. and Ecu. reaches temp. zone. Text p. 291

7 BLACK-BREASTED PUFFLEG *Eriocnemis nigrivestis*. Like a very dark Glowing P., but female (**b**) lacks tawny on breast. Male **a**. Pichincha area in Ecu. Text p. 272

8 SAPPHIRE-VENTED PUFFLEG *Eriocnemis luciani*. Dark glittering green; blackish blue tail fairly long and deeply forked (less in *marcapatae*). **8a** male *luciani*, **b** male *sapphiropygia*, **c** fledgling. Humid forest of Ecu. and Peru. Text p. 273

9 GLOWING PUFFLEG *Eriocnemis vestitus*. Both sexes with purplish blue throat-patch, female (**b**) partly tawny on throat and breast. Male **a**. Locally in Andes of Col. to extreme n Peru. Text p. 272

10 COPPERY-BELLIED PUFFLEG *Eriocnemis cupreoventris*. Unlike Glowing P. lacking purplish throat-patch, and has golden coppery belly and more deeply forked tail. Mérida Ven. to E Andes, Col. Text p. 273

11 RAINBOW STARFRONTLET *Coeligena iris*. With brick-red posterior parts, more or less green or coppery foreparts. Ad. males have strongly glistening, multicolored caps, ad. males usually black napes; **a** *hesperis*, **c** *eva*, **d** *aurora*, **e** *fuliginiceps*. **11b** imm. *aurora*. Patches of humid forest in s Ecu. and nw Peru. Text p. 265

12 VIOLET-THROATED STARFRONTLET *Coeligena violifer*. Tail orange-buff, or whitish in birds from Ayacucho (**a**). Male with dark green throat with purplish gular spot, **a** unnamed ssp, **d** *osculans*, **e** the black-hooded ssp *violifer*. **12b** female *violifer*, **c** juv. Patches of humid montane forest in Peru and Bol. Text p. 264

13 GOLDEN-BELLIED STARFRONTLET *Coeligena bonapartei*. Tail golden buff to bronzy; ad. males of *bonapartei* (**a**) and *consita* black-hooded. Ssp *eos* with cinnamon rufous wing-patch (Plate LXIV 10). **13 b** female, **c** imm. male. E Andes Col. into Ven. Text p. 263

14 DUSKY STARFRONTLET *Coeligena bonapartei orina*. Blackish. Paramó Frontino in n W Andes, Col. Text p. 263

15 COLLARED INCA *Coeligena torquata*. Tail partly white. Throat white from Ven. to c Peru (**b** female *insectivora*, **c** male *eisenmanni*, **d** male *fulgidigula*) or tawny in Cuzco to Bol. (**a**, male *inca*). Humid montane forest. Text p. 261

16 BLACK INCA *Coeligena prunellei*. Black with white half-collar. Humid forest edge on w slope of E Andes, Col. Text p. 261

17 BUFF-WINGED STARFRONTLET *Coeligena lutetiae*. Dark tail, buff (transparent-looking) wingpatch. **17a** male, **b** female. Call nasal. Humid shrub from C Andes of Col. to extreme n Peru. Text p. 264

Text for 18 to 22 on previous page

30 per cent of nat. size

PLATE XXVIII

Hummingbirds; species with fairly long, straight bills

PLATE XXIX Hummingbirds; species with generally short bills
(see also plates XXVIII, XXX-XXXI, LXIII and LXIV)

1 PERIJÁ METALTAIL *Metallura iracunda*. Male blackish with red tail. Perijá mts. on Col./Ven. border.
Text p. 285

2 TYRIAN METALTAIL *Metallura tyrianthina*. Tail varies from violet furthest n through reddish purple (**c** male *oreopola*) and golden bronze (**b** male *quitensis*) to deep blue in the s (**e** male *smaragdinicollis*). Juv. (**a**, *quitensis*), imm. (**d**, male *tyrianthina*) and female are partly orange-cinnamon below. Widespread in humid montane shrub from Ven. to Bol.
Text p. 284

3 COPPERY METALTAIL *Metallura theresiae*. Deep purplish coppery with narrow green gorget. Tail bronzy shot with blue. **3a** male, **b** female *theresiae*. E Andes of n Peru.
Text p. 283

4 BLACK METALTAIL *Metallura phoebe*. Black tail coppery. **4a** male, **b** female. Mainly pacific slope in Peru.
Text p. 281

5 VIRIDIAN METALTAIL *Metallura williami*. Tail shot with blue. Males with throat glittering green (**c** *primolina*) or black centrally (**d** *atrigularis*). Females with spotted throats (**a** *primolina*, **b** *williami*). Treeline shrub locally in C Andes, Col., and Ecu.
Text p. 281

6 VIOLET-THROATED METALTAIL *Metallura baroni*. Dark bronzy with violet throat. **6a** male, **b** female. In Azuay, sw Ecu.
Text p. 282

7 NEBLINA METALTAIL *Metallura odomae*. Bronzy, throat reddish purple. **7a** male, **b** female. Extreme n Peru.
Text p. 282

8 FIRE-THROATED METALTAIL *Metallura eupogon*. Bronzy with rather narrow orange gorget. **8a** male, **b** female. C Peru.
Text p. 283

9 SCALED METALTAIL *Metallura aeneocauda*. Rather scaled with tan below. Males with golden green throat. Tail shot with blue above (*aeneocauda*, **b** female, **c** male) or coppery (*malagae*, **a** male). Yungas of se Peru and Bol.
Text p. 283

10 BLUE-MANTLED THORNBILL *Chalcostigma stanleyi*. Sooty glossed purplish, males (**a** northern *stanleyi*, **b** southern *vulcani*) with black head with glittering beard. Female (**c**) with disintegrated beard, and juv. (**d**) paler, male with unspotted throat. *Gynoxys/Polylepis* shrub, locally from Ecu. to Bol.
Text p. 287

11 OLIVACEOUS THORNBILL *Chalcostigma olivaceum*. Bronzy brown. Ad. male of ssp *olivaceum*. Rare in high mts of Peru to La Paz, Bol.
Text p. 286

12 MOUNTAIN AVOCETBILL *Opisthoprora euryptera*. Bill-tip up-turned. Resembles a female metaltail, but with more streaky effect and prominent white postocular spot. Rare from C Andes, Col. to n Peru.
Text p. 290

13 RUFOUS-CAPPED THORNBILL *Chalcostigma ruficeps*. Male (**a**) with chestnut cap. Female resembles coppery-tailed Tyrian Metaltail, differing by slightly smaller bill and narrower (0.7 cm) tail-feathers. Imm. (**b**) more diffusely bronzy below. Humid montane forest.
Text p. 286

14 PURPLE-BACKED THORNBILL *Ramphomicron microrhynchum*. With extremely small bill. Male (**a**, **b** imm.) deep metallic purple with greenish yellow V-shaped gorget. Female (**c**) green, spotted on white background below, and with white-tipped tail. Ssp *albiventer* shown. Treetops in humid treeline habitat.
Text p. 280

15 BRONZE-TAILED THORNBILL *Chalcostigma heteropogon*. Bronzy, coppery on nape and rump. Ad. male (**a**) with glittering beard, disintegrating in female (**b**), lacking in juv. (**c**). E Andes, Col., to nw Ven.
Text p. 287

16 RAINBOW-BEARDED THORNBILL *Chalcostigma herrani*. Note large white tail-tips. **12a** male, **b** female, **c** juv. *herrani*. Páramos from C Andes, Col., to n Peru.
Text p. 288

17 AMETHYST-THROATED SUNANGEL *Heliangelus amethysticollis*. Pectoral bar white in the n to buff in the s, belly more or less ochraceous. Tail rather broad and square, with bronzy c feathers, vent mostly white. **17a** male *clarisse*, **b** *amethysticollis*, **c** female *decolor*, **d** juv. Widespread in humid montane forest. Text p.269

18 TOURMALINE SUNANGEL *Heliangelus exortis*. Dark metallic green with emerald pectoral bar, vent white. **18a** male *micrastur* of se Ecu., the rest *exortis*, **b** female, **c** juv., **d** male. Humid montane forest of Col. to n Peru.
Text p. 270

19 ORANGE-THROATED SUNANGEL *Heliangelus mavors*. Ad. male with orange gorget. E Andes of Col., into nw Ven.
Text p. 268

20 GORGETED SUNANGEL *Heliangelus strophianus*. Differs from sympatric Amethyst-throated S. by whiter pectoral bar, more slaty underparts, and more forked steel-blue tail. **20 a** male, **b** female. W Ecu.
Text p. 269

21 PURPLE-THROATED SUNANGEL *Heliangelus viola*. Dark metallic green with emerald pectoral bar, but unlike Tourmaline S. with buff and green vent, and more forked tail. **21a** male, **b** female. Sw Ecu. into n Peru.
Text p. 271

22 ROYAL SUNANGEL *Heliangelus regalis*. **22a** male unmistakable blue-black, **b** female resembles a frail Amethyst-throated S. with buff pectoral bar, spotted vent, and long, totally blue-black tail. Local in extreme n Peru.
Text p. 271

30 per cent of nat. size

PLATE XXIX

Hummingbirds; species with generally short bills

PLATE XXX Hummingbirds; species with long or otherwise peculiar tails (see also plates XXVII-XXXIX, XXXI, LXIII and LXIV)

1 BUFF-TAILED SICKLEBILL *Eutoxeres condamini*. Bill strongly curved. Note streaking and buff underside of tail. Humid forest, reaching temp. zone in Peru. Text p. 244

2 PLANALTO HERMIT *Phaethornis pretrei*. Tawny below and on rump, tail-feathers tipped white. Andean slopes of Bol. to n Arg. Text p. 242

3 GREEN HERMIT *Phaethornis guy*. Female *apicalis*. Bill very long. Dark with buff face-stripes (unlike White-whiskered and Sooty-capped H., see text). Ven.-Peru. Text p. 243

4 TAWNY-BELLIED HERMIT *Phaethornis syrmatophorus*. Bill very long. Tawny of underside extends to tail. Ssp *columbianus* shown. Col.-Peru. Text p. 243

5 LONG-TAILED SYLPH *Aglaiocercus kingi*. Deeply forked tail of males green, the streamers deep blue at base, or violet. **5a** imm. male *margaretae*, **b** female (pale underparts more or less rufous), **c** ad. male *mocoa*. Widespread in montane forest. Text p. 290

6 MARVELLOUS SPATULETAIL *Loddigesia mirabilis*. **6a** ad. male unmistakable, **c** imm. male. **6b** female, resembling Booted Racket-tail, but colors subdued, and lacking leg-puffs. Utcubamba valley in n Peru. Text p. 292

7 RUFOUS-SHAFTED WOODSTAR *Chaetocercus jourdani*. Male. Remark the strongly red gorget. N Ven.- ne Col. Text p. 298

8 PURPLE-THROATED WOODSTAR *Philodice mitchellii*. Dark bronzy with tawny vent. **8a** ad. male, **b** eclipse. W Col.-Ecu. Text p. 294

9 PERUVIAN SHEARTAIL *Thaumastura cora*. Male with thin white streamers and small bill. Pacific slope of Peru-n Chile. Text p. 293

10 GORGETED WOODSTAR *Acestrura heliodor*. Tiny, male dark, with flared gorget and tiny tail. Locally from n Ven. to n Ecu. Text p. 297

11 OASIS HUMMINGBIRD *Rhodopis vesper*. Long curved bill and rufous rump. **11a** male *atacamensis* (eclipse), **b** *vesper* in breeding garb. Arid zone of w Peru-n Chile. Text p. 293

12 PURPLE-COLLARED WOODSTAR *Myrtis fanny*. Green rump and less extreme bill than Oasis Hummingbird, but not short bill of Peruvian Sheartail. **12a** ad., **b** imm. male *fanny*. Sw Ecu.-n and w Peru. Text p. 294

13 SLENDER-TAILED WOODSTAR *Microstilbon burmeisteri*. Flared gorget and deep gray belly. Andean slopes of Bol.- nw Arg. Text p. 294

14 BOOTED RACKET-TAIL *Ocreatus underwoodii*. Tiny. Tail forked and with white tips (juv., females) or with partly bare-shafted rackets (ad. males). Large leg-puffs of males white (Ven.-Ecu.) or tawny (Peru-Bol.). **14a** female, **b** male *underwoodii*, **c** imm. male, **d** female *melanantherus*, **e** ad. male *addae*, **f** *annae*, **g** *peruvianus*, **h** female *peruvianus*. Humid Andean forest. Text p. 276

15 GRAY-BREASTED COMET *Taphrolesbia griseiventris*. Stocky. Above bronzy, below gray with blue-spotted throat. **15a** ad., **b** juv. Rare in arid valleys of n Peru. Text p. 290

16 BRONZE-TAILED COMET *Polyonymus caroli*. Dark bronzy with forked but not very long tail. **16a** female, **b** male. Pacific slope and some Andean valleys of Peru. Text p. 279

17 RED-TAILED COMET *Sappho sparganura*. **17a** male *sparganura* (golden tail, ad. with black feather-tips), **b** female, **c** juv., lacking purple on back (ssp *sapho* with red tail). Bushy Andean slopes of Bol. and Arg. Text p. 278

18 GREEN-TAILED TRAINBEARER *Lesbia nuna*. Tail extremely forked, feather-tips glittering emerald green. Bill generally small and straight, gorget round. **18a** female, **b** male *gracilis*, **c** male *nuna*, **d** imm. male *nuna*. Bushy Andean slopes from Col. to Bol. Text p. 278

19 BLACK-TAILED TRAINBEARER *Lesbia victoriae*. Tail extremely forked, to 18 cm in male, black with slight bronzy feather-tips. Bill slightly curved, but in c Peru close to that of Green-tailed T. **19a** ad. male *aequatorialis*, **b** imm. males *berlepschi* (note downwards extension of the incomplete gorget), **c** female and **d** imm. male *aequatorialis*. Andean slopes of n Col.-se Peru. Text p. 277

20 BEARDED MOUNTAINEER *Oreonympha nobilis*. Dark hood demarcated white, and long forked tail mainly white laterally. **20a** male, **b** juv. *nobilis* of Cuzco; **c** male *albolimbata* of arid valleys of c Peru. Text p. 289

21 BEARDED HELMETCREST *Oxypogon guerinii*. See also Plates LXIII and LXIV. Bronzy, dark hood demarcated white. Medium-long tail notched but blunt-tipped. **21a** juv., **b** ad. male *guerinii*, with broad white stripe in outer tail-feathers. Ssp *strubeli* has buff outer tail-feathers. **21c** female *lindeni*. Páramos of Ven. and Col. Text p. 288

PLATE XXX

15 per cent of nat. size

Hummingbirds; species with long or otherwise peculiar tails

PLATE XXXI Hummingbirds; mainly species with curved bills (see also plates XXVII-XXX, LXIII and LXIV). Coraciiform birds

1 LAZULINE SABREWING *Campylopterus falcatus*. Fairly large with arched bill and truncate tail extensively chestnut. **1a** ad. male, **b** imm. male (ad. female similar with gray chin), **c** juv. Humid forest in the n Andes. Text p. 245

2 WHITE-BELLIED HUMMINGBIRD *Amazilia chionogaster*. Outer tail-feathers with inner web pale to base (**a** and **b** shows variants). Peru – n Arg. Text p. 249

3 GREEN-AND-WHITE HUMMINGBIRD *Amazilia viridicauda*. Tail-feathers dark (never white edge to base of outer tail-feathers). Very locally in humid premontane shrub in c Peru. Text p. 249

4 STEELY-VENTED HUMMINGBIRD *Amazilia saucerottei*. Dark with steel-blue tail. Female. Ven.-Col. Text p. 250

5 BLUE-TAILED EMERALD *Chlorostilbon mellisugus*. Tiny. Male glittering green with forked steel-blue tail. Plumage variation as in following species. Ssp *melanorhynchus* shown. Widespread in premontane zone. Text p. 247

6 NARROW-TAILED EMERALD *Chlorostilbon stenura*. Tail-feathers green, small and acute in male (**a**). **6b** female, **c** imm. male. Mainly premontane n Andes. Text p. 249

7 BLUE-HEADED SAPPHIRE *Hylocharis grayi*. Green with blue head and tail; bill red. Premontane n Andes. Text p. 249

8 GLITTERING-BELLIED EMERALD *Chlorostilbon aureoventris*. Male ssp *aureoventris*. Note red bill. To temp. zone in Bol. and Arg. Text p. 248

9 GREEN VIOLETEAR *Colibri thalassinus*. Subterminal tail-bar. Purple of ear-plumes does not continue to chin. **9a** ad., **b** juv. *thalassinus*. Widespread in humid forest. Text p. 245

10 WHITE-VENTED VIOLETEAR *Colibri serrirostris*. Lighter than previous species, and vent white. **10a** ad., **b** extremely pale juv. Andean slopes in Bol. and n Arg. Text p. 247

11 MOUNTAIN VELVETBREAST *Lafresnaya lafresnayi*. Bill curved. Tail extensively white (or buff in *tamae* and *lafresnayi*). **11a** male *saul*, **b** male *lafresnayi*, **c** juv. (female similar). Andes from nw Ven. to e Peru. Text p. 259

12 CHESTNUT-BREASTED CORONET *Boissonneaua matthewsii*. Rufous below, on wing-linings and tail. Premontane forest Col. to Peru. Text p. 267

13 BUFF-TAILED CORONET *Boissonneaua flavescens*. Wing-linings rufous, tail buff tipped bronzy; **a** male *flavescens*, **b** female *tinochlora*. Montane forest of the n Andes. Text p. 267

14 SPECKLED HUMMINGBIRD *Adelomyia melanogenys*. Dull. Note white supercilium and dark ear-patch. **14a** *cervina*, **b** southern *inornata*. Humid Andean forest. Text p. 251

15 WHITE-BELLIED WOODSTAR *Acestrura mulsant*. Tiny. **15a** male with breast and belly white, **b** juv. with rusty bars above. Female and juv. tawny and white below, with rounded tail. Col.- Bol. Text p. 296

16 GORGETED WOODSTAR *Acestrura heliodor*. Female. Minute with rump and tail rufous, below uniform cinnamon. 2-lobed tail. Locally in Andes from Ven. to Ecu. Text p. 297

17 SLENDER-TAILED WOODSTAR *Microstilbon burmeisteri*. Female. Below uniform cinnamon. Small tail square. Slopes of Bol. and Arg. Text p. 294

18 PURPLE-THROATED WOODSTAR *Philodice mitchellii*. Female, with breast-bars. Tiny tail slightly 2-lobed. W Col.- Ecu. Text p. 294

19 PURPLE-COLLARED WOODSTAR *Myrtis fanny*. Female ssp *fanny*. Unlike Oasis Hummingbird not rufous on rump. Sw Ecu.-n and w Peru. Text p. 294

20 RUFOUS-SHAFTED WOODSTAR *Chaetocercus jourdanii*. Female ssp *andinus*. With distinctly 2-lobed tail. N Ven.-ne Col. Text p. 298

21 OASIS HUMMINGBIRD *Rhodopis vesper*. With long curved bill. Female with small 2-lobed tail with white tips. Rump cinnamon. Arid zone of w Peru – n Chile. Text p. 293

22 SPARKLING VIOLETEAR *Colibri coruscans*. Fairly large. Tail with dark subterminal bar. Blue of ear-plumes continues as chin-strap. **22a** ad., **b** juv. Widespread. Text p. 246

23 MASKED TROGON *Trogon personatus*. **23a** male, **b** female ssp *temperatus*. Note barred underside of tail (Crested Quetzal female has white barring distally on black under-tail). Widespread in humid montane forest. Text p. 301

24 PYGMY KINGFISHER *Chloroceryle aenea*. 13 cm, with chestnut underside. **24a** male, **b** female. Casual near Bogotá. Text p. 303

25 RINGED KINGFISHER *Ceryle torquata*. 40 cm. **25a** male, **b** female of southern ssp *stellata*. Widespread mainly in lowlands. Text p. 302

26 TOUCAN BARBET *Semnornis ramphastinus*. Ssp *ramphastinus* shown. Metallic coos characteristic. Premontane w Col. and Ecu. Text p. 305

27 WHITE-FACED PUFFBIRD *Hapaloptila castanea*. Note white area around large bill. Premontane forest in w Col.-n Peru. Text p. 304

30 per cent of nat. size, but
25 only 20 per cent of nat. size

PLATE XXXI

**Hummingbirds; mainly species with curved bills
Coraciiform birds**

PLATE XXXII Toucans and Woodpeckers (see also plate XXXIII)

1 BLUE-BANDED TOUCANET *Aulacorhynchus coeruleicinctus*. Bill gray with pale tip. Throat white. Ssp *coeruleicinctus*. Humid forest at 1700-3000 m from c Peru to c Bol. Text p. 306

2 CHESTNUT-TIPPED TOUCANET *Aulacorhynchus haematopygius*. Bill black with red base and ridge. No chestnut on vent. Locally to 2750 m. S Col. to n Bol. Text p. 307

3 EMERALD TOUCANET *Aulacorhynchus prasinus*. Bill black with yellow upper ridge and broad white demarcation at base. Vent and tip of tail-coverts tawny. **3a** southern *atrigularis*, **b** northern *albivitta*, **c** juv., **d** *phaeolaemus*. Mainly humid premontane zones, but to the temp. zone in the Andes of Ven. to Ecu. Text p. 306

4 GROOVE-BILLED TOUCANET *Aulacorhynchus sulcatus*. With yellow eyebrow and lacking tawny tail-tip (vs. Emerald T.). Ssp *calorhynchus* shown. To lower edge of temp. zone in nw Ven. and Sta Marta mts, n Col. Text p. 307

5 YELLOW-BROWED TOUCANET *Aulacorhynchus huallagae*. Gray bill with pale tip and white demarcation at base. Throat white, vent yellow. At 2100-2600 m below Ongon in La Libertad n Peru. Text p. 307

6 GRAY-BREASTED MOUNTAIN-TOUCAN *Andigena hypoglauca*. 3-colored bill with black spot on the yellow base. Black hood demarcated by blue-gray collar. Ssp *hypoglauca* shown, **a** ad., **b** juv. At 2500-3400 m from sw Col. along e slope of Ecu. (and sw Ecu.) and Peru. Text p. 308

7 PLATE-BILLED MOUNTAIN-TOUCAN *Andigena laminirostris*. Bill black with buffy white plate and some red at base. Black cap. Side with yellow patch. At 1000-3000 m in sw Col. and w Ecu. Text p. 308

8 HOODED MOUNTAIN-TOUCAN *Andigena cucullata*. Very dark, with yellow base of mandible. Pale gray nape-collar. Not conspicuously yellow rump. At 2500-3300 m in the yungas of se Peru and Bol. Text p. 309

9 BLACK-BILLED MOUNTAIN-TOUCAN *Andigena nigrirostris*. Bill black (**c**, *nigrirostris*; or partly red in **b** *spilorhynchus* and **a** *occidentalis*). Below sky-blue with white throat. At 1600-3200 m in Andes of nw Ven. and Col. to e Ecu. Text p. 309

10 OLIVACEOUS PICULET *Picumnus olivaceus*. Like a small passerine bird with striped tail. Below streaked olivaceous. Premontane forest to 2500 m in nw Ven. and Col. Text p. 311

11 SMOKY-BROWN WOODPECKER *Veniliornis fumigatus*. Small, almost uniform olivaceous brown. **11a** female, **b** male *fumigatus*. At 600-3000 m in humid forest from n Ven. to Bol. (Arg.?), and at 4000 m in Cord. Blanca, Peru. Text p. 313

12 ACORN WOODPECKER *Melanerpes formicivorus*. Unmistakable. Male. Oak woods at 1400-3500 m in Andes of Col. Text p. 311

13 MAGELLANIC WOODPECKER *Campephilus magellanicus*. Only large woodpecker in s Arg. and Chile. **13a** female, **b** male. Text p. 318

14 CRIMSON-CRESTED WOODPECKER *Campephilus melanoleucos*. Large, with red 'hammerhead'. **14a** male, **b** female *malherbii*. Reaches temp. forest w of E Andes in Col., and maybe in Bol. Text p. 317

15 POWERFUL WOODPECKER *Campephilus pollens*. Large, mostly black, with white malar stripe. Pale rump and white wing-patch conspicuous in flight. **15a** male, **b** female ssp *pollens*. At 1300-3000 m or higher in montane forest of Col. to c Peru. Text p. 317

16 BAR-BELLIED WOODPECKER *Veniliornis nigriceps*. Olivaceous, with all underparts boldly barred. **16a** male, **b** female *pectoralis*, **c** female *nigriceps*. At 2000-4000 in humid shrub and bamboo from Col. to Bol. Text p. 312

17 CRIMSON-MANTLED WOODPECKER *Piculus rivolii*. Above crimson, side of head pale yellow. **17a** male *atriceps*, **b** male, **c** female *rivolii*. At 1000-3700 m in humid forest from nw Ven. to c Bol. Text p. 314

18 STRIPED WOODPECKER *Picoides lignarius*. Densely checkered black and white. **18a** juv., **b** ad. male. At 1600-4000 m in semiarid c Bol., and along border of s Arg. and Chile. Text p. 312

20 per cent of nat. size

PLATE XXXII

Toucans and Woodpeckers

PLATE XXXIII. Woodpeckers (see also Plate XXXII).
Woodcreepers and Ovenbirds (see also Plates XXXIV-XXXVIII)

1 CHILEAN FLICKER *Colaptes pitius*. Head pale with cap gray, body barred gray-brown and whitish. Pale rump conspicuous in flight. In open woodlands in s Arg. and Chile. Text p. 315

2 ANDEAN FLICKER *Colaptes rupicola*. Light cinnamon-buff with gray cap and whisker. Upperparts of body barred, but showing pale yellow lower back in flight. **2a** the northern *cinereicapillus* female (no red on nape). **b** *puna* male. Around 4000 m through Peru to nw Arg. P. 315

3 BLACK-NECKED FLICKER *Colaptes atricollis*. Colorful head with black throat; body olivaceous, barred. **3a** *peruvianus* female (crown gray); **b** *atricollis* male (crown red). N and w Peru, locally to 4000 m. Text p. 315

4 SPOT-CROWNED WOODCREEPER *Lepidocolaptes affinis*. Bill pale, slender, distinctly curved. Below with black-edged white streaks. **4a** *lacrymiger* of Ven. and E Andes, Col.: throat-feathers black-edged, streaks below rounded and black-tipped; **b** *aequatorialis* of sw Col. and Ecu.: throat white to buffy white, unscaled; **c** *bolivianus* of Bol.: more olivaceous, bill dull whitish except at base. Upper trop. to lower temp. forest in C Am. and the Andes from Ven. to Bol. Text p. 322

5 GREATER SCYTHEBILL *Campylorhamphus pucherani*. Bill long, slender, curved. Headside with 2 white streaks. Rare in humid montane forest from Col. to se Peru. Text p. 323

6 STRONG-BILLED WOODCREEPER *Xiphocolaptes promeropirhynchus*. Large, bill stout and curved. **6a** *promeropirhynchus* of nw Ven. and E Andes Col.; **b** *lineatocephalus* of se Peru to Bol. with throat and ear-coverts more streaked (some individuals almost lacking crown-streaks). Widespread, from C Am. to Arg., with several sspp. To 2850 m in humid forest. Text p. 320

7 BLACK-BANDED WOODCREEPER *Dendrocolaptes picumnus*. Told from 6 by smaller, straighter bill. **7a** *validus*, a tropical form resembling *multistrigatus* of higher elevations; **b** *casaresi* of nw Arg.: more rufous above, barring less distinct. Widespread, in humid forest and woodland to 2700 m, from C Am. to Arg. Text p. 321

8 OLIVACEOUS WOODCREEPER *Sittasomus griseicapillus*. Small, unstreaked. Ssp *viridis*. In humid forest and second growth, to 2300 m in Bol. Text p. 320

9 TYRANNINE WOODCREEPER *Dendrocincla tyrannina*. Bill straight. Almost uniform, with no dark moustache like lower-elevation Plain-brown W. At 1500-3000 m in humid forest from sw Ven. to se Peru. Text p. 320

10 OLIVE-BACKED WOODCREEPER *Xiphorhynchus triangularis*. Bill barely curved. Triangular spots below. At 1500-2700 m in humid Andean forest from Ven. to Bol. Text p. 322

11 RUSTY-WINGED BARBTAIL *Premnornis guttuligera*. Not scansorial. Tail chestnut, wings chestnut brown. **11a** ad. **b** juv., back with rufous-buff streaks. At 1300-2750 m in humid Andean forest from Ven. to Bol. Text p. 385

12 FULVOUS-DOTTED TREERUNNER *Margarornis stellatus*. Mostly rufous, lower throat and breast with fine black-bordered white streaks. At 900-2200 m in very humid, mossy forest in w Col. and nw Ecu. Text p. 384

13 PEARLED TREERUNNER *Margarornis squamiger*. Recognized by broad pale eyebrow, chestnut upperparts and tail, and pearl-spotted underside. **13a** juv.: crown and throat mottled; **b** *perlatus*: spots white; **c** *squamiger* in s of range: spots yellowish. Humid montane forest from Ven. to Bol. Text p. 384

14 SPOTTED BARBTAIL *Premnoplex brunnescens*. Dark, tail blackish, below rusty-spotted. **14a** juv., back spotted; **b** ad. *brunnescens*. At 600-2750 m in humid forest from Costa Rica and Ven. to Bol. Text p. 386

15 STREAKED TUFTEDCHEEK *Pseudocolaptes boissonneautii*. Conspicuous white feather-tuft on neck-side. **15a** ad. female, bill longer than in male; **b** juv., cap black, below heavily scaled; **c** imm., cap streaked, below heavily streaked. Mainly montane humid forest from Ven. to Bol. Text p. 386

16 WHITE-THROATED TREERUNNER *Pygarrhichas albogularis*. Bill long. Much white below. **16a** ad. **b** imm.: much darker above and on belly. *Nothofagus* forest of the southern Andes. Text p. 395

30 per cent of nat. size   PLATE XXXIII

Woodpeckers, Woodcreepers, and Ovenbirds

PLATE XXXIV. Miners, Earthcreepers, and Horneros

1 SLENDER-BILLED MINER *Geositta tenuirostris*. Bill long, slender, slightly curved. In flight extensive rufous in wing and tail. No white on rump or tail. **1a** typical ad.: with mottled breast; **b** ad., a bird from above Lima: breast distinctly streaked; **c** imm.: breast heavily streaked. Gentle highland slopes with short grass and bare earth. C Peru to nw Arg. Text p. 329

2 THICK-BILLED MINER *Geositta crassirostris*. Bill very stout; legs white. No white on rump or tail. Sometimes browner. 600-3000 m in cactus country of Pacific slope from Lima to Arequipa, Peru. Text p. 329

3 COMMON MINER *Geositta cunicularia*. Upper tail-coverts pale, tail based cinnamon. Breast streaked except in c Peru. Long rufous patch in closed wing, lesser and median wing-coverts irregularly mottled (vs. *G. antarctica*). **3a** imm.; **b** southern *cunicularia*; **c** *fissirostris* of c Chile: breast-streaks distinct; **d** *juninensis* of c Peru: breast-streaks obsolete or lacking, bill slender, base of tail whitish. Peru to Patagonia. Text p. 328

4 SHORT-BILLED MINER *Geositta antarctica*. Primaries project well beyond tertials. Wing-coverts with narrow, pale gray bars, not mottled. In flight tail looks white with triangular dark center, and wings have dull rufous gray (not rufous) pattern. Sandy or heavily grassed Fuegian plains, migrating n. Text p. 327

5 CREAMY-RUMPED MINER *Geositta isabellina*. Large; bill slightly curved. Rump and most of tail whitish. Unstreaked below, no rufous in closed wing. Arg.-Chile, very high up. Text p. 325

6 RUFOUS-BANDED MINER *Geositta rufipennis*. Bill straight. Tail and rump with no white, tail mainly bright rufous with black subterminal bar. In flight conspicuous bright rufous wingbar outlined in black. Breast and belly buff, unstreaked. 2500-4400 m (in Chile to coast) in arid open land and scrub from Bol. to s Arg. and Chile. Text p. 326

7 PUNA MINER *Geositta punensis*. Whitish at base of tail does not extend to rump. In flight with strongly rufous wingbar and broad black trailing edge. Small, round; tip of bill curved. Uniform below; prominent rufous patch in closed wing. **7a** juv.; **b** ad. At 3000-5000 m in very arid areas from extreme s Peru to Catamarca Arg. and Atacama Chile. Text p. 327

8 DARK-WINGED MINER *Geositta saxicolina*. Rump and base of tail buffy white, rest of tail dark. In flight without rufous wingbar. Sides of head and neck buffy cinnamon. At 3700-4900 in grassy or stony slopes in wc Peru. Text p. 325

9 GRAYISH MINER *Geositta maritima*. Flanks washed pinkish; in flight dark wings and tail. Coastal desert to 2600 (3500) m from Ancash Peru to Tarapacá Chile. Text p. 324

10 BUFF-BREASTED EARTHCREEPER *Upucerthia validirostris*. Wings with more rufous than in 11. Uniformly buff below; bill deeply curved. Above 2700 m from Jujuy to Mendoza Arg. Text p. 332

11 PLAIN-BREASTED EARTHCREEPER *Upucerthia jelskii*. Wing-coverts much like back or (*pallida*) with slight rufous tinge; below uniform. **11a** *jelskii* ad.; **b** juv. Highlands from Ancash c Peru to Salta nw Arg. and Tarapacá n Chile. Text p. 332

12 WHITE-THROATED EARTHCREEPER *Upucerthia albigula*. Wing-coverts dark rufous. At 2300-3550 (3900) m in desert scrub on Pacific slope from Ayacucho Peru to n Chile. Text p. 331

13 STRIATED EARTHCREEPER *Upucerthia serrana*. Distinctly streaked above and below. At 3000-4200 m in shrub and *Polylepis* woodlands from Cajamarca to Huancavelica Peru. Text p. 333

14 STRAIGHT-BILLED EARTHCREEPER *Upucerthia ruficauda*. Bill almost straight. Tail rufous streaked dark, cocked upright. Throat and breast white, lower underparts white streaked buff. At 3500-4500 m (lower in the s) in *Polylepis* shrub and among rocks, from Arequipa Peru to Chubut Arg. and Santiago Chile. Text p. 334

15 ROCK EARTHCREEPER *Upucerthia andaecola*. Bill almost straight. Fine blackish streaking below; tail uniform rufous. **15a** ad.; **b** juv., scaled below. At 3000-3500 (4500) m in bushy canyons of Bol. to nw Arg. and n Chile. Text p. 334

16 BOLIVIAN EARTHCREEPER *Upucerthia harterti*. Small. Below grayish buff, unstreaked. At 1500-3050 m in *Acacia* scrub in c Bol. Text p. 335

17 CHACO EARTHCREEPER *Upucerthia certhioides*. Small. Unstreaked, forehead and brow rufous. **17a** juv. almost lacking rufous on forehead; **b** ad. *luscinia*. Mainly lowlands, but to 1350 m in the n, from Parag. to Mendoza Arg. Text p. 335

18 SCALE-THROATED EARTHCREEPER *Upucerthia dumetaria*. Bill curved; breast distinctly scaled. **18a** ssp *saturatior*; **b** juv., entire underside scaled; **c** *dumetaria*. S Peru s-wards, in the n only in very cold parts, in the s to sea level. Text p. 330

19 BAND-TAILED EARTHCREEPER *Eremobius phoenicurus*. Bill long and straight; earcoverts rufous. Patagonian desert. Text p. 336

20 RUFOUS HORNERO *Furnarius rufus commersoni*. Almost uniformly dull rufous. To temp. zone in Cochabamba, Bol. Often near houses. Text p. 343

30 per cent of nat. size

PLATE XXXIV

Miners, Earthcreepers, and Horneros

PLATE XXXV. Tit-spinetails, Chilia, and Cinclodes

Tit-spinetails *Leptasthenura* resemble tit-mice of other continents, forage acrobatically in the outer foliage. Very long and slender tail graduated but fork-shaped.

1 PLAIN-MANTLED TIT-SPINETAIL *Leptasthenura aegithaloides*. Back unstreaked. The short outer tail-feathers whitish. **1a** *berlepschi* (distinct species?) of highlands of s Peru to nw Arg. washed buffy throughout, crown-streaks fairly broad; **b** *grisescens* from elevations below 2500 m in sw Peru to n Chile, with broad crown-streaks; **c** juv. *pallida* of c and s Arg. very pale. Arid country from s Peru to Isla Grande.
Text p. 348

2 ANDEAN TIT-SPINETAIL *Leptasthenura andicola*. Whitish supercilium distinct; heavily streaked above and below. **2a** ad. *andicola*.; **b** *peruviana* juv. (crown unstreaked, belly mottled ochraceous); **c** *peruviana* ad. At 3000-4000 m in humid páramo shrubbery from Ven. to Ecu., and at 3700-4500 in drier *Polylepis* shrub and *Lepidophyllum* brush from c Peru to n Chile.
Text p. 346

3 STREAKED TIT-SPINETAIL *Leptasthenura striata*. Almost unstreaked below; prominent rufous wingpatch. **3a** juv. *striata*; **b** ad. *superciliaris*; **c** imm. *striata*. Coast to 3500 (4000) m in arid scrub and *Polylepis* from c Peru to n Chile.
Text p. 347

4 WHITE-BROWED TIT-SPINETAIL *Leptasthenura xenothorax*. Checkered throat sharply set off from plain drab gray underside; plain rufous crown. Semihumid *Polylepis* woodlands in Vilcanota mts, Cuzco Peru.
Text p. 348

5 RUSTY-CROWNED TIT-SPINETAIL *Leptasthenura pileata*. No rufous wingpatch. Supercilium indistinct; throat distinctly spotted, gray breast and sides usually indistinctly streaked. **5a** *cajabambae* with crown streaked; **b** *pileata* with crown uniform. 2000-4000 m in *Polylepis* woods and bushy slopes from s Cajamarca to Ayacucho Peru.
Text p. 347

6 BROWN-CAPPED TIT-SPINETAIL *Leptasthenura fuliginiceps*. Uniform light gray-brown with darker cap. Wing-patch chestnut, tail rusty. **6a** juv.; **b** ad. At 1500-3250 (3900) m, in arid and semi-humid scrub in Bol. and n Arg.
Text p. 349

7 TAWNY TIT-SPINETAIL *Leptasthenura yanacensis*. Uniform orange-cinnamon. **7a** juv. **b** ad. High-elevation shrub (mainly *Polylepis*) locally in Ancash, Lima, Cuzco, and Puno Peru, at 3800-4600 m, and in Bol., where found down to 3200 m. Text p. 350

8 CRAG CHILIA *Chilia melanura*. Bill straight. Throat and breast white, tail black. **8a** ad., **b** juv. Coast to 2500 m, in c Chile, in semi-arid scrub among rocky outcrops.
Text p. 337

Cinclodes *Cinclodes*. Open country, usually near running water. Tail often cocked. In flight conspicuous wingbar.

9 GRAY-FLANKED CINCLODES *Cinclodes oustaleti*. Smaller than Dark-bellied C. and with slightly curved bill; also higher up and less near water. **9a** juv. **b** ad. Fuegian zone, and at 1500-4200 m n to Antofagasta Chile and Mendoza Arg. Migrates.
Text p. 339

10 DARK-BELLIED CINCLODES *Cinclodes patagonicus*. Large; dark. Bill long and straight. **10a** ad., **b** juv. Coast to 2500 m from Aconcagua Chile and Mendoza Arg. to Isla Grande.
Text p. 338

11 STOUT-BILLED CINCLODES *Cinclodes excelsior*. Bill stout. **11a** the dark *aricomae* (**ROYAL C.**, possibly a distinct species) of humid *Polylepis* woods from Cuzco Peru to La Paz Bol. (rare!); **b** *excelsior* of páramos in Ecu. Also in C Andes Col.
Text p. 337

12 OLROG'S CINCLODES *Cinclodes olrogi*. Small, dark. Extensive white wingbar; tail tipped dull rufous. Above 1600 m, down to 900 m in winter. Isolated mountains of Córdoba and San Luis, Arg.
Text p. 339

13 BAR-WINGED CINCLODES *Cinclodes fuscus*. Fairly small. Pale tip of outer rectrix extends up along outer web. Wingbar white, or rufous in n and s part of range. **13a** southern ssp *fuscus*; **b** *albiventris* of Peru and Bol.; **c** *albidiventris* of Ecu.; **d** *paramo* juv. Distributed throughout the Andes, at 3500-5000 m in the n, but to sea level in the s; only migrant to ranges of 12 and 14.
Text p. 341

14 SIERRAN CINCLODES *Cinclodes comechingonus*. Rufous wingbar very bright; base of mandible orange-yellow. At 1600-2800 m in sierras of Córdoba, and wintering (down to 1000 m) to Tucumán and S. del Estero, Arg.
Text p. 340

15 WHITE-WINGED CINCLODES *Cinclodes atacamensis*. Wingbar pure white. Tailfeathers broad and rounded, white tip of outer feather restricted. **15a** *atacamensis*, mainly at 2500-5000 m from Ancash Peru to c Arg. and Chile; **b** dark *schocolatinus* of Sierras Grandes de Córdoba Arg.
Text p. 342

16 WHITE-BELLIED CINCLODES *Cinclodes palliatus*. Very large; pure white below. C Peru, at 4000-5000 m on mineral-rich bogs in Lima, adjacent Junín and Huancavelica. Rare.
Text p. 342

30 per cent of nat. size

PLATE XXXV

Tit-spinetails, Chilia, and Cinclodes

PLATE XXXVI Spinetails, Rayadito, Wiretail, and Thistletails.

Cranioleuca Spinetails creep like woodcreepers and crawl acrobatically in epiphytes and foliage, usually well hidden. Chestnut tail graduated but looks forked.

1 CREAMY-CRESTED SPINETAIL *Cranioleuca albicapilla*. Noisy. Entire top of head creamy. **1a** *albicapilla* from Apurímac; **b** *albicapilla* from Huancavelica. At 2400-4050 m in montane woodlands and semihumid shrubbery locally from Junín to Cuzco. Text p. 360

2 LIGHT-CROWNED SPINETAIL *Cranioleuca albiceps*. Crown buff or white, bordered black. **2a** *discolor* of Cochabamba to St Cruz Bol.; **b** and **c** *albiceps* of Puno and La Paz. At 2200-3400 m in open cloud forest, bamboo and humid second growth. Text p. 360

3 MARCAPATA SPINETAIL *Cranioleuca marcapatae*. Possibly sspp of 2. **3a** *marcapatae* of Cord. Vilcabamba and Carabaya Cuzco; **b** *weskei* of Cord. Vilcabamba; **c** juv. *marcapatae*. At 2400-3500 m, in dense humid vegetation, esp. bamboo. Text p. 359

4 STRIPE-CROWNED SPINETAIL *Cranioleuca pyrrhophia*. Crown streaked; white supercilium conspicuous behind eye. **4a** ssp *pyrrhophia*, to 1500 m in most of Arg. and s Bol., inner remiges like back; **b** *striaticeps* imm. (juv. has uniform crown); **c** *rufipennis* from Tilo-Tilo La Paz, inner remiges dull rufous. Also at 600-3100 m in c Bol. Arid scrub-forest. Text p. 357

5 SOUTHERN LINE-CHEEKED SPINETAIL *Cranioleuca baroni*. Distinct supercilium and throat white. **5a** *zaratensis*. **b** juv. *zaratensis*, crown more or less gray. Birds of some populations twice as large. At 1700-4500 m in semiarid montane woodlands from s Cajamarca to Huánuco and Lima Peru. Text p. 358

6 NORTHERN LINE-CHEEKED PINETAIL *Cranioleuca antisiensis*. Supercilium white or buffy; white of throat extending to upper breast. Ssp *palamblae*. At 900-2800 m in humid montane woods of w Ecu. and nw Peru. Text p. 358

7 ASH-BROWED SPINETAIL *Cranioleuca curtata*. Supercilium gray. **7a** ad.; **b** juv., with ochraceous orange supercilium and underside; crown sometimes chestnut. Amongst dense foliage in humid subtrop. forest from Col. to Bol. Text p. 356

8 RED-FACED SPINETAIL *Cranioleuca erythrops*. Head mostly rufous. In humid subtrop. forest of wc and w Col. and w Ecu. Text p. 357

Rayaditos *Aphrastura* resemble titmice. In flocks. Restless.

9 THORN-TAILED RAYADITO *Aphrastura spinicauda*. Unmistakable. Note long buff supercilium. **9a** ad.; **b** juv.: back spotted. In noisy flocks in *Nothofagus* forest of the southern Andes. Text p. 344

10 WHITE-BROWED SPINETAIL *Hellmayrea gularis*. Small. Distinct supercilium and throat white. **10a** *gularis*, red phase; **b** *gularis* imm. gray phase (juv. is mottled below). At 2300-3900 m in dense, humid forest and shrub from Ven. to c Peru. Text p. 355

Synallaxis spinetails. With long tails. In pairs within dense and tangled shrubbery and bamboo.

11 RUFOUS SPINETAIL *Synallaxis unirufa*. Rufous; lores dark. **11a** *ochrogaster*; **b** juv. At 1200-3700 m, on humid Andean slopes from Ven. to Cord. Vilcabamba Peru. Text p. 354

12 RUSSET-BELLIED SPINETAIL *Synallaxis zimmeri*. Restricted range in Ancash (and Lima?) wc Peru, in scrub at 1900-3000 m. Text p. 354

13 APURÍMAC SPINETAIL *Synallaxis courseni*. Ampay massif in Apurímac c Peru, at 2450-3500 m, in humid and semi-humid shrubbery. Not sympatric with 14. Text p. 351

14 AZARA'S SPINETAIL *Synallaxis azarae*. Belly fairly dark plumbeous (scaled with white in the s). At 1250-3700 m, in humid shrub from c Peru to Bol. Not sympatric with 13 or 15. Text p. 352

15 ELEGANT SPINETAIL *Synallaxis elegantior*. Belly whitish, tail mostly rufous. **15a** *media*: lacking pale postocular streak; **b** *elegantior*; **c** juv. At (900) 1100-3800 m, in humid shrub from Ven. to nc Peru. Not sympatric with 14. Text p. 351

16 SILVERY-THROATED SPINETAIL *Synallaxis subpudica*. Voice *Cranioleuca*-like. Tail dusky. **16a** ad.; **b** juv.: rufous of crown absent or duller and barely suggested. At 1200-3200 m, in forest edge and second growth in Bogotá and Cundinamarca Col. Text p. 353

17 BUFF-BROWED SPINETAIL *Synallaxis superciliosa*. Distinct postocular streak buff. **17a** juv.; **b** ad. *superciliosa*. To 2900 m in thorny thickets from Cochabamba Bol. to Tucumán Arg. Text p. 351

18 DES MUR'S WIRETAIL *Sylviorthorhynchus desmursii*. Tiny. Tail enormously long and very decomposed. **18a** ad.; **b** juv., lacking rufous on forehead. In dense bamboo in s Chile and sw Arg. Text p. 343

19 THISTLETAILS *Schizoeaca* replace *Synallaxis* in humid treeline habitats at 2600-4000 m from Ven. to Bol. **19a** *S. perijana* of Perijá mts; **b** *S. f.fuliginosa* of sw Ven. to c Ecu.; **c** *S. fuliginosa plengei* of Huánuco c Peru; **d** *S. griseomurina* of s Ecu. and extreme n Peru; **e** *S. palpebralis* of Junín c Peru; **f** *S. vilcabambae ayacuchensis* of Ayacucho; **g** *S. v.vilcabambae* of Vilcabamba mts Cuzco; **h** *S. h.helleri* of Vilcanota and Carabaya mts Cuzco; **i** *S. helleri* ssp. from Sandia in Puno se Peru; **j** *S. h.harterti* of La Paz; **k** *S.harterti bejeranoi* of Cochabamba and Sta Cruz Bol. Text p.362

30 per cent of nat. size

PLATE XXXVI

Spinetails, Rayadito, Wiretail, and Thistletails

PLATE XXXVII. Canasteros, Great Spinetail (compare Plates XXXVI and XXXVIII).

Canasteros *Asthenes* are small, mostly terrestrial furnariids. Long graduated tails are sometimes cocked.

1 CANYON CANASTERO *Asthenes pudibunda*. Tail shorter and bill longer than in 2. At 2500-3700 m. Huancavelica and Pacific Peru. Text p. 373

2 RUSTY-FRONTED CANASTERO *Asthenes ottonis*. Tail long. 2750-3600 m from Huancavelica to Cuzco Peru. Text p. 374

3 MAQUIS CANASTERO *Asthenes heterura*. Tail long. 3000-4150 m in arid Andean scrub in La Paz and Cochabamba Bol. Text p. 374

4 LESSER CANASTERO *Asthenes pyrrholeuca*. Outer 3 tail-feathers wholly rufous. **4a** *pyrrholeuca* of Arg.; **b** juv. To 3000 m, in bushes, in winter often in swamps. Breeds c Arg. and Chile s to Straits of Magellan. Migrates n to s Bol. and Urug. Text p. 365

5 BERLEPSCH' CANASTERO *Asthenes berlepschi*. Resembles 6. Outer 2 rectrices rufous, 3rd partly so. At 2600-3700 m on Mt Illampu nw Bol. Text p. 370

6 CREAMY-BREASTED CANASTERO *Asthenes dorbignyi*. May involve 3 species. Most forms with rufous rump and vent, and mainly black tail. The *arequipae* group (**DARK-WINGED C.**) mainly inhabits *Polylepis* woods (vs. arid scrub-forest at lower elevations in other forms). Dark-winged; **a** unnamed very dark ssp from Ayacucho and s Lima; **b** *arequipae* of s Peru to extreme nw. Bol. and n Chile.
The *huancavelicae* group (**PALE-TAILED C.**): **c** *usheri* of Apurímac, with outer 3-4 rectrices whitish; **d** unnamed form from Ancash.
The *dorbignyi* group (**RUSTY-VENTED C.**): **6e** *consobrina* of La Paz; **f** *dorbignyi* of Cochabamba Bol. s to Mendoza Arg., wings rufous as in 6d. Text p. 365

7 CHESTNUT CANASTERO *Asthenes steinbachi*. Resembles sympatric Creamy-breasted C., but lacks rufous gular patch, and tail extensively rufous. 800-3000 m from Salta to Mendoza Arg. Text p. 370

8 CACTUS CANASTERO *Asthenes cactorum*. Bill long. Breast unstreaked; outer 2 tail-feathers wholly rufous, rest with black and rufous longitudinal pattern. 250-2400 m, in in Pacific Peru. Text p. 373

9 CORDILLERAN CANASTERO *Asthenes modesta*. Breast faintly streaked; each rectrix with black and rufous longitudinal pattern. **9a** *rostrata* of Cochabamba Bol.; **b** southern *navasi*; **c** juv. *navasi*. 3000-4600 m, lower in s. C Peru s to Straits of Magellan. Text p. 372

10 DUSKY-TAILED CANASTERO *Asthenes humicola*. Tail dark. **10a** *humicola* of Atacama to Maule Chile and n Mendoza Arg.; **b**: *polysticta* of s Maule to Malleco Chile. To 2200 m. Text p. 371

11 PATAGONIAN CANASTERO *Asthenes patagonica*. Breast plain gray; tail mostly dark. To base of Andes in sc Arg. Text p. 371

12 STREAK-THROATED CANASTERO *Asthenes humilis*. Foreparts appear dark, with streaked throat; tail dark. **12a** *robusta*; **b** juv. Puna grassland, often near water, from Cajamarca Peru to La Paz Bol. Text p. 378

13 STREAK-BACKED CANASTERO *Asthenes wyatti*. Outer tail-feathers rufous from base. **13a** northern *aequatorialis*; **b** southern *graminicola*. At 3000-5500 m, in bunchgrass, from Ven. to Puno Peru. Text p. 375

14 PUNO CANASTERO *Asthenes punensis*. Like 13, but outer 3 tail-feathers black at base. **14a** an intergrade between *punensis* and 13 from Puno Peru; **b** *punensis* from La Paz Bol. 4000 m in the n, down to 2150 m in the s, in bunchgrass from e Peru to La Rioja Arg. Text p. 376

15 CÓRDOBA CANASTERO *Asthenes sclateri*. With more prominent rufous wingbar than 14. At 2000-2900 m in Córdoba Arg. Text p. 377

16 AUSTRAL CANASTERO *Asthenes anthoides*. Distinctly streaked above; outer 3 rectrices with whitish outer web. **16a** ad.; **b** juv. Brush and adjacent tall grass along Andes from Neuquen Arg. and Concepción Chile to Isla Grande. Rare. Text p. 377

17 SCRIBBLE-TAILED CANASTERO *Asthenes maculicauda*. Streaked throughout. No chin-patch. **17a** bird from Puno Peru; **b** from Arg. At 2250-4300 m in humid bunchgrass near treeline, locally from Puno Peru to Catamarca Arg. Text p. 378

18 JUNÍN CANASTERO *Asthenes virgata*. Streaked; tail with some rufous. Larger and browner than 19, with rufous chin-patch, and more white on belly. At 3350-4800 m in humid bunchgrass, often near treeline, locally from Lima to Puno Peru. Text p. 379

19 MANY-STRIPED CANASTERO *Asthenes flammulata*. Streaked above and below; wings and tail with some rufous. **19a** *taczanowskii* of Peru, **b** *flammulata* of Ecu., **c** *quindiana* of C Andes Col. 2800-4500 m in humid grass and treeline shrub from E and C Andes Col. to Junín Peru. Text p. 380

20 LINE-FRONTED CANASTERO *Asthenes urubambensis*. Tail dark, feathers very pointed. Forehead striped; body dark, scarcely streaked above. **20a** richly colored *urubambensis* of Cuzco to Cochabamba; **b** *huallagae* of La Libertad to Pasco Peru. Edge of humid *Polylepis* wood. Text p. 380

21 GREAT SPINETAIL *Siptornopsis hypochondriacus*. 2450-2800 m in dense humid shrubbery in Marañón watershed n Peru. Rare. Text p. 361

30 per cent of nat. size

PLATE XXXVII

Canasteros and Great Spinetail

PLATE XXXVIII Thornbirds, Foliage-gleaners, and Treehunters

Thornbirds *Phacellodomus* inhabit scrub and are most often revealed by their large stick-nests. Stiff, pointed feathers of forehead rufous.

1 CHESTNUT-BACKED THORNBIRD *Phacellodomus dorsalis*. Large. Rufous above. In dry scrub with scattered trees from s Cajamarca to Ancash Peru.
Text p. 383

2 RUFOUS-FRONTED THORNBIRD *Phacellodomus rufifrons*. Wings and tail olive-brown (remiges occ. edged rufous). In arid subtrop. scrub and woods, disjunctly, in n Peru and Sta Cruz Bol. to Tucumán Arg.
Text p. 382

3 STREAK-FRONTED THORNBIRD *Phacellodomus striaticeps*. Resembles Creamy-breasted Canastero but more extensive rufous on tail and flanks. Ssp *striaticeps* shown. At 2800-5000 m, in arid scrub from Apurímac Peru to Catamarca Arg.
Text p. 382

4 RUSSET-MANTLED SOFTTAIL *Thripophaga berlepschi*. **4a** juv.; **b** ad.: mostly rufous. At 2450-3350 m in elfin forest in C Andes in Amazonas and La Libertad Peru.
Text p. 381

5 SCALY-THROATED FOLIAGE-GLEANER *Anabacerthia variegaticeps temporalis*. Spectacles buff. **5a** imm.: throat rusty; **b** ad. W slope of Andes in Col. and Ecu. Premontane forest.
Text p. 389

6 MONTANE FOLIAGE-GLEANER *Anabacerthia striaticollis*. Spectacles whitish. Mandible distinctly upturned. **6a** juv.; **b** ad. *striaticollis*; **c** *yungae*. At 950-2600 m in humid forest in n W Andes and e Andean slopes from Ven. to Bol.
Text p. 389

7 SPECTACLED PRICKLETAIL *Siptornis striaticollis*. Like a small short-tailed *Cranioleuca* spinetail. Head with varying amounts of white. Premontane humid forest E Andes Col. to n Peru.
Text p. 383

Foliage-gleaners forage actively along branches and in foliage in humid forest. *Philydor* feed in the subcanopy, often open to view; *Syndactyla* live in undergrowth, and are difficult to see; *Automolus* are intermediate.

8 RUFOUS-NECKED FOLIAGE-GLEANER *Automolus ruficollis*. *Syndactyla*-like. Neckside contrasting orange. At 400-2900 m in dense undergrowth from El Oro Ecu. to Cajamarca Peru.
Text p. 391

9 BUFF-BROWED FOLIAGE-GLEANER. *Syndactyla rufosuperciliata*. Back unstreaked; throat scaled. Bill rather slender and pointed, mandible upturned. **9a** juv.; **b** ad. *oleagineus*: brow buff; **c** *cabanisi*: faint brow ochraceous. At 1300-3000 m, above 10 in range of overlap, in humid forest undergrowth from extreme n Peru to La Rioja Arg.
Text p. 387

10 LINEATED FOLIAGE-GLEANER. *Syndactyla subalaris*. Streaked below and on mantle, with whitish in e part of range. **10a** *subalaris* of w Col. and Ecu.; **b** juv.: brow and underside orange. Undergrowth of humid premontane forest from Ven. to Cuzco Peru.
Text p. 388

11 CRESTED FOLIAGE-GLEANER. *Automolus dorsalis*. Throat and upper breast yellowish white. Humid terra firme forest to 2200 m from se Col. to se Peru.
Text p. 390

12 BUFF-FRONTED FOLIAGE-GLEANER. *Philydor rufus*. Headside and brow ochraceous orange. **12a** ad? *columbiana*; **b** juv. *riveti* of Pacific slope of Col. and nw Ecu.: bill long and slender; fore-crown brown. Humid premontane forest from C Am. to Arg.
Text p. 390

Treehunters *Thripadectes* are stocky strong-billed birds of the dense understory and second growth in humid regions.

13 BUFF-THROATED TREEHUNTER *Thripadectes scrutator*. Head and underparts streaked buff. At 2100-3650 m from e and s of the Marañón Peru to Cochabamba Bol.
Text p. 392

14 FLAMMULATED TREEHUNTER *Thripadectes flammulatus*. Heavily streaked above and below. Ssp *flammulatus* shown. At 800-3300 m from Ven. to nw and e Ecu. and extreme n Peru.
Text p. 392

15 STRIPED TREEHUNTER *Thripadectes holostictus*. Throat with pointed scales; streaks below not edged black. At 800-3100 m (below 13 and 14, above 16 and 17) from sw Ven. to w Ecu. and Cochabamba Bol.
Text p.393

16 STREAK-CAPPED TREEHUNTER *Thripadectes virgaticeps*. Only head and upper breast streaked. At 1000-2500 m, above 17, below 15, from Ven. to n Ecu.
Text p. 394

17 BLACK-BILLED TREEHUNTER *Thripadectes melanorhynchus*. Throat with pointed scales as in 15, but breast virtually unstreaked. In premontane forest on e-slope from Col. to Puno se Peru. Text p. 393

18 SHARP-TAILED STREAMCREEPER *Lochmias nematura*. On or near the ground along dark streams. Below dark spotted white. **18a** *sororia*, Ven. to n Peru; **b** *obscurata*, **c** Peru to Jujuy nw Arg. At 725-2780 m, widespread but somewhat local.
Text p. 396

19 WREN-LIKE RUSHBIRD *Phleocryptes melanops*. Rush-beds. Song endless series of monotonous *clicks*. Streaked above. **19a** *juninensis* of 4000-4300 m in Juañ Peru; **b** juv. In highlands from Puno Peru to Jujuy nw Arg., and also in lowland areas.
Text p. 345

30 per cent of nat. size

PLATE XXXVIII

Thornbirds, Foliage-gleaners, and Treehunters

PLATE XXXIX Antbirds (see also Plate XL)

Antshrikes, *Thamnophilus*, are stocky passerines with hooked bills.

1 VARIABLE ANTSHRIKE *Thamnophilus caerulescens*. **1a** male *aspersiventer* from Puno Peru to Cochabamba Bol.: cap black, wing-coverts tipped white. Northern males (*subandinus*) wholly black below or (*melanochrous*) with scaled vent and flanks. Southern birds (*dinellii*) buff below with grayish white throat and grayish olive back, wings, and tail. **1b** female *aspersiventer*: wingcoverts black tipped white. In northern females gray of throat extends to breast, cap mixed with black, and wing-coverts lack white tips. Southern birds have rufescent crown, whitish throat, and ochraceous buff remaining underparts. At 1300-2750 m in thickets and humid forest undergrowth from Amazonas C Andes of Peru to Cochabamba Bol., below 2300 m from Santa Cruz and s Cochabamba Bol. to n Arg. Also in e Brazil. Text p. 398

2 RUFOUS-CAPPED ANTSHRIKE *Thamnophilus ruficapillus*. Crown chestnut; wings rufous. **2a** *cochabambae* male, breast barred; **b** *cochabambae* female, breast unbarred or almost so; **c** *marcapatae* male, heavily barred below; **d** *marcapatae* female, back brown tinged gray, below ochraceous orange; **e** *jaczewski* male, heavily barred below, tail barred. In the Andes from Arg. to Peru. Locally at 2000-2600 m at edge of humid forest in se Cajamarca and C Andes of n Peru, and from Cuzco Peru to Cochabamba Bol. Ascends to 2300 m in semi-arid scrub from Sta Cruz Bol. to Tucumán Arg. Also se Brazil to ne Arg. Text p. 399

3 UNIFORM ANTSHRIKE *Thamnophilus unicolor*. **3a** male, uniform gray; **b** female, brown with gray face and upper throat. Both sexes with faint wingbar in the s. At 900-2700 m in dense undergrowth of humid forest from Col. to w Ecu. and Huánuco c Peru. Text p. 397

4 BARRED ANTTHRUSH *Chamaeza mollissima*. Terrestrial, rail-like. Brow black and white. **4a** *yungae* juv.; **b** *yungae* ad. spotted below; **c** *mollissima* barred below. At 1800-3100 m in humid forest from c Col. to Cochabamba Bol. Text p. 403

5 RUFOUS-BREASTED ANTTHRUSH *Formicarius rufipectus*. Crown and nape dark chestnut (depicted *carrikeri*; crown black in *thoracicus*); breast rufous chestnut. Call 2 similar notes. 1000-2400 (3100) m in humid forest from Costa Rica and Ven. to w Ecu. and Cuzco Peru. Text p. 403

Grallaria antpittas are compact, long-legged, and short-tailed birds of secretive habits. Mainly terrestrial, inhabiting humid montane forest and thickets.

6 GIANT ANTPITTA *Grallaria gigantea*. Large. Forehead rufous chestnut; undulated or barred rufous below. Ssp *hylodroma* shown. At 2000-3000 m. Stagnant pools in humid forest in s Col. and both slopes of Ecu. Rare. Text p. 405

7 UNDULATED ANTPITTA *Grallaria squamigera*. Song a long, slow, deep, and hollow trill. **7a** *canicauda* ad.; **b** *squamigera* juv. At (1800) 2600-3800 m near treeline in humid forest from Ven. to Cochabamba Bol. Text p. 404

8 GRAY-NAPED ANTPITTA *Grallaria griseonucha*. Song a c. 2 s long series of 12-18 notes; also series of 5 even notes with pause after 3rd. **8a** juv.; **b** *griseonucha* ad. (*tachirae* has pale throat and brownish olive back). At 2300-3000 m in humid forest of Mérida and Táchira Ven. Text p. 411

9 SCALED ANTPITTA *Grallaria guatimalensis*. Song a deep and hollow slow trill. White whisker. **9a** *regulus* ad.; **b** *regulus* juv., with whitestreaked underparts. Mainly upper trop. zone, but to 2400 m in nw Peru. Humid and semi-humid forest from C Am. and Ven. to nw Peru and Cochabamba Bol. Text p. 405

10 CHESTNUT-NAPED ANTPITTA *Grallaria nuchalis*. Song an accelerating rising series of notes. Below slate-colored to dusky brown; nape chestnut. **10a** *obsoleta*: only nape chestnut; **b** *ruficeps*: crown and nape chestnut; **c** *ruficeps* juv., orange-brown. At 1900-3150 m, mainly 2600-2900 m in wet forest, esp. in bamboo, from Col. to extreme n Peru. Text p. 408

11 PALE-BILLED ANTPITTA *Grallaria carrikeri*. Song 6 even notes with pauses after 1st and before last. Like 10, but bill pale. **11a** ad.; **b** juv. At 2350-2900 m in wet forest, esp. in bamboo, in C Andes in Amazonas and La Libertad n Peru. Text p. 408

12 RUFOUS-FACED ANTPITTA *Grallaria erythrotis*. Song of 3 faintly ascending notes. Headside rufous. At 2050-2900 m in humid forest from La Paz to Sta Cruz Bol. Text p. 412

30 per cent of nat. size

PLATE XXXIX

Antbirds

PLATE XL Antpittas and Tapaculos (see also Plates XXXIX and XLI)

Antpittas *Grallaricula* are smaller and less terrestrial than *Grallaria*.

1 RUSTY-BREASTED ANTPITTA *Grallaricula ferrugineipectus*. Song a rhythmic series of 16-18 soft notes. Crown like back; below rufous. White spot near eye. **1a** *leymebambae*; **b** juv. Humid forest and woodland at (250) 800-2200 m locally in Ven. and Col., and at 1750-3350 m from Piura Peru to La Paz Bol. Text p. 415

2 SLATE-CROWNED ANTPITTA *Grallaricula nana*. Song a soft descending series of slightly buzzy whistles. Crown gray in contrast to back. At 700-2930 m in humid forest from Ven. to extreme n Peru.
Text p. 416

3 CRESCENT-FACED ANTPITTA *Grallaricula lineifrons*. Streaked black below; forehead with white 'horns'. At 2500-3220 m in swampy humid forest in s Col. and ne Ecu. Text p. 416

4 OCHRE-BREASTED ANTPITTA *Grallaricula flavirostris*. Song a short accelerating ascending series of 5-8 soft whistles. Call a descending gargle. Scalloped below (ad. male has yellow mandible and sometimes most of maxilla). **4a** *mindoensis*; **b** *boliviana*. At (500) 1150-2750 m in humid forest and second growth from Col. to w Ecu. and Cochabamba Bol. Text p. 414

5 BROWN-BANDED ANTPITTA *Grallaria milleri*. Brown; lores, throat, and belly white. At 2700-3140 m in humid forest in Quíndio and Caldas in C Andes Col. Text p. 413

6 STRIPE-HEADED ANTPITTA *Grallaria andicola*. Song a faint, ascending series of buzzy notes. Heavily streaked. **6a** *andicola* juv.; **b** *andicola* ad.; mantle streaked; **c** *punensis*, back unstreaked. At 3000-4300 m in *Polylepis* woods from s Cajamarca Peru to La Paz Bol.
Text p. 407

7 CHESTNUT ANTPITTA *Grallaria blakei*. 1-note call (?). Rufous; abdomen gray and faintly barred. At 2135-2470 m in humid forest from c Amazonas to Huánuco (Pasco? Ayacucho?) Peru. Text p. 411

8 RUFOUS ANTPITTA *Grallaria rufula*. 2-note call. Song variable. **8a** *rufula* juv., in molt; **b** *rufula* ad.; **c** *saltuensis*, olivaceous, pale below; **d** *cochabambae*, olivaceous above, tawny below. At 2300-3650 m in humid forest, often in bamboo, from Ven. to Cochabamba Bol. Text p. 412

9 BICOLORED ANTPITTA *Grallaria rufocinerea*. Gives a long clear whistle and also a long whistle slurred down at end. Rufous; breast and belly gray. At 2100-3300 m in humid forest in C Andes and e Nariño Col. Text p. 407

10 BAY ANTPITTA *Grallaria capitalis*. 4-note song, last 3 lowest and alike. Chestnut, c belly paler. At (1525) 2600-3000 m in humid shrubbery in Huánuco and Pasco c Peru. Text p. 410

11 WHITE-BELLIED ANTPITTA *Grallaria hypoleuca*. 3-note song, last 2 highest and alike. Throat and belly white. Ssp *castanea* shown. 1400-2350 m at edge of humid forest and in second growth from Col. to extreme n Peru. Text p. 409

12 RUSTY-TINGED ANTPITTA *Grallaria przewalskii*. 3-note song, middle note lower. Crown gray. **12a** ad.; **b** juv. At 2200-2750 m in undergrowth of dense humid forest often with bamboo in C Andes in Amazonas and San Martín n Peru. Text p. 410

13 RED-AND-WHITE ANTPITTA *Grallaria erythroleuca*. 3(-4)-note song, last note drawn-out. Rufous; throat and c underparts white. At 2200-3000 m in bamboo and other humid second growth in Vilcabamba and Vilcanota mts Cuzco se Peru. Text p. 410

14 CHESTNUT-CROWNED ANTPITTA *Grallaria ruficapilla*. 3-note song, middle note lowest and longest. **14a** *albiloris* juv.; **b** *albiloris* ad. At 550-3600 m in both dry and humid shrub from Ven. to Lambayeque and San Martín n Peru. Text p. 406

15 TAWNY ANTPITTA *Grallaria quitensis*. Song of (2-)3 unclear notes, 1st loudest and usually highest. **15a** *quitensis*; **b** Peruvian *atuensis*, more decidedly spotted below; **c** *quitensis* juv. At 2200-4000 m in humid treeline shrub from Col. to La Libertad n Peru. Text p.413

Tapaculos, Family Rhinocryptidae, are secretive groundliving inhabitants of thickets. Large southern species shown here.

16 CRESTED GALLITO *Rhinocrypta lanceolata*. Call *chirrup*. Crown and headside streaked, sides chestnut. To 1800 m in brushland in chaco from e Bol. to Río Negro Arg. Text p. 420

17 HUET-HUET *Pteroptochos tarnii*. Name describes call. Chestnut; hindcrown, back, tail, headside, and throat slaty black (throat like breast in *castaneus*). **17a** *castaneus* juv., barred and spotted; **b** *tarnii*. Lowlands to 1600 m in dense undergrowth and bamboo in *Nothofagus* forest from Colchagua to Messier Channel, Chile, and adjacent Arg. from Neuquén to Sta Cruz.
Text p. 418

18 MOUSTACHED TURCA *Pteroptochos megapodius*. Voice distinctive, a gargling crake-like sound; brow and broad moustache white. To 3050 m in bushy hillsides, dense bamboo and undergrowth of humid forest from Atacama to Concepción Chile. Text p. 419

PLATE XL

30 per cent of nat. size

Antpittas and Tapaculos

PLATE XLI Tapaculos (continued)

1 ANDEAN TAPACULO *Scytalopus magellanicus*. Treeline habitats into tall grass. Small-billed. Dark patch suggested in front of eye. Bars on flanks straight and narrow when present. **1a** *canus* ad. (type shown) at 2800-3800 in C Andes Col. **b** *griseicollis* juv.; **c** *griseicollis* ad. at 2500-3200 m in E Andes Col. **d** *opacus* sub-ad., at 3100-4000 m in e Ecu. **e** *affinis* ad. at 2600-4000 m in nw Peru: **f** *altirostris*, imm. female, at 2450-3300 m in nc Peru; **g** *altirostris* ad., **h** unnamed ssp from 3500 m in Apurímac: **i** *simonsi* at 3000-4300 from Cuzco to Cochabamba Bol. **j** *zimmeri*, of s Bol. **k** *superciliaris*, of nw Arg.: **l** *magellanicus* ad. male, of southern Andes. **m** *fuscus* of lowlands of c Chile (to 4000 m in the n?) (apparently specifically distinct from Andean T.; large blackish specimen from Chile with white fore-crown may represent a 3rd species). Ven. to Isla Grande. Text p. 437-43

2 BROWN-RUMPED TAPACULO *Scytalopus latebricola*. Trilled song. Bright flanks with wavy bars. **2a** *meridanus* ad. of 2000-3650 m in Col. and Ven.: gives short fast trills and a long rattly trill; bill small; **b** *spillmanni*, (Cerro Chinguela), inhabiting bamboo at 2200-3200 m in Ecu. to n Peru. Text p. 433

3 BLACKISH TAPACULO *Scytalopus unicolor, latrans* group. Barks and gives low-pitched single whistle. Uniform blackish. **3a** ad. (no subad. plumage); **b** juv.: drab, only flanks barred. At 1500-3700 m from Ven. to San Martín (Huánuco?) Peru.
Text p. 424

4 UNICOLORED TAPACULO *Scytalopus unicolor, parvirostris* group. Gives long trills. **4a** *parvirostris* juv. of Peru-Bol.: scaled, flanks barred; **b** *parvirostris* imm.; **c** *parvirostris* subad. (type depicted): flanks more or less barred, belly often with silvery tips. At 1850-2500 m in c Peru, to 3200 m in the s. Text p. 426

5 TAPACULO, unnamed species *Scytalopus* sp. Call a 1-2 s series of 3-7 notes. Song *keek-krrr*. **5a** ad.: uniform dark gray; **b** imm. At 2675-3500 m in Huánuco, Pasco Peru. Text p. 427

6 TAPACULO, unnamed species *Scytalopus* sp. Long fast trills and short rising trills. Large. Not in bamboo. N Ecu., at 2000-2800 m; see text under **Nariño T.** Text p. 432

7 RUFOUS-VENTED TAPACULO *Scytalopus femoralis*. Repeated song of 2 notes n of the Marañón, 1 note s of it. Bill strong. Wavy flank-bars. **7a** *femoralis* ad. from Peru; **b** *femoralis* juv. (800) 1500-2000 (2950) m from s Col. to Cuzco Peru. Text p. 429

8 SOUTHERN WHITE-CROWNED TAPACULO *Scytalopus (femoralis?) bolivianus*. Long low slow trill. Premontane zone (1200-2150 m) of se Peru and Bol. Text p. 431

9 NORTHERN WHITE-CROWNED TAPACULO *Scytalopus (femoralis?) atratus*. Ssp *confusus* shown. Song irregular series of 1 to 10 notes. White forehead. 1050-1800 (2100) m from Ven. to Peru. Text p. 430

10 LARGE-FOOTED TAPACULO *Scytalopus macropus*. Song a low-pitched note repeated at 3 per s. Large, uniform blackish. **10a** subad., flanks dull brown with wavy bars; **b** ad. C Peru, at 2400-3500 m, often near streams. Text p. 428

11 ASH-COLORED TAPACULO *Myornis senilis*. In bamboo; song up to minute-long, of single well spaced notes and terminated with a hysterical trill. **11a** ad.; **b** and **c** juv. Fairly uniform, sometimes with white on face and belly. 2300-3950 m Ven. to c Peru. Text. p. 423

12 SLATY GNATEATER *Conopophaga ardesiaca*. Whistled calls. **12a** imm. male, female similar, but without postocular tuft and narrow buff tips of greater wing-coverts; **b** male, long white postocular tuft. 800-2400 m in humid forest from Cuzco Peru to Tarija Bol. Text p. 417

13 OCHRE-FLANKED TAPACULO *Eugralla paradoxa*. Song a staccato series of 3-8 notes. Maxilla elevated. **13a** ad.; **b** juv., densely barred throughout. S Chile and adjacent Arg. Text p. 423

14 OLIVE-CROWNED CRESCENTCHEST *Melanopareia maximiliani*. Black breastband; white brow and throat. Mainly 1700-2100 m, but reaches 3500 m in Cochabamba. Forest edge and semi-arid shrubbery from La Paz Bol. to Mendoza Arg.
Text p. 421

15 SANDY GALLITO *Teledromas fuscus*. Brow white, tail mostly black. Loud monotonous song, and a 4 s descending series of 14 notes. Low brush in arid parts from sw Salta to Río Negro Arg. Text p. 421

16 OCELLATED TAPACULO *Acropternis orthonyx*. Song a repeated, clear whistle. **16a** ad.; **b** juv. molting. At 2700-3900 m in humid mossy forest and *Polylepis* woods from Ven. to Cord Colán in Amazonas Peru. Text p. 444

17 CHUCAO TAPACULO *Scelorchilus rubecula*. To 1500 m, in bamboo and other undergrowth in undisturbed *Nothofagus* forest from (Colchagua) Bío-Bío to Aysén s Chile and adjacent Neuquén to Chubut Arg.
Text p. 420

18 WHITE-THROATED TAPACULO *Scelorchilus albicollis*. Brow and throat white; belly barred. To 1600 m on bushy rocky slopes from Atacama to Curicó Chile. Text p. 419

30 per cent of nat. size

PLATE XLI

Tapaculos

PLATE XLII Waxwing, Plantcutters, and Cotingas

1 CEDAR WAXWING *Bombycilla cedrorum*. Accidental visitor from N Am. to Col. Text p.563

Plantcutters *Phytotoma* inhabit open scrubby wood and frequent orchards. Song and calls a mechanical cracking.

2 RUFOUS-TAILED PLANTCUTTER *Phytotoma rara*. Tail rufous chestnut and black. **2a** flight; **b** ad. male; **c** ad. female. Scrub and forest clearings in Chile and sw Arg. Text p. 452

3 WHITE-TIPPED PLANTCUTTER *Phytotoma rutila*. Tail with conspicuous white terminal bar. **3a** flight; **b** female; **c** imm. male; **d** male. At 600-3600 m in n, to 1800 m in the s, in thorny *Prosopis* wood, orchards, brush, and fields in Bol. and Arg. s to Río Negro.
Text p. 452

Ampelion cotingas inhabit cloud and elfin forest. Perch conspicuously atop trees, usually in upright stance.

4 RED-CRESTED COTINGA *Ampelion rubrocristata*. Bill appears whitish at a distance. Note white in tail. **4a** displaying; **b** ad.; **c** juv. Common at 2500-3700 m in Andes from Ven. to Lima Peru and Cochabamba Bol. Text p. 445

5 WHITE-CHEEKED COTINGA *Ampelion stresemanni*. White cheeks. At 2700-4240 m in montane woods with mistletoes from La Libertad to w Ayacucho Peru. Text p. 446

6 CHESTNUT-CRESTED COTINGA *Ampelion rufaxilla*. Most of head rufous. Yellow underparts streaked black. At 1860-2740 m in wet forest from Col. to Bol. Rare. Text p. 446

7 BAY-VENTED COTINGA *Ampelion sclateri*. Dark; no white in tail. **7a** ad.; **b**: juv. At 2500-3500 m in elfin forest from La Libertad to Junín Peru. Recent sighting in s Ecu. Rare and local. Text p. 447

8 DUSKY PIHA *Lipaugis fuscocinereus*. Large; tail long. Solitary except at leks where it gives piercing song. At 1700-3000 in humid forest from Col. to extreme n Peru; in Bol. replaced by similar Scimitar-winged P. Text p. 450

Fruiteaters *Pipreola* sit motionless in fruiting trees, and are most often revealed by very high-pitched wheens.

9 GREEN-AND-BLACK FRUITEATER *Pipreola riefferii*. Tail uniform green above. **9a** *riefferii* male in most of Col.; **b** *occidentalis* male of sw Col. and w Ecu.; **c** *melanolaema* female of Ven.; **d** *chachapoyas* of nc Peru, male: with c belly unmarked; **e** *tallmanorum* of c Peru, male: small, almost unmarked below. At (900) 1500-3050 m, below 10 in zone of overlap. Humid forest from Ven. to Huánuco c Peru. Text p. 448

10 BAND-TAILED FRUITEATER *Pipreola intermedia*. Tail with black subterminal band and white tip; markings below extended to c underparts in the form overlapping with 9. **10a** *signata* of c Peru, male; **b** female; **c** *intermedia* of s part of range, male: markings below blacker. Peru and Bol. at 2000-3000 m, above 9 in zone of overlap. Humid forest from La Libertad Peru to Cochabamba Bol. Text p. 449

11 BARRED FRUITEATER *Pipreola arcuata*. Large; barred below. **11a** juv. male; **b** female; **c** male. At 1800-3500 m in humid forest from Ven. to Cochabamba Bol. Text p. 449

PLATE XLII

30 per cent of nat. size
(except 2a and 3a)

Waxwing, Plantcutters, and Cotingas

PLATE XLIII Tyrant Flycatchers; mainly terrestrial species

Muscisaxicola ground-tyrants are terrestrial birds of open country. Color of crown patch often distinctive.

1 SPOT-BILLED GROUND-TYRANT *Muscisaxicola maculirostris*. Small, brownish. **1a** *rufescens* juv.; **b** *maculirostris* ad. Near steep walls and cliffs. Col. to Straits of Magellan; at 2000-4000 m in the n to below 1500 m in the s. Text p. 508

2 CINEREOUS GROUND-TYRANT *Muscisaxicola cinerea*. Gray. Usually smaller than 3, and with shorter and narrower brow, and with wings tinged brown. **2a** *cinerea*; Peru, Bol., Chile; **b** *argentina* of nw Arg. (may be a ssp of 3). On rocky slopes. At 4000 m in s Peru and Bol. down to 2700 m in the s, in Talca Chile and Catamarca Arg. Text p. 512

3 PLAIN-CAPPED GROUND-TYRANT *Muscisaxicola alpina*. Bill short. **3a** *alpina* of Ecu.; **b** *grisea* of Bol., juv.; **c** *grisea* ad. Forages on level ground, at 2700-4800 m in the Andes from Col. to Bol. Text p. 512

4 WHITE-FRONTED GROUND-TYRANT *Muscisaxicola albifrons*. Largest of the genus. Brow short, very broad; long wings with conspicuous silvery edging. At 4000-5600 m from Ancash Peru to Oruro Bol. and adjacent n Chile. Text p. 513

5 WHITE-BROWED GROUND-TYRANT *Muscisaxicola albilora*. Hindcrown broadly tawny; differs from 6 also by long bill, dull pale gray underparts, less conspicuous wing-panel, and whiter lores. Breeds on barren rocky slopes at 1500-2500 m from Straits of Magellan to Neuquén Arg. and Aconcagua Chile, and winters at 3000-4000 in Bol., Peru, and Ecu. Text p. 511

6 PUNA GROUND-TYRANT *Muscisaxicola juninensis*. Most of crown dull cinnamon. From 5 also by smaller bill, faint buffy yellow tinge below, brighter wing-panel, and less white on lores. At (4000) 4200-5000 from c Peru to nw Arg. and n Chile. Text p.511

7 BLACK-FRONTED GROUND-TYRANT *Muscisaxicola frontalis*. Forehead and c crown black. Breeds above 2900 m in c Arg./Chile, winters above 3600 m from nw Arg. to s Peru. Rare. Text p. 514

8 DARK-FACED GROUND-TYRANT *Muscisaxicola macloviana*. Small; face dark. Breeds to 1200 m from Isla Grande to w Neuquén Arg. and Llanquihué Chile, and migrates to coast of Peru. Text p. 509

9 CINNAMON-BELLIED GROUND-TYRANT *Muscisaxicola capistrata*. Forehead black, crown rufous chestnut; **9a** ad.; **b** juv. Breeds extreme s Arg. and Chile. Winters at increasing elevations n-wards to s Peru, where at 4000 m. Text p. 509

10 RUFOUS-NAPED GROUND-TYRANT *Muscisaxicola rufivertex*. Long slender bill. Pale, cap chestnut to cinnamon. **10a** *occipitalis* of Peru to Cochabamba Bol. at 3000-4500 m; **b** *pallidiceps* from coast to 4000 m from sw Peru to Cordobá Arg. Peru to Mendoza c Arg., southern birds migratory. Text p.510

11 OCHRE-NAPED GROUND-TYRANT *Muscisaxicola flavinucha*. Large with long bill and very long wings. Hindcrown yellow-ocher. **11a** ad.; **b** imm. Breeds Chile and adjacent Arg.; winters at 3200-4700 m from Bol. to nc Peru. Text p. 513

12 RUSTY-BACKED MONJITA *Neoxolmis rubetra*. Back rufous brown; sides of rump and wing-edgings whitish (much smaller *salinarum* has more white on wing and rump). Breeds w Arg. from Mendoza to Chubut, wintering n to Tucumán. Text p.503

13 CHOCOLATE-VENTED TYRANT *Neoxolmis rufiventris*. Large. Distinctive wing-pattern. Breeds Fuegian zone, and winters n to s Brazil. Text p. 503

14 FIRE-EYED DIUCON *Xolmis pyrope*. Open woods and gardens from Neuquén Arg. and Atacama Chile to Isla Grande. Migrates in the s. Text p. 502

15 RUFOUS-WEBBED TYRANT *Polioxolmis rufipennis*. Dark gray. **15a** ad.; **b** hovering like a kestrel. 3050-4300 m in *Polylepis* woods and adjacent slopes from n Peru to Potosí Bol. Text p. 499

16 WHITE-TAILED SHRIKE-TYRANT *Agriornis andicola*. Longer-tailed, larger, and with much heavier bill than 19; mandible pale. Tail mostly white. Ssp *andicola* shown. Rare. Ecu. to n Arg. and Chile. High elevations. Text p. 505

17 GREAT SHRIKE-TYRANT *Agriornis livida*. Large. No white brow, tail mostly black, c belly cinnamon. **17a** imm.; **b** ad. From Atacama Chile and w Río Negro Arg. to Isla Grande. Text p. 506

18 MOUSE-BROWN SHRIKE-TYRANT *Agriornis murina*. To 2000 m. Breeds Chubut to Tucumán Arg., wintering n to w Parag. and c Bol. Text p. 507

19 BLACK-BILLED SHRIKE-TYRANT *Agriornis montana*. Much like 16, but smaller, shorter-tailed, and with smaller bill; most forms with less white in tail. Bill black except in juv. **19a** *solitaria* of s Col. and Ecu.; tail much like 16; **b** *leucura* in s part of range. At 2400-4300 m, lower in the s, from s Col. to Cautín c Chile and Sta Cruz Arg. Text p. 504

20 GRAY-BELLIED SHRIKE-TYRANT *Agriornis microptera*. Much like 17, but smaller, paler, with whitish brow. Also grayer wash on belly, esp. in sympatric form. **20a** male; **b** female or imm. Lowlands of s Arg., wintering n to s Bol. and Urug.; also breeds at 2900-5000 m in the Andes from Catamarca, nw Arg. to n Chile and adjacent s Peru. Text p. 506

30 per cent of nat. size, (except 15b)

PLATE XLIII

Tyrant Flycatchers; mainly terrestrial species

PLATE XLIV Tyrant Flycatchers (see also Plates XLIII to XLVII)

1 CLIFF FLYCATCHER *Hirundinea ferruginea*. Long wings rufous. Headside mottled. Sallies from cliffs. Lowlands, ascending into the cloud forest along roads, from Ven. to Bol., to 4000 m. Text p. 519.

2 STREAK-THROATED BUSH-TYRANT *Myiotheretes striaticollis*. Large. Throat and breast streaked; in flight triangular wings with much rufous. **1a** ad.; **b** in flight. At (600) 1700-3700 m in open country and gardens, in Andes from Ven. to nw Arg. Text p. 500

3 SMOKY BUSH-TYRANT *Myiotheretes fumigatus*. In flight shows rufous in wings but not in tail. Much like 5, but darker (esp. on throat) and dark trailing edge of wing wider (1.5 cm). **3a** *cajamarcae* ad.; **b** *fumigatus* juv. 1800-3600 m in humid forest and shrubbery from Ven. to Cuzco Peru. Text p. 501

4 RED-RUMPED BUSH-TYRANT *Cnemarcus erythropygius*. Large. White forehead and wing-patch; rump rufous. In flight no rufous in wings. At 3000-4300 m in humid shrubbery from Col. to Cochabamba Bol. Text p. 499

5 RUFOUS-BELLIED BUSH-TYRANT *Myiotheretes fuscorufus*. Much like juv. 3, but throat white; dark trailing edge of wing narrower (1 cm). At 2130-3550 m at edge of humid forest from Pasco Peru to Cochabamba Bol. Text p. 502

Chat-tyrants *Ochthoeca* are most often seen low, at edges or inside montane forest of mainly humid types. Make chat-like jerks.

6 D'ORBIGNY'S CHAT-TYRANT *Ochthoeca oenanthoides*. Cinnamon-rufous below; very long, broad white brow. **6a** *polionota* in n part of range; **b** *oenanthoides* in s of range. At 2800-4500 m, lower in the s, often near *Polylepis* groves. From n Peru to n Chile and Tucumán nw Arg. Text p. 497

7 BROWN-BACKED CHAT-TYRANT *Ochthoeca fumicolor*. Unlike 6 with brow narrow or washed ochraceous, and in more humid habitat. **7a** *superciliosa* juv.; **b** *brunneifrons*; **c** *berlepschi*. At (1800) 2400-4400 m from Ven. to Cochabamba Bol. Text p. 496

8 WHITE-BROWED CHAT-TYRANT *Ochthoeca leucophrys*. Rather slender and upright; perched conspicuously. Distinct white brow; below light gray. **8a** *leucophrys*; **b** *leucometopa*; **c** *tucumana*. Lowlands to 1000 m in fairly arid habitats from n Peru to n Chile and San Juan c Arg. Text p. 498

9 SLATY-BACKED CHAT-TYRANT *Ochthoeca cinnamomeiventris*. Drawn-out, thin wheen recalls a fruiteater. Slaty; short brow and under wing-coverts white. **9a** *nigrita* of Ven.; **b** *thoracica* of c Peru to Bol.; **c** *cinnamomeiventris* of Col. and Ecu. At 900-3000 m along streams in dark humid montane forest from Ven. to Sta Cruz Bol. Text p. 493

10 RUFOUS-BREASTED CHAT-TYRANT *Ochthoeca rufipectoralis*. Rufous breast; large white brow. **10a** *centralis*; **b** *rufipectoralis*. At 2000-3600 (4100) m in humid montane shrubbery from Col. to Cochabamba Bol. Text p.495

11 PIURA CHAT-TYRANT *Ochthoeca piurae*. From 8 mainly by being smaller and rounder, and by having broad rufous wingbars, white edges to tertials and outer primary, and rufescent rump. Retiring habits. At 1500-2800 m on shrubby hillsides from Piura to Ancash Peru. Text p. 498

12 CROWNED CHAT-TYRANT *Ochthoeca frontalis*. Protruding breast. Brow yellow in front of eyes (except in E Andes Col.), white and not narrowing posteriorly. **12a** *albidiadema*; **b** *frontalis* juv.: brow buffy; **c** *boliviana*. At (1300) 2100-3660 m in humid shrubbery. At higher elevation than 13, from Col. to Bol. Text p. 495

13 GOLDEN-BROWED CHAT-TYRANT *Ochthoeca pulchella*. Probably sspp of 15. Round. **13a** *similis* juv.: posterior brow washed buff; **b** *pulchella*. Difficult to see. At 1700-2750 m in understory of interior of humid forest from Amazonas Peru to c Bol. Text p. 494

14 CROWNED CHAT-TYRANT *Ochthoeca (frontalis) jelskii*. Like 12c, but lower back and rump more reddish brown. From sw Ecu. to Lima Peru. Text p. 495

15 YELLOW-BELLIED CHAT-TYRANT *Ochthoeca diadema*. Jizz and habits like 13. Below yellow, breast olive (least so in Peru). Brow narrow posteriorly. **15a** *gratiosa*; **b** *rubellula*; **c** *diadema* juv. At (800) 1750-3100 m in humid forest from Ven. to n Peru. Text p. 494

16 PATAGONIAN CHAT-TYRANT *Ochthoeca parvirostris*. Call high-pitched, drawn-out, and quavering. Dark earpatch. S and c Chile and sw Arg. Text p. 497

17 WHITE-WINGED NEGRITO *Lessonia oreas*. Terrestrial, in open marshes and lake shores. **17a** female: much darker than female 18; juv. lighter and told from 18 by light underwing; **b** male: black, back rufous; **c** wing-flap: exposing characteristic silvery white underwing. At (1000) 4000-4900 m on altiplanos of c Peru to n Coquimbo Chile and Catamarca Arg. Text p. 515

18 RUFOUS-BACKED NEGRITO *Lessonia rufa*. Like 17, but no white on underwing. **18a** female: much lighter than female 17, and with small cinnamon patch on underwing; **b** male: back rufous chestnut. Breeds from coast to 2000 m. Nw Arg. and n Chile s-wards to Isla Grande. Text p. 515

30 per cent of nat. size
(except 1 and 2b)

PLATE XLIV

Tyrant Flycatchers

PLATE XLV Tyrant Flycathers (see also Plates XLIII to XLVII)

1 FORK-TAILED FLYCATCHER *Tyrannus savana*. Unmistakable. **1a** ad.; **b** juv.: from 2 by brownish cap and forked tail without white tip. In open country with scattered arboreal growth almost throughout lowlands of S Am. Vagrant to the high Andes. Text p. 524

2 EASTERN KINGBIRD *Tyrannus tyrannus*. Forest and second growth. Cap black, tail white-tipped. N Am. migrant s to n Arg., casual in the high Andes. Text p. 524

3 OLIVE-SIDED FLYCATCHER *Contopus borealis*. Upright. Whitish line down c underside. Forest clearings. N Am. migrant to 400-3300 m in the Andes s to Bol. Rare. Text p. 487

4 VERMILION FLYCATCHER *Pyrocephalus rubinus*. Semi-arid country. **4a** male molting to 1st ad. plumage: darker above than ad. female; **b** ad. male: unmistakable, blackish and scarlet to vermilion; **c** female: throat and breast white, breast and sides streaked, belly vermilion. N Am. to n Arg. and n Chile, locally to 3000 m. Southern birds migratory. Text p. 492

5 YELLOW-BROWED TYRANT *Satrapa icterocephala*. Brow and underparts yellow. **5a** ad.: pale gray wingbar; **b** juv.: breast spotted olive, wingbar faint. In fairly open country. Lowlands of e Bol., s Brazil, and n Arg., locally to 2600 m on Andean slopes. Winters n to c Peru, rarely Ven. Text p. 519

6 TROPICAL KINGBIRD *Tyrannus melancholicus*. Perches on telephone-lines. Tail slightly forked, throat and upper breast light gray. Widespread, except in dense forest, locally to c. 3000 m, from N Am. to wc Peru and Río Negro Arg. Text p. 524

7 TROPICAL PEWEE *Contopus cinereus*. Loral feathers white basally; wings reach less than half-way down tail. Ssp *pallescens* shown. In semi-open arid country in much of S Am., ascending to 2600 m in wc Peru, Bol., and nw Arg. Text p. 489

8 WESTERN WOOD-PEWEE *Contopus sordidulus*. Wings reach over half-way down tail; loral feathers pale gray basally; only those ad.s which have all dark bill safely told from 9. Also more throaty call than 9. N Am. migrant crossing the Andes in Col. to winter in humid forest clearings and woodland at 80-1600 m s to Bol. Text p. 488

9 EASTERN WOOD-PEWEE *Contopus virens*. Averages paler than 8, and always has pale mandible; not safely distinguishable. N Am. migrant crossing the Andes in Col. to winter in humid forest clearings at 500-1300 m s to se Peru. Text p. 488

10 GREATER PEWEE *Contopus fumigatus*. Conspicuous, and with a distinctive *peeew* call. Large, dark, upright. **10a** juv.: ochraceous wingbars; **b** *zarumae* of sw Col. to nw Peru: uniformly dark gray; **c** *ardosiacus* of Ven. to se Peru: lighter and more smoky gray. To 3000 m in clearings in humid forest clearings from N Am. to Ven. and nw Arg. Text p. 487

11 GREAT KISKADEE *Pitangus sulphuratus*. Call distinctive: *kis ka-dee*. White brow and throat, yellow underparts. In clearings, gardens etc., throughout trop. and subtrop. zone, and at 1500-3300 m in Bol. (ssp *bolivianus* depicted). Text p. 522

12 PLUMBEOUS TYRANT *Knipolegus signatus*. Perched inconspicuously. Constantly vibrates tail from side to side. Eye red in ad. **12a** *cabanisi* of 1100-2500 m from se Peru to nw Arg., male: slaty gray with light belly; **b** *cabanisi* female: tail-feathers edged rufous; **c** *signatus* of n and c Peru, male: black, white patch on underwing less conspicuous than in 13; **d** display flight. Humid premontane forest and shrub. Text p. 516

13 WHITE-WINGED BLACK-TYRANT *Knipolegus aterrimus*. Arid regions (*aterrimus*, *heterogyna*) or semi-humid second growth (*anthracinus*). **13a** *aterrimus* of Cochabamba Bol. to Chubut s Arg., female: largest ssp, breast unstreaked, outer web of outer primary rufous, crown usually blackish; **b** *aterrimus* 2nd year male; **c** *anthracinus* Junín c Peru to La Paz Bol., female: much like 16, but more robust, and with dark eye, conspicuous pale loral region, pale c belly, and more rufous on rump; **d** male *heterogyna*: black; bill whitish; **e** male *aterrimus*: glossy; bill gray; **f** display-flight. At (250) 600-3250 (in Bol. 3700) m from n Peru to s Arg. Text p. 517

14 GOLDEN-CROWNED FLYCATCHER *Myiodynastes chrysocephalus*. Headside traversed by 2 pale lines; back unstreaked, no white wingbar. **14a** ad.; **b** juv. less streaked below, no crown-patch, upperparts grayer. Second growth in humid premontane zone (to 2800 m) from Panama to w Ecu. and s Bol. Text p. 522

15 SPECTACLED TYRANT *Hymenops perspicillata*. Reed-beds and marshes. **15a** male black with mostly white primaries, conspicuous bare orbital skin yellow; **b** female: breast distinctly streaked, rufous wing-patch. Breeds in lowlands marshes from e Bol. to Chubut s Arg., and from Atacama to Valdivia Chile; Chilean birds migrating across the Andes. Text p. 518

16 RUFOUS-TAILED TYRANT *Knipolegus poecilurus*. Rather slender, eye red (except in juv.), rump with less rufous than 13, flanks gray, c belly pale rufous. Ssp *peruanus* shown. Locally at 900-2500 (3100) m at humid forest edge and in second growth from Ven. to Bol. Text p. 517

17 BLACK PHOEBE *Sayornis nigricans*. Black with fairly long tail. White wing-panel and c belly. On rocks in premontane streams (occ. to 3000 m) in semi-arid to semi-humid zones locally from N Am. to nw Arg. Ssp *angustirostris* shown. Text p. 491

30 per cent of nat. size
(except 12d and 13f)

PLATE XLV

Tyrant Flycatchers

PLATE XLVI Tyrants (see also Plate XLVII)

1 **TUFTED TIT-TYRANT** *Anairetes parulus*. **1a** *patagonicus*; **b** juv.; **c** *aequatorialis*. At 1830-4200 (lower i s). S Col. to Isla Grande. Text p. 472

2 **ASH-BREASTED TIT-TYRANT** *Anairetes alpinus*. 4000-4600 m in semi-humid *Polylepis* woods in Ancash and Cuzco Peru (Bol.?). Rare. Text p. 468

3 **YELLOW-BILLED TIT-TYRANT** *Anairetes flavirostris*. **3a** *arequipae*; **b** juv. At (0) 1000-3650 m from nw Peru to n Chile and s Arg. Text p. 472

4 **PIED-CRESTED TIT-TYRANT** *Anairetes reguloides*. **4a** male; **b** female; **c** juv. 0-2900 m in n, to 3550 m in s. From wc Peru to n Chile. Text p. 471

5 **BLACK-CRESTED TIT-TYRANT** *Anairetes nigrocristatus*. **5a** juv.; **b** male; **c** female. At 2000-4200 m in shrub in nw to c Peru. Text p. 470

6 **UNSTREAKED TIT-TYRANT** *Anairetes agraphia*. **7a** *agraphia*; **b** juv.; **c** *squamigera*. 2700-3600 m in Peru. Text p. 469

7 **AGILE TIT-TYRANT** *Anairetes agilis*. **6a** ad.; **b** juv. At (1800) 2700-3500 m. Ven. to Ecu. Text p. 470

8 **YUNGAS TODY-TYRANT** *Hemitriccus spodiops*. 800-2450 m in Bol. Text p. 482

9 **BLACK-THROATED TODY-TYRANT** *Hemitriccus granadensis*. **9a** *pyrrhops;* **b** *granadensis;* **c** *intensus*. 1500-3300 m. Ven. to Bol. Text p. 482

10 **RUFOUS-CROWNED TODY-TYRANT** *Poecilotriccus ruficeps*. **10a** *ruficeps* of Ven. to se Ecu.; **b** *melanomystax* of nW and C Andes of Col.; **d** unnamed form of n Peru s of the Marañón; **c** *peruvianus* of extreme n Peru. At (1000) 1600-2750 m. Text p. 481

11 **OCHRE-FACED TODY-FLYCATCHER** *Todirostrum plumbeiceps*. Ssp *obscurum*. At 750-2600 m. Humid understory. Se Peru to nw Arg. Text p. 483

12 **MOTTLE-CHEEKED TYRANNULET** *Phylloscartes ventralis angustirostris*. Compare XLVII 19. 1125-2400 m. C Peru to nw Arg. Text p. 479

13 **RUFOUS-HEADED PYGMY-TYRANT** *Pseudotriccus ruficeps*. **1a** ad.; **b** juv.: from 14 by brighter rufous wings. At 1400-3600 m. Col. to Bol. Text p. 480

14 **HAZEL-FRONTED PYGMY-TYRANT** *Pseudotriccus simplex*. Se Peru to Cochabamba Bol. at 1670-2000 m. Text p. 480

15 **HANDSOME FLYCATCHER** *Myiophobus pulcher*. **15a** *pulcher* of w Col. and nw Ecu.; **b** *oblitus* of se Peru. At (800) 1400-3050 m. Text p. 485

16 **OCHRACEOUS-BREASTED FLYCATCHER** *Myiophobus ochraceiventer*. **16a** male; **b** female. 2800-3500 m. Peru and Bol. Text p. 485

17 **FLAVESCENT FLYCATCHER** *Myiophobus flavicans*. **17a** *flavicans* male; **b** female.1500-2700 m. Ven. to c Peru. Text p. 483

18 **UNADORNED FLYCATCHER** *Myiophobus inornatus*. 1800-2600 m. C Peru to Bol. Text p. 484

19 **ORANGE-BANDED FLYCATCHER** *Myiophobus lintoni*. 2250-2800 m. S Ecu. Text p. 484

20 **CINNAMON FLYCATCHER** *Pyrrhomyias cinnamomea*. **20a** *pyrrhoptera* of Ven. to n Peru; **b** *cinnamomea* of Peru to nw Arg. 670-3350 m. Text p. 486

21 **SUBTROPICAL DORADITO** *Pseudocolopteryx acutipennis*. **a** ad.; **b** juv. In marshes at 1500-3000 m in Bol. and nw Arg., migrating (breeding?) n to Col. Text p. 475

22 **WARBLING DORADITO** *Pseudocolopteryx flaviventris*. Lowlands of Bol., Chile, and Arg., Chilean birds migrating across the Andes. Text p. 476

23 **BEARDED TACHURI** *Polystictus pectoralis*. Male (?) of ssp *bogotensis* of Col. shown. Rare on rushy parts of savannas. Text p. 475

24 **ACADIAN FLYCATCHER** *Empidonax virescens*. N Am. migrant s to Ecu. Text p. 490

25 **ALDER FLYCATCHER** *Empidonax alnorum*. Resembles a pewee, but less conspicuous, shorter-winged. Wags tail. N Am. migrant s to nw Arg. Text p. 490

26 **MANY-COLORED RUSH-TYRANT** *Tachuris rubrigastra*. Unmistakable. **26a** ad.; **b** juv. Reed-beds. Peru (to 4100 m) s-wards. Text p. 473

27 **BARRED BECARD** *Pachyramphus versicolor*. **27a** male; **b** female. 1500-3500 m. Edge of humid forest. Costa Rica and Ven. to Bol. Text p. 525

28 **CRESTED BECARD** *Pachyramphus validus*. **a** male; **b** female. Premontane, depicted ssp *audax* to 3500 m in woodlands from s Peru to nw Arg. Text p. 526

29 **DUSKY-CAPPED FLYCATCHER** *Myiarchus tuberculifer*. Widespread in lowlands, *atriceps* at 1100-3000 (3300) m from s Ecu. to nw Arg. Text p. 520

30 **GREAT-CRESTED FLYCATCHER** *Myiarchus crinitus*. Base of mandible pale. N Am. migrant to woodlands and dry forest to Col. Text p. 521

31 **PALE-EDGED FLYCATCHER** *Myiarchus cephalotes*. 1125-2750 m. Ven. to Bol. Text p. 521

30 per cent of nat. size

PLATE XLVI

Tyrants

Plate XLVII Tyrannulets and Flycatchers (see also Plate XLVI)

1 SULPHUR-BELLIED TYRANNULET *Mecocerculus minor*. Locally from w Ven. to extreme n Peru, at 1700-2600 m. Text p. 465

2 BUFF-BANDED TYRANNULET *Mecocerculus hellmayri*. Alone in mixed-species flocks. Makes short upward sallies. At 1100-2700 m in humid shrubbery from extreme se Peru to nw Arg. Text p. 465

3 WHITE-TAILED TYRANNULET *Mecocerculus poecilocercus*. Strict perch-gleaner. 1400-3000 m. Humid shrubbery from Col. to se Peru. Text p. 464

4 WHITE-BANDED TYRANNULET *Mecocerculus stictopterus*. Broad white wingbars. Snarling call. 1800-3700 m from Ven. to Bol. Text p. 466

5 WHITE-THROATED TYRANNULET *Mecocerculus leucophrys*. **5a** *leucophrys;* **b** *rufomarginatus*. 1850-4250 m, Ven. to nw Arg. Text p. 463

6 TORRENT TYRANNULET *Serpophaga cinerea*. On rocks and banks in streams. **6a** ad.. **b** juv. Ven. to Bol. Subtrop. to lower temp. zone. Text p. 467

7 WHITE-BELLIED TYRANNULET *Serpophaga munda*. Small, delicate; very pale. In dry wood and scrub in Bol. to c Arg., to 2600 m. Text p. 467

8 GREATER WAGTAIL-TYRANT *Stigmatura budytoides*. Dry wood from e Brazil to 2600 m in valleys of Bol. Text p. 468

9 SLATY ELAENIA *Elaenia strepera*. **9a** male; **b** female. Premontane, in nw Arg. to Bol., migrating to Ven. Text p. 461

10 SMALL-BILLED ELAENIA *Elaenia parvirostris*. Not safely told from 13 b; bill smaller. At least to 2000 m in Bol. Text p. 460

11 SIERRAN ELAENIA *Elaenia pallatangae*. Conspicuous yellowish wingbars; most of underparts pale yellow. Col. to Bol., at 1400-3650 m. Text p. 462

12 HIGHLAND ELAENIA *Elaenia obscura*. Large, long-tailed, pot-bellied. Short-billed. Head rounded without crest or patch, appearing small. Extreme n Peru to nw Arg., at 1850-2500 m. Text p. 462

13 WHITE-CRESTED ELAENIA *Elaenia albiceps*. Above grayish (except in *chilensis*), below whitish. **13a** *chilensis* juv.: no crown-patch; **b** *chilensis* ad. of s Andes; **c** *modesta* of w Peru to n Chile, in winter to e Peru; **d** *griseogularis* of n Peru n-wards. Often in semi-arid conditions. Col. to Isla Grande, to 3300 m. Text p. 459

14 MOUNTAIN ELAENIA *Elaenia frantzii*. Small. Flutters within shrubbery and bushes. Ssp *pudica* shown. Ven. to Col., at 1800-2800 m. Text p. 461

15 SOUTHERN BEARDLESS TYRANNULET *Camptostoma obsoletum*. Small. Perch-gleans. Often raises crest and tail and spreads wings in a jerk. Characteristic call. Plumage often faded, forehead sometimes whitish. **15a** *bolivianum* of Bol. and nw Arg., to 3125 m; **b** *sclateri* of w Ecu. to extreme nw Peru. In open habitats with trees and scrubs throughout to nw Arg., generally below 2600 m. Text p. 458

16 ASHY-HEADED TYRANNULET *Phyllomyias cinereiceps*. Dark patch on earcoverts, crown slaty. Ven. to se Peru, at 1150-2750 m. Text p. 455

17 BLACK-CAPPED TYRANNULET *Phyllomyias nigrocapillus*. Difficult to observe. Ssp *nigrocapillus* depicted. At 1600-3400 m in humid forest canopy from Ven. to c Peru. Text p. 454

18 TAWNY-RUMPED TYRANNULET *Phyllomyias uropygialis*. Fairly easy to observe. 1500-3700 m in humid shrubbery from Ven. to Bol. Text p. 455

19 SCLATER'S TYRANNULET *Phyllomyias sclateri*. Canopy of humid forest. Much like XLVI 12, but cap gray and underparts less yellow. Southern ssp *sclateri* shown; ssp of La Paz and Peru less yellow. 1525-2400 m, from se Peru to nw Arg. Text p. 454

20 PALTRY TYRANNULET *Zimmerius villissimus*. Like sympatric 22b, but eyebrow whitish or pale yellowish white (white in ssp *tamae*); wing-coverts edged pale yellowish green. Ssp *improbus* shown. At 800-3000 m in Ven. and n Col. Text p. 456

21 BOLIVIAN TYRANNULET *Zimmerius bolivianus*. **21a** *bolivianus* of Bol., **b** *viridissimus* of se Peru. At 1100-2830 m from se Peru to Bol. Text p. 457

22 GOLDEN-FACED TYRANNULET *Zimmerius viridiflavus*. **22a** *viridiflavus* of c Peru; **b** *chrysops* of Ven. to n Peru: below pale; usually only chin and under tail-coverts pale yellow. Common to 2700 m from Ven. to c Peru. Text p. 457

23 INCA FLYCATCHER *Leptopogon taczanowskii*. Headside mottled, forming light bar across cheek. At 1350-2650 m in Peru. Text p. 478

24 RUFOUS-BREASTED FLYCATCHER *Leptopogon rufipectus*. Ssp *rufipectus* depicted. At 1600-2700 m in Ven. to n Peru. Text p. 477

25 OLIVE-STRIPED FLYCATCHER *Mionectes olivaceus meridae*. Faint buffy wingbar on tips rather than edges. Ven. and Perijá mts, to 3000 m, other sspp below 26. Text p. 477

26 STREAK-NECKED FLYCATCHER *Mionectes striaticollis*. **26a** *viridiceps* of sw Col. and w Ecu.; **b** *columbianus;* **c** *striaticollis;* Col. to Bol. At 1300-3000 m. Text p. 477

30 per cent of nat. size

PLATE XLVII

Tyrannulets and Flycatchers

PLATE XLVIII Swifts, Swallows, Pipits, and Larks

1 WHITE-COLLARED SWIFT *Streptoprocne zonaris*. Large. Black with white collar all around neck; closed tail slightly forked. **1a** *albicincta* of the premontane to lower montane parts of the Tropical Andes; **b** *altissima* of the higher parts: large, forehead pale. In large parts of S Am. (but not in Chile), mainly over montaneous country. Text p. 238

2 WHITE-TIPPED SWIFT *Aeronautes montivagus*. Mottled white on throat and c belly. Locally over grassy or wooded slopes at 500-2600 m from Ven. to Bol. Text p. 239

3 ANDEAN SWIFT *Aeronautes andecolus*. White collar, rump, and c underparts, incl. vent. **3a** *andecolus* of the e Andean slope of Bol. and Arg., **b** *parvulus* of the w slope in Peru and n Chile. Mainly at 2000-3500 m over semi-arid rocky or bushy slopes from w Peru to n Chile and c Arg. Text p. 240

4 CHIMNEY SWIFT *Chaetura pelagica*. Small, sooty, with short square tail. Rapid wingweats. Frequently soars. N Am. migrant along W coast to s Peru; rarely to 3600 m. Text p. 239

5 CHESTNUT-COLLARED SWIFT *Cypseloides rutilus*. Dark. **5a** ad. male: collar rufous, chin and upper throat dark; **b** juv. molting male (true juv. lacks rufous, and in the Andean zone also most females lack rufous). Premontane to lower temp. zone of C Am. to Bol. Text p. 239

6 PURPLE MARTIN *Progne subis*. Large martin with long, triangular wings. Often glides. **6a** ad. male: purplish black with small whitish patch on lower sides (see drawing); **b** ad. female: belly whitish (gray in Southern M.); **c** imm. female. N Am. migrant to n Arg. On migration to 4000 m in the Andes. Text p. 528

7 BROWN-CHESTED MARTIN *Phaeoprogne tapera*. Large. Breast-band and spots down c belly. Breeds in lowlands, from Col. to nw Peru and Chubut Arg., but southern migrants appear up to 4000 m in the Andes. Text p. 529

8 BLUE-AND-WHITE SWALLOW *Notiochelidon cyanoleuca*. **8a** southern ssp *patagonicus*: wing-linings pale; **b** northern *cyanoleuca*: wing-linings dark, vent extensively black; **c** juv. *cyanoleuca*. Lowlands to 4000 m from Costa Rica to Isla Grande, southern birds migratory. Text p. 531

9 BROWN-BELLIED SWALLOW *Notiochelidon murina*. Dark smoky gray below. At 2100-4300 m, mostly in humid and semi-humid regions, from Ven. to Cochabamba Bol. Text p. 530

10 PALE-FOOTED SWALLOW *Notiochelidon flavipes*. Throat pale rufous, sides dark, winglinings black. At (2000) 2800-3600 m in humid treeline habitats locally from Ven. to Bol. Text p. 531

11 ANDEAN SWALLOW *Petrochelidon andecola*. Stout with triangular wings and hardly forked tail. Throat gray-brown grading towards whitish belly. At (2500) 3100-4600 m in open arid country from Ancash Peru to Tarija Bol. Text p. 533

12 CLIFF SWALLOW *Petrochelidon pyrrhonota*. Stockily built. Back streaked, rump rufous. Depicted ssp *melanogaster* is a N Am. migrant to n Arg., uncommon in the High Andes. Text p. 533

13 BARN SWALLOW *Hirundo rustica erythrogaster*. Only swallow in the range with deep fork and white spots in tail. **13a** juv.; **b** ad. N Am. migrant to all parts of S Am., but regular in the high Andes only on passage. Esp. over fields and marshes. Text p. 532

14 CORRENDERA PIPIT *Anthus correndera*. Heavily streaked, legs fairly dark. Outer tail-feather mostly white. **14a** *chilensis* of Chile; penultimate tail-feather white along shaft; **b** *catamarcae* from 3350-4450 m from Antofagasta Chile and nw Arg.; palest form; penultimate tailfeather mostly white; **c** *calcaratus* of similar elevations from Lima Peru to Jujuy nw Arg. Tail like b; underparts more streaked. Text p. 565

15 SHORT-BILLED PIPIT *Anthus furcatus*. Pale pattern on head, back appears scaled, streaked breast sharply separated from white belly. Depicted ssp *brevirostris* lives at 3500-4250 m (in s down to 2300 m) from Lima Peru to Jujuy nw Arg. Also lowlands from s Brazil to c Arg. Text p. 563

16 HELLMAYR'S PIPIT *Anthus hellmayri*. Slightly less buffy and with more, darker, and broader streaks on breast than 17; wingbars usually paler than panel. **16a** *hellmayri* from Puno Peru to Tucumán nw Arg.; outer tail-feather mainly buffy white, next with tiny white dot at tip when fresh; **b** *dabbenei* from Neuquén to Chubut, migrating n to Tucumán, Arg.; always white tip on penultimate tail-feather, as in 17. At 1650-3400 m. Also lowlands from se Brazil to ne Arg. Text p. 564

17 PARAMO PIPIT *Anthus bogotensis*. Below buffy and very sparsely streaked. **17a** *shiptoni* of nw Arg. (Bol.?); supercilium unstreaked, underparts mainly white; **b** *immaculatus* of Peru to Bol.; more buffy, esp. below; much like 16, but wingbars slight and not in sharp contrast to panel. Forms further n more buffy throughout. At (1950) 2200-4500 m from Ven. to nw Arg. Text p. 565

18 HORNED LARK *Eremophila alpestris peregrina*. Terrestrial. **18a** juv.: heavily scaled; **b** ad.: distinct facial pattern. At 2500-3000 m on fields and lake shores in E Andes of Col. Text p. 527

30 per cent of nat. size, but
1, 6, and 7 only 15 per cent

PLATE XLVIII

Swifts, Swallows, Pipits, and Larks

PLATE IL Wrens, Mockingbirds, and Dippers

Most wrens skulk in dense forest undergrowth, but reveal their presence by their lout melodic songs.

1 GRAY-MANTLED WREN *Odontorchilus branickii*. High in canopy, restlessly moving along horizontal branches. White below. Fairly long tail barred. Humid premontane forest of Col. to Bol. Text p. 539

2 INCA WREN *Thryothorus eisenmanni*. Loud song in duet or chorus of 5-6 birds. Dense bamboo thickets at 1830-3350 m, Cuzco Peru. Text p. 543

3 PLAIN-TAILED WREN *Thryothorus euophrys*. Tail almost unbarred. **3a** *atriceps* of extreme n Peru; **b** *longipes* of e Ecu., juv: breast plain rufous; **c** *longipes* ad.: breast lightly spotted; **d** *euophrys* of sw Col. and w Ecu.: lower throat and breast distinctly spotted black. At 1850-3500 m in dense bamboo. Text p. 543

4 MOUSTACHED WREN *Thryothorus genibarbis consobrinus*. Distinct moustache. In dense, humid thickets (not bamboo), mainly in premontane zone (locally to 2800 m) from Ven. to Bol. Text p. 544

5 RUFOUS WREN *Cinnycerthia unirufa*. Wings and tail virtually unbarred. Distinct wheezy call. **5a** *unibrunnea*, **b** juv: cap grayish. At (1800) 2200-3800 m in cloud forest from Ven. to n Peru. Text p. 539

6 SEPIA-BROWN WREN *Cinnycerthia peruana*. Wings and tail finely barred. Older birds may get white feathers on face, s-wards from se Ecu. **6a** *olivascens* of w and c Col. to n Peru; **b** *peruana* of n to c Peru; **c** *fulva* of se Peru and Bol.: tiny, superciliated, and usually white-faced. In cloud forest, premontane in area of overlap with Rufous W., but to 3370 m in the s. Text p. 540

7 GRAY-BREASTED WOOD-WREN *Henicorhina l. leucophrys*. Song loud, each phrase repeated several times. In pairs. Mainly premontane, but locally to 3600 m, in humid forest, often along dripping wet mossy banks, from Ven. to Bol. Text p. 547

8 CHESTNUT-BREASTED WREN *Cyphorhinus thoracicus*. Distinctive voice (see text) and headshape. Breast rufous chestnut. Low in wet forest undergrowth, at 750-2650 m from Col. to se Peru. Text p. 548

9 GRASS WREN *Cistothorus platensis*. Tiny. Mouse-like. Tumbling low flight with spread tail. Sings with tail cocked high over streaked back. **9a** *polyglottus* at 500-3275 m in Ven. and Sta Marta and Perijá mts: supercilium faint or absent, flanks unbarred (vs. barred in Paramo W. Plate LXIV); **b** *aequatorialis* at 2240-4500 m in E and C Andes Col. to c Ecu.; **c** *graminicola* at 3500-4600 from s Ecu. to La Paz Bol.; **d** *hornensis* below 2400 m in Chile and s Arg.: large, heavily streaked above. In fens, edge of humid shrub, grassy bogs, and swampy grassland (in Bogotá area never in tule marshes). Widespread but local. Text p. 540

10 HOUSE WREN *Troglodytes aedon*. Back with or without bars. Brow, when present, not as broad and contrasting as in Mountain W. **10a** southern *chilensis*; **b** *tecellatus* of Pacific slope of n Chile and s Peru; **c** large cinnamon-washed ssp *puna* of Peruvian highlands. From sea level to 4500 m, in many habitats, not unbroken forest. Often near houses. Text p. 545

11 MOUNTAIN WREN *Troglodytes solstitialis*. Very high-pitched voice. Broad supercilium contrasts dark stripe through eye. Vent distinctly barred. **11a** large barred *monticola* of Sta Marta mts n Col. **b** *solitaria* of Ven. and most of Col.; **c** *solstitialis* of s Col. to n Peru; **d** *frater* of se Peru and Bol., **e** *frater* juv. In forest and forest clearing from Mexico to nw Arg., at 1700-3500 m (lower in Arg.). Text p. 546

Mockingbirds of the genus *Mimus* are conspicuous, fairly long-tailed birds of open country.

12 CHILEAN MOCKINGBIRD *Mimus thenca*. Large. Long, broad supercilium whitish. Streaked back, flanks, and mustache. Lowlands in Chile from Atacama to Valdivia. Text p. 551

13 WHITE-BANDED MOCKINGBIRD *Mimus triurus*. White wing-pattern unmistakable. To 2600 m, mainly in Arg., occ. Chile. Text p. 550

14 PATAGONIAN MOCKINGBIRD *Mimus patagonicus*. Back uniform, breast grayish. Short, broad eye-brow from above eye. Bill short. Tail slightly shorter and less broadly white-tipped than in 15. To 1800 m in Arg. and s Chile. Text p. 549

15 CHALK-BROWED MOCKINGBIRD *Mimus saturninus*. Above faintly streaked. Long, broad supercilium white, below whitish. Tail broadly white-tipped. To base of Andes in Bol. and n Arg. Text p. 549

16 BROWN-BACKED MOCKINGBIRD *Mimus dorsalis*. Large patch in wing and 4 outer tail-feathers white. **16a** ad., **b** juv., with breast streaked (like juv. of others in the genus). 1700-4200 m in Bol., nw Arg., and extreme n Chile. Text p. 551

17 LONG-TAILED MOCKINGBIRD *Mimus longicaudatus*. A northern representative of 12. Long tail broadly tipped white. To lower temp. zone on Pacific slope of Ecu. and Peru. Text p. 551

Dippers, frequent rivers and streams.

18 RUFOUS-THROATED DIPPER *Cinclus schulzi*. Streams in alder woods at 1500-2000 m from Tarija Bol. to Tucumán Arg. Text p. 538

19 WHITE-CAPPED DIPPER *Cinclus leucocephalus*. **19a** *leucocephalus* of Peru and Bol., **b** *leuconotus*. Ven. to Ecu.; **c** juv. 1500-3300 m. Text p. 537

30 per cent of nat. size

PLATE IL

Wrens, Mockingbirds, and Dippers

PLATE L Thrushes (compare blackbirds Plate LI)

1 SLATY NIGHTINGALE-THRUSH *Catharus fuscater*. Melodic song. Difficult to see. Small. Orange bill and eye-ring, pale blue iris. Throat unstreaked. **1a** *fuscater* of Ven., e Col. and Ecu.; **b** *mentalis* of se Peru and Bol.: darker and tinged brown below. At 600-3250 m, low in mossy forest from Ven. to Bol. Text p. 554

2 SPOTTED NIGHTINGALE-THRUSH *Catharus dryas maculatus*. Hard to see. Low in humid premontane forest from C Am. to Arg. Text p. 553

3 SWAINSON'S THRUSH *Catharus ustulatus swainsoni*. Short bill. Distinct pale lore and eye-ring. N Am. migrant wintering in woodlands and forest below 2400 m from Mexico to nw Arg. Text p. 555

4 GRAY-CHEEKED THRUSH *Catharus m. minimus*. From Swainson's T. by more distinct spotting below, and lack of pale lore and eye-ring. In clearings and thickets along streams. N Am. migrant to premontane zone s to n Peru, to 3000 m on migration. Text p. 556

5 ORANGE-BILLED NIGHTINGALE-THRUSH *Catharus aurantiirostris*. Elusive. Distinctive song (not given at dusk and dawn). Bill, legs, and narrow eye-ring orange. Below white with gray breast and sides. Ssp *phaeopleurus* shown; other sspp with brownish crown and headside. At 600-2900 m in second growth from C Am. to Ven. and Col. Text p. 554

6 GLOSSY-BLACK THRUSH *Turdus serranus*. Male sings from bromeliad high in canopy. **6a** male: much smaller and more glossy than Great T., iris dark; **b** female: brownish below, fore-neck, chest, and flanks with slight olivaceous or dull brownish wash (imm. warmer tinged above); **c** juv.: spotted. At 900-3000 m in humid forest from Ven. to nw Arg. Text p. 559

7 YELLOW-LEGGED THRUSH *Platycichla flavipes venezuelanus*. Song (from high in canopy) with much mimicry. Bill fairly short. **7a** male: gray with black foreparts, bill and eye-ring orange-yellow, legs yellow; **b** female: olivaceous brown, rump grayer, smoky gray underparts heavily washed clay on throat, breast, and sides. Under tail-coverts marked with whitish. Legs yellow. At 500-2500 m in humid forest and plantations in n Col. and Ven. Text p. 556

8 WHITE-EARED SOLITAIRE *Entomodestes leucotis*. Difficult to see, but unmistakable. Distinctive ringing song. At 900-3350 m on mid levels in cloud forest from c Peru to Bol. Text p. 552

9 ANDEAN SOLITAIRE *Myadestes ralloides*. Characteristic song of 'rusty' noes. Plump shape, short, broad-based bill. White in tail. In flight note white band on underside of wing. **9a** *venezuelanus* from Ven. along e slope to n Peru: mandible wholly pale; **b** *ralloides* of Peru and Bol. with some spotted juv. feathers. At 650-2900 m in humid forest from Ven. to Bol. P. 553

10 SLATY THRUSH *Turdus nigriceps*. Song unmusical. Small. **10a** male: blue-gray with bright yellow bill, and streaked throat. **10b** female: blue-gray replaced by olive-brown, bill dusky suffused with dark yellow, legs more brownish; **c** juv.: scaled below. In premontane woodlands (to 2550 m) in nw Peru and Bol. to nw Arg., migrating to e Peru and Ecu. Text p. 562

11 AUSTRAL THRUSH *Turdus falcklandii*. Rump gray. Crown and headside blackish, throat distinctly streaked, bill and legs yellow. Below buffy gray-brown. **11a** ad.; **b** juv.: spotted underparts vary from buff to whitish. Woodlands and parks in lowlands to 2150 m from Atacama Chile and w Neuquén to Isla Grande. Text p. 561

12 CREAMY-BELLIED THRUSH *Turdus amaurochalinus*. Eye-ring dark, most of underparts (incl. vent) whitish. May show narrow white crescent below streaked throat. **12a** male: bill yellow; **b** female: bill dark or with some yellow. In fairly dry woodlands and gardens in lowlands of Brazil, Arg. and Bol., where ascending to 2600 m in Cochabamba. Text p. 561

13 RUFOUS-BELLIED THRUSH *Turdus rufiventris*. Belly rufous. Groves in lowlands of Brazil, Arg., and Bol., where ascending to 2600 m in Cochabamba. Text p. 560

14 CHESTNUT-BELLIED THRUSH *Turdus fulviventris*. Head black (male) or dark brown (female). Breast gray, belly ochraceous orange. Female with lightly streaked throat. At 1300-2700 m in stunted humid forest and shrub from Ven. to n Peru. Text p. 560

15 CHIGUANCO THRUSH *Turdus chiguanco*. Smaller and with shorter tail than Great T. (see also text). **15a** *chiguanco* of Ecu. to nw Bol., ad. male: dark eye-ring, paler than coexisting Great T.; **b** juv.: spotted below; **c** *anthracinus* of Bol., Chile, and nw Arg.: darker than coexisting Great T.: ad. male with yellow eye-ring. Female distinguished by size only. Mainly at 2000-4300 m, lower in the s, and to coast of Peru. In woods and open, fairly dry country. Text p. 558

16 GREAT THRUSH *Turdus fuscater*. Larger and with longer tail than Chiguanco T., also darker in zone of overlap (except in Bol., where paler). Ad. male with yellow eye-ring. **16a** *ockendeni* of se Peru, male: blackish with unstreaked throat; **b** *gigantodes* of Ecu. to c Peru, ad. female: dark gray-brown, eye-ring dark; **c** *gigas* of Ven. and e Col., largest ssp., olive-brown; **d** *gigantodes* juv.: belly may appear barred; **e** *fuscater* of Bol., male: dark gray-brown, c belly pale gray to gray, bill orange, throat streaked; **f** *fuscater* imm.: bill dull yellow, streaks of throat continue onto breast and belly; **g** *fuscater* female: no eye-ring, throat virtually unstreaked. At 1500-4250 m, in woods and forest edge in area of overlap with Chiguanco T., but n-wards from n Ecu. also in open country. Ven. to Cochabamba Bol. Text p. 557

30 per cent of nat. size

PLATE L

Thrushes

PLATE LI Blackbirds

1 CRESTED OROPENDOLA *Psarocolius decumanus*. C tail-feathers blackish, rest bright yellow. Black (belly sometimes mixed with chestnut). Heavy, ivory white bill. Widespread in premontane zone, occ. to 2600 m in Col. and Bol. (*maculosus*). Text p. 572

2 RUSSET-BACKED OROPENDOLA *Psarocolius angustifrons*. Tail with blackish c pair, next pair with bright yellow outer web, rest bright yellow with dusky terminal half of outer web and tip of inner web. Some sspp with yellow on forehead. Widespread in trop. zone, occ. to lower temp. zone in Ven. (*neglectus*). Text p. 572

3 DUSKY-GREEN OROPENDOLA *Psarocolius atrovirens*. Smaller than 2. Tail with olive 2 c feather-pairs, following pairs yellow with olive tips, penultimate with entire outer web olive, and outer pair wholly olive. C Peru to Bol., in humid premontane forest along streams. Text p. 572

4 MOUNTAIN GRACKLE *Macroagelaius subalaris*. Tail long. Glossy black. Chestnut on underwing, rump-feathers with gray bases. Probably in flocks. Virtually unknown. At 1950-3100 m in humid forest in ne Col. Text p. 575

5 BOLIVIAN BLACKBIRD *Oreopsar bolivianus*. Colonial. Gives loud clear whistles. Larger than 10. Primaries brownish. Bill slightly larger and maxilla more curved. At 2400-3200 m in dry shrubby areas in valleys of c Bol. Text p. 575

6 SCRUB BLACKBIRD *Dives warszewiczi*. Large. Fairly long tail and long, conical bill. Loud melodious song. Thickets, fields and plantations in the arid region. Pairs or loose flocks. Pacific slope from w Ecu. to c Peru, southern ssp *kalinowskii* (depicted) to 3200 m. Text p. 574

7 AUSTRAL BLACKBIRD *Curaeus curaeus*. In flocks. Sexes alike. Larger than 10, and with longer bill. **7a** juv.: tinged fuscous, bill shorter than in ad., and with pale tip; **b** ad.: feathers of head stiff, narrow, and pointed. Open woods, heathland and agricultural areas to 1500 m in s Arg. and Chile. Text p. 574

8 YELLOW-BILLED CACIQUE *Cacicus holosericeus*. Bill yellow, iris yellowish white. Harsh calls. In pairs. At 1500-3500 m in dense cloud forest mixed with bamboo from Ven. to Bol. Text. p. 573

9 MOUNTAIN CACIQUE *Cacicus leucorhamphus*. In flocks. Frequent, loud, sometimes whistled calls. Lower back and rump golden yellow. **9a** *chrysonotus* of c Peru to Bol.: wings usually without yellow; **b** *leucorhamphus* of Ven. to n Peru: inner wing-coverts golden yellow. Concealed nape-collar white. At 1500-3300 m in humid montane forest from Ven. to Bol. Text p. 573

10 SHINY COWBIRD *Molothrus bonariensis*. In flocks, foraging on the ground. Bill longer than in Screaming C. (see text). **10a** female, pale; **b** female, dark; **c** female, blackish; **d** male. Almost throughout, where cattle are found, mainly in trop. and subtrop. zones, but to 3350 m in Bol. Text p. 570

11 BAY-WINGED COWBIRD *Molothrus badius*. Conical bill and rufous wings. See text for juv. Screaming C. Woodland near rangeland, sometimes in very open terrain; throughout the chaco, and at 1100-3350 m in Bol. (ssp *bolivianus*). Text p. 571

12 YELLOW-WINGED BLACKBIRD *Agelaius thilius*. Bill slender. **12a** male, fresh plumage; **b** female; **c** juv. Reed and tule marshes and adjacent fields, in southern lowlands and around 4000 m in s Peru and Bol. Text p. 576

13 YELLOW-BACKED ORIOLE *Icterus chrysater giraudii*. Sexes alike. Mainly in lowland shrub in C Am. to n Col. and Ven., but to 2900 m in Ven. P. 577

14 NORTHERN ORIOLE *Icterus galbula*. Most of tail orange. **14a** female: crown and back streaked or spotted; **b** male, 1st year: crown and back unmarked; **c** male ad. Woodland, thickets, gardens and farmland. N Am., wintering s to n Col. and nw Ven. Text p. 577

15 WHITE-BROWED BLACKBIRD *Leistes superciliaris*. Smaller, and with shorter bill than meadowlarks. Supercilium from above eye. **15a** male: under wing-coverts black, bill blackish; **b** female: under wing-coverts dark gray, bill horn. Savanna, marshes, and grasslands, occ. to 2600 m, generally in taller vegetation than 17, in se Peru to n Arg. Accidental in Chile. Text p. 578

16 PERUVIAN MEADOWLARK *Sturnella bellicosa*. No similar species in region: **a** male, **b** female. On fields, meadows, and in open arid shrub on Pacific slope (to 2600 m) of Ecu., Peru, and n Chile. Text p. 579

17 LONG-TAILED MEADOWLARK *Sturnella l. loyca*. Under wing-coverts white. Supercilium red in front of eye. **17a** male; **b** female. In open country to 2450 m, often near cattle, from Córdoba Arg. and Atacama Chile to Isla Grande. Text p. 578

18 EASTERN MEADOWLARK *Sturnella magna*. Open country from N Am. to Surinam, depicted ssp *meridionalis* at 1700-3500 m in Col. and Ven. Text p. 580

19 BOBOLINK *Dolichonyx oryzivorus*. Sparrow-like. Flight call a characteristic *pink*. Tail-feathers pointed. **19a** female in fresh winter dress; **b** worn winter female; **c** male, early spring molt; **d** male in fresh spring dress. N Am. migrant. s to Arg., occ. visiting temp. zone, in fields and reed-beds. Text p. 580

PLATE LI

30 per cent of nat. size, but
flying birds only 15 per cent

Blackbirds

PLATE LII Vireos and Warblers

1 OLIVACEOUS GREENLET *Hylophilus olivaceus*. Sometimes several in mixed-species flocks. Humid premontane forest of e Ecu. to c Peru. Text p. 569

2 BROWN-CAPPED VIREO *Vireo gilvus josephae*. 700-2800 m in second growth, gardens, and forest edge from C Am. to Bol. Text p. 569

3 RED-EYED VIREO *Vireo olivaceus*. Iris brown (depicted *chivi*), or red in ad. N Am. migrants (*olivaceus*). White supercilium outlined in black. In C Peru to n Arg., in woodlands and thickets; premontane, but ascending to 3000 m in Peru. Text p. 568

4 RUFOUS-BROWED PEPPERSHRIKE *Cyclarhis gujanensis*. Song a loud, pleasant. Bill partly pale (see text for Black-billed P.). **4a** *saturatus* of c Peru; **b** *dorsalis* of highlands of Bol. Lowlands from Mexico to c Arg., and at 1050-3200 m in second growth, forest clearings, and arid shrub from Peru to Bol. Text p. 567

5 WHITE-FRONTED WHITESTART *Myioborus albifrons*. From lower elevation 8 by white on face and yellow throat. At 2200-4000 m in cloud forest in Andes of Ven. (except Páramo de Tamá). Text p. 588

6 GOLDEN-FRONTED WHITESTART *Myioborus ornatus*. From 8 by yellow face and throat. **6a** *ornatus* of E Andes: earcoverts; **b** *chrysops* of C and W Andes: At (1800) 2400-3400 m in humid forest of Paramo de Tamá, Ven. and most of Col. Text p. 587

7 SPECTACLED WHITESTART *Myioborus melanocephalus*. From lower elevation 8 by yellow spectacles and throat, from 9 by black and yellow (rather than gray and white) face. **7a** *ruficoronatus* of s Col. and Ecu., female; **b** *ruficoronatus* male, variant with more saturated coloration; **c** *ruficoronatus* juv.; **d** *bolivianus* of se Peru and Bol. At 2000-3950 m in cloud forest shrubbery of s Col. to c Bol. Text p. 587

8 SLATE-THROATED WHITESTART *Myioborus miniatus*. 500-3000 m in second growth and humid forest edge from Mexico to Bol. Text p. 586

9 BROWN-CAPPED WHITESTART *Myioborus b. brunniceps*. Secretive. Rarely spreads tail. At 1400-3200 (3800) m on humid and semi-humid wooded slopes in Bol. and nw Arg. Text p. 589

10 BAY-BREASTED WARBLER *Dendroica castanea*. **10a** male; **b** juv.; **c** female. N Am. migrant s to Col. and Ven., mainly in the tropics. Text p. 584

11 BLACKBURNIAN WARBLER *Dendroica fusca*. **11a** ad. male; **b** juv. N Am. migrant to subtrop. and temp. woodlands s to Bol. Text p. 583

12 BLACKPOLL WARBLER *Dendroica striata*. **12a** juv.; **b** female; **c** male. N Am. migrant. P. 584

13 CERULEAN WARBLER *Dendroica cerulea*. **13a** ad.; **b** juv. N Am. migrants to Peru and Bol., mainly in the tropics, in semi-open country. Text p. 583

14 GOLDEN-WINGED WARBLER *Vermivora chrysoptera*. N Am. migrant s to Ecu. and n Ven. 1000-3000 m. Text p. 581

15 BLACK-AND-WHITE WARBLER *Mniotilta varia*. Creeps trunks, dropping like leaf. **a** ad.; **b** juv. N Am. migrant s to Peru and Ven. Humid premontane forest edge. Text p. 581

16 AMERICAN REDSTART *Setophaga ruticilla*. **16a** female; **b** imm. male; **c** ad. male. N Am. migrant s to Peru, mainly in trop. woodland. Text p. 586

17 CONNECTICUT WARBLER *Geothlypis agilis*. See text for Mourning and Macgillivray's W. On or near the ground. N Am. migrant s to Peru, mainly in trop. humid forest. Text p. 585

18 MASKED YELLOWTHROAT *Geothlypis aequinoctialis*. **18a** male: black mask not bordered above by pale line, belly yellow (see text for Common Y.); **b** female. Low in dense thickets, occ. near marshes. Almost throughout lowlands s to n Arg., entering lower temp. zone in Bol. and Arg. Text p. 585

19 TROPICAL PARULA *Parula pitiayumi*. Mexico to n Arg., mainly lowlands, humid forest edge, second growth, humid patches in dry woods. Text p. 582

20 BLACK-CRESTED WARBLER *Basileuterus nigrocristatus*. 2600-3950 m in humid shrubbery. Ven. to n Peru. Text p. 590

21 PALE-LEGGED WARBLER *Basileuterus signatus*. **21a** *signatus* of c Peru to Urubamba Valley; **b** *flavovirens* of se Peru to Arg. At 1800-3050 (3400) m, below Citrine W. in zone of overlap. In understory of humid forest from c Peru to nw Arg. (1 rec. from E Andes Col.). Text p. 591

22 CITRINE WARBLER *Basileuterus luteoviridis*. No eye-ring, but area below eye mottled with whitish. **22a** *luteoviridis* of Ven. to n Peru: no dark on crown; **b** *richardsoni* of W Andes Col.: **c** *euophrys* of se Peru to Bol.: **d** juv. At (1700) 2300-3400 m in understory of humid forest from Ven. to Bol. Text p. 590

23 THREE-STRIPED WARBLER *Basileuterus tristriatus*. **23a** *punctipectus* of Cochabamba Bol. **b** *tristriatus* of se Ecu. and most of Peru. At 300-2700 m in second growth and forest. Ven. to Bol. Text p. 592

24 RUSSET-CROWNED WARBLER *Basileuterus coronatus*. **24a** *coronatus* of c and se Peru: belly yellow; **b** *orientalis* of e Ecu.: belly whitish. At 1300-3100 in undergrowth of humid forest from Ven. to Bol. Text p. 593

30 per cent of nat. size

PLATE LII

Vireos and Warblers

PLATE LIII Honeycreepers (compare Plates LIV-LVI)

1 GIANT CONEBILL *Oreomanes fraseri*. Flakes bark. In groups. Call high-pitched *zit*, song a siskin-like twitter. **1a** and **b** ad., showing variation in facial pattern; **c** juv; **d** imm. Locally at 2700-4850 m in *Polylepis* woods from sw Col. to Bol. Text p. 598

2 BLUE-BACKED CONEBILL *Conirostrum sitticolor*. Black-chested. **2a** *sitticolor* of Andes of Ven. to c Peru; **b** *cyaneum* of se Peru to Bol. At 2450-3800 m in humid shrubbery and páramo. Text p. 597

3 RUFOUS-BROWED CONEBILL *Conirostrum rufum*. C forehead, supercilium, and entire underparts rufous. At 2700-3400 m in humid shrubbery in Sta Marta and E Andes Col. Text p. 597

4 WHITE-BROWED CONEBILL *Conirostrum ferrugineiventre*. Above slaty, supercilium white, underparts rufous. At 2000-3700 (4100) m at cloud forest edge and elfin forest of c Peru to Bol. Text p. 596

5 CINEREOUS CONEBILL *Conirostrum cinereum*. White or whitish supercilium, wingbar, and wing speculum. **5a** southern *cinereum*; **b** buff-tinged northern *fraseri*. From lowlands to 4500 m in arid to fairly humid shrub from Col. to n Chile and Bol. Text p. 595

6 TAMARUGO CONEBILL *Conirostrum tamarugense*. Often in flocks. Much like 5, but supercilium, throat, and breast rufous. **6a** ad.; **b** imm. female. At 3400-4050 m in *Polylepis* and *Gynoxys* shrub in s Peru and below 2000 m in leguminaceous trees in n Chile. Text p. 596

7 CAPPED CONEBILL *Conirostrum albifrons*. In pairs. **7a** *albifrons*, male of E Andes Col.: black and blue with white cap; **b** *atrocyaneum* male of w Col. to n Peru: with cap blue; **c** *lugens* male of Peru and Bol.: black with blue cap; **d** imm. male: greenish with gray foreparts of female, but has attained blue 'shoulders' of ad. male; **e** female. At 1500-3000 m (mainly below 2500) in humid forest and adjacent second growth in Andes from Ven. to Bol. Text p. 598

8 TIT-LIKE DACNIS *Xenodacnis parina*. Loud, liquid calls. Gleans *Gynoxys* leaves for sugary secretions. **8a** *parina* male of c Peru: blue; **b** *parina* female: blue-capped, cheeks and all underparts bright rufous; **c** *petersi*, of w Peru to Ecu.: deep blue with bright streaks; **d** *petersi* female: only forehead blue, and dull cheeks and sides; **e** juv. At 3000-4600 m in *Gynoxys* shrub, sometimes into adjacent cloud forest edge or *Polylepis* woodland.
Text p. 606

9 MOUSTACHED FLOWERPIERCER *Diglossa mystacalis*. Glossy black, rump tinged blue-gray. Distinct white moustache. Bill black. **9a** *pectoralis* of c Peru: lower half of breastband white; **b** *albilinea* of se Peru: lacking breastband, and with shoulders blue-gray; **c** *unicincta* of nc Peru: with breastband rufous. **9d** *unicincta* juv. Mandible pale. At 2500-3600 m in humid forest and shrubbery of c Peru to Bol. Text p. 602

10 CHESTNUT-BELLIED FLOWERPIERCER *Diglossa gloriossissima*. Glossy black, shoulder and rump slaty, lower breast and belly rufous chestnut. At 3150-3800, replacing Moustached F. in W Andes Col.
Text p. 601

11 GLOSSY FLOWERPIERCER *Diglossa lafresnayii*. Glossy black, shoulders slaty blue, bill black. At (2000) 2700-3750 m, replacing Moustached F. in Ven. to n Peru. Text p. 602

12 GRAY-BELLIED FLOWERPIERCER *Diglossa carbonaria*. **12a** juv, **b** ad. 2100-4300 m in arid to semi-humid montane scrub in Bol. Text p. 604

13 BLACK-THROATED FLOWERPIERCER *Diglossa brunneiventris*. **13a** ad., **b** im. At (1450) 2000-4000, replacing Gray-bellied and Black F. in n Bol. and Peru, and at n end of W and C Andes Col. Text p. 603

14 BLACK FLOWERPIERCER *Diglossa humeralis*. Bill black with gray base below. **14a** *aterrima* of s Col. to n Peru, imm.; **b** *aterrima*, ad.: black, glossy when fully ad; **c** *humeralis* of E Andes Col.: with gray shoulders and rump. At 1500-4000 m, replacing Black-throated F. in much of the northern Andes. Text p. 604

15 RUSTY FLOWER-PIERCER *Diglossa sittoides*. Mandible distinctly upturned. **15a** juv., **b** female; **c** male *dorbignyi* of Col.: narrow forehead and headside blackish; **d** male *decorata* of Ecu. and n Peru. At 600-3500 m (occ. to coast in Peru) from Ven. to nw Arg., in thickets and hedgerows in semi-arid open country, in the n also in humid habitats. Text p. 601

16 WHITE-SIDED FLOWERPIERCER *Diglossa albilatera*. Song a flat trill. Lengthened white axillars. Mandible distinctly upturned. **16a** male imm; **b** blackish male of *affinis* from most of Peru; **c** female *albilatera*, told from 15 by lacking pale edges to tertials; **d** blue-gray male *albilatera* of the northern Andes. At (1300) 1600-3300 m in bamboo and shrubbery at edge of humid forest and on bushy hillsides from Ven. to sc Peru.
Text p. 605

17 MASKED FLOWERPIERCER *Diglossopis cyanea*. Dark blue with extensive black mask. Iris glowing red. **17a** ad., **b** *dispar* juv.: dark gray below. At (1500) 2000-3700 m in humid forest and adjacent shrubbery from Ven. to Bol. Text p. 600

18 BLUISH FLOWERPIERCER *Diglossopis caerulescens*. Iris deep red. **18a** bird from Cutucú mts Ecu.; **b** *saturata* from Ven. and E Andes of Col.; **c** imm. At 1350-3200 m at edge of humid forest, possibly mainly on poor soil, from Ven. to Bol. Text p. 599

30 per cent of nat. size

PLATE LIII

Honeycreepers

PLATE LIV Tanagers; mainly species of mossy cloud forest

1 YELLOW-THROATED TANAGER *Iridosornis analis*. Throat yellow, breast pale buff (see text for very similar Purplish-mantled T. of Col. and Ecu.). At 1130-2600 m in mossy forest in se Ecu. and e Peru.
Text p. 616

2 GOLDEN-COLLARED TANAGER *Iridosornis jelskii*. Hindcrown and nape golden, face black, belly rufous. At 3100-3650 m in elfin forest in Peru and Bol.
Text p. 616

3 YELLOW-SCARFED TANAGER *Iridosornis reinhardti*. Purplish black with golden nuchal collar. At (1800) 2050-3400 m in humid shrubbery in Loja s Ecu., and s and e of the Marañón from Amazonas to Cord. Vilcabamba Peru. Text p. 617

4 GOLDEN-CROWNED TANAGER *Iridosornis rufivertex*. Purplish black with golden crown. **4a** *ignicapillus* of sw Col.: vent chestnut; **b** *caeruleoventris* of n Col.: vent dark blue (occ. with some chestnut). At (2000) 2300-3800 m in humid shrubbery from Col. to n Peru. Text p. 617

5 LACRIMOSE MOUNTAIN-TANAGER *Anisognathus lacrymosus*. Dull blue above, yellow below. **5a** *pallididorsalis* of Perijá mts: dullest ssp, usually without yellow 'tears' on headside; **b** *palpebrosus* of s Col. and e Ecu.: 2 yellow 'tears'; **c** *lacrymosus* in most of Peru: usually lacking rear 'tear'. At (1800) 2600-3500 m from Ven. to Cord. Vilcabamba Peru. Text p. 619

6 BLUE-WINGED MOUNTAIN-TANAGER *Anisognathus flavinuchus*. C crown and nape yellow. Wings edged turquoise. **6a** *somptuosus* of se Ecu. and most of Peru: rump dull dark olive-green; **b** *flavinuchus* of se Peru and Bol.: rump blue. At 900-2750 m in humid forest from Ven. to Bol. Text p. 620

7 SCARLET-BELLIED MOUNTAIN-TANAGER *Anisognathus igniventris*. Unmistakable. **7a** *erythrotis* of c Col. and Ecu.: vent black; **b** *igniventris* (juv.) of se Peru and Bol.: vent red, wings edged blue. At (2250) 2600-3950 m in humid shrub from Col. to Bol.
Text p. 618

8 BLACK-CHESTED MOUNTAIN-TANAGER *Buthraupis eximia*. Large. Black mask extends onto breast. **8a** *chloronota* of se Col. and n Ecu.: back and rump dark moss-green; **b** *eximia* of ne Col.: rump blue. At (2000) 2750-3800 m in mossy forest from Col. to n Peru. Text p. 622

9 HOODED MOUNTAIN-TANAGER *Buthraupis montana*. Large. In noisy single-species flocks. Iris red. Breast only black at sides. **9a** *cucullata* of Col. and Ecu.: enamel-like purplish blue above; **b** *montana* of Bol.: pale blue nuchal collar. In 1700-3500 m in humid forest from Ven. to Cochabamba Bol. Text p. 620

10 GRASS-GREEN TANAGER *Chlorornis riefferii*. Unmistakable. Large. Green with red bill and feet, and rufous face and vent. **10a** *boliviana* of Bol.; **b** *riefferii* of Col. to n Peru: chestnut only to upper throat. At (1500) 2000-2700 (3350) m in cloud forest with bamboo from Col. to Cochabamba Bol. (not in Cuzco).
Text p. 639

11 GOLDEN-COLLARED HONEYCREEPER *Iridophanes pulcherrima*. Bill slender, slightly curved. **11a** male: black hood and upper back contrasting golden nuchal collar, wings and tail blue; **b** female or juv.: olive, lighter below, wings blue-green. In humid premontane forest from Col. to se Peru. Text p. 606

12 GOLDEN-BACKED MOUNTAIN-TANAGER *Buthraupis aureodorsalis*. Largest tanager. Unmistakable. At 3000-3500 m in elfin forest from San Martín to Huánuco Peru. Local. Text p. 622

13 MASKED MOUNTAIN-TANAGER *Buthraupis wetmorei*. Crown and back dull golden olive, black of face bordered with yellow. **a** male; **b** female imm.: forehead olive. At 2900-3650 m in mossy elfin forest from s Col. to n Peru. Rare. Text p. 621

14 FAWN-BREASTED TANAGER *Pipraeidea melanonota*. Blue above, black on headside, pale buff to buff below. No black mustache. At 700-3000 (4000) in woodlands and humid forest from Ven. to Arg., and in lowlands from se Brazil to ne Arg. Text p. 609

15 CHESTNUT-BELLIED MOUNTAIN-TANAGER *Dubusia castaneoventris*. Blue above, black on headside, rufous brown below with black mustache. At 2150-3600 m in elfin forest from c Peru to Bol.
Text p. 623

16 BUFF-BREASTED MOUNTAIN-TANAGER *Dubusia taeniata*. Distinctive song. Appears dull yellow below. **a** *carrikeri* of Sta Marta mts.: mid-throat buff streaked black, brow light blue; **b** *stictocephala* of Peru: entire crown spangled blue; **c** *taeniata* of Ven. to n Peru: pale blue brow. At 2000-3600 m in humid shrub and forest from Ven. to se Bol. Text p. 623

PLATE LIV

30 per cent of nat. size

Tanagers; mainly species of mossy cloud forest

Plate LV Tanagers (see also Plate LIV and LVII)

Tanagers of the genus *Tangara* are usually seen with mixed-species flocks feeding in the canopy in humid forest.

1 SAFFRON-CROWNED TANAGER *Tangara xanthocephala*. Yellow head with small black mask. Back streaked. **1a** *lamprotis* of se Peru to Bol.: crown orange; **b** *venusta* of Ven. to c Peru: crown yellow; **c** juv. Humid forest at 1100-2600 m, occ. higher. Text p. 610

2 SILVERY TANAGER *Tangara viridicollis*. **2a** male *fulvigula*: black foreparts with golden throat and silvery back; **b** female: green with golden-brown throat. Humid premontane forest and deciduous woodland, occ. to 3050 m. S Ecu. to Puno Peru. Text p. 615

3 GOLDEN-NAPED TANAGER *Tangara ruficervix*. Blue with golden nape. **3a** *leucotis* of w Ecu., juv.; **b** *leucotis* ad.: blue-green, nuchal patch margined black; **c** *fulvicervix* of Bol.: ultramarine, no black on hindcrown. At 1100-2700 m in humid forest from Col. to Bol. Text p. 612

4 FLAME-FACED TANAGER *Tangara parzudakii*. Black markings on orange-yellow head, and with opalescent wing-coverts and rump. **4a** *lunigera* of w Col. and w Ecu.; **b** widespread ssp *parzudakii* (Ayacucho specimen): face deep orange; **c** *parzudakii* juv. At 1000-2625 m in humid forest from Ven. to Peru. Text p. 610

5 BLACK-CAPPED TANAGER *Tangara heinei*. Cap dark. **5a** male, **b** female. At 700-2700 m in humid forest from Ven. to Ecu. Text p. 614

6 BLUE-BROWED TANAGER *Tangara cyanotis*. Crown and headside black, brow silvery green. **6a** *cyanotis* of Bol.: back dark blue-green; **b** *lutleyi* of Col. to Peru: back black. At 1100-2600 m in humid forest. Text p. 611

7 GREEN-CAPPED TANAGER *Tangara meyerdeschauenseei*. Headside dark, crown old gold. Male (female greener). Semiarid premontane shrub in Puno Peru. Text p. 613

8 SCRUB TANAGER *Tangara vitriolina*. Headside dark, crown rufous. Mainly at 1000-2600 m in semiarid shrub in Col. and n Ecu. Text p. 612

9 BLUE-AND-BLACK TANAGER *Tangara vassorii*. **9a** juv. *atrocoerulea* of c Peru to Bol.: grayish, faintly streaked on breast; **b** ad. *atrocoerulea*: black-spotted below, nape-patch yellowish white; **c** ad. *vassorii* of Ven. to n Peru: back blue, underparts uniform, no nape-patch; **d** juv. *vassorii*: uniform grayish below. At (1300) 2400-3350 m in cloud forest. Text p. 614

10 METALLIC-GREEN TANAGER *Tangara labradorides*. Dull blue-green. Loral region and nape black. At (500) 1200-2750 m in humid forest from Col. to n Peru. Text p. 611

11 BERYL-SPANGLED TANAGER *Tangara nigroviridis*. Black spotted turquoise. Humid premontane forest, occ. to 3000 m, from Ven. to Bol. Text p. 613

12 SAYACA TANAGER *Thraupis sayaca*. Blue-gray. Wings and tail dull turquoise. In woodland and gardens from e Brazil to 3200 (3600) m in Bol. and nw Arg. (depicted ssp *obscura*). Text p. 625

13 BLUE-CAPPED TANAGER *Thraupis cyanocephala*. Crown blue, headside blackish, back olive, underparts gray to bluish. Under wing-coverts bright yellow. In humid second growth woodland mainly at 2000-3000 m from Ven. to Bol. Text p. 624

14 BLUE-AND-YELLOW TANAGER *Thraupis bonariensis*. **14a** *darwinii* of Ecu. to Bol., at 2000-4000 m, male: back olive, rump yellow; **b** *darwinii* female: nondescript, yellowish buff below; **c** *schulzei* of s Bol. to lowlands of nw Arg., female: lores dark; **d** *schulzei* male: back black, rump, and breast bright orange. In semiarid woodland, scrub, and gardens. Text p. 624

15 RED-HOODED TANAGER *Piranga rubriceps*. Yellowish with red head and breast (no red spots on yellow parts). Bill black. **15a** male: entire breast red; **b** female: only upper breast red. At 1700-3200 m in humid forest from Col. to c Peru. Text p. 627

16 SUMMER TANAGER *Piranga rubra rubra*. Bill mostly pale. **16a** male; **b** female (imm. male similar, but more or less mixed with red). N Am. migrant to premontane second growth habitats s to Peru. To 3000 m on migration. Text p. 627

17 HEPATIC TANAGER *Piranga flava flava*. Bill dark. **17a** male: darker than Summer T.; **b** female. Premontane semi-arid woodland and shrub, occ. to 3150 m. From Ven. to Bol. Text p. 626

30 per cent of nat. size

PLATE LV

Tanagers

PLATE LVI Tanagers (see also Plates LV and LVII)

1 BLUE-HOODED EUPHONIA *Euphonia musica*. **1a** *pelzelni* of sw Col. to Ecu., female; **b** *pelzelni* juv.; **c** *insignis* of Azuay s Ecu., male; **d** *pelzelni* male. Widespread in premontane zones, occ. to 3000 m. Arid to humid shrub and forest. Text p. 608

2 CHESTNUT-BREASTED CHLOROPHONIA *Chlorophonia pyrrhophrys*. **2a** female; **b** male. 1400-3600 m. Humid forest and clearings. Ven. to c Peru. Rare. Text p. 608

3 BLUE-NAPED CHLOROPHONIA *Chlorophonia cyanea*. **3a** juv.; **b** female; **c** *longipennis* of Perijá mts to Bol., male. Premontane humid forest from Ven. to Bol. and se Brazil. Text p. 608

4 GRAY-HOODED BUSH-TANAGER *Cnemoscopus rubrirostris*. **4a** *chrysogaster*, most of Peru: bill black; **b** *rubrirostris* of Ven. to n Peru: large, bill pink. At 1900-3350 m in humid forest. Text p. 633

5 RUFOUS-CHESTED TANAGER *Thlypopsis ornata*. **5a** ad., **b** juv. At 1800-3500 m in humid and semi-humid shrubbery from Col. to Peru. Text p. 628

6 BROWN-FLANKED TANAGER *Thlypopsis pectoralis*. **6a** imm.; **b** ad. At 2000-3100 m in semiarid shrub in c Peru. Text p. 629

7 RUST-AND-YELLOW TANAGER *Thlypopsis ruficeps*. Often in flocks. **7a** ad., **b** juv. At (1300) 1850-3700 m in humid second growth and riparian thickets from c Peru to nw Arg. Text p. 629

8 RUFOUS-CRESTED TANAGER *Creurgops verticalis*. **8a** male; **b** female. 1150-2700 m in humid forest from Ven. to c Peru. Text p. 628

9 SLATY TANAGER *Creurgops dentata*. **9a** female; **b** male. Humid premontane shrubby forest from se Peru to Bol. Text p. 628

10 COMMON BUSH-TANAGER *Chlorospingus ophthalmicus*. Most sspp with pale iris and some white behind eye, breastband yellow to olive, throat whitish to buffy, often speckled. **10a** *venezuelanus* of Lara to Táchira Ven.: **b** *phaeocephalus* of Ecu. to n Peru; juv.; **c** *peruvianus* of se Peru; **d** *bolivianus* of La Paz into Cochabamba Bol. In humid premontane forest and edge, locally to 3000 m. Mexico to Ven. and nw Arg. Text p. 630

11 DUSKY-BELLIED BUSH-TANAGER *Chlorospingus semifuscus*. Ssp *livingstoni* sometimes with white postocular spot. Humid premontane forest, occ. to 2750 m, in w Col. and Ecu. Text p. 632

12 WHITE-CAPPED TANAGER *Sericossypha albocristata*. In noisy flocks sailing through the air. **12a** male; **b** female (throat black in juv.). 1600-3200 m in humid forest. Local. Ven. to c Peru. Text p. 630

13 BLACK-BACKED BUSH-TANAGER *Urothraupis stolzmanni*. Flocks in humid shrub (2750-3950 m) in C Andes of Col. and e Ecu. Text p. 633

14 PARDUSCO *Nephelornis oneilli*. In flocks in elfin forest at 3000-3800 m from s San Martín to Huánuco Peru. Text p. 640

15 BLACK-HEADED HEMISPINGUS *Hemispingus verticalis*. Imm. has white throat. At 2350-3600 m, atop dense canopies of humid shrubbery from Col. to n Peru. Text p. 638

16 DRAB HEMISPINGUS *Hemispingus xanthophthalmus*. 2375-3525 atop dense canopies of humid shrubbery. Nc Peru to La Paz Bol. Text p. 639

17 OLEAGINOUS HEMISPINGUS *Hemispingus frontalis*. **17a** *frontalis* of Col. to se Peru; **b** *ignobilis* of s Lara to Táchira, Ven. 1300-2900 m in humid forest from Ven. to se Peru. Text p. 637

18 SUPERCILIARIED HEMISPINGUS *Hemispingus superciliaris*. **18a** *chrysophrys* of Ven.; **b** *maculifrons* of sw Ecu. and n Peru; **c** *nigrifrons* of Col. and Ecu.; **d** *urubambae* of se Peru and nw Bol.; **e** *leucogaster* of c Peru. At 1900-3600 m in humid shrubbery. Text p. 636

19 PARODI'S HEMISPINGUS *Hemispingus parodii*. Elfin forest in Cuzco Peru. Text p. 635

20 ORANGE-BROWED HEMISPINGUS *Hemispingus calophrys*. 2300-3500 m in dense bamboo and cloud forest understory in the Yungas of Puno Peru and Bol. Text p. 635

21 THREE-STRIPED HEMISPINGUS *Hemispingus trifasciatus*. Mainly at 3000-3350 m in cloud forest canopy from c Peru to Bol. Text p. 639

22 RUFOUS-BROWED HEMISPINGUS *Hemispingus rufosuperciliaris*. 2550-3500 m in cloud forest undergrowth in nc Peru. Text p. 634

23 BLACK-EARED HEMISPINGUS *Hemispingus melanotis*. Head incl. throat black, with supercilium indistinct, or long and white in depicted *piurae* of nw Peru. Below mainly rufous. Humid premontane shrub from Ven. to Bol., ascending to 2900 m in Col., and 3050 m in nw Peru. Text p. 637

24 BLACK-CAPPED HEMISPINGUS *Hemispingus atropileus*. Crown and most of headside black with long narrow supercilium. **a** *atropileus* of Ven. to n Peru; **b** *auricularis* of Peru. At (1800) 2300-3600 m in undergrowth of wet forest. Text p. 634

30 per cent of nat. size

PLATE LVI

Tanagers

PLATE LVII Cardinals, Sparrows, and Finches (compare Plates LVIII-LXII)

1 SLATY FINCH *Haplospiza rustica*. Bill rather long and conical. **1a** male: uniform slaty; **b** female: above brownish olive, below yellowish gray, obscurely streaked dusky, esp. on throat and breast. Inside dense bamboo from Mexico to Ven. and Bol., in the Andes at 1200-3300 (3500) m. Text p. 668

2 MASKED SALTATOR *Saltator cinctus*. Black breastband contrasts white belly. Bill more or less washed with red. Tail fairly long, graduated. At 1700-3000 m, mainly inside dense bamboo, from Ecu. to c Peru. Rare. Text p. 644

3 RUFOUS-BELLIED SALTATOR *Saltator rufiventris*. Gray with rufous belly and vent. Long narrow supercilium white or whitish. At 2800-3800 m in thickets with mistletoes in dry valleys in Bol. to Jujuy nw Arg. Text p. 644

4 GOLDEN-BILLED SALTATOR *Saltator aurantiirostris*. Orange bill. Blackish face to upper breast, with white throat well developed in the s, but semiconcealed in the n. Tail conspicuously white-tipped. **4a** *hellmayri* from Bol., imm. female: bill dark; **b** *hellmayri* male: supercilium pure white; **c** *nasica* of Arg., male: bill large, tail without white; **d** *albociliaris* of c Peru to nw Bol. and n Chile: much white in tail; **e** *nigriceps* of s Ecu. and nw Peru (a distinct species?): lacking postocular streak, belly gray; Mainly in semiarid shrub and copses, at 1350-4000 m in the n, low in the s. Text p. 643

5 YELLOW GROSBEAK *Pheucticus chrysopeplus*. No pure black on crown or sides. Black wings with large white spots. **5a** *chrysogaster* from s Col. to s Peru, male: entire head and underparts yellow; **b** *chrysogaster* female: crown and back streaked; **c** *chrysogaster* imm.: breast streaked. Mexico to Ven. and Peru, at 900-3500 m (but to coast in Ecu. and Peru). Semiarid regions. Text p. 645

6 ROSE-BREASTED GROSBEAK *Pheucticus ludovicianus*. No yellow in plumage. Dark wings with large white spots. **6a** male: breast rosy, rump white; **b** female: less white in wing, below whitish streaked dusky. N Am. migrant to humid premontane zones s to Peru, reaching 3800 m on migration. Text p. 646

7 GRASSHOPPER SPARROW *Ammodramus savannarum caucae*. Supercilium dark yellow above lore, pale posteriorly. Tail-feathers fairly short, narrow, and pointed. Song a grasshopper-like buzz. Rare in tall grass at 1000-2850 m in Cauca valley Col. and n Ecu. Text p. 679

8 WEDGE-TAILED GRASS-FINCH *Emberizoides herbicola*. Tail long, graduated, feathers narrow and quite pointed. Upperparts streaked. Tall grass and entangled shrub mainly in the chaco, but to 2300 m in Bol. (*herbicola*). Text p. 683

9 GREAT PAMPA-FINCH *Embernagra platensis*. Large. Tail relatively long, slightly graduated, feathers broad and blunt. Bill orange with dark culmen. Dull olive-gray above with contrasting grayish rump, wings olive. **9a** ad.; **b** juv.: streaked above and on breast and sides. Bushy terrain, fields, and sawgrass, from se Brazil to c Arg., ascending to 3500 m in Bol. Text p. 689

10 BLACK-BACKED GROSBEAK *Pheucticus aureoventris*. Black wings with large white spots and yellow shoulders. Birds from Ecu. and female and imm. best told from Yellow G. by darker headside and by blackish spots on side of breast. **10a** *aureoventris* of se Peru to nw Arg., male: feathers of vent whitish narrowly tipped black, no white spots on tertials; **b** *terminalis* of e Peru, male: vent yellow, tertials with small white spots; **c** *crissalis* of s Col. and Ecu. (female ?): much like male; **d** *crissalis* male: yellow c of throat and most of breast; **e**: *uropygialis* of Col., male; **f** *aureoventris* female: browner above, white in wing less extensive. In humid forest and adjacent farmland at 1750-3700 m (lower in the s) from Col. to nw Arg. Text p. 646

11 GRAY-CRESTED FINCH *Lophospingus griseocristatus*. Gray, lower belly and vent white. Tail blackish, outer 3 feathers with white terminal half. Head crested. **11a** ad., **b** juv. At 1500-3100 m in arid thorny scrub of Bol. to Salta nw Arg. Text p. 669

12 HOUSE SPARROW *Passer domesticus*. **12a** male: crown gray, napeside chestnut, headside whitish, throat extensively black in worn plumage; **b** female: fairly uniform on head and below, streaked on back. Near houses. Mainly in southern lowlands, but locally in the Andes, and n to Col. Text p. 695

13 RUFOUS-COLLARED SPARROW *Zonotrichia capensis*. Distinct rufous half-collar and blackish spot on breastside. **a** *australis* of s Chile s Arg., juv.: heavily streaked, very pale; **b** *australis* ad.: no black on crown; **c** ad. from Ecu.; **d** Cochabamba juv.: much darker than juv. *australis*, and plumage worn for much shorter time. Esp. near houses, but in all habitats except closed forest. Mexico to Isla Grande, n of Peru primarily montane. Text p. 682

30 per cent of nat. size

PLATE LVII

Cardinals, Sparrows, and Finches

PLATE LVIII. Finches (see also Plates LVII and LIX-LXII)

Sierra-finches of the genus *Phrygilus* are fairly large finches of shrub and open country mainly in the high Andes and Patagonia.

1 PERUVIAN SIERRA-FINCH *Phrygilus punensis*. Unlike in the Black-hooded S-f. not chestnut on back (sometimes a slight golden brown tinge), and wings grayer. **1a** *punensis* male, **b** *punensis* juv, **c** *punensis* female, **d** *chloronotus* female. Puna country with rocks and bushes of Peru into La Paz Bol. Text p. 662

2 GRAY-HOODED SIERRA-FINCH *Phrygilus gayi*. Belly and lower flanks white (yellow in Patagonian S-f.), primaries edged silvery. **2a** juv, **b** *gayi* female, **c** *gayi* male. Open country in the Andes of Arg. and Chile, at 1500-4000 m furthest n. Text p. 660

3 PATAGONIAN SIERRA-FINCH *Phrygilus patagonicus*. Belly and flanks yellow, primaries only narrowly pale-edged. **3a** juv., **b** female, **c** male, with back golden-brown. *Nothofagus* shrub and riparian thickets in s Arg. and Chile. Text p. 659

4 BLACK-HOODED SIERRA-FINCH *Phrygilus atriceps*. Belly and lower flanks yellow. **4a** and **b** males, back golden chestnut, wings black in w Bol.; **c** female imm., **d** juv. Open land and *Polylepis* wood at 2400-4500 m from nw Arg. to sw Peru. Text p. 661

5 ASH-BREASTED SIERRA-FINCH *Phrygilus plebejus*. In flocks. Small. **5a** female juv.: streaked; **b** ad.: weakly streaked above, pale gray below. At 2400-4900 m in usually sparsely vegetated country in the Andes from Ecu. to c Arg. Generally common. Text p. 665

6 PLUMBEOUS SIERRA-FINCH *Phrygilus unicolor*. Male (and female *inca*) uniform gray, female streaked. **6a** *grandis* male; **b** *grandis* female imm.; **c** *grandis* juv.; **d** *inca* female ad.: like male; **e** *inca* female imm.; **f** *inca* juv. In open country, often near bogs, from Ven. to Isla Grande, at 2700-5300 m in the n, to lowlands in the s. Text p. 663

7 RED-BACKED SIERRA-FINCH *Phrygilus dorsalis*. Small. Back reddish, breast gray. May interbreed with White-throated S-f. **7a** ad.; **b** imm.: faintly streaked above. At 3350-4500 m in rocky country and tola heath from n Chile and Bol. to nw Arg. Text p. 664

8 WHITE-THROATED SIERRA-FINCH *Phrygilus erythronotus*. Small. Gray above and on breast, throat and belly white, area below eye grizzled with white. Primaries often contrasting brownish. No white in wing. **8a** ad.; **b** juv.: faintly streaked above, inner remiges edged warm brown, vent tinged buff. At 3650-4750 m in barren rocky country in s Peru to n Chile and Bol. Text p. 665

9 MOURNING SIERRA-FINCH *Phrygilus fruticeti*. Large. 2 white wingbars. Rump conspicuously gray in flight (streaked in female). **9a** male (*peruvianus*): breast and upper belly more or less black, bill yellow in breeding season, otherwise brownish; **b** female (*fruticeti*): ochraceous, streaked; earcoverts reddish brown; **c** male non-breeding; **d** male of the almost black *coracinus* of the semidesert of w Bol. (female of this form resembles male of other forms). Bushy slopes from n Peru to Isla Grande, at 1500-4000 m in the n. Text p. 662

10 BAND-TAILED SIERRA-FINCH *Phrygilus alaudinus*. White band in spread tail. Bill longer than in smaller Band-tailed Seedeater (Plate LIX 5). **10a** ad. female (*alaudinus*) in worn plumage; **b** male (*alaudinus*) in fresh 2nd year plumage: ridge of bill dark; **c** male (*alaudinus*) in worn 2nd year plumage. **d** ad. male (*excelsus*) in worn plumage; **e** 1st year male (*bipartitus*); **f** juv. At 2500-4100 m (and to coast of sw Ecu. and Peru) on slopes and flats with sparse vegetation from Ecu. to c Chile and c Arg. Text p. 666

11 CARBONATED SIERRA-FINCH *Phrygilus carbonarius*. Superficially like Mourning S-f., but much smaller, and without wingbars. Male gets blacker with wear. **11a** female: no reddish brown on earcoverts, lightly streaked below; **b** male fresh plumage (head); **c** worn male. Open grassy and bushy country in c Arg. Text p. 667

12 YELLOW-BRIDLED FINCH *Melanodera xanthogramma*. **12a** female (juv.?), **b** male. Told from very similar Black-throated F. by golden olive forehead. Bleak rocky habitats in s Arg. and Chile, breeding above 1200 m. Text p. 668

13 BLACK-THROATED FINCH *Melanodera melanodera*. **13a** female (juv.?), **b** male. Told from similar Yellow-bridled F. by white outline of black face. Ad. male with mostly yellow wings. Grassland in s Arg. and n Isla Grande. Vanishing except on Islas Malvinas. Text p. 667

30 per cent of nat. size

PLATE LVIII

Finches

Plate LIX Finches (see also Plates LVII-LVIII and LX-LXII)

1 PLUSH-CAPPED FINCH *Catamblyrhynchus diadema*. Stubby bill. Slaty above. **4a** ad.: forehead yellow, underparts rufous; **b** juv.: unpatterned. At 1750-3500 m in bamboo thickets, from Ven. to nw Arg.
Text p. 641

Seedeaters are small finches with stubby bills. Females of *Sporophila* are usually not identifiable.

2 YELLOW-BELLIED SEEDEATER *Sporophila nigricollis*. Smaller and with smaller bill than 3, wings with barely visible speculum. Belly pale yellow. **2a** male, **b** female. Widespread in fields and at edges from arid to humid lowlands and premontane areas.
Text p. 648

3 BLACK-AND-WHITE SEEDEATER *Sporophila luctuosa*. Bill thick, maxilla curved. Wings with prominent white speculum. Male (in fresh plumage black covered by gray-brown feathertips). At 900-3600 m in second growth and fields from Ven. to Bol. Text p. 647

4 DOUBLE-COLLARED SEEDEATER *Sporophila caerulescens*. Dark throat and breastband. **4a** male; **b** male imm.; **c** female. Lowlands and shrubby drier premontane zones from Col. to Arg. Text p. 648

5 BAND-TAILED SEEDEATER *Catamenia analis*. Smaller with much stubbier bill than in Band-tailed Sierra-finch (Plate LVIII 10). Spread tail with white band. Male with yellow bill in breeding season. **5a** *schistaceifrons* of e and c Col., juv; **b** *schistaceifrons* female; **c** *schistaceifrons* male: no wing speculum; **d** *analis* of s Peru s-wards, male: conspicuous speculum, c abdomen white; **e** *analoides* of w Peru, male: no speculum; **f** *analis* male in flight; **g** *analis* juv. At 2500-3700 (4650) m in bushy country, hedgerows etc. from Col. to c Arg.
Text p. 648

6 PARAMO SEEDEATER *Catamenia homochroa*. From 7 by habitat, longer bill, pointed tail-feathers, and darker plumage. Ssp *homochroa* shown. **6a** juv.; **b** 2nd year female; **c** ad. female; **d** 2nd year male; **e** ad. male. At 2300-3800 m in humid páramo and jalca shrubbery from Ven. to Bol. Text p. 650

7 PLAIN-COLORED SEEDEATER *Catamenia inornata*. From 6 by drier habitat, stubbier bill, lighter color tone, and less pointed tail-feathers. Bill pink in breeding season. **7a** juv.; **b** *inornata* female; **c** *inornata* male. At (2000) 2900-4400 m on grassland and shrubby slopes from Ven. to Arg. and Chile. Text p. 649

Inca-finches genus *Incaspiza* inhabit arid regions of n and c Peru. Resemble juncos by jizz, much white in tail, and very high-pitched calls and song. Bill long, pointed, orange.

8 BUFF-BRIDLED INCA-FINCH *Incaspiza laeta*. Distinct pale buff mustache. **8a** juv., **b** ad. At 1500-2750 m in *Bombax* woodland with admixed cacti, and thorny underscrub.
Text p. 682

9 GREAT INCA-FINCH *Incaspiza pulchra*. No black on forehead, but large black throat bib. **9a** juv., **b** ad. At 1000-2750 (lowest in zone of overlap with 10) on arid slopes with thorny scrub with *Melocactus* in w Ancash and Lima.
Text p. 680

10 RUFOUS-BACKED INCA-FINCH *Incaspiza personata*. Black mask includes forehead, but only upper throat. From 11 by some rufous on back and wings. **10a** juv.; **b** ad. At 2700-4000 m in dry montane scrub and sometimes *Puya raimondii* in e La Libertad and from s Cajamarca to Pasco.
Text p. 681

11 GRAY-WINGED INCA-FINCH *Incaspiza ortizi*. Rather gray above. **11a** ad., **b** juv. At 2000-2300 m in thick, dense, arid brush around the Huancabamba drainage of n Peru. Rare.
Text p. 680

12 SHORT-TAILED FINCH *Idiopsar brachyurus*. Large, tail short, bill long. Gray, grizzled white below eye. **12a** juv.: legs brown, flanks with warm wash; **b** ad.: legs fleshy gray. At 3950-4600 m in barren, rocky country from Puno Peru to nw Arg. Uncomon.
Text p. 659

13 COMMON DIUCA-FINCH *Diuca diuca*. Gray with white throat and c belly, rufous flanks, and extensively white outer tail-feathers. Pale mandible. **a** juv., **b** ad. male. In fields, steppes, and open forest, from s Bol. to c Arg. and Chile.
Text p. 658

14 WHITE-WINGED DIUCA-FINCH *Diuca speculifera*. In pairs. Tame. Large. Gray with white in wing and below eye, and with white throat and c belly. Eye red. **a** juv.; **b** ad. Near bogs at 4500-5350 m from c Peru to Bol. and n Chile.
Text p. 658

30 per cent of nat. size

PLATE LIX

Finches

PLATE LX Yellow-finches

Yellow-finches genus *Sicalis*. Mainly olive-colored finch-like birds of shrub and grassland. Often wander in flocks.

1 PUNA YELLOW-FINCH *Sicalis lutea*. From Greenish Y-f. by being brighter yellow throughout. Bill stout. **1a** juv.: mandible orange-yellow; **b** male in worn plumage: above gray-brown mottled with olive-green, rump yellowish olive; **c** fairly worn female; **d** fresh male: bright golden green and yellow. Not in towns. At 3350-4100 m in the arid puna of se Peru to nw Arg.
Text p. 651

2 STRIPE-TAILED YELLOW-FINCH *Sicalis citrina*. Shy, usually flying far. Small. Bill distinctly smaller and more pointed, and tail with more white than in Grassland Y-f. **2a** female: breast and sides streaked; **b** male: forecrown (when worn) and rump bright golden olive. Crown unstreaked. 400-2800 m in n, 2100-3700 m in s. Locally from n Col. to Guiana and se Peru to nw Arg, and e Brazil. In semiarid savanna. Text p. 651

3 GREATER YELLOW-FINCH *Sicalis auriventris*. Large. Wingtip long (c. 3 cm beyond tertials). In fresh plumage most feathers are tipped pale gray-brown. **3a** female: above gray-brown finely streaked dark brown, rump tinged olive-yellow, c belly and throat yellow, flanks and sides often with faint dark streaks (covered by fresh feather-tips); **b** male: golden olive (back still with grayish feather-tips of fresh plumage), yellow on belly and vent. Wings grayish, lesser coverts like back. In open country, often near houses, breeding above 1800 m. Chile and Arg. Text p. 653

4 SAFFRON FINCH *Sicalis flaveola*. **4a** worn male from temp. zone of Cochabamba (*koenigi*?): forehead orange; **b** fresh Cochabamba male: orange more or less covered by gray feather-tips; **c** Cochabamba female: breast streaked, wing-panel olive-yellow (brown in Grassland Y-f.); **d** ad female of *pelzelni* from adjacent lowlands; **e** juv.: like female, but olive in wing less distinct, olive in tail sometimes lacking; **f** fresh *pelzelnii* male: above olive-yellow with dark shaft-streaks, feathers tipped pale gray-brown, rump and upper tail-coverts olive-yellow, feathers tipped gray. Gray feathertips wear off, leaving bright yellow underside and orange forehead, but not as extensive as in Cochabamba birds. Savanna, gardens etc., spottily from Ven. to Arg., in lowlands to 2500 m, but to 3200 m in Bol. (*koenigi*?). Text p. 656

5 GRASSLAND YELLOW-FINCH *Sicalis luteola*. Fairly small with distinctive facial markings. Dull-plumaged birds told from Saffron Finch by brown wing-panel, from 2 by stubbier bill and streaked crown. **5a** *bogotensis* of Ven. to s Peru, mainly highlands, male: outer tail-feathers with variable amount of whitish distally on inner web, but never as extensive as in Stripe-tailed Y-f. **5b-e** *luteiventris* from se Peru s-wards, **b** female: unstreaked below; **c** *luteiventris* juv.: much like 4e, but belly tinged yellow; **d** and **e** *luteiventris* males: note distinct facial pattern. Mainly at 2200-4000 m, but low in w Peru and from s Brazil to Isla Grande and Chile. In rush-beds (esp. *Juncus* and *Scirpus*) and adjacent pastures and fields. Text p. 657

6 CITRON-HEADED YELLOW-FINCH *Sicalis luteocephala*. Wing-tips short. Stubby bill. Primaries and outer tail-feathers edged yellow; throat and narrow c underparts yellow, extensively gray on sides. **6a** male; **b** female: browner above. At 2550-3800 m in dry valleys with sparse herbs and bushes in c Bol. Very local.
Text p. 653

7 BRIGHT-RUMPED YELLOW-FINCH *Sicalis uropygialis*. Tail fairly short, primaries fairly long. In all plumages color of back contrasts more citrine hue of rump (more gradual change in other Y-f.s). Bill smaller and less arched than in Puna and Greenish Y-f.s. **7a** *sharpei* of nc and c Peru, male: no gray on sides and flanks; **b** *sharpei* female; **c** southern *uropygialis*, male: posterior part of earcoverts, lores, and half-ring below eye gray, sides grayish; **d** *uropygialis* female; **e** *uropygialis* juv. At 3200-4800 m (lower in the s) on short grass and on rocky slopes, and sometimes in villages, in Peru to n Chile and nw Arg. Common. Text p. 652

8 RAIMONDI'S YELLOW-FINCH *Sicalis raimondii*. Distinctly grayer (esp. on sides) than Grassland Y-f. **8a** worn male; **b** slightly worn male; **c** fresh male; **d** imm. female. To 2000 (3450) m in w Peru, in rocky lomas with fog vegetation and scattered trees and cacti. Text p.658

9 PATAGONIAN YELLOW-FINCH *Sicalis lebruni*. Note long wingtip and whitish panel. **9a** fresh male; **b** worn male; **c** juv.: unstreaked above. To 1200 m on grasslands of s Arg. and Chile, where the only Y-f. regularly present. Text p. 655

10 GREENISH YELLOW-FINCH *Sicalis olivascens*. Maxilla somewhat arched. Wingtips rather short (1.5-2 cm beyond tertials). Generally less yellow than Puna Y-f. **10a** juv.; **b** female; **c** fresh male; **d** worn male. At 1500-3600 (4250) m in rocky ravines, scrubby slopes, fields, and towns (but not puna grassland), from Peru to c Arg. and Chile. Text p. 654

30 per cent of nat. size

PLATE LX

Yellow-finches

PLATE LXI Brush-finches

Brush-finches typically live in pairs near the ground in thick brush or montane forest.

1 MOUSTACHED BRUSH-FINCH *Atlapetes albofrenatus*. Citrine with white whisker. **1a** *albofrenatus* of ne Col., **b** *meridae* of Ven. Premontane, in humid forest and dry shrub. Text p. 670

2 PALE-NAPED BRUSH-FINCH *Atlapetes pallidinucha*. Orange-yellow forehead turning whitish on crown. Larger than Rufous-naped B-f., but with similar habits. **2a** *papallactae*; **b** juv. *pallidinucha*. Mainly at 2000-3500 m in humid shrub and stunted cloud forest from Col. to n Peru. Text p. 670

3 CHESTNUT-CAPPED BRUSH-FINCH *Atlapetes brunneinucha*. Slow, near or on the ground. Chestnut hind-crown, olive back, white throat. **3a** *frontalis* from Panama to se Peru: black breastband; **b** *inornatus* of cw Ecu.: breastband lacking or barely suggested. In tall humid premontane forest from Mexico to se Peru. Text p. 677

4 WHITE-RIMMED BRUSH-FINCH *Atlapetes leucopis*. Large. Fast, near the ground. Chestnut crown and distinct white rim behind eye. At 2350-3200 m in cloud forest in Col. and Ecu. Rare. Text p. 679

5 TRICOLORED BRUSH-FINCH *Atlapetes tricolor*. Crown tawny gold. Bill long and stout. In humid forest and adjacent second growth in w Col. to w Ecu. (300-2000 m) and in c Peru (1500-2650 m) (Rufous-naped B-f. absent in this latter area). Text p. 672

6 RUFOUS-NAPED BRUSH-FINCH *Atlapetes rufinucha*. Crown rufous, underparts yellow. **6a** *elaeoprorus* of c Col.: small supraloral yellow spot, and faint dark submalar streak, large wing speculum; **b** *spodionotus* of s Col. to n and w Ecu.: no speculum; **c** *latinuchus* of se Ecu. to nc Peru: wing speculum, yellow supraloral spot small or absent; **d** *baroni* of nw Peru: crown rufous, very pale on nape; **e** *latinuchus* juv.; **f** *melanolaemus* of se Peru into nw Bol., juv.: entire headside (incl. moustache) black; **g** *rufinucha* of Bol.: distinct supraloral spot and submalar streak; back, wings, and tail black. Mainly at 2000-3000 m, in humid shrubbery and forest edge, from Ven. to Bol. Text p. 671

7 OCHRE-BREASTED BRUSH-FINCH *Atlapetes semirufus*. Foreparts uniform rufous, hindparts olive. **7a** *semirufus*; **b** *zimmeri*. At 1200-3050 m in open humid forest of Ven. and ne Col. Text p. 673

8 WHITE-WINGED BRUSH-FINCH *Atlapetes leucopterus*. Wings with white speculum. Frequently shows partial albinism on head. Breast white or whitish. **8a** *leucopterus* of Ecu.; **b** *dresseri* of n Peru, male: extensively white around eye; **c** *dresseri* female, individual with strong partial albinism. At 600-2900 m on arid, bushcovered hillsides (in Ecu. also humid second growth). Text p. 675

9 RUSTY-BELLIED BRUSH-FINCH *Atlapetes nationi*. **9a** southern ssp *arequipae*: paler, and more albinistic on head than *nationi* of Lima (**b**); both forms rusty below; **c** *seebohmi* of nw Peru: white belly, crown bay, supraloral spot white; **d** *celicae* of s Ecu.: no supraloral spot, and crown light. At 1850-4300 on bushy and wooded slopes and in open woodlands in s Ecu. and w Peru. Text p. 676

10 YELLOW-STRIPED BRUSH-FINCH *Atlapetes citrinellus*. Broad supercilium yellow. At 1000-3100 m in shaded ravines, dense shrub, and *Podocarpus* forest in nw Arg. Text p. 674

11 FULVOUS-HEADED BRUSH-FINCH *Atlapetes fulviceps*. Chestnut head with yellow moustache and 'horn'. **11a** male: submalar streak blackish; **b** female; submalar streak fulvous. At 1850-3150 m, lower in winter, in second growth in s Bol. and nw Arg. Text p. 674

12 RUFOUS-EARED BRUSH-FINCH *Atlapetes rufigenis*. Large. Crown and headside mainly rufous. **12a** *forbesi* of sc to se Peru: black on face; **b** *rufigenis* of wc Peru; **c** *rufigenis* juv. At 2750-4000 m in understory of dry to semi-humid woodlands. Text p. 674

13 SLATY BRUSH-FINCH *Atlapetes schistaceus*. Dark with brown crown. Moves quickly along twigs and vines, often up to 5 m above ground. **13a** *castaneifrons* of Ven.: no speculum; **b** *canigenis* of se Peru: no white on head; **c** *schistaceus* of Col. and Ecu.: darker than White-winged B-f. Mainly at 2500-3000 m in humid forest and adjacent second growth from Ven. to se Peru. Text p. 676

14 TANAGER-FINCH *Oreothraupis arremonops*. Shy. Distinctive call. Mainly chestnut with black head striped whitish. On or near the ground in mossy, humid premontane forest. W Col. and nw Ecu. Text p. 669

15 STRIPE-HEADED BRUSH-FINCH *Atlapetes torquatus*. Large. Forages slowly on the ground. White throat often puffed. **15a** *poliophrys* of c to se Peru, imm.; **b** *torquatus* of Bol.; **c** *assimilis* of Col. to n Peru; **d** *borellii* of Bol. to nw Arg.; **e** *assimilis* juv. Mainly at 1650-3300 m in humid forest and montane shrubbery, from Costa Rica to Ven. and nw Arg. Text p. 678

30 per cent of nat. size

PLATE LXI

Brush-finches

PLATE LXII Siskins and Warbling-finches

Siskins genus *Carduelis* are small vivid and twittering social finches usually with a conspicuous yellow wingbar. Inhabits open woodland or open, often semi-cultivated land with herbs, bushes, or rocks.

1 LESSER GOLDFINCH *Carduelis psaltria*. Ad. with restricted white wingbar. Olive tips above wear off leaving back black and rump more or less white. **1a** worn male, **b** male in fresh plumage, **c** juv. N Am. to nw Peru, in shrub, plantations, and reed-marsh, from lowlands to 3100m locally. Text p. 694

2 YELLOW-BELLIED SISKIN *Carduelis xanthogastra*. Black above, and on foreparts to upper breast. **2a** male, **b** female. Costa Rica to Ven. and w Ecu., se Peru to Bol., at 800-3700 m. Text p. 692

3 BLACK SISKIN *Carduelis atrata*. Mostly black. **3a** male, **b** juv female. At 3500-4800 m (lower in s) in open terrain with rocks and low shrubs in c Peru to c Arg. and Chile and Arg. Text p. 693

4 YELLOW-RUMPED SISKIN *Carduelis uropygialis*. Larger than Hooded S. Entire neckside and breast black, black back-feathers tipped and edged olive. Rump yellow. **4a** ad. worn plumage, **b** ad. fresh plumage. Mainly at 2500-3500 m locally from c Peru to c Arg. and Chile. Text p. 693

5 BLACK-CHINNED SISKIN *Carduelis barbata*. Wings with 2 yellowish bars. **5a** male: crown, chin, and c throat black; **b** juv. female. From Atacama Chile and w Neuquen Arg. s-wards, in *Nothofagus* shrub, gardens, and fields. Text p.694

6 ANDEAN SISKIN *Carduelis spinescens*. Male with black cap. **6a** male of widespread *spinescens*: base of tail yellow; **b** male *nigricauda* of n end of C and W Andes, Col.; **c** *spinescens* female: under tail-coverts yellowish white; **d** *spinescens* juv. At 1500-4100 m in *Espeletia* páramo, elfin forest, and farmland in Ven., Col., and e Ecu. Text p. 690

7 HOODED SISKIN *Carduelis magellanica*. Back olive. Rump yellow. Yellow in tail. Some populations with black of throat extending to breast, but never to neck-side. **7a** assumed hybrid *urubambensis* x *C. uropygialis*, male; **b** *urubambensis* male; **c** *boliviana* male; **d** *capitalis* male; **e** *urubambensis* female; **f** male imm. In many habitats from sea level to 5000 m from Col. to Arg. Text p. 691

8 THICK-BILLED SISKIN *Carduelis crassirostris*. Much like Hooded S., but bill thicker, and with yellow patch on neckside, female much grayer. **8a** male of southern *crassirostris*; **b** male of dull green *amadoni*; **c** male of unnamed northern ssp; **d** *amadoni* female; **e** imm. female; **f** juv. Mainly around 4000 m, in *Polylepis* woodlands in c Peru to c Arg./Chile. Text p. 690

Warbling-finches genus *Poospiza* inhabit scrub of Andean slopes and valleys. Most have white in tail.

9 RINGED WARBLING-FINCH *Poospiza torquata*. With pectoral bar and white edges in wing. **9a** juv; **b** *torquata*; **c** female of southern *pectoralis*. Bol., at 1500-3800 m, and Arg. to lowlands. Text p. 685

10 RUFOUS-SIDED WARBLING-FINCH *Poospiza hypochondria*. **10a** juv, **b** ad. Common at 2550-4200 m in arid to semi-humid shrub, gardens, and hedges, from La Paz Bol. to Mendoza Arg. p. 684

11 BOLIVIAN WARBLING-FINCH *Poospiza boliviana*. White throat contrasts brick-red chest. Uncommon at 1700-3100 m, on dry scrubby hillsides and in thorny thickets, in Bol. Text p. 683

12 RUSTY-BROWED WARBLING-FINCH *Poospiza erythrophrys*. Conspicuous white wing-panel. At 1800-3150 m in semi-humid second growth and shrubbery in woodlands and patchily wooded, watered ravines. Cochabamba Bol. to nw Arg. Text p. 687

13 BLACK-AND-RUFOUS WARBLING-FINCH *Poospiza nigrorufa whitii*. Chestnut below. At 1200-2300 m in undergrowth of woodland and second growth bordering grassland, fields, and gardens. Bol. to Arg. Text p. 686

14 BLACK-CAPPED WARBLING-FINCH *Poospiza cinerea melanoleuca*. Thorny woodland and thickets in open country. Bol. below 2500 m to s Brazil and n Arg. Text p. 685

15 PLAIN-TAILED WARBLING-FINCH *Poospiza alticola*. No white in tail. In nc Peru, at 2900-4300 m in *Polylepis* and *Gynoxys* woodland and adjacent humid shrubbery. Uncommon. Text p. 684

16 CHESTNUT-BREASTED MOUNTAIN-FINCH *Poospiza caesar*. Semiarid valleys of Apurímac and Cuzco Peru. At 2900-3900 m in mosaics of shrubbery, small tree groups or woods, and small fields. Text p. 688

17 RUFOUS-BREASTED WARBLING-FINCH *Poospiza rubecula*. **17a** ad, **b** imm. At 2500-3500 m in shrubbery in wc Peru. Rare. Text p. 687

18 COCHABAMBA MOUNTAIN-FINCH *Poospiza garleppi*. **18a** ad, **b** juv. At 3000-3800 m in watered semi-arid ravines with dense shrubbery and *Polylepis* and *Alnus* trees in Cochabamba Bol. Text p. 689

19 TUCUMAN MOUNTAIN-FINCH *Poospiza baeri*. **19a** ad, **b** juv. At 2300-3000 m in impenetrable scrub in Jujuy, Salta, Tucumán, and Catamarca nw Arg. Local. Text p. 688

PLATE LXII

Siskins and Warbling-finches

PLATE LXIII Santa Marta endemics

Sierra Nevada de Santa Marta (5775 m) lies isolated from the main Andes near the Caribbean coast in n Col. The following species and subspecies are endemic to the forest and páramo habitats of the mountain.

1 SANTA MARTA CONURE *Pyrrhura viridicata*. 1700-2500 m. Text p. 206

2 SANTA MARTA WOODSTAR *Chaetocercus astreans*. Text p. 297

3 BLACK-BACKED THORNBILL *Ramphomicron dorsale*. 3a juv./female; b male ad. At 2000-4600 m. Text p. 280

4 BEARDED HELMETCREST *Oxypogon guerinii cyanolaemus*. At 3100-5200 m. Text p. 288

5 SANTA MARTA SABREWING *Campylopterus phainopeplus*. 5a male; b female. At 1300-2000 m. Text p. 245

6 WHITE-TAILED STARFRONTLET *Coeligena phalerata*. 6a male; b female. At 1400-3200 m. Text p. 262

7 BAND-TAILED GUAN *Penelope argyrotis colombiana*. At 800-3050 m. Text p. 137

8 EMERALD TOUCANET *Aulacorhynchus prasinus lautus*. At 1600-2600 m. Text p. 306

9 SANTA MARTA BUSH-TYRANT *Myiotheretes pernix*. At 2100-2900 m. Text p. 501

10 BLACK-CAPPED TYRANNULET *Phyllomyias nigrocapillus flavimentum*. At 1600-3400 m. Text p. 454

11 RUFOUS-BREASTED CHAT-TYRANT *Ochthoeca rufipectoralis poliogastra*. At 2000-3600 m. Text p. 495

12 CINNAMON FLYCATCHER *Pyrrhomyias cinnamomea assimilis*. At 670-3100 m. Text p. 486

13 ANDEAN TIT-SPINETAIL *Leptasthenura andicola extima*. At 3000-4500 m. Text p. 346

14 SPOTTED BARBTAIL *Premnoplex brunnescens coloratus*. At 1200-2600 m. Text p. 386

15 RUDDY FOLIAGE-GLEANER *Automolus rubiginosus rufipectus*. Lowlands to 2450 m. Text p. 391

16 STRONG-BILLED WOODCREEPER *Xiphocolaptes promeropirhynchus sanctaemartae*. At 1500-2850 m. Also in Perijá mts. Text p. 320

17 RUSTY-HEADED SPINETAIL *Synallaxis fuscorufa*. At (700) 2000-3000 m. Text p. 353

18 STREAK-BACKED CANASTERO *Asthenes wyatti sanctaemartae*. At 2400-5000 m. Text p. 375

19 PARAMO SEEDEATER *Catamenia homochroa oreophila*. Female imm. At 2200-3300 m. Text p. 650

20 GREAT THRUSH *Turdus fuscater cacozelus*. At 2000-4100 m. Text p. 557

21 BLUE-NAPED CHLOROPHONIA *Chlorophonia cyanea psittacina*. At 600-2100 m. Text p.608

22 BLACK FLOWERPIERCER *Diglossa humeralis nocticolor*. At 1500-4000 m. Also in Perijá mts. Text p. 604

23 SANTA MARTA MOUNTAIN-TANAGER *Anisognathus melanogenys*. At 1500-3200 m. Text p. 618

24 SANTA MARTA WARBLER *Basileuterus basilicus*. At (2100) 2300-3000 m. Text p. 592

25 SANTA MARTA WHITESTART *Myioborus flavivertex*. At (1500) 2000-3050 m. Text p. 589

26 STREAK-CAPPED SPINETAIL *Cranioleuca hellmayri*. Crown and headside streaked. At 1520-3000 m. Text p. 356

27 WHITE-CAPPED DIPPER *Cinclus leucocephalus rivularis*. At (100) 1500-3900 m. Text p. 537

28 MOUNTAIN WREN *Troglodytes solstitialis monticola*. At 1500-4500 m. Text p. 546

29 SANTA MARTA WHITE-CROWNED TAPACULO *Scytalopus (femoralis) sanctaemartae*. At 1350-1700 m. Text p. 431

30 BROWN-RUMPED TAPACULO *Scytalopus l. latebricola*. At 2150-3650 m. Text p. 433

31 GRAY-THROATED LEAFSCRAPER *Sclerurus albigularis propinquus*. At 1500-2100 m. Text p. 395

32 STRIPE-HEADED BRUSH-FINCH *Atlapetes torquatus basilicus*. At 600-2800 m. Text p. 678

33 SANTA MARTA BRUSH-FINCH *Atlapetes melanocephalus*. At (600) 1500-3200 m. Text p. 672

PLATE LXIII

30 per cent of nat. size
(except 7)

Santa Marta endemics

PLATE LXIV Mérida endemics

The Mérida mts in nw Ven. forms the northeastern extreme of the high Andes. The fauna of its complex of páramos is relatively well isolated from that of the E Andes of Col. and Táchira mts by the low and arid subtropical Táchira depression. Consequently, a number of species and sspp are endemic to the Mérida mts.

1 ROSE-CROWNED CONURE *Pyrrhura rhodocephala*. Text p. 207

2 TYRIAN METALTAIL *Metallura tyrianthina oreopola*. To the treeline. Text p. 284

3 BEARDED HELMETCREST *Oxypogon guerinii lindeni*. At 3100-5200 m. In Mérida and Trujillo mts. Text p. 288

4 MERIDA SUNANGEL *Heliangelus spencei*. **4a** female; **b** male. Rosy throat, white breastband. At 2000-3600 m. C Mérida mts. Text p. 268

5 COLLARED INCA *Coeligena torquata conradi*. At 1500-3000 m. Mérida, Táchira (not on Tamá) and adjacent Col. Text p. 261

6 RUSTY-FACED PARROT *Hapalopsittaca amazonina theresae*. At 2500-3000 m in c Mérida to n Táchira. Text p. 213

7 WHITE-THROATED SCREECH-OWL *Otus albogularis meridensis*. At 1300-3100 m in Mérida and Táchira. Text p. 225

8 FLAMMULATED TREEHUNTER *Thripadectes flammulatus bricenoi*. At 2700-3000 m. Text p. 392

9 ANDEAN TIT-SPINETAIL *Leptasthenura andicola certhia*. At 3500-4100 m. Text p. 346

10 SLATY-BACKED CHAT-TYRANT *Ochthoeca cinnamomeiventris nigrita*. At 1900-2900 m in Mérida and Táchira. Text p. 493

11 BROWN-BACKED CHAT-TYRANT *Ochthoeca fumicolor superciliosa*. At 2400-4200 m from n Táchira to Trujillo. Text p. 496

12 GOLDEN-BELLIED STARFRONTLET *Coeligena bonapartei eos*. At 1400-3200 m from Táchira to Trujillo. Text p. 263

13 COLLARED JAY *Cyanolyca viridicyana meridana*. 1650-3250 m. Text p. 535

14 COMMON BUSH-TANAGER *Chlorospingus ophthalmicus venezuelanus*. At 900-3000 m from Táchira to s Lara. Text p. 630

15 OLEAGINOUS HEMISPINGUS *Hemispingus frontalis ignobilis*. At 1600-2900 m from Táchira to s Lara. Text p. 637

16 SUPERCILIARIED HEMISPINGUS *Hemispingus superciliaris chrysophrys*. At 1900-3200 m from Táchira to Trujillo. Text p. 636

17 SLATY-BACKED HEMISPINGUS *Hemispingus goeringi*. Female has less developed supercilium. At 2600-3200 m from n Táchira to Mérida. Text p. 638

18 GRAY-CAPPED HEMISPINGUS *Hemispingus reyi*. At 1900-3200 m. Text p. 636

19 WHITE-FRONTED WHITESTART *Myioborus albifrons*. At 2200-4000 m. Text p. 588

20 THREE-STRIPED WARBLER *Basileuterus tristriatus meridanus*. At 800-2700 m. Text p. 592

21 BANANAQUIT *Coereba flaveola montana*. 1200-1900 m. Táchira, Mérida. Text p. 595

22 MERIDA FLOWERPIERCER *Diglossa gloriosa*. 2500-4150 m. Táchira to Trujillo. Text p. 604

23 PARAMO PIPIT *Anthus bogotensis meridae*. At 2200-4100 m in Mérida and Trujillo. Text p. 565

24 BAR-WINGED CINCLODES *Cinclodes fuscus heterurus*. 3800-5000 m. Text p. 341

25 CRESTED SPINETAIL *Cranioleuca s. subcristata*. At 700-2300 m. Coastal mts to Zulia and Andes in Barinas. Text p. 356

26 WHITE-BROWED SPINETAIL *Hellmayrea gularis cinereiventris*. At 2300-3200 m. Text p. 355

27 GRAY-NAPED ANTPITTA *Grallaria g. griseonucha*. At 2300-2800 m in c Mérida. Text p. 411

28 GRASS WREN *Cistothorus platensis alticola*. Shown for comparison with Paramo W. At 500-3275 m in Sta Marta Col., Perijá and coastal mts, Andes in Lara and Mérida, and the Tepuis. Text p. 540

29 PARAMO WREN *Cistothorus meridae*. At 3000-4100 m from Mérida to Trujillo. Text p. 542

30 GRAY-BREASTED WOOD-WREN *Henicorhina leucophrys meridana*. 900-3000 m. Text p. 547

31 ANDEAN TAPACULO *Scytalopus magellanicus fuscicauda*. 2500-3200 m. Lara, Trujillo. Text p. 443

32 OCHRE-BROWED THISTLETAIL *Schizoeaca coryi*. 3000-4100 m. Táchira to Trujillo. Text p. 362

33 SLATY BRUSH-FINCH *Atlapetes schistaceus castaneifrons*. At 2000-3800 m. Text p. 676

34 YELLOW GROSBEAK *Pheucticus chrysopeplus laubmanni*. 950-2000 m. Sta Marta, Perijá, and coastal mts., not in Mérida. Text p. 645

35 BLACK-BACKED GROSBEAK *Pheucticus aureoventris meridensis*. Rec. at 2500 m in c Mérida. Text p. 646

PLATE LXIV

30 per cent of nat. size

Mérida endemics

References continued from p. 702

Girard, P (1933) Notas sobre algunas aves de Tucumán. *Hornero* 5:223-225.

Gochfeld, M, S Keith & P Donahue (1980) Records of rare or previously unrecorded birds from Colombia. *Bull. Brit. Orn. Club* 100:196-201.

Goodfellow, W (1901) Results of an ornithological journey through Colombia and Ecuador. *Ibis* 1901: 300-319.

Goodfellow, W (1902) Results of an ornithological journey through Colombia and Ecuador. *Ibis* 1902: 56-67.

Gore, MEJ & ARM Gepp (1978) *Las Aves del Uruguay*. Montevideo: Mosca Hernandos.

Graham, GL, GR Graves, TS Schulenberg & JP O'Neill (1980) Seventeen bird species new to Peru from the Pampas de Heath. *Auk* 97:366-370.

Grant, CHB (1911) List of Birds collected in Argentina, Paraguay, Bolivia and Southern Brazil, with field-notes. Part I-III. *Ibis* 1911: 80-137, 317-350 and 459-478

Graves, GR (1985) Elevational correlates of speciation and intraspecific geografical variation in plumage in Andean forest birds. *Auk* 102: 556-579.

Graves, GR (1988) Linearity of geographical range and its possible effect on the population structure of Andean birds. *Auk* 105:47-52.

Greenway, JC (1967) *Extinct and vanishing Birds of the World*. Revised ed. New York: Dover Publications.

Grigera, DE (1976) Ecologia Alimentaria de Cuatro Especies de Fringillidae frequentes en la Zona del Nahuel Huapí (*sic.*). *Physis C* (Buenos Aires) 35, 91:279-92.

Gyldenstolpe, N (1945) A contribution to the ornithology of northern Bolivia. *Kungl. Svenska Vet.-Akad. Handl.* ser. 3, 23(1).

Haffer, J (1967) On birds from the northern Chocó region, NW-Colombia. *Veröff. Zool. St. Samml. München* 11:123-149.

Haffer, J (1970) Entstehung und Ausbreitung nord-Andiner Berg-Vögel. *Zool. Jahrb. Syst.* 97:301-337.

Haffer, J (1974) Avian speciation in tropical South America. *Publs. Nuttal ornithol. Club* 14:1-390.

Haffer, J (1975) Avifauna of northwestern Colombia, South America. *Bonn. zool. Monogr.* 7.

Haffer, J (1986) On the avifauna of the upper Patía Valley, southwestern Colombia. *Caldasia* 15:533-553.

Handford, PT (1983) Breves notas sobre las aves del valle de Tafí, Provincia de Tucumán. *Neotrópica* 29(81): 97-105.

Hargitt, E (1890) *Catalogue of the birds in the British Museum*, 18. Picidae. London.

Hargitt, E (1898) On a collection of birds from north-western Ecuador, collected by Mr. WFH Rosenberg. *Novit. Zool.* 5:477-505.

Hargitt, E (1901) On some birds from north-west Ecuador. *Novit. Zool.* 8:369-371.

Hargitt, E (1902) Some further notes on the birds of north-west Ecuador. *Novit. Zool.* 9:599-617.

Harris, M (1982) *A Field Guide to the Birds of Galapagos*. Revised edition. London: Collins.

Harrison, P (1983) *Seabirds. An identification guide.* Boston: Houghton Mifflin Company.

Hartert, E & A Goodson (1917a) Notes and descriptions of South American birds. *Novit. Zool.* 24:410-419.

Hartert, E & A Goodson (1917b) Further notes on South American birds. *Novit. Zool.* 24:494-501.

Hartert, E & C Hartert (1894) On a collection of hummingbirds from Ecuador and Mexico. *Novit. Zool.* 1:43-64.

Hartert, E & S Venturi (1909) Notes sur les oiseaux de la République Argentina. *Novit. Zool.* 16:159-267.

Hartlaub, P (1853) Bericht über eine Sendung von Vögeln, gesammelt um Valdivia im südlichsten Chile durch Dr. Philippi. *Naumannia* 3:207-222.

Haverschmidt, F (1968) *The Birds of Surinam*. Edinburgh: Oliver & Boyd.

Hellmayr, CE (1905) A contribution to the ornithology of western Colombia. *Proc. Zool. Soc. London* 1911:1084-1213.

Hellmayr, CE (1919) Ein Beitrag zur Ornithologie von Südost-Peru. *Arch. Naturgesch.* 85, Abt. A, no. 10, 131 pp.

Hellmayr, CE (1921) Review of the birds collected by Alcide d'Orbigny in South America. Parts I & II. *Novit. Zool.* 28:171-213, 230-276.

Hellmayr, CE (1929, 1934, 1935, 1936, 1937, 1938) Catalogue of birds of the Americas. Parts 6-11. *Field Mus. Nat. Hist Publ.* 266; 330; 347; 365; 381; 430.

Hellmayr, CE (1932) The birds of Chile. *Field Mus. Nat. Hist. Publ.* 308.

Hellmayr, CE & B Conover (1942, 1948a,b, 1949) Catalogue of birds of the Americas. Part 1, no. 1-4. *Field Mus. Nat. Hist. Publ.* 514; 615; 616; 634.

Herklots, GAC (1961) *The Birds of Trinidad and Tobago*. London: Collins.

Hilty, SL (1977) *Chlorospingus flavovirens* rediscovered, with notes on other pacific Colombian and Cauca Valley Birds. *Auk* 94:44-49.

Hilty, SL & WL Brown (1983) Range extensions of Colombian birds as indicated by the M. A. Carriker Jr. collection at the National Museum of Natural History, Smithsonian Institution. *Bull. Brit. Orn. Club* 103:5-17.

Hilty, SL & WL Brown (1986) *A Guide to the Birds of Colombia*. Princeton, New Jersey: Princeton University Press.

Hilty, SL & JR Silliman (1983) Puracé National Park, Colombia. *Amer. Birds* 37:247-256.

Holmberg, EL (1878a) Contribuciones para el conocimiento de la fauna de Salta. *Natural. Argent.* 1:18-31, 43-52, 83-69 (96), 118-122, 152-156, 315-320.

Holmberg, EL (1878b) Apuntes sobre una colección de aves formada por el Sr. D. Manuel Oliveira César en el Partido de Las Conchas (Abril, Mayo y Junio de 1878). *Natural. Argent.* 1:231-241.

Holmberg, EL (1939) Las Aves Argentinas. Reedición del capitulo *Aves de la Fauna Argentina* publicado en el 'Segundo Censo de la República Argentina' (1895). *Hornero* 7:142-233.

Horváth, L & G Topál (1963) The zoological results of Gy. Topál's collectings in South Argentina. 9 Aves. *Ann. Hist.-nat. Mus. Natn. Hungarici* 55:531-542.

Housse, R (1924) Apuntes sobre las aves de la isla La Mocha. *Rev. Chil. Hist. Nat.* 28:47-54.

Housse, R (1925a) Avifauna de San Bernardo y sus alrededores. *Rev. Chil. Hist. Nat.* 29:141-150.

Housse, R (1925b) Adición a los 'Apuntes sobre las aves de la isla La Mocha.' *Rev. Chil. Hist. Nat.* 29:225-227.

Housse, RPE (1945) *Las Aves de Chile: su Vida y Costumbres*. Santiago: Ediciones Universidad de Chile.

Housse, RPE (1948) *Les Oiseaux du Chili*. Paris: Crété.

Hoy, G (1968, 1971) Über Brutbiologie und Eier einiger Vögel aus Nordwest-Argentinien, 1 and 2. *J. Orn.* 109:425-433; 112:158-63.

Hoy, G (1969) Addendas a la avifauna Salteña. *Hornero* 11:53-56.

Hoy, G (1975) Notes et fait divers. *Oiseaux* 45(2):189-192.

Hoy, G (1976, 1980) Notas nidobiológicas del noroeste Argentino, 1 and 2. *Physis, secc. C.* Buenos Aires 35:205-209; 39:63-66.

Hoy, G, F Contino & ER Blake (1963) Addendas a la avifauna Argentina. *Bol. Acad. Nac. Cienc., Córdoba* 43:295-308.

Hudson, WH & PL Sclater (1872) On the birds of the Río Negro of Patagonia. *Proc. Zool. Soc. London* 1872: 534-550.

Hughes, RA (1970) Notes on the birds of the Mollendo district, southwest Peru. *Ibis* 112:229-241.

Hughes, RA (1980) Midwinter breeding by some birds in the high Andes of southern Peru. *Condor* 82:229.

Hughes, RA (1988) Nearctic migrants in southwest Peru. *Bull. Brit. Orn. Club* 108:29-43.

Humphrey, PS & D Bridge (1970) Apuntes sobre distribución de aves en la Tierra del Fuego y la Patagonia, Argentina. *Rev. Mus. Argent. Cienc. Nat. Bernadino Rivadavia* 10:251-265.

Humphrey, PS, D Bridge, PD Reynolds, & RT Peterson (1970). *Birds of Isla Grande, Tierra del Fuego*. Lawrence, Kansas: Univ. of Kansas Museum of Natural History for the Smithsonian Institution, Washington, D.C.

Iafrancesco V, GM, CM Mateus & G Oviedo P (1985-86) Contribución al estudio de los Passeriformes de Colombia. I, II. *Bol. Cient. Univ. La Salle* 1:25-65, 2:37-82.

Jaffuel, F & A Pirion (1928) Aves observadas en el valle de Marga-Marga. *Rev. Chil. Hist. Nat.* ('1927') 31:102-115.

Jardine, W (1849) Ornithology of Quito. *Jardine's Contrib. Orn.*: 41-45, 66, 67.

Jardine, W (1850) Ornithology of Quito. *Jardine's Contrib. Orn.*: 1-3.

Jardine, W (1855) Prof. W. Jameson's collections from eastern cordillera of Ecuador. *Edinburgh New Philos. Journ.*, new ser. 2:113-119.

Jardine, W (1856) Professor W. Jameson's collections from the eastern cordillera continued--expedition from Quito to the mountain Cayambe. *Edinburgh New Philos. Journ.*, new ser. 3:90-92.

Johansen, H (1966) Die Vögel Feuerlands (Tierra del Fuego). *Vidensk. Meddr Dansk Naturh. Foren.* 129:215-260.

Johnson, AW (1965, 1967, 1972) *The birds of Chile and adjacent regions of Argentina, Bolivia and Peru*. Vols. 1, 2 and supplement. Buenos Aires: Platt Establecimientos Gráficos.

Johnson, AW (1970) Aves observadas en Mamiña (Tarapacá) desde el 15 al 30 de Agosto de 1968 y 1969. *Bol. Ornitol.* 2:1-2.

Keast, A & ES Morton, eds. (1980) *Migrant Birds in the Neotropics: Ecology, Behavior, Distribution and Conservation*. Washington, D.C.: Smithsonian Inst. Press.

Keith, AR (1970) Bird observations from Tierra del Fuego. *Condor* 72:361-363.

Kempff M, N (1981) *Aves de Bolivia*. La Paz.

King, WB, ed. (1978-1979) *Red Data Book, Aves*. Morges, Switzerland: IUCN.

Koepcke, HW (1958) Über die Wälder an der Westseite der peruanischen Anden und ihre tiergeographischen Beziehungen. *Verh. Dt Zool. Ges. Graz* 1957, nr 4:108-119, 9 Abb.

Koepcke, HW (1961) Synökologische Studien an der Westseite der peruanischen Anden. *Bonn. Geogr. Abhand.* 29.

Koepcke, HW & M Koepcke (1953) Die warmen Feuchtluftwüsten Perus. *Bonn. Zool. Beitr.* 4:79-146.

Koepcke, M (1958) Die Vögel des Waldes von Zárate (Westhang der Anden in Mittelperu). *Bonn. Zool. Beitr.* 9:130-193.

Koepcke, M (1961) Birds of the western slope of the Andes of Peru. *Amer. Mus. Novit.* 2028.

Koepcke, HW (1963a) Probleme des Vogelzuges in Peru. *Proc. XIII. Int. Orn. Congr.* 396-411.

Koepcke, M (1963b) Anpassungen und geographische Isolation bei Vögeln der peruanischen Küstenlomas. *Proc. 13th Int. Orn. Congr.* 1195-1213.

Koepcke, M (1963c) Zur Kenntnis einiger Finken des peruanischen Küstengebietes. *Beitr. Neotrop. Fauna* 3:2-19.

Koepcke, M (1964) *The birds of the Department of Lima, Peru*. Wynnewood, Penn.: Livingston Publishing Company.

Koepcke, HW & M Koepcke (1963) *Las aves silvestres de importancia económica del Perú*. Lima.

Koslowsky, J (1895a) Faunas locales argentinas. II. Enumeración sistemática de las aves de Chilecito (provincia de La Rioja – República Argentina) coleccionada durante los meses de marzo á mayo de 1895. *Rev. Mus. La Plata* 6:277-287.

Koslowsky, J (1895b) Aves recogidas en la Provincia de Catamarca (República Argentina) durante los meses de marzo y abril de 1895. *Rev. Mus. La Plata* 6:289-292.

Krieg, H (1940) *Als Zoologe in Steppen und Wäldern Patagoniens*. Munich, West Germany: Bayerischen Landwirtschaftsverlag.

Lafresnaye, F De & A d'Orbigny (1837, 1838) Synopsis Avium. *Mag. Zool.* 7, cl. 2:1-88; 8, cl.2:1-34.

Land, HC (1970) *Birds of Guatemala*. Wynnewood, Penn.: Livingston.

Landbeck, L (1877) Bemerkungen über die Singvögel Chiles. *Zoologischer Garten* 18:233-261.

Lane, AA (1897) Field-Notes on the Birds of Chili. *Ibis* 1897: 8-51; 177-195; 297-317.

Lataste, F (1894) Minuscule contribution à l'ornithologie chilienne. *Act. Soc. Scient. Chili* 3, livr. 3, '1893',: 113-116.

Lataste, F (1895a) Liste d'oiseaux recueillis, en trois jours de chasse, à la fin du mois de décembre, dans la hacienda de Caillihue (dép. de Vichuquen). *Act. Soc. Scient. Chili* 5, livr. 1-3: 33-34.

Lataste, F (1895b) Liste d'oiseaux capturés à Llohué (Itata), du 8 au 13 avril, et à Junquillos (San Carlos), les 13 et 14 avril 1895; avec refléxions sur le vol à voile et le vol ramé. *Act. Soc. Scient. Chili* 5, livr. 1-3: 60-63.

Laubmann, A (1939, 1940a) *Die Vögel von Paraguay. Wissenschaftliche Ergebnisse der Deutschen Gran Chaco-Expedition*. 2 Vols. Stuttgart: Strecker und Schröder.

Laubmann, A (1940b) Beiträge zur Avifauna Patagoniens. *Verh. Orn. Ges. Bayern* 22:1-98.

Lehmann, FC (1957) Contribuciones al Estudio de la Fauna de Colombia XII. *Noved. Colomb.* 3:101-156.

Lehmann, FC (1960) Contribuciones al Estudio de la Fauna de Colombia XV. *Noved. Colomb.* 1(5):256-276.

Lemke, TO & P Gertler (1978) Recent observations on the birds of the Sierra de la Macarena, Colombia. *Condor* 80:453-455.

Lesson, RP (1828) Observations générales sur l'histoire naturelles des diverses contrées visitées par la corvette la Coquille, et plus particul'rement sur l'ornithologie de chacune d'elles. Zoologie par Lesson et Garnot. 1, Part 1, livr. 6:229-246 in LI Duperrey *Voyage autor du monde exécuté ... sur la corvette ... la Coquille, pendant les années 1822, 1823, 1824 et 1825*. Paris.

Lillo, M (1889) Apuntes sobre la fauna de la Provincia de Tucumán. Enumeración y descripción de las especies animales indígenas con las custumbres y daños ó beneficios que ocasionan las más caracteristicas. Vertebrados. *Bol. Oficina Quimica Tucumán* 2:73-94.

Lillo, M (1905) Fauna tucumána. Aves catálogo sistemático. *Rev. Letras y Cienc. Soc.* (Tucumán), 3:3-41.

Lillo, M (1909) Notas ornitológicas. *Apuntes Hist. Nat. Buenos Aires* 1:21-26, 41-44.

Lönnberg, E (1903) On a collection of birds from northwestern Argentina and the Bolivian Chaco. *Ibis* 1903: 441-471.

Lönnberg, E & H Rendahl (1922) A contribution to the ornithology of Ecuador. *Arkiv Zool.* 14(25): 1-87.

Lopez T, E (1977) Contribución al conocimiento

de la ornitofauna de las Lomas de Mollendo, Arequipa. *V Congreso Nac. Biol. Cuzco.*

Lynch Arribálzaga, E (1902a) Apuntes ornitológicos. I. Dos especies nuevas para la avifauna argentina. *Anal. Mus. Nac. Buenos Aires* 8(ser. 3, vol. 1, entraga 1): 151-157.

Lynch Arribálzaga, E (1902b) Apuntes ornitológicos. II. Aves del Chubut occidental colectadas ó observadas por el Sr. Guillermo F. Gerling. *Anal. Mus. Nac. Buenos Aires* 8(ser. 3, vol. 1, entraga 1): 158-168.

Læssøe, T (1983) Avifaunal notes from Ecuador. Unpublished.

MacFarlane, JRH (1887) Notes on Birds in the Western Pacific, made in H. M. S. 'Constance', 1883-5. *Ibis* 1887: 201-215.

Maclean, GL (1974) Ornithologische Beobachtungen in Patagoniens Steppen und Feuerlands Urwäldern. *Gefiederte Welt* 98:190-193.

Marchant, S (1960) The breeding of some s.w. Ecuadorian birds. *Ibis* 102:349-382.

Marín A, M, LF Kiff & L Peña G (1989) Notes on Chilean birds, with description of two new subspecies. *Bull. Brit. Orn. Cl.* 109:66-82.

Markham, BJ (1971) *Catálogo de los anfibios, reptiles, aves y mamíferos de la provincia de Magallanes, Chile*. Punta Arenas.

Masramón, DO de (1969-1980) Contribución al estudio de las aves de San Luis, 1-5. *Hornero* 11:33-45, 113-123, 309-311, 413-416; 12(1979): 59-68.

Mathew, GF (1873) Natural History notes from Coquimbo. *Zoologist* 2nd ser., 8:3578-3579.

Matos F, J (1977) Las aves del valle del Cuzco. *V Congreso Nac. Biol. Cuzco.*

Mayr, E & WH Phelps, Jr (1967) The origin of the bird fauna of the south Venezuelan highlands. *Bull. Amer. Mus. Nat. Hist. 136:269-328.*

Mayr, E & GW Cottrell, eds. (1979) *Checklist of birds of the World*, 1. 2nd ed. Cambridge, Mass.: Museum of Comparative Zoology.

Mayr, E & JC Greenway, eds. (1960) *Check-list of birds of the World*, 9. Cambridge, Mass.: Museum of Comparative Zoology.

Mayr, E & JC Greenway, eds. (1962) *Check-list of birds of the World*, 15. Cambridge, Mass.: Museum of Comparative Zoology.

Mayr, E & RA Paynter, Jr, eds. (1964) *Check-list of birds of the world*, 10. Cambridge, Mass.: Musuem of Comparative Zoology.

McFarlane, RW (1975) The status of certain birds in northern Chile. *ICBP Bull.* 12:300-309.

McNeil, R (1982) Winter resident repeats and returns of austral and boreal migrant birds in Venezuela. *J. Field Orn.* 53(2):125-132.

Ménégaux, A (1908) Étude d'une collection d'oiseaux de l'Equateur donnée au Museum d'Histoire Naturelle. *Bull. Soc. Philom. Paris*, ser. 9, 10:84-100.

Ménégaux, A (1909) Étude d'une collection d'oiseaux provenant des hautes plateaux de la Bolivie et du Pérou meridional. *Bull. Soc. Philom. Paris*, ser. 10, 1:205-229.

Ménégaux, A (1910a) Étude d'une collection d'oiseaux provenant des hautes plateaux de la Bolivie et du Pérou meridional. *Rev. Franc. Orn.* 1:132-138 (mis-numbered as pp. 122-128 in journal)

Ménégaux, A (1910b) Étude d'une collection d'oiseaux du Pérou, 1. *Rev. Franc. Orn.* 1:318-322.

Ménégaux, A (1911a) Étude d'une collection d'oiseaux du Pérou, 2. *Rev. Franc. Orn.* 2:8-11.

Ménégaux, A (1911b) Étude des oiseaux de l'Equateur, rapportés par le Dr. Rivet. *Arc. de Méridien Équatorial* 9(1).

Ménégaux, A & CE Hellmayr (1905) Étude des espèces critiques et des types du groupe des Passeraux trachéophones de l'Amerique tropicale appartenant aux collections du Muséum. I et II. Conopophagidés et Hylactidés. *Bull. Mus. Hist. Nat. Paris* 11, 6:372-381.

Ménégaux, A & CE Hellmayr (1906) Étude des espèces critiques et des types du groupe des Passeraux trachéophones de l'Amerique tropicale appartenant aux collections du Muséum. III. Dendrocolaptidés. *Mem. Soc. Hist. Nat. Autun* 19:43-126.

Meyer de Schauensee, R (1944a) Notes on Colombian parrots. *Notulae Naturae* 140.

Meyer de Schauensee, R (1944b) Notes on Colombian woodpeckers, with the description of a new form. *Notulae Naturae* 141.

Meyer de Schauensee, R (1944c) Notes on Colombian birds, with a description of a new form of *Zenaida*. *Notulae Naturae* 144.

Meyer de Schauensee, R (1945a) Notes on Colombian birds. *Proc. Acad. Nat. Sci. Philadelphia* 97:1-16.

Meyer de Schauensee, R (1945b) Notes on Colombian flycatchers, manakins and cotingas. *Proc. Acad. Nat. Sci. Philadelphia* 97:41-57.

Meyer de Schauensee, R (1945c) Notes on Colombian antbirds, ovenbirds, and woodhewers, with a description of a new form from Peru. *Notulae Naturae* 153.

Meyer de Schauensee, R (1946a) Colombian zoological survey, part II. Notes on Colombian crows, wrens, and swallows. *Notulae Naturae* 161.

Meyer de Schauensee, R (1946b) Colombian zoological survey, part III. Notes on Colombian birds. *Notulae Naturae* 163.

Meyer de Schauensee, R (1946c) Colombian zoological survey. Part IV. Further notes on Colombian birds with the description of new forms. *Notulae Naturae* 167.

Meyer de Schauensee, R (1947a) On the genera *Automolus* (Furnariidae) and *Myrmeciza* (Formicariidae) in Colombia. *Notulae Naturae* 186.

Meyer de Schauensee, R (1947b) New or little-known Colombian birds. *Proc. Acad. Nat. Sci. Philadelphia* 99:107-126.

Meyer de Schauensee, R (1948-1952) The birds of the Republic of Colombia, 1-5. *Caldasia* 5:251-1214.

Meyer de Schauensee, R (1950a) Colombian zoological survey, part V. New birds from Colombia. *Notulae Naturae* 221.

Meyer de Schauensee, R (1950b) Colombian zoological survey, part VII. A collection of birds from Bolívar, Colombia. *Proc. Acad. Nat. Sci. Philadelphia* 102:111-139.

Meyer de Schauensee, R (1951a) Notes on Ecuadorian birds. *Notulae Naturae* 234.

Meyer de Schauensee, R (1951b) Colombian zoological survey, part VIII. On birds from Nariño, Colombia, with the description of four new subspecies. *Notulae Naturae* 232.

Meyer de Schauensee, R (1952) Colombian zoological survey, part X. A collection of birds from Southeastern Nariño, Colombia. *Proc. Acad. Nat. Sci. Philadelphia* 102:1-33.

Meyer de Schauensee, R (1953) Manakins and cotingas from Ecuador and Peru. *Proc. Acad. Nat. Sci. Philadelphia* 105:29-43.

Meyer de Schauensee, R (1959) Additions to the 'Birds of the Republic of Colombia.' *Proc. Acad. Nat. Sci. Philadelphia* 111:53-75.

Meyer de Schauensee, R (1964) *The Birds of Colombia*. Narberth, Penn.: Livingston Publishing Company.

Meyer de Schauensee, R (1966) *The Species of Birds of South America and their Distribution*. Narberth, Penn.: Livingston Publishing Company.

Meyer de Schauensee, R (1970) *A guide to the birds of South America*. Edinburgh: Oliver & Boyd.

Meyer de Schauensee, R & WH Phelps, Jr (1978) *A Guide to the Birds of Venezuela*. Princeton, New Jersey: Princeton University Press.

Miller, AH (1947) The tropical avifauna of the Upper Magdalena Valley, Colombia. *Auk* 64:351-381.

Miller, AH (1960) Additional Data on the distribution of some Colombian birds. *Noved. Colomb.* 1,5:235-237.

Mischler, TC (1986) Die Avifauna von Río Lurín bei Cieneguilla, Dep. Lima, Peru. *Bonn. zool. Beitr.* 37:257-279.

Moore, RT (1934) The Mt. Sangay labyrinth and its fauna. *Auk* 51:141-156.

Moreno, FP (1879) *Viaje á la Patagonia Austral Emprendido Bajo los Auspicios del Gobierno Nacional 1876-77.* Vol. 1. La Nación, Buenos Aires.

Morrison, A (1939a) The birds of the department of Huancavelica, Peru. *Ibis* 1939: 453-486.

Morrison, A (1939b) Notes on the birds of Lake Junín, central Peru. *Ibis* 1939: 643-654.

Morrison, A (1940) Brief Notes on the Birds of South Chile. *Ibis* 1940: 248-256.

Morrison, A (1948a) Notes on the birds of the Pampas River Valley, south Peru. *Ibis* 90:119-126.

Morrison, A (1948b) A list of the birds observed at the Hacienda Huarapa, Department of Huánuco, Peru. *Ibis* 90:126-128.

Morrison, A (1948c) Notes on Peruvian birds. *Ibis* 90:130-132.

Moynihan, M (1979) Geographic variation in social behavior and in adaptation among Andean birds. *Publ. Nuttal Orn. Club* 18.

Munves, J (1975) Birds of a highland clearing in Cundinamarca, Colombia. *Auk* 92:307-321.

Narosky, S (1978) *Aves Argentinas*. Buenos Aires: Editorial Albatros, Saci.

Narosky, S, Fraga, R & de la Pena, M (1983) *Nidificación de las aves Argentinas (Dendrocolaptidae y Furnariidae)*. Buenos Aires: Asoc. Orn. del Plata.

Narosky, T y D Yzurieta (1987) *Guia para la identificación de las aves de Argentina y Uruguay*. Buenos Aires: Vazquez Mazzini Editores for Asoc. Orn. del Plata.

Nation, W (1885) Notes on Peruvian birds. *Proc. Zool. Soc. London*: 277-279.

Navas, JR (1959) Lista preliminar de las aves del Parque Nacional Comechingones. *In* M Dimitre, La protección de la flora en el noreste de San Luis. *Anal. Parq. Nac.* 8:77-79.

Navas, JR (1970) Nuevos registros de aves para la Patagonia. *Neotropica* 18:11-16.

Navas, JR (1971a) Notas sobre aves del Parque Nacional Nahuel Huapí (*sic*.). *Neotropica* 17:153-156.

Navas, JR (1971b) Estudios sobre la avifauna Andinopatagonica. 1. Generos *Upucerthia, Ochetorhynchus* Y *Eremobius*. *Rev. Mus. Arg. Cienc. Nat.*

'Bernadino Rivadavia' 7: 267-304.

Niceforo, H (1945) Notas sobre aves de Colombia. *Caldasia* 3: 367-395, 4: 317-377, 5: 201-210, 7: 173-175.

Niceforo, H & A Olivares (1964-1976) Adiciones a la avifauna Colombiana. I, *Bol. Inst. la Salle* 204(1964): 5-27. II, *Bol. Soc. Venez. Cienc. Nat.* 26(1965): 36-58. III, *Ibid.* 26(1966): 370-393. IV, *Hornero* 10(1967): 403-435. V, *Bol. Inst. la Salle* 208(1968): 271-291. VI, *Lozania* 19(1975): 1-16, 20(1976): 19-34 and 21(1976): 1-15.

Niethammer, G (1953, 1956) Zur Vogelwelt Boliviens. *Bonn. zool. Beitr.* 4:195-303, 7:84-150.

Nores, M & D Yzurieta (1980a) *Aves de ambientes acuáticos de Cordoba y centro de Argentina.* Córdoba, Arg.: Sec. Estado Agric. y Ganadería.

Nores, M & D Yzurieta (1980b) Nuevas aves de la Argentina. *Hist. Nat.* (Mendoza) 1:169-172.

Nores, M & D Yzurieta (1981-3) Nuevas localidades para aves Argentinas, 1-5. *Hist. Nat.* 2:33-42, 101-104, 151-152. 3:41-43, 159-160.

Nores, M & D Yzurieta (1983) Especiación en las sierras pampeanas de Córdoba y San Luis (Argentina), con descipción de siete nuevas subespecies de aves. *Hornero*, N° extrao.: 88-102.

Nores, M, D Yzurieta, & R Miatello (1983) Lista y distribución de las aves de Córdoba, Argentina. *Bol. Acad. Nac. Cienc., Córdoba* 56(1-2):9 + 114 pp.

Norton, DW, G Orces V, & E Sutter (1972) Notes on rare and previously unreported birds from Ecuador. *Auk* 89:889-894.

Norton, WJE (1975) Notes on the birds of the Sierra Nevada de Santa Marta, Colombia. *Bull. Brit. Orn. Club* 95:109-115.

Oberholser, HC (1902) Catalogue of a collection of hummingbirds from Ecuador and Colombia. *Proc. U.S. Natn. Mus.* 24:309-342.

Ogilvie-Grant, WR (1893) *Catalogue of the birds in the British Museum*, 22. Pterocletes, Gallinae, Opisthocomi, and Hemipodii. London.

Olivares, A (1959) Aves migratorias en Colombia. *Rev. Acad. Col. Cienc. Exactas Fisicas Nat.* 10:371-442.

Olivares, A (1962) Aves de la región sur de la Sierra de la Macarena, Meta, Colombia. *Rev. Acad. Col. Cienc. Exactas Fisicas Nat.* 11:305-345.

Olivares, A (1963) Notas sobre aves de los Andes Orientales en Boyacá. *Bol. Soc. Venez. Cienc. Nat.* 25:91-125.

Olivares, A (1969) *Aves de Cundinamarca.* Bogotá: Univ. Nac. Colombia.

Olivares, A (1971) Aves de la ladera oriental de los Andes Orientales, Alto Río Cusiana, Boyacá, Colombia. *Caldasia* 11:203-226.

Olivares, A (1974) Aves de la Sierra Nevada del Cocuy, Colombia. *Rev. Acad. Col. Cienc. Exactas, Fisicas Nat.* 54:39-48.

Olrog, CC (1948) Observaciones sobre la avifauna de Tierra del Fuego y Chile. *Acta Zool. Lilloana* 5:437-531.

Olrog, CC (1949) Breves notas sobre la avifauna del Aconquija. *Acta Zool. Lilloana* 7:139-159.

Olrog, CC (1959) *Las aves argentinas: una guía de campo.* Tucumán, Argentina: Inst. 'Miguel Lillo'.

Olrog, CC (1963a) *Lista y distribucion de las aves Argentinas.* Tucuman, Argentina.

Olrog, CC (1963b) Notas sobre aves Bolivianas. *Acta Zool. Lilloana* 19:407-478.

Olrog, CC (1969) *Las aves sudamericanas: una guía de campo.* Tucumán, Argentina: Inst. 'Miguel Lillo'.

Olrog, CC (1979) Nueva lista de la avifauna argentina. *Op. Lilloana* 27, Tucumán.

Olrog, CC (1984) *Las aves Argentinas. Una nueva guía de campo.* Buenos Aires: Admin. Parq. Nac.

Olrog, CC & F Contino (1970) Dos especies nuevas para la avifauna Argentina. *Neotrópica* 16:94-95.

O'Neill, JP (1969) Distributional notes on the birds of Peru, including twelve species previously unreported from the Republic. *Occ. Pap. Mus. Zool. Louisiana State Univ.* 37:1-11.

Orbigny, AD d' (1847) *Voyage dans l'Amerique Méridionale.* Vol. 4, pt. 3, Oiseaux. Strasbourg: Berger-Levrault.

Orcés V, G (1944) Notas sobre la distribución geográfica de algunas aves neotropicas. *Flora* 4:103-123.

Orcés V, G (1974) Notas acera de la distribución geográfica de algunas aves del Ecuador. *Ciencia y Naturaleza* 15(1):8-11.

Orejuela, JE, RJ Raitt & H Alvarez (1980) Differential use by North American migrants of three types of Colombian forests. Pp. 253-264 in A Keast & ES Morton, eds. *Migrant Birds in the Neotropics: Ecology, Behavior, Distribution and Conservation.* Washington, D.C.: Smithsonian Inst. Press.

Ortiz-Crespo, FI & S Valarezo-Delgado (1975) Lista de aves del Ecuador. *Publ. Soc. Ecuatoriana Francisco Campos de Amigos Naturaleza* 2; 37 pp.

Orton, J (1871) Contributions to the natural history of the valley of Quito, no. 1. *Amer. Naturalist* 5:619-626.

Oustalet, E (1891) *Mission Scientifique du Cap Horn.* 1882-1883. Zoologie. Oiseaux. Pp. B1-341. Paris.

Parker III, TA & JP O'Neill (1976a,b, 1981) An introduction to bird finding in Peru: parts I-III. *Birding* 8:140-144, 205-216; 13:100-106.

Parker III, TA & JP O'Neill (1980) Notes on little known birds of the upper Urubamba Valley, southern Peru. *Auk* 97:167-176.

Parker III, TA, JV Remsen Jr, & JA Heindel (1980) Seven bird species new to Bolivia. *Bull. Brit. Orn. Club* 100:160-162.

Parker III, TA & SA Parker (1982) Behavioural and distributional notes on some unusual birds of a lower montane cloud forest in Peru. *Bull. Brit. Orn. Club* 102:63-70

Parker III, TA, SA Parker, & MA Plenge (1982) *An annotated check-list of Peruvian birds*. Vermillion, S.Dakota: Buteo Books.

Parker III, TA & RA Rowlett (1984) Some noteworthy records of birds from Bolivia. *Bull. Brit. Orn. Club* 104:110-113.

Parker III, TA, TS Schulenberg, GR Graves, & MJ Braun (1985) The avifauna of the Huancabamba region, northern Peru. Pp. 169-197 *in* PA Buckley *et al.*

Parker III, TA & JV Remsen (1987) Fifty-two Amazonian bird species new to Bolivia. *Bull. Brit. Orn. Club* 107:94-107.

Parker III, TA, TS Schulenberg, M Kessler & W Wust (in prep.) Species limits, natural history, and conservation of some endemic birds in northwest Peru.

Parkes, KC (1975) Birds of the Sierra Nevada de Santa Marta, Colombia: corrections and clarifications. *Bull. Brit. Orn. Club* 95:173-175.

Partridge, WH (1953) Observaciones sobre aves de las Provincias de Córdoba y San Luis. *Hornero* 10:23-73.

Pässler, R (1922) In der Umgebung Coronel's (Chile) beobachtete Vögel. Beschreibung der Nester und Eier der Brutvögel. *J. Orn.*: 430-482.

Paynter, RA, Jr, ed. (1968) *Check-list of birds of the world*, 14. Cambridge, Mass.: Museum of Comparative Zoology.

Paynter, RA, Jr, ed. (1970) *Check-list of birds of the world*, 13. Cambridge, Mass.: Museum of Comparative Zoology.

Paynter, RA, Jr (1985) *Ornithological gazetteer of Argentina*. Cambridge Mass.: Museum of Comparative Zoology.

Paynter, RA, Jr (1987) *Check-list of birds of the world*, 16. Cambridge, Mass.: Museum of Comparative Zoology.

Paynter, RA, Jr & MA Traylor, Jr (1977) *Ornithological Gazetteer of Ecuador*. Cambridge Mass.: Museum of Comparative Zoology.

Paynter, RA, Jr & MA Traylor, Jr (1981) *Ornithological Gazetteer of Colombia*. Cambridge Mass.: Museum of Comparative Zoology.

Paynter, RA, Jr, MA Traylor, Jr & B Winter (1975) *Ornithological gazetteer of Bolivia*. Cambridge Mass.: Museum of Comparative Zoology.

Pearson, DL & MA Plenge (1974) Puna bird species on the coast of Peru. *Auk* 91:626-631.

Pearson, OP & CP Ralph (1978) The diversity and abundance of vertebrates along an altitudinal gradient in Peru. *Mem. Mus. Hist. Nat. 'Javier Prado'* 18:1-97.

Pelzeln, A von (1865) In *Reise der österreichischen Fregatte 'Novara' um die Erde in den Jahren 1857, 1858, 1859 etc*. Zoologischer Theil. I. Band, Wirbelthiere. 2, pp. 1-4, 1-176. *Wien*

Peña, LE (1961) Results of research in the Antofagasta Ranges of Chile and Bolivia. *Postilla* 49:3-42.

Pereyra, JA (1942) Avifauna Argentina (contribucion a la ornitologia). *Mems. Jard. Zool. La Plata* 10:172-274.

Pereyra, JA (1945) Las aves del territorio del Neuquén. *Anal. Mus. Patagonia 'Francisco P Moreno'* 1:61-99.

Pereyra, JA (1950, 1951, 1969) Avifauna Argentina. *Hornero* 9:178-241, 291-347, 11:1-19.

Peters, JL (1923) Notes on some summer birds of northern Patagonia. *Bull. Mus. Comp. Zool. Harvard* 65:277-337.

Peters, JL (1934, 1937, 1940, 1945, 1948) *Check-list of birds of the World*, 2, 3, 4, 5, 6. Cambridge, Mass.: Harvard University Press.

Peters, JL (1951) *Check-list of birds of the World*, 7. Cambridge, Mass.: Museum of Comparative Zoology.

Peters, JL & JA Griswold, Jr (1943) Birds of the Harvard Peruvian Expedition. *Bull. Mus. Comp. Zool. Harvard* 92:281-327.

Peterson, RT & EL Chalif (1973) *A Field Guide to Mexican Birds*. Revised edition. Boston: Houghton Mifflin Company.

Phelps, WH (1944b) Las aves de Perijá. *Bol. Soc. Venez. Cienc. Nat.* 56:265-338.

Phelps, WH (1945) Resumen de las colecciones ornitológicas hechas en Venezuela. *Bol. Soc. Venez. Cienc. Nat.* 9(1944): 325-444.

Phelps, WH (1961) Night migration at 4,200 meters in Venezuela. *Auk* 78:93-94.

Phelps, WH & WH Phelps, Jr (1950) Lista de las aves de Venezuela y su distribución. Parte 2.

Passeriformes. *Bol. Soc. Venez. Cienc. Nat.* 12:1-427.

Phelps, WH & WH Phelps, Jr (1952) Nine new birds from the Perijá Mountains and eleven extensions of ranges to Venezuela. *Proc. Biol. Soc. Washington* 65:89-105.

Phelps, WH & WH Phelps, Jr (1953b) Eight new birds and thirtythree extensions of ranges to Venezuela. *Proc. Biol. Soc. Washington* 66:125-144.

Phelps, WH & WH Phelps, Jr (1954) Notes on Venezuelan birds and descriptions of six new subspecies. *Proc. Biol. Soc. Washington* 67:103-113.

Phelps, WH & WH Phelps, Jr (1955) Five new Venezuelan birds and nine extensions of range to Colombia. *Proc. Biol. Soc. Washington* 68:47-68.

Phelps, WH & WH Phelps, Jr (1956) Five new birds from Río Chiquito, Táchira, Venezuela and two extensions of ranges from Colombia. *Proc. Biol. Soc. Washington* 69:157-166.

Phelps, WH & WH Phelps, Jr (1957) Descriptions of four new Venezuelan birds, extensions of ranges to Venezuela and other notes. *Proc. Biol. Soc. Washington* 70:119-127.

Phelps, WH & WH Phelps, Jr (1958a) Descriptions of two new Venezuelan birds and distributional notes. *Proc. Biol. Soc. Washington* 71:119-124.

Phelps, WH & WH Phelps, Jr (1958b) Lista de las aves de Venezuela con sus distribución. Tomo II. Parte I. No Passeriformes,. *Bol. Soc. Venez. Cienc. Nat.* 19: 1-317.

Phelps, WH & WH Phelps, Jr (1963) Lista de las aves de Venezuela con su distribución. Parte II. Passeriformes. 2nd. ed. *Bol. Soc. Venez. Cienc. Nat.* 24: 1-479.

Philippi, F (1886) Reise nach der Provinz Tarapacá. *Verh. dt. Wiss. Ver. Santiago* 1, 4:135-163.

Philippi, RA (1868) Catálogo de las aves chilenas existentes en el Museo Nacional de Santiago. *Anal. Univ. Chile* 31: 241-335.

Philippi, RA (1888) Ornis der Wüste Atacama und der Provinz Tarapacá. *Ornis* 4: 155-160.

Philippi, RA & L Landbeck (1863) Beiträge zur Fauna von Peru. *Arch. Naturg.* 29: 119-138.

Philippi, RA & L Landbeck (1864) Contribución a la ornitolojía de Chile. *Anal. Univ. Chile* 25: 408-439.

Philippi B, RA (1936) Aves de Arica y alredores (extremo norte de Chile). *Hornero* 6: 225-239.

Philippi B, RA (1964) Catálogo de las aves Chilenas. *Inv. Zool. Chilenas* 11: 1-179.

Philippi B, RA, AW Johnson, & JD Goodall (1944) Expedición ornitológicas al Norte de Chile. *Bol. Mus. Nac. Hist. Nat. Chile* 22: 65-120.

Philippi B, RA, AW Johnson, JD Goodall & F Bahn (1954) Notas sobre aves de Magallanes y Tierra del Fuego. *Bol. Mus. Nac. Hist. Nat. Chile* 26: 1-65.

Plenge, MA (1974) Notes on some birds in west-central Perú. *Condor* 76:326-330.

Pliego, PE & F Vuilleumier (1989) *Directorio de colecciones ornitológicas en los países de la America Neotropical*. Am. Mus. Nat. Hist. (Orn. Dept.).

Porter, CE (1912) Bibliografía Ornitolójica de Chile. *Bol. Mus. Nac. Chile* 4: 197-206.

Rasmussen, PC & N López H (1988) Notes on the status of some birds of Region X, Chile. *Bull. Brit. Orn. Club* 108:154-159.

Reed, EC (1877) Apuntes de la Zoolojía de la hacienda de Cauquenes, provincia de Colchagua. *Anal. Univ. Chile* 49:535-569.

Reed, EC (1893) Notes on the Birds of Chili. *Ibis* 1893: 595-596.

Reed, EC (1896) Catálogo de las Aves Chilenas. *Anal. Univ. Chile* 93:197-213.

Reed, CS (1916) *Las aves de la Provincia de Mendoza*. Mendoza: Mus. Educ.

Reed, CS (1919) Breves notas acerca de nidos y huevos de algunas aves de la Cordillera de Mendoza. *Hornero* 1:267-273.

Reed, TE (1977) *A guide to the birds of Patagonia*. Privately printed, 40 pp.

Reiss, W & O Finsch (1905) Zum Vogelzuge auf die Hochgebirge von Ecuador. *Aquila* 12:302-304.

Remsen, JV & RS Ridgely (1980) Additions to the avifauna of Bolivia. *Condor* 82:69-75.

Remsen, JV, Jr, TA Parker III & RS Ridgely (1982). Natural history on some poorly known Bolivian birds. *Gerfaut* 72:77-87.

Remsen, JV, Jr & MA Traylor (1983) Additions to the avifauna of Bolivia, part 2. *Condor* 85:95-98

Remsen, JV, Jr (1984a) Natural History on some poorly known Bolivian Birds, Part II. *Gerfaut* 74:163-179.

Remsen, JV, Jr (1984b) High incidence of 'leap-frog' pattern of geographic variation in Andean birds: implications for the speciation process. *Science* 224:171-173

Remsen, JV, Jr (1984c) Arboreal dead-leaf-searching birds of the Neotropics. *Condor* 86:36-41.

Remsen, JV Jr (1985) Community organization and ecology of birds of high elevation humid forest of the Bolivian Andes. Pp. 733-756 *in* PA Buckley *et al.*, eds.

Remsen, JV, MA Traylor & KC Parkes (1985) Range extensions for some Bolivian birds, 1. *Bull. Brit. Orn. Club* 105:124-130.

Remsen, JV, Jr, MA Traylor, Jr & KC Parkes (1986) Range extensions for some Bolivian birds, 2. *Bull. Brit. Orn. Club* 106:22-32.

Remsen, JV & MA Traylor (in press). *Checklist of the birds of Bolivia*. (Vermillion, South Dakota: Buteo Books ?).

Reynolds, PW (1932) Notes on the Birds of Woodcock and Snipe islands. *Ibis* 1932: 34-39.

Reynolds, PW (1934) Apuntes sobre aves de Tierra del Fuego. *Hornero* 5: 339-353.

Reynolds, PW (1935) Notes on the Birds of Cape Horn. *Ibis* 1935: 97-98.

Rhoads, SN (1912) Birds of the paramo of central Ecuador. *Auk* 29:141-149.

Ribera, MO & W Hanagarth (1982) Aves de la región altoandina de la Reserva Nacional de Ulla-Ulla. *Ecologia en Bolivia* 1:35-45.

Ridgely, RS (1976) *Birds of Panama*. Princeton, N.J.: Princeton University Press.

Ridgely, RS (1980) Notes on some rare or previously unrecorded birds in Ecuador. *Amer. Birds* 34:242-248

Ridgely, RS & SJC Gaulin (1980) The Birds of Finca Merenberg, Huila Department, Colombia. *Condor* 82:379-391

Ridgway, R (1901, 1902, 1904, 1907, 1911, 1914, 1916, 1919), and H Friedmann (1941, 1946, 1950) The Birds of North and Middle America. 11 vols. *Bull. U.S. Natn. Mus.* 50, parts 1-11.

Ripley, SD (1950) A small collection of birds from Argentine Tierra del Fuego. *Postilla* 3:1-11.

Robbins, MB, RS Ridgely, TS Schulenberg & FB Gill (1987) The avifauna of the Cordillera de Cutucú, Ecuador, with comparisons to other Andean localities. *Proc. Acad. Nat. Sci. Philadelphia* 139:243-259.

Rodríguez P & D Narváez (1978) *Panorama de la avifauna Colombiana*. Bogotá: Editorial Jeroglífico.

Roe, NA & WW Rees (1979) Notes on the puna avifauna of Azángaro Province, Department of Puno, southern Peru. *Auk* 96:475-482.

Romero Z, H (1977) Primer registro de quatro aves para Colombia. *Lozania (Acta Zool. Colombiana)* 25.

Romero Z, H (1978) Primer registro de doce aves para Colombia. *Lozania (Acta Zool. Colombiana)* 26.

Rylander MK (1983) Notes on the Birds of the Central Peruvian Puna. *Beitr. Vogelkd. Jena* 29:152-160.

Salvador, SA & S Narosky (1984) Nuevos nidos de aves argentinas *Muscisaxicola rufivertex, Catamenia inornata, Sicalis olivascens* y *Carduelis crassirostris*. *Hornero* 12(1983): 134-137.

Salvadori, T (1891) *Catalogue of the birds in the British Museum*, 20. Psittaci. London.

Salvadori, T (1893) *Catalogue of the birds in the British Museum*, 21. Columbae. London.

Salvadori, T (1895) *Catalogue of the birds in the British Museum*, 27. Crypturi and Ratitae. London.

Salvadori, T (1900) Contribuzione all'avifauna dell'America Australe (Patagonia, Terra del Fuego, Isola Degli Stati, Isole Falkland). *Annali Mus. Civ. Stor. Nat. Genova* ser. 2, 20:609-634.

Salvadori, T & E Festa (1899, 1900) Viaggo del Dr. Enrico Festa nell' Ecuador,1-3. *Boll. Mus. Anat. Comp. Torino* 15, nos. 357,362, 368.

Salvin, O (1874) A visit to the principal museums of the United States with notes on some of the birds contained therein. *Ibis* 1874: 305-329.

Salvin, O (1880) A list of birds collected by the late Henry Durnford during his last expedition to Tucumán and Salta. *Ibis* 1880: 351-364.

Salvin, O (1883) A List of the Birds collected by Captain A. H. Markham on the West Coast of America. *Proc. Zool. Soc. London*: 419-432.

Salvin, O (1895) On birds collected in Peru by Mr. O. T. Baron. *Novit. Zool.* 2:1-22.

Salvin, O & FD Godman (1879) On a collection of birds from the Sierra Nevada of Santa Marta, Columbia. *Ibis* 1879: 196-206.

Salvin, O & FD Godman (1880) On the birds of the Sierra Nevada of Santa Marta, Colombia. *Ibis* 1880: 114-125, 169-178.

Salvin, O & E Hartert (1892) *Catalogue of the birds of the British Museum*, 16. Upupae, Trochili, Cypselidae, Caprimulgidae, Podargidae, and Steatornithidae. London.

Saunders, H & O Salvin (1896) *Catalogue of the birds in the British Museum*, 25. Gaviae and Tubinares. London.

Schmitt, CG & DC Schmitt (1987) Extensions of range of some Bolivian birds. *Bull. Brit. Orn. Club* 107:129-134.

Schulenberg, RS (1986) Adiciones a la avifauna de Pampa Galeras. *Boletin de Lima* 48:89-90.

Schulenberg, TS (1987) New records of birds from western Peru. *Bull. Brit. Orn. Club* 107: 184-189.

Schulenberg, TS & TA Parker, III (1981) Status and distribution of some northwest Peruvian birds. *Condor* 83:209-216.

Schulenberg, TS & JV Remsen, Jr (1982) Eleven bird species new to Bolivia. *Bull. Brit. Orn. Club*

102:52-57.

Schulenberg, TS, SE Allen, DF Stotz, & DA Wiedenfeld (1984) Distributional records from the Cordillera Yanachaga, central Peru. *Gerfaut* 74:57-70.

Sclater, PL (1851) Ornithological observations. *Jardine's Contrib. Orn.*: 93-101,105-110.

Sclater, PL (1855) List of a collection of birds recieved by Mr. Gould from the province of Quijos in the republic of Ecuador. *Proc. Zool. Soc. London* 1854:109-115.

Sclater, PL (1858) Notes on a collection of birds recieved by M. Verreaux of Paris from the Rio Napo in the republic of Ecuador. *Proc. Zool. Soc. London* 1858: 59-77.

Sclater, PL (1859a) On some new or little-known birds from the Rio Napo. *Proc. Zool. Soc. London* 1858: 440-441.

Sclater, PL (1859b) List of birds collected by Mr. Louis Fraser, at Cuenca, Gualaquiza and Zamora, in the republic of Ecuador. *Proc. Zool. Soc. London* 1858:449-461.

Sclater, PL (1859c) On the birds collected by Mr. Fraser in the vicinity of Riobamba, in the republic of Ecuador. *Proc. Zool. Soc. London* 1858:549-556.

Sclater, PL (1859d) List of the first collection of birds made by Louis Fraser at Pallatanga, Ecuador, with notes and descriptions of new species. *Proc. Zool. Soc. London*: 135-147.

Sclater, PL (1860a) List of additional species of birds collected by Mr. Louis Fraser at Pallatanga, Ecuador, with notes and descriptions of new species. *Proc. Zool. Soc. London*: 63-73.

Sclater, PL (1860b) List of birds collected by Mr. Fraser in the vicinity of Quito, and during excursions to Pichincha and Chimborazo, with notes and descriptions of new species. *Proc. Zool. Soc. London* 1860: 73-83.

Sclater, PL (1860c) List of birds collected by Mr. Fraser in Ecuador, at Nanagal, Calacali, Perucho, and Puellaro, with notes and descriptions of new species. *Proc. Zool. Soc. London* 1860: 83-97.

Sclater, PL (1860d) List of birds collected by Mr. Fraser at Babahoyo in Ecuador, with descriptions of new species. *Proc. Zool. Soc. London* 1860: 272-290.

Sclater, PL (1867) Notes on the Birds of Chili. *Proc. Zool. Soc. London* 1867: 319-340.

Sclater, PL (1886a) *Catalogue of the birds in the British Museum*, 11. Passeriformes: Coerebidae, Tanagridae, and Icteridae. London.

Sclater, PL (1886b) List of a Collection of Birds from the Province of Tarapacá, Northern Chili. *Proc. Zool. Soc. London* 1886: 395-404.

Sclater, PL (1888) *Catalogue of the birds in the British Museum*, 14. Passeriformes: Tyrannidae, Oxyrhaphidae, Pipridae, Cotingidae, Phytotomidae, Philepittidae, Pittidae, Xenicidae, and Eurylaemidae. London.

Sclater, PL (1890) *Catalogue of the birds in the British Museum*, 15. Passeriformes: Dendrocolaptidae, Formicariidae, Conopophagidae, and Pteroptochidae. London.

Sclater, PL (1891) On a second Collection of Birds from the Province of Tarapacá, Northern Chili. *Proc. Zool. Soc. London* 1891: 131-137.

Sclater, PL & WH Hudson (1888-1889) *Argentine ornithology*. 2 vols. London: RH Porter.

Sclater, PL & O Salvin (1869) Second List of Birds collected, during the Survey of the Straits of Magellan, by Dr. Cunningham. *Ibis* 1869: 283-286.

Sclater, PL & O Salvin (1870) Third List of Birds collected, during the Survey of the Straits of Magellan, by Dr. Cunningham. *Ibis* 1870: 499-504.

Sclater, PL & O Salvin (1867-1876a) On Peruvian birds collected by Mr. Whitely. Parts 1-9. *Proc. Zool. Soc. London* 1867:982-991; 1868:173-178, 568-570; 1869:151-158, 596-602; 1873:184-187, 779-784; 1874:677-680; 1876:15-19.

Sclater, PL & O Salvin (1876b) On new species of Bolivian birds. *Proc. Zool. Soc. London* 1876: 352-358.

Sclater, PL & O Salvin (1878) Reports on the Collections of Birds made during the Voyage of H. M. S. 'Challenger,' - No. IX. On the Birds of Antarctic America. *Proc. Zool. Soc. London* 1878: 431-438.

Sclater, PL & O Salvin (1879a) On the birds collected in Bolivia by Mr. C. Buckley. *Proc. Zool. Soc. London* 1879: 588-645.

Sclater, PL & O Salvin (1879b) On the birds collected by the late Mr. T. K. Salmon in the state of Antioquia, United States of Colombia. *Proc. Zool. Soc. London* 1879: 486-550.

Sclater, PL & GE Shelley (1891) *Catalogue of the birds in the British Museum*, 19. Rhamphastidae, Galbulidae, Bucconidae, Indicatoridae, Capitonidae, Cuculidae, and Musophagidae. London.

Scott, DA & M Carbonell (1986) *A directory of Neotropical wetlands*. Cambridge - Slimbridge.

Scott, WED & RB Sharpe (1904-1915) *Reports of the Princeton University Expeditions to Patagonia, 1896-1899*. Vol. II, Ornithology, Part 1(1904): 1-112; Part 2(1910):113-344; Part 3(1912):345-

504; Part 4(1915):505-718.
Seebohm, H (1881) *Catalogue of the birds in the British Museum*, 5. Passeriformes: Turdidae. London.
Serna D, MA (1980) *Catálogo de Aves Museo de Historia Natural*. Medellín, Colombia: Colegio de San José.
Serrano, P & J Cabot (1982) Lista de las Aves de la Reserva Nacional de Ulla-Ulla con comentarios sobre su abundancia y distribución. Seria EE-42 INFOL, La Paz, Bol.
Sharpe, RB (1874) *Catalogue of the birds in the British Museum*, 1. Accipitres. London.
Sharpe, RB (1875) *Catalogue of the birds in the British Museum*, 2. Striges. London.
Sharpe, RB (1877) *Catalogue of the birds in the British Museum*, 3. Passeriformes: Corvidae, Paradiseidae, Oriolidae, Dicruridae, and Prionopidae. London.
Sharpe, RB (1881) Account of the Zoological Collections made during the Survey of H. M. S. 'Alert' in the Straits of Magellan and on the Coast of Patagonia. Aves. *Proc. Zool. Soc. London* 1881: 6-18.
Sharpe, RB (1885) *Catalogue of the birds in the British Museum*, 10. Passeriformes: Dicaeidae, Hirundinidae, Ampelidae, Mniotiltidae, and Motacillidae. London.
Sharpe, RB (1888) *Catalogue of the birds in the British Museum*, 12. Passeriformes: Fringillidae. London.
Sharpe, RB (1890) *Catalogue of the birds in the British Museum*, 13. Passeriformes: Artamidae, Sturnidae, Ploceidae, Alaudidae. Also Atrichidae and Menuridae. London.
Sharpe, RB (1894) *Catalogue of the birds in the British Museum*, 23. Rallidae and Heliornithidae. London.
Sharpe, RB (1896) *Catalogue of the birds in the British Museum*, 24. Limicolae. London.
Sharpe, RB & WR Ogilvie-Grant (1892) *Catalogue of the birds in the British Museum*, 17. Leptosomatidae, Coraciidae, Meropidae, Alcedinidae, Momotidae, Todidae, Coliidae, Bucerotes, and Trogones. London.
Sharpe, RB & WR Ogilvie-Grant (1898) *Catalogue of the birds in the British Museum*, 26. Plataleae, Herodiones, Steganopodes, Pygopodes, Alcae, and Impennes. London.
Short, LL (1975) A zoogeographic analysis of the South American chaco avifauna. *Bull. Amer. Mus. Nat. Hist.* 154:163-352.
Short, LL & JJ Morony, Jr. (1969) Notes on some birds of Central Peru. *Bull. Brit. Orn. Club* 89: 112-115.

Sibley CG & JE Ahlquist (1985a) Phylogeny and classification of New World suboscine passerine birds (Passeriformes: Oligomyodi: Tyrannides). Pp. 396-428 *in* PA Buckley *et al.*
Sibley, CG & JE Ahlquist (1985b) The phylogeny and classification of the passerine birds, based on comparisons of the genetic material, DNA. *Acta XVIII Congr. Int. Orn.*:83-121.
Sick, H (1984, 1985a) *Ornitologia Brasileira*. 2 vols. Brasilia: Editora Universidade de Brasilia.
Sick, H (1985b) Observations on the Andean-Patagonian component of southeastern Brazil's avifauna. Pp. 233-237 *in* PA Buckley *et al.*
Skutch, AF (1960) Life History of Central American Birds. Vol II. *Pacific Coast Avifauna* 34. Berkeley, Calif.: Cooper Orn. Soc.
Skutch, AF (1969) Life History of Central American Birds. Vol III. *Pacific Coast Avifauna* 35. Berkeley, Calif.: Cooper Orn. Soc.
Slud, P (1964) The birds of Costa Rica, distribution and ecology. *Amer. Mus. Nat. Hist. Bull.* 128.
Slud, P (1976) Geographic and climatic relationships of avifaunas with special reference to comparative distribution in the neotropics. *Smithsonian Contr. Zool.* No. 211.
Smyth, CH (1927, 1928) Descripción de una colección de huevos de aves argentinas, 1-2. *Hornero* 4:1-16, 125-152.
Snow, DW (1976) The relationship between climate and annual cycles in the Cotingidae. *Ibis* 118: 366-401.
Snyder, DE (1966) *The Birds of Guyana*. Salem, Mass.: Peabody Mus.
Steinbacher, J (1962) Beiträge zur Kenntnis der Vögel Paraguays. *Abh. Senckenber. Natur. Gesell.* 502:1-106.
Stephens, L & MA Traylor, Jr (1983) *Ornithological Gazetteer of Peru*. Cambridge, Mass.: Museum of Comparative Zoology.
Steullet, AB & EA Deautier (1935-46) *Catálogo systemático de aves de la República Argentina*. Buenos Aires: Obra del Cincuent. Mus. La Plata, Univ. Nac. La Plata.
Stiles, FG & PG Aquilar F (1985) *Primer symposio de ornitologia Neotropical*. As. Peruana para Conservacion de la Naturaleza, Lima, 126 pp.
Stone, W (1928) *Reports of the Princeton University Expeditions to Patagonia, 1896-1899*. Vol. II, Ornithology, Part 5:719-857. Princeton, New Jersey: Princeton University Press
Stresemann, E (1937a) Vögel von Monte Illiniza (Central-Ecuador). *Orn. Monatsb.* 45:13-15.
Stresemann, E (1937b, 1938) Ueber einige seltene

Vögel aus Ecuador, 1-2. *Orn. Monatsb.* 45:13-5; 46:115-8.

Sztolcman, J (1926) Revision des oiseaux néotropicaux de la collection du Musée Polonais d'Histoire Naturelle à Varsovie. 1. *Ann. Zool. Mus. Polon. Hist. Nat.* 5:197-235.

Taczanowski, L (1874) Liste des oiseaux recueillis par M. Constatin Jelski dans la partie centrale du Pérou occidental. *Proc. Zool. Soc. London*: 501-565.

Taczanowski, L (1877a) Liste des oiseaux recueillis au nord duPérou occidental par MM. Jelski et Stolzmann. *Proc. Zool. Soc. London*: 319-333.

Taczanowski, L (1877b) Supplément à la liste des oiseaux recueillis au nord du Pérou occidental par MM. Jelski et Stolzmann. *Proc. Zool. Soc. London*: 744-754.

Taczanowski, L (1879) Liste des oiseaux recueillis au nord du Pérou par MM. Stolzmann et Jelski en 1878. *Proc. Zool. Soc. London*: 220-245.

Taczanowski, L (1880) Liste des oiseaux recueillis au nord du Pérou par M. Stolzmann pendant les derniers mois de 1878 et dans les premiere moitié de 1879. *Proc. Zool. Soc. London*: 189-215.

Taczanowski, L (1882) Liste des oiseaux recueillis par M. Stolzmann au Pérou nord-oriental. *Proc. Zool. Soc. London*: 2-49.

Taczanowski, L (1884-86). *Ornithologie du Pérou.* 3 vols. Paris: Oberthur.

Taczanowski, L & H Berlepsch (1885) Troisième liste des oiseaux recueillis par M. Stolzmann dans l'Ecuadeur. *Proc. Zool. Soc. London*: 67-124.

Tallman, DA, TA Parker, GD Lester & RA Hughes (1978) Notes on two species of birds previously unreported from Peru. *Wilson Bull.* 90: 445-446.

Terborgh, J (1971) Distribution on environmental gradients: Theory and a preliminary interpretation of distributional pattern in the avifauna of Cordillera Vilcabamba, Peru. *Ecology* 52: 23-40.

Terborgh, JW (1977) Bird species diversity on an Andean elevation gradient. *Ecology* 58:1007-1019.

Terborgh, J & JS Weske (1969) Colonization of secondary habitats by Peruvian birds. *Ecology* 50:765-782.

Terborgh, JW & JS Weske (1975) The role of competition in the distribution of Andean birds. *Ecology* 56:562-576.

Todd, WEC (1942a) List of the tinamous in the collection of the Carnegie Museum. *Ann. Carnegie Mus.* 29:1-29.

Todd, WEC (1942b) List of the hummingbirds in the collection of the Carnegie Museum. *Ann. Carnegie Mus.* 29:271-370.

Todd, WEC (1943) Critical remarks on the trogons. *Proc. Biol. Soc. Washington* 56:3-15.

Todd, WEC & MA Carriker, Jr (1922) The birds of the Santa Marta region of Colombia: A study in altitudinal distribution. *Ann. Carnegie Mus.* 14:1-611.

Todd, WEC & MA Carriker, Jr (1927) Gnateaters and Antbirds from Tropical America with a revision of the genus *Myrmeciza* and its allies. *Proc. Biol. Soc. Washington* 40:149-178.

Tosi, JA (1960) *Zonas de Vida Natural en el Perú: Memoria Explicativa sobre el Mapa Ecológica del Perú.* Instituto Inter-americano de Ciencias Agrícolas de la OEA, Zona Andina, Proyecto 39, Programa de Cooperación Tecnica, Boletin Técnico No. 5. Washington, D.C.

Traylor, MA (1951) Notes on some Peruvian birds. *Fieldiana (Zool.)* 31: 613-21.

Traylor, MA (1952) Notes on birds from the Marcapata Valley, Cuzco, Peru. *Fieldiana (Zool.)* 34:17-23.

Traylor, MA (1958) Birds of northeastern Peru. *Fieldiana (Zool.)* 35:87-141.

Traylor, MA, ed. (1979) *Check-list of birds of the world*, 8. Cambridge, Mass.: Museum of Comparative Zoology.

Tschudi, JJ von (1844-1846) *Untersuchungen über die Fauna Peruana.* Ornithologie. St.Gallen, Switzerland: Scheitlin und Zollekofer.

Unzueta Q, O (1975) *Mapa ecologico de Bolivia. Memoria explicativativa.* La Paz.

Valencia, N & I Franke (1980) El bosque de Zárate y su conservación. *Boletin de Lima* 7/8:1-20

Varty, N, J Adams, P Espin & C J Hambler (1986) An ornithological survey of Lake Tota Colombia, 1982. *ICBP study rep.* 12.

Vaurie, C (1972) An ornithological gazetteer of Peru (based on information compiled by J.T. Zimmer). *Amer. Mus. Novit.* 2491.

Venegas, C & J Jory (1979) Guia de campo para las aves de Magallanes. *Publ. Inst. Pat. Punta Arenas. Monogr.* 11:1-253.

Venero G, JL & HP Brokaw (1980) Ornitofauna de Pampa Galeras, Ayacucho Peru. *Publ. Mus. Hist. Nat. 'Javier Prado' Ser. A* 26:1-32.

von Sneidern, K (1954) Notas sobre algunas aves del Museo de Historia Natural de la Universidad del Cauca, Popayán, Colombia. *Noved. Colombianas* 1:3-13.

von Sneidern, K (1955) Notas ornitológicas sobre la colección del Museo de Historia Natural de la Universidad del Cauca. *Noved. Colombianas* 2:35-44.

Vuilleumier, F (1967) Phyletic evolution in modern birds of the Patagonian forests. *Nature* 215:247-248.

Vuilleumier, F (1969) Field notes on some birds from the Bolivian Andes. *Ibis* 111:599-608.

Vuilleumier, F (1970) Insular biogeography in continental regions. I. The northern Andes of South America. *Amer. Nat.* 104:373-388.

Vuilleumier, F (1980) Speciation in Birds of the High Andes. *Acta XVII Congr. int. orn.*, Berlin 1978, Vol. 2: 1256-1261.

Vuilleumier, F (1984) Zoogeography of Andean birds: two major barriers; and speciation and taxonomy of the *Diglossa carbonaria* superspecies. *Natn. Geogr. Soc. Res. Rep.* 16:713-731.

Vuilleumier, F (1985) Forest birds of Patagonia: ecological geography, speciation, endemism, and faunal history. Pp. 255-304 *in* PA Buckley *et al.*

Vuilleumier, F (1988) Avian diversity in tropical ecosystems of South America and the design of National Parks. *Biota Bull.* 1:5-32.

Vuilleumier, F & DN Ewert (1978) The distribution of birds in Venezuelan páramos. *Bull. Am. Mus. nat. Hist.* 162:49-90.

Vuilleumier, F & M Monasterio, eds. (1986) *High altitude tropical biogeography*. New York-Oxford.

Vuilleumier, F & D Simberloff (1980) Ecology vs. history as determinants of patchy and insular distributions in high Andean birds. *Evolutionary Biology* 12:235-379.

Walker, B & DG Ricalde (1988) Aves de Machupicchu y alrededores. *Boletin de Lima* 58:69-79.

Waugh, E & F Lataste (1894) Quelques jours de chasse à Peñaflor durant les mois de janvier et de mars. *Act. Soc. Sci. Chili* 4, livr. 2:83-89.

Waugh, E & F Lataste (1895a) Une semaine de chasse, au mois de juin, dans la hacienda de San Alfonso (département de Quillota). *Act. Soc. Sci. Chili* 4, livr. 4, '1894':167-173.

Waugh, E & F Lataste (1895b) Addition à la liste des Oiseaux de Peñaflor. *Act. Soc. Sci. Chili* 5, livr. 1-3:59-60.

Weske, JS (1972) *The distribution of the avifauna in the Apurímac Valley of Peru with respect to environmental gradients, habitat, and related species.* Ph.D. thesis.

Norman, Oklahoma: University of Oklahoma.

Wetmore, A (1926a) Report on a collection of birds made by J. R. Pemberton in Patagonia. *Univ. California Publ. Zool.* 24:395-474.

Wetmore, A (1926b) Observations on the birds of Argentina, Paraguay, Uruguay, and Chile. *Bull. U.S. Natn. Mus.* 133:1-448.

Wetmore, A (1946) New birds from Colombia. *Smiths. Misc. Coll.* 106 (16):1-14

Wetmore, A (1953) Further additions to the avifauna of Colombia. *Noved. Colomb.* 2:45-47.

Wetmore, A (1965-1973) The birds of the Republic of Panama. Parts 1-3. *Smiths. Misc. Coll.* Vol 150, pts. 1-3.

Wetmore, A, R Pasquier, & S Olson (1984) *The Birds of the Republic of Panama*. Part 4. Washington, D. C.: Smithsonian Institution Press.

Whitely, H (1873) Notes on humming-birds collected in high Peru. *Proc. Zool. Soc. London*: 187-191, 784.

Wiedenfeld, DA, TS Schulenberg & MB Robbins (1985) Birds of a tropical deciduous forest in extreme northwestern Peru. Pp. 305-315 *in* PA Buckley *et al.*

Woods, RW (1988) *Guide to birds of the Falkland Islands*. P.O. Box 9, Oswestry, U.K.: Anthony Nelson Ltd.

Wyatt, CW (1871) Notes on some of the birds of the United States of Columbia. *Ibis*: 113-131, 319-335, 373-384.

Yarrell, W (1847) Descriptions of the Eggs of Some of the Birds of Chile. *Proc. Zool. Soc. London* 15:51-55.

Zimmer, JT (1930) Birds of the Marshall Field Peruvian expedition 1922-1923. *Field Mus. Nat. Hist., Zool. Ser.* 17: 231-493.

Zimmer, JT (1931-1955) Studies of Peruvian birds. *Amer. Mus. Novit.* 500, 509, 523, 524, 538, 545, 558, 584, 646, 647,668, 703, 728, 753, 756, 757, 785, 819, 860, 861, 862, 889, 893, 894, 917, 930,962, 963, 994, 1042, 1043, 1044, 1045, 1066, 1095, 1108, 1109, 1126, 1127, 1159, 1160, 1168, 1193, 1203, 1225, 1245, 1246, 1262, 1263, 1304, 1345, 1367, 1380, 1428, 1449, 1450, 1463, 1474, 1475, 1513, 1540, 1595, 1604, 1609, 1649, 1723.

Zotta, A (1940) Lista sobre el contenido estomacal de las aves argentinas. *Hornero* 7:402-411.

Addenda

The following data was recieved too late for inclusion in the main text, except for the alterations in the distribution maps. It consists mainly of data gathered by JF during field work in Peru from mid October to mid December 1989. Co-workers in the field were Thomas Valqui Haase, Barry P. Walker (**BPW**), Walter Wust, and Constantino Aucca Chutas. Some data were obtained from José Luis Venero (**LV**), whose bird collection in Calca (2926 m) in the Urubamba valley, Cuzco, was examined.

Taczanowski's Tinamou *Nothoprocta taczanowskii* (p. 58). In Bosque Ampay near Abancay, Apurímac, inhabits bushy areas and ecotones between forest and fields/grassland from 2700 to 4000 m.

Andean Tinamou *Nothoprocta pentlandii* (p. 60). Seen near Andamarca, c Ayacucho (ssp?).

Agami Heron *Agamia agami* (p. 78). Rec. at 2900 m in Urubamba valley of Cuzco (LV).

Bicolored Hawk *Accipiter bicolor* (p. 96). Remains (prey of *Bubo virginianus*) found at 4100 m near Chua E Abancay, Apurímac.

Mountain Caracara *Phalcoboenus megalopterus* (p. 108). Habits: may scrape large mounds of cameliid dung in search of insects, exactly like a chicken. Nestlings Dec (Apurímac).

Rosy-billed Pochard *Netta peposaca* (p. 131). Seen near Cuzco (D Ricalde).

Plumbeous Rail *Rallus sanguinolentus* (p. 143). Rec. near Andamarca in Ayacucho.

Paint-billed Crake *Neocrex erythrops* (p. 147). Rec. near Calca in Cuzco (LV).

Ochre-bellied Dove *Leptotila ochraceiventris*. A rare inhabitant of deciduous and humid forest of Pacific El Oro and Loja, sw Ecu. and Tumbes and Piura, nw Peru, mainly below 1800 m, but rec. to 2600 m at Ayabaca in Piura (C Clarke). 24 cm. Iris yellow (1 specimen). Crown tawny grading to whitish pink on forehead, nape with purple gloss, sides of head warm buff, back bronzy brown. Throat white, remaining underparts warm buff, upper breast tinged purplish. Two outer tail-feathers broadly white-tipped. Underwings chestnut as in congeners. Mainly encountered with larger numbers of White-tipped Dove (p. 197), but more partial to forest.

Scarlet-fronted and **Mitred Conures** *Aratinga wagleri* and *mitrata* (p. 202-3). Flocks of unidentified conures move up into the temperate zone of the valleys of c Peru seasonally, to feed on flowering *Erythrina* trees or ripe corn. BPW believes that Scarlet-fronted C. inhabits arid valleys, Mitred C. humid zones. This holds true near Abancay, Apurímac, but all specimens from the arid Urubamba valley (in Cuzco Univ. and collection of LV) are Mitred C. (apparently ssp *alticola*).

Rusty-faced Parrot *Hapalopsittaca amazonina*. Splitting in 3 species (Fuertes', Red-faced, and Rusty-faced P, see p. 213) and description of *H. amazonina velezi* from Nevado del Ruiz is published by Graves & Restrepo in *Wilson Bull.* 10 (1989):369-76.

Andean Potoo *Nyctibius (leucopterus) maculosus* (p. 232). Heard in Bosque Ampay near Abancay, Apurímac.

Common Potoo *Nyctibius griseus*. Much like Andean Potoo (p. 232), but no pale band in wing. Voice a series of slowly descending whistles. 2 rec. near Calca, Cuzco, shows casual wandering up the Urubamba valley (LV).

Hummingbird *Taphrospilus* sp. (p. 251). One specimen from Maraynioc in Junín (P Hocking).

Great Sapphirewing *Pterophanes cyanopterus* (p. 260). Ascends to 4500 m in semi-humid *Myrsinanthes*, *Escallonia*, and *Polylepis* woodlands in Apurímac. Here it feeds mainly from *Fuchsia* and *Salpichroa*.

Sword-billed Hummingbird *Ensifera ensifera* (p. 266). Ascends to 4100 m in Apurímac, Peru.

Olivaceous Thornbill *Chalcostigma olivaceum* (p. 286). 3 sightings 1989 in mts E Abancay, Apurímac.

Royal Cinclodes *Cinclodes excelsior aricomae*. Inhabits mts immediately E Abancay, Apurímac. Population estimate for this area 40 pairs. Always in shady places in dense, mossy *Polylepis* wood, seeking food by throwing aside moss and debris from the forest floor with such efficiency that it looks as if pigs have been rummaging through the habitat (drawing p. 338). More rarely flakes moss off rocks and thick branches. When alarmed flies up to a thick moss-clad branch and waits motionless. Calls *queet*. Song of possibly this form *tirrirrrrrr cucuqueeticucu*, occ. interspersed with *cu cu cu* notes..

White-browed Tit-spinetail *Leptasthenura xenothorax* (p. 348). Also inhabits mts immediately E Abancay, Apurímac.

Tawny Tit-spinetail *Leptasthenura yanacensis* (p. 350). In mts E Abancay, Apurímac.

Pale-tailed Canastero ssp *usheri* (p. 367). Fairly common in valley in Río Negro Mayo (Andamarca area) in Ayacucho. This area (and probably the adjacent Soras area) may be the most significant present tracts of fallow land with widespread occurrence of *Cylindropuntia* cacti, in which the birds nest.

Streak-throated Canastero *Asthenes humilis* (p. 378). Rec. on Nevado Ampay W and mts E Abancay, Apurímac.

Junín Canastero *Asthenes virgata* (p. 379). Observed near Lahuane E Abancay, Apurímac.

Russet-bellied Spinetail *Synallaxis zimmeri* (p. 354). Habitat dense shrubbery of *Croton*.

Chestnut Antpitta *Grallaria blakei* (p. 411). A specimen with barred belly is kept in University of Cuzco.

Andean tapaculo, unnamed ssp (p 440) occurs at 3000-4000 m in Bosque Ampay W Abancay, but at 4000-4600 m in mts E Abancay.

Ash-breasted Tit-tyrant *Anairetes alpinus* (p.468). Common in numerous small but dense *Polylepis* woods in mts E Abancay, Apurímac. Juv. with white wing-covert tips edged buff. Usually feeds among dense *Polylepis* foliage, in dense low regrowth as well as tall trees. Often sallies for passing insects, and occ. climbs on trunks. Voice much like other *Anairetes* species (*sensu stricto*). During interaction *titirri titirri titirrri* often culminating as a *trrrrrrrrrrr* (descending, sometimes with introductory notes *t t t*). Usual call a labored *hueee*. Juv. call repeatedly *trrrr* or *queeer*. Warning *wrr*; upon take-off often a *tuirrr*.

White-winged Diuca-finch *Diuca speculifera* (p. 658). In mts E Abancay, Apurímac.

White-throated Sierra-finch *Phrygilus erythronotus* (p. 665). Rec. from Pampa Galeras, Ayacucho, Peru (LV).

Black-crested Warbler

Index

Roman numerals refer to plates.

Aburria 138
aburri, Aburria 138
Accipiter 95
Accipitridae 91
Acestrura 296
Acropternis 444
(*Actitis*) 174
acuta, Anas 126, **VI** 3, **IX** 2ab
acuticauda, Aratinga 201, **XXIII** 5ab
acutipennis, Chordeiles 234, **XXVI** 6
acutipennis, Pseudocolopteryx 475, **XLVI** 21ab
(*acutirostris*), *Scytalopus* 428, (440)
adela, Oreotrochilus 255, **XXVII** 7ab
Adelomyia 251
aedon, Troglodytes 545, **IL** 10a-c
aegithaloides, Leptasthenura 348, **XXXV** 1a-c
Aegolius 231
aenea, Chloroceryle 303, **XXXI** 24ab
aeneocauda, Metallura 283, **XXIX** 9a-c
(*aequatorialis*), (*Ciccaba*) see *albogularis, Otus* 225
(*aequatorialis*), *Momotus* 303
(*aequatorialis*), *Rallus* 144
aequinoctialis, Geothlypis 585, **LII** 18ab
Aeronautes 239
(*aestiva*), *Dendroica* see *petechia* 582
affinis, Aythya 132, **VIII** 6ab
affinis, Lepidocolaptes 322, **XXXIII** 4a-c
(*affinis*), *Scytalopus* 441, **XLI** 1e
agami, Agamia 78, 846
Agamia 78
Agelaius 576
agilis, Anairetes 470, **XLVI** 7ab
agilis, Geothlypis 585, **LII** 17
Aglaeactis 257
Aglaiocercus 290
agraphia, Anairetes 469, **XLVI** 6a-c
Agriornis 504
Ajaja 84
ajaja, Ajaja 84
Alaudidae 527
alaudinus, Phrygilus 666, **LVIII** 10a-f
alba, Calidris 170, **XX** 25
(*alba*), (*Egretta*) 75
alba, Tyto 221, **XXVI** 1ab
albicapilla, Cranioleuca 360, **XXXVI** 1ab
(*albicauda*), *Agriornis* 506
albicaudatus, Buteo 103
albiceps, Cranioleuca 360, **XXXVI** 2a-c
albiceps, Elaenia 459, **XLVII** 13a-d
albicollis, Nyctidromus 235

albicollis, Saltator 644
albicollis, Scelorchilus 419, **XLI** 18
albicollis, Turdus 562
(*albicrissa*), *Heliomaster* see *longirostris* 292
albifrons, Conirostrum 598, **LIII** 7a-d
albifrons, Muscisaxicola 513, **XLIII** 4
albifrons, Myioborus 588, **LII** 5, **LXIV** 19
albigula, Buteo 102, **XII** 2ab
albigula, Upucerthia 331, **XXXIV** 12
albigularis, Sclerurus 395, **LXIII** 31
albilatera, Diglossa 605, **LIII** 16a-d
(*albilinea*), *Columba* 189, **XXII** 3ab
albilora, Muscisaxicola 511, **XLIII** 5
albitarsus, Ciccaba 228, **XXV** 6
(*albiventer*), *Phalacrocorax* 73
(*albivitta*), *Aulacorhynchus*, see *prasinus* 306
albocristata, Sericossypha 630, **LVI** 12ab
albofrenatus, Atlapetes 670, **LXI** 1
albogriseus, Pachyramphus 526
albogularis, (*Ciccaba*) see *Otus* 225
albogularis, Otus 225, **XXV** 4, **LXIV** 7
albogularis, Phalcoboenus 108, **XI** 3ab
albogularis, Pygarrhichas 395, **XXXIII** 16ab
(*albogularis*), *Schistes* see *geoffroyi* 291
albus, Casmerodius 75, **III** 3
Alcedinidae 301
(*alfredi*), *Psarocolius* see *angustifrons* 572
aliciae, Aglaeactis 258, **XXVIII** 20
(*aliciae*), *Catharus* see *minimus* 556
alinae, Eriocnemis 274
alnorum, Empidonax 490, **XLVI** 25
alpestris, Eremophila 527, **XLVIII** 18ab
alpina, Muscisaxicola 512, **XLIII** 3a-c
alpinus, Anairetes 468, **XLVI** 2, 847
(*alticola*), *Aratinga* 203
alticola, Charadrius 162, **XVIII** 6a-c, **XX** 10
alticola, Poospiza 684, **LXII** 15
(*altirostris*), *Scytalopus* 441, **XLI** 1fg
amaurochalinus, Turdus 561, **L** 12ab
Amazilia 249
amazilia, Amazilia 250
Amazon (see Parrot)
Amazona 215
amazona, Chloroceryle 302
amazonina, Hapalopsittaca 213, **XXIV** 3a-d, 846
ambiguus, Ramphastos 310
(*Amblycercus*) 574
americana, Anas 123, **IX** 11ab
americana, Fulica 154, **X** 2
americana, Mycteria 82
americanus, Coccyzus 217, **XIV** 13ab
amethysticollis, Heliangelus 269, **XXIX** 17a-d

Ammodramus 679
(*Amoropsittaca*) 209
Ampelioides 450
Ampelion 445
Anabacerthia 388
Anairetes 468
analis, Catamenia 648, **LIX** 5a-g
analis, Iridosornis 616, **LIV** 1
(*analoides*), *Catamenia* see *analis* 648
Anas 123
Anatidae 115
Anatini 122
andaecola, Upucerthia 334, **XXXIV** 15ab
andecola, Petrochelidon 533, **LXVIII** 11
andecolus, Aeronautes 240, **XLVIII** 3
andicola, Agriornis 505, **XLIII** 16
(*andicola*), *Bolborhynchus* 210
andicola, Grallaria 407, **XL** 6a-c
andicola, Leptasthenura 346, **XXXV** 2a-c, **LXIII** 12, **LXIV** 9
Andigena 308
andina, Gallinago 179, **XIX** 3ab, **XX** 17
andina, Recurvirostra 158, **XVIII** 1
andinus, Phoenicoparrus 86, **IV** 1a-d
andinus, Podiceps 69, **X** 4
(*andium*), *Anas* 124
(*angelae*), *Cyanolyca* see *viridicyana* 535
angustifrons, Psarocolius 572, **LI** 2
Anhinga 74
Anhinga 74
anhinga, Anhinga 74
Anhingidae 74
Ani, Greater 219, **XIV** 15
   Groove-billed 220
   Smooth-billed 219, **XIV** 14ab
ani, Crotophaga 219, **XIV** 14ab
Anisognathus 618
(*annae*), *Ocreatus* 276
Anser 134
anser, Anser 134
ANSERIFORMES 115
antarctica, Geositta 327, **XXXIV** 4
(*antarcticus*), *Rallus* 145, **XV** 6
Antbird, Gray-headed 402
   Long-tailed 400
Anthocephala 250
anthoides, Asthenes 377, **XXXVII** 16ab
(*Anthoscenus*) see *Heliomaster* 292
(*anthracinus*), *Turdus* see *chiguanco* 558
Anthus 563
antisianus, Pharomachrus 299
antisiensis, Cranioleuca 358, **XXXVI** 6
(*antisiensis*), *Pharomachrus* see *antisianus* 299
Antpitta, Bay 410, **XL** 10
   (Bay-backed) see Bay, Rusty-tinged,
      and White-bellied
   Bicolored 407, **XL** 9
   Brown-banded 413, **XL** 5
   Chestnut 411, **XL** 7, 847
   Chestnut-crowned 406, **XL** 14ab

   Chestnut-naped 408, **XXXIX** 10a-c
   Crescent-faced 416, **XL** 3
   Giant 405, **XXXIX** 6
   Gray-naped 411, **XXXIX** 8ab, **LXIV** 27
   Hooded 417
   Ochre-breasted 414, **XL** 4ab
   Pale-billed 408, **XXXIX** 11ab
   (Puno) see Stripe-headed
   Red-and-white 410, **XL** 13
   Rufous 412, **XL** 8a-d
   Rufous-faced 412, **XXXIX** 12
   Rusty-breasted 415, **XL** 1ab
   Rusty-tinged 410, **XL** 12ab
   Santa Marta 407
   Scaled 405, **XXXIX** 9ab
   Slate-crowned 416, **XL** 2
   Stripe-headed 407, **XL** 6a-c
   Tawny 413, **XL** 15a-c
   Undulated 404, **XXXIX** 7ab
   (Watkins') 406
   White-bellied 409, **XL** 11
   (Yellow-breasted) 409
Antshrike, (Marcapata) 400
   Rufous-capped 399, **XXXIX** 2a-e
   Uniform 397, **XXXIX** 3ab
   Variable 398, **XXXIX** 1ab
Antthrush, Barred 403, **XXXIX** 4a-c
   Rufous-breasted 403, **XXXIX** 5
   Rufous-tailed 403
Antwren, Rufous-rumped 400
Aphrastura 344
(*Apocryptornis*) see
   Grallaricula
Apodidae 237
APODIFORMES 237
apolinari, Cistothorus 542, **X** 12ab
aquila, Eutoxeres 244
Ara 200
Aratinga 201
araucana, Columba 190, **XXII** 4ab
(*Archiplanus*) see *Cacicus*
arcuata, Pipreola 449, **XLII** 11a-c
Ardea 81
Ardeidae 75
Ardeola 77
ardesiaca, Conopophaga 417, **XLI** 12ab
ardesiaca, Fulica 152, **XVI** 6a-e
Arenaria 167
(*arequipae*), *Asthenes* 367-8
(*argentea*), *Tangara* see *viridicollis* 615
argyrofenges, Tangara 616
argyrotis, Penelope 137, **XIV** 3, **LXIII** 7
(*aricomae*), *Cinclodes* 338, **XXXV** 11b
armata, Merganetta 121, **VII** 1a-h, **VIII** 11ab
(*armillata*), *Cyanolyca* 536
armillata, Fulica 150, **XVI** 5a-c
Arremon 679
arremonops, Oreothraupis 669, **LXI** 14
Asio 229
(*assimilis*), *Atlapetes* see *torquatus* 678

Asthenes 364
astreans, Acestrura 297
atacamensis, Cinclodes 342, **XXXV** 15ab
(aterrima), Diglossa see *humeralis* 604
aterrimus, Knipolegus 517, **XLV** 12a-f
Athene 230
Atlapetes 670
atrata, Carduelis 693, **LXII** 3ab
atratus, Coragyps 88, **XI** 7ab
(atratus), Scytalopus 430, **XLI** 9
atriceps, Phalacrocorax 73, **VIII** 16
atriceps, Phrygilus 661, **LVIII** 4a-c
atricilla, Larus 184
atricollis, Colaptes 315, **XXXIII** 3ab
atrifrons, Odontophorus 141
(atrigularis), Metallura see *williami* 281
(atrocyaneum), Conirostrum see *albifrons* 598
atropileus, Hemispingus 634, **LVI** 24ab
atrovirens, Psarocolius 572, **LI** 3
Attagis 181
augusti, Phaethornis 243
Aulacorhynchus 306
aura, Cathartes 89, **XI** 8ab
aurantiirostris, Catharus 554, **L** 5
aurantiirostris, Saltator 643, **LVII** 4a-e
(aureata), (Tanagra) see *musica, Euphonia* 608
aureliae, Haplophaedia 276
aureodorsalis, Buthraupis 622, **LIV** 12
aureoventris, Chlorostilbon 248, **XXXI** 8
aureoventris, Pheucticus 646, **LVII** 35
auriceps, Pharomachrus 300
auriculata, Zenaida 191, **XXII** 1a-d
aurifrons, Bolborhynchus 209, **XXIV** 8a-e
auriventris, Sicalis 653, **LX** 3ab
australis, Phalcoboenus 109
Automolus 390
autumnalis, Dendrocygna 116
Avocet, Andean 158, **XVIII** 1
Avocetbill, Mountain 290, **XXIX** 12
aymara, Bolborhynchus 209, **XXIV** 10
aymara, Metriopelia 196, **XXII** 8ab
Aythya 132
Aythyini 131
azarae, Synallaxis 352, **XXXVI** 14

badius, Molothrus 571, **LI** 11
baeri, Poospiza 688, **LXII** 19ab
bahamensis, Anas 127, **VI** 6, **IX** 3
(bairdi), Myioborus see *melanocephalus* 587
bairdii, Calidris 169, **XIX** 12, **XX** 28
balliviani, Odontophorus 142, **XVII** 2
Bananaquit 595, **LXIV** 21
bangsi, Grallaria 407
barbata, Carduelis 694, **LXII** 5ab
barbata, Penelope 138, **XIV** 2ab
Barbet, Toucan 305, **XXXI** 26
Barbtail, Rusty-winged 385, **XXXIII** 11ab
  Spotted 386, **XXXIII** 14ab, **LXIII** 13
(baritula), Diglossa 601
baroni, Cranioleuca 358, **XXXVI** 5ab

baroni, Metallura 282, **XXIX** 6ab
(barrali), Heliangelus 270
Bartramia 174
Basileuterus 590
basilicus, Basileuterus 592, **LXIII** 24
Becard, Barred 525, **XLVI** 17ab
  Black-and-white 526
  Crested 526, **XLVI** 28ab
bellicosa, Sturnella 579, **LI** 16ab
(Belonopterus) 159
berlepschi, Asthenes 370, **XXXVII** 5
berlepschi, Thripophaga 381, **XXXVIII** 4ab
bicolor, Accipiter 96, **XIII** 6a-d, 846
bicolor, Dendrocygna 116, **VIII** 10
bicolor, Tachycineta 535
Bittern, Least 80, **XI** 9ab
  Pinnated 80, **III** 7
Blackbird, Austral 574, **LI** 7ab
  Bolivian 575, **LI** 5
  (Melodious) 574
  (Red-breasted) 578
  Scrub 574, **LI** 6
  White-browed 578, **LI** 15ab
  Yellow-hooded 577, **X** 6a-c
  Yellow-winged 576, **LI** 12a-c
Black-tyrant, White-winged 517, **XLV** 12a-f
blakei, Grallaria 411, **XL** 7, 847
Blossomcrown 250
Bobolink 580, **LI** 9a-d
Bobwhite, Crested 140, **XVII** 1a-c
bogotensis, Anthus 565, **XLIII** 17ab, **LXIV** 23
Boissonneaua 267
boissonneautii, Pseudocolaptes 386, **XXXIII** 15a-c
Bolborhynchus 209
boliviana, Poospiza 683, **LXII** 11
bolivianus, Oreopsar 575, **LI** 5
(bolivianus), Scytalopus 431, **XLI** 8
bolivianus, Zimmerius 457, **XLVII** 21ab
bombus, Acestrura 297
Bombycilla 563
Bombycillidae 563
bonapartei, Coeligena 263, **XXVIII** 13a-c, 14
bonapartei, Nothocercus 56, **I** 6ab
bonariensis, Molothrus 570, **LI** 10a-d
bonariensis, Thraupis 624, **LV** 14a-d
borealis, Contopus 487, **XLV** 3
Botaurus 80
bougueri, Urochroa 252
(bourcieri), Geotrygon see *frenata* 199
brachyura, Synallaxis 353
brachyurus, Buteo 102, **XII** 3ab
brachyurus, Idiopsar 659, **LIX** 12ab
branickii, Leptosittaca 203, **XXIII** 3
branickii, Odontorchilus 539, **IL** 1
(branickii), Theristicus see *melanopis* 83
(brasilianum), Glaucidium 227
(brevipennis), Polystictus 475
Brilliant, Fawn-breasted 252
  Green-crowned 252
  Violet-fronted 252

Brotogeris 212
(brunneifrons), Ochthoeca see *fumicolor* 496
brunneinucha, Atlapetes 677, **LXI** 3ab
(brunneitorques), Cypseloides see *rutilus* 239
brunneiventris, Diglossa 603, **LIII** 13ab
brunnescens, Premnoplex 386, **XXXIII** 14ab, **LXIII** 13
brunniceps, Myioborus 589, **LII** 9
Brush-finch, (Bay-crowned) 677
    Chestnut-capped 677, **LXI** 3ab
    Dusky-headed 673
    Fulvous-headed 674, **LXI** 11ab
    Moustached 670, **LXI** 1
    Ochre-breasted 673, **LXI** 7ab
    Pale-naped 670, **LXI** 2ab
    Rufous-eared 674, **LXI** 12ab
    Rufous-naped 671, **LXI** 6a-g
    Rusty-bellied 676, **LXI** 9a-d
    Santa Marta 672, **LXIII** 33
    Slaty 676, **LXI** 13a-c, **LXIV** 33
    Stripe-headed 678, **LXI** 15a-e, **LXIII** 32
    Tricolored 672, **LXI** 5
    White-winged 675, **LXI** 8a-c
    White-rimmed 679, **LXI** 4
    Yellow-striped 674, **LXI** 10
(Buarremon) see *Atlapetes*
Bubo 226
(Bubulcus) 76
Bucconidae 304
(buckleyi), Columbina 193
budytoides, Stigmatura 468, **XLVII** 8
buffoni, Circus 95
burmeisteri, Microstilbon 294, **XXX** 13, **XXXI** 17
Bush-tanager, Ashy-throated 632
    Black-backed 633, **LVI** 13
    Common 630, **LVI** 10a-d, **LXIV** 14
    (Dusky) 632
    Dusky-bellied 632, **LVI** 11
    Gray-hooded 633, **LVI** 4ab
    Short-billed 631
    (Yellow-whiskered) 632
Bush-tyrant, (Jelski's) 517
    Red-rumped 499, **XLIV** 4
    Rufous-bellied 502, **XLIV** 5
    Santa Marta 501, **LXIII** 9
    Smoky 501, **XLIV** 3ab
    Streak-throated 500, **XLIV** 2ab
Buteo 100
Buthraupis 620
(Butorides) 77
Buzzard-eagle, Black-chested 98, **XI** 6a-c

(cabanisi), Knipolegus 517
Cacicus 573
Cacique, Mountain 573, **LI** 9ab
    Yellow-billed 573, **LI** 8
cactorum, Asthenes 373, **XXXVII** 8
cactorum, Melanerpes 312
(Caenotriccus) see *Pseudotriccus*
caerulea, Egretta 77, **III** 4ab
caerulescens, Diglossopis 599, **LIII** 18a-c
caerulescens, Sporophila 648, **LIX** 4a-c
caerulescens, Thamnophilus 398, **XXXIX** 1ab
caesar, Poospiza 688, **LXII** 16
Cairina 121
Cairinini 121
(cajabambae), Leptasthenura 347
Calidrinidae, 167
Calidris 168
californicus, Lophortyx 140
callinota, Terenura 400
calliparaea, Chlorochrysa 610
(Calliphlox) see *Philodice*
calliptera, Pyrrhura 207, **XXIII** 7
calophrys, Hemispingus 635, **LVI** 20
Campephilus 317
campestris, Colaptes 316
Camptostoma 458
Campylopterus 245
Campylorhamphus 323
(cana), Thraupis see *episcopus* 626
canadensis, Wilsonia 586
Canastero, Austral 377, **XXXVII** 16ab
    Berlepsch's 370, **XXXVII** 5
    Cactus 373, **XXXVII** 8
    Canyon 373, **XXXVII** 1
    Chestnut 370, **XXXVII** 7
    Cordilleran 372, **XXXVII** 9a-c
    Córdoba 377, **XXXVII** 15
    Creamy-breasted 365, **XXXVII** 6a-d
    (Dark-winged) 367, **XXXVII** 6ab
    Dusky-tailed 371, **XXXVII** 10ab
    Junín 379, **XXXVII** 18, 847
    Lesser 365, **XXXVII** 4ab
    Line-fronted 380, **XXXVII** 20ab
    Many-striped 380, **XXXVII** 19a-c
    Maquis 374, **XXXVII** 3
    (Pale-tailed) 366, **XXXVII** 6cd, 847
    Patagonian 371, **XXXVII** 11
    Puno 376, **XXXVII** 14ab
    Rusty-fronted 374, **XXXVII** 2
    (Rusty-vented) 369, **XXXVII** 6ef
    Scribble-tailed 378, **XXXVII** 17ab
    Streak-backed 375, **XXXVII** 13ab, **LXIII** 17
    Streak-throated 378, **XXXVII** 12ab, 847
(candidissima), Egretta see *thula* 77
candidus, Melanerpes 312
canigularis, Chlorospingus 632
(canus), Scytalopus 443, **XLI** 1a
(Canutus) see *Calidris*
canutus, Calidris 168
(Capella) 177
capensis, Zonotrichia 682, **LVII** 13a-d
capistrata, Muscisaxicola 509, **XLIII** 9ab
capitalis, Grallaria 410, **Pl.XL** 10
(capitalis), (Spinus) see *magellanica, Carduelis* 691
Capitonidae 304
Caprimulgidae 233
CAPRIMULGIFORMES 232
Caprimulgus 235
Caracara, Carunculated 107, **XI** 5a-c

Crested 109, **XI** 1ab
Chimango 110, **XI** 4
Mountain 108, **XI** 2ab, 846
Striated 109
White-throated 108, **XI** 3ab
Yellow-headed 110
carbonaria, Diglossa 604, **LIII** 12
carbonarius, Phrygilus 667, **LVIII** 11a-c
Carduelis 690
caripensis, Steatornis 232
caroli, Polyonymus 279, **XXX** 16ab
carolina, Porzana 146, **XV** 7
(carolinensis), Anas 124
carolinensis, Caprimulgus 235
carrikeri, Grallaria 408, **XXXIX** 11ab
carunculatus, Phalcoboenus 107, **XI** 5a-c
Casmerodius 75
castanea, Dendroica 584, **LII** 10a-c
castanea, Hapaloptila 304, **XXXI** 27
(castaneiceps), Basileuterus see *coronatus* 593
castaneoventris, Dubusia 623, **LIV** 15
(castaneus), Pteroptochos 418
(castanoptera), Pyriglena see *leuconota* 401
castelnaudii, Aglaeactis 258, **XXVIII** 19
(Catamblyrhynchidae) 642
Catamblyrhynchus 641
Catamenia 648
Cathartes 89
Cathartidae 88
Catharus 553
Catoptrophorus 174
caudata, Drymophila 400
caudatus, Theristicus 83
cayana, Piaya 218, **XIV** 10
(cayennensis), Vanellus 159
ceciliae, Metriopelia 195, **XXII** 9a-c
cedrorum, Bombycilla 563, **LXII** 1
(celicae), Atlapetes see *nationi* 676
(Centropelma) 67
cephalotes, Myiarchus 521, **XLVI** 31
(Cerchneis) see *Falco*
certhioides, Upucerthia 335, **XXXIV** 17ab
cerulea, Dendroica 583, **LII** 13ab
Ceryle 302
Chachalaca, Speckled, 139
(Chaemepelia) see *Columbina*
Chaetocercus 298, see also *Acestrura*
Chaetura 239
chalcopterus, Pionus 215
Chalcostigma 285
Chamaepetes 138
Chamaeza 402
CHARADRIIFORMES 155
Charadriidae 158
Charadrius 162
Chat-tyrant, Brown-backed 496, **XLIV** 7, **LXIV** 11
Crowned 495, **XLIV** 12a-c, 14
d'Orbigny's 497, **XLIV** 6ab
Golden-browed 494, **XLIV** 13ab
(Jelski's) 495, **XLIV** 14

Patagonian 497, **XLIV** 16
Piura 498, **XLIV** 11
Rufous-breasted 495, **XLIV** 10ab, **LXIII** 11
Slaty-backed 493, **XLIV** 9a-c, **LXIV** 10
White-browed 498, **XLIV** 8a-c
Yellow-bellied 494, **XLIV** 15a-c
(cheriway), Polyborus see *plancus* 109
chiguanco, Turdus 558, **L** 15a-c
chihi, Plegadis 82, **IV** 5
(chilensis), Accipiter 96, **XIII** 6a-d
chilensis, Phoenicopterus 85, **IV** 3a-d
(chilensis), Rollandia 65
chilensis, Vanellus 159, **XVIII** 3ab
Chilia, Crag 337, **XXXV** 8ab
Chilia 336
chimachima, Milvago 110
chimango, Milvago 110, **XI** 4
chimborazo, Oreotrochilus 253, **XXVII** 4a-e
chionogaster, Amazilia 249, **XXXI** 2ab
(chiriri), Brotogeris 212
(chivi), Vireo 569
(Chirocylla) 451
Chloephaga 118
Chloroceryle 302
Chlorochrysa 610
(Chloronerpes) see *Piculus*
Chlorophonia, Blue-naped 608, **LVI** 3a-c
Chestnut-breasted 608, **LVI** 2ab
Chlorophonia 608
(chloropoda), Phaetusa see *simplex* 187
chloropus, Gallinula 149, **XVI** 7a-e
Chlorornis 639
Chlorospingus 630
Chlorostilbon 247
choliba, Otus 223, **XXV** 5a-d
Chondrohierax 92
Chordeiles 234
chrysater, Icterus 577, **LI** 13
chrysocephalus, Myiodynastes 522, **XLV** 14ab
(chrysogaster), Pheucticus 645
(chrysonotus), Cacicus 573
chrysopeplus, Pheucticus 645, **LVII** 5a-c, **LXIV** 3
chrysops, Cyanocorax 537
(chrysops), Zimmerius 458
chrysoptera, Vermivora 581, **LII** 14
(Chrysoptilus) 315
Chubbia) 175, 176
Chuck-will's-widow 235
Ciccaba 228
Ciconia 82
Ciconiidae 81
CICONIIFORMES 75
Cinclidae 537
Cinclodes, Bar-winged 341, **XXXV** 13a-d
Dark-bellied 338, **XXXV** 10ab
Gray-flanked 339, **XXXV** 9ab
Olrog's 339, **XXXV** 12
(Royal) 338, **XXXV** 11b, 847
Sierran 340, **XXXV** 14
Stout-billed 337, **XXXV** 11ab

White-bellied 342, **XXXV** 15ab
White-winged 342, **XXXV** 16
Cinclodes 337
Cinclus 537
cinctus, Saltator 644, **LVII** 2
cinerea, Muscisaxicola 512, **XLIII** 2ab
cinerea, Poospiza 685, **LXII** 14
cinerea, Serpophaga 467, **XLVII** 6ab
cinereiceps, Phyllomyias 455, **XLVII** 16
cinereum, Conirostrum 595, **LIII** 5ab
cinereus, Circus 95, **XXI** 1a-d
cinereus, Contopus 489, **XLV** 7
cinnamomea, Pyrrhomyias 486, **XLVI** 20ab, **LXIII** 12
cinnamomeiventris, Ochthoeca 493, **XLIV** 9a-c, **LXIV** 10
(cinnamomeus), Furnarius see *leucopus* 343
Cinnycerthia 539
Circus 94
(cissiurus), Ocreatus see *underwoodi* 276
Cistothorus 540
citrea, Protonotaria 585
citrina, Sicalis 651, **LX** 2ab
citrinellus, Atlapetes 674, **LXI** 10
Claravis 194
(clarissae), Heliangelus 270
(claudia), Heliangelus 270
clypeata, Anas 130, **IX** 13ab
Cnemarcus 499
Cnemoscopus 633
Cnemotriccus 490
Coccyzus 217
Cochlearius 79
cochlearius, Cochlearius 79, **III** 6
cocoi, Ardea 81, **III** 5
(coelestis), Thraupis see *episcopus* 626
Coeligena 261
coeligena, Coeligena 261
Coereba 595
Coerebidae 594
coeruleicinctus, Aulacorhynchus 306, **XXXII** 1
Colaptes 315
Colibri 245
Colinus 140
collaris, Charadrius 163, **XX** 2
(colombiana), Merganetta 121
colombianus, Neocrex 147, **XV** 9
(colombianus), Otus 225
(Colorhamphus) 498
Columba 189
columbarius, Falco 112, **XIII** 8ab
columbianus, Odontophorus 141
(columbianus), Philydor see *rufus* 390
Columbidae 189
COLUMBIFORMES 189
(Columbigallina) 192
Columbina 192
(Colymbus) see *Tachybaptus* and *Podiceps*
comechingonus, Cinclodes 340, **XXXV** 14
Comet, Bronze-tailed 279, **XXX** 16ab
  Gray-breasted 290, **XXX** 15ab
  Red-tailed 278, **XXX** 17a-c

(Compsocoma) 620
(Compsospiza) 688
(Compsothlypis) see *Parula*
concolor, Xenospingus 680
condamini, Eutoxeres 244, **XXX** 1
Condor, Andean 90, **XI** 9a-c
Conebill, Blue-backed 597, **LIII** 2ab
  Capped 598, **LIII** 7a-d
  Cinereous 595, **LIII** 5ab
  Giant 598, **LIII** 1a-d
  Rufous-browed 597, **LIII** 3
  Tamarugo 596, **LIII** 6ab
  White-browed 596, **LIII** 4
Conirostrum 595
Conopophaga 417
(Conopophagidae) 417
conoveri, Leptotila 197
Contopus 487
Conure, Austral 207, **XXIII** 11
  Blue-crowned 201, **XXIII** 5ab
  (Brown-breasted) 207, **XXIII** 7
  Flame-winged 207, **XXIII** 7
  Golden-plumed 203, **XXIII** 3
  Green-cheeked 205, **XXIII** 9
  Maroon-tailed 206, **XXIII** 6a-c
  Mitred 203, **XXIII** 1a-c, 846
  Rose-crowned 207, **LXIV** 1
  Santa Marta 206, **LXIII** 1
  Scarlet-fronted 202, **XXIII** 2a-d, 846
  Slender-billed 208, **XXIII** 10
(conspicillatus), Basileuterus 594
conspicillatus, Forpus 212
cooperi, Accipiter 97, **XIII** 5
Coot, American 154, **X** 2
  Andean 152, **XVI** 6a-e
  Giant 151, **XVI** 1a-c
  Horned 152, **XVI** 2
  Red-fronted 150, **XVI** 3ab
  Red-gartered 150, **XVI** 5a-c
  (Slate-colored) 152
  White-winged 153, **XVI** 4a-c
cora, Thaumastura 293, **XXX** 9
CORACIIFORMES 299
Coragyps 88
(corallinus), Pionus see *sordidus* 214
Cormorant, Neotropic 72, **VIII** 15a-c
  (Olivaceous) 72, **VIII** 15a-c
correndera, Anthus 565, **XLVIII** 14a-c
cornuta, Fulica 152, **XVI** 2
coronatus, Basileuterus 593, **LII** 24ab
Coronet, Buff-tailed 267, **XXXI** 13ab
  Chestnut-breasted 267, **XXXI** 12
coruscans, Colibri 246, **XXXI** 22ab
Corvidae 535
coryi, Schizoeaca 362, **LXIV** 32
Coscoroba 117
coscoroba, Coscoroba 117, **VIII** 1ab
(costaricensis), Grallaricula see
  flavirostris 414
Cotinga, Bay-vented 447, **XLII** 7ab

Chestnut-crested 446, **XLII** 6
Red-crested 445, **XLII** 4a-c
White-cheeked 446, **XLII** 5
Cotingidae 444
courseni, *Synallaxis* 351, **XXXVI** 13
Cowbird, Bay-winged 571, **LI** 11
  Screaming 572
  Shiny 570, **LI** 10a-d
Cracidae 135
Crake, Black 146, **XV** 3a-d
  Colombian 147, **XV** 9
  Paint-billed 147, **XV** 8, 846
Cranioleuca 355
crassirostris, *Carduelis* 690, **LXII** 8a-f
crassirostris, *Geositta* 329, **XXXIV** 2
crecca, *Anas* 124, **IX** 7
Crescentchest, Olive-crowned 421, **XLI** 14
Crested-tinamou, Elegant 63, **I** 4ab
Creurgops 628
crinitus, *Myiarchus* 521, **XLVI** 30
(crissalis), *Pheucticus* see *aureoventris* 646
cristatus, *Colinus* 140, **XVII** 1a-c
(Crocethia) 170
Crotophaga 219
cruziana, *Columbina* 194, **XXII** 12
Crypturellus 57
Cuckoo, Black-billed 217
  Dark-billed 218, **XIV** 12
  Dwarf 217
  Gray-capped 218
  Guira 220, **XIV** 9
  Squirrel 218, **XIV** 10
  Striped 220, **XIV** 11
  Yellow-billed 217, **XIV** 13ab
Cuculidae 217
CUCULIFORMES 217
cucullata, *Andigena* 309, **XXXII** 8
(cucullata), *Buthraupis* see *montana* 620
cucullata, *Grallaricula* 417
cunicularia, *Athene* 230, **XXV** 11a-c
cunicularia, *Geositta* 328, **XXXIV** 3a-d
cupreoventris, *Eriocnemis* 273, **XXVIII** 10
cupripennis, *Aglaeactis* 257, **XXVIII** 18a-c
Curaeus 574
curaeus, *Curaeus* 574, **LI** 7ab
curtata, *Cranioleuca* 356, **XXXVI** 7ab
curvirostris, *Nothoprocta* 61, **I** 9ab
cyanea, *Chlorophonia* 608, **LVI** 3a-c, **LXIII** 20
cyanea, *Diglossopis* 600, **LIII** 17ab
cyaneus, *Circus* 94, **XXI** 2ab
cyanocephala, *Thraupis* 624, **LV** 13
Cyanocorax 536
(cyanolaemus), *Aulacorhynchus* see *prasinus* 306
(Cyanolesbia) see *Aglaiocercus*
cyanoleuca, *Notiochelidon* 531, **XLVIII** 8a-c
Cyanoliseus 204
Cyanolyca 535
cyanoptera, *Anas* 129, **VII** 2a-d, **IX** 9ab, **X** 3
cyanopterus, *Pterophanes* 260, **XXVII** 2ab, 846
cyanotis, *Tangara* 611, **LV** 6ab

(cyanotus), *Colibri* 246
Cyclarhis 567
Cygnus 117, 134
Cyphorhinus 548
Cypseloides 239

dabbenei, *Penelope* 136, **XIV** 5
Dacnis, Tit-like 606, **LIII** 8a-e
(Dafila) see *Anas*
(Darter) 74
Darters 74
(darwinii), *Thraupis* see *bonariensis* 624
darwinii, *Nothura* 61, **I** 2
(decipiens), *Leptotila* 197
(decolor), *Leptotila* 197
decumanus, *Psarocolius* 572, **LI** 1
deiroleucus, *Falco* 112
(delicata), *Gallinago* 178, **XIX** 1
(Delothraupis) 623
Dendrocincla 320
Dendrocolaptes 321
Dendrocolaptidae 319
(Dendrocopos) 312
Dendrocygna 116
Dendroica 582
dentata, *Creurgops* 628, **LVI** 9ab
derbianus, *Aulacorhynchus* 307, **XXXII** 2
derbyi, *Eriocnemis* 275, **XXVIII** 1ab
desmursii, *Sylviorthorhynchus* 343, **XXXVI** 18ab
diadema, *Catamblyrhynchus* 641, **LIX** 1ab
diadema, *Ochthoeca* 494, **XLIV** 15a-c
Diglossa 601, see also *Diglossopis*
Diglossopis 599
dignus, *Veniliornis* 313
(Diphlogaena) 266
(Diphogena) see *(Diphlogaena)*
Dipper, Rufous-throated 538, **IL** 18
  White-capped 537, **IL** 19a-c, **LXIII** 27
(discolor), *Dendrocygna* see *autumnalis* 116
discors, *Anas* 128, **VII** 3ab, **IX** 8ab
(dispar), *Chloephaga* 118
Diuca 658
diuca, *Diuca* 658, **LIX** 13ab
Diuca-finch, Common 658, **LIX** 13ab
  White-winged 658, **LIX** 14ab, 847
Diucon, Fire-eyed 502, **XLIII** 14
Dives 574
(dives), *Dives* 574
Dolichonyx 580
(Doliornis) 447
domesticus, *Passer* 695, **LVII** 12ab
dominica, *Oxyura* 133, **VII** 6ab, **VIII** 12
dominica, *Pluvialis* 161, **XVIII** 10, **XX** 7
dominicanus, *Larus* 186, **XXI** 4ab
dominicus, *Tachybaptus* 67, **II** 1a-c
Doradito, Subtropical 475, **XLVI** 21ab
  Warbling 476, **XLVI** 22
dorbignyi, *Asthenes* 365, (369), **XXXVII** 6a-f
dorsale, *Ramphomicron* 280, **LXIII** 3ab
dorsalis, *Automolus* 390, **XXXVIII** 11

dorsalis, Mimus 551, **IL** 16ab
dorsalis, Phacellodomus 383, **XXXVIII** 1
dorsalis, Phrygilus 664, **LVIII** 7ab
Doryfera 244
Dotterel, Rufous-chested 163, **XVIII** 7a-d, **XX** 5
   Tawny-throated 164, **XVII** 8a-c, **XX** 4
Dove, Eared 191, **XXII** 1a-d
   Gray-headed 198
   Large-tailed 198, **XXII** 5
   Ochre-bellied 846
   Tolima 197
   White-tipped 197, **XXII** 6ab
Dowitcher, Short-billed 170
(dresseri), Atlapetes see *leucopterus* 675
dryas, Catharus 553, **L** 2
Drymophila 400
(dubius), Heliangelus 270
Dubusia 623
Duck, Black-bellied Whistling 116
   (Bronze-winged) 126
   Comb 121, **VIII** 8a-c
   Crested 125, **VI** 1a-d, **VIII** 4
   Fulvous Whistling 116, **VIII** 10
   Lake 134, **VII** 8ab
   Masked 133, **VII** 6ab, **VIII** 12
   Muscovy 121, 135
   (Peposaca) 131, **VII** 4ab, **VIII** 10
   Ruddy 133, **VII** 7a-d, **VIII** 13ab, **X** 5ab
   Spectacled 126, **VI** 2ab, **VIII** 3
   Torrent 121, **VII** 1a-h, **VIII** 11ab
   White-faced Whistling 116
dumetaria, Upucerthia 330, **XXXIV** 18a-c
dumicola, Polioptila 566
(dyselius), Eriocnemis 273

Eagle, Black-and-chestnut 98, **XII** 9ab
   Solitary 97
Earthcreeper, Band-tailed 336, **XXXIV** 19
   Bolivian 335, **XXXIV** 16
   Buff-breasted 332, **XXXIV** 10
   Chaco 335, **XXXIV** 17ab
   Plain-breasted 332, **XXXIV** 11ab
   Rock 334, **XXXIV** 15ab
   Scale-throated 330, **XXXIV** 18a-c
   Straight-billed 334, **XXXIV** 14
   Striated 333, **XXXIV** 13
   White-throated 331, **XXXIV** 12
Egret, Cattle 76, **III** 1
   Great White 75, **III** 3
   Snowy 77, **III** 2
Egretta 75
eisenmanni, Thryothorus 543, **IL** 2
Elaenia, Highland 462, **XLVII** 12
   Mountain 461, **XLVII** 14
   (Peruvian) 460
   Sierran 462, **XLVII** 11
   Slaty 461, **XLVII** 9ab
   Small-billed 460, **XLVII** 10
   White-crested 459, **XLVII** 13a-d
   Yellow-bellied 461

Elaenia 459
Elanoides 93
Elanus 93
elegans, Eudromia 63, **I** 4ab
(elegans), Progne 529
elegantior, Synallaxis 351, **XXXVI** 15a-c
Emberizoides 683
Embernagra 689
Emerald, Blue-tailed 247, **XXXI** 5
   Coppery 248
   Glittering-bellied 248, **XXXI** 8
   (Green-and-white) 249, **XXXI** 3
   Narrow-tailed 249, **XXXI** 6a-c
   (Rufous-tailed) 250
   Short-tailed 248
   (Steely-vented) 250, **XXXI** 4
   (White-bellied) 249, **XXXI** 2ab
(Empidochanes) see *Knipolegus*
Empidonax 490
Empidonomus 523
Enicognathus 207
Ensifera 266
ensifera, Ensifera 266, **XXVII** 3ab, 847
Entomodestes 552
(eos), Coeligena 263, **XXVIII** 14
episcopus, Thraupis 626
(equifasciatus), Veniliornis see *nigriceps* 312
Eremobius 336
Eremophila 527
(Ereunetes) see *Calidris*
Eriocnemis 271
(Erismatura) 133
(erythrogaster), Hirundo see *rustica* 532
erythroleuca, Grallaria 410, **XL** 13
(erythromelas), Piranga see *olivacea* 627
(erythronemius), Accipiter 96
erythronotus, Phrygilus 665, **LVIII** 8ab, 847
erythrophrys, Poospiza 687, **LXII** 12
erythrophthalma, Netta 131, **VII** 4a-c, **VIII** 7ab
erythrophthalmus, Coccyzus 217
erythrops, Cranioleuca 357, **XXXVI** 8
erythrops, Neocrex 147, **XV** 8, 846
erythropygius, Cnemarcus 499, **XLIV** 4
erythrotis, Grallaria 412, **XXXIX** 8
estella, Oreotrochilus 253, **XXVII** 5a-h
(Euchlornis) see *Pipreola*
Eudocimus 82
Eudromia 63
(Eudromias) 164
Eugralla 423
Eulidia 296
(euophrys), Basileuterus 591
euophrys, Thryothorus 543, **IL** 3a-d
(Eupelia) 194
Euphonia, Blue-hooded 608, **LVI** 1a-d
Euphonia 608
eupogon, Metallura 283, **XXIX** 8ab
euryptera, Opisthoprora 290, **XXIX** 12
(euryzonus), Turdus see *fulviventris* 560
(Euscarthmus) see *Hemitriccus*

Eutoxeres 244
excelsior, Cinclodes 337, **XXXV** 11ab, 847
excelsior, (Geositta) see *Cinclodes*
excelsior, (Upucerthia) see *Cinclodes*
exilis, Ixobrychus 80, **XI** 9ab
eximia, Buthraupis 622, **LIV** 8ab
exortis, Heliangelus 270, **XXIX** 18a-d
falcatus, Campylopterus 245, **XXXI** 1a-c
falcklandii, Turdus 561, **L** 11ab
Falco 111
Falconidae 106
FALCONIFORMES 88
Falconinae 111
Falcon, Aplomado 112, **XIII** 10a-c
  Orange-breasted 112
  Peregrine 113, **XIII** 9a-e
  (Kleinschmidt's) 114
  (Pallid) 114
falklandicus, Charadrius 163, **XVIII** 5ab, **XX** 9
fanny, Myrtis 294, **XXX** 12ab, **XXXI** 19
fasciata, Columba 189, **XXII** 3ab
fasciatum, Tigrisoma 79, **III** 8
fasciatus, Myiophobus 486
femoralis, Falco 112, **XIII** 10a-c
femoralis, Scytalopus 429, **XLI** 7ab, 8, 9, **LXIII** 29
ferruginea, Hirundinea 519, **XLIV** 1
(ferruginea), Oxyura 133
ferrugineifrons, Bolborhynchus 211, **XXIV** 11
ferrugineipectus, Grallaricula 415, **XL** 1ab
ferrugineiventre, Conirostrum 596, **LIII** 4
ferrugineus, Enicognathus 207, **XXIII** 11
Finch, Black-throated 667 **LVIII** 13ab
  Gray-crested 669, **LVII** 11ab
  Plush-capped 641, **LIX** 1ab
  Saffron 656, **LX** 4a-f
  Short-tailed 659, **LIX** 12ab
  Slaty 668, **LVII** 1ab
  Slender-billed 680
  Yellow-bridled 668, **LVIII** 12ab
Finfoots 154
Firecrown, Green-backed 266, **XXVIII** 5a-c
Fire-eye, White-backed 401
(fissirostris), Geositta 328
Flamingo, Andean 86, **IV** 1a-d
  Chilean 85, **IV** 3a-d
  (James') 86, **IV** 2a-c
  Puna 86, **IV** 2a-c
flammeus, Asio 229, **XXVI** 2ab
flammulata, Asthenes 380, **XXXVII** 19a-c
flammulatus, Thripadectes 392, **XXXVIII** 14, **LXIV** 8
flava, Piranga 626, **LV** 17ab
flaveola, Coereba 595, **LXIV** 21
flaveola, Sicalis 656, **LX** 4a-f
flavescens, Boissonneaua 267, **XXXI** 13ab
flavicans, Myiophobus 483, **XLVI** 17ab
flavinucha, Muscisaxicola 513, **XLIV** 11ab
flavinuchus, Anisognathus 620, **LIV** 6ab
(flavipes), Hylophilus 569
flavipes, Notiochelidon 531, **XLVIII** 10
flavipes, Platycichla 556, **L** 7ab

flavipes, Tringa 172, **XIX** 9, **XX** 16
flavirostris, Anairetes 472, **XLVI** 3ab
flavirostris, Anas 124, **VI** 5a-d, **IX** 6
flavirostris, Arremon 679
flavirostris, Grallaricula 414, **XL** 4ab
flavirostris, Porphyrula 148
flavogaster, Elaenia 461
(flavotincta), Grallaria 409
flaviventris, Pseudocolopteryx 476, **XLVI** 22
flavivertex, Myioborus 589, **LXIII** 25
(flavoviridis), Vireo see
olivaceus 568
Flicker, Andean 315, **XXXIII** 2ab
  Black-necked 315, **XXXIII** 3ab
  Campo 316
  Chilean 315, **XXXIII** 1ab
floriceps, Anthocephala 250
(Florida) 77
Flower-piercer, Black 604, **LIII** 14a-c, **LXIII** 22
  Black-throated 603, **LIII** 13ab
  Bluish 599, **LIII** 18a-c
  (Carbonated) 604
  Chestnut-bellied 601, **LIII** 10
  Deep-blue 600
  Glossy 602, **LIII** 11ab
  Gray-bellied 604, **LIII** 12
  Masked 600, **LIII** 17ab
  Mérida 604, **LXIV** 22
  Moustached 602, **LIII** 9a-d
  Rusty 601, **LIII** 15a-d
  (Slaty) 601
  White-sided 605, **LIII** 16a-d
fluviatilis, Muscisaxicola 508
Flycatcher, Acadian 490, **XLVI** 24
  Alder 490, **XLVI** 25
  Bran-colored 486
  Cinnamon 486, **XLVI** 20ab, **LXIII** 12
  Cliff 519, **XLIV** 1
  Dusky-capped 520, **XLVI** 29
  Flavescent 483, **XLVI** 17ab
  Fork-tailed 524, **XLV** 1ab
  Fuscous 490
  Great-crested 521, **XLVI** 30
  Golden-crowned 522, **XLV** 14ab
  Handsome 485, **XLV** 15ab
  Inca 478, **XLVII** 23
  McConnell's 477
  Ochraceous-breasted 485, **XLVI** 16ab
  Olive-sided 477, **XLV** 3
  Olive-striped 477, **XLVII** 25
  Orange-banded 484, **XLVI** 19
  Pale-edged 521, **XLVI** 31
  Rufous-breasted 478, **XLVII** 24
  Slaty-capped 479
  Streak-necked 477, **XLVII** 26a-c
  Sulphur-bellied 523
  (Traill's) 490
  Unadorned 484, **XLVI** 18
  Variegated 523

Vermilion 492, **XLV** 4a-c
Foliage-gleaner, Buff-browed 387, **XXXVIII** 9a-c
   Buff-fronted 390, **XXXVIII** 12ab
   Crested 390, **XXXVIII** 11
   Lineated 388, **XXXVIII** 10ab
   Montane 389, **XXXVIII** 6a-c
   Ruddy 391, **LXIII** 15
   Rufous-necked 391, **XXXVIII** 8
   Scaly-throated 389, **XXXVIII** 5ab
forficatus, Elanoides 93, **XIII** 1
Formicariidae 397
Formicarius 403
formicivorus, Melanerpes 311, **XXXII** 12
Forpus 212
(*franklini*), *Larus* see *pipixcan* 184
frantzii, Elaenia 461, **XLVII** 14
(*fraseri*), *Conirostrum* see *cinereum* 595
fraseri, Oreomanes 598, **LIII** 1a-d
frenata, Geotrygon 199, **XXII** 14ab
Fringillidae 642
frontalis, Hemispingus 637, **LVI** 17ab, **LXIV** 15
frontalis, Muscisaxicola 514, **XLIII** 7
frontalis, Ochthoeca 495, **XLIV** 12a-c, 14
frontalis, Synallaxis 352
(*frontata*), *Aratinga* see *wagleri* 202
Fruitcrow, Red-ruffed 451
Fruiteater, Band-tailed 449, **XLII** 10a-c
   Barred 449, **XLII** 11a-c
   Green-and-black 448, **XLII** 9a-e
   Scaled 450
fruticeti, Phrygilus 662, **LVIII** 9a-d
(*furcata*), *Cranioleuca* 357
(*fuertesi*), *Hapalopsittaca* 214
fulgidus, Pharomachrus 300
Fulica 149
fulica, Heliornis 154
fuliginiceps, Leptasthenura 349, **XXXV** 6ab
fuliginosa, Schizoeaca 362, **XXXVI** 19bc
(*fulgidigula*), *Coeligena* see *torquata* 261
(*fulva*), *Cinnycerthia* 540
fulviceps, Atlapetes 674, **LXI** 11ab
fulviventris, Turdus 560, **L** 14
fumicolor, Ochthoeca 496, **XLIV** 7ab, **LXIV** 11
fumigatus, Contopus 487, **XLV** 10a-c
fumigatus, Myiotheretes 501, **XLIV** 3ab
fumigatus, Veniliornis 313, **XXXII** 11ab
furcatus, Anthus 563, **XLVIII** 15
Furnariidae 323
Furnarius 343
fusca, Dendroica 583, **LII** 11ab
fuscater, Catharus 554, **L** 1ab
fuscater, Turdus 557, **L** 16a-g, **LXIII** 20
fuscatus, Cnemotriccus 490
(*fuscescens*), (*Buteo*) 98
fuscescens, Catharus 556
(*fuscicauda*), *Scytalopus* 443, **LXIV** 31
fuscicollis, Calidris 168, **XIX** 13, **XX** 24
(*fuscocaerulescens*), *Falco* see *femoralis* 112
fuscocinereus, Lipaugus 450, **XLII** 8
fuscoolivaceus, Atlapetes 673

fuscorufa, Synallaxis 353, **LXIII** 17
fuscorufus, Myiotheretes 502, **XLIV** 5
fuscus, Cinclodes 341, **XXXV** 13a-d, **LXIV** 24
(*fuscus*), *Scytalopus* 438, **XLI** 1m
fuscus, Teledromas 421, **XLI** 15

galbula, Icterus 577, **LI** 14a-c
gallardoi, Podiceps 70, **II** 5ab
GALLIFORMES 135
Gallinago 175
gallinago, Gallinago 178, **XIX** 1
Gallinula 148
Gallinule, Azure 148
   Common 149, **XVI** 7a-e
   Purple 147
   Spot-flanked 148, **X** 8ab, **XV** 10ab
Gallito, Crested 420, **XL** 16
   Sandy 421, **XLI** 15
garleppi, Poospiza 689, **LXII** 18ab
gayi, Attagis 181, **XVII** 7a-c
gayi, Phrygilus 660, **LVIII** 2a-c
genibarbis, Thryothorus 544, **IL** 4
geoffroyi, Schistes 291, **XXVIII** 6ab
georgica, Anas 126, **VI** 4a-c, **IX** 1ab, **X** 1
Geositta 324
Geothlypis 585
Geotrygon 198
Geranoaetus 98
gibsoni, Chlorostilbon see *mellisugus* 247
gigantea, Fulica 151, **XVI** 1a-c
gigantea, Grallaria 405, **XXXIX** 6
gigas, Patagona 255, **XXVII** 1a-c
gilvus, Mimus 549
gilvus, Vireo 569, **LII** 2
(*Gisella*) see *Aegolius*
glauca, Diglossopis 600
Glaucidium 226
Glaucis 242
glaucopoides, Eriocnemis 275, **XXVIII** 2ab
gloriosa, Diglossa 604, **LXIV** 22
gloriossissima, Diglossa 601, **LIII** 10
(*glyceria*), (*Zodalia*) 277
Gnatcatcher, Masked 566
Gnateater, Slaty 417, **XLI** 12ab
godini, Eriocnemis 273, **XXVIII** 4ab
Godwit, Hudsonian 175, **XX** 13
goeringi, Hemispingus 638, **LXIV** 17
Goldfinch, (Dark-backed) 695
   Lesser 694, **LXII** 1a-c
Goose, Andean 118, **V** 4a-c
   Ashy-headed 119, **V** 2a-d
   (Graylag) 134
   Magellan 118, **V** 3a-f
   Ruddy-headed 120, **V** 1ab
   Orinoco 120
   (Upland) 118, **V** 3a-f
goudotii, Chamaepetes 138, **XIV** 1
Grackle, (Colombian) 576
   Mountain 575, **LI** 4
   Red-bellied 575

Grallaria 404
Grallaricula 414
(graminicola), Asthenes see *wyatti* 375
(granadense), (Idioptilon) 482
granadensis, Hemitriccus 482, **XLVI** 9a-c
granadensis, Picumnus 311
Grass-finch, Wedge-tailed 683, **LVII** 8
grayi, Hylocharis 249
Grebe, (Black-necked) 69
  Colombian 69, **X** 4
  Hooded 70, **II** 5ab
  Least 67, **II** 1a-c
  Great 68, **VIII** 14a-c
  Junín Flightless 70, **II** 7a-c
  Pied-billed 67, **II** 2a-c
  (Puna) 70, **II** 7a-c
  (Short-winged) 66, **II** 4a-d
  Silvery 69, **II** 6a-d
  Titicaca Flightless 66, **II** 4a-d
  White-tufted 65, **II** 3a-c
Greenlet, Olivaceous 569, **LII** 1 (Scrub) 569
griseicapillus, Sittasomus 320, **XXXIII** 8
(griseiceps), Catharus 555
griseiceps, Myrmeciza 402
(griseicollis), Scytalopus 443, **XLI** 1bc
(griseigularis), Elaenia see *albiceps* 459
griseiventris, Taphrolesbia 290, **XXX** 15ab
griseocristatus, Lophospingus 669, **LVII** 11ab
griseonucha, Grallaria 411, **XXXIX** 8ab, **LXIV** 27
griseomurina, Schizoeaca 362, **XXXVI** 19d
griseus, Limnodromus 170
griseus, Nyctibius 846
Grosbeak, Black-backed 646, **LVII** 10a-f, **LXIV** 35
  (Golden-bellied) 645
  Rose-breasted 646, **LVII** 6ab
  (Southern Yellow) 645
  Yellow 645, **LVII** 5a-c, **LXIV** 3
Ground-dove, (Bare-eyed) 195, **XXII** 10ab
  Bare-faced 195, **XXII** 9a-c
  Black-winged 196, **XXII** 7
  Common 192, **XXII** 11
  Croaking 194, **XXII** 12
  (Ecuadorean) 193
  Golden-spotted 196, **XXII** 8ab
  Maroon-chested 194, **XXIV** 13ab
  Moreno's Bare-faced 195, **XXII** 10ab
  Picui 193
  Plain-breasted 193
  Ruddy 193
  (Scaly-breasted) 192, **XXII** 11
Ground-tyrant, Black-fronted 514, **XLIII** 7
  Cinereous 512, **XLIII** 2ab
  Cinnamon-bellied 509, **XLIII** 9ab
  Dark-faced 509, **XLIII** 8
  Little 508
  Ochre-naped 513, **XLIII** 11ab
  Plain-capped 512, **XLIII** 3a-c
  Puna 511, **XLIII** 6
  Spot-billed 508, **XLIII** 1ab
  Rufous-naped 510, **XLIII** 9ab

  White-browed 511, **XLIII** 5
  White-fronted 513, **XLIII** 4
GRUIFORMES 143
gryphus, Vultur 90, **XI** 9a-c
Guan, Andean 136, **XIV** 4a-c
  Band-tailed 137, **XIV** 3, **LXIII** 7
  Bearded 138, **XIV** 2ab
  Crested 136
  Red-faced 136, **XIV** 5
  Sickle-winged 138, **XIV** 1
  Wattled 138
(Guara) see *Eudocimus*
guatimalensis, Grallaria 405, **XXXIX** 9ab
guerinii, Oxypogon 288, **XXX** 21a-e, **LXIII** 4, **LXIV** 3
Guira 220
guira, Guira 220, **XIV** 9
gujanensis, Cyclarhis 567, **LII** 4ab
gularis, Hellmayrea 355, **XXXVI** 10ab, **LXIV** 26
Gull, Andean 186, **XXI** 8a-e
  Brown-hooded 185, **XXI** 5a-c
  Franklin's 184, **XXI** 3ab
  Gray 185
  Kelp 186, **XXI** 4ab
  Laughing 184
guttata, Ortalis 139
guttuligera, Premnornis 385, **XXXIII** 11ab
guy, Phaethornis 243, **XXX** 3
haemastica, Limosa 175, **XX** 13
Haematopodidae 155
Haematopus 156
haematopygius, Aulacorhynchus 307
haliaetus, Pandion 91, **XII** 5
Hapalopsittaca 212
Hapaloptila 304
Haplophaedia 276
Haplospiza 668
Harpyhaliaetus 97
Harrier, Cinereous 95, **XXI** 1a-d
  (Hen) 94, **XXI** 2ab
  Long-winged 95
  Northern 94, **XXI** 2ab
harrisii, Aegolius 231, **XXV** 9
harterti, Schizoeaca 364, **XXXVI** 19jk
harterti, Upucerthia 335, **XXXIV** 16
Hawk, Bay-winged 106
  Broad-winged 101, **XII** 4ab
  Bicolored 96, **XIII** 6a-d, 846
  Cooper's 97, **XIII** 5
  (Harris') 106
  Puna 104, **XII** 8a-c
  Red-backed 103, **XII** 7a-c
  Roadside 100, **XIII** 4ab
  Rufous-tailed 106, **XII** 6
  Sharp-shinned 96, **XIII** 7a-g
  Short-tailed 102, **XII** 3ab
  Swainson's 103, **XII** 10
  (Variable) 104, **XII** 8a-c
  White-rumped 101, **XII** 1a-c
  White-tailed 103

White-throated 102, **XII** 2ab
heinei, Tangara 614, **LV** 5ab
(*Heliodytes*) see *Campylorhynchus*
Heliangelus 268
(*Helianthea*) see *Coeligena*
helianthea, Coeligena 263, **XXVIII** 22ab
(*Heliochera*) see *Ampelion*
heliodor, Acestrura 297, **XXX** 10, **XXXI** 16
Heliodoxa 252
Heliomaster 292
Heliornis 154
Heliornithidae 154
helleri, Schizoeaca 363, **XXXVI** 19hi
Hellmayrea 355
hellmayri, Anthus 564, **XLVII** 16ab
hellmayri, Cranioleuca 356, **LXIII** 26
hellmayri, Mecocerculus 465, **XLVII** 2
Helmetcrest, Bearded 288, **XXX** 21a-e, **LXIII** 4, **LXIV** 3
Hemispingus, Black-capped 634, **LVI** 24ab
    Black-eared 637, **LVI** 23
    Black-headed 638, **LVI** 15
    Drab 639, **LVI** 16
    Gray-capped 636, **LXIV** 18
    Oleaginous 637, **LVI** 17ab, **LXIV** 15
    Orange-browed 635, **LVI** 20
    Parodi's 635, **LVI** 19
    Rufous-browed 634, **LVI** 22
    Slaty-backed 638, **LXIV** 17
    Superciliaried 636, **LVI** 18a-e, **LXIV** 16
    Three-striped 639, **LVI** 21
Hemispingus 634
Hemitriccus 482
Henicorhina 547
(*hepatica*), *Piranga* 627
herbicola, Emberizoides 683, **LVII** 8
Hermit, Green 243, **XXX** 3
    Planalto 242, **XXX** 2
    Rufous-breasted 242
    Sooty-capped 243
    Tawny-bellied 243, **XXX** 4
    White-whiskered 243
herodias, Ardea 81
Heron, Agami 78, 846
    Boat-billed 79, **III** 6
    (Chestnut-bellied) 78
    Great Blue 81
    Green 78
    Little Blue 77, **III** 4ab
    Night 78, **III** 9a-c
    Striated 77, **III** 10ab
    White-necked 81, **III** 5
herrani, Chalcostigma 288, **XXIX** 16a-c
(*hesperus*), *Coeligena* see *iris* 265
heteropogon, Chalcostigma 287, **XXIX** 15a-c
(*Heteroscelus*) 173
heterura, Asthenes 374, **XXXVII** 3
(*hiaticula*), *Charadrius* 162
Hillstar, Andean 253, **XXVII** 5a-h
    Black-breasted 254, **XXVII** 6ab

    Chimborazo 253, **XXVII** 4a-e
    Wedge-tailed 255, **XXVII** 7ab
    White-sided 254, **XXVII** 8ab
    White-tailed 252
Himantopus 157
(*himantopus*), *Himantopus* 157
himantopus, Micropalama 170, **XX** 26
hirsuta, Glaucis 242
hirundinacea, Sterna 187, **XXI** 7a-c
Hirundinea 519
Hirundinidae 528
Hirundo 532
holosericeus, Cacicus 573, **LI** 8
holostictus, Thripadectes 393, **XXXVIII** 15
homochroa, Catamenia 650, **LIX** 6a-e, **LXIII** 19
Honeycreeper, Golden-collared 606, **LIV** 11ab
Hornero, Pale-legged 343
    Rufous 343, **XXXIV** 20
huallagae, Aulacorhynchus 307, **XXXII** 5
(*huancavelicae*), *Asthenes* 367
(*hudsonicus*), *Numenius* see *phaeopus* 175
(*hudsonius*), *Circus* 94, **XXI** 2ab
Huet-huet 418, **XL** 17ab
humeralis, Diglossa 604, **LIII** 14a-c, **LXIII** 22
humicola, Asthenes 371, **XXXVII** 10ab
humilis, Asthenes 378, **XXXVII** 12ab, 847
Hummingbird, Amazilia 250
    Giant 255, **XXVII** 1a-c
    Green-and-white 249, **XXXI** 3
    Oasis 293, **XXX** 11ab, **XXXI** 21
    Rufous-tailed 250
    Speckled 251, **XXXI** 14ab
    Steely-vented 250, **XXXI** 4
    Sword-billed 266, **XXVII** 3ab, 847
    Wedge-billed 291, **XXVIII** 6ab
    White-bellied 249, **XXXI** 2ab
Hylocharis 249
(*Hylocichla*) see *Catharus*
Hylophilus 569
Hymenops 518
hyperythrus, Odontophorus 141
hypochondria, Poospiza 684, **LXII** 10ab
hypochondriacus, Siptornopsis 361, **XXXVII** 21
hypoglauca, Andigena 308, **XXXII** 6
hypoleuca, Grallaria 409, **XL** 11
Hypopyrrhus 575
(*Hypoxanthus*) see *Piculus*

ibis, Egretta 76, **III** 1
Ibis, (Bare-faced) 82
    Black-faced 83, **IV** 4a-d
    Buff-necked 83
    Puna 83, **IV** 6a-c
    Scarlet 82
    Whispering 82
    White-faced 82, **IV** 5
(*Ibycter*) see *Phalcoboenus*
Icteridae 570
icterocephalus, Agelaius 577, **X** 6a-c
icterophrys, Satrapa 519, **XLV** 5ab

icterotis, Ognorhynchus 204, **XXIII** 4
Icterus 577
Ictinia 94
Idiopsar 659
(Idioptilon) 482
(Idiospiza) see *Catamenia*
igniventris, Anisognathus 618, **LIV** 7ab
ignobilis, Turdus 561
imperialis, Gallinago 176, **XIX** 7
Inca, Black 261, **XXVIII** 16
   Bronzy 261
   Collared 261, **XXVIII** 15a-d, **LXIV** 5
(inca), Coeligena 261
Inca-finch, Buff-bridled 682, **LIX** 8ab
   Great 680, **LIX** 9ab
   Gray-winged 680, **LIX** 11ab
   Rufous-backed 681, **LIX** 10ab
incana, Tringa 173, **XX** 21
Incaspiza 680
infuscatus, Phimosus 82
ingens, Otus 224, **XXV** 2a-c
ingoufi, Tinamotis 64, **I** 16ab
(inornatus), (Buarremon) see *brunneinucha, Atlapetes* 677
inornatus, Catamenia 649, **LIX** 7a-c
inornatus, Myiophobus 484, **XLVI** 18
(insignis), (Tanagra) see *musica, Euphonia* 608
intermedia, Pipreola 449, **XLII** 10a-c
interpres, Arenaria 167, **XX** 1
(iolata), Colibri see *coruscans* 246
(Ionornis) see *Porphyrula*
iracunda, Metallura 285, **XXIX** 1
Iridophanes 606
(Iridoprocne) 535
Iridosornis 616
iris, Coeligena 265, **XXVIII** 11a-e
(isaacsonii), Eriocnemis 275
isabellina, Geositta 325, **XXXIV** 5
isidorii, Oroaetus 98, **XII** 9ab
Ixobrychus 80

Jabiru 82
Jabiru 82
Jacana 155
jacana, Jacana 155, **XV** 1ab
Jacana, (Northern) 155
   Wattled 155, **XV** 1ab
Jacanidae 155
jacula, Heliodoxa 252
(jamaicensis), Buteo 106
jamaicensis, Laterallus 146, **XV** 3a-d
jamaicensis, Oxyura 133, **VII** 7a-d, **VIII** 13ab, **X** 5ab
jamesi, Phoenicoparrus 86, **IV** 2a-c
(jamesoni), Gallinago 175, **XIX** 6ab
jardinii, Glaucidium 226, **XXV** 10a-c
(jardinii), (Spodiornis) see
   rustica, Haplospiza 668
Jay, (Black-collared) 536
   Collared 535, **XIV** 7ab, **LXIV** 13
   Green 536, **XIV** 6ab
   Plush-crested 537

   Turquoise 536, **XIV** 8
   (White-collared) 536
jelskii, Iridosornis 616, **LIV** 2
(jelskii), Ochthoeca 495, **XLIV** 14
jelskii, Upucerthia 332, **XXXIV** 11ab
jourdanii, Chaetocercus 298, **XXX** 7, **XXXI** 20
jubata, Neochen 120
julius, Nothocercus 56, **I** 5
juninensis, Muscisaxicola 511, **XLIII** 6
kalinowskii, Nothoprocta 59, **I** 13
Kestrel, American 111, **XIII** 11a-h
Killdeer 162, **XX** 3
Kingbird, Eastern 525, **XLV** 2
   Tropical 524, **XLV** 6
Kingfisher, Amazon 302
   Green 302
   Pygmy 303, **XXXI** 24ab
   Ringed 302, **XXXI** 25ab
kingi, Aglaiocercus 290, **XXX** 5a-c
Kiskadee, Great 522, **XLV** 11
Kite, (Everglade) 93
   Hook-billed 92, **XII** 11a-d
   Plumbeous 94
   Snail 93, **XIII** 3ab
   Swallow-tailed 93, **XIII** 1
   White-tailed 93, **XIII** 2
Knipolegus 516
Knot, Red 168
(koepckei), Otus 223
(kreyenborgi), Falco 114, **XIII** 9b

labradorides, Tangara 611, **LV** 10
lacrymosus, Anisognathus 619, **LIV** 5a-c
laeta, Incaspiza 682, **LIX** 8ab
Lafresnaya 259
lafresnayi, Diglossa 602, **LIII** 11ab
lafresnayi, Lafresnaya 259, **XXXI** 11a-c
laminirostris, Andigena 308, **XXXII** 7
Lancebill, Green-fronted 244
lanceolata, Rhinocrypta 420, **XL** 16
Lapwing, Andean 160, **XVIII** 4a-d
   (Chilean) 159
   Southern 159, **XVIII** 3ab
Laridae 184
Lark, Horned 527, **XLVIII** 18ab
   (Shore) 528
Larus 184
latebricola, Scytalopus 433, **XLI** 2ab, **LXIII** 30
Laterallus 146
(Lathria) see *Lipaugus*
(laticlavius), Heliangelus see *amethysticollis* 269
(latinuchus), Atlapetes see *rufinucha* 671
(latrans), Scytalopus 424
(latreillii), Attagis see *gayi* 181
leadbeateri, Heliodoxa 252
Leafscraper, Gray-throated 395, **LXIII** 31
(Leaftosser) see Leafscraper
lebruni, Sicalis 655, **LX** 9a-c
Leistes 578
Lepidocolaptes 322

Leptasthenura 346
Leptopogon 478
leptorhynchus, Enicognathus 208, **XXIII** 10
Leptosittaca 203
Leptotila 197
Lesbia 277
Lessonia 515
(*Leucippus*) 249
leucocephalus, Cinclus 537, **IL** 19a-c, **LXIII** 27
(*leucogenys*), *Merganetta* 121
(*Leuconerpes*) see *Melanerpes*
leuconota, Pyriglena 401
(*leuconotus*), *Cinclus* see *leucocephalus* 537
leucophrys, Henicorhina 547, **IL** 7, **LXIV** 30
leucophrys, Mecocerculus 463, **XLVII** 5ab
leucophrys, Ochthoeca 498, **XLIV** 8a-c
(*leucophrys*), *Vireo* 569
leucopis, Atlapetes 679, **LXI** 4
leucopleurus, Oreotrochilus 254, **XXVII** 8ab
leucopodus, Haematopus 156, **XVIII** 9ab
leucops, Platycichla 556
leucoptera, Fulica 153, **XVI** 4a-c
leucoptera, Henicorhina 547
leucopterus, Atlapetes 675, **LXI** 8a-c
(*leucopterus*), *Nyctibius* 232, 846
leucopus, Furnarius 343
leucopyga, Tachycineta 534
leucorhamphus, Cacicus 573, **LI** 9ab
leucorrhous, Buteo 101, **XII** 1a-c
leucotis, Entomodestes 552, **L** 8
leucurus, Elanus 93, **XIII** 2
lignarius, Picoides 312, **XXXII** 18
limicola, Rallus 144, **XV** 4a-c
Limnodromus 170
Limosa 175
linearis, Geotrygon 199
lineifrons, Grallaricula 416, **XL** 3
lineola, Bolborhynchus 211, **XXIV** 12ab
lintoni, Myiophobus 484, **XLVI** 19
Lipaugus 450
livia, Columba 191
livida, Agriornis 506, **XLIII** 17ab
Lochmias 396
Loddigesia 292
longicauda, Bartramia 174, **XX** 18
longicaudatus, Mimus 551, **IL** 17
(*longipennis*), *Chlorophonia* see *cyanea* 608
longirostris, Caprimulgus 235, **XXVI** 10a-d
longirostris, Heliomaster 292
(*Lophonetta*) 125
Lophortyx 140
Lophospingus 669
lorata, Sterna 186
loweryi, Xenoglaux 228, **XXV** 3
loyca, Sturnella 578, **LI** 17ab
(*luchsi*), *Myiopsitta* 208
luciani, Eriocnemis 273, **XXVIII** 8a-c
luctuosa, Sporophila 647, **LIX** 3
ludoviciae, Doryfera 244
ludovicianus, Pheucticus 646, **LVII** 6ab

lugens, Haplophaedia 276
(*luminosus*), *Heliangelus* 270
(*lunigera*), *Tangara* see *parzudakii* 610
(*lunulata*), (*Poecilothraupis*) see
    igniventris, Anisognathus 618
Lurocalis 233
lutea, Sicalis 651, **LX** 1a-d
luteiventris, Myiodynastes 523
luteocephala, Sicalis 653, **LX** 6ab
luteola, Sicalis 657, **X** 10ab, **LX** 5a-e
luteoviridis, Basileuterus 590, **LII** 22a-d
lutetiae, Coeligena 264, **XXVIII** 17ab
(*lutleyi*), *Tangara* see *cyanotis* 611
lyra, Uropsalis 237, **XXVI** 9

Macaw, Military, 201
    Red-fronted, 201
macconnelli, Mionectes 477
macloviana, Muscisaxicola 509, **XLIII** 8
Macroagelaius 575
(*Macropsalis*) see *Uropsalis*
macropus, Scytalopus 428, **XLI** 10ab
(*macrorhyncha*), *Dendrocincla* 320
macularia, Tringa 174, **XIX** 11, **XX** 31
(*maculata*), (*Pisobia*) see *melanotos, Calidris* 168
maculicauda, Asthenes 378, **XXXVII** 17ab
maculipennis, Larus 185, **XXI** 5a-c
maculirostris, Muscisaxicola 508, **XLIII** 1ab
maculosa, Columba 191, **XXII** 2
maculosa, Nothura 62, **I** 3
maculosus, Nyctibius 232, **XXVI** 7, 846
magellanica, Carduelis 691, **LXII** 7a-d
magellanica, Gallinago 179, **XIX** 2ab
magellanicus, Campephilus 318, **XXXII** 13ab
magellanicus, Scytalopus 437, **XLI** 1a-m, **LXIV** 31
(*magellanicus*), (*Spinus*) see
    magellanica, Carduelis
magna, Sturnella 580, **LI** 18
maguari, Ciconia 82
magnirostris, Buteo 100, **XIII** 4ab
major, Crotophaga 219, **XIV** 15
major, Podiceps 68, **VIII** 14a-c
Mallard 135
malouinus, Attagis 181, **XVII** 4ab
marcapatae, Cranioleuca 359, **XXXVI** 3a-c
(*marcapatae*), *Thamnophilus* 400
Margarornis 384, see also *Premnoplex* and
    Premnornis
(*Marila*) see *Netta* and *Aythya*
maritima, Geositta 324, **XXXIV** 9
marshalli, Otus 224, **XXV** 8
Martin, Brown-chested 529, **XLVIII** 7
    Purple 528, **XLVIII** 6a-c
    (Sand) 533
    Southern 529
martinica, Porphyrula 147, **XV** 11
matthewsii, Boissonneaua 267, **XXXI** 12
mavors, Heliangelus 268, **XXIX** 19
maximiliani, Melanopareia 421, **XLI** 14
Meadowlark, Eastern 580, **LI** 18

Long-tailed 578, **LI** 17ab
Peruvian 579, **LI** 16ab
(Peruvian Red-breasted) 580
Mecocerculus 463
megalopterus, Phalcoboenus 108, **XI** 2ab, 846
megalura, Leptotila 198, **XXII** 5
(Megapicos) see *Campephilus*
megapodius, Pteroptochos 419, **XL** 18
melacoryphus, Coccyzus 218, **XIV** 12
melambrotos, Cathartes 90
melancholicus, Tyrannus 524, **XLV** 6
melancoryphus, Cygnus 117, **VIII** 2ab
Melanerpes 311
melanocephalus, Atlapetes 672, **LXIII** 33
melanocephalus, Myioborus 587, **LII** 7a-d
Melanodera 667
melanodera, Melanodera 667, **LVIII** 13ab
melanogaster, Oreotrochilus 254, **XXVII** 6ab
melanogenys, Adelomyia 251, **XXXI** 14ab
melanogenys, Anisognathus 618, **LXIII** 23
(melanoleuca), Poospiza 686, **LXII** 14
melanoleuca, Tringa 172, **XIX** 8, **XX** 12
melanoleucos, Campephilus 317, **XXXII** 14ab
melanoleucus, Geranoaetus 98, **XI** 6a-c
melanonota, Pipraeidea 609, **LIV** 14
Melanopareia 421
melanopis, Schistochlamys 640
melanopis, Theristicus 83, **IV** 4a-d
melanops, Gallinula 148, **X** 8ab, **XV** 10ab
melanops, Phleocryptes 345, **XXXVIII** 19ab
melanoptera, Chloephaga 118, **V** 4a-c
melanoptera, Metriopelia 196, **XXII** 7
(melanorhynchus), Chlorostilbon see *mellisugus* 247
melanorhynchus, Thripadectes 393, **XXXVIII** 17
melanotis, Hapalopsittaca 212, **XXIV** 2ab
melanotis, Hemispingus 637, **LVI** 23
melanotos, Calidris 168, **XIX** 14, **XX** 27
melanotos, Sarkidiornis 121, **VIII** 8a-c
melanura, Chilia 337, **XXXV** 8ab
melanura, Pyrrhura 206, **XXIII** 6a-c
mellisugus, Chlorostilbon 247, **XXXI** 5
mercenaria, Amazona 216, **XXIV** 1ab
Merganetta 121
meridae, Cistothorus 542, **LXIV** 29
(meridanus), Scytalopus 434, **XLI** 2a
Merlin 112, **XIII** 8ab
Metallura 281
Metaltail, Black 281, **XXIX** 4ab
  Coppery 283, **XXIX** 3ab
  Fire-throated 283, **XXIX** 8ab
  Neblina 282, **XXIX** 7ab
  Perijá 285, **XXIX** 1
  Scaled 283, **XXIX** 9a-c
  Tyrian 284, **XXIX** 2a-c, **LXIV** 2
  Violet-throated 282, **XXIX** 6ab
  Viridian 281, **XXIX** 5a-d
Metriopelia 194
(mexicana), Coereba see *flaveola* 595
mexicanus, Himantopus 157, **VIII** 2a-c, **XX** 15
meyerdeschauenseei, Tangara 613, **LV** 7

(micrastur), Heliangelus 270
Micropalama 170
microptera, Agriornis 506, **XLIII** 20ab
microptera, Rollandia 66, **II** 4a-d
microrhynchum, Ramphomicron 280, **XXX** 14a-c
(Microspingus) 639
Microstilbon 294
militaris, Ara 201
(militaris), Leistes 578
(militaris), (Pezites) 579
milleri, Grallaria 413, **XL** 5
Milvago 10
Mimidae 548
Mimus 549
Miner, Creamy-rumped 325, **XXXIV** 5
  Common 328, **XXXIV** 3a-d
  Dark-winged 325, **XXXIV** 8
  Grayish 324, **XXXIV** 9
  Puna 327, **XXXIV** 7ab
  Rufous-banded 326, **XXXIV** 6
  Short-billed 327, **XXXIV** 4
  Slender-billed 329, **XXXIV** 1a-c
  Thick-billed 329, **XXXIV** 2
miniatus, Myioborus 586, **LII** 8
minimus, Catharus 556, **L** 4
minor, Chordeiles 234, **XXVI** 5
minor, Mecocerculus 465, **XLVII** 1
minuta, Columbina 193
minutilla, Calidris 169, **XX** 29
Mionectes 476
mirabilis, Eriocnemis 274
mirabilis, Loddigesia 292, **XXX** 6a-c
mitchellii, Phegornis 165, **XVII** 10a-c, **XX** 20
mitchellii, Philodice 294, **XXX** 8, **XXXI** 18
mitrata, Aratinga 203, **XXIII** 1a-c, 846
Mniotilta 581
(mocinno), Pharomachrus 299
Mockingbird, Brown-backed 551, **IL** 16ab
  Chalk-browed 549, **IL** 15
  Chilean 551, **IL** 12
  Long-tailed 551, **IL** 17
  (Northern) 549
  Patagonian 549, **IL** 14
  Tropical 549
  White-banded 550, **IL** 13
(mocoa), Aglaiocercus see *kingi* 290
modesta, Asthenes 372, **XXXVII** 9a-c
(modesta), Elaenia 460
modesta, Progne 529
modestus, Charadrius 163, **XVIII** 7a-d, **XX** 5
modestus, Larus 185
molinae, Pyrrhura 205, **XXIII** 9
mollissima, Chamaeza 403, **XXXIX** 4a-c
Molothrus 570
Momotidae 303
Momotus 303
momota, Momotus 303
monachus, Myiopsitta 208, **XXIII** 8
mondetoura, Claravis 194, **XXIV** 13
Monjita, (Mouse-brown) 507, **XLIII** 18

Rusty-backed 503, **XLIII** 12
montagnii, Penelope 136, **XIV** 4a-c
montana, Agriornis 504, **XLIII** 19ab
(montana), Anabacerthia see *striaticollis* 389
montana, Buthraupis 620, **LIV** 9ab
montana, Geotrygon 199, **XXII** 13ab
(monticola), Grallaria see *quitensis*
(monticola), Troglodytes 546, **LXIII** 28
montivagus, Aeronautes 239, **XLVIII** 2
(Moorhen) 149, **XVIII** 7a-e
morenoi, Metriopelia 195, **XXII** 10ab
moschata, Cairina 121, 135
mosquera, Eriocnemis 274, **XXVIII** 3
Motacillidae 563
Mountain-finch, Chestnut-breasted 688, **LXII** 16
    Cochabamba 689, **LXII** 18ab
    Tucumán 688, **LXII** 19ab
Mountain-tanager, (Black-cheeked) 619
    Black-chested 622, **LIV** 8ab
    Blue-winged 620, **LIV** 6ab
    Buff-breasted 623, **LIV** 16a-c
    Chestnut-bellied 623, **LIV** 15
    Golden-backed 622, **LIV** 12
    Hooded 620, **LIV** 9ab
    Lacrimose 619, **LIV** 5a-c
    Masked 621, **LIV** 13ab
    Santa Marta 618, **LXIII** 23
    Scarlet-bellied 618, **LIV** 7ab
Mountain-toucan, Black-billed 309, **XXXII** 9a-c
    Gray-breasted 308, **XXXII** 6
    Hooded 309, **XXXII** 8
    Plate-billed 308, **XXXII** 7
Mountaineer, Bearded 289, **XXX** 20a-c
mulsant, Acestrura 296, **XXXI** 15ab
munda, Serpophaga 467, **XLVII** 7
murina, Agriornis 507, **XLIII** 18
murina, Notiochelidon 530, **XLVIII** 9
murina, Phaeomyias 456
Muscisaxicola 507
(*Muscivora*) 525
(*musculus*), *Troglodytes* see *aedon* 545
musica, Euphonia 608, **LVI** 1a-d
Myadestes 553
Mycteria 82
mycteria, Jabiru 82
Myiarchus 520
Myioborus 580
(*Myiochanes*) see *Contopus*
Myiodynastes 522
Myiophobus 483
Myiopsitta 208
(*Myiosympotes*) see *Pseudocolopteryx*
Myiotheretes 499, see also
    Knipolegus and *Polioxolmis*
(*Myiothlypis*) see *Basileuterus*
Myornis 423
Myrmeciza 401
(*Myrmoderus*) see *Myrmeciza*
Myrtis 294
mystacalis, Diglossa 602, **LIII** 9a-d

(*mystacalis*), *Thryothorus* see *genibarbis* 544

naevia, Tapera 220, **XIV** 11
nana, Grallaricula 416, **XL** 2
nanum, Glaucidium 227, **XXV** 7ab
nationi, Atlapetes 676, **LXI** 9a-d
Negrito, Rufous-backed 515, **XLIV** 18ab
    White-winged 515, **XLIV** 17a-c
nematura, Lochmias 396, **XXXVIII** 18ab
Neochen 120
Neocrex 147
Neoxolmis 503
Nephelornis 640
Netta 131
(*Nettion*) see *Anas*
(*niceforoi*), *Anas* 127, **X** 1
Nighthawk, Common 234, **XXVI** 5
    Lesser 234, **XXVI** 6
    (Rufous-bellied) 234
    Short-tailed 233, **XXVI** 4ab
Night-heron, (Black-crowned) 78,
    III 9a-c
Nightingale-thrush, (Gray-headed) 555
    Orange-billed 554, **L** 5
    Slaty-backed 554, **L** 1ab
    Spotted 553, **L** 2
Nightjar, Band-winged 235, **XXVI** 10a-d
    Lyre-tailed 237, **XXVI** 9
    Swallow-tailed 236, **XXVI** 8ab
nigra, Rynchops 188
nigricans, Rallus 144
nigricans, Sayornis 491, **XLV** 17
(*nigricapillus*), (*Tyranniscus*) see
    nigrocapillus, Phyllomyias 454
(*nigriceps*), *Saltator* 643
nigriceps, Turdus 562, **L** 10a-c
nigriceps, Veniliornis 312, **XXXII** 16a-c
(*nigricollis*), *Podiceps* 69
nigricollis, Sporophila 648, **LIX** 2ab
nigrirostris, Andigena 309, **XXXII** 9a-c
nigrirostris, Cyclarhis 568
nigrivestis, Eriocnemis 272, **XXVIII** 7ab
nigrocapillus, Nothocercus 56, **I** 8ab
nigrocapillus, Phyllomyias 454, **XLVII** 17, **LXIII** 10
nigrocristatus, Anairetes 470, **XLVI** 5a-c
nigrocristatus, Basileuterus 590, **LII** 20
nigrorufa, Poospiza 686, **LXII** 13
nigroviridis, Tangara 613, **LV** 11
nobilis, Gallinago 177, **XIX** 4ab
nobilis, Oreonympha 289, **XXX** 20a-c
(*Nomonyx*) see *Oxyura*
Nothocercus 56
Nothoprocta 58
Nothura 61
Nothura, Darwin's 61, **I** 2
    Spotted 62, **I** 3
Notiochelidon 530
nuchalis, Grallaria 408, **XXXIX** 10a-c
Numenini 174
Numenius 175

nuna, Lesbia 278, **XXX** 18a-d
(Nunbird), White-faced 304, **XXXI** 27
(Nuttalornis) 487
Nyctibiidae 232
Nyctibius 232
Nycticorax 78
nycticorax, Nycticorax 78, **III** 9a-c
Nyctidromus 235

obscura, Elaenia 462, **XLVII** 12
obsoletum, Camptostoma 458, **XLVII** 15ab
obsoletus, Crypturellus 57, **I** 7
occidentalis, Pelecanus 74
occipitalis, Podiceps 69, **II** 6a-d
(Ochetorhynchus) 335
ochraceiventris, Leptotila 846
ochraceiventris, Myiophobus 485, **XLVI** 16ab
(ochraceus), Hemispingus see *melanotis* 637
(Ochthodiaeta) see *Myiotheretes*
Ochthoeca 493
(ockendeni), Turdus see *fuscater* 557
Ocreatus 276
odomae, Metallura 282, **XXIX** 7ab
Odontophorus 141
Odontorchilus 539
oenanthoides, Ochthoeca 497, **XLIV** 6ab
Ognorhynchus 204
Oilbird 232
(oleagineus), Veniliornis see *fumigatus* 313
(olivacea), (Pachysylvia) see *olivaceus, Hylophilus* 569
olivacea, Piranga 627
(olivacea), (Vireosylvia) see *olivaceus, Vireo* 568
olivaceum, Chalcostigma 286, **XXIX** 11, 847
olivaceus, Hylophilus 569, **LII** 1
olivaceus, Mionectes 477, **XLVII** 25
olivaceus, Phalacrocorax 72, **VIII** 15a-c
olivaceus, Picumnus 311, **XXXII** 10
olivaceus, Vireo 568, **LII** 3
(olivascens), Cinnycerthia see *peruana* 540
olivascens, Sicalis 654, **LX** 10a-d
olor, Cygnus 134
olrogi, Cinclodes 339, **XXXV** 12
oneilli, Nephelornis 640, **LVI** 14
(opacus), Scytalopus 442, **XLI** 1d
ophthalmicus, Chlorospingus 630, **LVI** 10a-d, **LXIV** 14
Opisthoprora 290
(Oporornis) 585
(oreophila), Catamenia 650, **LXIII** 19
ornatus, Myioborus 587, **LII** 6ab
(Orochelidon) see *Notiochelidon*
(Orodynastes) see *Myiotheretes*
orbygnesius, Bolborhynchus 210, **XXIV** 9
orbignyianus, Thinocorus 182, **XVII** 5a-d
oreas, Lessonia 515, **XLIV** 17a-c
Oreomanes 598
Oreonympha 289
(Oreopelia) 199
Oreopholus 164
Oreopsar 575
Oreothraupis 669

Oreotrochilus 252
orina, Coeligena 263, **XXVIII** 14
Oriole, Northern 577, **LI** 14a-c
    Yellow-backed 577, **LI** 13
ornata, Nothoprocta 59, **I** 10ab
ornata, Thlypopsis 628, **LVI** 5
Oroaetus 98
(Oropezus) see *Grallaria*
Ortalis 139
orthonyx, Acropternis 444, **XLI** 16ab
ortizi, Incaspiza 680, **LIX** 11ab
Ortygonax see *Rallus*
Oropendola, Crested 572, **LI** 1
    Dusky-green 572, **LI** 3
    Russet-backed 572, **LI** 2
(ortoni), (Zodalia) 277
oryzivorus, Dolichonyx 580, **LI** 19a-d
Oscines 527
Osprey 91, **XII** 5
(Ostinops) see *Psarocolius*
ottonis, Asthenes 374, **XXXVII** 2
Otus 223
oustaleti, Cinclodes 339, **XXXV** 9ab
Owlet, Long-whiskered 228, **XXV** 3
Owl, Barn 221, **XXVI** 1ab
    Buff-fronted 231, **XXV** 9
    Burrowing 230, **XXV** 11a-c
    Great Horned 226, **XXVI** 3ab
    Rufous-banded 228, **XXV** 6
    Rufous-legged 228, **XXV** 12
    Short-eared 229, **XXVI** 2ab
    Stygian 229, **XXV** 1
(Oxyechus) see *Charadrius*
Oxypogon 288
Oxyura 133
Oxyurini 132
Oystercatcher, American 156
    Magellanic 156, **XVIII** 9ab

(Pachosylvia) see *Hylophilus*
Pachyramphus 525
pallatangae, Elaenia 462, **XLVII** 11
palliatus, Cinclodes 342, **XXXV** 16
palliatus, Haematopus 156
pallidinucha, Atlapetes 670, **LXI** 2ab
(pallidior), Mecocerculus 464
palmarum, Thraupis 626
palpebralis, Schizoeaca 363, **XXXVI** 19e
(palpebrosa), (Poecilothraupis) see
    lacrymosus, Anisognathus 619
pamela, Aglaeactis 259, **XXVIII** 21ab
Pampa-finch, Great 689, **LVII** 9ab
Pandion 91
Pandionidae, 91
papa, Sarcoramphus 91
Parabuteo 106
paradoxa, Eugralla 423, **XLI** 13ab
(paraguaiae), Gallinago 179
Parakeet (see also Conure)
    Andean 210, **XXIV** 9

Barred 211, **XXIV** 12ab
Canary-winged 212
Gray-hooded 209, **XXIV** 10
Monk 208, **XXIII** 8
Mountain 209, **XXIV** 8a-e
(Orange-winged) 212
Rufous-fronted 211, **XXIV** 11
(*Pardirallus*) 144
Pardusco 640, **LVI** 14
parina, Xenodacnis 606, **LIII** 8a-e
parodii, Hemispingus 635, **LVI** 19
Parrot, Alder 215, **XXIV** 4
Black-winged 212, **XXIV** 2ab
Bronze-winged 215
Burrowing 204, **XXIII** 12a-d
(Fuertes's) 214
Plum-crowned 214, **XXIV** 7ab
Red-billed 214, **XXIV** 6ab
(Red-faced) 214
(Red-spectacled) 216
Rusty-faced 213, **XXIV** 3a-d, **LXIV** 6, 846
Scaly-naped 216, **XXIV** 1ab
(Speckle-faced) 215
(Tucumán) 215
White-capped 215, **XXIV** 5
Yellow-eared 204, **XXIII** 4
Parrotlet, Spectacled 212
Parula 582
Parula, Tropical 582, **LII** 19
Parulidae 581
parulus, Anairetes 472, **XLVI** 1a-c
parvirostris, Chlorospingus 631
parvirostris, Elaenia 460, **XLVII** 10
parvirostris, Ochthoeca 497, **XLIV** 16
(*parvirostris*), *Scytalopus* 426
parzudakii, Tangara 610, **LV** 4a-c
Passer 695
Passeres 527
PASSERIFORMES 319
passerina, Columbina 192, **XXII** 11
patachonicus, Tachyeres 120, **V** 5a-e, **VIII** 5
Patagona 255
patagonica, Asthenes 371, **XXXVII** 11
(*patagonica*), *Notiochelidon* see *cyanoleuca* 531
patagonicus, Cinclodes 338, **XXXV** 10ab
patagonicus, Mimus 549, **IL** 14
patagonicus, Phrygilus 659, **LVIII** 3a-c
patagonus, Cyanoliseus 204, **XXIII** 12a-d
Pauraque 235
pectoralis, Polystictus 475, **XLVI** 23
pectoralis, Thlypopsis 629, **LVI** 6ab
pelagica, Chaetura 239, **XLVIII** 4
Pelecanidae 74
PELECANIFORMES 72
Pelecanus 74
Pelican, Brown 74
(*pelzelni*), *Pseudotriccus* 480
(*pelzelni*), *Sicalis* 656
Penelope 136
pennata, Pterocnemia 53

pentlandii, Nothoprocta 60, **I** 11a-d, 846
pentlandii, Tinamotis 63, **I** 15ab
peposaca, Netta 131, **VII** 5ab, **VIII** 9, 846
Peppershrike, Black-billed 568
Rufous-browed 567, **LII** 4ab
perdicaria, Nothoprocta 60, **I** 12
peregrina, Vermivora 582
peregrinus, Falco 113, **XIII** 9a-e
perijana, Schizoeaca 362, **XXXVI** 19a
(*perlata*), *Margarornis* see *squamiger* 384
pernix, Myiotheretes 501, **LXIII** 9
(*personata*), *Diglossa* see *cyanea, Diglossopis*
personata, Incaspiza 681, **LIX** 10ab
personatus, Trogon 301, **XXXI** 23ab
perspicillata, Hymenops 518, **XLV** 15ab
peruana, Cinnycerthia 540, **IL** 6a-c
(*peruanus*), (*Spinus*) see
magellanica, Carduelis 691
(*peruvianus*), *Ocreatus* 276
peruvianus, Rallus 145, **XV** 5
petechia, Dendroica 582
(*petersi*), *Xenodacnis* see *parina* 606
(*petersoni*), *Otus* 225
Petrochelidon 533
Pewee, Eastern Wood- 488, **XLV** 9
Greater 487, **XLV** 10a-c
(Smoke-colored) 488, **XLV** 10a-c
Tropical 489, **XLV** 7
Western Wood- 488, **XLV** 8
(*Pezites*) 579
Phacellodomus 381, see also *Thripophaga*
(*phaeocephalus*), *Chlorospingus* see *ophthalmicus* 630
Phaeomyias 456
Phaeoprogne 529
phaeopus, Numenius 175, **XX** 14
Phaethornis 242
Phaetusa 187
phainopeplus, Campylopterus 245, **LXIII** 5ab
(*Phaiolaima*) see *Heliodoxa*
Phalacrocoracidae 72
Phalacrocorax 72
Phalarope, Wilson's 171, **XIX** 15ab, **XX** 19
Phalaropodinae 171
Phalaropus 171
Phalcoboenus 107
phalerata, Coeligena 262, **LXIII** 6ab
Pharomachrus 299
Phasianidae 139
Phegornis 165
Pheucticus 645
(*Pheugopedius*) see *Thryothorus*
philadelphia, Geothlypis 585
phillipsi, Tangara 615
Philodice 294
Philydor 389, see also *Anabacerthia* and *Syndactyla*
Phimosus 82
Phleocryptes 345
(*Phloeoceastes*) 317
Phoebe, Black 491, **XLV** 17
(White-winged) 491, **XLV** 17

phoebe, Metallura 281, **XXIX** 4ab
Phoenicoparrus 86
Phoenicopteridae 84
Phoenicopterus 85
phoenicotis, Chlorochrysa 610
phoenicurus, Eremobius 336, **XXXIV** 19
Phrygilus 659
Phyllomyias 454
Phylloscartes 479
Phytotoma 452
Phytotomidae 451
Piaya 218
Picidae 310
Picoides 312
PICIFORMES 304
picta, Chloephaga 118, **V** 3a-f
picui, Columbina 193
Piculet, Grayish 311
 Olivaceous 311, **XXXII** 10
Piculus 314
Picumnus 311
picumnus, Dendrocolaptes 321, **XXXIII** 7ab
Pigeon, Band-tailed 189, **XXII** 3ab
 Chilean 190, **XXII** 4ab
 Feral 191
 Ruddy 191
 Spot-winged 191, **XXII** 2
Piha, Dusky 450, **XLII** 8
 Scimitar-winged 451
pileata, Leptasthenura 347, **XXXV** 5ab
(pileatus), Accipiter see *bicolor* 96
pinnatus, Botaurus 80, **III** 7
Pintail, (Bahama) 127, **VI** 6, **IX** 3
 Northern 126, **VI** 3, **IX** 2ab
 (Niceforo's) 127
 White-cheeked 127, **VI** 6, **IX** 3
 Yellow-billed 126, **VI** 4a-c, **IX** 1ab, **X** 1
Pionus 214
Pipit, Correndera 565, **XLVIII** 14a-c
 Hellmayr's 564, **XLVIII** 16ab
 Páramo 565, **XLVIII** 17ab, **XLIV** 23
 Short-billed 563, **XLVIII** 15
pipixcan, Larus 184, **XXI** 3ab
Pipraeidea 609
Pipreola 447
(Pipromorpha) 478
Piranga 626
(Pisobia) see *Calidris*
Pitangus 521
pitiayumi, Parula 582, **LII** 19
pitius, Colaptes 315, **XXXIII** 1ab
(Pitylus) see *Saltator*
piurae, Ochthoeca 498, **XLIV** 11
plancus, Polyborus 109, **XI** 1ab
Plantcutter, Rufous-tailed 452, **XLII** 2a-c
 White-tipped 452, **XLII** 3a-d
platalea, Anas 130, **VI** 7a-c, **IX** 12ab
platensis, Cistothorus 540, **IL** 9a-d, **LXIV** 28
platensis, Embernagra 689, **LVII** 9ab
Platycichla 556

(Platypsaris) 527
platypterus, Buteo 101, **XII** 4ab
platyrhynchos, Anas 135
plebejus, Phrygilus 665, **LVIII** 5ab
Plegadis 82 Ploceidae 695
Plover, American Golden 161, **XVIII** 10, **XX** 7
 (Black-bellied) 161, **XVIII** 11, **XX** 6
 Collared 163, **XX** 2
 Gray 161, **XVIII** 11, **XX** 6
 (Lesser Golden) 161
 Magellanic 166, **XVII** 9a-c, **XX** 11
 Puna 162, **XVIII** 6a-c, **XX** 10
 (Ringed) 162
 Semipalmated 162, **XIX** 8, **XX** 8
 Two-banded 163, **XVIII** 5ab, **XX** 9
plumbea, Ictinia 94
plumbeiceps, Leptotila 198
plumbeiceps, Todirostrum 483, **XLVI** 11
Pluvialis 161
Pluvianellus 166
Pochard, Rosy-billed 131, **VII** 5ab,
 VIII 9, 846
 Southern 131, **VII** 4a-c, **VIII** 7ab
Podiceps 68
podiceps, Podilymbus 67, **II** 2a-c
PODICIPEDIFORMES 65
Podicipedidae 65
Podilymbus 67
poliocephala, Chloephaga 119, **V** 2a-d
poecilocercus, Mecocerculus 464, **XLVII** 3
poecilochrous, Buteo 104, **XII** 8a-c
(Poecilothraupis) see *Anisognathus*
Poecilotriccus 480
poecilurus, Knipolegus 517, **XLV** 16
Polioptila 566
Polioptilidae 566
Polioxolmis 499
(pollens), Agriornis see *andicola* 505
pollens, Campephilus 317, **XXXII** 15ab
Polyborinae 107
Polyborus 109
(polyglottus), Mimus 549
Polyonymus 279
polyosoma, Buteo 103, **XII** 7a-c
Polystictus 475
poortmanni, Chlorostilbon 248
Poospiza 683
(Poospizopsis) 688
(Porphyriops) 148
porphyrocephala, Iridosornis 616
Porphyrula 147
Porzana 146
Potoo, Andean 232, **XXVI** 7, 846
 Common 846
 (White-winged) 232
(Psalidoprymna) see *Lesbia*
prasinus, Aulacorhynchus 306,
 XXXII 3a-d, **LXIII** 8
Premnoplex 385
Premnornis 385

(pretrei), Amazona 216
pretrei, Phaethornis 242, **XXX** 2
Prickletail, Spectacled 383, **XXXVIII** 7
(primolina), Metallura see *williami* 281
Progne 528, see also *Phaeoprogne*
promeropirhynchus, Xiphocolaptes 320,
    XXXIII 6ab, **LXIII** 16
Protonotaria 585
prunellei, Coeligena 261, **XXVIII** 16
przewalskii, Grallaria 410, **XL** 12ab
psaltria, Carduelis 694, **LXII** 1a-c
Psarocolius 572
Pseudocolaptes 386
Pseudocolopteryx 475
(Pseudospingus) 638
Pseudotriccus 480
(Psilopogon) 210
Psittacidae 200
PSITTACIFORMES 200
(Psittacula) see *Forpus*
(Psittospiza) see *Chlorornis*
Pterocnemia 53
Pterophanes 260
Pteroptochos 418
(Ptiloscelis) 160
pucherani, Campylorhamphus 323, **XXXIII** 5
pudibunda, Asthenes 373, **XXXVII** 1
Puffbird, White-faced 304, **XXXI** 27
Puffleg, Black-breasted 272, **XXVIII** 7ab
    Black-thighed 275, **XXVIII** 1ab
    Blue-capped 275, **XXVIII** 2ab
    Colorful 274
    Coppery-bellied 273, **XXVIII** 10
    Emerald-bellied 274
    Greenish 276
    Glowing 272, **XXVIII** 9ab
    Golden-breasted 274, **XXVIII** 3
    Hoary 276
    Sapphire-vented 273, **XXVIII** 8a-c
    Turquoise-throaed 273, **XXVIII** 4ab
pulchella, Ochthoeca 494, **XLIV** 13ab
pulcher, Myiophobus 485, **XLVI** 15ab
pulcherrima, Iridophanes 606, **LIV** 11ab
pulcherrima, (Tangara) see *Iridophanes*
pulchra, Incaspiza 680, **LIX** 9ab
pumilus, Coccyzus 217
puna, Anas 128, **VI** 9a-d, **IX** 5
punensis, Asthenes 376, **XXXVII** 14ab
(punensis), Contopus see *cinereus* 489
punensis, Geositta 327, **XXXIV** 7ab
(punensis), Grallaria 407
punensis, Phrygilus 662, **LVIII** 1a-d
purpurascens, Penelope 136
(purpureicauda), (Metallura) 277
(pusillum), Camptostoma see *obsoletum* 458
Pygarrhichas 395
Pygmy-owl, Andean 226, **XXV** 10a-c
    Austral 227, **XXV** 7ab
Pygmy-tyrant, Hazel-fronted 480, **XLVI** 14
    Rufous-headed 480, **XLVI** 13ab

(Pygochelidon) see *Notiochelidon*
Pyriglena 401
Pyrocephalus 492
Pyroderus 451
pyrohypogaster, Hypopyrrhus 575
(Pyrope) see *Xolmis*
pyrope, Xolmis 502, **XLIII** 14
pyrrholeuca, Asthenes 365, **XXXVII** 4ab
Pyrrhomyias 486
pyrrhonota, Petrochelidon 533, **XLVIII** 12
pyrrhophia, Cranioleuca 357, **XXXVI** 4a-c
pyrrhophrys, Chlorophonia 608, **LVI** 2ab
(pyrrhops), (Euscarthmus) see
    granadensis, Hemitriccus 482
(pyrrhops), Hapalopsittaca 213
Pyrrhura 205
Quail, California 140
Quail-dove, Lined 199
    Ruddy 199, **XXII** 13ab
    White-throated 199, **XXII** 14ab
(Querquedula) see *Anas*
Quetzal, Crested 299
    Golden-headed 300
    White-tipped 300
quitensis, Grallaria 413, **XL** 15a-c

Racket-tail, Booted 276, **XXX** 14a-g
Rail, (Austral) 145, **XV** 6
    Blackish 144
    Bogotá 145, **X** 11
    (Carolina) 146, **XV** 7
    Lesser 144, **XV** 4a-c
    Peruvian 145, **XV** 5
    Plumbeous 143, **XV** 2a-d, 846
    (Virginia) 144, **XV** 4a-c
raimondii, Sicalis 658, **LX** 8a-d
Rallidae 143
ralloides, Myadestes 553, **L** 9ab
Rallus 143
Ramphastidae 305
ramphastinus, Semnornis 305, **XXXI** 26
Ramphastos 310
Ramphomicron 280
rara, Phytotoma 452, **XLII** 2a-c
Rayadito, Thorn-tailed 344, **XXXVI** 9ab
Recurvirostra 158
Recurvirostridae 157
Redstart (see also Whitestart)
    American 586, **LII** 16a-c
regalis, Heliangelus 271, **XXIX** 22ab
reguloides, Anairetes 471, **XLVI** 4a-c
reinhardti, Iridosornis 617, **LIV** 3
resplendens, Vanellus 160, **XVIII** 4a-d
reyi, Hemispingus 636, **LXIV** 18
Rhea, Lesser 53
    (Puna) 53
Rheidae 53
RHEIFORMES, 53
Rhinocrypta 420
Rhinocryptidae 418

(Rhinorchilus) see *Cyphorhinus*
rhodocephala, *Pyrrhura* 207, **LXIV** 1
Rhodopis 293
Rhynchotus 57
(richardsoni), Contopus see
    sordidulus 488
ridgwayi, *Plegadis* 83, **IV** 6a-c
riefferii, *Chlorornis* 639, **LIV** 10ab
riefferii, *Pipreola* 448, **XLII** 9a-e
Riparia 533
riparia, *Riparia* 533
rivolii, *Piculus* 314, **XXXII** 17a-c
(roboratus), Otus 224
rolland, *Rollandia* 65, **II** 3a-c
Rollandia 65
Rostrhamus 93
(Rosybill) 131, **VII** 5ab, **VIII** 9
(rothschildi), Heliangelus 270
rubecula, *Poospiza* 687, **LXII** 17ab
rubecula, *Scelorchilus* 420, **XLI** 17
ruber, *Eudocimus* 82
(ruber), Phoenicopterus 85
rubetra, *Neoxolmis* 503, **XLIII** 12
rubidiceps, *Chloephaga* 120, **V** 1ab
rubiginosus, *Automolus* 391, **LXIII** 15
rubiginosus, *Piculus* 314
rubinoides, *Heliodoxa* 252
rubinus, *Pyrocephalus* 492, **XLV** 4a-c
rubra, *Piranga* 627, **LV** 16ab
rubriceps, *Piranga* 627, **LV** 15ab
rubricollis, *Campephilus* 318
rubrigastra, *Tachuris* 473, **XLVI** 26ab
rubrirostris, *Cnemoscopus* 633, **LVI** 4ab
rubrocristatus, *Ampelion* 445, **XLII** 4a-c
rubrogenys, *Ara* 201
rufa, *Lessonia* 515, **XLIV** 18ab
rufaxilla, *Ampelion* 446, **XLII** 6
rufescens, *Rhynchotus* 57, **I** 1ab
ruficapilla, *Grallaria* 406, **XL** 14ab
ruficapillus, *Thamnophilus* 399, **XXXIX** 2a-e
ruficauda, *Chamaeza* 403
ruficauda, *Upucerthia* 334, **XXXIV** 14
ruficeps, *Chalcostigma* 286, **XXIX** 13ab
ruficeps, *Poecilotriccus* 481, **XLVI** 110a-d
ruficeps, *Pseudotriccus* 480, **XLVI** 13ab
ruficeps, *Thlypopsis* 629, **LVI** 7ab
(ruficervix), (Systellura) see
    longirostris, *Caprimulgus* 235
ruficervix, *Tangara* 612, **LV** 3a-c
ruficollis, *Automolus* 391, **XXXVIII** 8
ruficollis, *Oreopholus* 164, **XVII** 8a-c, **XX** 4
ruficollis, *Stelgidopteryx* 532
(ruficoronatus), Myioborus 588
rufifrons, *Fulica* 150, **XVI** 3ab
rufifrons, *Phacellodomus* 382, **XXXVIII** 2
rufigenis, *Atlapetes* 674, **LXI** 12ab
rufinucha, *Atlapetes* 671, **LXI** 6a-g
rufipectoralis, *Ochthoeca* 495, **XLIV** 10ab, **LXIII** 11
rufipectus, *Formicarius* 403, **XXXIX** 5
rufipectus, *Leptopogon* 478, **XLVII** 24

rufipennis, *Geositta* 326, **XXXIV** 6
rufipennis, *Polioxolmis* 499, **XLIII** 15ab
rufipes, *Strix* 228, **XXV** 12
(rufiventris), Lurocalis 234
rufiventris, *Neoxolmis* 503, **XLIII** 13
rufiventris, *Saltator* 644, **LVII** 3, 847
rufiventris, *Turdus* 560, **L** 13
rufivertex, *Iridosornis* 617, **LIV** 4ab
rufivertex, *Muscisaxicola* 510, **XLIII** 9ab
rufoaxillaris, *Molothrus* 572
rufocinerea, *Grallaria* 407, **XL** 9
rufosuperciliaris, *Hemispingus* 634, **LVI** 22
rufosuperciliata, *Syndactyla* 387, **XXXVIII** 9a-c
rufula, *Grallaria* 412, **XL** 8a-d
rufum, *Conirostrum* 597, **LIII** 3
rufus, *Furnarius* 343, **XXXIV** 20
rufus, *Philydor* 390, **XXXVIII** 12ab
(rufus), (Platypsaris) 527
rumicivorus, *Thinocorus* 183, **XVII** 6ab, **XX** 23
rupicola, *Colaptes* 315, **XXXIII** 2ab
(Rupornis) 100
Rushbird, Wren-like 345, **XXXVIII** 19ab
Rush-tyrant, Many-colored 473, **XLVI** 26ab
russatus, *Chlorostilbon* 248
rustica, *Haplospiza* 668, **LVII** 1ab
rustica, *Hirundo* 532, **XLVIII** 13ab
ruticilla, *Setophaga* 586, **LII** 16a-c
rutila, *Phytotoma* 452, **XLII** 3a-d
rutilans, *Xenops* 394
rutilus, *Cypseloides* 239, **XLVIII** 5ab
(rutilus), Xenops see rutilans 394
Rynchops 188
(rytirhynchos), Rallus 143

Sabrewing, Lazuline 245, **XXXI** 1a-c
    Santa Marta 245, **LXIII** 5ab
(salinarum), Neoxolmis 503
(salmoni), Tigrisoma see *fasciatum* 79
Saltator 643
Saltator, (Black-cowled) 643
    Golden-billed 643, **LVII** 4a-e
    Masked 644, **LVII** 2
    Rufous-bellied 644, **LVII** 3, 847
    Streaked 644
(sanctaemartae), Scytalopus 431, **LXIII** 29
Sanderling 170, **XX** 25
Sandpiper, Baird's 169, **XIX** 12, **XX** 28
    Buff-breasted 170, **XX** 30
    Least 169, **XX** 29
    Pectoral 168, **XIX** 14, **XX** 27
    Solitary 173, **XIX** 10, **XX** 22
    Spotted 174, **XIX** 11, **XX** 31
    Stilt 170, **XX** 26
    Upland 174, **XX** 18
    White-rumped 168, **XIX** 13, **XX** 24
Sandpiper-plover, Diademed 165,
    XVII 10a-c, **XX** 20
sanguinolentus, *Rallus* 143, **XV** 2a-d, 846
(sapho), Sappho 278
Sapphire, Blue-headed 249

Sapphirewing, Great 260, **XXVII** 2ab, 846
Sappho 278
Sarcoramphus 91
Sarkidiornis 121
Satrapa 519
(*saturatior*), *Upucerthia* 331
saturninus, *Mimus* 549, **IL** 15
saucerottei, *Amazilia* 250, **XXXI** 4
savana, *Tyrannus* 524, **XLV** 1ab
savannarum, *Ammodramus* 679, **LVII** 7
saxicolina, *Geositta* 325, **XXXIV** 8
sayaca, *Thraupis* 625, **LV** 12
Sayornis 491
(*scapularis*), *Jacana* see *jacana* 155
Scaup, Lesser 132, **VIII** 6ab
Scelorchilus 419
schulzi, *Cinclus* 538, **IL** 18
Schistes 291
Schistochlamys 640
Schizoeaca 362
sclateri, *Ampelion* 447, **XLII** 7ab
sclateri, *Asthenes* 377, **XXXVII** 15
sclateri, *Phyllomyias* 454, **XLVII** 19
Sclerurus 395
Scolopacidae 166
Scolopacinae 175
Screech-owl, Cinnamon 225
    Colombian 225
    Cloud-forest 224, **XXV** 8
    Rufescent 224, **XXV** 2a-c
    Savanna 223, **XXV** 5a-d
    (Tropical) 223, **XXV** 5a-d
    (West Peruvian) 224
    White-throated 225, **XXV** 4, **LXIV** 7
scrutator, *Thripadectes* 392, **XXXVIII** 13
scutatus, *Pyroderus* 451
Scytalopus 424
Scythebill, Greater 323, **XXXIII** 5
(*seebohmi*), *Atlapetes* 677
Seedeater, Band-tailed 648, **LIX** 5a-g
    Black-and-white 647, **LIX** 3
    Double-collared 648, **LIX** 4a-c
    Páramo 650, **LIX** 6a-e, **LXIII** 19
    Plain-colored 649, **LIX** 7a-c
    (Santa Marta) 650, **LXIII** 19
    Yellow-bellied 648, **LIX** 2ab
Seedsnipe, Gray-breasted 182, **XVII** 5a-d
    Least 183, **XVII** 6ab, **XX** 23
    Rufous-bellied 181, **XVII** 7a-c
    White-bellied 181, **XVII** 4ab
segmentata, *Uropsalis* 236, **XXVI** 8ab
Seiurus 584
semifuscus, *Chlorospingus* 632, **LVI** 11
semipalmatus, *Catoptrophorus* 174
semipalmatus, *Charadrius* 162, **XIX** 8, **XX** 8
semiplumbeus, *Rallus* 145, **X** 11
semirufus, *Atlapetes* 673, **LXI** 7ab
semitorquatus, *Lurocalis* 233, **XXVI** 4ab
Semnornis 305
senilis, *Myornis* 423, **XLI** 11a-c

seniloides, *Pionus* 215, **XXIV** 5
sephaniodes, *Sephanoides* 266, **XXVIII** 5a-c
Sephanoides 266
Sericossypha 630
Serpophaga 467
serrana, *Upucerthia* 333, **XXXIV** 13
serranus, *Larus* 186, **XXI** 8a-e
serranus, *Turdus* 559, **L** 6a-c
serrirostris, *Colibri* 247, **XXXI** 10ab
Setophaga 586
Shag, (Blue-eyed) 73, **VIII** 16
    Imperial 73, **VIII** 16
Sheartail, Peruvian 293, **XXX** 9
Shoveler, Northern 130, **IX** 13ab
    Red 130, **VI** 7a-c, **IX** 12ab
Shrike-tyrant, Black-billed 504, **XLIII** 19ab
    Gray-bellied 506, **XLIII** 20ab
    Great 506, **XLIII** 17ab
    Mouse-brown 507, **XLIII** 18
    White-tailed 505, **XLIII** 16
sibilatrix, *Anas* 123, **VI** 10a-c, **IX** 10
Sicalis 651
Sicklebill, Buff-tailed 244, **XXX** 1
    White-tipped 244
Sierra-finch, Ash-breasted 665, **LVIII** 5ab
    Band-tailed 666, **LVIII** 10a-f
    Black-hooded 661, **LVIII** 4a-c
    Carbonated 667, **LVIII** 11a-c
    Gray-hooded 660, **LVIII** 2a-c
    Mourning 662, **LVIII** 9a-d
    Patagonian 659, **LVIII** 3a-c
    Peruvian 662, **LVIII** 1a-d
    Plumbeous 663, **LVIII** 6a-f
    Red-backed 664, **LVIII** 7ab
    White-throated 665, **LVIII** 8ab, 847
signatus, *Basileuterus* 591, **LII** 21ab
(*signatus*), *Chlorospingus* see *canigularis* 632
signatus, *Knipolegus* 516, **XLV** 12a-d
(*simonsi*), *Atlapetes* see *nationi* 676
(*simonsi*), *Scytalopus* 439, **XLI** 1i
simplex, *Phaetusa* 187, **XXI** 6
simplex, *Pseudotriccus* 480, **XLVI** 14
Siptornis 383
Siptornopsis 361
Siskin, Andean 690, **LXII** 6a-d
    Black 693, **LXII** 3ab
    Black-chinned 694, **LXII** 5ab
    Hooded 691, **LXII** 7a-d
    Thick-billed 690, **LXII** 8a-f
    Yellow-bellied 692, **LXII** 2ab
    Yellow-rumped 693, **LXII** 4ab
(*Silvicultrix*) 493
Sittasomus griseicapillus 320
sitticolor, *Conirostrum* 597, **LIII** 2ab
sittoides, *Diglossa* 601, **LIII** 15a-d
Skimmer, Black 188
(Slaty-thrush) see Thrush
Snipe, (Andean) 175, **XIX** 6ab
    Banded 176, **XIX** 7
    Cordilleran 175, **XIX** 5ab, 6ab

(Imperial) 176, **XIX** 7
Magellanic 179, **XIX** 2ab
Noble 177, **XIX** 4ab
North American 178, **XIX** 1
(Paramo) 177, **XIX** 4ab
Puna 179, **XIX** 3ab, **XX** 17
sociabilis, Rostrhamus 93, **XIII** 3ab
socialis, Pluvianellus 166, **XVII** 9a-c, **XX** 11
(sodestromi), Eriocnemis 275
Softtail, Russet-mantled 381, **XXXVIII** 4ab
(solitaria), Agriornis see *montana* 504
solitaria, Tringa 173, **XIX** 10, **XX** 22
solitarius, Harpyhaliaetus 97
Solitaire, Andean 553, **L** 9ab
    White-eared 552, **L** 8
solstitialis, Troglodytes 546, **IL** 11a-e, **LXIII** 28
(somptuosa), (Compsocoma) see
    flavinuchus, Anisognathus 620
Sora 146, **XV** 7
sordidulus, Contopus 488, **XLV** 8
sordidus, Pionus 214, **XXIV** 6ab
sparganura, Sappho 278, **XXX** 17a-c
Sparrow, Grasshopper 679, **LVII** 7
    House 695, **LVII** 12ab
    Rufous-collared 682, **LVII** 13a-d
    Saffron-billed 679
sparverius, Falco 111, **XIII** 11a-h
Spatuletail, Marvellous 292, **XXX** 6a-c
(Spodiornis) see *Haplospiza*
(speciosus), Heliangelus 270
specularioides, Anas 125, **VI** 1a-d, **VIII** 4
specularis, Anas 126, **VI** 2ab, **VIII** 3
speculifera, Diuca 658, **LIX** 14ab, 847
spencei, Heliangelus 268, **LXIV** 4ab*(Speotyto)* 230
(spillmanni), Scytalopus 435, **XLI** 2b
spinescens, Carduelis 690, **LXII** 6a-d
Spinetail, Apurímac 351, **XXXVI** 13
    Ash-browed 356, **XXXVI** 7ab
    Azara's 352, **XXXVI** 14
    Buff-browed 351, **XXXVI** 17ab
    Creamy-crested 360, **XXXVI** 1ab
    Crested 356, **LXIV** 25
    Elegant 351, **XXXVI** 15a-c
    (Fork-tailed) 357
    Great 361, **XXXVII** 21
    Light-crowned 360, **XXXVI** 2a-c
    (Line-cheeked) 358
    Marcapata 359, **XXXVI** 3a-c
    Northern Line-cheeked 358, **XXXVI** 6
    Red-faced 357, **XXXVI** 8
    Rufous 354, **XXXVI** 11ab
    Russet-bellied 354, **XXXVI** 12, 847
    Rusty-headed 353, **LXIII** 17
    Silvery-throated 353, **XXXVI** 16ab
    Slaty 353
    (Sooty) 353
    Sooty-fronted 352
    Southern Line-cheeked 358, **XXXVI** 5ab
    Streak-capped 356, **LXIII** 26
    Stripe-crowned 357, **XXXVI** 4a-c

    White-browed 355, **XXXVI** 10ab, **LXIV** 26
spinicauda, Aphrastura 344, **XXXVI** 9ab
(spinosa), Jacana 155
(Spinus) see *Carduelis*
(Spiziornis) see *Anairetes*
(spodionotus), Atlapetes see *rufinucha* 671
spodiops, Hemitriccus 482, **XLVI** 8
(Spodiornis) 669
Spoonbill, Roseate 84
Sporophila 647
(Sporothraupis) see *Thraupis*
squamiger, Margarornis 384, **XXXIII** 13a-c
squamigera, Grallaria 404, **XXXIX** 7ab
(squamigularis), Heliangelus 270
(Squatarola) see *Pluvialis*
squatarola, Pluvialis 161, **XVIII** 11, **XX** 6
stanleyi, Chalcostigma 287, **XXIX** 10a-d
Starfrontlet, Blue-throated 263, **XXVIII** 22ab
    Buff-winged 264, **XXVIII** 17ab
    Dusky 263, **XXVIII** 14
    Golden-bellied 263, **XXVIII** 13a-c, **LXIV** 12
    Rainbow 265, **XXVIII** 11a-e
    Violet-throated 264, **LXIII** 12a-d
    White-tailed 262, **LXIII** 6ab
Starthroat, Long-billed 292
Steamer-duck, Flying 120, **V** 5a-e, **VIII** 5
Steatornithidae 232
Steatornis 232
(Steganopus) 171
steinbachi, Asthenes 370, **XXXVII** 7
Stelgidopteryx 532
stellatus, Margarornis 384, **XXXIII** 12
stenura, Chlorostilbon 249, **XXXI** 6a-c
Sterna 187
stictopterus, Mecocerculus 466, **XLVII** 4
Stigmatura 468
Stilt, Black-necked 157, **XVIII** 2a-c, **XX** 15
stolzmanni, Urothraupis 633, **LVI** 13
Stork, Maguari 82
    Wood 82
Streamcreeper, Sharp-tailed 396,
    XXXVIII 18ab
strepera, Elaenia 461, **XLVII** 9ab
Streptoprocne 238
stresemanni, Ampelion 446, **XLII** 5
striata, Ardeola 77, **III** 10ab
striata, Dendroica 584, **LII** 12ab
striata, Leptasthenura 347, **XXXV** 3a-c
striaticeps, Phacellodomus 382, **XXXVIII** 3
striaticollis, Anabacerthia 389, **XXXVIII** 6a-c
striaticollis, Mionectes 477, **XLVII** 26a-c
striaticollis, Myiotheretes 500, **XLIV** 2ab
striaticollis, Siptornis 383, **XXXVIII** 7
(striatipectus), Saltator see *albicollis* 644
striatus, Accipiter 96, **XIII** 7a-g
stricklandii, Gallinago 175, **XIX** 5ab, 6ab
Strigidae 222
STRIGIFORMES 221
Strix 228
strophianus, Heliangelus 269, **XXIX** 20ab

strophium, Odontophorus 141, **XVII** 3
Sturnella 578
stygius, Asio 229, **XXV** 1
subalaris, Macroagelaius 575, **LI** 4
subalaris, Syndactyla 388, **XXXVIII** 10ab
subcristata, Cranioleuca 356, **LXIV** 25
(*subcristata*), *Serpophaga* 468
subis, Progne 528, **XLVIII** 6a-c
Suboscines 319
subpudica, Synallaxis 353, **XXXVI** 16ab
subruficollis, Tryngites 170, **XX** 30
subvinacea, Columba 191
sulcatus, Aulacorhunchus 307, **XXXII** 4
sulcirostris, Crotophaga 220
sulphuratus, Pitangus 522, **XLV** 11
Sunangel, Amethyst-throated 269, **XXIX** 17a-d
    (Glistening) 270
    (Golden-throated) 270
    Gorgeted 269, **XXIX** 20ab
    (Longuemare's) 270
    Mérida 268, **LXIV** 4ab
    (Olive-throated) 270
    Orange-throated 268, **XXIX** 19
    Purple-throated 271, **XXIX** 21ab
    (Rothschild's) 270
    Royal 271, **XXIX** 22ab
    Tourmaline 270, **XXIX** 18a-d
Sunbeam, Black-hooded 259, **XXVIII** 21ab
    Purple-backed 258, **XXVIII** 20
    Shining 257, **XXVIII** 18a-c
    White-tufted 258, **XXVIII** 19
Sungrebe 154
superciliaris, Hemispingus 636, **LVI** 18a-e, **LXIV** 16
superciliaris, Leistes 578, **LI** 15ab
superciliaris, Leptopogon 479
(*superciliaris*), *Scytalopus* 438,
    XLI 1(j)k
superciliosa, Synallaxis 351, **XXXVI** 17ab
swainsonii, Buteo 103, **XII** 10
Swallow, Andean 533, **XLVIII** 11
    Bank 533
    Barn 532, **XLVIII** 13ab
    Blue-and-white 531, **XLVIII** 8a-c
    Brown-bellied 530, **XLVIII** 9
    Chilean 534
    Cliff 533, **XLVIII** 12
    (Cloud-forest) 532
    Pale-footed 531, **XLVIII** 10
    Southern Rough-winged 532
    Tree 535
Swan, Black-necked 117, **VIII** 2ab
    Coscoroba 117, **VIII** 1ab
    Mute 134
Swift, Andean 240, **XLVIII** 3
    Chestnut-collared 239, **XLVIII** 5ab
    Chimney 239, **XLVIII** 4
    White-collared 238, **XLVIII** 1ab
    White-tipped 239, **XLVIII** 2
Sylph, Long-tailed 290, **XXX** 5a-c
Sylviorthorhynchus 343

Synallaxis 350, see also *Hellmayrea*
Syndactyla 387
syrmatophorus, Phaethornis 243, **XXX** 4

Tachuri, Bearded 475, **XLVI** 23
Tachuris 473
Tachybaptus 67
Tachycineta 534
Tachyeres 120
taczanowskii, Leptopogon 478, **XLVII** 23
taczanowskii, Nothoprocta 58, **I** 14ab, 846
taczanowskii, Podiceps 70, **II** 7a-c
taeniata, Dubusia 623, **LIV** 16a-c
talpacoti, Columbina 193
tamarugense, Conirostrum 596, **LIII** 6ab
Tanager, Beryl-spangled 613, **LV** 11
    Black-capped 614, **LV** 5ab
    Black-faced 640
    Blue-and-black 614, **LV** 9a-d
    Blue-and-yellow 624, **LV** 14a-d
    Blue-browed 611, **LV** 6ab
    Blue-capped 624, **LV** 13
    Blue-gray 626
    Brown-flanked 629, **LVI** 6ab
    Fawn-breasted 609, **LIV** 14
    Flame-faced 610, **LV** 4a-c
    Glistening-green 610
    Golden-collared 616, **LIV** 2
    Golden-crowned 616, **LIV** 4ab
    Golden-naped 612, **LV** 3a-c
    Grass-green 639, **LIV** 10ab
    Green-capped 613, **LV** 7
    Green-throated 615
    Hepatic 626, **LV** 17ab
    (Highland Hepatic-) 627
    Metallic-green 611, **LV** 10
    Orange-eared 610
    Palm 626
    Purplish-mantled 616
    Red-hooded 627, **LV** 15ab
    Rufous-chested 628, **LVI** 5
    Rufous-crested 628, **LVI** 8ab
    Rust-and-yellow 629, **LVI** 7ab
    Saffron-crowned 610, **LV** 1a-c
    Sayaca 625, **LV** 12
    Scarlet 627
    Scrub 612, **LV** 8
    (Silver-backed) 615
    Silvery 615, **LV** 2ab
    Sira 615
    Slaty 628, **LVI** 9ab
    (Straw-backed) 615
    Summer 627, **LV** 16ab
    White-capped 630, **LVI** 12ab
    Yellow-scarfed 617, **LIV** 3
    Yellow-throated 616, **LIV** 1
Tanager-finch 669, **LXI** 14
(*Tanagra*) see *Euphonia*
(*Tanagrella*) see *Tangara*
Tangara 610

Tapaculo, Andean 437, **XLI** 1a-m, **LXIV** 31, 847
    Ash-colored 423, **XLI** 11a-c
    Brown-rumped 433, **XLI** 2ab, **LXIII** 30
    Chucao 420, **XLI** 17
    Large-footed 428, **XLI** 10ab
    Nariño 432, **XLI** 6
    Ocellated 444, **XLI** 16ab
    Ochre-flanked 423, **XLI** 13ab
    Rufous-vented 429, **XLI** 7ab, 8, 9, **LXIII** 29
    Unicolored 424, **XLI** 3ab, 4a-c
    (White-browed) 443, **XLI** 1jk
    White-throated 419, **XLI** 18
Tapera 220
tapera, Phaeoprogne 529, **XLVIII** 7
Taphrolesbia 290
Taphrospilus 251, 846
(tarapacensis), Pterocnemia 53
tarnii, Pteroptochos 418, **XL** 17ab
Tattler, Wandering 173, **XX** 21
Teal, (Andean) 124
    Blue-winged 128, **VII** 3ab, **IX** 8ab
    Cinnamon 129, **VII** 2a-d, **IX** 9ab, **X** 3
    Green-winged 124, **IX** 7
    Puna 128, **VI** 9a-d, **IX** 5
    Silver 128, **VI** 8, **IX** 4
    Speckled 124, **VI** 5a-d, **IX** 6
    (Yellow-billed) 124
Teledromas 421
(temperatus), Trogon 301
(temporalis), Anabacerthia 389, **XXXVIII** 5ab
tenuirostris, Geositta 329, **XXXIV** 1a-c
Tephrolesbia (see *Taphrolesbia*)
Terenura 400
Tern, Large-billed 187, **XXI** 6
    Peruvian 187
    South American 187, **XXI** 7a-c
(testacea), Piranga see *flava*
(thagus), Pelecanus 74
thalassinus, Colibri 245, **XXXI** 9ab
Thamnophilus 397
(thaumasta), (Zodalia) 277
Thaumastura 293
thenca, Mimus 551, **IL** 12
theresiae, Metallura 283, **XXIX** 3ab
Theristicus 83
thilius, Agelaius 576, **LI** 12a-c
Thinocoridae 180
Thinocorus 182
Thistletail, (Ayacucho) 363, **XXXVI** 19f
    Black-throated 364, **XXXVI** 19j
    (Cochabamba) 364, **XXXVI** 19k
    Eye-ringed 363, **XXXVI** 19e
    Mouse-colored 362, **XXXVI** 19d
    Ochre-browed 362, **LXIV** 32
    Perijá 362, **XXXVI** 19a
    (Peruvian) 363
    (Plenge's) 363, **XXXVI** 19c
    (Sandia) 364, **XXXVI** 19i
    Vilcabamba 363, **XXXVI** 19g
    White-chinned 362, **XXXVI** 19b

Thlypopsis 628
thoracicus, Cyphorhinus 548, **IL** 8
(thoracicus), Formicarius see *rufipectus* 403
Thornbill, Black-backed 280, **LXIII** 3ab
    Bronze-tailed 287, **XXIX** 15a-c
    Blue-mantled 287, **XXIX** 10a-d
    Olivaceous 286, **XXIX** 11, 847
    Purple-backed 280, **XXX** 14a-c
    Rainbow-bearded 288, **XXIX** 16a-c
    Rufous-capped 286, **XXIX** 13ab
Thornbird, Chestnut-backed 383, **XXXVIII** 1
    Rufous-fronted 382, **XXXVIII** 2
    Streak-fronted 382, **XXXVIII** 3
Thraupidae 607
Thraupis 624
Threskiornithidae 82
Thripadectes 392
Thripophaga 381, see also *Asthenes*
Thrush, (Andean Slaty-) 562
    Austral 561, **L** 11ab
    Black-billed 561
    Chestnut-bellied 560, **L** 14
    Creamy-bellied 561, **L** 12ab
    Chiguanco 558, **L** 15a-c
    Glossy-black 559, **L** 6a-c
    Gray-cheeked 556, **L** 4
    Great 557, **L** 16a-g, **LXIII** 20
    Pale-eyed 556
    Rufous-bellied 560, **L** 13
    Slaty 562, **L** 10a-c
    Swainson's 555, **L** 3
    (Veery) 556
    White-necked 562
    Yellow-legged 556, **L** 7ab
(Thryophilus) see *Thryothorus*
Thryothorus 543
thula, Egretta 77, **III** 2
Tiger-heron, Fasciated 79, **III** 8
Tigrisoma 79
Tinamidae 55
TINAMIFORMES 55
Tinamotis 63
Tinamou, Andean 60, **I** 11a-d, 846
    Brown 57, **I** 7
    Chilean 60, **I** 12
    Curve-billed 61, **I** 9ab
    Highland 56, **I** 6ab
    Hooded 56, **I** 8ab
    Kalinowski's 59, **I** 13
    Ornate 59, **I** 10ab
    Patagonian 64, **I** 16ab
    Puna 63, **I** 15ab
    Red-winged 57, **I** 1ab
    Taczanowski's 58, **I** 14ab, 846
    Tawny-breasted 56, **I** 5
Tit-spinetail, Andean 346, **XXXV** 2a-c,
    LXIII 12, **LXIV** 9
    Brown-capped 349, **XXXV** 6ab
    Plain-mantled 348, **XXXV** 1a-c
    Rusty-crowned 347, **XXXV** 5ab

Streaked 347, **XXXV** 3a-c
Tawny 350, **XXXV** 7ab, 847
White-browed 348, **XXXV** 4, 847
Tit-tyrant, Agile 470, **XLVI** 7ab
    Ash-breasted 468, **XLVI** 2, 847
    Black-crested 470, **XLVI** 5a-c
    Pied-crested 471, **XLVI** 4a-c
    Tufted 472, **XLVI** 1a-c
    Unstreaked 469, **XLVI** 6a-c
    Yellow-billed 472, **XLVI** 3ab
Todirostrum 482
Tody-flycatcher, Ochre-faced 483, **XLVI** 11
Tody-tyrant, Black-throated 482, **XLVI** 9a-c
    Rufous-crowned 481, **XLVI** 10a-d
    Yungas 482, **XLVI** 8
tolmiei, Geothlypis 585
torquata, Ceryle 302, **XXXI** 25ab
torquata, Coeligena 261, **XXVIII** 15a-d, **LXIV** 5
torquata, Poospiza 685, **LXII** 9a-c
torquatus, Atlapetes 678, **LXI** 15a-e, **LXIII** 32
(*Totanus*) see *Tringa*
Toucanet, Blue-banded 306, **XXXII** 1
    Chestnut-tipped 307, **XXXII** 2
    Crimson-rumped 307
    Emerald 306, **XXXII** 3a-d, **LXIII** 8
    Groove-billed 307, **XXXII** 4
    Yellow-browed 307, **XXXII** 5
Trainbearer, Black-tailed 277, **XXX** 19a-d
    Green-tailed 278, **XXX** 18a-d
(*traillii*), *Empidonax* 490
(*traviesi*), *Coeligena* 262
Treehunter, Black-billed 393, **XXXVIII** 17
    Buff-throated 392, **XXXVIII** 13
    Flammulated 392, **XXXVIII** 14, **LXIV** 8
    Streak-capped 394, **XXXVIII** 16
    Striped 393, **XXXVIII** 15
Treerunner, Fulvous-dotted 384, **XXXIII** 12
    Pearled 384, **XXXIII** 13a-c
    White-throated 395, **XXXIII** 16ab
triangularis, Xiphorhynchus 322, **XXXIII** 10
trichas, Geothlypis 585
(*Trichopicus*) see *Melanerpes*
tricolor, Atlapetes 672, **LXI** 5
tricolor, Phalaropus 171, **XIX** 15ab, **XX** 19
trifasciatus, Basileuterus 593
trifasciatus, Hemispingus 639, **LVI** 21
Tringa 172
Tringini 172
tristriatus, Basileuterus 592, **LII** 23, **LXIV** 20
triurus, Mimus 550, **IL** 13
Trochilidae 241
Troglodytes 545
Troglodytidae 539
Trogon 301
Trogon, Masked 301, **XXXI** 23ab
Trogonidae 299
(*Trogonurus*) see *Trogon*
(*Trupialis*) see *Sturnella*
Tryngites 170
tschudii, Ampelioides 450

tuberculifer, Myiarchus 520, **XLVI** 29
tucumana, Amazona 215, **XXIV** 4
Tuftedcheek, Streaked 386, **XXXIII** 15a-c
tumultuosus, Pionus 214, **XXIV** 7ab
Turca, Moustached 418, **XL** 18
turcosa, Cyanolyca 536, **XIV** 8
Turdidae 552
Turdus 557, see also *Platycichla*
Turnstone, Ruddy, 167, **XX** 1
Tyrannidae 453
tyrannina, Dendrocincla 320, **XXXIII** 9
(*Tyranniscus*) see *Phyllomyias* and *Zimmerius*
Tyrannulet, Ashy-headed 455, **XLVII** 16
    Buff-banded 465, **XLVII** 2
    Black-capped 454, **XLVII** 17, **LXIII** 10
    Bolivian 457, **XLVII** 21ab
    Golden-faced 457, **XLVII** 22ab
    Mottle-cheeked 479, **XLVI** 12
    Mouse-colored 456
    (Olrog's) 454
    Paltry 456, **XLVII** 20
    Sclater's 454, **XLVII** 19
    Southern Beardless 458, **XLVII** 15ab
    Sulphur-bellied 465, **XLVII** 1
    Tawny-rumped 455, **XLVII** 18
    Torrent 467, **X** 7, **XLVII** 6ab
    White-banded 466, **XLVII** 4
    White-bellied 467, **XLVII** 7
    (White-crested) 467
    White-tailed 464, **XLVII** 3
    White-throated 463, **XLVII** 5ab
Tyrannus 523
(*tyrannus*), (*Muscivora*) 525
tyrannus, Tyrannus 525, **XLV** 2
Tyrant, Chocolate-vented 503, **XLIII** 13
    (Patagonian) 497, **XLIV** 16
    Plumbeous 516, **XLV** 12a-d
    Rufous-tailed 517, **XLV** 16
    Rufous-webbed 499, **XLIII** 15ab
    Spectacled 518, **XLV** 15ab
    Yellow-browed 519, **XLV** 5ab
tyrianthina, Metallura 284, **XXIX** 2a-c, **LXIV** 2
Tyto 221
Tytonidae 221
tzacatl, Amazilia 250

uncinatus, Chondrohierax 92, **XII** 11a-d
underwoodii, Ocreatus 276, **XXX** 14a-g
unicinctus, Parabuteo 106
unicolor, Phrygilus 663, **LVIII** 6a-f
unicolor, Scytalopus 424, **XLI** 3ab, 4a-c
unicolor, Thamnophilus 397, **XXXIX** 3ab
unirufa, Cinnycerthia 539, **IL** 5ab
unirufa, Synallaxis 354, **XXXVI** 11ab
Unnamed species *Scytalopus* 427, **XLI** 5ab
Upucerthia 330
Urochroa 252
(*Uromyias*) 469
Uropsalis 236
uropygialis, Carduelis 693, **LXII** 4ab

uropygialis, Lipaugus 451
uropygialis, Phyllomyias 455, **XLVII** 18
uropygialis, Sicalis 652, **LX** 7a-e
Urothraupis 633
(urubambae), Scytalopus 440
urubambensis, Asthenes 380, **XXXVII** 20ab
(urubu), Coragyps see *atratus* 88
(usheri), Asthenes 367, **XXXVII** 6c, 847
ustulatus, Catharus 555, **L** 3

validirostris, Upucerthia 332, **XXXIV** 10
validus, Pachyramphus 526, **XLVI** 28ab
Vanellus 159
varia, Mniotilta 581, **LII** 15ab
variegaticeps, Anabacerthia 389, **XXXVIII** 5ab
varius, Empidonomus 523
vassorii, Tangara 614, **LV** 9a-d
Veery 556
Velvetbreast, Mountain 259, **XXXI** 11a-c
Veniliornis 312
(ventralis), Accipiter 96
ventralis, Buteo 106, **XII** 6
ventralis, Phylloscartes 479, **XLVI** 12
Vermivora 581
verreauxi, Leptotila 197, **XXII** 6ab
versicolor, Anas 128, **VI** 8, **IX** 4
versicolor, Pachyramphus 525, **XLVI** 27ab
versicolorus, Brotogeris 212
verticalis, Creurgops 628, **LVI** 8ab
verticalis, Hemispingus 638, **LVI** 15
(verticalis), Myioborus see *miniatus* 586
vesper, Rhodopis 293, **XXX** 11ab, **XXXI** 21
(Vestipedes) see *Eriocnemis* and *Haplophaedia*
vestitus, Eriocnemis 272, **XXVIII** 9ab
vicinior, Scytalopus 432, **XLI** 6
victoriae, Lesbia 277, **XXX** 19a-d
viduata, Dendrocygna 116
(vigua), Phalacrocorax see
olivaceus 72
vilcabambae, Schizoeaca 363, **XXXVI** 19fg
villissimus, Zimmerius 456, **XLVII** 20
viola, Heliangelus 274, **XXIX** 21ab
Violetear, Green 245, **XXXI** 9ab
  (Mountain) 246, **XXXI** 9ab
  Sparkling 246, **XXXI** 22ab
  White-vented 247, **XXXI** 10ab
(violicollis), Heliangelus 269
violifer, Coeligena 264, **XXVIII** 12a-d
virens, Contopus 488, **XLV** 9
(virenticeps), Cyclarhis see *gujanensis*
Vireo 568
Vireo, (Brown-capped) 569
  Red-eyed 568, **LII** 3
  Warbling 569, **LII** 2
Vireonidae 567
(Vireosylvia) see *Vireo*
virescens, Ardeola 78
(virescens), Brotogeris see *versicolorus* 212
virescens, Empidonax 490, **XLVI** 24
virgata, Asthenes 379, **XXXVII** 18, 847

virgaticeps, Thripadectes 394, **XXXVIII** 16
virginianus, Bubo 226, **XXVI** 3ab
(virginianus), Chordeiles see *minor* 234
(virginianus), Rallus see *limicola* 144
viridicata, Pyrrhura 206, **LXIII** 1
viridicauda, Amazilia 249, **XXXI** 3
viridicollis, Tangara 615, **LV** 2ab
viridicyana, Cyanolyca 535, **XIV** 7ab
viridiflavus, Zimmerius 457, **XLVII** 22ab
vitriolina, Tangara 612, **LV** 8
vittata, Oxyura 134, **VII** 8ab
vociferus, Charadrius 162, **XX** 3
Vultur 90
Vulture, Black 88, **XI** 7ab
  Greater Yellow-headed 90
  King 91
  Turkey 89, **XI** 8ab
Vultures, New World 88

wagleri, Aratinga 202, **XXIII** 2a-d, 846
Wagtail-tyrant, Greater 468, **XLVII** 8
Warbler, Black-and-white 581, **LII** 15ab
  Bay-breasted 584, **LII** 10a-c
  Black-crested 590, **LII** 20
  Blackburnian 583, **LII** 11ab
  Blackpoll 584, **LII** 12ab
  Canada 586
  Cerulean 583, **LII** 13ab
  Citrine 590, **LII** 22a-d
  Connecticut 585, **LII** 17
  Golden-winged 581, **LII** 14
  Macgillivray's 585
  Mourning 585
  Pale-legged 591, **LII** 21ab
  Prothonotary 585
  Russet-crowned 593, **LII** 24ab
  Santa Marta 592, **LXIII** 24
  Tennessee 582
  Three-banded 593
  Three-striped 592, **LII** 23ab, **LXIV** 20
  (White-lored) 594
  Yellow 582
Warbling-finch, Black-and-rufous 686, **LXII** 13
  (Black-capped) 686, **LXII** 14
  Bolivian 683, **LXII** 11
  (Chestnut-breasted) 688
  (Cinereous) 686
  (Cochabamba) 689
  Gray-and-white 685, **LXII** 14
  Plain-tailed 684, **LXII** 15
  Ringed 685, **LXII** 9a-c
  Rufous-breasted 687, **LXII** 17ab
  Rufous-sided 684, **LXII** 10ab
  Rusty-browed 687, **LXII** 12
  (Tucumán) 688
(warscewiczi), Lepidocolaptes see *affinis* 322
warszewiczi, Dives 574, **LI** 6
Waterfowl, 115
Waterthrush, Northern 584
(watkinsi), Grallaria 406

Waxwing, Cedar 563, **XLII** 1
wetmorei, Buthraupis 621, **LIV** 13ab
Whimbrel 175, **XX** 14
Whitestart, Brown-capped 589, **LII** 9
   (Chestnut-crowned) 588
   Golden-fronted 587, **LII** 6ab
   Santa Marta 589, **LXIII** 25
   Slate-throated 586, **LII** 8
   Spectacled 587, **LII** 7a-d
   White-fronted 588, **LII** 5, **LXIV** 19
   (Yellow-crowned) 589
(*whitii*), *Poospiza* 686, **LXII** 13
Wigeon, American 123, **IX** 11ab
   Chiloe 123, **VI** 10a-c, **IX** 10
   (Southern) 123, **VI** 10a-c, **IX** 10
Willet 174
williami, Metallura 281, **XXIX** 5a-d
Wilsonia 586
Wiretail, DesMurs' 343, **XXXVI** 18ab
Woodcreeper, Black-banded 321, **XXXIII** 7ab
   Olivaceous 320, **XXXIII** 8
   Olive-backed 322, **XXXIII** 10
   Spot-crowned 322, **XXXIII** 4a-c
   Strong-billed 320, **XXXIII** 6ab, **LXIII** 16
   Tyrannine 320, **XXXIII** 9
Woodpecker, Acorn 311, **XXXII** 12
   Bar-bellied 312, **XXXII** 16a-c
   (Black-necked) 315, **XXXIII** 3ab
   Crimson-crested 317, **XXXII** 14ab
   Crimson-mantled 314, **XXXII** 17a-c
   Golden-olive 314
   Magellanic 318, **XXXII** 13ab
   Powerful 317, **XXXII** 15ab
   Red-headed 318
   Smoky-brown 313, **XXXII** 11ab
   Striped 312, **XXXII** 18
   White 312
   White-fronted 312
   Yellow-vented 313
(Wood-pewee) see Pewee
Wood-quail, Black-fronted 141
   Chestnut 141
   Gorgeted 141, **XVII** 3
   Stripe-faced 142, **XVII** 2
   Venezuelan 141
Woodstar, Chilean 296
   Gorgeted 297, **XXX** 10, **XXXI** 16
   Little 297
   Purple-collared 294, **XXX** 12ab, **XXXI** 19
   Purple-throated 294, **XXX** 8, **XXXI** 18
   Rufous-shafted 298, **XXX** 7, **XXXI** 20
   Santa Marta 297, **LXIII** 2
   Slender-tailed 294, **XXX** 13, **XXXI** 17
   White-bellied 296, **XXXI** 15ab
Wood-wren, Bar-winged 547
   Gray-breasted 547, **IL** 7, **LXIV** 30
Wren, Apolinar's Marsh 542, **X** 12ab
   Chestnut-breasted 548, **IL** 8
   Grass 540, **IL** 9a-d, **LXIV** 28
   Gray-mantled 539, **IL** 1
   House 545, **IL** 10a-c
   Inca 543, **IL** 2
   (Marsh) 541
   (Mérida) 543
   Mountain 546, **IL** 11a-e, **LXIII** 28
   Moustached 544, **IL** 4
   Páramo 542, **LXIV** 29
   Plain-tailed 543, **IL** 3a-d
   Rufous 539, **IL** 5ab
   (Santa Marta) 546, **LXIII** 28
   Sepia-brown 540, **IL** 6a-c
   (Short-billed Marsh) 541
wyatti, Asthenes 375, **XXXVII** 13ab, **LXIII** 17

xanthocephala, Tangara 610, **LV** 1a-c
xanthogastra, Carduelis 692, **LXII** 2ab
xanthogramma, Melanodera 668, **LVIII** 12ab
(*Xanthomyias*) 454
xanthophthalmus, Hemispingus 639, **LVI** 16
(*Xanthoura*) 537
(*Xenicopsoides*) see *Anabacerthia*
(*Xenoctistes*) see *Syndactyla*
Xenodacnis 606
Xenoglaux 228
Xenops, Streaked 394
Xenospingus 680
xenothorax, Leptasthenura 348, **XXXV** 4, 847
Xiphocolaptes 320
Xiphorhynchus 322
Xolmis 502, see also *Agriornis* and *Polioxolmis*
yanacensis, Leptasthenura 350, **XXXV** 7ab, 847
yarrellii, Eulidia 296
yaruqui, Phaethornis 243
Yellow-finch, Bright-rumped 652, **LX** 7a-e
   Citron-headed 653, **LX** 6ab
   Grassland 657, **X** 10ab, **LX** 5a-e
   Greater 653, **LX** 3ab
   Greenish 654, **LX** 10a-d
   Patagonian 655, **LX** 9a-c
   Puna 651, **LX** 1a-d
   Raimondi's 658, **LX** 8a-d
   (Saffron) 656, **LX** 4a-f
   Stripe-tailed 651, **LX** 2ab
Yellowlegs, Lesser 172, **XIX** 9, **XX** 16
   Greater 172, **XIX** 8, **XX** 12
Yellowthroat, Common 585
   Masked 585, **LII** 18ab
yncas, Cyanocorax 536, **XIV** 6ab
(*Zamelodia*) see *Pheucticus*
(*Zaratornis*) 447
(*zeledoni*), *Cochlearius* see *cochlearius* 79
Zenaida 191
(*Zenaidura*) 191
(*zimmeri*), *Scytalopus* 439, **XLI** 1j
zimmeri, Synallaxis 354, **XXXVI** 12, 847
Zimmerius 456
(*Zodalia*) 277
zonaris, Streptoprocne 238, **XLVIII** 1ab
(*Zonibyx*) 163
Zonotrichia 682

Plants illustrated on the color plates

Among the branches, leaves, and flowers shown on the color plates, some are identifiable. Thus, the plates can be used, in a small and incomplete scale, as a guide for learning some of the plants mentioned in the text. Several of them are important vegetational elements in the high Andes. Roman numerals refer to plates, numbers to birds with which the plant is illustrated.

Acacia sp., Plates XXIV 8d, LI 17b, LIII 5ab.
Alnus acuminata, Plates XLV 12a, LII 4a.
Anacardiaceae, Plate XXX 18 cd

Barnadesia sp., Plates XXVII 7a, XXXV 6b, LV 14cd, LXII 18b.
Beloperone sp., Plate XXVIII 21ab.
Berberis sp., Plates XXXIV 18a, XXXVII 4b, XLIII 17b, XLVI 1a, LI 7b.
Bidens laevis, Plate X before 11.
Blechnum sp., Plate XXIX 5a-d, XXXII 15b.
Bomarea spp., Plates XXVIII 11e 12e, LXII below 10c.
Brachyotum sp., Plate XXIX 17d 22b.
Bromeliaceae, indet., Plates XIV 3, XXIV 7, XXXII 7,8, LIV 13a.
Buddleia incana, Plate XIII 11c and XXX 19c.

Cavendishia sp., Plates LIII 2b, LXIII 22.
Cecropia sp., drawing p. 305
Chuquiragua insignis, Plate XXVII 4e.
C. spinosa, Plate XXVII 6a.
Chusquea bamboo, Plates XXIV 12 13, XXX 14b-f, XXXIX behind 6, XLI 5a 11c, XLVI 6c, LI 9b, LII 21b 22, LVI 12, LVII 1,2, LIX 1ab, LXI 6g, LXIII 21
Cladonia sp., Plate XXXVI 19j.
Cleta sp., Plate LIX 6e.
Clusia sp., Plates XIV 1, XXIX 7ab.
Colletia sp., Plates XXXV 1a, XLVI 3a.
Coriaria sp., Plate XLVII 11 and drawing p. 461.
Cortaderia sp., Plate LIX 4c, LX 5a.

Dipsacus sp., Plate XLV 16.

Ephedra sp., Plate LXII 3b.
Escallonia myrtilloides XXVIII 20, XLIV 7b, LIII 2a.
Espeletia spp., Plate XXX 20, XXXV 2a.
Eucalyptus Plate XIII 10a, XXXI 22ab, LVII 4b.
Eupatorium sp. XLVI 4a, LXII 13.

Fuchsia sp., XXVIII 5b 12a.

Gynoxys sp., Plate XXIX 10c, LIII 8abc.

Heliconia leaf-tip, Plate XXX 4, flowers p. 242.
Hypericum sp., Plate XXIX 1, XLIV 7c.

Lantana sp., Plate XXX 13.
Lepidophyllum quadrangulare, Plate XXXV 1b, LVIII 10b.
Limnobium stoloniferum Plate X, by 8b and behind 11a.

Melastomataceae, indet., Plates XLVI 10b, LXI 6b, LXII 9d.
Mutisia acuminata Plate XXVII 1c
Margarocarpus sp., Plate IL 10c.
Miconia sp., Plate LIII 3.
Myzodendron mistletoe, Plate XLIV 16.

Nicotiana glauca Plate XXX 17a-c.
Nothofagus spp., Plates XXIII 11, XLIV 16, XLVII 6, LVIII 3a-c, LXII 5ab.

Oreopanax, Plate XXXIII 2c, and drawing p. 202.
Oxalis, Plate XLVI 15ab

Passiflora flower bud, Plate XXVIII 13b.
Pentacalia sp., Plate XXXVII 19b
Peperonia sp., Plate XLVII 10.
Podocarpus sp., Plates XXXVI 1a, XLVII 2, LXI 12b.
Polylepis spp., Plates XXIX 10b, XXXV 2c 3c 4 5b 7b 11a, XXXVII 2 3 6ab 5 20a, XLVII 5b, LIII 1a-c 4 6ab, LXII 8bc 15.
Polypodium sp., Plate XXV 8.
Prosopis sp., Plates XLII 3b-d, LX 4a-c, LXII 3b-d 14.

Rubus sp., Plate XLVII 14

Salpichroa sp., Plate XXVII 3b.
Schinus molle, Plate XLVII 8.
Schinus sp., Plates XLII 19b, XLIII 20a.
Scirpus californicus Plate X 6a-c 10 ab, XLVI 21 22 26, LI 12.
Selaginella sp., Plate XXXIX 4c.
Siphocampylos sp., Plate XXVIII 8b.
Stipa vicugnarum, Plate XLVIII 16ab.

Tillandsia usneoides, Plates XXIII 2a, XXIV 9 10, XXXVII 21, XLV 2, LIII 15a-d, LV 14a.
Trichocereus cactus, Plates XXXIII 3b, XXXV 1c, XXXVII 7.
Tristerix mistletoe, Plates XLIV 2a, XLVII 21ab, LXII 5.

Vallea stipularis, Plate XXIX 13.

Verbena tridentata, Plate XXXVII 16a.

Weinmannia sp., Plates LV 9ab, LVI 14, LXII 7.

Zea mais, Plate LVII 5c.

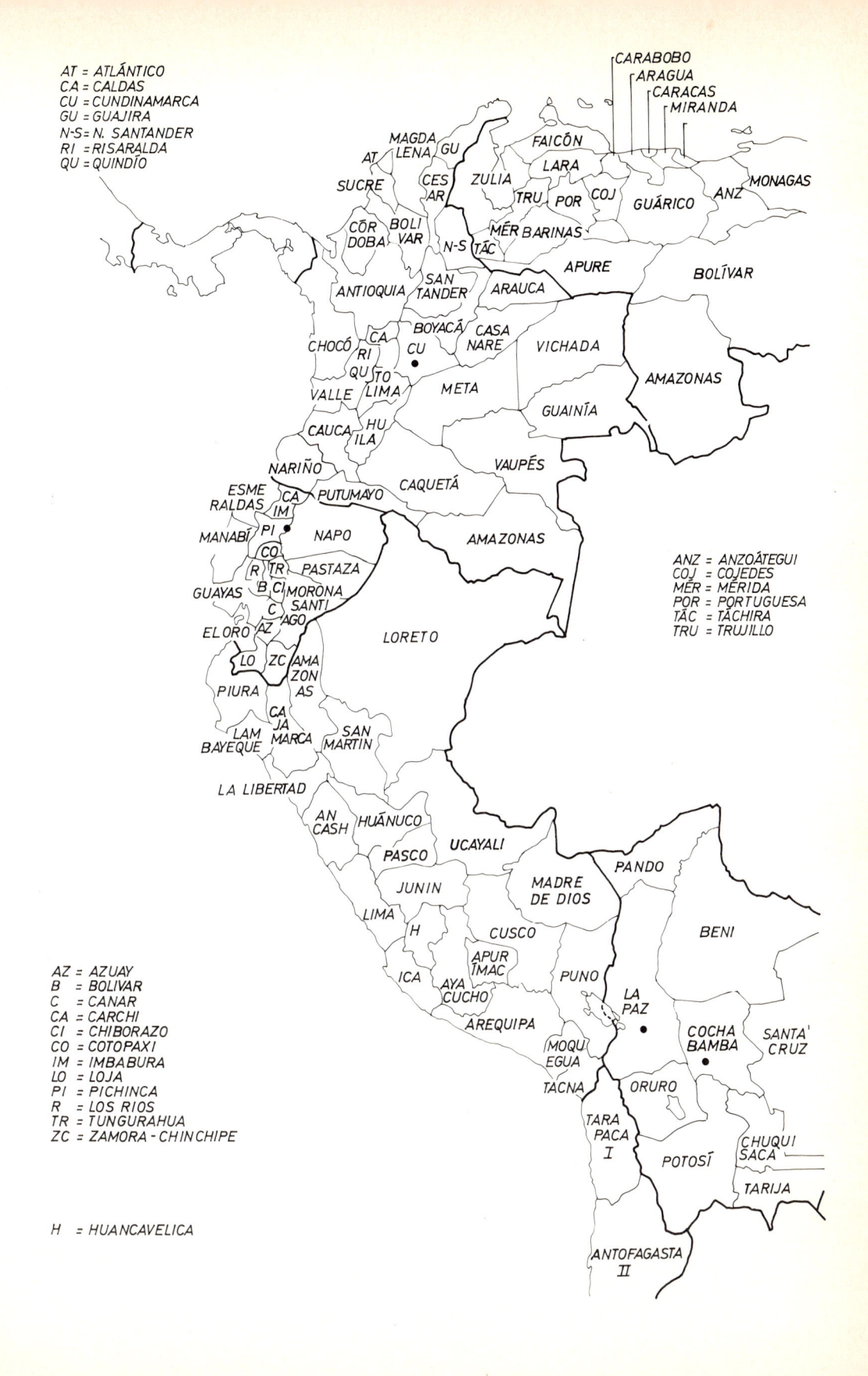

AT = ATLÁNTICO
CA = CALDAS
CU = CUNDINAMARCA
GU = GUAJIRA
N-S = N. SANTANDER
RI = RISARALDA
QU = QUINDÍO

ANZ = ANZOÁTEGUI
COJ = COJEDES
MÉR = MÉRIDA
POR = PORTUGUESA
TÁC = TÁCHIRA
TRU = TRUJILLO

AZ = AZUAY
B = BOLIVAR
C = CAÑAR
CA = CARCHI
CI = CHIBORAZO
CO = COTOPAXI
IM = IMBABURA
LO = LOJA
PI = PICHINCA
R = LOS RIOS
TR = TUNGURAHUA
ZC = ZAMORA-CHINCHIPE

H = HUANCAVELICA

CARABOBO
ARAGUA
CARACAS
MIRANDA
MAGDALENA
GU
FALCÓN
SUCRE
AT
ZULIA
LARA
MONAGAS
CES AR
TRU POR COJ GUÁRICO ANZ
CÓR DOBA BOLIVAR
MÉR BARINAS
N-S TÁC
ANTIOQUIA SAN TANDER ARAUCA APURE BOLÍVAR
CHOCÓ CA BOYACÁ CASA NARE VICHADA
RI CU
QU TO
VALLE LIMA META AMAZONAS
CAUCA HU ILA GUAINÍA
NARIÑO VAUPÉS
ESME RALDAS CA PUTUMAYO CAQUETÁ
IM
MANABÍ PI NAPO
CO
R TR PASTAZA AMAZONAS
GUAYAS B CI MORONA
C SANTI
EL ORO AZ AGO
LO ZC AMA ZONAS LORETO
PIURA
LAM CA JA SAN
BAYEQUE MARCA MARTIN
LA LIBERTAD
AN HUÁNUCO
CASH UCAYALI PANDO
PASCO
JUNIN MADRE DE DIOS BENI
LIMA
H CUSCO
ICA APUR ÍMAC
AYA PUNO
CUCHO LA PAZ
AREQUIPA COCHA SANTA
MOQU BAMBA CRUZ
EGUA
TACNA ORURO
TARA CHUQUI
PACA SACA
I POTOSÍ
TARIJA
ANTOFAGASTA II

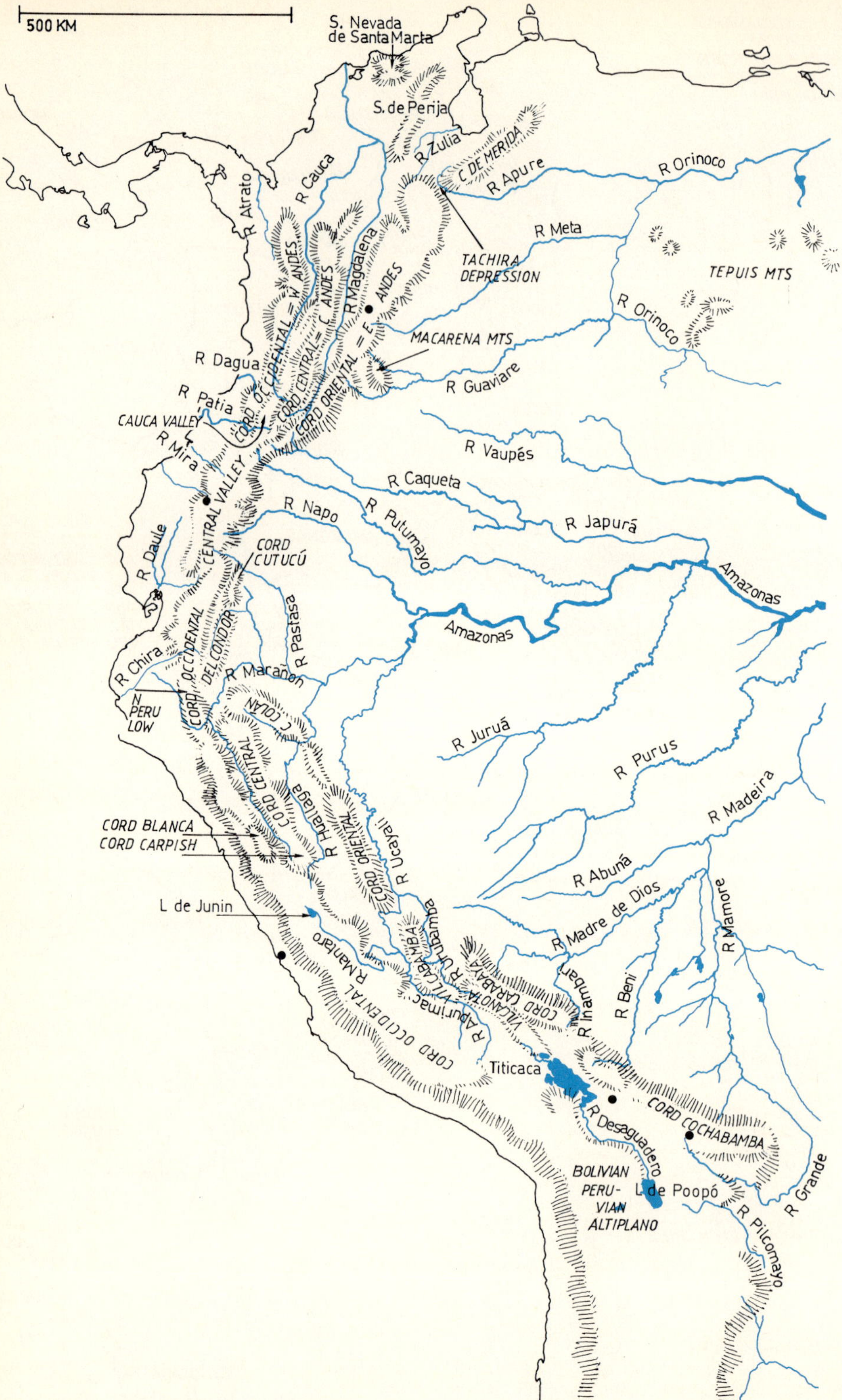
500 KM
S. Nevada de Santa Marta
S. de Perija
R Zulia
C. DE MERIDA
R Apure
R Orinoco
R Atrato
R Cauca
TACHIRA DEPRESSION
R Meta
TEPUIS MTS
R Magdalena
CORD OCCIDENTAL = W. ANDES
CORD CENTRAL = C. ANDES
CORD ORIENTAL = E. ANDES
R Orinoco
MACARENA MTS
R Dagua
R Guaviare
R Patia
CAUCA VALLEY
R Mira
R Vaupés
R Caqueta
CENTRAL VALLEY
R Napo
R Putumayo
R Japurá
R Daule
CORD CUTUCÚ
Amazonas
Amazonas
R Pastasa
CORD OCCIDENTAL
DEL CONDOR
R Chira
R Marañon
N PERU LOW
C. COLAN
CORD CENTRAL
R Juruá
R Purus
R Huallaga
CORD ORIENTAL
R Ucayali
R Madeira
CORD BLANCA
CORD CARPISH
R Abuná
R Urubamba
R Mamore
L de Junin
R Madre de Dios
R Mantaro
VILCABAMBA
CORD CARABAYA
VILCANOTA
R Inambari
R Beni
R Apurimac
CORD OCCIDENTAL
Titicaca
CORD COCHABAMBA
R Desaguadero
BOLIVIAN PERUVIAN ALTIPLANO
L de Poopó
R Pilcomayo
R Grande

ROMAN NUMERALS FOLLOWING NAMES OF CHILEAN PROVINCES REFER TO REGIONS. NAMES OF REGIONS COVERING MORE THAN ONE PROVINCE ARE: VALPARAISO (V), O'HIGGINS (VI), MAULE (VII), BIO-BIO (VIII), LA ARAUCANIA (IX), LOS LAGOS (X).

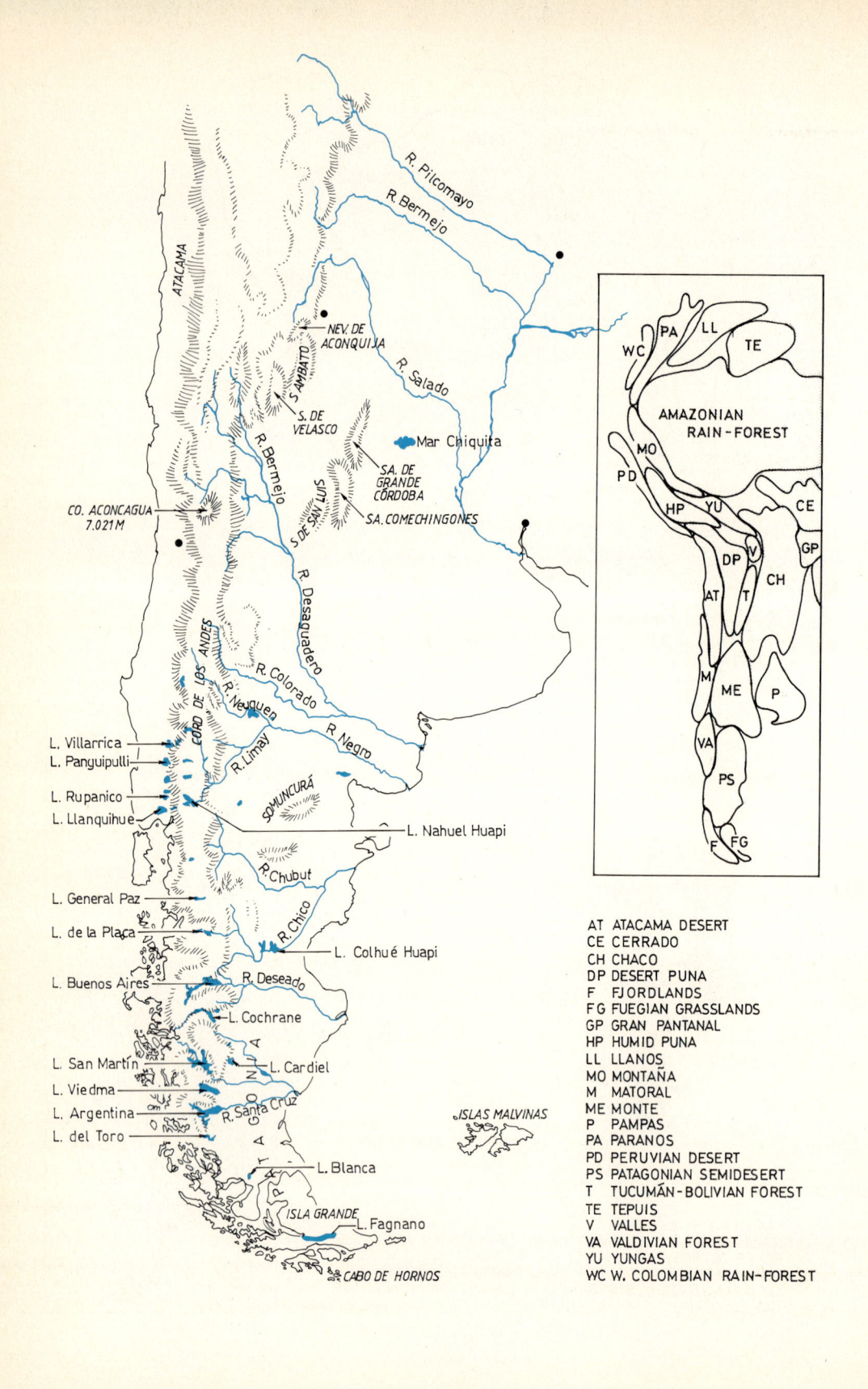

AT ATACAMA DESERT
CE CERRADO
CH CHACO
DP DESERT PUNA
F   FJORDLANDS
FG FUEGIAN GRASSLANDS
GP GRAN PANTANAL
HP HUMID PUNA
LL LLANOS
MO MONTAÑA
M   MATORAL
ME MONTE
P   PAMPAS
PA PARANOS
PD PERUVIAN DESERT
PS PATAGONIAN SEMIDESERT
T   TUCUMÁN-BOLIVIAN FOREST
TE TEPUIS
V   VALLES
VA VALDIVIAN FOREST
YU YUNGAS
WC W. COLOMBIAN RAIN-FOREST

Left:
Interior of undisturbed *Polylepis* forest at 4300 meters elevation in Quebrada Balcón in the mountains southeast of Abancay in Apurímac, Peru. The flowering creeper is *Salpichroa* sp. Giant Hummingbird *Patagona gigas* chasing White-tufted Sunbeam *Aglaeactis castelnaudii*. Above an Ash-breasted Tit-tyrant *Anairetes alpinus*.

Right:
Bunchgrass terrain degraded by sheep-grazing on Meseta de las Viscachas in Santa Cruz in southern Argentina. Fox *Dusicyon griseus* and White-bellied Seedsnipes *Attagis malouinus*.